Global Conference on Polymer and Composite Materials 2015

(PCM2015)

IOP Conference Series: Materials Science and Engineering
Volume 87

Beijing, China
16-18 May 2015

ISBN: 978-1-5108-0976-5
ISSN: 1757-8981

Printed from e-media with permission by:

Curran Associates, Inc.
57 Morehouse Lane
Red Hook, NY 12571

Some format issues inherent in the e-media version may also appear in this print version.

Printed by Curran Associates, Inc. (2015)

For permission requests, please contact the Institute of Physics
at the address below.

Institute of Physics
Dirac House, Temple Back
Bristol BS1 6BE UK

Phone: 44 1 17 929 7481
Fax: 44 1 17 920 0979

techtracking@iop.org

Additional copies of this publication are available from:

Curran Associates, Inc.
57 Morehouse Lane
Red Hook, NY 12571 USA
Phone: 845-758-0400
Fax: 845-758-2634
Email: curran@proceedings.com
Web: www.proceedings.com

Table of contents

Volume 87

2015 Global Conference on Polymer and Composite Materials (PCM2015)
16–18 May 2015, Beijing, China

Accepted papers received: 10 June 2015
Published online: 16 July 2015

Preface

Papers

IOP Conf. Series: Materials Science and Engineering **87** (2015) 011001 doi:10.1088/1757-899X/87/1/011001

Preface

The *2015 Global Conference on Polymer and Composite Materials (PCM2015)* sponsored by Wuhan Advanced Materials Society and the Ningbo Adhesives and Products Industry Association was held from May 16-19 2015 in Beijing, the capital of the People's Republic of China and one of the most populous cities in the world. The technical program consisted of seven international keynote speakers, oral presentations, a Mini-Workshop: "Writing a Scientific Publication" and a poster session.

The present issue of IOP Conference Series: Materials Science and Engineering (MSE) records the proceedings of PCM2015 and contains 117 specially selected manuscripts submitted to the PCM2015 conference. The electronic submission and handling of manuscripts via the conference website, including the selection of reviewers and evaluation of manuscripts, were identical to the procedures applied to manuscripts submitted as regular contributions for publication.

The organization of this conference and the preparation of the proceedings volume would have been impossible without the tremendous efforts and dedication of many individuals, especially from Ms. Lan Mengguo, who oversaw the organization of the conference and the program; and a large team of reviewers with their timely submission of quality reports. We express our sincere thanks to all authors and presenters for their contributions. We also thank our sponsors very much for their generous support.

The 2016 Global Conference on Polymer and Composite Materials (PCM2016) will be held from June 17-20 2016 in Shanghai, the largest city in China and one of the largest urban areas in the world; all participants are welcome to the renewed and vibrant PCM2016.

Prof. Dr. Esteban Broitman
Linköping University, Sweden
The Chairman of PCM2015

Peer review statement

All papers published in this volume of *IOP Conference Series: Materials Science and Engineering* have been peer reviewed through processes administered by the proceedings Editors. Reviews were conducted by expert referees to the professional and scientific standards expected of a proceedings journal published by IOP Publishing.

IOP Conf. Series: Materials Science and Engineering **87** (2015) 012001 doi:10.1088/1757-899X/87/1/012001

Extinction performance of microwave by core-shell spherical particle

Youpeng Wu[1,3], Jinxing Cheng[1,2], Wenping Zhou[1], Fengtao Zhao[1], Weiwei Wen[1], Junhui Liu[1] and Yuxin Hu[1]

[1]Kadinuo Science and Technology (Beijing) Co., Ltd., Beijing 100013, China
[2]Institute of Nuclear and New Energy Technology, Tsinghua University, Beijing 100084, China

E-mail: pengpengstudent@163.com

Abstract. A new method was presented for calculating extinction cross-section of electromagnetic wave by the core-shell spherical granule. An equivalent method was used to calculate the equivalent permittivity of the core-shell spherical granule. A concentric core-shell spherical granule can be replaced by a solid granule possessing the same radius with the original coating shell, and the equivalent permittivity is equal to that of the product of equivalent coefficient and the kernel permittivity. Based on the electromagnetic wave theory and the equivalent method, the scattering and absorption characteristics of the electromagnetic wave by a core-shell spherical particle were analyzed under the condition that wavelength was far greater than the particle diameter. The influence of geometrical and dielectric parameters on the scattering and absorption of microwave was discussed numerically in detail.

1. Introduction

Particle with core-shell structure, a kind of functional composite material, possesses the properties of both the shell layer and the core material, meanwhile it also has different physical and chemical characteristics from the shell layer and the core material. Therefore, studies on the core-shell granules have attracted extensive concern due to their preparation and their electrical, magnetic, optical as well as catalytic properties [1-5]. Composites with core-shell granules owning electromagnetic absorbing function have been widely applied not only in the military stealth technology, but also in the civil domain for reducing disturbance of electromagnetic wave to microelectronic communication systems, eliminating the radiation of electromagnetic wave to human body as well as controlling environmental pollution of electromagnetic wave [6-10]. The outer shell can change and endue the nucleus granule with special optical, electrical and magnetic properties. Thus a solid granule can be upgraded to a core-shell one, and then its electromagnetic loss and frequency dispersion characteristics can be adjusted by the way of altering the shell layer, which is an effective method to develop composites with excellent absorbing capacity, thin thickness and little weight.

This paper presents an equivalent method to investigate the effective permittivity of core-shell particle. A complex model of permittivity of two-phase composites is established and utilized to

[3] Address for correspondence: Youpeng Wu, Kadinuo Science and Technology (Beijing) Co., Ltd., Beijing 100013, China. E-mail: pengpengstudent@163.com.

simplify the composite system with core-shell granules into the one with solid granules. The concentric core-shell spherical particle randomly mixed with matrix is replaced by solid particle mixed with the same matrix. The permittivity of the equivalent solid particle, which possesses the same radius with the original coating shell, is equal to that of the product with equivalent coefficient and permittivity of the original nucleus. Then the electromagnetic scattering and absorption characteristics of spherical core-shell particle are analyzed. In the condition that the particle diameter is much smaller than the wavelength of incident electromagnetic wave, the formulas are deduced for calculating extinction coefficient of core-shell particle. The influence of the structure and dielectric parameters on the scattering and absorption performance is investigated numerically. The results are beneficial to the theoretical direction on the design of absorbing coating materials with core-shell granules.

2. Extinction model of core-shell particle

2.1. Effective theory of composite with core-shell spherical particle

An equivalent method, integrating the shell layer and the core particle as a "complex particle", is adopted to investigate the effective permittivity of the core-shell particle. Here a model with three-phase components is established. Usually, the filler granules can be considered as spherical ones. For simplicity, the permittivity of the core is assumed to be unchangeable. It is supposed that a and b are the radii of core particle with permittivity ε_1 and shell with permittivity ε_2 of the complex particle, respectively. The permittivity of matrix is ε_m, and the thickness of the shell layer is $t=b-a$. Under quasi-static approximation, when an electromagnetic field $\vec{E}_0$ is perpendicularly incident to the complex particle, the electric potential in each component of the composite is given by the Laplace equation. The overall electrical dipole moment of the spherical core-shell particle $\vec{p}$ is calculated through the electric potential generated by the electrical dipole moment in the space [11].

$$\vec{p} = \frac{\gamma\varepsilon_1 - \varepsilon_m}{\gamma\varepsilon_1 + 2\varepsilon_m} 4\pi\varepsilon_m b^3 \vec{E}_0 \quad (1)$$

Where $\gamma = [\beta(1+2\beta) + 2\alpha\beta(1-\beta)]/[(1+2\beta) - \alpha(1-\beta)]$ and $\alpha = R_1^3/R_2^3$ are called equivalent coefficient and structure parameter, respectively, and β represents the ratio $\varepsilon_1/\varepsilon_2$.

Solid spherical granules with permittivity ε_1 and radius b embedding in the same homogeneous matrix can be made up of another composite. Illuminated perpendicularly by electromagnetic field E_i, the overall electrical dipole moment of solid particle is calculated with the following formula.

$$\vec{p} = \frac{\varepsilon_1 - \varepsilon_m}{\varepsilon_1 + 2\varepsilon_m} 4\pi\varepsilon_m b^3 \vec{E}_0 \quad (2)$$

Comparing equations (1) with (2), it is shown that the difference in the two equations is just to replace $\gamma\varepsilon_1$ in equation (1) with ε_1 in equation (2). Therefore spherical particle with interface equates another spherical one without shell layer, and this leads to a result that the composite system can be replaced by solid sphere with equivalent permittivity $\gamma\varepsilon_1$ and equivalent radius b mixed with the same matrix. Consequently, the particle is simplified from binary components to the single component.

2.2. Scattering and absorption of electromagnetic wave by core-shell particle

When the wavelength of the incident electromagnetic wave is much larger than the scale of a scatterer, the scattering effect of the scatterer on electromagnetic field can be accurately processed by the Mie scattering theory. By this means, the scattering cross-section (C_{sca}), the absorption cross-section (C_{abs}) and the extinction cross-section (C_{ext}) can be expressed as follows [10, 12].

$$C_{sca} = \frac{8}{3}\pi k^4 a^6 \left|\varepsilon_r - 1\right|^2 / \left|\varepsilon_r + 2\right|^2 \quad (3)$$

$$C_{abs} = 12\pi k a^3 \varepsilon_r'' / |\varepsilon_r + 2|^2 \quad (4)$$

$$C_{ext} = C_{sca} + C_{abs} \quad (5)$$

Where a is the radius of spherical particle with relative permittivity $\varepsilon_r = \varepsilon_r' - i\varepsilon_r''$, $k=2\pi/\lambda$ serves as the propagation constant, and λ is the wavelength of incident wave in the background. It is shown in equation (4) that extinction properties of single particle are determined by the granule diameter a, permittivity ε_r, as well as wavelength λ of incident wave.

Now the scattering and absorption characteristics of dielectric core-shell particle to electromagnetic wave are investigated. The parameters of the core-shell particle are the same as the front signs a, b, ε_1 and ε_2. According to the above equivalent method, core-shell particle can be considered as symmetrical solid particle with permittivity $\gamma\varepsilon_1$ and radius b. To make the results more reasonable, extinction properties are replaced by extinction parameters per unit volume. Thereby, the three cross-sections per unit volume are expressed as follows.

$$C'_{sca} = 2k^4 b^3 |\gamma\varepsilon_1 - 1|^2 / |\gamma\varepsilon_1 + 2|^2 \quad (6)$$

$$C'_{abs} = 9k \, \mathrm{Im}[\gamma\varepsilon_1] / |\gamma\varepsilon_1 + 2|^2 \quad (7)$$

$$C'_{ext} = C'_{sca} + C'_{abs} \quad (8)$$

Where $\mathrm{Im}[\gamma\varepsilon_1]$ denotes taking the imaginary part of $\gamma\varepsilon_1$. The above formulas are also suitable to multilayer core-shell particle. Based on the deduction mentioned above, several investigations are carried out to explore the impact of particle parameters on the scattering and absorption properties.

3. Results and discussion

Usually the main operating frequency range of radar is between 2 GHz and 18 GHz, so the incident electromagnetic wave with wavelength $\lambda=3\times10^{-2}$m is selected as representative. It is supposed that a=50nm and ε_1=20-5i are the radius and permittivity of core particle, respectively, b and ε_2 are the radius and permittivity of shell layer, respectively, and then $\beta=\varepsilon_2/\varepsilon_1$. The thickness of shell is $d=b-a$, and the structure parameter $\alpha=a^3/b^3$. Based on equations (6)-(8), the effects of shell parameters on scattering and absorption properties are analyzed.

Firstly, the influence of core-shell structure parameter on scattering and absorption capacity is explored. It is given that the ratio β is 0.1 and 2.0, respectively. Figure 1 presents the relationship between the extinction cross-section and the shell thickness. Figures 1(a) and 1(b) respectively display the variational tendency of absorption cross-section and the scattering cross-section along with the shell thickness.

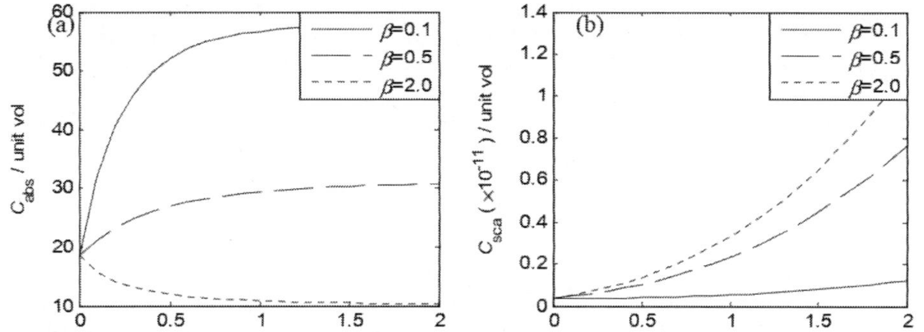

Figure 1. Extinction cross-section versus ratio d/a.

IOP Conf. Series: Materials Science and Engineering **87** (2015) 012001 doi:10.1088/1757-899X/87/1/012001

As shown in these two figures, the absorption cross-section is much greater than the scattering one, where their difference achieves more than 10 orders of magnitude. Therefore the impact of scattering on extinction can be ignored usually. The numerical results indicate that scattering cross-section enhances along with the increase of shell thickness when other parameters are identical. Absorption cross-section is fairly sensitive to the shell permittivity, which makes the variational tendency of absorption cross-section versus different shell thickness evident. Along with the increasing shell thickness, absorption cross-section rises when the value of β is relatively great and decreases when the value of β is relatively small. Absorption cross-section changes rapidly when the value d/a is lower than 0.5 (especially for higher β), and then mildly goes to a fixed value.

Secondly, the influence of core-shell dielectric parameter on scattering and absorption capacity is explored. It is given that the ratio d/a is 0.02, 0.1 and 0.2, respectively. Figure 2 presents the relationship between the extinction cross-section and the shell permittivity. Figures 2(a) and 2(b) respectively display the variational trends of the absorption cross-section and the scattering one along with the shell permittivity. The numerical results indicate that scattering cross-section is enhanced straight along with the increase of shell permittivity while other parameters are identical, whereas absorption cross-section shows a reverse tendency. When $\beta<1$, a variety of shell permittivities put significant impact on absorption cross-section. It seems easier to obtain higher absorption cross-section by decreasing shell permittivity and increasing shell thickness simultaneously.

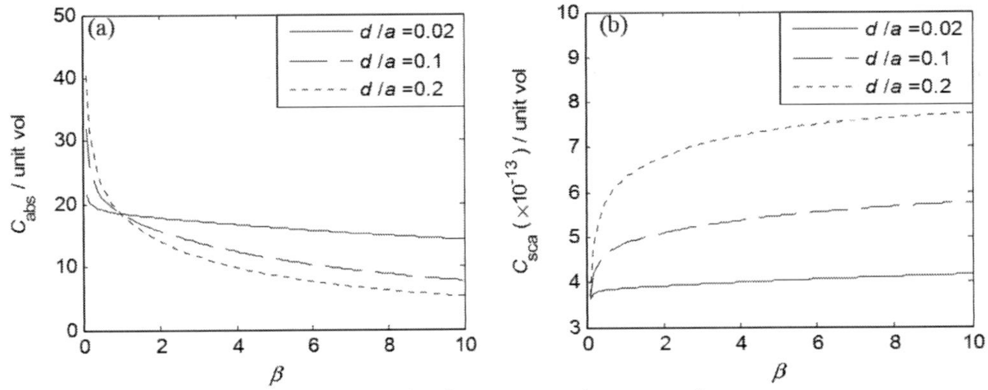

Figure 2. Extinction cross-section versus β.

Finally, the difference of extinction characteristics between solid particle and hollow particle is discussed. It is given that shell radius is b=50 nm, and shell permittivity ε_2=20-5i. The frequency range of incident wave is between 2 GHz and 18 GHz. Figure 3 displays the relationship between the extinction cross-section and the frequency. As shown in figures 3(a) and 3(b), the extinction cross-sections of both solid particle and hollow particle increase along with the enhancing the frequency. Being identical in other conditions, the scattering cross-section of solid particle is superior to that of hollow particle, and increasing shell thickness is in favour of raising the scattering capacity to the electromagnetic wave. Yet, absorption capacity exhibits an opposite feature. This phenomenon occurs due to the following reasons. As shown in equation (4), extinction properties of single particle are determined by its permittivity ε_r while the granule radius and wavelength are fixed. Given that $\varepsilon_r = \varepsilon'_r - i\varepsilon''_r$, absorption cross-section per unit volume can be expressed as follows.

$$C'_{abs} = 9k\varepsilon''_r/|\varepsilon_r+2|^2 =_r 9k\varepsilon''/\left[(\varepsilon'_r+2)^2+\varepsilon''^2_r\right] \tag{9}$$

The extreme values of equation (9) happen at $\varepsilon''_r = \varepsilon'_r + 2$. As shown in this equation, absorption cross-section decreases along with the enhancing ε'_r, and ε''_r deviates further from the optimal value with a smaller absorption cross-section. As given ε_2=20-5i, when the permittivity of the hollow particles is closer to the optimal values, higher absorption cross-section can be achieved. Therefore, the absorption capacity of the core-shell particle can be improved by adjusting the structural parameters and dielectric parameters, which would also help reduce the weight of the compound particle.

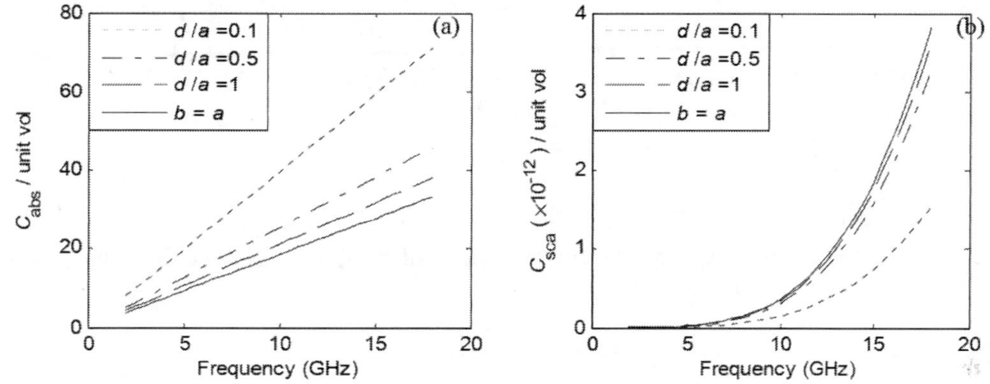

Figure 3. Extinction cross-section versus frequency.

4. Conclusion

In summary, this work puts emphasis on the effect of core-shell particle parameters on the scattering and absorption properties. An equivalent method, replacing the concentric core-shell spherical particle with equivalent solid particle, is used to reduce the complicacy of composite system. The equivalent solid particle possesses the same radius with the original shell layer, and its equivalent permittivity is equal to that of the product with equivalent coefficient and nucleus permittivity. Then formulas for scattering, absorption and extinction cross-sections of core-shell particle are deduced. In the condition that the particle diameter is far less than the wavelength of the incident wave, the impact of the geometry and dielectric parameters on the scattering and absorption properties is investigated. Numerical analysis shows that scattering cross-section increases along with the increasing thickness and permittivity of shell layer. Increasing the shell permittivity and thickness cannot certainly enhance the value of absorption cross-section, indicating that there exist a group of optimum values. The absorption capacity of the core-shell particle can be improved by adjusting the structural parameters and dielectric parameters simultaneously. The results provide theoretical foundation and guidance for the design of core-shell structure composites to optimize some desired electromagnetic properties.

References

[1] Orlowska S, Beroual A and Fleszynski J 2004 *J. Phys. D: Appl. Phys.* **37** 758-63
[2] Schlamp M C, Xiao G and Alivisators A P 1997 *J. Appl. Phys* **82** 5837-42
[3] Clementia L A, Vegaa J R, Gugliottaa L M and Quirantes A 2012 *J. Quant. Spectrosc. Ra.* **113** 2255-64
[4] Jarriel G W, Riggs L S and Baginski M E 1997 *IEEE T Electromagn. C* **39** 324-31
[5] Liu X X, Wu Y P and Zhang Z Y 2010 *Physica B* **405** 4393-6
[6] Yang S Y, Lozano K , Lomeli A, Foltz H D and Jones R 2005 *Compos. Part A-Appl. S.* **36** 691-7
[7] Wang D K, Huang H, Yu K, Zhang X F and Dong X L 2009 *J. Inorg. Mater.* **24** 340-4
[8] Ge F D, Zhu J and Chen L M 1996 *Acta Electronica Sinica* **24** 82-5
[9] Tian H Y and Wang X Z 2006 *Journal of Lanzhou University (Natural Sciences)* **42** 135-9
[10] Xu Q J and Zhang S Y 2008 *The Journal of Light Scattering* **20** 223-6
[11] Gu L P and Gao L 2007 *Journal of Applied Sciences* **25** 91-5
[12] Jiang C Y 1996 *Chinese Journal of Radio Science* **11** 65-88

IOP Conf. Series: Materials Science and Engineering **87** (2015) 012002 doi:10.1088/1757-899X/87/1/012002

Classification of acoustic emission sources produced by carbon/epoxy composite based on support vector machine

Peng Ding[1], **Qin Li and Xunlei Huang**

Standard and Quality Control Research Institute, Ministry of Water Resources, P.R.C, Hangzhou 310012, China

E-mail: qddp2008@126.com

Abstract. Carbon/epoxy specimens were made and stretched to fracture. In the process, acoustic emission (AE) signals were collected and their parameters were set as the input parameters of the neural network. Results show that using support vector machine (SVM) network can recognize the difference of AE sources more accurately than using the BP neural network. In addition, the accuracy of the SVM increases when the number of the training set increases. It is proved that using AE signal parameters and SVM network can recognize the AE sources' pattern well.

1. Introduction

Carbon fiber reinforced composites (CFRP) have been widely used in fields of civil aviation, medical care as well as architecture because of their excellent properties such as high strength, light quality, corrosion resistant and high temperature resistant [1]. However, due to the high price of CFRP, it is quite necessary to find ways to discover and prevent the defects as early as possible. Acoustic emission (AE) testing is a novel non-destructive testing method which can be used online, and support vector machine (SVM) is a general feedforward network which can be used for pattern recognition [2]. Therefore, combining AE and SVM can well solve the pattern recognition problems of CFRP and prevent the loss.

2. Basic theories

2.1. Acoustic emission

AE is defined as "the class of phenomena whereby transient elastic waves are generated by the rapid release of energy from localized sources within a material or the transient waves so generated" [3, 4]. AE is generated by the material itself. In material, AE sources can be from dislocation movement, cracks, fractures, and even corrosion. The waveform of the AE sources is shown in figure 1 and the features of the waveform are presented as below.

In figure 1, amplitude means the peak voltage of the largest excursion attained by the signal waveform from an emission event. In other words, peak amplitude is the highest point of the signal. Counts means the number of times the AE signal exceed a preset threshold during any selected portion

[1] Address for correspondence: Peng Ding, Standard and Quality Control Research Institute, Ministry of Water Resources, P.R.C, Hangzhou 310012, China. E-mail: qddp2008@126.com.

of a test. Rise time is defined as the time between the start of the AE signal (Time of Hit) and its peak amplitude. Duration is the time from the start to the end of the AE signal. The energy is contained in a detected AE burst signal, with units usually reported in joules and values which can be expressed in logarithmic from (dB) [5]. Energy is defined as:

$$E = \frac{1}{2}\int_{t_1}^{t_2} f_+^2(t)dt - f_-^2(t)dt$$

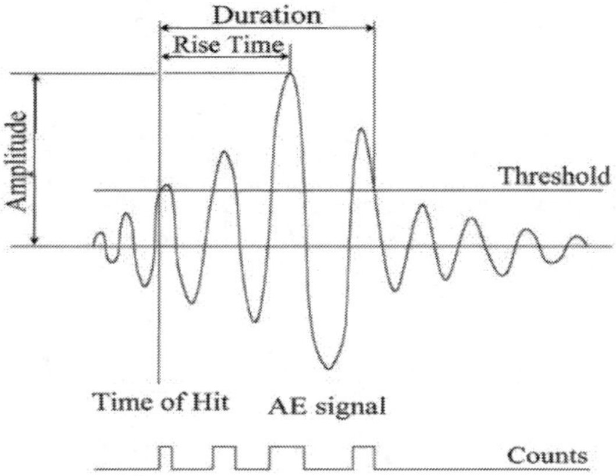

Figure 1. Signal features of the AE source.

2.2. Support vector machine

SVM, first proposed by Vapnik, is a typical neural network based on study algorithm [6-8]. The algorithm of SVM is as follows:

Firstly, the input vectors are mapped into high dimensional feature space through a nonlinear transform Φ.

Secondly, the maximal α_{0p} of the objectives of equation (1) is solved under the constraint condition $\sum_{p=1}^{P} \alpha_p d^p = 0$, $0 \le \alpha_p \le C(\text{or}\alpha_p \ge 0)$, p=1, 2, …, P.

$$Q(\alpha) = \sum_{p=1}^{P} \alpha_p - \frac{1}{2}\sum_{p=1}^{P}\sum_{j=1}^{P} \alpha_p \alpha_j d^p d^j \Phi^T(X^p)\Phi(X^j) \tag{1}$$

Thirdly, the optical weight can be calculated through equation (2):

$$W_0 = \sum_{p=1}^{P} \alpha_0 d^p \Phi(X^p) \tag{2}$$

Fourthly, given an unclassified model X, the classification discriminant function is obtained through equation (3):

$$f(X) = \text{sgn}[\sum_{p=1}^{P} \alpha_{0p} d^p \Phi^T(X^p)\Phi(X) + b_0] \tag{3}$$

The X's classification is judged from the value of f(x).

3. AE signals acquisition

CFRP specimens respectively with 0, 1, 5, 9 bundles of carbon fiber and carbon-cloth/epoxy specimens were made and drawn to fracture at a speed of 2 mm/min. PCI-2 AE system was used to collect AE signals in the process. The sampling rate of the AE system was set at 2 MHz while the hit length was 1k. The threshold was set at 30 dB in order to remove the environmental noise and mechanical noise. The preamplifier gain was set at 20 dB and the analog filter was 2 kHz-3 MHz. The sensor was R15α with Vaseline as couplant. The construction and the geometry size of the specimen were shown in figure 2.

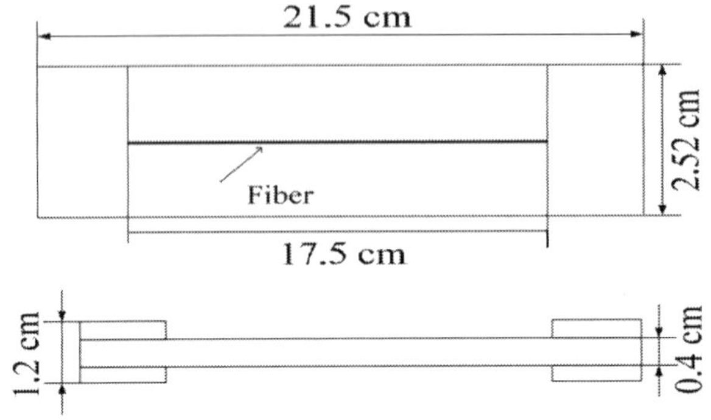

Figure 2. Construction and geometry of specimen.

Table 1. Input data of the neural network.

			Rise Time (μs)	Counts	Energy (mV^2)	Duration (μs)	Amplitude (dB)
Matrix Crack	**Training (10 data)**	**Sum**	547	165	128	2779	584
		Average	54.7	16.5	12.8	277.9	58.4
	Testing (20 data)	**Sum**	398	280	126	3142	1145
		Average	19.9	14	6.3	157.1	57.25
Delaminations	**Training (10 data)**	**Sum**	531	621	1566	9670	791
		Average	53.1	62.1	156.6	967	79.1
	Testing (20 data)	**Sum**	7819	1705	6182	23826	1574
		Average	390.9	85.25	309.1	1191.3	78.7
Fiber Breakage	**Training (10 data)**	**Sum**	205	2241	1154	3910	783
		Average	20.5	224.1	115.4	391	78.3
	Testing (20data)	**Sum**	461	3772	7188	7241	1537
		Average	23.05	188.6	359.4	362.05	76.85

4. Result and discussion

In this tensile experiment, the AE signals were divided into three main groups through the waveform and frequency spectrum, that is, matrix crack, delaminations and fiber breakage. It is revealed that the rise time, duration, amplitude and energy of matrix crack signals are low while those of delaminations are high, the rise time and duration of fiber breakage are low and amplitude and energy of fiber breakage are high. Rise time, counts, energy, duration and amplitude were set as the input of the neural network. Besides, 30 data of these AE signal features were chosen for each group of the AE signals, 10 for training and 20 for testing. Sum and average of these signals features of the three groups were shown in table 1.

After the training data were respectively trained for the BP neural network and SVM network, the testing data were put into the two neural networks to test the veracity of the two neural networks. Their classification results were shown in figures 3 and 4, respectively.

The details of the contrast between BP neural network and SVM network were shown in table 2.

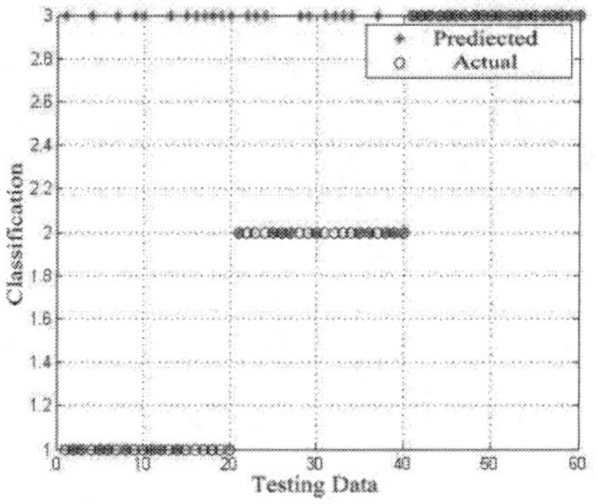

Figure 3. Classification of BP neural network.

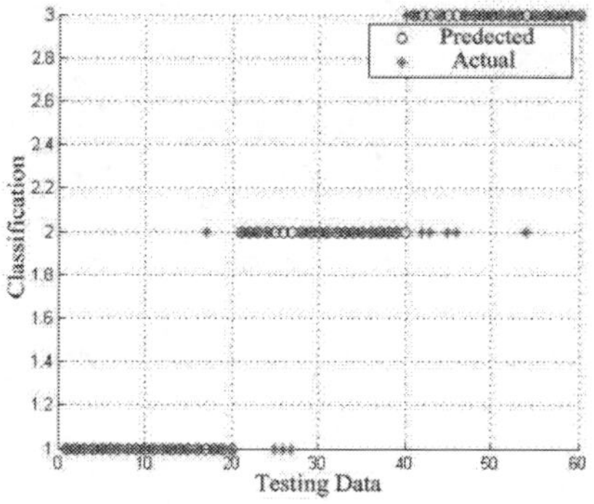

Figure 4. Classification of SVM network.

Table 2. Result of the neural network.

		Matrix Crack	Delaminations	Fiber Breakage	Total Accuracy
BP	**Accurate number**	8	10	20	63.33%
	Accurate rate	40%	50%	100%	
SVM	**Accurate number**	19	16	15	83.33%
	Accurate rate	95%	80%	75%	

It reveals that classification of the AE signals based on SVM network is more accurate than that based on BP neural network from figures 1 and 2 and table 1, except for classification of the fiber breakage signals. When the training data of SVM neural network increased to 15 from 10 and the testing data remained 15, the classification result was shown in figure 5.

It can be concluded that the classification result is more accurate when then training increases to 15 as demonstrated by figures 4 and 5. Table 3 shows the results contrast of the SVM network with 15 and 10 training data.

From figures 4 and 5 and table 3, it is revealed that when the training data increase the accuracy of the SVM network also increases dramatically. It can be concluded from table 3 that the accurate number and rate of matrix crack, delaminations, fiber breakage all increase much, especially the accurate rate of fiber breakage.

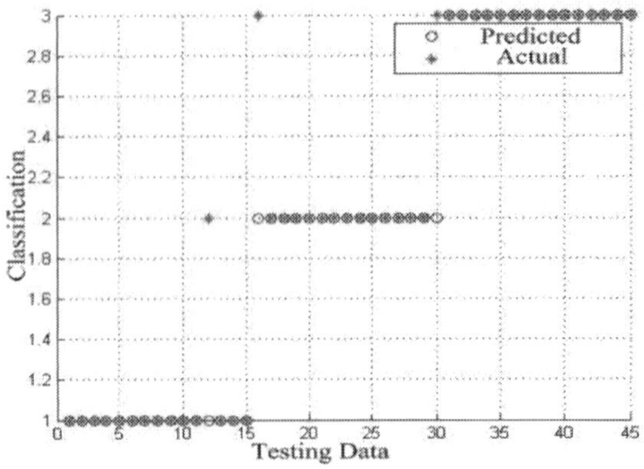

Figure 5 Classification of the SVM training data increased

Table 3. Result of the SVM with different training data.

		Matrix Crack	Delaminations	Fiber Breakage	Total Accuracy
15 Training data	Accurate number	14	13	15	93.33%
	Accurate rate	93.33%	86.67%	100%	
10 Training data	Accurate number	19	16	15	83.33%
	Accurate rate	95%	80%	75%	

5. Conclusion

It is concluded from this paper that the AE signal features and SVM network can be used for pattern recognition of the AE sources. The AE sources' pattern recognition based on AE signal features and SVM network is more accurate than that based on BP neural network. What's more, with the increase of the training data, the accuracy of pattern recognition based on SVM also increases. Therefore, adding the training data without impacting the arithmetic speed can improve the effect of SVM network on the AE signal sources' recognition.

Acknowledgments

This work is supported by the Graduate Innovation Base of Jiangxi Province. We also gratefully acknowledge the National Natural Science Foundation of China (Nos. 61062010) and Aviation Science Foundation (Nos. 2007ZF56013) for their support of this work.

References

[1] Edward Arnold 2003 *Acoustic Emission Evaluation of FRP Composite Specimens in Tension and Bonding* (West Virginia University)
[2] Subcommittee: E07.92. 2008 *Standard Terminology for Nondestructive Examinations* (Philadelphia: American Society for Testing and Materials)
[3] Christian U Grosse and Masayasu Ohtsu 2008 *Acoustic Emission Testing* (Berlin Heidelberg: Springer Publications)
[4] Shawn Allen Carey 2008 *Acoustic Emission Acousto-Ultrasonic Signature Analysis of Failure Mechanisms in Carbon Fiber Reinforced Polymer* (University of South Carolina)
[5] Adrian A. Pollock 2005 *User's Manual* (Physical Acoustics Corporation)
[6] Cortes C and Vapnik V 1995 Support Vector net-work *Machine Learning* **20** 273-9

[7] Zhao Qun and Prncipe J C 2001 Support vector machines for SAR automatic target recognition *IEEE Transactions on Aerospace and Electronic Systems* **37** 643-5
[8] Han Liqun 2008 Recognition of the part of growth of flue-cured tobacco leaves based on support vector machine *7th World Congress on Intelligent Control and Automation* (Chong Qing, China) pp 3624-7

Design of metal matrix composite with particle reinforcement produced by deep cryogenic treatment

Seyed Ebrahim Vahdat[1, 3] **and Keyvan Seyedi Niaki**[2]

[1]Department of Engineering, Ayatollah Amoli Branch, Islamic Azad University, Amol, Iran
[2]Department of Mechanical Engineering, Iranian Research Organization for Science and Technology (IROST), Tehran, Iran

E-mail: e.vahdat@iauamol.ac.ir

Abstract. One mechanism for strengthening and increasing toughness of composites is through debonding mechanism. Interface strength between reinforcement particles and matrix and the effective surface of debonding greatly affect toughness and strength of the composites. In this study, a model was proposed to estimate the effect of interface and matrix strength of composites on increasing tensile toughness and strength. Then, interface strength and its effect on increasing tensile toughness and strength were calculated in a case study on composites containing particle reinforcement in matrix of deep cryogenic treated steel. Using the methods for estimating strength and toughness of composites reduces trial-and-error time in the design process.

1. Introduction

High-performance steels possess high strength, hardness, abrasion resistance and toughness along with reasonable price. In general, hardness and strength are almost inversely related to toughness, which limits the application of steels. In order to overcome this problem, engineers have attempted to produce composites with hardness and abrasion resistant reinforcements. Factors such as adequate strength for the interface between reinforcement and matrix have challenged the successful production of this type of material. Furthermore, uniformity of properties in these materials is a requirement for using very fine reinforcements that are usually finer than one micron and are highly dispersed, which leads to problems like agglomeration of reinforcement [1]. Similar to behavior of composites, scanning electron microscopy (SEM) image of debonding $M_{23}C_6$ particles from matrix of 45WCrV7 [2, 3] and AISI D2 [4] steels, and that of breaking down M_7C_3 particles have been observed [4]. In addition, in force-displacement (F-ΔL) curve of deep cryogenic treated 1.2542 steel, the elastic part contains several slopes [5-7].

Using the methods for calculating strength and toughness of composites, trial-and-error rate was reduced in experimental production and designing these types of materials. Many models have been presented for estimating strength and toughness of composites with metal matrix and particle reinforcement. For instance, for the composite with aluminum matrix reinforced with 10 and 20% volume of alumina in similar conditions, fracture toughness was reported to be in proportion to the distance between the particles [8]. In other words, in a particular composite with similar interface in

[3] Address for correspondence: Seyed Ebrahim Vahdat, Department of Engineering, Ayatollah Amoli Branch, Islamic Azad University, Amol, Iran. Email: e.vahdat@iauamol.ac.ir.

which volume fraction of particles is constant, the finer the particles are , the less the distance between the particles will be; therefore, fracture toughness ratio would be reduced. Nardone and Prewo estimated strength of composites with particle reinforcement by proposing an improved shear-lag model [9]. Shen et al. used finite element methods for composites with particle reinforcement, demonstrating that particle form had no significant impact on its tensile strength. They indicated that, in a composite with aluminum matrix (3.5% age hardened copper containing 20% volume of reinforcement) in similar conditions, interface strength decreased with cylindrical, spherical, defective cylindrical and two truncated cones reinforcement, respectively [10]. In Hahn and Rosenfield's model, fracture toughness was directly related to the size of particles, Young's modulus and yield strength of composite and had inverse relationship with volume percent of particles [11]. In Garrett and Knott's model, fracture toughness was directly related to work hardening, Young's modulus and yield strength of composite [11].

The above-mentioned models [8-11] are not based on microstructure, i.e., these models ignore the effect of particles population density and interface strength on load transfer. Therefore, interface strength between reinforcement particles and matrix and also effective surface of debonding particles are the research variables in the present work. By determining interface strength between reinforcement particles and matrix in certain operating conditions and controlling population density, size and content of reinforcement, desirable toughness and strength could be designed.

The focus of this study was the interface strength between reinforcement of $M_{23}C_6$ particles and matrix of AISI S1 steel so that this method could be generalized. Moreover, in case that debonding mechanism was active, the effect of this mechanism on increasing strength and tensile toughness was also investigated.

2. Materials and methods

Figure 1 presents a model for strength and tensile toughness of the composites with particle reinforcement in which debonding mechanism is dominant. If particles are broken down, the maximum effective surface will exceed the particle's diameter, which is equal to the circle area in spherical particles (see figure 1(a)). According to figure 1(b), if particles debond from the matrix (without being broken down), effective surface of the particle will be equal to its perimeter, and to the surface of the sphere in spherical particles. In other words, when debonding mechanism is dominant, effective surface of particles will increase for strengthening because circle area (πr^2) of each spherical particle is smaller than sphere surface ($4\pi r^2$).

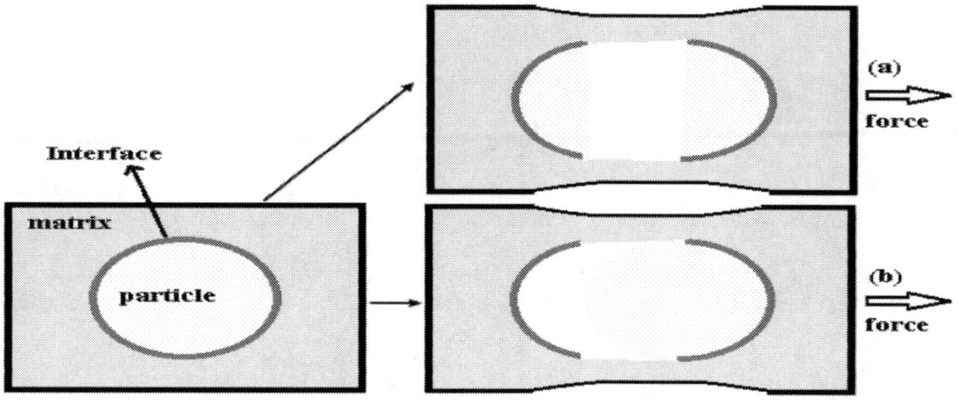

Figure 1. Effective surface for strengthening (a) when particles are broken down (b) when particles are debonding.

Interface has less strength; therefore, as long as particles debond from the matrix, the matrix along with debonding mechanism of particles plays a role in strengthening. Therefore, strength of composite is obtained using equations (1), (2) and (3). When the particles debond from the matrix, it will only resist until the composite is broken down.

$$A_{CO} = A_{P,ef} + A_{M,ef} \rightarrow A_{M,ef} = A_{CO} - PD \times \pi r^2 \quad (1)$$

$$\sigma_{CO,UTS} \times A_{CO} = (\sigma_{P,PO} \times A_{P,PO}) + (\sigma_{M,UTS} \times A_{M,ef}) \quad (2)$$

$$A_{P,PO} = PD \times 4\pi r^2 \quad (3)$$

By substituting equations (1) and (3) in equation (2), equation (4) can be simplified as.

$$\sigma_{CO,UTS} = (\sigma_{P,PO} \times PD \times 4\pi r^2) + (\sigma_{M,UTS} \times (A_{CO} - PD \times \pi r^2)) \quad (4)$$

Toughness is the amount of work per volume unit of material before rupturing. Considering equation (5), it could be regarded as equivalent to the product of multiplying the force required for deformation by the length of pathway through which the particle is drawn in the matrix. If force is in Newton ($Force = \sigma \times A$) and the pathway is in meter, tensile toughness unit will be in Joule. Note that, length of the pathway through which the particle is drawn in the matrix ($2\pi r$) is in proportion to particle size.

$$U_T \equiv (\sigma_{P,PO} \times A_{P,PO} \times 2\pi r) + (\sigma_{M,UTS} \times A_{M,ef} \times \Delta L_{CO}) \quad (5)$$

Table 1. Chemical composition of AISI S1.

Element	(W%)	Element	(W%)	Element	(W%)	Element	(W%)	Element	(W%)	Element	(W%)
C	0.4800	Ni	0.1280	S	0.0250	W	1.5700	Mo	0.0281	Fe	Rest
Si	0.9950	V	0.0148	Cr	1.1200	Mn	0.3360	P	0.0567		

Table 2. Content, average size and population density of M_7C_3 and $M_{23}C_6$ particles for AISI S1.

Code	M_7C_3 content	$M_{23}C_6$ content	M_7C_3 average size μm	$M_{23}C_6$ average size μm	population density of M_7C_3 mm^{-2}	population density of $M_{23}C_6$ mm^{-2}
241	0.42V%	2.18V%	0.5 (0.3 to 1)	0.22 (0.065to0.5)	62000	660000
242	0.47V%	2.42V%	0.55 (0.3 to 1)	0.23 (0.065to0.5)	65000	630000
243	0.37V%	3.73V%	0.7 (0.6 to 0.9)	0.28 (0.065to0.7)	60000	600000
361	0.57V%	4.69V%	0.65 (0.5 to 0.8)	0.30 (0.065to1)	64000	894000
362	0.60V%	6.92V%	0.7 (0.4 to 1.7)	0.35 (0.065to0.7)	63000	750000
363	0.34V%	8.91V%	0.7 (0.4 to 1.5)	0.40 (0.065to0.6)	62000	726000
481	0.35V%	10.04V%	0.7 (0.3 to 1.4)	0.24 (0.065to0.7)	65000	707000
482	0.25V%	12.66V%	0.7 (0.4 to 1.4)	0.5 (0.065to1)	62000	650000
483	0.24V%	12.87V%	0.7 (0.4 to 2)	0.52 (0.065to1)	65000	620000

By substituting equations (1) and (3) in equation (5), equation (6) is simplified for calculating tensile toughness.

$$U_T \equiv (\sigma_{P,PO} \times PD \times 8\pi^2 r^3) + (\sigma_{M,UTS} \times (A_{CO} - PD \times \pi r^2) \times \Delta L_{CO}) \quad (6)$$

In equal conditions, tensile toughness of composites described above is higher than that of composites shown in figure 1(a). This difference increases with the increasing effective surface area of the debonding particles, surface area of interface ($A_{P,PO}$) and interface strength ($\sigma_{P,PO}$).
The first terms in equation (6) ($\sigma_{P,PO} \times PD \times 8\pi^2 r^3$) and equation (4) ($\sigma_{P,PO} \times PD \times 4\pi r^2$) are in fact effects of reinforcement debonding mechanism on increasing tensile toughness and strength, respectively.

In this study, AISI S1 steel was utilized to calculate interface strength of reinforcement of $M_{23}C_6$ particles in matrix of steel. Accordingly, presenting a model for designing toughness and strength of AISI S1 steel in impact loading conditions is important in practical terms. Its chemical composition is listed in table 1.

To calculate interface strength, 9 sets of specimens, each of which included three specimens, were used for providing sufficient data for desirable conclusion and discussion. In order to determine microstructure characteristics, cylindrical specimens were 12 mm in diameter and 15 mm in length. TESCAN MIRA II device with EDS was used to obtain SEM images. In addition, OLYSIA m3 metallographic software which was calibrated for 2048 × 1536 pixels was utilized. For calculating each phase, at least 5 SEM images with magnification of 10^4 from one region were used. Mean of the data is reported in table 2.

Table 3. Results of tensile test at room temperature for AISI S1.

Specimen code	$\sigma_{CO,UTS}$ MPa	ΔL_{CO} mm	U_T (MJ)= $2/3\times\sigma_{M,UTS}\times\Delta L_{CO}/50\times100$
241	2279±21	2.4±0.37	72.2
242	2265±31	1±0.5	30.2
243	2137±53	3±0.75	85.5
361	2268±65	3.5±0.75	*105.9*
362	2201±65	2.5±0.75	73.4
363	2245±65	2.5±0.75	74.8
481	2244±64	3.1±0.15	*92.8*
482	2206±65	3.8±0.75	*110.3*
483	2249±28	3.1±0.4	*93.0*

In accordance with BS EN 10002-1 standard [12], specimens of tensile test were prepared in dumbbell shape with diameter (d) of 5 mm, initial base length (L) of 25 mm and total length (L_t) of 15 cm; results are shown in table 3. Tensile test was performed at a strain rate of 0.00166 s^{-1}. Machining of the specimens was implemented using a CNC milling device.

3. Results and discussion

Figure 2 shows SEM images of the composition, morphology and size of carbides. Figure 3 shows SEM image of the necking area of specimen 482 using secondary electron gun. It shows pull-out of carbide of 0.96 μm in diameter that produced micro-void during tensile test because of carbide pull-out.

M_7C_3 particles were broken down while $M_{23}C_6$ particles were debonding; therefore, effects of M_7C_3 particles on tensile toughness and strength based on debonding mechanism were neglected. Additionally, volume fraction of $M_{23}C_6$ particles was much greater than that of M_7C_3 particles (see table 2). On the other hand, condition of figure 1(a) never happened. In condition of figure 1(b), four factors contributed to increased tensile toughness and strength of the studied composite, which included matrix strength, interface strength, matrix surface and interface surface. The amount of debonding surface with a role in strengthening was calculated for $M_{23}C_6$ particles using equation (3) per square meter of composite surface ($A_{CO}=1m^2$), as listed in table 4.

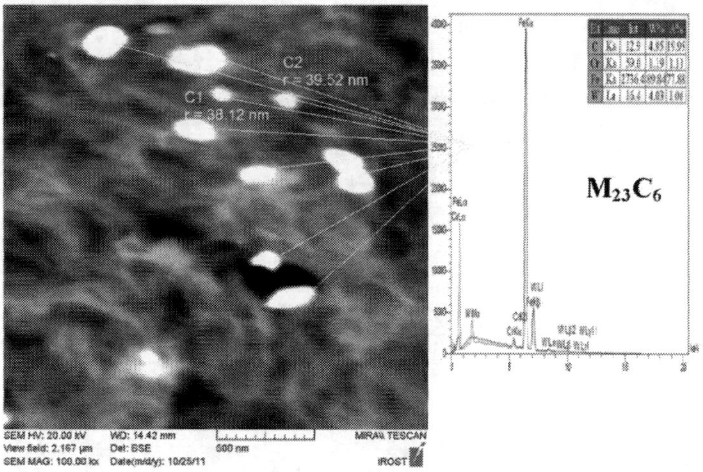

Figure 2. Composition, morphology and size of $M_{23}C_6$ carbides.

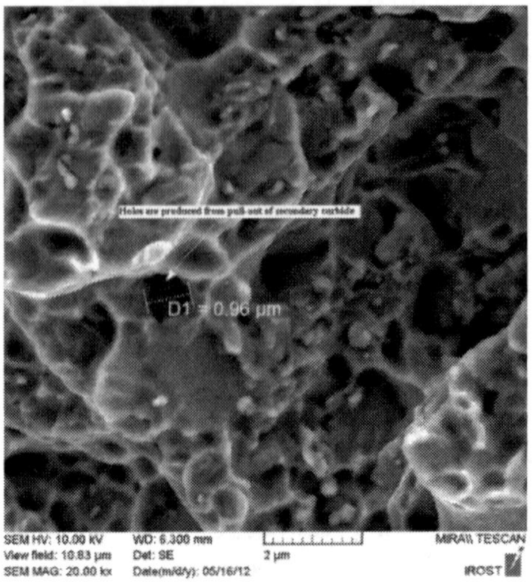

Figure 3. SEM image of the necking area of specimen 482, ×20000, after ultrasonic cleaning.

Effective surface of the matrix was obtained by subtracting composite surface (1 m^2) from total surface of particles, equation (1), as given in table 4. Accordingly, values of two factors were determined. According to the data in tables 2 and 3, solving equations (4) and (6) in a system of two equations with two unknowns led to determining two other factors of matrix strength ($\sigma_{M,UTS}$) and interface strength ($\sigma_{P,PO}$) based on equations (7) and (8), the values of which are listed in table 4.

$$\sigma_{P,PO} = (\sigma_{CO,UTS} - \sigma_{M,UTS} \times (1 - PD \times \pi r^2)) / (PD \times 4\pi r^2) \quad (7)$$

$$\sigma_{M,UTS} = ((U_T/14.7) - 2\pi r \times \sigma_{CO,UTS}) / ((\Delta L_{CO} - 2\pi r) \times (1 - PD \times \pi r^2)) \quad (8)$$

Table 4. Debonding surface, matrix surface, strength of matrix, strength of interface and effect of debonding mechanism on increasing strength and tensile toughness.

Code	$A_{P,PO}$ mm^2	$A_{M,ef}$ mm^2	Matrix strength	Interface strength	Effect of reinforcement debonding mechanism in increasing tensile toughness, the first term of equation (6) ($\sigma_{P,PO} \times PD \times 8\pi^2 r^3$)	Effect of reinforcement debonding mechanism in increasing strength, the first term of equation (4) ($\sigma_{P,PO} \times PD \times 4\pi r^2$)
241	100304	974924	2099	2317	--------------	--------------
242	104647	973838	2109	2012	2238J (0.007%)	211MPa (9.3%)
243	147706	963074	2012	1344	2570J (0.003%)	199MPa (9.3%)
361	252644	936839	2195	834	2922J (0.003%)	211MPa (9.3%)
362	288488	927878	2151	709	3309J (0.005%)	205MPa (9.3%)
363	364742	908814	2240	572	3857J (0.005%)	209MPa (9.3%)
481	429788	892553	2280	485	4241J (0.005%)	209MPa (9.3%)
482	510250	872438	2293	402	4738J (0.004%)	205MPa (9.3%)
483	526415	868396	2349	397	5024J (0.005%)	209MPa (9.3%)

Effect of debonding mechanism of $M_{23}C_6$ particles in matrix of AISI S1 steel on increasing tensile toughness, i.e. the first term of equation (6), was calculated (see table 4), which was negligible since particle reinforcements had small effective surface for tensile toughness ($PD \times 4\pi r^2 \times 2\pi r$) whereas fiber reinforcements had much larger effective surface for tensile toughness ($PD \times 2\pi r L/2 \times L/2$). For example, in equal conditions, if fiber length was at least 300 times of fiber radius ($L=300 \times r$), then effective surface for tensile toughness would be 1800 ($\approx 300^2/16\pi$). However, effects of debonding mechanism of $M_{23}C_6$ particles in matrix of AISI S1 steel on increasing strength, i.e. the first term in equation (4) (in accordance to table 4), would be 9.3%, which was considerable.

As presented in figure 2, in all the specimens except 241, the strength of interface was less than that of the matrix. This issue provided the field for debonding particles; i.e. it was in agreement with the initial assumption that governed the condition of figure 1(b). On the other hand, interface strength was being reduced because, size of $M_{23}C_6$ particles became larger (according to table 3); larger particles decreased coherency of interface, which reduced strength of the interface.

According to table 4, effect of debonding mechanism on increasing strength was the same (9.3%) for all the specimens since (as demonstrated in table 3) all of them had almost equal strength.

According to figure 4, interface strength (from 2012 to 397 MPa) was smaller than matrix strength (from 2012 to 2349 MPa) in all the specimens, except specimen 241. Therefore, conditions were appropriate for activating debonding mechanism. Moreover, according to table 2, in the studied steel, debonding mechanism of the reinforcement particles had a negligible effect (0.003 to 0.007%) on increasing tensile toughness; this effect was considerable in the case of tensile strength (9.3%).

According to table 2, the amount of secondary carbide constantly increased. Thus, the matrix metal around the carbide had poor carbon and alloying elements. Accordingly, the matrix which was poor in carbon and alloying elements could have an effective role in increasing tensile toughness.

For strength, the above finding was confirmed in figure 4, where strength of metal matrix is almost constant. Therefore, strengthening of debonding mechanism had significant effects on strength of composite.

This method can be generalized to the composite with fiber reinforcement. In such a state, equations (7) and (8) can be changed as equations (9) and (10).

$$\sigma_{fiber,PO} = (\sigma_{CO,UTS} - \sigma_{M,UTS} \times (1 - PD \times \pi r^2)) / (PD \times \pi DL/2) \quad (9)$$

$$\sigma_{M,UTS} = ((U_T/K) - L/2 \times \sigma_{CO,UTS}) / ((\Delta L_{CO} - L/2) \times (1 - PD \times \pi r^2)) \quad (10)$$

Whereas D is fiber diameter, L is fiber length and K is equality coefficient of equation (6).

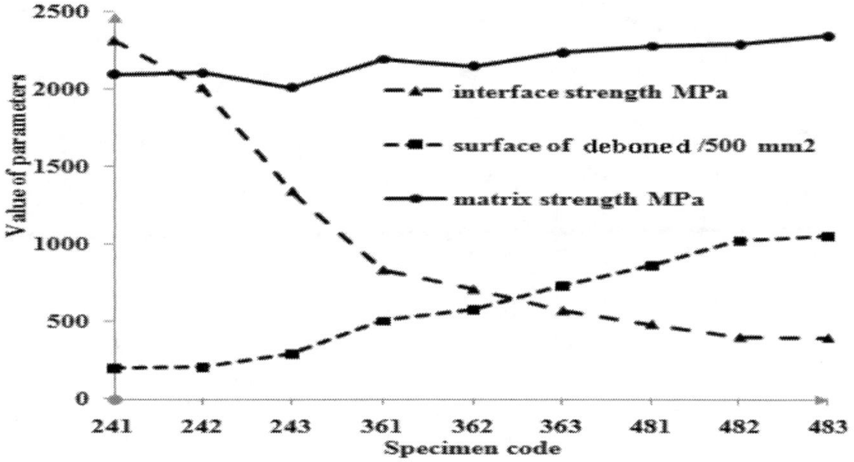

Figure 4. Comparing debonding surface, matrix strength and interface strength for 9 different specimens.

4. Conclusion

In this study, a method is proposed for designing tensile toughness and strength of composites based on reinforcement debonding mechanism. In this method, simultaneous increase of tensile strength and tensile toughness occurs when interface strength is less than matrix strength and reinforcement strength. This method is utilized for 9 specimen sets of AISI S1 steel with particle reinforcement. Considering mean diameter of spherical particles, it can be concluded that:

- Debonding mechanism has a negligible effect on increasing tensile toughness of composite with particle reinforcement.
- Debonding mechanism has a considerable effect on increasing tensile strength of composite with particle reinforcement.

References

[1] Cory A S 2000 Discontinuous reinforcement for metal-matrix composites *Composites* **21** 131
[2] Vahdat S E, Nategh S and Mirdamadi Sh 2013 *Mater. Sci. Eng. A* **585** 444
[3] Vahdat S E 2014 *Int. J. Steel Struc.* **14** 571
[4] Das D and Ray K K 2012 *Mater. Sci., Eng. A* **541** 45
[5] Farhani F, Niaki K S, Vahdat S E and Firouzi A 2012 *Mater. Des.* **42** 279
[6] Vahdat S E, Pournaghi A, Mohammadynia A and Hajili M 2014 *Procedia Mater. Sci.* **6** 202
[7] Vahdat S E, Nategh S and Mirdamadi Sh 2014 *Int. J. Preci. Eng. Manu.* **15** 497
[8] Chawla N and Allison J E 2001 *Fatigue of Particle Reinforced Materials* In *Encyclopedia of Materials: Science and Technology* 2nd Edi (Elsevier Oxford) **1** 2967
[9] Nardone V C and Prewo K M 1986 *Scripta Metallurgica* **20** 43
[10] Shen,Y L, et al. 2000 *Scripta Materialia* **42**427
[11] Bhaskar S and Majumdar S 2000 Engineering mechanics and analysis of metal matrix composite *Composites* **21** 975
[12] British Standard 10002–1 2001 *Metallic Materials Tensile Testing Part 1: Method of Test at Ambient Temperature* (UK: British Standards Institution)

IOP Conf. Series: Materials Science and Engineering **87** (2015) 012004 doi:10.1088/1757-899X/87/1/012004

Towards enhancing peel strength of adhesively bonded joints

W Ameen, S M Darwish and A AlSamhan

Industrial Engineering Department, Advanced Manufacturing Institute King Saud University, Saudi Arabia

E-mail: Darwish@ksu.edu.sa

Abstract. The evaluation of strength as well as failure behavior of adhesively bonded structures is essential to further improvement of their applicability. Since adhesively bonded structures are very poor in resisting peel loading, designers of these joints usually avoid subjecting these structures to peel loadings. However, in some circumstances, designers fail to avoid peel stressing in their designed structures. The aim of the present work is to enhance peel resistance of adhesively bonded structures in order to confront inevitable situations that bonded structures are subject to peel loading, without adding extra weight to the design.

1. Introduction

It is found that strength and failure behavior of adhesively bonded structures depend much on mechanical properties of adhesive material and stress concentration in bonded structures [1]. Usually, adhesive manufacturers offer a wide range of adhesives with variable mechanical properties. However, the selection of an adhesive depends on the materials to be bonded (adhered), surface treatment methods, bonded structures, working environment and types/magnitude of applied loads. Moreover, stress distribution plays a significant role in achieving a strong adhesive bonded structure.

Peel testing allows users to understand the strength of adhesive structures and evaluate joint strength, such as maximum and minimum peeling forces and average peel force. Peel resistance, which refers to the required load to peel bonded adherent parts, can be calculated as average peel load per bond line width dimension [2]. According to ASTM D1876-01 Standard Test Method [2], average peeling force per unit width is issued to measure the peel strength. The average peeling is taken from the force-displacement diagram after the initial peak reading. It has been found that adding fillers to the adhesive mastics results in a change in the composite material properties (mechanical, physical and chemical properties). Nowadays, Nano fillers become more attractive to many researchers, due to its great improvement in composite material properties compared with other fillers due to their small size and large surface area. Nano-reinforcements are able to provide unique combination of properties, which cannot be reached for conventional fillers in micrometer range size. Hence, it is possible to design the adhesive in the Nano scale by adding nanoparticles such as nanoclays, carbon nanotubes (CNT) or carbon nano fibers (CNF) and others in the adhesive matrix [3]. Due to the large surface area of the nano-sized particles, only a small amount can lead to significant changes in the resulted properties of the nano-composite adhesives [3]. The effect of adding nanoclay to epoxy adhesive on shear strength was investigated, using single lap joints. Different percentages (1%, 3% and 5%) of nanoclay filled epoxy adhesive were tested, under both static (tensile) and dynamic (impact) loadings. The results showed that the adhesive joints with 1% nanoclay particles had the maximum shear loadings. It was found 7% strength enhancement compared with plain adhesive results [4].

Experiments were conducted to study the enhancement of joint strength by adding zirconia nanoparticles to epoxy matrix [5]. Three percentages of nanoparticles are considered, i.e., 0.5%, 1% and 1.5%. The tests showed that the mechanical properties improved with the percentage of zirconia nanoparticles increasing up to 1% and declined with the percentage further increasing to 1.5%. Similar studies were carried out for the strength enhancement by adding alumina nanoparticles to epoxy adhesives [6] using four percentage levels (0.5%, 1%, 1.5% and 2%). The results found that the alumina nanoparticles improved the shear strength of the bonded joints by 60%, at optimal filler percentage of 0.5%. In order to investigate the effect of adding carbon nanotubes on the strength of epoxy resins, different studies have been conducted. Bonded similar substrates of carbon/carbon (C/C) and carbon/carbon–siliconcarbide (C/C–SiC) composites using pure epoxy resin and 3% filled adhesive of multiwall carbon nanotubes (MWCNTs) were used to study the strength enhancement of overlap joint. The experimental results showed that the MWCNT filled epoxy resin bonded C/C–C/C andC/C–SiC–substrates have a higher strength than those bonded with plain epoxy adhesive. Furthermore, it has been reported that the overlap joints having dissimilar adherents (C/C–SiC) provide higher bonding strength compared with similar adherent substrates (C/C) for both plain and MWCNTs filled adhesives, since particles loaded composite materials increase the toughness and resist the formation ofcracks [7]. The effect of MWCNTs filled adhesive Epoxy 862 with four levels of percentages on joint strength is also studied. The four levels of percentage are 0.1, 0.2, 0.3 and 0.4%. From the overall experimental results, it is clearly observed that the joint strength reaches the maximum with 0.3% of MWCNTs with 26% strength enhancement [8].

According to the above literature, it was observed that most of the research was carried out for enhancing only the shear resistance of single overlap joints. For this reason, the current study will focus on enhancing T-peel strength using MWCNTs filled adhesive, in order to prepare adhesively bonded joints to cope with inevitable peel loadings.

2. Strength of adhesively bonded T-peel joints

Several research works have been conducted to investigate the effect of T-peel joint geometry (e.g. adhesive thickness, adherend thickness, flange radius, and spew fillet) on enhancing the peel strength. Matsui [9] compared the peel strength of T-peel joint with different bonding line adhesive thickness. The results pronounced an increase in strength of the peel joint with the increasing bond line thickness (adhesive layer thickness).

Numerically the effect of joint geometry and properties (adhesive fillet, adherend thickness, flange radius and adherent material properties) on different Von Misses stress distribution within the adhesive layer were considered in the study [10]. The parametric investigation showed that Von Misses, peel and axial stress concentration increase with the decrease of adherend thickness, adhesive fillet and adherend material stiffness. In addition, it has been found that the effect of flange radius is less pronounced [10].

The finite-element analysis was conducted to evaluate the effect of adherend thickness dissimilarity on the peel strength of spot-welded and weld-bonded T-peel joint. The finite element (FE) model assumes isotropic adhesive materials, cohesive failure mode, while neglecting the adhesive contact boundary condition. The results show that dissimilarity of thickness magnifies the stress concentration located at the boundaries of the weld nugget. Furthermore, the stress concentration located in the thinner adherends of the joint is higher than that in the thicker adherends. The introduction of an adhesive layer between dissimilar thickness joints, with the spot weld nugget results in the improvement of the peel strength of spot-welded through reducing the associated level of stress concentration (by nearly 84%) [11].

Numerical methods were also used to study the effect of spew fillet ratio ($(f/RO)\times 100$) (0, 25, 50, 75 and 100%, on the T-peel joint stiffness. The results show that the spew fillet parameter of T-peel joint has a significant influence on the T-peel stiffness. For example, it has been found that the T-peel stiffness increases by 11 times for 100% spew fillet ratio compared with non- spew fillet model using modified shell solid element [12]. Based on the literature review, the following conclusions can be drawn:

- Peel resistance of the bonded and weld-bonded joint is very low.
- Little attention has been paid to strengthening the peel strength for bonded joints.
- Mixing nano particles or nano carbon tubes with plain adhesive results in the enhancement of shear strength of bonded joint.

The current research focuses on the enhancement of peel strength of adhesively bonded 2024 aluminum alloy joints. The technique covers introducing carbon nano-tubes to plain adhesive for performance enhancement of T-peel bonded joints.

3. Experimental work

3.1. Tee peel specimens

The peel specimens of the present work were manufactured according to ASTM [D1876-01], as shown in figure 1.

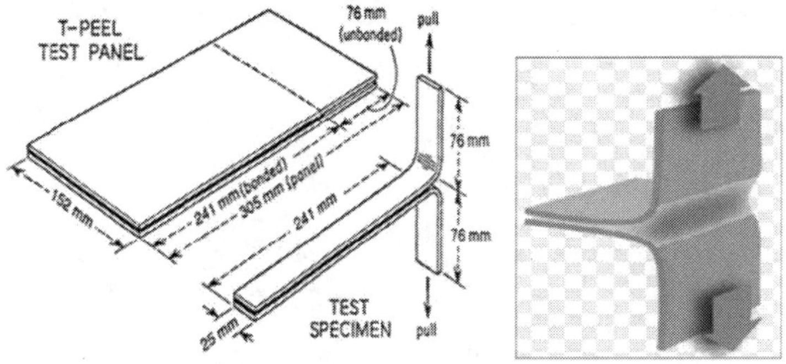

Figure 1. T-peel specimen ASTM [D 1876-01].

3.2. Peel joint substrates preparation

Aluminum alloy of grade 2024-T3 was used for the experimental study. This alloy provides high ratio of the strength to weight and good fatigue performance, so it is often used for aircraft body structure especially fuselage and wing structures [13]. The chemical composition of the specimens is obtained using spectra machine as listed in table 1.

Table 1. Chemical composition of aluminum alloy (2024-T3).

Composition of aluminum alloy(2024-T3)					
Component	Al	Cu	Mg	Fe	Other
Wt.%	93.1	4.45	1.45	0.136	0.864

The T-peel joint adherents were manufactured with thickness of 1.6mm and flange radius of 6mm [1].

3.3. Plain adhesive bonded specimen preparation

The specimens having unfilled adhesive were prepared by washing the bonding contact surfaces with distilled water, and then polished with 200 and 400 grade abrasive papers, and finally wiped by an acetone agent to remove any pollutants [14, 15]. Epoxy adhesive (type Araldite 2011) was selected for the current study based upon its performance [15-17]. The adhesive was prepared by manually mixing the resin and hardener with weight ratio of 100:80, respectively [15]. In order to ensure a uniform thickness (0.3 mm) of applied adhesive layer, a special fixture (adhesive applying fixture, see figure 2) was designed and manufactured using CNC milling machine.

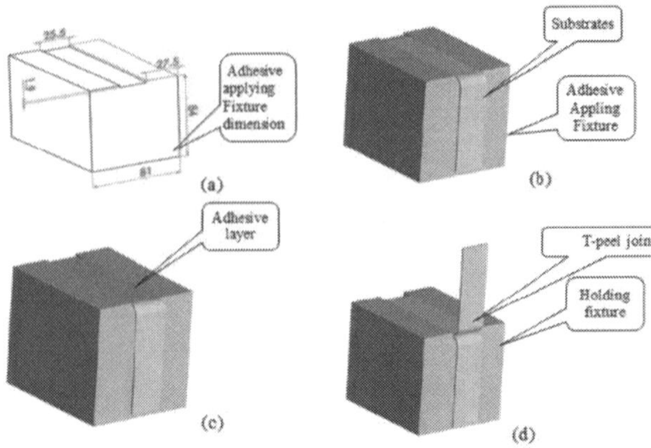

Figure 2. Adhesive applying fixture.

After applying the adhesive on the surfaces of the metal substrates, the specimens were held in the suitable position until the adhesive bonded joint was safe for handling (2 hours) using holding fixture.

Finally, the specimens were transferred to the oven for curing purpose. The curing temperature and time were set according to the manufacturer recommendations, that is, 120 ℃ and 60 minutes, respectively [15, 17], and figure 3 shows the prepared T-peel joint.

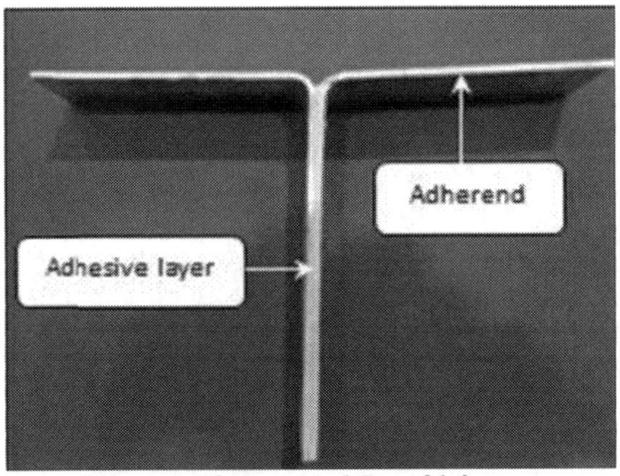

Figure 3. The prepared T-peel joint.

3.4. MWCNT filled adhesive specimen preparations

In the present work, MWCNT was used to enhance the strength of T-peel bonded joints with varying percentages (0, 0.1, 0.4, 0.5, 0.6 and 1%) of MWCNT adhesive, and tensile test facility was utilized to prepare and test them in order to allocate the optimum percent of MWCNT adhesive that ensures the highest joint strength. Tensile test was carried out for Aluminum alloy 2024-T3, specimen having plain epoxy adhesive and MWCNT filled adhesive in order to obtain the mechanical properties.

MWCNTs were obtained from Hanwha Nanotech Corporation with average diameter of 9.5 nm and length of 1.5 μm of 99% purity. Different percentages of MWCNTs were respectively mixed with adhesives to obtain the optimum strength enhancement of T-peel joint. Figure 4 shows the scanning electron microscope (SEM) image of MWCNTs.

The adhesive resin is mixed with MWCNT, for 5 minutes using electric shock device in order to ensure uniform distribution of the MWCNTs on the surface of resin. Furthermore, to achieve a homogenous mixture, a magnetic stirrer was used to steer the mixture for 10 hours, at 200 rpm at a

IOP Conf. Series: Materials Science and Engineering **87** (2015) 012004 doi:10.1088/1757-899X/87/1/012004

temperature of 50℃. It is worth noting that the mixing procedure with MWCNTs was established when adhesive temperature increased to 50℃.

The hardener resin was added to it and then mixed manually to obtain the required adhesive mixture to be used for the T-peel joints. Optical microscope was used to inspect the homogeneity of MWCNTs distribution on adhesive (see figure 5).

Based upon the literature review [4, 8], MWCNT mixing percentages were selected as follows: 0.1%, 0.4%, 0.5%, 0.6% and 1% of adhesive weight.

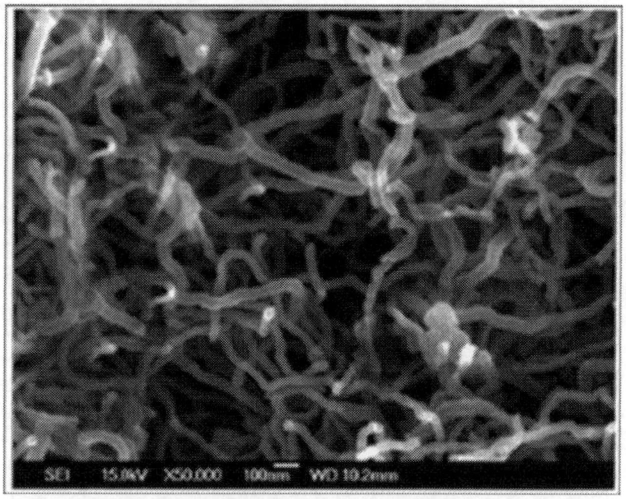

Figure 4. SEM image of the MWCNTs.

Figure 5. Optical image of MWCNTs filled adhesive.

3.5. Adherends and adhesive materials properties

The mechanical properties of both adherends and adhesive materials are evaluated in order to measure their specifications. The mechanical properties of sheet metal 2024 Aluminum were obtained through tensile testing according to ASTM standard E8-81 (cutting along rolling direction) in order to be tested on Instron Tensile test Machine (see figure 6 and table 2).

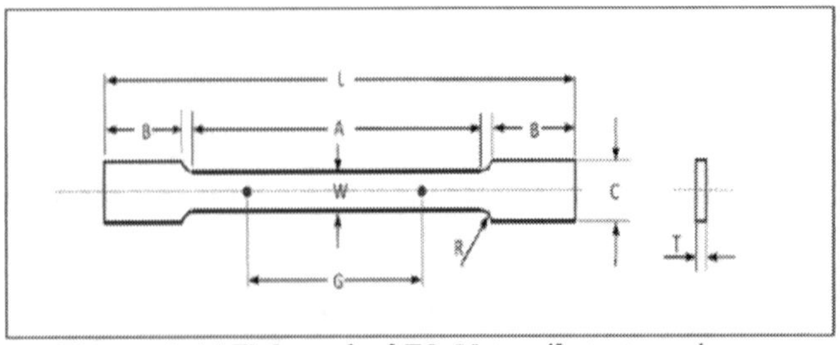

Figure 6. ASTM standard E8-81 tensile test specimens.

Table 2. Tensile test specimen dimensions.

Symbols	Description	Dimension
L	Overall length,	min 200
A	Length of reduced section,	min 57
G	Gage length	50±0.1
W	Width	12.5 6 0.2
B	Length ofgrip section,	min 50 [2]
C	Width of grip section	20 [0.750]
R	Radius of fillet,	min12.5 [0.500]
T	Thickness	thickness of material

Five standard specimens were prepared and tested to obtain the average mechanical properties values. The experimental stress strain curves and measured material properties are shown in figure 7 and table 3, respectively.

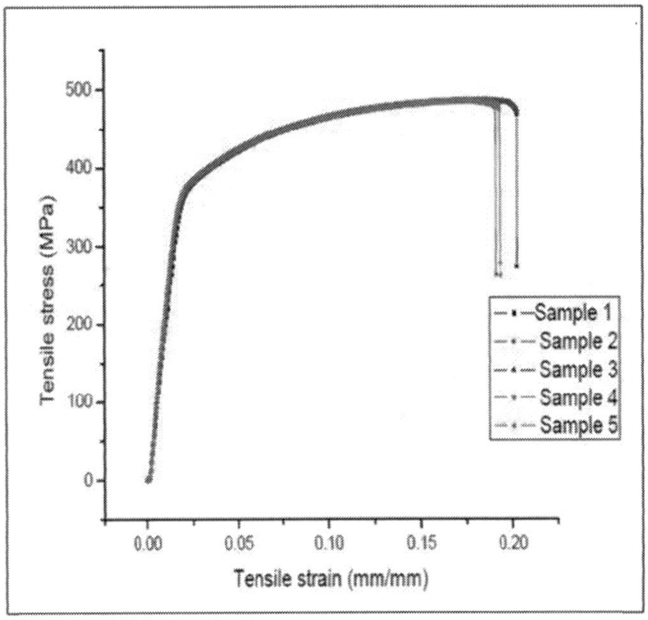

AA 2024-T3

Figure 7. The stress strain curves for five samples.

Table 3. Measured mechanical properties of aluminum alloy2024-T.

Material Code	Peak Load (KN)	E-Modulus (MPa)	Yield Point (MPa)	Ultimate Tensile Stress(MPa)	Mass Density g/cm3	Tangent Modulus MPa Et
AA2024-T3	9.72	73.100	358.16	486.12	2.78	830.0

3.6. Mechanical properties of plain and MWCNTs filled epoxy adhesive materials
In order to measure the properties of the used adhesive materials (Araldite 2011), standard ASTM D3763 tensile test specimens with overall dimensions (127×12.7×4 mm and gauge length 25 mm) were fabricated using specimen casting method, where a dumbbell-sharp cutter was used for manufacturing the mold cavity from wax (see figure 8).

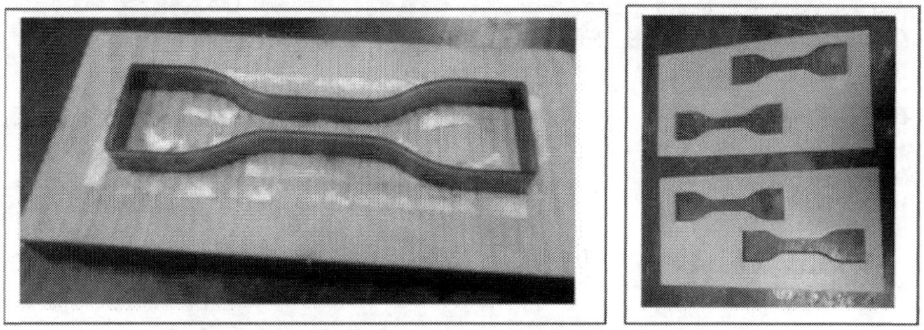

Figure 8. Dumbbell-sharp cutter metal tool and wax cavity.

The cast specimens were then placed in an oven in order to melt the wax and to achieve the curing time required by adhesive material. The cured specimens were then subject to tensile testing in order to obtain the mechanical properties of the adhesive; the results are shown in figure 9 and table 4.

Figure 10 shows the tensile test specimen for MWCNTs filled adhesive with 0.5% percent composition. The results obtained from Instron tensile machine are shown in figure 11 and table 5.

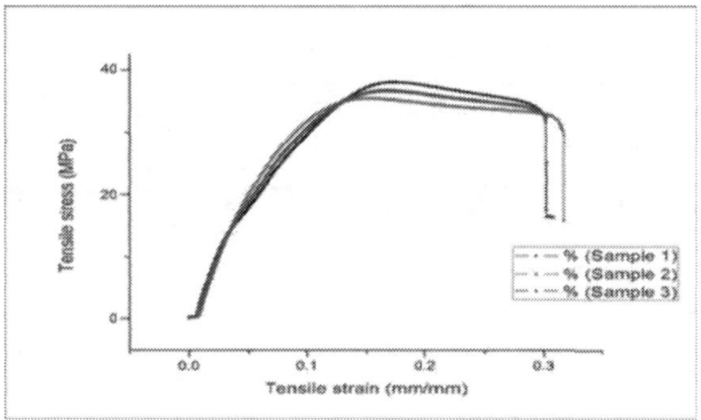

Figure 9. Stress strain curve of plain adhesive obtained from Instron machine.

Table 4. Mechanical properties of plain adhesive.

Peakload (KN)	E- Modulus (MPa)	Yield Point (MPa)	Ultimate Tensile Stress(MPa)	Mass Density g/cm^3	Tangent Modulus MPa E_t
1.16	437.5917	16	36.73	1.0350	188.2

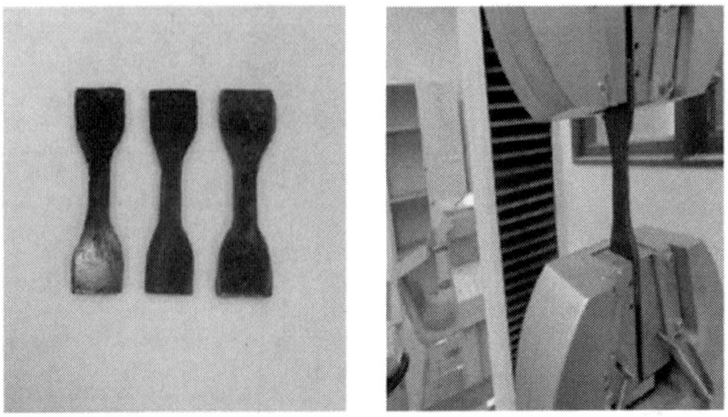

Figure 10. Tensile test specimen of MWCNTs filled adhesive (0.5% MWCNTs filled).

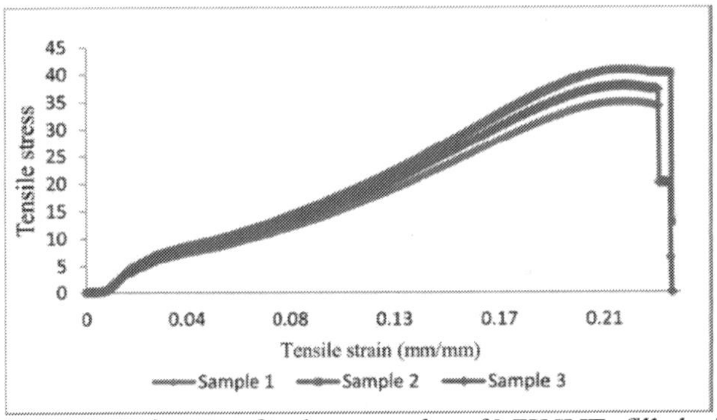

Figure 11. Stress strain curve for three samples of MWCNTs filled adhesives.

Table 5. Mechanical properties of MWCNTs filled adhesive (0.5%).

PeakLoad (KN)	E- Modulus (MPa)	Yield Point (MPa)	Ultimate Tensile Stress(MPa)	Mass Density g/cm^3	Tangent Modulus MPa Et
1.97	185.22	34.633	37.3	1.0309	88.66

The peak peel load increased from 1.16 KN to 1.97 KN in case of MWCNTs filled adhesive compared with plain adhesive results. Furthermore, it was reported that the yield strength also increased from 16.0 MPa in case of plain adhesive to 34.63 MPa in case of MWCNTs filled adhesive.

3.7. Strength measurement of the plain and MWCNT filled joints

To ensure the consistency of the results, tests were repeated three times, for each case, Instron Tensile testing machine (Series 3369H) with load capacity up to 50 KN was used for peel testing of all experiments. The tests were carried out at a crosshead speed of 5 mm/min as recommended [13]. Figure 12 shows the T-peel joint test setup on the Instron Machine and the experimental results analysis with provided software (Bluehill® 2).

Figure 13 shows the cohesive failure of T-peel bonded joint, while figure 14 shows the load displacement plots for three tests specimens.

4. Results and discussions

From figure 14, it can be observed that the average of 3 peak loads of the three samples is 300 N. The average peeling load is 52.3 N and the peel strength is 2.08 N/mm. (After being divided by bonding line width, is used to obtain the T-peel strength. The same procedures and preparation conditions were followed to evaluate MWCNTs filled adhesive T-peel bonded joints.

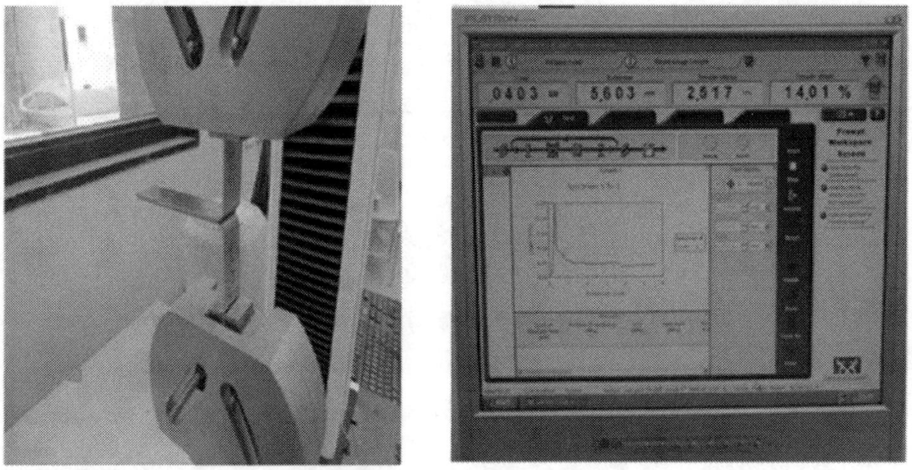

Figure 12. T-peel test setup.

Figure 13. Tested specimen after peel test.

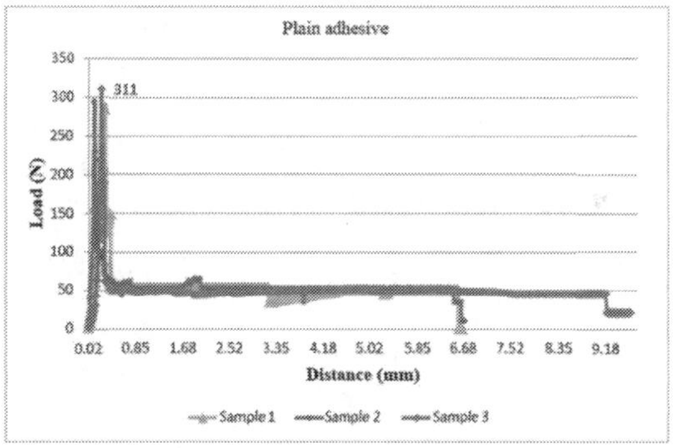

Figure 14. Load displacement curves for three plain adhesive bonded peel joints.

Table 6 shows the recorded values of the T-peel strength for unfilled and filled specimens with different MWCNTs percentages.

One-way ANOVA was used to show whether different MWNCTs percentages have significant effects on the peel strength or not. The data has been tested and successfully verified for normality as shown in figure 15. From ANOVA results (figure 16), it has been observed that MWNCTs percentages significantly affect peel strength (p-value = $0 < 0.05$).

Table 6. Peel strength T-peel joints with MWCNTs filled adhesive.

MWCNTs	**T-peel joint strength N/mm**		
Percentage (%)	**Sample 1**	**Sample 2**	**Sample 3**
0(Plain adhesive)	2.084	2.144	2.016
0.1	2.268	2.192	2.220
0.4	2.300	2.336	2.364
0.5	2.596	2.548	2.564
0.6	2.504	2.492	2.484
1	2.380	2.292	2.388

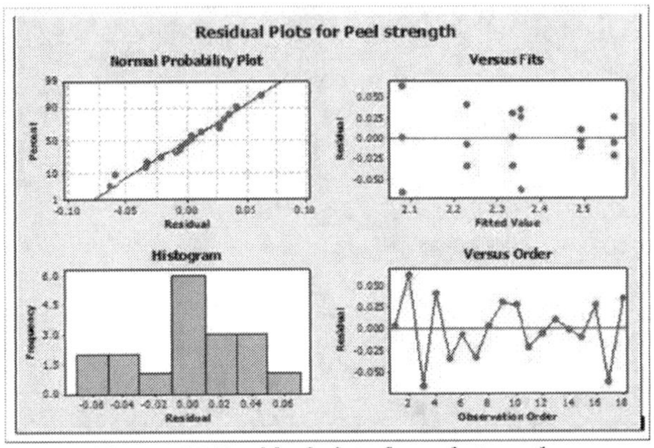

Figure 15. Residual plots forpeel strength.

```
One-way ANOVA: Peel strength versus MWCNTs percentage (%)

Source                DF        SS        MS      F      P
MWCNTs percentage (%)  5   0.46809   0.09362  55.37  0.000
Error                 12   0.02029   0.00169
Total                 17   0.48838

S = 0.04112   R-Sq = 95.85%   R-Sq(adj) = 94.11%

                              Individual 95% CIs For Mean Based on
                              Pooled StDev
Level  N    Mean   StDev  ------+---------+---------+---------+---
0.0    3  2.0813  0.0640  (---*--)
0.1    3  2.2267  0.0384            (--*---)
0.4    3  2.3333  0.0321                   (---*--)
0.5    3  2.5693  0.0244                                 (--*---)
0.6    3  2.4933  0.0101                            (--*---)
1.0    3  2.3533  0.0533                    (---*--)
                          ------+---------+---------+---------+---
                             2.10      2.25      2.40      2.55
```

Figure 16. ANOVA test results.

In order to determine the optimum percentage of MWNCTs that provides the highest peel strength, average strengths were calculated and plotted as shown in figure 17. It can be observed from this figure that T-peel strength increases with the increase of MWCNTs percentage up to 0.5%, and then decreases when the MWCNTs percentage further increases. It is worth mentioning that the trend obtained in the experimental results is identical to what was reported in literature [8].

It can be also observed that addition of 0.5% of MWCNTs has resulted in 23% increase in peel strength.

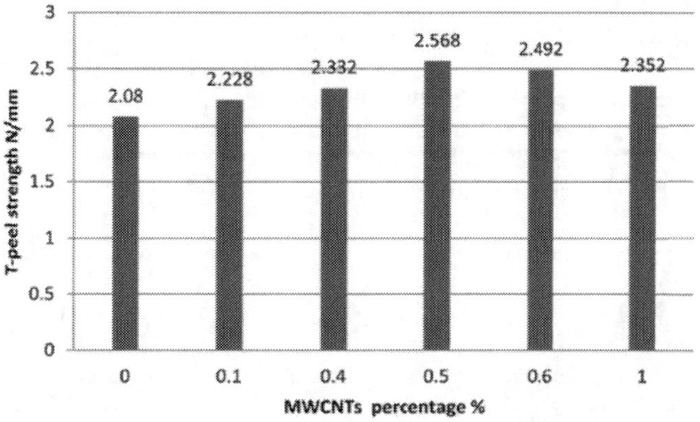

Figure 17. Average peel strength of samples with the different MWCNTs percentages filled adhesive.

5. Conclusion

The current research focuses on the enhancement of peel strength of bonded joints of 2024 aluminum alloy. The method covers application of carbon nano-tubes filled adhesive on T-peel bonded joint. It is worth noting, that mixing 0.5% MWNT with adhesives enhances the strength of the T-peel joint without increasing the structure weight. The average-peeling load for the MWCNTs filled adhesive T-peel joint increases by 23 % compared with plain adhesive joints.

Acknowledgments

The project was financially supported by King Saud University, Vice Deanship of Research Chairs.

References

[1] Mohd Afendi T and Hairul Bin Bakri 2011 Strength prediction of epoxy adhesively bonded scarf joints of dissimilar adherends *International Journal of Adhesion & Adhesives* **31** 402-11

[2] ASTM Standard 2008 Standard test method for peel resistance of adhesives (T-Peel Test) *Active Standard ASTM D1876* **15.06** doi: 10.1520/D1876-08R15E01

[3] Silvia G Prolongo, María R Gude and Alejandro Ureña 2010 *Nano Reinforced Adhesives, Nano Fibers* Ed Ashok Kumar (Madrid: University Rey Juan Carlos Spain)

[4] Khalili S, Tavakolian M and Sarabi A 2010 Mechanical properties of nanoclay reinforced epoxy adhesive bonded joints made with composite materials *Journal of Adhesion Science and Technology* **24** 1917-28

[5] Dorigato A, Pegoretti A, Bondioli F and Messori M 2010 Improving epoxy adhesives with zirconia nanoparticles *Composite Interfaces* **17** 873-92

[6] Dorigato A and Pegoretti A 2011 The role of alumina nanoparticles in epoxy adhesives *Journal of Nanoparticle Research* **13** 2429-41

[7] Srivastava V 2011 Effect of carbon nanotubes on the strength of adhesive lap joints of C/C and C/C–SiC ceramic fibre composites *International Journal of Adhesion and Adhesives* **31** 486-9

[8] Zhou Y, Pervin F, Lewis L and Jeelani S 2008 Fabrication and characterization of carbon/epoxy composites mixed with multi-walled carbon nanotubes *Materials Science and Engineering: A* **475** 157-65

[9] Matsui K 1990 Effects of size on nominal ultimate tensile stresses of adhesive-bonded circular or rectangular joints under bending or peeling load *International Journal of Adhesion and Adhesives* **10** 90-8

[10] Hinopoulos G and Broughton W 1999 Evaluation of the T-peel joint using the finite element method *NPL Report CMMT (A) 207*

[11] Darwish S 2004 Analysis of weld-bonded dissimilar materials *International Journal of Adhesion and Adhesives* **24** 347-54
[12] Beevers A, Steidler S, Durodola J and Coackley M 2001 Analysis of stiffness of adhesive joints in car bodies *Journal of Materials Processing Technology* **118** 95-100
[13] Lacroix L, Blanc C, Pébère N, Thompson G, Tribollet B and Vivier V 2012 Simulating the galvanic coupling between S-Al2CuMg phase particle sand the matrix of 2024 aerospace aluminum alloy *Corrosion Science* **64** 213-21
[14] Darwish S M H 2011 Science of weld: Adhesive joints *Hybrid Adhesive Joints* (Springer) pp 1-36
[15] Huntsman Corp. 2007 Technical data sheet of structural adhesive Araldite 2011 *Huntsman Advanced Martials*
[16] Ghosh P 2003 Weld bonding of stainless steel *International - the Iron and Steel Institute of Japan Internationa* **43** 85-94
[17] Herwan J 2010 *Load-Displacement Curve Prediction of Weld Bonded Stainless Steel Using Finite Element Method* (Saudi Arabia, Riyadh: King Saudi University, Mechanical Engineering Department)

IOP Conf. Series: Materials Science and Engineering **87** (2015) 012005 doi:10.1088/1757-899X/87/1/012005

Combining 3D printed forms with textile structures – mechanical and geometrical properties of multi-material systems

L Sabantina, F Kinzel, A Ehrmann[1] and K Finsterbusch

Niederrhein University of Applied Sciences, Faculty of Textile and Clothing Technology, Mönchengladbach, Germany

E-mail: andrea.ehrmann@fh-bielefeld.de

Abstract. The 3D printing belongs to the rapidly emerging technologies which have the chance to revolutionize the way products are created. In the textile industry, several designers have already presented creations of shoes, dresses or other garments which could not be produced with common techniques. 3D printing, however, is still far away from being a usual process in textile and clothing production. The main challenge results from the insufficient mechanical properties, especially the low tensile strength, of pure 3D printed products, prohibiting them from replacing common technologies such as weaving or knitting. Thus, one way to the application of 3D printed forms in garments is combining them with textile fabrics, the latter ensuring the necessary tensile strength. This article reports about different approaches to combine 3D printed polymers with different textile materials and fabrics, showing chances and limits of this technique.

1. Introduction

Several methods enable the additive production of 3D forms, such as stereo-lithography, selective laser sintering or fused deposition modelling (FDM) [1]. The FDM technique is based on resistively heating filaments in an extruder nozzle; the molten material is deposited on the printing bed line by line and then hardens there. The second layer is printed after the first one is finished and the printing bed is lowered [2].

FDM printers can handle filaments produced from several materials, such as Acrylonitrile butadiene styrene (ABS), polyactic acid (PLA), polyamide, polycarbonate, polyethylene, polypropylene, or wax [3]. Additionally, special materials are available, e.g. wood or brick, which are especially useful for model makers in architecture. Soft materials can cause problems in the transport mechanism [4]. One possibility to overcome these problems is using special filaments, such as Lay Tekkks, produced by Kai Parthy from CC-Products (Cologne). The parts printed with this hard filament can be put into warm water for a certain time, which dissolves the hard parts and softens the sample [4].

[1] Address for correspondence: A Ehrmann, Niederrhein University of Applied Sciences, Faculty of Textile and Clothing Technology, Mönchengladbach, Germany. E-mail: andrea.ehrmann@fh-bielefeld.de.

Since the FDM technique is used in the most inexpensive 3D printers, this method is the most interesting one for small companies, allowing for testing the possibilities of this technology without the necessity of high investments. While previous experiments always concentrated on the possibilities to create textile-based structures from different polymeric materials by FDM printing [4] and to insert fibrous materials in 3D printed forms [5], this article examines possibilities to print 3D forms directly on textile fabrics.

2. Experimental

The 3D forms shown here were designed using the free CAD software SketchUp. The 3D elements were exported as standard template library (STL) files and imported in the repetier host which was used to slice the form into a layer model. This sliced result was sent to the 3D printer.

For the models depicted in this article, the FDM printer X400, created by German RepRap, was used.

The 3D printed additions to garments should be soft and not too brittle. Thus, instead of the commonly used ABS, which has shown insufficient mechanical properties if used in open structures [6], PLA soft was chosen as filament material. The filament diameter was 3.0 mm.

The extruder temperature was set to 211°C and the printing bed temperature to 63°C. The layer height was 0.3 mm. For filling the samples, a linear pattern was chosen.

3. Results

In a first test series, a pattern of 55 mm x 75 mm lateral dimension with 1.2 mm height (equal to 4 layers) was designed and printed on different textile fabrics. Figure 1 shows the results of printing PLA on cotton, wool, viscose, and a polyester net (from left to right), respectively.

Figure 1. Floral pattern printed on cotton, wool, viscose fabric and polyester net (from left to right).

While the printed forms can be easily stripped from the cotton and viscose fabrics, the adhesion on wool increases slightly due to the fabric's rough surface. However, only for the polyester net, the printed material is really fixed on the fabric, since here the molten PLA flows around the single threads and encloses them, resulting in a mechanical connection of both parts.

To increase the adhesion strength of the printed pattern on a textile fabric, several tests using high temperature after treatments were performed. For this, a VEIT Kannegiesser fixation machine was used as well as an industrial steam iron VEIT HP 2003, allowing for combinations of pressure and temperature treatment. Since PLA has a softening temperature of about 50°C and should be processed at about 195-220°C, different temperatures between 120°C and 200°C were used in these tests. For an after-treatment at 200°C, a slight increase of the connection strength could be reached; however, the printed pattern could still be easily stripped from the textile fabric.

Thus, the next test concentrated on the idea of increasing the mechanical connection using "open" textile structures, such as different nets. Figure 2 shows the strip pattern which was used for the following examinations, with different strip heights (the largest one in the middle) and constant strip width. The textile layer was introduced between the first and the second PLA layer.

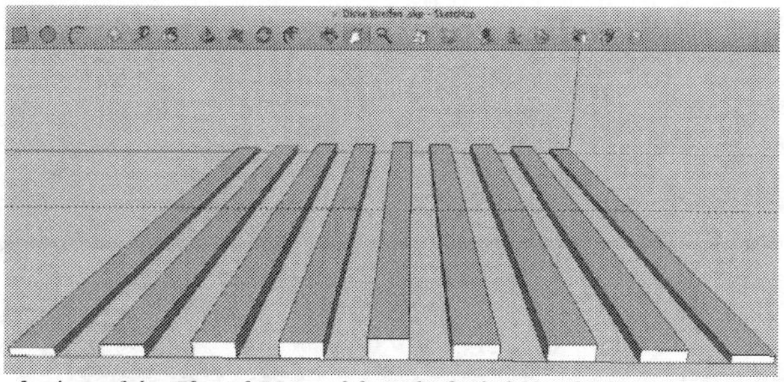

Figure 2. Strip pattern, designed in SketchUp, with strip heights of 0.9 mm, 1.2 mm, 1.5 mm, 1.8 mm, and 2.1 mm, a constant strip width of 2.5 mm and a length of 90 mm. The textile net is placed on the first layer.

This test sample was firstly tested using the Martindale Abrasion Tester Method according to standard ISO 12947-1. In this test, the abrasion of the sample under examination is tested by rubbing it against a worsted wool cloth abradent using a pressure of 9 kPa, and continuously changing the direction in a Lissajous-like figure.

As can be seen in figure 3, the polyester net fabric was not strong enough to stand the Martindale test, and was destroyed after 4,500 Martindale tours. The PLA printed pattern, however, was not influenced by the Martindale test – neither the surface changed visibly, nor the strips were pulled off from the base fabric. Apparently in this combination the connection between the textile and the surrounding PLA printed form is stronger than the textile net itself.

Next, the same PLA pattern was printed again, with one textile net between the first and the second layer, and another textile net between the second and the third layer. On this sample, a separation test according to standard ISO DIN 53530 was performed. For this, the polyester net layers were secured in the grip of a tensile testing machine and pulled apart with constant speed. In all cases – independent of the direction of the printed PLA pattern with respect to the grips – the test ended with the net fabric being destroyed (figure 4) for separation forces of ~ 2,600-3,700 cN. No separation between the PLA parts and the included net fabrics was detected.

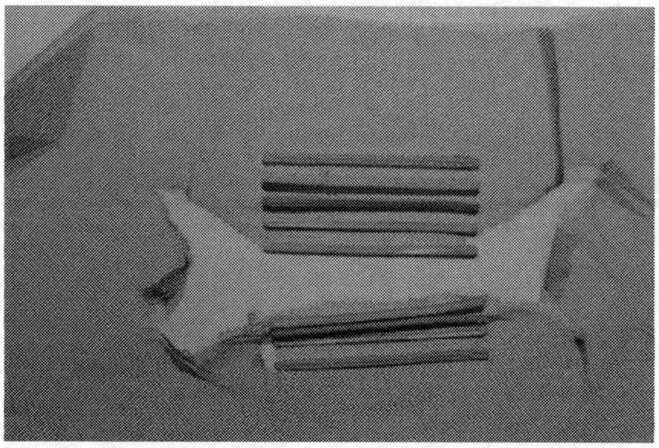

Figure 3. Test pattern, printed with PLA on a polyester net fabric, after 4,500 Martindale tours.

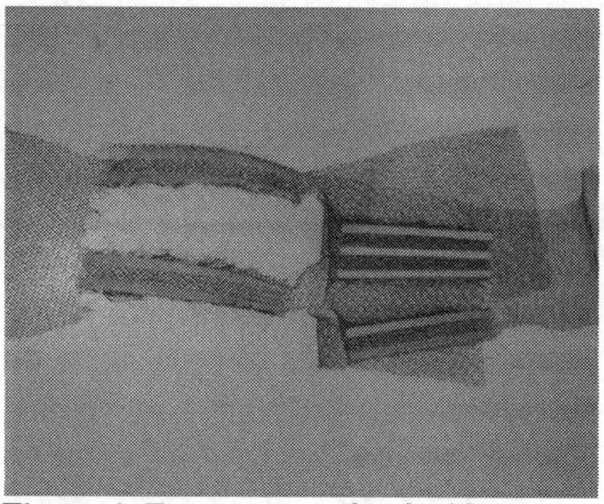

Figure 4. Test pattern, printed with PLA on polyester net fabric, after test according to standard DIN 53530.

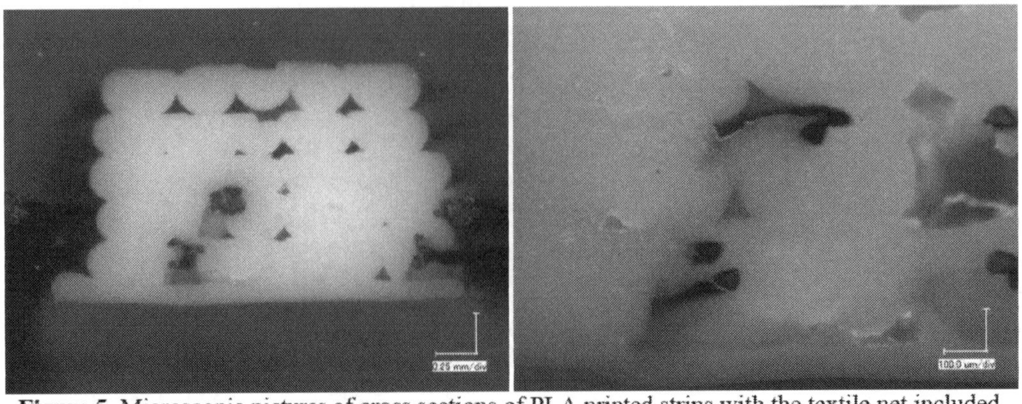

Figure 5. Microscopic pictures of cross sections of PLA printed strips with the textile net included.

To examine the connection between the printed layers and the included textile fabrics, cross sections of the printed PLA strips were prepared and inspected using a digital optical microscope. As figure 5 shows, the black net threads are well included in the white PLA matrix, building a reliable connection between both materials. The cross sections, however, also depict a typical problem of several inexpensive FDM printers: While the very first layer is printed on the hot printing bed, allowing for the first printed lines to fuse together and create one closed layer, the following layers still consist of clearly distinguishable lines whose connections seem to be reduced continually when the printed form becomes higher. This problem is connected with an undesired lateral shift of the layers with respect to their neighbours.

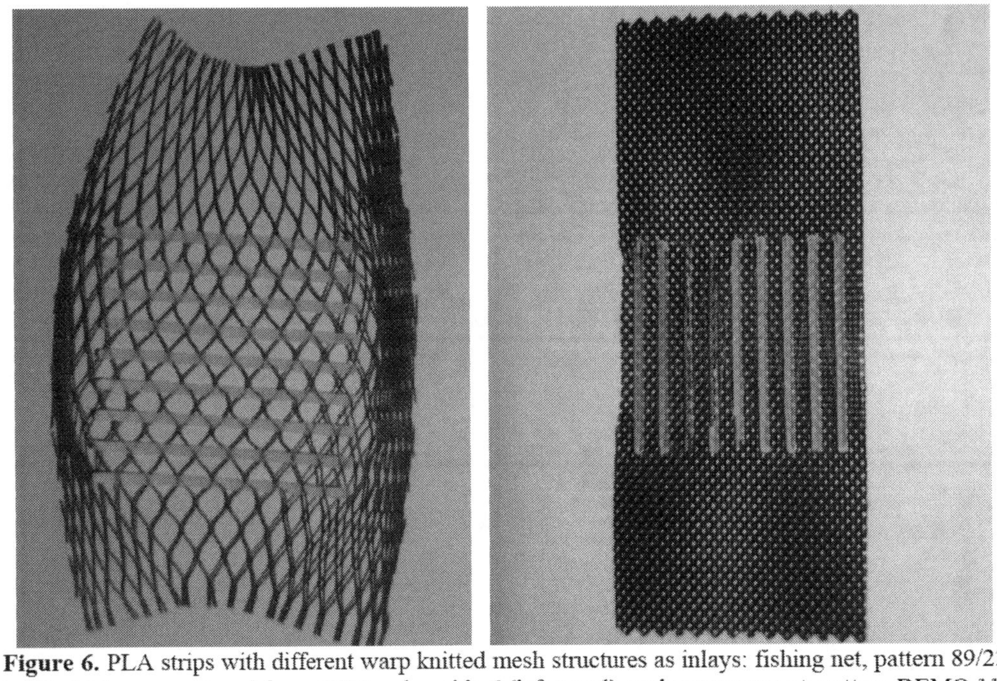

Figure 6. PLA strips with different warp knitted mesh structures as inlays: fishing net, pattern 89/232, by Karl Meyer, produced from 100% polyamide 6 (left panel), and sunscreen net, pattern DEMO 1134, by Karl Meyer, produced from 100% polyester (right panel).

In order to overcome these problems, future experiments will focus on adjusting the printing parameters, such as nozzle temperature, printing bed temperature, printing velocity, but also environment temperature etc., to enhance the layer-layer adhesion as well as the line-line adhesion.

Due to the insufficient mechanical properties of the polyester net used for previous experiments, several other textile fabrics with relatively open structures were tested as possible base layers. Figure 6 shows results of PLA strips, now all with an identical height of 1.5 mm, in which different textile mesh structures are inlayed.

While these textile structures are significantly stronger than the previously used fine polyester net, the connection between PLA strips and textile base materials now depends on the mesh-opening-diameter-to-thickness ratio of the net structures. Structures with large mesh openings (e.g. figure 6, left panel) can be fixed very well in the PLA strips, while thicker materials with smaller mesh openings (e.g. figure 6, right panel) cannot be completely interfused by the molten PLA due to its high viscosity. Thus, the latter is not perfectly surrounded by a PLA matrix, allowing for the PLA to be separated from the textile base layer by relatively low forces. Figure 6 (right panel) shows the results of a test to pull the strips off by hand, while the connection between PLA strips and the open-structure fishing net (figure 6, left panel) could not be destroyed by hand.

4. Conclusion

To combine the tensile strength of common textile fabrics with the creative freedom of printing different 3D forms, composites of different textile mesh structures with PLA printed matrices have been examined due to their mechanical properties. For several textile nets, the connection between both materials, generated by the printed material surrounding the single textile threads, is found to be sufficient for utilization in garments and appears promising for use in technical textiles.

Future tests will concentrate on enhancing the adhesion between the printed lines, with and without textile net structures embedded, by optimizing the printing parameters.

References

[1] Chua C K, Leong K F and Lim C S 2003 *Rapid Prototyping: Principles and Applications, 2nd Edition* (Singapore: World Scientific Publishing Co. Pte. Ltd.)

[2] Novakova-Marcincinova L 2012 Application of fused deposition modeling technology in 3D printing rapid prototyping area *Manuf. and Ind. Eng.* **11** 35-7

[3] Noorani R 2005 *Rapid Prototyping: Principles and Applications* (New Jersey: John Wiley & Sons)

[4] Melnikova R, Ehrmann A and Finsterbusch K 2014 3D printing of textile-based structures by Fused Deposition Modelling (FDM) with different polymer materials *IOP Conf. Ser.: Mater. Sci. Eng.* **62** 012018

[5] Richter C, Schmülling S, Ehrmann A and Finsterbusch K 2015 FDM printing of 3D forms with embedded fibrous materials, *Proc. Int. Conf. on Design, Manufacturing and Mechatronics*(Wuhan), accepted

[6] Passlack B, Ehrmann A and Finsterbusch K 2013 *Melliand Textilberichte* **94** 224

Preparation and characterization of Phase change material microcapsules by a core-shell-like emulsion polymerization method

Li-ming Ding and Guang-ling Pei [1]

School of Materials Science and Engineering, Beijing Institute of Fashion Technology, Beijing 100029, China

E-mail:clypgl@bift.edu.cn

Abstract. Phase change material microcapsules (MicroPCMs) were synthesized by a core-shell-like emulsion polymerization method. Styrene and methylacrylic acid copolymer (PS-MAA) was used as a wall material, and paraffin was used as a core material in order to prepare spherical, high resistance and high enthalpy MicroPCMs. Scanning Electron Microscope (SEM), laser particle size analyzer, Fourier Transform Infrared Spectroscopy (FTIR), Thermogravimetry (TG) and Differential Scanning Calorimeter (DSC) were employed to characterize the MicroPCMs. The results indicated that the average particle size of MicroPCMs was 42.29 μm, and the content of paraffin within microcapsules was 57.6%. The melting temperature and crystallization temperature were 30.7℃ and 25.2℃.The melting enthalpy and crystallization enthalpy were -84.1 J/g and 91.3 J/g, respectively.

1. Introduction

Phase change materials (PCMs) are thermal energy storage materials that store and release heat energy at a constant temperature by undergoing a phase change [1]. PCM is a kind of interesting material for latent heat storage due to its high energy storage density and small temperature variation from storage to recovery [2]. Therefore, PCMs have attracted considerable interest in the fields of energy storage during the past decades. However, the direct application of PCMs is not easily due to their low thermal conductivity, flammability, instability and leakage problems [3]. A solution for those problems is microencapsulation of PCMs. Microencapsulation means that individual particles or droplets of solid or liquid material (the core) are surrounded or coated by a continuous film of polymeric material (the shell) to produce capsules from the micrometer to millimeter range [4]. Microencapsulation has many advantages, such as preventing the reaction of PCMs with the outside environment and the leakage of liquid PCMs, increasing heat-transfer area, providing constant volume [5-7].

Phase change material microcapsules (MicroPCMs) not only have the advantages of PCMs but also have the advantages of microcapsules. Therefore, MicroPCMs have drawn greater attention than traditional PCMs. Many methods have been used to prepare the MicroPCMs. Fang et al. [8] prepared nano-encapsulated n-tetradecane as phase change material for thermal energy storage by in situ polymerization. Alay et al. [9] prepared poly (methyl methacrylate)/n-hexadecane microcapsules using

[1] Address for correspondence: Guang-ling Pei, School of Materials Science and Engineering, Beijing Institute of Fashion Technology, Beijing 100029, China. E-mail:clypgl@bift.edu.cn.

emulsion polymerization and studied the applicability to the textiles. Qiu et al. [10] fabricated microcapsules of n-octadecane with different n-butyl methacrylate as copolymer shells by a suspension-like polymerization method. Malekipirbazari et al. [11] microencapsulated paraffin with gelatin/gum Arabic as the wall by the complex coacervation method.

The aim of this paper is to synthesize MicroPCMs by a core-shell-like emulsion polymerization method. The paraffin is used as a core material and methylacrylic acid copolymer (PS-MAA) is chosen as the wall material.

2. Experimental

2.1. Materials

Styrene (St, A.R., Tianjin Fuchen Chemical Reagent Co., Ltd.), Methyl acrylic acid (MAA, A.R., Sinopharm Chemical Reagent Co., Ltd), Azodiisobutyronitrile (AIBN, C.R., Sinopharm Chemical Reagent Co., Ltd) and Sodium dodecyl sulfate (SDS,C.R., Sinopharm Chemical Reagent Co., Ltd) are used directly without further treatment. Paraffin is distilled from crude wax, and Water is purified by distillation before using.

2.2. Preparation of MicroPCMs

A core-shell-like emulsion polymerization method was used to fabricate the MicroPCMs. First, the paraffin was added to an aqueous solution of SDS and the mixture was heated to melt the paraffin. The mixture was emulsified under the stirring speed of 10000 rpm for 1min after the paraffin was melted and the paraffin emulsion was obtained. Then the paraffin emulsion was transferred to a four-necked flask equipped with a stirrer and a condenser and maintained for 10min at room temperature and at a stirring speed of 1000 rpm. Next, added the mixture of St, MAA and AIBN to the paraffin emulsion and dispatched the stirring speed to 400 rpm. The polymerization process was carried at the temperature of 75℃ under the protection of N_2 for 6h rs. The reaction products were separated by vacuum filtering and were washed with boiling water for three times. The purified MicroPCMs were obtained after being dried in an oven at 50℃ for 24 hrs. The experiment repeated three times, weighing the quality of the dried MicroPCMs and calculating the productive rate.

2.3. Characterization of MicroPCMs

Laser particle size analyzer (Mastersizer 2000, Malvern Instruments Ltd.,UK) was used to determine the size and size distribution of MicroPCMs. The MicroPCMs were dispersed in distilled water and the pump speed set as 2000rpm. Scanning Electron Microscopes (SEM, JSM-IT300 and JEOL JSM-7500, Japanese electronics Co., Ltd.,Japan) were used to observe the morphologies and structures of MicroPCMs. All samples were coated with a layer of gold prior to the observation. Fourier Transform Infrared Spectroscopy (FTIR, Bruker Tensor27, Bruker Co., Ltd., Germany) was used to determine the FTIR spectra of paraffin, PS-MAA and MicroPCMs. The samples were mixed with KBr and pressed into a pellet. The FTIR spectra in absorbance mode were recorded within the range of 500-4000cm-1. Thermogravimetry (TG, Netzsch TG209F1, Netzsch Instruments Ltd., Germany) was used to characterize the thermal stabilities of paraffin, PS-MAA and MicroPCMs. The heating was in the range of 30-600℃ at the rate of 5℃/min in a nitrogen atmosphere. Differential Scanning Calorimeter (DSC, EXSTAR DSC6200, NSK LTD., Japan) was used to measure the thermal storage properties of paraffin, PS-MAA and MicroPCMs. The tests were in the range of 0-80℃ at a heating or cooling rate of ±5℃/min in a nitrogen atmosphere.

3. Results and discussion

3.1. Preparation of MicroPCMs

This article is to explore whether MicroPCMs can be successfully synthesized by a core-shell-like emulsion polymerization method. On the one hand, the composition of the product should be characterized; on the other hand, the productive rate also should be considered. The productive rate of MicroPCMs can be calculated by the following formula:

$$\text{productive rate} = \frac{m_{MicroPCMs}}{m_{core} + m_{shell\ monomer}} \times 100 \quad (1)$$

Where $m_{MicroPCMs}$, m_{core} and $m_{shell\ monomer}$ are the weight of dried MicroPCMs, core and shell monomer, respectively. The mean productive rate is 81.8%, which is lower than the productive rate of PS-MAA [12].

3.2. Size and size distribution of MicroPCMs

The size and size distribution of the MicroPCMs are shown in figure 1. The average particle size of MicroPCMs is 42.29 μm and the distribution index is 2.19. As the figure 1 indicates that the particle size mainly distributed at 10 μm. Combining with the photographs of SEM in figure 2, we can deduce that the particles with larger size are uncoated paraffin. As seen from the figure 1, the distribution of the MicroPCMs is wide. Maybe this is caused by the inadequate emulsion of paraffin.

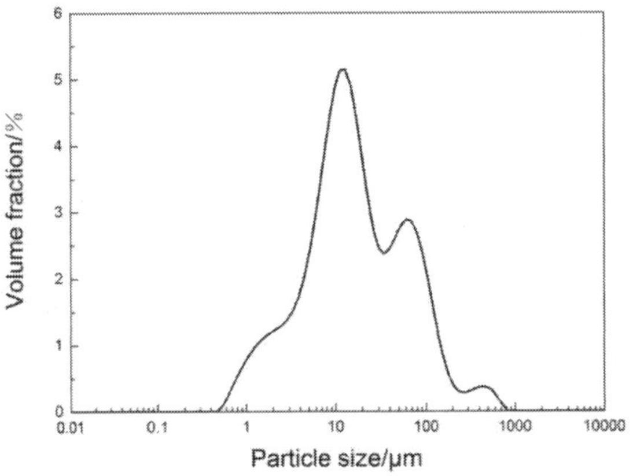

Figure 1. Diameter distribution of the MicroPCMs.

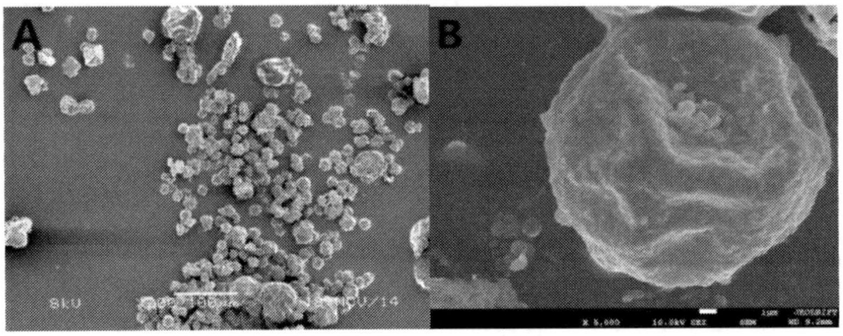

Figure 2. Morphology of the MicroPCMs.

3.3. Morphology of MicroPCMs

It can be seen from SEM image A, the sizes of the particles are mainly between 5 μm and 25 μm. This is well in accordance with that shown in figure 1. SEM image B shows that the MicroPCMs have a spherical profile. However, there are many dimples on the surface of MicroPCMs. This could be attributed to the generation of reserved expansion space in the MicroPCMs fabrication process [13]. On one hand, the density of the polymer is higher than that of the monomers, on the other hand, the volume of paraffin reduces as its phase changes from melting state to crystal state. Both factors lead to a generation of reserved expansion space during the progress of the MicroPCMs preparation, thus to the shrinkage of the shell [14-15].

3.4. Chemical characterization

Figure 3 shows the FTIR spectra of paraffin, PS-MMA and MicroPCMs. FTIR spectrum of paraffin has a typical multiple strong absorption peaks between 2850 cm^{-1} and 2950 cm^{-1}. The peaks are the symmetric and asymmetric stretching vibration of CH_2 and CH_3. The absorption peak appears at 1467 cm^{-1} represents the unsymmetrical bending vibration of CH_3 and scissors bending vibration of CH_2 in aliphatic chain. The absorption peak appears at 1378cm^{-1} corresponding to the symmetric stretching vibration of and CH_3 in aliphatic chain.

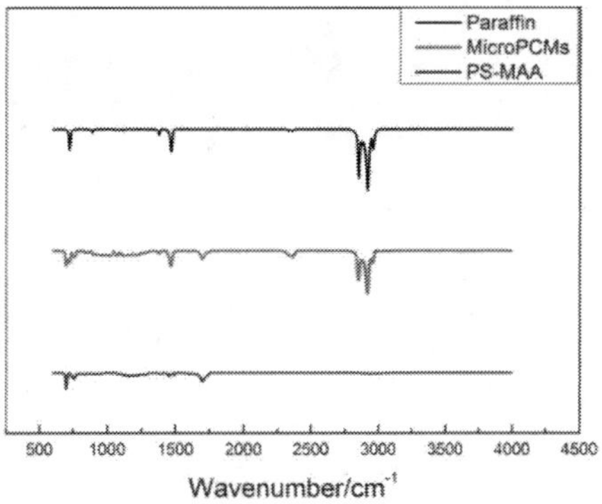

Figure 3. FTIR spectra of paraffin, PS-MMA and MicroPCMs.

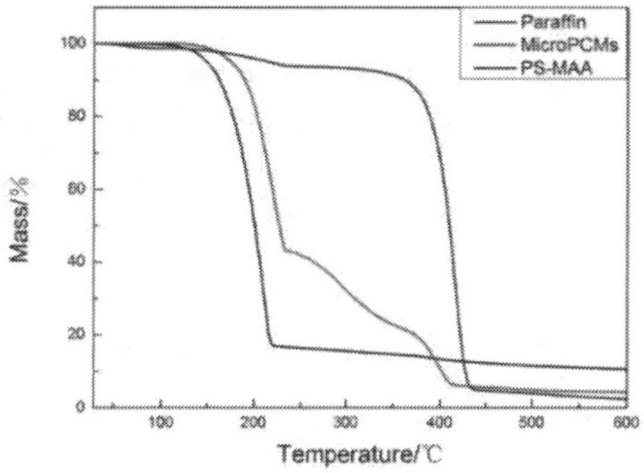

Figure 4. TG curves of paraffin, PS-MMA and MicroPCMs.

The strong absorption peak appears at about 720 cm^{-1} corresponds to the in-plane bending vibration and characterizes that the alkane chains has more than methylene. FTIR spectra of PS-MAA have a typical multiple weak absorption peaks between 3100 cm^{-1} and 2900 cm^{-1}. The peak indicates unsaturated C-H stretching vibration on the benzene ring. The strong absorption peaks appear at 699cm^{-1} and 758 cm^{-1} are out-plane deformation vibration of H on benzene ring, which is the characteristic peak of mono-substitution benzene ring. The strong absorption peak appears at 1698 cm^{-1} corresponding to the blue shift stretching vibration of C=O, and the weak absorption peak appears at 1400 cm^{-1} corresponding to stretching vibration of OH. These peaks indicate the existence of carboxyl group. From the above analysis we could surely come to the conclusion that the shell is the copolymer between St and MAA.

The FTIR spectrum of MicroPCMs not only includes the characteristic peak of paraffin but it also has characteristic peak of PS-MAA. So the MicroPCMs consists of core materials and wall materials. There are no addition peaks appear on the FTIR spectrum of MicroPCMs, which indicates that there are no chemical reaction between core materials and shell materials.

3.5. Thermal stability of MicroPCMs

Figure 4 shows TG curves of paraffin, PS-MMA and MicroPCMs. As is shown in the figure 4, paraffin started to lose its weight at approximately 147℃, and it fast lost its weight from 147℃ to 212 ℃. The TG curve of paraffin is sharp because it is composed of linear alkane molecules. The shell material PS-MAA began to lose weight at approximately 366℃ and it completely lost its weight at about 437℃. However, TG curve of MicroPCM mainly shows two steps of mass lose. The first mass lost occurred between 165℃ and 225℃ resulted from the weight loss of paraffin. The second mass lost happened from 366℃ to 420℃ caused from the degradation of PS-MAA. The results indicate encapsulation of paraffin can delay the volatilizing.

3.6. Thermal properties of MicroPCMs

The content of PCM in microcapsules is an important factor for MicroPCMs, and it directly determines the phase change enthalpy and energy storage efficiency of MicroPCMs. The DSC curves of paraffin, PS-MMA and MicroPCMs are shown in figure 5 and the specific data are listed in table 1.

As figure 5 shows, the PS-MAA has no phase change behavior in the test range. Therefore, the content of paraffin in MicroPCMs can be calculated by the following formula [16]:

$$\text{paraffin content} = \frac{\Delta H_{MicroPCMs}}{\Delta H_{PCMs}} \times 100 \tag{2}$$

Where $\Delta H_{MicroPCMS}$ and ΔH_{PCMS} are the enthalpies of MicroPCMs or pure PCMs, respectively.

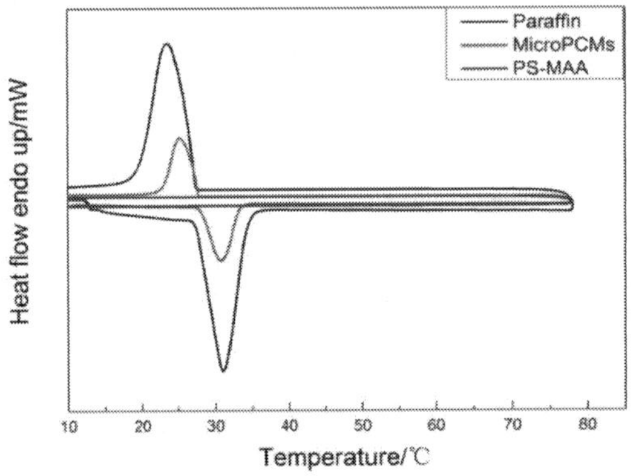

Figure 5. DSC curves of paraffin, PS-MMA and MicroPCMs.

Table 1 Thermal property of MicroPCMs.

Sample	Tpm/℃	ΔHm（J/g）	Tpc/℃	ΔHc（J/g）	PCM content/%
paraffin	30.9	-146	23.3	167	100
MicroPCMs	30.7	-84.1	25.2	91.3	57.6.

[a]Tpm is melting temperature,
[b]ΔHm is melting enthalpy;
[c]Tpc is crystallization temperature;
[d]ΔHc is crystallization enthalpy.

It can be seen from figure 5 and table 1 that the melting temperature and crystallization temperature of paraffin have no obvious change before and after the microencapsulation. So the microencapsulation doesn't affect the thermal properties of paraffin. The paraffin content in MicroPCMs is about 57.6%. The melting temperature and crystallization temperature are 30.7℃ and 25.2℃.The melting enthalpy and crystallization enthalpy are -84.1 J/g and 91.3, respectively.

4. Results and discussion

MicroPCMs using paraffin as core materials are successfully prepared by core-shell-like emulsion polymerization. The average particle size of the MicroPCMs is 42.290 μm, and the content of paraffin is 57.6%. The melting temperature and crystallization temperature are 30.7℃ and 25.2℃.The melting enthalpy and crystallization enthalpy are -84.1 J/g and 91.3 J/g, respectively. The MicroPCMs prepared in this paper have potential application in the field of energy storage.

References

[1] Farid M M, Khudhair A M and Razack S A K 2004 *Energy Conversion and Management* **45** 1597-615
[2] Chen C, Liu S, Liu W, Zhao Y and Lu Y 2012 *Solar Energy Materials and Solar Cells* **96** 202-9.
[3] Ma Y, Chu X, Tang G and Yao Y 2013 *Materials Letters* **91** 133-5
[4] Vidhyalakshmi R, Bhakyaraj R and Subhasree R S 2009 A*dvances in Biological Research* **3** 96-103
[5] Sarıer N and Onder E 2007 *Thermochim. Acta* **452** 149-60
[6] Onder E, Sarıer N and Cimen E 2008 *Thermochim. Acta* **467** 63-72
[7] Kakarla Raghava Reddy,Han Mo Jeong,Youngil Lee and Anjanapura Venkataramanaiah Raghu 2009 *Journal of Polymer Science:Part A: Polymer Chemistry* **48** 1477-84
[8] Fang G Y, Li H, Yang F, Liu X and Wu S M 2009 *Chemical Engineering Journal* **153** 217-21
[9] Sennur Alay, Cemil Alkan and Fethiye Gode 2011 *Thermochimica Acta* **518** 1-8
[10] Qiu X L, Lu L X, Wang J, Tang G Y and Song G L 2014 *Solar Engergy Materials & Solar Cells* **128** 102-11
[11] Malekipirbazari Milad, Sadrameli S M, Dorkoosh Farid and Sharifi Hamed 2014 *Int. J. Energy Res.* **38** 1492-500
[12] Wang Y X and Zhao Y T 1995 *Journal of East China University of Science and Technology* **21** 702-7
[13] Shan X L, Wang J P, Zhang X X and Wang X C 2009 *Thermochimica Acta* **494** 104-9
[14] You M, Zhang X and Wang J 2009 *Journal of Materials Science* **44** 3141-7
[15] Li W, Song G, Tang G, Chu X, Ma S and Liu C 2011 *Energy* **36** 785-91
[16] Cemil Alkan, Ahmet Sarl and Ali Karaipekli 2011 *Energy Conversion and Management* **52** 287-92

Gaseous detonation synthesis and characterization of nano-oxide

Honghao Yan[1], Linsong Wu, Xiaojie Li and Xiaohong Wang

Department of Engineering Mechanics, Dalian University of Technology, No. 2 Linggong Road, Dalian 116024, P. R. China

E-mail: 923309973@qq.com

Abstract. Gaseous detonation is a new method of heating the precursor of nanomaterials into gas, and integrating it with combustible gas as mixture to be detonated for the synthesis of nanomaterials. In this paper, the mixed gas of oxygen and hydrogen is used as the source for detonation, to synthesize nano TiO_2, nano SiO_2 and nano SnO_2 through gaseous detonation method, characterization and analysis of the products, it was found that the products from gaseous detonation method were of high purity, good dispersion, smaller particle size and even distribution. It also shows that for the synthesis of nano-oxides, gaseous detonation is universal.

1. Introduction

The concept of nanomaterials was originally proposed by Gleiter of Germany, who had artificially prepared nanocrystals [1] in the 1980s for the first time. When the structure of a material enters the range of nano-scale features, the electronic structure and crystal structure of the crystal surface of the materials have undergone great changes, such as small size effect, surface effect, macroscopic quantum tunneling effect, quantum size effect and dielectric limit domain effects, which shows its broad application prospects in aerospace, chemistry, medicine, biology, electronics and other fields [2-4].

Detonation synthesis is the reaction procedure using the instantaneous high temperature, high pressure and high detonation velocity generated by the detonation, under these extreme conditions, the precursor undergoes pyrolysis, phase transition or chemical reaction with the explosion products. This destroys the structure of the original substance and enables the synthesis of new materials [5, 6]. Detonation was first used to synthesize nanodiamond [7] and was then extended to synthesize a variety of nanomaterials, including nanoscale TiO_2 and Al_2O_3 and carbon-encapsulated metal, etc. [8-10]. Gaseous detonation method directly used gas precursor to turn the precursor into gas, such as using plasma, laser evaporation, electron beam heating, etc, and then detonate the precursor in gas state with the combustible gas, and finally condense and grow into ultra-fine powder during the expansion cooling process of the products; its advantages are that it can reasonably choose the initial condition such as detonation system, regulate the reactive gas species, change the pressure, temperature and density of gas, and choose a strong detonation methods, etc., the nano-powder can be obtained with high purity, good dispersion, even distribution of particle size and small aggregation. In

[1] Address for correspondence: Honghao Yan, Department of Engineering Mechanics, Dalian University of Technology, No. 2 Linggong Road, Dalian 116024, P. R. China. E-mail: 923309973@qq.com.

this paper, chlorides were selected as the metal precursor, and H_2 and O_2 were chosen as the detonation medium to synthesize nano TiO_2, nano SiO_2 and nano SnO_2. The crystalline structure and phase component were obtained on powder X-ray diffraction (XRD) and the evolution of the morphology and particle size of the nano oxide powders was examined by transmission electron microscopy (TEM). As there is no carbon source in gas detonation system, the carbon pollution to nanomaterials which exists in detonation synthesis of conventional explosive was solved.

2. Experimental

2.1. Experimental equipment

Figure 1 is Temperature-control Gaseous Detonation Tube, which is the main experimental equipment for preparing nanoparticles by gaseous detonation method. It is a cylindrical closed container consisting of titanium tube with inner diameter of 100 mm and titanium flange, while being equipped with detonating point, filling port, gas valves and outer loop temperature control system, etc.

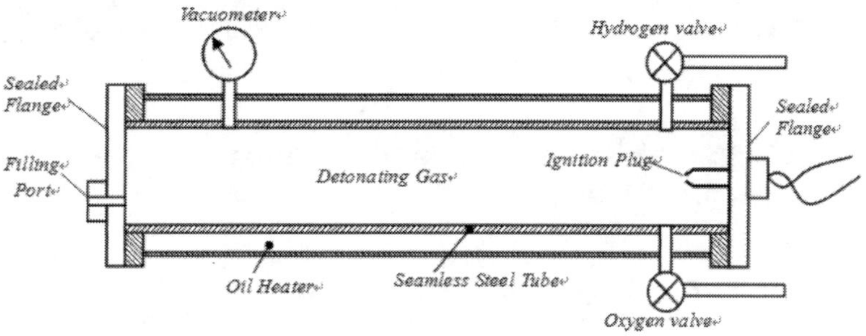

Figure 1. Schematic of detonation tube.

2.2. Experimental procedure

The experimental procedures for the preparation of nano-oxides through gaseous detonation method are basically the same; below is an overview of experimental procedure for preparation of nano TiO_2 by gaseous detonation as an example.

Firstly, heat the gaseous detonation tube to 423.15 K to turn the precursor into gas completely by outer loop temperature control system (boiling point of $TiCl_4$ is 409.55 K), vacuumize the tube with a vacuum pump, later inject a certain amount of $TiCl_4$ solution, then observe the pointer changes of the vacuum meter on the detonation tube, and determine the extent for the evaporation of the solution according to variation of the pointer. Subsequently, hydrogen and oxygen are introduced, and their amount can be calculated from the value of the pressure gauge according to Dalton's Law of partial pressure and the Ideal Gas Law.

Let it stand for 3 to 5 minutes, so that the gas can be completely mixed. Then ignite the mixed gas with detonator and collect the product 15 to 20 minutes later. For the crystal structure of the collected powder particles, XRD-6000 X-ray diffractometer is utilized to carry out polymorph analysis, and the morphology and size of the powder are observed by Tecnai $G^2$20 S-Twin TEM.

3. Results and discussion

3.1. Preparation of Nano SiO_2

With hydrogen and oxygen as detonation energy, $SiCl_4$ (A Johnson Matthey Company, purity 99%, boiling point: 331.75 K) as precursor, the mixture is heated to 393.15 K and ignited within the detonation tube. Nanoparticles of nano SiO_2 are formed with the reaction equation as follows:

$$SiCl_4(g)+O_2(g)+2H_2(g)\rightarrow SiO_2\ (s)+4HCl(g) \tag{1}$$

Reactant molar ratio is 1:1:2, and XRD is utilized to carry out analysis for detonation products. The parameters measured are: Cu target (Ka, $\lambda = 0.15406$ nm), scanning speed 5 °/min, scan range 2θ: 10°-90°.

Figure 2a shows the XRD curve of nano SiO_2 powder, and the product spectrum appears dispersion peak near the diffraction angle of 22°; it is typical amorphous structure, so the product is amorphous particles.

As can be seen from figure 3a, the crystal structure of the product is spherical, mostly in the particle size of 100nm with good dispersibility.

3.2. Preparation of Nano TiO_2

With hydrogen and oxygen as detonation energy, $TiCl_4$ (sinopharm Chemical Reagent, purity 99% boiling point: 409.55 K) as precursor, the mixture is heated to 423.15 K and ignited within the detonation tube. Nanoparticles of nano TiO_2 are formed with the reaction equation as follows:

$$TiCl_4(g)+O_2(g)+2H_2(g)\rightarrow TiO_2(s)+4HCl(g) \tag{2}$$

Reactant molar ratio is 1:1:2, and XRD is utilized to carry out analysis for detonation products. The parameters measured are: Cu target (Ka, $\lambda = 0.15406$ nm), scanning speed 10°/min, scan range 2θ: 10°-80°.

Figure 2b is the XRD curve of nano TiO_2 powder. By comparing with ASTM card, the product prepared is identified as the TiO_2 powder with mixed crystals of anatase phase and rutile phase, where the mass ratio of anatase phase to rutile phase is 45.8:54.2. The particle size is calculated according to Scherrer Equation: $D=K\lambda/\beta\cos\theta$, where D is the crystallite size, λ is the wavelength of X-ray radiation (0.154 nm), β is the full width at half maximum (FWHM), θ is the diffraction angle, and K is usually taken as 0.89. The calculated result is anatase phase being 26.24 nm and rutile phase being 36.77 nm.

As can be seen from figure 3b, the product crystal is spherical with particle size mostly of 80-100 nm, and there are also particles with diameter of 150 nm, but the products are of good dispersibility.

3.3. Preparation of Nano SnO_2

With hydrogen and oxygen as detonation energy, $SnCl_4$ (Sinopharm Chemical Reagent, purity: 99%, boiling point: 387.25 K) as precursor, the mixture is heated to 393.15 K and ignited within the detonation tube. Nanoparticles of nano SnO_2 are formed with the reaction equation as follows:

$$SnCl_4(g)+O_2(g)+2H_2(g)\rightarrow SnO_2(s)+4HCl(g) \tag{3}$$

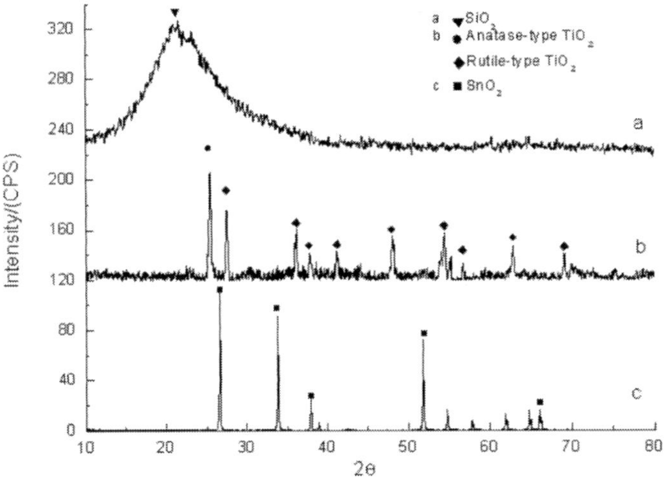

Figure 2. XRD diagrams (a) SiO_2; (b) TiO_2; (c) SnO_2.

Reactant mole ratio is 1:1:2, and XRD is utilized to carry out analysis for detonation products. The parameters measured are: Cu target (Ka, λ= 0.15406 nm), scanning speed 5°/min, scan range 2θ: 10°-70°.

Figure 2c is the XRD diagram of the collected product, where the main component is found to be SnO_2. All the peaks of SnO_2 can be observed in figure 2c, and the small peaks of Cl is also found, which may be due to the incompletely reacted liquefied $SnCl_4$. According to Scherrer Equation, the average particle size of detonation products is calculated to be 8.2 nm.

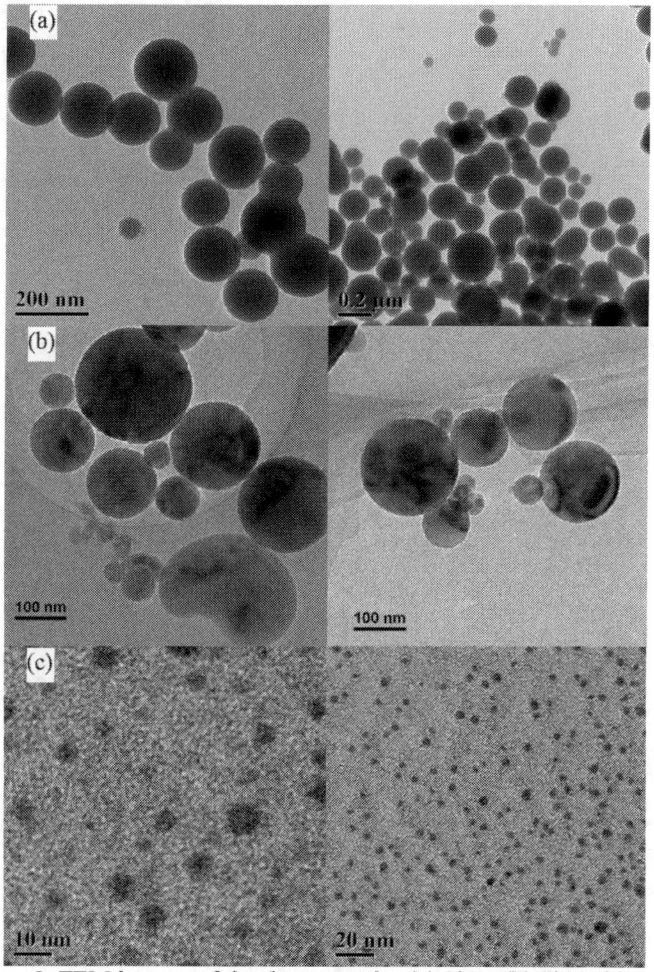

Figure 3. TEM images of the three samples (a) SiO_2; (b) TiO_2; (c) SnO_2.

Figure 3c is the topography observed from TEM. It can be seen that the product particles are spherical, of even size ranging between 1-10 nm, which is in line with the XRD analysis. The lattices crystal grains are clearly visible, without obvious aggregation, and the product particles are of good dispersibility.

4. Conclusion

This paper has prepared nano SiO_2, nano TiO_2 and nano SnO_2 by gaseous detonation method, the products' particles are spherical; the size is even and they show good dispersibility. It is concluded that using gaseous detonation method to synthesize nano-oxides is feasible and universal.

Acknowledgments
The project was financially supported by the National Science Foundation of China (No. 10872044, No. 10602013 & No. 10972051 & No. 10902023) and the Natural Science Foundation of Liaoning Province of China (No. 20082161).

References

[1] Birringer R, Gleiter H, Klein H P and Marquardt P 1984 *Phys. Lett. A* **102** 365-9
[2] Stockle R, Fokas C, Deckert V, Zenobi R, Sick B, Hecht B and Wild U P 1999 *Appl Phys Lett* **75** 160-2
[3] Bui J D, Zelles T, Lou H J, Gallion V L, Phillips M I and Tan W H 1999 *J Neurosci Methods* **89** 9-15
[4] Almeida P D, Deelen J V and Catry C 2003 *Applied Physics A* **26** 1007-12
[5] Staver A M and Gubareva N A 1984 *Fizika Goreniia I Vzryva* **20** 100-7
[6] Beloshapk A G, Bukaemskii A A and Staver A M 1990 *Combust. Explo. Shock Waves* **26** 457-61
[7] Graham R A, Morosin B and Venturini E L 1986 Ann. *Rev. Mater. Sci* **16** 315-41
[8] Li X J, Ouyang X and Yan H H 2008 *Advanced Materials Research* **32** 13-16
[9] Li R Y, Li X J and Xie X H 2006 *Combust. Explo. Shock waves* **42** 607-10
[10] Yan H H, Wu L S, Li X J and Wang X H 2013 *Rare Metal Materials and Engineering* **42** 1325-7

Effects of preparation process on performance of rubber modified asphalt

Hanbing Liu, Guobao Luo, Xianqiang Wang and Yubo Jiao[1]

College of Transportation, Jilin University, Changchun 130025, China

E-mail: jiaoyb@jlu.edu.cn

Abstract. The rational utilization of waste rubber tire is essential for the environmental protection. Utilizing rubber particles to modify asphalt can not only improve asphalt performance, but also help the recycling of waste materials. Considering the effect of different preparation process parameters on the performance of rubber modified asphalt, this paper analyzes the effects of the shear temperature, shear time and shear rate on the performance of rubber modified asphalt, and provided a reference for its preparation.

1. Introduction

The increasing traffic load and the more frequent extreme weather conditions put higher requirements on the performance of the road construction materials. In recent years, road damage phenomena such as pavement rut, bleeding and bulling have happened very frequent and become increasingly serious. Therefore, how to improve the performance of asphalt mixture and prolong the service life of pavement becomes an important research direction of the road building materials [1].

The amount of waste rubber tire has been increasing along with the number of cars growing in China. Relevant data show that there are new 250 million waste rubber tires totaling 520 tons in 2010. The existing recycling level of waste rubber tire is low and large numbers of tires are disposed with fire, which has aggravated the environmental pollution. In fact, waste can be turned into treasure as the effective components contained in waste rubber tire can be used for asphalt modification. Studies on rubber modified asphalt have attracted broad attention of scholars all over the world [1, 2].

Waste rubber tire is an ideal asphalt modifier, including natural rubber, synthetic rubber, carbon black and other ingredients. It's shown in some research papers that adding rubber particles into asphalt can significantly improve the performance of asphalt and its mixture, such as enhancing durability of pavement, increasing the toughness and elasticity recovery of asphalt, improving the aging resistance and anti-rutting ability, and reducing the pavement noise etc. [3, 4].

The process of rubber modified asphalt mainly consists of dry and wet processes, and the wet process has been widely applied in real-world engineering. Considering the effect of different preparation process parameters on the performance of rubber modified asphalt, this paper analyzed the effects of the shear temperature, shear time and shear rate on the performance of the rubber modified asphalt, which provided a reference for its preparation [5-10].

2. Materials

[1] Address for correspondence: Yubo Jiao, College of Transportation, Jilin University, Changchun 130025, China. E-mail: jiaoyb@jlu.edu.cn.

The 90 # matrix asphalt was selected in the experiment, whose properties were shown in table 1.

The rubber particles used in the experiment were obtained through normal temperature crushing method, and the pass rate of 30 mesh was 69.8%, 40 mesh 33.5% and 50 mesh 20.8%.

Table 1. Test results of matrix asphalt performance.

Test properties	Unit	Standard requirements	Test results
Penetration (25℃, 100g, 5s)	0.1mm	80~100	96
Ductility (15℃)	cm	≥100	>100
Softening point (ring and ball method)	℃	≥42	56

3. Experimental results and discussion

3.1. *Effect of shear temperature on the performance of modified asphalt*

The shear temperature will affect the performance of rubber modified asphalt. The low shear temperature will reduce asphalt fluidity and lead to the high consistency of asphalt, which is bad for its adsorption and swelling, while high shear temperature is easy to cause the aging of asphalt. Therefore, shearing temperature was set as a single variable in this experiment to study its effect on the performance of rubber modified asphalt. In this experiment, the dosage of rubber particles was 15%, shear time 45 min, shear rate 5000r/min, swelling time 60 min (180℃) and the shear temperature 150℃, 160℃, 170℃, 180℃ and 190℃. The penetration softening point and ductility at 5℃ index were chosen as evaluation indexes. Test results at different shear temperatures were shown in table 2 and the curve of each parameter with shear temperature was shown in figure 1.

Table 2. Test results of properties at different shear temperatures.

Shear temperature (℃)	Ductility (5℃) (cm)			Average (cm)	Penetration 25℃ (100g, 5s, 0.1mm)			Average (0.1mm)	Softening point (℃)		Average (0.1mm)
	1	2	3		1	2	3		1	2	
150	11.52	11.39	11.53	11.48	61.1	60.5	60.5	60.7	51.2	51.4	51.3
160	11.98	12.05	11.85	11.96	60.2	61.3	60.6	60.7	51.8	52.2	52.0
170	12.19	12.31	12.34	12.28	60.3	61.1	61.3	60.9	52.3	52.9	52.6
180	12.62	11.80	12.54	12.32	62.7	61.5	59.2	61.1	53.1	53.9	53.5
190	12.45	11.68	12.62	12.25	61.2	62.1	61.2	61.5	53.4	52.4	52.9

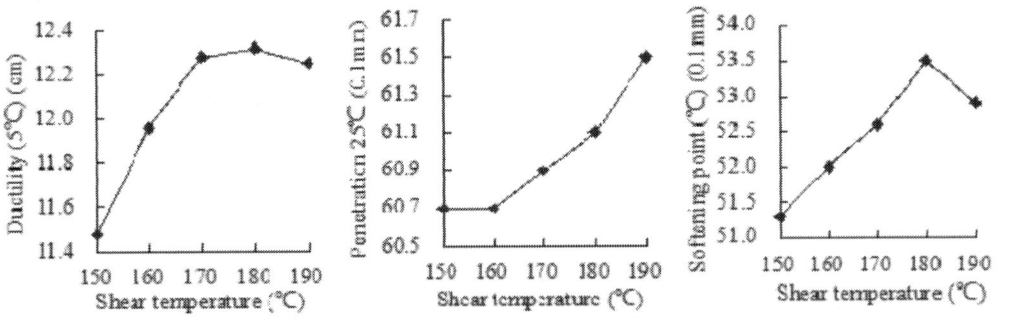

Figure 1. The curve of each parameter with shear temperature.

The conclusion can be drawn based on the experiment data that the penetration value increases but only in a small extent when the shear temperature increases. The softening point increases with the shear temperature increasing from 150℃ to 180℃ and reaches its maximum at 180℃, then decreases with the temperature further increasing. Thus, 180℃ could be taken as the optimal shear temperature for softening point index. Ductility shows a significant increase at 5℃ index when the shear temperature increases from 150℃ to 170℃. However, as the shear temperature continues to increase, the ductility only increases slightly and then decreases. The peak value of ductility shows up at 180℃. Therefore, the range from 170℃ to 180℃ is chosen to be the optimal shear temperature for ductility at 5℃ index.

3.2. Effect of shear time on the performance of modified asphalt

The shear time is another important factor contributing to the preparation quality of rubber asphalt. A short shear time would not allow rubber particles to shear adequately, while a long shear time is easy to cause the aging of asphalt. So the shearing time was set as a single variable to study its effect on the performance of modified asphalt in this paper. In this experiment, the dosage of rubber particles was 15%, shear temperature 180℃, shear rate 5000 r/min, swelling time 60min(180℃) and shear time 15 min, 30 min, 45 min and 60 min. The penetration softening point and ductility at 5℃ index were chosen as evaluation indexes. The test results at different shear time were shown in table 3 and the curve of each parameter with shear time was shown in figure 2.

Table 3. Test results of properties at different shear time.

Shear time (min)	Ductility (5℃) (cm)			Average (cm)	Penetration 25℃ (100g, 5s, 0.1mm)			Average (0.1mm)	Softening point (℃)		Average (0.1mm)
	1	2	3		1	2	3		1	2	
15	12.23	12.15	12.13	12.17	61.9	63.2	61.8	62.3	51.8	52.6	52.2
30	12.24	12.35	12.19	12.26	62.3	62.3	61.1	61.9	52.6	53.2	52.9
45	12.62	11.80	12.54	12.32	62.7	61.5	59.2	61.1	53.1	53.9	53.5
60	12.19	12.41	12.33	12.31	62.5	61.8	60.5	61.6	53.1	54.1	53.6

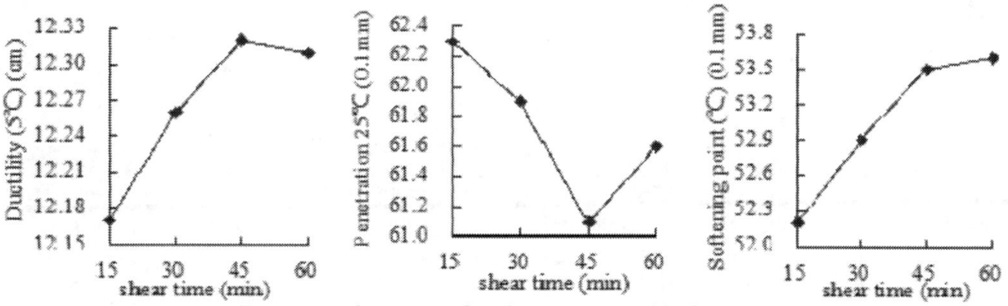

Figure 2. The curve of each parameter with shear time.

From the table and the figure it can be concluded that the penetration of rubber modified asphalt has a minimum value when the shear time is 45min, which can be used as the shear time for penetration index at 25℃. The softening point of rubber modified asphalt is relatively stable in the range of 45 min to 60 min, and this eigenvalue interval of the shear time is suitable for softening point index of rubber modified asphalt. The ductility of rubber modified asphalt at 5℃ is increasing as the shear time increases from 15min to 45min. The value of ductility at 60 min is close to that at 45min, so that 45 min is chosen as the shearing time for the ductility index at 5℃.

3.3. Effect of shear rate on the performance of modified asphalt

IOP Conf. Series: Materials Science and Engineering **87** (2015) 012008 doi:10.1088/1757-899X/87/1/012008

The shear rate is another factor with the same importance as shear temperature and shear time in affecting the performance of rubber modified asphalt. The low shear rate will lead to a bad shear result, nevertheless if the shear rate is too high, the temperature of rubber particles will rise rapidly and cause waste of resources. So in this project the shearing rate was set as a single variable in the experiment to study its effect on the performance of modified asphalt. In this experiment, the dosage of rubber particles was 15%, shear temperature 180℃, shear time 45 min, swelling time 60 min (180℃) and shear rate 3000 r/min, 4000 r/min, 5000 r/min, 6000 r/min and 7000 r/min. The penetration softening point and ductility at 5℃ were chosen as the evaluation indexes. Test results at different shear rates were listed in table 4 and the curves of each parameter with shear rate were shown in figure 3.

Table 4. Test results of properties at different shear rate.

Shear rate (r/min)	Ductility (5℃) (cm)			Average (cm)	Penetration 25℃ (100g, 5s, 0.1mm)			Average (0.1mm)	Softening point (℃)		Average (0.1mm)
	1	2	3		1	2	3		1	2	
3000	12.22	12.18	12.23	12.21	62.1	61.4	61.9	61.8	51.8	52.4	52.1
4000	12.23	12.25	12.30	12.26	62.1	60.8	61.3	61.4	52.3	52.9	52.6
5000	12.62	11.80	12.54	12.32	62.7	61.5	59.2	61.1	53.1	53.9	53.5
6000	12.31	12.34	12.28	12.31	61.2	60.8	61.6	61.2	53.4	54.0	53.7
7000	12.32	12.31	12.36	12.33	61.2	62.4	60.9	61.5	53.0	53.6	53.3

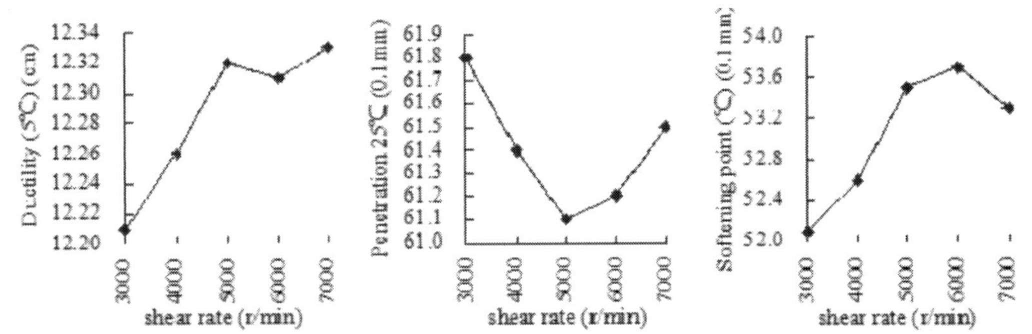

Figure 3. The curve of each parameter with shear rate.

According to experimental data, the following conclusions can be drawn. The penetration of rubber modified asphalt is decreasing until the shear rate reaches 5000 r/min, but the change of penetration rate is only in a small range. The peak value of the softening point appears when the shear rate is in the range of 5000 r/min to 6000 r/min, and such eigenvalue range of shear rate is suitable for the softening point index of rubber modified asphalt. The ductility of rubber modified asphalt at 5℃ shows an increasing trend with the increasing shear rate, but the change is not obvious after 5000 r/min, So 5000 r/min is applicable to 5℃ ductility index of rubber modified asphalt as shearing rate.

4. Conclusion

Considering the effect of different preparation process parameters on the performance of the rubber modified asphalt, this paper selects the shear temperature, shear time and shear rate as the research objects and analyzes the effect of these process parameters on the performance of rubber modified asphalt. From the test results it can be concluded that all preparation process parameters have effects on the performance of the rubber modified asphalt. Comprehensively considering these effects, this paper recommends the shear temperature 180℃, shear time 45 min and shear rate 5000 r/min as the best preparation process parameters for the rubber modified asphalt. The process parameters determined by the institute in this paper can provide a reference for preparation of the rubber modified asphalt.

Acknowledgments
The authors express their appreciation for the financial support of National Natural Science Foundation of China (51408258, 51378236, 51278222), China Postdoctoral Science Foundation funded project (2014M560237), Fundamental Research Funds for the Central Universities and Science (JCKY-QKJC06) & Technology Development Program of Jilin Province (20140203002SF).

References
[1] Presti D L 2013 Recycled tyre rubber modified bitumen for road asphalt mixtures: A literature review *J. Construction and Building Materials* **49** 863-81
[2] Li X, Jian C and Guo H Q 2009 Microstructure and performance of crumb rubber modified asphalt *J. Construction and Building Materials* **23** 3586-90
[3] Cao W D 2007 Study on properties of recycled tire rubber modified asphalt mixtures using dry process *J. Construction and Building Material* **21** 1011-5
[4] Moreno F, Rubio M C and Martinez-Echevarria M 2012 The mechanical performance of dry-process crumb rubber modified hot bituminous mixes: The influence of digestion time and crumb rubber percentage *J. Construction and Building Material* **26** 466-74
[5] Liu H Y, Chen Z J, Wang W, Wang H N and Hao P W 2014 Investigation of the rheological modification mechanism of crumb rubber modified asphalt (CRMA) containing TOR additive *J. Construction and Building Materials* **67** 225-33
[6] Kök B V and Çolak H 2011 Laboratory comparison of the crumb-rubber and SBS modified bitumen and hot mix asphalt *J. Construction and Building Materials* **25** 3204-12
[7] Zhang F and Hu C B 2015 The research for structural characteristics and modification mechanism of crumb rubber compound modified asphalts *J. Construction and Building Materials* **76** 330-42
[8] Liu S T, Cao W D, Fang J G and Shang S J 2009 Variance analysis and performance evaluation of different crumb rubber modified (CRM) asphalt *J. Construction and Building Materials* **23** 2701-8
[9] Mull M A, Stuart K and Yehia A 2002 Fracture resistance characterization of chemically modified crumb rubber asphalt pavement *J Mater Sci.* **37** 557-66
[10] González V, Martínez-Boza F, Gallegos C, Pérez-Lepe A and Páez A 2012 A study into the processing of bitumen modified with tire crumb rubber and polymeric additives *J. Fuel Process Technology* **95** 137-43

IOP Conf. Series: Materials Science and Engineering **87** (2015) 012009 doi:10.1088/1757-899X/87/1/012009

Plasmonic nanoparticles: Towards the fabrication of biosensors

Hui Shen[1]

Anhui Sanlian University, No. 47 He'an Road, Hefei, Anhui, China

E-mail: Shenhuii@mail.ustc.edu.cn

Abstract. Au and Ag nanoparticles are mainly employed in the fabrication of biosensors owing to their unique optical properties compared to other noble metal nanoparticles. Many biosensors are fabricated for the rapid detection of different analytes such as organic and inorganic molecules, biomolecules like DNA, proteins, biotoxins and pathogens. In this mini review we mainly discuss on the usage of Au and Ag nanoparticles for the fabrication of colorimetric, SERS and two photon based photoluminescence biosensors.

1. Introduction

In recent years much attention is paid towards the synthesis of plasmonic nanoparticles owing to their unique optical, electronic and catalytic properties compared to their bulk counterparts [1-7]. A strong localized surface plasmon resonance (LSPR) phenomenon can be seen in Au and Ag nanoparticles due to collective oscillations of conduction electrons induced by the interaction of electromagnetic radiations [8]. LSPR frequencies of Au and Ag nanoparticles (NP) are mainly depending upon on size, shape, inter particle distance and composition [8, 9]. A strong electromagnetic enhancement can be attained through the excitation of their LSPR and this peculiar property attracted for the usage of these NP in the fabrication of biosensors [9]. The fabrication of biosensors becomes crucial for the improvement of health and environmental sectors for the early rapid detection of analytes and diseases. In recent years several biosensors such as colorimetric [10-12], SERS [13], two photon based photoluminescence [14], electrochemical and refractive index based sensors are fabricated for the real time monitoring of different analytes and biotoxins present in water and food samples. It is still a challenging task to fabricate low-cost based biosensors [1]. In this report, we mainly summarize on the usage of Au and Ag nanoparticles in fabrication of colorimetric, SERS and two photon based photoluminescence biosensors towards the detection of analytes related to biological, environmental and clinical sectors.

2. Colorimetric based sensors

In colorimetric based sensors the analytes are mainly detected based on the colour change in NP solution. In presence of specific analytes the NP aggregate to give different colour intensity and measuring the change in colour can quantify the presence of analytes in the given samples [15]. The detection method is very simple, selective, portable and instrumentation free which even can judge color change based on the naked eye [1]. For the first time Mirkin et al. developed a colorimetric assay

[1] Address for correspondence: Hui Shen, Anhui Sanlian University, No. 47 He'an Road, Hefei, Anhui, China. E-mail: Shenhuii@mail.ustc.edu.cn.

for the detection of DNA by using Au Nps [16]. In their method two different batches of gold nanoparticles of size 13 nm were attached with thiol modified DNA and the other batch with its complimentary DNA molecules. On mixing these two batches of Au Nps aggregate with each other due to binding affinity between the DNA molecules and developed a color change from red to blue. The binding interaction between DNA molecules can be controlled by its thermal denaturation [16]. After this work several similar strategies were developed for the analysis of RNA [17], proteins [18] cancerous cells [19], organic [20] and inorganic molecules [21]. Xia, et al. developed a label free detection of DNA, proteins and ions molecules by using unmodified Au Nps in presence of positive electrolytes [15]. In their approach, single stranded DNA molecules in the presence of positive electrolytes got aggregated and color change from red to blue whereas double stranded DNA molecules are stabilized without aggregation and thus no color change. This approach can be used for the detection of different DNA molecules by simply changing the probe DNA. Lu et al. developed a low cost biosensor for the detection of melamine in raw milk. In their approach 1-(2-mercaptoethyl)-1,3,5-triazinane-2,4,6-trione (MTT) stabilized gold nanoparticles were synthesized and in the presence of melamine, MTT shows affinity with melamine due to hydrogen bonding and thus aggregation of gold nanoparticles takes place to give different color as shown in figure 1 [20].

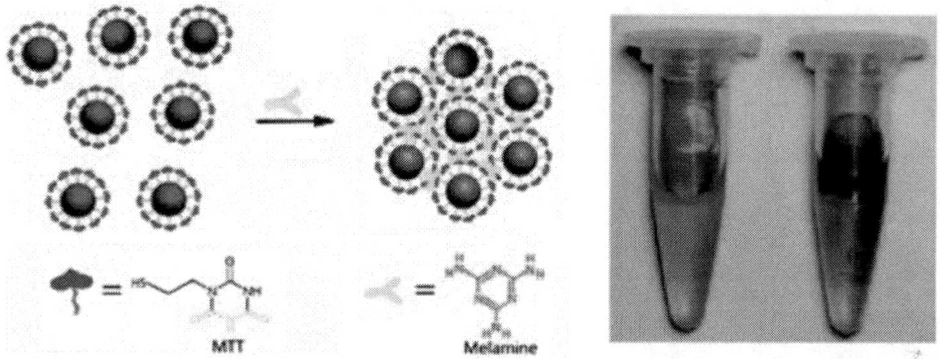

Figure 1. (A) Visual color transition of the MTT-stabilized gold nanoparticles from red to blue upon addition of melamine (from left to right: 0, 1.5 μM) [20].

3. Surface-enhanced Raman scattering (SERS) based biosensors

Raman scattering spectroscopy, the powerful technique which can provide information about molecules based on their vibrations. Raman spectrum is unique and act as fingerprint for the analysis of molecules. But this technique is based on inelastic scattering which is week compared to elastic scattering and thus generates very low Raman signal [1]. For the enhanced Raman signals Au and Ag nanoparticles were used and this technique is known as Surface-enhanced Raman scattering (SERS) [22, 23]. Several strategies were developed for SERS activity and this mainly depends upon size, shape, inter-particle distance and composition of NPs [24, 25]. Aggregation of NPs can improve SERS activity compared to individual NPs. Gary Braun et al. developed a label free DNA detection approach using Ag nanoparticles as shown in figure 2 [26]. In their method, hot spots were created by self-assembly tethering of Ag nanoparticles on a thin Ag film which gives an enhanced Raman signal for the detection of DNA molecules.

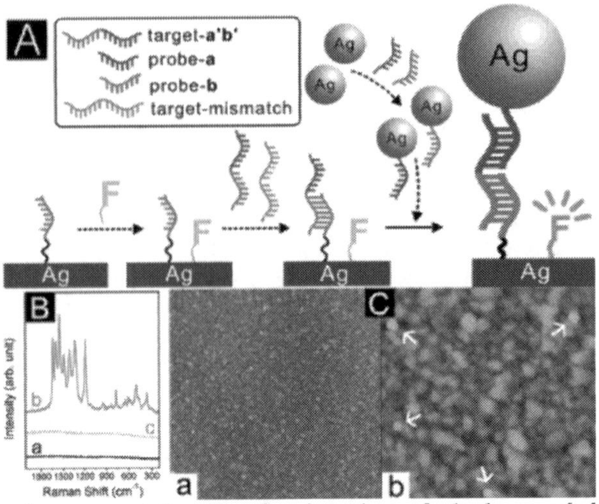

Figure 2. (A) Schematic representation for the detection of single-stranded DNA with SERS technique by using bio-functionalized Ag NP. (B) SERS spectra before and after detection of probe molecule. (C) Representative AFM images of (a) single stranded DNA coated Ag NP; (c) after detection of target DNA probe. Arrows point indicates the AgNPs on the surface [26].

Feng shao, et al. reported a similar strategy for the detection of animal viruses as shown in figure 3 [27]. In their method 3D SERS substrates were synthesized by depositing Ag islands on a biomimetic scaffold chitin with sputtering technique. After deposition of Ag islands developed a hierarchical structure with hot spots and this can be used as SERS for label free detection of animal viruses. Several similar strategies were developed for the detection of biomolecules, metal ions and organic molecules.

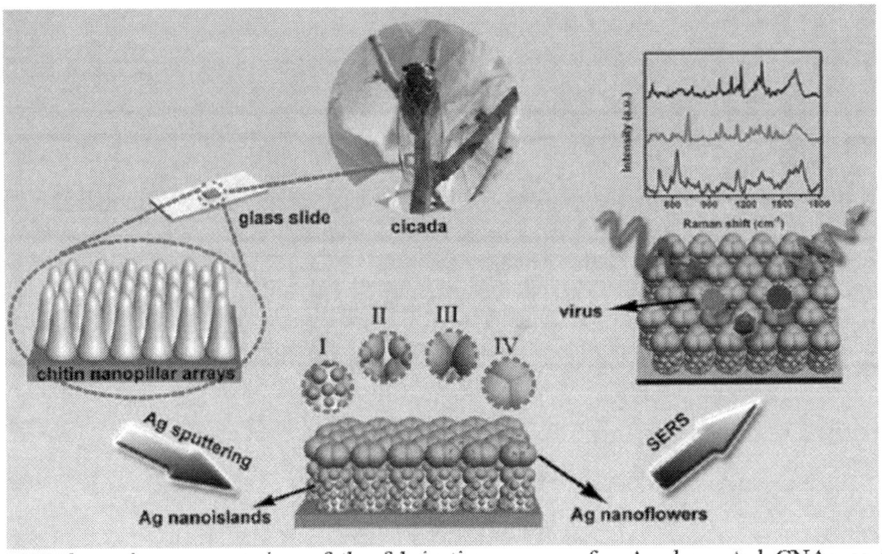

Figure 3. Schematic representation of the fabrication process for Ag-decorated CNAs as the 3D biomimetic substrate applied in label-free animal virus detection. By using sputtering technique, Ag-NIs and Ag-NFs are formed simultaneously on the side surfaces and top ends of chitin nanopillars with hierarchical nanogaps, respectively [27].

IOP Conf. Series: Materials Science and Engineering **87** (2015) 012009 doi:10.1088/1757-899X/87/1/012009

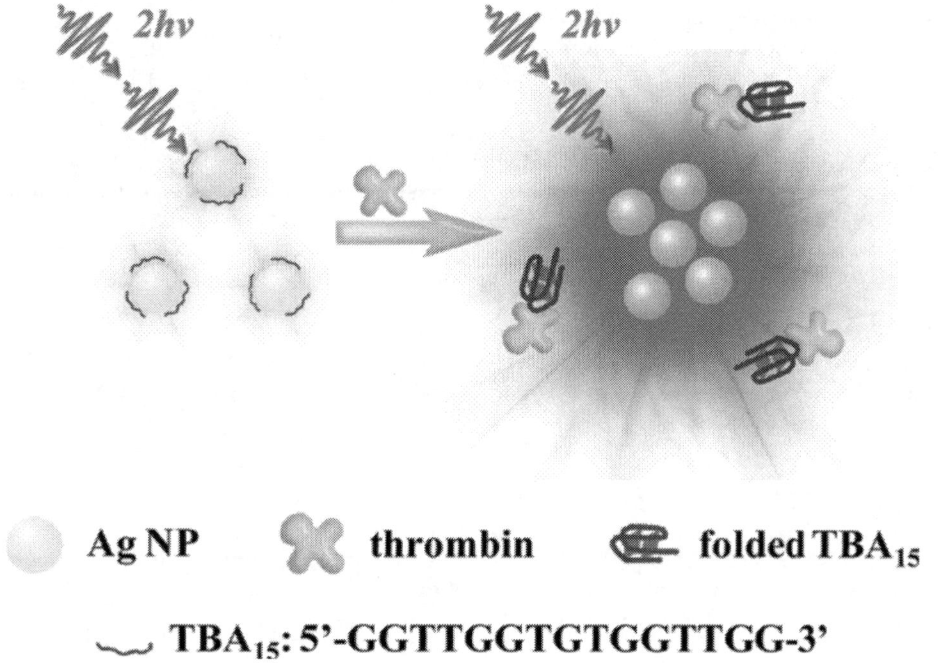

Figure 4. Schematic representation of two-photon sensing of thrombin using Ag NPs and TBA15 [29].

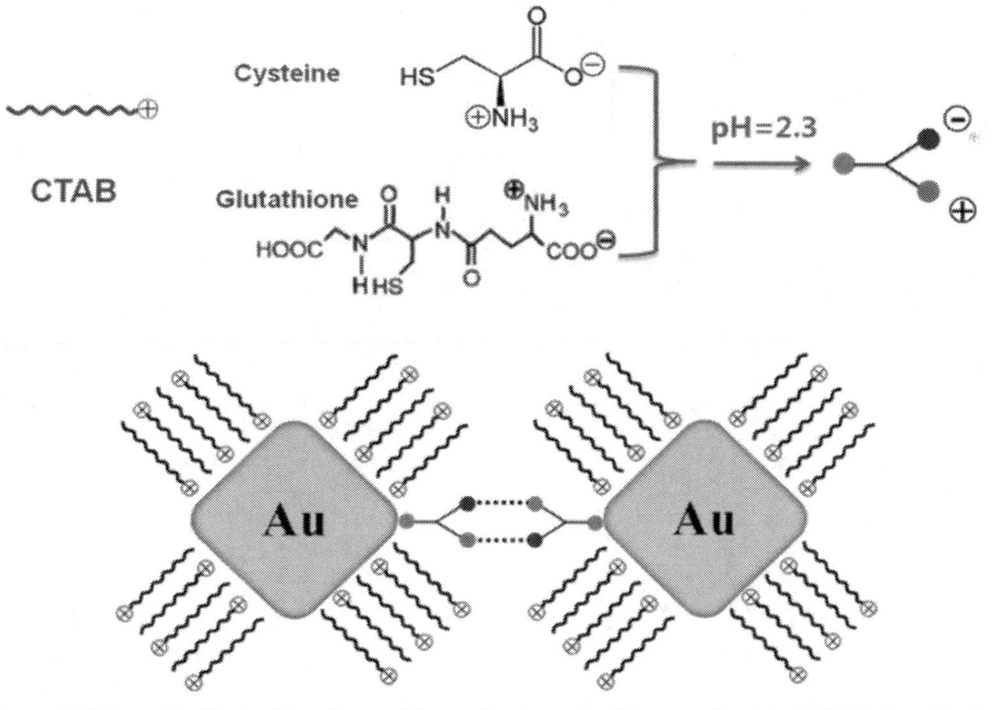

Figure 5. Schematic illustration of assembly mechanism of gold nanocubes (Au NCs) induced in the presence of Cysteine/Glutathione aminoacids [30].

4. Two-photon photo luminescence based biosensors (TPPL)

The two photon photo excitation has better advantages compared to single photon excitation such as quenching of auto-fluorescence, high penetration depth and reduced photo bleaching. Au and Ag nanoparticles show enhanced TTPL and strongly depend on LSPR frequencies and surrounding medium. For an example a Au nanoparticle shows nearly 58 times enhanced TPPL compared to conventional single organic dye (Rhodamine) [14]. TPPL intensity mainly depends on the size of Au and Ag nanoparticles [28]. As compared to SERS technique which requires an additional Raman tags as probe molecules may not require in TPPL technique since the plasmonic nanostructures itself act as TPPL probe molecules. Several biosensors were fabricated based on this strategy. For example Cuifeng Jiang et al. developed TPPL based biosensor using Ag nanoparticles for the detection of thrombin protein as shown in figure 4 [29].

Zhenping Guan, et al. followed a similar strategy and developed a biosensor for the detection of cysteine/glutathione amino acid using Au nanocubes as shown in figure 5 [28]. The developed method has good potential to be used for in vivo biosensing and imaging. This approach can also be applied for the detection of metal ions.

5. Conclusion

In this report we briefly describe about the usage of plasmonic nanostructures such as Au and Ag nanoparticles for the fabrication of biosensors for the detection of different analytes based on colorimetric, SERS and TPPL techniques. Though several biosensors were developed using nanoparticles but the real challenge lies in the fabrication of portable low cost based label free biosensors for the error free real time detection of different analytes.

References

[1] Polavarapu L, Pérez-Juste J, Xu Q H, Liz-Marzán L M 2014 Optical sensing of biological, chemical and ionic species through aggregation of plasmonic nanoparticles *Journal of Materials Chemistry C* **2** 7460-76

[2] Polavarapu L and Liz-Marzán L M 2013 Towards low-cost flexible substrates for nanoplasmonic sensing *Physical Chemistry Chemical Physics* **15** 5288-300

[3] Wang Z and Ma L 2009 Gold nanoparticle probes, *Coordination Chemistry Reviews* **253** 1607-18

[4] Rosi N L and Mirkin C A 2005 Nanostructures in biodiagnostics *Chemical Reviews* **105** 1547-62

[5] Ghosh S K and Pal T 2007 Interparticle coupling effect on the surface plasmon resonance of gold nanoparticles: From theory to applications *Chemical Reviews* **107** 4797-862

[6] Vilela D, González M C and Escarpa A 2012 Sensing colorimetric approaches based on gold and silver nanoparticles aggregation: Chemical creativity behind the assay- A review *Analytica Chimica Acta* **751** 24-43

[7] Lin M, Pei H, Yang F, Fan C and Zuo X 2013 Applications of gold nanoparticles in the detection and identification of infectious diseases and biothreats *Advanced Materials* **25** 3490-6

[8] Liz-Marzán L M 2006 Tailoring surface plasmons through the morphology and assembly of metal nanoparticles *Langmuir* **22** 32-41

[9] Alvarez-Puebla R, Liz-Marzán L M, García de and Abajo F J 2010 Light Concentration at the nanometer scale *The Journal of Physical Chemistry Letters* **1** 2428-34

[10] Lin Y W, Huang C C and Chang H T 2011 Gold nanoparticle probes for the detection of mercury, lead and copper ions *Analyst* **136** 863-71

[11] Zamborini F P, Bao L and Dasari R 2012 Nanoparticles in measurement science *Analytical Chemistry* **84** 541-576

[12] Liu D, Wang Z and Jiang X 2011 Gold nanoparticles for the colorimetric and fluorescent detection of ions and small organic molecules *Nanoscale* **3** 1421-33

[13] Alvarez-Puebla R A and Liz-Marzán L M 2012 SERS detection of small inorganic molecules

and ions *Angewandte Chemie International Edition* **51** 11214-23

[14] Wang H, Huff T B, Zweifel D A, He W, Low P S, Wei A and Cheng J X 2005 In vitro and in vivo two-photon luminescence imaging of single gold nanorods *Proceedings of the National Academy of Sciences of the United States of America* **102** 15752-6

[15] Xia F, Zuo X, Yang R, Xiao Y, Kang D, Vallée-Bélisle A, Gong X, Yuen J D, Hsu B B Y, Heeger A J and Plaxco K W 2010 Colorimetric detection of DNA, small molecules, proteins, and ions using unmodified gold nanoparticles and conjugated polyelectrolytes *Proceedings of the National Academy of Sciences of the United States of America* **107** 10837-41

[16] Mirkin C A, Letsinger R L, Mucic R C and Storhoff J J 1996 A DNA-based method for rationally assembling nanoparticles into macroscopic materials *Nature* **382** 607-9

[17] Alhasan A H, Kim D Y, Daniel W L, Watson E, Meeks J J, Thaxton C S and Mirkin C A 2012 Scanometric MicroRNA array profiling of prostate cancer markers using spherical nucleic acid-gold nanoparticle conjugates *Analytical Chemistry* **84** 4153-60

[18] Kim Thanh N T and Rosenzweig Z 2002 Development of an aggregation-based immunoassay for anti-protein a using gold nanoparticles *Analytical Chemistry* **74** 1624-8

[19] Medley C D, Smith J E, Tang Z, Wu Y, Bamrungsap S and Tan W 2008 Gold nanoparticle-based colorimetric assay for the direct detection of cancerous cells *Analytical Chemistry* **80** 1067-72

[20] Ai K, Liu Y and Lu L 2009 Hydrogen-bonding recognition-induced color change of gold nanoparticles for visual detection of melamine in raw milk and infant formula *Journal of the American Chemical Society* **131** 9496-7

[21] Xue X, Wang F and Liu X 2008 One-step, room temperature, colorimetric detection of mercury (Hg 2+) using DNA/nanoparticle conjugates *Journal of the American Chemical Society* **130** 3244-5

[22] Jeanmaire D L and Van Duyne R P 1977 Surface Raman spectroelectrochemistry Part I. Heterocyclic, aromatic, and aliphatic amines adsorbed on the anodized silver electrode *Journal of Electroanalytical Chemistry* **84** 1-20

[23] Moskovits M 2013 Persistent misconceptions regarding SERS *Physical Chemistry Chemical Physics* **15** 5301-11

[24] Guerrero-Martínez A, Barbosa S, Pastoriza-Santos I and Liz-Marzán L M 2011 Nanostars shine bright for you, Colloidal synthesis, properties and applications of branched metallic nanoparticles *Current Opinion in Colloid and Interface Science* **16** 118-27

[25] Liu X Y, Cheng F, Liu Y, Liu H J and Chen Y 2010 Preparation and characterization of novel thermoresponsive gold nanoparticles and their responsive catalysis properties *Journal of Materials Chemistry* **20** 360-8

[26] Braun G, Lee S J, Dante M, Nguyen T-Q, Moskovits M and Reich N 2007 Surface-enhanced raman spectroscopy for DNA detection by nanoparticle assembly onto smooth metal films *Journal of the American Chemical Society* **129** 6378-9

[27] Shao F, Lu Z, Liu C, Han H, Chen K, Li W, He Q, Peng H and Chen J 2014 Hierarchical nanogaps within bioscaffold arrays as a high-performance SERS substrate for animal virus biosensing *ACS Applied Materials and Interfaces* **6** 6281-9

[28] Guan Z, Gao N, Jiang X F, Yuan P, Han F and Xu Q H 2013 Huge enhancement in two-photon photoluminescence of au nanoparticle clusters revealed by single-particle spectroscopy *Journal of the American Chemical Society* **135** 7272-7

[29] Jiang C, Zhao T, Li S, Gao N and Xu Q H 2013 Highly sensitive two-photon sensing of thrombin in serum using aptamers and silver nanoparticles *ACS Applied Materials and Interfaces* **5** 10853-7

[30] Guan Z, Li S, Cheng P B S, Zhou N, Gao N and Xu Q-H 2012 Band-selective coupling-induced enhancement of two-photon photoluminescence in gold nanocubes and its application as turn-on fluorescent probes for cysteine and glutathione *ACS Applied Materials & Interfaces* **4** 5711-6

IOP Conf. Series: Materials Science and Engineering **87** (2015) 012010 doi:10.1088/1757-899X/87/1/012010

Ultrasonic and dielectric studies of polymer PDMS composites with ZnO and onion-like carbons nanoinclusions

V. Samulionis[1,3], J. Macutkevic[1], J. Banys[1], J. Belovickis[1] and O. Shenderova[2]

[1]Physics Faculty of Vilnius University, Sauletekio al. 9/3, Vilnius 10222, Lithuania
[2]International Technology Center, Raleigh, NC 27715, USA

E-mail: vytautas.samulionis@ff.vu.lt

Abstract. The ultrasonic and dielectric temperature investigations were performed in polydimethylsiloxane (PDMS) with zinc oxide (ZnO) and onion-like carbon (OLC) nanocomposites. In the glass transition region, the ultrasonic velocity dispersion and large ultrasonic attenuation maxima were observed. The positions of ultrasonic attenuation peaks were slightly shifted to higher temperatures after doping PDMS with OLC and ZnO nanoparticles. The ultrasonic relaxation was compared to that of dielectric and such behaviour was described by Vogel-Fulcher law. The upshift of the glass transition temperature with addition of nanoparticles was confirmed by both methods. The additional increase of ultrasonic attenuation in composites doped with OLC and ZnO was observed at room temperature and such behaviour we attributed to ultrasound–nanofiller interaction in polymer matrix.

1. Introduction

In recent years, polymer nanocomposites and understanding their physical and chemical properties have attracted great attention [1, 2]. The presence of nanoparticles in polymer allows both properties from nanoparticles material and polymer matrix to be combined, thus obtaining advanced polymer nanocomposites with improved mechanical, electrical, and optical properties of the materials. Especially, nanoscale inclusions such as carbon nanotubes or onion like carbons have been used to reinforce polymer materials, metal oxide ZnO nanoparticles doped polymers have been studied as alternative materials for optical applications [3]. Polymers, such as polyurea elastomers and PDMS have been proved to be suitable matrices in the development of composite structures due to their ease of production and processing, good adhesion with reinforcing elements. Earlier, using ultrasonic method, we have studied relaxation processes that determine the nanocomposite elastic behaviour of polyurea elastomers with inorganic nanofillers [4, 5]. The PDMS/OLS composites were studied by dielectric and ultrasonic methods [6, 7]. In this contribution we present further ultrasonic and dielectric studies of PDMS with ZnO and OLC nanofilers. The nanofiller content dependent dielectric losses, ultrasonic attenuation and velocity dispersion have been observed above the glass transition temperature. It is shown that the maximum of dielectric losses and ultrasonic attenuation in PDMS nanocomposite can be described according to the Vogel-Fulcher law and the glass transition temperature increases with nanofiller concentration. The addition of ZnO and OLC nanoparticles in

[3] Address for correspondence: V. Samulionis, Physics Faculty of Vilnius University, Sauletekio al. 9/3, Vilnius 10222, Lithuania. E-mail: vytautas.samulionis@ff.vu.lt.

polymer resulted in the increase of ultrasonic velocity in low temperature region and the increase of attenuation at room temperature.

2. Experimental procedure

The OLC samples which were used to fabricate OLC/PDSM composites were prepared according procedure described in [6]. NanoDiamonds with appropriate aggregate size were heated in vacuum (10−2 Torr) at 850°C for 3 h providing the production of OLC with corresponding size of aggregates. Fractions of ZnO with aggregate sizes 30 nm were used as filler similarly as it was described in the literature [6]. The polydimethylsiloxane, Sylgard, was purchased from Dow-Corning as a two part material. When forming PDMS/OLC or PDMS/ZnO composites, an intermediate solvent was employed that served as a dispersion medium for the nanoparticles prior to mixing with the polymer matrix. The nanoparticles were dispersed in the solvent and sonicated to break up large agglomerates, then the suspension was mixed with uncured PDMS. The solvent was subsequently removed in vacuum. The PDMS-nanoparticle mixture was cured at 60°C for 2 hrs and 40°C overnight. Thick films with continuous nanoparticle distribution were obtained. The complex dielectric permittivity of PDMS/OLC or ZnO nanocomposites was measured as a function of frequency and temperature using a bridge setup with LCR meter HP4284A. The ultrasonic studies were carried out by means of automatic computer controlled pulse-echo ultrasonic system as described in [8]. The large dynamic range and large input ultrasonic power allowed the large ultrasonic attenuation values to be measured. The procedure of attenuation and velocity measurement in polymer composites was presented earlier [7]. Samples of dimensions 4 mm × 4 mm × 2-3 mm were cut from larger thick films. We used pulse transmission method, therefore the sample was held confined between two buffer rods. The temperature dependencies of longitudinal ultrasonic velocity and attenuation in pure PDMS and composites with nanoparticles were measured changing temperature (rate of 0.1 K/min) in heating and cooling cycles.

3. Results and discussion

The temperature dependencies of longitudinal ultrasonic attenuation at 10 MHz frequency for PDMS/ZnO composites are shown in figure 1 in cooling cycle. These graphs demonstrate that there

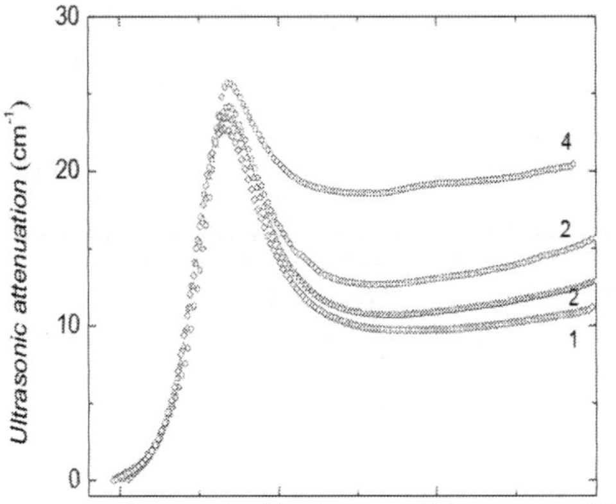

Figure 1. Temperature dependences of ultrasonic attenuation in composites PDMS /ZnO: 1 – 0, 2 – 2, 3 – 5, 4-10 wt% of ZnO nanoinclusions.

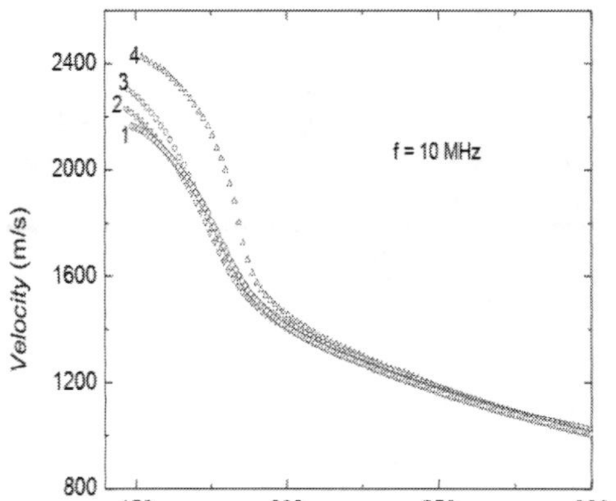

Figure 2. Temperature dependences of ultrasonic velocity in composites PDMS /ZnO: 1 – 0, 2 – 2, 3 – 5, 4 - 10 wt% of ZnO nanoinclusions.

IOP Conf. Series: Materials Science and Engineering **87** (2015) 012010 doi:10.1088/1757-899X/87/1/012010

are attenuation peaks characteristic for relaxation phenomena. Low temperature onset of the attenuation correlate with the glass transition temperature $T_g \approx$ 140–160 K, known from earlier publications [6, 7]. The temperature dependences of ultrasonic velocity in the composites and reference sample are shown in figure 2. These dependencies were obtained from variation of delay time in mechanical system after subtracting the known part of delay variation in the quartz buffers. One can see that temperature dependencies of ultrasonic velocity in PDMS are also dependent on adding of ZnO nanoparticles. The steepest velocity dispersion region corresponds to attenuation maxima. The shift of attenuation and velocity curves to higher temperatures ($\Delta T \approx$ 2-3 K) is clearly seen especially in the low temperatures attenuation wing. Such behaviour is related to a shift of the glass transition temperature in the composites with ZnO. An interesting hysteresis of ultrasonic properties we observed in composites with ZnO nanoparticles. This temperature hysteresis is merely noticeable in pure PDMS but it is more evident for nanocomposites with larger ZnO amount (figures 3 and 4). Such hysteresis can be attributed to the melting phase transition from semi-crystalline to pure amorphous phase which had been observed in PDMS near 220 K [9]. We suppose that ZnO nanoparticles serve as crystallization centres and at low temperatures the crystalline phase in PDMS with ZnO occupies higher volume. Therefore at the first order melting transition ultrasonic attenuation and velocity anomalies are more pronounced in composites with higher degree of crystallinity.

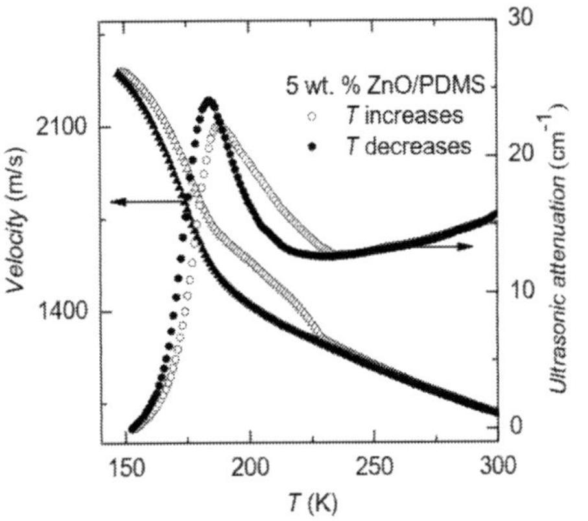

Figure 3. The ultrasonic attenuation and velocity temperature hysteresis in PDMS/ZnO composite with 5 wt % ZnO.

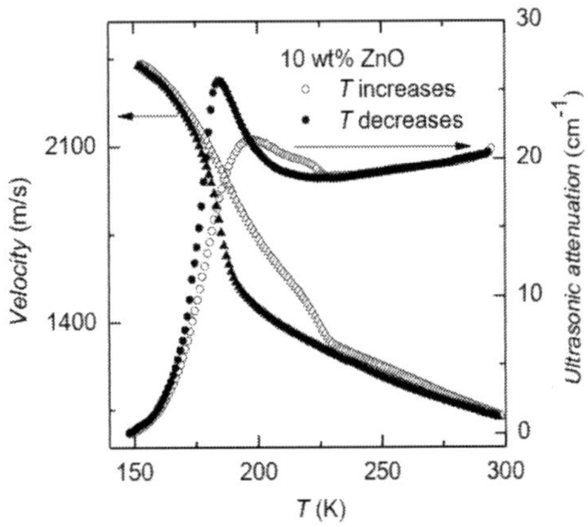

Figure 4. The ultrasonic attenuation and velocity temperature hysteresis in PDMS/ZnO composite with 10 wt % ZnO.

It is necessary to note that in PDMS/ZnO composites after addition of nanoparticles ultrasonic velocity increases, especially at low temperatures, and it is consequence of the reinforcement of PDMS polymer, because elastic coefficients of zinc oxide material are considerably larger. Dielectric measurements of PDMS/ZnO composites in the frequency range 500 Hz – 1 MHz have shown that dielectric relaxation appears in the same temperature region above glass transition and is similar to that of ultrasonic relaxation presented above. Temperature dependences of imaginary part of complex permittivity attributed to dielectric losses for pure PDMS and composites with 1 wt% and 5 wt% and 10 wt% of ZnO inclusions at frequency 1 MHz are shown in figure 5. It is clearly seen that the peak of dielectric loss maximum (T_{max}) also shifts to higher temperature values with increasing the concentration of ZnO nanoparticles in the same way as ultrasonic attenuation peaks (see figure 1). These frequency dependent maxima are caused by alpha relaxation as in pure PDMS matrix [10]. The temperatures T_m corresponding to dielectric loss peaks are also frequency f dependent (figure 6) and can be described by Vogel –Fulcher relationship [11]:

$$f = f_0 \exp(-E_f / k(T_m - T_0)) \tag{1}$$

where k is the Boltzmann constant, f_0 is the frequency in limit T_m - ∞, E_f is the parameter as activation energy, and T_0 is the glass transition temperature. Obtained parameters are summarized in table 1.

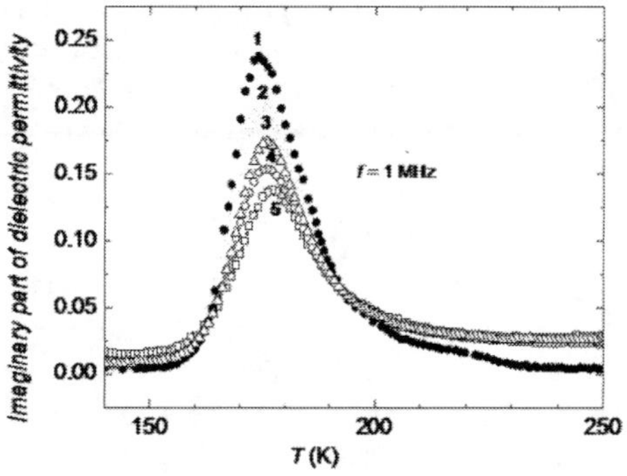

Figure 5. Temperature dependences of imaginary part of dielectric permittivity in composites PDMS/ZnO: 1 – 0, 2 – 1, 3 – 2, 4 – 5, 5-10 wt% of ZnO nanoinclusions.

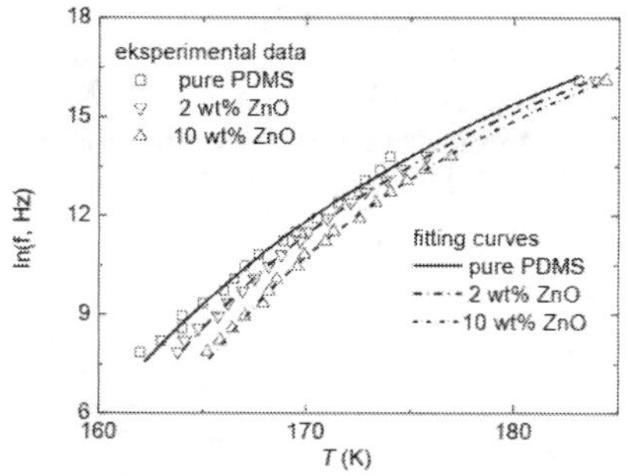

Figure 6. Measured frequencies f versus T_m of PDMS/ZnO composites 1 – 0, 2 – 2, 3 -10 wt% of ZnO nanoinclusions.

It is clearly seen in figure 5 that at low (up to 160 K) and high (210 K- 250 K) temperatures dielectric losses of pure polymer are lower than those of doped with ZnO nanoparticles. Moreover, the higher is the concentration of ZnO in the nanocomposite the higher is the imaginary part of the complex dielectric permittivity. Such differences in losses can be attributed to the impact of oxygen vacancies induced by ZnO inclusions on dielectric properties. Similar phenomenon in bulk $SrTiO_{3-x}$ and its local properties due to the gradient in the concentrations of oxygen vacancies from surface to surface has been observed and reported previously [12-14]. Hence, oxygen vacancies in the interface between ZnO particles and the ambient matrix can change the whole composite dielectric and electric properties.

The glass transition temperature is increasing with the ZnO filler concentration. It should be noted, that relaxation time $\tau = 1/2\pi f$ obtained from dielectric measurements was used to describe ultrasonic attenuation using equation for single relaxation time [5]. After fitting procedure the analogous to T_0 values of T_g and increase of glass transition temperature were verified, but parameters f_0 and E_f were different for elastic relaxation.

Table 1. Vogel- Fulcher fit parameters of dielectric relaxation in ZnO/PDMS nanocomposites.

Material	f_0 (THz)	E_a (eV)	T_0 (K)
Pure PDMS	9.0	0.063	129.4
2 % ZnO/PDMS	0.82	0.046	136.8
10 % ZnO/PDMS	0.89	0.046	140.8

Therefore, the most significant finding of the present investigation on ZnO/PDMS nanocomposites are the upshift of the glass transition temperature T_g with addition of nanoparticles. This upshift of T_g was observed earlier in polyurea elastomer composites with inorganic nanoinclusions and in PDMS/OLC nanocomposites [6, 15]. Molecular dynamics simulations have shown [16] that the presence of nanoscale structures in the polymer network affects the glass transition temperature due to a change of the polymer dynamics. In the case of attractive interactions between the polymer and the nanoparticles, the chain relaxation time is increased and leads to a higher T_g relative to the pure system. In our case, an attractive interaction between polymer chains and the ZnO nanoparticles can be then proposed to explain the observed increase in T_g relative to the pure PDMS polymer matrix. Moreover, according to another theoretical calculation, the density of composite could be higher than the pure polymer density, and an increasing of the glass transition temperature with the filler concentration could be observed [17].

Earlier in polyurea elastomers, we explained the ultrasonic attenuation and velocity dispersion simply by relaxation process with single temperature dependent relaxation time described by Arrhenius equation [5]. In case of PDMS with ZnO and OLC nanoinclusions the attenuation peak is more asymmetric and at higher temperatures the ultrasonic damping is considerably higher than at T = 130 K. The high temperature wing is strongly dependent on doping.

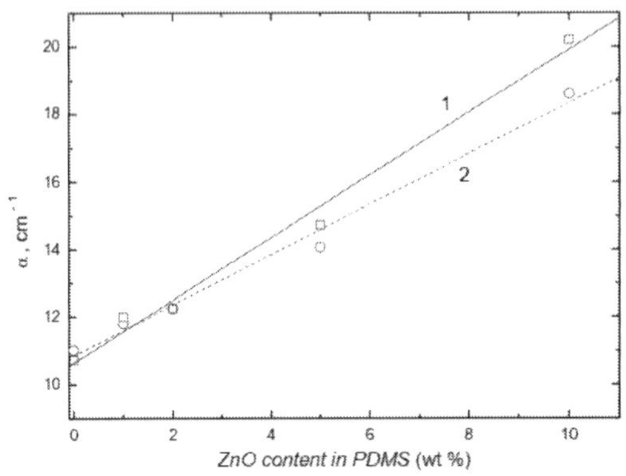

Figure 7. The ultrasonic attenuation dependences on OLC (1) and ZnO (2) content in PDMS composites. T = 290 K.

Figure 7 shows the longitudinal ultrasonic attenuation dependence of OLC and ZnO content at room temperature. It demonstrates that there is another mechanism involved in elastic dissipation process. Such behavior can be related to the ultrasound nanoparticle interaction in amorphous state of PDMS.

4. Conclusions

The temperature dependencies of longitudinal ultrasonic attenuation and velocity in polydimethylsiloxane/zinc oxide composites were measured. Near 180 K the ultrasonic attenuation relaxation maxima and velocity dispersion determined by glass transition in PDMS have been observed. Additionally, the ZnO content dependent ultrasonic attenuation has been observed at the high temperature region above T_g. At low temperatures longitudinal ultrasonic velocity increases with ZnO concentration and it shows that these PDMS polymers are reinforced by nanoparticles.

Acknowledgment

This research is funded by the European Social Fund under the Global Grant measure.

References

[1] Godovsky D Y 2000 *Advances in Polymer Science* **153** 163-205

[2] Kasim F A, Mahdi M A, Hassan J J, J Al-Ani S K and Kasim S J 2012 *International Journal of Nanoelectronics and Materials* **5** 57-66

[3] Tachikawa S, Noguchi A, Tsuge T, Hara M, Odawara O and Wada H 2011 *Materials* **4** 1132-443

[4] Samulionis V, Banys J, Sánchez-Ferrer A and Mezzenga R 2011 *Sensors & Transducers* **12**, 66-70

[5] Samulionis V, Svirskas Š, Banys J, Sánchez-Ferrer A, Jecg Chin S and McNally T 2013 *Phys. Status Solidi.* **A 210**, 2348-52
[6] Macutkevic J, Kranauskaite I, Banys J, Moseenkov S, Kuznetsov V, and Shenderova O 2014 *J. Appl. Phys.* **115** 213702
[7] Samulionis V, Macutkevic J, Banys J, Belovickis J, Shenderova O 2014 *IEEE IUS Proceedings* 1987-8
[8] Valevicius V, Samulionis V, Banys J and Brilingas A 1996 *J. de Physique IV* **6** C8-405-9
[9] Del Valle-Carrandi L, Alegria A, Colmenero 2010 *Eur. Phys. J. Special Topics* **189** 257-61
[10] Fragiadakis D, Pissis P, Bokobza L 2005 *Polymer* **46** 6001-8
[11] Fogel H 1921 *Z. Phys.* **22** 645-6, Vulcher G. S 1925 *J. Am. Ceram. Soc.* **8** 339-55
[12] Liu Z Q, Leusink D P, Wang X, Lu W M, Gopinadhan K, Annadi A, Zhao Y L, Huang X H, Zeng S W, Huang Z, Srivastava A, Dhar S, Venkatesan T and Ariando 2011 *Phys. Rev. Letters* **107** 146802
[13] Liu Z Q, Leusink D P, Lu W M, Wang X, Yang X P, Gopinadhan K, Lin Y T, Annadi A, Zhao Y L, Roy Barman A, Dhar S, Feng Y P, Su H B, Xiong G, Venkatesan T and Ariando 2011 *Phys. Rev. B* **84** 165106
[14] Liu Z Q, Lu W, Zeng S W, Deng J W, Huang Z, Li C J, Motapothula M, Lu W M, Sun L, Han K, Zhong J Q, Yang P, Bao N N, Chen W, Chen J S, Feng Y P, Coey J M D, Venkatesan T and Ariando 2014 *Adv. Mater. Interfaces* **1** 1400155
[15] Fuith A, Reinecker M, Sanchez-Ferrer A, Mezzenga R, Mrzel A, Knitte M, Aulika I, Dunce M, Schranz W 2011 *Sensors & Transducers* **12** 71-9
[16] Starr F W, Schroder T B and Glotzer S C 2002 *Macromolecules* **35** 4481-92
[17] Wei Ch, Srivastava D and Cho K 2002 *Nano Letters* **2** 647-50

Thermal degradation and tensile strength of sansevieria trifasciata-polypropylene composites

H Abral[1] and E Kenedy

Mechanical Engineering, Andalas University, Padang, Indonesia

E-mail: abral@ft.unand.ac.id

Abstract. The paper exhibits thermal degradation and tensile strength of Sansevieria Trifasciata (ST) fibers and polypropylene (PP) composites. Thermal degradation of ST fibers PP composites was conducted by using thermogravimetry (TGA) instrument, meanwhile tensile strength of the composite was done by using tensile equipment. The results show that the thermal resistance of ST fibers PP composites was higher than that of virgin PP only. Increases in volume fraction of fibers in the composites enhance the tensile strength. Scanning Electron Microscope (SEM) observation exhibits good interface bonding between ST fibers and PP matrix.

1. Introduction

Natural fibers still provided a great attention of researcher communities to be developed being a new material having interesting properties. Combination between natural fibers and polymers resulted in some kinds of alternative substance. Even, the compounds of both materials provided better properties in comparison to other conventional materials like steel. Natural fibers have special advantages in cases of light weight, renewability, low price, environmental friendly [1].

Indonesia as tropical country provided huge types of plants. One of those is ST plant growing very easy and fast. They have interesting properties like high mechanical properties, low density and small diameters [2]. As reported by last study that characterization of natural fibers depends on growing location like temperature, humidity, elements in ground [3]. Aim of the study is to research properties of individual ST fibers growing in local area, Indonesia, as well as their effect on PP matrix composites.

2. Methods

2.1. Extraction the ST fibers and composites

The ST fibers were extracted from leaves of local ST plant growing in Padang, Indonesia. Steps the extraction of the fibers from a ST leaf, in detail, were reported in Final Project Kenedy E [4]. The extracted ST fibers were then treated by 5% NaOH solution for 30 min. The treated, cleaned, dried ST fibers were cut, 5 mm long, and distributed between PP sheets with fibers fraction of 2%. Composite samples were made by hot press machine at 150°C, for 20 min by 20 ton in load. Preparation the samples for tensile testing was based on ASTM D638.

[1] Address for correspondence: H Abral, Mechanical Engineering, Andalas University, Padang, Indonesia. E-mail: abral@ft.unand.ac.id.

2.2. Thermogravimetric

Equipment used for measuring degradation of composites while increasing temperature was TGA Q500, a product of TA Instrument. Temperature rate was 20°C per min. from range of temperature 20 up to 1000°C. Nitrogen rate flown during experiment was 40 ml per min.

2.3. Tensile testing

Equipment Com-Ten testing machine of 95T Series was used for measuring tensile strength of composite sample. Speed and temperature during testing were 3 mm/min. and room temperature, respectively.

2.4. SEM observation

Fracture surface of ST-PP composite was observed using SEM photographs. Kind of SEM was a Hitachi 3400 N series equipment.

3. Results and analysis

3.1. Thermal properties

Figure 1 shows thermal degradation of research sample against increase in temperature. Sample in figure 1(a) was pure PP degraded while raising temperature. Meanwhile, Figure 1(b) exhibited thermal properties of 2%ST-PP composites. It is clear to see that both figures showed different characterization. Virgin PP sample has no higher resistance on increasing temperature in comparison to ST-PP composites. Similar result of last study was also reported by Chen, et al [5] that pure PP was only in one stage, meanwhile composites samples exhibits thermal behavior in two steps as raising temperature. The different step of both samples was due to different chemical composition [5].

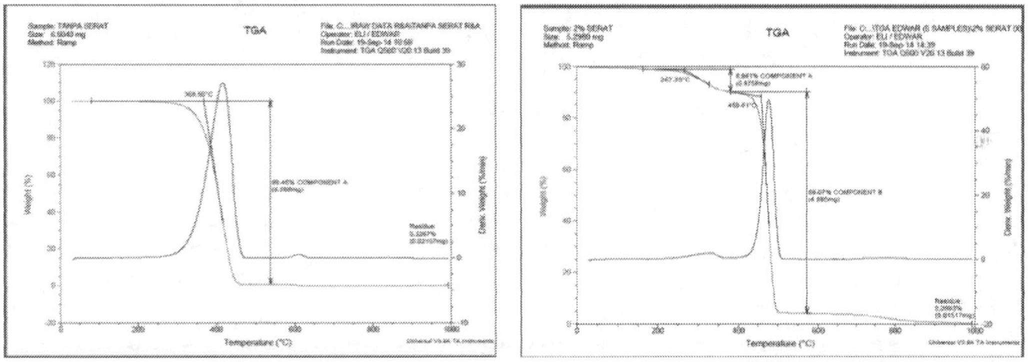

(a) (b)

Figure 1. (a) Thermal degradation for PP only; (b) Thermal degradation for 2%ST - PP matrix.

3.2. Tensile strength of ST-PP composites

Figure 2 shows effect of filler on tensile strength the composites. ST fibers were applied as filler in between PP sheet as matrix component. ST fibers have been treated with sodium hydroxide of 5% for 30 min. In this paper only limited ST fibers fraction was reported. The results can be observed in figure 2 in which tensile strength of the composites increases in rising ST weight. As reported in last studies, enhancement of the mechanical properties was due to external load distributed affectively to ST fibers [6].

3.3. SEM photographs

Figure 3 shows SEM observation on fracture surface of pure PP as well as composites. In the figure 3(a), there is no void, micro crevice observed on the surface, hence, all thin sheets of PP was well fused. Figure 3(b) exhibited fracture surface of the PP-ST composites in which the fibers were bonded by PP matrix. No gap between ST fibers and PP sheets was seen as indication better interface bonding both substances. This result was a good agreement with last studies [7, 8]. Therefore, a good bonding from these components in making composites provided increasing tensile strength as shown in figure 2.

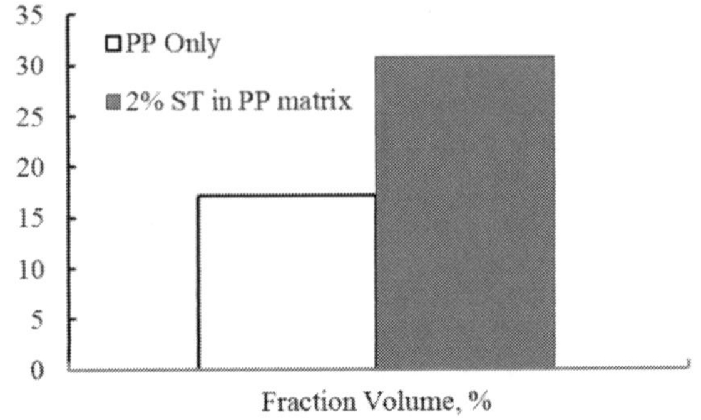

Figure 2. Tensile strength in MPA of PP only as well as 2% ST-PP composites.

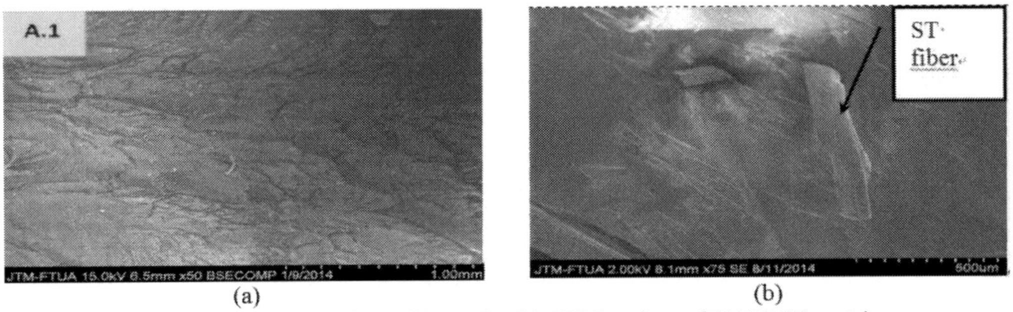

Figure 3. (a) SEM surface of PP only; (b) SEM surface of 2%ST-PP matrix.

4. Conclusion

The paper informed the thermal properties of PP only as well as tensile strength of 2%ST-PP composites. Some conclusions may be explained that 2% ST fibers in PP matrix have increased thermal properties in comparison to PP only. PP was degraded in one step in increasing temperature, meanwhile, composites with 2% ST fibers showed two steps in thermal degradation. Tensile strength of PP composed by 2% ST fibers is higher than that of PP only.

Acknowledgments

Acknowledgement is addressed from the authors to Directorate General of Higher Education, Ministry of National Education, Indonesia, with project "International Research Collaboration and Publication, 2015, for supporting research finance.

References

[1] Abral H, Gafar M F, Andriyanto H, Ilhamdi, Sapuan S M, Ishak M R and Evitayani 2012 *Polym-Plast Technol Eng.* **51** 12–8
[2] Munawar S S, Umemura K and Kawai S 2007 *J Wood Sci* **53** 108–13
[3] Rout J, Misra M, Tripathy S S, Nayak S K and Mohanty A K 2001 *Compos Sci Techn.* **61** 1303–10
[4] Kenedy E 2014 *Report of Final Project* (Andalas University)
[5] Chen M, Ma Y, Xu Y, Chen X, Zhang X and Lu C 2013 *Polym-Plast Technol Eng.* **52** 1566–73
[6] Abral H, Andriyanto H, Samera R, Sapuan S M and Ishak M R 2012 *Polym-Plast Technol Eng.* **51** 500–6

[7] Vilay V, Mariatti M, Taib R M and Todo M 2008 *Compos Sci Techn.* **68** 631–8
[8] Abral H, Putra H, Sapuan S M and Ishak M R 2013 *Polym-Plast Technol Eng.* **52** 446–51

Plasma-modified graphene nanoplatelets and multiwalled carbon nanotubes as fillers for advanced rubber composites

M. Sicinski[1, 5], T. Gozdek[1], D. M. Bielinski[1, 2], H. Szymanowski[3], J. Kleczewska[1] and A. Piatkowska[4]

[1]Institute of Polymers and Dyes Technology, Faculty of Chemistry, Technical University of Łódź, Stefanowskiego 12/16 Street, 90-924 Łódź, Poland
[2]Institute for Engineering of Polymer Materials and Dyes, Division of Elastomers and Rubber Technology, Harcerska 30 Street, 05-820 Piastów, Poland
[3]Institute of Materials Engineering, Faculty of Mechanical Engineering, Technical University of Łódź, Stefanowskiego 1/15 Street, 90-924 Łódź, Poland
[4]Institute of Electronic Materials Technology, Wólczyńska 133 Street, Warsaw, Poland

E-mail: mariusz.sicinski@p.lodz.pl

Abstract. In modern rubber industry, there still is a room for new fillers, which can improve the mechanical properties of the composites, or introduce a new function to the material. Modern fillers like carbon nanotubes or graphene nanoplatelets (GnP), are increasingly applied in advanced polymer composites technology. However, it might be hard to obtain a well dispersed system for such systems. The polymer matrix often exhibits higher surface free energy (SFE) level with the filler, which can cause problems with polymer-filler interphase adhesion. Filler particles are not wet properly by the polymer, and thus are easier to agglomerate. As a consequence, improvement in the mechanical properties is lower than expected. In this work, multi-walled carbon nanotubes (MWCNT) and GnP surface were modified with low-temperature plasma. Attempts were made to graft some functionalizing species on plasma-activated filler surface. The analysis of virgin and modified fillers' SFE was carried out. MWCNT and GnP rubber composites were produced, and ultimately, their morphology and mechanical properties were studied.

1. Introduction

Carbon materials are commonly used as fillers in rubber industry, and their major role is to improve the mechanical properties of the vulcanizates. Too low SFE of the filler can dramatically decrease the adhesion between filler and the polymer matrix, but on the other hand – too high SFE can intensify the agglomeration effect [1, 2]. The possibility of controlling the dispersive part of carbon filler's SFE is strongly limited, so the improvement can only be achieved via changing its polar part. Currently applied matrix – filler interphase compatibilization methods seem not to be effective in many cases, thus some attempts were made to adapt low-temperature plasma to filler surface modification [3, 4].

[5] Address for correspondence: M. Sicinski, Institute of Polymers and Dyes Technology, Faculty of Chemistry, Technical University of Łódź, Stefanowskiego 12/16 Street, 90-924 Łódź, Poland. E-mail: mariusz.sicinski@p.lodz.pl.

Graphene and carbon nanotubes, since their structures were discovered, attracted much interest of many scientific groups from different sectors [5]. Due to their unique mechanical, electrical and thermal properties, they immediately became one of the most promising materials for advanced composites fillers [6]. Unfortunately it was quickly found, that there is a major problem with achieving a satisfactory dispersion in polymer matrix.

It seems that inactive, thin amorphous carbon layer on GnP and MWCNT surface, can significantly reduce expected polymer-filler interactions, which, along with agglomeration (especially talking about CNT which form strongly entangled micro-sized structures) can result in much lower mechanical properties of the composite than expected.

One of well-known chemical methods of surface layer activation is solvent treatment, with generation of covalently bonded acidic, amine or fluorine groups [7]. The process is effective, but unfortunately also has a few major disadvantages–e.g. reagent's harm on environment, and generation of dangerous solvent waste. Besides, more attempts were made for surface properties modification with alternative methods, such as low-temperature plasma treatment. The application of plasma techniques in polymer industry is very promising, especially due to environment protection issues – plasma treatment does not generate waste, is fast, and energetically advantageous process.

Low-temperature plasma can be observed as a discharge between electrodes of low-pressure reactor chamber. The discharge is maintained in a presence of a process gas, which, for example, can be Ar_2, O_2, H_2, N_2, acetylene, methane, or simply air. Depending on a type of gas by controlling the process parameters, plasma treatment can be used for cleaning of the material's surface (so called "micro-sandblasting"), or etching of some functional groups via material's reaction with ionized gas particles. For different material and purposes, a variety of generator frequencies can be applied – from kHz up to MHz. The process can be repeated, and the effect is stable for a few minutes up to months depending on the material (e.g. plasma activated polypropylene can be further processed after several weeks). The most important advantages of plasma technique can be pointed as follows: no solvents have to be used, explosion protection is not necessary, staff is not exposed to toxic chemicals, modified materials do not have to be dried, running costs are very low, and last but not least - all types of solid materials can be treated with plasma.

The process and general mechanism of MWCNT oxygen plasma activation were described. There are a few stages of modification, and their time depends on virgin MWCNT properties, as well as on applied plasma power. The modification stages can be described as follows:

- during the first few minutes of treatment, oxygen plasma activates the surface layer of the material, and the concentration of oxygen increases;
- as a consequence of ion and electron bombardment the amorphous carbon layer covering nanotubes loosens, and the material slightly expands;
- as the bombardment continues, amorphous layer is oxidized and removed–this process proceeds the proper modification of clean nanotubes;
- further process running provides etching of different oxygen groups to material's surface – mainly nanotubes edges and active centers where some structural defects are present (sp^3 hybridization makes modification much easier than in the case of stable sp^2 structure).

Some works consider oxygen plasma treatment as a simple, quick and cheap method for MWCNT cleaning–an inactive, amorphous carbon layer is eliminated in this way. In this work, attempts to modify the commercial GnP and MWCNT were made to obtain improved filler dispersion in elastomer matrix, which was expected after the increasing of filler's SFE values. To achieve a homogenously modified powdery filler, samples were treated in a rotating reactor chamber. Based on previous experience, the process was carried out only in a discharge power of 100 W [8].

2. Experimental

2.1. Materials

The materials treated and used as fillers for rubber composites were:

- GnP Grade 4-Research grade, (CheapTubes, USA)
- MWCNT, purity wt. 95%, (CheapTubes, USA)

As an elastomer matrix, a styrene-butadiene rubber (KER 1500, Synthos S.A., Polska) was used.

2.2. Composites production
Rubber mixes were prepared with a Brabender Plasticorder laboratory micro mixer (Brabender, Germany), operated with 45 rpm, for 30 min. Their composition is presented in table 1. The only one variable was the type of modified mineral filler.

Table 1. Composition of the rubber compounds studied.

Component	Quantity[phr]
SBR (KER 1500)	100
Carbon Black N550	40
MWCNT/GnP	10
ZnO	3,5
Stearine	1
Polnox (antioxidant)	2
TMTD (tetramethyl thiuram disulfide)	0,2
TBBS (N-tert-butyl-2-bezothiazole sulfenamide)	1,4
S	1

2.3. Modification of fillers
Fillers studied were modified with a Zepto plasma reactor (Diener, Germany). The reactor was operated with the frequency of 40 kHz and the maximum discharge power of 100 W. Two different system set-ups were used – one for simple filler activation, and the other for activation, followed by transportation of the modifying agent vapours to the reactor chamber.

Carbon fillers were subject to the oxygen plasma treatment during various time. Efficiency of process gas flow was 20 - 60 cm^3/min, and the pressure in the reactor chamber was maintained at 30-80 Pa. Symbols of the modified fillers were as follows:

- CNT_0/GnP_0 – virgin material
- CNT_1/ GnP_1 - 16 min. oxygen plasma treatment
- CNT_2/ GnP_2 - 32 min. oxygen plasma treatment
- CNT_3/ GnP_3 – 48 min. oxygen plasma treatment
- CNT_4/ GnP_4 - 64 min. oxygen plasma treatment

3. Methods

3.1. Tensiometric measurements of SFE
Effectiveness of the modification is represented by changes in the filler's SFE and its components – polar and dispersive. SFE was examined with a K100 MKII tensiometer (KRÜSS GmbH, Germany). Contact angle was determined using polar (water, methanol, ethanol) and non-polar (n-hexane, n-heptane) solvents. SFE and its components were calculated by the method proposed by Owens-Wendt-Rabel-Kaeble [9].

3.2. Micromorphology of fillers
Micromorphology of the filler was studied with an AURIGA (Zeiss, Germany) scanning electron microscope (SEM). Secondary electron signal (SE) was used for surface imaging. Accelerating voltage of the electron beam was set to 10 keV.

3.3. Mechanical properties of elastomer composites

Mechanical properties of the vulcanizates studied were determined with a Zwick 1435 universal mechanical testing machine (Germany). Tests were carried out on "dumbbell" shape, 1.5 mm thick and 4 mm width specimens, according to PN-ISO 37:1998 standard. The following properties of the materials were determined: elongation at break (Eb), stress at elongation of 100% (SE100), 200% (SE200), 300% (SE300) and tensile strength (TS).

3.4. Micromechanical properties of composites

Mechanical properties of the surface layer of the samples were determined with a NanoTest 600 instrument (MicroMaterials Ltd., UK) – figure 1. The Berkovich diamond penetrated the sample with a normal load of P = 0.5 mN, applying loading/unloading rate of dP/dt = 0.01 mN/s, in 20 different points of the surface. All the experiments were run under the controlled conditions of temperature (T = $23^{\pm 2}$ °C) and relative humidity (60^{+5} %).

4. Results

4.1. MWCNT surface layer modification

The tensiometric analysis revealed (table 2), that a virgin material was oxidized. Sixteen minutes' modification of MWCNT caused the significant decrease in SFE polar part, what stays in conformity with second and third stage of cleaning process described by Xu [10]. Extending the plasma treatment time to 32 minutes causes very significant SFE value increase, which is connected with the sudden change of its polar part form 13.3 up to 54.5 mJ/m^2. Longer modification brings almost no effect on SFE and its parts – the total value stays at the level of 73 mJ/m^2 after 48 minutes, and 75.6 mJ/m^2 after 64 minutes.

Sample (oxygen plasma treatment time)	SFE [mJ/m^2]	polar part [mJ/m^2]	dispersive part [mJ/m^2]
$MWCNT_0$	39.3	18.2	21.2
$MWCNT_1$(16 min.)	31.6	18.3	13.3
$MWCNT_2$(32 min.)	72.9	18.4	54.5
$MWCNT_3$ (48 min.)	73.3	18.3	55.0
$MWCNT_4$(64 min.)	33.6	18.0	15.6

Table 2. Total SFE of vulcanizates containing MWCNT with its components (L – total surface free energy, Ld – dispersive part, Lp – polar part).

For GnP, there is a lack of information on plasma treatment in the literature, as rather mono- and double- layer structures are investigated [11, 12]. In this case, the situation is completely different – although the process was also carried out for time up to 64 minutes, the obtained maximum SFE value was 30 mJ/m^2, with a polar part changing from 4.7 for virgin material, to 12.7 mJ/m^2 for 64 minutes

Table 3. Results of the analysis of total SFE of vulcanizates containing graphene with its components (L – total surface free energy, Ld – dispersive part, Lp – polar part).

Sample (oxygen plasma treatment time)	SFE [mJ/m^2]	polar part [mJ/m^2]	dispersive part [mJ/m^2]
GnP_0	22.9	4.7	18.2
GnP_1(16 min.)	24.2	5.9	18.3
GnP_2(32 min.)	22.8	4.5	18.3
GnP_3 (48 min.)	26.1	7.7	18.4
GnP_4(64 min.)	30.0	12.7	17.3

treatment (table 3). It seems that also for GnP, some equilibrium of oxidation and amorphous carbon layer elimination can be observed, as SFE values for particular treatment time vary from 4.5 to 7.7 mJ/m^2 before the adequate surface oxidation starts(between 48 and 64 minutes). It also has to be underlined, that also in case of commercial GnP, the virgin material demonstrates the presence of oxides layer (polar part of SFE – 4.7 mJ/m^2).

It is easily seen that the modified nanotubes form a bit different structures from virgin material (figures 1A and 1B). Plasma-treated agglomerates seem to be loosened, what is confirmed by micrographs (figures 1C and 1D), where a lot of "molten" edges, and shorter nanotubes are visible. The loosened structure may be achieved due to the amorphous phase elimination stage during MWCNT plasma modification. It seems that in a consequence of long-time oxygen plasma treatment, some of MWCNT tend to break and form shorter structures. Such phenomena can be reflected by better filler distribution in polymer matrix, and ultimately better mechanical properties of a composite.

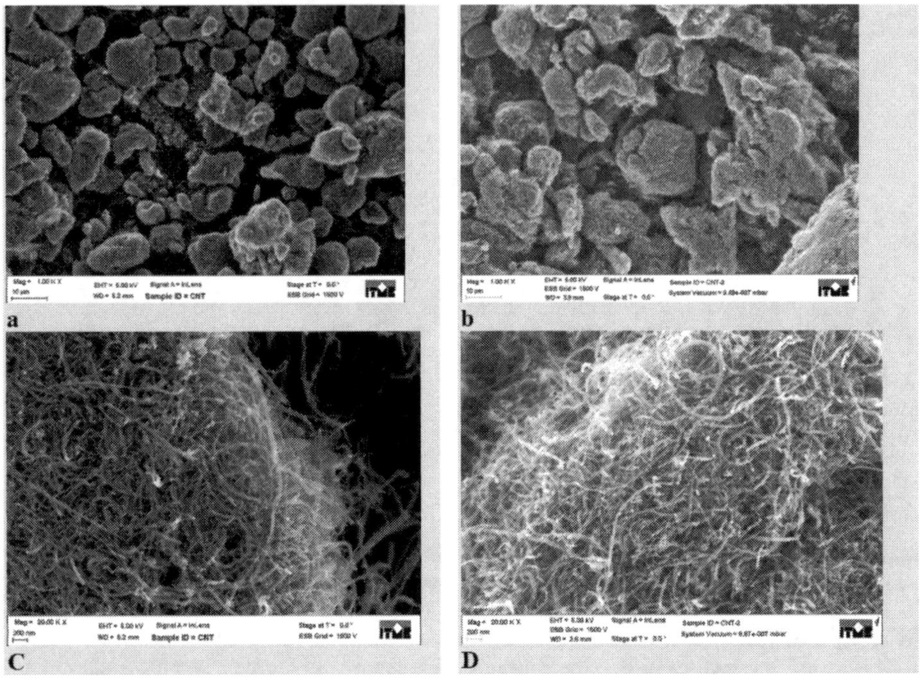

Figure 1. Morphology of carbon nanotubes studied: A, C,– not modified MWCNT, B, D,– modified MWCNT.

Results of SFE analysis for GnP suggested, that its mechanism is completely different from that for MWCNT. GnP surface, probably very stable, almost free of structure defects, is much harder to modify compared with carbon nanotubes. Relatively small changes in SFE values are reflected by SEM analysis – only slight differences can be observed between virgin and plasma treated GnP – for modified materials, some delayered graphene galleries were observed, whereas such structures were not found for virgin samples (figure 2).

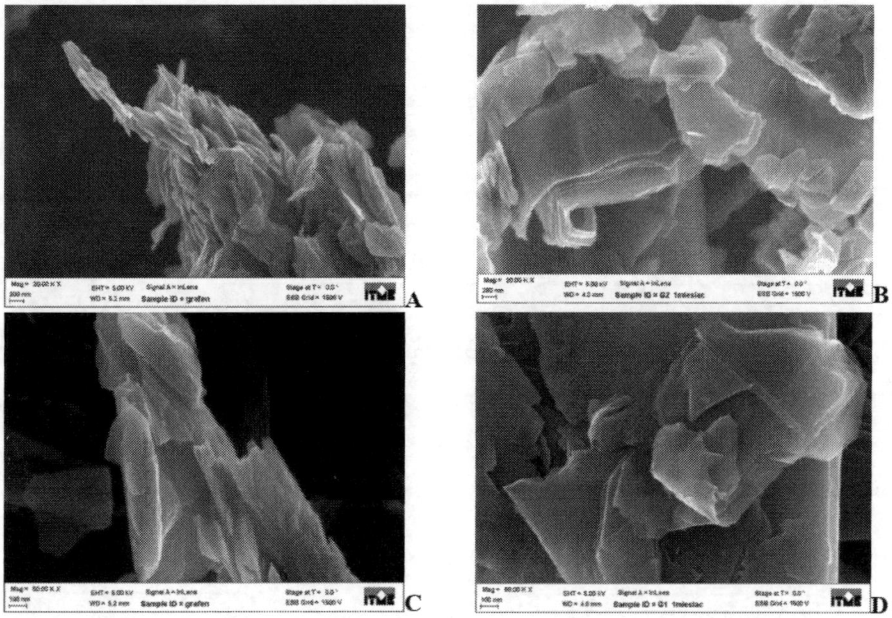

Figure 2. Morphology of graphene samples studied: A, C – not modified graphene, B, D – modified grapheme.

4.2. Mechanical properties of composites

The results of mechanical properties of the composites are consistent with SEM observations and SFE analysis. In case of MWCNT, plasma modification causes significant increase of composite tensile strength. For virgin samples, the obtained value was 7.2 MPa, 12 MPa for 16 min. modification, 15.2 MPa for 32 min. modification, and approximately 18 MPa for 48 min. and 64 min. treatment (table 4). The results of mechanical test confirmed that plasma treatment positively affects the composite tensile strength, probably due to improved filler dispersion in polymer matrix.

Table 4. Mechanical properties of elastomer composites filled with carbon nanotubes.

	SBR/CNT_0	SBR/CNT_1	SBR/CNT_2	SBR/CNT_3	SBR/CNT_4
SE_{100}, MPa	2.3	4.1	4.5	4.6	4.7
SE_{200} ,MPa	4.2	8.5	9.9	9.4	8.8
SE_{300} ,MPa	5.9	8.9	14.7	13.6	13.8
TS ,MPa	7.2	12	15.2	17.9	18.3
E_b ,%	366	312	310	393	361

Such an effect was not observed for GnP containing composites (table 5). The vulcanizates with a modified filler showed lower tensile strength compared with the virgin sample, and only the 64 min. GnP treated sample presented similar parameters to virgin samples. However, what is interesting is Sample SBR/GnP_1, where significant increase of elongation at break was observed. Probably short modification makes the graphene galleries loosened, enabling the relative movement of structures in this way during material stretching.

4.3. Micromechanical properties of composites surface layer

The composites microhardness reaches the maximum value for samples containing 32 min. treated filler, and it doesn't depend on the filler type (table 6). The extension of a modification time to 48 or

64 min. is reflected by significant microhardness decrease. Plastic deformation of the composites is inversely proportional to the microhardness values.

Table 5. Mechanical properties of elastomer composites filled with grapheme.

	SBR/GnP_0	SBR/GnP_1	SBR/GnP_2	SBR/GnP_3	SBR/GnP_4
SE_{100}, MPa	4.2	2.8	3.7	3.3	4.5
SE_{200}, MPa	9.2	6.7	8.5	7.5	9.8
SE_{300}, MPa	13.3	10.4	12.7	11.2	14.3
TS, MPa	19.3	17	16.1	15.6	20.5
E_b, %	413	469	374	418	435

Table 6. Micromechanical properties of the surface layer of the composites studied.

	Microhardness, MPa	Plastic deformation, μm
SBR/CNT_0	2,6 ±0,1	2,9 ±0,3
SBR/CNT_1	3,1 ±0,1	2,6 ±0,2
SBR/CNT_2	4,5 ±0,1	2,2 ±0,2
SBR/CNT_3	3,3 ±0,1	2,4 ±0,2
SBR/CNT_4	2,2 ±0,1	3,1 ±0,3
$SBR/grafen_0$	3,2 ±0,1	2,6 ±0,4
$SBR/grafen_1$	4,1 ±0,1	2,3 ±0,2
$SBR/grafen_2$	5,7 ±0,1	1,9 ±0,1
$SBR/grafen_3$	3,1 ±0,1	2,6 ±0,2
$SBR/grafen_4$	2,3 ±0,1	3,1 ±0,5

5. Conclusion

The plasma modification effect strongly depends on a filler type and character, which is confirmed by SFE measurements and mechanical tests. The oxygen plasma activation seems to be reasonable only for MWCNT, as in this case process causes a significant SFE increase reflected by the increased mechanical performance of the composite. During plasma treatment MWCNT, agglomerates tend to expand and form loosened structures, what probably makes the dispersion better. The micromechanical tests have shown some similar tendency for MWNCT and GnP. The measurement indicates that different phenomena are present - visible at micro- and nano-level, and negligible when the composite is tested in a macro-scale.

Acknowledgments

This work is supported by a project grant from National Science Center, Poland (No. DEC-2012/05/B/ST8/02922).

References

[1] Bieliński D M, Dobrowolski O and Przybytniak G 2008 Morphological aspects of rubber fracture and wear *Journal of Applied Polymer Science* **110** 55-60

[2] Wolff S and Wang J 1992 Filler-elastomer interactions, part IV. the effect of the surface energies of Fillers on elastomer reinforcement *Rubber Chemistry and Technology* **65** 329-42

[3] Dierkes W K, Guo R, Mathew T, Tiwari M, Datta R N, Talma A G, Noordemeer J W M and van Ooij W J 2011 A key to enhancement of compatibility and dispersion in elastomer blends *Kautschuk Gummi Kunststoffe* **64** 28-35

[4] Chityala A and van Ooij W J 2000 Plasma deposition of polymer films on pmma powders using vacuum fluidisation techniques *Surface Engineering* **16** 299-302

[5] Dresselhaus M and Avouris P 2001 *Carbon Nanotubes: Synthesis, Properties and Application* (Berlin, Heidelberg: Springer Verlag)

[6] Gong Q M, Li Z, Bai X D, Li D, Zhao Y and Liang J 2004 Thermal properties of aligned carbon nanotube/carbon nanocomposites *Materials Science and Engineering: A* **384** 209-14
[7] Valentini L, Macan J, Armentano I, Francesco M and Kenny J M 2006 Modification of fluorinated single-walled carbon nanotubes with aminosilane molecules *Carbon* **44** 2196-201
[8] Bieliński D, Parys G and Szymanowski H 2012 Plazmochemiczna modyfikacja powierzchni sadzy jako napełniacza mieszanek gumowych *Przemysł Chemiczny* **91** 1508-12
[9] Owens D K and Wendt R C 1969 Estimation of the surface free energy of polymers *Journal of Applied Polymer Science* **13** 1741-7
[10] Xu T, Yang J, Liu J and Fu Q 2007 Surface modification of multi-walled carbon nanotubes by O_2 plasma *Applied Surface Science* **253** 8945-51
[11] Nourbakhsh A, Cantoro M, Vosch T, Pourtois G, Clemente F, van der Veen MH, Hofkens J, Heyns MM, De Gendt S and Sels BF 2010 Bandgap opening in oxygen plasma-treated grapheme *Nanotechnology* **21** 435203
[12] Felten A, Eckmann A, Pireaux J J, Krupke R and Casiraghi C 2013 Controlled modification of mono- and bilayer graphene in O_2, H_2 and CF_4 plasmas *Nanotechnology* **24** 35570

Application of nano material for shielding power-frequency electromagnetic field

Hualiang Li[1], Li Li and Jiawen Liu

Electric Power Research Institute of Guangdong Power Grid Corporation, Guangzhou, China

E-mail: li-hualiang@163.com

Abstract. Only limited data are available on shielding electromagnetic field exposure in professional work. In our paper, we studied the electromagnetic field intensity in 500 kV substations, and explored influence of nano material in high voltage laboratory simulation. Moreover, the results of nano-fabrics material for shielding power frequency electromagnetic field indicated that, both shielding fabrics can almost completely shield the electric field, but have weak shielding effectiveness against magnetic field.

1. Introduction

Increasing exposure to power frequency electromagnetic fields (EMF, 50 Hz), generated by power lines and electric appliances, raises concern about potential adverse health effects of EMF. Early epidemiological studies indicated a correlation between EMF exposure and the development of childhood leukemia [1, 2], which has been confirmed in other studies [3-5]. The power worker are interested in knowing the sources of professional magnetic and electric fields to help determine whether exposure to these fields causes ill health and in particular [6-8], to provide a measure for reducing the health risk of the electromagnetic fields.

In professional environments, there are various publications that report on measurements of electromagnetic fields inside high voltage substations (110 kV, 220 kV, or 500 kV), and some researcher [9, 10], including us [11] claimed weak relation between electromagnetic and health. While, so far, less information is available on the measure, especially shielding the electromagnetic fields. In the measures of reducing potential health effects of electromagnetic fields, shielding persists as the principal challenge. In this paper, a variety of experiments have been used to assess the nano material for shielding EMFs.

2. Materials and methods

2.1. Experimental measurements

One epresentative open type air substation (500 kV) belonging to the Guangdong Power Grid Corporation (CSG) were selected for achieving the real exposure data. The electrical equipments were

[1] Address for correspondence: Hualiang Li, Electric Power Research Institute of Guangdong Power Grid Corporation, Guangzhou, China. E-mail: li-hualiang@163.com.

installed in outdoor. Circuit-breaker, switch mechanism, current transformer (CT), potential transformer (PT), transformer and others were installed independently. All measuring spots were set near the electric equipments mention above. The measurements were performed using PMM 8053A electromagnetic radiation analyzer meter with the isotropic EPH-50B probe ((PMM, Italy) [11].

2.2. Electromagnetic shielding experiment

The nano silver fiber woven fabric and nano alloy coated fabric (Kangyi, Shenzhen) were selected as the shielding textile fabric. The electromagnetic shielding experiment was tested in high voltage laboratory simulating the real electromagnetic field. The characteristics of two kinds of textile fabric are depicted in table 1.

Table 1. Comparison of the two shielding fabrics.

Material	Structure	Character	Useage
Nano alloy coated fabric	Firm, hard film	The permeability and antioxidant capability are good	the lining or pockets in protective radiation suit
Nano silver fiber woven fabric	Firm, hard, antioxidant film	The product has functions of antibiotic and deodorize.	wearring close to the skin

The following formula can be used to compute the shielding effectiveness (SE).

$$SE_{E(H)} = 20\lg\frac{E_0(H_0)}{E_s(H_s)} \tag{1}$$

Where E0 (H0) and Es(Hs) are the inducing electric (magnetic) field intensity and the shielding electromagnetic field intensity in fabric.

2.3. Exposure limits to electromagnetic fields

Both the testing and evaluating directives have been transposed in the Chinese national standard. The testing method is the measurement of physical agents in workplace part 3: power frequency electric field (GBZ/T 189.3-2007). The values of electromagnetic field intensity were evaluated according to the occupational exposure limits for hazardous agents in the workplace part 2: physical agents (GBZ 2.2-2007) and the labour environment monitoring technological specification of electric power industry part 7: monitoring of power frequency electromagnetic fields (DL/T 799.7-2010). Considering the 50 Hz power frequency, the professional exposure to the electric field is limited to 5 kV/m, and the professional exposure to the magnetic field is limited to 500μT, which is more rigorous than the guidelines for limiting exposure of the International Commission on Non-Ionizing Radiation Protection (ICNIRP) [10].

3. Results and discussion

3.1. Electromagnetic field distribution in open type air substations

The electromagnetic field distribution was tested in 500 kV substations. As can be seen in table 2, there are 2 measuring spots in which the electric field intensity were more than 5 kV/m. there are all below the busbar. In relative high voltage regions (500 kV regions), the nearest distance measuring spots is 22 m, which largely due to the higher voltage.

Furthermore, there are also 2 measuring spots in which the magnetic field intensity exceed the professional exposure limit 500 μT. The maximum value of magnetic fields intensity in shunt reactor regions is 1490 μT. The 35 kV region contains a variety of devices (such as a capacitor, reactor), higher currents can induce high magnetic flied intensity.

Table 2. The electromagnetic field distribution of open type air 500kV substation.

location	Distance (m)	Electric fields V/m	Magnetic fields μT
high voltage wire of 1# transformer A phase	below	268	22.3
low voltage wire of 1# transformer A phase	below	1269	49.4
high voltage wire of 2# transformer A phase	below	274	24
low voltage wire of 3# transformer A phase	below	1738	65.2
500kV 1# busbar A phase	below	7424.0	3.6
500kV 1# busbar B phase	below	3389.0	2.4
500kV 2# busbar A phase	below	5988.0	15.0
2579 circuit breaker of Transmission Lines	below	269.0	22.3
35 kV #3 shunt reactor #7 A phase	1	493.0	1233.0
35 kV #3 shunt reactor #7 A phase	3.5	797.5	489.7
35 kV #3 shunt reactor #4 C phase	1	263.2	1490.0
50227 circuit breaker of 2M busbar	2	4701	3.604
35 kV 4M busbar A phase	below	966.3	52.65
35 kV 2M busbar C phase	below	1266	25.28
35 kV #3 shunt reactor C phase	2	1963	427.6

3.2. The effect of shielding material

It has been very clearly asserted that the electric and magnetic fields may cause physiological effects due to the currents induced into the human tissues exceeding certain levels [12, 13]. Nowadays, many positively techniques have conducted in order to face the requirements of the governmental documents regarding the professional and residential exposure to the electromagnetic fields. Wearing protective clothing and expanding working distance were the main two ways to reduce the adverse effect of electromagnetic fields.

Two shielding textile fabric were selected to test in a power frequency electric field environment. Figure 1 shows the effect of two shielding fabrics on the electric field. In table 3, the comparison of the shielding effectiveness is shown.

As the test voltage increasing, the electric field intensity were from 2.069 kV/m to 14.59 kV/m, the magnetic field intensity were from 44 μT to 115 μT. After coated nano alloy coated fabric, the electric field intensity was not higher than 6 V/m, and the shielding effectiveness was 66.28 dB~70.69 dB.

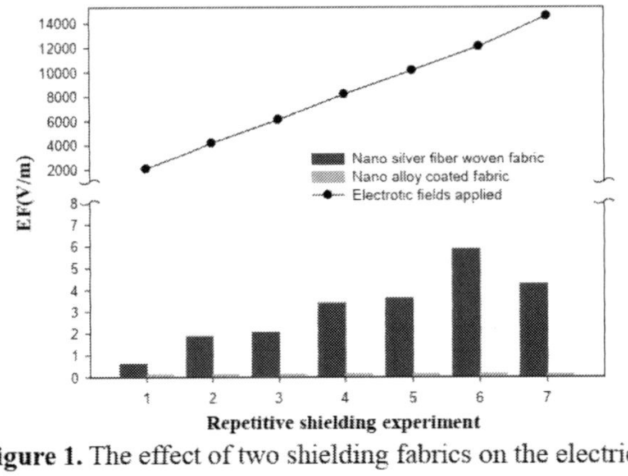

Figure 1. The effect of two shielding fabrics on the electric field.

Table 3. Comparison of the shielding effectiveness using the two shielding fabrics.

experimental value		shielding effectiveness (dB)	
		nano silver fiber woven fabric	Nano alloy coated fabric
electric field intensity (V/m)	2069	70.58	84.45
	4191	67.07	90.94
	6117	69.5	94.08
	8186	67.64	96.25
	10130	68.94	97.97
	12090	66.28	98.18
	14590	70.69	101.99

accordingly, the electric field intensity after coated the nano silver fiber woven fabric were not higher than 1 V/m, and the shielding effectiveness was 84.45 dB~101.99 dB, by which the shielding effect is better. The effects of both were higher than the previous literature [14]. That is to say the nano silver fiber woven fabric and nano alloy coated fabric can effectively shield the power frequency electric field.

However, we found that both materials can hardly shield the power frequency magnetic field in the shielding experiments. The the power frequency magnetic field intensity were nearly the same to magnetic field intensity applied [15, 16].

Nano alloy coated fabrics has good function of anti-static, anti-ultraviolet radiation and so on, but the other function of bad softness and easy breakage meant it was not suitable for the shielding cloth. On the contrary, with the characteristic of anti-bacterial, deodorant, anti-static, good absorption and permeability, the nano silver fiber woven fabric can be used for personal wearing and have good application prospects.

4. Conclusions

The electromagnetic field distribution of an open type air 500 kV substation was measured. Two kinds of shielding material were studied for their shielding effectiveness. The results showed that the power frequency electric field intensity exceed the professional limits in the spots of below busbar. It should be noted that the power frequency magnetic field were low except 35 kV reactor regions. The maximum magnetic fields intensity is 1490 μT. On the other hand, the results of shielding experiment showed that two kinds of material can effectively reduce the external power frequency electric field intensity. The shielding effectiveness were 66.28 dB~70.69 dB and 84.45 dB~101.99 dB, respectively. However both materials had no serious effect on the shielding power frequency magnetic field.

References

[1] Wertheimer N and Leeper E 1979 *Am. J. Epidemiol.* **109** 273-84
[2] Li CY, Lee W C and Lin R S 1998 *J. Occup. Environ. Med.* **40** 144-7
[3] Ahlbom A, Day N, Feychting M, Roman E, Skinner J and Dockerty J 2000 *Br. J. Cancer.* **83** 692-8
[4] Greenland S, Sheppard A R, Kaune W T, Poole C and Kelsh M A 2000 *Epidemiology* **11** 624-34
[5] Hardell L and Sage C 2008 *Biomed. Pharmacother* **62** 104-9
[6] Balcavage W X, Alvager T and Swez J 1996 *Biochem. Bioph. Res. Co.* **222** 374-8
[7] Li P, McLaughlin J and Infante-Rivard C 2009 *Cancer Cause Control* **20** 945-55
[8] Bolte J F, Baliatsas C, Eikelboom T and Kamp I 2015 *Environ. Pollut.* **196** 224-9
[9] IARC 2002 *IARC Monogr Eval Carcinog Risks Hum.* **80** 1-395
[10] ICNIRP 2010 *Health Phys.* **96** 818-36
[11] Li L, Defu X, Jiawen L, Zixin L, Guangcheng Z and Hualiang L 2014 *Australas Phys. Eng. Sci. Med.* **37** 37-44

[12] Olden K, Booman G A and Bemheim N J 1999 *Niehs report on health effects from exposure to power-line frequency electric and magnetic fields* (Bethesda: NIH Publication) pp 99-4493
[13] Auger N, Park AL, Yacouba S, GoneauM and Zayed J 2012 *Occup. Environ. Med.* **69** 147-9
[14] Chung D D L 2001 *Carbon* **39** 279-85
[15] Robert K Adair 1991 *Phys. Rev.* A **43** 1039-D
[16] Chung D D L 2000 *J. Mater. Eng. Perform.* **9** 350-4

PCM/ graphite foam composite for thermal energy storage device

C X Guo[1], X L Ma and L Yang

School of Chemical and Energy Engineering, Zhengzhou University, Zhengzhou, 450001, China

E-mail: guochaxiu@zzu.edu.cn

Abstract. Numerical studies are proposed to predict and investigate the thermal characteristics of a thermal storage device consists of graphite foam matrix saturated with phase change material, PCM. The composite (graphite foam matrix saturated with PCM) is prepared by impregnation method under vacuum condition, and then is introduced into a cylindrical shell and tube device while it experiences its heat from an inner tube fluid. The two-dimensional numerical simulation is performed using the volume averaging technique; while the phases change process is modelled using the enthalpy porosity method. A series of numerical calculations have been done in order to analyze the influence of fluid operating conditions on the melting process of the paraffin/ graphite foam. The results are given in terms of temperature or liquid fraction time history in paraffin/ graphite foam composite, which show that the heat transfer rate of the device is effectively improved due to the high thermal conductivity of graphite foams. Therefore, paraffin/ graphite foam composite can be considered as suitable candidates for latent heat thermal energy storage device.

1. Introduction

High Efficiency thermal storage device is an important requirement for heat recovery, solar energy utilization and seasonal storage since it can bridge the time gap between energy requirements and energy use.

There are sensible heat and latent heat (i.e., phase change) stores in the thermal energy storage device. Thermal storage utilizing solid–liquid phase change materials (PCMs) is one of the most attractive thermal storage techniques due to its inherent advantage of simplicity and reliability. The major advantages of phase change stores are their large heat storage capacity and their isothermal behavior during the charging and discharging process. Therefore, how to select a suitable PCM should be considered in the high efficient latent thermal storage device. Among the various organic or inorganic PCMs, normal paraffin is an excellent PCM for low temperature thermal energy storage due to its large latent heat, good stability, no subcooling degree and no toxicity [1]. Moreover, various paraffin waxes that both the melting point and the heat of fusion increase with their increasing chain length, so it's available in a selection of melting point ranges to make a good match between melting range and system operating temperature.

[1] Address for correspondence: C X Guo, School of Chemical and Energy Engineering, Zhengzhou University, Zhengzhou, 450001, China. E-mail: guochaxiu@zzu.edu.cn.

Although paraffin waxes exhibit desirable properties as PCMs, they present a low thermal conductivity below 0.4 W/(m K). This property increases the thermal resistance of the growing layer of the molten/ solidified medium and thereby may cause the surface heat flux decreases, reducing the rate of heat storage further in the thermal storage device.

In order to offset the low thermal conductivity of the PCM, the device must be designed with an adequate heat transfer area and the material must have suitable heat transfer coefficient. Therefore, the structures of the device have been researched, such as spherical capsule [2], unfinned or finned cylinder [3], which leads to reduced effective energy storage capacity. The low thermal conductivity of paraffin wax can also be enhanced, such as like inserting fins [4], metal matrices [5].

Some authors have put forward to disperse high-conductivity particles inside PCM, such as expanded graphite [6], metal nanoparticles [7], to prepare composite PCMs and measure the effect of containing particles on the thermal conductivity of the composite PCM. The thermal conductivity of the composite PCM was found to increase with the loading of metal nanoparticles. Nevertheless, these enhanced methods add significant weight and cost to the storage devices.

Some other authors have used paraffin supported within the porous structure of silica catalyst [8] or activated carbon [9]. The results have found that the mean pore size of the support could be a critical parameter: too small pore size could decrease the latent heat storage capacity. On the other hand, if the pores are too big, there will not be enough capillary forces to retain the wax liquid phase inside the matrix. However, these two porous supports did not improve the effective thermal conductivity [10].

In addition to the aforementioned methods, inserting PCMs into a porous media appears to be an attractive choice to enhance heat transfer. A number of experimental and theoretical investigations on this problem have been conducted in the past two decades. For example, Lafdi et al. experimentally investigated the phase-change heat transfer within a PCM composite permeated with aluminum foam [11]. Li M et al. [12] gave one of the preparation, structure and thermal energy storage property of capric-palmitic acid/attapulgite composite PCMs; Zhao et al. [13] experimentally and numerically studied the paraffin melting in metallic foam. The results showed that the paraffin embedded in metal foam (copper or aluminum) was found to be more effective in improving the heat conduction of PCM.

In recent years, development of high thermal conductivity, graphite foams opened a new horizon to thermal storage and thermal management applications. Graphite foams present advantages as the bulk thermal conductivities varying correspondingly with density from 40 to 150 W/m K, densities ranging from 0.2 to 0.6 g/cm^3, and good mechanical properties and chemical inertness. It's good compatibility with many surfaces and has a porous structure with a high porosity, which can reach 95% of the volume consist of void spaces and provides a high ratio of surface area to volume. So Graphite foam can be filled with various materials. On the other hand, graphite foam is known to be superior to porous metallic like aluminum, copper or nickel due to having higher thermal conductivity than conductive metals. The above advantages make graphite/PCMs composites desirable for latent heat thermal storage applications. Py et al. [14] proposed a supported PCM made of paraffin impregnated by capillary forces in a graphite matrix. They found that the thermal conductivity of the composite was equal to that of the sole porous graphite matrix. Moreover, the composite presented the same anisotropy with respect to the compression axis. Zhong et al. [15] characterized the thermal performance of paraffin wax/graphite foam composites using experimental measurements of thermal diffusivity and latent heat, experimental results indicated noticeable improvements in thermal diffusivity of the composite compared to that of pure PCM, especially with lower porosity of the foam. A numerical study has been carried out by Lafdi et al. [16] to investigate the thermal performance of graphite foam with different porosities saturated with the PCM (paraffin wax). The energy absorption rate for graphite/PCM composite was compared with that of pure PCM and found a significant improvement in the energy absorption rate.

Thermal characteristics of a thermal protection system consisted of graphite foam saturated with PCM were investigated [17-19]. The results illustrated that the higher the porosity the more stability of the thermal performance of the matrix composite.

Considering the above-reviewed papers, it can be concluded that utilizing highly-porous, highly-conductive graphite foam is one of the most effective methods of improving thermal conductivity and heat transfer of phase change processes.

In the present work, we focus on the shell-and-tube thermal energy storage device, which the PCM occupies the space between the tube and the shell, and the heat is transferred from the heat transfer fluid (HTF) which flows through a single tube, is a highly attractive option in the energy storage applications. In the last few years, several authors numerically or experimentally analyzed the paraffin wax melting process in a shell-and-tube type storage device, for example, in [20], the temperature distribution and melting times of the pure paraffin wax were studied. Here, a new kind of PCM/graphite composite prepared by commercially available graphite foam under the brand name POCO-HTC™ being infiltrated with paraffin wax is introduced as heat storage material in the device. The thermal characteristics of the proposed device are investigated numerically. The numerical model is based on volume averaging technique and the enthalpy porosity method is used to model the phase change process in the shell-and-tube type storage device. Moreover, parametric study is performed to analyze the effects of influential factors on thermal performance of the proposed device.

2. Experimental

PCM used in the experiments is a kind of commercial paraffin. Thermophysical properties of the paraffin wax are measured and listed in table 1. The latent heat is 173.4 kJ//kg, the thermal conductivity is 0.21 W/mK as solid phase and 0.12 W/mK as liquid phase, and the specific heat is 2890 J/kg K. the melting process starts at 46.45°C (319.6K) and ends at 48.68°C (321.83k).

Table 1. Physical properties of PCM and porous materials.

Properties	*PCM1*	*Paraffin*	*Cupper foam*	*Graphite foam*
Density (kg m^{-3})	1000	747	8920	500
Specific Heat (J Kg^{-1}K^{-1})	1828	2890	390	700
Thermal conductivity (W m^{-1}K^{-1})	1.568(solid) 1.458(liquid)	0.21(solid) 0.12(liquid)	401	150
Latent heat (kJ kg^{-1})	220	173.4		
Melting temperature (K)	331-333	319.6-321.83		

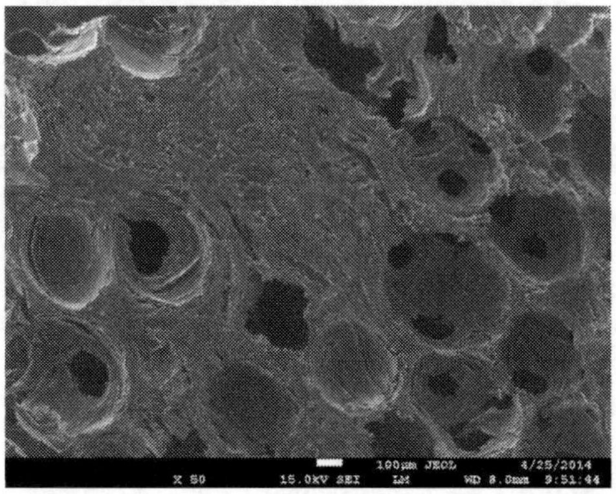

Figure 1. SEM image of graphite foam.

Figure 2. SEM image of PCM/graphite foam composite.

The graphite foam used in these experiments is POCO-HTC™ open pore graphite foam provided by POCO company in USA, which has the average pore diameter of 400 ms, total porosity of 72.8%, density of 0.5 g/cm^3, heat capacity of 0.7 kJ/kg·K and the average heat conductivity coefficient of 150 W/m·K.

Two graphite foam cylindrical disks are fabricated to be introduced as the solid matrices of the PCM/graphite foam composites. The graphite foam cylinders have 150 mm height, 126 mm and 25 mm in outer diameter and inner diameter, respectively. Each one of the graphite foam is heated under vacuum in a molten paraffin wax to achieve maximum wax absorption.

Each one of the graphite foam and paraffin wax are filled into a beaker, then is heated to 335K (above the melting point of the paraffin wax 321.83 K) in a vacuum oven for about 5h, after the graphite foam is submerged in the melted paraffin wax to achieve maximum wax absorption, the vacuum oven is cooled until the paraffin wax is solidified under normal atmosphere. Mass ratio of the paraffin wax in the composites is about 51.2%. The morphology of the graphite and PCM / graphite foam composites are observed by means of scanning electron microscopy (SEM, JSM-7500F) at room temperature. As illustrated in figure 1, there are some unconnected areas between pores in graphite foam. It can be known from figure 2 that some of the PCMs have been impregnated to the pores of the graphite foam.

The two composites are introduced into a vertical cylindrical concentric tube, which has inside radius (r) of 12.5mm and outside radius (R) of 63 mm, and the height (H) of 300 mm. The outer tube is well insulated using a suitable layer of Styrofoam. As presented in figure 3, the heat transfer fluid (HTF), water, flows through the inner tube and exchanges heat with PCM of the composite. During charging process hot water heats the PCM, and when PCM melts, the heat is stored. During discharging process, the PCM solidifies and the stored heat is delivered to the cold fluid, which is the working principle of thermal energy storage device.

3. Physical model

The thermal energy storage device used for the present analyze is shown in figure 3. It consists of an inner tube, outer tube and an annulus space filled with paraffin wax/graphite foam composites. The outer tube is insulated. Due to the symmetry, only a half of this domain will be considered in numerical simulation, as shown in figure 4, and the device is formulated in two-dimensional axi-symmetric coordinates. The volume averaged energy equation with phase change is used to model the heat transfer process inside the PCM/graphite composite with phase change.

The enthalpy porosity theory is employed to predict the phase change process of PCM/graphite foam in the numerical model. The thermal properties of the paraffin and graphite foam are listed in table 1.

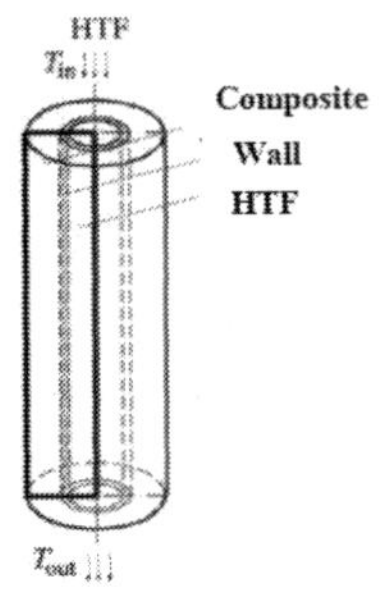

Figure 3. Schematic of the device.

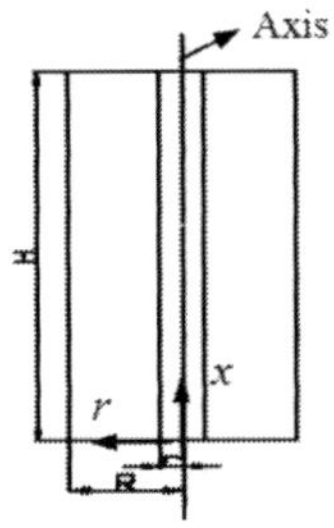

Figure 4. 2D Physical domain.

The governing equations describing the melting of phase change material inside graphite foam are obtained from volume averaging of the main conservation equations of mass, momentum and energy.

In order to simplify the numerical model, the following assumptions are made:

- Newtonian fluid, and the flow is laminar, inlet velocity and inlet temperature of the HTF are both constant;

- Thermophysical properties of the HTF, the tube wall and the PCMs are independent of the temperature;
- Natural convection in the liquid phase of PCM is caused by density variation, which is based on the Boussinesq assumption;

Accordingly, the governing equations used here are as follows:

- for the HTF

Continuity:

$$\frac{\partial(\rho_f u)}{\partial r}+\frac{\partial(\rho_f v)}{\partial x}=0 \tag{1}$$

Where u,v is the flow velocity.

Momentum:

$$\frac{\partial(\rho_f u_i)}{\partial t}+\frac{\partial}{x_i}(\rho_f u_i u_j)=\frac{1}{x_j}\left[\mu\left(\frac{\partial u_i}{\partial x_i}+\frac{\partial u_j}{\partial x_j}\right)\right]-\frac{\partial p}{\partial x_i}+S_i \tag{2}$$

Where p is an effective pressure, and μ is the viscosity. Subscribe i and j represents r and x coordinates. S_i is source term.

Energy:

$$\frac{\partial(\rho_f T_f)}{\partial t}+\frac{\partial(\rho_f u T_f)}{\partial x}=\frac{\partial}{\partial x}\left(\frac{k_f}{c_f}\frac{\partial T_f}{\partial x}\right)+S_T \tag{3}$$

For the numerical solution of the phase change, the enthalpy-porosity model developed by Voller [21], which implemented in FLUENT, is applied. In this model, the entire domain is considered as a porous medium where γ takes the value of 1 in the liquid phase, 0 in the solid phase and $0<\gamma<1$ in the mushy region between solid and liquid. When PCM and porous media are mixed as composite materials, the flow pressure loss due to the solid PCM and the porous media can be determined by the momentum equations as equation (2), but the source terms are expressed as:

$$S_x=-\rho g\beta(T-T_{ref})+Av-\left(\frac{\mu}{\alpha}v+C_2\frac{1}{2}\rho v|v|\right) \tag{4}$$

$$S_r=Au-\left(\frac{\mu}{\alpha}u+C_2\frac{1}{2}\rho u|u|\right) \tag{5}$$

Where β is a thermal expansion coefficient, and α and C_2 are the permeability and the inertia coefficient of graphite foam, respectively. The function A is the "porosity function", is written as [22]

$$A=-\frac{C(1-\gamma)^2}{(\gamma^3+\omega)} \tag{6}$$

Where ω= 0.001 is a small computational constant used to avoid division by zero, and C is a constant reflecting the morphology of the melting front. This constant is a large number, usually 10^4–10^7. The value of $C=10^5$ has been used in the present study.

The equilibrium thermal model is applied to simulate the heat transfer process of PCM in graphite foam, which treats that the liquid PCM and the graphite foam have the same temperature. The energy equation for the device can be described as:

$$\frac{\partial(\rho_{eff,n}h)}{\partial t}=\frac{1}{r}\frac{\partial}{\partial r}\left(k_{eff,n}\frac{\partial T}{\partial r}\right)+k_{eff,n}\frac{\partial^2 T}{\partial x^2}-\varepsilon\rho_{p,l}L\frac{\partial\gamma}{\partial t} \tag{7}$$

Where ε is the porosity of graphite foam and L is the latent heat of the PCM, h is the sensible enthalpy and can further be written as:

$$h = h_{ref} + \int_{T_{ref}}^{T} c_{eff,n} dT \tag{8}$$

The effective thermal conductivity k_{eff} and effective density ρ_{eff} in equation (7) are calculated as the volume average of the conductivities and density of porous material and PCM, respectively, which are denoted by:

$$k_{eff,n} = k_s - (k_s - k_{p,n}) \cdot \varepsilon \tag{9}$$

$$\rho_{eff,n} = \rho_s (1 - \varepsilon) + \rho_{p,n} \cdot \varepsilon \tag{10}$$

Where ρ_s and ρ_p are densities of graphite foam and PCM, respectively; k_s and k_p are thermal conductivities of graphite foam and PCM, respectively. Subscribe n represents liquid or solid state of PCM.

The initial and boundary conditions are:

The initial temperature of PCM is at melting temperature Tm, and the HTF enters the inner tube at a constant temperature which is a little higher than Tm and causes melting of the PCM in the graphite foam. So the velocity of HTF is specified at the inlet, and pressure outlet condition is set at the outlet. The outer tube is insulated.

4. Computational methodology

The model for the numerical study is created using pre-processor software GAMBIT 2.3.16. Meshing of the numerical model is generated and the boundary is applied at appropriate surfaces. Equations of the HTF and the PCM, with initial and boundary conditions have been solved as one domain. The computational domain is resolved with a fine unstructured mesh near the wall to resolve the boundary layer and an increasingly coarser mesh in the rest of the domain in order to reduce the computational time. The GAMBIT model is then exported to FLUENT for problem solving. Time step of 0.05s is conducted to provide a stable convergent solution and accuracy of results. The FIRST ORDER UPWIND differencing scheme has been utilized for solving the momentum and energy equations, whereas the PRESTO scheme is adopted for the pressure correction equation. The PISO algorithm has been applied for the velocity-pressure coupling. The under-relaxation factors for the velocity components, pressure correction and thermal energy were 0.5, 0.3 and 1 respectively. The converged results are assumed to be reached when the maximum relative changes of all velocity, temperature and enthalpy values between consecutive iterations are less than 10^{-6}. Grid-dependence test is carried out by 10000 elements, 18000 elements and 25000 elements and the maximum difference of melt fraction at an identical time is within 0.01% between using 18000 elements and 25000 elements with a time step of 0.05s. Therefore, 18000 elements with a time step of 0.05 s are used in the computations considering both accuracy and computing time are suitable for the accuracy of solution.

5. Numerical model validation

The proposed numerical model has been checked by comparison between numerical results and existing experimental data obtained by [23]. In this experiment, a shell and tube thermal energy storage device, similar to that proposed in the present work as presented in figure 3, with PCM1 and copper foam with porosity of 95% filling the annual space and a HTF flowing through the inner tube, is used.

The thermal properties of the PCM and porous materials are given in table 1. The melting of the PCM1 is studied experimentally for HTF inlet temperatures above the initial temperature of the PCM, $T_{f,in}$ = 343 K. In this experiment, temperature was measured at representative locations point 2 (r=24 mm, x= 215 mm). Figure 5 shows the comparison of measured and calculated PCM1 temperature at point 2. As can be seen in figure 5, a good agreement between present predicted and experimental results is observed. Minor discrepancies between predictions and measurements were observed. The reason for this can be ascribed to the measurements uncertainties and model simplifications. Therefore, the present numerical model can be accurately used for the study of the shell-and-tube thermal energy storage device proposed.

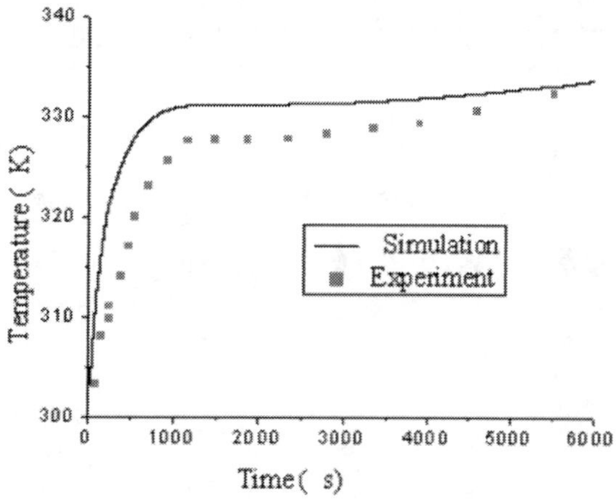

Figure 5. Comparison of experimental and calculated PCM temperature at point 2.

6. Results and discussion

After the validity of the numerical model has been checked, the numerical analysis of the heat transfer during charging of the shell-and-tube thermal energy storage device has been performed. A series of numerical calculations have been carried out in order to analyze the transient heat transfer and the effect of graphite foam on the heat transfer of PCM. Initially, PCMs are solids and their temperatures are assumed to 303 K, PCM is paraffin with the properties presented in table 1, and graphite foam has the porosity of 0.728. Analysis has been performed for various HTF operating conditions: the HTF inlet velocities from 0.35 to 0.5 m/s, and inlet temperatures of the HTF varied from 330 to 345 K.

6.1. Temperature distribution and liquid fraction

In Comparison with the device with pure PCM, figure 6 gives the temperature distribution at 500 s, 1500 s and 2500 s for the case of HTF temperature of 345 K and inlet velocity of 0.35 m/s.

In the melting process of PCM, Temperature distributions at 500 s, 1500 s and 2500 s in shell and tube thermal storage device with pure PCM and PCM/graphite foam composites are given in figures 6(a) and 6(b), respectively. It can be seen for pure paraffin, isothermals is almost vertical and temperature distribution is not uniform, which is much higher near the inner tube wall than that of outer tube wall, due to its low thermal conductivity leading to slow heat conduction. However, for the PCM/graphite foam composite, its isothermals of top are tilted to the outside wall gradually with the increase of melting time because liquid phase of paraffin of high temperature rises with the melting of paraffin wax and solid phase of paraffin wax of low temperature sinks resulting from the gravity and natural convection, and temperature distribution of phase transition zone tend to be homogeneous with time increase. In the melting process, paraffin of upper and near the inner tube wall melt first and that of lower and away from the inner tube wall melt finally.

Figure 7 shows that liquid fraction of PCMs at 500 s, 1500 s and 2500 s in melting process. It also can be learn that heat transfer can be greatly enhanced by graphite foam compared with the melting of pure paraffin wax in shell-and-tube thermal energy storage device. Influence of the HTF inlet temperature on the melting time is shown in figure 8. It can be seen that the melting time decreases with the increase of HTF inlet temperature.

IOP Conf. Series: Materials Science and Engineering **87** (2015) 012014 doi:10.1088/1757-899X/87/1/012014

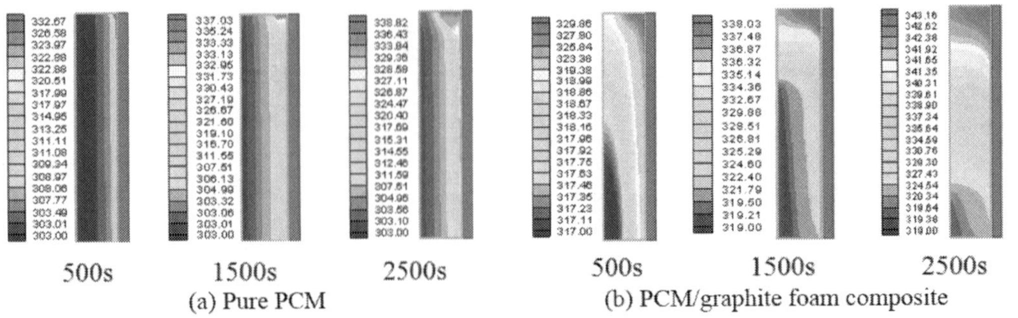

Figure 6. Temperature distributions in device.

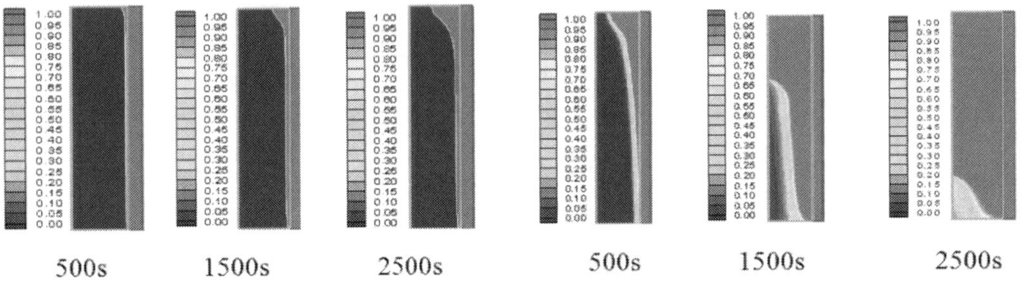

Figure 7. Liquid fraction of PCM in device.

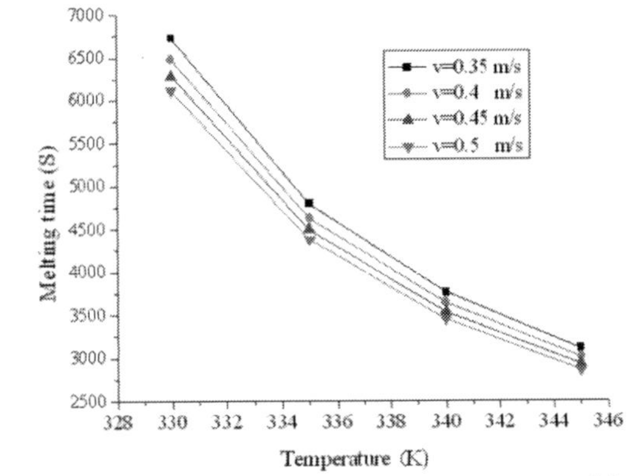

Figure 8. Influence of HTF temperature on melting time.

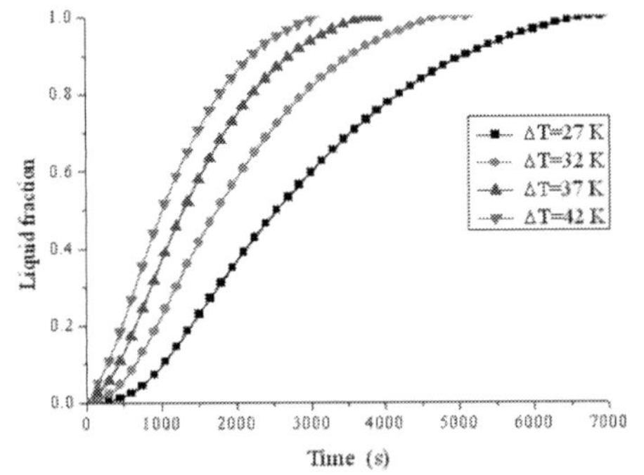

Figure 9. Influence of temperature difference on liquid fraction.

Influences of the different HTF inlet velocity on the melting time of the PCM are also presented in figure 8, the melting time decreases with the flow velocity increase when the HTF inlet temperature is constant. when HTF inlet temperature is 330 K, the melting time is 6725, 6484.85, 6283.85 and 6111.7 s corresponding HTF inlet velocity of 0.35, 0.4, 0.45 and 0.5 m/s, respectively, which means the melting time difference between adjacent velocity is 240.15, 201, and 172.15s, while when HTF inlet temperature is 345 K, the melting time is 3112.5, 3011.35, 2927.85 and 2857.3 s for the corresponding flow velocity of 0.35, 0.4, 0.45, 0.5 m/s, and the melting time difference of adjacent velocity is 101.15, 83.5 and 70.55s, so the influence of velocity on melting time decreases gradually with the increase of HTF inlet temperature.

6.2. Influence of the temperature difference on melting of the composites

The temperature difference between HTF inlet temperature and PCM initial temperature, ΔT, on liquid fraction of PCM is illustrated in figure 9. It can be observed that the liquid fraction of PCM/graphite composite changes over time at different ΔT, and the bigger the temperature difference causes the faster variation of liquid fraction while HTF inlet velocity is equal to 0.35 m/s. It can also be noted that increasing the temperature difference results in decreasing of the melting time of paraffin/graphite composite in the thermal energy storage device.

Figure 10 shows the transient thermal response of the point (r=25.25 mm, x=157.5 mm) in pcm/graphite composite to the different temperature differences. It is observed that thermal response is identical despite of the temperature different. In each case, the temperature rises quickly at the beginning due to the higher heat transfer temperature between the HTF and PCM/graphite composite, then gradually to a steady-state temperature of 320 K where PCM begins to melt by absorbing heat. After all the PCM become liquid state, the temperatures of the composites begin to rise again by absorbing heat from the HTF, until to the heat balance between the HTF and the composites. However, with the temperature difference decreases, phase change point is lagged and time of phase change is prolonged, so the melting time of phase change is longer. As given in figure 10, for example, time of phase change range is from 625 s to 1750 s when the temperature difference is 42 K. while time difference is 32 K, it's from 875s to 2695 s, that is 1820s for melting of PCM.

6.3. Influence of HTF inlet velocity on melting of the composites

Figure 11 gives the liquid fraction of PCM changes with time under different flow velocity when HTF inlet temperature is set as 345 K. Results show that the bigger flow velocity is, the faster melting rate of paraffin wax is. But the influence of HTF inlet velocity on liquid fraction is not obvious.

Figure 12 shows temperature changes of five points in the composites under the conditions of HTF velocity of 0.35 m/s and inlet temperature of of 345 K. Five points are selected as: point 1 (25.25 mm, 157.5 mm), point 2 (31.5 mm, 157.5 mm), point 3 (19 mm, 157.5 mm), point 4 (25.25 mm, 236.25 mm) and point 5 (25.25 mm, 78.75 mm). It can been seen from figure 12 that the temperature of composites rise and heat stores in the form of sensible heat before the composites reaching the melting point; on the contrary, temperature of composites keep at melting point and heat stores in the form of latent heat; while all PCMs melting completely, it again stores heat by means of sensible heat, so the temperature continue to increase. Results show that temperatures of these points are identical, but temperature of point 4 increases first, then point 3, point1and point 2 and finally temperature of point 5 rises. Because point 4 locates in the upper of phase change region and absorbs heat from HTF first. Therefore, it's proved that paraffin wax of the composite near the top and the inner tube wall melts first, then melting front gradually to other areas.

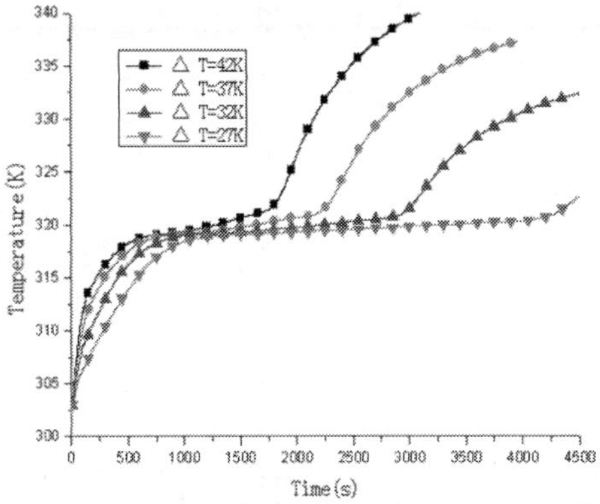

Figure 10. Transient thermal response of a point in PCM/graphite foam composites

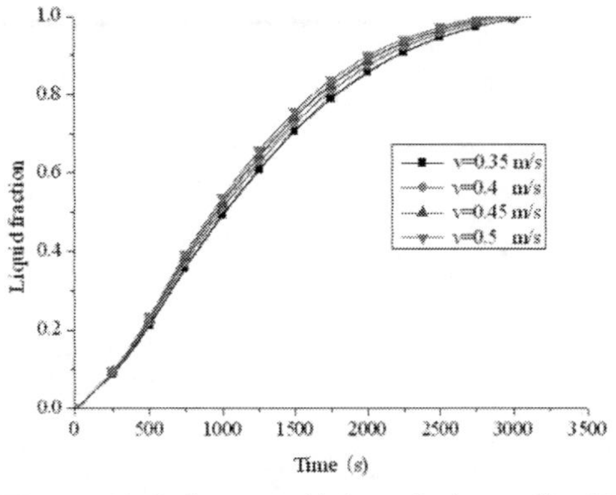

Figure 11. Influence of inlet velocity on liquid fraction of PCM.

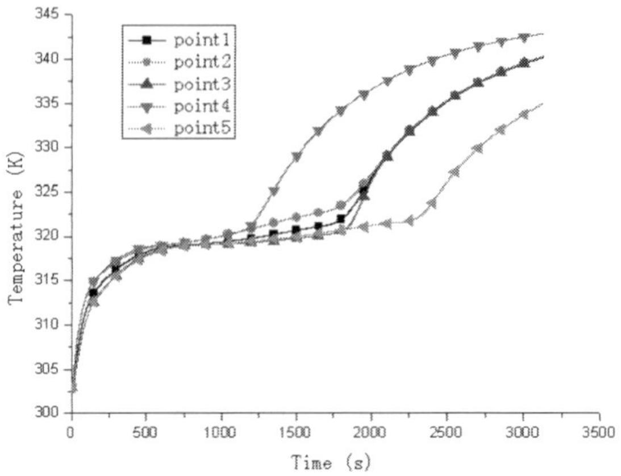

Figure12. Variation of temperature over time at five points.

7. Conclusion

Graphite foam are introduced as the porous medium while paraffin wax was introduced as the occupied PCM into the matrix pores. The PCM/graphite foam composite is prepared and introduced into a tube and shell thermal energy storage device, where paraffin/graphite foam composites as the PCMs fill the shell side and water as HTF circulating inside the tube, has been numerical analyzed in this paper. Compared with the results of the pure PCM, the phase change heat transfer can be greatly enhanced by using graphite foam in device. Moreover, several simulation calculations have been conducted in order to study the effects of operating parameters: HTF inlet temperature and velocity on melting process of the composites. It can be concluded that HTF inlet temperature plays a significant role for reducing the melting time and liquid fraction, but the influence of flow velocity on melting process is small although increasing velocity can reduce the melting time.

Acknowledgments

The authors are thankful for financial contributions from the National Natural Science Foundation of China (No. 51176173) and the Key Programs of the Science and Technique of Henan Education Committee (No. 14A480002).

Nomenclature

A	[-]	Porosity function	Special characters:		
C	[-]	Constant	α	[m^2]	Permeability
c	[$J\ kg^{-1}\ K^{-1}$]	Specific heat	ρ	[$kg\ m^{-3}$]	Density
C_2	[-]	Inertia coefficient	β	[-]	Thermal expansion coefficient
g	[$m\ s^{-2}$]	Gravity acceleration	ε	[-]	Porosity of foam material
h	[$J\ kg^{-1}$]	Sensible enthalpy	γ	[-]	Liquid fraction of PCM
k	[$W\ (mK)^{-1}$]	Thermal conductivity	μ	[$kg\ (ms)^{-1}$]	Dynamic viscosity
L	[$J\ kg^{-1}$]	Latent heat of fusion			
p	[Pa]	Pressure	Subscripts:		
u	[$m\ s^{-1}$]	Velocity in r-direction	eff		Effective expression
v	[$m\ s^{-1}$]	Velocity in x-direction	f		Fluid (Phase change material)
t	[s]	Time	in		Inlet
T	[K]	Temperature	ref		Reference value
Tm	[K]	Melting temperature	s		Graphite foam
x	[m]	Cartesian axis direction			
r	[m]	Radius			

References

[1] Oro E, de Gracia A, Castell A, Farid M M and Cabeza L F 2012 *J. Appl Energy* **99** 513-33
[2] Bedecarrats J P, Strub F, Falcon B and Dumas J P 1996 *Int. J. Refrig.* **19** 187-96
[3] Ryu H W, Hong S A, Shin B C and Kim S D 1991 *Energy* **16** 727-37
[4] Shatikian V, Ziskand G and Letan R 2005 *Int. J. of Heat and Mass Transfer* **48** 3689-706
[5] Mesalhy O, Lafdi K, Elgafy A and Bowman K 2005 *Energy Conver. and Manag.* **46** 847-86
[6] Marin J M, Zalba B, Cabeza L F and Mehling H 2005 *Int. J. of Heat and Mass Transfer* **48** 2561-70
[7] Valan A A and Mujumdar S 2012 *Int. Commun. in Heat and Mass Transfer* **39** 8-16
[8] Erk H F and Dudukovic M P 1996 *AIChE J.* **42** 791-808
[9] Chapotard C and Tondeur D 1983 *Chem. Eng. Commun.* **24** 183-204
[10] Cacciola G, Restuccia G and Mercadante L 1995 *Carbon* **33** 1205-10
[11] Lafdi K, Mesalhy O and Shaikh S 2007 *J. Appl. Phy.* **102** 083549
[12] Li M, Wu Z and Kao H 2011 *Appl. Energy* **88** 3125-32
[13] Zhao C Y, Lu W and Tian Y 2010 *Sol. Energy* **84** 1402-12
[14] Py X, Olives R and Mauran S 2001 *Int. J Heat Mass Transfer* **44** 2727-37
[15] Zhong Y, Guo Q, Li S, Shi J and Liu L 2010 *J. Sol. Energy Mater. Sol. Cells* **94** 1011-14
[16] Lafdi K, Mesalhy O and Elgafy A 2008 *J. Carbon* **46** 159-68
[17] Mesalhy O, Lafdi K and Elgafy A 2006 *Carbon* **44** 2080-88
[18] Wirtz R, Fuchs A, Narla V, Shen Y, Zhao T and Jiang Y 2003 *AIAA Paper* 0513
[19] Chintakrinda K, Weinstein R D and Fleischer A S 2011 *Int. J. of Thermal Sci.* **50** 1639-47
[20] Rosler F and Bruggemann D. 2011 *Heat Mass Transfer* **47** 1027-33
[21] Voller V R and Pankrash C 1987 *Int. J. Heat Mass Transfer* **30** 1709-20
[22] Brent A D, Voller V R and Reid K J 1988 *Numer. Heat Transfe* **13** 297-318.
[23] Liu F Q 2010 *Numerical simulation and experimental study on enhancement of foam metals to phase change heat storage* (China: Hebei Science and Technology University) Master Thesis

IOP Conf. Series: Materials Science and Engineering **87** (2015) 012015 doi:10.1088/1757-899X/87/1/012015

Mechanical properties of GFRP tube confined recycled concrete under axial compression

Xiaogang Wang[1], **Chaofeng Liang, Zechenglong Zhou, Lanqi Dong, Kewei Ding and Jialun Huang**

Department of Civil Engineering, Shaoxing University, Shaoxing, China, 312000

E-mail: 550886527@qq.com

Abstract. This article outlines the recycled aggregate replacement rate and thick-diameter rate of GFRP tube confined in recycled concrete, which has an important impact on the material's compressive strength. Overall, under the same conditions of using recycled concrete, the bearing capacity of short concrete columns can be improved by using broader GFRP tubes. There is a four-fold increase in the bearing capacity of short concrete columns compared to the short column without the restriction of a GFRP tube. The bearing capacity of a short column crafted by recycled coarse aggregate is much lower (about 30%). than those made by common concrete column Additionally, the bearing capacity of short columns made by recycled fine aggregates is also lower than those made by common concrete (approximately 20%). Finally, we find that there is no significant difference between experimental and theoretical data.

1. Introduction

GFRP (short glass fiber reinforced plastic) is characterized by having a high strength, strong corrosion resistance, and thermal expansion coefficient of concrete material; its winding mold made of GFRP tube at a certain angle, and pouring concrete in it can form the GFRP pipe concrete column [1]. So, the GFRP tube can reduce the influence of environment on the concrete, improving the durability of the structure, but also greatly improves the carrying capacity of the concrete column. Also, it can be applied as a template, and pouring concrete in concrete pouring is completed. Meanwhile, concrete is the modern civil engineering material, and it is difficult for people to handle a social problem such as the waste of it. Recycled concrete can easily solve the current problem, and concrete can be used responsibly again. As a result of these arguments, recycled concrete has become more popular in modern civil engineering [2].

2. The experiment

2.1. Specimen preparation

This study was conducted by a30 GFRP tube recycled concrete short column test; the specimen diameter is at 150 mm, height is 300 mm, and short columns are made from ordinary Portland cement, C32.5. GFRP tube is provided by Shanghai Ning Trading Company, while recycled concrete material is crushed by the test of C25 concrete beam, after crushing screening to the experimenter. The

[1] Address for correspondence: Xiaogang Wang, Department of Civil Engineering, Shaoxing University, Shaoxing, 312000, China. Tel.: +8618257557508; E-mail: 550886527@qq.com.

thickness of the GFRP tube in the coarse aggregate replacement ratio and the fine aggregate content is 3 mm, and the thickness diameter ratio on the recycled concrete axial compression performance is recorded at 3 mm, 6 mm, 9 mm respectively. The data from this study can be seen in table 1.

Table 1. Test parameters.

Specimen number	Wall thickness (mm)	Replace rate of recycled coarse aggregate Ñ (%)	Replace rate of recycled fine aggregate P (%)	f_c (Mpa)
1	3	0	100	76.40
2	3	50	100	61.15
3	3	100	100	52.6
4	3	100	0	66.1
5	3	100	50	58.9
6	3	100	100	54.2
7	0	100	100	12.9
8	3	100	100	53.7
9	6	100	100	70.4
10	9	100	100	118.1

The specimen of GFRP tube and the short concrete column height are the same; in the GFRP tube strain gauge and tile, according to the figure 1, below.

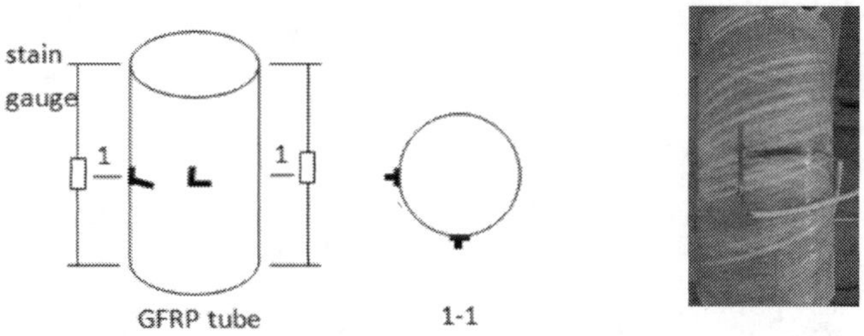

Figure 1. Strain gauge arrangement.

2.2. Test methods

In the middle of the test piece and GFRP pipeline flank placed on horizontal and vertical strain gauges then connect the static resistance strain gauges, measure test concrete short column in the process of each phase of the transverse and longitudinal strain. Specimens were loaded on 300 KN pressure machine, and the loading speed is recorded at 0.5 Mpa/s slow continuous loading, until the specimen damaged.

Specimen loaded figure as showed in figure 2.

Figure 2. The loading diagram and experimental process diagram.

3. The thickness-diameter ratio of GFRP tube on the axial compression bearing capacity of recycled concrete

The thickness to diameter ratio on the influence of the axial compression bearing capacity of recycled concrete short column is reflected in the ultimate compressive strength of GFRP tube and GFRP tube binding of recycled concrete short columns.

According to the loading characteristics of the GFRP tube's recycled concrete short column, there are six basic assumptions:

- The radial restraint stress of the recycled concrete is evenly distributed;
- Regardless of the slippage between recycled concrete and GFRP tube;
- Concrete is non-linear elastic isotropic material;
- By the radial stress of the GFRP tube is far less than its axial and hoop stress, it can be ignored. Consider with thin GFRP tube [3];
- Consider the GFRP pipe is thin, we think axial stress and hoop stress is evenly distributed along the thickness direction;
- Considering the short column is shorter, we do not consider the stability coefficient.

By the force balance of the GFRP tube recycled concrete column analysis reveals the vertical axis pressure by the core of recycled concrete compression and the compression of GFRP tube share to occur at the same time because of the short column inferior height, therefore we don't consider the effect of the stability coefficient, axial compression bearing capacity of GFRP tube recycled concrete column by the static equilibrium condition analysis, the design equation of the axial compressive strength of recycled concrete with the constraint, the compressive strength of GFRP tube, and GFRP tube for the core of the recycled concrete constraint.

Therefore, we apply the equation:

$$N_u = N_0 = f_{cc}A_c + f_{fy}A_f \tag{1}$$

In the equation, N_0 is the core of the GFRP tube recycled concrete ring to restraint stress, A_c and A_f is the recycled concrete, and the cross sectional area of the GFRP tube; f_{fy} is the compressive strength of GFRP tube; f_{cc} is restrained by GFRP tube of the compressive strength of recycled concrete.

f_{cc} derivation equation [3]:

$$f_{cc} = f_{ck} + \zeta_{cf} f_{fy} \tag{2}$$

In the equation, f_{cc} is constrained axial compressive strength of recycled concrete, f_{ck} is the ultimate compressive strength of GFRP tube, and ζ_{cf} is the GFRP pipe ring enhancement coefficient to constraints.

Enhancement coefficient of GFRP pipe ring [4] to the constraints ζ_{cf}:

$$\zeta_{cf} = \frac{f_{hl}}{f_{ck}} \tag{3}$$

$$f_{hl} = \frac{2f_{hf}t}{d} \tag{4}$$

In the equation, f_{hl} is the core of GFRP tube's recycled concrete ring to restraint stress. f_{hf} is the GFRP materials used to attain the tensile strength, while t and d are the GFRP tube wall thickness and diameter, respectively.

The ultimate compressive strength of GFRP tube is represented by f_{fy} [4]:

$$f_{fy} = \frac{(1+\mu_f v_{f\theta_x})E_f \varepsilon_{\theta_u}}{(1-v_{f\theta_x} v_f)\mu_f} \tag{5}$$

In the equation: $v_{f\theta_x}$ is ring of GFRP tube to the Poisson ratio, v_f is GFRP tube axial poisson's ratio, μ_f is GFRP tube axial compression poisson's ratio of concrete reinforcement phase, ε_{θ_u} is GFRP tube to tensile strain limit cycles, E_f is the GFRP tube axial compressive modulus of elasticity.

Upon examination, we utilize the following equation: $N_u = f_{cc}A_c + f_{fy}A_f$.

The theoretical equation for calculating the bearing capacity calculation of the theoretical equation is consistent with the experimental data. Specific data are shown in table 2.

Table 2. Comparison table theoretical values and the measured values.

Specimen number	Wall thickness (mm)	f_c (Mpa)	N (Kn)	N_u (Kn)	N_u/N
8	3	53.7	1136.69	1230	1.08
9	6	70.4	1599.51	1575	0.98
10	9	118.1	2086.58	2190	1.04

4. The parameter analysis

During the experiments, we continued the study of the recycled concrete short column GFRP tube's mechanical properties. According to the test data, we analyse the whole loading process of short column, and in analysing we adopt the following basic assumptions [5]:

- There is no relative slip between GFRP and recycled concrete;
- During the process of the stress of the short column, a deformation compatibility condition is met, is=ε_f ; is recycled concrete strain, and GFRP tubes area strain.

We analysed recycled coarse aggregate content, the recycled fine aggregate super test with a testing program, then drew the corresponding strain-time curve chart.

4.1. Coarse aggregate replacement ratio effect on strain - time curve [6]

In the study, we analysed the recycled coarse aggregate content, which was found to be at 0%, 50%, and 100%; three cases of strain - time curve, it can be seen that the ultimate bearing capacity of the short column decreased as the recycled coarse aggregate content increased, but the short column of ultimate compressive strain did not decrease with the rise of recycled coarse aggregate content. However, when the recycled coarse aggregate replacement ratio is 50% (the results were shown in figures 3 and 4), the ultimate tensile strain is a larger aggregate. There is the recycled coarse aggregate and mortar between old and recycled coarse aggregates, which form the relatively weak section of the

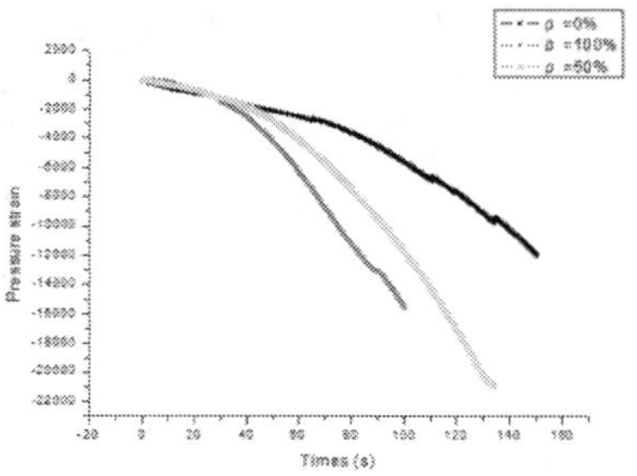

Figure 3. Strain coarse aggregate content - time curve.

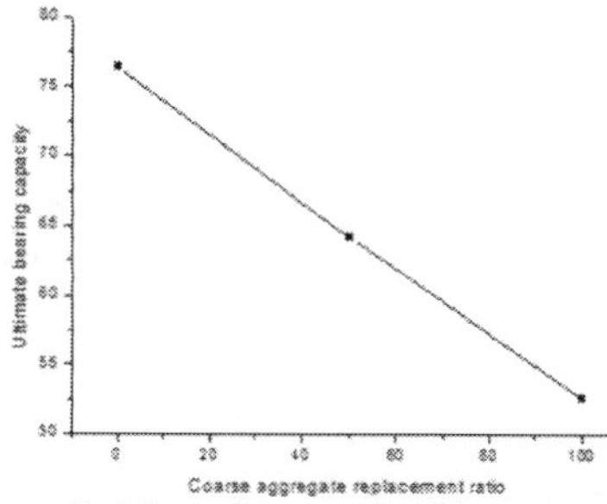

Figure 4. The ultimate bearing capacity in elation to the coarse aggregate content.

transition between the new mortar strains as a result of the increase being too large.

4.2. Fine aggregate replacement ratio effect on strain - time curve

We analyzed the rho recycled fine aggregate, which was found to be at 0%, 50%, 100%, three cases of strain - time curve, it can be seen that the ultimate bearing capacity of the short column decreased with increasing of recycled fine aggregate. When the recycled coarse aggregate replacement ratio is 50%, the overall strain is more significant than the fine aggregate to replace at a rate of 0% and 100% recycled coarse aggregate content (the results were shown in figures 5 and 6). Similarly, the ultimate 50% strain of this group of components is greater than the concrete with the aggregate replacement ratio of recycled concrete.

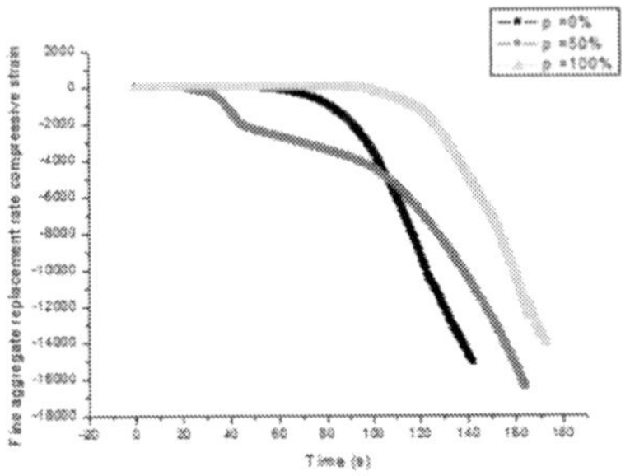

Figure 5. Strain fine aggregate content - time curve.

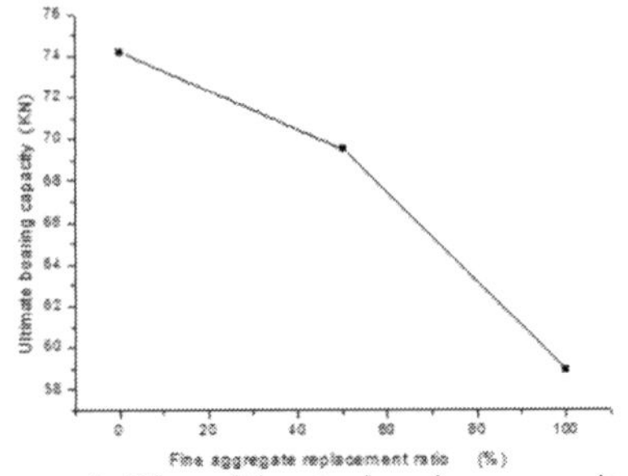

Figure 6. The ultimate bearing capacity in relation to the fine aggregate content.

Based on these results, we can initiate a hypothesis stating that when components both have new and recycled aggregates, new and old concrete of components of the new and old mortar gelation and the development of micro cracks are caused by the results. Therefore, these components' deformation is better than that of concrete with aggregate replacement ratio of recycled concrete.

5. Conclusion

- Based on the influence of GFRP tube recycled concrete short columns, axial compression performance analysis shows that based on GFRP tube recycled concrete short column constraints, recycled concrete short column's axial compression performance can enhanced. Additionally, the thick diameter ratio on the bearing capacity of the axial compression short column has obvious promotional qualities, however due to the limited number of specimens, more tests are needed in order to represent the overall truth.
- The replacement ratio of recycled concrete has a profound effect on short column's axial bearing capacity. Compared with natural coarse aggregates, recycled aggregates contain more old mortar, therefore the strength is lower than a natural coarse aggregate. With the increase of aggregates, the bearing capacity has fallen dramatically; but when the component is contained in the new and old aggregate, the deformation of such components is better than that of concrete with an aggregate replacement ratio of recycled concrete.
- In this experiment, a single source of recycled aggregates, strength grade is C30, the replace rate type is less, however in advanced engineering, the source of the recycled aggregates and their replace rate is more complex. Additionally, the performance is vastly different, so the recycled aggregates still need supplementary study.

Acknowledgments

This work was financially supported by National Nature Science Foundation (51308340), the College Students Scientific and Technological Innovation Programs of Zhejiang Province (2013R426013), and Nation (201310349014)

References

[1] Xiao,Y 2014 *From steel tubed columns to FRP tubed columns* (Washington, D.C.: Proc. ASCE Structural Congress)

[2] Yu Qing 2000 Charactor and application of FRP composites in civil engineering *Journal of Harbin University of Civil Engineering and Architecture* **3** 26-30

[3] Chen Bailing, Qin Guopeng and Wang Lianguang 2010 Study on bearing capacity of axially-loaded GFRP tube filled with steel-reinforced concrete short colums *Journal of Northeastern University* **31** 1035

[4] Wu Donghui, Yang Song, Tang Guiyun and Zhou Ting 2006 Axial ultimate compressive strength of FRP-tube reinforced concrete struture *Fiber Composites* **1** 15

[5] Yuan Jifeng, Wang Wei and Lu Yongtao 2014 Study on impacts of water cement ratio and RCA replacement ratio on recycled concrete strength *Technology and Materi* **13** 53

[6] Qin Guopeng, Wang Lianguang and Zhou Le 2011 Study on bearing capacity of axially-Loaded GFRP tube composite columns filled with concrete *Building Structure* **47** 102

IOP Conf. Series: Materials Science and Engineering **87** (2015) 012016 doi:10.1088/1757-899X/87/1/012016

Carbon footprint of automotive ignition coil

Huey-Ling Chang[1,4], Chih-Ming Chen[2], Chin-Huang Sun[3] and Hung-Di Lin[1]

[1]Department of Chemical and Materials Engineering, National Chin Yi University of Technology, Taichung, Taiwan (ROC)
[2]Department of Mechanical Engineering, National Chin Yi University of Technology, Taichung, Taiwan (ROC)
[3]Department of Mechanical Engineering, Fuzhou Polytechnic, Fuzhou 350108, Fujian, China

E-mail: hlchangtc@gmail.com

Abstract. In recent years, environmental issues, such as climate change and global warming due to the excessive development of industry, have attracted increasing attention of citizens worldwide. It is known that CO_2 accounts for the largest proportion of greenhouse gases. Therefore, how to reduce CO_2 emissions during the life cycle of a product to lessen its impact on environment is an important topic in the industrial society. Furthermore, it is also of great significance to cut down the required energy so as to lower its production costs during the manufacturing process nowadays. This study presents the carbon footprint of an automotive ignition coil and its partial materials are defined to explore their carbon emissions and environmental impact. The model IPCC GWP_{100a} calculates potential global greenhouse effect by converting them into CO_2 equivalents. In this way, the overall carbon footprint of an ignition coil can be explored. By using IPCC GWP_{100a}, the results display that the shell has the most carbon emissions. The results can help the industry reduce the carbon emissions of an ignition coil product.

1. Introduction

Carbon footprint can be defined as the carbon dioxide emissions directly or indirectly caused by a product, person or other things. As the international community is increasingly valuing the environmental issues in the modern society, many organizations have been established to monitor the effect of carbon emissions on environment. Regulations on carbon emissions also become clearer and more definite.

As consumers are more concerned about greenhouse gas emissions, this means more cost loading for the industry in terms of a production cost. Therefore, the establishment of carbon footprint of a product can not only save costs, but also do the work for environmental protection.

Sinha and Kutnar [1] conducted life cycle assessment for different window frame materials to explore the environmental impact by using different materials, namely, aluminum, plastic (PVC), and wood for window frames. The result showed that wood window frames has the lowest impact on the environment, as the carbon footprint of aluminum and PVC frames is four times and twice more than

[4] Address for correspondence: Huey-Ling Chang, Department of Chemical and Materials Engineering, National Chin Yi University of Technology, Taichung, Taiwan (ROC). E-mail: hlchangtc@gmail.com.

that of wood frames, respectively. Therefore, wood material is a good selection for low carbon emissions material and thus the environmental selection.

Kim [2] directed a comparison of life cycle assessment of existing and alternative personal computer (PC). The result showed that the greenhouse gas emission of an alternative PC is 15 percent less than an existing computer. However, the existing PCs are still in the usage phase and will generate huge greenhouse gas emissions. How to reduce carbon emissions of the existing PCs is necessary.

Matthews [3] explained the importance of boundary conditions in the analysis of carbon footprint. The study divided carbon emissions into three categories, namely direct emissions, indirect emissions, and emissions in addition to the previous two kinds in the supply chain caused by other industries. Direct emissions of greenhouse gases arise from the industry itself which are the combustion of natural gas and oil during a manufacturing process. Indirect emissions of greenhouse gases are a discussion of carbon emissions generated from the energy purchase process.

Vink et al. [4] performed a life cycle assessment of materials for polylactic acid ester (polylactide, PLA). A PLA material produced by Doyle Company will consume 54 MJ/kg of fossil fuels. The production process was improved by Doyle Company to reduce to 7 MJ/kg of fossil fuels consumption after five to eight years. The greenhouse gas emissions were also reduced from +1.8 to -1.7 carbon dioxide equivalents.

2. Global warming potential

The United Nations Intergovernmental Panel on Climate Change (IPCC) was co-founded by the World Meteorological Organization and the United Nations Environment Program in 1988, specializing in climate change caused by human's activities. IPCC published four official climate change assessment reports in 1900, 1995, 2001, and 2007, respectively. The SimaPro software referring to Global-warming potential (referred as GWP) IPCC presents the idea that greenhouse gas emissions can be converted into carbon equivalent in order to calculate the carbon footprint.

The GWP_{100a} value of CO_2 is calculated by referring to 100 years from the time of the calculation to reflect the cumulative effects of radiation of CO_2. For example, in 2007, the GWP of methane is 25, which means that one ton of methane after 100 years from the computing time giving the cumulative effects of radiation is equivalent to the cumulative equivalent of 25 tons of CO_2 in the current time [5].

The method of IPCC GWP_{100a} in this study only applied the procedure of characterization. Its characterized factor is the direct global warming potential of air emissions. The considerations are as follows:

- Do not include the indirect emissions of nitrogen caused by the formation of nitrous oxide.
- Do not calculate the radiation effects of nitrogen oxides, water and sulfuric acid emissions arising from the troposphere to the stratosphere.
- Do not consider indirect effects.
- Include the carbon dioxide caused by the CO emissions.
- Consider the offset of bio-absorption of carbon dioxide.

The formulation is as follows:

$$\mathbf{E_{CO2e}} = \mathbf{M_i} \times \mathbf{GWP_i} = \mathbf{EF_i} \times \textstyle\sum_j \mathbf{A_j^T} \quad (1)$$

Where, $\mathbf{Eco_2e}$ is the overall emissions of carbon dioxide equivalent, **Mi** is for each amount of greenhouse gases, $\mathbf{GWP_i}$ is the global warming potential of all greenhouse gases, $\mathbf{EF_i}$ refers to the ith greenhouse gas emission, $\mathbf{A_j^T}$ refers to the activities information in the manufacturing process.

3. Study procedure

The detailed study procedure is as follows. First, identify this study's purposes. This study discusses the carbon emissions for the materials that are composed of automotive ignition coil. Ignition coils represent the so-called "power plant" of the ignition process in the petrol engine. As part of the ignition system, they are charged with the task of providing the spark plug with the high voltage required to generate an ignition spark between the center and earth electrode of the spark plug and ignite the air-fuel mixture.

For this purpose, the ignition coils draw current from the battery through the low-voltage connection and depending on the type, multiply it to up to 45 kilovolts. Then collect and organize data for analysis. The SimaPro software is utilized as the main analytical tool, in different assessment models for analysis and evaluation. The software also can produce a tree diagram to clearly show the environmental impact, as the tree diagram clearly exhibit the branch of respective input energy and the material. In the tree-branch subsystem, the system in a measurable way, based on an expression like a thermometer, quickly determines the materials and the impact of the energy on the environment. The filling material of automotive ignition coil is epoxy resin which is a liquid insulating resin. The elements of an ignition coil will be divided into three parts, i.e. polymer materials, metal materials, and package material. Table 1 lists the material composition, and the weight of an ignition coil (typeAS-944).

Table 1. Materials used for an ignition coil.

Component	**Packaging Materials**	**Material composition**	**weight**
Terminal Block and cover		PBT (Polybutylene terephthalate)	11.0 g
Primary spools		PBT	7.0 g
Secondary spools		PPS(Polyphenylene Sulfide)	29.8 g
Shell		PBT	30.0 g
Insulating resin		Epoxy resin	68.6 g
Insulating body		PBT	110.7 g
Terminal		copper	2.6 g
Coil		copper	32.8 g
Secondary terminal		copper	0.2878 g
High-voltage connection		copper	1.5 g
High-voltage terminal		Iron	6.5 g
Core		Iron	111.5 g
Ground sheet		stainless	3.0 g
	Waterproof rubber	rubber	2.0 g
	Carton	paper	52.9 g
	plastic bags	PE (Polyethylene)	4.72 g

4. Results and discussion

Carbon footprint calculates the amount of greenhouse gas emissions caused by a particular activity or entity, commonly also referred to as global warming potential (GWP). It is measured in kilograms of carbon dioxide equivalent (CO_2eq.).The net diagram analysis was illustrated in figures 1-3. In this study, carbon footprint of different elements was calculated (table 2).

For the categories of polymer materials (figure 1): insulating resin, insulating body, secondary spools, and shell are 0.185 kg, 0.534 kg, 0.164 kg, and 0.145 kg of carbon dioxide equivalent emissions respectively. For the categories of metal materials (figure 2), the biggest emission source is core (0.102 kg CO_2 eq.), followed by coil (0.0619 kg CO_2 eq.). For the categories of package materials (figure 3): waterproof rubber, carton, and plastic bags are 0.00386kg, 0.0778kg, and 0.00976kg of carbon dioxide equivalent emissions, respectively.

From the analysis (table 2), it can be known that each production of an AS-944 type ignition coil will generate a total of 1.934 kg carbon dioxide equivalent (kg CO_2 eq.). The biggest emission source was Polybutylene terephthalate (54.89%), followed by epoxy resin (13.28%), Polyphenylene Sulfide (11.77%), Iron (8.47%), and paper (5.58%).

The most amount of polybutylene terephthalate is used in insulating body and shell. The paper is packaging material which produces carbon emissions by up to 0.0778 kg CO_2 eq.

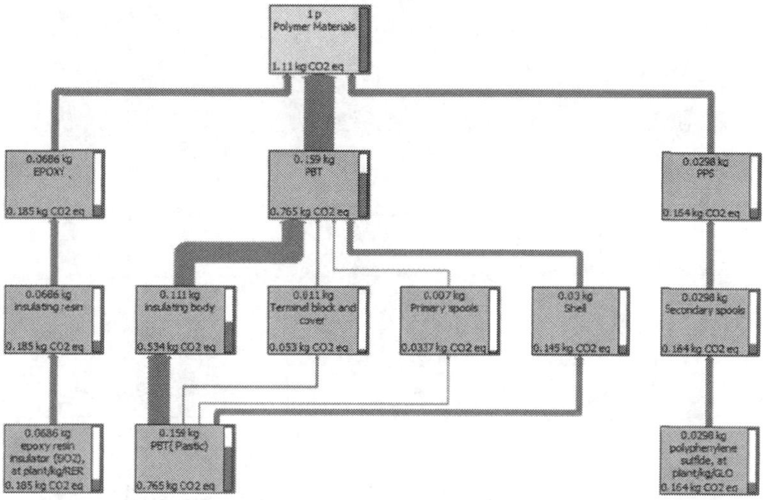

Figure 1. The polymer materials net analysis diagram by IPCC GWP.

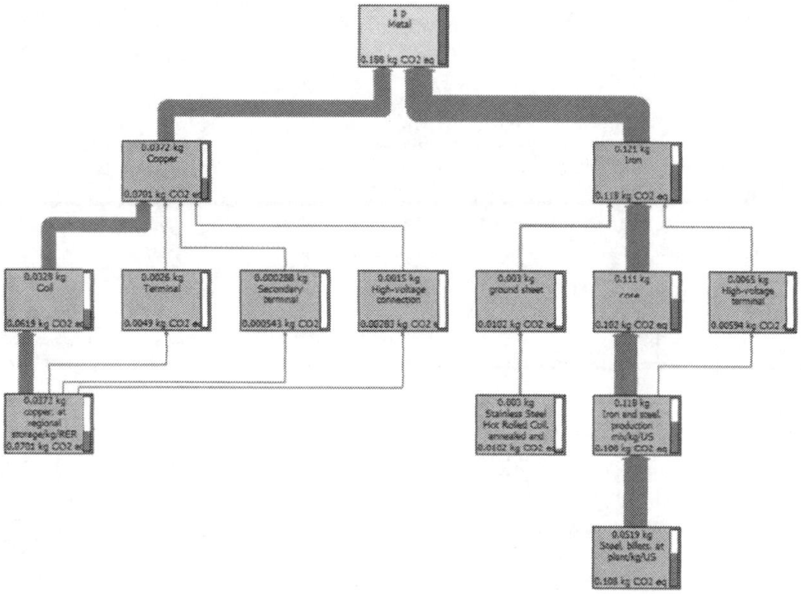

Figure 2. The metal materials net analysis diagram by IPCC GWP.

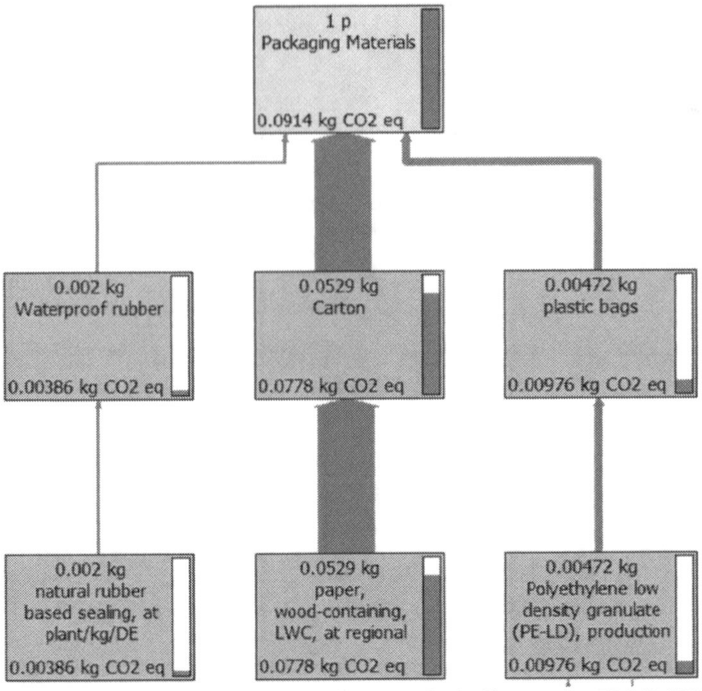

Figure 3. The package materials net analysis diagram by IPCC GWP.

Table 2. The carbon footprint of an ignition coil product.

Emission source	Total use weight (g)	Carbon footprint ($kg\ CO_2eq$)	Emission source carbon footprint (%)
Epoxy resin	68.6	0.185	13.28
PBT	159	0.765	54.89
PPS	29.8	0.164	11.77
copper	37.2	0.0701	5.03
Iron	121	0.118	8.47
rubber	2.0	0.00386	0.28
paper	52.9	0.0778	5.58
Polyethylene	4.72	0.00976	0.7

5. Conclusion

In the study of carbon footprint, IPCC utilizes global warming as conversion factor to calculate CO_2 equivalent. From the material composition of ignition coil, one can understand that each material used corresponds to how many equivalents of carbon dioxide emissions. It is found that polybutylene terephthalate is used in insulating body and shell produces the most emissions equivalents of carbon dioxide. Therefore, one should pay special attention to the selection and development of the materials. This study presents the carbon footprint and environmental impact of automotive ignition coil. The results can help the industry reduce the emission source carbon footprint in an ignition coil product.

Acknowledgments
The research was supported by Ministry of Science and Technology of the Republic of China. The researchers hereby express the gratitude.

References
[1] Sinha A and Kutnar A 2012 Carbon footprint versus performance of aluminum, plastic, and wood window frames from cradle to gate *Buildings* **2** 542-53
[2] Sun-Wook Kim, Ik-Won Choi, Chun-Youl Baek and Kun-Mo Lee 2012 Comparison of product carbon footprint between existing and alternative product: A case study of personal computer *Design for Innovative Value Towards a Sustainable Society* (Berlin: Springer Netherlands) 988-9
[3] Matthews H S, Chris T. Hendrickson and Christopher L. Weber 2008 The Importance of Carbon Footprint Estimation Boundaries *Environ. Sci. Technol.* **42** 5839-42
[4] Riva A, D'Angelosante S and Trebeschi C 2006 Natural gas and the environments results of life cycle assessment *Energy* **31** 138-48
[5] Intergovernmental Panel on Climate Change (IPCC) Climate change 2007 The physical science basis *Contribution of Working Group I to the Fourth Assessment Report* eds Solomon S D and Qin M (Geneva, Switzerland: IPCC)

IOP Conf. Series: Materials Science and Engineering 87 (2015) 012017 doi:10.1088/1757-899X/87/1/012017

Electrochemical research on corrosion behavior of A3 steel in compound sodium molybadate and organic inhibitor solution

C X Sun[1], Y M Chen, H W Xu, M Zhang, M Chen, M Xue, J Y Wu and C S Huang

Institute of Energy Research, Henan Academy of Science, No. 29 Garden Road, Zhengzhou 450008, R.P. China

E-mail: caixiasun@alum.imr.ac.cn

Abstract. The electrochemical corrosion behavior of A3 in compound sodium molybdate and organic inhibitor solution was tested by the electrochemical workstation method. The concentration of the compound inhibitor set to range 250 mg/L to 3000 mg/L. The polarization curve results of A3 in different concentration inhibitor solutions show that the inhibitor markedly represses the anodic processes. The EIS has two time constant. The extreme concentration is 1500 mg/L.

1. Introduction

Inhibitors play an important part in preventing metal used in recycling cooling water from corroding. Sodium molybdate has low toxicity to environment and biological as a kind of inorganic inhibitor that can influence electrochemical corrosion reaction [1, 2].

ATA as an organic inhibitors can forme a SMS (self-assembled monolayers) on the surface of metal matrix through the chemical bond. ASM technology has the advantage of higher inhibition efficiency at the low dosage compared with the traditional method [3-6].

This study meanly used organic inhibitors ATA formed a SMS on the surface, meanwhile with the help of inorganic inhibitors sodium molybdate and ZnSO4 fill the void of the SMS, by the two methods cooperating with each other the compound inhibitors can form an impact and uniform membrane on the surface of the A3.

2. Expreimental

2.1. Materials

Sodium Molybdate 10 mg/L, ATA (3-Amino-1, 2, 4-triazole) 1 mg/L, POCA (Copolymer of Phosphono and Carboxylic Acid) 2.5 mg/L, $ZnSO_4$ 8 mg/L, PESA (Sodium of Polyepoxysuecinic Acid) 10 mg/L. The concentration set to range from 250 mg/L to 3000 mg/L.

2.2. Inhibitors preparations

[1] Address for correspondence: C X Sun, Institute of Energy Research, Henan Academy of Science, No. 29 Garden Road, Zhengzhou 450008, R. P. China. E-mail: caixiasun@alum.imr.ac.cn.

Firstly, dilute the ATA at 60℃ and then cool it to room temperature; secondly add molybdate POCA and PESA to the molybdate solution waiting for them dissolved, and then stir it well; Thirdly, add $ZnSO_4$ to the compound solution and stir well for use.

2.3. Instrument
Potentiostat/galvanostat CS 2305, JSM-6510 scanning electron microscopy (SEM), BT224s Electronic Balance. Magix Pro PW2440 X-ray photoelectron spectrocopy (XPS).

2.4. Experimental procedure

2.4.1. Electrochemical test of A3. A3 electrode was sealed in epoxy resin except the exposed cross-section as working surface. The A3 electrode was embedded in epoxy resin with an exposed cross-sectional area of 1 cm^2. The counter electrode in electrochemical test was platinum while saturated calomel electrode (SCE) was used as the reference electrode. The electrochemical studies were operated on the Potentiostat/galvanostat CS2305. Linear polarization measurement were carried out with a potential perturbation of ±100 mV around the open circuit potential (OCP), Electrolyte was prepared by adding compound sodium molydbate and organic inhibitor solution with the concentration ranging between 250 mg/L~3000 mg/L in tap water. The electrchemical impedance spectra (EIS) of the A3 electrode system were measured at 10mV disturbing sinusoidal wave and the frequency was 0.1 Hz to 10 kHz.

2.4.2. SEM test. The A3 plate with size of 50 mm x 25 mm x 2 mm were immersed in the compound inhibitor solutions after 72 h at 45℃, then washing it by water and alcohol in turn. The operation parameters were at 25 kV acceleration voltage.

2.4.3. XPS test. The A3 The A3 plate with size of 50 mm x 25 mm x 2 mm were immersed in the compound inhibitor solutions after 72 h at 45℃, then then washing it by water and alcohol in turn. The operation parameters were at 3 kV focus voltage, 3 kV energy voltage. The XPS use magnesium as target and operated at 0.1 mpa vacuum degree.

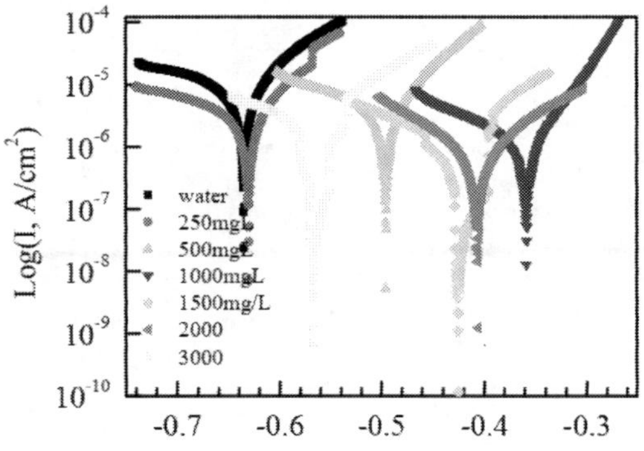

Figure 1. Polarization curves of A3 electrode in different concentration inhibitor solutions.

3. Results and discussions

3.1. Electrochemical test results

3.1.1. Effect of concentration on the polarization curves of A3 electrode. Figure 1 shows the polarization curves of A3 electrode in different concentration sodium molybdate and organic inhibitor solutions. From figure 1 a markedly represses on anodic processes and a positive shift of the corrosion potential E_{corr} can been found. Furthermore the anodic branch of the polarization curves is noted that have a higher slope than that in blank solution (without additive). The electrochemical parameters of A3 derived from polarization measurement were given in table 1. From table 1 that the R_p and I_0 value decrease with the concentration of the compound sodium moly- bdate and organic inhibitor increase when it below 1500 mg/L. The R_p and I_0 value increase with the concentration of the compound sodium molybdate and organic inhibitor increase when it above 1500 mg/L. From the change of R_p and I_0 that the film formed on the A3 by the compound sodium molybdate and organic inhibitor become wore thicker with the increase of inhibitors. The change of I_0 also indicate that the compound inhibitors has composite effect on the anode and cathode reaction.

Table 1. Electrochemical parameters of A3 derived from polarizational measurement.

	R_p ($\Omega\cdot cm^2$)	B_a (mV)	B_c (mV)	I_0 ($\mu a/cm^2$)	EI (%)
0mg/L	2422.0	61.27	110.38	6.60	
250mg/L	5725.5	57.22	119.95	2.99	55.05
500mg/L	15273.0	54.88	131.70	1.31	83.03
1000mg/L	13041.0	52.67	146.43	1.32	85.18
1500mg/L	100129.0	69.30	105.26	0.34	94.70
2000mg/L	58288.0	61.30	87.77	0.73	89.04
3000mg/L	23543.0	34.37	-	1.09	83.60

3.1.2. Effect of concentration of inhibitor on electrochemical impedance spectra. Figure 2 shows the Nyquist plots of A3 electrode in different concentration compound sodium molybdate and organic inhibitor concentration solutions. From figure 2 the radius of the nyquist plots of A3 change bigger with the increase of the inhibitor concentrations below 1500 mg/L while it change smaller with the increase of the inhibitor concentration above 1500 mg/L. The Nyquist curve of the compound sodium molybdate and organic inhibitor resume a single capactive with no sign of inductive impedance when the concentration set below1000 mg/L which meaning that the adsorption layer on the A3 electrode surface is stable.

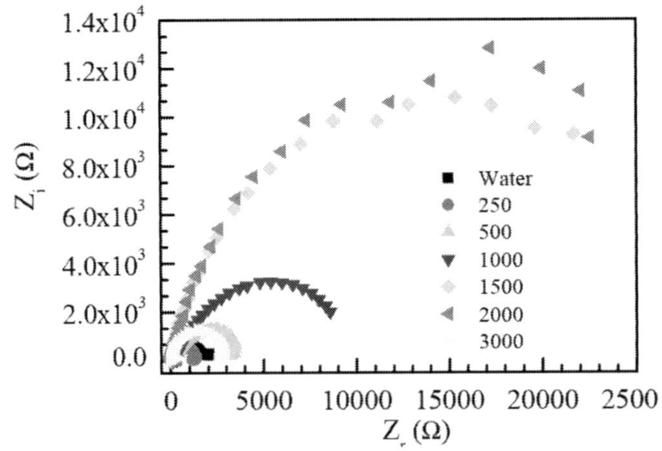

Figure 2. Nyquist spectra of electrode in different concentration inhibitor solutions.

Figure 3 shows the bode spectra of A3 plate electrode in different concentration compound sodium molybdate and organic inhibitor solutions. From figure 3a the introduction of the compound sodium molybdate and organic inhibitor solutions make the changes increase at high frequency. The change of phase angle at high frequency is discontinuous. The absolute value of Z at low frequency indicates the polarization resistance [7, 8]. From figure 3b the change of absolute value of Z at low frequency is discontinuous which indicate the introduction of the compound sodium molybdate and organic inhibitor have an extreme value to protect A3 from corrosion.

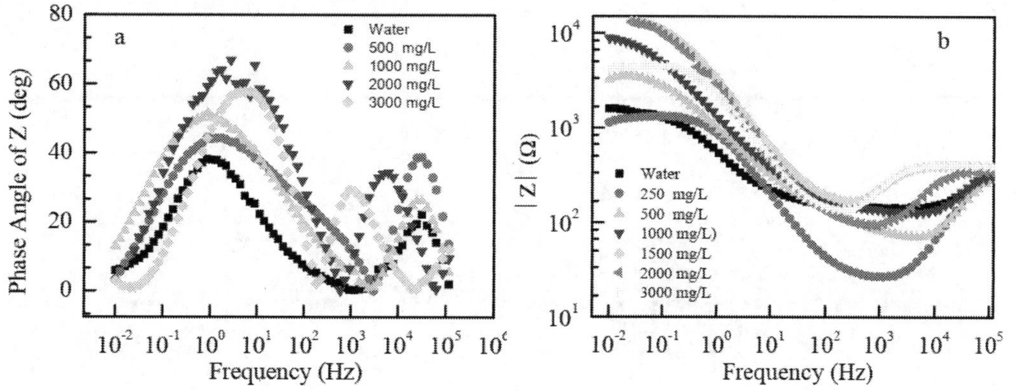

Figure 3. Bode spectra of A3 plate electrode in different concentration compound molybdate inhibitor solutions (a) phase angle plots, (b) absolute value of Z.

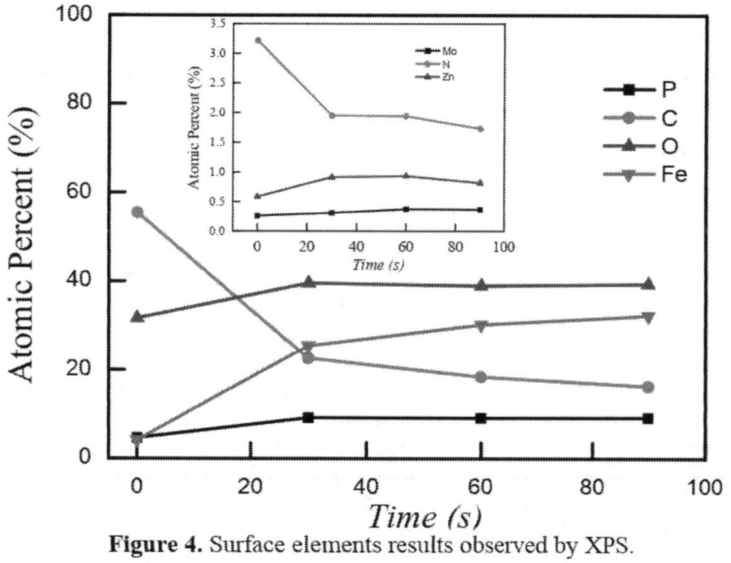

Figure 4. Surface elements results observed by XPS.

3.2. XPS results of A3 plate after test

Figure 4 shows the XPS results of A3 plate which immersed in the compound sodium molybdate and organic inhibitor solutions for 72h. The results show that the surface was composed of C、O、P、Fe、Mo、N、Zn elements. The C contents of A3 plate on the surface were higher than that after sputtering. The reason leading to the C content in surface higher is the C surface adsorption result. The Mo content still can be tested after sputtering 90s which illustrate the compound sodium molybdate and organic inhibitor formed a thin membrane on the A3 plate surface. The membrane formed by the compound sodium molybdate and organic inhibitor solution was comprised of C、O、P、Fe、Mo、N、Zn elements.

3.3. SEM results of A3 plate

Figure 5 shows the SEM results of A3 plate which immersed in the compound sodium molybdate and organic inhibitor solutions for 72 h. From figures 5a and 5b a uniform and compact membrane was formed on the surface of the A3 plate after after immersed in the compound sodium molybdate and organic inhibitor solutions. The membrane on the surface can protect A3 used cooling water system from corrosion.

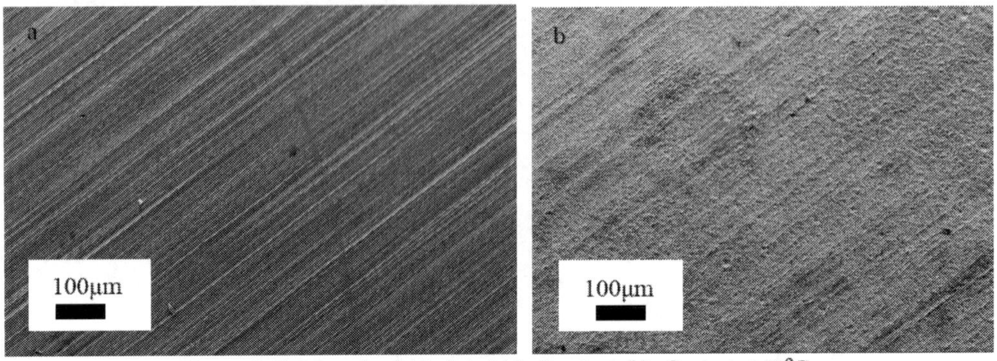

Figure 5. SEM result of A3 plate (a) before test, (b) after test at 45°C.

4. Summary

The researched compound sodium molybdate and organic inhibitor can protect the A3 from corrosion by markedly impact the anode reaction. The electrochemical parameters derived from polarization curves and the EIS result shows the 1500mg/L is an extreme concentration. The bode phase has two time constance.

The XPS results show that the membrane formed on the surface of A3 composed of C/O/P/Fe/Mo/N/Zn elements.

The SEM results show that the membrane formed on the surface of A3 after immersed in the compound sodium molybdate and organic inhibitor solutions is uniform and impact.

Acknowledgments

This work was financially supported by the Henan province key project of science and technology (132102210238) and plan for scientific innovation talent province (144200510025).

References

[1] Diamantino T C, Guilhermino L, Almeida E and Soares A M V M 2000 *Ecotox. Environ. Safe* **45** 253-9
[2] Araminezhaad M K, Ravari F B, Dadgrynezhad A and Kavianpoor Y 2007 *J. Chem* **19** 297-306
[3] Cao C N 2004 *Principles of Erosive Electrochemistry* (Beijing: Chemical Industry Press) p 232
[4] Yin R H, Ji X B, Wang H J, Zhang L and Shi X H 2007 *J. Acta. Metall. Sin* **43** 273
[5] Lininis P E and Whitesides G M 1992 *J. Am. Chem. Soc* **114** 9022
[6] Wan Z Y, Zhang L, Yin R H, Xu Q J, Chen H and Zhu L J 2008 *J. Acta. Metall. Sin* **44** 203
[7] Zhang D Q, Gao L X and Zhou G D, 2004 *Corros. Sci* ***46*** 3031
[8] Du H Y, Lu M X, Wu Y S and Wu W M 2006 *J. Acta. Metall. Sin* **42** 533-6

Micro/nanoscale continuous printing: direct-writing of wavy micro/nano structures via electrospinning

Feiyu Fang, Zefeng Du, Jun Zeng, Ziming Zhu, Xin Chen, Xindu Chen, Yuanjun Lv and Han Wang[1]

Guangdong Provincial Key Laboratory of Micro-nano Manufacturing Technology and Equipment, Guangdong University of Technology, Guangzhou, 510006, China

E-mail: wanghangdut@126.com

Abstract. Micro/nanofibers that are created by direct-writing using an electrospinning (ES) technique have aroused much recent attention, owing to their intriguing physical properties and great potential as building blocks for micro/nanoscale devices. In this work, a wavy direct-writing (WDW) process was developed to directly write wavy micro/nano structures suitable for the fabrication of micro/nanoscale devices. The low voltage WDW technique is anticipated to be useful for a broad range of applications including flexible/stretchable electronics, micro optoelectronics, nano-antennas, microelectromechanical systems (MEMS), and biomedical engineering.

1. Introduction

The efficiency and low-cost production of micro/nanostructures by electrospinning (ES) via a direct-writing process has gained popularity recently, owing to its enormous potential for application in many fields.ES is considered to be a straightforward, low-cost technique to fabricate ultra-thin fibers with diameters ranging from tens of nanometers to several micrometers. In particular, having the ability to deposit functional materials directly and without contact on a variety of substrates, including flexible substrates, ES can be used to directly print large-area organic semiconducting nanowire arrays on device substrates, enabling sophisticated large-area nanowire lithography for nanoelectronics [1, 2]. Thus, ES is considered to be a promising alternative to traditional lithography technology to produce the microelectronic devices in an ingenious and cost-effective way. However, owing to the bending instability of a charged jet under coupled multi-field forces, the ES process is unstable and almost uncontrollable. Serpentine/wavy structures show enormous potential for applications in flexible/stretchable electronics owing to their excellent mechanical and electrical properties and performance [3-5]. In order to get fibers to assume a specific serpentine/wavy structure, several exploratory studies have been carried out, yielding remarkable results and very long wave-shaped fibers have been successfully made [6-11]. Although the above work shows promise in fabrication of novel structures, challenges remain in stably producing the micro/nano scale serpentine patterns which frequency, amplitude and wavelength can be control precisely. In previous research, we supplemented a high voltage AC applied to auxiliary electrodes (with an input voltage exceeding 1 kV) to produce

[1] Address for correspondence: Han Wang, Guangdong Provincial Key Laboratory of Micro-nano Manufacturing Technology and Equipment, Guangdong University of Technology, Guangzhou, 510006, China. E-mail: wanghangdut@126.com.

IOP Conf. Series: Materials Science and Engineering **87** (2015) 012018 doi:10.1088/1757-899X/87/1/012018

wavy fibers [12]. However, the resulting fiber pattern was not smooth enough and the scale was not small enough to meet the requirements for manufacture of micro/nano devices. Furthermore, owing to the large whipping angle and amplitude, the pattern of fibers exhibited waveform distortion.

In this study, we applied lower voltages to the auxiliary electrodes; the AC voltage was set to below 300 V and the DC voltage set to less than 1.5 kV. This method differs from traditional ES in that the jetted liquid fiber is pulled by a combination of the DC and AC electrical field forces, while the low voltage AC-DC coupling electric field constrains the bending of the jet to make it controllable. This low voltage wavy direct-writing (WDW) process can be effectively controlled by adjusting the substrate speed, as well as the magnitude and frequency of the applied AC field. WDW permits direct-writing of small size complex serpentine patterns, because the formerly unstable jet can be precisely oriented and positioned. What's more, unlike conventional drop on-demand-type jet printing techniques [13], WDW can print the serpentine patterns in continuous jet mode. This method is anticipated to be useful for a broad range of applications including flexible/stretchable electronics, micro optoelectronics, nano-antennas, MEMS, and biomedical engineering.

2. Experimental

2.1. Preparation

Polyethylene oxide (PEO) with an average molecular weight of 2 000 000 (Aladdin, Shanghai, China) was chosen for the preparation of the solutions. PEO fibers were electrospun using 3%-8% (w/w) concentrations of PEO in deionized water with 4 hours stirring at 20°C. The ground collector was made of chromium-plated glass. All samples were stored at 20°C.

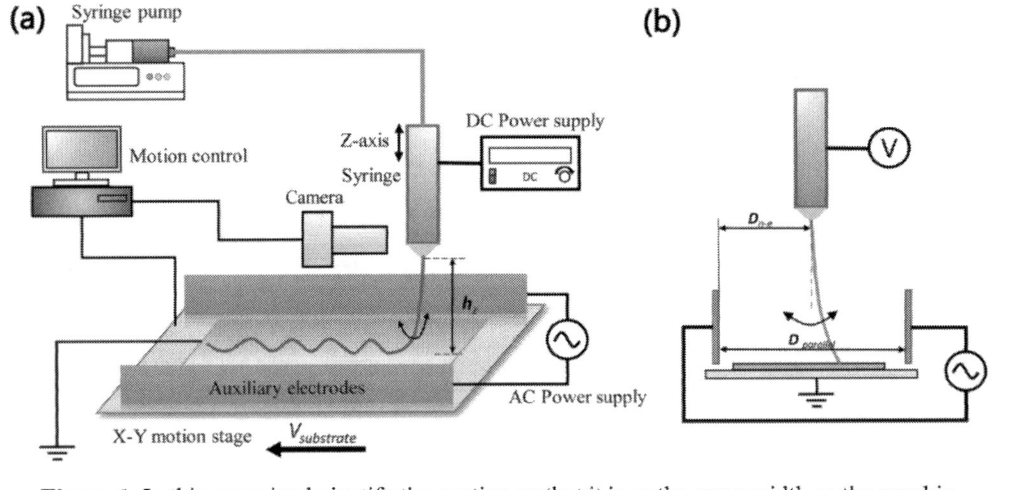

Figure 1. In this case simply justify the caption so that it is as the same width as the graphic.

2.2. Apparatus

The schematic diagram of the wavy direct-writing (WDW) apparatus is shown in figure 1. The PEO solution was delivered with a syringe pump (Lange, Inc., China). A stainless steel nozzle (inner diameter 260 μm and external diameter 520 μm) was adopted as an electrode, and the ground collector was a Cr-coated glass plate fixed to a moving stage (Suruga, Japan). A high voltage, generated by a direct current power supply (DW-P403, Dongwen Inc., China), was applied between the nozzle and the collector to generate a Taylor cone to assist in pulling out the jet from the nozzle. The nozzle-to-collector distance was adjusted to vary from 0-50 mm, and the moving speed of the substrate was adjusted from 0-400 mm s^{-1}. Furthermore, two parallel auxiliary electrodes made of copper foil were stationed beside the collector and connected with an AC variable-frequency power supply (Sunan SNP-605, China).

2.3. Characterization
The microstructure was characterized by an image measurement instrument (Rational VMS-3020H, China) and scanning electron microscope (SEM) (Hitachi TM-3030).

3. Results and discussion

In this work, low voltage wavy direct-writing (WDW) is employed to achieve formation of the required serpentine patterns. Two parallel auxiliary electrodes made of copper foil are placed beside the collector and connected to an AC power supply, which can adjust the driving force acting on the fiber during the dropping process, as shown in figure 1b. The low voltage AC-DC coupling electric field constrains the bending of the jet, making it to become controllable. Therefore, WDW is able to directly write high-resolution serpentine patterns (figure 2). In the WDW process, the frequency of serpentine patterns is approximately equal to the frequency of AC electric field and the amplitude of the patterns depends linearly on the applied AC voltage value. The pattern wavelength can be controlled precisely by changing the speed of the collector.

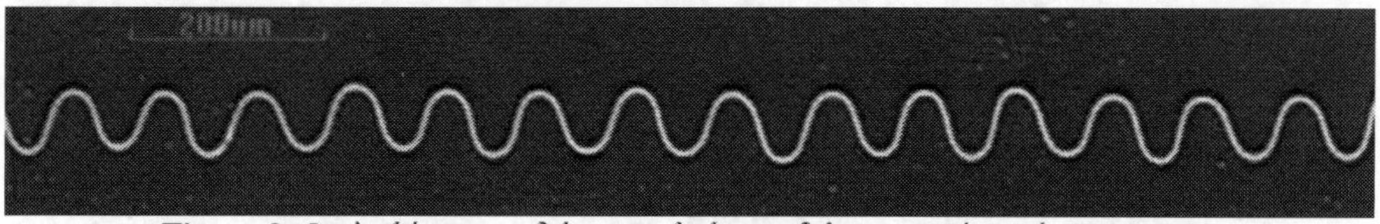

Figure 2. Optical images of the morphology of the serpentine microstructure.

What's more, we found that the formation behaviour of the serpentine structures depended greatly on the strength of the AC electricfield and the force status of the charged jet. With optimal DC voltage and nozzle height, the controllability of the whipping jet instability mainly correlates with the spacing of the two auxiliary electrodes, $D_{parallel}$, and $D_{n\text{-}e}$, the horizontal distance between the nozzle and auxiliary electrodes. When the two auxiliary electrodes are far apart or the $D_{n\text{-}e}$ is large, the AC electric field has almost no effect on the charged jet, and repulsive electric forces between the uniformly-distributed charges play the leading role in influencing electrical bending, as shown in figure 3(a). This AC electric field effect becomes more and more apparent after a decrease in the distance between the nozzle and auxiliary electrodes, as shown in figures 3(b) and 3(c) .When $D_{n\text{-}e}$ falls below a specific distance, d, the AC electric field force is dominant and the path of the charged jet can be precisely controlled. The near-perfect serpentine pattern direct-write by WDW is shown in figure 2, but it was influenced by many possible factors and currently it is not possible to define the stability critical condition.

When $D_{parallel}$ and $D_{n\text{-}e}$ are both optimal values, the nozzle-to-substrate distance, h_z, will greatly impact the movement of the charged jet. However, when h_z is lower than a specific height (in our experiment, 2 mm), WDW converts to NFES and the AC-DC coupling electric field barely influences the charged jet swing, as shown in figures 4(a) and 4(b).When h_z is huge, the system behaves as a conventional far-field ES with uncontrollable bending instability.

Additionally, solution concentration and collector speed also play important roles in the formation behaviour of the serpentine structures. When concentration is just lower than the critical value, bending superposition is observed in the direction of the jet path of motion, as shown in figures 4(e), 4(f) and 4(g). Under high substrate speed conditions, WDW converts to Mechano-Electrospinning (MES) [10, 14] and the AC-DC coupling electric field has almost no effect on controlling the charged jet swing. In particular, thinner and better-aligned nanofibers result from MES than NFES in an AC-DC coupling electric field, as shown in figures 4 (c) and 4(d).

IOP Conf. Series: Materials Science and Engineering **87** (2015) 012018 doi:10.1088/1757-899X/87/1/012018

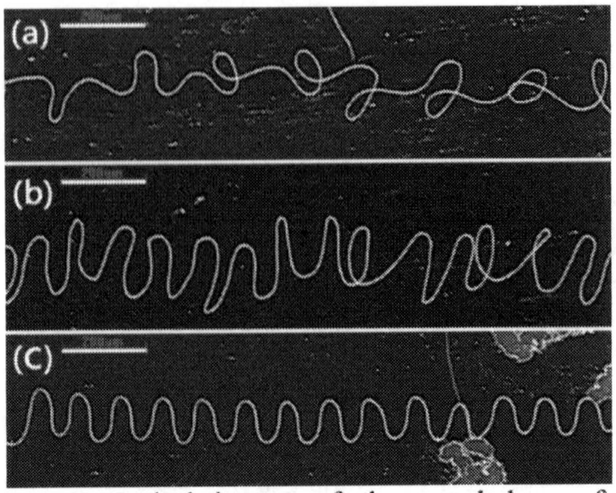

Figure 3. Optical images of the morphology of the microstructure with different distance between the nozzle and auxiliary electrodes: (a) 11 mm; (b) 9 mm; (c) 7.5 mm.

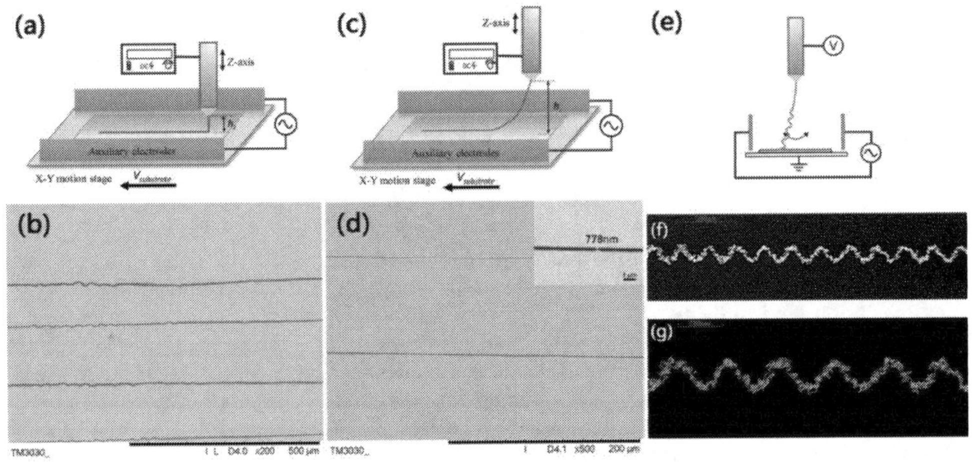

Figure 4. (a)(c)(e) Schematic diagrams of WDW in different condition. (a)(b) WDW converts to NFES when the hz is lower than a specific height. The nanofibers arrays deposited in the condition: AC voltage 1 KV, DC voltage 1.5 KV, 4% PEO, nozzle-to-substrate distance 2 mm and substrate speed 10mms^{-1}; (c)(d) Under high substrate speed conditions, WDW converts to MES. The parallel nanofibers arrays deposited in the condition: AC voltage 1 KV, DC voltage 1.5 KV, 4% PEO, nozzle-to-substrate distance 5mm and substrate speed 30mms^{-1}; (e)-(g) when concentration is just lower than the critical value, bending superposition is observed and the patterns deposited are coarse. The fiber patterns deposited in the condition: 3% PEO, DC voltage 1.5 KV, AC voltage 300 V, 50 Hz, substrate speed 12 mms^{-1} and nozzle-to-substrate distance 5 mm.

4. Conclusion

In this study, we have developed the low voltage WDW process to directly write serpentine micro/nano structures suitable for the fabrication of micro/nanoscale devices and demonstrated the ability of low voltage WDW to precisely pattern nanofibers on substrates. This fabrication ability will permit the use of electrospun nanofibers based wiring of structural and functional components in microelectronics, MEMS, sensor, micro optoelectronics devices and biological apparatus.

Acknowledgments
This work was financially supported by National Natural Science Foundation of China (51305084), Project on the Integration of Industry, Education and Research of Dongguan City (No. 2013509109101), Guangdong Innovative Research Team Program (No. 201001G0104781202), Guangdong Provincial Key Laboratory of Micro-nano Manufacturing Technology and Equipment ((GDMNML2013-1, Guangdong Provincial Key Laboratory Construction Project of China (Grant No. 2011A060901026)), Key Joint Project of National Natural Science Foundation of China (Grant No. U1134004), China Post-Doctoral Science Foundation (125032).

References

[1] Min S Y, Kim T S, Kim B J, Cho H, Noh Y Y, Yang H, Cho J H and Lee T W 2013 *J. Nat. Commun.* **4** 1773
[2] Sun B, Long Y-Z, Chen Z-J, Liu S-L, Zhang H-D, Zhang J-C and Han W-P 2014 *J. J. Mater. Chem. C* **2** 1209-19
[3] Song Y M, Xie Y, Malyarchuk V, Xiao J, Jung I, Choi K J, Liu Z, Park H, Lu C, Kim R H, Li R, Crozier K B, Huang Y and Rogers J A 2013 *J. Nature* **497** 95-9
[4] Kim D H, Lu N, Ma R, Kim Y S, Kim R H, Wang S, Wu J, Won S M, Tao H, Islam A, Yu K J, Kim T I, Chowdhury R, Ying M, Xu L, Li M, Chung H J, Keum H, McCormick M, Liu P, Zhang Y W, Omenetto F G, Huang Y, Coleman T and Rogers J A 2011 *J. Science* **333** 838-43
[5] Xu S, Zhang Y, Cho J, Lee J, Huang X, Jia L, Fan J A, Su Y, Su J, Zhang H, Cheng H, Lu B, Yu C, Chuang C, Kim T I, Song T, Shigeta K, Kang S, Dagdeviren C, Petrov I, Braun P V, Huang Y, Paik U and Rogers J A 2013 *J. Nat. Commun.* **4** 1543
[6] Xin Y and Reneker D H 2012 *J. Polymer* **53** 4254-61
[7] Yu J, Qiu Y, Zha X, Yu M, Yu J, Rafique J and Yin J 2008 *J. Eur. Polym. J.* **44** 2838-44
[8] Han T, Reneker D H and Yarin A L 2007 *J. Polymer* **48** 6064-76
[9] Kessick R and Tepper G 2004 *J. Appl. Phys. Lett.* **84** 4807
[10] Huang Y, Duan Y, Ding Y, Bu N, Pan Y, Lu N and Yin Z 2014 *J. Sci. Rep.* **4** 5949
[11] Sun B, Long Y Z, Liu S L, Huang Y Y, Ma J, Zhang H D, Shen G and Xu S 2013 *J. Nanoscale* **5** 7041-5
[12] Zhu Z M, Chen X D, Zheng J W, Liang F, Li M H and Wang H 2014 *Proceedings of the 1st International Conference on Progress in Additive Manufacturing* (Singapore: Nanyang Executive Centre) pp 352-7
[13] Park J U, Hardy M, Kang S J, Barton K, Adair K, Mukhopadhyay D K, Lee C Y, Strano M S, Alleyne A G, Georgiadis J G, Ferreira P M and Rogers J A 2007 *J. Nat. Mater.* **6** 782-9
[14] Bu N, Huang Y, Wang X and Yin Z 2012 *J. Mater. Manuf. Processes* **27** 1318-23

IOP Conf. Series: Materials Science and Engineering **87** (2015) 012019 doi:10.1088/1757-899X/87/1/012019

Catalyzed oxidative degradation of methyl orange over Au catalyst prepared by ionic liquid-polymer modified silica

Y Wang[1] **and J S Guo**

Key Laboratory of Textile Science and Technology, Ministry of Education, College of Textiles, Donghua University, 2999 North Renmin Road, Shanghai 201620, P. R. China

Abstract. A new type of hybrid material was prepared by grafting an ionic liquid monomer, 1-(p-vinylbenzyl)-3-methylimidazolium chloride, on the surface of the porous silica which was synthesized via sodium silicate hydrolysis. The as-synthesized products were characterized by scanning electron microscope, nitrogen physisorption experiment, thermogravimetric analysis and Fourier transform infrared spectra. A catalyst with Au was prepared using the hybrid material as carrier. The experimental results show that the catalyst exhibits a better catalytic effect of hydrogen peroxide on the degradation of methyl orange. The reason may be that the metal component of the catalyst facilitated the dissociation of hydrogen peroxide to produce abundant highly active free radicals which can rapidly ruin the structure of methyl orange molecules in water. Finally, a probable catalytic degradation mechanism based on diffusion was discussed.

1. Introduction

Methyl orange, an azo dye, which is extensively applied in various industries, is known to be carcinogenic and mutagenic. Thus, the degradation of methyl orange has attracted particular attention in the last decade. The degradation of methyl orange can be achieved through an intricate oxidation process with multiple auxiliary means. However, some oxidants such as hydrogen peroxide possess the ability to disrupt methyl orange molecules through their dissociation and yet the oxidant is usually stable in the presence of methyl orange. In order to degrade methyl orange thoroughly and quickly, a few auxiliary measures are necessary. Inorganic or organic components in some compounds have been proved to be efficient activators for hydrogen peroxide to generate a variety of free radicals and intermediates [1-5]. Xu et al found that bicarbonate anion can stimulate the dissociation of hydrogen peroxide to liberate active oxygen species which disrupt methyl orange molecules [6]. Ember et al studied the mechanism of Mn(Ⅱ) activating hydrogen peroxide in the presence of bicarbonate anion and found that the active species formed in the reactions such as the Mn(Ⅳ)=O intermediate and hydroxyl radicals are the motivation of the degradation of methyl orange [7]. In addition, Fenton and Fenton-like methods are usually used in the degradation of methyl orange as important means [8-11].

Ionic liquids are typical green chemicals due to their special physicochemical properties. In this paper, we grafted an ionic liquid monomer onto the surface of the porous silica via surface-initiated polymerization, obtaining modified silica. A novel catalyst containing Au was prepared by means of

[1] Address for correspondence: Y Wang, Key Laboratory of Textile Science and Technology, Ministry of Education, College of Textiles, Donghua University, 2999 North Renmin Road, Shanghai 201620, P. R. China. E-mail: 1109001@mail.dhu.edu.cn.

the adsorption of the metal ions with the modified silica. This catalyst has prominent catalytic activity for the degradation of methyl orange in the presence of hydrogen peroxide.

2. Experimental

2.1. Materials

4-(chloromethyl) styrene (90%), 1-methylimidazole (99%), chloroauric acid ($HAuCl_4$, 48-50% based on Au), 2-bromoisobutyryl bromide (98%), γ-aminopropyl triethoxysilane (APS, 98%), triethylamine (TEA, 99%), 2, 2'-bipyridyl (BiPy, AR, 99%) and sodium silicate (47% based on SiO_2) were purchased from Aladdin. Ethyl acetate (AR, ≥99.5%), sulfuric acid (AR, 95-98%), acetone (AR, ≥99.5%), ethanol (AR, ≥95%), sodium borohydride ($NaBH_4$, ≥96.0%), hydrogen peroxide (H_2O_2, AR, ≥30%), N, N-Dimethylformamide (DMF, AR, ≥99.5%) and methyl orange (IND) were purchased from Sinopharm Chemical Reagent Co., Ltd. Toluene (AR, ≥99.5%) was purchased from Sinopharm Chemical Reagent Co., Ltd and dehydrated with 4A molecular sieve prior to use. Copper (I) chloride (CuCl, ≥99.99%) was purchased from Sigma-Aldrich. Other reagents were of analytical grades and used as received.

2.2. Characterization

Scanning electron microscope (SEM) measurements were carried out on a FEI (company) Quanta 200F SEM. Nuclear magnetic resonance (NMR) spectra were carried out on a Bruker AVANCE III spectrometer. Fourier transform infrared (FT-IR) spectra were recorded on a Bruker TENSOR27 spectrometer using KBr pellets. Thermogravimetric analysis(TGA) was performed on a NETZSCH STA449F3 thermogravimetric analyzer at a heating rate of 10℃/min under nitrogen atmosphere. Ultraviolet-Visible (UV-Vis) spectra were recorded on a SHIMADZU 2550 spectrometer. The nitrogen physisorption experiment was performed on a QUANTACHROME QUADRASORB SI gas sorption system in static measurement mode at 77 K. The Au loading of the catalyst was determined by an inductively coupled plasma (ICP) spectrometer (Prodigy, Leeman Corporation).

2.3. Preparation of ionic liquid monomer

Ionic liquid monomer, 1-(p-vinylbenzyl)-3-methylimidazolium chloride ([VBMI][Cl]), was prepared according to a previous literature [12]. Typically, a mixture of 4-(chloromethyl) styrene (4.58 g, 30 mmol) and 1-methylimidazole (2.05 g, 25 mmol) was stirred under N_2 at 42℃ for 24 h (figure 1). The reaction mixture was washed with ethyl acetate (3×45 mL) and then dried overnight under vacuum at ambient temperature. ^{1}H NMR (400 MHz, δ, ppm, DMSO-d_6): 3.92 (s,3H,-N-CH_3), 5.30 (d,1H, *J*=11.0 Hz, C*H*H=CH-), 5.58 (s,2H,Ph-CH_2-N-), 5.90 (d,1H, *J*=17.7 Hz, CH*H*=CH-), 6.72-6.79 (m,1H, CH_2=CH-), 7.42-8.02 (m,6H, -N-CH-CH-N- and Ph), 9.81 (s,1H, -N-CH-N-).

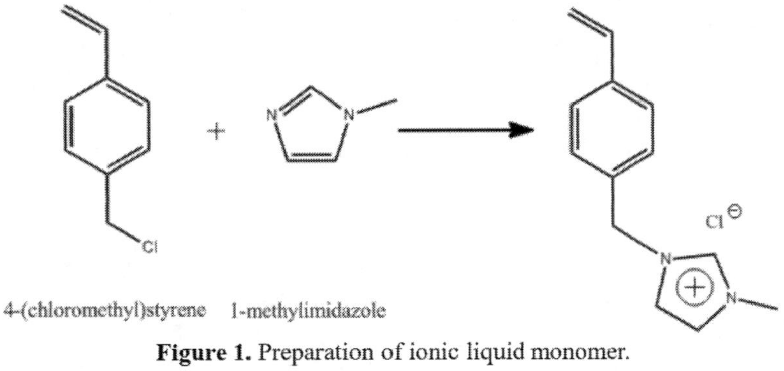

Figure 1. Preparation of ionic liquid monomer.

2.4. Preparation of porous silica

Porous silica was prepared via sodium silicate hydrolysis. In a typical process, to a solution of sodium silicate (5.0 g) in deionized water (50 mL) was added with stirring an aqueous solution of sulfuric acid (1.8 M, 20 mL). The reaction mixture was stirred at 75℃ for 5 h to form silica gel and then hydrothermally treated at pH=10 at 100 ℃ for 24 h. The crude product obtained was washed many times with deionized water and dried under vacuum conditions at 120℃ for 12 h.

2.5. Synthesis of initiator 2-bromo-2-methyl-N-(3-(triethoxysilyl) propyl) propanamide (BTPAm)
BTPAm was synthesized via the reaction of 2-bromoisobutyryl bromide and APS as described previously [13]. A solution of 2-bromoisobutyryl bromide (1.0 mL,8 mmol) in toluene (100 mL) was added dropwise to a cold solution of APS (1.8 mL,8 mmol) in dry toluene(100 mL) with TEA (1.2 mL, 8 mmol) at 0℃. The mixture was magnetically stirred for 3 h at 0℃ and then for another 10 h at ambient temperature (figure 2). The precipitate in the mixture was removed by filtration and the filtrate was evaporated under reduced pressure. ^{1}H NMR (400 MHz, δ, ppm, $CDCl_3$): 4.58 (t, 2H, *J*=7.0 Hz, -Si-CH_2-), 5.16 (t, 9H, *J*=6.2 Hz, CH_3-CH_2-O-Si-), 5.57-5.59 (m, 2H, -Si-CH_2-CH_2-), 5.87 (s, 6H, CH_3-C-Br), 7.15 (m, 2H, -CH_2-NH-), 7.74-7.78 (m, 6H, CH_3-CH_2-O-Si-), 8.70 (s, 1H, -NH-).

2.6. Immobilization of the initiator onto the porous silica [14]
Porous silica (1.0 g) was added to a solution of BTPAm (0.1 mL) in anhydrous toluene (20 mL) and the mixture was stirred under N2 at room temperature for 24 h (figure 3). The solid was separated from the mixture by filtration and washed with toluene, acetone, and ethanol and then dried under vacuum conditions at room temperature for 12 h, affording BTPAm modified silica (silica-BTPAm).

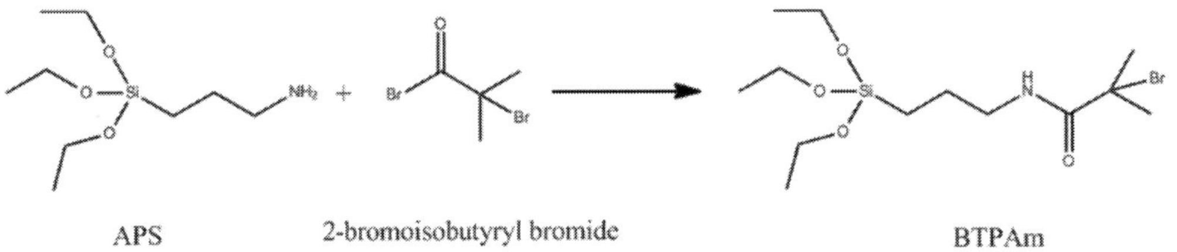

Figure 2. Synthesis of initiator BTPAm.

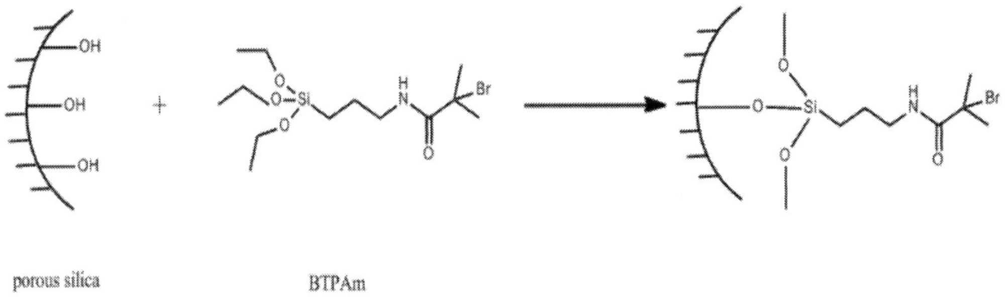

Figure 3. Immobilization of the initiator onto the porous silica.

2.7. Preparation of the porous silica grafted by poly ([VBMI][Cl]) via surface-initiated polymerization
Silica-BTPAm (2.0 g) was added to a round-bottom flask containing a solution of CuCl (9.9 mg, 0.1 mmol) and BiPy (31.2 mg, 0.2 mmol) in DMF (40 mL) and the mixture was degassed and then filled back with N_2. A degassed solution of [VBMI] [Cl] (2.5 g) in DMF (10 mL) was added to the flask. After the reaction mixture was stirred at 60℃ for 24 h (figure 4), the solid was separated from the reaction system by filtration and washed many times with DMF and water. Finally, the product was

dried under vacuum at 80℃ for 12 h, obtaining polymer-modified silica (Silica-PIL). It is shown that Silica-PIL can adsorb gold and palladium ions in water, but could not adsorb methyl orange in water.

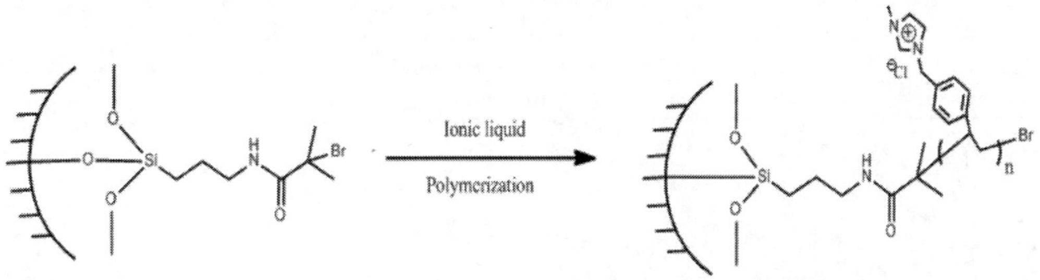

Figure 4. Preparation of porous silica grafted by poly ([VBMI][Cl]) via surface-initiated polymerization.

2.8. Preparation of catalyst.

Silica-PIL (0.1 g) was added into a vessel containing an aqueous solution of chloroauric acid (7 mM, 4.0 mL) and the mixture was sonicated in water bath and then stirred until the solution was discolored. The solid was separated from the mixture by filtration and dried under vacuum at 60℃ for 4 h. To a flask containing the above solid and deionized water(10 mL) was added with vigorous stirring a cold aqueous solution of sodium borohydride (0.1 M, 1.0 mL) at 0℃. The product was obtained by filtration and dried under vacuum at 40℃ for 4 h, affording the catalyst loaded with Au with a slightly ruby color (Silica-PIL-Au, Au: 4.52%, ICP).

2.9. Oxidative degradation of methyl orange by hydrogen peroxide in the absence of the catalyst

To a vessel containing an aqueous solution of methyl orange (100 mg/L, 3 mL) was added an aqueous solution of H_2O_2 (30%, 3 mL). The reaction mixture was kept at room temperature for 16 h without stirring.

2.10. Oxidative degradation of methyl orange by hydrogen peroxide in the presence of Silica-PIL

To a vessel containing the mixture of an aqueous solution of methyl orange(100 mg/L, 3 mL) and Silica-PIL(0.1 g) was added an aqueous solution of H_2O_2 (30%, 3 mL). The reaction mixture was kept at room temperature for 16 h without stirring.

2.11. Oxidative degradation of methyl orange by hydrogen peroxide in the presence of the catalyst

To a vessel containing the mixture of an aqueous solution of methyl orange(100 mg/L, 3 mL) and the above catalyst (0.1 g) was added an aqueous solution of H_2O_2 (30%, 3 mL). The reaction mixture was kept at room temperature for 16 h without stirring.

3. Results and discussion

3.1. Morphology and physicochemical properties of porous silica

Morphology of the porous silica was investigated by SEM observation and shown in figure 5. It can be observed that the surface of the sample is scraggly and found many alveolate holes. This special structure may adsorb copious catalytic substances under certain conditions and further provide many active sites for chemical reactions.

The specific surface area and pore volume of the porous silica were calculated from the nitrogen physisorption isotherm (figure 6) using Brunauer- Emmett-Teller (BET) method. The pore size of the porous silica was determined through the above SEM image. Some parameters on the properties of the porous silica are shown in table 1.

3.2. Thermal stability of Silica-PIL

TGA, an effective tool to investigate thermal stabilities of various materials, was carried out for Silica-PIL from 30 to 850℃ under nitrogen atmosphere (figure 7). For Silica-PIL, the weight loss around 100℃ is due to the desorption of the physic-adsorbed water. In the range of 200-700℃, a weight loss of 37.38% is mainly ascribed to the decomposition of the organic substances, such as polymeric ionic liquids grafted onto the surface of the sample.

Figure 5. Morphology of the porous silica.

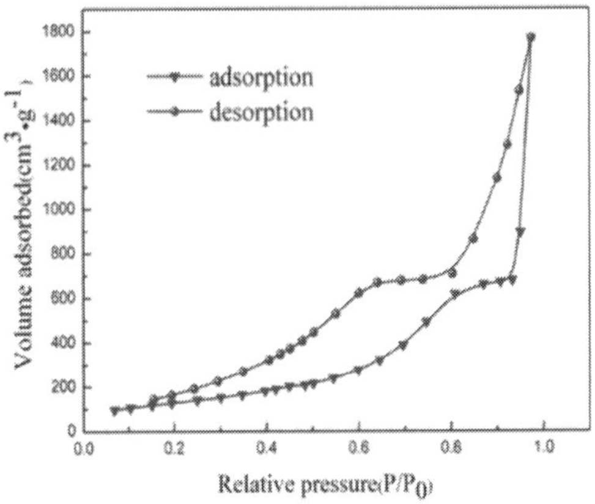

Figure 6. Nitrogen physisorption isotherm for the porous silica.

Table 1. Parameters on the properties of the porous silica.

	specific surface area (m^2 g^{-1})	pore volume (mL g^{-1})	pore size (nm)
porous silica	479	2.73	20.0

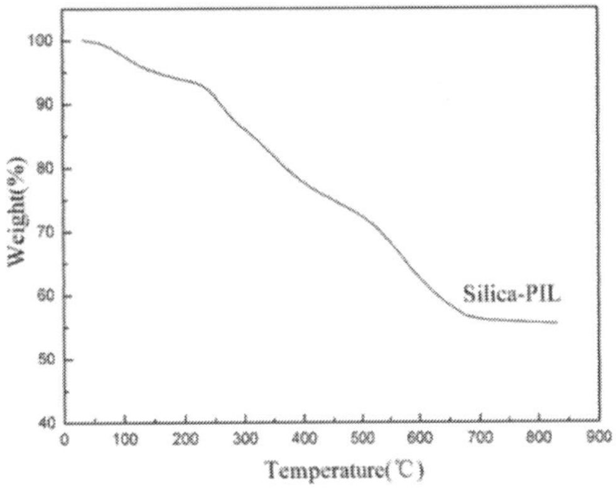

Figure 7. TGA curve of Silica-PIL.

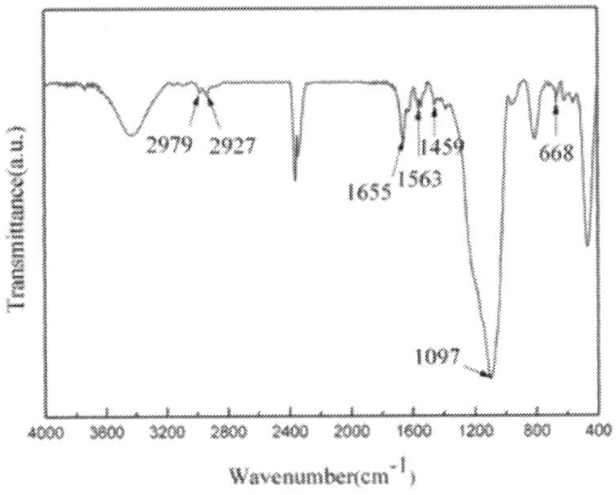

Figure 8. FT-IR spectrum of Silica-PIL.

3.3. The structural analysis of Silica-PIL

The structure of ionic liquid-polymer modified porous silica was investigated by means of FTIR spectrum (figure 8). The FTIR spectrum exhibits a strong band at 1097 cm^{-1} due to Si-O-Si stretching vibration [15, 16]. The bands at 2979 and 2927 cm^{-1} are attributed to C-H stretching vibration [17, 18]. The spectrum also shows other characteristic absorption bands at 1655 and 1563 cm^{-1} for imidazole ring vibration [19-21] and 1459 cm^{-1} for benzene ring vibration [15, 16]. The band at 668 cm^{-1} can be assigned to C-H bending vibration [15, 16].

3.4. Evaluation of catalytic activities

The color of the aqueous solution of methyl orange oxidated by hydrogen peroxide was changed from orange to yellow in the absence of the catalyst and kept this color even after 14 days (image (a) in figure 9). The result shows that the effect of the degradation of methyl orange only over hydrogen peroxide without any catalyst is poor and the thorough degradation of methyl orange in water could not be achieved under the condition. The probable reason is that hydrogen peroxide is stable in the presence of methyl orange [22] and could not produce abundant free radicals to decompose methyl orange. Similar result was obtained for Silica-PIL, suggesting that Silica-PIL has no the catalytic activity for methyl orange.

However, the aqueous solution of methyl orange oxidated by hydrogen peroxide catalyzed by Silica-PIL-Au was decolorized and became clarified after 16 h (image (b) in figure 9), which indicated that the thorough degradation of methyl orange was achieved.

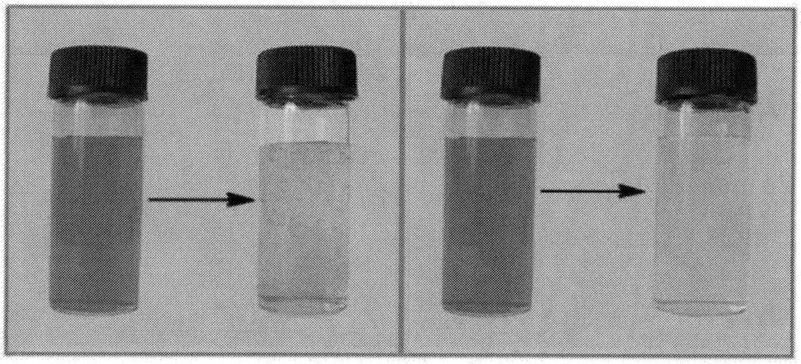

(a) (b)

Figure 9. Comparison of the degradation of methyl orange without or with the catalyst.

UV-Vis spectra also demonstrated the complete degradation of methyl orange in water oxidated by hydrogen peroxide in the presence of the catalyst (figure 10). The band at wavelength of 463 nm due to the conjugated structure constructed via azo bond [23, 24] disappeared after the degradation, indicating that the azo structure of methyl orange molecule was damaged. Benzene ring of methyl orange molecule was also damaged after the degradation owning to the disappearance of the band at 273 nm assigned to π-π^* transition of a benzene molecule [25].

Based on the experimental process and results, a probable mechanism is proposed. Obviously, Au immobilized on the catalysts is able to stimulate the decomposition of hydrogen peroxide to form abundant free radicals, such as hydroxyl radicals [10]. These free radicals can rapidly degrade the methyl orange molecules in water (figure 11). And the action continued through the diffusion of the methyl orange molecules in the reaction system toward the surface of the catalyst.

4. Conclusions

The synthesis of a novel ionic liquid-polymer modified porous silica was achieved via surface-initiated polymerization. The composite was used as a kind of catalyst after it was loaded with Au. The results showed that the degradation of methyl orange was significantly accelerated and thoroughly achieved under the action of the catalyst compared with that without the catalyst. The probable reason is that Au immobilized on the catalyst facilitates hydrogen peroxide decomposition to produce abundant free radicals which can rapidly damage the molecular structure of methyl orange. Our work has significant implications for preparation of novel catalysts and degradation mechanism of methyl orange.

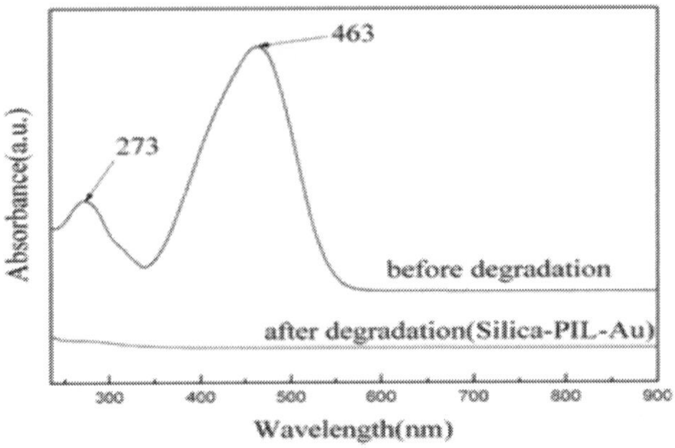

Figure 10. UV-Vis spectra of methyl orange before/after the degradation.

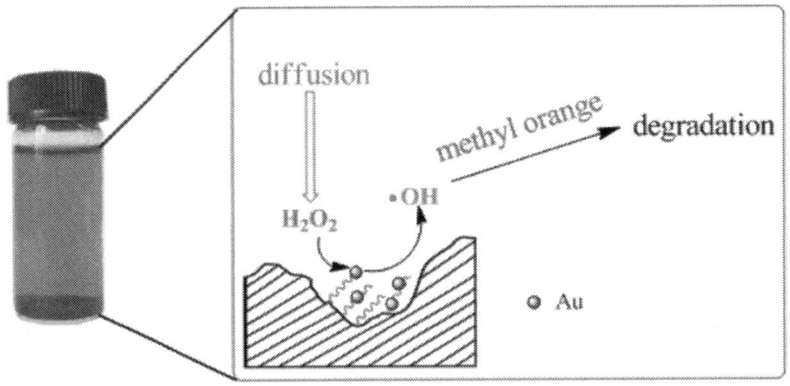

Figure 11. Schematic diagram of a possible mechanism of the catalyzed oxidative degradation of the methyl orange in the reaction system.

References

[1] Thi D N, Ngoc H P, Manh H D and Kim T N 2011 *J. Hazard. Mater.* **185** 653-61
[2] Panda N, Sahoo H and Mohapatra S 2011 *J. Hazard. Mater.* **185** 359-65
[3] Xu H Y, Prasad M, Qi S Y and Li Y 2010 *Sci. China. Technol. Sc.* **53** 3014-19
[4] Ran J W, Li X Y, Zhao Q D, Qu Z P, Li H, Shi Y and Chen G H 2010 *Inorg. Chem. Commun.* **13** 526-8
[5] Devi L G, Kumar S G, Reddy K M and Munikrishnappa C 2009 *J. Hazard. Mater.* **164** 459-67
[6] Xu A H, Li X X, Xiong H and Yin G C 2011 *Chemosphere* **82** 1190-95
[7] Ember E, Gazzaz H A, Rothbart S, Puchta R and van Eldik R 2010 *Appl. Catal. B-Environ.* **95** 179-91
[8] Ali M E M, Gad-Allah T A, Elmolla E S and Badawy M I 2014 *Desalin. Water. Treat.* **52** 4538-46
[9] Pham A N, Xing G W, Miller C J and Waite T D 2013 *J. Catal.* **301** 54-64
[10] Lin T Y and Wu C H 2005 *J. Catal.* **232** 117-26
[11] Shah V, Verma P, Stopka P, Gabriel J, Baldrian P and Nerud F 2003 *Appl. Catal. B-Environ.* **46** 287-92
[12] He X Y, Yang W and Pei X W 2008 *Macromolecules* **41** 4615-21
[13] Sun Y B, Ding X B, Zheng Z H, Cheng X, Hu X H and Peng Y X 2007 *Eur. Polym. J.* **43** 762-72

[14] Schepelina O and Zharov I 2008 *Langmuir* **24** 14188-94
[15] Hu F X, Neoh K G, Cen L and Kang E T 2006 *Biomacromolecules* **7** 809-16
[16] Li C M, Yang J, Wang P Y, Liu J and Yang Q H 2009 *Micropor. Mesopor. Mat.* **123** 228-33
[17] Nasef M M, Saidi H, Dessouki A M and El-Nesr E M 2000 *Polym Int* **49** 399-406
[18] Castillejos E, Debouttiere P J, Roiban L, Solhy A, Martinez V, Kihn,Y, Ersen O, Philippot K, Chaudret B and Serp P 2009 *Angew. Chem. Int. Edit.* **48** 2529-33
[19] Dan Y M, Yu H G, Long Q, Hou A, Liu Y and Qu S S 2004 *Thermochim. Acta.* **419** 169-72
[20] Wu J P, Wang M J and Stark J P W 2006 *J. Quant. Spectrosc. Ra.* **102** 228-35
[21] Nanbu N, Sasaki Y and Kitamura F 2003 *Electrochem. Commun.* **5** 383-7
[22] Cheng L, Wei M Y, Huang L H, Pan F, Xia D S, Li X X and Xu A H 2014 *Ind. Eng. Chem. Res.* **53** 3478-85
[23] Chen Y P, Liu S Y, Yu H Q, Yin H and Li Q R 2008 *Chemosphere* **72** 532-6
[24] Zhang A P and Fang Y 2006 *Chem. Phys.* **331** 55-60
[25] Takeuchi M, Hidaka M and Anpo M 2014 *Res. Chem. Intermediat.* **40** 2315-25

Solidification behavior of high-density polyethylene (HDPE) during injection molding: Correlation between crystallization kinetics and thermal gradient field

Bin Yang[1,4], Yan-Li Deng[1], Gui-Jing Li[1,2], Ji-Bin Miao[1], Ru Xia[1,5], Jia-Sheng Qian[1], Peng Chen[1] and Jing-Wang Liu[3]

[1]College of Chemistry & Chemical Engineering and the Key Laboratory of Environment-Friendly Polymeric Materials of Anhui Province, Anhui University, Hefei 230601, P.R. China

[2]Department of Materials Science & Engineering, Zhejiang University, Hangzhou 310058, P.R. China

[3]College of Polymer Science & Engineering and the State Key Laboratory of Polymer Materials Engineering, Sichuan University, Chengdu 610065, P.R. China

E-mail: yangbin@ahu.edu.cn; xiarucn@sina.com

Abstract. This work mainly investigated the effect of thermal field on the crystallization kinetics of high-density polyethylene (HDPE) during injection molding (IM) process. The thickness $X = 0.4$ was found to be a crucial location heavily influenced by thermal conduction. The temperature decay tended to be stable, with limited variation of the crystallization rate when $X > 0.4$. It was observed that the crystallization rate was in good proportion to the cooling rate (ϕ). Our experimental finding showed that the consequence of relative crystallinity (χ) was in agreement with that of the secondary temperature difference (STD). This study is practically significant to the further investigation on the relationship among "processing-structure-property" of polymeric materials.

1. Introduction

The transient heat conduction problems (e.g., the ice formation), also referred to as the "Stefan problems" or "moving boundary problems", were initially proposed by J. Stefan in 1981 [1]. Nowadays, issues of this kind still have practical significance in both engineering and industrial fields. Numerous methods have been proposed to numerically solve the multi-dimensional phase-change heat conduction problems, such as, the variational method [2, 3], the moving heat source method [4], the finite element method [5], the enthalpy method [6], the variable space-grid method [7, 8], the nodal

[4] Address for correspondence: Bin Yang, College of Chemistry & Chemical Engineering and the Key Laboratory of Environment-Friendly Polymeric Materials of Anhui Province, Anhui University, Hefei 230601, P.R. China. E-mail: yangbin@ahu.edu.cn.

[5] Address for correspondence: Ru Xia, College of Chemistry & Chemical Engineering and the Key Laboratory of Environment-Friendly Polymeric Materials of Anhui Province, Anhui University, Hefei 230601, P.R. China. E-mail: xiarucn@sina.com.

integral method [9], and so forth.

Injection molding (IM) is one of the most commonly used fabricating techniques, accounting for about one third of all plastics processing [10, 11]. Generally speaking, an IM cycle consists of four stages, namely, filling, packing/holding, cooling and demolding (ejection) [12]. The product quality will also be considerably influenced by the operational parameters, mixing with another material, etc [13-15]. In spite of the significant improvement on modeling/simulation of the IM process of thermoplastic polymers over the past few decades, a reliable prediction of the viscoelastic response and microstructure development during real processing operation still remains challenging so far. In this study, the cooling stage of IM process was numerically studied using the enthalpy transformation methodology (ETM), which has been recently proved to be an efficient method to treat the phase-change heat conduction issues of crystalline polymers.

The degree of crystallinity is a vital parameter that basically dictates many end-use properties of polymers, including dimensional stability, mechanical, optical and thermal properties [16-18]. For instance, the difference in density between crystalline and non-crystalline regions leads to different refractive indices. The heat resistance always increases with the improvement of crystallinity. Besides, many defects of the injection-molded articles (e.g., warpage, shrinkage, dimensional instability, etc.) are usually associated with crystallinity, primarily owing to the improper selection of processing variables. As a result, the acquisition of desired material performance could also be realized by regulating the crystallinity via the optimization of process parameters, such as, melt temperature, mold temperature, injection speed, holding pressure, etc. [19-21]. Besides, recent findings have shown that the mold temperature has a tremendous influence on the product life-time [22].

In light of the importance of crystallinity to polymeric materials, a lot of literature has been already presented on the simulation and experimental studies of the polymer crystallization process [23-31]. Among them, the Avrami Equation is a widely established relationship between crystallinity and crystallization time especially under quiescent isothermal crystallization conditions. On this basis, a number of mathematical models and modifications have also been proposed for non-isothermal crystallization circumstances [23, 27-29]. The theoretical analyses generally showed good agreement with their resultant experimental data. As is known, real processing conditions (e.g., extrusion, blow molding, injection molding, etc.) are considerably different from the quiescent crystallization investigations (either isothermal or non-isothermal) since the polymers are actually subject to the varying mechanical and thermal fields simultaneously. More importantly, the liberation of latent heat due to polymer crystallization could remarkably alter the in-cavity temperature fields [32]. Nevertheless, the correlation between solidification kinetics and polymer crystallization is still unclear up to now.

Based on our previous work [10, 33-37], the objective of the present study is to explore the effect of the initial temperature difference (ITD) and the secondary temperature difference (STD) on the solidification kinetics of HDPE, in an attempt to disclose the physical nature behind the phase-change heat conduction during injection molding process. Crystallization behaviors under different processing conditions were compared, with the degree of crystallinity obtained using the ETM coupled with the material P-V-T relationship [1, 8, 34]. The present work is of fundamental importance to the optimization of the process variables as well as the further investigation on the relationship among "processing-structure-property" of crystalline polymers.

2. Experimental

2.1. Material

The material used in present experiments was high-density polyethylene (HDPE), Model: 5000S, which was kindly supplied by the *Daqing Petroleum Chemical Co.*, China, with a number-averaged molecular weight (M_n) of 5.30×10^5, and a melt flow rate (MFR) of 1.0 g/10 min (2.16 kg/190°C) according to ASTM D1238. Detailed material thermal parameters of HDPE had been presented elsewhere [10, 35].

IOP Conf. Series: Materials Science and Engineering **87** (2015) 012020 doi:10.1088/1757-899X/87/1/012020

2.2. Rheological property characterization
Rheological behavior was examined using the melting index (MI) tester, Model: ZRZ-1452, provided by the *SANS Instrument Co*., China. The capillary diameter is 2.09 mm, with a capillary length of 8.0 mm. The logarithmic apparent viscosity (η_a) of the material used shows good linearity with the increase of the inverse absolute temperature (T) under a constant load (2.16 kg) at various temperatures ranging from 150 to 230°C with an increment of 20°C, as demonstrated in figure 1. The melt flow activation energy (E_a) of the HDPE resin was 26.6 kJ/mol, evaluated using the Arrhenius Equation [38].

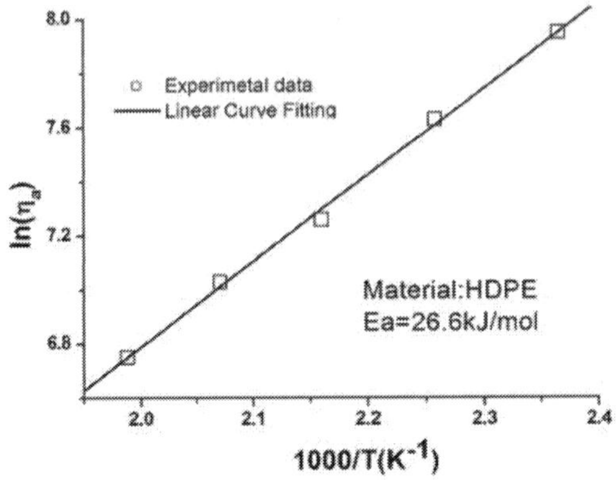

Figure 1. Measured rheological curve of the HDPE resin used.

2.3. In-cavity temperature measurement during IM process
IM experiments were carried out on a reciprocating screw precision injection-molding machine (Model: HYF-350, *Haiying Plastic Machinery Manufacturing Co*., China). The mold cavity utilized had its dimensions as follows: 175.0 mm in length, 12.0 mm in width, and 4.0 mm in thickness. The mold temperature was regulated through the use of circulating water. The injection time and packing time were 1.5 sec and 3.0 sec, respectively. Two armored copper-constantan micro-thermocouples (Model: TK-247, *Anthone Electronics Co*., China), with a 0.5 mm diameter sensor tip and measuring range from 35 to 350°C, were installed to monitor the temperature decays at specific positions. The insert depth of both thermocouples was adjustable so as to measure the in-cavity temperature profiles at different positions from the mold surface to the central plane of the cavity. The operation has been detailed in our previous studies [39]. The temperature changes throughout the injection molding cycles were recorded by a *Keithley*-2700 *Data Acquisition System* with a sampling time of 0.1 sec. In the present work, the injection molding experiments were conducted according to the processing conditions, as listed in table 1. For better comparison, four cooling conditions used in this study were denoted as Case A (T_0=190°C, T_w=20°C), Case B (T_0=210°C, T_w=40°C), Case C (T_0=210°C, T_w=60°C) and Case D (T_0=230°C, T_w=50°C), where T_0 and T_w represent the melt and mold temperatures, respectively.

Table 1. Processing conditions used in injection molding experiments.

Processing parameters	Values
Melt temperature (°C)	190, 210, 230
Mold temperature (°C)	20, 40, 50, 60
Injection pressure (MPa)	50.0
Packing pressure (MPa)	35.0
Injection time (sec)	1.5
Packing time (sec)	3.0
Cooling time (sec)	120.0

3. Results and discussion

Figure 2 shows the temperature profiles obtained via ETM through the part thickness under the processing conditions of Cases A-D from $ln\ t$=0 to $ln\ t$=4, where t is the cooling time. Two horizontal imaginal lines of T=110°C and T=120°C were portrayed so as to indicate the region representing the mushy zone, which consists of both liquid and solid phases [33]. For simplicity, the thickness is presented in the form of normalized distance (X), which is defined as $X=x/b$ with b denoting the reference length (i.e., b=2 mm in the present work) [36]. It can be seen from the comparison among figures 2a-2d that the temperature at X=0 (i.e., the position closest to the mold wall) falls sufficiently rapidly to the wall temperature (T_W) from the initial melt temperature (T_0), suggesting the existence of a large cooling rate. At a specific time, the cooling rate decreases gradually with the increasing X from the mold wall surface. As compared with the outer parts of the injection-molded article, the inner parts maintain relatively high temperature. This could be due to the fact that the latent heat released during the melt crystallization could not be transferred away timely in the central area. The vertical distance between two neighboring curves at a specific location can be adopted to be an estimate of the cooling rate. For instance, at X=0.5, the time when the temperature drops from T_0 to the range of mushy zone T_1~T_2 [34, 39] (i.e., T_1=115.1°C and T_2=118.2°C for the present case), can be employed to compare the cooling rate under various cooling conditions. For brevity, ITD and STD were defined as ITD=T_0-T_w and STD=T_f-T_w, respectively, with the reference phase transition temperature (T_f) given by T_f =$(T_1+T_2)/2$ [34, 36]. The values of both ITD and STD are presented in table 2. The cooling rate (ϕ) was estimated in the following order: $\phi|_{\text{Case A}} > \phi|_{\text{Case B}} > \phi|_{\text{Case D}} > \phi|_{\text{Case C}}$, which is consistent with the consequence of STD [39]. Both Cases A and B have the same ITD of 170°C but $T_{W|\text{Case A}} < T_{W|\text{Case B}}$, which also indicated that mold temperature exerted greater influence on the controlling of the cooling rate [11, 33]. It was shown that at a given location, when $X < 0.4$, the cooling speed increased firstly and then decreased; while $X \geq 0.4$, the cooling speed increased continuously. At X = 1.0, it was not until $ln\ t$ = 2 that the temperature began to change, suggesting that it took some time for heat to be transferred out.

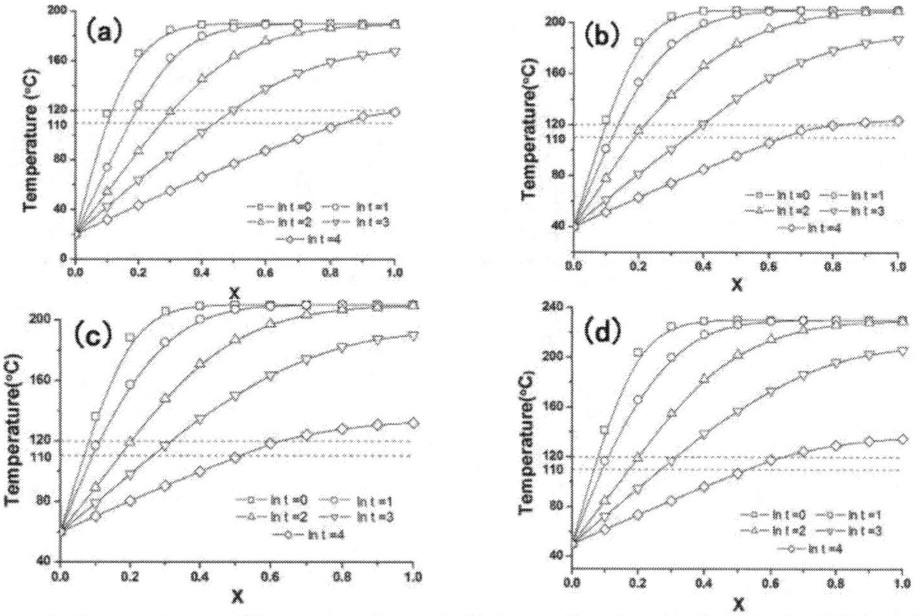

Figure 2. Temperature profiles across the part thickness direction (evaluated via ETM) under various cooling conditions: (a) T_0=190 °C, T_W=20 °C, (b) T_0=210 °C, T_W=40 °C, (c) T_0=210 °C,

T_W=60 °C and (d) T_0=230 °C, T_W=50 °C.

Table 2. Comparison of ITD and STD values under various cooling conditions (unit: °C).

Temperature Difference	Case A	Case B	Case C	Case D
ITD	170.0	170.0	150.0	180.0
STD	96.7	76.7	56.7	66.7

For better comparison, the dimensionless temperature (θ) was also introduced by $\theta=(T-T_W)/(T_0-T_W)$ [33, 37]. The phase transition occurred at θ = 0.56~0.58, 0.44~0.46, 0.50~0.52, 0.36~0.38, respectively, under various cooling conditions from Case A to Case D. Two horizontal lines (also given in dimensionless form) were depicted in figure 3 to distinguish the area that underwent the phase-transition process. It can be observed that no phase-transition platform was displayed at X = 0.1 under Case C. The onset time when phase-change platform occurs at the locations of X = 0.1 and X = 0.3 ranks as, Case C < Case A < Case B < Case D, which is generally in accordance with the rank of ITD. At X=0.3, the length of phase-transition plateau (L_{pc}) ranks as $L_{pc|\text{Case C}} < L_{pc|\text{Case B}} < L_{pc|\text{Case D}} < L_{pc|\text{Case A}}$, which is different from our previous finding that L_{pc} is merely indicated by STD [39]. It was thus suggested that L_{pc} is more relevant to the initial melt temperature (T_0) instead of the STD. It is obvious that the closer to the center, the slower cooling rate (that is to say, the longer time the polymers stay within the crystallization temperature range T_1~T_2). The cooling rates at the occurrence of the phase-change transition are summarized in table 3. At X=0.7, the sequence of cooling rate is consistent with that of ITD, indicating that when X > 0.4 the influence of various factors (such as, shear force, mold temperature, etc.) tends to be stable. But at X = 0.3, Cases D and A are on the contrary, since Case A has relatively low mold temperature as compared to Case D, which also suggests that the mold temperature displays greater influence on the cooling rate [33].

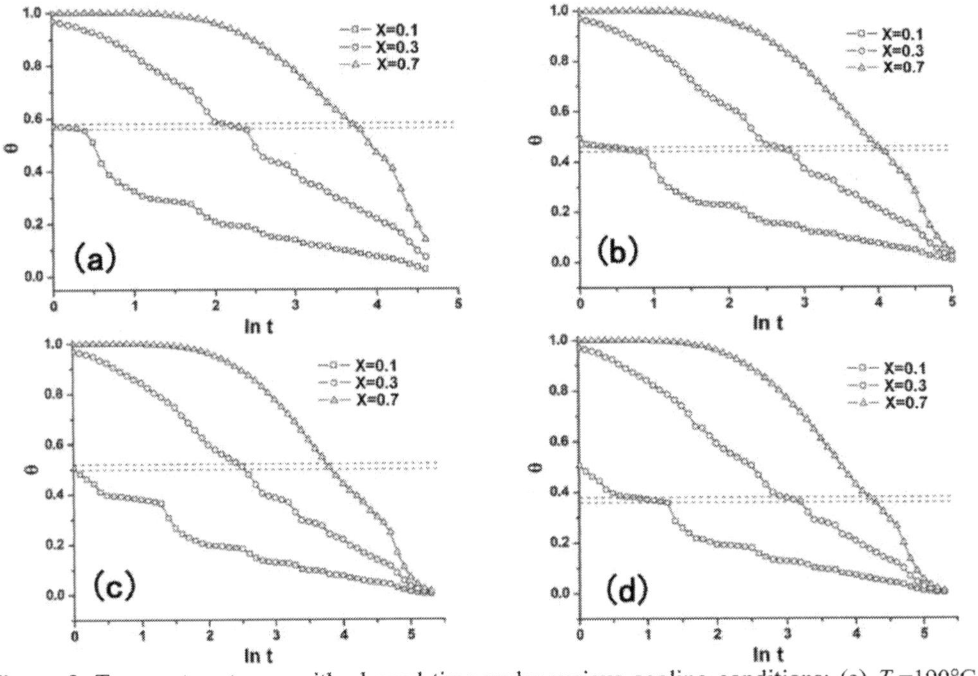

Figure 3. Temperature traces with elapsed time under various cooling conditions: (a) T_0=190°C, T_W=20°C, (b) T_0=210°C, T_W=40°C, (c) T_0=210°C, T_W=60°C and (d) T_0=230°C, T_W=50°C.

Table 3. Comparison of cooling rates at the occurrence of phase transition under various cooling conditions (unit: °C/s).

Location	Cooling rate at the occurrence of phase transition			
	Case A	Case B	Case C	Case D
X=0.3	1.35	1.04	2.84	0.65
X=0.7	0.91	0.70	1.29	0.51

As can be seen in figure 4, the length of the phase transition (Lpc) close to the mold surface is much shorter than that far away from it. At all positions in the cavity, L_{pc} ranks as $L_{pc|\text{ Case A}} < L_{pc|\text{ Case B}} < L_{pc|\text{ Case D}} < L_{pc|\text{ Case C}}$, indicating that the cooling speed ranks as, Case A > Case B > Case D > Case C. That is to say, the larger the L_{pc} value, the slower the cooling rate, which is consistent with the result from STD. In Case A, at X = 0.4~0.6, the L_{pc} value approximately remains constant. Generally speaking, the length of phase transition platform tends to share similar value when $X \geq 0.4$, for instance, Case B: X= 0.4~0.7, Case C: X= 0.5~0.7, and Case D: X= 0.6~0.7.

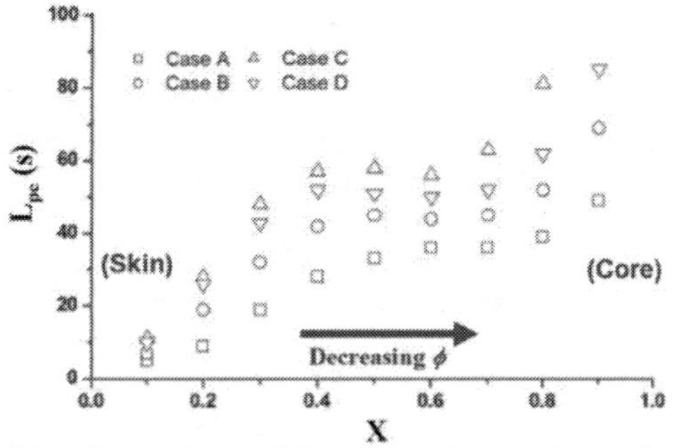

Figure 4. Position dependence of the phase-change plateau length (L_{pc}).

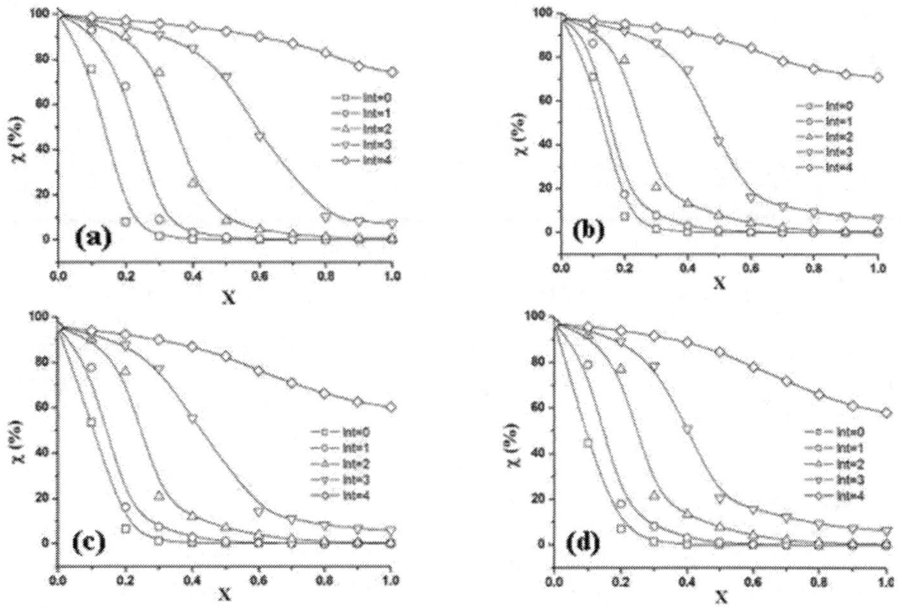

Figure 5. Distribution of crystallinity (χ) across the part thickness direction: (a) T_0=190 °C,

T_W=20 °C, (b) T_0=210 °C, T_W=40 °C, (c) T_0=210 °C, T_W=60 °C and (d) T_0=230 °C, T_W=50 °C.

Figure 5 shows the variation of crystallinity (χ) of injection-molded articles as a function of position. It should be mentioned that the degree of crystallinity (χ) can be calculated by using the following equation:

$$\chi = \frac{\rho_c}{\rho} \cdot \frac{\rho - \rho_a}{\rho_c - \rho_a} \times 100\% \frac{V_a - V}{V_a - V_c} \times \quad 10 \tag{1}$$

where ρ_c and V_c are the density and specific volume of the crystalline phase, respectively; while ρ_a and V_a are the density and specific volume of the amorphous phase of polymers, respectively. According to literature [40], ρ_c = 1.014 g/cm^3, ρ_a = 0.854 g/cm^3, for HDPE. With the aid of the 2d-*Tait* equation (see the Appendix), the specific volume can be estimated when the temperature and pressure are known. Then, the crystallinity (χ) can be readily calculated using equation (1).

The crystallinity of the polymer decreased from the mold surface to the core, as shown in figure 5. As the cooling time elapsed, the crystallinity of the polymer increased. The change of crystallinity at the central region was more obvious than that at the cortical region where the crystallinity hardly varied. Since there exist multi-scale structures as well as amorphous fractions within the macromolecules (unlike the ideal small molecular substances), polymer crystallization is quite difficult to achieve one hundred percent at all positions across the part thickness direction. In figure 6, the red curve is a master curve obtained via non-linear curve fitting technique, with the fitting curve being expressed as $t_{1/2}^{-1} = 0.0061 \cdot X^{-2.188}$. It can be readily seen that the experimental data under various cooling conditions have a relatively good fit to it. Since the value of $t_{1/2}^{-1}$ reflects the average crystallization rate [41, 42], the larger the value of $t_{1/2}^{-1}$, the greater the average crystallization rate of polymers. It can be seen that the crystallization rate considerably slows down with the decrease of cooling rate.

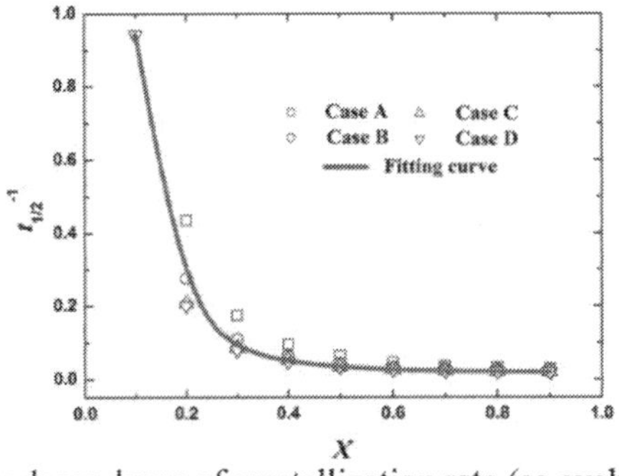

Figure 6. Position dependence of crystallization rate (as evaluated by $t_{1/2}^{-1}$).

Figure 7a suggests that the values of crystallinity at log t=0 (namely, $\chi|$ log t = 0) are quite similar under various cooling conditions for a given position (esp. in the inner locations). At X = 0.1, when *log t* = 0, the crystallinity ranks as $\chi|_{\text{Case A}} > \chi|_{\text{Case B}} > \chi|_{\text{Case C}} > \chi|_{\text{Case D}}$, which could be primarily dictated by the melt temperature (T_0). Under different conditions, the evolution of crystallinity with time generally takes an "*S*" shape, which is also in accordance with the non-isothermal crystallization kinetic studies [23, 42, 43]. However, a remarkable distortion on the kinetic curves is observed roughly at X_t = 80%, and the phenomenon could possibly be associated with the existence of shear flows during the melt post-filling stage. The start-stop time of phase transition can be readily acquired from the first derivation of the χ vs. *log t* curves. From the comparison, it can be concluded that the start time at X=0.1 is earlier than that at X=0.3 and X=0.7. At the position quite close to the mold surface, there exists large thermal gradient, which acts as the driving force of heat transfer [38], leading to the occurrence of phase transformation at the polymer/mold interface. Moreover, due to the thermal protection effect of the outer polymer (allowing for the poor thermal conductivity of polymers), the inner parts of the injection-molded article show relatively slower response to heat, thus these kinetic curves change slightly at the beginning of crystallization.

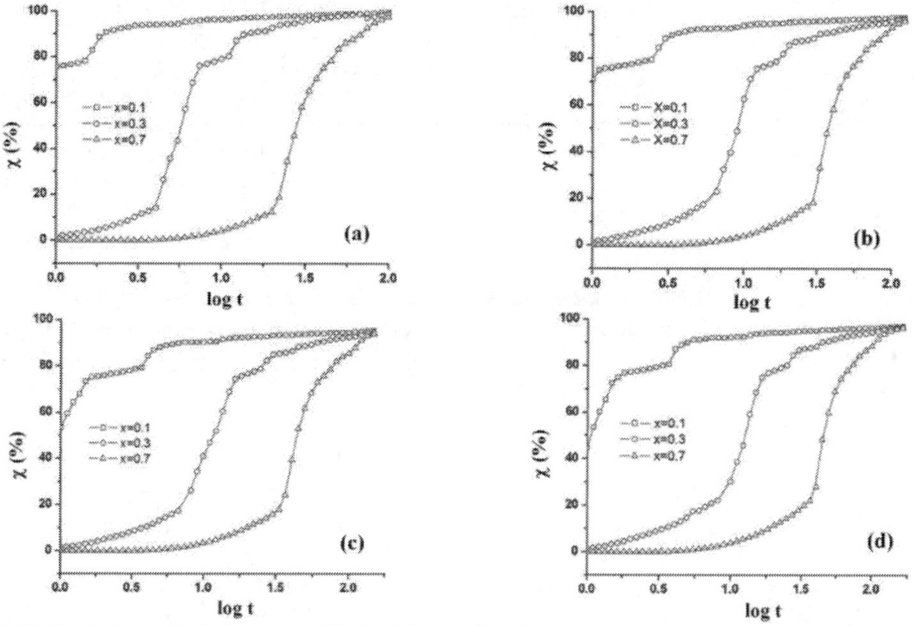

Figure 7. Evolution of relative crystallinity (χ) as a function of cooling time under various processing conditions: (a) T_0=190°C, T_W=20°C, (b) T_0=210°C, T_W=40°C, (c) T_0=210°C, T_W=60°C and (d) T_0=230 °C, T_W=50°C.

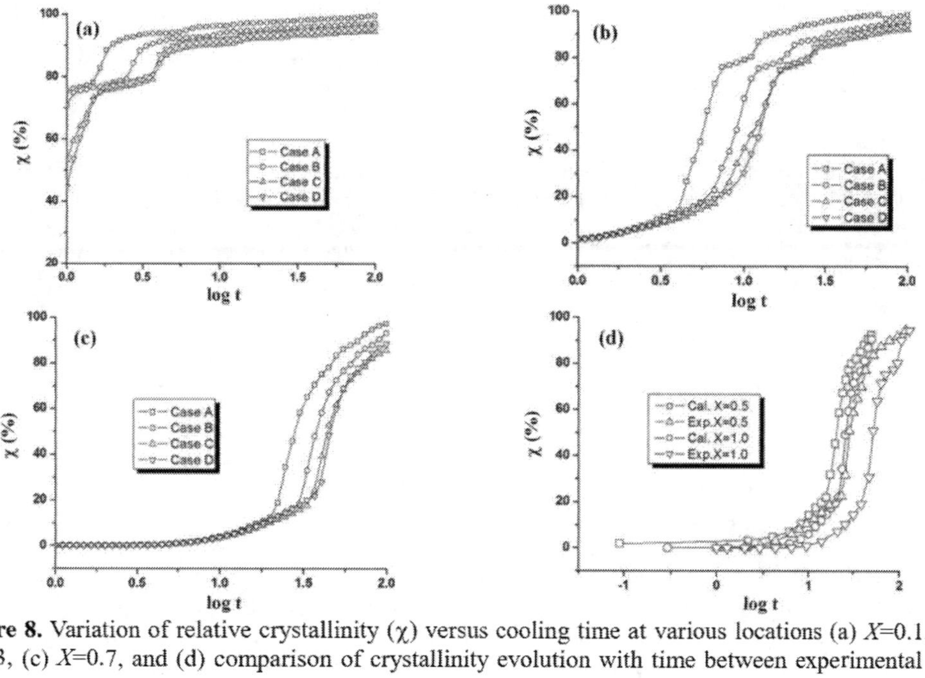

Figure 8. Variation of relative crystallinity (χ) versus cooling time at various locations (a) X=0.1, (b) X=0.3, (c) X=0.7, and (d) comparison of crystallinity evolution with time between experimental and

calculated data at X=0.5 and X=1.0 (under Case D).

Figure 8 shows that under the four different cooling conditions, the crystallinity (χ) ranks as, $\chi|_{\text{Case A}} > \chi|_{\text{Case B}} > \chi|_{\text{Case C}} > \chi|_{\text{Case D}}$ at the beginning; while $\chi|_{\text{Case A}} > \chi|_{\text{Case B}} > \chi|_{\text{Case D}} > \chi|_{\text{Case C}}$ at the later stage of the IM cooling process. The exchange of sequence in crystallinity under Cases C and D can be explained by the crystallinity being primarily influenced by T_0 at the beginning (high T_0 is unfavorable to the crystal nucleus formation). Subsequently, the crystallinity shows the same trends with STD. The variation of crystallinity versus location ranks as, $\chi|_{X=0.1} > \chi|_{X=0.3} > \chi|_{X=0.7}$, which indicates that the position close to the mold wall achieves higher crystallinity. Consistent with figure 7, the curve becomes complicated when the crystallinity is close to 80%, which might be caused by the effect of shear stress. When $log\ t = 0$, an obvious degree of crystallinity (χ) can be obtained only at X=0.1; while its value is close to zero at the locations of X=0.3 and X=0.7. Through the comparison among figures 8a-c, it is seen that there exists an induction time before the jump of crystallinity, the length of which depends heavily on the melt temperature, since the melt temperature is too high to form the crystal nucleus. Figure 8d shows the theoretical and experimental kinetic curves at X=0.5 and X=1.0 under Case D, where the trends of both experimental curves are generally in reasonable agreement with those obtained from theoretical prediction using the *Avrami* Equation. Anyway, investigation on quantitative relationship between crystallization kinetics and cooling rate is an ongoing subject in our group.

4. Conclusion

The work further explores the correlation between crystallization kinetics and thermal gradient field on the basis of our previous studies. It is found that the thickness X = 0.4 is a crucial position heavily influenced by the thermal conduction. When X > 0.4, the temperature decay tends to be stable, and the phase-change plateau lengths share quite similar values within a specific range (*cf.* figures 2 and 4). In addition, the crystallization rate, which is in proportion to the cooling rate, hardly changes after X = 0.4 (*cf.* figure 6). The crystallization rate is higher at the location close to the mold wall than that in the core region during the IM cooling process, primarily due to the temperature difference between the melt temperature and the mold temperature. At a later stage of melt cooling, the consequence in relative crystallinity is in agreement with that in STD (*cf.* figure 8). The present study is of practical significance to the optimization of cooling parameters and the further investigation on the correlation among "processing-structure-property" of polymers and their composites.

Acknowledgments

The authors are grateful to the National Natural Science Foundation of China (No. 51203002 and 51273001), the Key Research Project of Anhui Provincial Department of Education (No. KJ2012A011 and KJ2011z015), the Doctoral Program of Higher Education of China (No. 20113401110003), and the "211 Project " of Anhui University for the financial support to this work.

Appendix

The P-V-T relation of polyethylene is presented as below:

$$V(T,P) = V_0(T)\cdot[1 - C_0\cdot\ln(1+\frac{P}{B(T)})] + V_t(T,P);$$

$$V_0(T) = b_{1m} + b_{2m}\cdot(T-b_5); B(T) = b_{3m}\cdot\exp[-b_{4m}\cdot(T-b_5)]; V_t(T,P) = 0, T > T_t(P)$$

$$V_0(T) = b_{1s} + b_{2s}\cdot(T-b_5); B(T) = b_{3s}\cdot\exp[-b_{4s} - (T-b_5)];$$

$$V_t(T,P) = b_7\cdot\exp[b_8\cdot(T-b_5) - b_9\cdot P], T < T_t(P)$$

where $T_t(P) = b_5 + b_6\cdot P$ and $C_0 =$ 0.0894

The values of the material parameters are listed as follows: b_{1m}=0.001274 $m^3\cdot kg^{-1}$; b_{1s}=0.001075 $m^3\cdot kg^{-1}$; b_{2m}=1.026×10^{-6} $m^3\cdot kg^{-1}\cdot K^{-1}$; b_{2s}=2.077×10^{-7} $m^3\cdot kg^{-1}\cdot K^{-1}$; b_{3m}=9.263×10^7 Pa; b_{3s}=3.324×10^8 Pa; b_{4m}=0.004941 K^{-1}; b_{4s}=2.46×10^{-6} K^{-1}; b_5=414.5 K; b_6=1.543×10^{-7} $K\cdot Pa^{-1}$; b_7=0.0001872 $m^3\cdot kg^{-1}$;

b_8=0.05158 K^{-1}; b_9=1.023×10^{-8} Pa.

References

[1] Cao Y, Faghri A, and Chang W S 1989 A numerical analysis of Stefan problems for generalized multidimensional phase-change structures using the enthalpy transforming model *Int. J. Heat Mass. Transfer* **32** 1289-98
[2] Bondarev V A 1997 Variational method for solving non-linear problems of unsteady-state heat conduction *Int. J. Heat Mass Transf.* **40** 3487-716
[3] Gorla R S R, Canter M S and Pallone P J 1998 Variational approach to unsteady flow and heat transfer in a channel *Heat Mass Transf.* **33** 439-42
[4] Westerberg K W, Wiklof C and Finlayson B A 1994 Time-dependent finite-element models of phase change problems with moving heat sources *Numer. Heat Transf. Part B: Fund.* **25** 119-43
[5] Zhang S, Cao W, Zheng G, Jia Z and Shen C 2011 Model and numerical simulation for the second penetration in water-assisted injection molding *Int. Polym. Proc.* **26** 560-8
[6] Jana S, Ray S and Durst F 2007 A numerical method to compute solidification and melting processes *Appl. Math. Model.* **31** 93-119
[7] Savovic S and Caldwell J 2009 Numerical solution of stefan problem with time-dependent boundary conditions by variable space grid method *Thermal Sci.* **13** 165-74
[8] Sadoun N, Si-Ahmed E K, Colinet P and Legrand J 2012 On the boundary immobilization and variable space grid methods for transient heat conduction problems with phase change: Discussion and refinement *Comptes Rendus: Mecanique* **340** 501-11
[9] Mitchell S L 2011 An accurate nodal heat balance integral method with spatial subdivision *Numer. Heat Transf., Part B: Fund.* **60** 34-56
[10] Yang B, Fu X R, Yang W, Huang L, Yang M B, Feng J M 2008 Numerical prediction of phase-change heat conduction of injection-molded high density polyethylene thick-walled parts via the enthalpy transforming model with mushy zone *Polym. Eng. Sci.* **48** 1707-17
[11] Chen W C, Fu G L, Tai P H and Deng W J 2009 Process parameter optimization for MIMO plastic injection molding via soft computing *Expert Systems Appl.* **36** 1114-22
[12] Liu S J and Su P C 2009 An experimental setup to measure the transient temperature profiles in water assisted injection molding *Int. Polym. Proc.* **24** 234-41
[13] Chen C C A and Chang S W 2008 Shrinkage analysis on convex shell by injection molding *Int. Polym. Proc.* **23** 65-71
[14] Iannaccone A, Amitrano S and Pantani R 2013 Rheological and mechanical behavior of ethyl vinyl acetate/low density polyethylene blends for injection molding *J. Appl. Polym. Sci.* **127** 1157-63
[15] Sannen De, Munck M, Van Puyvelde P and De Keyzer J 2012 Water penetration behavior in water-assisted injection molding (WAIM): A study of product quality for different process and material parameters *Int. Polym. Proc.* **67** 602-16
[16] Tsai C C, Wu R J, Cheng H Y, Li S C, Siao Y Y, Kong D C and Jang G W 2010 Crystallinity and dimensional stability of biaxial oriented poly (lactic acid) films *Polym. Degrad. Stab.* **95** 1292-8
[17] Okada K N, Washiyama J, Watanabe K, Sasaki S, Masunaga H and Hikosaka M 2010 Elongational crystallization of isotactic polypropylene forms nano-oriented crystals with ultra-high performance *Polym. J.* **42** 464-73
[18] Anssens V, Block C, Van Assche G, Van Mele B and Van Puyvelde P 2010 RheoDSC analysis of hardening of semi-crystalline polymers during quiescent isothermal crystallization *Int. Polym. Proc.* **25** 304-10
[19] Villmow T, Pegel S, Potschke P and Wagenknecht U 2008 Influence of injection molding parameters on the electrical resistivity of polycarbonate filled with multi-walled carbon nanotubes *Compos. Sci. Technol.* **68** 777-89
[20] Wu C.H and Liang W J 2005 Effects of geometry and injection-molding parameters on weld-line strength *Polym. Eng. Sci.* **45** 1021-30

[21] Cheng W S, Chen C S and Chien R D 2009 Investigation of the effects of injection molding processing parameters on conductive polymeric composites for electromagnetic interference shielding effectiveness, polymer-plastics technology and engineering *Polym. Plast. Technol. Eng.* **48** 216-20

[22] Engels T A P, van Breemen L C A, Govaert L E and Meijer H E H 2009 Predicting the long-term mechanical performance of polycarbonate from thermal history during injection molding *Macromol. Mater. Eng.* **294** 829-38

[23] Kim K H, Isayev A I and Kwon K 2005 Flow-induced crystallization in the injection molding of polymers: A thermodynamic approach *J. Appl. Polym. Sci.* **95** 502-23

[24] Battegazzore D, Bocchini S and Frache A 2011 Crystallization kinetics of poly (lactic acid)-talc composites *Express Polym. Lett.* **5** 849-58

[25] Tronci S, Grosso M L, Baratti R and Romagnoli J A 2011 A stochastic approach for the prediction of PSD in crystallization processes: Analytical solution for the asymptotic behavior and parameter estimatio *Comput. Chem. Eng.* **35** 2318-25

[26] Ojwang J G O, van Santen R, Kramer G J, van Duin A C T and Goddard W A 2008 Predictions of melting, crystallization, and local atomic arrangements of aluminum clusters using a reactive force field *J. Chem. Phys.* **129** 244506-19

[27] Das D and Langrish T A G 2012 An activated-state model for the prediction of solid-phase crystallization growth kinetics in dried lactose particles *J. Food Eng.* **109** 691-700

[28] Falk R F, Marziano I, Kougoulos T and Girard K P 2011 Prediction of agglomerate type during scale-up of a batch crystallization using computational fluid dynamics models *Organic Process Res. Develop.* **15** 1297-1304

[29] Lin JX, Wang C Y and Zheng Y Y 2008 Prediction of isothermal crystallization parameters in monomer cast nylon 6 *Comput. Chem. Eng.* **32** 3023-9

[30] De Santis F, Pantani R and Titomanlio G 2011 Nucleation and crystallization kinetics of poly (lactic acid) *Thermochimic Acta* **522** 128-34

[31] Cheng W S, Chen C S, Chen S C and Chien R D 2009 Investigation of the effects of injection molding processing parameters on conductive polymeric composites for electromagnetic interference shielding effectiveness *Polym. Plast. Technol. Eng.* **48** 216-20

[32] Yang B, Xia R, Miao J B, Qian J S, Yang M B and Chen P 2013 Probing solidification kinetics of high-density polyethylene during injection molding using an in-situ measurement technique *Polym. Test.* **32** 202-8

[33] Yang B, Lin J-Z, Xia R, Su L-F, Miao J-B, Qian J-S, Chen P, Liu J-W and Deng S-Q 2014 Elucidating correlation between solidification behavior and melt crystallization kinetics of isotactic polypropylene (*i*PP) during injection molding *J. Macromol. Sci. B: Phys.* **53** 462-73

[34] Yang B, Fu X R, Yang W, Liang S P, Sun N, Hu S and Yang M B 2009 Simulation of phase-change heat transfer during cooling stage of gas-assisted injection molding of high-density polyethylene via enthalpy transformation approach *Polym. Eng. Sci.* **49** 1234-42

[35] Yang B, Fu X R, Yang W, Liang S P, Sun N, Hu S, Yang M B and Feng J M 2010 Prediction of heat conduction with phase-change effects during cooling stage of injection molding of high-density plyethylene: approximate integral approach *J. Macromol. Sci. Part B: Phys.* **49** 734

[36] Liang S P, Yang B, Fu X R, Yang W, Sun N, Hu S and Yang M B 2010 A simple method for forecast of cooling time of high-density polyethylene during gas-assisted injection molding *J. Appl. Polym.* Sci. **117** 729-35

[37] Wang L, Yang W, Huang L, Yang B, Sun N and Yang M B 2010 Effect of thermal gradient field with phase change on crystal morphologies of HDPE during GAIM process *Plast. Rubber Compos.* **39** 385-91

[38] Deng S Q, Yang B, Miao J B, Xia R, Qian J S, Chen P and Yang M B 2012 Study on the solidification kinetics of high-density polyethylene during thin-walled injection molding process *J. Polym. Eng.* **32** 355-63

[39] Wang L, Wang J-H, Yang B, Wang Y, Zhang Q-P, Yang M-B and Feng J-M 2013 A novel

hierarchical crystalline structure of injection-molded bars of linear polymer: co-existence of bending and normal shish–kebab structure *Colloid Polym. Sci.* **291** 1503-11

[40] Isayev AI, Lin T H, and Kwon K 2010 Frozen-in birefringence and anisotropic shrinkage in optical moldings: II. Comparison of simulations with experiments on light-guide plates *Polymer* **51** 5623-39

[41] Jiang X L, Luo S J, Sun K and Chen X D 2007 Crystallization behavior of pet materials *Express Polym. Lett.* **1** 245

[42] Cai J L, Yu Q, Han Y, Zhang X Q and Jiang L S 2007 Thermal stability, crystallization, structure and morphology of syndiotactic 1, 2-polybutadiene/organoclay nanocomposite *Euro. Polym. J.* **43** 2866-81

[43] Sun N, Yang B, Wang L, Feng J M, Yin B, Zhang K and Yang M B 2012 Crystallization behavior and molecular orientation of high density polyethylene parts prepared by gas-assisted injection molding *Polym. Int.* **61** 622-30

IOP Conf. Series: Materials Science and Engineering **87** (2015) 012021 doi:10.1088/1757-899X/87/1/012021

Preparation and characterization of titania/silicone nanocomposite material

Y Shen, L Wang[1], H Zhang, T Wu and H Y Pan

College of Optics and Electronic Science and Technology, China Jiliang University, Hangzhou 310018, China

E-mail: 05a0402031@cjlu.edu.cn

Abstract. The preparation and properties of high refractive index nanocomposite material were studied. The TiO_2 nanoparticles were synthesized by sol-gel method using acetic acid as a chelating ligand. The nanoparticles were dispersed directly into the polymer matrix to prepare transparent high refractive index nanocomposite thin films. The refractive index of films will be enhanced with the increase of titania contents. The particles were characterized by X-ray diffraction (XRD), Transmission Electron Microscope (TEM), and Fourier Transform Infrared Spectroscopy (FTIR), respectively. The results showed that all samples with different amounts of TiO_2 exhibit good optical transparency. Furthermore, the pattern of the TiO_2 NPs shows a pure anatase phases. From TEM image, the TiO_2 has little agglomeration. The FT-IR spectrum indicated that acetate ions and titanium ions show good chelation.

1. Introduction

Polymer–inorganic nanocomposites have attracted much attention due to the excellent properties in optics, mechanics, and physics [1-6]. Especially in the light-emitting diode (LED) field, a high refractive index and transparency is needed to increase light-extraction efficiency. The light-extraction efficiency will double increase, when the encapsulant refractive index rises from 1.5 to 2.0 [7]. However, the refractive indices of most organic materials do not exceed 1.5. On the other hand, some of the polymers have a greater refractive index than 1.7, but they usually have a high optical dispersion and strong absorption in the visible region. Therefore, if a high refractive index of inorganic material is incorporated into polymers, the problems will be solved, and the polymer can adjust refractive indices over a wide range.

In recent years, there have been a large number of reports about polymer–inorganic hybrid materials with high refractive indices [3, 4, 8-20]. There are many kinds of inorganic particles, and TiO_2 has been studied a lot [10, 11 , 16-18]. For example, Joseph et al. [4] reported the preparation of acetic acid-modified TiO_2 nanoparticles by sol–gel synthesis method, and the refractive index of hybrid nanocomposite thin film reached to 2.38. Liu et al. [10] synthesized anatase titania nanocrystals (NCS), which was then incorporated into (3-glycidyloxypropyl) trimethoxysilane organic silica sol followed by a sol–gel process. The refractive index at 632.8 nm increased from 1.502 to 1.663 as the TiO_2 weight content went from 10% to 50%; Tao et al. [14] prepared grafting polymer chains onto anatase TiO_2 nanoparticles by a combination of phosphate ligand engineering and alkyneazide "click"

[1] Address for correspondence: L Wang, College of Optics and Electronic Science and Technology, China Jiliang University, Hangzhou 310018, China. E-mail: 05a0402031@cjlu.edu.cn.

chemistry. The refractive index of the composite material increased linearly from 1.5 up to 1.8 by increasing the loading of TiO_2 particles to 30 vol%.

In this study, acetic acid was utilized as a chelated agent to synthesize TiO_2 NPs, then directly incorporated into silicone to prepare TiO_2/silicone nanocomposite. Acetic acid can improve the dispersion of TiO_2 in the organic matrix, and reduce the agglomeration of nanoparticles. The particles were characterized by XRD, TEM, and FTIR, respectively. The results showed that all samples with different amounts of TiO_2 exhibit good optical transparency. Furthermore, the pattern of the TiO_2 NPs shows pure anatase phases. From TEM image, the TiO_2 had little agglomeration. The FT-IR spectrum indicated that acetate ions and titanium ions showed good chelation.

2. Experimental

2.1. Material
Tetrabutyl titanate (98%, Sinoreagent) was the precursor of TiO_2 in the sol–gel process. Other materials are listed as follows: acetic acid (HAc, 99.5%, Sinoreagent), nitric acid (65-68%), dimethylacetamide (DMAC, 99%), deionized water, silicone resin and curing agent.

2.2. Synthesis of TiO_2 NPs
In the experiment, 17 g tetrabutyl titanate and 6 g HAc were mixed for 10 min. at room temperature. Then, mixture was rapidly poured into deionized water with stirring. Further, a few drops of nitric acid were added into the solution, and heated to 80℃, maintained under reflux for 90min. The as-synthesized TiO_2 sol was centrifuged and washed three times with DMAC. Finally, TiO_2 sol was dispersed into DMAC through ultrasonication for 1 h. The crystal structure of the nanoparticles was characterized by XRD with Cu Kα radiation. TEM was used to observe the agglomeration of TiO_2.

2.3. *Preparation of high-refractive-index hybrid films*
First, 1 g of silicone and 1 g of curing agent were mixed for 10min, then various amounts of TiO_2 solution(10, 20 and 30 wt.% of TiO_2 respectively) were added, and stirred for 1 h. Then the mixture was spin-coated on glass or silicon wafer. The sample was put in a vacuum oven for 80℃ until the DMAC solvent had been evaporated, and then cured at 120℃ for 4 h.

3. Results and discussion
Figure 1 shows the XRD pattern of the TiO_2 NPs dominant peaks with values of 2θ being located at 25.3 (101), 38 (004), 48 (200), 54 (105) and 62.4 (204). It belongs to an anatase phase. Thus, the as-prepared TiO_2 NPs was pure anatase phase. Figure 2 shows the particle size of the TiO_2 NPs, which was about 10 to 20 nm.

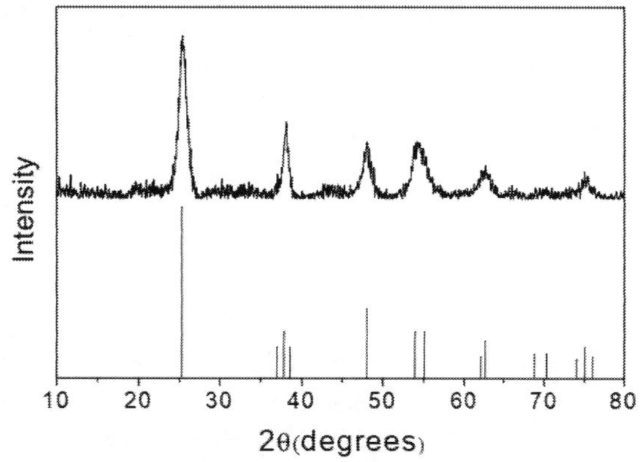

Figure 1. XRD pattern of TiO_2 nanoparticles.

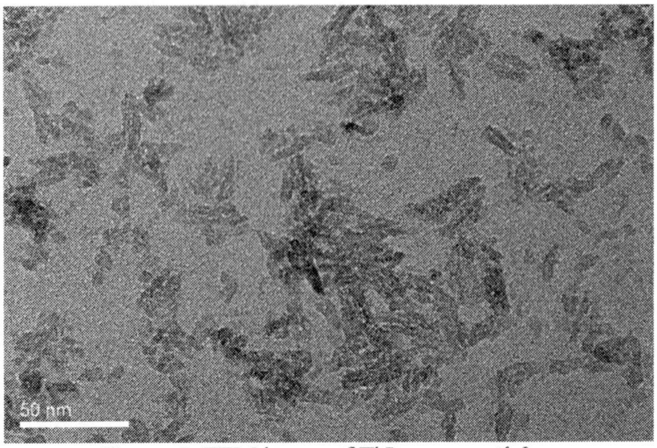

Figure 2. TEM image of TiO_2 nanoparticles.

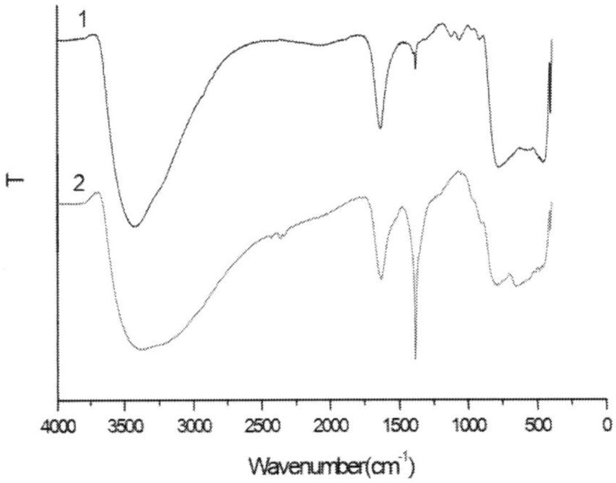

Figure 3. FTIR spectra of (1) TiO_2 and (2) the TiO_2-HAc NPs.

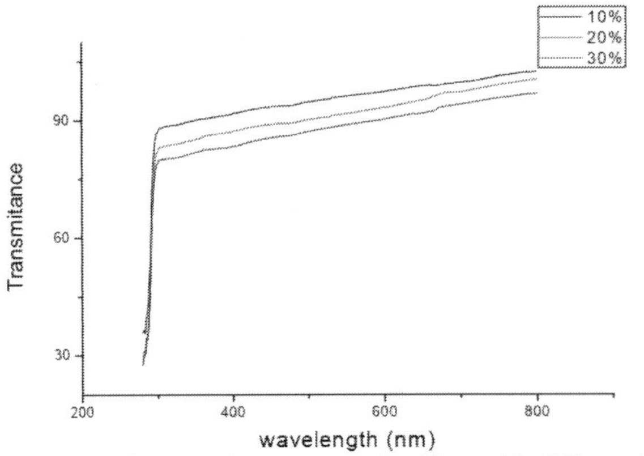

Figure 4. Variation of the transmittance of nanocomposite films with different TiO_2 concentrations.

Figure 3 presents FTIR spectra of TiO_2 and TiO_2-HAc NPs; both exhibit an absorption at 3200–3000 cm^{-1} for –OH units, a peak near 1633 cm^{-1} for H-O-H bending vibrations, and Ti-O bond at 800-450 cm^{-1}. Line 2 reveals the peak of C-H stretching vibration at 3000-2800 cm^{-1}. New peak appears at 1500 cm^{-1}, and 1430cm^{-1} peak is strengthened, the acetate ions and titanium ions show good chelation. Figure 4 shows the transmittance of the nanocomposite films with different concentrations of TiO_2 which exhibits good optical transparency. The lowest transmittance exceeds 80%.

4. Conclusion

The TiO_2 nanoparticles, modified by acetic acid, are prepared by the sol–gel method. The modified nanoparticles have well dispersion in epoxy resin. The hybrid thin film exhibits excellent optical transparency in the visible region, so the nanocomposite material can be used in optical components potentially.

References

[1] Baah D, Tigner J, Bean K, Britton B, Walker N, Henderson G and Floyd-Smith T 2011 Preparation of planar graded refractive index nanocomposites using microfluidics *Materials Science and Engineering: B* **176** 883-8

[2] Baghayeri M, Zare E N and Lakouraj M M 2014 Novel superparamagnetic PFu@ Fe_3O_4 conductive nanocomposite as a suitable host for hemoglobin immobilization *Sensors and Actuators B: Chemical* **202** 1200-8

[3] Chatterjee A and Islam M S 2008 Fabrication and characterization of TiO_2–epoxy nanocomposite *Materials Science and Engineering: A* **487** 574-85

[4] Chau J L H, Lin Y-M and Li A-K 2007 Transparent high refractive index nanocomposite thin films Materials *Letters* **61** 2908-10

[5] Chung P T, Yang C T, Wang S H, Chen C W, Chiang A S T and Liu C-Y 2012 ZrO_2/epoxy nanocomposite for LED encapsulation *Materials Chemistry and Physics* **136** 868-76

[6] Kang M and Kang S 2011 Influence of Al_2O_3 additions on the crystallization mechanism and properties of diopside/anorthite hybrid glass-ceramics for LED packaging materials *Journal of Crystal Growth* **326** 124-7

[7] Lakouraj M M, Zare E N and Moghadam P N 2014 Synthesis of novel conductive poly (p - phenylenediamine)/Fe_3O_4 nanocomposite via emulsion polymerization and investigation of antioxidant activity *Advances in Polymer Technology* **33** 21385

[8] Liu B-T and Li P-S 2013 Preparation and characterization of high-refractive-index polymer/inorganic hybrid films containing TiO_2 nanoparticles prepared by 4-aminobenzoic acid *Surface and Coatings Technology* **231** 301-6

[9] Liu B-T, Tang S-J, Yu Y-Y and Lin S-H 2011 High-refractive-index polymer/inorganic hybrid films containing high TiO_2 contents *Colloids and Surfaces A: Physicochemical and Engineering Aspects* **377** 138-43

[10] Liu Y, Lü C, Li M, Zhang L and Yang B 2008 High refractive index organic–inorganic hybrid coatings with TiO_2 nanocrystals *Colloids and Surfaces A: Physicochemical and Engineering Aspects* **328** 67-72

[11] Mont F W, Kim J K, Schubert M F, Schubert E F and Siegel R W 2008 High-refractive-index TiO_2-nanoparticle-loaded encapsulants for light-emitting diodes *Journal of Applied Physics* **103** 083120

[12] Nakayama N and Hayashi T 2007 Preparation and characterization of TiO_2 and polymer nanocomposite films with high refractive index *Journal of applied polymer science* **105** 3662-72

[13] Nazarzadeh Zare E, Lakouraj M M and Baghayeri M 2015 Electro-Magnetic Polyfuran/Fe_3O_4 Nanocomposite: Synthesis, characterization, antioxidant activity, and its application as a biosensor *International Journal of Polymeric Materials and Polymeric Biomaterials* **64** 175-83

[14] Tao P, Li Y, Rungta A, Viswanath A, Gao J, Benicewicz B C, Siegel R W and Schadler L S 2011 TiO_2 nanocomposites with high refractive index and transparency *Journal of Materials Chemistry* **21** 18623-9

[15] Yan L, Ziyin L, Xueying Z, Kyoung-sik M, Sehoon Y, Choi J and Wong C P 2013 High refractive index and transparency nanocomposites as encapsulant for high brightness LED packaging. *2013 Electronic Components and Technology Conference* (Las Vegas, USA) pp 553-6

[16] Yan L, Ziyin L, Xueying Z, Sehoon Y, Kyoung-sik M and Wong C P 2012 ZnO quantum dots-filled

encapsulant for LED packaging *2012 Electronic Components and Technology Conference* (San Diego, USA) pp 2140-4

[17] Yang J M, Shih C H, Chang C N, Lin F H, Jiang J M, Hsu Y G, Su W Y and See L C 2003 Preparation of epoxy–SiO_2 hybrid sol–gel material for bone cement *Journal of Biomedical Materials Research Part A* **64** 138-46

[18] Yi L, Yangyang S, Lingbo Z, Jianwen X, Jiongxin L, Hongjin J, Dong H and Wong C P 2006 Nanocomposite for Advanced Packaging of Microelectronics *Proc. Int. Conf. on Electronic Materials and Packaging*(Hong Kong, China) pp 1-16

[19] Yuwono A H, Xue J, Wang J, Elim H I, Ji W, Li Y and White T J 2003 Transparent nanohybrids of nanocrystalline TiO_2 in PMMA with unique nonlinear optical behavior *Journal of Materials Chemistry* **13** 1475-9

[20] Zare E N, Lakouraj M M, Moghadam P N and Azimi R 2013 Novel polyfuran/functionalized multiwalled carbon nanotubes composites with improved conductivity: Chemical synthesis, characterization, and antioxidant activity *Polymer Composites* **34** 732–9

Magnetic-luminescent properties of Fe_3O_4@Gd_2O_3:Eu^{3+} core-shell nanocomposites

T Wu, L Wang[1], H Zhang, Y Shen and H Y Pan

College of Optics and Electronic Science and Technology, China Jiliang University, Hangzhou 310018, China

E-mail: 05a0402031@cjlu.edu.cn

Abstract. A simple Urea Homogeneous Precipitation (UHP) method was employed to synthesize bifunctional core-shell nanoparticles of Fe_3O_4@Gd_2O_3:Eu^{3+}. X-ray diffraction (XRD), scanning electron microscopy (SEM), photoluminescence spectroscopy (PL) and vibrating sample magnetometer (VSM) were used to characterize the structure, morphology, luminescence and magnetic properties of the composites, respectively. Results show that the nanocomposites possess both luminescent and magnetic properties, thus the synthesized nanoparticles may find potential application in the area of biomedicine, such as drug targeting, fluorescence label and cell separation.

1. Introduction

In recent years, increasing attention has been paid to bifunctional core-shell nanocomposites, which have been gradually applied in biomedical applications, for example, magnetic-resonance imaging, biological labeling, biochemical separation, targeted drug and the like [1-5]. Iron oxide (for instance Fe_3O_4) is identified as an ideal material in biomedicine applications, which possesses unusual magnetic performances, nontoxicity, good biocompatibility [6]. Luminescent substances, for instance, quantum dots (QDs) and organic dyes, have been extensively used for fluorescence labeling in bio-applications [7]. Nevertheless, these nanoparticles have intrinsic limitations; organic dyes typically display rapid photobleaching and a low fluorescence quantum yield. Likewise, QDs show unstable chemical properties, potential toxicity, and fluorescence intermittence as well [8, 9]. Therefore, these disadvantages limit their application in the area of biomedicine, particularly in the human medicine to some extent [10]. Compared with organic dyes and QDs, lanthanide-doped nanoparticles are now becoming widely popular being expected to represent a new kind of fluorescent biomedicine label, because they promise large Stokes shifts, high quantum yields, and superior photostability etc. [11, 12]. Relative to other lanthanide-doped nanoparticles, Gd_2O_3 is an important host crystal for preparation of fluorescent materials. For instance, Gd_2O_3:Eu^{3+} provides many merits: stronger fluorescence properties, stronger charge transfer bands and so on [13].

To the best of our knowledge, the preparation of bifunctional nanocomposites has been used in several different routes. For instance, Shi et al. adopted a sol–gel method to synthesize Fe_3O_4@C@YVO_4:Eu^{3+} magnetic-luminescent nanocomposites [14]. Lu et al. reported a modified Stober method to prepare core-shell Fe_3O_4@SiO_2@$Eu(DBM)_3\cdot 2H_2O$ microsphere [15]. Yang et al.

[1] Address for correspondence: L Wang, College of Optics and Electronic Science and Technology, China Jiliang University, Hangzhou 310018, China. Tel.: 0571 86875686; E-mail: 05a0402031@cjlu.edu.cn.

reported the Fe_3O_4@$nSiO_2$@$mSiO_2$@YVO_4:Eu^{3+} nanoparticles by the sol-gel method [16]. Nevertheless, the preparation process in these methods was relatively complex.

In this article, bifunctional core-shell nanocomposites of Fe_3O_4@Gd_2O_3:Eu^{3+} were prepared through Urea Homogeneous Precipitation (UHP) approach. The preparation process is shown in figure 1. The synthetical nanocomposites have strong magnetic and fluorescence properties. Thus, this nanocomposite has potential bio-applications, which includes target-drug delivery and luminescence label.

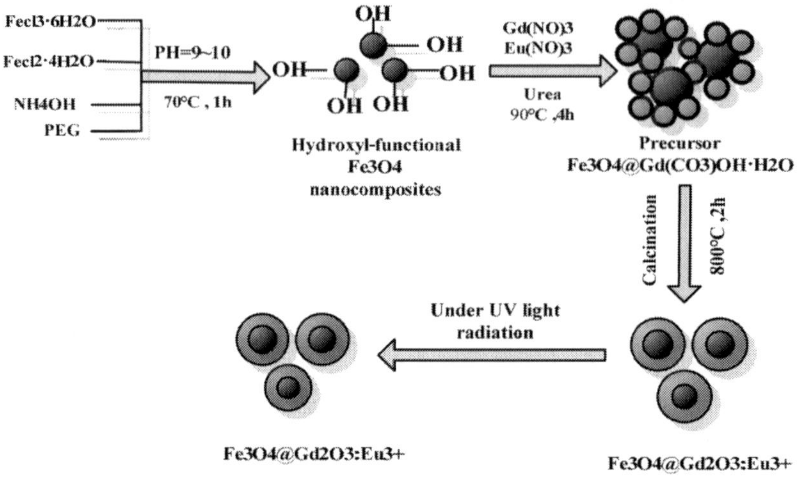

Figure 1. The preparation process of spherical Fe_3O_4@Gd_2O_3:Eu^{3+} core-shell nanocomposites.

2. Experimental

2.1. Materials

Ferric chloride hexahydrate ($Fecl_3 \cdot 6H_2O$, purity:99%), ferrous chloride tetrahydrate ($Fecl_2 \cdot 4H_2O$), polyethylene glycol (PEG, MR=4,000), ammonium hydroxide (28–30%NH_3 basis), gadolinium oxide (Gd_2O_3, 99.9%), europium oxide (Eu_2O_3, 99.99%), nitric acid (HNO_3, purity:70%), urea ($(NH_2)_2CO$, purity: 99–100.5%) and ethanol (C_2H_5OH) were bought from Sinopharm Chemical Reagent Limited Company. Reagents are used without further refinement. Deionized (DI) water was used over all the experimental process.

2.2. Preparation of hydroxyl-functional Fe_3O_4 nanoparticles

Hydroxyl-functional Fe_3O_4 nanoparticles were synthesized through the reported co-precipitation method [17]. $Fecl_3 \cdot 6H_2O$ (1.8 g) and $Fecl_2 \cdot 4H_2O$ (0.6 g) at a 2:1 molar ratio were dissolved in 60 ml DI water at 70°C, followed by the addition of PEG (1 g). Solution was stirred for 30 min until complete dissolution. Next the mixture solution was added slowly to a NH_4OH solution with vigorous stirring for 45 min, and then cooled to room temperature naturally. Fe_3O_4 nanocomposites were separated, rinsed six times using DI water and ethanol. Precipitates were put at a vacuum drying oven at 75°C for 12h, and Fe_3O_4 nanoparticles were finally acquired.

2.3. Preparation of Fe_3O_4@Gd_2O_3:Eu^{3+} nanocomposites

Gadolinium oxide (Gd_2O_3) luminescent shell was deposited on the magnetic Fe_3O_4 nanocomposites by UHP method, with minor modifications [18]. Briefly, first 0.3444 g (0.95 mmol) of gadolinium oxide (Gd_2O_3) and 0.0176 g (0.05 mmol) of europium oxide (Eu_2O_3) were dissolved in nitric acid respectively, then the solution was sonicated for 20 min, and the pH was adjusted to 2~3 through ammonium hydroxide (NH_4OH). Next solutions were mixed with DI water (60 ml) and urea (6.0 g), and stirred vigorously for 30 min to form clear solution, after that Fe_3O_4 (20 mg) nanoparticles were injected in solution. Mixtures were sonicated for 30 min, heated to 90°C and maintained for 4 h under vigorous stirring. Resulting suspensions were centrifuged at 8000rpm for 10min. The products were separated, rinsed six times with DI water and ethanol; the precipitates were put at a vacuum drying oven at 75°C for 12 h. Finally, the particles were calcinated at 800°C for 2 h to obtain bifunctional Fe_3O_4@ Gd_2O_3:Eu^{3+} nanocomposites.

2.4. Characterization

The phase of prepared products was characterized by XRD (D2 PHASER) using Cu-Ka (λ=0.15405 nm) radiation within the range of 15–80° 2θ. The structure of products was observed by SEM (ZEISS SUPRASS). The photoluminescence properties were investigated by Hitachi F-4500 spectrofluorimeter. The magnetic properties were measured by VSM at 300 K. Measurements were examined at room temperature (300 K).

3. Results

3.1. Phase and morphology

Phase of the samples was examined by XRD. Figure 2 shows XRD patterns of Fe_3O_4, Fe_3O_4@Gd_2O_3:Eu^{3+} nanoparticles. XRD peaks for Fe_3O_4 are matched with the face-centered(Fd-3m space group)cubic-single crystal of Fe_3O_4 (PDF#75-1610) [19]. As for Fe_3O_4@Gd_2O_3:Eu^{3+} nanocomposites, besides the diffractions of cubic-single crystal Fe_3O_4 (labeled with*), the diffraction peaks at 2θ =28.6°, 33.1°, 47.5° and 56.4° can be indexed to the Gd_2O_3 diffraction peaks (PDF#86-2477, marked with°). No additional peaks from the doped components are detected owing to the low concentration of doping ions.

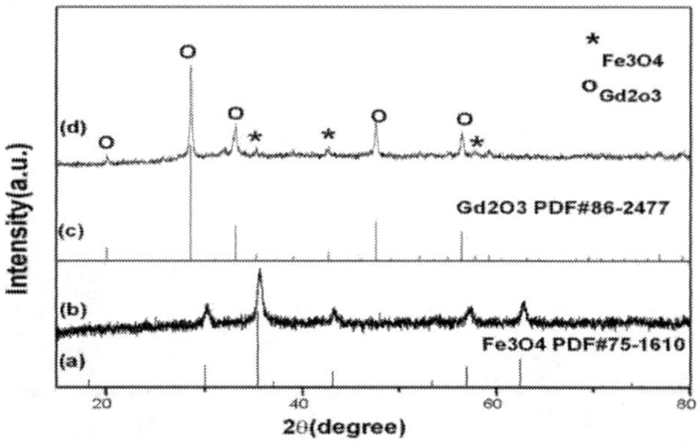

Figure 2. XRD patterns of pure Fe_3O_4 (b), Fe_3O_4@Gd_2O_3:Eu^{3+} (d).

The morphology of products was observed by SEM. In figure 3A, SEM image of pure Fe_3O_4 showed that the obtained Fe_3O_4 particles have a spherical shape with an average size of about 12 nm. Additionally, these nanocomposites are non-aggregated with narrow size distribution. Figure 3B shows SEM of the prepared Fe_3O_4@Gd_2O_3:Eu^{3+} nanocomposites. The particle size of core-shell nanocomposites under an 80,000 magnification is observed as about 80 nm, the shape of nanocomposites is almost spherical and there are some agglomerates. It is indicated that the Gd_2O_3:Eu^{3+} is deposited on the core Fe_3O_4 nanocomposites.

3.2. Optical properties of Fe_3O_4@Gd_2O_3:Eu^{3+} nanocomposites

Through the excitation and emission spectra, the photoluminescence properties of the pure Gd_2O_3:Eu^{3+}, Fe_3O_4@Gd_2O_3:Eu^{3+} nanoparticles were further characterized. In excitation spectra (figure 4A), the excitation spectra of pure Gd_2O_3:Eu^{3+} nanoparticle, monitored wavelength at 612 nm, contained strong,

IOP Conf. Series: Materials Science and Engineering **87** (2015) 012022 doi:10.1088/1757-899X/87/1/012022

broad band with a maximum at 256 nm. No obvious peak of Gd_2O_3 host at 228 nm (near 230 nm) was detected. Compared with the excitation of pure $Gd_2O_3:Eu^{3+}$ nanoparticle, one can see that the excitation of $Fe_3O_4@Gd_2O_3:Eu^{3+}$ nanoparticle consists of a broad and strong excitation at 248 nm and an excitation peak at 228 nm.

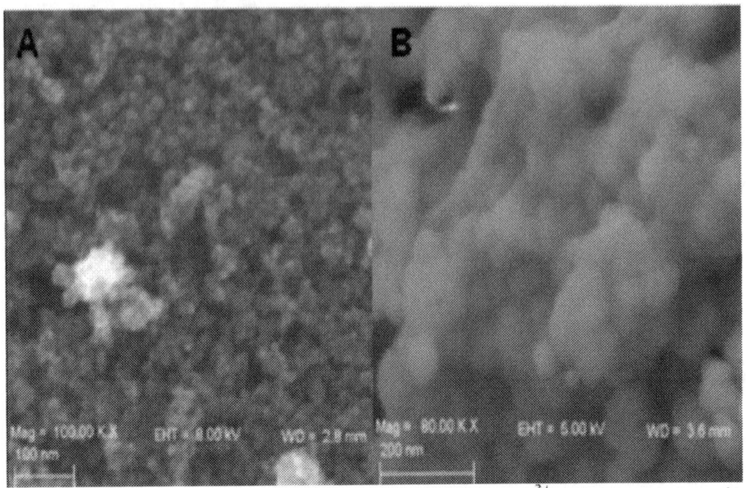

Figure 3. SEM images of Fe_3O_4 (A), $Fe_3O_4@Gd_2O_3:Eu^{3+}$ (B) nanocomposites.

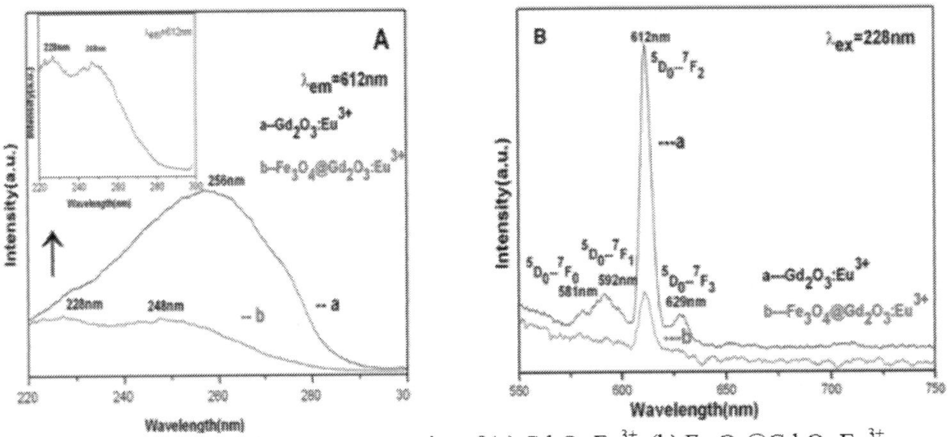

Figure 4. Photoluminescence properties of (a) $Gd_2O_3:Eu^{3+}$, (b) $Fe_3O_4@Gd_2O_3:Eu^{3+}$.

In emission spectra (figure 4B), the monitoring wavelength is 228 nm, a dominant peak at 612 nm is the characteristic luminescence of Eu^{3+}, which corresponds to the $^5D_O-^7F_2$ electric dipole transition of Eu^{3+}. There are other several minor peaks corresponding to $^5D_O-^7F_o$ (581 nm), $^5D_O-^7F_1$ (592 nm) and the $^5D_O-^7F_3$ (629 nm) transitions of Eu^{3+} [20]. As shown in figures 4A and 4B, the intensities of both the excitation and emission spectra of $Fe_3O_4@Gd_2O_3:Eu^{3+}$ (figure 4B) decrease compared with pure $Gd_2O_3:Eu^{3+}$ (figure 4A), as we know, magnetic Fe_3O_4 can absorb UV energy of $Gd_2O_3:Eu^{3+}$, resulting in the strong quenching effect on fluorescence material. However, these results indicate that nanocomposites still have good fluorescence properties, in spite of partial quenching by black magnetite.

3.3. Magnetic performances of $Fe_3O_4@Gd_2O_3:Eu^{3+}$ nanoparticles

Magnetic properties were tested by VSM at room temperature (300 K). Figure 5 shows the magnetic hysteresis loops of Fe_3O_4 and $Fe_3O_4@Gd_2O_3:Eu^{3+}$, the saturation magnetization (Ms) value of Fe_3O_4 is 62.99 emu/g. After $Gd_2O_3:Eu^{3+}$ coating, the Ms value decreases to 1.29 emu/g, the prepared $Fe_3O_4@Gd_2O_3:Eu^{3+}$ shows coercivity Hc of 60Oe and a remanence of 0.01emu/g at room temperature. The decrease of saturation magnetization, in comparison to the value of Fe_3O_4, can be attributed to the presence of nonmagnetic contents of coated $Gd_2O_3:Eu^{3+}$ layer surrounding the Fe_3O_4 [21].

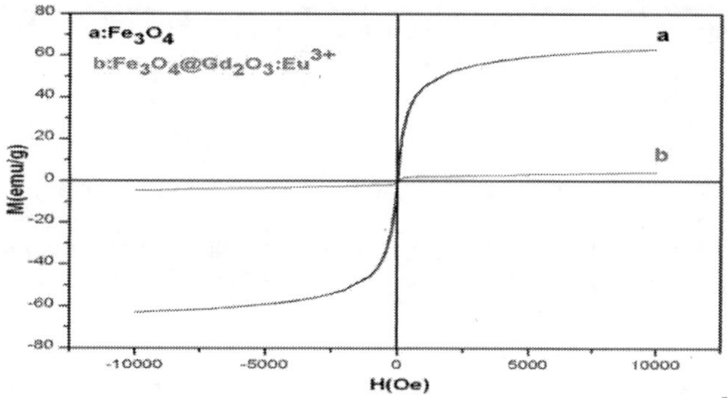

Figure 5. Magnetic properties of Fe_3O_4 (a) and $Fe_3O_4@Gd_2O_3:Eu^{3+}$ (b).

4. Conclusion

In short, bifunctional $Fe_3O_4@Gd_2O_3:Eu^{3+}$ nanocomposites are synthesized successfully using a simple urea UHP method. It is indicated that the morphology of nanocomposites is almost spherical shape with an average size of about 80nm, and the bifunctional nanocomposites display strong red emission peak stemming from $^5D_O-^7F_2$ of Eu^{3+} ions at 612 nm, and high saturation magnetization Ms of 1.29 emu/g, coercivity Hc of 60Oe and a remanence Mr of 0.01 emu/g. Therefore, bifunctional magnetic-luminescent core-shell nanocomposites have well potential application at the area of biomedicine, such as drug targeting, fluorescence label and cell separation.

References

[1] Dosev D, Nichkova M, Dumas R K and Kennedy I M 2007 *Nanotechnology* **18** 055102
[2] Yu X, Wan J, Shan Y, Chen K and Han X 2009 *Chemistry of Materials* **21** 4892-8
[3] Pradhan A, Bah R and Sellmyer D 2008 *Journal of Applied Physics* **103** 07F704-07F-3
[4] Guardia P, Labarta A and Batlle X 2010 *The Journal of Physical Chemistry C* **115** 390-6
[5] Sobha K, Surendranath K, Meena V and Latha K 2010 *J. Biotech. Mol. Bio. Rev* **5** 001-12
[6] Leng Y, Sato K, Shi Y and Kamiya H 2009 *The Journal of Physical Chemistry C* **113** 16681-5
[7] Zhang Y, Ren Z, Fu Y and Zhai H 2009 *Journal of Physics and Chemistry of Solids* **70** 505-9
[8] Jie G, Wang L and Zhang S 2011 *Chemistry-A European Journal* **17** 641-8
[9] Wang G and Su X 2011 *Analyst* **136** 1783-98
[10] Sun P, Zhang H, Liu C, Fang J, Wang M, Chen J, Zhang J and Xu S 2009 *Langmuir* **26** 1278-84
[11] Cho N-H, Cheong T-C, Min J H and Seong S-Y 2011 *Nature Nanotechnology* **6** 675-82
[12] Li C and Lin J 2010 *Journal of Materials Chemistry* **20** 6831-47
[13] Vetrone F, Naccache R and Capobianco J A 2009 *Advanced Functional Materials* **19** 2924-9
[14] Shi J, Tong L, Liu D and Yang H 2012 *Journal of Nanoparticle Research* **14** 1-9
[15] Lu P, Zhang J-L, Liu Y-L, Sun D-H, Liu G-X, Hong G-Y and Ni J-Z 2010 *Talanta* **82** 450-7
[16] Yang P, Quan Z, Hou Z, Li C, Kang X, Cheng Z and Lin J 2009 *Biomaterials* **30** 4786-95
[17] Vu H H T and Atabaev T S 2014 *Journal of Sol-Gel Science and Technology* **71** 391-5
[18] Li J-G, Li X, Sun X and Ishigaki T 2008 *The Journal of Physical Chemistry C* **112** 11707-16
[19] Peng H, Liu G, Dong X, Xu J and Yu W 2011 *Journal of Alloys and Compounds* **509** 6930-4
[20] Li J-G, Zhu Q, Li X, Sun X and Sakka Y 2011 *Acta Materialia* **59** 3688-96
[21] Wang Q, Yang X, Yu L and Yang H 2011 *Journal of Alloys and Compounds* **509** 9098-104

Fiber breakage phenomena in long fiber reinforced plastic preparation

Chao-Tsai Huang[1,3], Huan-Chang Tseng[1], Jiri Vlcek[2] and Rong-Yeu Chang[1]

[1]CoreTech System (Moldex3D) Co., Ltd., Hsinchu, Taiwan
[2]Compuplast International, Inc., Zlin, Czech Republic

E-mail: cthuang@moldex3d.com

Abstract. Due to the high demand of smart green, the lightweight technologies have become the driving force for the development of automotives and other industries in recent years. Among those technologies, using short and long fiber-reinforced plastics (FRP) to replace some metal components can reduce the weight of an automotive significantly. However, the microstructures of fibers inside plastic matrix are too complicated to manage and control during the injection molding through the screw, the runner, the gate, and then into the cavity. This study focuses on the fiber breakage phenomena during the screw plastification. Results show that fiber breakage is strongly dependent on screw design and operation. When the screw geometry changes, the fiber breakage could be larger even with lower compression ratio.

1. Introduction

Short and long fiber-reinforced plastics (FRPs) are now commercially available materials. FRPs can greatly improve the mechanical properties of injection molding products in automotive. This advantage is helpful for reducing costs and vehicle weight of FRP manufacturing. The important fiber microstructure variables, including orientation, length, and concentration, determine the degree of enhancement of mechanical properties. Furthermore, it is critical that the microstructure influences final shrinkage/warpage of anisotropy [1-3]. On the other hand, little attention has been paid to the material preparation inside the plastification unit. For example, the properties of the final product may be acceptable while poor mixing of color additives in the material makes the appearance unacceptable. Another example is when the process conditions are such that the material prior to injection has a high variation in temperature. Even with a perfect simulation of the filling stage, the results may not correspond to the reality. It is because the basic assumption for the filling simulation (that the material temperature is uniform, or the fiber length is the same as that before going through the screw) is not satisfied. In addition, there is potential for even more problems such as excessive shear heating of the melt or break the fiber length [4-6].

Since the microstructures of fiber inside plastic matrix are very complex, it is not easy to visualize them. Besides, there are too many combinations for material, process conditions and the dynamic features of microstructures (such as fiber length breakage, fiber length distribution, fiber orientation, fiber density) are too complicated to manage and control during the injection molding through the screw, the runner, the gate, and then into the cavity. This study focuses on the fiber microstructures.

[3] Address for correspondence: Chao-Tsai Huang, CoreTech System (Moldex3D) Co., Ltd., Hsinchu, Taiwan. E-mail: cthuang@moldex3d.com.

More specifically, most of our attention is paid to the fiber breakage prediction during screw plastification. Results show that fiber breakage is strongly dependent on screw design and operation. When the screw geometry changes, the fiber breakage could be larger even with lower compression ratio.

2. Numerical theory and assumption

2.1. *Model for screw plastification*

The plastification unit in the injection molding simulation was first studied by Potente at al. [4-6]. For an Injection Molding Plastification Unit, this also corresponds to the relative motion of the barrel moving forward. Consequently, there is another velocity component for the relative motion of the barrel against the screw. As is shown in figure 1, this velocity component contributes to the down channel material transportation but at the same time it reduces the material rotation inside the channel. This means that it reduces the mixing efficiency. The simulation means that the velocity components set as boundary conditions for the velocity down the channel and across the channel also have a contribution to the barrel shifting.

$$v_{ch} = V_b \cos(\theta) + V_s \sin(\theta) \quad (1)$$

$$v_r = V_b \sin(\theta) - V_s \cos(\theta) \quad (2)$$

where vch is the down channel component, vr is the rotation velocity component, Vb is the barrel surface rotation velocity and Vs is the barrel shift velocity.

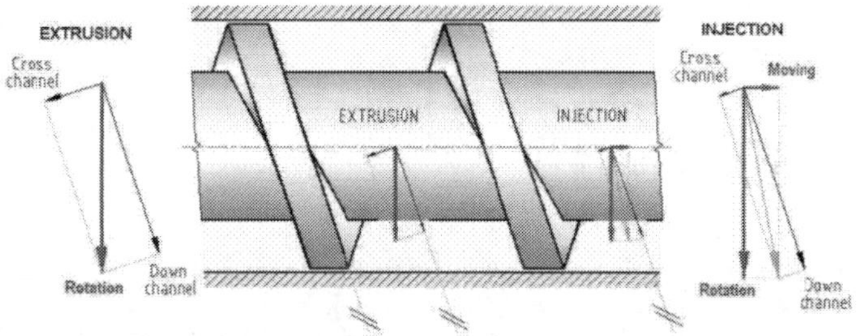

Figure 1. A comparison of velocity components in an extrusion and injection molding process [4].

Assume that there is no conduction in the channel direction, the heat transfer can be solved as a planer problem. The solid and molten materials are taken as they are at the end of the back motion on every cross-section. The time dependent energy balance equation is:

$$\rho C_p \frac{\partial T}{\partial t} = \lambda \left(\frac{\partial^2 T}{\partial x^2} + \frac{\partial^2 T}{\partial y^2} \right) + E(x, y) \quad (3)$$

where T is the temperature, x and y are the coordinates, t is the time, ρ , C_p and λ are the density, heat capacity and conductivity either of the melt or the solid based on the position, respectively. E is the energy consumption inside the solved field.

2.2. *Model for fiber breakage*

The plastification unit in the injection molding simulation was first studied by Potente at al. [4-6]. For an Injection Molding Plastification Unit, this also corresponds to the relative motion of the barrel moving forward.

A mean fiber length is described by number average L_N and weight average L_W, as follows:

$$L_N = \sum_{i=1}^{N} N_i l_i / \sum_{i=1}^{N} N_i \tag{4}$$

$$L_W = \sum_{i=1}^{N} N_i l_i^2 / \sum_{i=1}^{N} N_i l_i \tag{5}$$

where N_i represents the expected value of the number of fibers of length l_i in a sample taken from a small region. Note that $l_i = il$. l is the discretized length segment, namely, $l = L/N$; L is the initial length; N is the number of segment. Phelps et al. [7-8] describe the change rate of the number of fibers with length l_i, called as the fiber length attrition model:

$$\dot{N}_i = -P_i N_i + \sum_{k=1}^{N} R_{ik} N_k \tag{6}$$

$$R_{ik} = N_{norm}(l_i, \frac{l_k}{2}, Sl_k) \tag{7}$$

$$R_{ki} = R_{(i-k)i} \tag{8}$$

$$\sum_{k=1}^{N} R_{ji} = 2P_i \tag{9}$$

where P_i is a breakage probability for parent fibers of length l_i at time Δt; R_{ik} is the rate of production of child fibers of length l_i by breaking parent fibers of length l_k, thus with the mass conservation law.

R_{ik} is given by the normal distribution function, as follows:

$$N_{PDF}(x, \mu, \sigma) = \frac{1}{\sqrt{2\pi\sigma^2}} e^{-\frac{(x-\mu)^2}{2\sigma^2}} \tag{10}$$

where x is variable, μ is the mean, and σ is the variance.

3. Investigation model and information

3.1. Original screw design

The original screw geometry design is shown in figure 2. For conventional screw, there are three zones, i.e., feeding zone (10D), compression zone (8D), and metering zone (4D), respectively. Here 10D means the length is 10 times of the diameter. Initially, the channel depth of the feeding zone is 6.4 mm, and that of the metering zone is 2.3 mm (shown as Screw (1-1)). The ratio of the screw length to diameter is L/D = 22. The diameter of screw is 36 mm. The compression ratio is about 2.8. Then the screw is lightly revised with 2.1 mm metering zone (shown as Screw (1-2)).

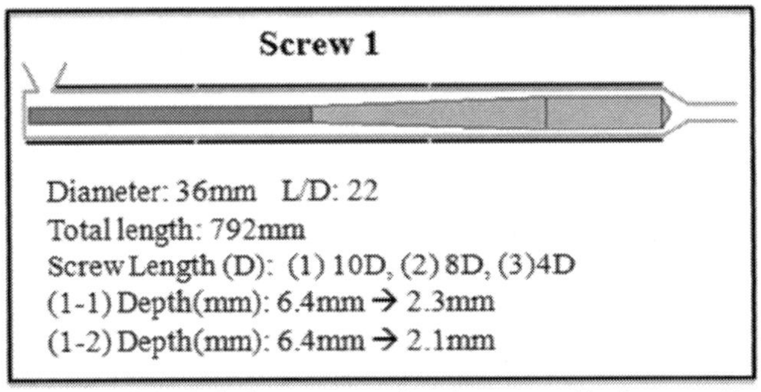

Figure 2. Original design: conventional screw with three sections, compression ratio = 6.4/2.3 for screw (1-1).

For injection molding process setting is as follows: filling 0.74 sec, packing 3 sec, cooling 10.6 sec, and mold open and close 5 sec. Cycle time is 19.3 sec. Material is SABIC ABS Cycolac CGF20 (20%

IOP Conf. Series: Materials Science and Engineering **87** (2015) 012023 doi:10.1088/1757-899X/87/1/012023

glass fiber). The barrel is set up with three sections. The temperature is set to 225°C, 235°C, and 245°C. Screw operation is 100 RPM. Screw stoke is 34.91 mm. The original fiber length is 13 mm.

3.2. New screw design

To test the screw design effect on the fiber breakage, the original design is modified as the new one as shown in figure 3. The new screw has feeding zone (6D), compression zone (8D), and metering zone (8D). The channel depth of the feeding zone is 7 mm, and that of the metering zone is 3 mm. The compression ratio is about 2.3. Compared with the original design, the compression ratio is lower. Other parameters are kept the same.

To cover full injection molding process, the screw plastification is simulated using Moldex3D R13 ScrewPlus module. Fiber microstructures, including fiber length, fiber orientation, and fiber density distribution, are predicted using Moldex3D software[4].

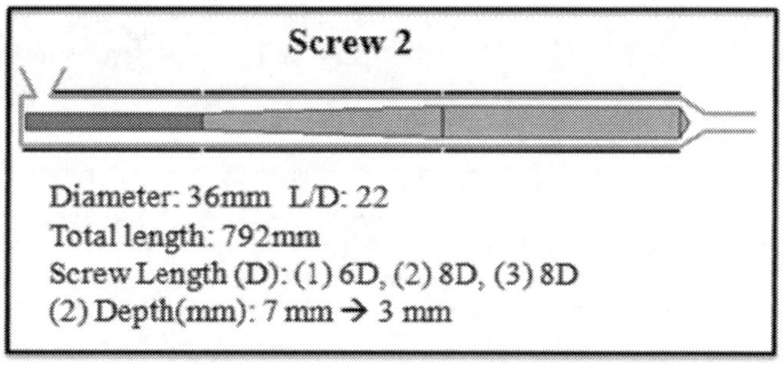

Figure 3. New design: conventional screw with different three sections, compression ratio = 7/3.

4. Results and discussion

4.1. Fiber breakage investigation

Figure 4 shows the shear rate variation along the screw performance for Screw 1. For Screw 1-1, from feeding zone to metering zone, the highest shear rate happens around the end of compression zone. For Screw 1-2, since the channel depth of metering zone is reduced to 2.1 mm, the shear rate increases significantly. Figures 5 and 6 show the fiber length variation along the screw direction during the operation. In figure 5, the number average of fiber length is reduced to around 2 mm. No significant effect on fiber breakage is observed when the metering zone is reduced to the channel depth. Similar behavior can be obtained based on weight average of fiber length as shown in figure 6.

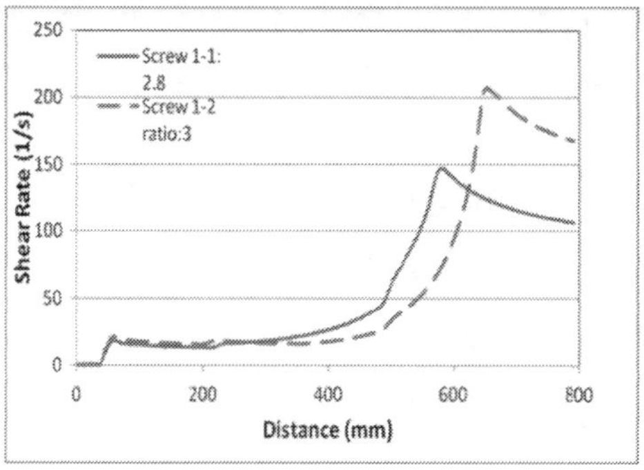

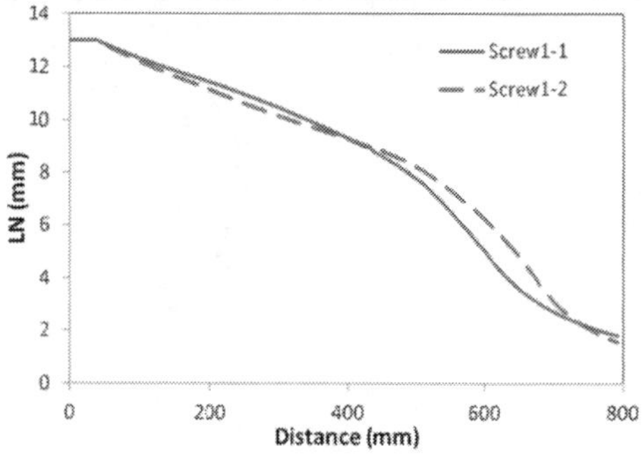

[4] Moldex3D is a CAE software for injection molding simulation. It is created by CoreTech System Co. Ltd from Taiwan.

Figure 4. Shear rate variation along the screw performance for Screw 1.

Figure 5. Number average of fiber length along the screw direction during processing.

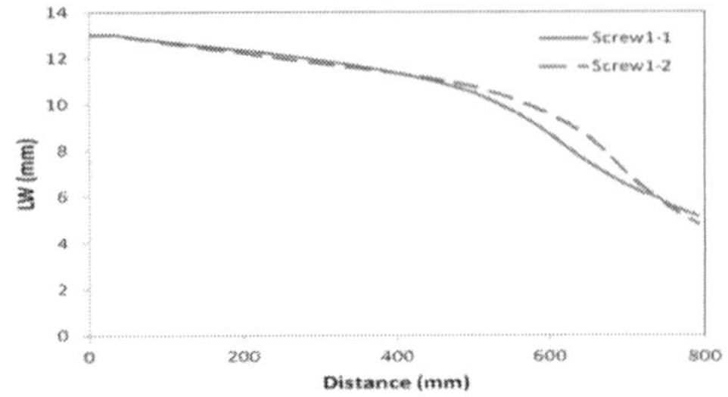

Figure 6. Weight average of fiber length along the screw direction during processing.

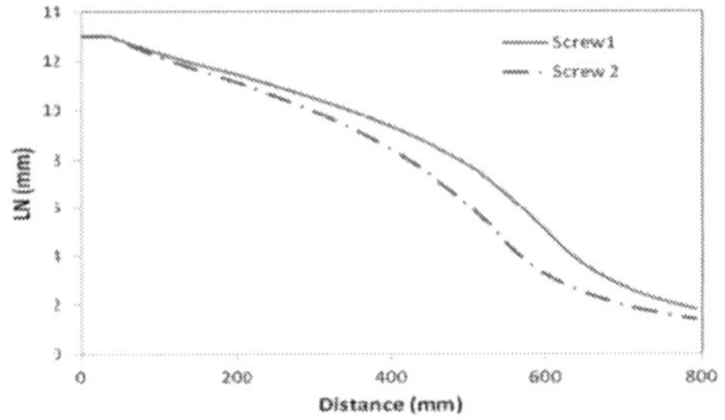

Figure 7. Comparison between new and original screw design: number average of fiber length along the screw direction during processing.

Furthermore, when the screw is modified with some new structures (as shown in figures 2 and 3), the fiber length breakage history is very different. In figure 7, although Screw 2 has lower compression ratio, the long fiber is broken earlier and becomes shorter at the end of tip of screw based on number average of fiber length. The result is similar based on the weight average of fiber length in figure 8. To compare the final fiber length at the end of screw tip, figure 9 shows Screw 2 has the shortest fiber length because the higher shear stress happens earlier in screw operation.

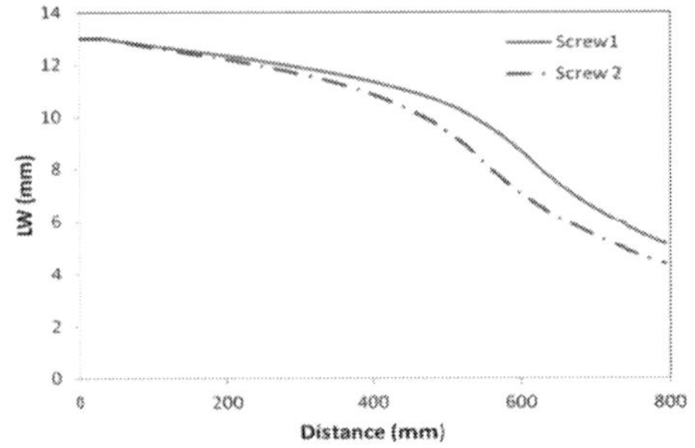

Figure 8. Comparison between new and original screw design: weight average of fiber length along the screw direction during processing.

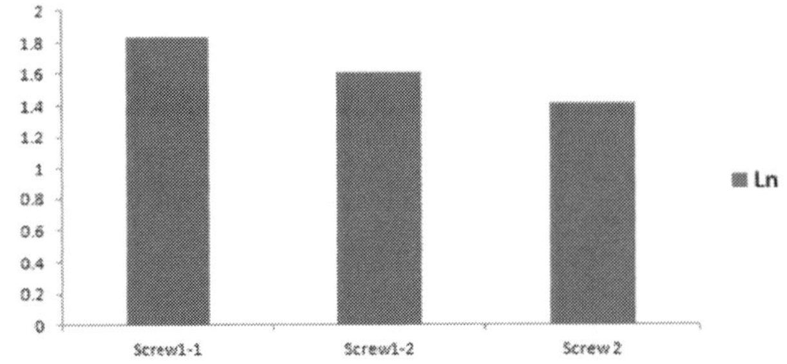

Figure 9. Average fiber length at the exit of screw with various screw designs.

IOP Conf. Series: Materials Science and Engineering **87** (2015) 012023 doi:10.1088/1757-899X/87/1/012023

4.2. Validation of fiber breakage

In the study conducted by Patcharaphun and Opaskornkul [9], the fiber length distribution (FLD) was measured along the screw channels, and then the molten composite was injected through the nozzle of an injection unit. Their material was a polypropylene filled with 40 wt% of long glass fibers. The initial fiber length was 7 mm. Generally, the injection molding machine screw has three zones: feeding, compression, and metering. Based on their experimental data, a comparative numerical analysis was performed with results of fiber breakage under a screw speed condition being kept at 12 RPM. Figure 10 shows fiber degradation in number average length at different screw sections. The predicted fiber length distribution is in good agreement with that of the experimental results [8].

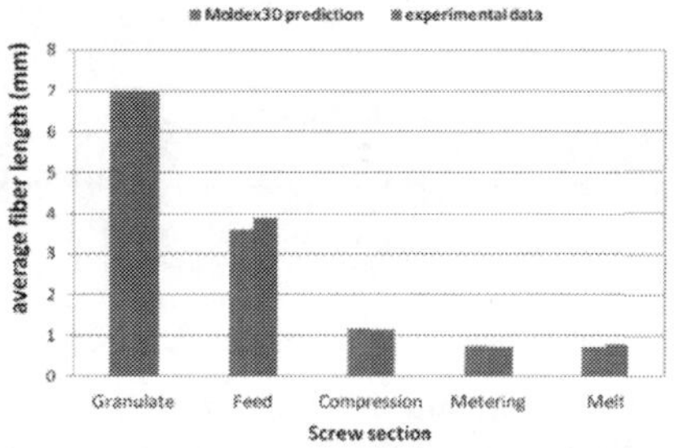

Figure 10. Fiber degradation in number-average length with various screw sections.

5. Conclusion

This study integrates the screw plastification with injection molding for fiber microstructures investigation. More specifically, most of our attention is paid to fiber breakage prediction during screw plastification. Results show that fiber breakage is strongly dependent on screw design and operation. When the screw geometry changes, even the compression ratio is lower, the fiber breakage could be larger.

References

[1] Advani S G 1994 *Flow and Rheology in Polymer Composites Manufacturing* (New York: Elsevier)

[2] Tseng H-C, R-Y Chang and C-H Hsu 2013 Phenomenological improvements to predictive models of fiber orientation in concentrated suspensions *J. Rheol.* **57** 1597

[3] Tseng H-C, R-Y Chang and C-H Hsu 2013 Method and computer readable media for determining orientation of fibers in a fluid *U.S. Patent No. 8571828*

[4] Vlcek J, Miller L and Huang C Simulation of Screws for Injection 2009 *Molding SPE Annual Tech Meeting* pp 2478-81

[5] Potente H, Schulte H and Effen N 1993 Simulation of injection molding and comparison with experimental values *International Polymer Processing* **8** 224-35

[6] White J L and Potente H 2002 *Screw Extrusion: Science and Technology*: Chap 4 Technology of Single Screw Extrusion with Reciprocating Screws (Munich: Hanser)

[7] Phelps J H, Abd El-Rahman A I, Kunc V, Charles L and Tucker I 2013 *Composites Part A: Applied Science and Manufacturing* **51** 11-21

[8] Phelps J H *Processing-Microstructure Models for Short- and Long-Fiber Thermoplastic Composites*, Ph.D. Thesis, University of Illinois at Urbana-Champaign (2009).

[9] Patcharaphun S and Opaskornkul G 2008 Characterization of fiber length distribution in short and long-glass-fiber reinforced polypropylene during injection molding process *Kasetsart J. (Nat. Sci.)* **42** 392-7

Effects of vermiculite on the tribological behavior of PI-matrix friction materials

X L Xu[1], X Lu, D L Yang and E Zhang

College of Materials Science and Engineering, Dalian Jiaotong University, Dalian 116028, China

E-mail:xxiulin@tom.com

Abstract. Effects of vermiculite as a friction modifier on brake friction performance of PI-matrix materials have been tested. PI-matrix friction materials with different vermiculite content (0%, 5%, 10%, 15%, 20% in wt) were prepared by the heat-press molding method. The friction tests were carried out on a pin-on-disk tester in the speed range of 20~120 km/h. The experimental results indicate that the effects of vermiculite on the friction and wear performance of the materials are closely related to the friction speed. At low friction speeds (between 20~40 km/h), the friction coefficient increases with the increasing of vermiculite content. At high friction speeds (between 60~120 km/h), the friction coefficient increases at first when vermiculite content below 10 wt% and then decreases with vermiculite content increasing. The wear rate decreases by adding 5 wt% vermiculite, and then increases. As the friction speeds above 100 km/h, the wear rate changes dramatically.

1. Introduction

The friction materials used in the brake system should have stable and reliable coefficient of friction (COF) at wide range of braking load, friction speed, service temperature, humidity, and so on [1]. These characteristics accompanied with high wear resistance, low noise, light weight, acceptable cost and easy processing make a friction material ideal for the brake system. The polymer-based composite materials have been known to be the best candidate to serve as friction materials in the brake system.

The performance of the friction material is strongly affected by the selection of the ingredients. The number of ingredients used for formulation of such composites may exceeds 700, and they can be mainly classified into four prime classes; namely binder, fibers, fillers, and friction modifiers [2–4]. The binder (such as phenolic resin and rubber) binds the ingredients firmly so that they can perform the desired function in the friction materials [3]. Ideally, there should be no significant deterioration in the function of the binder when the brake is operated under diverse conditions. The fibers play a major role in maintaining strength, stiffness, thermal stability, and frictional properties of the composite such as organic and mineral fibers (aramid pulp, oxidized polyacrylonitrile fiber, acrylic, cellulose, carbon, glass and slag fibers, wollastonite). The fillers are low-cost minerals (such as barite and clay) whose primary role is to increase the volume of the friction material and thus to cut the overall cost of a

[1] Address for correspondence: X L Xu, College of Materials Science and Engineering, Dalian Jiaotong University, Dalian 116028, China, Email: xxiulin@tom.com

IOP Conf. Series: Materials Science and Engineering **87** (2015) 012024 doi:10.1088/1757-899X/87/1/012024

composite on a volume basis. The friction modifier's roles is the alteration of the friction behavior, which are classified into two kinds: abrasives, for example aluminum oxide (Al_2O_3), zircon ($ZrSiO_4$), zirconium oxide (ZrO_2), and silicon carbide (SiC) with Mohs hardness values of about 7–9 and lubricants, such as graphite, coke, MoS_2, and Sb_2S_3 and so on [5].

Vermiculite is one of important minerals with lamellar structure composed of two silica tetrahedral sheet. Vermiculite has the advantages of low density, low hardness, low thermal conductivity and good thermal stability, strong sound-absorbing. Vermiculite can take place of asbestos in the field of friction materials. As the cohesive force of vermiculite between interlayer is weak, and the structural characteristics of infinite lax and dispersion, vermiculite can closely be connected with the matrix. Thus the friction materials containing vermiculite have high bonding strength and resistance to shear stress [6]. The hardness of vermiculite is low (Mohs hardness values between 1 ~ 1.5). However, vermiculite can increase the friction coefficient under certain conditions for the lamellar structure. Cao et al. [7, 8] developed a new friction composite reinforced with flake vermiculite, and the results indicated the flake vermiculite has good reinforcing properties; Zeng et al. [9] confirmed the feasibility of vermiculite taking place of asbestos fiber.

The friction process is always accompanied by the development of debris (the third body) [10], which adheres to the rubbing couple. As a result, the characteristic friction layers form on the friction surface, which determines friction performance. The characterization of friction layer differed with the friction conditions and the component of the friction materials. The purpose of this study is to evaluate the effects of vermiculite as friction modifier on friction performance and friction layers of PI-matrix materials. Some insights into the friction and wear mechanism of the friction materials are provided.

2. Experimental

2.1. Fabrication of the samples

The investigated friction materials contain different ingredients including resin (table 1), reinforcing fibers (wollastonite 10 um), solid lubricant (graphite, 314~400 um), abrasive (iron powder EPA-512), fillers (barite 44 um) and friction modifier (vermiculite 44 um and friction powder FB-80. All the ingredients were well mixed, then PI-matrix materials were prepared by the heat-press molding method at heat-press temperature 220℃ under heat-press pressure of 8 KN for 40 minutes. The obtained samples were cylindrical of ϕ16 mm×20 mm. Then the friction materials were post-treated at 120℃ for 60 minutes, at 150℃ for 60 minutes, and at 180℃for 120 minutes.

Table 1. The compositions of the designed samples (in wt %).

ingredients	Samples				
	1	2	3	4	5
PI	10	10	10	10	10
Wollastonite	8	8	8	8	8
Graphite	18	18	18	18	18
Friction powder	10	10	10	10	10
Iron powder	7	7	7	7	7
vermiculite	0	5	10	15	20
barite	37	32	27	22	17
others	10	10	10	10	10

2.2. Tests of friction performance

The tribological properties were tested on a pin-on-disk tester (see in figure 1). The disk is made of H13 steel, and the diameter of the friction is 150 mm. The friction interface pressure is 0.57 MPa, The friction rotating velocity is 200~3000 r/min (11~170 km/h), the friction time was 90s under each friction speed; the microstructures were observed optical microscopy (OLS 3000) and JSM−6360LV scanning electron microscopy (SEM).

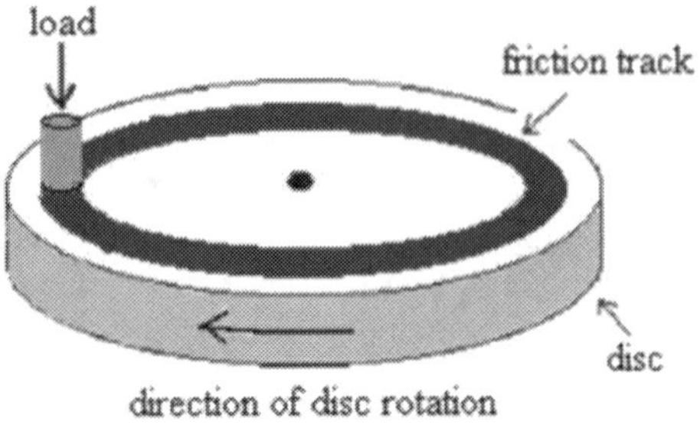

Figure 1. Schematic of the pin-on-disk tester.

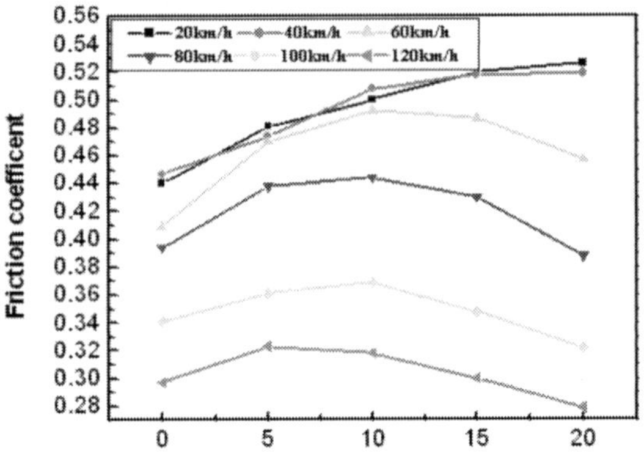

Figure 2. Effect of vermiculite content of friction coefficient of the friction materials.

3. Results and discussion

3.1. Effect of vermiculite content on the friction coefficient of the friction materials

The dependence of the vermiculite content, on the average friction coefficient of PI-matrix friction materials at different friction speed are shown in figure 2. The friction coefficient increases with vermiculite content increasing at the friction speeds between 20~40 km/h. When the friction speeds between 60~120 km/h, the friction coefficient increases to the maximum value at the content of vermiculite between 5~10 wt% and then decreases.

The friction force results from the mechanical action and intermolecular force between the two friction surfaces. A large amount of micro peaks or valleys on the friction surface are unavoidable. The micro peaks are generally referred to as asperities. The mechanical force included actions, such as the micro peaks and valleys meshed with each other, leading to deformation, shearing, breaking, and the asperities embedded into the dual surface as a result of ploughing on the friction surfaces, and so on. As the cohesive force of vermiculite between layer is weak, and the structural characteristics of infinite lax and dispersion, vermiculite can be closely bond with the matrix when the amount to be optimized because of the interlayer structure of vermiculite. The friction torque increases and the friction coefficient increases with the increasing of vermiculite content (below 10 wt. %).when the concentration of vermiculite is above 10 wt%, the cohesive force between PI and vermiculite decreases, stripping and drawing cause the friction torque to decrease. As a result, the friction coefficient decreases.

With friction speed increasing, the impact and shear force between asperities in the dual friction surface rapidly increase, so that a lot of debris are formed by shearing and ploughing. Under the effect of friction pressure, the debris fill the gap between peaks and valleys, thereby the ploughing effects of the asperities are reduced. As the friction speed increases, the temperature on the friction surface is likely to be very high, which induced the matrix closed to the friction surface to be oxidized and carbided, and the compressive and inter-laminar shear strength closed to the friction surface to be lowered, which make more debris be produced, leading a increased areas of friction surface covered by the debris. Finally, a continuous friction film is formed on the friction surface, resulting in the decreasing of friction coefficient. Schematic illustration of the filled debris and the formed friction film is shown in figure 3 [11].

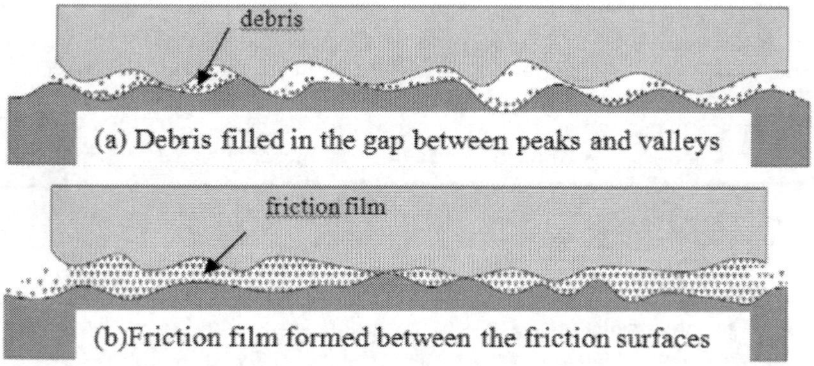

Figure 3. Schematic illustration of the debris filled and the friction film formed.

3.2. Effect of vermiculite content on wear rate of the friction materials

The relationships between the wear rate and the content of vermiculite at different friction speed are shown in figure 4. The wear rate increases slowly with the increasing of friction speed below 80 km/h. However, the wear rate increases rapidly as the friction speed above 80 km/h. There are possibly two reasons: 1, high friction speed results in a great impact and shear force between asperities in the friction surface, and then a great deal of debris are produced, which increases the wear rate; 2, the temperature on the friction surface is likely to be very high, which induce the matrix closed to the friction surface to be oxidized and carbided, and the compressive and inter-laminar shear strength closed to the friction surface to be lowered, which make the wear be aggravated with the speed increasing [12].

Figure 4 shows that the wear rate decreases at first and then increases with the increasing of vermiculite content. The effects of vermiculite content on wear rate of the friction materials are related to the friction speed. At the low friction speed, the wear rate changes a little. As the friction speeds above 100 km/h, the wear rate changes dramatically. When the content of vermiculite is 5 wt%, the wear rate is lowest at all friction speed. Because the cohesive force between vermiculite and PI matrix is high; the specific surface area of layered vermiculite is large; the layered vermiculite has lots of pore, which can adsorb lots of water and small molecular compounds decomposed from PI- matrix of the friction materials [9]. As a result, the wear rate decreases. When the content of vermiculite is above 10wt%, the PI-matrix cannot bond all the sheet of vermiculite well. Vermiculite is easily peeled off from the matrix, and the wear rate increases rapidly.

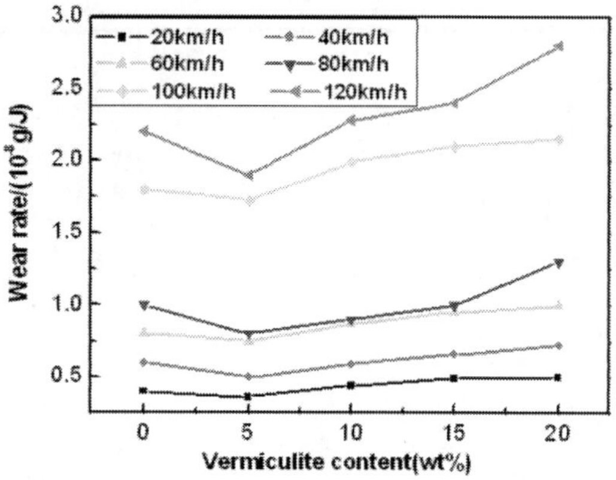

Figure 4. Effect of Vermiculite contents on wear rate of the friction materials.

3.3. The morphology of friction surface

In order to study the role of vermiculite in PI- matrix of the friction materials, the friction film and debris under the friction speeds 20km/h and 120km/h have been shown in figures 5 and 6. Character of the friction layer determines the friction performance. It can be seen from figures 5(a) and 6(a) that the worn surface is uneven with many grooves and wear debris distributed on it at the speed of 20 km/h. Figure 6(b) shows the morphology of debris collected at counter friction surface at the friction speed of 20 km/h, which are small and equiaxial particles. At the friction speed of 120 km/h, it can be seen from figures 6 and 7 that a continuous friction film by the main constitution of vermiculite is formed on the friction surface, and the friction surface is flat, as shown in figures 5(b) and 6(c). Figure 6(d) shows that the debris collected on the counter surface are big sheets, which are quite different from the figure 6(b) at 20 km/h. At low friction speed, the friction force partly comes from the force shearing the flake vermiculite into smaller particles, so the friction force increases as the rising content of vermiculite. As a result the friction coefficient increases at low friction speed. With the braking speed increasing, the impact and shear force between asperities in the dual friction surface increase rapidly. The lamellar vermiculite peels off, which need to overcome the cohesive force of vermiculite between interlayer, so the friction coefficient of friction materials containing vermiculite between 5~10wt% is higher compared with the non-vermiculite friction material. When the content of vermiculite up to 10 wt%, the friction coefficient decreases rapidly with the friction speed rising. Because the impact and shear force increase continuously and the temperature maybe high with braking speed increasing, the PI-matrix cannot bond all the sheet of vermiculite well, vermiculite is easily peeled off from the matrix. And the hardness of vermiculite is low. The vermiculite peeled off can play the role of lubrication, so the friction coefficient decreased.

4. Conclusions

The effects of vermiculite content on friction performance were comprehensively evaluated. The effects of Vermiculite content on the friction and wear performance of the materials are closely related to the friction speed.

- At low friction speeds (between20~40km/h), the friction coefficient increases with the increasing of Vermiculite content.
- At high friction speeds (between 60~120km/h), the friction coefficient increases and then decreases with the increasing of Vermiculite content. The friction coefficient increases when the content of vermiculite between 5wt%~10wt%. With friction speed increasing, the impact and shear force between asperities in the dual friction surface increase rapidly. So that the flake vermiculite drops off, which need to overcome the cohesive force of vermiculite between interlayer. When the content of vermiculite is up to10 wt%, the friction coefficient

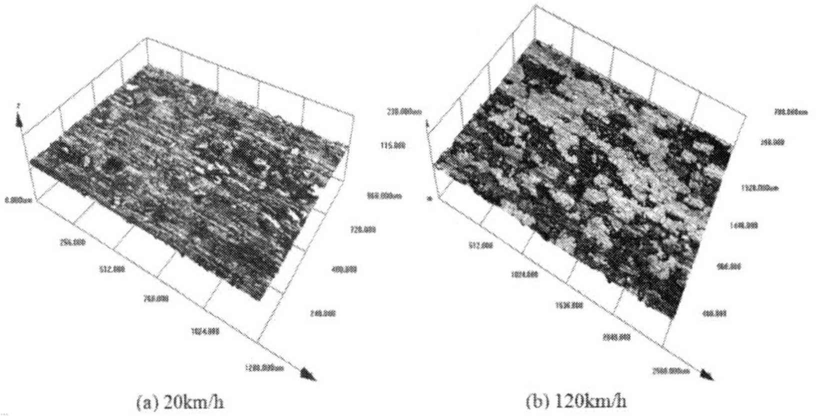

(a) 20km/h (b) 120km/h

Figure 5. The 3−D morphology of friction surface at different friction speeds.

IOP Conf. Series: Materials Science and Engineering **87** (2015) 012024 doi:10.1088/1757-899X/87/1/012024

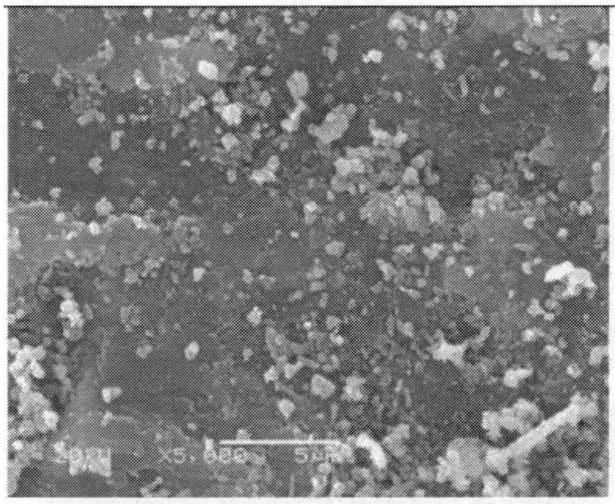

(a) friction surface feature 20 km/h

(b) morphologies of debris 20 km/h

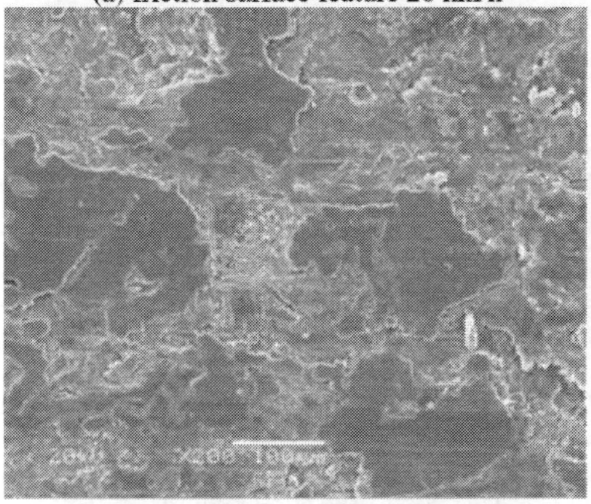

(c) friction surface feature 120 km/h

(d) morphologies of debris 120 km/h

Figure 6. Morphologies of friction surfaces and debris at different friction speeds.

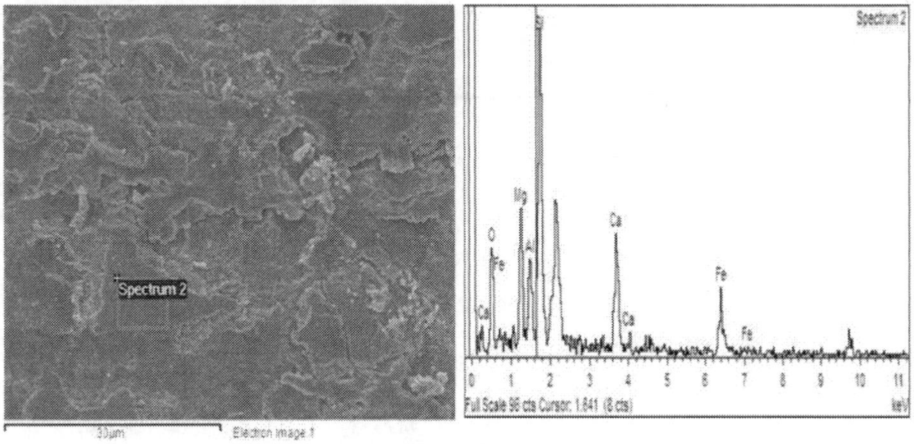

Figure 7. The spectrum of the continuous film at friction speed of 120 km/h.

decreases rapidly with the friction speed rising, because the PI-matrix cannot bond the vermiculite well.

- The wear rate decreases at first and then increases with the increasing of vermiculite content. As the friction speeds above 100 km/h, the wear rate changes dramatically
- The content of Vermiculite is about 5wt. %, The friction coefficient of the friction material is relatively stable, and the wear rate is low, the friction material shows good friction and wear performance.

References

[1] Eriksson M 2002 On the nature of tribological contact in automotivebrake *J. Wear.* **252** 26-36
[2] Kim S J, Jang H 2000 Friction and wear of friction materials containing two different phenolic resins reinforced with aramid pulp *J. Tribol.* **33** 477-84
[3] Gurunath P V and Bijwe J 2007 Friction and wear studies on brake-pad materials based on newly developed resin *J. Wear.* **263** 1212-19
[4] Kumar M and Bijwe J 2010 Role of different metallic fillers in non-asbestos organic (NAO) friction composites for controlling sensitivity of coefficient of friction to load and speed *J. Trio. Int.* **43** 965-74
[5] Vladimír Tomášek,Gabriela Kratošová, Rongping Yun, Yanli Fan and Yafei Lu 2009 Effects of alumina in nonmetallic brake friction materials on friction performance *J. Mater Sci.* **44** 266-73
[6] Gao H M, 2007 *Mineral Composite Friction Material* (Beijing: Chemical Industry Press) chapter 4 pp183-6
[7] Cao X K, Lian L and Wen D J 2000 Study on the mechanism of vermiculite flake reinforced modified phenol resin in friction composites *J. Engineering Plastics Applications* **28** 1-3
[8] Cao X Q, Yao A Y and Wen D J 2000 Effects of volume fraction diameter-thickness ratio of flake reinforcements on performance of friction composites *J. Journal of Wuhan University of Technology* **22** 18-20
[9] Zeng Z B and Yao A Y 1997 The study of vermiculite friction materials *J. Non-Metallic Mineral.* **118** 68-71
[10] Filip P 2002 On friction layer formation in polymer matrix composite materials for brake applications *J. Wear.* **252** 189-98.
[11] Fan S W, Zhang L T, Cheng L F, Tian G L and Yang S J 2010 Effect of braking pressure and braking speed on the tribological properties of C/SiC aircraft brake materials *J. Composites Science and Technology* **70** 959–65
[12] Kukutschovaa J 2009 Wear mechanism in automotive brake materials, wear debris and its potential environmental impact *J. Wear.* **267** 807-17

Study of mould design and forming process on advanced polymer-matrix composite complex structure

S J Li[1], L H Zhan[1, 2, 3], H M Bai[1], X P Chen[1] and Y Q Zhou[1]

[1]Light Alloys Research Institute, Central South University, Changsha 410083, China
[2]High-performance Complex Manufacturing State Key Laboratory of Central South University, Changsha 410083, China

E-mail: yjs-cast@csu.edu.cn

Abstract. Advanced carbon fibre-reinforced polymer-matrix composites are widely applied to aviation manufacturing field due to their outstanding performance. In this paper, the mould design and forming process of the complex composite structure were discussed in detail using the hat stiffened structure as an example. The key issues of the mould design were analyzed, and the corresponding solutions were also presented. The crucial control points of the forming process such as the determination of materials and stacking sequence, the temperature and pressure route of the co-curing process were introduced. In order to guarantee the forming quality of the composite hat stiffened structure, a mathematical model about the aperture of rubber mandrel was introduced. The study presented in this paper may provide some actual references for the design and manufacture of the important complex composite structures.

1. Introduction

As the lightweight and high-performance material, advanced carbon fibre-reinforced polymer-matrix composites are widely applied to aviation manufacturing field due to their high specific strength and stiffness, anisotropy and good designability [1, 2]. Currently, complex composite structures are mainly used in some bearing structures, such as the large wing plates of airfoil and empennage. The wing plates usually employ the long truss as the stiffened structure, and kinds of types such as T type, L type, J type and hat type are often consisted in the long truss stiffened structure, as shown in figure 1. Thus it can be seen that the complex composite structures play an important role in the process of composites manufacturing. Therefore, how to fabricate the complex structures accurately so as to ensure the forming precision and internal quality has become the key issue for the composites manufacturing. In response to this issue, many scholars have done a lot of relative studies in this field. Li [3] fabricated the T-stiffened skins integrally by different tool assembly schemes in autoclave, and discussed the influence of tool assembly schemes and integral molding technologies on the compaction of T-stiffened skins in detail. Huang [4] studied the co-cured process to form the composite panels with T-type stiffeners. Vijayaraju [5] used three configurations to study the failure mechanism and the failure progression of T-stiffened skin. Sala [6] studied the hollow composite components manufactured by the treated elastomeric tooling by analytical, numerical and

[3] Address for correspondence: L H Zhan, Light Alloys Research Institute, Central South University, Changsha 410083, China. E-mail: yjs-cast@csu.edu.cn.

experimental approaches, and the results demonstrated the need to design the mandrel accurately. Huybrechts [7] proposed a theory governing the tooling behavior of hybrid tooling and expansion block tooling, and the theory was validated by experimental data. Williams [8] studied J- and blade-stiffened panels, comparing how well experimental data could be correlated with analysis results for buckling behavior. However, the study on forming and manufacturing process of more complex composite structures were seldom or not systematically introduced. As the best representative composite structure in the aircraft, hat stiffened structures employ a more complex configuration, and the stiffened structures of approximately 80% are hat stiffened structure. Thus, in this paper, the mould design and forming process of the hat stiffened structure were discussed in detail. The hat stiffened structure was taken as a typical example to introduce the manufacture process of complex composite structures. And it may provide some actual reference for the design and manufacture of the important complex structures.

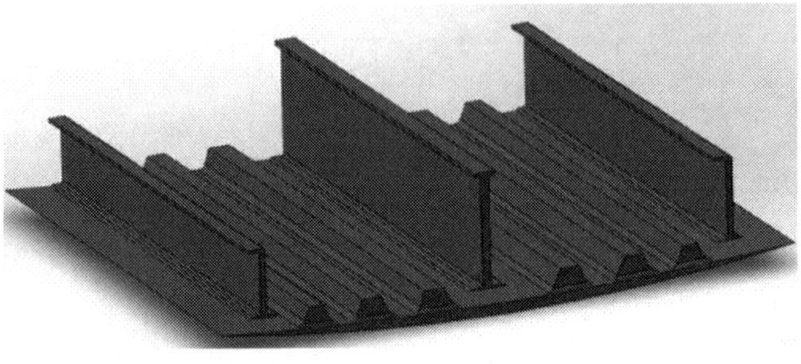

Figure 1. Kinds of stiffened structures in aircraft.

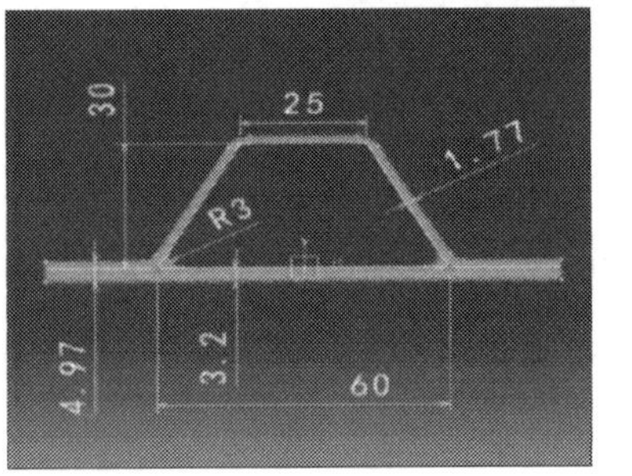

Figure 2. Typical size of the hat stiffened structure.

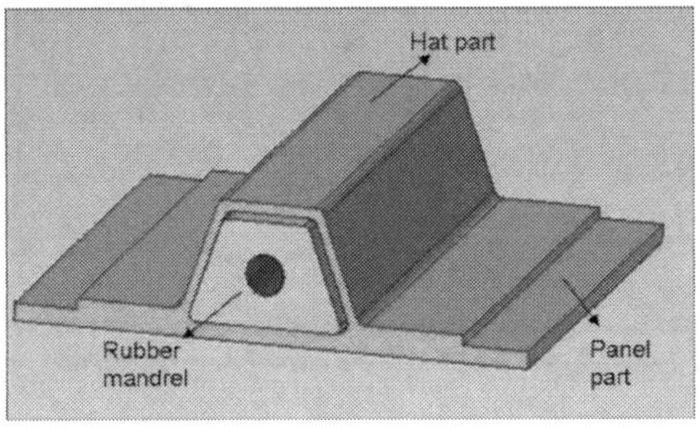

Figure 3. The assembly relationship of the hat stiffened skin.

2. Mould design

2.1. Design of composite covering plate

The composite covering plate is made by a kind of composite material as similar as the preform, and formed by the autoclave process. There are two advantages using the composite covering plate. Firstly, it plays a key role to locate the long truss with hat stiffener in the skin-stiffener structure. Secondly, because the coefficient of thermal expansion (CTE) is close to the preform, the forming quality of the composite preform can be better ensured. Thus, the composite covering plate may be seen as an indemnity of the external manufacturing environment for the composite hat stiffened structure. The typical size of the hat stiffened structure and the assembly relationship were shown in figures 2 and 3. According to the size and assembly relationship, the mould of the composite covering plate was designed, as shown in figure 4. It can be seen from figures 2 and 3 that two aspects should be considered during the process of mould design for the composite covering plate. On the one hand, the temperature uniformity of the mould interface is pretty important, which may influence the forming precision of the covering plate. Thus, the modest size and uniform distribution of the air vent should be taken into account. On the other hand, owing to the reason that the preform is covered by the composite covering plate, the size of the mould groove should be larger than the preform. In the common of hat stiffened structures, the thickness of the hat stiffened structures is usually about 1.77 mm. Therefore, the clearance compensation for the groove is often selected about 2 mm.

Figure 4. Mould of the composite covering plate.

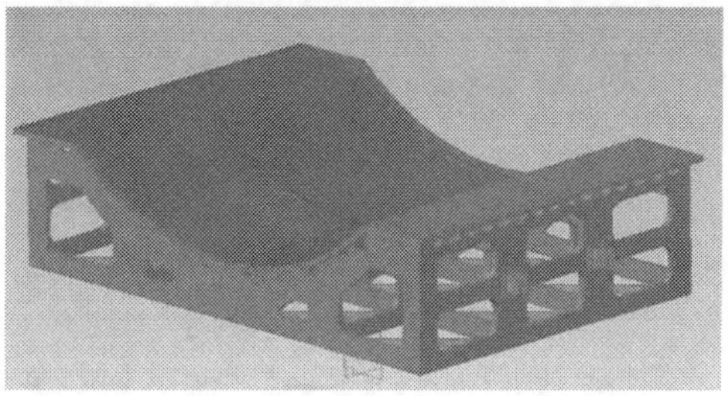

Figure 5. Typical mould design of stiffened skin panel.

2.2. Mould design of skin panel

In the curing process, the complex skin panel comes into direct contact with the tooling of skin panel, and the heat is transferred to the skin panel through the mould of skin panel. Generally, as the large bearing composite structure, the size of skin is also pretty big. Thus, it is difficult to achieve an absolute uniform temperature field on the surface of the skin panel. But in the manufacturing process of the complex composite structure, the uniform temperature can make sure to achieve an ideal degree of cure and slight curing deformation, so that the curing quality of skin panel fairly rely on the temperature field. For the complex structure of hat stiffened skin, in order to obtain a relatively uniform temperature, often the mould design and selection of mould material are the key points to focus on. Therefore, enough vent holes and evened-temperature holes are requested in the mould design, which can ensure to achieve a relatively uniform distribution of temperature. In view of the difficulty of manufacturing the composite mould due to the complex structure, the invar steel is often selected as the mould material, which is also based on the approximate CTE between the skin panel and mould material. In addition, the designing accuracy of the mould surface is also very important and noteworthy. The typical mould design of stiffened skin panel is shown in figure 5.

2.3. Mould design of mandrel structure for the hat stiffened structure

Composite hat stiffened structure is a semi-enclosed structure, which needs an internal mandrel to support the hat section so as not to be crushed by the external pressure. In actual production, vacuum bag and rubber mould usually are used as the internal mandrel. However, the vacuum bag method gradually is abandoned due to the reasons such as easy to leak, poor security, so the rubber mould is more and more selected as the preferred tooling to assisted forming the composite hat stiffened structure. The rubber mould is formed by pouring the liquid rubber to a cavity of the hat type and then solidified. Thus, the mould used to pour the rubber mould is needed to design. During the mould design, there are also several issues need to be considered. Firstly, the size of mould cavity and the size of the rubber mould should stay the same. And in order to avoid the excessive liquid rubber pouring into the mould cavity so as not to influence the forming accuracy, a diversion trench about depth of 1mm should be machined on the mould surface. Secondly, in the curing process, the rubber mould expands gradually as the temperature increases. If the inflation is too large, the forming quality may be hard to ensure. To settle the issue, a prefabricated hole can be designed in the rubber mould, and a metallic solid shaft is designed to assembly in the middle of the hat cavity accordingly. Under this condition, the prefabricated hole can be made, and if the prefabricated hole cannot adapt the manufacturing environment, it can be replaced by changing the shaft diameter to meet the product requirements. Thirdly, all the rounding chamfers are degrees of 3~5, this can be conducive to demould and to obtain a perfect surface of mould. According to the three aspects mentioned above, the exploded views of the mandrel mould were shown in figure 6.

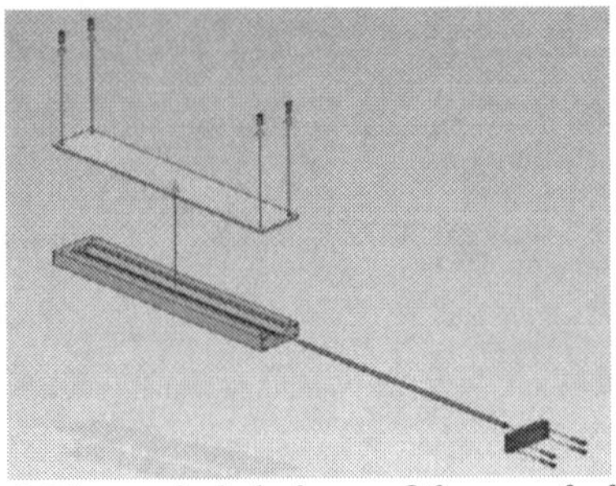

Figure 6. The exploded views of the mandrel mould.

3. The forming process of composite hat stiffened structure

3.1. Determination of material and stacking sequence

In view of the severe environment that plane suffered, the fibre strength and the resin content should be taken into account firstly, and the volume fraction of fibres is also an important factor that needs to be considered. Recently, the fibres with high tensile strength and modulus as well as the zero-bleeding resin are the main constituent parts of preperg. High tensile strength and modulus ensure the design strength, while the zero-bleeding resin makes sure the volume fraction of fibres and internal curing quality. For example, the common prepreg tape P2352W-19 and woven fabric FM6673-37KC. In addition, the rubber mandrel material is obtained by curing the RHODORSIL RTV 3248 A and B in accordance with the ratio of 10:1. Besides the determination of material, the stacking sequence of prepregs is another key point. The carbon fibre-reinforced polymer-matrix composite is a kind of anisotropy material, and the mechanical property of each part in the complex structure can be improved by changing the stacking sequence. For the composite hat stiffened structure, the stacking sequence of hat part and panel part are often as follows: the stacking sequence of the hat part is (±45/0/90/0/90/0/90/0/±45) and the stacking sequence of the panel part is $(\pm45/\pm45/90/-45/\pm45/45/0/45/\pm45/-45/45/90)_{2s}$.

3.2. The autoclave co-curing process

The co-curing process of autoclave is often employed to fabricate the large complex bearing component owing to the higher curing quality, such as the composite hat stiffened panel. In the curing process of autoclave, the temperature and pressure are the two important manufacturing factors, which can make sure the prepregs cured and compacted ideally. Figure 7 showed the typical curing process of complex composite structure. Before temperature was heated, vacuuming to 0.095 MPa at the room temperature firstly, it was mainly aimed at pre-compressing the prepregs and excluding the excrescent particles trapped in the prepregs. Then the temperature was raised to 180℃ at a rate of 1.5℃/min and maintained for 150 minutes under the curing pressure of 0.6 MPa. In the end, pressure relief and air cooling were employed when the temperature was cooled to 70℃ with the same rate of 1.5℃/min.

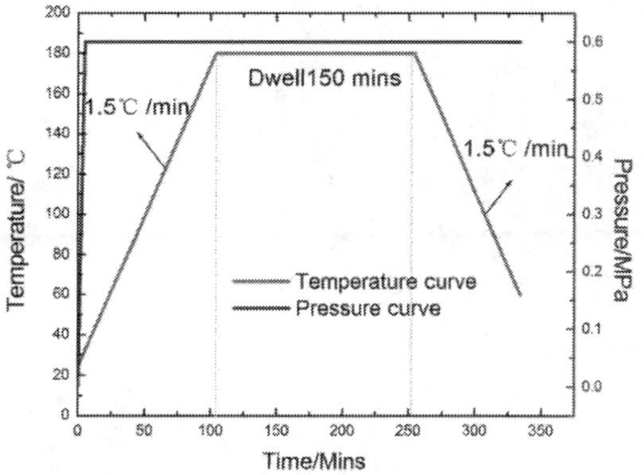

Figure 7. Typical curing process of composite structure.

3.3. Selection of mandrel structure

The hat stiffener part and panel part is consisted in the hat stiffened structure, thus during the co-curing process, a closed trapezoid cavity will be formed. The stiffener is very soft when it is uncured, if the load is only applied on the surface of the hat stiffener, the hat stiffened structure may collapse. Therefore, the mandrel is requested as a support in the middle of the hat stiffened structure, and silicone rubber mandrel is often chosen as the important support. However, in the curing process, the inflation of rubber mandrel is unavoidable due to the synchronous temperature increase of the mandrel and preform. If the design of mandrel structure is unreasonable, the hat stiffened structure is not enough to keep the shape during the process of raising temperature. If the inflation or the stiffness of the mandrel structure is too small, the mandrel may not hold out the trapezoid cavity, and the defects would follow. In engineering practice, in order to solve the problem, optimizing and selecting proper mandrel structure becomes an effective way to go. In general, a prefabricated adjusting hole is set in the middle of the rubber mandrel to solve the difficulty. It is mainly based on the principle that the inflation of the mandrel generated in the curing process can exactly be absorbed by the prefabricated adjusting hole. Thus, how to determine the size of the prefabricated adjusting hole becomes the key point.

Table 1. The material property parameters.

Parameters	Value
Elasticity modulus/ MPa	3.552
CTE/ (m.℃$^{-1}$)	9e^{-4}
Heat conductivity/(W^{-1}.m.K^{-1})	0.23
Poisson ratio	0.48

To determine the optimal size of the hole is quite time-consuming through specific experiment, while the finite element modelling (FEM) method provides an efficient approach to ensure the optimal size of hole. The boundary condition was set according to the actual conditions, the material property parameters were shown in table 1. Tetrahedral mesh was taken as the meshing scheme of the rubber mandrel, and the typical meshing scheme and the result of the thermal deformation nephogram for the rubber mandrel were shown in figures 8 and 9, respectively.

Through a lot of modelling and simulation, the mathematical model between the size of prefabricated adjusting hole and other technical parameters was established as follows:

The volume of the prefabricated adjusting hole:

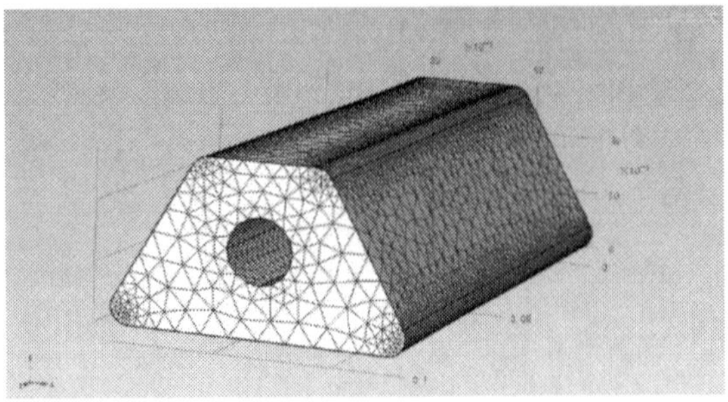

Figure 8. Typical meshing scheme of the rubber mandrel.

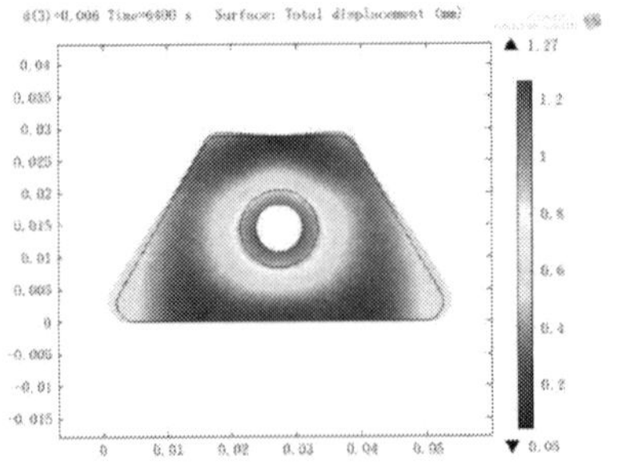

Figure 9. Nephogram of the thermal deformation for rubber mandrel.

$$V=\frac{1}{4}\pi d^2 L \tag{1}$$

The total volume inflation of mandrel:

$$\Delta V=\alpha\Delta T V_0-\frac{3P(1-2\upsilon)}{E}V_0 \tag{2}$$

Let, $V=\Delta V\zeta$, there the mathematical model of diameter of prefabricated adjusting hole is as follows:

$$d=\left\{\frac{4V_0}{\pi L}\left[\alpha\Delta T-\frac{3P(1-2\upsilon)}{E}\zeta\right]\right\}^{\frac{1}{2}} \tag{3}$$

Where d is the diameter of the prefabricated adjusting hole, V is the volume of the prefabricated adjusting hole, L is the length of the rubber mandrel, V_0 is the initial volume of the rubber mandrel, ΔV is the change of the rubber mandrel, α is the coefficient of thermal expansion, E is the elastic modulus, υ is the Poisson's ratio, T is temperature, P is the actual curing pressure and ζ is the correction factor. Finally, multiple regression tests were employed to the optimal correction factor by FEM method, and ζ of 0.658~0.778 was determined, which can provide an academic and practical reference for the selection of the mandrel structure.

4. Conclusions

In this paper, the mould design and forming process of the bearing complex composite structure were introduced using the hat stiffened structure as an example. Some conclusions by this study were as follows.

(1) In the process of mould design for the hat stiffened structure, the mould of composite covering plate, panel and rubber mandrel become the three important factors to take into account. For the mould of covering plate, the vent air should distributed uniformly inner the mould so as to ensure the uniform temperature field, and the allowance of the groove is often selected about 2 mm. For the mould of panel, enough vent holes and evened-temperature holes are requested. The invar steel is often selected as the mould material, and the surface precision should also be ensured. For the mould of rubber mandrel, the size of mould cavity and the size of the rubber mould should stay the same, a diversion trench about depth of 1 mm should be machined on the mould surface, and the diameter variable shaft should be designed to meet the different request of products.

(2) For the forming process of composite hat stiffened structure, the autoclave co-curing process is often employed to fabricate the complex composite structure. Besides the material and stacking sequence of the prepregs, the structure of rubber mandrel is also crucial. By FEM method, the mathematical model between the size of prefabricated adjusting hole and other technical parameters was established, and finally correction factor ζ of 0.658~0.778 was determined.

Acknowledgments
This work was supported by the National Program on Key Basic Research Project (2014CB046502), the Fundamental Research Funds for the Central Universities of Central South University (2014zzts036) and sponsored by Hunan Postdoctoral Scientific Program (2014RS4013). The authors would like to gratefully acknowledge the support of Composites Centre of Shenyang Aircraft Industry Group.

References

[1] Davies L W, Day R J, Bond D, Nesbitt A, Ellis J and Gardon E 2007 *Compos. Sci. Technol.* **67** 1895
[2] Li S J, Zhan L H, Chen R, Peng W F, Zhang Y A, Zhou Y Q and Zeng L R 2014 *Fiber Polym.* **15** 2404
[3] Li M, Wang X M, Xie F Y and Zhang Z Z 2010 *J. Reinf. Plast. Comp.* **29** 1311
[4] Huang C K 2003 *Compos. Part A-Appl. S.* **34** 403
[5] Vijayaraju K, Mangalgiri P D and Dattaguru B 2004 *Compos. Struct.* **64** 227
[6] Sala G 1996 *Mater. Design* **17** 33
[7] Huybrechts S M, Meink T E, Wegner P M and Ganley J M 2002 *Compos. Part A-Appl. S.* **33** 155
[8] Williams J K and Stein M 1976 *AIAA J.* **14** 1618

IOP Conf. Series: Materials Science and Engineering **87** (2015) 012026 doi:10.1088/1757-899X/87/1/012026

Forming strengthening nanoparticles in the metal matrix of plasma deposited powder alloys coatings

D L Alontseva[1] and N V Prokhorenkova

East-Kazakhstan State Technical University, 69 Protozanov Str, Ust-Kamenogorsk, 070004, Kazakhstan

E-mail: dalontseva@mail.ru

Abstract. This paper presents the generalized results of investigation of the structure-phase compositions of thick coatings on the base of Ni and Co deposited by plasma-jet on steel substrates. Research methodology included transmission and scanning electron microscopy with energy dispersive analysis, X-ray structure phase analysis, mechanical testing. The phase structures and morphology of precipitation of strengthening nanoparticles from solid solution are defined; factor k is defined in the Hall-Petch equation for the coating materials; the coating structure model was developed. It was found that the exposure to plasma jet during the coating deposition leads to the evolution of the structural-phase state and to substantial improvement of microhardness of modified surfaces. The reason of doing this research is the necessity of the study of the structural and phase structure of these coatings and understanding what phase changes exactly are desirable in order to improve the structural behavior of the coatings at modifying treatment. Based on the experimental study, we plan to develop a scheme of the coating structure for the further use it in modeling processes taking place during additional irradiation treatment, and eventually to give evidence-based recommendations on the selection of modes of further processing.

1. Introduction

The method of plasma powder coating is a modern and promising method to deposit a coating of refractory metal materials or ceramics on various substrates [1, 2]. The formation of nanostructures in material of the coatings deposited using this method is predictable, as it is known. In particular, effective amorphous or nanocrystalline states are achieved at high heating rates, high pressure and short-term exposure to high temperatures [3]. Since coatings are deposited at high plasma temperatures, it is possible to expect the formation of a thermally stable coating, unlike nanocomposite films prepared by magnetron sputtering [4, 5].

The papers [6-10] point out that the intermetallic compounds formed in the Ni-based plasma-detonation coatings have improved corrosion resistance and strength. It is necessary to note that the carbidic strengthening phase in Ni and Co-based alloys at the temperatures of 700-800°C coagulates much faster than the intermetallic one, which leads to quick loss of alloy strength [11]. Therefore for employing in high temperature environments the alloys with intermetallic hardening are more

[1] Address for correspondence: D L Alontseva, East-Kazakhstan State Technical University, 69 Protozanov Str, Ust-Kamenogorsk, 070004, Kazakhstan. E-mail: dalontseva@mail.ru.

preferable. Due to considerable depth of the plasma-detonation coating (100-500 μm) there is no problem of bonding of a high-duty and brittle surface film with the main material with much lower strength and high plasticity. The process of deformation takes place in a consistent structure mode through all the coating and the intermediate layer between the coating and the substrate.

Ni, Cr, and Co are the basic materials for development of solid, corrosion- and wear-resistant coatings [2, 12, 13]. As well as alloys based on Ni, the Co-based alloys can have an fcc lattice, which gives a technological advantage in comparison with the alloys based on Cr with a bcc lattice. In particular, the alloys with the fcc lattice are better welded, they retain strength at higher temperatures, are less prone to grain increase during heating, and do not lose ductility at low temperatures [12].

Some of the main problems of plasma detonated thick coatings (100-500 μm thick) are their porosity, lack of homogeneity on account of poor agglomeration of powder particles, high roughness of surface, and low adhesion to substrate. These result in insufficient corrosion and wear resistance of such coatings [1, 2]. To eliminate these drawbacks the coatings may be irradiated by electron beam in vacuum or re-treated by direct current pulse plasma jet on the surface without powder coating in air (duplex treatment). Practical experience of the use of combined technologies of coating deposition by plasma detonation with the subsequent modification by e-beam or plasma jet allows proves that the mechanical properties of such coatings of metals and alloys (microhardness, nanohardness, wear resistance, corrosion resistance) are highly desirable [2, 9, 10]. To scientifically justify the modes of additional irradiation we need to predict the structure of coatings before irradiation. There are not enough published TEM-data about the structure-phase composition of coatings deposited by plasma detonation. We need to provide the choice of duplex treatment modes according to the energy of a plasma jet or electron beam, the time of exposure of the surface, etc. Therefore we need a correct model of a structure-phase composition of a coating before additional treatment. On the basis of this model we may recommend additional irradiation modes.

One of the aims of this research is establishing structure-phase composition and mechanical properties of the plasma deposited coatings formed in result of plasma jet exposure during the coatings deposition. The other is developing a model of the coating composition on the basis of the experimental data analysis.

2. Materials and methods

An "Impulse-6" plasma detonation unit was used to form 150 -300 μm thick protective coatings of powder alloys on stainless Steel 3 substrate. For the coatings we used the Co and Ni- based powder alloys with additives of Cr: AN-35 (Co-based) and the PG-19N-01 (Ni-based). The electric current density in a plasma jet can vary from 1 to 7 A/cm^2, and the heat transmission rate in the sample varies in the $q = (0.1\ldots5)\ 10^6\ W/cm^2$ depending on the electric current density. The average temperatures of the plasma flow at the nozzle exit of the installation reach the $T = (10\ldots15)\ 10^3\,°C$. The average diameter of a plasma jet on a sample is 25 mm; the pulse duration is of the order of 10 μs. Propane, oxygen and air were used as combustible and orifice gases. Mo was selected as a plasma-jet eroding electrode material. The coating was deposited and the additional treatment was produced at the Sumy Institute for Surface Modification (Sumy, Ukraine).

The experimental methods of structural-phase analysis are presented in table 1. For a more detailed analysis of the coating, it was mechanically cut off the surface of the substrate to examine the structure at different depth from the coating surface. We used the method of arbitrary secant line to define the volume fraction of phases according to the TEM-images and data [14] and CrystalMaker software to define the parameter of the crystal-lattice according to the TEM-diffraction pattern. The foils for TEM were prepared by the Ar ion sputter etching method using the Precision Ion Polishing System PIPS – M-69 ("Gatan", USA).

The selection of the additional exposure modes is based on modeling of temperature profiles in the coatings and the substrate during irradiation. Mathematical modeling of the temperature profiles in the double-layer absorbers is described in Krasavin et al. and Alontseva et al.'s work [15-18]. We took such modes of irradiation at which the melting temperature is achieved on the surface the Ni or the Co, and on the border between the coating and the substrate the temperature is high enough to significantly accelerate the diffusion processes and to increase the width of the diffusion zone from the coating to the substrate, but not to lead to coating melting. The optimal modes were assumed those that would lead to homogenization of the coating and improve its adhesion to the substrate due to the acceleration of diffusion processes during irradiation. Microhardness was measured at the angle laps of coatings by PMT-3 microhardness meter (LOMO, Russia) with an indentation load of 2, 5, and 10 N. We also carried out the corrosion and wearability tests as described in our work [6], however for the present research the results of microhardness tests are more important since they are directly correlated with the content of the reinforcing phases.

Table 1. The experimental methods of structural-phase analysis.

Method	Equipment	NB
X-ray diffraction (XRD)	X'Pert PRO ("PANalytical", the Netherlands)	For studying the samples' phase composition
Scanning Election Microscopy (SEM) with Energy Dispersive Spectrometry (EDS)	JSM-6390LV ("JEOL", Japan) with EDS ("Oxford Instruments", Great Britain)	The investigation of surface structure and morphology and for elemental analysis
Transmission Electron Microscopy (TEM)	JEM-2100 ("JEOL", Japan)	For studying the nanostructure

3. Results and discussion

The TEM images and microelectron-diffraction pattern surfaces confirm the presence of an amorphous layer on the surface (figure 1). The coatings have differences in the phase structure with depth. We noted the reduction of γ-phase parameter a in the coatings and relative increase of γ-phase peaks (220), (422), (440) intensity with depth. In general, the γ-phase peaks spread. In the coating layers that contact the substrate the volume concentration of Fe–based phases rises (Fe is the basic component of a substrate), and oxides disappear.

According to the data of the XRD the material of the substrate contains a Fe-based bcc-lattice phase (cubic, I m-3 m, 229) with the parameter a=2, 8662 Å.

It was proven by TEM that the base through-thickness layer of plasma detonation coatings is a mixture of crystallographically differently oriented nanograins of austenitic γ-phase (figure 2(b)) with the size of 1-2 nanometers, and lamellas of intermetallic phases up to 50 nanometers long (figures 2(a), 3(a) and 4(a)). Every indexed reflex of intermetallic phases (figures 3(b) and 4(b)) was tested by the dark field method (figures 3(c) and 4(c)).

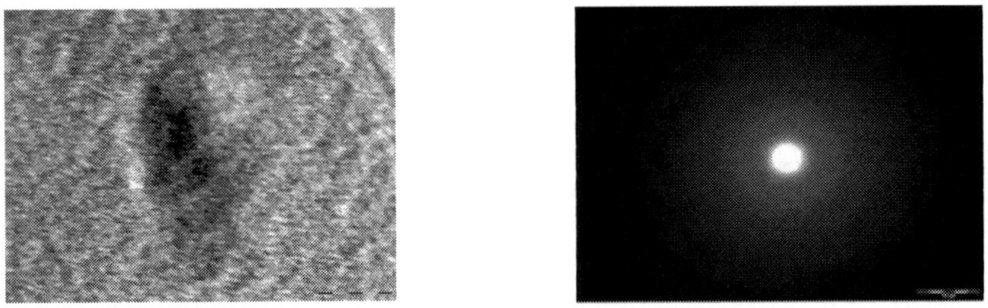

(a) (b)

Figure 1. TEM-image of the surface of AN-35 plasma-detonation coating (a); electron-diffraction pattern of the surface of AN-35 plasma-detonation coating (b).

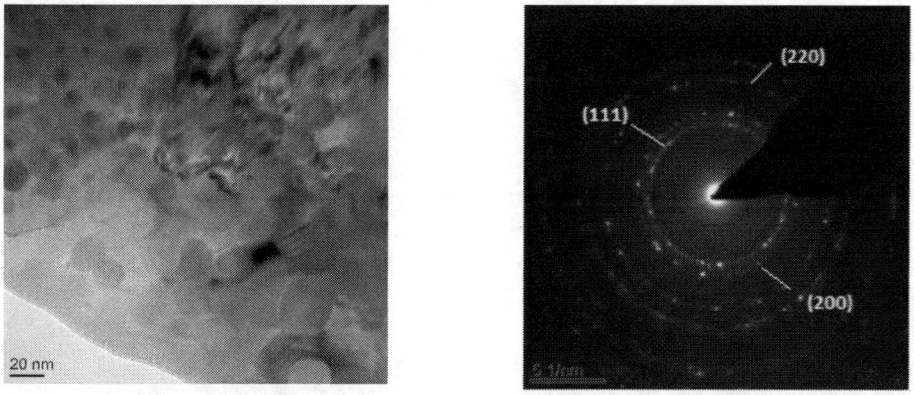

(a) (b)

Figure 2. TEM - images of PG-19N-01 powder coating deposited by plasma-detonation onto the Steel 3 substrate: the area of nanograins with different crystal-lattice orientation (a); the electron diffraction pattern of Ni-based matrix with indexes of corresponding crystallographic plates (b).

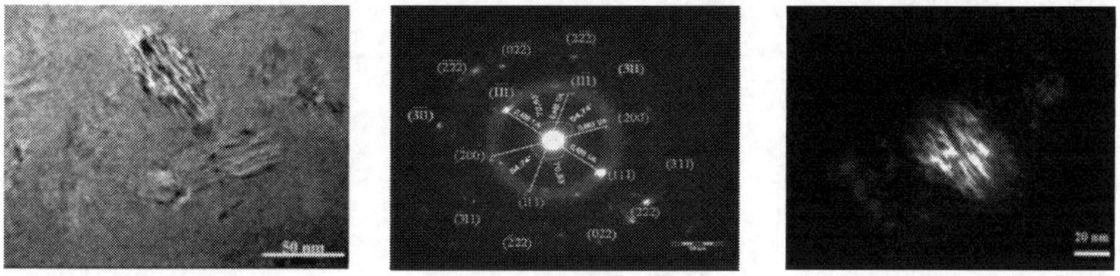

(a) (b) (c)

Figure 3. TEM - images of PG-19N-01 coating: the $CrNi_3$ particle (bright field) (a); the electron diffraction pattern of a $CrNi_3$ particle, the zone axis is $[0\bar{1}1]$(b); the $CrNi_3$ particle (dark field) shot in point reflex (111) (c).

(a) (b) (c)

Figure 4. TEM - images of AN-35 coating: the $Co_{0.8}Cr_{0.2}$ particle (bright field) (a); the electron diffraction pattern of a $Co_{0.8}Cr_{0.2}$ particle, the zone axis is [001] (b); the $Co_{0.8}Cr_{0.2}$ particle (dark field) shot in point reflex $(0\bar{1}0)$ (c).

According to the XRD data the PG-19N-01 coating contains a Ni-based fcc-lattice phase with the parameter a=3.525…3.540 Å. In accordance with the estimated electron-diffraction pattern (figure 2(c)) this parameter makes 3.53 Å. The estimated fcc-lattice parameter of $CrNi_3$ (figure 3) in the coating made a= 3.58 Å (the interplane distance is 2.05 Å), which is also very close to the XRD data. The volume ratio of $CrNi_3$ in the coating PG-19N-01, as defined by TEM images, makes about 20%. In accordance with the estimated electron-diffraction pattern (figure 4(b)) the parameters of $Co_{0.8}Cr_{0.2}$-

phase are: $a = в = 2.5$ Å and $c = 4.0$ Å. The volume ratio of $Co_{0.8}Cr_{0.2}$ in the coating AN-35, as defined by TEM images, makes about 30 %. These nanosize intermetallic phases may be called strengthening as the highest microhardness of a coating corresponds to those places where the volume concentration of these phases is highest (figure 5). Along with the microhardness measurement we measured the sizes of corresponding grains.

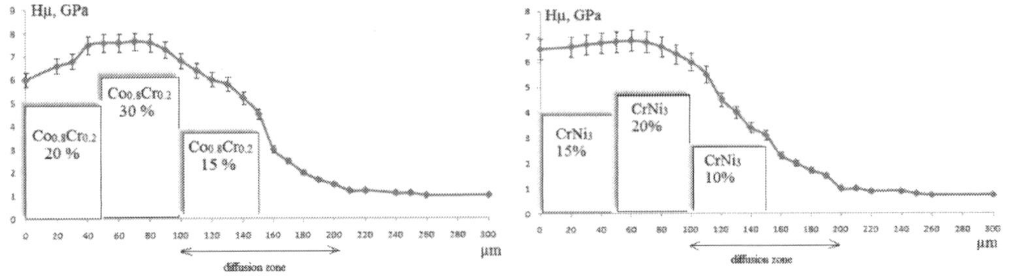

(a) (b)

Figure 5. Coating microhardness variation according to distance from the surface and to the volume concentration of intermetallic phase: AN-35 (a). Coating microhardness variation according to distance from the surface and to the volume concentration of intermetallic phase: PG-19N-01 (b).

It is particularly important that there is a transition (diffusion) zone between the coating and the substrate, where the Ni-based or Co-based γ-solid solution turns into the Fe-based solid solution with the bcc – structure, and compounds of Co or Ni with Fe evolve. The width of the diffusion zone for the samples before irradiation is estimated in average as 100 micron: 50 micron in the coating and 50 micron in the substrate (by processing the data on the distribution of microhardness across the depth from the surface, and according to the X-ray phase analysis data).

The analysis of dependence of microhardness at some distance from the surface shows that the coating microhardness is considerably higher than that of the steel substrate. The steel microhardness on the average is 1.4 GPa. The maximum microhardness of AN-35 is 8.0 GPa, and of PG-19N-01 is 7.0 GPa. It is observed in the coatings at the depth of 60 μm from the surface. The width of the transitional from a coating to a substrate layer with the increased microhardness is estimated as 100 microns, the analysis of its structurally-phase composition allows defining it as a diffusion zone.

Due to a considerable depth of the coating (150 μm) there is no problem of conjunction of high-duty and brittle surface film with the main material that possesses much lower strength and high plasticity. The process of deformation has a consistent structure rate in all the coating and the substrate intermediate layer. We think that the structure-phase state of a coating is defined by the following factors: deformational impact of a plasma jet, temperature profile distribution in the coating material, and inhomogeneous concentration of elements in the coating. The formed structures are stable at room temperature, no reduction of strength properties is observed. We explained the dependence between the linear sizes of grains and microhardness by the Hall-Petch equation (1)

$$H = H_0 + kd^{-\frac{1}{2}} \qquad (1)$$

where H is microhardness, actually the strength property; k is the constant, connected with the transfer of deformation through the grain boundary; H_0 is the strain connected with the energy dissipation in motion dislocation in an infinitely large grain; d is the size of a grain. This interpretation is true because of numerous well-known empirical dependences of strength σ on hardness H; and the similar explanation of dependence was used in the paper by Musil [4] devoted to research of superhard thin films. We have defined constants in this equation for AN-35 coating: H_0=4.6GPa, k=1.4 GPaμm$^{1/2}$ and for coating PG-19N-01: H_0=4.1 GPa, k=0.6 GPaμm$^{1/2}$ for a material with the grain microstructure. If we believe that the maximum microhardness corresponds to grains with intermetallic lamellas (figure 6), the constant k in the Hall-Petch equation will be negative for the material with the nanograin structure. Such an effect was observed by researchers in another paper by Musil [5] devoted to research of nanostructure films.

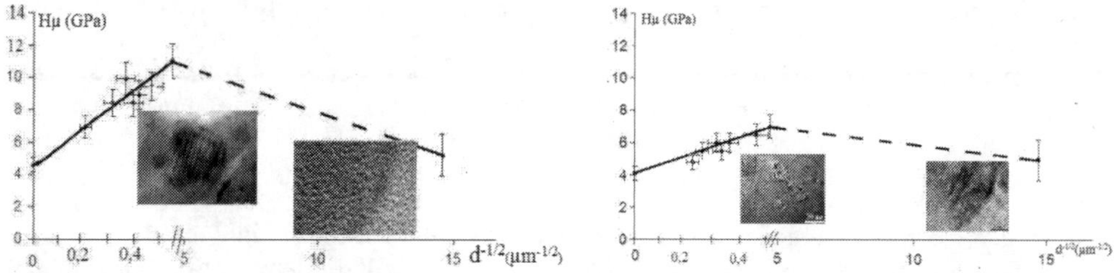

(a) (b)

Figure 6. Variation of the Hall-Petch curve trend: correlation between the grain size $d^{-1/2}$ ($\mu m^{-1/2}$) and microhardness Hμ (GPa) for coating AN-35 (a) and for coating PG-19N-01 (b).

We consider that the nanocrystalline patterns found in these coatings are distinctive for all the coatings deposited by the plasma detonation method, and partially characteristic for the substrate layer next to the coating. Some of the reasons are high micro structure defectiveness conditioned by the plasma jet impact action on the surface and steep temperature gradient in the coating, which may lead to a great deformation in a coating. As a result, in order to remove the strains in a coating there is formed a substructure of nanograins of different crystal lattice orientation with high and continuous disorientation, which is proved by distinctive ring electron-diffraction patterns. When the foils in a goniometer are oriented randomly, there are no characteristics of polycrystals features of diffraction contrast, namely, the changes of its intensiveness on the boundaries necessary for defining grain edges. We assume that we observe the pattern similar to the one of crystal polygonization that is a fragmented pattern with fragments being disoriented nanograins. The validation of this assumption for the coatings lies in some diffusion of the peaks and lowering of their intensiveness in the X-Ray diffractograms. Though in electron diffraction pattern all the reflexes of face-centered cubic lattice are observed, the planes with the axis <111> zone give brighter reflexes (figure 7(a)). It is indicative of fiber texture (figure 7(b)). The axis of the fiber texture is perpendicular to the coating surface along the heat current which develops at heating and the subsequent cooling of the coating during deposition by a plasma jet. Next to the substrate layer the coatings are deformed (figure 7(c)). Basing on the results of research of microstructure, the phase structure, and microhardness of plasma detonation coatings at depth from the surface, we developed the scheme of the coating structure (figure 8). This scheme was used at working out the model of the temperature profile distribution in these coating under irradiation by a plasma jet or an e-beam.

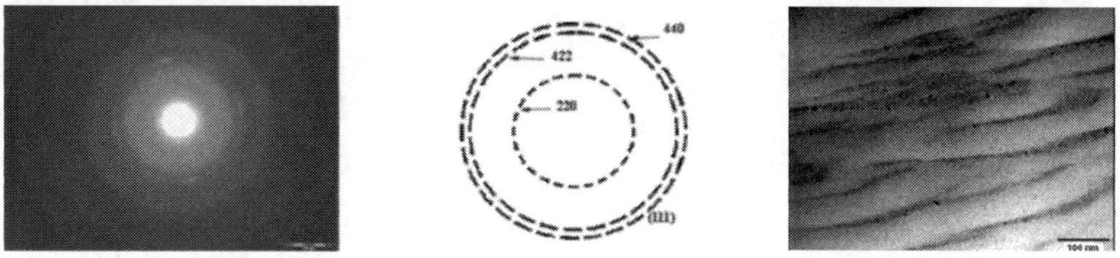

(a) (b) (c)

Figure 7. The electron-diffraction pattern of AN-35 coating with the texture with an axis type <111> zone (a) and (b); the defects in a contact layer of coating PG-19N-01with substrate (c).

The temperature interval of $CrNi_3$-phase in Ni-Cr solid solution (Cr 20-40 at.%) is 500-600°C according to the data of Nash [19]; the existence of the temperature interval of the $Co_{0.8}Cr_{0.2}$ phase is not specified in references. Blake et al. [20] note the existence of an intermediate $CrCo_3$–phase in the Fe-Co system. But the temperature intervals of the advent of phases in a non-equilibrium open system, like an irradiated and complex in structure coating, may not coincide with the data of equilibrium phase diagrams.

The formation of phases on the Cr basis caused some concern because of the known effect of embrittlement of austenite materials at chromiferrous phase precipitation on the boundaries of austenitic grains (σ - a phase with a tetragonal lattice or carbides) [21]. On the other hand, austenite alloys with high Cr content are known for the effect of their hardening by disperse intermetallic γ'-phase (fcc) or η – phase (hcp) due to the mechanism of cellular precipitation [22]. In case of our investigation it is obvious that precipitation of disperse nanosized lamellas of intermetallic compounds raises coating durability; the highest microhardness is observed in those layers of coatings where the volume fraction of intermetallic compounds is maximal (see figure 5).

The scheme of the structure of coatings on Ni - Cr and Co–Cr based powder alloys deposited by plasma-detonation on the steel substrates is presented in figure 8. Figure 8(a) displays the nanograins and intermetallic phase precipitation with their electron diffraction pattern, where: 1- amorphous layer with oxides and carbides; 2-textured layer (solid solution on Co-base with intermetallic phase) and unmelted particles of coating powder; 3-intermediate coating layer (coating-substrate) with deformed and broken particles of coating; 4-intermediate substrate layer with fine-grained microstructure; 5-substrate with large grains. The strokes show precipitations of lamellas of intermetallic phase.

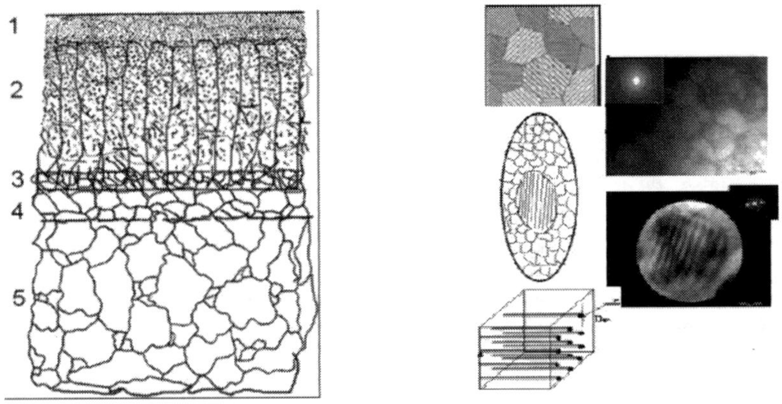

(a) (b)

Figure 8. The scheme of the structure of the Ni - Cr and Co–Cr based powder alloy coatings deposited by plasma-detonation on the steel substrates (a) the scheme of nanograin structure and the scheme of cubic grain with lamellas of an intermetallic phase, and their TEM – images with corresponding electron diffraction patterns (b).

The coatings need additional irradiation for reduction of surface roughness and the increase of their homogeneity. On the basis of the experimental structural and phase analysis of coating microstructure we developed the model of structure for plasma-detonation deposited Co and a Ni- based coating. The applied value of the model was used to calculate the mode of additional irradiation for modifying the coatings.

4. Conclusions

The study of the features of the plasma-deposited powder Ni and Co-based coatings allowed establishing their structural-phase composition and define the phase structures and morphology of precipitation of strengthening nanoparticles from solid solution. Factor k was defined in the Hall-Petch equation for the coating materials. The authors developed the model of structure for the coatings. It was found that the exposure of plasma jet during the coatings deposition leads to the evolution of the structural-phase state and substantial improvement of performance properties of modified surfaces. The authors also identified the mechanism of surface improvement. In general the structure-phase changes are represented by the forming of nanosized intermetallic phase volume concentration and existing of the diffuse zone in the coating layer contacting the substrate.

Acknowledgments
This research is funded by the Scientific Committee of the Ministry of Education and Science of the Republic of Kazakhstan for the project No 29-417-15 "Automated precision deposition of powder coating multifunctional systems".

References
[1] Misaelides P, Hatzidimitou A, Noli F, Pogrebnjak A, Tyurin Yu N and Kosionidis S 2004 Preparation, characterization and corrosion behavior of protection coatings on stainless steel deposited by plasma detonation *Surface and Coatings Technology* **180-181** 290-6
[2] Kadyrzhanov K, Komarov F, Pogrebnyak A, Russakov V and Turkebaev T 2005 *Ion-Beam and Ion-Plasma Modification of Materials* (Moscow: Moscow State University) p 640
[3] Poate J, Foti G and Jacobson D 1983 *Surface Modification and Alloying by Laser, Ion, and Electron Beams* (New York, USA) p 424
[4] Musil J 2000 Hard and superhard nanocomposite coatings *Surface and Coatings Technology* **125** 322-30
[5] Musil J 2005 *Physical and Mechanical Properties of Hard Nanocomposite Films Prepared by Reactive Magnetron Sputtering* (NewYork: Nanostructured Hard Coating) chapter 10 pp 407-63
[6] Alontseva D, Bratushka S, Il'yashenko M, Makhmudov N, Prokhorenkova N, Onanchenko E, Novgorodtsev A and Rogoz V 2012 Formation of stable nanostructured phases in plasma-jet-treated Ni-Cr powder coating *Physics of Metals and Metallography* **113** 785-94
[7] Alontseva D and Russakova A 2013 The structure-phase compositions and properties of plasma-detonation Ni and Co-based powder alloys coatings *Advanced Materials Research* **702** 94-9
[8] Alontseva D, Missevra S and Russakova A 2013 Characteristics of structure and properties of plasma-detonated Ni-Cr and Co-Cr based powder alloys coatings *Journal of Materials Science and Engineering* **3** 41-9
[9] Alontseva D, Pogrebnjak A, Kolesnikova T and Russakova A 2013 Modeling of processes in Co-based coatings exposed to plasma jet irradiation *Medziagotyra* **19** 277-82
[10] Alontseva D, Krasavin A, Pogrebnjak A and Russakova A 2013 Modification of Ni-based plasma detonation coatings by a low-energy DC E-Beam *Acta Physica Polonica A* **123** 867-70
[11] Hassanein A, Kulsinski G and Wolfer W 1981 Vaporization and melting of materials in fusion devices *Journal of Nuclear Materials* **103 -104** 321-6
[12] Burns R and Bradley W 1967 *Protective Coatings for Metals* (New York, USA) p 735
[13] Elsea A, Westerman A and Manning G 1949 The Co-Cr binary system *Trans. AIME Metals Technology* **180** 579
[14] Egerton R F 2010 *Physical Principles of Electron Microscopy. An Introduction to TEM, SEM and AFM* (Moscow) p 304
[15] Krasavin A, Alontseva D and Denisova N 2013 *Certificate of Authorship No. 0010558 of the Republic of Kazakhstan for the Computer Program*
[16] Alontseva D, Krasavin A, Kolesnikova T and Russakova A 2012 Computer simulation of temperature profiles of a two-layer sample during heating by the electron beam *Proceedings of the 7th International Symposium on Applied Informatics and Related Areas (AIS)* 53-56
[17] Alontseva D, Krasavin A, Russakova A and Sadvakasova N 2013 Computer simulation of temperature profiles for DC pulse plasma jet modification of Co-based plasma detonation

coatings *Scientific Journal Proceedings of the International Conference Nanomaterials: Applications and Properties* **2** 11-4
[18] Alontseva D, Krasavin A, Russakova A and Bektasova G 2014 Modeling of irradiation induced processes of heat transfer in bilayer metal absorbers *Materials of 5th International Conference "Radiation Interaction with Materials: Fundamentals and Applications"* 210-213
[19] Nash P 1986 The Cr–Ni (Chromium–Nickel) system *Bulletin of Alloy Phase Diagrams* **27** 466-7
[20] Blake F, Lord J and Focke A 1987 Solid solution of chromium in nickel and of iron in nickel *Physical Revew 2nd Series* **29** 206-7
[21] Naohara T 1983 The formation of s-phase in binary alloys *Scr. Met* **17** 111-4
[22] Sukhovarov V F 1983 *The Cellular Precipitation in Alloys* (Novosibirsk: Nauka) p 164

Dynamic mechanical properties and thermal stability of furfuryl alcohol and nano-SiO_2 treated poplar wood[1]

Youming Dong, Xiaoyan Shen, Shifeng Zhang[2], Jianzhang Li

MOE Key Laboratory of Wooden Material Science and Application; Beijing Key Laboratory of Wood Science and Engineering; MOE Engineering Research Center of Forestry Biomass Materials and Bioenergy, College of Materials Science and Technology, Beijing Forestry University, Beijing, 100083 China

E-mail: zhangshifeng2013@126.com

Abstract. Wood polymer nanocomposites (WPNC) were prepared from the furfuryl alcohol and nano-SiO_2 using a method of vacuum impregnation. Dynamic mechanical properties in storage modulus and mechanical loss factor, as well as the thermal stability of the WPNC were evaluated. The interface interaction between the organic and inorganic compounds was also studied by the scanning electron microscope and energy dispersive X-ray spectrometer. The dynamic mechanical analysis showed the improvement in the storage modulus and mechanical loss factor of WPNC as a result of the strong interfacial interaction between the organic and inorganic matrix. Additionally, with an increase in nanoparticles content in the composites, the thermo-stability of WPNC improved significantly.

1. Introduction

As one of the most relevant raw materials and renewable resources in the world, wood is a continual, durable, beautiful, versatile, and cost-effective material, which has been used extensively for furniture and construction. Interest has recently peaked regarding the development of novel wood materials with comprehensive functions [1]. For example, wood polymer composites (WPCs) fabricated by the impregnation and polymerization of monomer or oligomer have an emerged as an important family of green composites [2]. The preparation of wood polymer nanocomposites (WPNC) via nano-treatment would be a hot point to combine the characteristics of organic polymers with those of inorganic nanoparticles.

As a commercially used WPC, furfurylation endows wood with high dimensional stability, stellar biological resistance, and environmental friendliness via the in-situ polymerization of furfuryl alcohol (FA) monomer [3-5]. FA is a hydrophilic organic liquid derived from furfural, which can be obtained from hydrolysis products of pentose-rich biomass [6]. Therefore, a variety of studies have been conducted to explore FA as an applicable hydrophilic nanoparticle dispersant and an in-situ polymerization precursor for preparing bio-based nanocomposites [6-8]. Generally, researchers

[1] Youming Dong and Xiaoyan Shen contributed to this work equally.

[2] Address for correspondence: Shifeng Zhang, MOE Key Laboratory of Wooden Material Science and Application; Beijing Key Laboratory of Wood Science and Engineering; MOE Engineering Research Center of Forestry Biomass Materials and Bioenergy, College of Materials Science and Technology, Beijing Forestry University, Beijing, 100083 China. E-mail: zhangshifeng2013@126.com.

IOP Conf. Series: Materials Science and Engineering **87** (2015) 012027 doi:10.1088/1757-899X/87/1/012027

postulated that the cross-link reactions of PFA via the electrophilic addition or Diels-Alder cycloaddition. During wood modification, PFA also forms covalent bonds with lignin [9].

Nano-SiO_2 is a form of common inorganic nanoparticles used regularly for reinforcing polymer properties such as high mechanical properties, flame resistance, and thermal stability [10, 11]. WPNC with nano-SiO_2 and urea-formaldehyde resin were utilized to augment wood properties [12]. The effect of nano-SiO_2 on the wood's properties and styrene acrylonitrile/nanoclay composites was evaluated. The results proved the hardness, thermal stability, and UV stability had improved by incorporating nanoparticles [10]. Hence, the combination of FA and nano-SiO_2 had the potential to use the virtues of FA and nano-SiO_2 for preparation of an innovative and eco-friendly high-performance WPNC. This organic-inorganic system would also have a synergistic effect on the wood's properties. WPNC were prepared by the in-situ polymerization of FA utilizing nano-SiO_2 as the particulate phase within poplar wood. The effect of nanoparticles on the physical-mechanical properties, morphologies, and crystallinity of the composites were evaluated in our previous study [13]. More detailed studies are needed for a true understanding of the nano-SiO_2's effect on WPNC. In this work, the dynamic mechanical properties and thermal stability of WPNC are characterized in detail.

2. Experimental

2.1. *Materials*

Furfuryl alcohol (chemical purity 98.0%; Sinopharm Chemical Reagent Co., Ltd., China), nano-SiO_2 (particle diameter: 15 ± 5 nm; Shanghai Maikun Chemical Co., Ltd., China), maleic anhydride, and disodium tetraborate (analytically pure, Beijing Chemical Works, China) were used during these studies.

Obtained from Guangxi province of China, the untreated and treated wooden specimens of the sound sapwood of fast-growing poplar were cut into a size of 3 (Radial) × 8 (Tangential) × 60 (Longitudinal) mm3. The specimens were oven-dried at 105°C to achieve a consistent weight. Afterward, they were sorted into four groups: A, B, C and D. Group A, B, and C were treated with 30 wt% FA solutions containing 0, 1.0 and 2.0 wt% nano-SiO_2, respectively. Group D was untreated, and used as the control group.

2.2. *Preparation of wood polymer nanocomposites*

The nano-SiO_2 swelled in an FA water solution (30%) and sonicated for 20 minutes. Then 2 wt% maleic anhydride and 4 wt% di-sodium tetraborate was added to the solution to complete the impregnation. Next, the specimen were immersed into solutions under a vacuum pressure of 0.095 MPa for 30 minutes, and soaked under the atmospheric pressure for 2 hours. After the impregnation, the specimens were wrapped with aluminum foils, placed at the room temperature for 12 hours, and then oven-cured at 100°C for another 12 hours. The aluminum foil was removed, and the specimens were then oven-dried at 105°C until a constant weight was obtained. After the treatment, the wood percentage gain (WPG) was calculated using to the following equation:

$$WPG\ (\%) = (W_1 - W_0)/\ W_0 \times 100 \qquad (1)$$

In the equation, W_0 and W_1 are the oven-dried weights of a specimen before and after treatment, respectively.

2.3. *Measurements*

The morphological features of WPNC and the distribution of nanoparticles were examined using the Hitachi S-4800 scanning electron microscope, which was equipped with an energy dispersive X-ray spectrometer (SEM-EDX, Japan). The tangential sections of treated and untreated wood samples were mounted on conductive adhesives, sputter-coated with gold, then observed at an accelerating voltage of 15 kV.

The Dynamic Mechanical Analysis (DMA) of the untreated and treated specimens was conducted using the NETZSCH DMA 242 (Germany) in the dual cantilever-bending mode. The specimens were scanned over a range of 40 to 225°C, at an increased rate of 5°C/minute, and a frequency of 2 Hz. Five replicates were conducted.

The thermal stability of the untreated versus treated specimens were evaluated by TGA by using a Q50 TGA analyzer (TA Instruments, USA) at a constant heating rate of 10°C/minute from a room temperature to 600°C under the flowing nitrogen atmosphere.

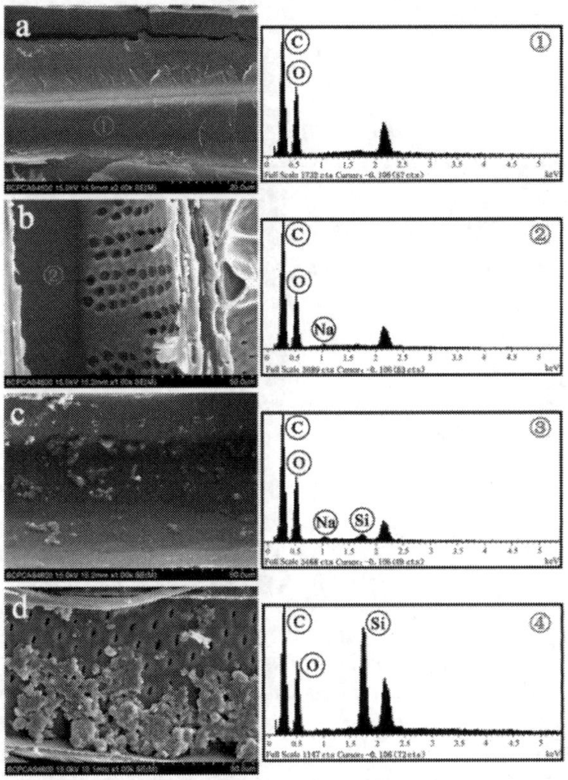

Figure 1. FE-SEM observations of (a) untreated wood, (b) FA-treated wood, (c) FA -treated wood with 1.0% SiO_2, and (d) FA -treated wood with 2.0% SiO_2 and their relative EDX spectrum of the elements.

3. Results and discussion

3.1. SEM-EDX analysis

The prepared WPNC was dark brown due to the conjugated structure derived from the aliphatically linked poly-FA. The average WPG was 78.11%. The morphological features of WPNC examined by SEM-EDX are shown in figure 1. It was noticed that, in the untreated wood, the avulsed lamellar structure of the cell wall can clearly be discerned (figure 1a). However, these details depicting the cell wall's division was absent in the FA-treated wood (figure 1b) due to the polymerization of FA on the vessel cell walls. In the FA/SiO_2-treated wood (figure 1c), the cluster aggregation of nano-SiO_2 was invisible. Additionally, most of the nanoparticles were clearly covered by the FA polymers, exhibiting a highly desirable combination of organic and inorganic compounds that could have an effect on the properties of the WPNC. When SiO_2 was increased, the cluster aggregation would become larger (figure 1d). In the EDX spectrum, no silicon was present in untreated poplar wood and FA-treated wood, while the distribution of silicon was obvious in the FA/SiO_2-treated wood. In the third part in figure 1c, the EDX spectrum suggested that the Si element evenly distributed onto the wood's cell walls. In addition to the silicon, a sodium signal in the FA-treated wood and FA/SiO_2-treated wood was present, which had resulted from the addition of disodium tetraborate in the furfurylation process.

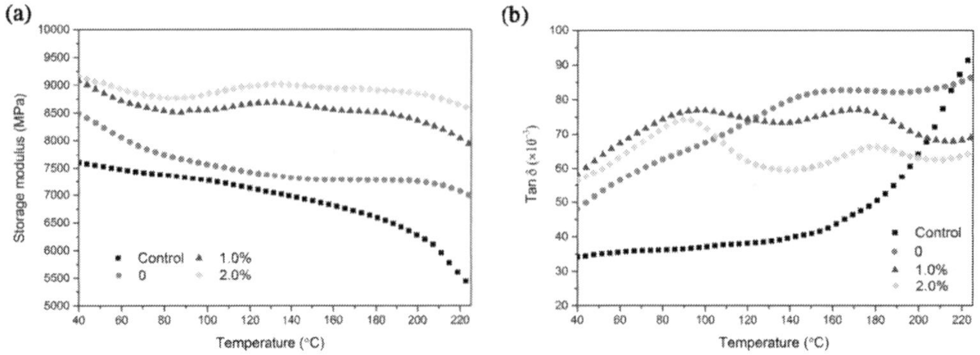

Figure 2. Dynamic mechanical analysis of untreated wood (Control) and treated wood with FA and 0, 1.0%, and 2.0% nano-SiO_2. (a): storage modulus (E'); (b): mechanical loss factor (tan δ).

3.2. *Dynamic mechanical analysis*

After the specimen were treated with FA and 0, 1.0%, and 2.0% nano-SiO_2, the dynamic mechanical properties such as the storage modulus (E') and mechanical loss factor (tan δ) as a function of temperature are shown in figure 2. Figure 2(a) depicts E' of the specimens. With the increase in temperature, E' of control group was found to decrease due to the heightened chain mobility of the wood cell wall components [14]. The steep decrease appeared when E' neared the glass transition temperatures of wood (~150°C). It was noticed that the higher E' was represented by the specimens treated with FA. The first steep decrease of E' value appeared at about 60 ~ 80°C, due to the glass transition temperature of PFA is 80°C [6]. The onset of the E' value increased according to the fluctuation in SiO_2 content. When the content of SiO_2 was 1.0%, the E' values improved (9256 MPa) considerably compared to the control (7602 MPa) and FA-treated wood (8658 MPa). This improvement in E' values for FA/SiO_2-treated wood samples may be related to the better interface interaction among PFA, nano-SiO_2, and wood components, which allowed better stress transfer at the interphase between organic and inorganic compounds [14]. Additionally, WPNC became more rigid when SiO_2 was added due to the rigidity property of the nano-SiO_2 [15]. Overall, a similar changing trend of both 1.0% and 2.0% SiO_2-treated wood samples was observed.

The ratio of the loss modulus in relation to the storage modulus is measured as the mechanical loss factor (tan δ). Figure 2(b) depicts the variation of tan δ as a function of temperature for both the control and treated wood fragments. When compared to the control groups, the tan δ variation of FA-treated wood samples changed significantly. Due to the higher WPG, PFA played the most vital role in the wood polymer composites. It is interesting to realize that the FA- and FA/SiO_2-treated wood samples displayed two distinct peaks, however they cannot be detected for the control samples in the tan δ curves. The peaks of tan δ for FA-treated wood samples were about 80°C and 150°C, while the peaks for FA/SiO_2-treated wood samples appeared at about 90°C and 175°C. It was indicated that the peaks shifted to the higher temperatures. Due to the slip of PFA, the molecular chains decreased with the nano-SiO_2 content. As a result of the barrier between nanoparticles, this phenomenon also implied the strong interphase interaction between the nanoparticles and PFA [14, 15].

3.3. *Thermal stability analysis*

The thermal degradations of the controlled and treated wood samples were examined by the thermogravimetry (TGA). Figure 3 shows the TGA scans of each sample. The TGA data confirmed a lowered decomposition temperature and an increased char formation for the treated wood samples, particularly for the treated wood sample with 2.0% SiO_2. The temperatures at the 20%, 40%, 60%, and

80% weight losses occurred, and the residual weights at 600°C as representative of thermo-stability are listed in table 1. It was apparent that the thermal stability of the FA- and FA/SiO_2-treated samples were improved when compared against the control samples. Specifically, the thermal stability of the nano-SiO_2 incorporated samples increased with the rise of the nanoparticle content. When the content of nanoparticles was 2.0%, the residual weight at 600°C was subsequently increased by 18.45% when compared with that of the FA-treated wood sample. This result indicates that the thermo-stability of the composites was significantly improved by the incorporation of nanoparticles. The main reason for the spike in thermal stability is the fact that the inorganic matrix could promote the formation of the self-insulating char on the wood surfaces and the heat transfer is inhibited [16]. Lastly, this result also suggested that a strong interface interaction between organic and inorganic matrix existed.

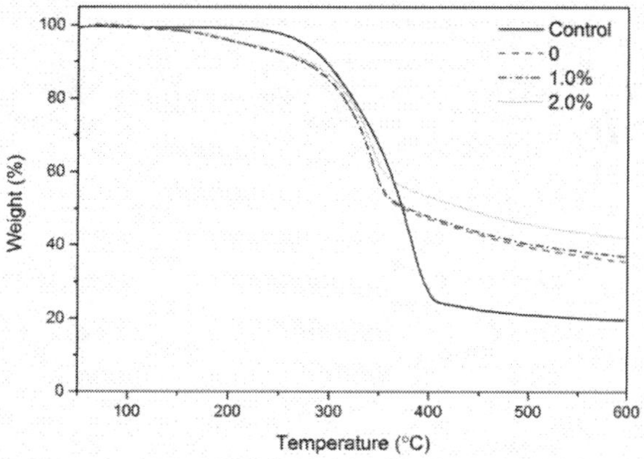

Figure 3. Thermogravimetric curves of control and treated wood samples with FA and 0, 1.0, and 2.0% SiO_2.

Table 1. Thermal analysis of the WPNCs.

Sample	Temperature of decomposition in °C at different weight loss (%)				RW% at 600 °C[a]
	20%	40%	60%	80%	
Control	321.78	361.50	384.20	564.28	19.60
FA	314.39	346.50	489.39	--	35.51
FA+1.0%SiO_2	314.45	346.95	509.47	--	36.91
FA+2.0%SiO_2	319.85	354.87	--	--	42.06

[a]RW refers to the residual weight.

4. Conclusion

The dynamic mechanical properties and thermal stability of WPNC treated with FA/SiO_2 were decidedly improved due to the secure interface interaction between PFA and nanoparticles. The storage modulus and softening temperature of WPNC rose with the increase of the SiO_2 content. When the content of SiO_2 was 1.0%, the onset of the storage modulus could be improved by 21.8% compared to that of the control sample. Additionally, the inorganic nanoparticles in WPNC can potentially improve its thermo-stability by fostering the self-insulating char formation in wood, and inhibiting the heat transfer. The combination of organic and inorganic materials can improve the thermal properties of wood and expand the wood's application.

References

[1] Qin C and Zhang W 2012 *Mater. Lett.* **89** 101
[2] Deka B K and Maji T K 2012 *Polym. Bull.* **70** 403
[3] De Vetter L, Pilgård A, Treu A, Westin M and Van Acker J 2009 *Wood Mater. Sci. Eng.* **4** 30

[4] Lande S, Westin M and Schneider M 2008 *Mol. Cryst. Liq. Cryst.* **484** 367
[5] Venås T M and Rinnan Å 2008 *Chemometrics Intellig. Lab. Syst.* **92** 125
[6] Guigo N, Mija A, Zavaglia R, Vincent L and Sbirrazzuoli N 2009 *Polym. Degradation Stab.* **94** 908
[7] Pranger L and Tannenbaum R 2008 *Macromolecules* **41** 8682
[8] Ahmad E E M, Luyt A S and Djoković V 2012 *Polym. Bull.* **70** 1265
[9] Nordstierna L, Lande S, Westin M, Karlsson O and Furo I 2008 *Holzforschung* **62** 709
[10] Devi R R and Maji T K 2013 *J. Taiwan Inst. Chem. Eng.* **44** 505
[11] Li J Z, Furuno T, Zhou W R and Yu Z M 2005 *J. Wood Chem. Technol.* **25** 257
[12] Shi J, Li J, Zhou W and Zhang D 2007 *Frontiers of Forestry in China* **2** 104
[13] Dong Y, Yan Y, Zhang S and Li J 2014 *BioResources* **9** 6028
[14] Islam M S, Hamdan S, Talib Z A, Ahmed A S and Rahman M R 2012 *Composites Sci. Technol.* **72** 1995
[15] Peng Z, Kong L X and Li S D 2005 *Synth. Met.* **152** 25
[16] Merk V, Chanana M, Keplinger T, Gaan S and Burgert I 2015 *Green Chem.* **17** 1423

Synthesis and characterization of nitrogen-phosphorus-based fire retardants modified by boride/propanetriol flyeidyl ether complex

Haijiao Kang, Linrong Ma, Shifeng Zhang[1] and Jianzhang Li[2]

MOE Key Laboratory of Wooden Material Science and Application; Beijing Key Laboratory of Wood Science and Engineering; MOE Engineering Research Center of Forestry Biomass Materials and Bioenergy, College of Materials Science and Technology, Beijing Forestry University, Beijing 100083, China

E-mail: zhangshifeng2013@126.com; lijianzhang126@126.com

Abstract. A Boride/propanetriol glyeidyl ether (B/PTGE) complex was employed to intensify the fire resistance capabilities of nitrogen-phosphorus (NP) fire retardants by reacting with phosphoric acid and urea to yield nitrogen-phosphorus-boron-PTGE fire retardants. The effects of NPB-PTGE fire retardants on wooden properties were characterized by limit oxygen index (LOI), cone calorimetry, X-ray Diffraction (XRD) and scanning electron microscopy (SEM). The results depict that the fire resistance of the B/PTGE complex modified by NP-based fire retardants was improved significantly. The PTGE was at 10% boride at 2%, and the treated wood has the LOI of 52%, which is 11.46% higher compared with woods treated with NP fire retardant.

1. Introduction

Wood products have been regularly used for centuries in many industries, particularly the building and construction sector. The advantages wood's worldwide availability, exemplary mechanical properties, and economical effectiveness make the flammability of wooden materials an increasingly worrisome topic [1]. The fire resistance of wood is limited. As no wooden material is completely fireproof, fire retardant treatment becomes a key precautionary tactic to prevent the combustion of structures [2]. To avoid fires, wood and wood-based materials can be treated using many methods, the most promising of which is fire retardant treatment.Nitrogen-phosphorus (NP) fire retardant has received considerable attention because it protects structures from fire with a minimum amount of associated health hazards, and is also convenient when enhancing the fire performance of wood-based materials. Under the heat

[1] Address for correspondence: Shifeng Zhang, MOE Key Laboratory of Wooden Material Science and Application; Beijing Key Laboratory of Wood Science and Engineering; MOE Engineering Research Center of Forestry Biomass Materials and Bioenergy, College of Materials Science and Technology, Beijing Forestry University, Beijing 100083, China. E-mail: zhangshifeng2013@126.com.
Jianzhang Li, MOE Key Laboratory of Wooden Material Science and Application; Beijing Key Laboratory of Wood Science and Engineering; MOE Engineering Research Center of Forestry Biomass Materials and Bioenergy, College of Materials Science and Technology, Beijing Forestry University, Beijing 100083, China. E-mail: lijianzhang126@126.com.
Haijiao Kang and Linrong Ma contributed to this work equally.

of a young fire, the retardancy is achieved by forming a charred layer (serving as a physical barrier) that deters the heat and mass transfer of flame between gas and the condensed phases [3, 4]. A study focused on the fire propagation of particle boards treated with three commercial fire retardants showed that diammonium phosphate was preferred for reducing heat release, followed by monoammonium phosphate and BP (mixture of 27-33% boric acid, 67-73% guanylurea phosphate and 0.0-4.2% phosphoric acid) [5]. Due to various fire retardant mechanisms, utilization of a single fire retardant is unable to achieve ideal results, and thus the growing industry trend is to modify the basic concept of a fire retardant. In a comprehensive thermal study on the main components in fire-retardant formulas (such as pentaerythritol, melamine, and ammonium phosphate), the transformation mechanism of these substances during thermal degradation was described [6]. Moreover, melamine, melamine-formaldehyde resin, urea–formaldehyde resin (UF), and melamine-modified UF (MUF) mixed with NP fire retardant or the modified fire retardants significantly improved the fire retardancy, such as increasing oxygen index, prolonging ignition time, reducing heat release rate (HRR) and total heat release (THR), and decreasing mass loss rate [7-10].

Boron has the potential to decompose B_2O_3 and fashion a glassy protective layer, which promotes carbon formation and holds back volatile combustibles [11]. A modern form of phosphorus-nitrogen-boron (PNB) wood fire retardant consisting of the optimum formula of phosphoric acid, boric acid, and dicyandiamide was applied when treating wood [12]. High-boron, high-solubility wood preservatives such as SBB and $(NH_4)_2HPO_4$ were used to examine the overall fire retardant performance and test the swelling property, strength, and fire resistance of the treated wood [13,14]. The study's results showed that the synergetic effect of $(NH_4)_2HPO_4$ and SBB improved the fire retardant's performance [13, 14]. Wood fire retardant FRW, primarily including high purity amidinourea phosphate (GUP), boric acid, and a small amount of additives was developed, and the fire retardant mechanism of FRW was proposed [15, 16]. The results show that boric acid and GUP were both effective at varying temperatures by different fire retardant mechanisms. The two complement each other as synergistic additives [15, 16].

Although such systems have been studied extensively, many problems exist. These include the complex synthesis of fire retardants, the high costs, the use of poisonous melamine and dicyandiamide, and the overall effect of some formulas is not ideal. For improving the fire retardant efficiency and reducing the cost in this study, a boride/propanetriol glyeidyl ether (B/PTGE) complex was employed to produce NPB- PTGE fire retardants, which can be compared to the boride or PTGE alone modified NP-based fire retardant. The fire resistant performance of the NPB fire retardant was evaluated by an LOI test and cone calorimetry, while the morphology and crystal structure of fire retardant- treated wood samples were investigated using XRD and SEM.

2. Experimental

2.1. Materials

Urea (Henan, ZhongYuan DaHua Company), 85% phosphate (Beijing Chemical Factory), Borax (Beijing Chemical Factory), and Propanetriol Glyeidyl Ether (PTGE) (Beijing Chemical Factory) were used during the study. Additionally, wooden poplar specimens ($150\times6\times3$ mm^3, $100\times100\times10$ mm^3) were used for the LOI and CONE experiments.

2.2. Synthesis of modified fire-retardants

Into a four-neck round-bottom flask:

- add pure phosphoric acid heat to 45°C under frequently stirring;
- add urea and heat to 90°C under stirring;
- add a given amount of catalysts and borax or PTGE or B/PTGE complex (seeing details in table 1);

after, heat to 135°C, hold for about 10 minutes, then pour the mixture out and obtain the fire retardant.

Table 1. Nitrogen and phosphorus fire retardant compound experiment with B/PTGE.

	NP/ (%)	Boride/ (%)	PTGE/ (%)
A	15	-	10
B	15	2	-
C_1	15	2	5
C_2	15	2	10
C_3	15	2	15
C_4	15	2	20

2.3. Fire-retarding treatment of wood samples
The fire-retardants were dissolved in water, with the mass concentration of 15 % for preparing the fire retardant woods. The wood specimens (150×6×3mm^3 and 100×100×10 mm^3) were immersed in the four fire-retardants/aqueous solutions for 8 hours, then oven-dried at 105±2°C. The control group was set simultaneously. The samples were then stored in desiccators for tests.

2.4. LOI text
LOI is the minimum oxygen percentage required for maintaining a subject's flaming combustion under specific laboratory conditions. LOI was measured by an LFY-606 LOI instrument (Shandong textile academy of sciences), and tests were conducted according to the method described in the National Standard of People's Republic of China (GB/T2406-93).

2.5. CONE test
The CONE tests (ISO 1993) were conducted using an FTT cone calorimeter (East Grinstead, UK). The specimens were then encased in aluminum foil and horizontally exposed to a 35 kW/m^2 external heat flux to 750°C. The specimens were then ignited with a spark igniter. The parameters of combustion were later calculated based on basic data derived from the CONE test on Microsoft Excel.

2.6. XRD analysis [7]
XRD examination of untreated vs. treated wood samples was performed on an X-ray 6000 (Shimadzu, Japan) machine. All of the samples were initially grounded to pass an 80- mesh sieve, and air-dried prior to analysis. The X-ray beam was Cu-Ka (k = 0.1540 nm) radiation, operated at 40 kV and 30 mA. The scanning rate was 2.0°/minute, and 2θ ranged from 5° to 50° with the rotation speed of 30 r/minute.

2.7. SEM analysis
Observations of the morphologies of untreated and treated wood specimen after the CONE test was performed by a Philips S-3400 SEM analyzer (Hitachi, Japan) using an acceleration voltage of 10 kV.

3. Results and Discussion

3.1. LOI of fire retardant treated wood
The LOI results are depicted in figure 1. Clearly, the LOIs increase significantly after being treated with fire retardants. As shown in our previous study [17], when the levels of Boride came to reach 2%, the fire retardant property is enhanced. The oxygen index of the fire retardant increases first, then decreases as the amount of PTGE increases, which is respectively 49%, 52%, 45% and 38%, and when compared with NP fire retardant processing material, the oxygen indexes are increased by 32.43%, 40.54%, 40.54% and 32.43%. When PTGE is at 10% and Boride lies at 2%, the oxygen index reaches its peak at 52%. This occurs primarily because the interactions between the boride and the fire

retardant. Additionally, the epoxy groups in PTGE can form chemical bonds with hydroxyl groups in the wood, ensuring the fixation.

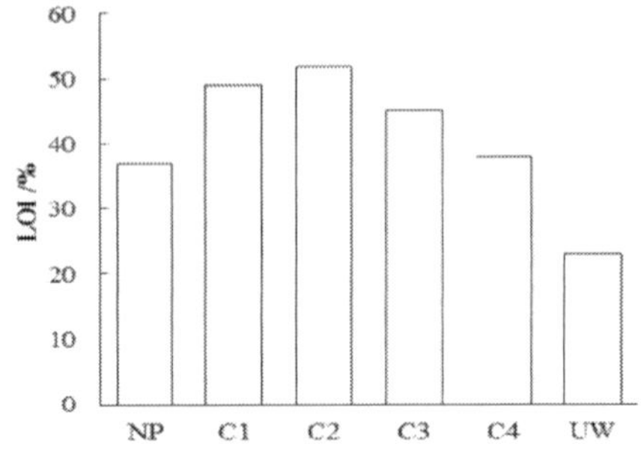

Figure 1. Oxygen index of wood treated with a composite fire retardant.

3.2. CONE test of fire retardant treated wood

Figure 2 presents the curves of the heat release rate (HRR) and the mass loss rate (MLS) of the untreated and treated woods. The typical HRR curve of wood initially increases to a peak, then drops to a steady state. This is followed by a second peak as the final portion of the specimen is consumed [18]. The precise numbers are expressed in table 2. All analyzed samples displayed two separate HRR peaks during combustion, which indicated that gradual burning occurred after the formation of initial charred layers [19].

Table 2. The main fire retardant parameters of different fire retardant for wood.

	LOI/%	HRR/KW·m^{-2}		THR/MJ·m^{-2}	TTI/s	FPI/s·m^2·KW^{-1}
		pk_1-HRR	pk_2-HRR			TTI/pk_1-HRR
NP	37.3	57.14(35s)	119.49(340s)	26.73(45.42%	15	0.263
A_2	46	58.83(25s)	93.90(340s)	22.82(53.40%	15	0.255
B_2	49	67.75(75s)	86.46(310s)	23.81(51.39%	10	0.147
C_2	52	51.12(30s)	92.38(335s)	21.30(56.5%	12	0.235
UW	25	138.67(55s)	170.75(275s)	48.97	21	0.151

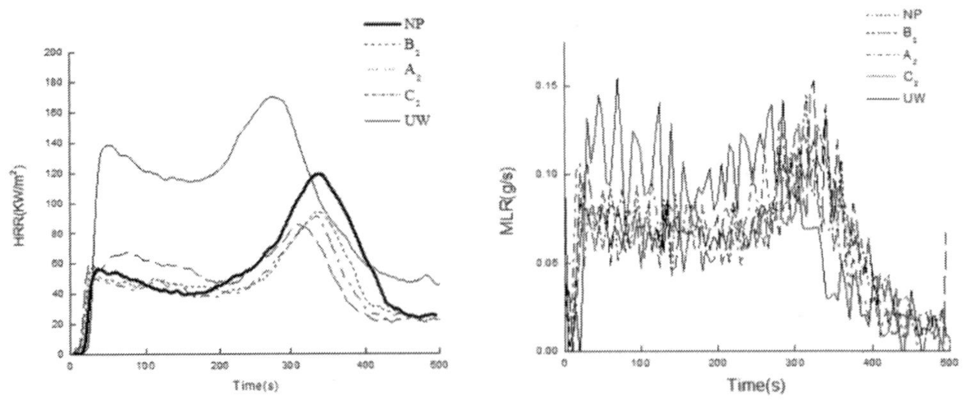

Figure 2. Heat release rates (left) and mass loss rate of treated and untreated wood (right).

The HRR of fire retardant treated wood (and the heat release rate peak) are lower than that of the untreated wood, indicating that fire retardants effectively inhibit the heat release from wood combustion. When compared to NP fire retardants, the PTGE, Boride and B/PTGE modified fire retardant complex perform more effectively, according to the heat release rate peaks at 21.4%, 27.6% and 22.7%.The first drastic HRR peak seen in the treated samples has been ahead in the A_2 and C_2 curve, which reveals that modified fire retardants can catalyze wood pyrolysis to a degree. The second sharp HRR peak of B_2 curve appears later than others, which is more effective to the fire retardant properties [20].

According to the mass loss rate (MLR) curve of different fire retardant samples, the treated and untreated wood play a big different, the MLR value of the untreated wood is initially higher than the treated samples before 320 s, while, after 320s the MLR value of the treated wood is higher than the untreated one's both under the continually decrease course. Which is nearly consistent with the HRR curve, therefore, the higher LOI value of the treated wood is due to their lower MLR and HRR value compared to the untreated ones especially at the first 300 seconds during their burning process.

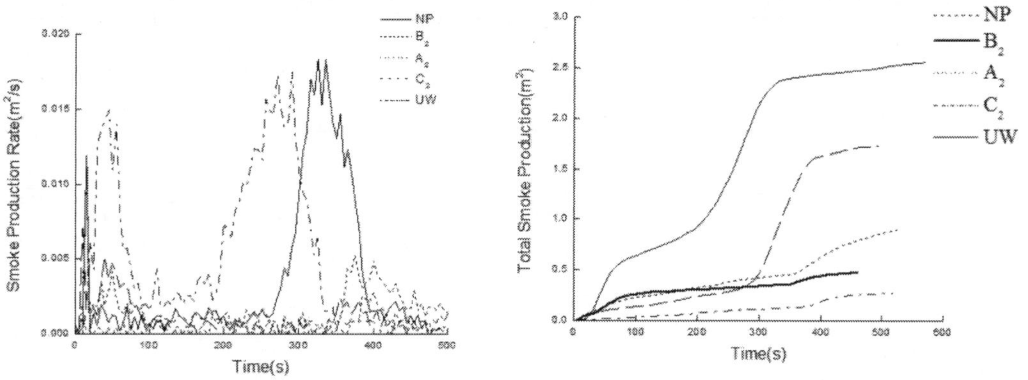

Figure 3. Smoke production rate (left) and total smoke production (right) of treated and untreated wood samples.

Smoke generated rate (SPR) is a dynamic value of the smoke generation, and the area beneath the curve represents the total amount of smoke production (TSP). As depicted in figure 3, the untreated wood appears in two peaks in the SPR curve, and similarly with the HRR. This declares the heat releasing process is accompanied by heavily smoke release. However, the peak of the NP fire retardant treated wood only has one higher smoke peak at 0.0178 m^2/s with a time delay compared to the untreated ones. The reason may be due to NP fire retardant's ability to reduce combustion at the beginning of a burn to control a lower smoke production rate. And then with time extending, the NP fire retardant inhibits the charcoal burning to incompletely combust, emitting a lot of smoke.

The smoke production rate curve of the modified fire retardants vary from the former ones. The Boride and PTGE modified fire retardants respectively at 10 seconds and 14 seconds get the SPR peak, which suggests the wood released heavily smoke at the time of initial lighting. The first broad peak appears at approximately 30 seconds, while the second peak of the modified NP-based fire retardant (Boride, PTGE, and B/PTGE complex modified NP-based fire retardant numbered A_2, B_2, and C_2) show different tendency: the A_2 curve at the 370 second mark come out the second mild peak about 0.005 m^2/s; the second peak in B_2 curve is reduced to disappear; the SPR of C_2 keeps in a low state around 0.0015 m^2/s.

We can also see that the TSP curves differ from the treated and untreated wood samples. While the untreated wood sample has the largest amount of 2.5 m^2 with a rapid increasing rate, the NP fire retardant, A_2, B_2, and C_2 are shockingly lower than the untreated one, being 1.72 m^2, 0.87 m^2, 0.49 m^2 and 0.27 m^2. Additionally the TSP of NP fire retardant is higher than A_2, B_2, and C_2, meaning that the

NP fire retardant can only manage the smoke release, not change its burning mechanism. And the results of C_2 are fulfilled, showing the synergy effect of boride and PTGE in the smoke release control.

3.3. XRD analysis of fire retardant treated woods

LOI is the lowest oxygen percentage appropriate for maintaining a specimen's flaming combustion under specified laboratory conditions. In the study, LOI was measured by an LFY-606 LOI instrument (Shandong Textile Academy of Sciences) according to the technique communicated in the National Standard of People's Republic of China (GB/T2406-93). Figure 4 shows the XRD patterns of the untreated and treated wood. The characteristic peaks for the untreated woods appear at 2θ=16°, 22° and 35°, which corresponds to the crystal plane diffraction peaks of (101), (002) and (040), respectively. On the diffractograms of the treated woods, peak (101) were enhanced and three additional peaks around 2θ=23.5°, 33.6° and 45° were registered in the XRD curves, and are related to the crystalline structure of fire retardants.

The crystallization of the wood was enhanced to 38.55% (NP), 40.18% (A_2), 39.96% (B_2) and 41.14% (C_2). PTGE can improve the wood's cellulose crystallinity, because it can react with the hydroxyl groups on cellulose in the amorphous region while rearranging the microfibril, and drag them to the crystalline region to increase the cellulose crystallinity. Therefore, the B/PTGE complex modified NP-based fire retardant is the optimal for wood modification [21].

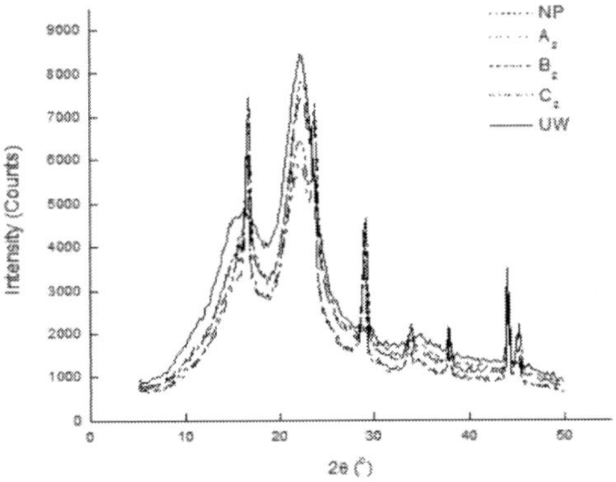

Figure 4. XRD of untreated wood and treated wood.

3.4. The morphologies of the fire retardant treated woods residuals after combustion

The morphology of the char residue that was formed after the CONE test was observed through SEM. As shown in figure 5, the char residue generated from the untreated wood reveals a gray sprawled structure (seen in figures 5a), while those generated from the treated wood show a dense form (figures 5b-5e). Additionally, the fiber structures of the Boride, PTGE or B/PTGE complex modified by the NP fire retardant treated wood samples retain relatively complete, and a dense carbon layer is formed at the surface. The justification may be that the NP fire retardant has the capability to thermally decompose acid material, which can promote the PTGE decomposing into a viscous polymer containing massive N/P elements. Therefore, the carbon layer structure and its properties may be regulated to a denser one, which can create a non-flammable coating, beneficial to the improvement of fire retardant capabilities on the surface of the burnt wood.

4. Conclusions

The fire retardant properties of wood were significantly enhanced after being treated with the B/PTGE complex modified NP-based fire retardant. When the PTGE levels were at 10% and Boride at 2%, the treated wood maintained the LOI of 52%, which is 11.46% higher compared with woods treated with NP fire retardant. The HRR and MLR curves show the wood treated with fire retardants had undergone obvious improvements.

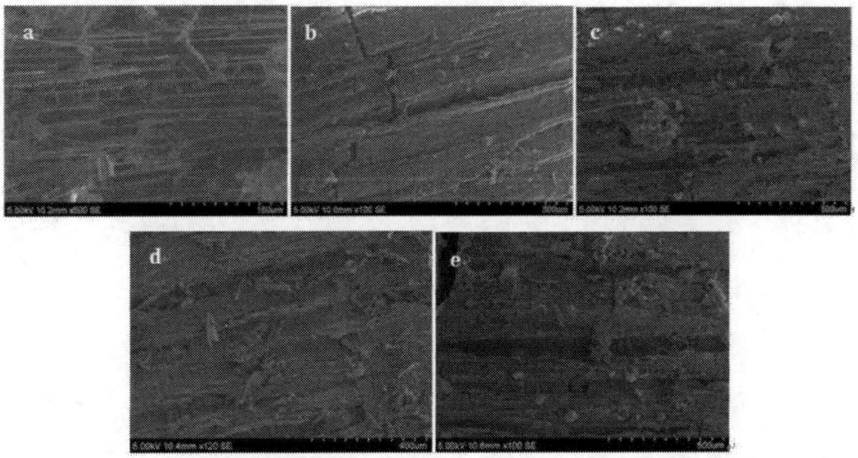

Figure 5. SEM images of untreated SAMPLE (a), NP treated sample (b), NP-PTGE treated sample (c), NPB treated sample (d), and NPB-PTGE treated sample (e).

XRD analysis indicated that the main crystalline structure of wood was not affected by the fire retardant treatment. The wood treated with the B/PTGE complex modified NP-based fire retardant has higher crystallization, which is optimal for wood modification.

According to SEM analysis, a dense structure was observed in treated woods, which is beneficial to improving fire retardant properties of wood.

An optimization of the NP-based fire retardant modified by the B/PTGE complex is achieved in this study, as to the formation of a protective cover on the wood surface, which validly inhibits the wood burning process.

References

[1] Seefeldt H and Braun U 2012 *Macromolecular Materials and Engineering* **297** 814-20
[2] Chuang C S, Tsai K C, Wang M K, Ou C C, Ko C H and Shiau I L 2008 *Wood Sci Technol* **42** 593-607
[3] Pedieu R, Koubaa A, Riedl B, Wang X M and Deng J 2012 *Eur. J. Wood Prod.* **70** 191-7
[4] Li Q, Zhong H F, Wei P and Jiang P K 2005 *J. Appl. Polym. Sci.* **98** 2487
[5] Izran K, Zaidon A, Rashid A M A, Abood F, Saad M J, Thirmizir M Z, Masseat K and Rahim S 2009 *Asian Journal of Applied Sciences* **2** 446-55
[6] Zybina O A, Varlamov A V, Chernova N S and Mnatsanov S S 2009 *Russian Journal of Applied Chemistry* **82** 1542-6
[7] Ma X X, Wu Y Z and Zhu H L 2013 *J Wood Sci.* **59** 419-25
[8] Kawarasaki M, Kikuchi S, Hiradate R and Hirabayashi Y 2004 *J Hokkai For Prod. Res. Inst.* **18** 16-22
[9] Jiang J X, Li J Z, Fan D B and Song F 2009 *Journal of Beijing Forestry University* **31** 39-42
[10] Chen L, Su H, Zhang Y and Xie G 2009 *Guangdong Forestry Science and Technology* **25** 14-7
[11] Yao C, Qing Y and Wu Y 2010 *China Forestry Science and Technology* **24** 91-3
[12] Qu H, Wu W, Wu H, Xie J and Xu J 2011 *J Therm Anal Calorim* **103** 935-42
[13] Liao H, Liu L, Xie G, Yun H and Guo Q 2008 *Guangdong Forestry Science and Technology* **24** 22-5
[14] Liao H, Liu L, Xie G, Yun H and Deng T 2008 *Journal of South China Agricultural University* **29** 120-4
[15] Wang Q and Li J 2005 *Scientia Silvae Sinicae* **41** 123-6
[16] Wang Q, Li J and Winandy J E 2004 *Wood Sci Technol* **38** 375-89

[17] Ma L, Zhang S, Gao Q, Li J J, Li J Z and Li L 2014 *China Wood Industry* **28** 30-3
[18] Ayrilmis N, Candan Z and White R 2007 *Holz Roh Werkst* **65** 449-58
[19] Sun Q F, Lu Y, Zhang H M, Yang D J, Xu J S, Li J, Liu Y X and Shi J T 2010 *Materials Research Innovations* **16** 326-31
[20] Fang G and Liu W 2000 *Journal of Northeast Forestry University* **28** 28-30
[21] Lang Q, Bi Z, Zheng X, Miao X and Pu J 2013 *Transactions of the Chinese Society of Agricultural Engineering* **29** 267-75

IOP Conf. Series: Materials Science and Engineering **87** (2015) 012029 doi:10.1088/1757-899X/87/1/012029

Multi-model modeling and its application of urban sewage treatment based on clustering analysis

Y U Ping[1]

College of Electrical and Information Engineering, Lanzhou University of Technology, 730050 Lanzhou, Gansu, China

E-mail: lutyup@163.com

Abstract. For urban sewage treatment processes, a single model suffers from heavy burden calculation and poor accuracy. A modeling method founded on an advanced supervised k-means clustering algorithm is proposed. The method introduced in this paper is the cluster center initialization idea of CCIA algorithm into a classical k-means clustering algorithm is applied to corral the data into clusters or second clustering by judging a pre-set threshold value, and the least squares method is applied to construct the ARX sub-models, and the system model is then constructed by weighing all ARX sub-models. The proposed method is used to determine the ammonia concentration model for wastewater treatment system benchmark and the actual WWTP practice. Simulation and application research results show that the proposed method can be effectively used to fit the nonlinear characteristics of the system with high precision.

1. Introduction

Biological sewage treatment process system has a complex mechanism because of its non-linear, uncertainties characters and other factors. It is very difficult to establish an individual model for these nonlinear mechanism systems. Researchers often use the input and output data from the system to model identification 1. For nonlinear systems with a vast scope of operating conditions, its creation single global model is often difficult to meet the necessary modeling accuracy. However, if we can decompose the nonlinear system into a plurality of partial scope of work conditions, then we would improve the modeling accuracy. This decomposition principles based approach is called multi-model modeling method. It has received widespread attention in nonlinear system modeling and control [1, 2]. A clustered approach is an advanced technique of data classification. Additionally, the use of clustering data classified multi-modeling has also become a subject worthy of study [3-7].

2. Methods

2.1. K-means clustering

K-means clustering is one of the simplest unsupervised data clustering algorithms with fast convergence, and it is suitable for large-scale data set classification [8]. *K*-means clustering uses k as

[1] Address for correspondence: Y U Ping, College of Electrical and Information Engineering, Lanzhou University of Technology, 730050 Lanzhou, Gansu, China. E-mail: lutyup@163.com.

the vital parameter to divide the data set into "k" clusters, so that the data points within the higher similarity between clusters of data points in the cluster.

The algorithm randomly selects the initialization cluster center, c_k, and the distance of each of the data points with k cluster centers is then calculated by an equation (1). Next, it is determined that every data point belongs to the certain class by equation (2).

$$dist(d_i, c_m) = \sum_{j=1}^{p} (d_{ij} - c_{mj})^2, j \in [1, p] \tag{1}$$

$$dist(d_i, c_j) = \min_{j \in [1,k]} \{dist(d_i, c_1), dist(d_i, c_2), \cdots, dist(d_i, c_k)\} \tag{2}$$

In the equation, j represents the data dimension.

When all remaining data points are assigned to the nearest cluster, the algorithm will re-calculate the average of each cluster in order to identify the cluster center using the equation (3).

$$c_k = \frac{\sum d_i}{|S_k|}, d_i \in S_k \tag{3}$$

Here, $|S_k|$ represents the number of data points belonging to the k-th cluster.

The above process is repeated until the cluster center is no longer shows any change at the end of the clustering operation.

2.2. *Supervised k-means clustering multi-model modeling improvement*

The k-means clustering algorithm is a simple and quick convergence, however it contains many deficiencies. For example, the clustering result relies on the selection of k's value; therefore the proper value of k is difficult to select when there is an absence of a clear understanding surrounding the data characteristics. The initial value of the cluster centers is randomly selected, which would make the clustering optimize process ended with local optimal condition or empty cluster. The clustering quality is sensitive to the isolated data points; unsupervised mechanisms data classification only considers the difference between the input data, without taking factors such as output into account. However, in multi-model modeling processes, the final modeling errors cannot be reflected in the classification process, so it will prone to large, long-term corruption.

In this paper, we focus on the initial point selection of the k-means clustering algorithm, and discuss how the supervision mechanism will be added into the classical algorithm to improve its performance in multi-model modeling applications.

We can see from the above analysis that k-means clustering is primarily the method to obtain the initial point (which is randomly selected), and that the k-means algorithm is an iterative algorithm for different initial values that may result in different clustering groups even if there is no complete solution, only when the initial value is close to the final classification results may better clustering results be obtained. Of course, we can also use multiple computing methods, and select objects as the initial points with a big difference (as far as possible) to improve the algorithm; however the efficiency of this would not be very high.

Khan and Ahmad studied in the initial value selection of the k-means clustering algorithm, and proposed the CCIA (cluster center initialization algorithm) algorithm [9]. The CCIA algorithm consists of two parts, one for the initialization of the cluster centers, and the second for density-based multi-scale data condensation algorithm (DBMSDC) [10]. The CCIA algorithm is an advanced equation based on the fact that the each variable property found in the data set will affect the spatial distribution of the sample data. It assumes that each dimension of the variable properties (in line with the normal distribution and data) can be divided into k clusters. This means each dimension corresponding to the normal distribution curve is divided into k parts with the equal area. Then, by selecting equal diversion points as interval points, we can ensure the differences of each cluster as large as possible. However, because the algorithm is based on the DBMSDC algorithm, the parameters required to be set by the user are very sensitive to the self-sufficiency (such as the setting of the number of core point). The algorithm has been improved, and is expressed in this paper. We use the main idea of the CCIA algorithm to initialize the cluster center, and then re-cluster the results in accordance with the classical k-means clustering algorithm. This is done with a class threshold value for judgment by performing the cluster again, and examining the termination conditions for the clustering process. If the spatial distance between clusters is larger than the threshold value, then the end of the clustering has been reached, however if the distance does not exceed the threshold value, there is a secondary cluster.

The basic steps of the above-mentioned improved clustering algorithm can be summarized as follows:

Step 1: Determine the number of categories for k, and create a set of sample data D ($n \times m$ -dimensional);

Step 2: For the sample data of D, initialize the cluster center, c_j, by selecting the equal diversion points as the interval points, and get pattern-strings (s_j) of the each sample data, where j represents the data attributes, $j \in [1, m]$.

Step 3: Repeat step 2 to obtain pattern-strings (s with $n \times m$ -dimensional) for D, then place each of the sample data with the same s into one cluster, and calculate the number of the current cluster (q), $k \leq q \leq k^m$.

Step 4: If $q > k$, it symbolizes that the current classification results have a re-cluster possibility, therefore proceed to the next step. Otherwise, output the current clustering result as the final clustering results, and turn to Step 8.

Step 5: Individually calculate the data in each cluster for its cluster center value and the spatial distance between the clusters. Then, analyzing the relationship between the spatial distance and the threshold value (we selected the half of the largest distance between two clusters spatial distance between the clusters as the threshold value). If the value is greater than the threshold, the algorithm will automatically move to the next step. Otherwise, it will retain the current classification, and set $k=q$. If this occurs, turn to Step 8.

Step 6: Re-clustering the two classes with the least amount of difference is calculated from equations (1) and (2), and the merged cluster center is calculated by equation (4), which is shown below:

$$c_p^{'} = \frac{n_i \times c_{ip} + n_j \times c_{jp}}{N}, p = 1, 2, \cdots, m \tag{4}$$

In this equation, c_i, and c_j represent the two class centers of the clusters about to merge. Additionally, n_i, and n_j represent the number of samples contained by the corresponding class; subsequently, N represents the number of all samples in these two classes.

Step 7: $q = q - 1$, return to Step 4.

Step 8: Use the k-means clustering algorithm to classify all data samples with the corresponding number of the cluster center get from above as the initial value of the k-means clustering, and output the final classification results.

Additionally, data classification based on unsupervised classification purposes only considers the differences between the input data without considering the output factors; it does not reflect the final modeling error found in the classification process, and will inevitably produce significant modeling errors.

Here, an improved supervised multi-model modeling method is proposed. The purpose is for clustering data points in each category. If the error is too significant for the corresponding parameters of the model, we put this point to the other cluster to make the model error smaller; then, re-identify the model's parameter after the operation.

Ultimately, the process of the new clustering-based supervision multi-model modeling algorithm is as follows:

Step 1: Add the designed appropriate excitation signal to the multi-input and multi-output systems to incentive it sufficiently, then withdraw the input and output data for identification during simulation;

Step 2: Preprocessing the input and output data, for example, by mapping the data to the same scale space by normalization.

Step 3: Select the ARX (auto-regressive exogenous) model, then order and divide the data into two types of modeling data and test data.

Step 4: Use the improved *k*-means clustering algorithm mentioned above to group the modeling data into *k* classes.

Step 5: Use the least squares method for the ARX model fitting to obtain the initial parameters of the models for each type of data.

Step 6: Calculate the error of data points for each model, the data points are classified into corresponding categories according to the principle of maximum error. When all data points move neither algorithm, proceed to Step 8.

Step 7: Re-fit the new clustered data to obtain the parameters of the sub-model, then return to Step 6.

Step 8: If all data points move neither, use the current model set as the best multi-model fitting set.

Step 9: Use the test data for the model fitting test, if the test error is in the allowed range, consider this group of models as the best expression of a multi-model, and otherwise return to Step 4.

3. Results and discussion

3.1. Simulation results

The wastewater treatment Benchmark BSM1 (Benchmark Simulation Model No. 1) is developed by the COST (the European Co-operation in the Field of Scientific and Technical Research) group 682, 624, and IWA (International Water Association) [11, 12], and is based on ASM1 (Activated Sludge Model No. 1). ASM1 is focused on carbon and nitrogen removal, and it also includes the system process description, the simulation model, the simulation steps and the evaluation standards. The advent of Benchmark makes up the standard measurement, and the same simulation environment for the performance of different control strategies. It is conducive for the sewage treatment experts and researchers to select the optimal control scheme.

Using the Benchmark BSM1 model built-in literature [12] to simulate and produce large volumes of data, and model the concentration of ammonia nitrogen using this paper's proposed method. Select 900 sets of data for modeling, and 250 sets of data for verification testing. In this situation, the output variable is the ammonia nitrogen concentration (S_{NH}), and the input variables are the external carbon source flow (Q_C). The set point of the dissolved oxygen concentration ($S_{O,set}$); while the measurable disturbance is the ammonia nitrogen concentration in the source water ($S_{NH,IN}$). The variables are written in y, u_1, u_2 and d.

The second order model structure described as follows:

$$y(k) = a_1 y(k-1) + a_2 y(k-2) + b_{11}u_1(k-1) + b_{12}u_1(k-2) + b_{21}u_2(k-1) + b_{22}u_2(k-2) + c_1 d(k-1) + c_2 d(k-2) \tag{5}$$

The modeling results are then shown below in figure 1.

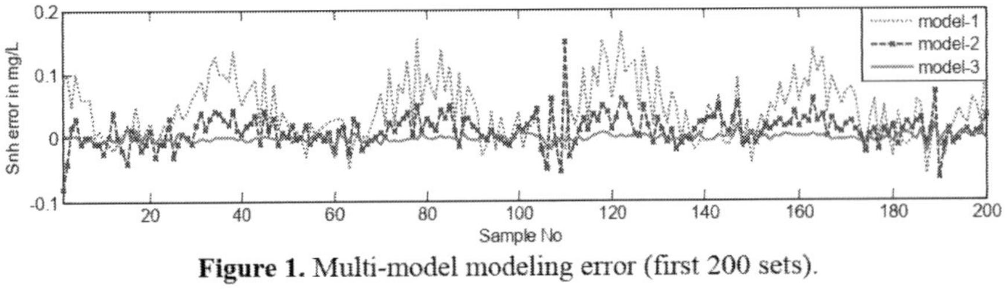

Figure 1. Multi-model modeling error (first 200 sets).

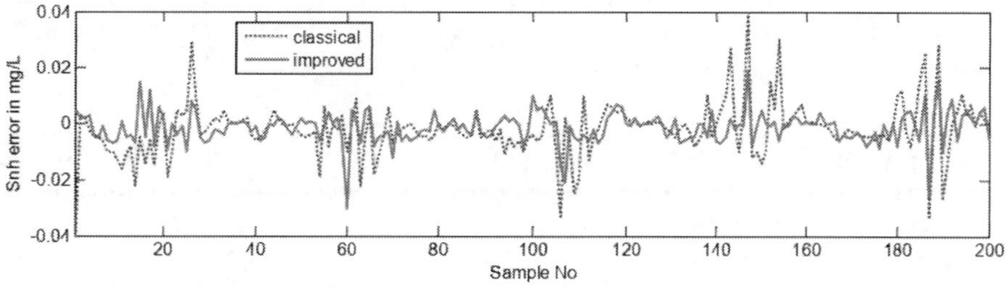

Figure 2. Compared results of minimum error of classical *k*-means algorithm and method in this paper.

Additional validation of the effectiveness of this strategy, is the comparison of the basic *k*-means clustering method (k=3) with the improved strategy in this section for clustering analysis. The compared results are displayed above in figure 2. Obviously, the smallest error of the improved strategy is less than the classical *k*-means clustering algorithm.

By using the verified data to test the above two method's multi-model modeling errors, the parity error is shown in figure 3.

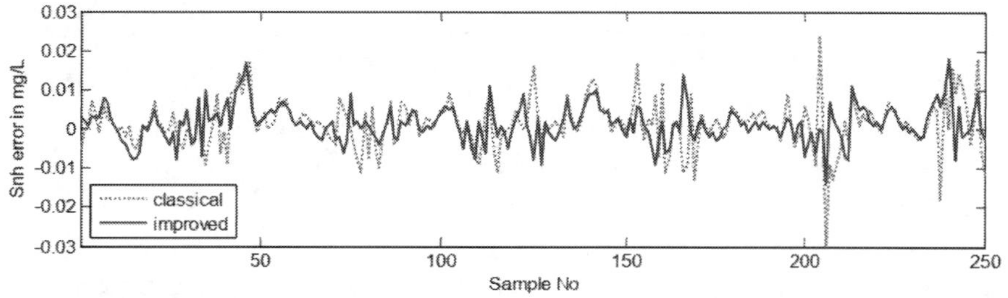

Figure 3. Compared results of verify error of classical *k*-means algorithm and method in this paper.

Also, we can use the standard deviation formula (found in equation (6) and the maximum absolute error formula (found in equation (7) to calculate the standard deviation and the maximum absolute error in the modeling and verification process. These results are shown below in table 1. In the equations, N represents the number of data points.

$$\sigma = \sqrt{\frac{\sum_{k=1}^{N}(y(k)-\hat{y}(k))^2}{N}} \tag{6}$$

$$MAXE = \max_{k\in[1,N]}\left|y(k)-\hat{y}(k)\right| \tag{7}$$

Table 1. Compared results of these two methods.

error	**classical *k*-means algorithm**	**method in this paper**
standard deviation of the modeling error	0.010927	0.005027
standard deviation of the verify error	0.0067942	0.0050152
maximum absolute error of modeling process	0.04151	0.02705
maximum absolute error of verify process	0.03024	0.02432

The data seen in table 1 further illustrates that the proposed method in modeling and verify error is smaller than the classical *k*-means method. The proposed method contains a higher level of precision, and offers better fitting performance of the system of non-linear characteristics.

3.2. Application results

The Sewage Treatment Plant of Pingliang City in Gansu Province was built in 2006. The designed processing capacity is 50,000 m^3 per day. It is designed for nitrogen and phosphorus removal by the Carrousel oxidation ditch process of live sewage, and a minute amount of industrial waste water, which benefits a population of about 220,000 people. In order to secure a better effect from the biological nitrogen and phosphorus removal process, modification to the original design is begun at the start of 2011. The initial oxidation ditch will be transformed to aerobic section with a bottom aeration of A_2/O process. It is put into operation at the end of 2012.

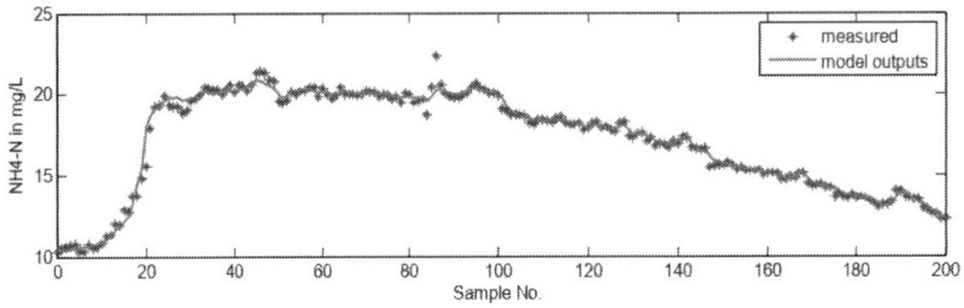

Figure 4. Ammonia concentration model outputs and measured values.

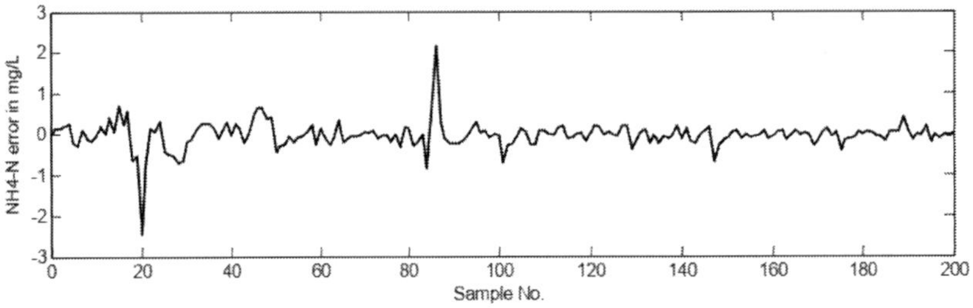

Figure 5. Verify error of ammonia concentration.

Off-line multi-model modeling and verification of the plant processes is conducted with 1,200 sets actual data obtained from the sewage treatment plant in January 2011, by means of the proposed method in this paper. 1,000 sets data for modeling, and 200 sets for verifying was used to conduct the study. Figure 4 shows the ammonia nitrogen concentration outputs of the model and the actual measured values, as well as the error between the model output and actual measurement values, which are shown in figure 5.

4. Conclusions

In this paper, an improved clustering algorithm is proposed and applied in the multi-model modeling process of the urban sewage activated sludge process. This study is conducted under Benchmark simulation and actual plant application. The initial value selection method and supervised mechanism are improved sufficiently throughout the study. Additionally, we offer the specific implementation steps for the algorithm. This method can be used to determine the initial model parameters of the model adaptive control. However, the multi-model obtained by this method is the high precision of its best matching sub-model output error. Therefore, the method of selecting the best matched sub-model in the control process (multi-model switching or weighted) is need to continue research in the future.

Acknowledgments
This study is supported by the Excellent Young Teachers Program of Lanzhou University of Technology (No. Q201408), and the Natural Science Foundation of Gansu Province (No. 1212RJYA031).

References

[1] Li Q L, Lei H M, Shao L, et al. 2010 *Control and Decision* **12** 1866-7
[2] Huang Y R and Zhang S D 2012 *Automation & Instrumentation* **2** 10-11
[3] Cong Q M, Zhao L J and Chai T Y 2010 *J. Northeastern Univ. (Natural Science)* **9** 1221-2
[4] Xu H X, Liu G H, Zhou D W, et al. 2009 *Chinese Journal of Scientific Instrument* **10** 2226-7
[5] Li W, Yang Y P and Wang N 2008 *Control and Decision* **5** 560
[6] Zhang Y, Liu G H, Wei H F, et al. 2012 *Control and Decision* **7** 1117
[7] Frey B J and Dueck D 2007 *Science* **5814** 972-5
[8] Likas A, Vlassis N and Verbeek J J 2003 *Pattern Recognition* **2** 451
[9] Khan S S and Ahmad A 2004 *Pattern Recognition Lecture* **11** 1293
[10] Mitra P, Murthy C A and Pal S K 2002 *IEEE Transaction on Pattern Analysis and Machine Intelligence* **6** 734-6
[11] Alex J, Benedetti L, Copp J, et al. 2008 *Benchmark Simulation Model No. 1 (BSM1)* (London: IWA Publishing House)
[12] Du X J, Hao X H, Li H J, et al. 2011 *Asian Journal of Chemistry* **10** 4457-9

Design and analysis of the glass fiber composite truss beam structure

Lina Chen[1,2] **and Zhi Li**[3,4]

[1]School of Science, Wuhan University of Technology, Hubei, China
[2]Hubei ChuAn Construction Co. Ltd, Hubei, China
[3]China Shipbuilding Industry Corporation No. 701 Institute, Hubei, China

E-mail: 64346585@qq.com

Abstract. To exert the corrosion-resistant characteristics of the composite materials, which have overcome the characteristics of low modulus of the truss structure design.We designed a new kind of beam-slab structure and it can be applied to the desulfurization and denitration tower.The kingpost is undertook by four successive FRP profiles and nodes connected by hough; both ends are connected by a rod with the thread structure.This research analyzed the structure of the finite element, and the result comes out that the stress, the strain and distortion of the whole kingpost, are meet regulatory requirements as well.

1. Introduction

With the development of economy and science and technology, people are becoming more concerned about environmental issues. National legislation and other means have used to control the enterprise production damage to the environment by our country, it requires industrial waste gas, produced by enterprises, such as steel, cement and chemical, must achieve purification standard through processing before released into the atmosphere[1-2]. The treatment of desulfurization and denitration method, is always being used to treat the exhaust gas, and completed in the desulfurization, denitrification tower[3-4]. The desulfurization equipment and packing accordingly are placed in the desulfurization tower, and to support these loads relies mainly on the reinforced concrete beam and slab structure or composite beam slab structure with their characteristics respectively, the concrete structure has good performance in temperature resistant but limited tower life for susceptibility to corrosion, and the composite material which corrosion resistance is well with low modulus of rigidity and large deformation[5-6]. This paper aims to make use of composite materials in a girder grillage structure, overcome composite defects and improve the structure of the overall life.

2. Overall FEM calculations of the trussed beam

Due to the truss beam node is not the real form of articulate, those member bars, which belong to the girder construction, cannot simply regard as a two-force member in the girder grillage structure. So we chose the BEAM188 element for considering because it can integrallty reflect the flexural torque and shearing action in the member bar and more close to the actual structure. The girder which is similar to

[4] Address for correspondence: China Shipbuilding Industry Corporation No. 701 Institute, Hubei, China. Email: 64346585@qq.com

the girder grillage structure, combined with member bar of fiber glass reinforced plastics, its length of span is 15 m and the arrangement shows in figure 1.

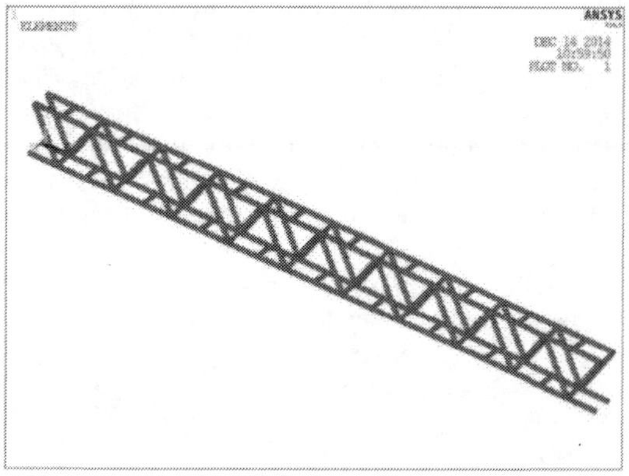

Figure 1. FEM calculation of trussed beam.

As it shows in the picture, the length of each arm is 15 m, and the length of the short bars, which connected between two-element girder grillage is 0.5 m. Beam with both ends built-in the wall of the tower, with the form of freely supported beam to consolidate entirety. The total load that acts on the girder is 166335 kN, and the upper limit load of the girder is 11089 N/m can obtain.

The member bar is designed with a ring-type area of section. It can ensure the member bar has a greater flexural rigidity and makes the structure safer while the area of section is confirmed. The member bar of the trussed beam is made of pultruded material; there are four long bars that are located up and down are connecting link whose length are 15 m. The oblique stand bar and the crosswise is connecting link connected to the main pole in the form of threaded connection. The type of construction of the connected node as shown in figure 2. The node contains two pieces of the upper and lowers which wrapped into one by unidirectional fiber when adjusted. To pour some clay-like substance into the model, which is made of chopped fibers mixed with resin, after it freezing the thread can be obtained, and so as the thread on the member bar. Type of thread on the ends of member bar is shown in figure 3. The thread in the form of trapezoidal thread, the pitch is 3 mm. To determine the parameter of each size of the thread, according to the standard of "GB 57961-2005 trapezoidal thread Part 1 dentoform", the length of the part of the thread is 15 cm.

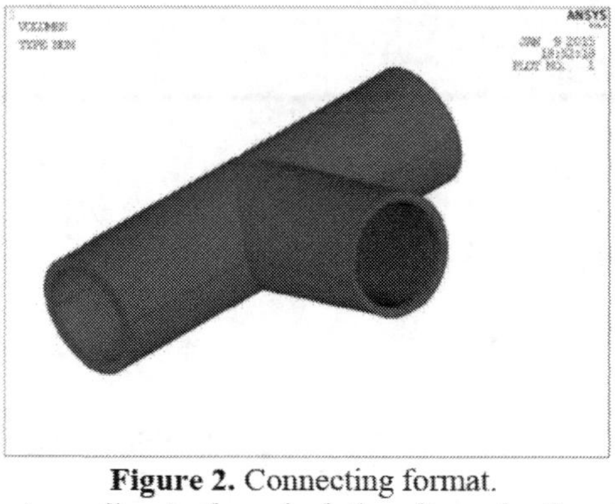

Figure 2. Connecting format.

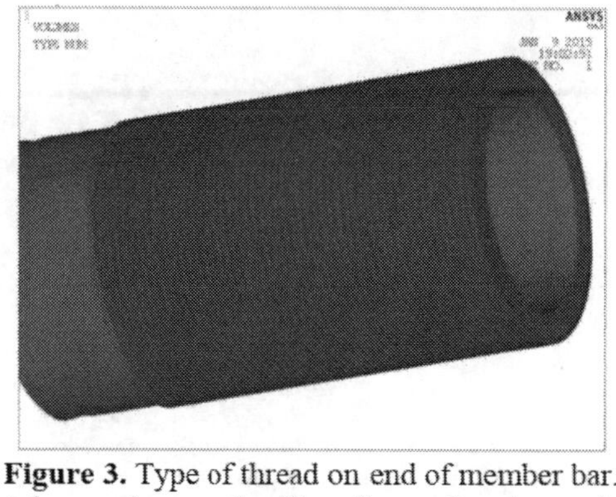

Figure 3. Type of thread on end of member bar.

According to the calculation, it can be found that the maximum of pulling force of each member bar in the girder did not have much change with changing area of section while the load is certained. The value stabilized at around 120 kN. To facilitate the optimal design, we assumed the maximum value of pulling force, which the member bar can bear, is approximate 120 kN. The axial force is mainly kind of stress that acts on the member bar of the girder grillage structure. Though the connected type of the member bar designed in this paper is not a girder grillage strictly, the force and momental conduct that between those member bars are similar to the girder grillage. So the stress that produced by flexural torque is much lower, compared with an axial force that is considered as the main factors during the designation of the area of the section of member bar. For pultruded fiberglass material, assume the allowable stress of 45 MPa, then determine the area of the section of member bar is about 2667 mm^2. Considering the entity deflection of the girder is the minimize, while the area of the section is confirmed, and the inner diameter and the external diameter of the member bar is 80 mm and 100 mm, then we can calculate the boom displacement with the influence of working load and dead load. The displacement shows in figure 4.

As it shows in the picture that the maximum displacement is 4.3 cm, scratch span ratio, which is satisfied with the standard. Then the designed sectional dimension is deemed to be reasonable.

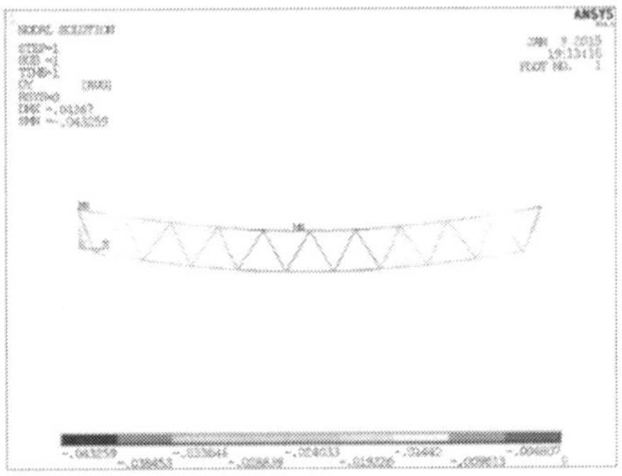

Figure 4. Vertical displacement nephogram of truss beam.

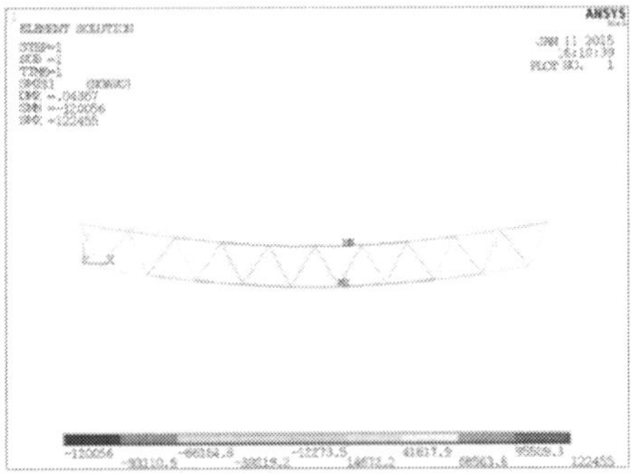

Figure 5. Axial force diagram.

3. The design of the node connection in the beam

As it shows in the figures 4 and 5, the deformation and the axial force of the girder comes out the symmetric distribution along the middle of the beam, and the maximum of the axial force and flexural torque of the girder bear can be obtained while the node locates in the middle of the girder bear. Consider to calculate the stress and strain distribution of the node with two forms that the middle.

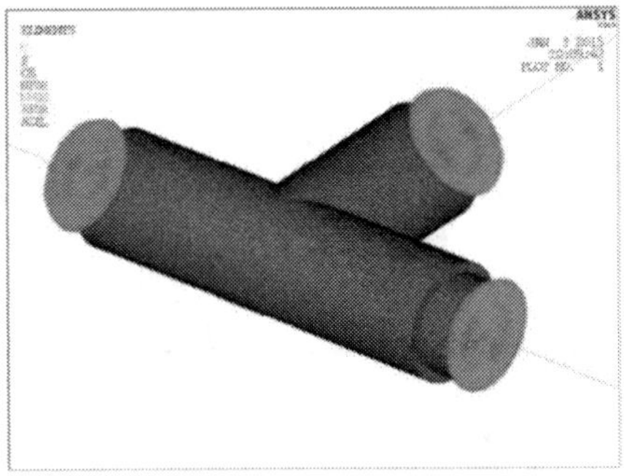

Figure 6. Form of node and the girder element coupling.

The approach of connecting node has described as previously, using the SOLID185 element to do the calculation. The connecting link is TARGE170 and CONTA174 between the connect node and member bar, and so as the connection of thread. The contact element produced both on the outer of the member bar and inner of the gusset plate. Does not consider about pretightenin force while connecting in the form of a thread, and then chosen part of the node to calculate, and connected with another part by the method of a degree of freedom coupling. The nodal point of entity element, which locates in the connected position, was chosen to couple with the node of girder element. Then the cross section, that entity element girder element connected, is formed with the district of stiffness. It will affect the calculation results of member bar within a certain scope, but, it will not cause important effect to the entity node, while the imitative range this nearby the district of those two elements connecting. In the figure 6, here shows the connected form of entity element and girder element.

3.1. Analysis of stress along length direction of member bar

Figure 7 shows nephogram of the stress that along the length direction of the beam bar. From this picture, it can be found that there are considerable stress focuses in the location where the two elements connected on the member bar and the scope is about 20mm, whereas the stress of other positions is normal. During the process of analysis the stress of member bar, while not chosen the end part element, then the stress nephogram can be obtained, as shown in figures 8, 9 and 10.

From the pictures above, we can find that the maximum of stress in the direction of radial, circumferential and axial of member bar is 3.8 MPa, 3.19 MPa, and 50.3 MPa, respectively, from which we can assure that the stress condition is satisfied with the strength of the designed material.

3.2. Analysis of stress of member bar along the vertical direction of member bar

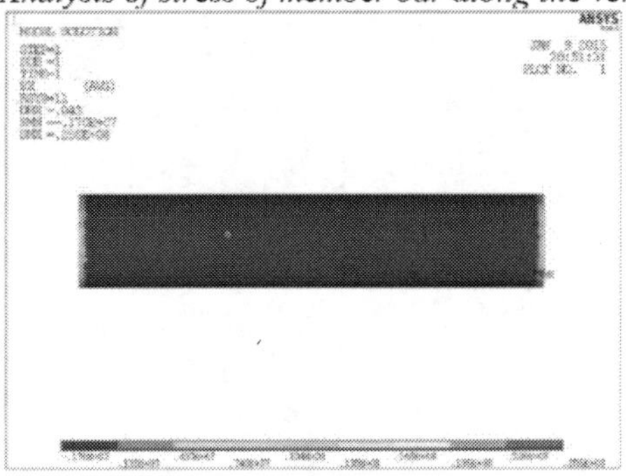

Figure 7. Stress nephogram in the radial direction of member bar.

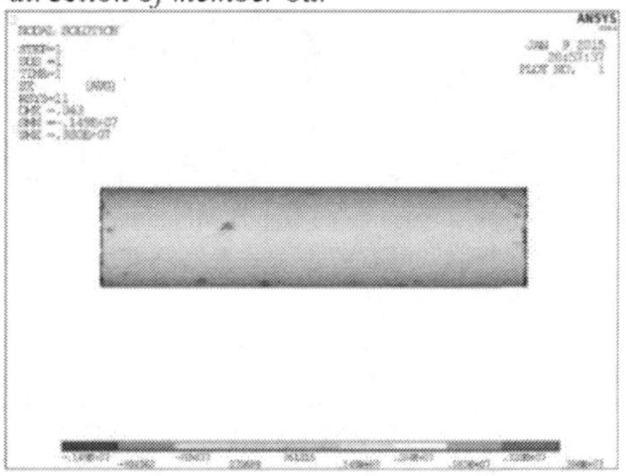

Figure 8. Stress nephogram in the radial direction of member bar.

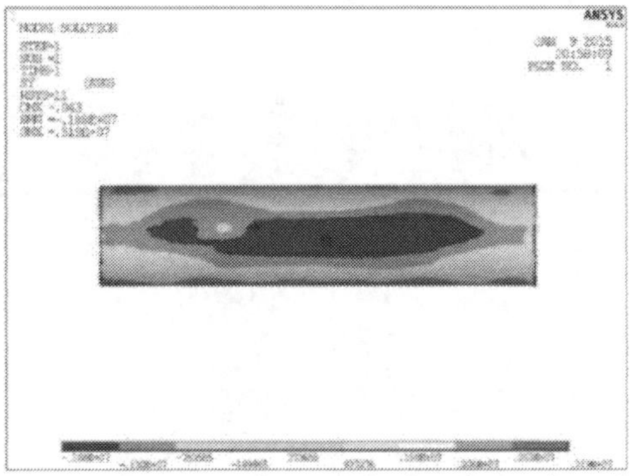

Figure 9. Stress nephogram in the circumferential direction of member bar.

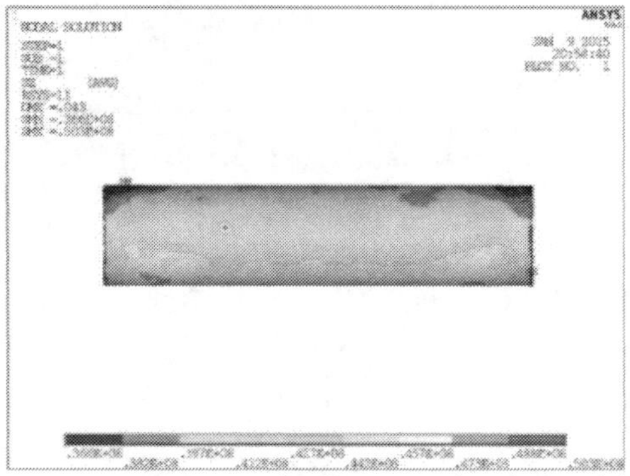

Figure 10. Stress nephogram in the axial direction of the member bar.

Figures 11-13 show part of stress nephogram of the thread of the crosswise connected member bar. And at the end of thread, it can finds out that there is a large stress, for squeeze that generates end with the casing wall which caused by the affect of flexural torque entity, and also comes out that in the position that nearby the upper, the direct stress decreased or even negative stress. However, the large stress that appears in the smooth segment of member bar, may due to the big value of the elastic ratio in the district where member bar and loop bar connected. Under the condition of the same deformation, large strain turned out at the root department of the member bar with short thread and will due to it broken down. It can find out that the maximum stress in three directions happened at the end of thread. So we must be avoided in the process of designation. We also can get that the maximum of stress is 4.46 MPa. In spite of the stress at the end position may oversize, the force in the crosswise connected member bar is tiny and the stress is low correspondingly, which is satisfied with the design strength of the material. The thread of member bar with oblique poling, requires capsulation and it does not list in this paper, and must be paid remarkable attention during designation.

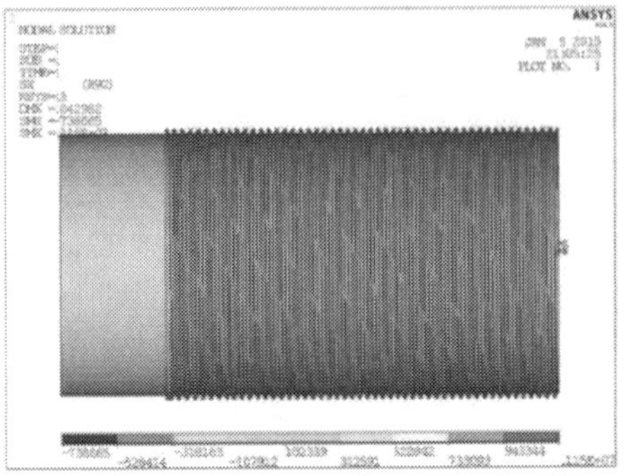

Figure 11. Stress nephogram in the radial direction of crosswise member bar.

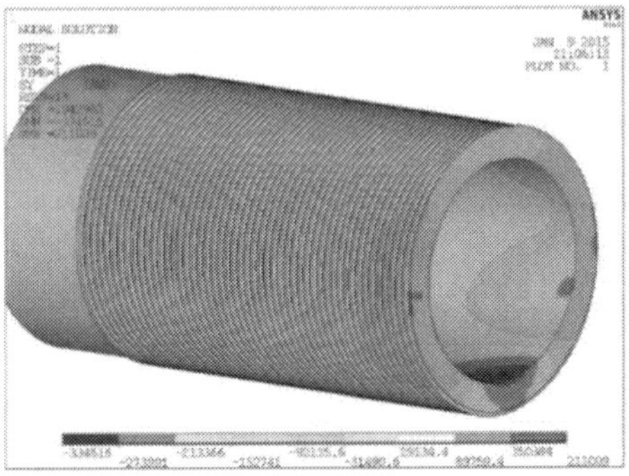

Figure 12. Stress nephogram in the circumferential direction of crosswise member bar.

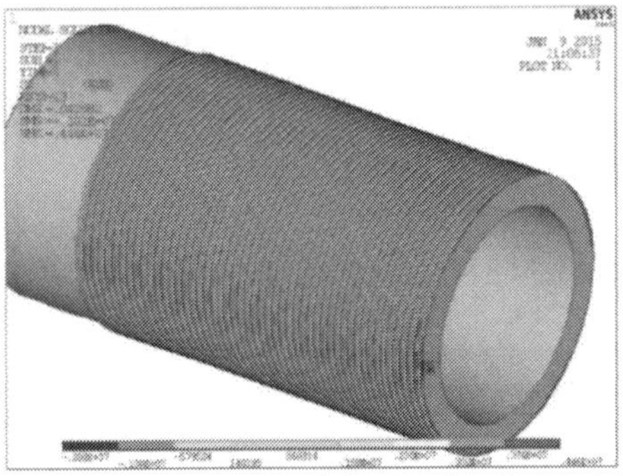

Figure 13. Stress nephogram in the axial direction of crosswise member bar.

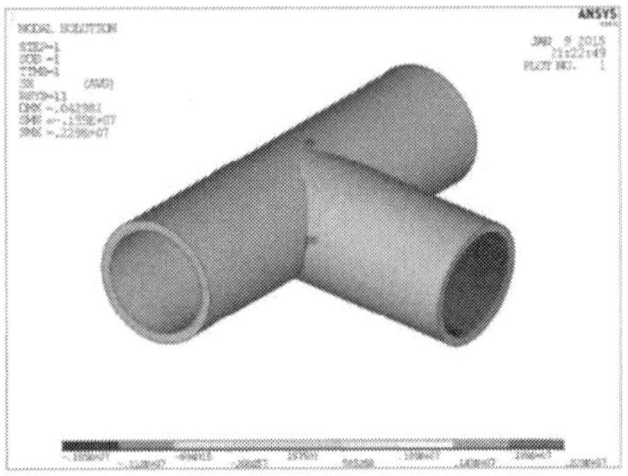

Figure 14. Stress nephogram in the radial direction of T-node.

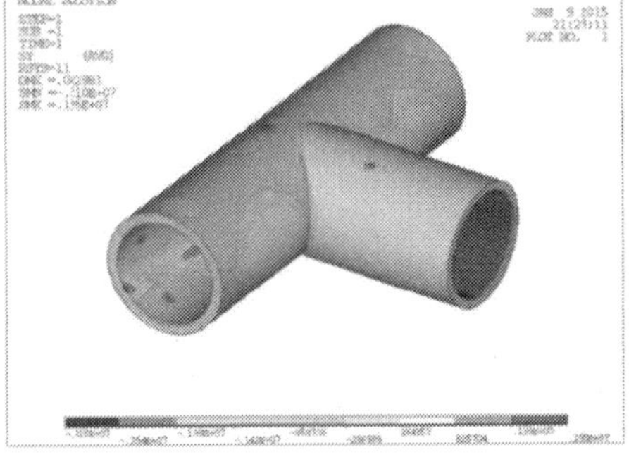

Figure 15. Stress nephogram in the circumferential direction of T-node.

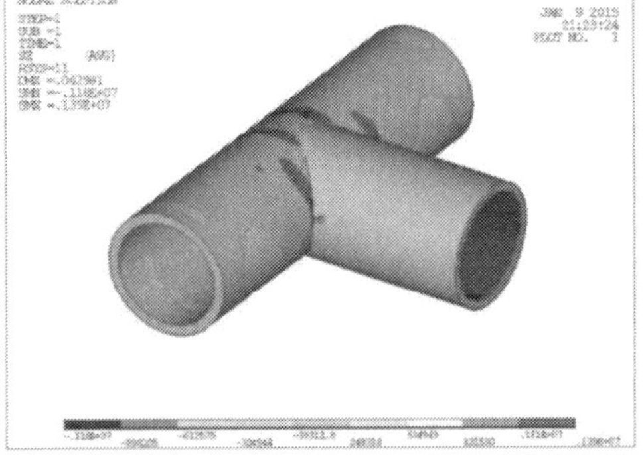

Figure 16. Stress nephogram in the axial direction of T-node.

3.3. Analysis of the stress on barrel with thread

According to the designation standards of straight thread connecting, the carrying power in barrel is 1.2 fold of that in the member bar, the thickness of the barrel is 15 mm. As a result, the size of the inner diameter is 100 mm while the external diameter is 130 mm.

The thread barrel plays a decisive role to the effective of member bar connecting, the figures 14-16 shows the barrel of T-node, which is used to connect some different parts of the member bar, and it is an important guarantee for the mechanical properties of the overall structure. As we can see from the figure 14-16, that the stress of barrel mainly concentrated at the connection point, and the maximum stress in radial direction is 2.29 MPa, occurred inside of the barrel and root of the thread, as we have analyzed before; while the maximum stress in circumferential direction, which is compression stress, brought out inside the thread, is 3.1 MPa; the maximum of stress in axial direction is 1.39 MPa, which happened on the position that the T-node connected. During the process of constructing the darrel, some treatments of consolidation should put on the connected position, to avoid the situation that the materials would be broken down and lose working ability. Then, we know that the stress on the node is small, that is suit with the design strength.

4. Conclusions

During the process of trussed beam designation, various complicated factors, combined with the calculation and analysis in the former part of the article. While in the case of the stress conditions have been fit, we should consider about the usefulness of economic as well. The difficulties of designation are the treatment of the member bar connecting, while good design could improve the force and decrease the requirements of stress strength, as a result, it can improve the life of the structure in the case of the same sectional dimension.

References

[1] Liu Jin, Gao Bingjun and Zhao Huilei 2008 FEM for strength and stability analysis of desulfurizer *Journal of Liaoning Tech-nical University (Natural Science)* **27** 104-6

[2] Zhang Bin, Zhang Hairong and Chen Xiaowei 2012 Reinforcement of concrete hollow slab beam and its finite element analysis *Journal of Guangxi University (Nature Science Edition)* **37** 630-5

[3] Samyn F and Bourbigot S 2008 Characterisation of the dispersion in polymer flame retarded nanocomposites *European Polymer Journal* **44** 1631-41

[4] Keller T, Bai Y and Vallée T 2007 Long-term performance of a glass fiber-reinforced polymer truss bridge *Journal of Composites for Construction* **11** 99-108
Peng Feng All FRP and FRP-concrete hybrid components for bridges: Experiments, theories and case study *3rd Asia-Pacific Conference on FRP in Structures* (Hokkaido, Japan) **11** 99-108.

[5] Bai Y and Keller T 2008 Modal parameter identification for a GFRP pedestrian bridge *Composite Structures* **82** 90-100

IOP Conf. Series: Materials Science and Engineering **87** (2015) 012031 doi:10.1088/1757-899X/87/1/012031

Catalytic ozone aqueous decomposition of methylene blue using composite metal oxides

Xuan Liu[1], Yongjiang Hou[1, 3], Jie Guo[1], Yaquan Wang[2], Qian Zuo[1] and Chunyu Wang[1]

[1]Environmental Science and Engineering, Hebei University of Science and Technology, Shi Jiazhuang 050018, P. R. China
[2]Chemical Engineering, Tianjin University, Tian Jin 30072, P. R. China

E-mail: 421127827@qq.com

Abstract. By using the method of co-precipitation, Fe-Mn, Al-Mn, Al-Mg composite metal oxides were prepared. Using X-ray diffractometer (XRD), the phases of catalysts synthesized were observed to be Mn_3O_4 and Fe_2O_3.With the increase of the calcination temperature, Mn_3O_4 was gradually transformed into Mn_2O_3. The experimental results show that: the best catalyst is the Fe-Mn composite metal oxide with the calcination temperature of 650℃. In this experimental condition, when the ozone amount was 1.92mg/min, and the dosage of catalyst was 0.5g, the removal rate of methylene blue (MB) was the best. The decolorization rate can reach to 100%. Compared with the effect of ozonation alone, the total organic carbon removal rate increased from 29.19% up to 65.78% after adding catalysts.

1. Introduction

Dyes are widely used in textile, printing and dyeing, photography, photochemistry and other industries. It is estimated that in the dyeing process, the textile industry has 10-15% of dye loss into the sewage discharge. Dye is not onlt refractory, but also has low biodegradability. These characteristics pose a great threat to human health and ecosystem.

Dye for extensive degradation products, the application of the carcinogenicity and low biodegradability, pose a great threat to human health and ecosystem [1]. Only ozone method is widely used to treatment of printing and dyeing waste water. Ozone dosage and the amount of azo groups have a great impact on the treatment results [2]. Ozone oxidation has selective degradation to organic matter and organic compounds. In order to enhance the ability of ozone oxidation of refractory organic pollutants, researchers put forward a series of advanced oxidation process to produce hydroxyl radicals [3].

The advantage of metal oxide catalytic ozone oxidation technology is not the introduction of other energy to the process, but does not adding chemicals to the water in the complex. At the same time, the use of metal oxides in catalytic ozone oxidation technology solves the problem of the separation of the catalyst. With reasonable cost and the simple operation, the metal oxide catalytic is used in large-scale water treatment [4]. Li Xin [5-7] research group studied $CuO-MnO_2/Al_2O_3$, $NiO-MnO_2/Al_2O_3$, NiO-CuO composite metal catalyst was used for catalytic ozonation of Songhua River, Harbin, China.

[3] Address for correspondence: Yongjiang Hou, Environmental Science and Engineering, Hebei University of Science and Technology, Shi Jiazhuang 050018, P. R. China. E-mail: 421127827@qq.com

They achieved good experimental results, but few were reported relative to the complex oxides for catalytic ozonation dye waste water.

Azo dyes can release with aromatic amine compounds which can increase the risk of cancer [8]. Azo dyes containing azo dyes, azo dyes and human contact process, can release with aromatic amine compounds [8] cancer risk. Nevertheless, azo dyes still occupy the chief status of in the field of chemical industry [9]. Taking a typical azo dye methylene blue (MB) as the research object, its structure is shown in figure 1. The preparation method of influence on catalytic effect was through the research system. In order to find efficient catalytic ozone oxidation catalyst, the best conditions of catalytic ozonation reaction is explored, influence factors is studied, catalytic mechanism is analyzed, which provides a theoretical basis for the industrial application of catalytic ozonation. A treatment method with high efficiency, economy and design basis was proposed. This has great significance on water resource recycling, energy saving and emission reduction.

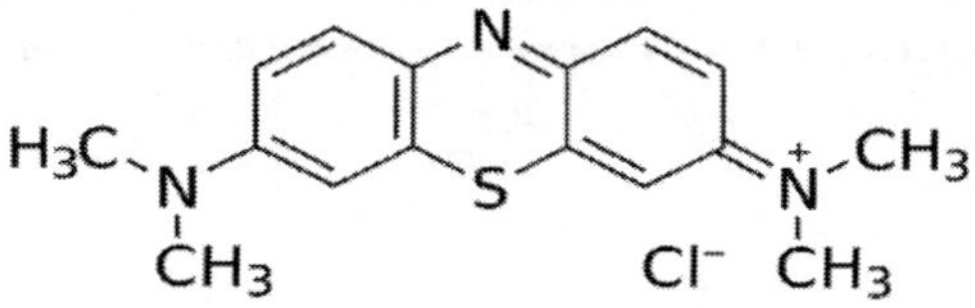

Figure 1. Schematic diagram of the MB molecular structure.

2. Experimental materials and methods

2.1. Main experimental reagent and instrument

Experimental reagents: methylene blue, Tianjin City Standard Technology Co. Ltd; manganese nitrate solution 50%, Tianjin Bodi Chemical Co. Ltd.; potassium iodide, Tianjin City Yongda Chemical Reagent Co. Ltd.; sodium thiosulfate, Tianjin City Yongda Chemical Reagent Co. Ltd.; ferric nitrate, Tianjin City Yongda Chemical Reagent Co. Ltd.; magnesium nitrate, Tianjin Bodi Chemical Limited by Share Ltd; aluminum nitrate, Tianjin City Yongda Chemical Reagent Co., Ltd; Tianjin city, sodium hydroxide, chemical reagent factory, above all is pure analysis.

Experimental instruments: SX-4-10 type muffle furnace, Tianjin Test Instrument Co. Ltd.; 101-1AB type drying oven, Tianjin city test instrument Co. Ltd.; DF-101S type heat collection type constant temperature heating stirrer, Gongyi Yuhua Instrument Co. Ltd.; UVProbe-2550 UV-VIS spectrophotometer, Japan SHIMADZU. FDX-2 type ozone generator, Dalian Beida Purifying Equipment Co. Ltd. AB204-E electronic balance Beijing international Science and Technology Co. Ltd..

Self-made experimental reactor is three neck flask (1L). The ozone reactor bottom porous cloth gas head was utilized to form small bubbles with the water solution full contact. The reaction condition for the pressure was 0.06 MPa, and generator working current was 50 mA. Emissions were exhausted by potassium iodide solution after absorption. Combined with a variety of laboratory test conditions and reaction system, the concentration of methylene blue solution was determined 100 mg/L.

2.2. Catalyst preparation and experimental method

2.2.1. Catalyst preparation

By mixing aqueous solution of Na_2CO_3, $Mn(NO_3)_2$ and $Fe(NO_3)_39H_2O$, the pH of the solution mixture was adjusted to about 10 with 1M NaOH. The mixed liquid was transferred to the autoclave with Teflon lined. The aging temperature were fixed at 120℃ for 24 h. Finally the reaction product was collected by filtration, washed with de-ionized water, and then air-dried at 80℃overnight. The solid was roasted at high temperature for 3h to obtain Fe-Mn composite metal oxide samples. Changing the species, the process was repeated to obtain Al-Mg, Al-Mn composite metal oxide [10].

2.2.2. Experiments and analysis method
Catalytic ozonation reaction of MB was carried out in a 1000 mL glass flask. In a typical catalytic reaction, a certain amount of the catalyst and 500 mL of MB solution with a concentration of 100 mg/L were stirred in the water bath (30±0.2°C) with a magnetic stir, while O_3 is added. And 5 ml sample solution was collected at different time during the catalytic reactions. The degradation of MB was determined from the absorption intensity at its maximum visible absorption wavelength (664 nm) by UV–vis spectrophotometer. The mineralization degree of sample solutions was analyzed using a total organic carbon analysis system.

3. Results and discussion

3.1. Characterization of XRD
The XRD images of Fe-Mn composite metal oxide calcined in different temperature were shown in figure 2. Diffraction peaks corresponding to the Fe-Mn composite metal oxide (550℃) were observed at 18°, 24°, 19°, 32°, 36°and 60°in the XRD. Compared with the standard card, the formed complexes were shown as Mn_3O_4 and Fe_2O_3. When the calcination temperature was increasing, the characteristic diffraction peak at 24 became weakened, while the others were enhanced. It shows that Mn_3O_4 was gradually transformed into Mn_2O_3 with the increase of calcination temperature.

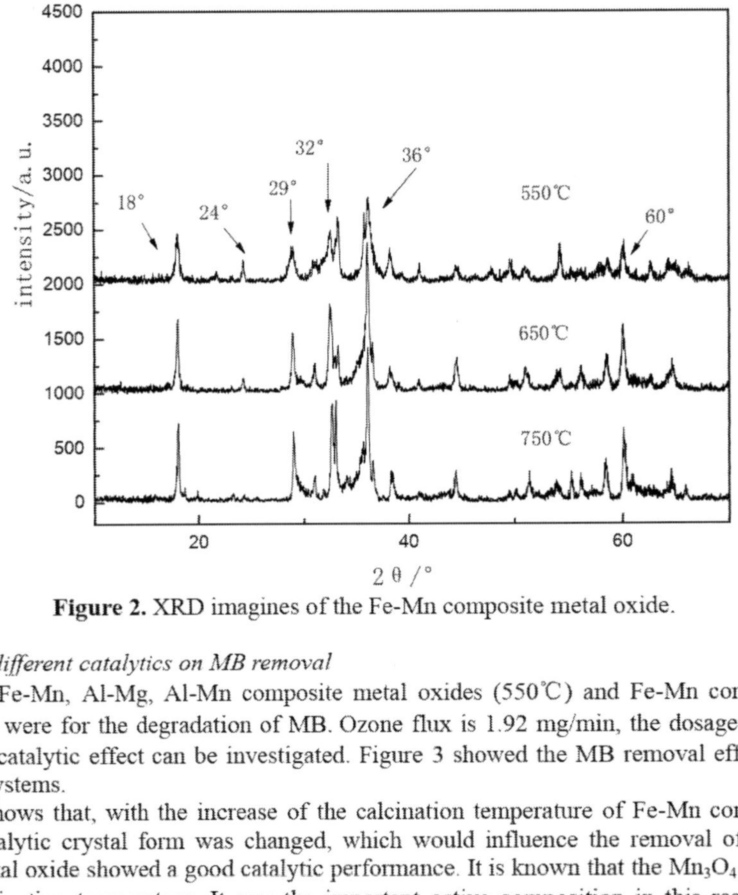

Figure 2. XRD imagines of the Fe-Mn composite metal oxide.

3.2. Effect of different catalytics on MB removal
Respectively, Fe-Mn, Al-Mg, Al-Mn composite metal oxides (550℃) and Fe-Mn composite metal oxide (650℃) were for the degradation of MB. Ozone flux is 1.92 mg/min, the dosage of catalyst is 0.5 g, and its catalytic effect can be investigated. Figure 3 showed the MB removal efficiency about the different systems.

Figure 3 shows that, with the increase of the calcination temperature of Fe-Mn composite metal oxide, the catalytic crystal form was changed, which would influence the removal of MB. Fe-Mn composite metal oxide showed a good catalytic performance. It is known that the Mn_3O_4 was produced at 550℃ calcination temperature. It was the important active composition in this reaction system.

With the increase of calcination temperature, Mn_3O_4 began to transform gradually to Mn_2O_3, and catalytic performance was also gradually reduced as well.

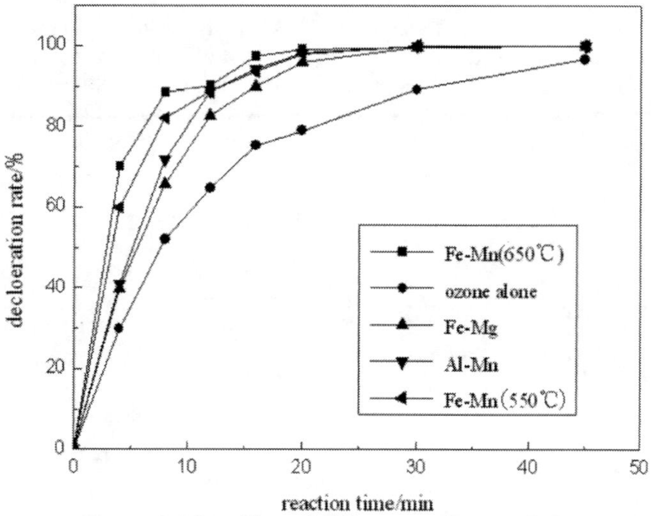

Figure 3. The effect of different catalysts on MB.

3.3. Influence of ozone flow on MB removal

Catalytic ozonation in water is three-phase reaction, including gas liquid and solid. The amount of catalyst is 0.5 g. The ozone flux is changed to research the effect on MB degradation.

Figure 4 showed that, with the increasing of ozone flux, the effect of the removal of MB has improved obviously. Because in the presence of the catalyst; hydroxyl radical increased with the increase of the concentration of ozone, which accelerated the degradation of MB [11]. But when the ozone flux kept increasing (2.88 mg/min), the removal of MB became decreased obviously. This is because the utilization rate of ozone was decreased, and ozone flux wasn't the main factor affecting the removal rate of MB.

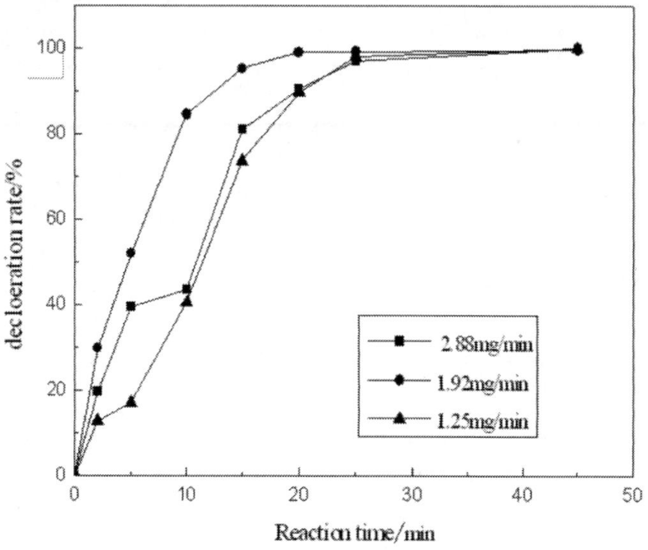

IOP Conf. Series: Materials Science and Engineering **87** (2015) 012031 doi:10.1088/1757-899X/87/1/012031

Figure 4. The effect of MB removal.

3.4. Influence of the dosage of catalyst on MB removal

Figure 5 showed that MB removal rate increased with the increasing catalyst dosage. The higher the dosage of catalyst, the larger for catalytic ozonation apparent rate [12]. The decomposition of ozone was accelerated under the high dose of catalysts. Catalytic reaction produced more hydroxyl radicals. They accelerated the degradation of MB. Although the low dose of catalyst resulted in a low removal rate, the degradation efficiency was higher than that using ozonation alone.

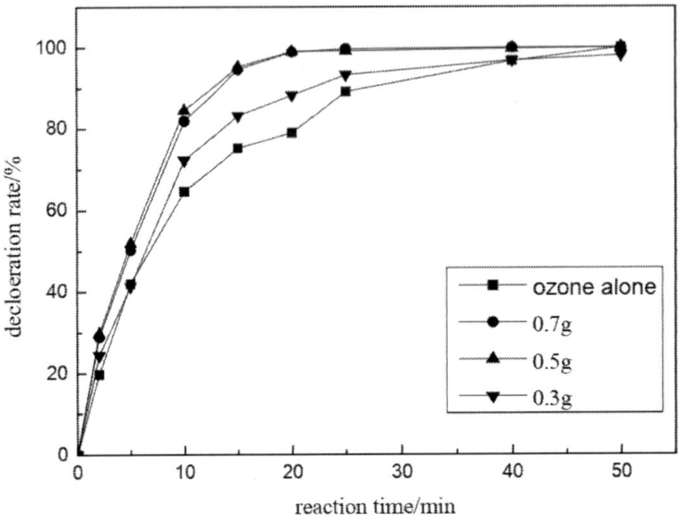

Figure 5. Effects of different catalyst dosage on removal of MB.

3.5. Mineralization degree of MB

The MB decoloration was investigated by UV-Visual, but MB degradation degree cannot be completely reflected. Therefore, the mineralization degree of MB (illustrated by TOC removal) was explored in this study. After the reaction, through the comparison of removal rate of total organic carbon (TOC) of MB, the catalytic performance of different catalysts was investigated.

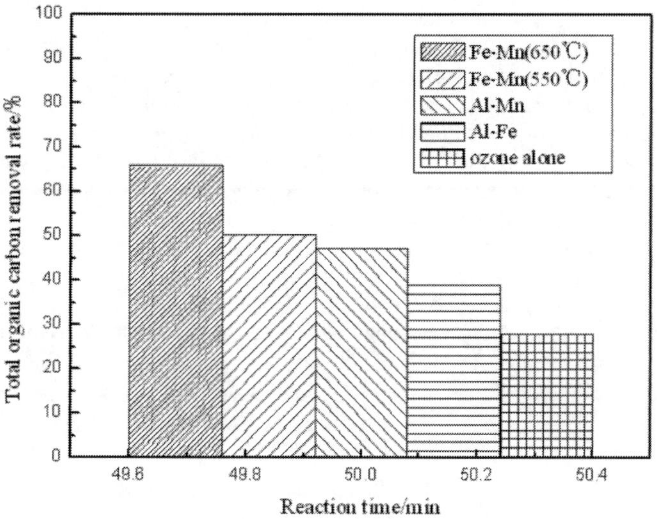

Figure 6. The mineralization degree of MB.

Figure 6 showed that the TOC removal rate is 65.8% in the presence of the Fe-Mn composite metal oxide (650℃), while it is only 31.2% in the single O_3 system at 50 min. Therefore, it can be concluded that Fe-Mn composite metal oxide added contribute a lot to higher TOC removal percentages. The decomposition of organic matter is more thorough than the single O_3 system.

4. Conclusion

In this paper, the high activity catalyst Fe-Mn composite metal oxide is successfully prepared with catalytic oxidation MB dye solution in the presence of O_3. The study reveals that the O_3/Fe-Mn composite metal oxide system has an obvious degradation effect on MB solution. The results show that catalyst calcination temperature, the dosage of catalyst and ozone flux are important influence factors of MB degradation in a certain flux of O_3. And the mineralization degree of MB is much higher (65.8%) than the pure ozone system (31.2%). Fe-Mn composite metal oxide, as promising water treatment materials, has a good application prospect.

Acknowledgments

This work was supported by the National Nature Science Foundation of China (No. 21276183) and the Natural Science Foundation of Hebei Province (No. B2012208086).

References

[1] Kasprzyk H B and Raczyk S U 2006 Catalytic ozonation of natural organic matter on alumina *Appl. Catal.* **62** 345-58

[2] Ma J and Graham D 1999 Degradation of atrazine by manganese-catalyzed ozonation: Influence of humic substances *Wat Res.* **3** 785-93

[3] Andreozzi R, Canterino M, Marotta R, et al. 2005 Antibiotic removal from wastewaters: The ozonation of amoxicillin *J. Hazard. Mater.* **122** 243-50

[4] Beltran F J, Rivas F J and Montero-de-Espinosa R. 2002 Catalytic ozonation of oxalic acid in an aqueous TiO_2 slurry reactor *Appl. Catal. B.* **39** 221-31

[5] Li W W, Qiang Z M, Zhang T, et al. 2010 Efficient degradation of pyruvic acid in water by catalytic ozonation with PdO/CeO_2 *J. Mol. Catal. A: Chem.* **348** 70-6

[6] Andreozzi R, Insola A, Caprio V, et al. 1996 The use of manganese dioxide as a heterogeneous catalyst for oxalic acid ozonation in aqueous solution *Appl. Catal. A.* **138** 75-81

[7] Aenreozzi R, Insola A, Caprio V, et al. 1992 The kinetics of Mn(II)-catalyzed ozonation of

oxalic-acid in aqueous-solution *Water Res.* **26** 917-21

[8] Rosal R, Gonzalo S, Santiago J, et al. 2011 Kinetics and mechanism of catalytic ozonation of aqueous pollutants on metal oxide *Ozone Sci. Eng.* **33** 434-40

[9] Zhao Lei, Sun Zhizhong and Ma Jun 2007 Research progress on the mechanism of metal catalytic ozonation in water treatment application *Modern Chemical Industry* **27** 78-82

[10] Liu Xiufang and Fan Binbin 2013 C/MgAl hydrotal citecomposite material as catalyst precursor forcitral acetone condensation reactions in the catalytic performance *Chinese Journal of Inorganic Chemistry* **29** 345-9

[11] WangYan, Zhang Hui, Zhang Jianhua et al. 2011 Degradation of tetracycline in aqueous media by ozonation in an internal looplift reactor *J .Journal of Hazardous Materials* **192** 35-43

[12] Buxton G V, Greenstock C L and Helman W 1988 Petal Critical review of rate constants for reactions of hydrated electrons hydrogen atoms and hydroxyl radicals in aqueous solution *J. Phys. Chem.* **17** 813-7

The optical and electrical characteristics of PMMA film prepared by spin coating method

H Q Zhang[1, 2], **Y Jin**[1, 2, 3] **and Y Qiu**[1]

[1]School of Physics and Optoelectronic Engineering, Dalian University of Technology, Dalian 116024, China
[2]Key Laboratory for Micro/Nano Technology and System of Liaoning Province, Dalian University of Technology, Dalian 116024, China

E-mail: jinye0721@163.com

Abstract. Thin films of poly (methyl methacrylate) (PMMA) were prepared on n-Si substrate and indium tin oxide (ITO) glass was prepared by spin coating. High frequency capacitance-voltage (C-V) and current-voltage (I-V) characterization were carried out on the Al/PMMA/n-Si and Al/PMMA/ITO glass structures, with the films as the insulator layer to evaluate the electrical properties. For the PMMA, the dielectric constant value obtained was about 3.9 at 1 MHz. The breakdown field strength of PMMA can be influenced by the impurities and surface charges or the interface states in the Si substrate. The average transmittance value of the PMMA/ITO glass layer was above 80% in the visible range.

1. Introduction

In recent years, the organic dielectric film has been of great interest for their potential applications. The polymer of Poly (methyl methacrylate) (PMMA) is considered as an option for the gate dielectric due to its high resistivity (>2×10^{15} Ω·cm), suitable dielectric constant similar to SiO_2 [1]. PMMA's chemical resistance, thermal stability, mechanical flexibility, low processing temperatures and thin film obtained on large areas by the sol-gel process make it an ideal candidate as a dielectric layer in organic electronic devices [2, 3]. Moreover, PMMA can be also used as an insulator layer in the field-effect transistor, which has been widely studied recently and presents good electrical performances [4-8]. For example, Puigdollers et al. fabricated pentacene thin film transistor using PMMA as gate dielectrics, indicating that transistors with PMMA showed better electrical characteristics than those with SiO_2. In this article, two structures, Al/PMMA/n-Si and Al/PMMA/ITO glass, were fabricated by spin coating technique to study their electrical properties respectively. Moreover, the transparency of PMMA fabricated on ITO glass was also investigated.

2. Experimental

The n-type (100) silicon wafers and ITO glasses were used as the substrates to form the metal-insulator-semiconductor (MIS) and metal-insulator-metal (MIM) structures. The silicon wafers came from Zhejiang Lijing Silicon Material Company and the ITO glasses were from Xiamen Yida. They were all ultrasonically cleaned by methylbenzene, acetone, ethyl alcohol and deionized water. The

[3] Address for correspondence: Y Jin, School of Physics and Optoelectronic Engineering, Dalian University of Technology, Dalian 116024, China. E-mail: jinye0721@163.com.

silicon wafer substrates were then immersed in hydrofluoric acid solution to remove the native SiO_2 from the surface.

The PMMA at 10 wt% was dissolved in glacial acetic acid (CH_3COOH). The acetic acid was purchased from Liaoning Fuyuan.The solution was deposited by spin-coating at 2500 rpm on different cleaned substrates to form dielectric coating, and these samples were baked in an electro thermostatic blast oven at 60°C for 2 h. The bottom metal contact was evaporated (Al) on the back side of the Si substrate and annealed at 550°C for 10 min in N_2 ambient. After annealing, the Al electrode in the Si substrate was divided into two parts with a knife, for ohmic measurement. The top metal contact was formed by evaporating Al over the PMMA film surface by using a suitable metal mask (dot size, 0.5×10^{-4} cm^2 for Si; 1.18×10^{-4}cm^2 for ITO). The corresponding schematic diagrams of both samples are shown in figure 1.

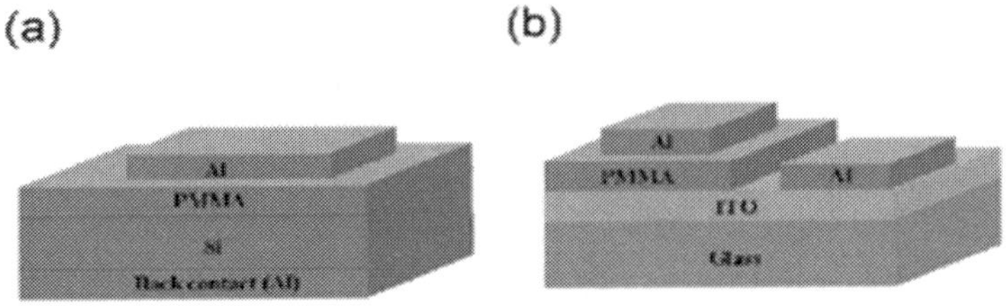

Figure 1. Schematic diagrams of MIS structure (a) and MIM structure (b) with PMMA as gate dielectric.

For studying the electrical characteristics of the PMMA film, current-voltage (I-V) and high frequency capacitance-voltage (HF C-V) electrical characterizations were performed on both MIS and MIM structures by using the Keithley 4200 semiconductor parameter analyzer (SCS). The bottom Al and Si contact was measured by SCS. The PMMA/ITO glass was used to study the optical transparency by UV-Vis. XRD was used to measure the chemical composition structure of the PMMA. Measurement of the layer thickness was done by Atomic-Profiler.

3. Results and discussion

The XRD spectrums for the MIS and MIM structure are shown in figures 2(a) and 2(b), respectively. As shown in figure 2(a), the peaks at 2θ=38.06°, 44.10°, 64.81° and 77.86° are caused by Al electrode. The peaks at 2θ=33.10° and 61.77° come from the Al_2O_3 in the back contact due to the process of Al allying. Referring to the XRD spectrum of Si, at 2θ=69.13°, the peak of Si substrate is broadened due to PMMA. In figure 2(b), the peaks of the Al electrode are at 2θ= 38.32°, 44.36°, 63.84° and 77.84°. At around 2θ=15°, the broad hump caused by the glass substrate is also broadened by PMMA referring to the XRD spectrum of ITO glass. The peak of ITO film is so weak and then vanishes. The first broad diffraction slop from 2° to 10° in the two graphs may be caused by the measurement instrument. There are no prominent peaks of PMMA in the spectra indicating its amorphous nature [1].

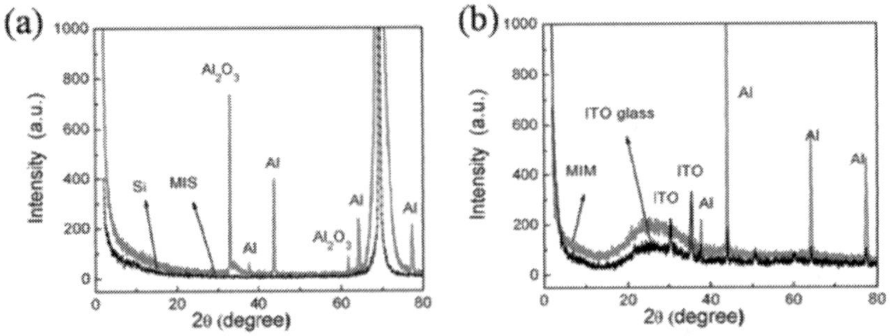

Figure 2. XRD spectrum for MIS (a) and MIM (b) structure.

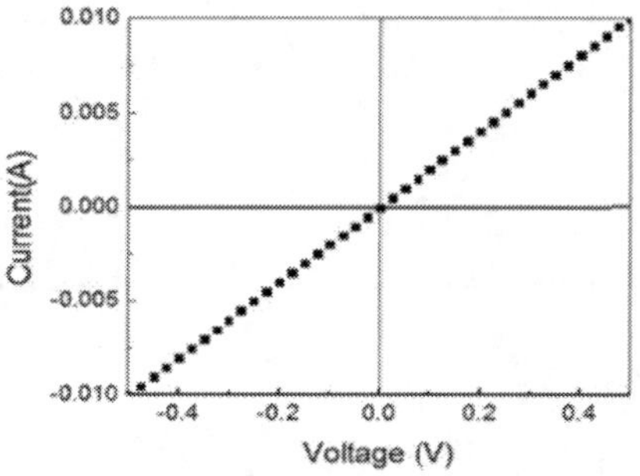

Figure 3. I-V characteristic curve for ohmic contact.

Figure 3 shows the I-V characteristic of Si and Al back electrode. The Al-Si contact has a linear ohmic characteristic. The C-V characteristic curves for MIS and MIM structures measured at a frequency of 1MHz are shown in figure 4. The curves are recorded by scanning from -30 V to 30 V at room temperature. For the MIS structure, the shape of the curve in figure 4(a) is similar to the typical C-V characteristic curve of MIS devices with n-Si substrate. At positive bias, it shows maximum capacitance with accumulation of holes in the n type silicon at the polymer–silicon interface. As the voltage decreases, depletion is clearly observed with the decrease of the capacitance because the total capacitance comes from the serial coupling of the capacitance from the polymer dielectric and the silicon layer. The MIM structure can be considered as a parallel plate capacitor and its C-V characteristic curve shows a constant capacitance value.

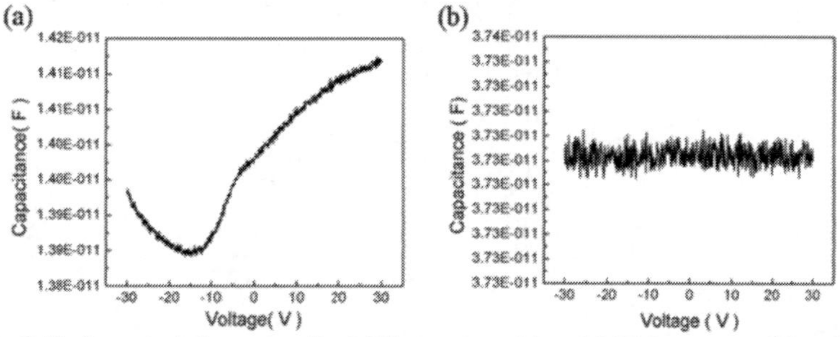

Figure 4. C-V characteristic curves for MIS structure (a) and MIM structure (b) at 1MHz.

The dielectric constant value (ε_i) of the PMMA film is determined using equation (1) for a parallel plate capacitor, where ε_0 is the permittivity in vacuum, A is the area of the capacitor, t is the thickness of the PMMA film and c is the measured capacitance.

$$\varepsilon_i = \frac{c}{\varepsilon_0} \times \frac{t}{A} \tag{1}$$

The film thickness of the MIM structure is 1.120 μm, the dielectric constant value (ε_{MIM}) is 3.99 at 1 MHz, which is determined from the constant capacitance value from the C-V measurement of the MIM structure. For the MIS structure, the layer thickness is 1.187 μm. At positive voltage, it shows the maximum capacitance with accumulation of electrons in the n type silicon at the polymer–silicon interface. The dielectric constant of the polymer film can be obtained from this capacitance using equation (1). The dielectric constant value (ε_{MIS}) calculated from the maximum capacitance is 3.78 at 1 MHz for MIS structure. ε_{MIS} is smaller than ε_{MIM}, which may be attributed to the fact that the maximum capacitance value of the MIS structure has not been saturated. The dielectric constant of PMMA film at 1MHz is similar to literature value which is in the range from 2 to 4 [1, 3, 4].

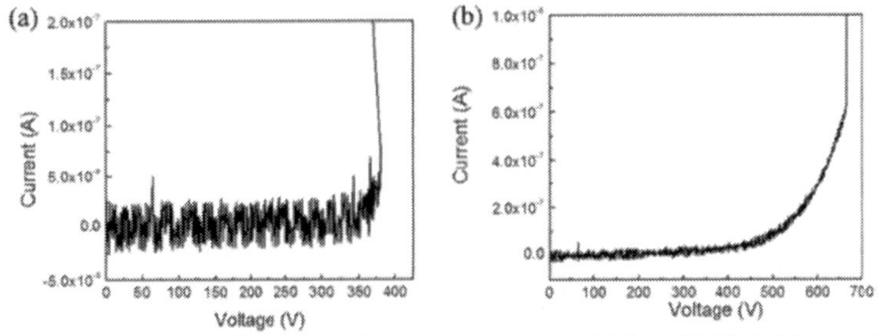

Figure 5. Current-voltage curve for MIS (a) and MIM (b) structures.

For the I-V characteristic measurements, because the breakdown voltage of PMMA is higher than the measurement range of SCS and the noise has a larger effect on the measurement with Keithley 2410, SCS is used to measure the leakage current and the Keithley 2410 to measure the breakdown voltage. From the I-V characteristics of MIS and MIM devices, the currents of the two structures are -2×10^{-10} A at -200 V and 2×10^{-10} A at 200 V. Due to the limit of the SCS measurement accuracy, the current cannot reach smaller magnitude, and the leakage current density is estimated to be about 10^{-6} Acm^{-2}. The breakdown voltage of the MIS devices is about 380 V as shown in figure 5(a). However, for the MIM devices, the current increases gradually and until 650 V the devices eventually break down (figure 5(b)). The breakdown field strength of the PMMA obtained from the MIS and MIM structure is 3.2 $MVcm^{-1}$ and 5.8 $MVcm^{-1}$, respectively. The breakdown voltage of the MIS structure is smaller than that of the MIM structure, which may be attributed to the impurities and surface charges or interface states existed in the Si substrate [9, 10].

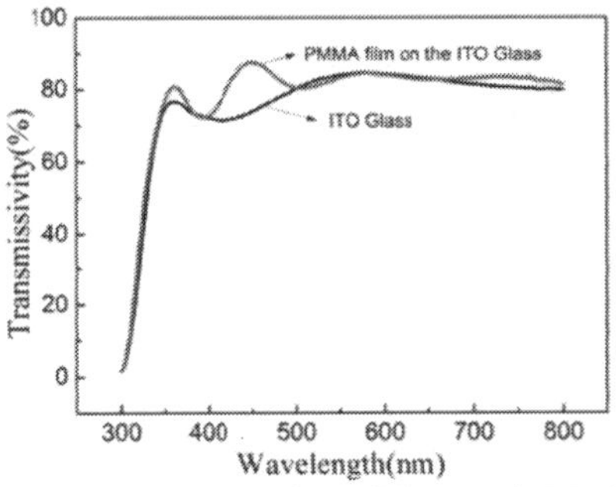

Figure 6. Transmittance spectrum of the fabricated PMMA/ITO glasslayer in the visible range.

Furthermore, the optical transparency of the fabricated PMMA/ITO glass layer in the visible range is also studied, as shown in figure 6, The ITO glass and the PMMA/ITO glass were measured respectively. The curve drops rapidly at approximately 340 nm due to the absorption of the glass substrate. The ITO glass with PMMA film has better transmittance in some regions, and it shows that the decrease of the reflected light is due to the PMMA film. The refractive index of PMMA film (n_{PMMA} is about 1.49 at room temperature) is lower than that of ITO film (n_{ITO} is about 1.9 in 450 nm region at room temperature). The wave of oscillation in the visible range may be caused by optical interference phenomenon [11, 12]. According to the optical path difference of the refractive light,

$$(2k+1)\times\frac{\lambda}{2}=2nd,(k=0,1,2,3...) \tag{2}$$

When the refractive light is weakened, the transmittance of the fabricated PMMA/ITO glass layer is increased. The change of waveform may be due to the heterogeneous thickness of the film and different absorption coefficients for different wavelengths [12].

4. Conclusion

In this work, the dielectric films PMMA with the structure of MIS and MIM are obtained by spin coating technique. The XRD spectrum shows the amorphous nature of PMMA. The dielectric constant of PMMA in the two structures obtained through the high frequency C-V electrical characterization of is similar, about 3.9 at 1 MHz. The impurities and surface charges or interface states existed in the Si substrate may contribute to the decrease of the breakdown voltage of PMMA. And the leakage current density of PMMA is about 10^{-6} A cm^{-2}. The transmittance spectra show that the transmittance of PMMA/ITO glass is above 80%, and PMMA improves the transparency of this structure in the visible range.

Acknowledgments
This work is supported by the fundamental Research Funds of the Central Universities (DUT14LK35).

References

[1] Na M and Rhee S W 2006 *Org Electron* **7** 205-12
[2] Gross S, Camozzo D, Di Noto V, Armelao L and Tondello E 2007 *European Polymer Journal* **43** 673-96
[3] Morales-Acosta M D, Quevedo-Lopez M A, Gnade B E and Ramirez-Bon R 2011 *J Sol-Gel Sci Techn* **58** 218-24
[4] Puigdollers J, Voz C, Martin I, Orpella A, Vetter M and Alcubilla R 2004 *Journal of Non-Crystalline Solids* **338** 617-21
[5] Horowitz G 1998 *Advanced Materials* **10** 365-77
[6] Veres J, Ogier S, Lloyd G and De Leeuw D 2004 *Chemistry of Materials* **16** 4543-55
[7] Horowitz G, Deloffre F, Garnier F, Hajlaoui R, Hmyene M and Yassar A 1993 *Synthetic Met* **54** 435-45
[8] Estrada M, Mejia I, Cerdeira A and Iniguez B 2008 *Solid State Electron* **52** 5253-9
[9] Kolodzey J, Chowdhury E A, Adam T N, Qui G, Rau I, Olowolafe J O, Suehle J S and Chen Y 2000 *IEEE Transactions on Electron Devices* **47** 121-8
[10] Ando T, Fowler A B and Stern F 1982 *Reviews of Modern Physics* **54** 437
[11] Swanepoel R 1983 *J. Phys. E: Sci* **16** 1214
[12] Manifacier J C, Gasiot J and Fillard J P 1976 *J. Phys. E: Sci* **9** 1002

Immobilization of laccase on a novel ZnO/SiO_2 nano-composited support for dye decolorization

Wei-Xun Li, Huai-Yan Sun and Rui-Feng Zhang[1]

Faculty of Materials Science and Chemical Engineering, Ningbo University, Ningbo, Zhejiang 315211, PR China

E-mail: zhangruifeng@nbu.edu.cn

Abstract. ZnO nanowires were introduced into macroporous SiO_2 by means of in situ hydrothermal growth. The obtained nano-composite was then used to immobilize laccase (secured from *Trametes versicolor*) through the process of static adsorption. The average loading amount was as high as 193.4 $\mu mol \cdot g^{-1}$. The immobilized laccase was proven to be an effective biocatalyst in the decolorization of two dyes: Remazol Brilliant Blue B, and Acid Blue 25. The decolorization percentage of Remazol Brilliant Blue B and Acid Blue 25 reached 93% and 82%respectively. The immobilized laccase exhibited enhanced thermal stability and pH adaptability compared to free laccase. After ten recycles, the immobilized laccase retained 42% decolorization catalytic activity.

1. Introduction

Every year, the textile industry generates large volumes of liquid effluent pollutants [1, 2]. Synthetic dyes are the largest of all textile dyestuffs produced, and their residuals left in wastewater have been a primary source of environmental contamination [3, 4]. Synthetic dyes contain azo bonds (-N=N-), aniline, phenol, or other aromatic compounds. Several physical and chemical approaches have been proposed to withdraw dyes from industrial wastewaters. Such examples are coagulation, adsorption, chemical oxidation, or photolysis [5-7].

The enzymatic treatment of pollutants is elegant process because of the mild reaction conditions they produce, without generating toxic sludge. Oxidative enzymes are one enzyme which maintains the capability for decolorizing dyes. Oxidative enzymes such as iron, which-contains chloroperoxidase [8] and horseradish peroxidase [9], or manganese, which-contains manganese peroxidase [10], have been applied to the decolorization of synthetic dyes. Laccases (EC 1.10.3.2) are copper-containing oxidases, and were first discovered in the Japanese lacquer tree, *Rhus vernicifera* [11, 12]. This enzyme has been applied to the decolorization of dyes, as stated in the previous report [13]. The reaction is easy to operate because it consumes O_2, at room temperature [14, 15]. Due to the susceptibility of enzymes to deactivate in the presence of other chemicals, and it is difficult to separate and recycle, which limits its application. So using immobilized laccase is separating expediently and recycling easily.

Generally, the nature of the support heavily influences the properties of the immobilized laccase.

[1] Address for correspondence: Rui-Feng Zhang, Faculty of Materials Science and Chemical Engineering, Ningbo University, Ningbo, Zhejiang 315211, PR China. Tel.: +86 13805843294; E-mail: zhangruifeng@nbu.edu.cn.

Polymer supports such as polyacrylamide [16] and alginate [17] are commonly used for enzyme immobilization, and are easy to be prepared, but short of thermal and chemical stability. Inorganic support maintains quality chemical stability, high mechanical strength, and low cost. Mesoporous materials particularly have large pore volume, tunable pore diameter, and a high specific surface area; these characteristics make them to be explored extensively as supports for immobilization of enzymes. The enzyme immobilization on different kinds of mesoporous materials: MCM [18, 19], SBA-15[20, 21], CNS [22], MCF [23, 24], has been extensively researched. Macroporous materials such as diatomite [25], ceramic [26] have also been used to immobilize enzymes, however their low specific surface area creates a disadvantage in regards to the loading amount of enzymes.

A new kind of large-sized macroporous SiO_2 developed by our group recently. It offers enhanced mechanical stability, low density, and high porosity (93%) [27, 28]. To solve the issue of low specific surface area found in this macroporous material, a special *in situ* hydrothermal growth technology was established. The technology was created to introduce ZnO nanowires in the macroporous SiO_2 support [29]. The composite has a higher specific surface area compared to the pure macroporous silica, and the ZnO nanowires are usually positive-charged, which is very useful to immobilize negative-charged enzymes through electrostatic interaction. In our previous study, ZnO nanowires/macroporous SiO_2 composite support has been used to immobilize lipase through the method of static adsorption [30, 31]. In this study, the negative-charged laccase (IP is 3.7, working pH is about 5.5) was immobilized on ZnO nanowires/macroporous SiO_2 composite supports in an aqueous solution. Two dye substrates (Remazol Brilliant Blue B (RBBR) and Acid Blue 25, which is shown in figure 1) were decolorized under the catalysis of both free and immobilized laccases. The loading amount and stability of adsorption were measured by using a standard method. The reaction conditions such as pH and temperature were then investigated and optimized. Finally, the storage stability and reusability were tested.

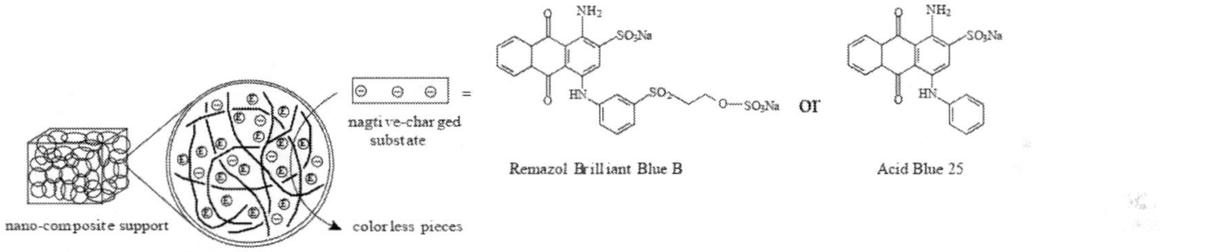

Figure 1. Scheme for composite support-laccase degradation two dyes.

2. Materials and methods

2.1. Materials

Laccase from *Trametes versicolor* was purchased from Aldrich, 2-Azino-bis (3-ethylben-zothiazoline-6-sulfonic acid) diammonium salt (ABTS), buffer salts, Remazol Brilliant Blue B and Acid Blue 25 are purchased from Aladdin (Shanghai).

2.2. Synthesis composite supports

A mixture of Epoxy resin (16.0 g) and polyethylene glycol (28.0 g of PEG1000 and 4.0 g of PEG2000) were heated and stirred to form a clear solution, 5.0 g of diethylenetriamine was then quickly added with strong stirring. The reaction system was kept at 70 °C for 3 hours. After curing the obtained the white product was immersed in pure water to completely remove the PEG resulting in a 3D skeletal polymer. Then, the white 3D skeletal polymer was naturally dried for 2 days and then immersed in pure ethyl silicate for 3 h. The hydrolysis of infiltrated ethyl silicate was carried out by exposing the samples in $NH_3 \cdot H_2O$ atmosphere at 40 °C for 12 h in a sealed container. After that the silica/polymer composites were dried at 60 °C for 2 h to remove the produced ethanol and the adsorbed $NH_3 \cdot H_2O$.

Calcinations of the composites were carried out at 5 °C·min^{-1} in a muffle furnace up to a temperature of 820 °C for 30 minutes. The white 3D silica has density of less than 0.2 g/mL [27].

Introduction of ZnO nanowires in macroporous silica was carried out by two steps. The first step was the formation of ZnO crystal seeds on the wall of silica support: A mixture of zinc acetate, PEG600 and water was used as the precursor solution, in which the SiO_2 supports were immersed 3 h and then dried at 100 °C for 1 h to remove most of the water. Firstly the samples were heated in a muffle furnace at 200 °C for 1 h, and then the temperature was raised to 650 °C at a rate of 7.5 °C/min and maintained for another 20 min. The second step was the *in-situ* growth of ZnO nanowires: zinc nitrate was dissolved in ammonia solution to form 1M solution, in this solution the seeds-contained supports were immersed for 6 h, and then the samples were placed in a sealed container equipped with ammonia-collecting material. The hydrothermal reaction was carried out in two stages: maintaining 90 °C for 6 h in the first stage and 100 °C for 12 h in the second stage. The obtained products were washed with pure water to completely remove the soluble inorganic salts and dried in vacuum at room temperature for 24 h [28-29].

2.3. *Loading amount assay*

Both of the supports are cut into small particles (~ 8 mm^3). In the adsorption experiment, 0.3 g support is mixed with a 10 ml laccase solution in acetate buffer (pH 6.5), and the mixture continues to be shaken at 25°C for 6 hours. After that, the material was recovered by centrifugation for 5 minutes, then washed with an acetate buffer until no activity detected in the supernatant. The final enzyme preparation was kept in acetate buffer.

The amount of immobilized laccase (P_0) was determined according to the Bradford method [32], and the P_0 was calculated from the below equation:

$$P_0 = \frac{(C_1 - C_2)V_1 - C_3V_2}{m} \quad (1)$$

The leaching of laccase from the supports described in the previous [30]. The calculation equation was as follows:

$$P_t = P_0 - \frac{C_tV}{m} \quad (2)$$

2.4. *Enzymatic dye decolorization*

Reactions were carried out at room temperature in 1 mL reaction mixtures containing two dyes (0.1 mol/L) and an acetate buffer pH 3-8, then added free and immobilized enzymes into it respectively. For all reactions, the concentration of the two dyes was 0.1mol/L, and the concentration of free and immobilized enzyme was maintained at 0.1 mmol/L. The decolorization efficiency was determined by monitoring the maximum absorbance (λ_{max}) of dyes at UV-Vis spectrophotometer, Remazol Brilliant Blue B at 550 nm and Acid Blue 25 at 600 nm respectively:

$$\text{Decolorization efficiency (\%)} = \frac{A_0 - A_t}{A_0} \times 100\% \quad (3)$$

2.5. *Effect of pH, temperature and thermal stability on catalytic activity*

The effect of pH and temperature on the catalytic activity of the free and immobilized laccase against the dyes was determined spectrophotometrically. This was done by measuring the changes in absorbance at the wavelength of maximum absorption of each dye at different pH balances and temperatures, keeping the constant temperature at 25℃ and the pH=5.5.

The thermal stability of the free and immobilized laccases was determined by measuring the residual activity of the enzyme exposed to 80℃ in acetate acid buffer pH 6.5 for 120 min. Every 20 min, a sample removed and assayed for enzymatic activity. The free and immobilized laccase activity was measured by monitoring the absorbance change of 2, 2-Azino-bis (3-ethylben-zothiazoline-6-sulfonic acid) diammonium salt (ABTS) during its oxidation with a UV–visible spectrophotometer at 420 nm.

2.6. Storage stability

The immobilized laccase was kept at 25℃ for different times (0-100 days), and was measured by monitoring the absorbance change of 2, 2-Azino-bis (3-ethylben-zothiazoline-6-sulfonic acid) diammonium salt (ABTS) during its oxidation with a UV–visible spectrophotometer at 420 nm.

2.7. Reusability of an immobilized laccase

The reusability of the immobilized laccase was assayed in ten sequential cycles for the decolorization of RBBR. The immobilized laccase was recovered by performing centrifugation after each reaction cycle, and was placed in a fresh reaction media. The residual activity data was normalized to the first cycle assay, which was considered to be at 100% activity.

3. Results and discussion

3.1. Structural characterization of the supports

Figures 2 and 3 display the SEM images of the macroporous SiO_2 support and the composite with ZnO nanowires. The structural characterization of the supports is described in the previous literature [30].

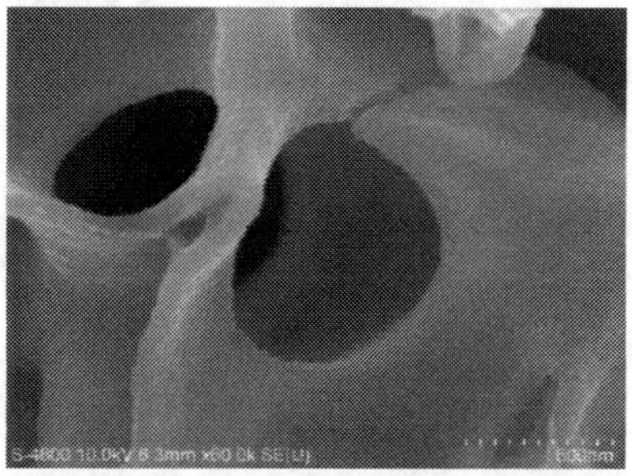

Figure 2. SEM image of macroporous SiO_2.

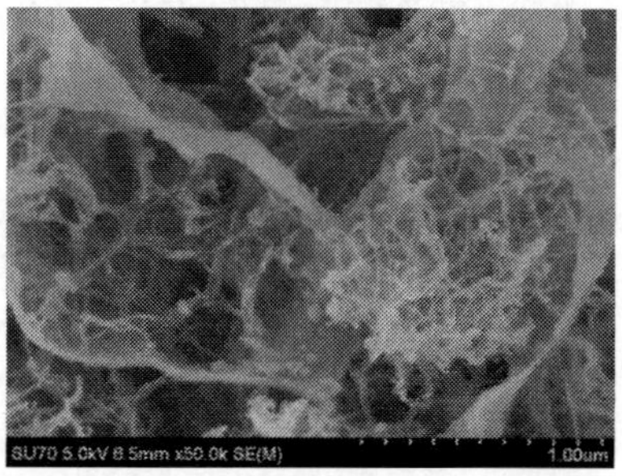

Figure 3. SEM image of the ZnO nanowires/macroporous SiO_2 composite.

3.2. Loading amount and adsorbing stability of the supports

The loading amount is described in table 1, and is about 193.4 nmol/g and pure macroporous SiO_2 is 43.7 nmol/g. It can be seen that the composite support is about 5 times the loading amount of macroporous SiO_2., and the loading amount of composite support is high than the silica-based mesoporous material [22].The enhancement of adsorption and stability should be a result of the increase of surface area and the electrostatic interaction between positive-charged ZnO nanowires and a negatively-charged laccase. Therefore, the interaction of a laccase with composite support is much faster than that with pure macroporous SiO_2.

Table 1. The loading amount of laccase immobilized on different supports.

Soaking time/h	Composite support Loading/μmol·g^{-1}	Macroporous SiO_2 Loading/μmol·g^{-1}
0	193.4	43.7
12	189.9	32.4
24	164.8	14.8
36	139.5	9.6
48	107.1	6.7

3.3. Effect of pH, temperature on the catalytic activity of free and immobilized laccases

3.3.1. pH. Figures 4 and 5 express the pH effect of immobilized and free enzymes on decolorization of the RBBR and Acid Blue 25 respectively. It is apparent that the decolorizations of both dyes under catalysis of immobilized and free enzymes are highly sensitive to the associated pH value, and that the optimum pH for free laccase is 5, and for the immobilized one is 5.5. Free laccase received the highest decolorization of 80% and 56% for RBBR and Acid Blue 25 respectively. The immobilized laccase provided the highest decolorization at 93% and 82% for RBBR and Acid Blue 25 respectively. The results indicate that the catalytic activity of laccase could be improved after immobilization in this way. Furthermore, the free laccase retained only 6% (RBBR) and 3% (Acid Blue 25) of decolorization activity at pH=8, while the immobilized laccase still has 32% (RBBR) and 25% (Acid Blue 25) of decolorization activity at the same pH value. It has been reported that supporting materials usually alter the effect of pH on the catalytic profile of some enzymes [33].

3.3.2. Temperature. The temperature of reaction is another important parameter affecting the activity of the enzyme. Figures 6 and 7 depict the temperature's effect on decolorization efficiency. The decolorization process was carried out under different temperatures varying from 20℃ to 80℃. The results are illustrated in figures 6 and 7, both free and immobilized laccases maintained high activities as the temperature maintained at 20～30℃. As the temperature above 40℃, the free laccase showed a rapid decrease in activity as a result of the denaturation of enzyme at high temperature [34]. The immobilized laccase kept catalytic activities much higher than free laccase at all the range of temperature. For example, the immobilized laccase still have over 50% of catalytic activity at 60℃, 32% of catalytic activity at 80℃. The effect of supports protecting enzymes from deactivation at high temperature has also been found in other reports [35-37].

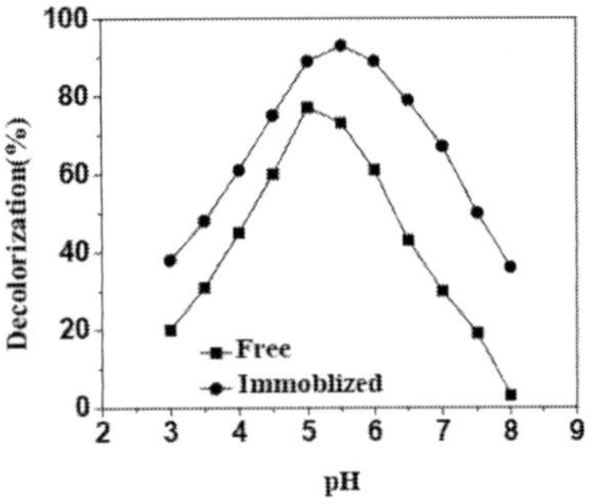

Figure 4. Effect of pH on the decolorization of RBBR by free and immobilized laccase.

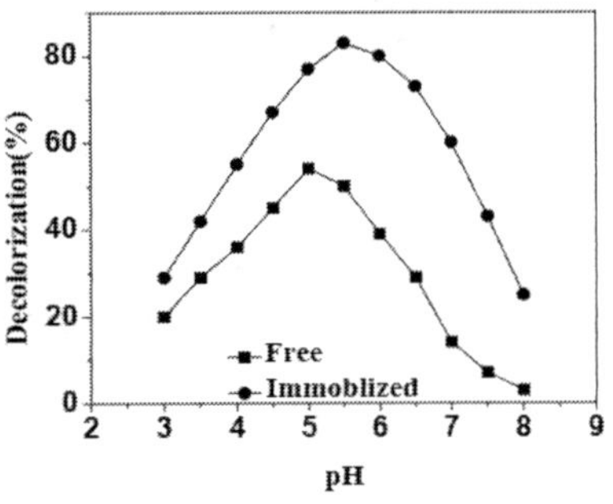

Figure 5. Effect of pH on the decolorization of Acid Blue 25 by free and immobilized laccase.

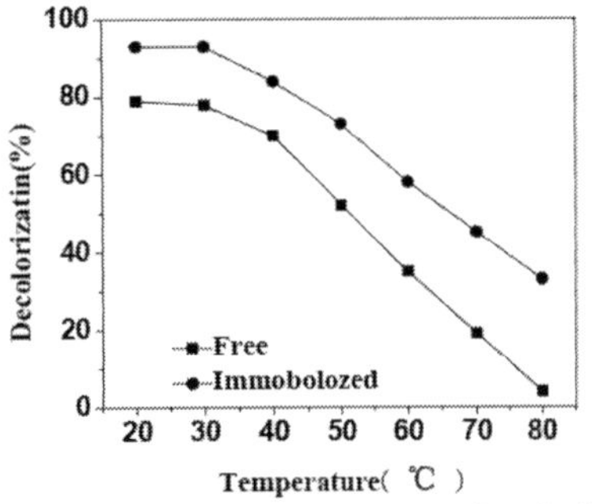

Figure 6. Effect of temperature on decolorization

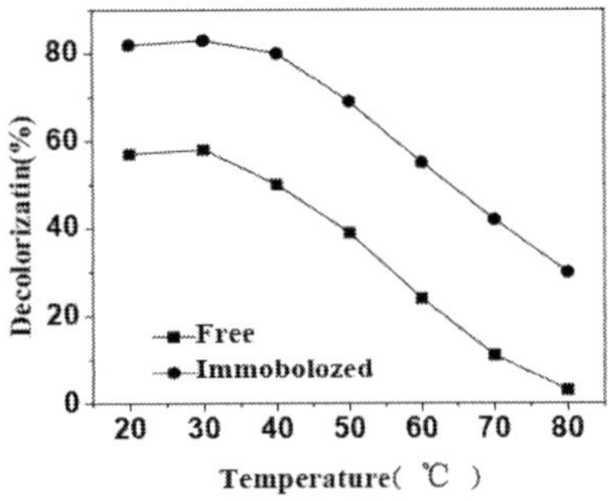

Figure 7. Effect of temperature on

of RBBR by free and immobilized laccase. decolorization of Acid Blue 25 by free and immobilized laccase.

3.3.3. Thermal stability. Figure 8 shows the heat inactivation curves for both the free and immobilized laccases. After an incubation period of 120 minutes at 80℃,the immobilized laccase retained 70% of its initial activity, while the free enzyme retained only 40%. Obviously, the nano-composite support is very effective at protecting laccase from inactivation at high temperatures. The electrostatic interaction improved the thermal stability of the immobilized enzyme, or, in other words, the immobilized laccase has an enhanced heat-resistance compared to the free laccase. The immobilized laccase could work in harsh environmental conditions with less activity loss as compared to the free laccase.

3.4. Storage stability

Figure 9 shows the decrease of relative activity in the immobilized laccase during incubation at 25℃ for many days. The immobilization laccase retained approximately 80% of its initial activity after three months of storage. That means the laccase immobilized on the nano-composite support is a stable biocatalyst, which may have practical applications in the processes of dye-containing waste water.

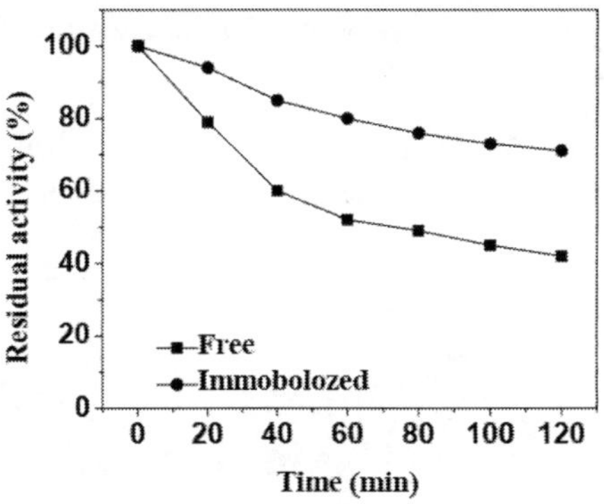

Figure 8. Thermal stability of the free and immobilized laccase at 80℃.

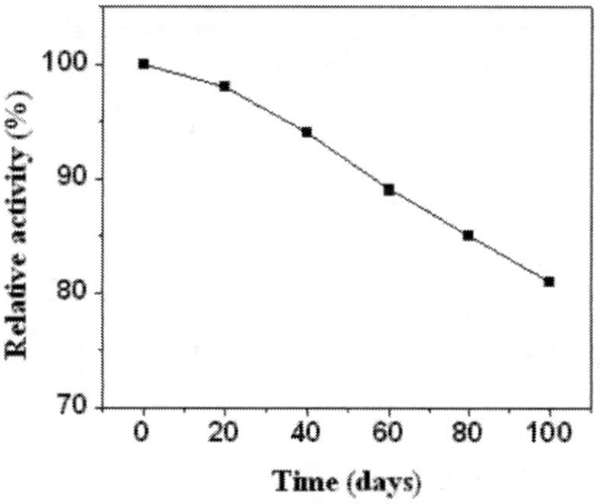

Figure 9. Storage stability of immobilized laccase at 25℃.

3.5. Reusability of the immobilized laccase

The reusable qualities of the immobilized laccase were tested by repeating catalysis of the decolorization of RBBR. The decolorization percentage was recorded as the relative activity of laccase in each cycle. As shown in figure 10, the relative activity showed a slight drop as the cycle number going from 1 to 10, 40% of initial activity maintained after ten times of cycle. It has never been reported that laccase immobilized on epoxy-activated carriers retained only 20% of initial catalytic activity after six cycles [37]. According to the comparison, it can be concluded that the laccase immobilized on the nano-composite support exhibits enhanced reusability.

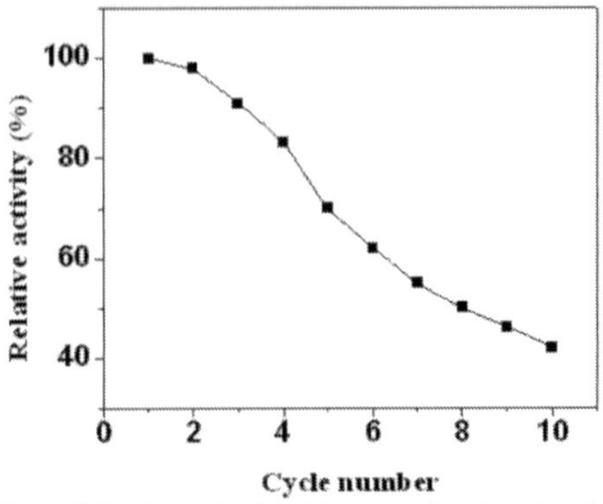

Figure 10. Reusability of the immobilized laccase for the decolorization of RBBR.

4. Conclusion

- The ZnO nanowires into macroporous SiO_2 resulted in a new composite support for the immobilization of laccase. The support contains 233 m^2/g of specific surface area, and 193.4μmol/g of adsorption capacity. The immobilized laccase exhibited excellent resistance to temperature and pH inactivation, high thermal stability, and reusability. After ten recycles, the immobilized laccase still maintained 40% of its initial activity.
- The optimal reaction conditions for 93% the decolorization of (RBBR) and 82% of the decolorization of Acid Blue 25 are as follows: molarity of two dyes (0.1 mol/L), molarity of free laccase and immobilized enzymes (0.1 mmol/L), reaction temperature (25℃), and pH (5.5); reaction time (1 h).

References

[1] Sun J H, Shi S H, Lee Y F and Sun SP 2009 *Chem. Eng. J.* **155** 680
[2] Bokare A D, Chikate R C and Paknikar K M 2008 *Appl. Catal. B: Environ.* **79** 270
[3] Telke A, Kalyani D, Jadhav J and Govindwar S 2008 *Acta Chim. Slov.* **55** 320
[4] Vandevivere P C,Bianchi R and Verstraete W 1998 J. *Chem. Technol. Biotechnol.* **72** 289
[5] Colar L A, Ilinoiu E C, Manea F and Pode R 2012 *Environ Eng Manag J* **11** 61
[6] Zucca P, Rescigno A, Olianas A, Maccioni S and Sollai F 2011 *J Mol Catal B Enzym.* **68** 216
[7] Pakshirajan K and Jaiswal S 2011 *J Sci Ind Res* **70** 987
[8] Zhang J, Feng M Y, Jiang Y C, Hu M C, Li S N and Zhai Q G 2012 *Chem. Eng. J.* **191** 236
[9] Kim G Y, Lee K B, Cho S H, Shim J and Moon S H 2005 *J. Hazard. Mater.* **126** 183
[10] Cheng X B, Jia R, Li P S, Tu S Q, Zhu Q, et al. 2007 *Enzyme Microb. Technol.* **41** 258
[11] Giardina P, Faraco V, Pezzella C and Piscitelli A 2010 *Cell Mol Life Sci* **67** 369
[12] Morozova O V, Shumakovich G P, et al. 2007 *Appl Biochem Microbiol* **43** 523
[13] Rodríguez Couto S and Toca Herrera J L 2006 *Biotechnol Adv* **24** 500
[14] Zamora P P, Pereira C M, Tiburtiusm E R L, et al. 2003 *Appl. Catal. B: Environ* **42** 131
[15] Zille A, Tzanov T, Gübitz G M and Paulo A C 2003 *Biotechnol. Lett* **25** 1473
[16] Dayaram P and Dasgupta D 2008 *J Environ Biol* **29** 831
[17] Lu L, Zhao M and Wang Y 2007 *World J Microbiol Biotechnol* **23** 159
[18] Wang Y, Zheng X and Zhao M 2008c *J Chem Eng Chin Univ* **22** 83
[19] Xu X H, Lu P, Zhou Y M, Zhao Z Z and Guo M Q 2009 *Mater. Sci. Eng. C* **29** 2160
[20] Salis A, Pisano M, Monduzzi V and Solinas E 2009 *J. of Mol. Catal. B: Enzym.* **58** 175
[21] Bautista L F, Morales G R and Sanz R 2010 *Bioresour. Technol.* **101** 8541
[22] Forde J, Tully E, Vakurov A, Gibson T D, et al. 2010 *Enzyme Microb.Technol.* **46** 430
[23] Schmidt P, Lukens W W, Zhao D Y, Yang P D, et al. 1999 *J Am Chem Soc.* **121** 254
[24] Xue P, Xu F and Xu L D 2008 *Appl Surf Sci.* **255** 1625
[25] Yang Y X, Zhang J B, Yang W M, Wu J D and Chen R S 2003 *Appl Surf Sci.* **206** 20

[26] Huang L and Cheng Z M 2008 *Chin J Catal* **29** 57
[27] Zhang R F, Long N B and Zhang L L 2009 *Thin Solid Films* **517** 6677
[28] Long N B and Zhang R F 2009 *Chin J Inorg Chem.* **25** 1153
[29] Li X F and Zhang R F 2013 *Chin J Mater Res.* **27** 526
[30] Shang C Y, Li W X and Zhang R F 2014 *Enzyme Microb. Technol.* **61** 28
[31] Shang C Y, Li W X and Zhang R F 2015 *J. of Mol. Catal. B: Enzym.* **113** 9
[32] Bradford M M 1976 *Analytical Biochemistr.* **72** 248
[33] Wang F, Guo C, Yang L R and Liu C Z 2010 *Bioresour. Technol.* **101** 8931
[34] Antczak S M, Kubiak A, Antczak T and Bielecki S 2009 *Renew Energ.* **34** 1185
[35] Takahashi H, Li B, Sasaki T, Miyazaki C, et al. 2001 *Micropor Mesopor Mat.* **755** 44
[36] Takahashi H, Li B, Sasaki T, Kajino T and Inagaki S 2000 *Chem. Mater.* **12** 3301
[37] Kunamneni A, Ghazi I, Camarero S, Ballesteros A, et al, 2008 *Process Biochem.* **43** 169

Whey protein isolate modified by transglutaminase aggregation and emulsion gel properties

Weiwei Qi, Chong Chen, Mujun Liu, Guoping Yu[1], Xinghang Cai, Peipei Guo, Yuxiu Yao and Sijie Mei

College of Food Science, Northeast Agricultural University, No. 59 Mucai Street, Xiang Fang District, Harbin 150030, China

E-mail:neauqiww@163.com

Abstract. Whey protein isolate and commercial soybean salad oil were used to produce the WPI emulsion dispersions. The properties of TG-catalyzed emulsion gelation produced from WPI emulsion dispersions were investigated by the amount of TG, temperature, pH and reaction time. Specifically, the texture properties (hardness and springiness), water-holding capacity and rheological properties (G 'and G ") were assessed. The result of Orthogonal tests showed WPI emulsion can form better hardness and springiness gel when the ratio of TG and WPI was 20U/g, pH 7.5, treatment temperature and time were 50℃ and 3 h, respectively. The microstructure of TG emulsion gels was more compact, gel pore is smaller, distribution more uniform, the oil droplets size smaller compared with untreated emulsion gels. Compared to the control of rheological properties, G ' and G " were significantly increased and G ' > G ", results showed that the gel was solid state, and TG speeded up the process of gelation.

1. Introduction

Whey protein isolate (WPI) is an economically important by-product of the cheese manufacturing process, used as an ingredient by the food industry because of its high nutritional value and functional attributes. Whey proteins comprise of a mixture of proteins, dominated by β-lactoglobulin (β-Lg) and α-lactalbumin with minor amounts of proteose-peptones, serum albumin and immu-noglobulins [1]. β-Lg is the major protein present in bovine milk whey. It exists, as a dimer with a molecular weight of 36,400 Da. Two disulfide bonds and a free thiol group exist within β-lactoglobulin [2]. Due to whey protein small molecular weight and more spherical shapes, it has the ability to form a stable functional properties and good emulsion gelation, therefore can prepare emulsified gel [3], and the gel matrix can retain a lot of water and other non-protein components, and can effect on texture and sensory properties of different food [4].

As the active factors, droplets can improve the emulsion gel strength and water-holding capacity. Yogurt, fresh cheese, gelatin, pudding, milk, snacks, hot dogs and other food products belong to these emulsion gel. Fat plays an important role to improve the texture characteristics of emulsion gel food. Rheological properties of such composite gel are strongly dependent on the volume fraction of filler

[1] Address for correspondence: Guoping Yu, College of Food Science, Northeast Agricultural University, No. 59 Mucai Street, Xiang Fang District, Harbin 150030, China. Tel.: 0451-55190637; E-mail: yuguopingneau@hotmail.com.

particles and the interaction with the gel matrix and oil droplets. Therefore, the research about emulsion gel to improve the textural properties of the food is necessary. In the present work, there are few papers published concerning with whey protein emulsion gel.

To be competitive in food ingredient markets, the functionality of whey proteins must be continuously improved and designed for specific end uses. Chemical and physical methods are commonly used. Food proteins can have their functionality altered by temperature and other chemical means. Specific functional attributes could be obtained by enzymatic polymerization of proteins. Here the enzymatic reaction could be controlled by time to enhance the functionality to the desired level [5]. The enzyme transglutaminase (EC 2.3.2.13) catalyzes an acyl transfer reaction between the γ-carboxyamide group of peptide-bound, glutamine residues as acyl donors and primary amines as receptors [6]; when transglutaminase acts on protein molecules, intra- and inter-molecular ε-(γ-glutamyl) lysine crosslinks are formed by the enzyme reaction [7]. The inter-molecular cross-linking, which increases the weight average molecular weight of the protein, may improve the gelation and emulsifying properties [8]. Many studies have been carried out the TG-induced polymerization of modified whey protein to obtained different functional properties of gels, such as high hardness [9], the changes temperature of the gel point [10], the reduction of syneresis cracking of gel catalyzed by TG [11], but the properties of whey protein emulsified gel catalyzed by transglutaminase rarely reported. The effect of the amount of TG, pH, temperature, heating time on emulsion gel were observed by texture analysis, water-holding capacity and rheology, and provide the reference about theoretical studies to improve whey protein gel emulsion properties.

2. Materials and methods

2.1. Materials

The whey protein isolate powder used in this study was a commercial product provided by Columbia company (USA), with an approximate composition of >90% protein, <5.1% moisture, <1.5% minerals as per experiment's data. Soybean salad oil (JIUSAN Oil Industry Group Co. Ltd. Harbin, China). Ca^{2+}-independent microbial transglutaminase (TG-B) was provided by Yiming Biological Products Co. Ltd. (Jiangsu, China). The enzyme (composed of lactose, maltodextrin and transglutaminase) presented mean enzymatic activity of 100 U/g, as per manufacturer's data. The enzyme was used in the original form without further purification. All other chemicals were of analytical grade.

2.2. WPI emulsion dispersions preparation

Preparation of 8% (w/v) whey protein isolate emulsion dispersions. WPI solutions were prepared by dispersing the WPI powder in distilled water adjusted to 7.0 pH using either 1.0 M HCl or 1.0 M NaOH, followed by mechanically stirring at room temperature for 120 min, then left overnight at 4℃ to equilibrate and ensure completely hydration. Commercial soybean salad oil was added to whey protein isolate solution, and the ratio of oil is 15% (v/v). This system was stirred and premixed for 15 min at 60℃. Emulsions were obtained using an YQ-3 high speed homogenizer (Scientific research machinery work, Jiangsu, China) at 10000 rpm for 120 s. Emulsions were cooled and stored at 4℃ until further analyses. All samples were prepared in triplicate.

2.3. The effects of TG on the characteristics of emulsion gel

The basic parameters were setting for the temperature of 45℃, cross-linking of time 3 h, pH of 7.5, TG-B for 30 U/g. The sample was heated at 90℃ for 30min, and then gel was rapidly cool to room temperature in ice-water bath. Finally, left overnight at 4℃ before the analysis.

2.3.1. Single factor experimental design. In order to investigate the effects of TG on emulsion gel characteristics, texture properties and water-holding capacity were determined. The amount of TG (0,

5, 10, 15, 20, 25, 30 U/g protein), pH of the reaction (5.5, 6.0, 6.5, 7.0, 7.5, 8.0, 8.5), the temperature (35, 45, 55, 65, 75℃), reaction time (1, 2, 3, 4, 5, 6, 7 h) were presented.

2.3.2. The orthogonal design. According to Univariate tests, four variables the amount of TG, pH, temperature, reaction time were selected to carry out L_9 (3^4) orthogonal test in table 1. Gel hardness, gel elasticity, water-holding capacity as index. Assistant II by orthogonal design software for analysis to determine the transglutaminase catalyzed gel emulsion optimal conditions. The form of factors and levels are shown as followed:

Table 1. The factors and levels of orthogonal test.

Number	Amount of TG A(U/g)	pH B	T C(℃)	Time D(h)
1	15	7.0	40	2
2	20	7.5	45	3
3	25	8.0	50	4

2.4. Texture analysis of emulsion gel

The texture of emulsion gel was determined according to the method of Xin Gu [12] with minor modification by TA-XT plus Texture Analyzer (Stable Micro System, England). The gel (25 mm in height and 50 mm in diameter) was equilibrated at room temperature for 30min before measured. Using P/0.5 cylindrical probe, which moves downward twice during measuring. setting forward speed: 10 mm/s, measuring speed: 10 mm/s; retreating speed: 10 mm/s, compression depth: 10 mm. Three measurements were conducted for each replication, and there were three replications in all treatments. Means and standard deviations were calculated from nine data.

2.5. Water-holding capacity

The water-holding capacity (WHC) of emulsion gel was determined according to the method of Vale'rie Leung Sok Line [13] with minor modification (centrifugal machine, Medicine centrifuge works, Beijing). The WPI emulsion gel was weighed and placed into 50 mL centrifugal tubes, and centrifuged at 9000 rpm for 20 min, then measured the weight of overflow water. Values are means of three replications with duplicate measurements. Water-holding Capacity equation is as follows:

$$\text{WHC}\ \% = (W_t - W_r)/W_t \times 100$$

Where W_t (g) is the weight of total moisture of gel and W_r (g) is the weight of overflow water after centrifuging.

2.6. Scanning electron microscopy

The gel was spiced into squares of 3×3×3 mm^3, and were determined according to the method of Chin K B [14]. Accelerating voltage 5 KV, observing and photographing the sample particle morphology with Japan JME-100CXII TEM.

2.7. Rheology

The Rheology of emulsion gel was determined according to the method of Li [15] with slight changes by MALI038384 Rotational rheometer (Malvern Instruments Ltd. England). Rheological measurements carried out using parallel-plate with 20 mm diameter and 0.3 mm gap. In order to prevent evaporation of moisture during the test, the sample was sealed around paraffin oil. The gelation conditions were the same as those previously described in Section 2.3. Heating temperature from 30℃ to 85℃ by 3℃/min constant heating rate and maintain 10 min. Oscillation frequency of 1

Hz, the stress amplitude of 0.122, which is the linear viscoelastic region, as measured by the initial trial. The parameter G' (storage modulus) and G" (loss modulus) were measured.

2.8. Statistical analysis

Each experiment was repeated three times. The data are the average of three times, standard deviation of the error term; Analysis of variance was used ($p<0.05$) and when the effect of the factors was significant. Data were statistically analyzed using SPSS13.0 software.

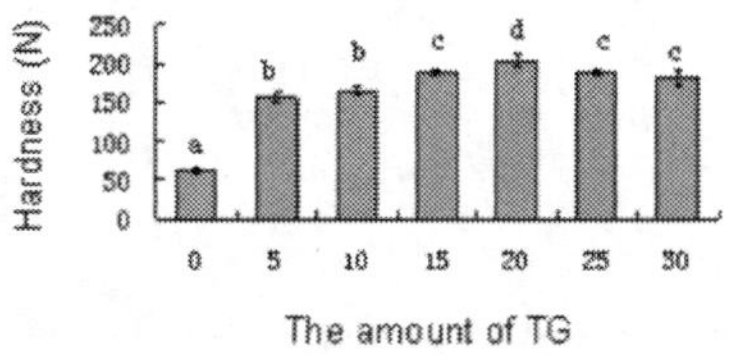

Figure 1. Effect of enzyme dosage on hardness of TG-WPI emulsion gels.

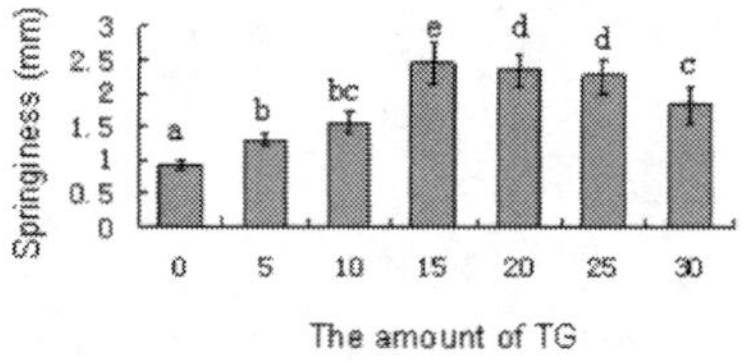

Figure 2. Effect of enzyme dosage on springiness of TG-WPI emulsion gels.

3. Results and discussion

3.1. The effects of the amount of TG on texture and water-holding capacity

The gel strength increased significantly at TG dosage (0-20 U/g).The emulsion gel strength reached maximum with the amount of enzyme 20 U/g; over 20 U/g, the gel strength decreased with the amount of TG increasing (figure 1). Figure 2 showed that amount of TG was 0-15 U/g, the gel springiness significantly enhanced; and then decreased when the dosage was over 15 U/g.

The α-lactalbumin structure has 8 glutamine residues and 12 lysine residues; while β-lactoglobulin has 16 glutamine residues and 15 lysine residues in its protein chain [16]. However, not all of these residues are available for enzymatic reaction with transglutaminase, due to the globular structure of whey proteins. The rate of crosslinking by transglutaminase is dependent on the macromolecular structure of each protein substrate; reactive glutamine residues, i.e., reside in flexible regions of the polypeptide chain or in regions with reverse turns [17]. Nieuwenhuizen et al. [18] investigated the accessibility of the lysine and glutamine residues of a-lactalbumin to the microbial transglutaminase reaction and showed that a maximum of five lysines and five glutamines can be modified by transglutaminase depending on the temperature, pH, and the presence or absence of calcium (Ca^{2+}).

Transglutaminase concentrations above 20 U/g of protein may be higher than the quantities required for the reaction with the available residues of α-lactalbumin and β-lactoglobulin, saturation of the protein substrate may occur, without altering, even decreasing the hardness and springiness of the gel at transglutaminase concentrations higher than 20 U/g of protein, this probable due to excessive crosslinks. Similar results were observed by Han Cui-ping et al. [19]. The polymerization of whey proteins, the gel strength reached maximum at the amount of 10 U/g of TG, over 10 U/g the strength decreased.

The water-holding rate was high at 94-98% at the of 5-30 U/g, Figure 3 showed when TG dosage was 0-5 U/g, the water-holding capacity of gel was significantly increased; subsequently slowly improved with 5-15U /g; enzyme dosage is 15 U/g, the WHC was maximum.As a consequence, when TG dosage was a range of 5-20 U/g, emulsion gel strength, springiness and WHC were at optimum values.

3.2. The effects of pH on texture and WHC of emulsion gel

Emulsion gel strength increased with increasing pH.The gel strength significantly increased at pH 5.0-7.5, and reached maximum when the pH is 7.5, whereas when the pH continues increasing, gel hardness declined (figure 4).Tang Chuan-he [20] also confirmed the optimum pH of TG catalyzed β-

IOP Conf. Series: Materials Science and Engineering **87** (2015) 012034 doi:10.1088/1757-899X/87/1/012034

lactoglobulin was 7.5, however the pH over 7.5, the catalysis of TG was weakened, TG was almost inactivation, when pH was over 8.5. Figure 5 described the effect of pH on gel springiness of TG-WPI emulsion gels. Like hardness, gel elasticity increases with increasing pH. Gel elasticity increases slowly at pH values 5.0-6.5, but at pH 6.5-7.5, the elasticity increases significantly and reached maximum at pH 7.5, the treatment of pH exceeded 7.5 was not found to significantly increase.

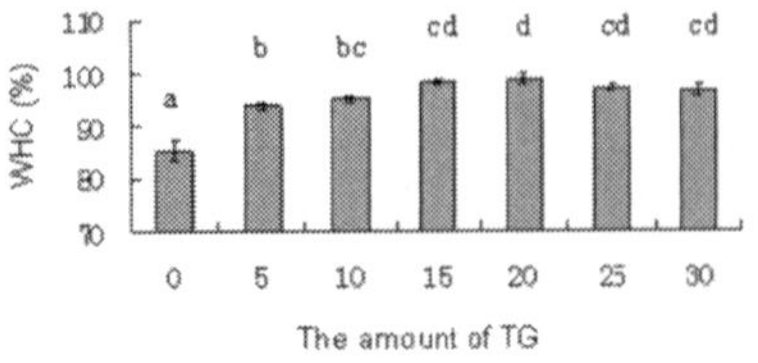

Figure 3. Effect of enzyme dosage on water-holding capacity of TG-WPI emulsion gels.

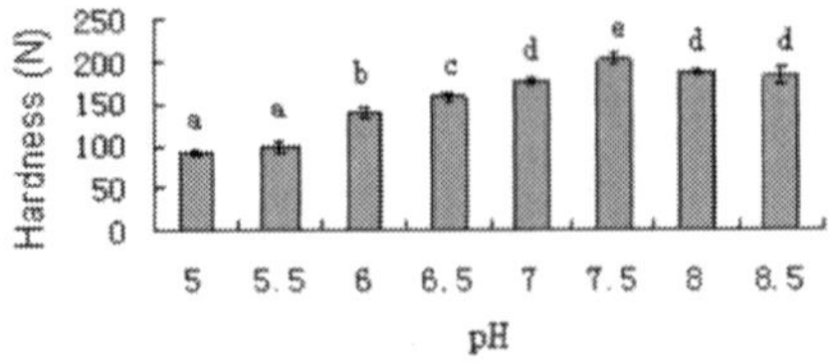

Figure 4. Effect of pH on gel hardness of TG-WPI emulsion gels.

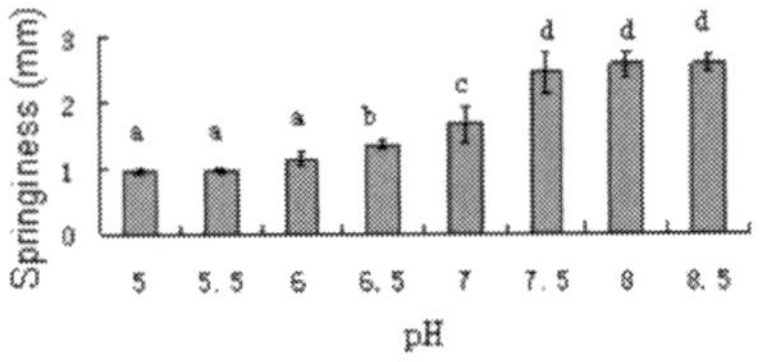

Figure 5. Effect of pH on gel springiness of TG-WPI emulsion gels.

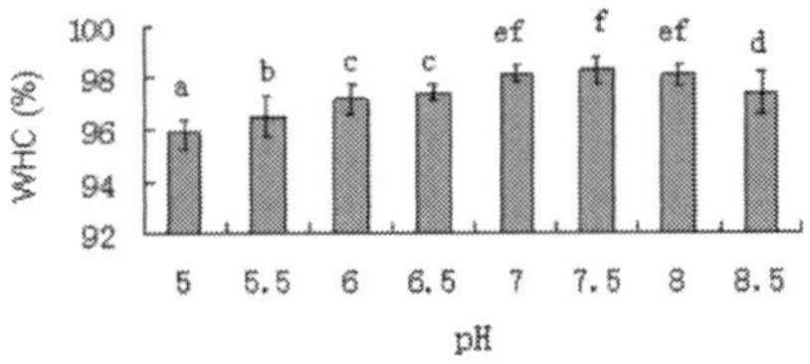

Figure 6. Effect of pH on water-holding capacity of TG-WPI emulsion gels.

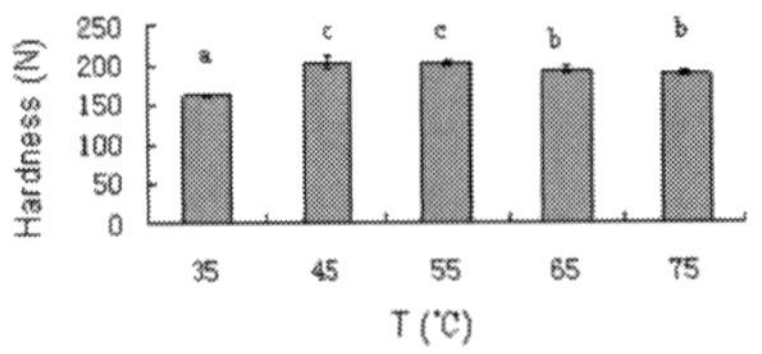

Figure 7. Effect of temperature on hardness of TG-WPI emulsion gels.

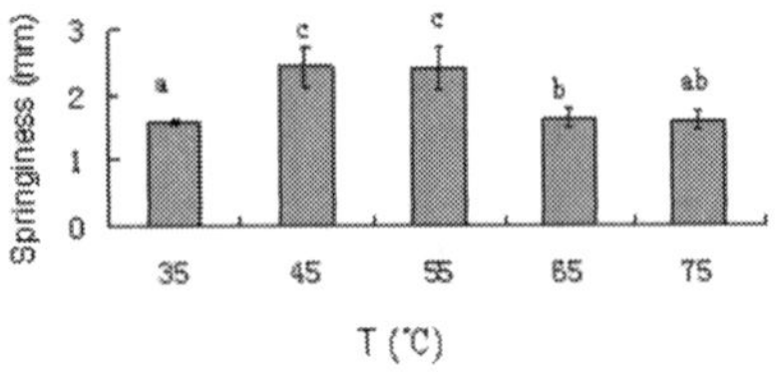

Figure 8. Effect of temperature on springiness of TG-WPI emulsion gels.

As shown in figure 6, WHC of emulsion gel significantly increased at pH 5.0-7.0. When the pH was 7.0-8.0, the WHC was not significant increased. Overall, the emulsion gel strength, elasticity and water-holding capacity are at optimum values at pH 7.0-7.5.

3.3. The effects of temperature on texture and WHC of emulsion gel

Emulsion gel strength increased with increasing temperature. Specially at 35-45℃, the strength is significantly increased. Each of enzymes itself has optimal temperature and pH. The optimal temperature of TG is at the range of 45-55℃, when the temperature is higher 55℃, the TG is passivated even inactivated. Therefore, as the temperature continues to rise, the emulsion gel strength is slightly decreased (figure 7).

As shown in figure 8, Emulsion gel elasticity increases with increasing temperature, the gel elasticity is significantly improved at 35-45℃; however, the elasticity has no significant change at 45-55℃; as the temperature continues to rise, the elastic gel is significantly reduced. The effect of temperature on WHC is the same like to hardness. When the temperature is 45℃, WHC of emulsion gel is maximum (figure 9). According to above analysis, the optimum values of gel strength, elasticity, WHC are at the range of 45-55℃.

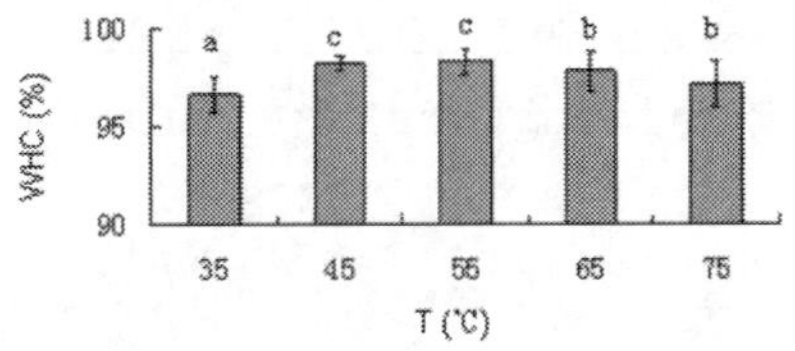

Figure 9. Effect of temperature on water-holding capacity of TG-WPI emulsion gels.

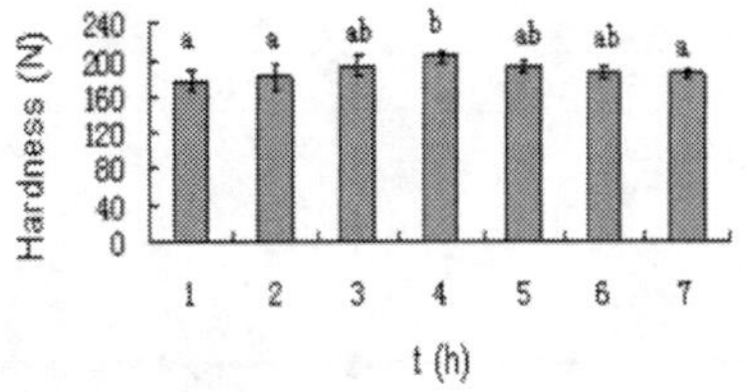

Figure 10. Effect of time on hardness of TG-WPI emulsion gels.

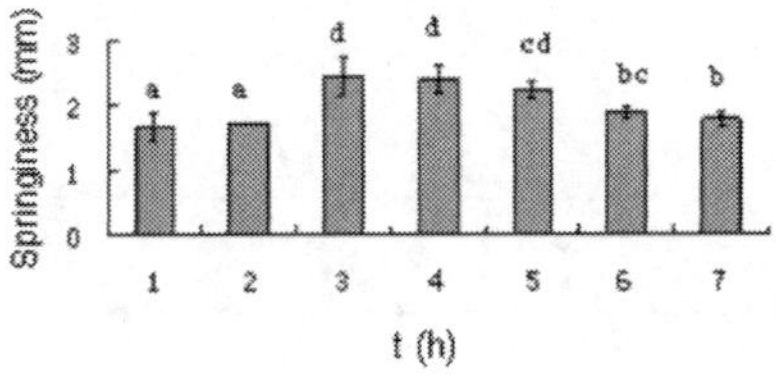

Figure 11. Effect of time on springiness of TG-WPI emulsion gels.

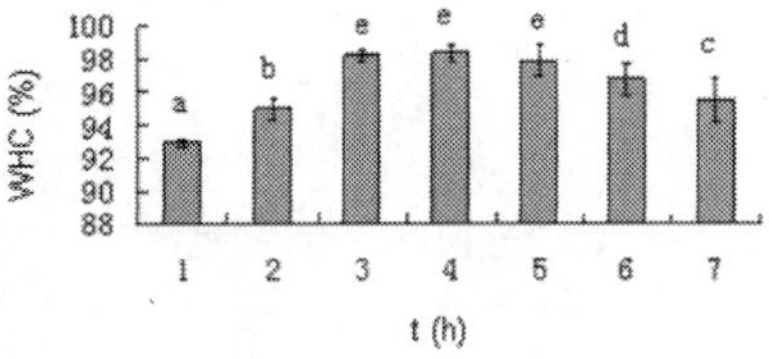

Figure 12. Effect of time on water-holding capacity of TG-WPI emulsion gels.

3.4. The effects of heating-time on texture and WHC of emulsion gel

The emulsion gel strength showed no significant change as heating-time extended, gel strength was maximized when heated 3-4 h (figure 10). As shown in figure 11, gel elasticity significantly increased when the reaction 2-3 h; at the time of 5-7 h, the emulsified gel elasticity decreased. As reaction time extended, the WHC increased firstly, then decreased, when the reaction was at 3-4 h, WHC reached the maximum (figure 12).This could be related that with heating-time extending, whey protein occured so excessive crosslinks that it is not prone to aggregate and form gel. Another reason maybe the mechanism of TG catalysis, the TG enzyme can catalyze and decomposite glutamine at the present of H_2O, thereby as the reaction extended, TG probably promotes reverse reaction and parts of aggregation are resolved.

3.5. Result of orthogonal design

According to single factor experimental results, L_9 (3^4) orthogonal test was carried out by optimum levels of the amount of enzyme added (A), pH (B), temperature (C), reaction time (h) in table 2 and table 3. The characteristics of gel were evaluated by lining-up score method to determine the optimum process conditions of the emulsion gel of whey protein isolate.

Table 2. L_9 (3^4) Orthogonal experiment design.

Test number	TG Amount	pH	T	Time
	A(U/g)	B	C(℃)	D(h)
1	1(15)	1(7.0)	1(40)	1(2)
2	1(15)	2(7.5)	2(45)	2(3)
3	1(15)	3(8.0)	3(50)	3(4)
4	2(20)	1(7.0)	2(45)	3(4)
5	2(20)	2(7.5)	3(50)	1(2)
6	2(20)	3(8.0)	1(40)	2(3)
7	3(25)	1(7.0)	3(50)	2(3)
8	3(25)	2(7.5)	1(40)	3(4)
9	3(25)	3(8.0)	2(45)	1(2)

Table 3. Data of $L_9(3^4)$ orthogonal experiment.

Number	Hardness (g)	springiness	WHC (%)
1	180.891±8.091	1.921±0.291	96.321±1.029
2	190.221±5.093	1.322±0.383	94.928±4.039
3	179.391±3.023	2.019±0.198	94.338±8.043
4	193.981±9.004	1.823±0.284	95.271±5.093
5	211.169±4.907	2.113±0.377	98.029±4.972
6	201.135±5.095	2.531±0.163	97.832±3.672
7	198.327±6.784	2.939±0.185	95.021±1.091
8	200.223±2.907	2.312±0.330	96.406±3.903
9	195.147±4.096	2.011±0.023	95.987±5.001

Table 4. Result data of lining-up score.

Number	Hardness scores	Springiness scores	WHC scores	comprehensive scores
1	1.441	1.000	5.875	8.316
2	4.080	4.889	5.054	14.023
3	1.000	3.856	1.720	6.576
4	5.055	5.411	3.578	14.044
5	10.000	7.733	10.00	27.733
6	7.012	10.000	5.898	22.910
7	6.371	6.517	3.161	16.049
8	6.907	4.844	5.875	17.626
9	5.566	7.900	1.000	14.466

Table 5. The range value analysis of $L_9(3^4)$ Orthogonal experiment.

	Factors			
	TG Amount A(U/g)	pH B	T C(℃)	Time D(h)
K1	28.915	38.409	48.852	50.515
K2	64.687	59.382	42.533	52.982
K3	48.141	43.952	50.358	38.246
k1	9.638	12.803	16.284	16.838
k2	21.562	19.794	14.178	17.661
k3	16.047	14.651	16.786	12.749
Range	11.924	6.991	2.608	4.912

Table 6. The variance value analysis of $L_9(3^4)$ Orthogonal Experiment.

Factors	DEVSQ(sum of square of deviations)	DF(degree of freedom)	FR(F ratio)	FCV(F critical value)
TG Amount	213.672	2	2.474	4.46
pH	78.742	2	0.912	
T	11.492	2	0.133	
Time	44.529	2	0.481	
Range	345.440	8		

Lining-up score method [21]: For the No. i (i = 1,2, ..., 9) test, Xmax represents the best, in the first place, marked 10 scores; Xmin means the worst, in the last, marked 1 score; for the remaining number of other indicators points, according to proportion of their degree of difference between the outstanding value of the index, namely:

$$\Delta X = (X_{max} - X_{min}) / 9;\ A\ (score) = (X_{max} - X_i);\ final\ score = 10\text{-}A;$$

Then the sum of each number test score for all indicators was comprehensive scores in table 4.

TG has the potential to improve gelation through the formation of covalent intra-and intermolecular bonds. Covalent bonds, such as disulfide and ε-(γ-Glu)-Lys bonds, can restrict the flow of protein chains, thereby enhancing the elasticity and hardness of the network [22]. Orthogonal test results in table 5 showed primary and secondary relationship of the four factors on the impact of the emulsion were A> B> D> C, that is the amount of TG impacts the most, and temperature the least. Optimal conditions established by the test results for the main parameters $A_2B_2C_3D_2$, namely transglutaminase catalyzed emulsion gel processing as follows: enzyme dosage 20 U/g, pH 7.5, temperature 50℃, reaction time 3 h. By variance analysis in table 6, the main factor was enzyme dosage, consistent with the range analysis in table 5. Be verified experimentally, the measured value of the gel strength of 211.927±2.433 g; gel elasticity was 2.018± 0.234, WHC was 98.357± 5.903%.

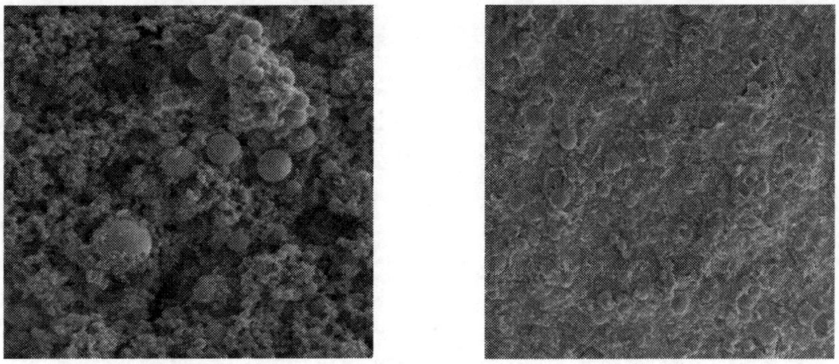

(a) the control (b) TG

Figure 13. Microstructure of TG-WPI emulsion gel observed by scanning electron microscopy. a: 8% (w/v) whey protein isolate emulsions, pH 7.5, heated at 90℃ for 30 min b:enzyme additives 20 U/g, pH 7.5, temperature 45℃ and heating-time 3 h, then heated at 90℃ for 30 min.
[a]Both the micrographs was 4000×magnified.

3.6. The microstructure of TG emulsion gels

Electron micrographs were taken using SEM in order to get insight into the structure of emulsion gel. The optimum conditions were tested rather than the full experimental design. Scanning electron microscopy (SEM) showed that there were differences in the microstructure of the control and experimental gel (figures 13(a) and 13(b)). The differences were mainly associated with the compactness of the protein matrix and the size of the void spaces containing the aqueous portion of the gel. Compared to the control (figure 13(a)), the network structure is more compact, led to emulsion gels with higher density and less porosity, more uniform and smoothly distribution, smaller particle size of droplets.

3.7. Rheology

Rheological properties of gels can be described by the storage modulus (G ') and loss modulus (G' '). When the G ' value is much greater than G' ' values, indicating the status of the sample exhibits solid-like behaviour with high elasticity; the performance characteristics similar to liquid (G' '>G '), on the contrary, has a high viscosity [23].

The control WPI emulsion gel (figure 14(a)) the maximum G 'values and G "values were 12670.6 Pa and 7840.8 Pa. TG catalyzed WPI forming emulsion gel (figure 14(b)), G ' value and G " values were increased by 334.85% and 338.81%, which implied that TG-WPI emulsion gels were more compact than WPI emulsion gels, this agreed with the micrographs (figure 13).

IOP Conf. Series: Materials Science and Engineering **87** (2015) 012034 doi:10.1088/1757-899X/87/1/012034

The gel point was about at 70℃ (figure 14a), probable due to the denatured temperature of whey protein was over 65℃. Although not the true definition, the gel point represented quite accurately the transition point from what was perceived to be a liquid to a more solid-like material. Gel point temperatures were lower for samples that had more extensive cross-linking [24]. The control of G ' increased rapidly at 60℃(figure 14(b)), the gel point decreased by 10℃ compared to figure 14(a), which showed the gel point temperature was decreased, indicating that TG enhanced the ability of the emulsion solutions forming gel.

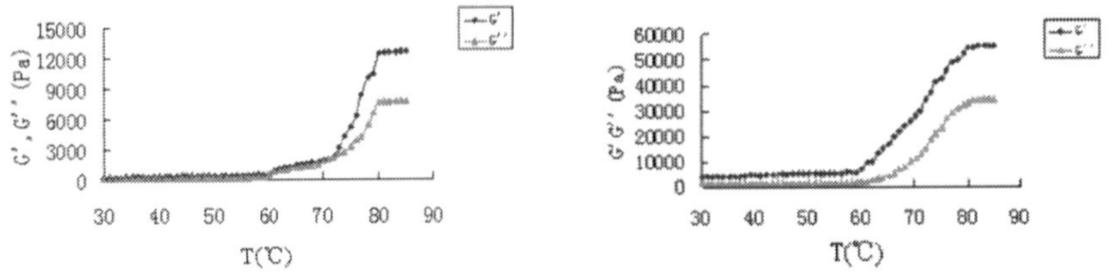

(a) control (b) TG

Figure 14. The temperature ramps of WPI emulsion gel. a: 8% (w/v) whey protein emulsions, pH 7.5; b:enzyme additives 20U/g, pH 7.5, temperature 50℃ and heating-time 3 h.

4. Conclusion

This study demonstrates that the optimum condition of forming emulsion gels catalyzed TG, enzyme dosage 20 U/g, pH 7.5, temperature 50℃, reaction time 3 h. Emulsion gels with high G ' and G " and good water-holding capacity were obtained. SEM show TG leads to changes in the structure of emulsion gels from a particulate to a mixed-type gel, where both fine-stranded and random aggregates are found. TG-WPI emulsion results in emulsion gels of unique properties that can be tailored to specific food applications.

Acknowledgment

The authors gratefully acknowledge the Ministry of Science and Technology of China (Project #2013BAD18B07-03) for funding this work. Thanks are also due to Yiming Biological Products Co. Ltd for providing enzyme samples.

References

[1] Schokker E P, Singh H and Creamer L K 2000 Heat-induced aggregation of β-lactoglobulin A and B with α-lactalbumin *International Dairy Journal* **10** 843-53
[2] Sawyer L and Kontopidis G 1999 β-lactoglobulin–A three-dimensional perspective. *International Journal of Food Science and Technology* **34** 409-18
[3] Han Xue and Sun Bing 2003 The functions and main applications of whey protein *China Dairy Industry* **3** 28-30
[4] Singh H 1991 Modification of food proteins by covalent cross linking *Trends in Food Science and Technology* **2** 196-200
[5] Mangino M E 1992 Gelation of whey protein concentrates *Food Technology* **46** 114-8
[6] Motoki M and Seguro K 1998 Transglutaminase and its use for food processing *Trends Food Sci Technol* **9** 204-10
[7] Kuraishi C, Yamazaki K and Susa Y 2001 Transglutaminase: Its utilization in the food industry *Food Rev Int* **17** 221-46
[8] Liu, M X and Damodaran S 1999 Effect of transglutaminase-catalyzed polymerization of β casein on its emulsifying properties *Journal of Agricultural and Food Chemistry* **47** 1514-9

[9] Eissa A S, Bisram S, Khan S A 2004 Polymerization and gelation of whey protein isolates at low pH using transglutaminase enzyme *J Agric Food Chem.* **52** 4456-64

[10] Truong V D, Clare D A, Catignani G L and Swaisgood H E 2004 Cross-linking and rheological changes of whey proteins treated with microbial transglutaminase *J Agric Food Chem.* **52** 1170-6

[11] Gauche C, Tomazi T, Barreto P L M and Bordignon-Luiz M T 2009 Physical properties of yoghurt manufactured with milk whey and transglutaminase *LWT–Food Science and Technology* **42** 239-43

[12] Xin G U, Lydia J. Campbell and Stephen R Euston 2009 Influence of sugars on the characteristics of glucono-δ-lactone-induced soy protein isolate gels *Food Hydrocolloids* **23** 314-26

[13] Vale´rie Leung Sok Line, Gabriel E. Remondetto and Muriel Subirade 2005 Cold gelation of β-lactoglobulin oil-in-water emulsions *Food Hydrocolloids* **19** 269-78

[14] Chin K B, Goa M Y and Xiong Y L 2009 Effect of soy protein substitution for sodium caseinate on the transglutaminate-induced cold and thermal gelation of myofibrillar protein *Food Research International* **4** 1-8

[15] Li J, Ould Eleya M M and Gunasekaran S 2006 Gelation of whey protein and xanthan mixture: Effect of heating rate on rheological properties *Food Hydrocolloids* **20** 678-86

[16] Morr C and Ha E W 1993 Whey protein concentrates and isolates processing and functional: Properties critical reviews *Food Sci Nutr.* **33** 431-76

[17] Dickinson E 1997 Enzymatic crosslinking as a tool for food colloid rheology control and Interfacial stabilization *Trends Food Sci Technol.* **8** 334-49

[18] Nieuwenhuizen W F, Dekker H L, Koning L J, Gro neveld T, Koster C G and Jong G A H 2003 Modification of glutamine and lysine residues in holo and apo-a-lactalbumin with microbial transglutaminase *J Agric Food Chem.* **51** 7132-9

[19] Han Cui-ping, Huo Gui-cheng, Duan Cui-cui 2009 Study on texture characteristic of whey protein polymerized by MTGase with different temperature and amount of MTGase *China Dairy Industry* **33** 28-30

[20] Tang Chuan-he,Yang Xiao-quan, Peng Zhi-ying 2002 Polymerization of whey protein catalyzed by microbial transglutaminase *China Dairy Industry* **30** 11-4

[21] Yuan Yufeng 2005 Analysis of multi-target orthogonal experiment *Journal of Hubei Automotive Industries Institute* **19** 53-6

[22] Chanyongvorakul Y, Matsumura Y, Nonaka M, Motoki M and Mori T 1995 Physical properties of soy bean and broad bean 11S globulin gels formed by transglutaminase reaction *Journal of Food Science* **60** 483-8

[23] Vale´rie Leung Sok Line, Gabriel E. Remondetto and Muriel Subirade 2005 Cold gelation of β-lactoglobulin oil-in-water emulsions *Food Hydrocolloids* **19** 269-78

[24] Wilcox C P and Swaisgood H E 2002 Modification of the rheological properties of whey protein isolate through the use of an immobilized microbial transglutaminase *J. Agric. Food Chem.* **50** 5546–51

Enhancement of impact strength of poly (methyl methacrylate) with surface fine-tuned nano-silica

Bin Wen, Yixiao Dong, Lili Wu, Chao Long and Chaocan Zhang[1]

School of Materials Science and Engineering, Wuhan University of Technology, Wuhan 430070, PR China

E-mail: polymers@whut.edu.cn

Abstract. Highly dispersible nanoparticles in organic solvent always receive wide interests due to their compatibility with polymer materials. This paper reported a kind of isopropanol alcohol silica dispersion which obtained using a method of azeotropic distillation. The isopropanol alcohol dispersed silica (IPADS) were treated with coupling agents to fine-tune their surface properties. Polymethyl methacrylate (PMMA) was then used as a research object to test the compatibility between IPADS and polymer. UV-vis spectra indicate that IPADS would reach its high compatibility with PMMA if coupling with trimethoxypropylsilane (PTMS). Followed experiments on PMMA proved that the high compatibility can prominently enhance the impact strength about 30%. The results may provide reference both for nano-silica modification and better understanding of nano-enhanced materials.

1. Introduction

Nanomaterials had been broadly applied in polymer composites to improve their performance, such as strength, thermal stability, electrical properties and optical properties [1-3]. Nano-silica is one of the most widely used nanomaterials because of its low cost and potential application in different fields including paints, catalysis, etc. Mostly, the surface of silica was modified by varied types of coupling agent to improve their compatibility with polymers, like oleic acid, methyltriethoxysilane, etc [4-6]. The coupling agents also help nanosilica to form complex with polymers hence increase their performance [7-10]. This also provide an important step to stop agglomerates and lower the intersurface energy between silica and polymer [5, 11-13]. For this purpose, different modification methods were highlighted in recent literature [14].

PMMA complexed with SiO_2, ZnO, CeO_2 had been studied to improve performances of high transparency, better mechanical strength, ultraviolet resistance, and the like [3, 15-20]. However, many existed works didn't fully get rid of the defects of agglomerate while fabricating a composite. The core problem is low interface compatibility. Herein nano-silica was surface modified with three kind of silane coupling agents in order to improve their interface compatibility with PMMA. The results were further test by UV-vis spectrum and the impact strength of PMMA. The water dispersed nano-silica was firstly transferred into isopropyl alcohol phase by a simple method of azeotropic distillation. Three kinds of coupling agent were selected to improve surface property of nano silica. The final test of silica modified PMMA bulk confirmed that nano-silica can effectively improve the

[1] Address for correspondence: Chaocan Zhang, School of Materials Science and Engineering, Wuhan University of Technology, Wuhan 430070, PR China. E-mail: polymers@whut.edu.cn.

performance of PMMA in light transmission and impact strength.

2. Experimental section

2.1. Materials

Isopropyl alcohol (IPA, ≥ 99%), Benzoyl peroxide(BPO, ≥ 99%) was purchased from Sinopharm Chemical Reagent Co., Ltd. Aqueous silica sol (25 wt% in water) was purchased from Jiangyinguolian Chemical Industry Co., Ltd. Hydrochloric acid (36%~38%) was purchased from Xinyang Chemical Reagent Co., Ltd. Methyl methacrylate (MMA, ≥99%) was purchased from Tianjin Kermel Chemical Reagent Co., Ltd. 3-chloropropylmethyldimethoxysilane (CPMDMS, ≥ 95%), 3-chloropropyltrimethoxysilane (CPTMS, ≥98%), trimethoxypropylsilane (PTMS, ≥ 97%) was purchased from Aladdin Chemical Co., Ltd. All chemicals were used as received without further purification.

2.2. Preparation of IPA dispersed silica nanoparticles (IPA-DSN) and its surface modification

The pH of 20g aqueous silica was adjusted to 2.0 by 5 N HCl and added to a 500 mL distillation flask which equipped with a magnetic stir bar. Then silica sol mixed with 180 g IPA for azeotropic distillation in order to remove water. When the boiling point of distillate raised from 80.4°C to 82.4°C, the transition finished, keeping on the distillation until the left silica sol in flask was about 45 g. The distillation product was subsequently purified by 220 nm millipore filter.

A round bottom flask which equipped a reflux condenser and a magnetic stir bar was charged with 40 g IPA-DSN (10 wt% is silica) and required silane coupling agents (about 20 wt% to silica mass). The mixture was heated to 75°C for 4h to finish nanoparticles surface modification.

2.3. Preparation of nano-enhanced poly (methyl methacrylate)

The four- necked flask containing MMA (50 g) and BPO (initiator, 0.25 g) was heated to 85°C in oil bath for 20min under nitrogen. Modified IPA-DSN was added under vigorous stirring to form a clear dispersion. Then the four- necked flask was rapidly taken to water bath under 40°C. The dispersion was applied a vacuum environment to remove residual IPA until the boiling point of the dispersion reached 100°C. At last, the PMMA serum was cured at 65°C for 2 h and 50°C for 12 h.

2.4. Sample characterization

Dynamic light scattering (DLS) data was record by Malvern Instruments ZEN1600 series; FTIR data was collected on Lambda 750 S which was purchased from PerkinElmer; Transmission electron microscopy (TEM) was performed with JEM-2100F STEM/EDS working at a voltage of 160 kV; TG was taken on Power-Compensation Differential Scanning Calorimeter which was purchased from PE; UV / visible spectroscopy were measured by UV-8800 series which was produced by Jiangsu Skyray Instrument Co., Ltd; Impact strength test was performed on XJJ-50.

3. Results and discussion

FTIR spectra in figure 1 confirm the silane coupling agent is partly grafted on the surface of nano silica. Detailed analysis is described as follows. Firstly, there is a strong absorption band at 800 cm^{-1}, which corresponds to the C-Si stretching vibration or Si-O-Si stretching vibration in silica sol. The absorption band in 1630 cm^{-1} corresponds to the O-H stretching vibrations of water, and 2960 cm^{-1} indicates the stretching vibration of asymmetric C-H bond. After treated by CPMDMS(spectrum b) or CPTMS(spectrum c), two new absorption peaks appear at 698 and 1415 to 1440 cm^{-1}, corresponding to the C-Cl stretching vibrations of halogenated hydrocarbon and alkyl group in coupling agent. A strong absorption band is enhanced at 2960 cm^{-1}, indicating that there are alkyl groups on the surface of silica sol, which has been modified by PTMS (spectrum d).

However, FTIR is not sufficient to confirm our aimed product since some peaks are relatively weak. So the modified IPA-DSN was test by thermal gravimetric (TG) analysis, additionally. The TG plot (figure 2a) showed an initial weight loss of 1.38% up to 89.5°C due to the adsorbed water molecules in silica which had been modified by CPMDMS. The weight loss increased significantly near 300°C, indicating the decomposition of organic part on silica surface. The followed TG plots (figures 2b and 2c) present similar results. Above all, it can be concluded that CPMDMS, CPTMS and PTMS are successfully grafted onto silica sol.

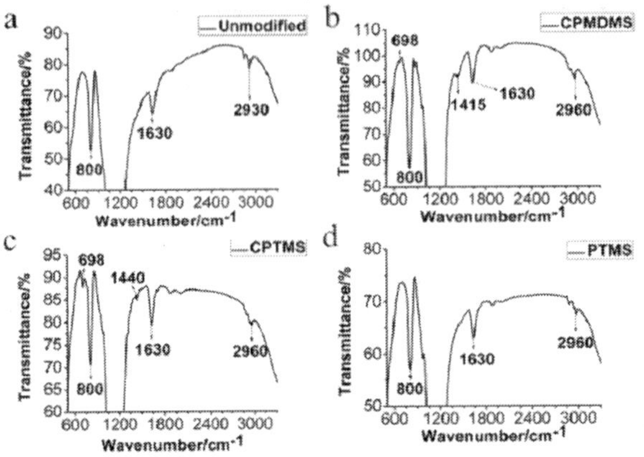

Figure 1. FTIR spectra of nano-silica: (a) Unmodified, (b) modified by CPMDMS, (c) modified by CPTMS, (d) modified by PTMS.

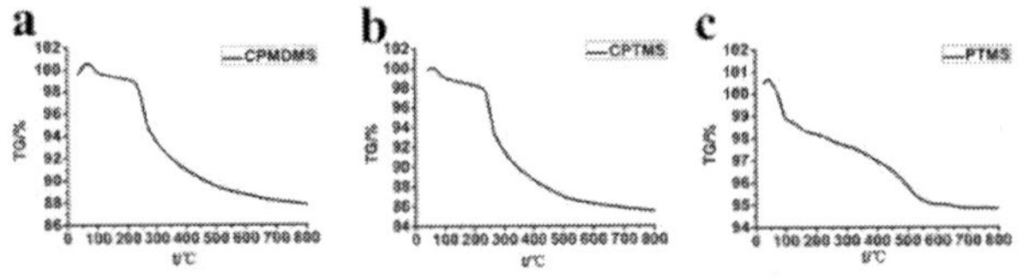

Figure 2. TG curves of nano-silica: (a) modified by CPMDMS, (b) modified by CPTMS, (c) modified by PTMS.

We also did a quantitative calculation of the grafting rate based on TG data (table 1), which had listed the number of grafting rate per square nanometer. The grafting rate could be calculated by equation (1),

$$\eta = \left(\frac{\alpha}{M-60} \cdot N_A\right)\left[\left(\frac{4\pi}{3}\left(\frac{d}{2}\right)^3 \rho\right)^{-1} \pi d^2\right]^{-1} \quad (1)$$

where α is the weight loss of the sample (measured by TG), N_A is the Avogadro constant, d is average diameter of nanosilica particles, ρ is density of silica and M is the molecular mass of coupling agents.

Table 1. Grafting rate of polymer on nano-silica surface.

Coupling Agent	CPMDMS	CPTMS	PTMS
Weight loss	10.17%	12.04%	5.01%
Grafting Rate(N/nm^2)	3.73	3.90	2.16

Furthermore, the quality of original silica dispersion is highly essential for nanocomposites so that we test it with DLS (table 2). It is obvious that the particle size (ca 17 nm) has not been much affected after dealing with silane coupling agent, so was its size distribution (demonstrated by PDI data).

Table 2. The DLS data of IPA-DSN.

Nano-silica Sample	Unmodified	CPMDMS	CPTMS	PTMS
Average Size	17.2	18.1	17.6	16.6
PDI	0.23	0.228	0.235	0.187

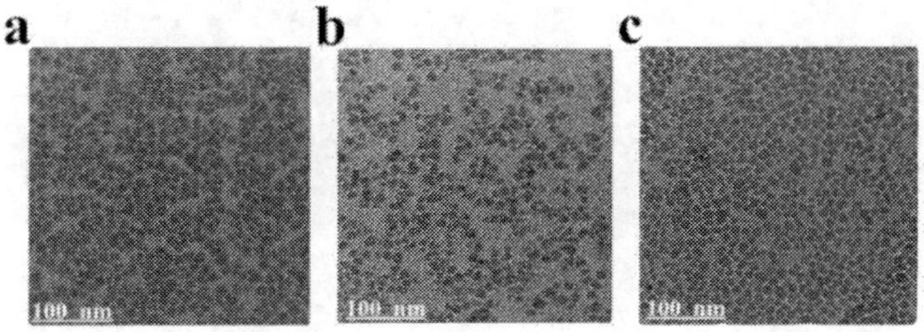

Figure 3. TEM image of (a) water dispersed nanosilica, (b) IPA-DSN. (c) modified silica sol by CPMDMS.

Images of transmission electronic microscopy (TEM, figure 3) also present the same results, of which with the original particle size of water dispersed nanosilica was about 10nm. Its size didn't have much change after transferred into IPA and modified by silane coupling agent (figures 3b and 3c). The result was coincident with DLS data and demonstrates the high quality of our nano silica materials.

Because PMMA is a kind of ultra transparent materials, it is very convenient to observe agglomeration of nano silica via transmittance. For this reason, UV / visible spectrum was applied as one of methods to judge the compatibility (figure 4). It is obvious that the light transmittance of PMMA modified by PTMS present the highest transmittance among the three, which means silane coupling agent which contain propyl groups have better compatibility with MMA.

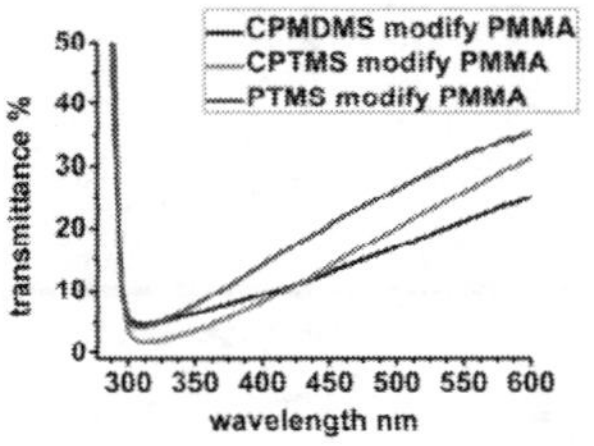

Figure 4. UV-vis spectrum for nanosilica modified-PMMA.

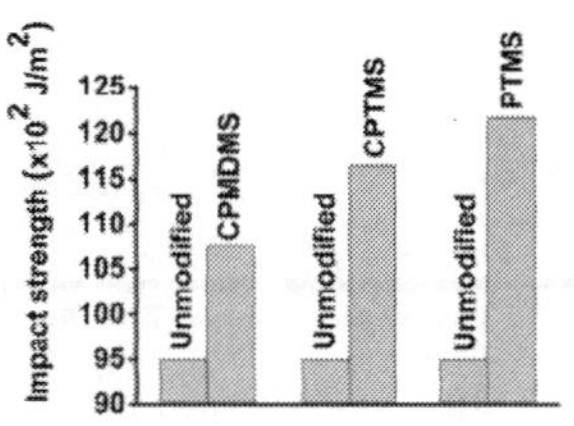

Figure 5. Impact strength of prepared PMMA

Additionally, the impact strength of PMMA further confirm the result acquire by UV-vis spectrum (figure 5). The most transparent sample which tailored by PTMS receive best mechanical performance. Nevertheless, the sample which was tailored by other two silane coupling agents also have great enhancement on impact strength compared to the unmodified one.

4. Conclusions

In this work, we have made three main contributions. 1) We have successfully transferred nanosilica into IPA phase without further aggregation. 2) Three kinds of silane coupling agent were applied to improve surface properties of IPA-DSN. The grafting effeciency was demonstrated by IR and TG measurement. 3) It is clear that IPA-DSN which has been modified by different coupling agent can improve the impact performance of PMMA without much changing of its transmission. There was also a comparison of three kind of as-used silane agents. The propyl group can improve the compatibility between PMMA and nanosilica better than other two functional groups. Above all, this study may provide reference for future work of fabricating nano doped composite materials.

References

[1] Castelvetro V and De Vita C 2004 *Advances in Colloid and Interface Science* **108** 167-85
[2] Yu Y Y and Chen W C 2003 *Materials Chemistry and Physics* **82** 388-95
[3] Kos T, Anžlovar A, Pahovnik D, Zagar E, Orel Z C and Zigon M 2013 *Macromolecules* **46** 6942-8
[4] Bauer F, Gläsel H J, Hartmann E, Langguth H and Hinterwaldner R 2004 *Int J Adhes Adhes* **24** 519-22.
[5] Ding X, Zhao J, Liu Y, Zhang H and Wang Z 2004 *Materials Letters* **58** 3126-30
[6] Gu S, Kondo T and Konno M 2004 *Journal of Colloid and Interface Science* **272** 314-20
[7] Zhang J, Yang J, Wu Q, Wu M, Liu N, Jin Z and Wang Y 2010 *Macromolecules* **43** 1188-90
[8] Wang Q, Luo Y, Feng C, Yi Z, Qiu Q, Kong L and Peng Z 2012 *Journal of Nanomaterials* **2012** 5
[9] Liu C H, Chiu L K, Yeh J Y and Tsiang R C C 2012 *Journal of Nanomaterials* **2012** 2
[10] Lin T and Zheng Y 2013 *Micro & Nano Letters Iet* **8** 217-20
[11] Bourgeat-Lami E, Espiard P and Guyot A 1995 *Polymer* **36** 4385-9
[12] Liu Y L, Hsu C Y, Wang M L and Chen H S 2003 *Nanotechnology* **14** 813-20
[13] Liu Y L, Hsu C Y and Hsu K Y 2005 *Polymer* **46** 1851-6
[14] Zhang H, Li C, Guo J, Zang L and Luo J 2012 *Journal of Nanomaterials* **2012** 3
[15] Tadano T, Zhu R, Muroga Y, Hoshi T, Sasaki D, Yano S and Sawaguchi T 2014 *Chem Lett* **43** 705-7
[16] Japić D, Djerdj I, Marinšek M and Orel Z C 2014 *Acta Chimica Slovenica* **60** 797-808
[17] J. Hu, Y. Zhou and X. Sheng 2015 *Journal of Materials Chemistry C* 3 2223-2230
[18] Dong Y, Zhang C, Wu L, Chen Y and Hu Y 2014 *Macromolecular Rapid Communications* **35** 1943-8
[19] Fielding L A, Tonnar J and Armes S P 2011 *Langmuir* **27** 11129-44
[20] Tunusoğlu O Z and Demir M M 2013 *Industrial & Engineering Chemistry Research* **52** 13401-10

IOP Conf. Series: Materials Science and Engineering **87** (2015) 012036 doi:10.1088/1757-899X/87/1/012036

Flexural analysis of palm fiber reinforced hybrid polymer matrix composite

G. Venkatachalam[1], A. Gautham Shankar[2], Dasarath Raghav[2, 3], R. Santhosh Kiran[2], Bhargav Mahesh[2] and Krishna Kumar[2]

[1]School of Mechanical and Building Sciences, Vit University, Vellore, India
[2]Mechanical Engineering, School of Mechanical and Building Sciences, Vit University, Vellore, India

E-mail: dasarath94@hotmail.com

Abstract. Uncertainty in availability of fossil fuels in the future and global warming increased the need for more environment friendly materials. In this work, an attempt is made to fabricate a hybrid polymer matrix composite. The blend is a mixture of General Purpose Resin and Cashew Nut Shell Liquid, a natural resin extracted from cashew plant. Palm fiber, which has high strength, is used as reinforcement material. The fiber is treated with alkali (NaOH) solution to increase its strength and adhesiveness. Parametric study of flexure strength is carried out by varying alkali concentration, duration of alkali treatment and fiber volume. Taguchi L9 Orthogonal array is followed in the design of experiments procedure for simplification. With the help of ANOVA technique, regression equations are obtained which gives the level of influence of each parameter on the flexure strength of the composite.

1. Introduction

Environmental Pollution, which is mostly caused by non-biodegradable plastics, is driving researchers all over the world to produce a bio-degradable substitute. Bio composite is one of the most promising solutions to environmental pollution. There has been an increase in the production of materials made from natural resins worldwide [1]. Practical applications of Bio composites include interior automotive components and housings for notebook computers [2]. The growth in the use of natural fibers has led to a decrease in the carbon dioxide in the atmosphere [3]. Bio degradability and disposability are the most important aspects of hybrid composites. But, Renewable Polymers being more sensitive to moisture provide poor gas barrier properties which leads to reduction in the mechanical properties of the composite [4].

Synthetic fibers are costly and the production of these fibers is generally associated with large carbon footprint. Natural fillers are more economic and eco-friendly alternative. The main disadvantages of using natural fibers as reinforcement are the moisture absorption and poor adhesion between the fiber which is hydrophilic and the matrix which is hydrophobic. Poor adhesion between the fibers and the matrix leads to reduction of mechanical properties [5] .Various treatments that have

[3] Address for correspondence: Mechanical Engineering, School of Mechanical and Building Sciences, Vit University, Vellore, India. E-mail: dasarath94@hotmail.com

been employed to improve the fiber matrix adhesion are bleaching, peroxide treatment, de-waxing, vinyl grafting, alkali treatment, isocyanate treatment, acetylation and treatment with coupling agents [6].

Low concentration alkali treatment has led to the increase in mechanical, physical and thermal properties of jure fibers. Alkali treatment also increased the hydrophobicity of the fibers which lead to better fiber matrix adhesion [7].Alkali treatment has also led to an increase in the tensile and flexural strengths of Kenaf fibre reinforced polypropylene composites [8].

Goulart et al. studied the mechanical behaviour of palm fiber reinforced polypropylene composites and concluded that the reinforcement of palm fibers to the matrix increased the flexural strength and modulus when compared to pure polypropylene [9].

The tensile, flexural, thermal stability and biodegradation properties increased up to 50% of the fiber weight and then started to drop [10]. Venkatachalam et al. studied the effect of reinforcement of natural fibers such as jute and aloe Vera on the mechanical properties of the composite and concluded that the reinforcement of natural fibers increases the tensile and flexural strength of the polymer hybrid matrix composite [11-13]. They have used Taguchi L9 array in the design of experiments procedure for simplification. Also Venkatachalam et al investigated the influence of various factors on flexural strength of coir fibre reinforced hybrid polymer matrix composite [11].

In this work, hybrid polymer is used as matrix. The primary constituent of the hybrid polymer is General purpose resin and the secondary constituent is Cashew Nut Shell Liquid. Palm fiber which is known for its high strength is used as the reinforcement material. The fiber is treated with NaOH solution. The parameters which have been varied are alkali concentration, alkali treatment duration and fiber volume. The flexural test is carried out with the help of INSTRON Testing machine. Analysis of variance (ANOVA) is a statistical technique which is employed to find the influence of parameters such as soak time, alkali treatment and fiber volume on the flexural strength of the composite. The main advantages of ANOVA are random variability and robust design. ANOVA analysis is performed using MINITAB software.

2. Experimental procedure

In this work, naturally occurring fibers on the branches of palm tree are used as reinforcement. The fibers are treated with Sodium Hydroxide solution (shown in figure 1). The NaOH concentrations used are 5%, 15% and 25%. The duration of treatment considered are 6, 12 and 24 hours. 3 sets of hybrid polymers are prepared with CNSL concentration being 5%, 15% and 25%. The prepared samples are shown in figure 2.

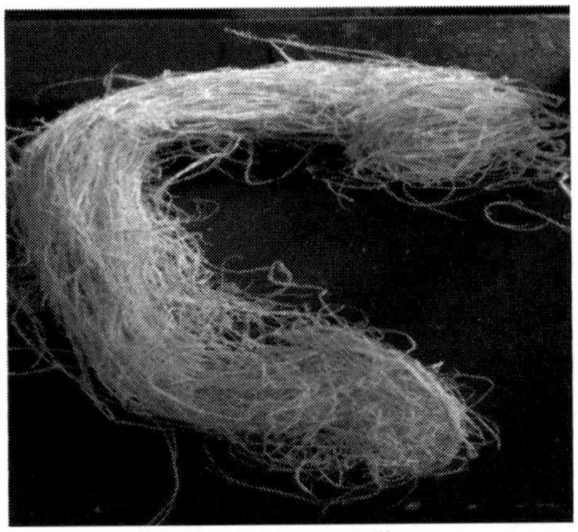

Figure 1. Treated fibers.

Figure 2. Prepared samples.

Table 1 shows the Taguchi L9 array for design of experiments procedure. 9 samples are prepared for each concentration of CNSL (5%, 15%, and 25%). Flexural test is performed on all 27 specimens

IOP Conf. Series: Materials Science and Engineering **87** (2015) 012036 doi:10.1088/1757-899X/87/1/012036

using INSTRON TESTING MACHINE. Figure 3 illustrates the sample undergoing flexural test and figure 4 presents the tested samples.

Table 1. Taguchi L9 array for design of experiments.

Sample No.	NaOH Concentration (%)	Fibre volume (%)	Soak Time (hrs.)
1	5	0.444	6
2	5	0.89	12
3	5	1.34	24
4	10	0.89	6
5	10	1.34	12
6	10	0.444	24
7	15	1.34	6
8	15	0.444	12
9	15	0.89	24

Figure 3. Sample undergoing flexural test.

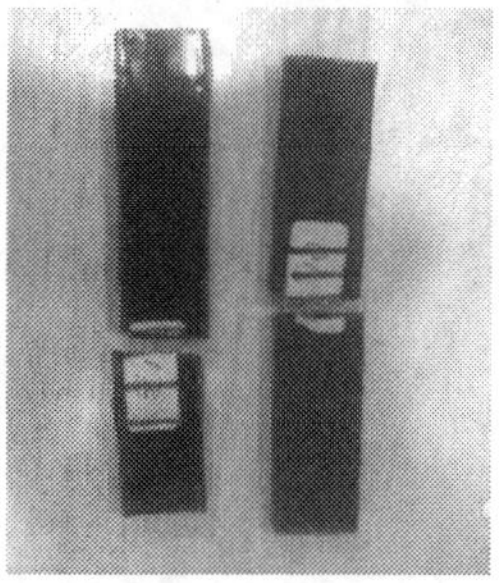

Figure 4. Samples after completion of test.

3. Results and discussions

Table 2 elucidates the Ultimate Flexural Strength of the 27 specimens.

Table 2. Ultimate flexural stress (MPa).

Sample No	Ultimate Flexural Stress (MPa)		
	5% CNSL	15% CNSL	25% CNSL
1	24.73	19.53	3.42
2	25.59	20.47	4.4
3	26.47	21.47	3.74
4	25.45	19.37	2.98
5	26.03	21.19	3.52
6	24.6	19.38	3.39
7	26.26	21.33	3.66
8	24.34	19.09	3.38
9	25.69	20.35	3.41

3.1. Main effects plot

The main effects plots are shown in figures 5-7 which illustrate the level of influence of different parameters on the Ultimate Flexural Stress of the composite. The fiber volume is directly proportional to the Ultimate flexural stress and also has the highest influence over other parameters. The optimum fiber volume for Ultimate Flexural stress is 1.34%.

For Composites with 5% CNSL Concentration, there is a reduction in the Ultimate Flexural Stress when the alkali concentration is increased from 5% to 15%. The increase in the Flexural Stress when the alkali concentration is further increased to 25% is negligible. The optimum alkali concentration is

IOP Conf. Series: Materials Science and Engineering **87** (2015) 012036 doi:10.1088/1757-899X/87/1/012036

5%. There is a reduction in the Ultimate Flexural Stress when the alkali treatment duration is increased from 6 to 12 hours. However, there is a considerable increase in the Ultimate Flexural Stress when the alkali treatment duration is further increased to 24 hours, hence the optimum alkali duration is 24 hours.

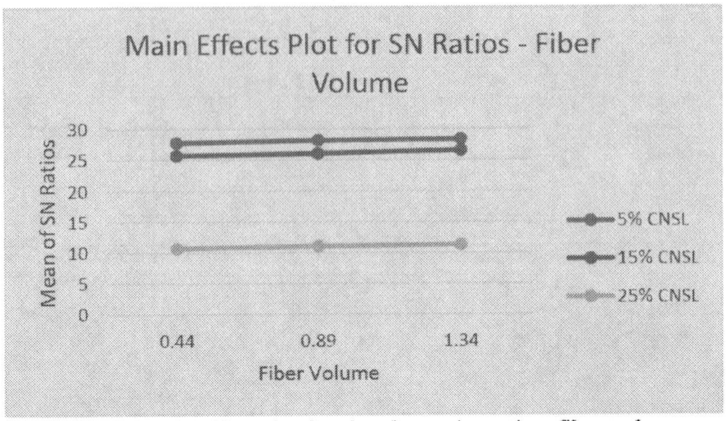

Figure 5. Main effect plot for signal to noise ratio - fiber volume.

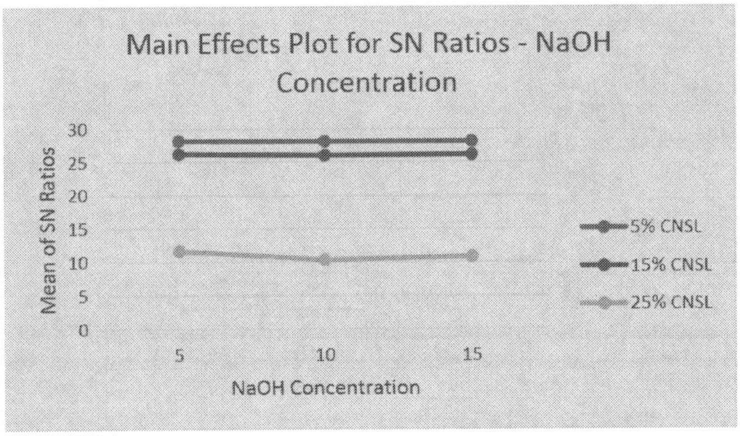

Figure 6. Main effect plot for signal to noise ratio - NaOH concentration.

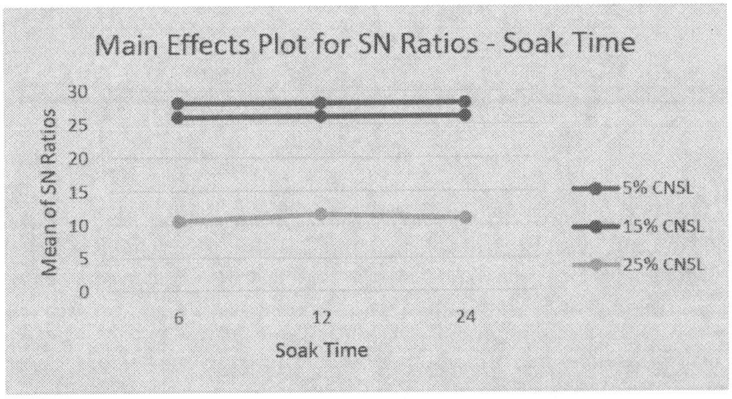

Figure 7. Main effect plot for signal to noise ratio – soak time.

On the other hand for hybrid matrix composites with 15% and 25% CNSL concentration, it has been observed that fiber volume is directly proportional to the Ultimate Flexural stress and the optimal fiber volume is 1.34%. When the NaOH concentration is increased from 5% to 10% there is a decrease in the Ultimate Flexural Stress. However, there is an increase in the Flexural Stress when alkali concentration is increased to 15%. Therefore, the optimum alkali concentration is 5%.

In the case of composites with 15% CNSL concentration the Ultimate Flexural Stress is directly proportional to soak time. Thus, the optimum soak time is 24 hours. On the contrary for 25% CNSL concentration composites, there is an increase in the Ultimate Flexural Stress when the soak time is increased from 6 hours to 12 hours. There is a substantial decrease in the Ultimate Flexural Stress when the alkali treatment duration is increased to 24 hours. Therefore the optimum soak time in this case is 12 hours.

3.2. Regression equation

Regression equations are obtained which enable us to find the Ultimate Flexural Stress values for different values of parameters such as alkali concentration, fiber volume and treatment duration. Equations (1)-(3) present the regression equations for 5%, 15% and 25% CNSL concentration composite respectively.

$$\text{Flexural Stress (MPa)} = 23.8 - 0.0167\ \text{NaOH}\ \% + 1.89\ \text{Fiber Volume} + 0.00825\ \text{Soak time} \quad (1)$$
$$\text{Flexural Stress (MPa)} = 18.2 - 0.0233\ \text{NaOH}\ \% + 2.23\ \text{Fiber Volume} + 0.0172\ \text{Soak time} \quad (2)$$
$$\text{Flexural Stress (MPa)} = 3.61 - 0.0370\ \text{NaOH}\ \% + 0.271\ \text{Fiber Volume} + 0.0046\ \text{Soak time} \quad (3)$$

NaOH-Concentration of alkali solution; Fiber Volume-Volume percentage of palm fibers; Soak time-Duration of alkali treatment in hours

4. Conclusion

In a series of studies on mechanical behaviour of palm fibre reinforced hybrid polymer matrix composite, authors made an attempt to study the flexural strength of the composites. ANOVA technique is employed to study the level of influence of fibre parameters on flexural strength. The three regression equations obtained from ANOVA analysis gives the level of influence of these fibre parameters on flexural strength.

On an average there is an 8.5% increase in the overall Ultimate Flexural Strength, when the fibre volume is increased from 0.44% to 1.34%. Likewise, when the duration of alkali treatment is varied, a small increment of 0.5664% in the Ultimate Flexural Stress is observed. On the other hand when the alkali concentration is varied, there was a negligible decrease of 0.633% in the Ultimate flexural stress is perceived. It is concluded that Fiber volume has the highest influence on the Ultimate Flexural Stress whereas the duration of alkali treatment and alkali concentration have a lesser impact on the Ultimate Flexural Stress.

References

[1] Omar Faruka, Andrzej K Bledzki, Hans-Peter Fink and Mohini Sain 2012 Biocomposites reinforced with natural fibers: 2000–2010 *Progress in Polymer Science* **37** 1552-96

[2] Caroline Baillie and Randika Jayasinghe 2004 *Green Composites - Polymer Composites and the Environment* (England: Woodhead Publishing) pp 1-50

[3] Yoshihiko Arao, Takayasu Fujiura, Satoshi Itani and Tatsuya Tanaka 2015 Strength improvement in injection-molded jute-fiber-reinforced polylactide green-composites *Composites Part B: Engineering* **68** 200-6

[4] Cao Y, Shibata S and Fukumoto I 2006 Mechanical properties of biodegradable composites reinforced with bagasse fibre before and after alkali treatments *Composites Part A: Applied Science and Manufacturing* **37** 423-9

[5] Saheb D N and Jog J P 1999 Natural fiber polymer composites: A review *Advances in Polymer Technology* **18** 351-63

[6] Mohanty A K, Misra M and Drzal L T 2001 Surface modifications of natural fibers and performance of the resulting biocomposites: An overview *Composite Interfaces* **8** 313-43

[7] Aparna Roy, Sumit Chakraborty, Sarada Prasad Kundu, Ratan Kumar Basak, Subhasish Basu Majumder and Basudam Adhikari 2012 Improvement in mechanical properties of jute fibres through mild alkali treatment as demonstrated by utilisation of the Weibull distribution model *Bioresource Technology* **107** 222-8

[8] Asumani O M L, Reid R G and Paskaramoorthy R 2012 The effects of alkali–silane treatment on the tensile and flexural properties of short fibre non-woven kenaf reinforced polypropylene composites *Composites Part A: Applied Science and Manufacturing* **43** 1431-40

[9] Oliveira T A, Teixeira A, Miléo P C and Mulinari D R 2011 ICM11 mechanical behaviour of polypropylene reinforced palm fibers composites goulart, S.A.S *Procedia Engineering* **10** 2034-9

[10] Hamdy Ibrahim, Mahmoud Farag, Hassan Megahed and Sherif Mehanny 2014 Characteristics of starch-based biodegradable composites reinforcedwith date palm and flax fibers *Carbohydrate Polymers* **101** 11

[11] Venkatachalam G, Gupta A and Gautham Shankar A 2015 Flexural analysis of coir fiber reinforced hybrid polymer matrix composites *Composites: Mechanics, Computations, Applications* **6** 105-12

[12] Vimalanand S, Venkatachalam G, Gautham A, Rohit A, Gupta G and Anantrao U S 2014 Investigations on tensile and flexural strengths aloevera fiber reinforced hybrid polymer matrix composite materials *Souvenir on INCCOM* (India: Thiruvananthapuram) p 5A6

[13] Viswanath S, Venkatachalam G, Ragavender K and Ashka 2014 Investigation on tensile behavior of coir fiber made hybrid polymer matrix composite *Proceedings of IMEC* (India, Tiruchirappalli) pp 311-4

IOP Conf. Series: Materials Science and Engineering **87** (2015) 012037 doi:10.1088/1757-899X/87/1/012037

A review on the electrochemical treatment of the salty organic wastewater

Xianjun Du[1]

College of Electrical and Information Engineering, Lanzhou University of Technology, 730050 Lanzhou, Gansu, China

E-mail: 8843079@qq.com

Abstract. Electrochemical technologies have proved to be useful for the treatment of wastewater, and recent years, there are growing interests in electrochemical treatment of the salty organic wastewater. The aim of this paper is to mainly present the source of the salty organic wastewater, the mechanism of direct and indirect oxidation process, and the research advances of electrochemical technologies in the salty organic wastewater by literature reports review.

1. Introduction

The processes used for wastewater treatment can be grouped into three major categories: chemical, physical, or biological treatment processes. It is possible for treatment processes to be a combination of these three major categories. The main contaminants of concern in wastewater are organic compounds, suspended solids, nutrients, and pathogens [1].

One of the most commonly used biological wastewater treatment processes is the activated sludge process. The activated sludge process is a suspended-growth process that recycles a portion of the solids from the secondary clarifier back into an aerated bioreactor in order to maintain a high population of microorganisms (biomass) within the bioreactor. The microorganisms convert biodegradable organic matter and some inorganic compounds, such as nitrogen and phosphorus, into new biomass and other products that are used in their metabolic processes. Biomass is separated from the treated wastewater in the secondary clarifier with a portion of the biomass being recycled and a portion of the biomass being wasted. Many municipal wastewater treatment plants use the activated sludge treatment process because it is highly flexible, reliable, and effective.

Because of the microbial activity is limited in high salt conditions by the traditional biological method in wastewater treatment [2]. Recent years, with the development of the industry, environmental protection become a hot focus of academe and government, and new laws demanding more strict environmental protection are being approved, and so, the search for more effective methods, such as low cost, high efficiency, multifunction, easy operation and pollution free, is increasing [3, 4]. Among the different techniques for wastewater treatment [5, 6], electrochemical methods have achieved a relevant place. In particular, the oxidation method of organic pollutants is anodic oxidation using different electrodes [7-10] or cathodic generation of hydrogen peroxide [11-13].

[1] Address for correspondence: Xianjun Du, College of Electrical and Information Engineering, Lanzhou University of Technology, 730050 Lanzhou, Gansu, China. E-mail: 8843079@qq.com.

2. Treatment mechanism

The process of electrochemical degradation of pollutants can be divided into direct and indirect oxidation process. In the direct oxidation process, pollutants are adsorbed by the surface of the anode at first, and then through the electron transfer reaction of hydroxyl radical formation (•OH); strong oxidative free radical damage of pollutant molecular structure, decomposition to CO_2. However, competitive reaction of oxygen evolution exists for this process consume part of free radical, reduce the oxidation efficiency. Therefore, the development of high oxygen evolution potential of the electrode is the main way to improve the efficiency.

In the indirect electrochemical oxidation process, to produce strong oxidizing agents such as hypochlorous acid/chlorine, ozone, H_2O_2, etc., oxidant pollutants subsequent generation and electrochemical oxidation occurs in solution. The oxidant will be generated and used all of it momently. Shortcoming of the chlorine oxidation method is concerns in the electrolytic process may produce organic chlorides increase the toxicity of the wastewater.

The hydrogen peroxide (H_2O_2) the advantages of easy to store, transported and its safety procedure process with high efficient, it is widely used as an oxidative agent. With the technical developments of the catalytic component materials of active coal, silica textures, zeolites, nafion membranes, structured clay and goethite formed Fe.

The Advanced Oxidation Processes (AOPs) fully or partially used in wastewater treatment aimed to oxidative degradation of the persistent organic contaminants are successful, and usually, it is using a combination of oxidants being more worthwhile to transform them into biodegradable aliphatic carboxylic acids followed by a biological process. With the very high oxidative potential, HO is the decision maker of the oxidation process by different mechanisms [14, 15].

3. Research and development of electrode materials

As mentioned above, the electrode material is one of the effects of electrochemical degradation of organic matter limit factors. Traditional insoluble anode materials, such as graphite electrode, platinum electrodes, still get a wide range of uses in wastewater treatment. In order to improve the efficiency of electrochemical processing, researchers began to study and develop anode materials which have higher catalytic activity, higher over potential and higher stability.

Three dimensional shape-stable anode (DSA) is a kind of metal Ti as catalyst coating substrate (for example: RuO_2 and IrO_2) electrode, precipitation on the chlorine and oxygen have a very high catalytic activity, electrochemical process has long been used in chlor alkali industry. Because the DSA electrode and releasing oxygen over potential is relatively low, the current efficiency of direct oxidation process of organic matter degradation of low [16, 17], however, in the presence of chloride ion in the solution conditions, the electrode surface is indirect oxidation of chlorine evolution, will promote the degradation of organic matter, the practical wastewater treatment, so this type of electrode is especially suitable for high NaCl, such as olive oil, printing and dyeing, tanning wastewater [18-20].

PbO_2 electrode has the advantages of low price, easy preparation, good electrical conductivity, high chemical stability and large surface area, high oxygen evolution over-potential. It is an excellent electrode with the experimental detection to the electrode surface producing a large number of •OH. PbO_2 electrode has been applied to the electrochemical degradation of salty organic wastewater, and gradually extended to the practical wastewater treatment, such as printing and dyeing wastewater, indole production wastewater, pesticides and herbicides wastewater, landfill leachate, tanning wastewater and anionic surface active agent [21-24].

Others have their own advantages such as graphite electrode, activated carbon fiber anode, SnO_2 electrode, boron doped diamond (BDD) electrode, are also widely used to electrochemical treatment of the organic wastewater [25-27]. Graphite electrode is cheap, large surface area, it can be combined with the use of adsorption and electrochemical degradation; activated carbon fiber (ACF) is a new type of porous carbon materials, it is numerous pores, large surface area and very strong adsorption; SnO_2 electrode materials are widely used in organic matter degradation; diamond membranes has high hardness, high thermal conductivity, high chemical stability and other excellent properties, and the

BDD membranes has good net conductive properties and semiconductor properties, it is get more and more application in wastewater treatment.

Composites material electrode research also had an advance in wastewater biological treatment recent years. Jiang [28] prepared three nano-composites, LDH/lysozyme, LDH/GO and LDH/GO/FeP, with the MgAl-NO_3-LDH as the carrier. Since it is a novel biological technologies, Jiang and her team were tried to apply to wastewater treatment. During their experimental study, the composites materials can continue to be used without replacement. Therefore, it would become a potential new disinfectant in the enzymatic treatment of sewage because of its good bactericidal activity and no pollution. In Liu's [29] studies, with carbon nanotube as the template, the polypyrrole/carbon nanotube (PPy/CNT) composites were prepared in ethanol solution through chemical oxidation method. The effects of preparation conditions on properties of PPy/CNT were studied to obtain the optimum preparation conditions. Besides these, Liu and her team also carried out to synthesize polypyrrole/polyaniline (PPy/PANI) composites in PANI dispersion. In Yang's [30] studies, in order to change the tactic pattern of tradition electrode matrix material, by considering the internal composition structure of the electrode composites material, Yang and her team decided to use the aluminum as the in-core, with the outer layer by the titanium gable layered composite construction matrix material, which in the electrode the core as aluminum for fine electrical conductivity, to play an important part of reducing the interface resistance and the equalization current distribution, and the outer layer using the titanium gable, to still maintain the original electrochemistry nature. Experimental results indicate that the change of the electrode matrix material not only improved electrode performance, but also attained the purpose of saving energy and reducing consumption.

4. Research and development of electrochemical reactor

Traditional electrolytic reactor is adopted in 2D planar electrode, the reactor has the advantages of small volume, small amount of body processing unit, but its low current efficiency and mass transfer efficiency make it restricted in electrochemical use. The main problem of the electrochemical wastewater treatment practical application is how the laboratory batch reactor will turn into continuous zoom processing unit in practice. Scholars have carried out a series of research on the new type of electrochemical reactor recent years.

Korbahti et al. [31] designed for processing simulation of printing and dyeing wastewater by electrochemical tubular continuous reaction, using iron as anode, stainless steel as cathode, investigation under the optimized conditions, researched the relationship between residence time and COD, color removal, turbidity, pH changes. The results showed that, the residence time increases to promote the removal of COD and chroma. Also, this processing device was used in coating wastewater [32].

Sakalis et al. [33] using a continuous reactor, NaCl and sodium sulfate as electrolyte, using Pt/Ti as anode material, carbon felt as cathode, studied the electrochemical degradation of dye wastewater. Under the optimal conditions, the removal efficiency of dye reached 94.4%, BOD_5 and COD were decreased by 35% and 45%, COD/BOD_5 value down from 4.3 to 3.6. This reactor is suitable for pretreatment of biological pretreatment. Szpyrkowicz [34] designed a continuous unidirectional filtration electrochemical reactor, provides a computational model for printing and dyeing wastewater treatment in different velocity, and the study of electrochemical reactor for treatment of industrial wastewater of direct and indirect oxidation process through the fluid mechanics.

5. Conclusions

Electrochemical oxidation technology has a very strong degradation ability of the salinity organic wastewater treatment, and it has a good prospect in field of salty organic wastewater treatment with the development of the theory of electrochemistry, the electrode materials, the electrochemical reactor technology and the joint use application with other methods.

Acknowledgments

This paper is supported by the open foundation of the Key Laboratory of Gansu Advanced Control for Industrial Processes (No. XJK201511), the Excellent Young Teachers Program of Lanzhou University of Technology (No. Q201408) and the Natural Science Foundation of Gansu Province (No. 1212RJYA031).

References

[1] Tchobanoglous G, Burton F L and Stensel H D 2003 *Wastewater Engineering: Treatment and Reuse 4th Ed* (New York: Metcalf & Eddy, Inc.)
[2] L'Amoura R J A, Azevedo E B, Leite S G F, et al. 2008 *Sep. Purif. Technol.* **2** 142-3
[3] Anastas P T and Lankey R L 2002 *ACS Symp. Ser.* **23** 1-4
[4] Lancaster M 2002 *Green Chemistry, An Introductory Text* (London: Royal Society of Chemistry)
[5] Rajeshwar K and Ibanez J 1997 *Environmental Electrochemistry Fundamentals and Applications in Pollution Abatement* (San Diego, CA: Academic Press Inc)
[6] Simonsson D 1997 *Chem. Soc. Rev.* **3** 181-3
[7] Brillas E, Cabot P, Casado J 2003 *Environ. Sci. Pollut.* **26** 235-6
[8] Iniesta J, Expósito E, González-García J, et al. 2002 *J. Electrochem. Soc.* **5** D57-8
[9] Iniesta J, González-García J, Expósito E, et al. 2001 *Water Res.* **14** 3291-2
[10] Jiang J, Chang M and Pan P 2008 *Environ. Sci. Technol.* **8** 3059-60
[11] Expósito E, Sánchez-Sánchez C M and Montiel V 2007 *J. Electrochem. Soc.* **8** E116
[12] Sánchez-Sánchez C M, Expósito E, Casado J, et al. 2007 *Electrochem. Commun.* **1** 19-20
[13] Brillas E and Casado J 2002 *Chemosphere* **3** 241
[14] Pera-Titus M, Garcia-Molina V, Banos M A, et al. 2004 *Appl. Catalysis B: Environ.* **47** 219-20
[15] Wang H and Wang J 2007 *Appl. Catalysis B; Environ.* **77** 58-9
[16] Li X Y, Cui Y H, Feng Y J, et al. 2005 *Water Res.* **10** 1972
[17] Feng Y J and Li X Y 2003 *Water Res.* **10** 2399
[18] Chatzisymeon E, Dimou A, Mantzavinos D, et al. 2009 *J. Hazard Mater.* **1-3** 268
[19] Vaghela S S, Jethva A D, Mehta B B, et al. 2005 *Environ. Sci. Technol.* **8** 2848-50
[20] Costa C R, Botta C M R, Espindola E L G, et al. 2008 *J. Hazard Mater.* **1-2** 616
[21] Andrade L S, Ruotolo L A M, Rocha R C, et al. 2007 *Chemosphere* **11** 2035-6
[22] Keech P G and Bunce N J 2003 *J. appl. Electrochem.* **1** 79
[23] Panizza M, Sires I and Cerisola G 2008 *J. Appl. Electrochem.* **7** 923
[24] Martinez-Huitle C A, De Battisti A, Ferro S, et al. 2008 *Environ. Sci. Technol.* **18** 6929
[25] Polcaro A M, Palmas S, Renoldi F, et al. 1999 *J. Appl. Electrochem.* **2** 147-8
[26] Yi F Y and Chen S X 2008 *J. Porous Mater.* **5** 565-7
[27] Liang L C, Chang J E and Wen T C 1995 *Water Res.* **2** 671
[28] Jiang F J 2014 *Application of Layered Double Hydroxide Composites in the Biological Treatment of Wastewater* (Ji'nan, China: Qilu University of Technology)
[29] Liu Q 2012 *Preparation and Evaluation on PPy Composite Electrode Materials for Capacitive Deionization Modules* (Tianjin, China: Tianjin University)
[30] Yang X Q 2010 *Preparation and Evaluation of the Layered Composites Electrode Materials of Ti-Al* (Kunming, China: Kunming University of Technology)
[31] Korbahti B K and Tanyolac A 2009 *J. Hazard Mater.* **2-3** 771-3
[32] Korbahti B K and Tanyolac A 2009 *Chem. Eng. J.* **2-3** 444
[33] Sakalis A, Mpoulmpasakos K, Nickel U, et al. 2005 *Chem. Eng. J.* **1** 63-4
[34] Szpyrkowicz L 2006 *J. Chem. Technol. Biotechnol.* **8** 1375-7

The preparation of a kind of phenylephrine intermediates

Pan Jiang and Feng Liu[1]

School of Perfume and Aroma Technology, Shanghai Institute of Technology, 100 Haiquan Rd, Shanghai 201418, P. R. China.

E-mail: liufeng@sit.edu.cn

Abstract. A two-step method to accomplish the synthesis of the important intermediate α-(N-methyl-N-benzylamino) -3-hydroxy acetophenone hydrochloride was discussed, including a brominaion and the following amination. Starting from the commercial available raw material, the target compound could be gained with 62% yield. This procedure involves cheap and easily obtained raw materials, simple operation, low cost, low waste production.

1. Introduction

Phenylephrine is an α1-adrenergic receptor agonist of the phenethylamine class, and is primarily used as a decongestant[1] for bronchiectasis and vascular contraction nasal mucosa. It can reduce nasal and sinus congestion relieve nasal mucosal hyperemia or swelling, and may also relieve symptoms of nasal congestion. This supplement can also be used to treat shock and supraventricular tachycardia, or used to maintain blood pressure during anaesthesia and mydriatic check[2,3]. Phenylephrine initially by the Frederick Stearns company, was successfully developed in 1934. For a long time, the use of the compounds by Aspirin, Phenylpropanolamine Hydrochloride, dextral ephedrine, and some other related compounds. The dosages of the pharmaceuticals in our country are relatively small, for many years in the domestic production technology research and preparation research basically is in a state of stagnation.

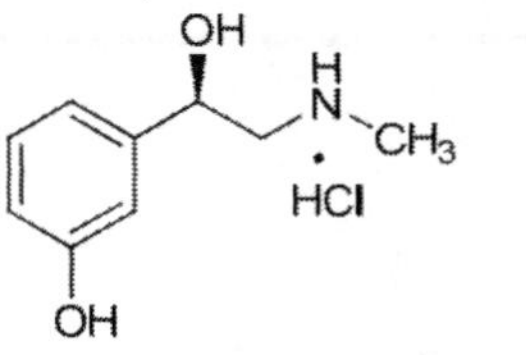

(R)-phenylephrine hydrochloride

a

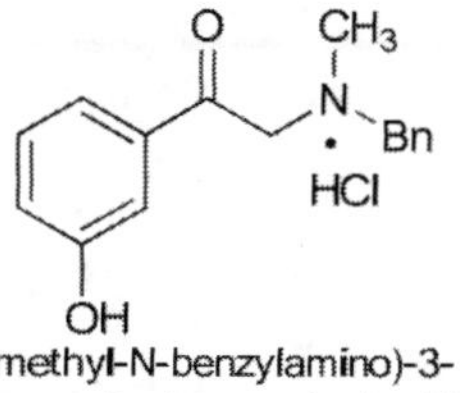

α-(N-methyl-N-benzylamino)-3-hydroxy acetophenone hydrochloride

b

Figure 1. Phenylephrine and it's precursor

There is vast social significance in our country pushing to catalyze the related research and application of phenylephrine, instead of continuing to use similar compounds, especially in the

[1] Address for correspondence: Feng Liu, School of Perfume and Aroma Technology, Shanghai Institute of Technology, 100 Haiquan Rd, Shanghai 201418, P. R. China. E-mail: liufeng@sit.edu.cn.

treatment of flu and related diseases of drug use to improve the level of people's rational drug use, and inhibit drug related crimes.

Various synthesis studies and findings regarding phenylephrine have been reported [4-6], but only a few are of industrial value. (R)-phenylephrine hydrochloride (figure 1) is the common form of phenylephrine in the pharmaceutical industry. α-(N-methyl-N-benzylamino)-3-hydroxy acetophenone hydrochloride (b) is a key intermediate to accomplish the synthesis of (R)-phenylephrine hydrochloride. Magnificently, after an asymmetric hydrogenation process, (R)-phenylephrine hydrochloride can be produced. There are many preparation methods for the intermediate or similar intermediates. In this study, we have developed a simple operation, promoting an energy saving synthesis method on the basis of previous studies, for a total of two steps bromination and amination.

2. Bromine generation reaction to obtain α-bromo-3-acetoxy acetophenone

Traditional bromine generation reaction[7-9], is a free radical substitution reaction between bromine and the substrate. Considering hydrogen bromide generated in bromine generation reaction could be oxidized into bromine again, the new generated bromine will attend the bromination reaction again, so that we can improve the utilization rate of bromine to reduce costs.

Table 1. Screening of reaction conditions for bromine generation reaction.

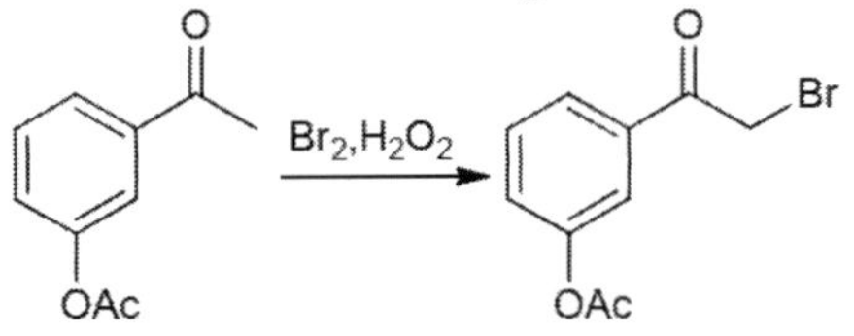

Entry	Br_2	HBr	H_2O_2	solvent	T/℃	Yield[%][b]
1	1.1	0	0	n-Butyl acetate	-5	0
2	1.1	0	0	n-Butyl acetate	0	83.7
3	1.1	0	0	n-Butyl acetate	10	75.9
4	1.1	0	0	n-Butyl acetate	20	73.8
5	1.1	0	0	ethyl acetate	0	76.2
6	1.1	0	0	isopropyl acetate	0	73.7
7	1.1	0	0	CH_2Cl_2	0	68.4
8	1.1	0	0	THF	0	62.5
9	1.1	0	0	n-Butyl acetate	30	65.9
10	0	1.3	1.3	n-Butyl acetate	0	0
11	0.2	1.0	2.0	n-Butyl acetate	0	18.9
12	0.4	0.4	0.8	n-Butyl acetate	0	39.7
13	0.6	0.2	0.7	n-Butyl acetate	0	62.3
14	0.7	0.2	0.6	n-Butyl acetate	0	76.5
15	0.7	0	0.6	n-Butyl acetate	0	75.7
16[a]	0.8	0	0.5	n-Butyl acetate	0	82.2
17	0.8	0	0.5[c]	n-Butyl acetate	0	73.6

[a]Reaction conditions: Step 1) c (1.0 mmol), Br_2 (0. 8 mmol), solvent 20 mL,1h; then; Step 2) H_2O_2 (0.5 mmol) 4h.
[b]Isolated yield, including 3-(2,2-dibromoacetyl) phenyl acetate which could also be transformed into the target product in the next reactions. [c]$KMnO_4$ is instead of H_2O_2.

First, we began by screening the reaction temperature and solvent, as illustrated below in table 1. It can be seen that the freezing point and n-Butyl acetate is at the most suitable conditions (entry1-8). The presence of hydrogen peroxide can reduce the necessary usage amount of bromine (entry 2 and 16), however only hydrogen bromide and hydrogen peroxide can't cannot cause bromination (entry 10). Other antioxidants can be tried, perhaps to the cause of the low dissolvability, however the product yield wasn't improved with the presence of $KMnO_4$ (17).

3. Amination reaction for α-(N-methyl-N-benzylamino)-3-hydroxy acetophenone hydrochloride

The reaction included two steps: the first amination, and the second the addition of acid to generate the salt. The whole process does not need to a chromatography column. The was generated by salt precipitation from the organic solvent, then (through recrystallization) can be the product as pure as a white powder. It's the better to choose an inorganic base to absorb acid (entries 1-6 in table 2). The reaction prefers aprotic solvents consistent with the previous reaction, therefore we chose n-butyl acetate as the solvent. The yield is exemplary at 20℃ to 40℃, so the operation is convenient.

Table 2. Screening of reaction a conditions for α-(N-methyl-N-benzylamino)-3-hydroxy acetophenone hydrochloride.

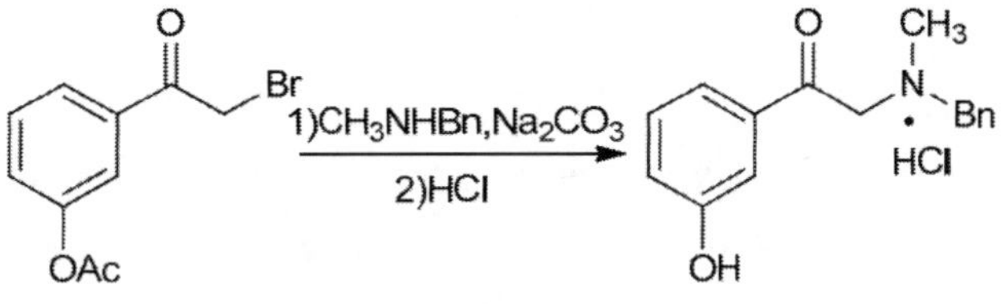

Entry	Base	Solvent	T/℃	Yield[%][b]
1	Na_2CO_3	n-Butyl acetate	30	73.5
2	K_2CO_3	n-Butyl acetate	30	58.2
3	$NaHCO_3$	n-Butyl acetate	30	63.7
4	NaOH	n-Butyl acetate	30	66.8
5	Et_3N	n-Butyl acetate	30	43.6
6	pyridine	n-Butyl acetate	30	32.1
7	Na_2CO_3	ethyl acetate	30	70.3
8	Na_2CO_3	isopropyl acetate	30	72.0
9	Na_2CO_3	CH_2Cl_2	30	68.5
10	Na_2CO_3	THF	30	62.3
11	Na_2CO_3	n-Butyl acetate	20	72.8
12	Na_2CO_3	n-Butyl acetate	40	71.4
13	Na_2CO_3	n-Butyl acetate	50	70.4

[a]Reaction conditions: Step 1) d(1.0 mmol), Na_2CO_3 (2.0 mmol),CH3NHBn (1.1 mmol), solvent (20 mL), 2 h; then Step 2) concentrated hydrochloric acid (3 mmol), 2 h.
[b]Isolated yield from c to b .Et_3N= triethylamine.

4. Preparation method of phenylephrine

Results of asymmetric hydrogenation of α-(N-methyl-N-benzylamino)-3-hydroxy acetophenone hydrochloride(b) to get (R)-phenylephrine hydrochloride (a)[10] by using different catalysts has been reported, however here, we introduce the reaction using Ru(Ⅱ)-(R)-BINAP as the catalyst, which could be bought from any market. The reaction was conducted in an Argonaut Technologies, α-(N-methyl-N-benzylamino)-3-hydroxy acetophenone hydrochloride(b), and a catalyst was placed in a 100ml Parr autoclave with an overhead stirrer, and a heating jacket under the nitrogen atmosphere in figure 2. Methanol was added through the injection port, and the reaction was purged five times by pressurizing to 30 bar hydrogens and releasing the pressure while stirring. The reaction was then pressurized to 30 bar hydrogens and stirred at room temperature for over 45 hours. Then, the reaction system was filtered, and evaporated under reduced pressure which gave off a white solid residue. After, recrystal from methanol the pure product was secured. The details were as follows: Purity: 98%, enantiomeric purity: 75%ee.

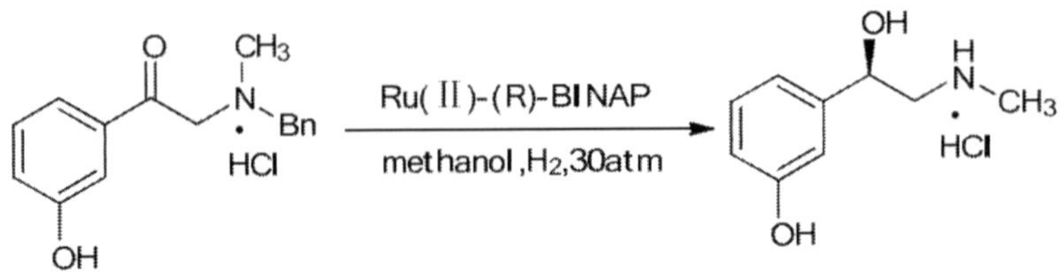

b a

BINAP= (±)-2,2'-Bis(diphenylphosphino)-1,1'-binaphthalene

Figure 2. Asymmetric hydrogenation to get (R)-phenylephrine hydrochloride

5. Summary

A method was developed to synthesize the α-(N-methyl-N-benzylamino)-3-hydroxy acetophenone hydrochloride; the important intermediate included two steps. In the first step, we used H_2O_2 oxidize HBr to reduce the consumption of expensive bromide. The two steps are merged into one after adding a concentrated hydrochloric acid product. The product precipitated from the organic phase in high yield. Then, the product could be purified by recrystallization .

6. Experimental section

d: Br_2 (128 mg, 0. 8 mmol) was slowly dropped into a solution of c(1.78 g, 1.0 mmol) in n-Butyl acetate (20mL) in a 50 mL flask, which was equipped with a stirrer bar. The reaction mixture was stirred for 1 hour in an ice-water bath. Then, 30% H_2O_2 (56.8 mg,0. 5 mmol) was dropped into the solution slowly. After stirring for another 4 hours in the ice-water bath. The resulting mixture was quenched with water and extracted with EtOAc. The combined organic layers were dried over Na2SO4 and concentrated under reduced pressure to produce d as a crude product of yellow oil (2.80g, >100%).

b: To the solution of d (2.8 g) in n-Butyl acetate (20 mL) the solution of Na_2CO_3 (212 mg, 2 mmol) in water (5 mL) was successively dropped, then the CH_3NHBn (233 mg, 1.1 mmol) was dropped into the solution. The reaction mixture was stirred for 2 hours at room temperature. The resulting mixture was extracted with n-Butyl acetate (10 mL) and cleansed with saturated sodium chloride. Then, to the combined organic layers concentrated hydrochloric acid (0.25 mL, 3 mmol) was dropped. Combined acid water layers and washed with n-Butyl acetate. Acid water was heated up to 40 ~ 45℃, then the temperature was kept for 2 hours, and cooled to the temperature of 0℃. The product was precipitated as a white solid, after recrystallization in acetone obtained the pure product,b (143g,62%).

^1H NMR (500 MHz, $CDCl_3$) δ 8.05(s, 1H), 7.53 (s, 1H), 7.43 (s, 1H), 7.37 (s, 1H), 7.29 – 7.17 (m, 5H), 7.01 (s, 1H), 5.66 (s, 1H), 4.32 (s, 1H), 3.89 (s, 1H), 3.57 (s, 1H), 3.48 (s, 1H), 2.34 (s, 3H). ESI-MS: 256.3.

Acknowledgment

This work was supported by the National Natural Science Foundation of China (Grant No. 21302126) and the Shanghai Engineering Technology Research Center of Fragrance and Flavor (12DZ2251400).

References

[1] 2006 *The Merck Index* 14th ed (NJ, USA: Merck and Co., Inc.: Whitehouse Station)
[2] John F McGarrity 2010 *Tetrahedron: Asymmetry* **21** 2479-86
[3] Helmut Legerlotz 1933 Beta-alkyl-amino componds Of Mono-hydroxy-phenyl-ethanols and process of producing same *US 1932347* Oct 24
[4] Hukki Jaakko and Honkanen Erkki 1959 *Acta Chemica Scandinavica.,* **13** 329-32
[5] H. Takeda, T. Tachinami, M. Aburatani, H. Takahashi, T. Morimoto, K. Achiwa.1989 *Tetrahedron Lett.* **30** 367-70
[6] S. SAKURABA, H. TAKAHASHI, H. TAKEDA, K. ACHIWA,1995 *Chem. Pharm. Bull.* **43** 738-47
[7] Helmut Legerlotz 1934 Process of manufacturing optically active mono-hydroxyphenyl methylamino-ethanols-1 *US 1954389* Apr 10
[8] Divi Murali Krishna Prasad 2012 Process for resolution Of 1-(3-hydroxyphenyl)-2-methylamino ethanol, *US2012/0108848* May 3
[9] Breuer Michael and Pletsch Andreas 2010 Method for producing L-phenylephrine using an alcohol dehydrogenase of aromatoleum aromaticum Ebn1 (azoarcus Sp. Ebn1) *WO2010/031776 A3* Sep. 10
[10] Feng Wenhua, Guo Fangchao and Yu Xiaoli 2011 Preparation methods Of L-phenylephrine hydrochloride, *WO2011131027* Oct. 27

Degradation of acid scarlet 3R with CuO/SiO_2 hollow sphere catalyst

F Xie[1, 2, 3], J Zhong[1], L Wang[1], K Wang[2] and D X Hua[2]

[1]School of Chemistry and Chemical Engineering, Tianjin University of Technology, Tianjin, China
[2]College of Science, China University of Petroleum (East China), Qingdao, China

E-mail: xiefei2013@foxmail.com

Abstract. Silica-supported copper catalyst materials have been synthesized via an incipient wetness impregnation. The resulting samples were characterized using X-ray diffraction (XRD) and Scanning electron microscope (SEM). The heterogeneous Fenton-like oxidation of reactive azo dye solutions by this catalyst was also investigated. The effects of various operating conditions on decolorization performance were evaluated, namely hydrogen peroxide dosage, initial pH, catalyst loading and initial dye concentration. The results indicated that by using 34 mmol/L of H_2O_2 and 6.0 g L^{-1} of the catalyst at 60°C, pH 3.5, 97% of decolorization efficiency was achieved within 90 min. CuO/SiO_2 hollow sphere is shown a promising catalyst for degradation of azo dye aqueous solution by Fenton-like processes.

1. Introduction

In the world, more than 700,000 t of about 10,000 different types of dyes on pigments are produced every year [1]. Unfortunately about 10%~20% dyes are lost or discharged as aqueous effluents during dying and finishing processes in the textile industry. The dye wastewater has a heavy colority, which is difficult to be treated. The traditional methods of dying wastewater treatment, such as membrane separation, ion-exchange, or activated carbon adsorption, just transfer the pollutants to another phase, and the biological method is not suitable because of the high COD value of the wastewater [2].

The Fenton process is an effective method for industrial wastewater treatment. In this process, powerful oxidant hydroxyl radical (·OH) is produced from H_2O_2 in the presence of ferrous cation, which can degrade many refractory compounds including dye [3]. However, homogeneous Fenton process suffer from some famous disadvantages such as narrow pH range of 2.5 - 3, generation of secondary pollutants and requirement of additional separation and disposal.

To overcome these disadvantages, heterogeneous Fenton catalysts have been developed. In heterogeneous Fenton process, the iron ions can be immobilized in the structure of solid supports, in which the catalysts might produce hydroxyl radicals from hydrogen peroxide and prevent the precipitation of iron hydroxide [4]. Besides iron, copper, manganese, zinc and so on, also have the effect to exciting the hydrogen peroxide and generating radicals. Results of some research show that among the transition metal oxides, iron and copper proved to be more active [5].

[3] Address for correspondence: F Xie, School of Chemistry and Chemical Engineering, Tianjin University of Technology, Tianjin, China. E-mail: xiefei2013@foxmail.com.

On the other hand, hollow micro-/nano-structured materials have received increasing attention due to their high chemical and thermal stability, low density and high specific surface [6]. These materials have potential applications in various fields such as catalysts, chemical sensors, photonic crystals etc [7]. The main purpose of the present study was to decolorize a non-biodegradable azo dye, Acid Scarlet 3R by heterogeneous Fenton-like process using silica-supported copper catalysts. The efficiency of the heterogeneous catalyst was evaluated by color removal and COD removal. The influence of temperature, pH values, catalyst loading and H_2O_2 dosage on the oxidation process was also discussed.

2. Experimental

2.1. Materials

The reactive dye Acid Brilliant Scarlet 3R was supplied by Cangzhou Lingang Kilda Dyestuffs Co., Ltd and used without purification. H_2O_2 solution (30%) and other reagents were obtained from Sinopharm Chemical Reagent Co., Ltd. These chemicals were analytical grade and used in the same state as they were received. Deionized water was used in the whole experiment process. 0.1mol/L H_2SO_4 was used to adjust the pH value of the dye solution to examine the effect of pH value on degradation reaction.

2.2. Catalyst preparation

Carbon microspheres were synthesized through the polycondensation reaction of glucose under hydrothermal conditions according to the method reported in the literature [8].

In the process of preparing silica hollow spheres, 26 mL of C_2H_5OH was firstly mixed with 14 mL of H_2O, followed by adding 0.1g of CTAB into the solution under vigorous stirring until dissolved. Subsequently, the core templates, carbon spheres were uniformly dispersed in the above solution under ultrasound irradiation. After that, a total of 4g of TEOS was added dropwise [9]. The controlled hydrolysis of TEOS around the carbon microspheres was processed under acidic conditions (pH=2.0) at 40°C for 24 h. The obtained solid product was separated by filtration and washed with water. After that, it was put in an oven and dried at 80°C for 6h. Calcination was performed at 350°C in air for 1 h (temperature increase = 1°C/min), and then at 550°C for 6 h.

The CuO/SiO_2 catalyst was prepared by impregnation method. Certain amount of SiO_2 hollow sphere powder was added to 40mL of 10wt. % copper nitrate aqueous solution, depositing at r.t. for 12 h. The deposit was then dried at 80°C for 6h, followed by calcination in air at 550°C for 4 h.

2.3. Catalyst Characterization

The surface morphology and hollow structure of the samples were observed using a Hitachi S-4800 Scanning Electron Microscopy (SEM) analyzer. X-ray diffraction (XRD) patterns were recorded on a diffractometer (Philips X-Pert) with CuKα radiation (0.15405 nm) equipped with a graphite monochromator in 2θ ranges between 0° and 75°.

2.4. Fenton-like reactions

The catalytic properties of the material were tested via oxidation of the reactive dye Acid Scarlet 3R. The experiments were performed in a round-bottom flask fitted with a magnetic stirrer. The initial concentration of the dye is 0.4 g/L. When 50mL of dye solution in the flask attains a fixed temperature, the required amount of catalyst and 30% H_2O_2 were added. The mixture was kept under constant agitation. Afterwards, samples were withdrawn at fixed time and analyzed by a UV-Vis spectrophotometer with the maximum absorbance for the dye at 510 nm. The oxygen demand (COD) was analyzed using dichromate method.

The performance of the catalysts for the decolorization of Acid Scarlet 3R was evaluated under different conditions. The temperature, catalyst dosage and initial H_2O_2 concentration were varied between 40 and 80°C, 2-12 g L^{-1} and 27.2-40.8 mmol, respectively. At selected reaction time, aliquots

IOP Conf. Series: Materials Science and Engineering **87** (2015) 012039 doi:10.1088/1757-899X/87/1/012039

of the solution were collected. After the catalyst was separated from the reaction mixture by centrifugation, the solution was analyzed for COD removal and decolorization efficiency. In all the experiments, other parameters were kept constant and samples were withdrawn at regular time for the analysis. The decolorization efficiency of Acid Scarlet 3R was calculated using the following equation:

$$Decolorization\ efficiency = \left(\frac{C_0 - C_t}{C_0}\right) \times 100\%$$

where C_0 is the initial concentration of the dye and C_t the concentration at reaction time t.

3. Results and discussion

3.1. Catalyst characterization

SEM images of the hollow SiO_2 spheres are shown in figure 1. It clearly shows the hollow structure of the SiO_2 spheres.

Figure 1. SEM image of SiO_2 hollow spheres.

The powder diffraction pattern of the catalyst is depicted in figure 2. It is seen that the spectrum of the catalyst shows the characteristic peaks of bulk CuO at 2θ = 35.5°, 38.7° and 48.8°. The widen diffraction peak around 23°, was characterized as amorphous silica, which proved the hollow silica spheres were amorphous silica.

3.2. Effect of reaction temperature

The influence of reaction temperature on the decolorization and COD removal efficiency of Acid Scarlet 3R was studied by changing temperature from 40°C to 80°C. It can be seen (figure 3) that raising the temperature has a positive impact on the decolorization of Acid Scarlet 3R. The COD removal increased from 13% to 90% as the temperature increased from 40 to 60°C. That's because the elevated temperature can increase the reaction rate of hydrogen peroxide and catalysts, thereby increasing the production rate of hydroxyl radicals or high-valence copper species [10]. Furthermore, increasing temperature makes the reactant molecules have more energy to overcome the activation energy of the reaction [11]. While raising temperature up to 60°C does not show significant improvement in COD removals maybe because too high temperature will promote the decomposition of hydrogen peroxide to oxygen and water [12].

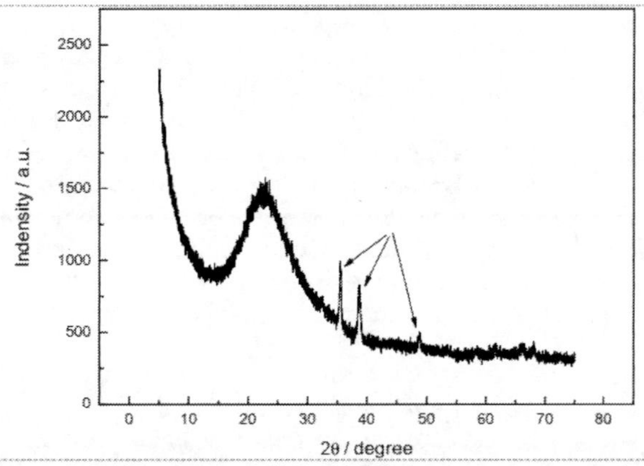

Figure 2. XRD patterns of CuO/SiO_2 heterogeneous catalyst.

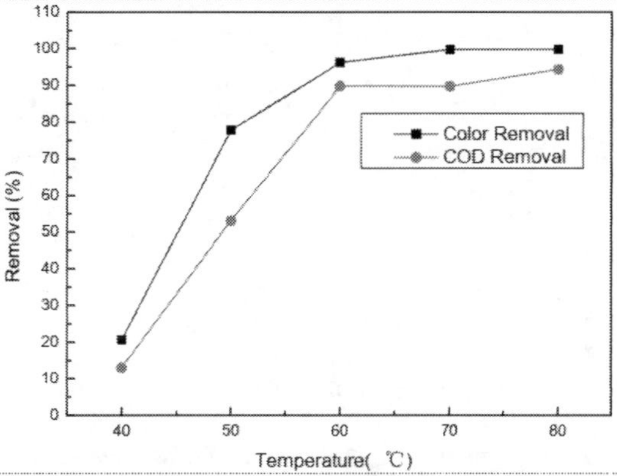

Figure 3. Effect of temperature on catalytic activity, (●) COD and (■) colority. Reactions conditions: $[dye]_o$ = 400 mg/L, catalyst = 6.0 g/L, $[H_2O_2]_o$ = 34 mmol/L, pH = 3.5.

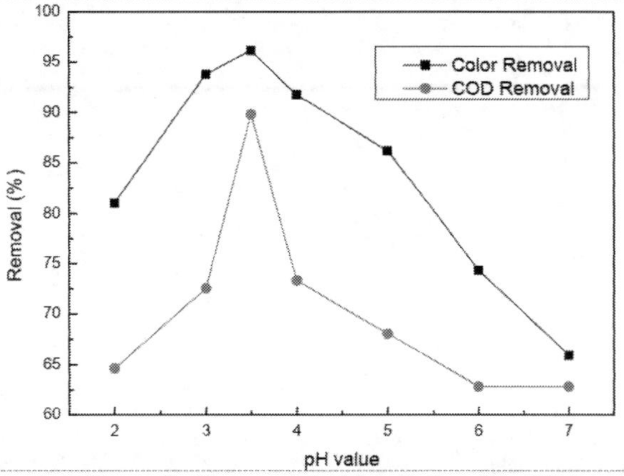

Figure 4. Effect of pH on catalytic activity, (●) COD and (■) colority. Reactions conditions: $[dye]_o$ = 400 mg/L, catalyst = 6.0 g/L, $[H_2O_2]_o$ = 34mmol/L, temperature = 60°C.

3.3. Effect of initial pH values on dye degradation

pH value is one of the most important parameters that influence the catalytic degradation. The effects of pH from 2 to 7 (adjusted by H_2SO_4, the initial pH was 9.18) on the catalytic decolorization are shown in figure 4. The figure shows that adjusting pH value of the dye solution from 2 to 3.5 increased the decolorization, but thereafter the decolorization decreased. At low pH (pH<3.0), more H_2O_2 molecules will be solvated by reaction with H^+ to form oxonium ion ($H_3O_2^+$), which results in their lower activity, or recombination of hydroxyl radicals becomes prominent resulting in inhibition of the catalytic reaction [13]. COD removal coincided with the color removal, but the curve of COD removal was below down the curve of color removal. That's because the hydroxyl radicals (·OH) mainly attacked the chromophores first, thus the colority of the solution slowed down fast.

3.4. Effect of H_2O_2 dosage

The results of COD and color removal for 400 mg/ L of Acid scarlet 3R for determination of optimum H_2O_2 dosage are illustrated in figure 5. It can be seen that color removal efficiency increased by increasing H_2O_2 dosage from 68% to 110% of stoichiometric amount (34 mmol/L). 110% of stoichiometric amount of H_2O_2 can be considered as the optimum dosage, which will produce 86% color removal and 67% COD removals. When the dosage of H_2O_2 increased to 125% of stoichiometric amount, COD removal didn't increase any more, but dropped to 65%. This is probably because hydroxyl radicals were produced from H_2O_2 with catalyst promoting and lower H_2O_2 dosage was not adequate to produce enough hydroxyl radicals. Furthermore, higher H_2O_2 concentration make hydroxyl radicals have more chance to react with each other, thereby reducing the concentration of them [14].

3.5. Effect of catalyst dosage

Figure 6 shows the influence of catalyst dosage on Acid scarlet 3R decolorization and COD removal. The results indicated that higher catalyst amount was associated with increased decolorization, and it was found that 99% decolorization efficiency was achieved with catalyst dosage of 10 g/L. Increased catalyst amount would lead to the formation of more copper sites on the catalyst surface, which would accelerate the decomposition of H_2O_2 to produce more ·OH. However, at low catalyst dosage, the color removal can also reach a high value if given adequate time. So the catalyst dosage can accelerate the reaction rate.

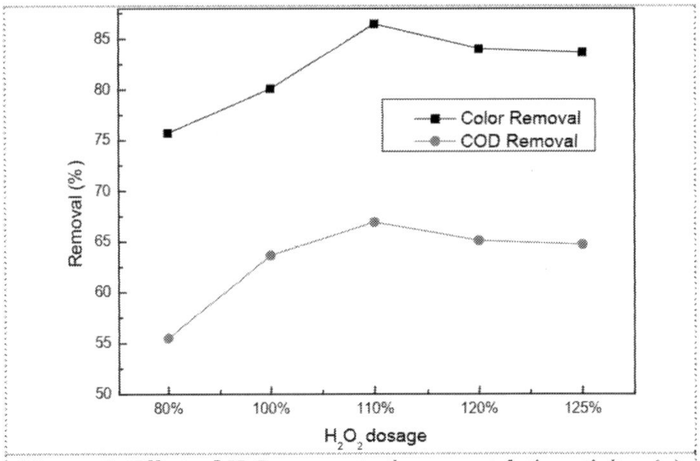

Figure 5. Effect of H_2O_2 concentration on catalytic activity, (•) COD and (■) colority. Reactions conditions: $[dye]_o$ = 400 mg/L, catalyst = 6.0 g/L, pH = 3.5, temperature = 60°C.

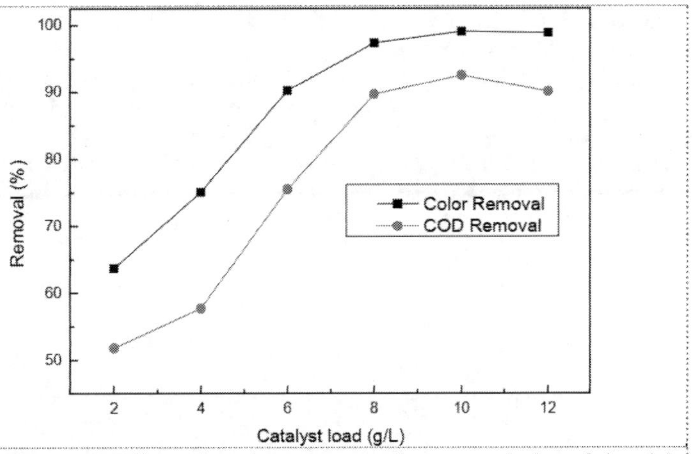

Figure 6. Effect of catalyst concentration on catalytic activity, (●) COD and (■) colority. Reactions conditions: $[H_2O_2]_o$ = 34 mmol/L, pH = 3.5, temperature = 60°C, reaction time = 90 min.

4. Conclusions

This study investigated the oxidation of azo dye using the heterogeneous catalyst and H_2O_2 under mild conditions. When temperature is 60°C, the dosage of H_2O_2 is 34 mmol L^{-1}, initial dye concentration is 400 mg L^{-1} and pH value of reaction solution is 3.5, the decolorization efficiency of Acid Scarlet 3R reached 97% within 90 min.

References

[1] Papic S, Vujevic D, Koprivanac N and Sinko D 2009 Decolourization and mineralization of commercial reactive dyes by using homogeneous and heterogeneous Fenton and UV/Fenton processes *J. Hazard. Mater.* **164** 1137-45

[2] Han F, Kambala V S R, Srinivasan M, Rajarathnam D and Naidu R 2009 Tailored titanium dioxide photocatalysts for the degradation of organic dyes in wastewater treatment: A review *Appl. Catal. A.* **359** 25-40

[3] Dükkancı M, Gündüz G, Yılmaz S and Prihod'ko R V 2010 Heterogeneous Fenton-like degradation of rhodamine 6G in water using CuFeZSM-5 zeolite catalyst prepared by hydrothermal synthesis *J. Hazard. Mater.* **181** 343-50

[4] Hassan H and Hameed B H 2011 Fe–clay as effective heterogeneous Fenton catalyst for the decolorization of reactive blue 4 *Chem. Eng. J.* **171** 912-18

[5] Xu X R, Li H B, Wang W H and Gu J D 2004 Degradation of dyes in aqueous solutions by the Fenton process *Chemosphere* **57** 595-600

[6] Yu J, Liu S and Yu H 2007 Microstructures and photoactivity of mesoporous anatase hollow microspheres fabricated by fluoride-mediated self-transformation *J. Catal.* **249** 59-66

[7] Wang C H, Chu X F and Wu M M 2007 Highly sensitive gas sensors based on hollow SnO_2 spheres prepared by carbon sphere template method *Sens Actuators B* **120** 508-13

[8] Sun X and Li Y 2004 Colloidal carbon spheres and their core/shell structures with noble-metal nanoparticles *Angew. Chem. Int. Edit.* **43** 597-601

[9] Shen W, Zhu Y, Dong X, Gu J and Shi J 2005 A new strategy to synthesize TiO_2 hollow spheres using carbon spheres as template *Chem. Lett.* **34** 840-41

[10] Sun J H, Sun S P, Wang G L and Qiao L P 2007 Degradation of azo dye Amido black 10B in aqueous solution by Fenton oxidation process *Dyes Pigments* **74** 647-52

[11] Xu H-y, Prasad M and Liu Y 2009 Schorl: A novel catalyst in mineral-catalyzed Fenton-like system for dyeing wastewater discoloration *J. Hazard. Mater.* **165** 1186-92

[12] Benitez F J, Acero J L, Real F J, Rubio F J and Leal A I 2001 The role of hydroxyl radicals for the decomposition of p-hydroxy phenylacetic acid in aqueous solutions *Water Res.* **35** 1338-43

[13] Chang M C, Shu H Y and Yu H H 2006 An integrated technique using zero-valent iron and UV/H_2O_2 sequential process for complete decolorization and mineralization of CI Acid Black 24 wastewater *J. Hazard. Mater.* **138** 574-81

[14] Tekbaş M, Yatmaz H C and Bektaş N 2008 Heterogeneous photo-Fenton oxidation of reactive azo dye solutions using iron exchanged zeolite as a catalyst *Micropor. Mesopor. Mat.* **115** 594-602

Optimized simulation of vortex jet mill in waste rubber grinding technology by LNG

Yuemei Han[1]

Institute of Mechanical Engineering Dalian University, Xuefu Street 10th, Dalian Economic Development Zone, Dalian, Liaoning, China

E-mail: hanyuemei@dlu.edu.cn

Abstract. Frozen rubber powder has excellent qualities and application value, and it can be achieved from waste rubber after being crushed at low temperature used liquefied natural gas (LNG) as cryogen. Vortex jet mill was the key equipment to further crush the rubber particles which the pressure-air was jet into in the basic LNG technological process. After confirming the structure and size of the jet nozzle, the Height (H) between the nozzle and the bottom of the mill, the incident angle α and the initial size of the rubber particles were changed then the continuous phase and the track of single particle were optimized in order to gain more excellent crushing effect. The results showed: the jetting gas were spiral rising in the mill and the speed of it was reduced, so the particle was graded by the gas. The impact and collision could reduce the particle diameter and crush them but the result was influenced by the initial size of the particle. The size of the original rubber particles must not be more than 110μm. The simulation was helpful and leading for the experiment.

1. Introduction

The basic principle of crushing waste rubber at low temperatures was that the waste rubber blocks were quickly chilled to a temperature below its brittle cold with refrigerant and then were put into a mill. The fine rubber powder by this way maintains the material qualities and its value in use increases so this technology at low temperatures will be the direction and trend of the comprehensive utilization of waste rubber. More and more research on the method of LNG (liquefied natural gas) as the refrigerant in waste rubber crushed was carried out recent years. The research about crush waste rubber at low temperatures with LNG was still limited in conceptual stage although it was a new and promising technology for the higher requests on the equipment and operating safety. There was no completely equipment or experiment practice reported now.

2. The basic flow

The basic flow of waste rubber crushing technology at low temperature was shown in figure 1. Waste rubber blocks were chilled to -90℃ with cold energy supplied by LNG at first and then the rubber particles reached a 20-40 mesh after a simple mechanical crushing. The particles are put into a crushing equipment- vortex jet mill to be crushed further by the cold pressure air in the last step. The vortex jet mill was the key to smash of rubber particles in the whole flow. The parameter of the cold

[1] Address for correspondence: Yuemei Han, Institute of Mechanical Engineering Dalian University, Xuefu Street 10th, Dalian Economic Development Zone, Dalian, Liaoning, China. E-mail: hanyuemei@dlu.edu.cn.

air and the rubber particles that entered into the mill must be controlled strictly. Size of particles entering must be between 20 and 40 mesh for too small particles would influence the classification efficiency and too large ones would result to decrease the smash efficiency. The pressure and flux of the cold air entering the mill must be controlled. The inlet pressure was about 0.4 MPa and the inlet flux was between 0.3 and 1 m^3/min adjusted by the inlet air valve [1].

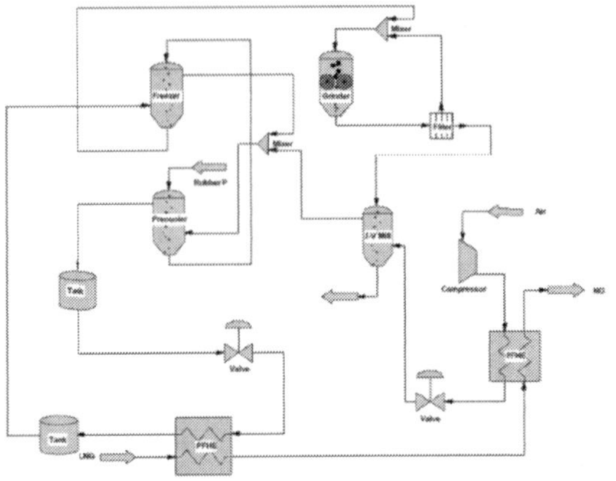

Figure 1. The basic flow by LNG.

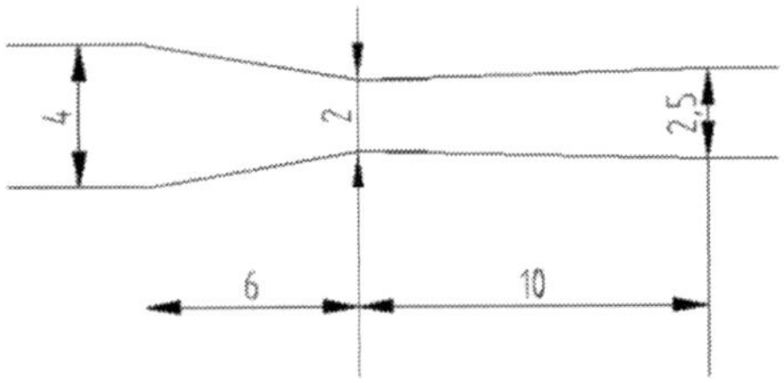

Figure 2. Structure and size of the nozzle.

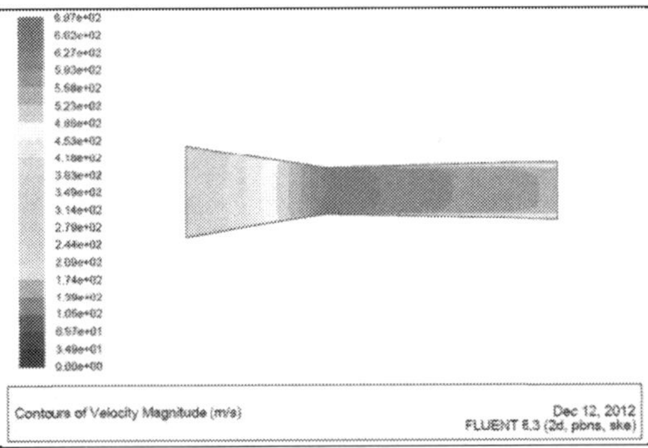

Figure 3. Velocity simulation profile in the nozzle.

3. Analog computations

3.1. Computations of the nozzle

In order to calculate the last speed of the cold air after going through the nozzle the cold air's motion in the nozzle was simulated. The inlet air pressure was set to 0.4 MPa and the outlet pressure was Atm. The equal diameter segments of inlet and outlet were ignored. The structure and parameters was showed on figure 2 while the computations results were shown on figure 3. The speed of cold air at the outlet was more than the speed of sound and the average speed was 550 m/s according to the results simulated by the software [2]. Then the data would be the boundary conditions in the following simulation of the grind.

Table 1. The model parameters in simulation.

No.	1	2	3	4	5	6	7	8	9
H (mm)	15	15	15	21	21	21	27	27	27
A (°)	30	35	40	30	35	40	30	35	40

3.2. Simulation of the continuous phase

The model parameters in simulation were shown in table 1 was the distance between the bottom and the nozzle and α was the angle between the nozzle line and the intersection point of tangential direction.

The Continuous phase was simulated at first without considering the rubber particles in the mill. There were 6 nozzles on the wall of the mill and the speed of the air inlet was 550 m/s while the outlet air was Atm. The 3D-model built based the data No. 4 in table 1 with software was shown in figure 3. The speed along the positive direction of Z axis and pressure inside of the grind was studied [3]. The results was shown that the maximum value of speed appeared at the wall of the mill along the positive direction of Z axis and the equal Z axis speed lines along the radial direction were in banding structure as figure 4.

Figure 4(c) was simulating results about the velocity distribution along Z axis at the longitudinal section in the mill. The velocity along the radial direction and the area near the outlet was upward while there was a small symmetrical area between the outlet and the wall in which the velocity along Z axis was down, so the air flow had a round trip between the top and bottom walls.

The air flow inside of the mill was keeping spiral upward movement and the radius became gradually smaller when the value of the speed became faster and reached to the outlet at last [4]. The results were shown on figure 4(d). The simulation result about air pressure was shown on figure 5. The pressure along the radial direction distributed in banding and there was negative pressure in the bottom centre area along Z axis. This phenomenon confirmed that the vacuum at the outlet was very necessary while the mill was working. The fastest speed of air along Z axis was 20 m/s according to the results above. Only data No. 4 was chosen to show out because the similar results were achieved under other data.

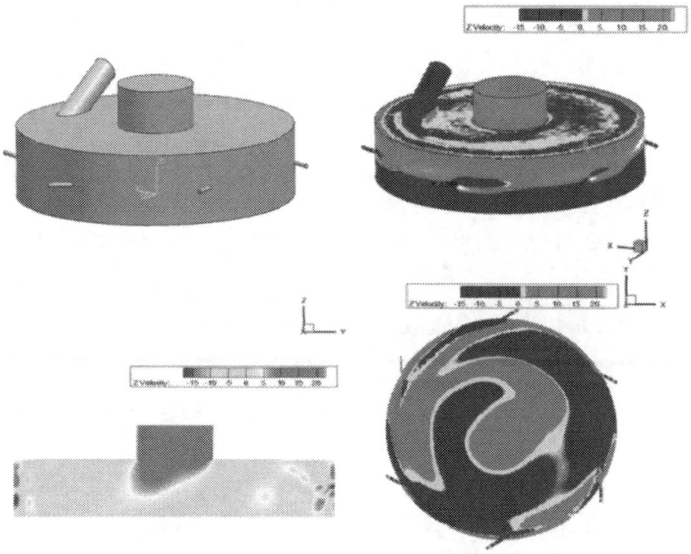

Figure 4. The simulation result of No. 3 model. a) 3D model; b) Profile of velocity along Z axis; c) along the Z-axis velocity profile at longitudinal section; d) Profile of velocity along Z axis at the nozzle level plane.

Table 2. The size of the feed rubber particles.

Size (um)	10	40	60	75	80	110	150	250	400	850
Feed rate (g/s)	1	1	1	1	1	1	1	1	1	1

3.3. *Simulation of the particles' track*

Because the particles' track can be simulated accurate only when the solid volume fraction particles was set to be 1g/s and the rate of the inlet particles to be 30 m/s. The solid volume fraction was below 10% for its density was 1000 kg/m^3. The simulation was basic on the model parameter of No. 3 with 10 nozzles. The particle phase's influence on the continuous phase was ignored and only the influence from continuous phase on the particle was considered. The size of rubber particles was shown in table 2.

Table 3. The statistics result of the rubber particles at exit.

Size(μm)	850	400	250	150	110	80	75	60	40	10
Track	216	216	216	216	216	216	216	216	216	216
flee	0	0	0	0	79	168	183	216	216	216
undone	216	216	216	216	137	48	33	0	0	0

The simulation result was shown in table 3. There were some rubber particles escaping from the exit whose size was 110 um under the model condition. All the rubber particles can escape from the exit when their size was not bigger than 60 um. The efficiency of classification of the vortex jet mill was about 50% according to the simulation. The efficiency in the actual situation would be larger than this number for the rubber particles would be smaller by the impact collision in the mill. There would be more rubber particles escaped from the exit and the efficiency of classification would be improved because of the smaller average size of the particles in the actual situation.

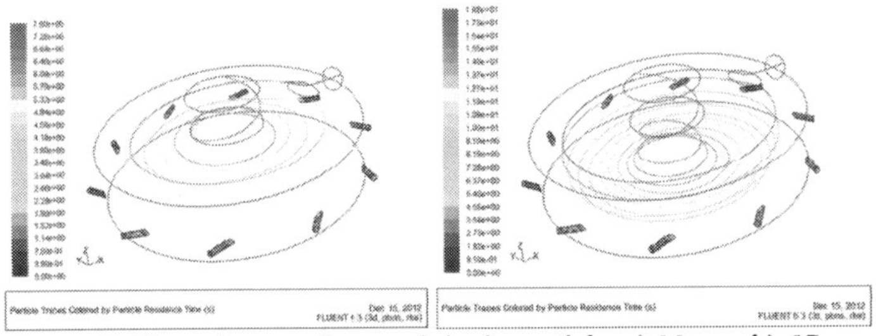

Figure 5. The trajectory figure of single particle. a) 10 um; b) 65 um.

From the trajectory figure of single particle we can see that rubber particles spiralled downward at first and became smaller and smaller simultaneously then spiralled upward to exit and escaped from it. The larger particles would spiral downward further and their radius of gyration would be larger than the small ones. Figure 5 was shown the trajectory of particle of 10 um and 65 um. From figure 5 we can see the 65 um particles spiralled downward at first and reached the bottom of mill so the larger particles had more probability to impact the bottom and the wall of the mill [6]. The gather of particles larger than 110 um near the bottom and circumferential wall increased the probability of collision and improved the crushing effect.

4. Conclusion

The simulation was carried out with different H (the distance between nozzles and the grind bottom, angle of incidence α and the size of inlet particle after the size of nozzles was conformed. The simulation result about continuous phase airflow showed that the jetting air spiralled upward and the rate of it became lower simultaneously so the airflow was very effective for classification. The trajectory of single particle showed that at first the actual situation spiralled downward and then spiralled upward after got into the mill. The larger particle spiralled downward further and had more collision probability than the smaller one so they can be crushed more completely. The efficiency of classification was about 50% according to the simulation while the efficiency would be higher in the actual situation for the big particles could be crushed by collision. The size of the original rubber particles must not be more than 110 um for they would gather in the mill and cannot escape from it.

References

[1] Kazumi Kozawa, Takafumi Seto and Yoshio Otani 2011 Development of a spiral flow jet mill with improved classification performance Advanced Powder Technology 23 601-6

[2] Shuli Teng, Peng Wang, Linjie Zhu, Ming-Wan Young, Costas G. Gogos Experimental and numerical analysis of a lab-scale fluid energy mill *Powder Technology* **195** 31.

[3] Avi Levy and Haim Kalman 2007 Numerical study of particle motion in jet milling particulate *Science and Technology* **25** 197-204

[4] Avi Katz and Haim Kalman 2007 Preliminary experimental analysis of a spiral jet mill performance *Part. Syst. Charact.* **24** 332-8

[5] Eskin D, Voropayev S and Vasilkov O 1999 Simulation of jet milling *Powder Technology* **105** 257-65

[6] Ishito K, Akiyama S and Tanaka Z 2002 Improvement of grinding and classifying performance using a closed-circuit system *Advances in Powder Technology* **13** 363-75

IOP Conf. Series: Materials Science and Engineering **87** (2015) 012041 doi:10.1088/1757-899X/87/1/012041

Laccase catalysed grafting of phenolic onto xylan to improve its applicability in films

Jicheng Pei[1], Bing Wang, Fangdong Zhang, Zhongyang Li, Yunbei Yin and Dongxu Zhang

Tianjin Key Laboratory of Pulp & Paper, Tianjin University of Science &Technology, Tianjin 300457, China

E-mail: jcpei@tust.edu.cn

Abstract. Xylan can be tailored for various value-added applications. However, its use in aqueous systems is hampered by its complex structure, and small molecular weight. This research aimed at improving the xylan molecular weight and changing its structure. Laccase-catalysed oxidation of 4-coumaric acid (PCA), ferulic acid (FA), syringaldehyde (SD), and vanillin (VA) onto xylan was grafted to study the changes in its structure, tensile properties, and antibacterial activities.

A Fourier transform infrared (FTIR) spectrum analyser was used to observe the changes in functional groups of xylan. The results showed a band at 1635 cm^{-1} corresponding to the stretching vibration of conjugated carbonyl carboxy hemoglobin and a benzene ring structure were strengthened; the appearance of a new band between 1200 cm^{-1} and 1270 cm^{-1} corresponding to alkyl ethers on the aryl C-O stretching vibration was due to the fact that during the grafting process, the number of benzene ring structures increased and covalent connections occurred between phenols and xylan.

The reaction mechanism for the laccase-catalysed oxidation of phenol compounds onto xylan was preliminary explored by ^{13}C-NMR. The results showed that PCA-xylan, FA-xylan graft poly onto xylan by C_7 ester bond, SD-xylan graft poly onto xylan by ether bond and an ester bond, and VD-xylan graft poly onto xylan by ether bond.

The film strength of xylan derivatives has been significantly increased, especially for the PCA-xylan derivative. The increases in tensile stress at break, tensile strength, tensile yield stress, and Young's modulus were: 24.04%, 31.30%, 55.56%, and 28.21%, respectively. After laccase/phenolics were modified, xylan had a good antibacterial effect to *E. coli*, *Corynebacterium glutamicum*, and *Bacillus subtilis*. The SD-xylan, FA-xylan, and PCA-xylan showed a greater efficacy against *E. coli*, *Corynebacterium glutamicum*, and *Bacillus subtilis*, respectively.

1. Introduction

Lignocellulosic biomass is one of the most widely available, renewable, resources: it represents a promising low-cost raw material for the production of biofuel, bioenergy, and value-added

[1] Address for correspondence: Jicheng Pei, Tianjin Key Laboratory of Pulp & Paper, Tianjin University of Science &Technology, Tianjin 300457, China. Tel.: +86 13352068064; E-mail: jcpei@tust.edu.cn.

Fund project: National Natural Science Foundation of China (31170548); Tianjin University of Science &Technology Scientific Foundation of China (20130117).

biomaterials [1]. The main components of lignocellulosic biomass are: cellulose, hemicellulose, and lignin. Hemicelluloses, the second most abundant class of polysaccharides found in nature after cellulose, comprises roughly one-fourth to one-third of most plant materials [2]. In the pulp and paper making process, significant quantities of hemicellulose are dissolved and burned as fuel. The portion of hemicellulose burned can be roughly estimated at about 40 million tonnes per year [3]. How to make better use of the hemicellulose has become a hot topic among researchers. However, its use in aqueous systems is hampered by its complex structure and low molecular weight [4-6].

Hemicellulose contains many hydroxyl groups in the molecule. Many studies on hemicelluloses have been carried out by chemical modification (grafting, esterification, sulphonyl reaction, etherification, etc.). An active functional group is introduced to improve the film formation properties (such as solubility, hydrophilicity, barrier properties, and crystallisation) [7]. Chemical modification of hemicelluloses requires harsh reaction conditions, and will produce certain harmful by-products, while the use of bio-enzymatic modification, with its milder reaction conditions, can be more environmentally friendly. Thus, in recent years, using biological enzymes to catalyse biological graft-modified hemicellulose has become another focus of research attention. Some teams use laccase catalysed oxidation of aromatic groups for grafting onto galactoglucomannan, and the cross-linking treatment resulted in significant improvement of the mechanical properties of barrier films made with spruce galactoglucomannan [8]. However, the aromatic group best suited for grafting onto xylan remains to be studied. In addition, the main product for the modern paper-manufacturing industry is mechanical pulp, and its main raw materials are hardwood and grasses, which are important sources of hemicellulose.

Xylan is the main component of hemicellulose from grasses and hardwood: it comprises complex molecular polysaccharides based on the β-1, 4 xylosidic bond connecting D-xylose residues for its main chain [7]. In the presence of oxygen, a phenolic hydroxyl group can be oxidised to phenolic oxygen free radicals by laccase, and phenoxy radical mutual condensation generates macromolecules. Thus, laccase catalysing phenols were used as agents to be grafted onto xylan in this work.

In this research, phenols were oxidised to phenolic oxygen free radicals by laccase, and reacted with xylan by graft copolymerisation to improve the film-forming properties thereof: this also endowed the material with good antibacterial properties, thus expanding its potential areas of application. Infrared radiation (IR), and nuclear magnetic resonance (NMR), analyses proved the feasibility of the proposed graft modification method.

2. Materials and methods

2.1. Raw materials and reagents

Straw was sourced from Shandong; laccase, catalogue item number NOVOZYME 51003, with an enzyme activity of 1072 u/mL was provided by the Novozyme Company; phenols were purchased from the Sigma-Aldrich Corporation; two types of xylan were used: a) commodity-grade xylan: purchased from Shanghai Jing Chun Biochemical Technology Co., Ltd, with a molecular weight of about 30 kDa (200 sugar units); b) homemade xylan: laboratory preparation; other reagents were purchased from Tianjin Chemical Reagent Factory.

2.2. Experimental methods

2.2.1. Preparation of xylan. Straw powder was soaked in 60 g/L of NaOH solution with a solid-liquid ratio of 1:15, and the reaction took place at 100°C for 2 h, the filtrate was then obtained after filtering and pH adjustment to 7.0, the filtrate was then precipitated with the ratio of filtrate to ethanol being 0.8. After washing with anhydrous ethanol, the precipitate was obtained for use as the xylan sample [9].

2.2.2. Xylan with grafted phenols reaction. First, 90 ml of acetic acid buffer solution (pH 4.8) was added to 1 g of homemade xylan in a beaker; then 10 ml of 5 mmol/L methanol solution of phenols was added to the beaker; thereafter, the beaker was stored in water at 40°C for 5 min and oxygen was supplied using an oxygen pump, then 200 μl of laccase was added and the reaction continued for 4 h in water at 40 °C. After the completion of this grafting reaction, absolute ethyl alcohol of a volume of three times of the reaction solution volume was added, and the resulting precipitate observed. After 24 h, the refined precipitate was obtained through repeated washing in absolute ethyl alcohol. The precipitate was then dried in vacuum at 60 °C for 24 h and the sample of xylan with grafted phenol was obtained.

2.2.3. FT-IR analysis. The VECTOR FTIR spectrometer (Brooke Instrument Company, Germany) was used on the xylan and its derivatives over 16 scans, capturing spectra between 500 and 4000 cm^{-1}, at a resolution of 4 cm^{-1}: samples were previously kept in the dark and dried for 24 h at room temperature in a desiccator.

2.2.4. NMR analysis. NMR spectra were captured by a Swiss Brook Baiesibin 400 M superconducting NMR spectrometer: 20 mg of xylan and graft product were dissolved in 0.6 mL of deuterated D_2O, the sampling conditions were as follows: sampling time 3.98 s, relaxation time 2.0 s, and accumulated 128 times, with a concentration of 4.70 ppm (using the solvent peak pattern) for correction. ^{13}C spectra were taken from 100 mg samples dissolved in 0.6 mL of D_2O under the following sampling conditions: 30° (using the pulse sequence), sampling time 1.36 s, relaxation time 1.89s, and accumulated 30,000 times [10].

2.2.5. Preparation of the film from xylan and grafted product. Eight 0.6 g xylan or graft-modified product samples, and 0.4 g carboxymethyl cellulose sodium were each added separately to 35 ml of deionised water and stirred for 60 min at 80 °C, and then poured into a glass mould at room temperature (23°C, 50% humidity) for casting, and dry film manufacture [8, 10].

2.2.6. Tensile properties of film. A computer-controlled electronic universal testing machine (CMT4503, Shenzhen sans Testing Machine Co., Ltd, China) was used for film testing. The dried film was cut into 80 mm × 15 mm spindle-shaped samples, then their thickness was measured. The prepared film strip was placed on a microcomputer-controlled electronic universal testing machine for testing in accordance with Chinese Standard GB/T 1040.3-2006. The cross-head speed was 50 mm / min, and the test gauge length was 25 mm.

2.2.7. Antibacterial properties. The inhibition zone method was used for detecting the antibacterial activity of the film. First, indicator bacteria cultures were taken whose bacteria concentration was 105 to 106 CFU / mL, each with a volume of 0.1 mL which was added dropwise to beef extract peptone medium plates (diameter 90 mm), and the coating stick was then used to paint them evenly onto the flat surface before natural drying for 30 min. Then sterilised film sample wafers (diameter 5 mm) were placed evenly on the plate. The dishes were placed in an incubator for 24 h as 37°C, for measurement of the inhibition zone diameter. Each experiment was repeated three times per sample, and the results averaged.

3. Results and discussion

3.1. Xylan and grafted product FT-IR analysis

3.1.1. FT-IR characterisation of xylan. In the experiments, straw powder was used as a raw material in the alkali extraction method for preparing xylan. Firstly, compared with homemade xylan, the infrared absorption peaks of factory produced xylan are shown in figure 1.

Figure 1 shows that the infrared spectrum of homemade xylan had two distinct peaks at 1043 cm^{-1}

IOP Conf. Series: Materials Science and Engineering **87** (2015) 012041 doi:10.1088/1757-899X/87/1/012041

and 3396 cm^{-1}: these were from a sugar unit within a hydroxyl group and an ether linkage (C-O-C) absorption peak [8]. Spectrum (1) was the IR spectrum of commodity-grade xylan: it revealed that the band between 1170 cm^{-1} and 1000 cm^{-1} contained the typical absorption peaks of xylan, the bands at 1382 cm^{-1}, 1237 cm^{-1}, and 1249 cm^{-1} were -C-H stretching vibration, or –OH, or -C-O bending vibrations, the bands at 1157 cm^{-1} and 989 cm^{-1} were typical arabinose absorption peaks; and the band at 1045 cm^{-1} was a -C-C stretching vibration and hemicellulose -C-OH mixed-mode vibrations [10]. Spectrum (2) contained typical xylan absorption characteristics (1157.08 cm^{-1} and 1043.30 cm^{-1} peaks) that showed that homemade xylan contained the main structural elements of commercial-grade xylan: *i.e.* its pyran ring structure [11]. By this analysis, the structure of homemade xylan was shown to be consistent with commodity-grade xylan.

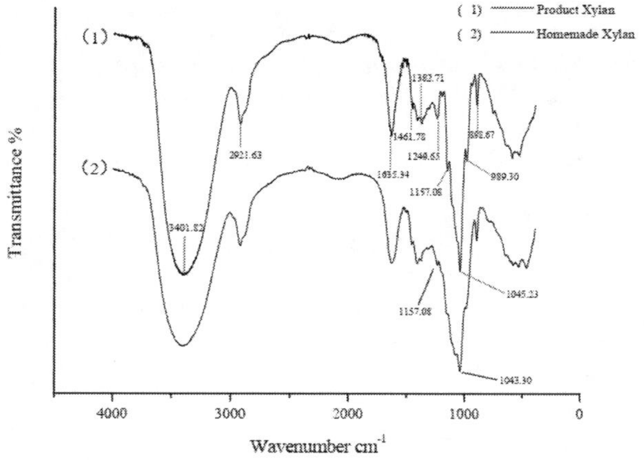

Figure 1. Commodity-grade xylan and homemade xylan blank FT-IR comparative analysis.

3.1.2. FT-IR characterisation of grafted product. The study aimed to improve the molecular weight of xylan and change its structure. 4-coumaric acid (PCA), ferulic acid (FA), syringaldehyde (SD), and vanillin (VA) were grafted onto xylan by laccase-catalysed oxidation. FTIR was used to observe the changes in functional groups, as shown in figure 2.

As can be seen from figure 2, a band at 1635 cm^{-1} came from conjugated carbonyl carbon and oxygen stretching vibration and the benzene ring structure absorption peak. This showed that the homemade xylan contained side groups with a benzene ring structure. A strong absorption peak near 1043 cm^{-1} came from a stretching vibration of the carbonyl carbon and oxygen in the xylan: this further explained that the hemicellulose prepared was, indeed, xylan.

The xylan after being laccase grafted, had a band at 1635 cm^{-1} corresponding to stretching vibration of conjugated carbonyl carboxyhemoglobin and a benzene ring structure which had been strengthened [12, 13]; a new band appeared between 1200 cm^{-1} and 1270 cm^{-1} corresponding to those alkyl ethers on the aryl C-O stretching vibration which showed that, during the grafting process [11], the benzene ring structure was stiffened and covalent connections occurred between phenols and the xylan.

Furthermore, in the infrared spectra of 4-coumaric acid-xylan and ferulic acid-xylan, new bands appeared at 1245 cm^{-1} and 1395 cm^{-1}, these two absorption peaks arose from carboxylic acid C=O stretching vibrations, and-OH bending vibrations, respectively [8]. In the infrared spectra of ferulic acid-xylan, syringaldazin-xylan, and vanillin-xylan, a new band appeared at 1384 cm^{-1}, this was the bending vibration of -CO-CH_3. By FTIR analysis, the phenols used were covalently grafted onto the xylan, and its benzene ring structure was stiffened.

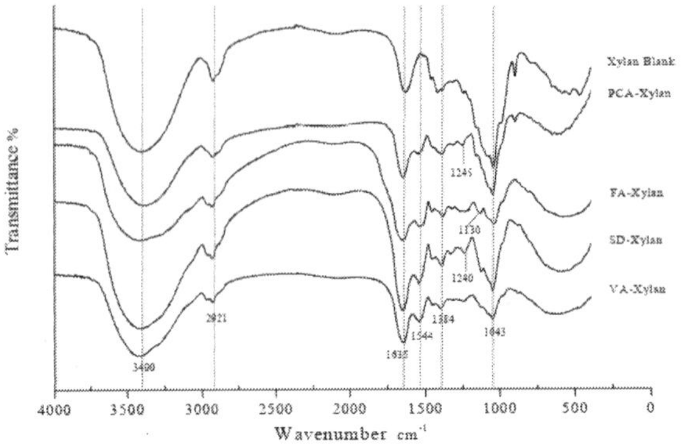

Figure 2. Homemade xylan and phenol-xylan FT-IR comparative analysis.

3.2. Xylan and grafted product 13C-NMR analysis

To observe the xylan and grafted product functional group structure, ^{13}C-NMR analysis (see figure 3) was used.

Xylan was the repeating structural unit represented by the inter-polymers of the monosaccharides which were mainly linked to each other by covalent bonding, hydrogen bonding, an ether bond, and an ester bond. Figure 3(a) shows anomeric carbon located at peak 1 (at 101.90 ppm) which showed that the xylan was based on β glycosidic linkage for its main chain [9].

As seen from the comparison chart of figures 3(a) and 3(b), in figure 3(b), peak 13 (at 23.48 ppm), and 14 (at 18.37 ppm) was the methyl absorbent of the acetyl group [11]; a weak absorption peak appeared at 168.41 ppm which was the C-4 of the hydroxyphenyl; in addition, there was a weak absorption signal at 171.25 ppm which was the O-C=O absorption of the acetyl group and Cγ absorption of the cinnamic acid [14]. The hydroxyphenyl ring structure was found on the grafted xylan: the PCA was therefore grafted onto the xylan by C_γ ester bonding.

In figure 3(c), the peaks at 26.25 ppm, and 19.84 ppm were the methyl absorbent of the acetyl group; a weak absorption peak appeared at 145.11 ppm which was the C_α absorption of ferulic acid [11]; figure 3(b) showed strong absorption signals at 180 ppm and 173.89 ppm which were ester bond carbonyl and ferulic acid Cγ signals, respectively [14]. That showed that FA was grafted onto xylan by C_γ ester bonding.

In figure 3(d), an absorption peak appeared at 51.67 ppm which was the $-OCH_3$ absorption peak; a strong absorption band appeared at 75 ppm which formed the C2, C3, and C5 absorption signal for xylan, and the C3 and C4 absorption signal of a xylose structure; in addition, SD-xylan had a C_α peak(connected with the carbohydrate ether bond), etherified C_α (β-O-4) absorption signals at 83.12 ppm, and 85.20ppm, an aromatic ring structure from C3/C5 (etherified) absorption at 154.75 ppm, and ester absorption signals at 171.19 ppm and 173.84 ppm [14]. That showed that SD-xylan grafted onto xylan by ether bonding and an ester bond.

Figure 3(e) shows an absorption peak at 38.04ppm from the γ- methyl, and α and β- methylene absorption signals from lignin propane side chains; in addition, VA-xylan had a C_α peak (from carbohydrate ether bonding) at 82.19 ppm [11]; and a hydroxyphenyl C-4, vanillin C_α (etherified) absorption signal at 168.57 ppm [15]. That showed that VA-xylan was grafted onto xylan by ether bonding.

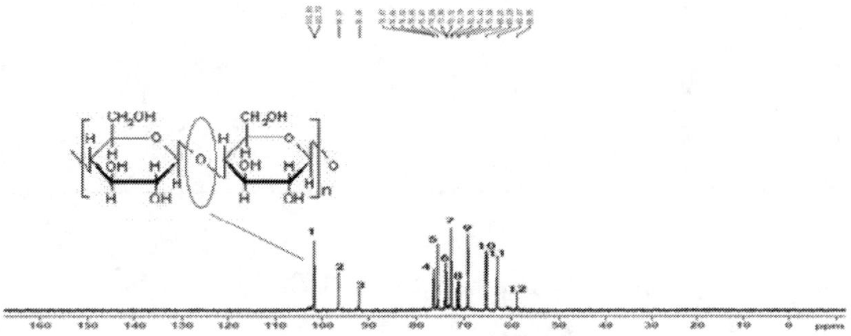

(a) Xylan blank ^{13}C-NMR

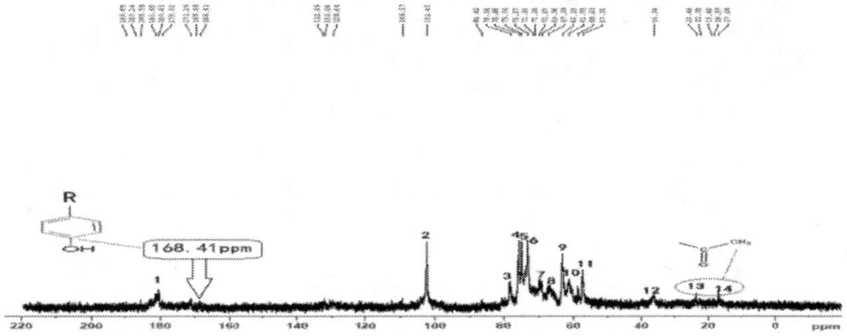

(b) 4- coumaric acid-xylan 13C-NMR

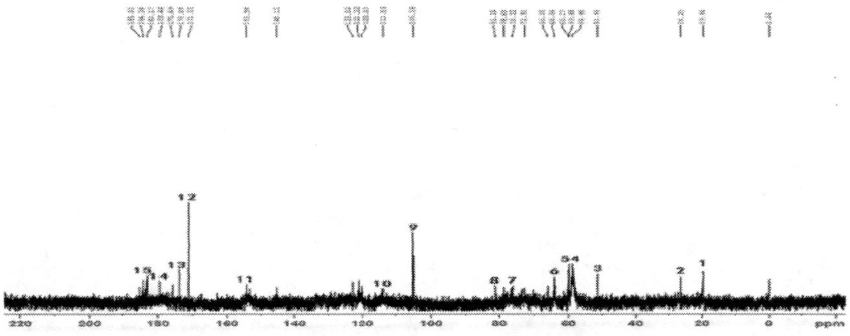

(c) Ferulic acid-xylan 13C-NMR

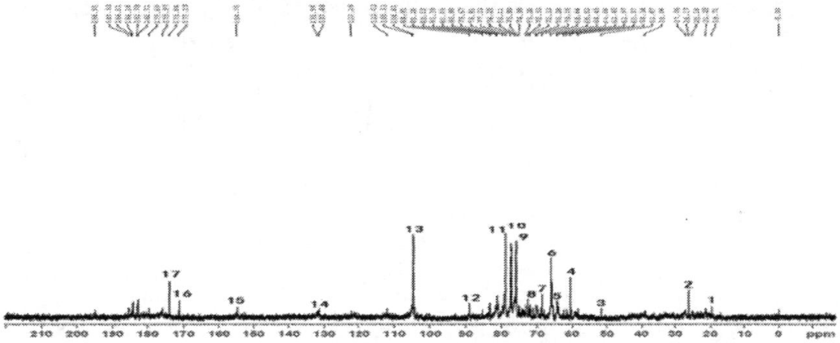

(d) Syringaldazin-xylan 13C-NMR

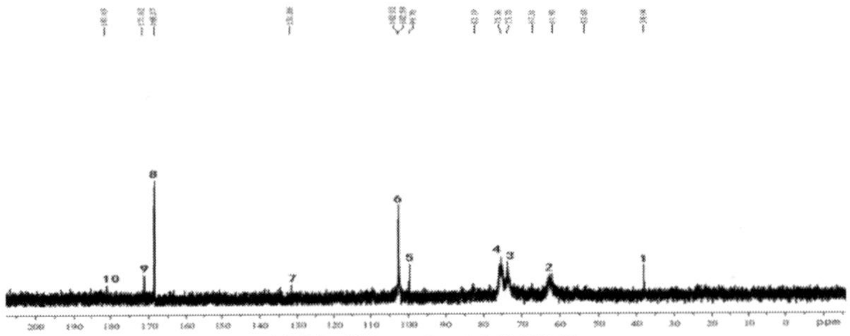

(e) Vanillin-xylan 13C-NMR

Figure 3. 13C NMR spectrogram of homemade xylan and graft product.

3.3. Tensile properties of the film

We tested the tensile stress of the xylan and the grafted xylan films to characterise their mechanical strength, tensile yield stress, and Young's modulus (see results in table 1).

Table 1. Homemade xylan and its graft product film tensile properties.

Sample	Tensile Stress (σB) /MPa	Tensile Strength (σM) /MPa	Tensile yield stress (σy) /MPa	Young's modulus (E) /MPa
Xylan	1.04	1.15	0.27	0.663
PCA-Xylan	1.29	1.51	0.42	0.850
FA-Xylan	1.21	1.43	0.33	0.771
SD-Xylan	1.13	1.47	0.35	0.758
VA-Xylan	1.18	1.33	0.37	0.761

Table 2. Changing material properties: homemade xylan compared to its graft products.

Sample	Increase in tensile stress/ %	Increase in tensile strength/ %	Increase in tensile yield stress/ %	Increase in Young's modulus/ %
PCA-Xylan	24.04	31.30	55.56	28.21
FA-Xylan	16.35	24.35	22.22	16.29
SD-Xylan	8.65	27.83	29.63	14.33
VA-Xylan	13.46	15.65	37.04	14.78

We measured the increase in strength and stiffness of the grafted xylan film (see table 2). The films produced with grafted xylan derivatives improved significantly with regards their tensile stress, tensile strength, tensile yield stress, and Young's modulus. Phenols were grafted onto the side chains of the molecular chain, the side groups induced steric hindrance, molecular chains of xylan underwent a relative separation, and reduced aggregation of molecular chains during film formation to improve its film-forming properties [16]. For testing, the grafted products selected were: PCA-xylose film, FA-xylose poly-film, SD-xylose poly-film, and VA-polyethylene film with xylose. The tensile stress increases at break were: 24.04%, 16.35%, 8.65%, and 13.46% respectively; the tensile strength increases were: 31.30%, 24.35%, 27.83%, and 15.65%, respectively; the tensile yield stress increases were: 55.56%, 22.22%, 29.63%, and 37.04% respectively; the Young's modulus increases were: 28.21%, 16.29%, 14.33%, and 14.78%, respectively. The results showed that the strength of films produced by xylan derivatives improved significantly, especially for the PCA-xylan derivative.

3.4. Antibacterial properties

In these experiments, 4-coumaric acid (PCA), ferulic acid (FA), syringaldehyde (SD), and vanillin (VA) were grafted onto xylan by laccase-catalysed oxidation, these phenolic compounds were commonly used in food preservatives, and their structures were similar to the structural phase in lignin (as found in paper). During the experiment, we mainly selected *E. coli*, *Corynebacteriumglutamicum*, and *Bacillus subtilis* for antibacterial property evaluation.

Table 3. Homemade xylan and its graft product film antibacterial performance test results.

Sample	Bacterial inhibition zone diameter（mm）		
	E. coli	*Corynebacteriumglutamicum*	*Bacillus subtilis*
Xylan	5	5	5
PCA-Xylan	10.0	10.5	10.0
FA-Xylan	11.0	11.5	8.5
VA-Xylan	9.0	10.0	7.5
SD-Xylan	12.0	11.0	8.0

Table 3 shows the antibacterial effect of xylan and grafted xylan on *E. coli*, *Corynebacteriumglutamicum*, and *Bacillus subtilis*.As could be seen from the table, the untreated xylan film did not have any antibacterial activity, but after laccase/phenolic modification, xylan had good antibacterial effect on *E. coli*, *Corynebacteriumglutamicum*, and *Bacillus subtilis*. The xylan grafted with different polyphenols exhibited different inhibitory effects. The SD-xylan showed a greater efficacy against *E. coli*; the FA-xylan showed a greater efficacy against *Corynebacteriumglutamicum*; and the PCA-xylan had a greater efficacy against *Bacillus subtilis*.

Laccase catalysed the oxidation of phenols and grafted it onto xylan: after laccase/phenolic modification xylan had good antibacterial activity which was related to the phenolic itself exerting an inhibitory effect. After polyphenols had been grafted onto the side chains of the xylan, the intermolecular hydrogen bonds appeared to reduce in number, the re-stabilisation phenomenon began to weaken between molecules, at the same time, hydrophobic groups were introduced onto the modified xylan, and affinity between xylan and bacterial cells was enhanced [17], antibacterial activity centres increased in number, thereby enhancing the antimicrobial activity of xylan. Thus, xylan derivatives exhibited strong antibacterial properties under the synergistic effect of the phenolics.

4. Conclusion

This article revealed that grafting between phenols and xylan by laccase catalysis and oxidation represented a new approach to improving the molecular weight of xylan, changing the structure thereof, and to improving the applicability of xylan when used in barrier films.

Infrared spectra showed that, during the grafting process, benzene ring structures increased in stiffness and covalent connections occurred between the phenols and the xylan. ^{13}C-NMR analysis showed that PCA-xylan, and FA-xylan, were grafted onto xylan by C_γ ester bonding, SD-xylan was grafted onto xylan by ether bonding and an ester bond, and VD-xylan was grafted by ether bonding.

The strength of films produced by xylan derivatives improved significantly, especially for the PCA-xylan derivative. After laccase/phenolic modification, xylan had good antibacterial effects on *E. coli*, *Corynebacterium glutamicum*, *Bacillus subtilis*, and SD-xylan, FA-xylan, and PCA-xylan showed a greater efficacy against *E. coli*, *Corynebacterium glutamicum*, and *Bacillus subtilis*, respectively.

References

[1] Jing B, Feng P, Pai P, et al. 2010 Isolation and fractionation of hemicelluloses by graded ethanol precipitation from Caraganakorshinskii *Carbohydrate Research* **345** 802-9

[2] Ren J L, Sun R C, Liu C F, et al. 2007 Synthesis and characterization of novel cationic SCB hemicelluloses with a low degree of substitution *Carbohydrate Polymers* **67** 347-57

[3] Abdus S, Richard A V, Joel J P, et al. 2011 Crosslinked hemicellulose citrate–chitosan aerogel foams *Carbohydrate Polymers* **84** 1221-9

[4] Egues I, Eceiza A and Labidi J 2013 Effect of different hemicelluloses characteristics on film forming properties *Industrial Crops and Products* **47** 331-8
[5] Ren Junli, Sun Runcang and Liu Chuanfu 2006 Advances in chemical modification of hemicelluloses *Modern Chemical Industry* **26** 68-71
[6] Yu Ziping, Peng Hong, Lin Da, et al. 2011 The structure characteristic of hemicellulose: A review *Polymer Bulletin* **6** 48-54
[7] Li Heping and He Lixia 2009 Advances in chemical modification of xylan and application of their derivatives *Chemical Industry and Engineering Progress* **28** 1955-81
[8] Oinonen P, Areskogh D and Henriksson G 2013 Enzyme catalyzed cross-linking of spruce galactoglucomannan improves its applicability in barrier films *Carbohydrate Polymers* **95** 690-6
[9] Peng Yuanhua, Lu Hongmei and Zeng Xiangqin 2007 Optimization of preparation conditions hemicellulose *Environmental Science and Technology* **13** 44-6
[10] Peng Xinwen 2012 *Efficient Synthesis and Applications of Functional Hemicellulose* (South China University of Technology)
[11] Gu Ruijun and Xie Yimin 2001 Studies on lignin and carbohydrate complex of monolignols (I)—synthesized DHP from holocellulose and coniferins *Paper Science and Technology* **20** 1-6
[12] Bourbonnais R and Paice M G 1992 De me thylation and delignification of kraft pulp by Coriolus Versicolor Laccase in the presence of 2, 2-azino-bis-(3-ethybenz-thiaoline-6-sulphonate) *Appl Microbiol Biotechnol* **36** 823-7
[13] Liu Haitang, Pei Jicheng, Hu Huiren, et al. 2010 Spectra analysis of lignin small molecular guaiacylconiferyl alcohol biological modification treated by laccase *Spectroscopy and Spectral Analysis* **30** 1469-73
[14] Yang Haitao and Xie Yimin 2007 Study on the chemical structure of xylan-lignin complex *China Pulp and Paper* **22** 1-4
[15] Shi Zhengjun 2013 Structural characterization of hemicelluloses and lignin, and their inter-linkages in Dendrocalam us brandisii and Dendrocalam us sinicus (Beijing Forestry University)
[16] Mikkonen K S, Rita H, Helén H, et al. 2007 Effect of polysaccharide structure on mechanical and thermal properties of galactomannan-based films *Biomacromolecules* **8** 3198-205
[17] Sudharshan N R, Hoover D G and Knorr D 1992 Antibacterial action of chitosan *Food Biotechnol* **6** 257-72

IOP Conf. Series: Materials Science and Engineering **87** (2015) 012042 doi:10.1088/1757-899X/87/1/012042

Preparation of PMMA/graphene oxide microcellular foams using supercritical carbon dioxide

Meijuan Li[1, 2], Ping Cheng[1,2], Ruizhi Zhang[2], Guoqiang Luo[2, 3], Qiang Shen[2] and Lianmeng Zhang[2]

[1]Department of Chemistry, School of Chemistry, Chemical Engineering and Life Sciences, Wuhan University of Technology, Wuhan, Hubei 430070, China
[2]State Key Laboratory of Advanced Technology for Materials Synthesis and Processing, Wuhan University of Technology, Wuhan, Hubei 430070, China

E-mail: luogq@whut.edu.cn

Abstract. Microcellular foams have a widely applications in many industries due to their superior properties. In this paper, the polymethymethacrylate (PMMA)/graphene oxide (GO) microcellular foams were prepared by supercritical carbon dioxide as a friendly foaming agent. The effect of graphene oxide amount on cellular structure and mechanical strength of foams had been investigated. Microstructure characterization bases on scanning electron microscopy (SEM). The cell size and cell density were calculated via image analysis. It was found that the average foam cell size was decreased from 20.1 μm to 2.2 μm and the cell density was increased from 2.8×10^{8} to 3.3×10^{10} when 1.5wt.% GO sheets were added into PMMA matrix. In addition, the compression strength of polymeric foams was increased from 13 MPa to 39 MPa.

1. Introduction

Microcellular foams, typically characterized by cell sizes smaller than 10 μm and cell densities larger than 10^9 cell/cm^3 [1], have been attracted a great attention because of their great properties such as lightweight, thermal insulation, high strength and cost reduction [2]. Microcellular foams have potential applications in building, packaging, automotive and aircraft industries [3]. In the past studies, many kinds of polymer have been used to product microcellular foams such as polymethymethacrylate (PMMA) [4], polystyrene (PS) [5], polycarbonate (PC) [6], polypropylene (PP) [7] and polylactic acid (PLA) [8]. In recent years, with the development of microcellular foaming technology, polymer nanocomposite microcellular foams also have received widespread attention due to a small amount of nanoparticles can significantly improve a variety of properties without sacrificing the properties of polymer matrices [9]. These nanoparticels include nanosilica [10], carbon nanotube [11], grapheme [4, 5] and nanoclay [12]. Graphene, as a new multifunctional material, has been attracting enormous attention since it was first exfoliated mechanically from graphite in 2004 [13]. For a variety of excellent properties such as large theoretical specific surface area, high thermal conductivity and strength [13], graphene can be produced multifunctional nanocomposite microcellular foams. In this

[3] Address for correspondence: Guoqiang Luo, State Key Laboratory of Advanced Technology for Materials Synthesis and Processing, Wuhan University of Technology, Wuhan, Hubei 430070, China. E-mail: luogq@whut.edu.cn.

study, we fabricated PMMA/GO microcellular foams using supercritical carbon dioxide ($ScCO_2$). GO was used as a heterogeneous nucleation agent and a reinforcement. The effect of graphene oxide on cell sizes and cell density to microcellular foams was investigated.

2. Experimental

2.1. Raw materials
PMMA pellets were commercially available from Nantong Rayon Chemical Co. Ltd., China. The density of the PMMA is 1.19 g/cm^3 and the glass transition temperature (T_g) is 107°C. Graphene oxide (GO) was obtained from Nanjing XFNANO Materials Tech Co. Ltd. The foaming agent is high-pressured CO_2 with a purity of 99.9%.

2.2. Preparation of PMMA/GO nanocomposites
The PMMA/GO nanocomposites were prepared by solution blending. Firstly, the GO sheets were dispersed in N, N-dimthylformamide (DMF) by ultrosonication for 4 h. The amounts of GO sheets were 0.5, 1.0 and 1.5 wt.% relative to PMMA. Furthermore, the PMMA particles were dissolved in DMF by heating. Then, the PMMA solution was mixed with the GO solution by ultrosonication for 4 h continually. Secondly, the solution blend was subsequently dropped into large amount of ethanol with strong agitation. Finally, the samples were dried in vacuum freeze-drying machine for 24 h to remove the solvent and then pressed into a flake by hot-pressing at 170°C.

2.3. Supercritical CO_2 foaming process
The PMMA/GO microcellular foams were prepared by the batch foaming process. We also used molds to constrain foaming process (mold-limited foaming). The samples were placed in a mold with an inner height 3 mm and then the mold was placed in a stainless steel vessel filled with CO_2. The saturation conditions were 16 MPa, 80°C and 8 h. After completed saturation, the pressure was quickly released within 1 s. The foam was then fixed using ice and water mixture bath after 30 s for foaming.

2.4. Characterization
The microstructures of PMMA/GO foam were characterized by scanning electron microscopy (SEM) with an accelerating voltage of 5 kV. The cell size and cell density were calculated via image analysis using Nano-Measure (Department of Chemistry, Fudan University) and more than 100 cells were tested. Then the cell density (N_f) was calculated according to [14]:

$$N_f = \left(\frac{nM^2}{A}\right)^{3/2} \quad (1)$$

where, A is the area of the SEM micrograph (cm^2), n is the number of cells in the micrograph and M is the magnification factor.
The bulk density of PMMA/GO foam was measured based on drainage method by an analytical balance with an accuracy of 0.1 mg. The bulk density (ρ) was calculated as:

$$\rho = \frac{m_0}{m_0\text{-}m_1} \times \rho_{water} \quad (2)$$

where, m_0 is the mass of the sample in atmosphere and m_1 is the mass of the sample in water and ρ_{water} is the density of water at room temperature.

The compression strength of PMMA/GO foam was measured by Universal Testing Machine (QJ-210A) at the loading speed was 0.5 mm/min. The typical dimensions of the specimens were 3 mm × 3 mm × 7 mm.

3. Results and discussion

3.1. Morphology of PMMA/GO foams

IOP Conf. Series: Materials Science and Engineering **87** (2015) 012042 doi:10.1088/1757-899X/87/1/012042

It is known that the nanoparticles can act as very effective heterogeneous nucleation sites due to the lowered energy barrier for nucleation. Nanoparticles especially suited for the production of microcellular foams of high a cell density and a small cell size [15]. In order to investigate the amounts of GO on the morphology of the microcellular foams, we produced microcellular foams with the GO amounts of 0 wt.%, 0.5 wt.%, 1.0 wt.% and 1.5 wt. %. The results showed in figure 1.

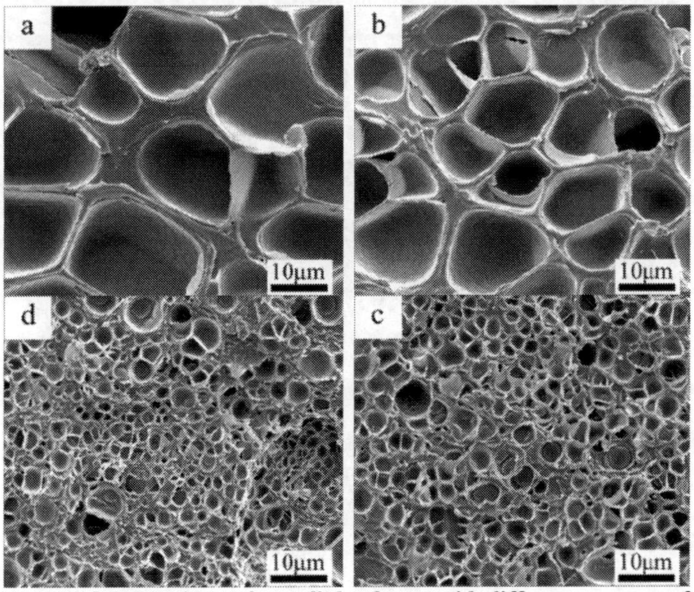

Figure 1. SEM images of PMMA/GO microcellular foams with different amounts of GO: a. 0 wt.%, b. 0.5 wt.%, c. 1.0 wt.%, d. 1.5 wt.%.

From the SEM images, it was found that average cell size decreased and cell density increased with the increasing GO content. It could be explained according to heterogeneous nucleation mechanisms [15]. GO sheets are heterogeneous nucleation sites, and more GO sheets could supply more nucleation sites and result in a higher nucleation rate. As it can be seen in figures 2 and 3, the average cell size were decreased from 20.1 μm to 2.2 μm and the average cell density were increased from 2.8×10^8 to 3.3×10^{10} with the 1.5 wt.% GO sheets compared with pure PMMA microcellular foam.

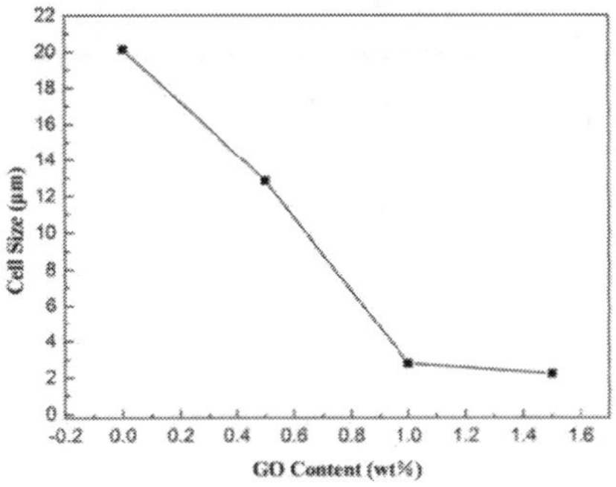

Figure 2. Variations of average cell size of PMMA/GO microcellular foam with different amounts of GO.

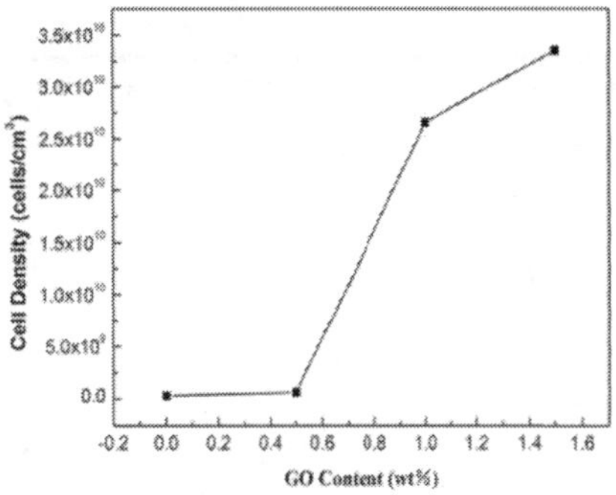

Figure 3. Variations of average cell density of PMMA/GO microcellular foam with different amounts of GO.

3.2. Compression properties of PMMA/GO foams

It is known to all that the problem of stress cracking could limit the application of PMMA. In order to investigate the effect of cell structure on mechanical properties of PMMA/GO foams, the compression properties of foams were tested. The result showed in figure 4. It was found that there is an obvious yield platform in the stress-strain curve of PMMA/GO foams. It showed that the foams have a great compression resistance capability due to the large number of cells with small size and high cell density. These cells could absorb a huge amount of energy when they were compressed. Moreover, the compressive strength has improved from 13 MPa to 39 MPa when the content of GO in PMMA matrix increased to 1.5wt.%. The density of PMMA/GO foams was showed in figure 5. It showed that the density of foams was increased with the increasing content of GO. The densities of foams also had an impact on compression strength. For the foams, the density increased means that the volume fraction of cell walls in pore structure increased. So it has strong bearing capacity to reply the compression.

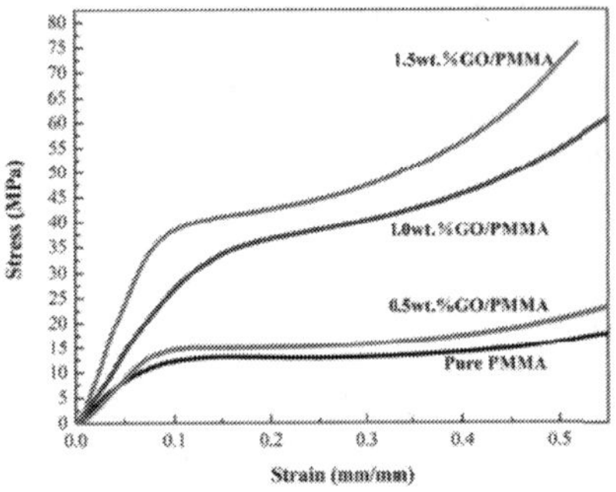

Figure 4. Compression stress-strain curves for PMMA/GO foams with different amounts of GO.

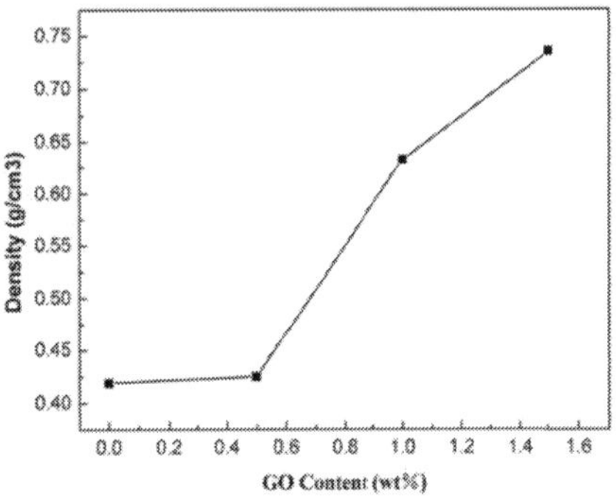

Figure 5. Variations of relative density of PMMA/GO foams with different amounts of GO.

4. Conclusion

In this work, the PMMA/GO microcellular foams were successfully prepared by superficial carbon dioxide foaming. The addition of GO sheets decreased the cell size and increased cell density in PMMA microcellular foam. With the increasing GO content, the average foam cell size was decreased from 20.1 μm to 2.2 μm and the cell density was increased from 2.8×10^8 to 3.3×10^{10} when 1.5wt.% GO sheets were added into PMMA matrix. The compression strength of polymeric foams was increased from 13 MPa to 39 MPa, compared with the pure PMMA foam.

Acknowledgment

This work is financially supported by the natural science foundation of Hubei Province, China (No. 2014CFB258 and No. 2014CFB257).

References

[1] Martini-Vvedensky J E, Suh N P and Waldman F A, 1984 Microcellular closed cell foams and their method of manufacture *US Patent*

[2] Yang G H, Su J J, Gao J, Hu X, Geng C Z and Fu Q 2013 Fabrication of well-controlled porous foams of graphene oxide modified poly (propylene-carbonate) using supercritical carbon dioxide and its potential tissue engineering applications *J. Supercrit. Fluids* **73** 1-9

[3] Xing Z, Wang M H, Du G H, Xiao T Q, Liu W H, Dou Q and Wu G Z 2013 Preparation of microcellular polystyrene/polyethylene alloy foams by supercritical CO_2 foaming and analysis by X-ray microtomography *J. Supercrit. Fluids* **82** 50-5

[4] Zhang H B, Yan Q, Zheng W G, He Z X and Yu Z Z 2011 Tough graphene-polymer microcellular foams for electromagnetic interference shielding, *ACS Appl. Mater. Interfaces* **3**

918-24

[5] Yang J T, Wu M J, Chen F, Fei Z D and Zhong M Q 2011 Preparation, characterization, and supercritical carbon dioxide foaming of polystyrene/graphene oxide composites *J. Supercrit. Fluids* **56** 201-7

[6] Chen H, Guan R, Zhao J Z, Jiang S F, Ke Z and Zha S W 2013 Microcellular foaming of plasticized thin PC sheet I. Effects of processing conditions *J. Wuhan Univ. Technol. Mater. Sci. Ed.* **28** 235-42

[7] Zhang P, Zhou N Q and Li. B 2007 Effects of process variables on microcellular structure and crystallization of polypropylene foams with supercritical CO_2 as the foaming agent-a study of microcellular foaming of polypropylene *Polym. Plast. Technol. Eng.* **46** 885-91

[8] Zhai W T, Ko Y, Zhu W L, Wong A S and Park C B 2009 A study of the crystallization, melting, and foaming behaviors of polylactic acid in compressed CO_2 *Int. J. Mol. Sci.* **12** 5381-97

[9] Lee L J, Zeng C C, Cao X, Han X M, Shen J and Xu G J 2005 Polymer nanocomposite foams, *Compos. Sci. Technol.* **65** 2344-63

[10] Ji G Y, Zhai W T, Lin D P, Ren Q, Zheng W G and Jung D W 2013 Microcellular foaming of poly (lactic acid)/silica nanocomposites in compressed CO_2: Critical influence of crystallite size on cell morphology and foam expansion *Ind. Eng. Chem. Res.* **52** 6390-8

[11] Zeng C C, Hossieny N, Zhang C and Wang B 2010 Synthesis and processing of PMMA carbon nanotube nanocomposite foams *Polym.* **51** 655-64

[12] Zeng C, Han X, Lee L J, Koelling K W, Tomasko D L 2003 Polymer-clay nanocomposite foams prepared using carbon dioxide *Adv. Mater.* **15** 1743-7

[13] Novoselov K S, Geim A K, Morozov S V, Jiang D, Zhang Y, et al. 2004 Electric field effect in atomically thin carbon films *Sci.* **306** 666-9

[14] Kumar V and Suh N P 1990 A process for making microcellular thermoplastic parts *Polym. Eng. Sci.* **30** 1323-9

[15] Han X, Lee L J and Tomasko D L 2005 CO_2 foaming of polymer nanocomposite blends Aust. *J. Chem.* **58** 492-503

Anti-aging cosmetics and its efficacy assessment methods

Xiang Li[1]

School of Perfume and Aroma Technology, Shanghai Institute of Technology, Shanghai 201418, China

E-mail: lixiang0902@hotmail.com

Abstract. The mechanisms of skin aging, the active ingredients used in anti-aging cosmetics and evaluation methods for anti-aging cosmetics were surmised in this paper. And the mechanisms of skin aging were introduced in the intrinsic and extrinsic ways. Meanwhile, the anti-aging cosmetic active ingredients were classified in accordance with the mechanism of action. Various evaluation methods such as human evaluation, in vitro evaluation were also summarized.

1. Introduction

As skin ages, it loses its natural elasticity and becomes thinner, more fragile and laxer, taking on a wrinkled appearance. The aging process of the skin generally has two kinds of forms: natural aging and light aging. The former mainly refers to the growth with the age, aging from within the body caused by such as genetic, gravity, endocrine and immune function of irresistible factors; while the latter is mainly due to the role of ultraviolet light, a series of changes in the weather and environmental pollution etc. which aggravates the natural aging process.

According to the mechanism of the senescence of skin, three kinds of methods can be summarized, basically including: protecting the skin from external environmental stimuli; scavenging intracellular free radicals; repair, supplement nutrition to skin cells. The use of some categories of personal care products to improve the appearance of aging is necessary.

The mechanisms of skin aging, the active ingredients used in anti-aging cosmetics and evaluation methods for anti-aging cosmetics were surmised in this paper. And the mechanisms of skin aging were introduced in the intrinsic and extrinsic ways.

2. The mechanisms of skin ageing

Aging of the skin has been attributed to two processes referred to as intrinsic or extrinsic processes. The intrinsic process

2.1. The intrinsic process

The intrinsic process occurs through the passage of time, and appears as fine wrinkles on the skin. The endogenous mechanism of skin aging is very complex, and representatives of the aging theory mainly

[1] Address for correspondence: Xiang Li, School of Perfume and Aroma Technology, Shanghai Institute of Technology, Shanghai 201418, China. E-mail: lixiang0902@hotmail.com.

include: the free radical theory, theory of genetic gene, mitochondrial DNA damage theory and other theories.

2.1.1. The theory of free radical. The free radical theory was introduced by Harman in 1956, its core includes: free radicals are common intermediates in the life of organisms with high chemical activity; On the normal physiological conditions, generation and elimination of free radicals are in dynamic equilibrium in the body; Once this balance is disrupted, resulting in excessive production of free radicals; The excessive free radical toxicity can cause the aging body by attacking on the chromosome, mitochondria, cell membrane and connective tissue and other biological tissues [1].

2.1.2. The theory of genetic gene. The theory of genetic gene, mainly analyzes the aging phenomenon from the chromosome and gene level. The doctrine said that, aging is the phenomenon of life determined by a genetic program. Along with the growth of the age, the modified gene loss, DNA methylation and phosphorylation reaction decreases, telomere shortens, DNA self-repair ability drops, oncogene and tumor suppressor gene regulation can result in chromosome mutation, normal cell differentiation [2, 3], all of these can lead to skin aging.

2.1.3. The theory of mitochondrial DNA damage. In 1980, Miquel and Cowiker [4] proposed the hypothesis of aging mitochondrial, they thought that mitochondrial DNA damage is the molecular basis of cell aging and death. At the perspective of cell biology, 90% of the energy is produced by mitochondria, which is needed for cells life. While mitochondrial DNA is damaged, the energy producing is reduced, affecting the energy supply of cells, causing a recession of cell, tissue, organ function. Therefore, mitochondria degeneration, rupture and leakage are important reasons for the aging.

2.2. The extrinsic process
The extrinsic process is often referred to as the effects of the environment (such as sun) and other exposures (such as weather) on the skin.

3. The active ingredients in anti-aging cosmetics
According to their functions or effects, anti-aging cosmetics can be divided into three categories: moisturizing cosmetics, antioxidant cosmetics and the biological activity of cosmetics. Active ingredients of different kinds of cosmetics are different. As the study on the bioactive components of knowledge continues to improve, the active ingredients used in anti-aging cosmetics are increasing. The ingredients of scavenging free radicals, cell repairing, moisturizing and UV absorption are discussed here.

3.1. The ingredients of scavenging free radicals
Oxidation is the greatest threat to skin ageing, unhealthy diet, sun, stress, environmental pollution, can make skin free radicals overflow, resulting in looking bleak, and water lack. Therefore, it's necessary to pay attention to the antioxidant in daily life for the health skin.
Vitamin E, vitamin C, superoxide dismutase, coenzyme Q10, zinc sulfate, ferulic acid, green tea, idebenone, polyphenols and carotenoids are the types of scavenging free radicals ingredients.

Djerassi D [5] found that vitamin E can penetrate into the skin and continue to accumulate, so that the skin becomes soft and smooth. Vitamin E is a good active substance to prevent light aging. The effect of sunscreens with vitamin E is more effective, and when the vitamin E and vitamin C are used in combination, the effect is better. In addition, vitamin E also has effects on anti-aging, inhibiting erythema solare, reducing wrinkles and anti-inflammatory.

Pinnell S R [6] conducted the skin penetration absorption experiment and found that any form of vitamin C can be identified and play a role in the body after being converted to ascorbic acid. Hoppe U, Bergemann J et al. [7] found that a concentration of 0.3% ubiquinone can significantly reduce eye wrinkles depth. There are also documents said, by UV ray irradiation, coenzyme Q10 can inhibit human fibroblast collagenase expression.

3.2. The ingredients of cell repairing

The decrease of dermal collagen will lead to the decline of skin elasticity and the appearance of wrinkles. Therefore, promoting the growth of collagen can also alleviate the aging of the skin. Vitamin A, retinoic acid, tartaric acid, cell growth factor, collagen protein, beta hydroxy acid, beta glucan are this kind of materials.

Oblong JE, Bissett D L [8] mentioned in their book, retinol and retinyl propionate can effectively reduce facial wrinkles at low concentrations. Duell E A [9] through the skin permeation experiments found, retinol skin permeation effect is better than retinol esters, even better than acid. Djerassi D also found that vitamin A can enhance the skin elasticity, improve complexion and skin texture. Study of Kafi R, Kwak H S et al. [10] shows that, vitamin A has a good repair capacity of natural aging skin. Yaar M, Gilchrest B A [11] proved that all trans retinoic acid can improve the fine wrinkles, pigmentation and depigmentation, facial nevus. Rabe J H, Mamelak A J, McElgunn P J et al. [12] said that the reason that retinoic acid has the efficacy may be due to the fact they have the ability to promote the synthesis of collagen.

Bermann P E [13] found that retinoic acid may contribute to the retention of water to increase the skin. Robert L [14] proved that, long-term use of alpha hydroxy acids (such as glycolic acid, lactic acid, citric acid, malic acid and so on) can improve the skin wrinkles, roughness and color due to light, and may also increase the thickness of epidermis, dermis, collagen synthesis induced by the increased water retention, thereby improving the elasticity of the skin and appearance.

3.3. The ingredients of moisturizing

The decrease of moisture content of the skin will lead to the decline of protease activity, causing the skin peeling, so moisture is an essential part of anti-aging.

The ingredient of moisturizing is a kind of agent which can lead to the water being attracted to the horny layer so as to increase the skin moisture content. Moisturizing agents currently used in anti-aging cosmetics mainly include: pyrrolidone carboxylic acid salt, sorbitol, glycerol, polyethylene glycol, cholesterol and oleic acid, hyaluronic acid, lactic acid and some microbial fermentation metabolites. Hyaluronic acid is one of the most widely used.

3.4. UV absorption

The effect of UV radiation on skin can lead to DNA damage directly, the apoptosis or growth arrest. The light aging and natural aging together, cause the skin barrier degradation and the formation of wrinkles and pigmentation, which may cause malignant change and other changes. Therefore, a sunscreen for anti-aging is very necessary. Using ultraviolet light scattering agent or UV absorber can reduce skin damage caused by the sun.

4. The efficacy evaluation of anti-aging cosmetics

4.1. In vitro evaluation

Determination of free radical scavenging capacity and fibroblast proliferation ability of detection are in vitro evaluation.

According to the free radical damage theory, the free radical is the main cause of natural aging and photo-aging skin. Therefore, it's one of the most important indexes that whether the anti-aging cosmetics (raw material) have the ability of scavenging free radicals. At present, the free radical scavenging ability evaluation indexes mainly include: the ability of cleaning 2, 2-diphenyl-1-picrylhydrazyl hydrate (DPPH), the ability of scavenging hydroxyl free radical and superoxide anion.

Can Xin, Zhenyu Wang et al. [15, 16] measured black water extract and Lonicera edulis ethanol elution effect on scavenging superoxide anion free radical, hydroxyl radical, DPPH radical, respectively, and confirmed that black bean water extract and ethanol elution of Lonicera edulis have a good ability of scavenging free radicals, and thus can be used in cosmetics.

4.2. In vivo evaluation
Methods of human evaluation include: expert evaluation, high-quality image analysis, transepidermal water loss, horny protein changes and clinical data of erythema and dryness and so on [17].

Fu J J J and Hillebrand G [18] through an 8-week, randomized, parallel-group study to compare the efficacy of a cosmetic moisturizer regimen vs. a prescription regimen with 0.02% tretinoin for improving the appearance of facial wrinkles, they found that the cosmetic regimen significantly improved wrinkle appearance after 8 weeks relative to tretinoin, with comparable benefits after 24 weeks. The cosmetic regimen was significantly better tolerated than tretinoin through 8 weeks by all measures.

Watson S, Ogden L F [19] conducted a randomized controlled experimental study on photo-aging effect of anti-aging cosmetics on skin using a baseline assessment, showing that, 43% of the volunteers who used testing products for 6 consecutive months, have an improvement on facial wrinkles. The study also shows that, fibrinogen is the effective biomarkers to repair the leather damaged by the light.

Liu Xiaoying [20] verified that the brown algae extract have effects on improving wrinkles and skin elasticity through the methods of using a high resolution UVA camera measuring the optical properties of skin, the cutometry instrument measuring the elasticity, smoothness, the appearance of wrinkles of skin .

Lerong Zhao [21] used a cream with green plum-blossom extract for 30 volunteers for 4 weeks, by measuring the roughness and elasticity of the skin, and found that cream with green plum-blossom extract can decrease the roughness and the depth of shallow wrinkles of the skin, and strengthen skin elasticity.

5. Conclusion
As our knowledge and technology continue to grow, more and more active ingredients are found and used in cosmetics. The mechanism of anti-aging cosmetics according to its active ingredients can be roughly classified into four kinds, moisturizing, removing free radicals, cell repair and UV absorption type. For evaluation of the efficacy of anti-aging cosmetics method, it mainly includes in vitro evaluation and in vivo evaluation. Determination of free radical scavenging capacity and fibroblast proliferation ability of detection are in vitro evaluation. Methods of human evaluation include: expert evaluation, high-quality image analysis, transepidermal water loss, horny protein changes and clinical data of erythema and dryness, etc. It will be an important task to improve the existing evaluation methods and to develop the new model in the field of anti-aging cosmetics in the future.

References
[1] Bernstein E F 2002 *Dermatol Surg* **28** 132-5
[2] Lin Fang 2010 *People's Military Surgeon* **53** 149-52
[3] Zou Y, Song E and Jin R 2009 *Skin Res. Technol.* **15** 399-406
[4] Juhong Wang 2003 *Journal of Hexi University* **2** 94-8
[5] Djerassi D 2001 *Cosmetics and Toiletries Manufacturers Worldwide* **10** 22-25
[6] Pinnell S R 2001 *Dermatol Surg* **27** 137-42
[7] Hoppe U, Bergemann J and Diembeck W 1999 *Biofactors* **9** 371-8
[8] Oblong J E and Bissett D L 2005 *Procedures in Cosmetic Dermatology* **2** 35-42
[9] Duell E A 1997 *J Investig Dermatol* **109** 301-5
[10] Kafi R, Kwak H S and Schumacher W E 2007 *Arch Dermatol* **143** 06-12
[11] Rabe J H, Mamelak A J and McElgunn P J 2006 *J. Am. Acad. Dermatol.* **55** 1-19
[12] Bermann P E 2007 *NursClin North Am* **42** 485-500
[13] Robert L, Bronaugh, Linda M, et al.2010 *Textbook of Aging Skin* (Berlin: Springer) 1063-8

[14] Can Xin 2012 *Journal of Hebei United University (Natural Science Edition)* **34** 115-8
[15] Zhenyu Wang 2012 *Science and Technology of Food Industry* **33** 163-74
[16] Griffiths T W 2010 *British Journal of Dermatology* **162** 469-71
[17] Fu J J J, Hillebrand G and Raleigh P 2010 *British Journal of Dermatol* **162** 47-54
[18] Watson S and Ogden L F 2009 *British Journal of Dermatology* **16** 419-26
[19] Xiaoying Liu, et al. 2012 *Detergent & Cosmetics* **35** 33-7
[20] Lerong Zhao 2012 *Flavor Fragrance Cosmetics* **4** 33-6

Microstructures controlled by nanoparticles in complicated star terpolymers

Ling Zhou[1]

School of Science, Nantong University, Nantong, 226019, China

E-mail: ZL7103@163.com

Abstract. Self-consistent field and density functional theories are applied to investigate microstructure transitions in $A(BC)_4$ star terpolymers due to the addition of nanoparticles when B block is a minority. With increasing nanoparticles that have a preferential affinity to B block, a variety of order-to-order transitions are observed, such as the transitions from a hexagonal structure to a core-shell hexagonal structure, from a hexagonal structure with beads at the interface to a core-shell hexagonal structure, from a lamellar structure with beads at the interface to a three-phase lamellar structure, and from a two-phase lamellar structure to a three-phase lamellar structure. The results are helpful in the fabrication of controllable functional materials through the role of nanoparticles.

1. Introduction

The self-assembly of copolymers provides a lot of opportunities to fabricate novel functional materials [1]. The components and their interaction energies, the entropy effect caused by the extension of chains and the system incompressibility play an important role in determining the phase behavior of copolymers [2, 3]. Recently, more and more researches indicate that the addition of few nanoparticles can dramatically change structures and properties of copolymers [4-6]. Based on electrostatic interactions, hydrogen bond, hydrophobic interactions and biological recognition between nanoparticles and copolymers, the mixed systems can realize all kinds of self-assemblies, which open up a new developing space by using nanoparticels as sensors and vastly widen the range of application. In this paper, we add several nanoparicles into $A(BC)_4$ star terpolymers. The combined self-consistent field and density functional theories [7] are used to investigate the influence of few nanoparticles on the microstructure transitions. It is well known that the increase of monomer and the change of topological structures may greatly increase the complexity of the self-assembled microstructures. Especially the star constraint of star copolymers can promote the formation of singular microstructures [8]. The $A(BC)_4$ star terpolymers are just this kind of copolymers whose complicated topological architecture leads to the appearance of various complex microstructures that cannot be observed in simple copolymers [9]. The motivation of this work is to study whether nanoparticles can effectively control the microstructure transitions in such complicated terpolymers and what the main factors influencing the microstructures are. Finally four kinds of microstructure transitions are observed.

2. Model and method

[1] Address for correspondence: Ling Zhou, School of Science, Nantong University, Nantong, 226019, China. E-mail: ZL7103@163.com.

We select a nanoparticle/terpolymers mixture as our model system. In this model, *A(BC)*$_4$ star terpolymers consist of one homopolymer arm *A* and four copolymer arms *BC*. The ends *A* and *B* of the five chains are attached to a core. Blocks *A*, *B* and *C* have N_A, N_B and N_C segments, respectively. The statistical length *b* of each segment is same. Therefore each segment has a common volume ρ_0^{-1}. In addition, we assume that the length of blocks *B* and *C* on four arms are same. Therefore the polymerization index of *A(BC)*$_4$ star terpolymers $N = N_A + 4N_B + 4N_C$. We use M_i to denote the monomer fraction of block *i* (*i*=*A*, *B*, *C*). Due to the incompressibility constraint of the system, we have $M_A = N_A / N$, $M_B = N_B / N$, $M_C = N_C / N = 1 - M_A - M_B$. Once nanoparticles are added into the *A(BC)*$_4$ star terpolymers in the volume *V*, the free energy *F* of the system can be written as:

$$\frac{NF}{\rho_0 k_B TV} = -\phi_{Ter}\ln(\frac{Q_{Ter}}{V\phi_{Ter}}) - \frac{\phi_n}{\alpha_n}\ln(\frac{Q_n\alpha_n}{V\phi_n}) + \frac{1}{V}\int dr(\chi_{AB}N\varphi_A(\vec{r})\varphi_B(\vec{r}) + \chi_{AC}N\varphi_A(\vec{r})\varphi_C(\vec{r}) + \chi_{BC}N\varphi_B(\vec{r})\varphi_C(\vec{r})$$
$$+\chi_{An}N\varphi_A(\vec{r})\varphi_n(\vec{r}) + \chi_{Bn}N\varphi_B(\vec{r})\varphi_n(\vec{r}) + \chi_{Cn}N\varphi_C(\vec{r})\varphi_n(\vec{r}) - W_A(\vec{r})\varphi_A(\vec{r}) - W_B(\vec{r})\varphi_B(\vec{r}) - W_C(\vec{r})\varphi_C(\vec{r})$$
$$-W_n(\vec{r})\rho_n(\vec{r}) - P(\vec{r})(1 - \varphi_A(\vec{r}) - \varphi_B(\vec{r}) - \varphi_C(\vec{r}) - \varphi_n(\vec{r})) + \rho_n\Psi(\bar{\varphi}_n)). \quad (1)$$

In above equation, k_B denotes the Boltzmann's constant and *T* denotes the temperature. ϕ_n is the nanoparticle average volume fraction, and ϕ_{Ter}(=1-ϕ_p) is the terpolymer average volume fraction. $\varphi_i(\vec{r})$ and χ_{ij} (= χ_{ji}, *i*, *j*=*A*, *B*, *C*, *n*) are the local volume fraction and the interaction parameter of components *i* and *j*. $W_i(\vec{r})$ $(i = A, B, C, n)$ is a mean chemical potential field that acts on the component *i*. $Q_{Ter}(= \int d\vec{r} q_A q_A^+ = \int d\vec{r} q_{BC} q_{BC}^+)$ is a partition function of a single teropolymer chain. $Q_n(= \int d\vec{r}\exp[-W_n(\vec{r})])$ is a nanoparticle partition function. According to the architecture of *A(BC)*$_4$ star terpolymers, we should introduce four distribution functions $q_A(\vec{r},s)$, $q_A^+(\vec{r},s)$, $q_{BC}(\vec{r},s,l)$, $q_{BC}^+(\vec{r},s,l)$ to determine Q_{Ter}. These four functions satisfy the following modified diffusion equations:

$$\partial q_A / \partial s = (Na^2/6)\nabla^2 q_A - W_A(\vec{r})q_A; \quad (2)$$
$$\partial q_A^+ / \partial s = -(Na^2/6)\nabla^2 q_A^+ + W_A(\vec{r})q_A^+; \quad (3)$$
$$\partial q_{BC} / \partial s = (Na^2/6)\nabla^2 q_{BC} - W_B(\vec{r})q_{BC} \quad \text{if } 0 < s < M_B/4; \quad (4)$$
$$\partial q_{BC}^+ / \partial s = -(Na^2/6)\nabla^2 q_{BC}^+ + W_B(\vec{r})q_{BC}^+ \quad \text{if } 0 < s < M_B/4; \quad (5)$$
$$\partial q_{BC} / \partial s = (Na^2/6)\nabla^2 q_{BC} - W_C(\vec{r})q_{BC} \quad \text{if } M_B/4 < s < (M_B + M_C)/4; \quad (6)$$
$$\partial q_{BC}^+ / \partial s = -(Na^2/6)\nabla^2 q_{BC}^+ + W_C(\vec{r})q_{BC}^+ \quad \text{if } M_B/4 < s < (M_B + M_C)/4. \quad (7)$$

The denotation *s* is the segment that changes continuously from one end to another along the arms. Equations (2)-(7) are solved from the initial conditions: $q_A(\vec{r}, M_A) = 1$, $q_{BC}^+(\vec{r}, (M_B + M_C)/4, l) = 1$ (l = 1,2,3,4), $q_A(\vec{r},0) = \prod_{l=1}^4 q_{BC}^+(\vec{r},0,l)$, $q_{BC}(\vec{r},0,l) = q_A(\vec{r},0)\prod_{J=1,J\neq l}^4 q_{BC}^+(\vec{r},0,J)$. The denotation *l* refers to the *l*th arm of *BC* copolymers. $\alpha_n(= \frac{4\pi R^3\rho_0}{3N})$ is a volume ratio of the spherical nanoparticles to terpolymers. *R* is the radius of spherical nanoparticles. $\varphi_n(\vec{r})(= \frac{\rho_0}{N}\int_{|\vec{r}'|<R} d\vec{r}'\rho_n(\vec{r} + \vec{r}'))$ is the local volume fraction of nanoparticles, where ρ_n denotes the nanoparticle center distribution. $\bar{\varphi}_n(\vec{r})(= \alpha_n / V_{2R} \int_{|\vec{r}'|<2R} d\vec{r}'\rho_n(\vec{r} + \vec{r}'))$ is the weighted density of nanoparticles. $\Psi(\bar{\varphi}_n)(= (4\bar{\varphi}_n - 3\bar{\varphi}_n^2)/(1 - \bar{\varphi}_n)^2)$ is the excess free energy per nanoparticle. V_{2R} is the spherical volume around $\vec{r}$ with radius 2*R*. $P(\vec{r})$ is a potential field enforcing the system incompressibility.

The final microstructures of the system are obtained through minimizing the free energy *F*, with respect to $W_A(\vec{r})$, $W_B(\vec{r})$, $W_C(\vec{r})$ and $W_n(\vec{r})$, $\varphi_A(\vec{r})$, $\varphi_B(\vec{r})$, $\varphi_C(\vec{r})$, $\varphi_n(\vec{r})$ and $P(\vec{r})$. A set of self-

IOP Conf. Series: Materials Science and Engineering **87** (2015) 012044 doi:10.1088/1757-899X/87/1/012044

consistent equations are obtained and solved with the periodic boundary conditions in a two-dimensional space. By the self-consistent iteration, the final local volume fractions φ_A, φ_B, φ_C and φ_n at the equilibrium are determined. In the calculation, all of the lengths are expressed in units of b.

3. Results and discussion

The parameters in the calculation are chosen as $N = 600$, $R = 2b$, $M_B = 0.1$. The nanoparticles have a preferential affinity to B block. The interactions between different components are asymmetrical and are set as $\chi_{AB}N = \chi_{An}N = 130$, $\chi_{BC}N = \chi_{Cn}N = 160$, $\chi_{AC}N = 80$, $\chi_{Bn}N = 0$.

First we study the microstructures of the terpolymer melts without nanoparticles. M_A is controlled from 0.1 to 0.7. The increment of M_A is set to be 0.1. Since M_B is fixed to 0.1, M_C changes from 0.8 to 0.2 correspondingly. Four kinds of microstructures changing with (M_A, M_B, M_C) are observed, which are a hexagonal structure (hex) when $(M_A, M_B, M_C) = (0.1, 0.1, 0.8)$, hexagonal structures with beads at the interface (h+b) when $(M_A, M_B, M_C) = (0.2, 0.1, 0.7)$, $(0.3, 0.1, 0.6)$, $(0.4, 0.1, 0.5)$, lamellar structures with beads at the interface (l+b) when $(M_A, M_B, M_C) = (0.5, 0.1, 0.4)$ and $(0.6, 0.1, 0.3)$, and a two-phase lamellar structure (l2) when $(M_A, M_B, M_C) = (0.7, 0.1, 0.2)$. The detailed distributions of different blocks are shown in figures 1(a), 1(c), 1(e) and 1(g).

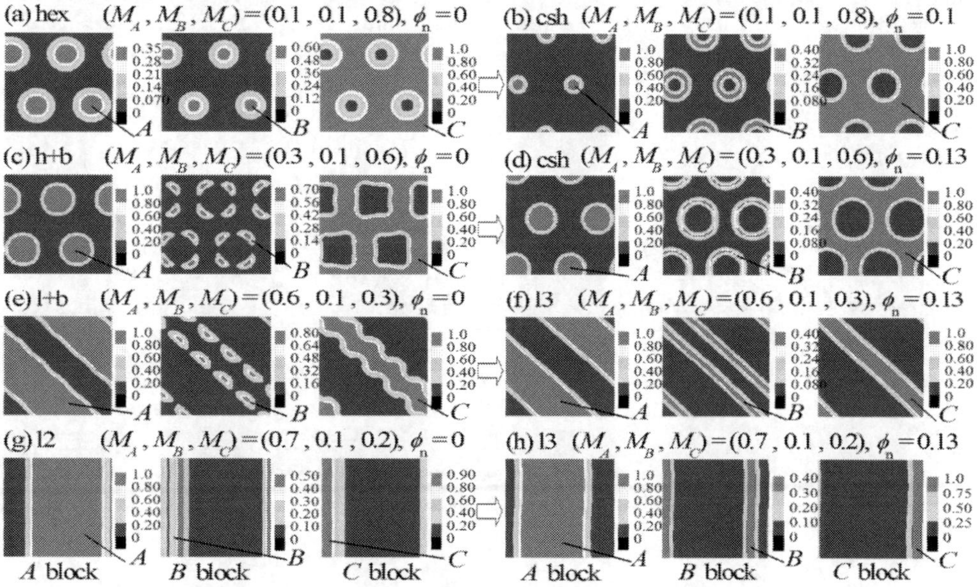

Figure 1. The local volume fractions of blocks A, B and C (color online).

Then we consider the influence of nanoparticles. Figures 1(b), 1(d), 1(f) and 1(h) illustrate the microstructures due to the addition of few nanoparticles. The color scale on the right of each figure gives the range of local volume fractions of the corresponding blocks A, B or C. Figure 1(b) gives the core-shell hexagonal structure (csh) when $(M_A, M_B, M_C) = (0.1, 0.1, 0.8)$ and $\phi_n = 0.1$. However, the morphology without nanoparticles is a hexagonal structure (hex) as shown in figure 1(a), where the hexagonal cores consist of A and B blocks, and the matrix consists of the C block. With increasing nanoparticle concentration, the further phase separation occurs between A and B where the block B and nanoparticles form the shells and the block A stays in the hexagonal cores. Therefore the original hexagonal structure (hex) is substituted by the core-shell hexagonal structure (csh) due to the addition of nanoparticles. Figure 1(c) shows the hexagonal structure with beads at the interface (h+b) when

$(M_A, M_B, M_C) = (0.3, 0.1, 0.6)$ and the nanoparticles are absent. In this case, the hexagonal cores consist of *A* blocks, four bead-shape domains of *B* blocks distribute symmetrically around the *A* cores, and the remaining part consists of *C* blocks. Similar microstructures are also found when $(M_A, M_B, M_C) = (0.2, 0.1, 0.7)$ and $(0.4, 0.1, 0.5)$. The only difference of the microstructures is that the numbers of the bead-shape domains. When $(M_A, M_B, M_C) = (0.2, 0.1, 0.7)$, $(0.3, 0.1, 0.6)$ and $(0.4, 0.1, 0.5)$, three, four and six bead-shape domains of *B* blocks appear in turn. The addition of nanoparticles makes the isolated *B* domains change to a shell surrounding the cores *A*. The core-shell hexagonal structure (csh) when $(M_A, M_B, M_C) = (0.3, 0.1, 0.6)$ and ϕ_n=0.13 is shown in figure 1(d). Figure 1(e) shows the lamellar structure with beads at the interface (l+b) when $(M_A, M_B, M_C) = (0.6, 0.1, 0.3)$. The blocks *A* and *C* separate into two parallel layers, and the blocks *B* shrink into bead-shape domains staying at the interface of two layers when $(M_A, M_B, M_C) = (0.6, 0.1, 0.3)$ and nanoparticles are absent. The same microstructures also can be found when $(M_A, M_B, M_C) = (0.5, 0.1, 0.4)$. With the gradually increased nanoparticles, the domains of block *B* increase and finally form the lamellar structure between layers *A* and *C*. Therefore a three-phase lamellar structure (l3) is observed. Such three-phase lamellar structure (l3) when $(M_A, M_B, M_C) = (0.6, 0.1, 0.3)$ and ϕ_n=0.13 is shown in figure 1(f). Therefore the addition of nanoparticles successfully realizes the microstructure transition from a l+b structure to a l3 structure. Figure 1(g) shows a two-phase lamellar structure (l2) without nanoparticles when $(M_A, M_B, M_C) = (0.7, 0.1, 0.2)$. Due to the relatively short chains *B* and *C*, and the relatively weak interaction $\chi_{BC}N$, the blocks *B* and *C* form a mixed layer and the block *A* occupies another layer. The addition of nanoparticles leads to the further phase separation of blocks *B* and *C*. Finally block *B* and block *C* completely separate and form two lamellae, respectively. Such three-phase lamellar structure (l3) is shown in figure 1(h) when $(M_A, M_B, M_C) = (0.7, 0.1, 0.2)$ and ϕ_n=0.13.

4. Conclusion

In summary, we theoretically study the microstructures controlled by nanoparticles in $A(BC)_4$ star terpolymers. We find four kinds of microstructures in the terpolymer melts when $\chi_{AB}N = 130$, $\chi_{BC}N = 160$, $\chi_{AC}N = 80$ and $M_B = 0.1$. With the increase of M_A, hex, h+b, l+b and l2 structures are observed in turn. In order to capture the microstructure transitions, we add few nanoparticles into $A(BC)_4$ star terpolymers. The interactions between nanoparticles and different blocks are $\chi_{An}N = 130$, $\chi_{Bn}N = 0$ and $\chi_{Cn}N = 160$. The calculations show that the order-to-order transitions, such as hex$\rightarrow$csh, h+b$\rightarrow$csh, l+b$\rightarrow$l3 and l2$\rightarrow$l3, can be realized. The strategy in the present study is helpful in the fabrication of controllable functional materials triggered by nanoparticles.

Acknowledgments

This work was supported by the Major National Scientific Instrument and Equipment Development Project of China (No. 2011YQ03013403), the National Natural Science Foundation of China (Nos. 61371057 and 11204143) and the Natural Science Foundation for pre-research of Nantong University (Nos. 11ZY005, 10ZY001).

References

[1] Xu W Q, Jiang K, Zhang P W and Shi A C 2013 *J. Phys. Chem. B* **117** 5296-305
[2] Subbotin A, Markov V and ten Brinke G 2010 *J. Phys. Chem. B* **114** 5250-6
[3] Shin D O, Mun J H, Hwang G, Yoon J Moon, Kim J Y, Yun J M, Yang Y, Oh Y, Lee J Y, Shin J, Lee K J, Park S, Kim J U and Kim S O 2013 *Acsnano* **7** 8899-907
[4] Zhou L and Ma Y Q 2008 *J. Phys.: Condens. Matter* **20** 095006
[5] Zhang L S and Lin J P 2009 *Macromolecule* **42** 1410-4
[6] Pan Q Y, Tong C H and Zhu Y J 2011 *ACS Nano* **5** 123-8

[7] Thompson R B, Ginzburg V V, Matsen M W and Balazs A C 2002 *Macromolecules* **35** 1060-71
[8] Tang P, Qiu F, Zhang H D and Yang Y L 2004 *J. Phys. Chem. B* **108** 8434-8
[9] Zhou L, Liu Y, Peng J and Fang J H 2013 *J. Polym. Res.* **20** 102

IOP Conf. Series: Materials Science and Engineering **87** (2015) 012045 doi:10.1088/1757-899X/87/1/012045

Preparation and characterization of aqueous polyurethane oil/polyacrylate latex interpenetrating polymer network

M M Zhou, L L Ma, J Du, F Cao and J J Xiao[1]

School of Materials Science and Engineering, Hebei University of Science and Technology, 050018 Shijiazhuang, P.R. China

E-mail: jijunxiao@aliyun.com

Abstract. A series of aqueous polyurethane oil (network I) /polyacrylate (network II) latex interpenetrating polymer networks (LIPNs) were synthesized via the technology of latex interpenetrating polymer network combined seed emulsion polymerization process. Fourier transform infrared (FTIR) spectroscopy, laser particle size distributing analyzer and universal tension machine were utilized to characterize the bulk structures and mechanical properties of LIPNs. For used as damping material, the damping performance of LIPNs were analyzed by dynamic mechanical analysis (DMA). It was found that the damping temperature region of LIPN was wider than those of aqueous polyurethane oil, the temperature region with greater tanδchanged with the TPGDA content and hard-/soft-segment mass weight ratio (m_{MMA}/m_{BA}) and the glass transition temperature (T_g) of the network I and network II in LIPN occurred within shift each other, even overlap with increasing m_{MMA}/m_{BA} value. The results show that LIPNs synthesized through the combined process have greater tanδ and wider damping temperature region, which is suitable for the use of damping coatings.

1. Introduction

In recent years, the research of technologies reducing vibration noise and materials has been paid more attention because of the serious noise pollution problem. The polymer damping coating is a kind of functional materials, which can reduce the vibration and noise of various mechanical vibrations [1, 2]. The polymer has higher damping values in the glass transition temperature (T_g) range. Generally, the glass transition temperature region of a single polymer is only 20℃-30℃, while the change of environmental temperature region is bigger than the T_g region of polymers. In order to meet the requirements of different environments, people have need to synthesis polymers with T_g more than 60℃. The technology of interpenetrating polymer networks (IPNs) could solve noise pollution problem [3-5]. The IPNs are a novel type of polymer blend composed of cross-linked polymer, combining the advantages of each component forming networks [6, 7]. In order to reduce the cost and the danger of organic solvent volatilization, the rapid development of waterborne damping coatings have been obtained, especially in the Latex interpenetrating polymer network (LIPN). LIPN is a unique type of polymer blend prepared by seed emulsion polymerization [8]. LIPN exhibits excellent properties superior to those of each component, being attributed to their morphological structure [9]. In this study, LIPN was composed of waterborne polyurethane oil referred to as network I and the

[1] Address for correspondence: J J Xiao, School of Materials Science and Engineering, Hebei University of Science and Technology, 050018 Shijiazhuang, P.R. China. E-mail: jijunxiao@aliyun.com.

IOP Conf. Series: Materials Science and Engineering **87** (2015) 012045 doi:10.1088/1757-899X/87/1/012045

polymer derived from methyl methacrylate and butyl acrylate as network II.

2. Experimental

2.1. Materials
The raw materials used were polyurethane oil (PUO, made by ourselves), methyl methacrylate (MMA, Shanghai Huayi Company), n-butyl acrylate (BA, Beijing Eastern Petrochemical Company), tripropylene glycol diacrylate (TPGDA, Ji'nan Venus Chemical Industry), potassium persulfate (Tianjin Yongda Chemical Reagent Company Limited), potassium bicarbonate (Tianjin Yongda Chemical Reagent Company Limited).

2.2. Synthesis of latex interpenetrating polymer network (LIPN) of aqueous polyurethane oil/polyacrylate
Polyurethane oil was put into a 500 mL four-flask with a mechanical stirrer, a thermometer and a reflux condenser, and stirred gradually heated to 70℃. Then the solution of potassium persulfate and sodium bicarbonate and the mixture of MMA, BA and TPGDA were added to the system by dropping under nitrogen for 2 h, and heated to 80℃. Subsequently the system was maintained for 1 h at approximately 80℃. After cooling to room temperature, the LIPN was obtained.

2.3. Measurements
FTIR spectra were collected on a Nicolet 6700/FTIR-Raman modules spectrometer. The particle size of LIPN was measured by dynamic light scattering analyzer (MALVERN NANO-S90). Dynamic mechanical analysis (DMA) were performed on a Q800 dynamic mechanical analyzer with film tensile-mode in a temperature range from -60℃~80℃ at a frequency of 2 Hz and a heating rate of 5°C/min. The tensile tests were performed in a universal testing machine (CMT7104) with a stretching rate of 500 mm/min.

3. Results and discussion

3.1. FTIR analysis
The FTIR spectra of polyurethane oil (Network I) and LIPN (solid content of 30%, hard-/soft-segment mass weight ratio (m_{MMA}:m_{BA})=1:1, the amount of crosslinking agent of 0.2% relative to monomer) are shown in figures 1 and 2, respectively.

In figure 1, a strong stretching band located at 3336 cm^{-1} corresponding to characteristic absorption peak of the N-H group, a stretching vibration band centered around 1719.4 cm^{-1} corresponding to one of the C=O group, and a strong stretching band located at 1241 cm^{-1} due to one of the C-O-C group are observed, indicating the existence of NHCOOR in network I. The results suggest that the network I was polyurethanes. It can also be observed that there was no absorption bands at 811 cm^{-1} (H-C=C) in figure 2, indicating that the network II have formed via radical polymerization of monomers. These results suggest that LIPN was successfully prepared.

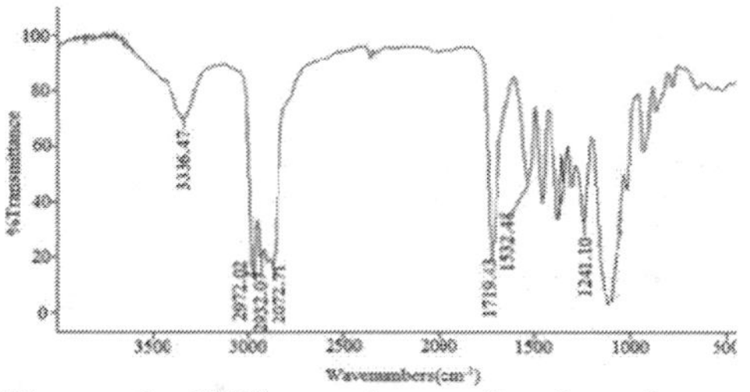

Figure 1. FTIR spectrum of polyurethane oil (network I).

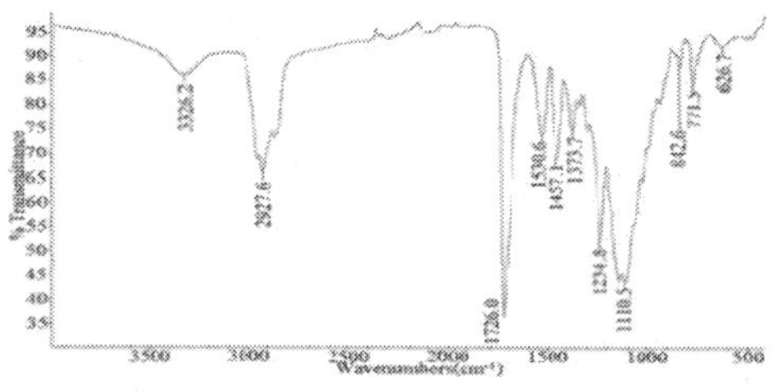

Figure 2. FTIR spectrum of LIPN.

3.2. Influence of various factors on latex particle size for LIPN

3.2.1. Effect of mMMA/mBA value on particle size of the LIPN. Figure 3 shows the influence of m_{MMA}/m_{BA} on the of average particle size of LIPN. The results shown that average particle size of LIPN decreases with the increasing m_{MMA}/m_{BA} value, indicating the greater the soft monomer BA, the softer the particles and more easy to coagulate to form larger particles.

3.2.2. Effect of the TPGDA content on particle size of the LIPN. As seen from figure 4, the average particle size of latex particle for LIPN initially increases with the increasing TPGDA content. As the TPGDA content is more than 0.4%, that of LIPN decreases. Because a small amount of crosslinking agent is equivalent of monomer which can not develop the role of crosslinking agent. While the amount of the crosslinking agent exceeds 0.4%, the crosslinking agent plays the role of crosslinking, and the maromolecular chains were cross-linked together to obtained smaller particles.

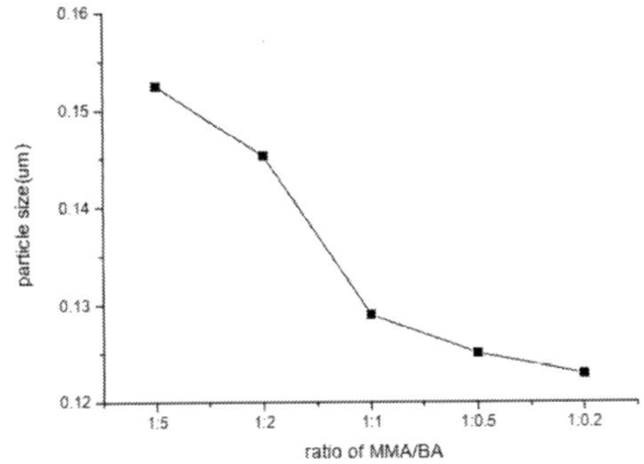

Figure 3. Effect of m_{MMA}/m_{BA} on LIPNs particle size.

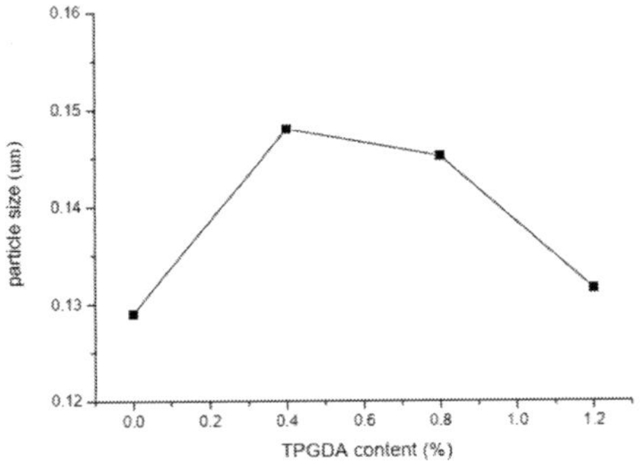

Figure 4. Effect of TPGDA content on LIPNs particle size.

3.3. Dynamic mechanical analysis (DMA)

3.3.1. DMA curve of LIPN compared as network 1. The loss factor (tanδ) represents the damping properties of polymers. In general, the greater tanδ, the better damping properties of polymer. As shown figure 5, there were two peaks of tanδ. In figure 5(a), one of peaks was corresponding to T_g (-49.54℃) of soft segment for aqueous polyurethane oil, the other was T_g (64.25℃) of hard segment. While in figure 5(b), one was corresponding to T_g (-23.35℃) of network II, the other was T_g (57.49℃) of aqueous polyurethane oil in LIPN. It can be seen from figure 5, the damping temperature region of LIPN was wider than those of aqueous polyurethane oil.

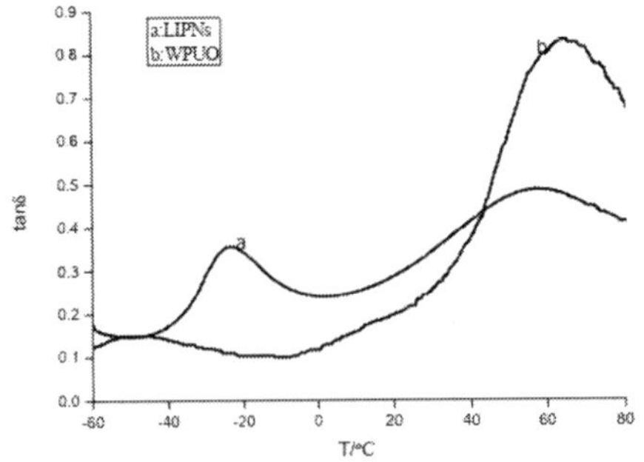

Figure 5. DMA curves of aqueous polyurethane oil and LIPN.

3.3.2. Effect of mMMA/mBA value on glass transition temperature (T_g) for LIPNs. The loss factor (tanδ) of LIPNs are shown in figure 6. According to FOX equation, the T_g of network II corresponding to the m_{MMA}/m_{BA} of 1:5, 1:2, 1:1, 1:0.5 were -39.7℃, -20.8℃,1.4℃, 28.4℃,respectively. It can be seen from Figure 6, The T_g of network I in LIPN was observed at approximately 54°C, and shifted to lower temperature compared with the T_g (65℃) of pure network I. The T_g of network II corresponding to the m_{MMA}/m_{BA} of 1:5, 1:2, 1:1, 1:0.5 were -25℃, 8℃, 27℃, 56℃, respectively, shifted to higher temperature with increasing m_{MMA}/m_{BA} value and greater than corresponding calculated value. The T_g of the network I and network II in LIPN occurred within shift each other, even overlap with increasing m_{MMA}/m_{BA} value. This result may be attributed to the physical entanglement between the network I and network II.

3.3.3. Effect of TPGDA content on T_g of LIPNs. Figure 7 shows the influence of crosslinking agent TPGDA content on tanδ and T_g of LIPNs. As can be seen, with increasing TPGDA content, the two peaks of LIPNs move within each other, the valley became smaller and gradually tend to be flat, or even overlap, Those results may be attributed to higher physical entanglement and interpenetrating of network I and network II at interface, resulting in lower phase separation and greater forced compatibility.

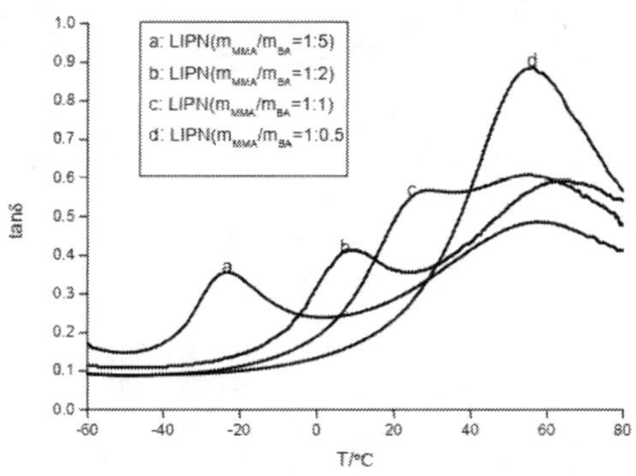

Figure 6. Effect of MMA/BA on T_g of LIPNs.

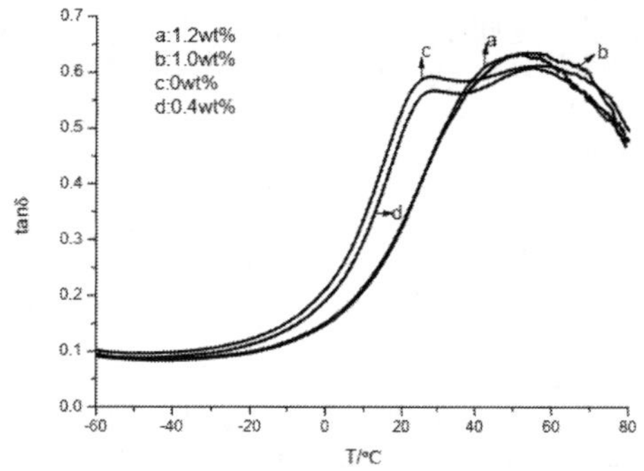

Figure 7. Effect of TPGDA content on T_g of LIPNs.

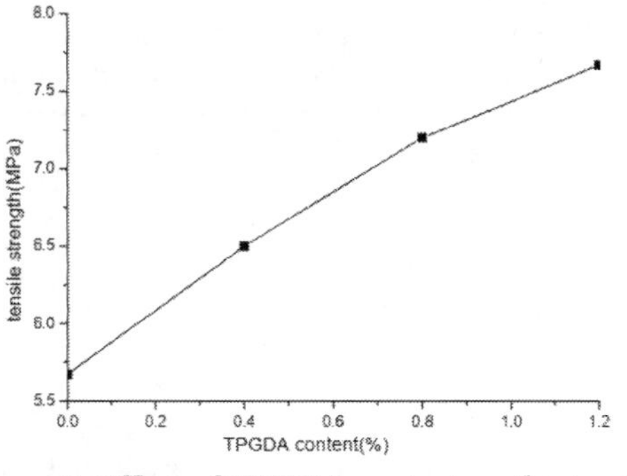

Figure 8. Effect of TPGDA content on the tensile strength of LIPNs.

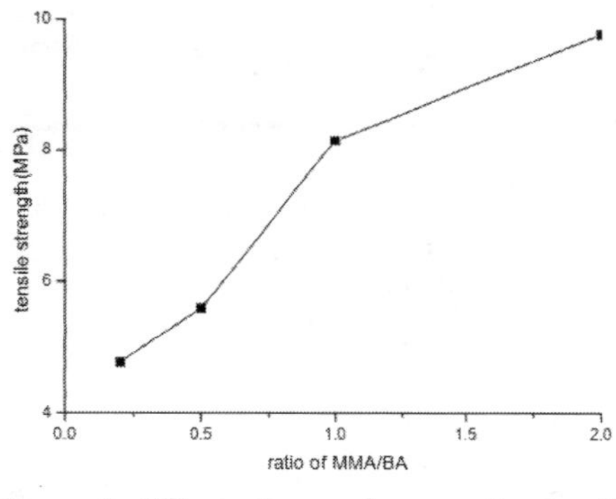

Figure 9. Effect of m_{MMA}/m_{BA} on the tensile strength of LIPNs.

3.4. Influence of various factors on the tensile properties for LIPN films

3.4.1. Effect of TPGDA content on the tensile strength for LIPN films. The influence of TPGDA content of LIPNs on tensile strength was shown in figure 8. It can be seen that the tensile strength of LIPN films increased with increasing TPGDA content. The results shown the interpenetrating degree of two networks in latex particle was widened, and improve the tensile strength of LIPN film.

3.4.2. Effect of m_{MMA}/m_{BA} on the tensile strength for LIPN films. The effect of m_{MMA}/m_{BA} on tensile strength for LIPN films was investigated. It can be found from figure 9 that the tensile strength of the LIPN films increase with the increasing m_{MMA}/m_{BA} value. Because MMA as hard segment could increase the strength of the LIPN film; While BA as soft segments could reduce the strength of the LIPN film.

4. Conclusions

A series of aqueous polyurethane oil/polyacrylate interpenetrating polymer networks (LIPNs) have been successfully synthesized via the technology of latex interpenetrating polymer network combined seed emulsion polymerization process. It shows good damping performances while the mass weight ratio of hard-/soft-segment (m_{MMA}/m_{BA}) was 1:2. The peak of damping temperature for LIPNs shifted gradually towards within each other, the T_g corresponding to maximal loss factor ($\tan\delta_{max}$) increased and the damping temperature range narrowed with increasing m_{MMA}/m_{BA} and TPGDA content. It indicated that the T_g and temperature range of damping materials would be satisfied by changing the TPGDA content and m_{MMA}/m_{BA}. This result implied that this type of LIPNs in our experiments is potential application in aqueous damping material with wider temperature region.

References

[1] Chen L and Chen S 2004 Latex interpenetrating networks based on polyurethane, polyacrylate and epoxy resin *Progress in Organic Coatings* **49** 252-8

[2] Patsiasa S, Saxtona C and Shiptonb M 2004 Hard damping coatings: An experimental procedure for extraction of damping characteristics and modulus of elasticity *Materials Science and Engineering: A* **370** 412-6

[3] Frounchi M, Burford R P and Chaplin R P 1994 Interpenetrating polymer networks of poly (diethyleneglycol bis (allylcarbonate) and poly (urethane acrylates) *Polymer* **35** 5073-8

[4] Kuo S W, Kao H C and Chang F C 2003 Thermal behavior and specific interaction in high glass transition temperature PMMA copolymer *Polymer.* **44** 6873-81

[5] Jia Q M , Zheng M, Zhu Y C, Li J B and Xu C Z 2007 Effects of organophilic montmorillonite on hydrogen bonding, free volume and glass transition temperature of epoxy resin/polyurethane interpenetrating polymer networks *European Polymer Journal* **43** 35-42

[6] Lynda Merlin D and Sivasanka B 2008 Synthesis and characterization of semi-interpenetrating polymer networks using biocompatible polyurethane and acrylamide monomer *European Polymer Journal* **45** 165-70

[7] Frisch H L and Frisch K C 1979 Polyurethane-epoxy interpenetrating polymer networks-barrier and surface properties *Progress in Organic Coatings* **7** 105-11

[8] Yu H B, Peng J, Zhai M L, Li J Q, Wei G S and Qiao J L 2007 Synthesis and characterization of poly (n-butyl acrylate)-poly (methyl methacrylate) latex interpenetrating polymer networks by radiation-induced seeded emulsion polymerisation *Radiation Physics and Chemistry* **76** 1746-50

[9] Tang E J, Tian B Y and Fu C Y 2009 Apparent kinetic study of latex interpenetrating polymer networks P(ethyl acrylate)/P(styrene) and P(styrene)/P(ethyl acr ylate) by two stage emulsion polymerization *Colloids and Surfaces A: Physicochemical and Engineering Aspects* **333** 7-8

Investigation on critical strain of dynamic recrystallization for SiCp/Al-Cu composite

Yali Sun, Jingpei Xie[1,4], Shiming Hao[3], Aiqin Wang[1,2], Pei Liu[1] and Min Li[1]

[1]School of Materials Science and Engineering, Henan University of Science and Technology, Luoyang 471023, China
[2]Collaborative Innovation Center of Non-Ferrous Materials of Henan Province, Luoyang 471023, China
[3]School of Physics and Engineering, Henan University of Science and Technology, Luoyang 471023, China

E-mail: Xiejp@haust.edu.cn

Abstract. Using Gleeble-1500D simulator, the critical strain of dynamic recrystallization of SiCp/Al-Cu composite were investigated at 623-773 K with strain rate of 0.01-10 s^{-1}. The results show that the main softening mechanism is the dynamic recrystallization through the stress-strain curves of SiCp/Al-Cu composite, the activation energy is 272.831 KJ/mol. $\ln\theta$-ε and $(-\partial(\ln\theta)/\partial\varepsilon)-\varepsilon)$ curves were plotted based on the experimental data to further obtain the critical strain and steady strain of DRX. Then the critical strain model and the steady-state strain model of DRX were set up. Furthermore, the dynamic recrystallization volume fraction of SiCp/Al-Cu composite was investigated.

1. Introduction

Aluminum matrix composites (AMCs) reinforced with SiCp have received great attention because of their high specific stiffness, improved wear resistance and high specific strength. The composites which have a higher content of SiC have more enhanced characteristics. Material deformation and microstructure change with technological parameters, material deformation region is divided into stability region and instability region, Stability region was the region where the material can be machine normally. But instability region often appear failure mode such as particle cracking, separation and wedge cracking. Dynamic recrystallization (DRX) was an important mechanism belong to stability region for the microstructure control during hot deformation, DRX not only can fine grains, but also can controlling mechanical properties during processing [1-3]. But using microstructure or stress-strain curves to determine the time when DRX begun was difficultly. So, it is necessary to study the critical strain of dynamic recrystallization. The purpose of this article is to study the critical strain of dynamic recrystallization and recrystallization fraction of SiCp/Al-Cu composite.

2. Experimental

As the experimental material, the preparation method of SiCp/Al-Cu composite was powder

---

[4] Address for correspondence: Jingpei Xie, School of Materials Science and Engineering, Henan University of Science and Technology, Luoyang 471023, China. Tel.:0379-64277396; E-mail: Xiejp@haust.edu.cn.
Foundation item: Project (No. 51371077) supported by the National Natural Science Foundation of China.

metallurgy. 30% SiCp particles with the average size of 3.5 microns which the shape is irregular sharp polygonal blocks, and 70% 2024Al powders that belonged to Al-Cu alloy which with main chemical compositions 3.8wt.% Cu, 1.5wt.% Mg and Al balance, and the average diameter is 10 μm, were mixed using a QQM/B roller mixer for 24 h, and subsequently sintered at 580℃ by vacuum hot-pressing. Figure 1 shows the optical micrograph of the sintered composite. Organization is relatively dense, no obvious gap, the gray phase is the Al matrix, the particles appear gray black is SiC particles, and the white phase is Al-Cu alloy phase that scattered in the matrix.

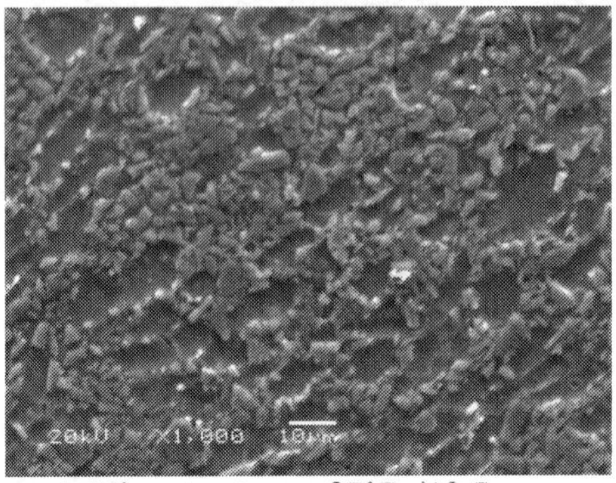

Figure 1. Microstructures of SiCp/Al-Cu composite.

The size of Cylindrical specimens that used for compressive is $\phi 8\times 12$ mm. Machine oil mixed with graphite powder were used for lubrication during hot deformation so that the friction between the specimen and die surface could be minimized. The isothermal constant strain rate deformation was achieved by Utilizing Gleeble-1500D simulator in the temperature range of 623-773 K with the strain rate of 0.01-10 s^{-1}. After straining, specimens were quenched within two seconds to preserve the hot worked microstructures.

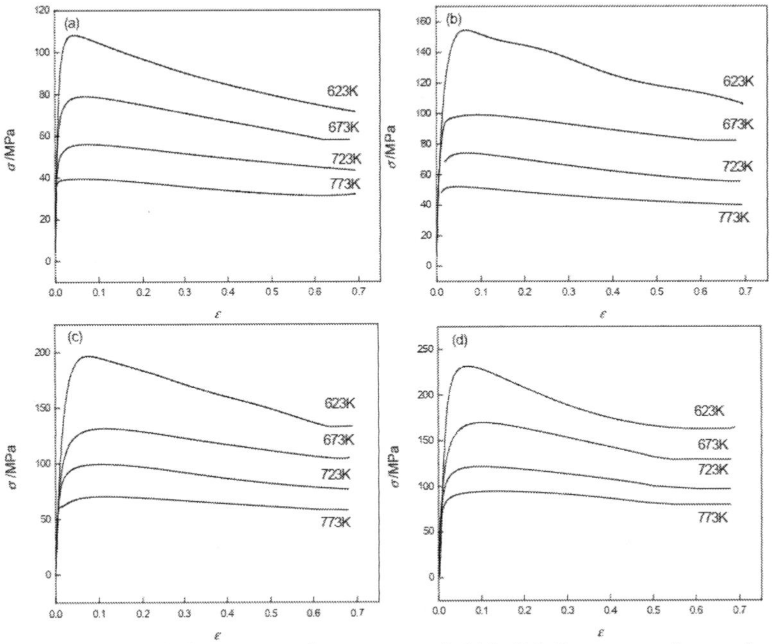

Figure 2. Typical stress-strain curves of SiCp/Al-Cu composite under

temperature of (a) $\dot{\varepsilon}$=0.01s^{-1}; (b) $\dot{\varepsilon}$=0.01 s^{-1}; (c) $\dot{\varepsilon}$=0.01s^{-1}; (d) $\dot{\varepsilon}$=0.01s^{-1}.

3. Results and analysis

3.1. Stress-strain curves

Figure 2 shows the true stress-strain curves of SiCp/Al-Cu composite in different deformation conditions. The stress-strain curves exhibit a typical dynamic recrystallization behavior. The flow stress revealed a trend of increase quickly to the maximum and then decrease slowly during the whole deformation process. This is an emergent property resulting from interactions among the work hardening, dynamic recovery and recrystallization softening. The strain hardening plays a leading role in the initial stages of deformation, and the flow stress increase rapidly. Peak stress appearance in the stress-strain curves when the work hardening and softening balance, recrystallization softening become the main mechanism subsequently, stress decreases gradually and reached a stable value finally.

3.2. Determination of constitutive equations

Determine the peak stress in the material deformation can help choose appropriate processing equipment and then improve the machining efficiency. Phenomenological power-Arrhenius was widely used to describe the constitutive equations, and the formula can be written as [4]:

$$\dot{\varepsilon} = A_1\sigma^{n'}\exp\left(\frac{-Q}{RT}\right) \tag{1}$$

$$\dot{\varepsilon} = A_2\exp(\beta\sigma)\exp\left(-\frac{Q}{RT}\right) \tag{2}$$

$$\dot{\varepsilon} = A[\sinh(\alpha\sigma)]^n\exp\left(-\frac{Q}{RT}\right) \tag{3}$$

Where σ is the flow stress, $\dot{\varepsilon}$ is the strain rate, T is the deformation temperature, Q is the activation energy for deformation, R is the ideal gas constant, A_1, A_2, A, n and α ($\alpha=\beta/n'$) are material constants. Zener-Hollomon parameter (Z) is widely used to state the hot deformation behavior, which correlates with the strain rate, deformation temperature, and activation energy by the expression [5]:

$$Z = \dot{\varepsilon}\exp\left(\frac{Q}{RT}\right) = A[\sinh(\alpha\sigma)]^n \tag{4}$$

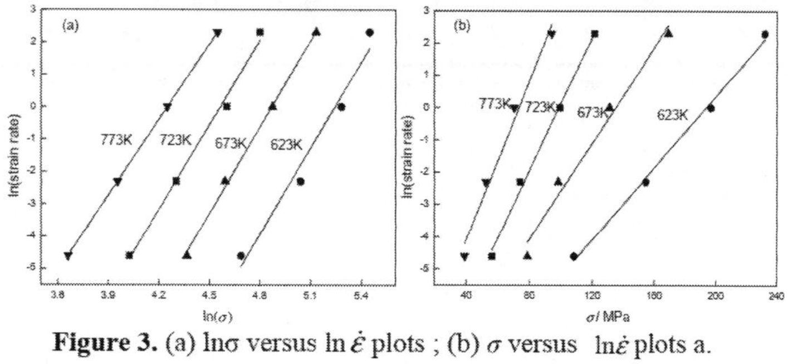

Figure 3. (a) lnσ versus ln$\dot{\varepsilon}$ plots ; (b) σ versus ln$\dot{\varepsilon}$ plots a.

From figures 3(a) and 3(b), the values of n' and β were evaluated, which n' and β are taken as the average values of the slopes of the $\ln\dot{\varepsilon}-\ln\sigma$ and $\ln\dot{\varepsilon}$ -σ plots [6], and the value of α was 0.01166 for the SiCp/Al-Cu composite.

The activation energy Q could be calculated and shown by the following equation through differentiating equation (3)

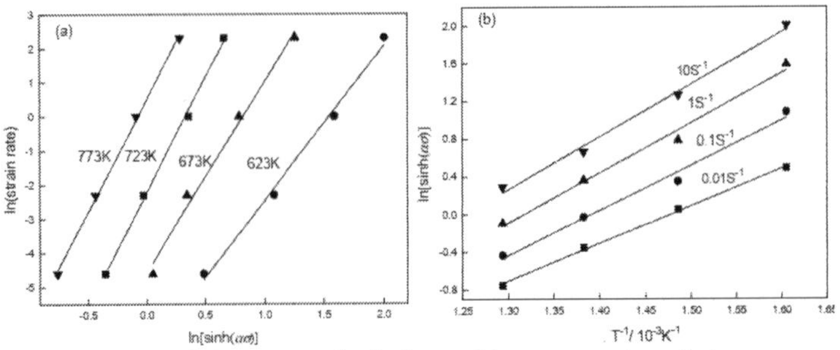

Figure 4. (a) lnσ versus $\ln\dot{\varepsilon}$ plots ; (b) σ versus $\ln\dot{\varepsilon}$ plots a.

$$Q = R\left.\frac{\partial \ln[\sinh(\alpha\sigma)]}{\partial(1/T)}\right|_{\dot{\varepsilon}}\left.\frac{\partial \ln\dot{\varepsilon}}{\partial \ln[\sinh(\alpha\sigma)]}\right|_{T} \quad (5)$$

Where n is the mean slope of $\ln\dot{\varepsilon}-\ln[\sinh(\alpha\sigma)]$ plots (figure 4(a)), and S is the mean slope of the $\ln[\sinh(\alpha\sigma)]-1/T$ plots (figure 4(b)). The value of activation energy of SiCp/Al-Cu composite is calculated to be Q=RnS=272.831 kJ/mol.

Figure 5 shows the relationship between ln[sinh(ασ)] and ln$Z$, the intercept of the straight line is ln$A$=44.5. Then the hot deformation equation of SiCp/Al-Cu composite can be written as

$$\dot{\varepsilon} = e^{44.5}[\sinh(0.01166\sigma)]^{6.2917}\exp(-\frac{272.831}{RT}) \quad (6)$$

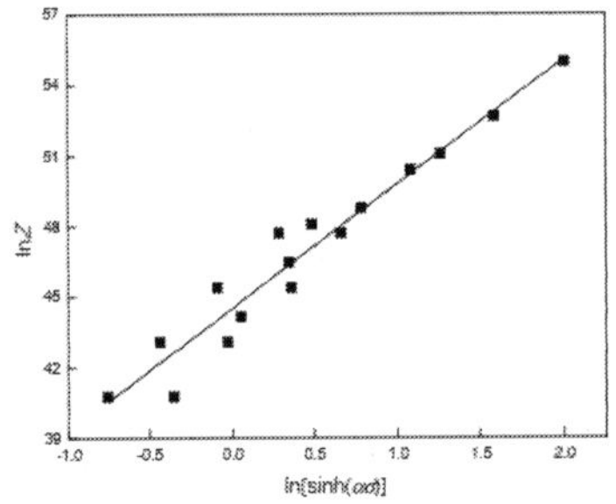

Figure 5. Relationships between peak stress and Z parameter of SiCp/Al-Cu composite.

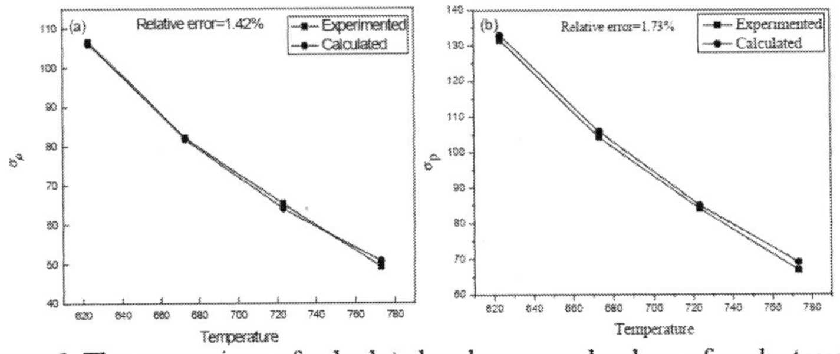

Figure 6. The comparison of calculated and measured values of peak stress for SiCp/Al-Cu composite at (a) 0.1s^{-1} ; (b) 1s^{-1}.

Using thermal simulation test data to verify the exactness of constitutive equation. Figure 6 is the comparison of calculated and measured valued of peak stress at 0.1s^{-1}, 1s^{-1} respectively. The relative error are 1.42% and 1.73%. The relative error is small, and the difference of the values in different strain rate is very small. It expressed that the results between calculated and measured had good agreement, indicating that the establishment of constitutive equation can provide theoretical basis for practical production.

3.3. *Hot deformation characteristics*

The true stress-strain curve is the outward manifestation of microstructure during hot deformation, and the appearance of peak stress indicates the onset of dynamic recrystallization. But it is difficult to reflect directly when and where the dynamic recrystallization is initiated for the true stress-strain. So this experiment adopts the proposed by Poliak and Jonas [7, 8] to determine the SiCp/Al-Cu composites dynamic recrystallization critical strain:

$$\frac{\partial}{\partial\sigma}\left(\frac{\partial\theta}{\partial\sigma}\right)=0 \qquad \left(\frac{\partial\theta}{\partial\sigma}\right)_{\dot{\varepsilon}}=\left(\frac{\partial\ln\theta}{\partial\varepsilon}\right)_{\dot{\varepsilon}} \tag{7}$$

According to the above equation, the characteristic of dynamic recrystallization is that not only the θ-σ curve but also the $\ln\theta$-ε curve presents an inflection point. The $\ln\theta$–ε and the $(-\partial(\ln\theta)/\partial\varepsilon)$–$\varepsilon$ curve can be drawn based on figure 2. Using the relationship of $-\partial^2(\ln\theta)/\partial\varepsilon^2=0$, the critical strain of dynamic recrystallization can be obtained directly.

Figure 7 shows the relationship between strain hardening rate and strain in different deformation conditions. The derivative of the strain hardening rate as a function of the strain in different conditions was calculated according to figure 7. Figure 8 are the $(-\partial(\ln\theta)/\partial\varepsilon)$–$\varepsilon$ curves. The critical strain ε_c is the point at which the strain value is minimum in the plot.

The Sellars is widely used to build the prediction model of critical strain of SiCp/Al-Cu composite [9]:

$$\varepsilon_c=kZ^m \tag{8}$$

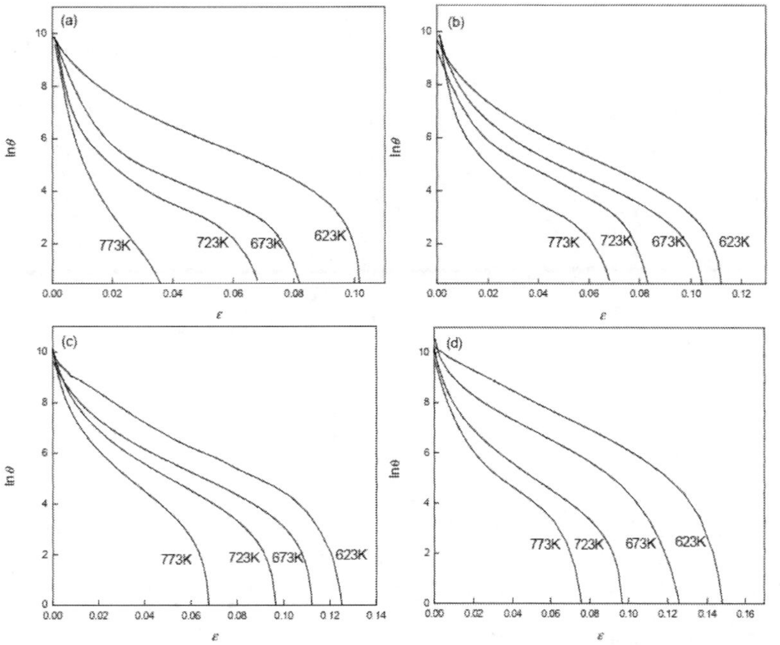

Figure 7. The $\ln\theta$–ε curves of SiCp/Al-Cu composite under different deformation conditions(a) $\dot{\varepsilon}$=0.01 s^{-1}; (b) $\dot{\varepsilon}$=0.1 s^{-1}; (c) $\dot{\varepsilon}$=1 s^{-1}; (d) $\dot{\varepsilon}$=10 s^{-1}.

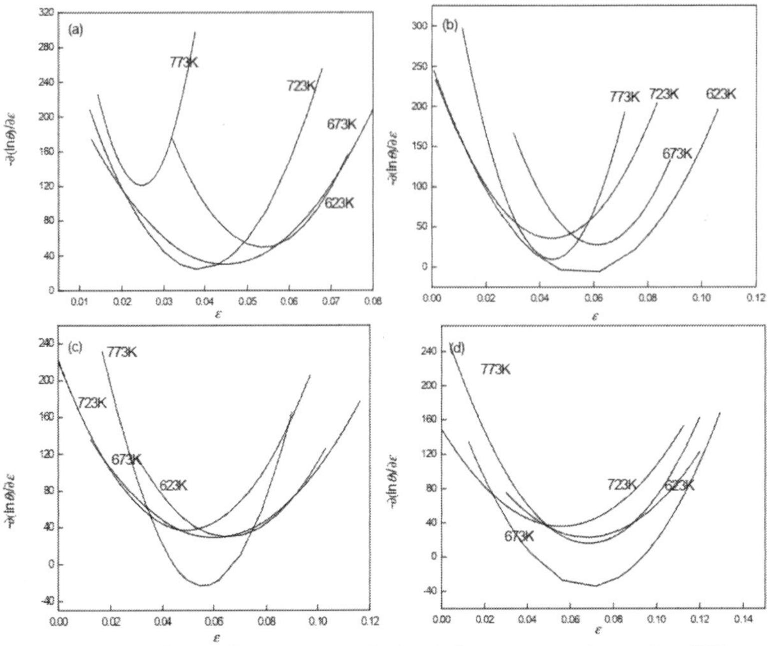

Figure 8. The $-\partial(\ln\theta)/\partial\varepsilon$–$\varepsilon$ curves of SiCp/Al-Cu composite under different deformation conditions (a) $\dot{\varepsilon}$=0.01 s^{-1}; (b) $\dot{\varepsilon}$=0.1 s^{-1}; (c) $\dot{\varepsilon}$=1 s^{-1}; (d) $\dot{\varepsilon}$=10 s^{-1}.

Where k and m are constants. Figure 9 illustrates the relationship between natural logarithm of the calculated strain and natural logarithm of the Zener-Hollomon parameter. It is clear that the value of $\ln\varepsilon_c$ increases with the increase of Zener-Hollomon parameter [10]. and $\ln\varepsilon_c$ shows higher linear relationship with lnZ , Therefore, the equation is determined as follows: $\ln\varepsilon_c$=0.079ln$Z$-6.62. The equation for critical strain is given by: $\varepsilon_c=1.32\times10^{-3}Z^{0.079}$.

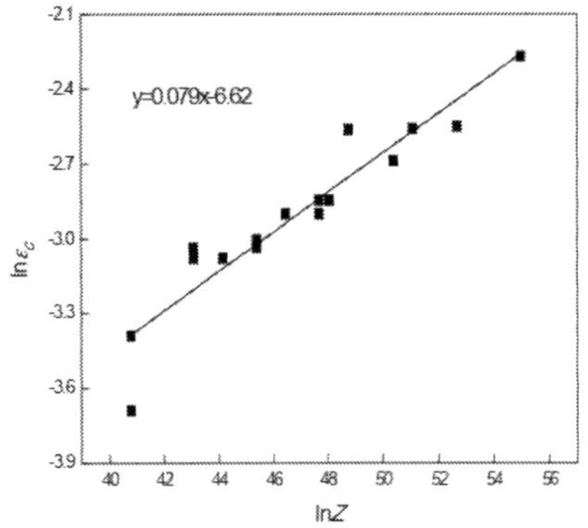

Figure 9. $\ln\varepsilon_c$ versus ln$Z$ plot.

The steady -state strain can be obtained from figure 10 which shows the relationship between the strain hardening rate (θ) and strain (ε), these plots reduce quickly and then rise slowly, and the minimum points are below zero and represent the position where the dynamic dynamic softening is maximized. After this minimum point the work hardening rate increases and in the steady-state strain, the strain hardening rate reaches zero [11].

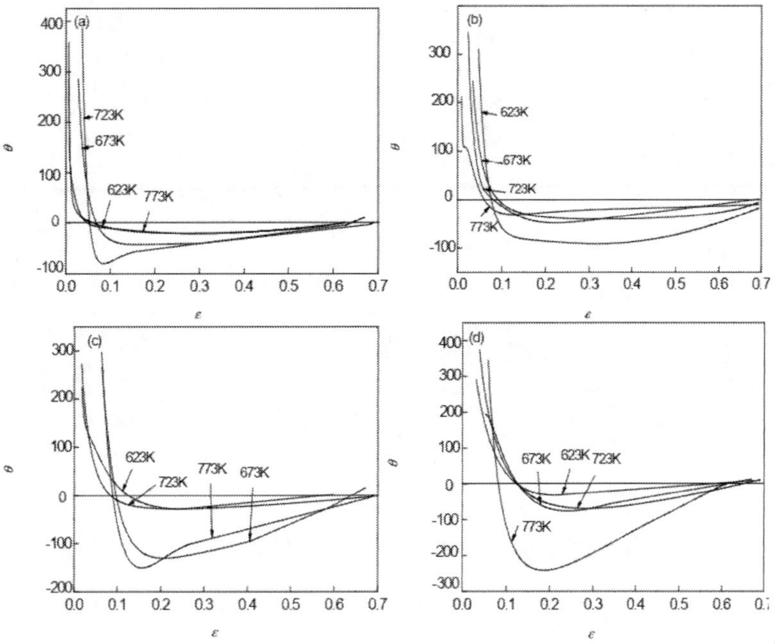

Figure 10. Strain hardening rate *vs.* strain for SiCp/Al-Cu composite ($\dot{\varepsilon}$=0.01 s^{-1}).

3.4. DRX kinetics

The critical strain is the point that dynamic recrystallization starts to happen, but it is difficultly to understand the change of recrystallized fraction from the above calculation. X_{DRX} represent the recrystallized fraction. The Avrami equation can directly conduct the relationship between X_{DRX} and ε [12-15]:

$$X_{DRX} = 1 - \exp\left[-B\left(\frac{\varepsilon - \varepsilon_c}{\varepsilon_{0.5}} \right)^n \right] \tag{8}$$

Where, B is the Avrami constant, n is the Avrami exponent, ε_c is the critical strain and $\varepsilon_{0.5}$ is the strain that the value of X_{DRX} is 50%. In order to avoid the interference of human factors in traditional observation of metallographic specimen, equation (9) is employed in the paper to determine the X_{DRX} [9]:

$$X_{DRX} = \frac{(\sigma_{drvx})^2 - (\sigma_{drxx})^2}{(\sigma_{drvss})^2 - (\sigma_{drxss})^2} \tag{9}$$

Where, σ_{drvss} and σ_{drxss} are the saturation and steady stresses, respectively. σ_{drvx} is the flow stress when DRV plays the important role in the softening mechanism. σ_{drxx} is the flow stress when DRX plays the important role in the softening mechanism.

X_{DRX} versus ε with various strain rates at different temperatures of SiCp/Al-Cu composite is shown in figure 11. From figure 11, the X_{DRX} increases with the increasing of true strain, and when the value of X_{DRX} reaches 1 indicate correspond strain by this point is steady-state strain. Furthermore, curves show a "S" shaped growth trend. The tendency of the curves is increased slowly-increased rapidly-increased slowly. In addition, when the strain is a constant, the recrystallized fraction decreases with the increase of strain rate [16].

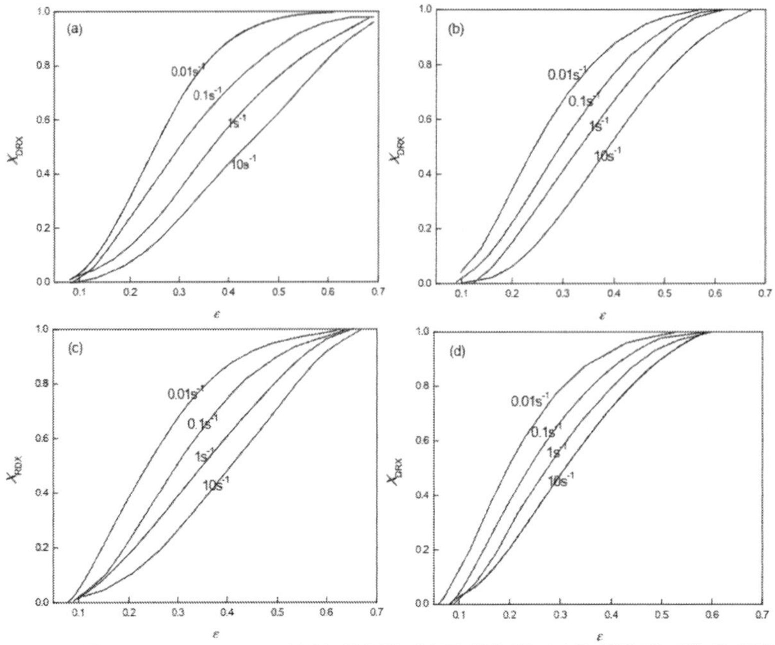

Figure 11. X_{DRX} versus ε curves(a)T=623 K; (b)T=673 K; (c)T=723 K; (d) T=773 K.

Using the relationship between strain hardening rate and strain that shows in figure 10, the position of the onset of steady-state strain can be determined, and this point is usually appear on the end of the curves where the strain hardening rate reaches zero. In figure 12, the strain hardening rate and recrystallized fraction are both plotted against strain at a strain rate of 0.01 s^{-1} and a temperature of 623 K. Figure 12 confirms the accuracy of the model used to predict the dynamically recrystallized volume fraction. And the point at which the strain is steady-state strain coincides with the volume fraction of dynamic recrystallization is unity. So the model of DRX kinetic can be illuminated by the Avramy-type expression.

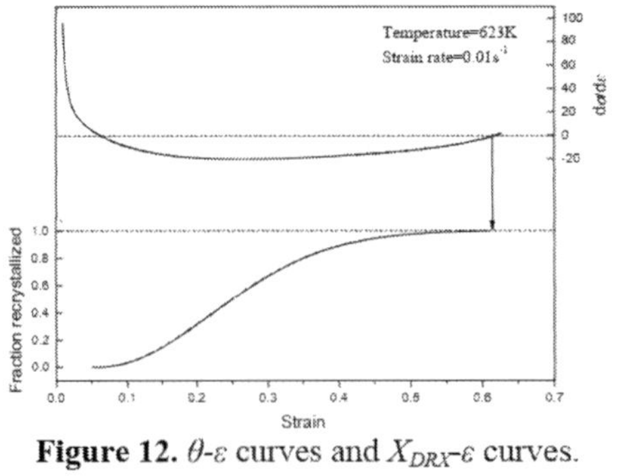

Figure 12. θ-ε curves and X_{DRX}-ε curves.

4. Conclusions

Hot compression tests on SiCp/Al-Cu composite were performed at various strain rates and deformation temperatures and got the following conclusions:

- The main softening mechanism of SiCp/Al-Cu composite during the thermal deformation is dynamic recrystallization mechanism. The activation energy is 272.831 KJ/mol.

- The critical strain can be identified from the inflection point on the $\ln\theta$–ε and $-\partial(\ln\theta)/\partial\varepsilon$–$\varepsilon$ curves. There is linear relationship between critical strain and peak strain, i.e., $\varepsilon_c/\varepsilon_p$=0.652.
- The initial critical strain of dynamic recrystallization can be calculated by Sellars model setting, and there is a relationship between ε_c and Z, $\varepsilon_c=1.32\times10^{-3}Z^{0.079}$. The steady-state strain can be determined using θ-ε curve.
- The dynamic recrystallization volume fraction of SiCp/Al-Cu composite is investigated. The model of DRX kinetic can be illuminated by the Avramy-type expression.

References

[1] Frommert M and Gottstein G 2009 *Mater. Sci. Eng. A* **506** 101-10
[2] Ebrahimi G R, Keshmiri H, Maldar A R and Momeni A 2012 *J. Mater. Sci. Technol.* **28** 467-73
[3] Mirzadeh H, Cabrera J M, Prado J M and Najafizadeh A 2011 *Mater. Sci. Eng. A* **528** 3876-82
[4] Meysami M and Mousavi S A A A 2011 *Mater. Sci. Eng. A* **528** 3049-55
[5] Lin Y C and Chen Xiao-min 2011 *Mater. Des.* **32** 1733-59
[6] Xiao X, Liu G Q, Hu B F, Zheng X, Wang L N, Chen S J and Ullah A 2012 *Comp. Mater. Sci.* **62** 227-37
[7] Poliak E I and Jonas J J 2003 *ISIJ Int.* **43** 684-91
[8] Jonas J J and Poliak E I 2003 *Mater. Sci. Forum* **426** 57-66
[9] Xu Yao-wen, Tang Di, Song Yong and Pan Xiang-gang 2012 *Mater. Des.* **36** 275-8
[10] Wei Hai-lian, Liu Guo-quan, Xiao Xiang and Zhang Ming-he 2013 *Mater. Sci. Eng. A* **573** 215-21
[11] Mirzadeh H, Cabrera J M, Najafizadeh A and Calvillo P R 2011 *Mater. Sci. Eng. A* **528** 3876-82
[12] Wang Meng-han, Li Yu-feng, Wang Wen-hao, Zhou Jie and Chiba Akihiko 2013 *Mater. Des.* **45** 384-92
[13] Wang Jian, Xiao Hong, Xie Hong-bao, Xu Xiu-mei and Gao Ya-nan 2012 *Mater. Sci. Eng. A* **539** 294-300
[14] Zeng Zhou-yu, Chen Li-qing, Zhu Fu-xian and Liu Xiang-hua 2011 *J. Mater. Sci. Tec.* **27** 913-9
[15] Xu Dong, Zhu Miao-yong, Tang Zheng-you and Sun Chao 2013 *J. Wuhan Uni. Tec.-Mater. Sci. Ed* **28** 819-24
[16] Shanban M and Eghbali B 2010 *Mater. Sci. Eng. A* **527** 4320-5

IOP Conf. Series: Materials Science and Engineering **87** (2015) 012047 doi:10.1088/1757-899X/87/1/012047

Preparation and characterization of electrospun poly(3-hydroxybutyrate-co-3-hydroxyvalerate) fibrous mats

Yongjing Xu, Liming Zou[1], Hongwei Lu and Zailing Chen

State Key Laboratory for Modification of Chemical Fibers and Polymer Materials, College of Materials Science and Engineering, Donghua University, 2999 North People Road, Shanghai 201620, People's Republic of China

E-mail: lmzou@dhu.edu.cn

Abstract. Poly (3-hydroxybutyrate-co-3-hydroxyvalerate) (PHBV), as a biodegradable polyester, was electrospun to obtain defect-free fibers with high surface-area-to-volume ratio. Several parameters such as solvent ratio, polymer concentration, applied voltage, flow rate, and tip-to-target distance were optimized to achieve defect-free morphology. The average diameter of the PHBV fibers was 1400 nm. In order to evaluate the final properties of PHBV nanofibers, the following characterization techniques were employed: scanning electron microscopy (SEM), Fourier transform infrared attenuated total reflectance (FTIR-ATR) spectroscopy, uniaxial tensile tests and dataphysics instruments.

1. Introduction

Electrospinning, which was introduced in early 1930s, is known as a novel and effective fabiration method for producing ultra fine fiber of nanometer to micrometer scale. Generally, electrospun nanofibrous mats with controllable diameters, high specific surface and porosity and three dimensional network structures, have the ability to closely mimic the biological function and structure of nature extracellular matrix (ECM). And electrospun nanofibrous scaffolds served as a three-dimensional template could provide favorable environment for directing cell adhesion, migration proliferation and differention, and, thus, promote tissue regeneration in vitro as natural extracelluar matrix (ECM) does in vivo. As a consequence, electrospinning has been widely used for nanofibrous scaffold fabrication and have shown potential as effective tissue engineering scaffolds [1, 2]. Therefore, electrospun nanofibers are being explored as scaffolds for nanodevices, biosensors, drug delivery, and tissue engineering applications. Various polymers including synthetic ones such as PCL, polylactic acid (PLA), polyglycolic acid (PGA), PLGA, polystyrene, polyurethane (PU), polyethylene terephthalate, poly (L-lactic acid)-co-poly (e-caprolactone) (PLACL) and biological materials such as collagen, gelatin, chitosan have been successfully electrospun to obtain fibers with diameters ranging from 3 nm to 5 μm [3-6].

Poly (3-hydroxybuty-rate-co-3-hydroxyvalerate) (PHBV), as a member of polyhydroxyalkanoate (PHA), is a well-known biodegradable, biocompatible, nontoxic, thermoplastic polyester produced by bacteria [7]. Besides that, PHBV possesses good oxygen permeability and mechanical strength, and

[1] Address for correspondence: Liming Zou, State Key Laboratory for Modification of Chemical Fibers and Polymer Materials, College of Materials Science and Engineering, Donghua University, 2999 North People Road, Shanghai 201620, People's Republic of China. E-mail: lmzou@dhu.edu.cn.

the ultimate degradation product is (R)-3-hydroxybutyric acid which is a normal constituent of human blood. Therefore, PHBV has been extensively explored as biomaterials for in vitro and in vivo studies, and used for various biomedical devices such as sutures, prosthetic devices, drug delivery systems and surgical applications.

According to the investigations, Nonwoven electrospun PHBV fibrous mats have already been produced and revealed (in vitro) a better performance for chondrocytes and fibroblasts attachment, growth and proliferation when compared to their 2D film counterparts [8, 9]. In additon, electrospun PHBV fibers loaded with an angiogenesis factor have shown to accelerate the wound healing process [10].

Electrospinning of PHBV fibers has been previously reported using solvents like hexafluoroisopropanol (HFIP) [11], trifluoroethanol (TFE) [12], chloroform [13], and chloroform with DMF [14]. Many studies have reported PHBV fiber diameters between 1 and 4 μm using chloroform as solvent [13]. In this paper, the objective of this study was to obtain electrospun poly(3-hydroxy-butyrate-co-3-hydroxyvalerate) (PHBV) fibers used dichloromethane as the solvents, which has low toxicity and residue. The study was aimed to make comparison between PHBV fibers and 2-D films, and then based on the conclusion of the experiment, the study may be explored furthur for culturing human skin fibroblast cells, which made PHBV fibers can be used for the skin tissue engineering.

It is important to underline that, to the best of our knowledge, dichloromethane as the solvents, which has not been related stdies by far. Meanwhile, such full characterization of electrospun PHBV has not been reported yet.

2. Materials and methods

2.1. Materials

Poly (3-hydroxybutyrate-co-3-hydroxyvalerate) (PHBV) with 1.09 mol% 3-hydroxyvalerate (HV) (Mw = 2.67×10^5), was purchased fromTianan Biologic Material CO., dichloromethane (CH_2Cl_2, 99.5%) was supplied by Shanghai Lingfeng Chemical Reagent CO. All reagents and solvents were used as-purchased without any further purification.

2.2. Fabrication of electrospun PHBV nanofibers and 2-D films

PHBV fibers and films were prepared by electrospinning and solvent casting, respectively. Electrospun PHBV was fabricated as follows. PHBV powder was dissolved in CH_2Cl_2 with a magnetic stirrer for 12 h to prepare homogeneous solution, the overall polymer amount was kept 10% wt/v. Electrospinning was carried out in air at room temperature, the solution was filled in a 10ml syringe equipped with a 22 G metallic needle, fixed in a digitally controlled syringe pump (KD scientific, MA, USA), and the flow rate was maintianed at 1mL/h. A rectangular steel plate covered with aluminum foil was placed 14 cm away from the needle tip. Meanwhile, the tip of the needle was connected to a high voltage of 15 KV (GAMMA, ES, USA). All electrospun fibers collected were dried under vacuum for 24 h and stored in a dessicator for further characterization.

Two-dimensional PHBV films were cast by dissolving the PHBV in CH_2Cl_2 and kept overnight at room temperature and further vacuum-dried for 24 h.

2.3. Surface morphology

The surface morphology of the electrospun PHBV fibers and 2-D films was analyzed using scanning electron microscopy (FE-SEM, JSM 6701F, JEOL, Tokyo, Japan) at an accelerating voltage of 3 kV at various magnifications. The samples were sputter-coated with platinum prior to imaging. The fiber diameters were measured from different locations, and the fiber diameter versus number of fibers was plotted.

2.4. Porosity analysis

The percentage porosity of the electrospunPHBVfibers and the 2-D film were measured by using the following equations [15], the thickness of the fibrous mats was measured by a micrometer.

$$\text{Porosity } (\%) = (1\text{-}d/D) \times 100\%$$

d: mat apparent density (g/cm^3)

D: bulk density of PHBV (g/cm^3)

2.5. *Fourier transform infrared spectroscopy*

Infrared (IR) spectra of all mats were recorded on a Nicolet 8700 spectrophotometer, by ATR reflection method. The spectrum was acquired in thewavelength range of 650–4000 cm^{-1}, spectral resolution 4 cm^{-1}, scans 128.

2.6. *X-ray diffractometry*

X-ray diffraction analyses (XRD, Philips X'Pert) were performed in the following conditions: CuKαλ=1.5402 Å, 2θ range 0°–50°, step size 0.020°, time per step 4s, scan speed 0.03°/s.

2.7. *Thermogravimetric analysis (TGA)*

Thermogravimetric measurements (TG) were carried out by simultaneous thermogravimetric and differential thermal analyses (TGA, Netzsch209 F1) in the following conditions: sample weight about 7 mg, N_2 flow 20 cm^3/min, temperature range 20–600°C, heating rate 10°C/min. Residual mass and maximum degradation temperature (T_d) were measured.

2.8. *Tensile properties*

The tensile properties of the PHBV fibers and 2-D films were evaluated using Universal Material Testing Machine. Fibers of 10 mm × 80 mm (n=5) and 2-D films of 10 mm×80 mm (n = 5) were used in this study. The ends of the samples were mounted on the gripping units of the tensile tester, and a load of 500Nat an extension rate of 5mm/min was applied until failure.

2.9. *Measure of water contact angle*

For evaluation of wettability, the water contact angle of the electrospun PHBV fibers and 2-D films was measured at room temperature using a Kruss GmbH DSA 100 Mk 2 goniometer (Hamburg, Germany) followed by image processing of sessile drop with DSA 1.8 software. A droplet of deionized water (2.0 μL) was placed on the surface of a film at room temperature. More than five measurements were carried out for a single sample, and the resulting values are the averaged results.

3. Results and discussion

3.1. *Morphology of electrospun mats and 2-D films*

The structure and surface morphology of PHBV nanofibers and 2-D film were observed under a scanning electron microscope. Figure 1B showed that PHBV nanofibers have a uniform structure without any sign of beads formation. Figure 1C illustrated that the average diameter of the fibers obtained at 10% (wt/v) polymer concertration was 1400 nm. Besides that, the nanofibrous scaffold had a pore size 4.2 μm and a porosity of 90%, whereas the 2-D film presented a pore size 75.42 nm and a porosity of 24.57%. In this study, it can be concluded that PHBV fibers have much higher porosity and pore size than 2-D films. As for the cell culture, controlled pore size and high porosity are essential for the transport of oxygen and nutrient supply to the cells and cellular growth. 2-D films may be allowed the nutrients and metabolic wastes pass through by the nanosized porosity, but, they seem too small to provide enough space for cell ingrowth. Most importantly, the average porosity of skin tissue engineering is 70%-80%. Therefore, PHBV fibers are more suitable for the potential skin tissue engineering. In addition, it should be noted that PHBVfibers with different fiber diameter, porosity and pore size can be collected by only varying the process parameters.

IOP Conf. Series: Materials Science and Engineering **87** (2015) 012047 doi:10.1088/1757-899X/87/1/012047

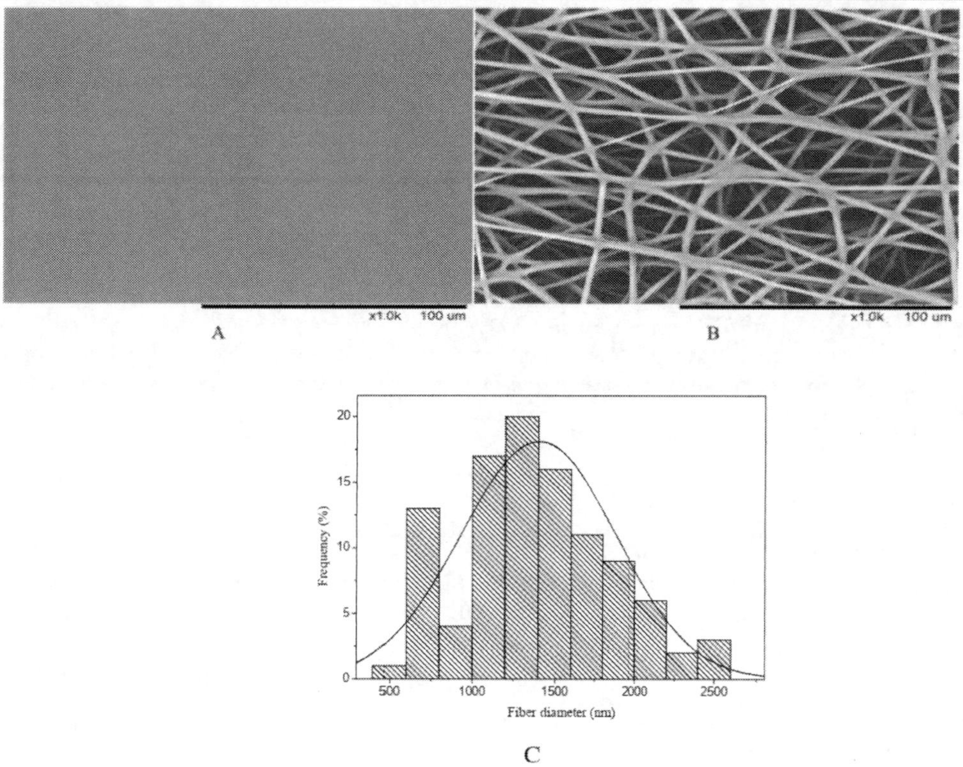

Figure 1. Scanning electron micrograph showing the surface morphology of (A) solvent cast PHBV 2-D film, (B) electrospun nonwoven PHBV fibers, and (C) histogram showing the PHBV fiber size distribution.

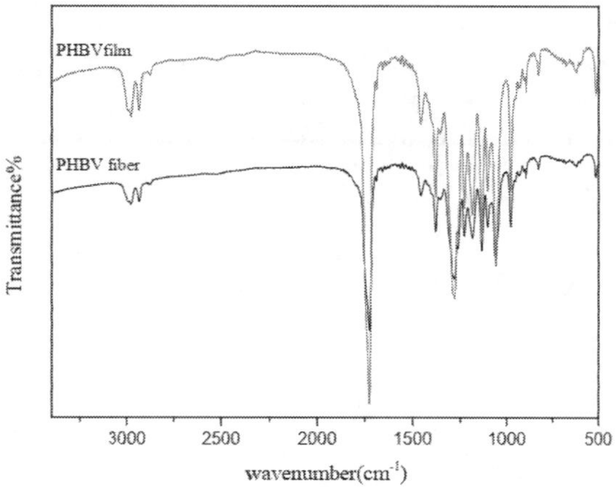

Figure 2. ATR spectra of electrospun PHBV nanofiber and solvent cast PHBV 2-D film.

3.2. ATR spectroscopy of electrospun mats and 2-D films

The spectra of the PHBV film and PHBV fiber were compared in figure 2, which confirmed the coincidence between 2-D film and nanofibers. Specially, all the characteristic PHBV vibrational modes were detected, in agreement with the related studies. Figure 2 showed the prominent characteristic of the ester carbonyl (C=O) of PHBV at 1720cm^{-1}, meanwhile, the absorption peak at 1282 cm^{-1} indicated the strong C-O stretching band, which further confirmed the presence of the ester carbonyl (C=O) group. In addition, the absorption peak observed at 1058-1282 cm^{-1} can be assigned to the stretching vibration of C-O-C and bending vibration of –CH, –CH_2. The bands centered at 1184 cm^{-1}, 1229 cm^{-1} were shown the crystalline phase of PHBV, revealling the high crystallinity of PHBV nanofiers and films.

3.3. XRD analyses of electrospun mats and 2-D films

In figure 3 diffraction patterns of electrospun mats and 2-D films are reported. It can be obtianed that the crystallinity of electrospun mats and 2-D films are 66.36%, 55.74% respectively, using the Analysis Software of JADE.

In figure 3, PHBV fibers and 2-D films presented diffraction peaks at around 2θ 14°,17°, 21°, 22°, 26°, 27.5° and 30.5°, ascribable to the helix lamellar α-form conformation of the PHBV and to the (020), (110), (101),(111), (121), (040) and (002) crystallographic planes, respectively. But in constrast, PHBV 2-D films showed relatively weak diffraction peaks compared with fibers, which can be explained that the crystal structure of 2-D films was imperfect owning to the untrolled volatilerate rate of solvent that hinder the formation of nuclei and the growth of crystal grain. It has been reported that PHBV exhibits isodimorphism due to the comparable chain conformation of poly (hydroxybutirate) (PHB) and poly (3-hydroxyvalerate) (PHV).[16] Because of the low HV content in the present work, it can be expected to crystallizes in the PHB lattice. Therefore, as to the fibers, the presence of β-form crystals indicates a high level of molecular stretching in the amorphous region between the α-crystalline lamellae, which might develop first. Wang et al. stated that significant chain orientation of the PHB molecules is likely to take place in the entrance region from the Taylor cone to the ejecting jet, where a severe jet stretching is induced by the convergent flow and electric repulsion [17]. The resulting orientation readily induces the crystallization of the α-form lamellae, and the interlaced chains between the crystalline lamellae are further elongated during jet whipping to develop the planar zigzag β-form crystals.

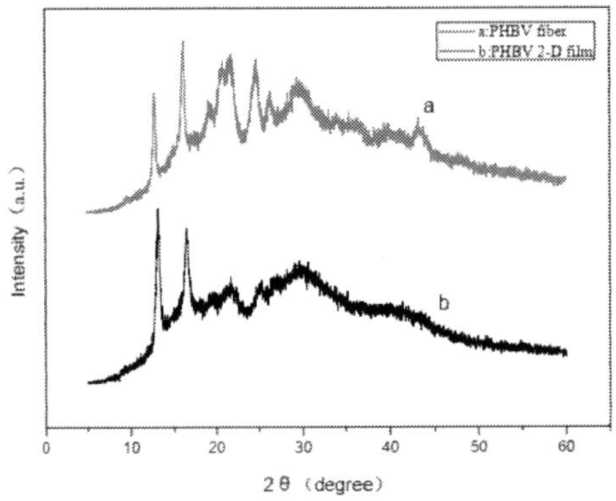

Figure 3. XRD patterns of PHBV, fibers (a) and 2-D films (b).

3.4. Thermogravimetric analysis (TGA) of electrospun mats and 2-D films

The TG and DTG curves of the electrospun mats and 2-D films were showen in figures 4A and 4B, respectively, showing that the thermal degradation of PHBV occurred in one step.

In figure 4A, it demonstrated that PHBV fibers had higher onset degradation temperature than 2-D films obviously. In figure 4B, the two derivative curves presented two distinctly distinguishable thermal decomposition (Td) peaks, in which the degradation temperature of PHBV fibers and 2-D films were 215.5℃~279.8℃ and 199.7℃~247.7℃, the maximum decomposing temperature were 268.1℃and 235.7℃ respectively. The reasons can be interpreted as follows, PHBV fibers were stretched in the process of electrospinning, which led to the internal structure of PHBV more regular and stable, thus, more energy should be needed to make the molecular chain start moving, and then decomposing. For the 2-D films, the fast and uncontrolled volatilization of dichloromethane made the molecular chain in a mixed and disorder arrangement, as a result, the decomposing temperature was lower compared with PHBV fibers. It was known that the decomposing temperature was high when the polymer had high crystallinity, which was corroborating the XRD results (figure 3). In short, PHBV fibers were equiped with excellent thermostability in constract with 2-D flims.

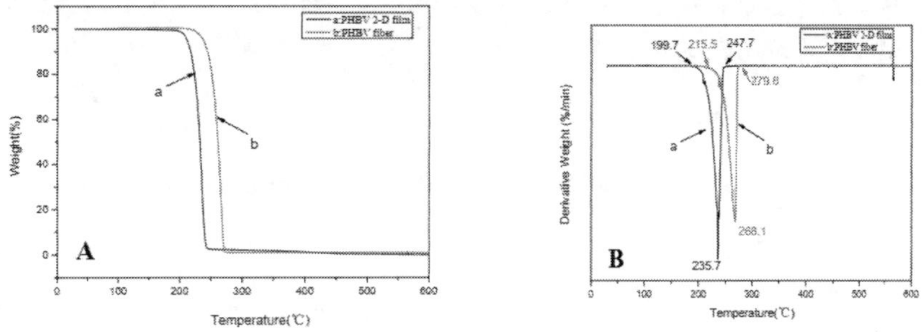

Figure 4. TG (A) and DTG (B) curves of electrospun mats and 2-D films.

3.5. Mechanical properties of electrospun mats and 2-D films

Table 1 showed that PHBV fibers exhibited higher tensile strength but lower Tensile modulus compared with 2-D film. The result can be explained as follows, the PHBV fiber was collected by rotating frame cylinder, therefore, the fibers obtained were equipped with orientation partly. The fast volatility of CH_2Cl_2, leading to 2-D films with low crystallinity when preparing the 2-D PHBV film may be the other reason. The increase in Young's modulus for the PHBV film may be due to its greater thickness (0.2 mm) and lower porosity (24.57%) when compared with the fibers (0.12 mm and 90%). Moreover, the strain at failure (ε) suggested that PHBV fibers were more elastic than the 2-D films, which were suitable for skin tissue engineering. Venugopal et al. have previously shown the efficacy of elastic poly (caprolactone) and poly (caprolactone)-collagen nanofibrous coating for skin regeneration applications when compared with less elastic matrices [18]. Hence, the perfect tensile stress and elasticity made PHBV fibers posess more advantages applied in the skin tissue engineering.

Table 1. Mechanical properties of PHBV fiber and film.

material	σ (MPa)	E (MPa)	ε (%)
PHBV fiber	2.91	76	18.25
PHBV 2-D film	2.68	89	14.75

σ: Tensile stress.
E: Tensile modulus.
ε: Strain at failure.

3.6. Water contact angle of electrospun mats and 2-D films

The static WCA valued are shown in figure 5, which confirmed that PHBV films (5B) exhibited a less hydrophilic surface with 128.2° and the PHBV fibers (5A) with 111.1°. The phenomenon for PHBVfilms may be due to its low porosity (24.57%) when compared with PHBV fibers (90%), moreover, there existed three-dimensional network structure for the PHBV fibers, the capillary effect may lead to the higher WCA. But for the skin tissue engineering, the hydrophilic scaffolds are beneficial to direct cell adhesion, migration proliferation and differention. The hydrophilcity of PHBV fibers had slight improvement compared with 2-D films, but the PHBV fibers need further modification which can make it serve as potential artificial skin scaffolds.

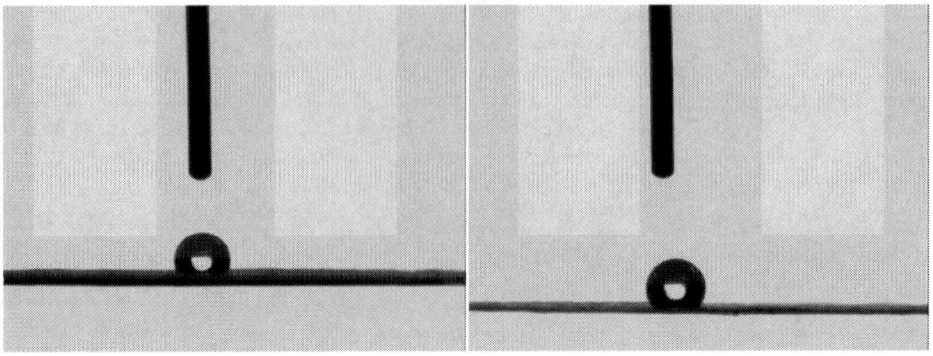

A B

Figure 5. The static water contact angles (WCA) of PHBV fibers (A) and solvent cast PHBV 2-D film (B).

4. Conclusions

Electrospun PHBV fibers were optimized to obtain defect-free morphology with an average fiber diameter of 1400 nm, and possess marked similarities to the natural ECM. Regarding the mechanical properties, the PHBV fibers presented excellent values (2.91 Mpa, 76 Mpa, 18.25%) with respect to the PHBV 2-D films (2.68 Mpa, 89 Mpa, 14.75%). On the ground of experimental fings, it was possible to conclude that the obtained PHBV fibers featured excellent properties with respect to the PHBV 2-D film. In particular, the SEM and WCA values made it possible to develop PHBV fibers with a potential skin tissue engineering application.

Acknowledgments

The authors would like to acknowledge the research center for analysis & measurement of Donghua University. This work is supported by Fundamental Research Program of Science and Technology Commission of Shanghai Municipality (13NM1401502).

References

[1] Pham Q P. Sharma U and Mikos A G 2006 *Tissue Eng.* **12** 1197
[2] Barnes C P, Sell S A, Boland E D, Simpson D G and Bowlin G L 2007 *Adv. Drug Deli Very Re V.* **59** 1413
[3] Venugopal J, Sharon L, Choon A T and Ramakrishna S J 2008 *Biomed Mater Res B Appl Biomater* **84B** 34-48
[4] Ma P X and Choi J W 2001 *Tissue Eng* **7** 23-33
[5] Zhang Y Z, Wang X, Feng Y and Lim C T 2006 Biomacromolecules **7** 1049–57
[6] Reneker D H and Chun I 1996 Nanotechnology **7** 216-23
[7] Mitomo H, BarhamP J and Keller A 1987 Polym. J. **19** 1241
[8] Lee I S, Kwon O H and Meng W 2004 *Macromolecular Research* **12** 3748
[9] Suwantong O, Waleetorncheepsawat S and Sanchavanakit N 2007 *International Journal of Biological Macromolecules* **40** 217-23
[10] Kuppan P, Vasanthan K S, Sundaramurthi D and Krishnan U M 2011 *Biomacromolecules* **12** 3156-65

[11] Meng W, Kim S, Yuan J, Kim J C, Kwon O H, Kawazoe N, Chen G, Ito Y and Kang I J 2007 *Biomater. Sci. Polym. Ed.* **18** 81-94
[12] Lee I S and Kwon O H 2004 *Macromol. Res.* **12** 374-8
[13] Sombatmankhong K, Suwantong O, Waleetorncheepsawat S, Supaphol P J 2006 *Polym. Sci. Part B: Polym. Phys.* **44** 2923-33
[14] Ndreu A, Nikkola L, Ylikauppila H, Ashammakhi N and Hasirci V 2008 Nanomedicine **3** 45-60
[15] He W, Ma Z W, Yong T, Teo W E and Ramakrishna S 2005 *Biomaterials* **26** 7606
[16] Furuhashi Y, Imamura Y, Jikihara Y and Yamane H 2004 *Polymer* **45** 5703-12
[17] Wang C, Hsu C H and Hwang I H 2008 *Polymer* **49** 4188-95
[18] Venugopal J, Ma L L and Ramakrishna S 2005 *Tissue Eng.* **11** 847-54

IOP Conf. Series: Materials Science and Engineering **87** (2015) 012048 doi:10.1088/1757-899X/87/1/012048

The preparation of lysine modified multi-walled carbon nanotubes and the study of its dispersion properties

Hongwei Lu, Liming Zou[1], Yizhe Wei, Xinlong Ling and Yongjing Xu

State Key Laboratory for Modification of Chemical Fibers and Polymer Materials, College of Materials Science and Engineering, Donghua University, 2999 North People Road, Shanghai 201620, People's Republic of China

E-mail: lmzou@dhu.edu.cn

Abstract. The poor dispersion in aqueous solution limited the application of carbon nanotubes (CNTs) in biological field. Here we used DCC/DMAP as Catalysis to prepare lysine modified multi-walled carbon nanotubes (MWNTs). FT-IR and TGA demonstrated that lysine have been successfully grafted to MWNTs, EA showed that lysine graft rate up to 23.4%. The dispersion of lysine modified MWNTs was investigated by direct visual inspection and microscope observation, the result showed that lysine modified MWNTs can be dispersed in aqueous solution and keep stable for long time.

1. Introduction

CNTs had drawn wide spread attention as soon as it was discovered by Iijima in 1991 [1]. Numerous investigation about CNTs have been made. Researchers focus on the structure and properties of CNTs at the initial stage. The unique structure of CNTs endue excellent properties such as high mechanical strength, high electrical conductivity and chemical & thermal stability. CNTs have been widely used in hydrogen storage materials [2], field emission devices [3], supercapacitors [4], bio-medical materials [5] and other fields.

However, the poor dispersion in solvents and aqueous solution limited the application of CNTs in many fields. CNTs are easily aggregated in solvents for the strong π-π stacking interactions between tubes, covalent modifications and noncovalent modifications were taken to solve these problems [6]. Although the noncovalent modification could not disrupted the structure of CNTs, the noncovalent interaction between wrapping molecules and the CNTs is not as strong as the covalent bond caused by covalent modification. Covalent modified CNTs could be dispersed in solvent more stability than noncovalent method [7]. The water-solubility and biocompatibility of CNTs improved a lot after attached by amino acid and peptide, which lead to the high potential application of CNTs in biomaterials [8].

In this paper, we grafted lysine to the surface of MWNTs by chemical function. Dicyclohexylcarbodiimide (DCC)/ 4-dimethylaminopyridine (DMAP) were used as coupling agents to

[1] Address for correspondence: Liming Zou, State Key Laboratory for Modification of Chemical Fibers and Polymer Materials, College of Materials Science and Engineering, Donghua University, 2999 North People Road, Shanghai 201620, People's Republic of China. E-mail: lmzou@dhu.edu.cn.

efficiently realize this reaction. Lysine modified MWNTs show excellent dispersion in water, in addition, the presence of amino improve reaction activity of CNTs. CNTs can be reacted with different functional groups to realized the applications in many fields.

2. Materials and method

2.1. Materials
The MWNTs (lengths 1.5 μm, diameters<10nm, purity 90%) were purchased from Shenzhen Nanotech Port Co., Ltd and used as received. Lysine, DCC, DMAP, N, N-dimethylformamide (DMF), Potassium permanganate ($KMnO_4$) were purchased from Sinopharm ChemicalReagent Co., Ltd. (China) and used as received. Sulfuric acid and hydrochloric acid were purchased from shanghai darui fine-chemical Co., Ltd. (China).

2.2. Characterization
Fourier transform infrared spectral (FT-IR) in the range of 400–4000 cm^{-1} were obtained by a Nicolet 8700 Fourier transform infrared spectrometer (Thermo Electron, American) with KBr pellets of MWNTs, and under the averaging of 32 scans at a resolution of 4 cm^{-1}. Thermogravimetric analyses (TGA) were performed by use of a TG 209 F1 Iris thermogravimetric analyzer (Netzsch-GeraetebauGmbH, Germany) under air flow. In each experiment, the MWNTs samples (5 mg) were heated from 30℃ to 900℃ at a rate of 20 K/min^{-1}. Elemental analyzes (EA) were got by a Vario EL III Elemental analyzer (Germany). Here we used as carrier gas, and the accuracy of result is (C, H, N) $\leq 0.1\%$. The dispersion of MWNTs were used a TK-C921EC metallurgical microscope (Japan, JVC) to take image at 40×10 shot.

2.3. Preparation of lysine modified MWNTs
MWNTs have been purified and carboxylation before we used [9]. 50 mg carboxylated MWNTs were added to a three-necked flask, 500 mg DCC and 10 mg DMAP followed, 40 ml DMF were used as solvent. The mixture were treated under a refluxed condenser for 5 h at 120℃. After filtration we got the Intermediates, DMF were used to wash it. Then we added 350 mg lysine and 40 ml DMF mixed with the Intermediates, treated under a refluxed condenser for 24 h at 120℃. Finally, we used ethanol and distilled H_2O rinsed the product and filtration for 3 times, and this lysine modified MWNTs were dried in a vacuum oven at 60℃ over night.

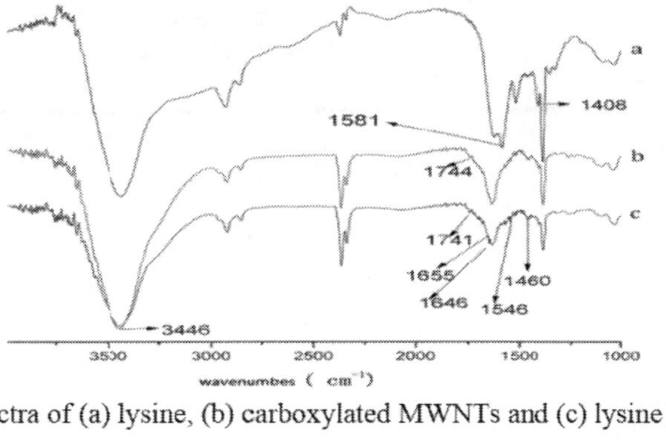

Figure 1. FT-IR spectra of (a) lysine, (b) carboxylated MWNTs and (c) lysine modified MWNTs.

3. Results and discussion

3.1. FT-IR spectra of lysine, carboxylated MWNTs and lysine modified MWNTs

IOP Conf. Series: Materials Science and Engineering **87** (2015) 012048 doi:10.1088/1757-899X/87/1/012048

The FT-IR spectra (figure 1) were used to characterize the functional groups of MWNTs before and after modification. There are double stretching vibration of backbone C=C at 1655 cm^{-1} and 1460 cm^{-1} both in the spectrum of carboxylated MWNTs and lysine modified MWNTs, and also a stretching vibration for C=O at 1740 cm^{-1}. The weak peaks at 1646 cm^{-1} and 1546 cm^{-1} are the vibration of amide I and amide II in the spectrum of lysine modified MWNTs, but there is no same peak appeared in the spectrum of carboxylated MWNTs. It means that lysine has been successfully grafted to the surface of MWNTs.

3.2. TGA of carboxylated MWNTs and lysine modified MWNTs
TGA is a thermal analysis that can be used to distinguish the component of products for the different thermal stability. Figure 2 shows the TGA of different products, we can see that lysine have a significant mass loss at the temperature from 200℃ to 500℃ (figure 2a), and this temperature range is the decomposition temperature of lysine. Lysine mixed with MWNTs or grafted to MWNTs both have a mass loss at the temperature from 200℃ to 500℃ (figure 2b). It can be used to prove that lysine has been grafted to MWNTs, and the different of mass loss due to the method lysine attached to MWNTs.

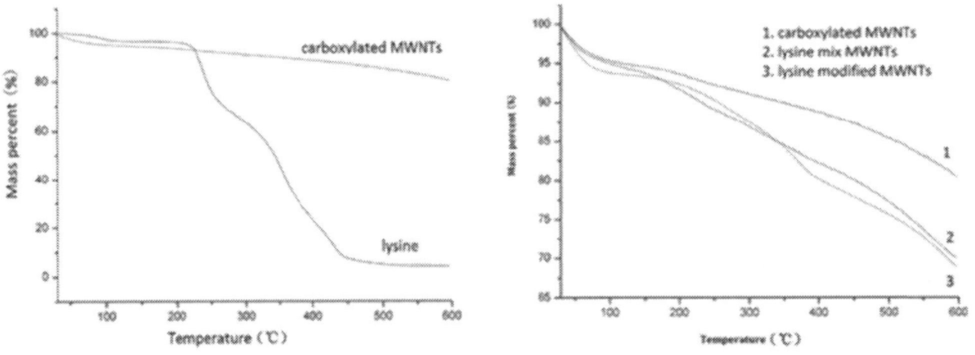

a b

Figure 2. TGA curves of lysine, carboxylated MWNTs, lysine mixed MWNTs, lysine modified WNTs.

3.3. EA of carboxylated MWNTs and lysine modified MWNTs
EA was used to analyze the element component of different kind MWNTs. There is no N in carboxylated MWNTs, but 4.5% N in Lysine modified MWCNTs (table 1), content of lysine which grafted to MWNTs was calculated to be 23.4% by calculating.

Table 1. C, H, N content of carboxylated MWNTs and lysine modified MWNTs.

	C (%)	H (%)	N (%)	Lysine (%)
carboxylated MWNTs	83.150	1.196	0.000	0.000
Lysine modified MWCNTs	75.520	3.077	4.515	23.381

3.4. Dispersion of Primitive MWNTs, carboxylated MWNTs and lysine modified MWNTs
Preparation solution of MWNTs, carboxylated MWNTs and lysine modified MWNTs at concentration of 0.1 mg/ml, ultrasonic 30 min, and settlement for 96 h.

Primitive MWNTs have fastest sedimentation rate, and lysine modified MWNTs keep good dispersion from the photo taken by digital camera (figure 3). All this due to that lysine have been grafted to MWNTs, it is a kind of water-soluble amino acid and contribute to improve the dispersion of MWNTs in water solution.

Figure 3. Digital photograph of different WWNTs solution taken by camera after settlement 96 h. MWNTs, (b) carboxylated MWNTs and (c) lysine modified MWNTs.

Image of microscope demonstrate the dispersion of MWNTs in the angle of microscopic (figure 4). After settlement 96 h, solutions of MWNTs have different degrees of aggregation and the size of aggregate reflect it. The image of different MWNTs solution taken by microscope at the same magnification (40×10) show that lysine modified MWNTs solution has the smallest size, followed by carboxylated MWNTs solution, then solution of MWNTs. The images taken by microscope in keeping with the photographs taken by camera, all this show that dispersion of MWNTs in aqueous solution improved after modified by lysine.

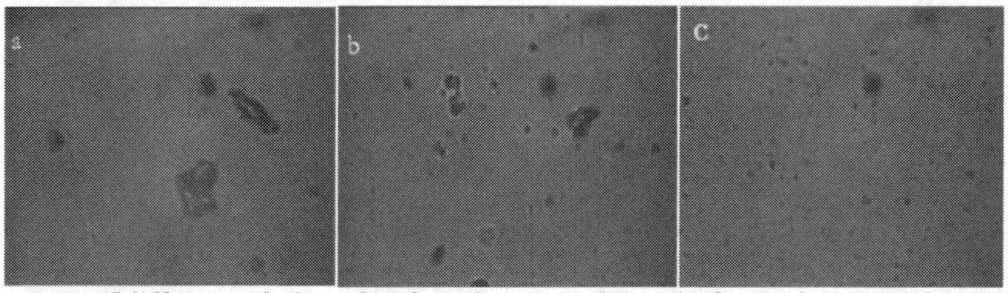

Figure 4. Image of different solution taken by microscope (40×10) after settlement 96 h. MWNTs, (b) carboxylated MWNTs and (c) lysine modified MWNTs.

4. Conclusions

DCC/DMAP Catalysis is an effective coupling agents to prepare lysine modified MWNTs, FT-IR and TGA demonstrated that lysine have been grafted to MWNTs, and the EA result Showed that lysine graft rate reach up to 23.4%. After modification MWNTs show excellent dispersion in water solution which made MWNTs have promising prospects in biological area.

Acknowledgments

The authors would like to acknowledge the research center for analysis & measurement of Donghua University. This work is supported by Fundamental Research Program of Science and Technology Commission of Shanghai Municipality (13NM1401502) and Innovation Program of Shanghai Municipal Education Commission (13ZZ052).

References

[1] Iijima S 1991 *Nature* **354** 56-8
[2] Pandyan R K, Seenithurai S, Kumar S V, et al. 2015 *Fullerenes, Nanotubes and Carbon Nanostructures* **23** 175-80
[3] Xu J Z, Xu P, Wei Q Y, et al. 2015 *Applied Physics Letters* **106** 073501-5
[4] Evan W and Islam M F 2015 *ACS Applied Materials & Interfaces* **7** 5612-8
[5] Singh S, Kumar A, Khare S, et al. 2014 *Applied Physics Letters* **105** 213701-4
[6] Dimitrios T, Nikos T, Alberto B, et al. 2006 *Chemical Reviews* **106** 1105-36
[7] Hu N T, Dang G D, Zhou H W, et al. 2007 *Materials Letters* **61** 5285-7.

[8] Marchesan S and Prato M 2015 *Chemical Communications* **51** 4347-59.
[9] Ling X L, Wei Y Z, Zou L M, et al. 2013 *Applied Surface Science* **276** 159-66

Improvement of toughness and water resistance of bioplastic based on wheat gluten using epoxidized natural rubber

S Hemsri[1], C Thongpin, P Somkid, S Sae-arma and A Paiykaew

Department of Materials Science and Engineering, Faculty of Engineering and Industrial Technology, Silpakorn University, Nakhon Pathom 73000, Thailand

E-mail: sudsirihemsri@yahoo.com

Abstract. Novel blends based on wheat gluten (WG) and epoxidized natural rubber (ENR) were fabricated with different ENR contents of 10, 20 and 30 wt% in an internal mixer. Sulfur vulcanization was used to crosslink the ENR phase in the blends. Comparatively, blends of WG and natural rubber (WG/NR) were prepared in the same condition as the WG/ENR blends. Tensile mechanical properties and impact strength of the WG/ENR blends were investigated and compared with the WG/NR blends as well as pure WG. Moreover, water absorption of pure WG and the WG/ENR blends was also tested. As investigated by scanning electron microscopy (SEM), the results revealed more compatibility between WG and ENR compared with NR. The elongation at break, impact strength and water resistance of the WG/ENR blends were found to remarkably increase with respect to the pure WG. Thus, incorporation of ENR into WG could improve toughness and water resistance of WG. Furthermore, the effect of adding glycerol acting as a plasticizer on the mechanical properties and impact strength of the WG/ENR blends was also studied. The blends with glycerol-plasticized WG (WG-Gly/ENR) showed more homogeneous morphologies and superior results in the mechanical properties and impact strength compared with the WG/ENR blends.

1. Introduction

During the past years, bio-based polymers from renewable natural resources have attracted much attention for academia and industry because of increasing environmental problems. Plant protein-based biopolymers can serve as an alternative to conventional petroleum-based plastics because of their low cost, large-scale availability, nontoxic, and environmentally friendly properties. Compared with other plant proteins, the wheat gluten (WG) has a relative cheap cost, good stability to heat, strong tensile strength and excellent barrier properties [1-3]. However, major drawbacks of WG-based plastics to prevent them from a wide range of applications are their inherent brittleness and high water absorption after being processed. This is due to high degree of the three-dimensional protein network formed by heat-induced covalent crosslinking through disulfide bonds (S-S) of thiol groups (-SH) of cysteine residues in WG peptides and due to hydrophilic nature (polar amino acids) of the protein [4]. To improve the mechanical properties and water resistance of WG plastics, there are various approaches to expand its usefulness in an even wider variety of applications. Plasticizers are widely

[1] Address for correspondence: S Hemsri, Department of Materials Science and Engineering, Faculty of Engineering and Industrial Technology, Silpakorn University, Nakhon Pathom 73000, Thailand. E-mail: sudsirihemsri@yahoo.com.

used to add into WG for improving flexibility of the WG plastics. Such plasticizers include glycerol, sorbitol, sucrose, diethanolamine and triethanolamine, and saturated fatty acid [5-8]. However, these plasticizers do not improve the water resistance of the WG, except the later plasticizer. Blending WG with other polymers is one practical way to achieve new polymeric materials with tailor-made properties. Many polymers have been reported to blend with WG in order improve mechanical properties of the protein, for instance, aliphatic polyester [9], poly (hydroxyl ester ether) [10], polycaprolactone [11] and maleic anhydride-modified polycaprolactone [12], poly (vinyl alcohol) [2, 13] and thiolated (vinyl alcohol) [14] and cassava starch [15].

Epoxidized natural rubber (ENR) is a commercial modified natural rubber which is prepared by reacting natural rubber (NR) latex with peroxy acid. ENR possesses excellent elastic properties and toughness, high strength, oil resistance and adhesive properties, and a high degree of damping [16]. Epoxy groups in ENR would improve the polarity of rubber, thereby promoting compatibility with polar polymers in polymer blends. ENR is a promising material acting as toughening agent for many polymers. Some studies on blending of ENR to improve flexibility and impact strength have been reported with polystyrene (PS) [17], poly (vinyl chloride) (PVC) [18], nylon 6 [19], poly (ethylene-co-acrylic acid) [20] and poly (lactic acid) (PLA) [21].

As noted above, many studies have reported blending WG with various thermoplastic polymers. However, based on our knowledge, there have been no reports on blending ENR elastomer into WG. Thus, the objective of this research is to fabricate a new WG/ENR blend which is expected to exhibit good flexibility, impact strength and water resistance with respect to pure WG plastic. In this work, mechanical properties, including tensile mechanical properties and impact strength, as well as water absorption of the blends were investigated. Comparison between WG/NR and WG/ENR blends was also described in terms of mechanical properties and morphology. Moreover, the effect of plasticization of WG by using glycerol as a plasticizer on mechanical properties, morphology and water resistance in the WG/ENR polymer blends is discussed in detail.

2. Experimental

2.1. Materials

Wheat gluten (WG) was supplied by Zhangjiagang HengFeng Starch Products Co. Ltd., China. Natural rubber (NR) of STR 5L grade (Standard Thai Rubber 5 L) was purchased from PI Industry Co. Ltd., Thailand. Epoxidized natural rubber containing 50 mol% of epoxidation (ENR50) was produced by Muang Mai Guthrie Public Co. Ltd., Thailand. Glycerol (Gly) used as a plasticizer, was obtained from Fisher Scientific. In this work, conventional vulcanization (CV) was used to vulcanize ENR phase to enhance tensile strength properties of polymer blends. All chemicals for the conventional vulcanization in ENR including zinc oxide (ZnO), stearic acid (SA), tetramethylthiuram disulfide (TMTD), *N*-tert-butyl-2-benzothiazolesulfenamide (TBBS) and sulfur were used as received. Zinc oxide and stearic acid were used as activators, TMTD and TBBS were added as accelerators, and sulfur was used as a curing agent to vulcanize the rubber phase.

2.2. Preparation of WG/ENR blends

WG was dried in an oven at 80 °C for 24 hr to remove moisture before use. ENR was masticated in an internal mixer (Brabender W50EHT, Germany) at 70°C for 15 min with a rotating speed of 70 rpm. WG and the masticated ENR with various ratios of 100/0, 90/10, 80/20 and 70/30 by weight were mixed together in the internal mixer at 70°C with a rotating speed of 70 rpm for 10 min to prepare premixed blends. To prepare pre-vulcanized WG/ENR blends, all chemicals for curing ENR including zinc oxide (5 phr), stearic acid (1.5 phr), TMTD (0.4 phr), TBBS (0.6 phr) and sulfur (2.5 phr) were added sequentially into the premixed WG/ENR blends for 5 min for each chemical Moreover, to compare with the WG/ENR blends, the WG/NR blends were prepared at the same condition as described above. In order to study effect of a plasticizer on mechanical properties of the WG/ENR blends, a 20% of glycerol by weight based on WG was homogeneously mixed with WG for 5 min in

Philips Blender HR2115 to obtain glycerol-plasticized WG (WG-Gly) first. The WG-Gly was mixed with the masticated ENR to prepare the premixed WG-Gly/ENR blends and then followed by adding the curative agents into the WG-Gly/ENR blends at the same condition as described above. All blends were compression- molded at 150°C for 5 min at pressure of 1,500 psi by using LabTech compression molding machine to prepare sample specimens for mechanical property and water absorption testing.

2.3. Mechanical property testing

2.3.1. Tensile testing. A universal testing machine (Instron 5969) was used to investigate tensile properties according to ASTM D638-10 type IV. The tensile tests were conducted under ambient condition with a 5 kN load cell and the crosshead speed of 50 mm/min. At least ten specimens were performed for each blend.

2.3.2. Izod impact testing. Izod impact testing, according to ASTM D256-10, was performed on notched impact specimens by using a pendulum impact tester (Zwick model B5102.202 Izod Pendulum 4 J). For each blend, fifteen specimens were tested and the average value of impact energy in kJ/m^2 was recorded.

2.4. Morphology

Morphology studies of WG and the WG/ENR blends with different ratios were carried out using a scanning electron microscope (SEM; Hitachi Model S3400N instrument). Cryogenic fractured surface of the polymer blends were used to study the morphology of the blends coated with platinum.

2.5. Water absorption testing

All samples of WG and WG/ENR as well as WG-Gly/ENR blends with dimensions of 64.0 x 12.60 x 2.00 mm were dried in an oven at 80°C for 24 hr to remove moisture in the specimens and the original weight of the samples was recorded. Then, the specimens were horizontally submerged in deionized water at a room temperature. At certain times of 5, 15, 25, 35, 50, 60, 120, 180, 240, 720, 1440 and 2880 min, the specimens were taken out from water. The excess surface water was removed with a tissue and weight of the specimen was immediately measured as a function of time. Three replicates were measured for each material at each time point. The percentage of water absorption (%WA) was calculated as shown in equation (1).

$$\% \text{ WA} = [(Ws-W_0)/W_0] \times 100 \quad (1)$$

where Ws and W_0 are weight of a specimen after submersion and before submersion, respectively.

3. Results and discussion

Generally, NR and ENR must be vulcanized to provide the better physical and mechanical properties for engineering application. From this reason, in this work, vulcanized NR and ENR by sulfur was used to improve tensile properties. Typically, rubbers can be vulcanized by two systems: sulfur-cured and peroxide-cured systems. The sulfur vulcanization provides high chain flexibility of the polymer network in the rubber phase due to formation of the mono, di and polysulphidic linkages (C-S-C, C-S-S-C and C-Sx-C, respectively), which more stable than rigid C-C linkages in the peroxide vulcanization [18]. Moreover, in this work, the temperature used for mixing all vulcanizing chemicals and the premixed WG/ENR or WG/NR blends to prepare the pre-vulcanized blends in the internal mixer was 70°C. At this temperature, which is lower than a range of use temperature of sulfur to vulcanize the rubber (120-150°C), possible crosslinking accidents of ENR or NR phase in the blends is avoided during the mixing step.

3.1. Comparison of WG/NR and WG/ENR blends

3.1.1. Morphology of the WG/NR and WG/ENR blends. Scanning electron micrographs of cryogenic fractured surfaces of the WG/NR and WG/ENR blends with ratios of 90/10 and 70/30 are given in figure 1. Regardless of the rubber types, obvious agglomeration of WG on the fractured surfaces and a continuous structure of rubber phase were observed in all blends (figures 1a-1d). Typically, in polymer blends containing rubber, the rubber phase is in form dispersed particles in a matrix after vulcanization. However, the rubber component in our blends was a continuous phase. This observation is similar to PLA/NR blends reported by Yuan et al [22]. This might be because, in fact, the WG mixed into the rubber phase was in powder form. The WG could not melt during the mixing step in the internal mixer, thus it formed agglomeration dispersed in the blends. Compared to the WG/ENR blends (figure 1b), the WG/NR blend with 10 wt% content of NR (figure 1a) showed clearly an appearance of a large number of cracks as well as holes between WG aggregates and NR phase. When NR content in the blends increased (30 wt% content of NR) as shown in figure 1c, some cracks and cavities between WG agglomerations and NR domain were observed. Comparatively, in the 70/30 WG/ENR blend (figure 1d) the size of WG aggregates became smaller than that of the 70/30 WG/NR blend as presented in figure 1c. Moreover, in the 70/30 WG/ENR blend, the ENR domain appeared smooth, while the rough NR domain on the fracture surface was observed in the corresponding WG/NR blend. These results indicated a better compatible morphology as well as a better interfacial adhesion in the WG/ENR blends relative to the WG/NR blends. This might be because the ENR is polar rubber due to a presence of epoxy groups in its structure, while the NR is non-polarity. Hence, blending polar wheat protein and non-polar NR led to incompatible blend due to a great difference in polarity of such polymers. In contrast, in the case of the WG/ENR blend, there is similarity in polarity between WG and ENR and there might be molecular interaction between epoxy groups of ENR and polar amino acids in WG, resulting in an enhancement in compatibility between WG and ENR.

(a) WG/NR (90/10)	(b) WG/ENR (90/10)	(c) WG/NR (70/30)	(d) WG/ENR (70/30)

Figure 1. SEM images of cryogenic fractured surfaces: (a) WG/NR (90/10), (b) WG/ENR (90/10), (c) WG/NR (70/30) and (d) WG/ENR (70/30) (x200 magnification).

3.1.2. Mechanical Properties of the WG/NR and WG/ENR blends. Tensile mechanical properties and impact strength of the WG/NR and WG/ENR blends prepared under the same condition are summarized in table 1 Obviously, the addition of either NR or ENR into WG decreased modulus and maximum tensile strength of WG due to a result of the low modulus of the rubbers. In contrast, elongation at break and impact strength of both blends were higher than those of pure WG. Considering the effect of weight percent of the rubbers on the mechanical properties of the blends, it can be seen that modulus and tensile strength continuously decreased but elongation and impact strength values significantly increased as the rubber content increased. However, the WG/ENR blends provided superior results for all tensile mechanical properties and impact strength compared to the WG/NR blends. This is attributed to more compatibility of WG and ENR as illustrated in the previous SEM images in figure 1.

As shown in table 1, incorporating the ENR component into the brittle WG dramatically enhanced elongation at break and also impact strength of the wheat protein. For instance, the WG/ENR blend at ratio of 80/20 provided remarkable improvement of elongation at break from 1.76 to 20.30% (around 12 times) and impact strength from 0.47 to 5.15 kJ/m^2 (around 11 times) with respect to pure WG.

Table 1. Tensile properties and impact strength of WG/NR and WG/ENR blends.

Polymer blends	Ratio (wt/wt)	Modulus (MPa)	Maximum Tensile strength (MPa)	Elongation at break (%)	Impact strength (kJ/m^2)
WG/NR	100/0	1668.20 ±142.15	37.70 ± 3.54	1.76 ± 0.20	0.47 ± 0.13
	90/10	415.32 ± 76.10	3.48 ± 0.65	1.08 ± 0.26	1.07 ± 0.29
	80/20	114.31 ± 2.29	2.63 ± 0.41	4.98 ± 0.57	2.95 ± 0.88
	70/30	24.87 ± 2.76	1.92 ± 0.34	18.38 ± 3.68	5.04 ± 0.54
WG/ENR	100/0	1668.20 ±142.15	37.70 ± 3.54	1.76 ± 0.20	0.47 ± 0.13
	90/10	792.16 ± 42.29	8.83 ± 0.97	1.90 ± 0.19	1.88 ± 0.34
	80/20	286.40 ± 37.95	6.71 ± 0.55	20.30 ± 2.31	5.15 ± 0.21
	70/30	49.17 ± 8.74	4.56 ± 0.63	41.10 ± 3.97	8.34 ± 0.65

3.2. Comparison of WG/ENR and WG-Gly/ENR blends

3.2.1. Morphology of the WG/ENR and WG-Gly/ENR blends. Figure 2 illustrates the SEM images of scanning electron micrographs of cryogenic fractured surfaces of the WG/ENR and WG-Gly/ENR blends with ratios of 90/10, 80/20 and 70/30.

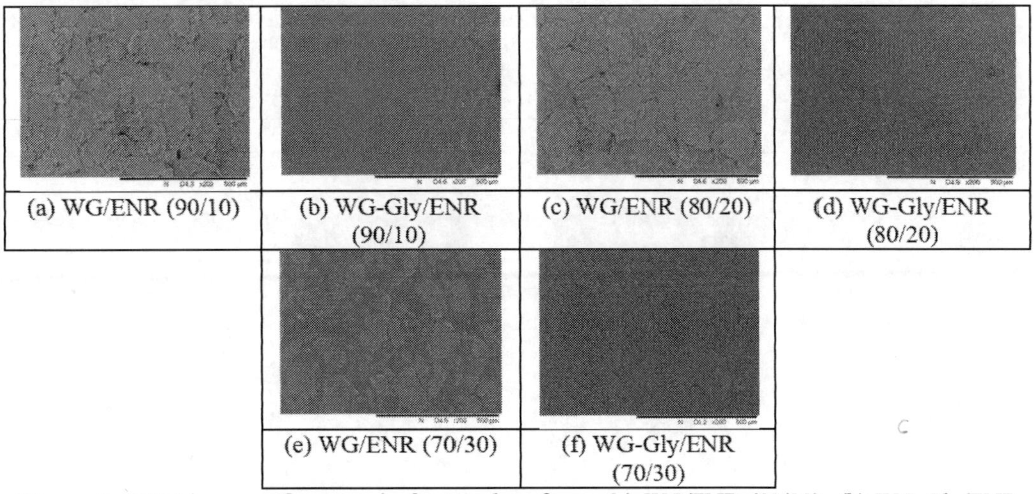

Figure 2. SEM images of cryogenic fractured surfaces: (a) WG/ENR (90/10), (b) WG-Gly/ENR (90/10), (c) WG/ENR (80/20), (d) WG-Gly/ENR (80/20), (e) WG/ENR (70/30) and (f) WG-Gly/ENR (70/30) (x2 00magnification).

Interestingly, it can see clearly that in the blends prepared from plasticized WG as shown in figures 2b, 2d and 2f, there is no appearance of large WG agglomeration compared to the blends prepared from WG without plasticization. The WG-Gly/ENR blend with ratio of 90/10 showed smooth surface whereas rough surfaces were observed in the blends with ratios of 80/20 and 70/30. No appearance of the large WG aggregates in the WG-Gly/ENR blends is because incorporation of glycerol acting as a plasticizer into WG led to not only a decrease in intermolecular forces, especially hydrogen bonding and electrostatic interactions, between protein chains but also an increase in polymer mobility in WG [23, 24], and thereby providing the better dispersion of WG in the ENR phase during the blending preparation. Therefore, plasticization of WG with glycerol is necessary in order to reduce size of WG aggregates in the blends, resulting in providing more homogeneous blends with respect to WG/ENR blends without plasticization.

3.2.2. Mechanical Properties of the WG/ENR and WG-Gly/ENR blends. The tensile modulus, maximum tensile strength, elongation at break and notch Izod impact strength of pure WG, plasticized WG (WG-Gly) and all blends including the WG/ENR and WG-Gly/ENR blends with various ratios is shown in table 2. As expected, in the WG-Gly, the modulus and maximum tensile strength reduced while the elongation at break and impact strength increased with respect to pure WG. The reduction of the modulus and tensile strength in WG-Gly can be attributed to the fact is that glycerol acting as a plasticizer decreases the glass transition temperature (T_g) of WG, resulting in higher chain mobility of WG. Comparatively, it was found that the all tensile mechanical properties and impact strength of the WG-Gly/ENR blends were higher than of the WG/ENR blends. For instance, the tensile strength, elongation and impact strength of the WG-Gly/ENR blends at low ENR content of 10 wt% increased around 2, 4 and 2.5 times, respectively with respect to the WG/ENR blends at the same rubber content. This result indicated that the plasticization of WG by glycerol significantly enhanced the mechanical and impact properties of the WG-Gly/ENR blends compared to the blends without plasticization. It was well known the mechanical properties of polymer blends depend largely on the resulting the phase structure and morphologies of such blends. Hence, the improvement of the mechanical and impact properties of the WG-Gly/ENR blends resulting from size reduction of WG aggregates in the blends due to the plasticization effect of glycerol in WG as shown in the aforementioned SEM micrographs in figure 2.

Table 2. Tensile properties and impact strength of WG/ENR and WG-Gly/ENR blends.

Polymer blends	Ratio (wt/wt)	Modulus (MPa)	Maximum Tensile strength (MPa)	Elongation at break (%)	Impact strength (kJ/m^2)
WG/ENR	100/0	1668.20 ±142.15	37.70 ± 3.54	1.76 ± 0.20	0.47 ± 0.13
	90/10	792.16 ± 42.29	8.83 ± 0.97	1.90 ± 0.19	1.88 ± 0.34
	80/20	286.40 ± 37.95	6.71 ± 0.55	20.30 ± 2.31	5.15 ± 0.21
	70/30	49.17 ± 8.74	4.56 ± 0.63	41.10 ± 3.97	8.34 ± 0.65
WG-Gly/ENR	100/0	601.20 ± 48.34	20.92 ± 4.46	3.68 ± 0.34	1.54 ± 0.69
	90/10	508.82 ± 12.31	18.10 ± 1.24	7.91 ± 1.30	4.68 ± 0.23
	80/20	428.40 ± 29.20	14.61 ± 0.87	13.9 ± 1.52	10.53 ± 0.87
	70/30	96.25 ± 15.48	9.54 ± 0.99	29.6 ± 3.04	15.04 ± 0.59

Figure 3 shows typical stress-strain curves of pure WG, glycerol-plasticized WG (WG-Gly) and the WG-Gly/ENR blends with different ENR contents. Obviously, compared to the WG-Gly/ENR blends, the WG-Gly showed more brittle fracture with high modulus and tensile strength but low elongation. In contrast, when the ENR was added into the plasticized WG, more ductility was observed. Considering areas under the stress-strain curves of all samples, it was found that the area under the curve, correlated to toughness of the sample increased as the concentration of ENR in the blends, indicating that incorporation of elastomeric ENR into WG remarkably enhanced toughness in the polymer blends.

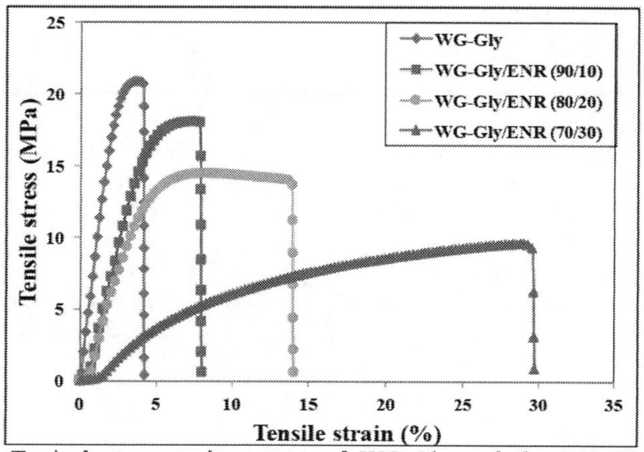

Figure 3. Typical stress-strain curves of WG-Gly and the WG-Gly/ENR blends with different ENR contents.

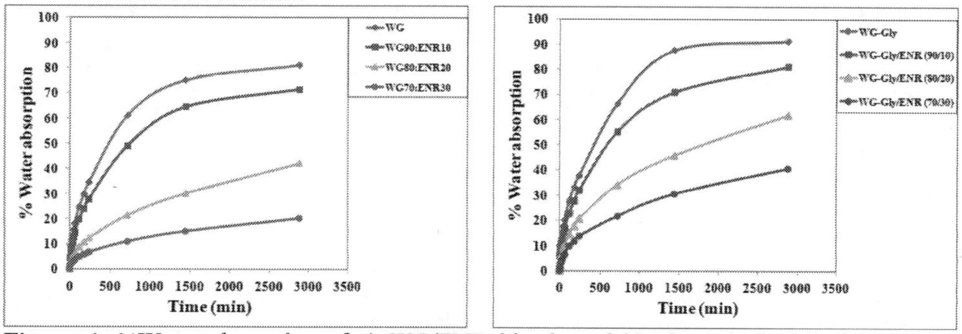

Figure 4. %Water absorption of a) WG/ENR blends and b) glycerol-plasticized WG/ENR blends.

3.3. Water absorption property

Water absorption of polymer blends is one of important properties for evaluating their stability on water affecting their long-term performance. The result of the percent water absorption (%WA) data of pure WG and the WG/ENR blends is illustrated in figure 4a, while %WA data of glycerol-plasticized WG (WG-Gly) and glycerol-plasticized blends (WG-Gly/ENR) is presented in figure 4b. From figure 4a, it can be clearly seen that pure WG absorbed much more water than did the WG/ENR blends. During 2 days of the testing, WG absorbed 80% of its original weight in water, indicating very poor water resistance of the WG plastic whereas %WAs of the blends were lower than 70%. Thus, incorporation of ENR rubber into WG significantly improved the water resistance of the blends compared with that of the pure WG because ENR is the hydrophobic natural rubber, while WG has hydrophilic nature due to polar amino acids in WG structure, resulting in a high degree of water absorption. Considering the WG/ENR blends, %WAs of the blends with ratios of 90/10, 80/20 and 70/30 were around 70, 40 and 20%, respectively. Thus, the water resistance of the blends increased with an increase of rubber content in the blends due to a presence of higher non-polar rubber content in the blends, thereby reducing water diffusion in the blends. In the case of the WG-Gly/ENR blends (figure 4b), a decrease in the water absorption as the rubber content was observed, similar to the result of the WG/ENR blends. For comparison of %WA of the WG/ENR and the WG-Gly/ENR blends, the WG-Gly/ENR blends provided higher water absorption with respect to the WG/ENR blend without the plasticizer. This might be because hydrophilic hydroxyl groups in glycerol molecules can absorb water, resulting in a high degree of water absorption in the plasticized WG-Gly/ENR blends.

4. Conclusions

The WG/ENR and WG/NR blends with various weight ratios of 90/10, 80/20 and 70/30 were fabricated in this work. The ENR provided more compatibility with WG than did NR as shown in the SEM results. More compatibility of the WG/ENR blends led to higher mechanical properties and impact strength of with respect to the WG/NR blends. Moreover, the influence of addition of ENR on mechanical and impact properties, morphologies and water absorption of the blends were studied compared with the pure WG. Incorporation of ENR tremendously enhanced the elongation at break, impact strength and water resistance of the blends compared with the pure WG. On the other hand, modulus and maximum tensile strength decreased as an increase in ENR content. Hence, addition of ENR into WG led to an enhancement of the flexibility and water resistance in the blends with respect to pure WG. The effect of adding glycerol plasticizer to WG on the tensile mechanical properties, toughness and water absorption were also investigated with respect to the blends without plasticization. From the SEM micrographs, compared to the WG/ENR blends, it was found that the aggregate size of WG in the blends dramatically decreased and more homogeneous morphologies was observed in the WG-Gly/ENR blends. This is due to a plasticization effect, which reduced intermolecular forces, such as hydrogen bonding and electrostatic interactions between the protein chains. From this result, the mechanical properties and toughness of the WG-Gly/ENR blends were higher than those of the WG/ENR blends. However, the WG-Gly/ENR blends showed lower the water resistance than the blends without glycerol plasticization due to a presence of hydrophilic glycerol in the blends.

Acknowledgments

This research was financially supported by Faculty of Engineering and Industrial Technology, Silpakorn University, Thailand. The author would like to thank Mr. Punnakit Nakkharat and for testing mechanical properties and Mr. Nontawat Kraitape for the SEM assistance.

References

[1] Bietz J A and Lookhart G L 1996 *Cereal Foods World* **41** 376
[2] Zhang X, Burgar I, Lourbakos E and Beh H 2004 *Polymer* **45** 3305
[3] Reddy N and Yang Y 2007 *Biomacromolecules* **8** 638
[4] Lagrain B, Thewissen B G, Brijis K and Delcour J A 2008 *Food Chem* **107** 753
[5] Redl A, Morel M H, Bonicel J, Vergnes B and Guilbert S 1999 *Cereal Chem.***76** 361
[6] Cherian G, Gennadios A, Weller C and Chinachoti P 1995 *Cereal Chem.* **72** 1
[7] Irissin-Mangata J, Bauduin G, Boutevin B and Gontard N 2001 *Eur. Polym. J.* **37** 1533
[8] Pommet M, Redl A, Morel M H and Guilbert S 2003 *Polymer* **44** 115
[9] Lim S W, Jung I K, Lee K H and Jin B S 1999 *Euro. Polym. J.* **35** 1875
[10] Wang C, Carriere C J and Willett J L 2002 *J. Polym. Sci. Part B: Polym. Phys.* **40** 2324
[11] Finkenstadt V L, Mohamed A A, Biresaw G and Willett J L 2008 *J. Appl. Polym. Sci.* **110** 2218
[12] John J, Tang J and Bhattacharya M 1998 *Polymer* **39** 2883
[13] Dicharry R M, Ye P, Saha G, Waxman E, Asandei A D and Parnas R S 2006 *Biomacromolecules* **7** 2837
[14] Dong J, Dicharry R, Waxman E, Parnas R S and Asandei A D 2008 *Biomacromolecules* **9** 568
[15] Hashimoto J M, Nabeshima E H, Cintra H S, Dias A R G, Bustos F M and Chang Y K 2002 *J. Sci. Food Agric.* **82** 924
[16] Gelling I R 1991 *J. Nat. Rubb. Res.* **6** 184
[17] Asaletha R, Kumaran M G and Thomas S 1999 *Eur. Polym. J.* **35** 253
[18] Jeerupun J, Wotthikanokkhan J and Phinyocheep P *2004 Macromol. Symp.* **219** 281

[19] Tanrattanakul V, Sungthong N and Raksa P 2008 *Polym. Testing* **27** 794
[20] Mohanty S, Nando G B, Vijayan K and Neelakanthan N R 1996 *Polymer* **37** 5387
[21] Zhang C, Wang W, Huang Y, Pan Y, Jiang L, Dan Y and Luo Y 2013 *Mater. Des.* **45** 198
[22] Yuan D, Chen K, Xu C, Chen Z and Chen Y 2014 *Carbohydr. Polym.* **113** 438
[23] Zhang H and Mittal G 2010 *Environ. Prog. Sustain. Energy* **29** 203
[24] Mekonnen T, Mussone P, Khalilb H and Bressler D 2013 *J. Mater. Chem. A* **1** 13379

The electronic structure and optical properties of Lu-doped ZnO nanotubes

Y Wu and S K Zheng[1]

College of Electronic and Informational Engineering, Hebei University, 071002, Baoding, China

E-mail: zhshk@126.com

Abstract. The electronic band structures, density of states, and optical properities of pure and Lu-doped zigzag single-walled (6, 0) ZnO nanotubes (SWZONT) were studied using the density functional method whithin generalized gradient approximation. For the pure SWZONT system, the Fermi level locates at the valence band maxmum, while it deviates to the conduction band, and exhibits metal-like charicteristics after the Lu atom is introduced into the system. The Lu-doped SWZONT system is stable, and the semiconductor type (N type) is enhanced. Additionally, the catalytic oxidation activity was improved. The optical properties indicate that the Lu-doped SWZONT materials can be applied to UV optoelectronic devices.

1. Introduction

Nano-materials (particularly the low-dimensional semiconductor materials) are booming in nanosenser, electrochemical capacitance storage, luminescent devices and many other fields [1-3]. After Iijima found Carbon nanotubes in 1991, research regarding nanotubes became competitive. In addition to Carbon nanotubes, ZnO nanotubes also have attracted the attention of scientific reports [4, 5]. Compared to other low-dimensional nanostructures of ZnO in the form of nanocrystalline, nanowire, nano-film and so on, ZnO nanotubes have a unique quantum structure and novel optoelectronic properties. Specifically, ZnO nanotubes possess the necessary porosity and high surface-to-volume ratio. These qualities make ZnO nanotubes the best choice for hydrogen storage devices, photovoltaic cells, electrochemical supercapacitors, and many other applications.

ZnO nanotubes that were initially reported are polycrystalline, or boast an amorphous structure. In recent years, more methods have been used for synthesizing ZnO nanotubes. Wu [6] synthesized ZnO nanotubes with the pyrolysis of zinc acetylacetonate method. Zhang [7] prepared aligned ZnO nanotubes by means of metal organic deposition, using a sapphire substrate. In his study, they found that only by changing the growth pressure (from 10 Torr down to 0.06 Torr), at low temperature (350-500°C) and under the condition without catalyst, could they secure modulate nanostructures forms, such as nano rods, nanotubes, or nano walls. These nanostructures are epitaxial growth, and have the same epitaxial growth connection with basement. The growth pressure of the nanotubes is described as P_v= 1-0.3Torr. Respectively, diameter and wall thickness of nanotube are about 500 nm and 100 nm.

Additionally, many methods regarding the preparation of ZnO nanotubes have been attempted, including thermal reduction [8], vapor phase growth [9], hydrothermal growth [10], gas-solid

[1] Address for correspondence: S K Zheng, College of Electronic and Informational Engineering, Hebei University, 071002, Baoding, China. E-mail: zhshk@126.com.

processes [11], sol-gel processes [12], Plasma aided molecular beam epitaxy growth [13], $Zn(NH_3)_4^{2+}$ precursor body heat decomposition [14], etc. All of the ZnO nanotubes gain a large value in regards to both diameter and thickness. Until now, a vast amount of effort was focused on the ultra-thin size of ZnO nanotubes. As for the rationality of their existence, and the possibility of the manufacturing of ZnO single-walled nanotubes, many theoretical explorations have been tested.

Single-walled ZnO nanotubes (SWZONTs) have not been experimentally reported, making theoretical analysis necessary. A vast number of theoretical simulation studies have predicted the feasibility of synthesis SWZONTs. Shen [15] calculated the energy of minute ZnO nanowires and single-walled nanotubes with the same number of atoms, using the first principles method. With the same size, the author discovered that a ZnO single wall nanotube is more stable than nanowires. Tu [16] introduced several characteristic vectors similar to single-walled carbon nanotubes. Chiral vector, C, is defined as two corresponding positions, with a pair of parameters (n, m), known as the index. However, the quantum size effect of SWZONTs [17] limits the energy band structure of ZnO nanotubes, and it is difficult to determine their exact potential optical and electrical properties.

Additionally, doping can alter the band structure and properties of the materials. This became a contraversial issue in the field of materials research, and provides a new, effective way for the potential performance of SWZONTs' development. It is reported that rare earth elements are widely used as dopant because of their 4f shell [18-19].

Pan [20] and An [21] calculated the electronic structure of ZnO nanotubes (6, 0) using first principles of the generalized gradient approximation. Although they gained different energy gaps, ZnO nanotubes were confirmed as a direct band gap semiconductor material. This finding was considered a big breakthrough. Combined, rare earth ions and semiconductor nanocrystals, and efficient light-emitting devices were formed in 2011 [22]. However, related theoretical research regarding doped SWZONTs is remains relatively small, and the rare earth element doped zigzag single-walled ZnO nanotubes of theoretical research has never been reported.

So, for this paper, we calculated the band structures, density of states, and optical absorption spectra of Lu doped SWZONTs and pure SWZONTs using first-principles based on density functional theory (DFT). By comparing the electronic structures' differences between the pure and doped system, we bring forward instructional advice about SWZONTs.

2. Material and methods

Firstly, we created an infinite zigzag single-walled (6, 0) C nanotube. Then, the C elements were replaced with Zn and O side-by-side. The bond length was set as 0.190 nm. Finally, the model of single-walled (6, 0) ZnO nanotube was formed. All of the theoretical calculations were completed at the a×a×c orthogonal supercell zone, which leads to the extension of the nanotubes along the direction of c. The pure SWZONT is generated from a Carbon nanotube crystal structure, which has the lattice parameters, as a=1.926 nm, c =1.710 nm, $\alpha=\beta=90^\circ$, $\gamma=120^\circ$. In order to have a significant understanding regarding the influences with Lu doped, one Zn atom is replaced by Lu. The zigzag single-walled (6, 0) ZnO nanotube is shown in figure 1.

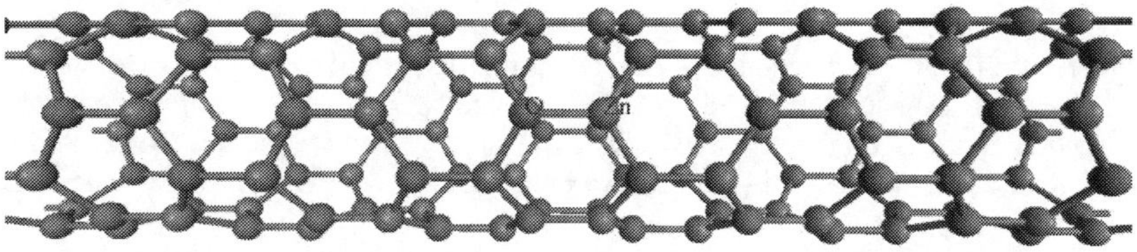

Figure 1.The Model of Pure Single-Walled (6, 0) ZnO Nanotube

The calculations are performed according to the density function theory (DFT), which is based on pseudo-potential technology. The interaction between ion core and valence electrons is described by ultrasoft pseudopotential. The Perdew-Burke-Ernzerh (PBE) of generalized gradient approximation (GGA) is employed to depict the exchange-correlation energy for valence electrons. The cutoff energy of plane-wave is determined to be 380 eV. The maximum root-mean-square convergent tolerance is less than 1×10^{-5} eV/atom, that is, the force imposed on each atom is not greater than 0.1 eV, and 0.1 GPa for stress. The Brillouin Zone's integration is approximated using the special *k*-points sampling scheme of Monkhorst-Pack, and the 1×1×4 *k*-points grid is used for the Brillouin zone integration along the nanotube axis. The valence electron configurations for Zn, O, Lu atoms are selected as: $3d^{10}4s^2$, $2s^22p^4$ and $4f^{14}5p^65d^16s^2$, respectively. All calculations are performed in reciprocal space.

3. Results and discussion

3.1. Band structure and density of states of SWZONT

The study of the electronic composition of the pure SWZONT system is essential to understand the influence caused by the dopant-Lu. For this reason, we calculated the band structure and the destiny of state of pure SWZONT, as shown below in figure 2.

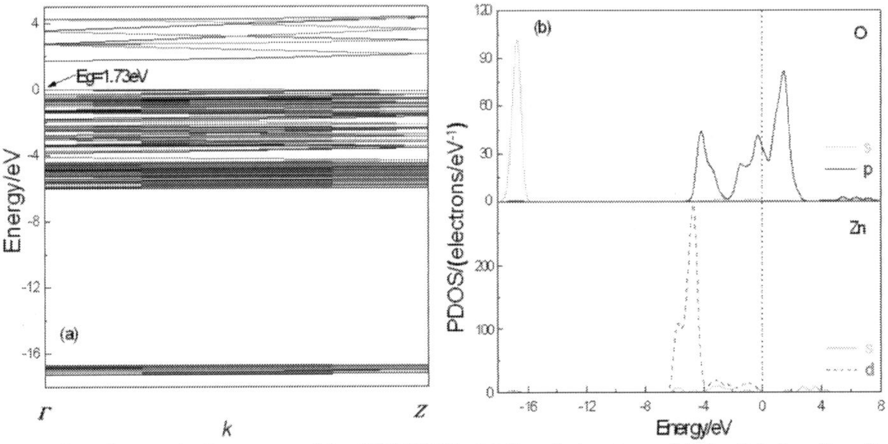

Figure 2. The Electronic Structure of the SWZONT. (a) Band structure; (b) Partial density of states.

It can be seen from the band structure that the valence band's top and the conduction band's bottom are at the same *k* point-*Γ*, which indicates that the SWZONT is a direct-gap semiconductor. Moreover, the energy band gap of SWZONT is approximately 1.75 eV, more approaching to Chen's experimental data 3.24 eV [23] than He's theoretical data 1.66 eV [24]. The most important reason for this is that the Perdew-Burke-Ernzerhof (PBE) generalized gradient approximation (GGA) to electronics and electronic exchange correlation function between the defect treatments [5]. The valence band structure of the SWZONT can be separated into three sections: the lower valence band from -6.10 ~ -4.00 eV, the higher valence band from -4.00 ~ 0eV and the valence band at -17 eV with a width about 0.5 eV. The conduction band structure of the SWZONT was mainly located in 1.75~4.33 eV.

From the figure 2(b), the lower valence band is primarily contributed by Zn-3d states, while the higher valence band mainly is contributed by O-2p, valence band at -17 eV mainly contributed by Zn-4s; the conduction band is composed of the hybridize influence caused by Zn-4s and O-2p. The conduction band minimum (CBM) and the valence band maximum (VBM) are composed of Zn-4s and O-2p, respectively.

3.2. The atomic structure of lu doped SWZONT

In order to examine the relationship between the dopant and the atomic structure directly, a model build is neutrally doped. The valence states of the dopant were not considered in the neutral doped model. Only substituted doping or vacancy doping was used, which means no consideration about the inherent oxygen vacancy or anion hole.

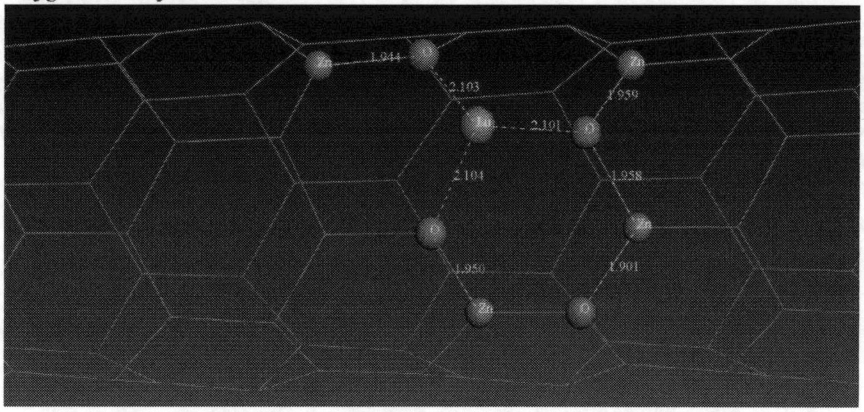

Figure. 3 Atomic Structure of Lu Doped SWZONT

The Lu doped SWZONT's structure after geometry optimization was conducted, and is depicted above in figure 3. It can be seen that the introduction of the impurity element, Lu, results in greater structural distortion of the doped model. In the intrinsic SWZONT, the bond length of Zn-O was set as 1.900 Å, but in figure 3, the Zn-O length round impurity element, Lu, is longer (at an average of 1.953 Å). Additionally, the length of Lu-O (2.103 Å) is much longer than that of the Zn-O. The phenomenon occurs due to the associated quantum chemistry; because of the bigger atomic radius of Lu (217 *pm*) than Zn (142 *pm*), the length of Lu-O is longer than that of Zn-O. But, as the distance of Zn-O and Lu grow farther, the length of Zn-O's gradual recovery proceeds to 1.900 Å. The change of the system's structure greatly encourages the formation of oxygen vacancy near the impurity element Lu, which improves the system's redox potential. Therefore, the catalytic ability has been improved as the introduction of Lu.

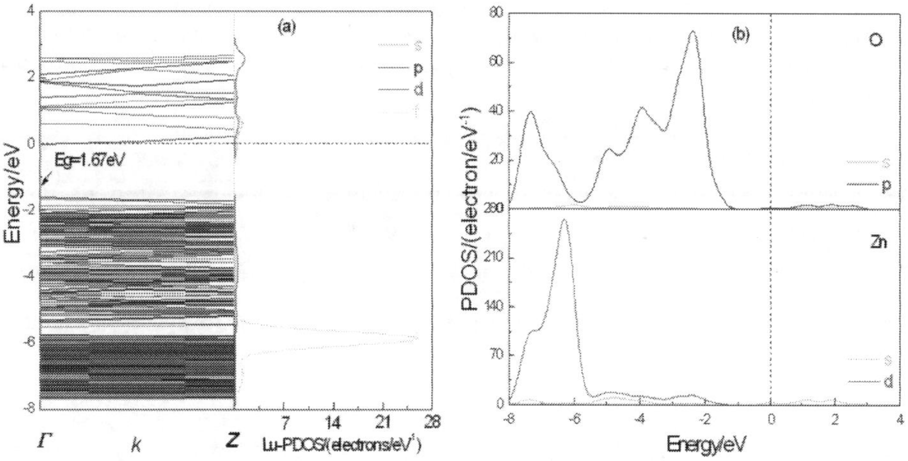

Figure 4. The Electronic Structure of the Lu-SWZONT. (a) Band Structure and DOS of Lu; (b) Partial Density of States.

3.3. Band structure and density of states of lu doped SWZONT

In order to analyze the influence of the impurity element, Lu, the band structure, and the partial density of states (PDOS) of Lu-SWZONT were plotted, which is shown in figure 4. Due to of the minute effect the semiconductor material made by the valence band at -17 eV, the graph's range is near the Fermi level. It is apparent that the conduction band minimum of Lu-SWZONT is at the Fermi level, which means that the strengthening of the semiconductor type (N type). The conduction band and valence band were both moving towards low energy end and the energy of the doped system diminishes, which indicates that the system becomes more stable. Based on that factor, we noticed that the distance of the conduction band minimum (CBM) movement was longer than that of valence band maximum (VBM), therefore the width of band gap decreases, which results in the enhancement of the metallicity and electrical conductivity of the Lu- SWZONT system. This predicts the red shift of the absorption spectrum.

We analysed the cause of the CBM failure from figure 4. The main reason was that the Lu-6s states' control of the CBM. In the case of pure SWZONT, Zn-4s states and O-2p states interact with each other to form the s-like anti-bonding and p-like bonding, and then form the energy gap. After Lu doping, Lu-6s states process the lower energy than Zn-4s states. Consequently, the s-like anti-bonding formed by Lu-6s and O-2p states in the Lu-SWZONT system contain low-energy than that formed by Zn-4s states and O-2p states in the intrinsic system. Thus, the CBM moved down. Simultaneously, the energy level between Lu-5d states and O-2p states is phenomenal; Compared to the Lu-SWZONT system, the p-d repulsion effect in the intrinsic system is much stronger when counteracting the p-p repercussion effect, which causes the VBM moved toward low energy end. Therefore, the p-p repercussion effect is stronger than the p-d repulsion effect, which results in the VBM moved toward low energy end at the Lu-SWZONT system.

Meanwhile, the conduction band is obviously broadening and becoming higher in destiny. Due to the repercussion effect from Lu-6s states and Lu-5d states, the conduction band grew corrugation fluctuation degree. This phenomenon reveals that the electron effective mass in conduction band digressed; and the non-local degree grew. There was an acceptor level intruded by Lu-5d states which decreased the necessary energy needed for electron transition. In total, Lu-4f leads to the separation of lower valence band and higher valence band obviously.

3.4. Optical properties

In regards to the zigzag single-walled (6, 0) ZnO nanotubes, optical properties are related to the electronic structure. So, the optical properties, dielectric function, absorption factors, and some other parameters can be deduced from the electronic composition. In the linear response range, the nanotubes macroscopic optical response characteristic can be described by the frequency dependent dielectric function [25]:

$$\varepsilon(\omega)=\varepsilon_1(\omega)+i\varepsilon_2(\omega) \tag{1}$$

Specifically, $\varepsilon_1=n^2-k^2, \varepsilon_2=2nk$. From the momentum matrix elements between the occupied and unoccupied electronic states, and the Kramers–Kronig transformation, the real part ε_1, the imaginary part ε_2, absorption coefficient, energy loss spectrum and others optical constants can be calculated. The real part ε_1, the imaginary part ε_2 and the absorption coefficient can be represented as follows:

$$\varepsilon_1(\omega)=1+\frac{8\pi^2e^2}{m^2}\cdot\sum_{V,C}\int_{BZ}d^3K\frac{2}{(2\pi)}\times\frac{|e\cdot M_{CV}(K)|^2}{[E_C(K)-E_V(K)]} \tag{2}$$

$$\varepsilon_2(\omega)=4\frac{8\pi^2}{m^2\omega^2}\cdot\sum_{V,C}\int_{BZ}d^3K\frac{2}{(2\pi)}\times|e\cdot M_{CV}(K)|^2\times\delta[E_C(K)-E_V(K)-\hbar\omega] \tag{3}$$

$$\alpha=\sqrt{2}\omega\left[\sqrt{\varepsilon_1^2(\omega)+\varepsilon_2^2(\omega)}-\varepsilon_1(\omega)\right]^{1/2} \tag{4}$$

In this equation, $\hbar$ is Planck constant, ω is angular frequency, k is extinction coefficient, and K is reciprocal lattice vector. Subscripts C and V represent the conduction and valence bands of SWZONT, respectively. BZ is first Brillouin zone, while $E_C(K)$ and $E_V(K)$ express the intrinsic level of the conduction and valence bands, respectively. We can deduce from (1)-(4) that the macro optical constants are closely related in microelectronic structure. The electronic structure is the theoretical foundation to analyse the optical properties.

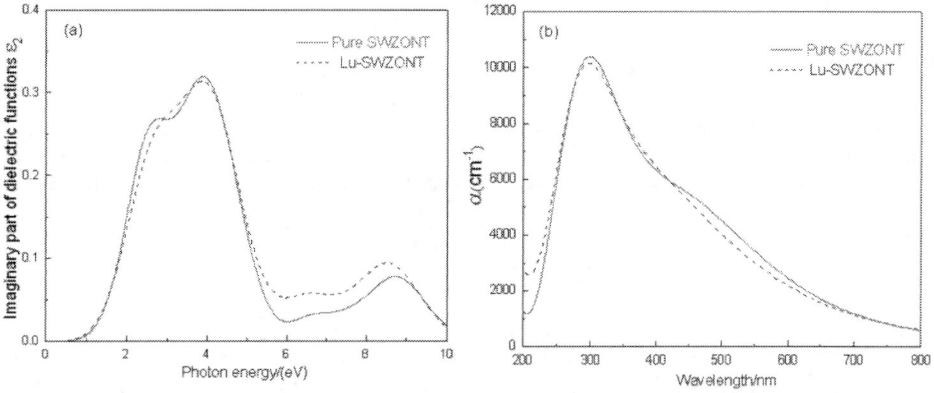

Figure 5. (a) Imaginary part of dielectric functions α_2; (b) Absorption coefficient.

The optical characteristics of the Lu-doped SWZONT are systematically discussed on the basis of the calculated dielectric functions and the absorption coefficient. Figure 5 displays the dielectric functions (ε_2), which are the pandect of the optical properties for the two different configurations of the SWZONT systems. At the low energy region of figure 5(a), there are no differences between the Lu-doped and intrinsic system; however there was an occurance at the high energy area, when the peak of the ε_2 increased. At this point, the value of Lu-doped system becomes twice as much as the pure system at 6 eV, which conform to in figure 4 that the energy band gap depress to 1.67 eV. This data indicates the introduction of the impurity element Lu would decrease the energy band gap effectively and improve the electrical conductivity simultaneously. On the other hand, we found that the area under the curve of Lu-doped is larger than that of the intrinsic, which suggests the width of dielectric functions are magnified. It also indicates that the introduction of the impurity element Lu would improve the dielectric functions width effectively.

Figure 5(b) presents the absorption coefficient of the SWZONT systems with two different configurations. It is clear that the absorption region is quite wide, and that the main absorption region still resides in the UV region. When compared with pure SWZONT, the additional absorption is observed at UV region for Lu-doped system, which is due to the decrease of energy band gap. More electrons in the valence band can be excited by the UV light to overcome the smaller energy. This declares that the doped materials can be used for UV optoelectronic devices.

4. Conclusions

The electronic structure and the optical properties of the Lu-doped zigzag single-walled (6, 0) ZnO nanotubes were studied by first-principles calculations, which are as follows:

- The Lu-doped SWZONT is still a direct-gap semiconductor; the introduction of Lu results in the decrease of the energy band gap and the improvement of the electrical conductivity.
- Lattice distortion was obvious in the Lu-doped system; the distortion encourages the formation of oxygen vacancy around the impurity Lu, which improves the system's redox potential. The catalytic oxidation activity of the Lu-doped SWZONT was therefore improved.

- The Lu-doped SWZONT system is stabilized because of the introduction of Lu, and the semiconductor type (N type) is enhanced in the Lu-doped system.
- Lu-doped SWZONT materials can be applied to UV optoelectronic devices.

Acknowledgments
This work was financially supported by National Students' Innovation and Entrepreneurship Training Program (201410075002) and CSC fellowship ([2014]-3012).

References

[1] Cui Y, Wei Q, Park H, et al. 2001 Nanowire nanosensors for highly sensitive and selective detection of biological and chemical species *Science* **293** 1289-92
[2] Cao L, Kong L B, Liang Y Y, et al. 2004 Preparation of novel nano-composite Ni (OH) 2/USY material and its application for electrochemical capacitance storage *Chem. Commun.* **14** 1646-7
[3] Xin H, Liu X, Shi X, et al. 2005 High efficiency light emitting device *U.S. Patent Application 11/196,856*
[4] Xing Y J, Xi Z H, Xue Z Q, et al. 2003 Optical properties of the ZnO nanotubes synthesized via vapor phase growth *Applied Physics Letters* **83** 1689-91
[5] Wu Y, Huang J C, Cui Y C, et al. 2014 First-principles study on optical properties of Cd-doped single-walled (8, 0) ZnO nanotube *Applied Mechanics and Materials* **668** 27-30
[6] Wu J J, Liu S C, Wu C T, Chen K H, Chen L C 2002 Heterostructures of ZnO-Zn coaxial nanocables and ZnO nanotubes *Applied Physics Letters* **81** 1312-4
[7] Zhang B P, Binh N T, Wakatsuki K, Segawa Y, Yamada Y, Usami N, Kawasaki M and Koinuma H 2004 Formation of highly-aligned ZnO tubes on sapphire (0001)substrates *Applied Physics Letters* **84** 4098-100
[8] Hu J Q, Li Q, Meng X M, Lee C S and Lee S T 2003 Thermal reduction route to the fabrication of coaxial Zn/ZnO nanocables and ZnO nanotubes *Chemistry Materials* **15** 305-8
[9] Xing Y J, Xi Z H, Xue Z Q, Zhang X D, Song J H, Wang R M, Xu J, Song Y, Zhang S L and Yu D P 2003 Optical properties of the ZnO nanotubes synthesized via vaporphase growth *Applied Physics Letters* **83** 1689-91
[10] Sun Y, Fuge G M, Fox N A, et al. 2005 Synthesis of aligned arrays of ultrathin ZnO nanotubes on a Si wafer coated with a thin ZnO film *Advanced Materials* **17** 2477-81
[11] Kong X Y, Ding Y and Wang Z L 2004 Metal-semiconductor Zn-ZnO core-shell nanobelts and nanotubes *Journal of the Physics Chemicals B* **108** 570-4
[12] Wu G S, Xie T, Yuan X Y, Li Y, Yang L, Xiao Y H and Zhang L D 2005 Controlled synthesis of ZnO nanowires or nanotubes via sol-gel template process *Solid State Communications* **134** 485-9
[13] Yan J F, Lu Y M, Liang H W, Liu Y C, Li B H, Fan X W and Zhou J M 2005 Growth and properties of ZnO nanotubes grown on Si (111) substrate by plasma-assisted molecular beam epitaxy *Journal of Crystal Growth* **280** 206-11
[14] Zhang J, Sun L D, Liao C S and Yan C H 2002 A simple route towards tubular ZnO *Chemical Communications* **3** 262-3
[15] Shen X, Allen P B, Muckrman J T, Davenport J W and Zheng J C 2007 Wire versus tube: Stability of small one-dimensional ZnO nanostructures *Nano Letters* **7** 2267-71
[16] Tu Z C and Hu X 2006 Elasticity and piezoelectricity of zinc oxide crystals, single layers, and possible singlewalled nanotubes *Physicals Review B* **74** 035434-40
[17] Mclaren A, Valdes-Solis T, Li G, et al. 2009 Shape and size effects of ZnO nanocrystals on photocatalytic activity *Journal of the American Chemical Society* **131** 12540-1
[18] Pan H and Feng Y P 2008 Semiconductor nanowires and nanotubes: Effects of size and surface-to-volume ratio *ACS Nano* **2** 2410-4
[19] An W, Wu X and Zeng X C 2008 Adsorption of O_2, H_2, CO, NH_3, and NO_2 on ZnO nanotube: A density functional theory study *The Journal of Physical Chemistry C* **112** 5747-55

[20] Minami T, Yamamoto T and Miyata T 2000 Highly transparent and conductive rare earth-doped ZnO thin films prepared by magnetron sputtering *Thin Solid Films* **366** 63-8

[21] Zhang Y G, Zhang G B and Wang Y X 2011 First-principles study of the electronic structure and optical properties of Ce-doped ZnO *Journal of Applied Physics* **109** 063510

[22] Luo W, Fu C, Li R, et al. 2011 Er^{3+}–doped anatase TiO_2 nanocrystals: Crystal–field levels, excited - state dynamics, upconversion, and defect luminescence *Small* **7** 3046-56

[23] Chen S J, Liu Y C, Shao C L, et al. 2006 Photoluminescence study of ZnO nanotubes under hydrostatic pressure *Applied Physics Letters* **88** 133127

[24] He A L, Wang X Q, Fan Y Q, et al. 2010 Electronic structure and magnetic properties of Mn-doped ZnO nanotubes: An ab initio study *Journal of Applied Physics* **108** 084308

[25] Shen X C 1992 *The Spectrum and Optical Property of Semiconductor* (Beijing: Science Press) p 76 (in Chinese)

IOP Conf. Series: Materials Science and Engineering 87 (2015) 012051 doi:10.1088/1757-899X/87/1/012051

Improving the quality of asphalt coating with carbon nanomodifiers

Urkhanova Larisa[1], Shestakov Nikolay[1,3], Semenov Aleksandr[2], Smirnyagina Natalya[2] and Semenova Irina[2]

[1]East Siberia State University of Technology and Management, KluchevskayaStr. 40b, Ulan-Ude,670013, Russia
[2]Institute of physical material sciences of Siberian division of Russian Academy of Science, SakhyanovoiStr. 6., Ulan-Ude,670047, Russia

E-mail: Nik.shestakov.1990@mail.ru

Abstract. This article deals with the possibility of modifying the binder by adding carbon nanomodifier to bitumen to improve the quality of asphalt. Addition of 0.05%-0.5% of nanomodifier significantly changes the properties of bitumen. Asphalt with this astringent has increased strength, heat resistance and shear resistance.

1. Introduction

One of the main reasons why pavement is breaking before its time is the poor quality of road bitumen. Bitumen does not have adhesion characteristics, which are necessary, because it agglutinates only mineral pieces of basic species. Beside that bitumen becomes fragile during severe winters in Russia. In sharply continental climate of the Republic of Buryatia the amplitude of the temperature fluctuations is very high, both annual and daily. Also cold and snowless winter may start very abruptly and maintain for a very long time. These circumstances lead to a formation of cavities on pavements, which, considering the low strength of asphalt, transform into potholes [1].

According to that, the problem of increasing the quality of road bitumen is becoming an urgent issue, and the solution of it will help to maintain terms of pavement usage and to increase efficiency of its building and reparation.

For managing the process of bitumen structure formation with specific parameters (distribution of different fractions of bitumen according to thickness of its adsorption layer) a variety of modifiers is being used, both organic and mineral compounds. These compounds are working in different ways: first, formation of the polymer "grid", which is responsible for increasing the interval of flexibility, or second, enlargement of quantity of active centers on mineral part surface, which can take part in chemical reacting with bitumen (chemisorption). The first method is provided by adding different organic compounds, such as rubber, polymeric additives, their combinations, etc. The second technique is implemented by other agents, such as phosphoric acid or hydrochloric acid. Different nano additives like nanotubes, fullerenes, astralenes and nanodiamonds compose a separate group of agents which performance is poorly investigated [2].

[3] Address for correspondence: Shestakov Nikolay, East Siberia State University of Technology and Management, Kluchevskaya Str. 40b, Ulan-Ude,670013, Russia. E-mail: Nik.shestakov.1990@mail.ru.

A number of scientists were studying the effects of adding carbon nanomodifiers into bitumen. Thus, in one of the works carbon nanotubes (CNT) were chosen as a modifier, in amount of 0.005% of bitumen weight. Oil bitumen modifying is implemented by using ultrasound. Increased strength, flexibility as well as water, heat and cold resistance are found as a result of adding the carbon nanomodifiers in asphalt concrete[3].

In the other work, as a modifier for asphalt concrete, carbon nanotubes were used in amount of 0.2%-10% of bitumen weight. Technical result of the work is obtainment of high wear resistance and durability of bituminous concrete[4].

Authors of another work applied "Taunit" as a modifier, which is dimensional nanoscale filamentous formations of polycrystalline carbon, represented as loose black powder, which was used in amount of0.01- 0.001% of bitumen weight. Using of this modifier makes positive impact on durability and stability of bituminous concrete[5].

Researchers are paying more attention to modifying of bitumen with CNT and nano-sized carbon black, when, at the same time, the impact of fullerenes on bitumen qualities is not well discovered.

The main goal of this work is to study the influence of carbon nanomodifiers (CNM) on road building materials exploited in sharply-continental climate.

This work deals with the asphalt concrete using bitumen, modified by CNM. The authors defined basic physical, mechanical and deformation properties of that concrete.

2. Material and methods

Synthesis of initial carbon nanomodifier were carried out on the special apparatus - a plasma-chemical reactor (figure1).

Figure 1. General view (plasma chemical reactor): 1 - camera for the synthesis 2-nitrogen trap, 3 - stocks, 4-block to meet the load, 5-carbon condensate storage device, 6-rack, 7-clips, 8-detent, 9-tap, 10 - the valve controlling the supply of cold water.

Apparatus were successfully tested previously for the synthesis of fullerene mixture at preparing targets, sputtered by ion beams and evaporated by electron beams. Targets were used in the synthesis of fullerene derivatives coatings [6].

The principle of the system is based on the erosion of graphite electrodes in the arc discharge plasma [7]. The discharge is initiated at a pressure of 105 Pa, by passing through the electrodes with current frequency 44 or 66 kHz. Erosion of rods occurs in a closed sealed volume, filled with helium.

The synthesis of plasma is taking place in the water-cooled chamber 1 of the apparatus. Towards the bottom and the top of the chamber nitrogen traps 2 are mounted through the flanges. There are inspection windows provided in the chamber for visual observation. The chamber caps are removable. Stocks 3 are attached to the removable bottom part of the chamber. There is a system which provides their movement according to rods combustion, along with coordination system with load 4.Synthesized carbon condensate is collected from the chamber walls in the store 5. Chamber is located on the rack 6. Electric power is supplied from the voltage generator with a rated capacity of 16 kW. Maximum amperage of erosion of graphite rods is 160 A. The average time of synthesis is 10 min. Fullerenes were selected by benzol from carboncondensatecontaining10-12% of fullerenes.CNM contains parts by weight: 0.8 - C60; 0.15 - C70; the rest of it - higher fullerenes and oxides C60O and C70O (figure 2).

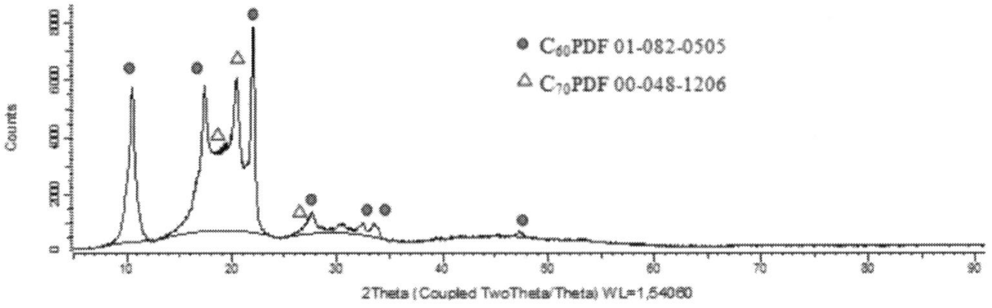

Figure 2. - XRD pattern of CNM.

Bitumen of grade 90/130 was used to determine the properties of the binder. For the preparation of bituminous mixtures the following materials were used: crushed granite, granite screenings of crushing with grain size of up to 5 mm, river sand with a fineness modulus Mk = 3 and mineral powder MP-1 - ground limestone.

Basic properties of the original and CNM modified bitumen were defined using standard test methods: the temperature of softening by the “ring and sphere” method was determined according to the StSt 11506-73, penetration was determined according to the StSt 11502-82.

The properties of original and CNM modified bitumen, BND 90/130, are shown in table 1.

Table 1. Properties of the original bitumen and the bitumen modified by CNM.

Name index bitumen	Initial bitumen BND 90/130	Bitumen + 0.05 % mass CNM	Bitumen + 0.1 % mass CNM	Bitumen + 0.25 % mass CNM	Bitumen + 0.5 % mass CNM
Penetrationat 25°C, 0.1mm	95	90	81	78	62
Temperature of softening, °C	46	45	43	42	40

In order to evaluate the impact of CNM on the quality of the binder, the samples of asphalt concrete mixtures were prepared according to the recipe in the table 2.

Table 2. Composition of the fine-grained asphalt mixes of the type B, brand 2 for a road-climatic zone1 (Siberian Region).

The name of asphalt component	Content by mass, %
Crushed stone from quarry "Vahmistrovo". Fractions 15-5 mm	44
Crushed sand, fractions 5-0mm	48
Mineral powder [a]	8
Organic binder [b] (Over 100 %)	5.6

[a]The limestone powder was used as a mineral powder.

[b]The organic binder – bitumen modified by additives used in this study.

The introduction was made by heating the bitumen to 140°C with mechanical stirring of CNM. In addition, the authors investigated the possibility of obtaining steady suspension of CNM in an organic solvent - toluene, with further introduction in bituminous concrete mixture. Moreover, the formation of bituminous concrete microstructure with strong aggregates adhesion with organic binder was taken into account. Determination of durability characteristics were carried out according to StSt 12801-98. Experimental bitumen compositions and methods of administration of CNM are presented in table 3. The results of CNM amount optimization studies are shown in tables 4, 5 and figures 3-6.

Table 3. Methods of dispositions of modified bitumen.

Composition	Distribution method CNM with binder	Quantity CNM by weight bitumen, %
1	-	0
2	Mechanical Mixing by 140°C	0.05
3	Mechanical Mixing by 140°C	0.1
4	Spreading CNM in toluene	0.1

3. Results and discussion

According to the table 1, it is clear that the introduction and increasing the number of CNM decrease the temperature of bitumen softening. Reducing the viscosity of bitumen in the process temperature promotes better wetting of the surface of the stone material - aggregate organic binder.

Possibly, resins, which are responsible for the bitumen plasticity, exhibit their properties much better by interacting with fullerenes. At the same time, the depth of the needle penetration is reduced, so bitumen turned up stronger. Hence, the asphaltenes, which determine the degree of bitumen hardness, also interact with fullerenes. That is to say, bitumen acquires new unique properties by administering CNM.

During the development of the asphalt composition using the modified bitumen the authors had a few tasks. They defined an optimal content and composition of CNM depending on the composition of electrode, used in the preparation of a modifying additive. Also the authors determined the method of introduction of CNM in bitumen, providing more evenly distributed nano-dispersed additive in the organic binder.

Table 4. Physical and mechanical parameters of organic compounds on the basis of bitumen, modified with CNM, identified by the StSt 12801-98 methods.

Number of the sample in the Table3	Average density, g/cm^3	Breaking point at compressing specimens, MPa			Water resistance index	Watersaturation in volume, %
		$R_{50°C}$	$R_{20°C}$	$R_{0°C}$		
1	2.37	0.91	2.53	4.31	0.94	3.22
2	2.38	1.05	2.5	4.13	0.95	2.98
3	2.39	1.53	2.75	4.37	0.98	2.31
4	2.37	1.21	2.53	4.55	0.97	2.29
Requirements of the StSt 9128-97		Least 0.9 MPa	Least 2.2MPa	Not more 10MPa	Least 0.9	1.5%-4.0%

IOP Conf. Series: Materials Science and Engineering **87** (2015) 012051 doi:10.1088/1757-899X/87/1/012051

Table 5. Thermal and mechanical parameters of organic compounds on the basis of bitumen, modified with CNM, identified by the StSt 12801-98 methods.

Number of the sample in the Table3	Heat resistance index, R_{50}/R_{20}	Temperature sensibility index, R_{50}/R_0	Calculated crack resistance, R_0/R_{50}	Shear resistance: Internal attrition index	Adhesiondisplacement,MPa
1	0.36	0.21	4.73	0.87	0.32
2	0.42	0.25	3.93	0.82	0.35
3	0.56	0.35	2.85	0.85	0.43
4	0.48	0.27	3.76	0.88	0.37

The analysis of the results showed that CNM modified asphalt has a higher streng that 20°Cand 50°Candincreasedwaterresistance. The compressive strength of specimens of asphalt concrete with the introduction of 0.1% of CNM at 20°C has increased by 10%, at 50°C –by 70%, with adding of 0.05% of CNM - 5% and 15% respectively. One can see a significant increase in strength at 50°C, especially by administering large amounts of nanomodifiers- 0.1%.

CNM dissolution in toluene showed no significant changes in the properties of asphalt, compared to the introduction of the original additive in hot bitumen.

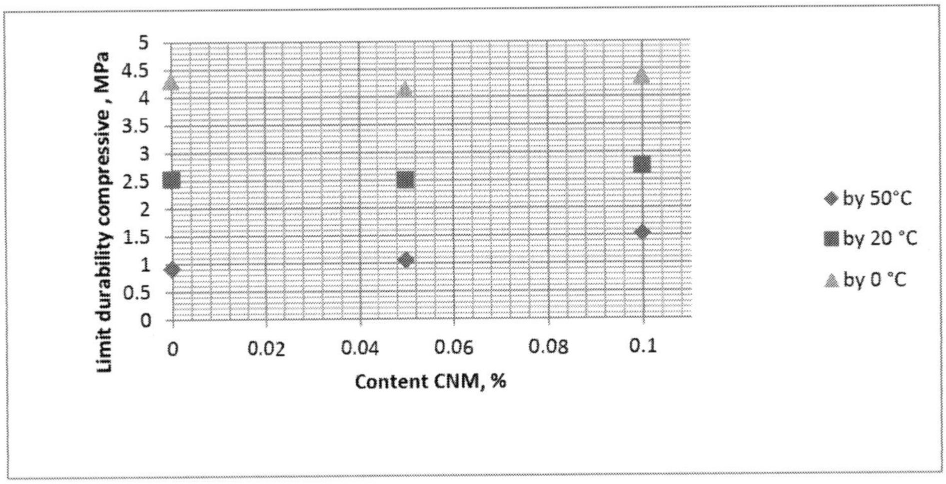

Figure 3. Characteristics durability bituminous concrete by different temperatures.

Increasing in strength can be explained by improving of bitumen structure, using carbon nanomodifers. Hence, binder can be more efficiently transferred from the bulk to the pellicle condition. Consequently, asphalt strength is increased at the 50°C and slightly decreased at 0°C. Therefore, the use of CNM modified bitumen will improve important performance characteristics such as heat resistance and shear resistance.

Analysis of the results showed that the optimum amount of CNM additive in the asphaltic mixture is equal to 0.1% by weight.

The temperature range, in which the asphalt remains in the elastic-plastic state, can be characterized by the temperature sensitivity index, which was calculated as the ratio of strength at 50°C to the strength at 0°C.

Results, presented on a figure 4, show the increasing of temperature sensitivity index as a result of the introduction of CNM: with the introduction of 0.05% - increasing by 19%, with the introduction of 0.1% - by 66%. This can be explained by decrease in the amount of bitumen in the bulk condition and more complete transfer of it into a structured condition, therefore asphalt will remain in the elastic-plastic state at greater temperature range, will be less exposed to cracking and will resist greater amount of temperature fluctuations.

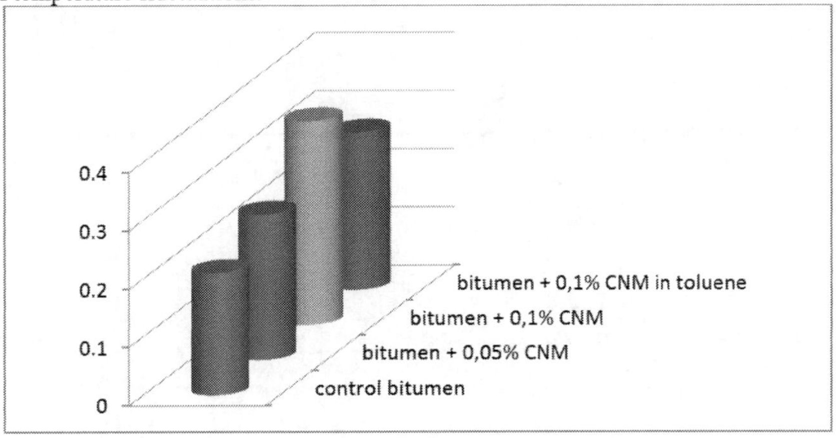

Figure 4. Change index temperature sensibility adding CNM.

The resistance of asphalt to stress at high temperatures is characterized by a heat resistance index. The results show a signifi cant increase of asphalt thermal resistance index. With the introduction of 0.05% CNM this parameter increased 1.16 times, the introduction of 0.1% - 1.55 times (figure5), which will reduce the risk of plastic deformations.

Resistance to the formation of waves, in fluxes and other shear deformations is characterized by the asphalt shear strength. This parameter is determined by the test of the Marshall method. The results, presented on figure 6, show the increase of asphalt shear strength using bitumen, modified with 0.1% CNM, by 34%. This is obviously duet enlarged cohesive strength hand viscosity of the modified bitumen. All this will create asphalt concrete with high strength and high shear resistance[8].

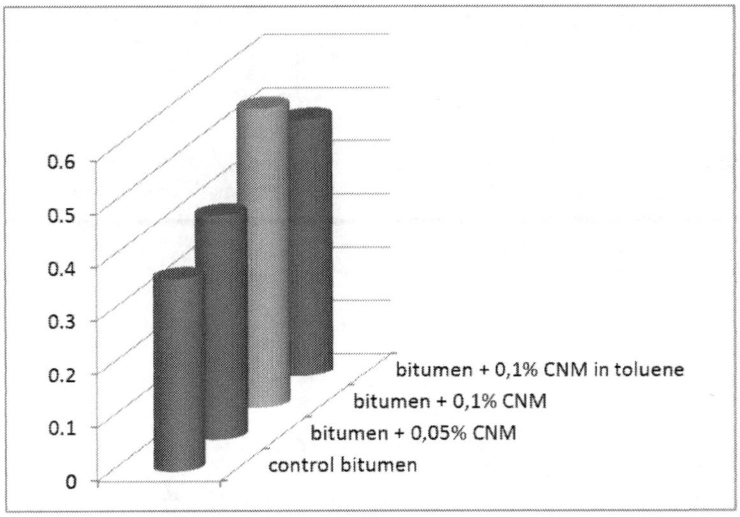

Figure 5. Change index heat resistance adding CNM.

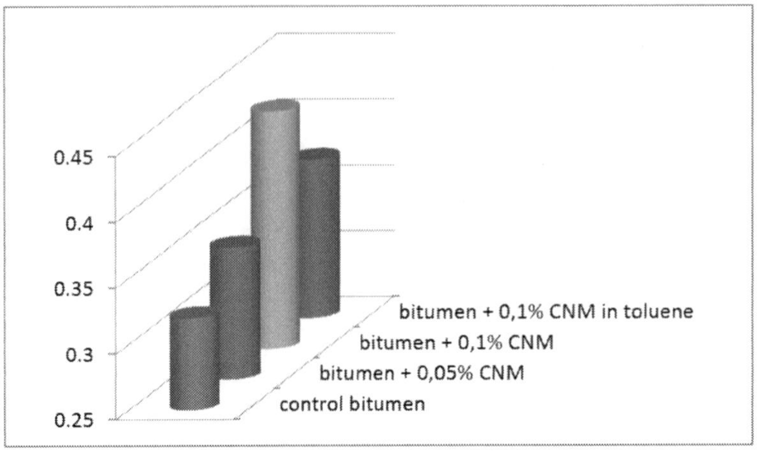

Figure 6. Change adhesion displacement bituminous concrete adding CNM.

From the literature it is known that asphaltenes make bitumen harder and heat resistant, resins determine its plasticity and flexibility, and oils have a plasticizing effect. Therefore, it can be assumed that aromatic and polycyclic structures of asphaltenes and resins, which include nitrogen heterocycles and sulfur, and have a π-bond and atoms with no divided electronic pairs, when interacting with CNM, can form strong structural clusters. Because of the spatial volume of the fullerene molecule cluster growth is happening throughout the entire volume of binder and, intertwining, forms a strong and rigid spatial network, which is increasing the hardness and heat resistance of bitumen.

The molecules of organic substances in bitumen are complex and contain a wide variety of functional groups. It can be as summed that the contact between the fullerene molecules and bitumen compounds will form strong bonds. In cases when there are several points at which the molecules are joined, even through the hydrogen bonds, to the surface, the desorption of substance is substantially irreversible [9]. At higher concentrations of CNM in bitumen there is a growth of clusters centers and, therefore, growth and density of the polymer matrix formation, which, in its turn, increases the strength characteristics.

4. Conclusions

With modifying bitumen by CNM the properties of a binder are changing, that leads to improved properties of the asphalt. Yet little studied and mastered, with small dosages modifiers significantly change the properties and structure of road materials and coatings based on them. Thus, CNM can be regarded as an effective modifier, and this direction of road-building materials science is extremely promising.

Acknowledgments

This work was supported by the Russian Academy of Sciences (Artica, Project No. 84).

References

[1] Gohman L M 2002 Polymer-bitumen binders based on SBS for road construction (PhD thesis, Informavtodor №4 , Russia) p 36

[2] Inozemtcev S S 2010 Structure and properties nanomodified stone mastic bitumen concrete with increased performance characteristics (PhD thesis, University of Moscow, Russia)p 184

[3] Zaporotckova I V and Sipliviy BN 2014 Method of strengthening asphaltic road coating by carbon nanomaterial *RU Patent* 2515007

[4] Kondratiev D N, GoldinV V, Mercelene N F 2011 Nanostuctured modifier for bitumen concrete *RU Patent* 2412126

[5] Yartcev V P, Polezhaeva E S, Yu. A Buchnev 2011 Influence of carbon nanoadditives (taunit) on physic-mechanical characteristics and durability of bitumen*Bulletin TGTU* **17** 1053
[6] Churilov G N, Bulina N V and Fedorov A S 2007 Fullerenes: synthesis and theory of formation (Monograph / Novosibirsk: Publishing House of Russian Academy of Sciences) p 227
[7] Semenov A P and Semenova I A 2014 Method of coating synthesis of fullerene derivatives *RU Patent* 2517706
[8] Asadullina Z U 2012 Mechanism of aging and plasticization of bitumen by tar *Construction Materials* **1** 51-3
[9] Yadykina V V 2004 Increasing of quality bitumen and cement concretes on technogenic raw materials given the state of its surface (PhD thesis, Belgorod State Technological University, Belgorod, Russia) p 455

Gel spinning of PVA composite fibers with high content of multi-walled carbon nanotubes and graphene oxide hybrids

Yizhe Wei, Dengpan Lai, Liming Zou[1], Xinlong Ling, Hongwei Lu and Yongjing Xu

State Key Laboratory for Modification of Chemical Fibers and Polymer Materials, College of Materials Science and Engineering, Donghua University, 2999 North People Road, Shanghai 201620, People's Republic of China

E-mail: lmzou@dhu.edu.cn.

Abstract. In this report, poly (vinyl alcohol) (PVA) composite fibers with high content of multi-walled carbon nanotubes and graphene oxide (MWCNTs-GO) hybrids were prepared by gel spinning, and were characterized by TGA, DSC, SEM, XL-2 yarn strength tester and electrical conductivity measurement. The total content of MWCNTs-GO hybrids in the PVA composite fibers, which is up to 25 wt%, was confirmed by TGA analysis. The DSC measurement shows that the melting and crystallization peaks decreased after the addition of nano-fillers. This is due to the reason that the motion of PVA chains is completely confined by strong hydrogen bonding interaction between PVA and nano-fillers. After the addtion of GO, the dispersibility of MWCNTs in composite fibers improved slightly. And the tensile strength and Young's modulus increased by 38% and 67%, respectively. This is caused by the increased hydrogen bonding interaction and synergistic effect through hybridization of MWCNTs and GO. More significantly, the electrical conductivity of PVA/MWCNTs/GO composite fibers enhanced by three orders of magnitude with the addition of GO.

1. Introduction

Carbon nanotubes (CNTs) have received great attention due to their unique mechanical, electrical and thermal properties since their discovery by Iijima [1]. They can find application in high performance nanocomposites [2], field-emission devices [3], hydrogen storage media [4] and sensors [5]. Incorporation of high mechanical and electrically conductive additives, carbon nanotubes, into polymer fibers can improve their mechanical properties and provide multi-functionalities needed for electrical energy storage, sensing and actuation [6]. In years past, PVA fibers containing a large fraction of carbon nanotubes were widely investigated because of their strong mechanical property, high toughness, and high electrical conductivity [7]. Vigolo proposed a new method by which fibers were spun by injecting the mixture of CNTs and SDS into a co-flowing PVA bath [8]. The fibers from this method contain large fractions of CNTs, that is, above 10 wt%. This led to fibers with large energy to failure, high electrical conductivity. However, it is particularly challenging to achieve this spinning process on large scale and in multifilament configuration. Meanwhile, much effort has been

[1] Address for correspondence: Liming Zou, State Key Laboratory for Modification of Chemical Fibers and Polymer Materials, College of Materials Science and Engineering, Donghua University, 2999 North People Road, Shanghai 201620, People's Republic of China. E-mail: lmzou@dhu.edu.cn.

made to directly spin composite fibers from mixture of PVA and CNTs in static coagulating baths. However, the final content of CNTs in composite fibers was low, which is insufficient to achieve significant properties. Moreover, large fraction of CNTs may result in agglomerates [9], and bad dispersion state reduces the mechanical and conductivity properties of the fibers.

In recent years, graphene, a two-dimensional, one-atom-thick material, has attracted great attention in various practical application areas because of its extraordinary electronic, thermal, and mechanical properties. Graphene has been considered as an ideal nanofillers in polymer fibers to improve the mechanical and electrical properties. GO which bearing various oxygen functional groups is hydrophilic and can be readily dispersed in water as individual sheets to form stable colloidal suspensions. Moreover, GO can serve as a superior dispersant to disperse pristine CNTs in aqueous media or polymer matrix through supramolecular interactions [10]. More significantly, the hybridization of graphene and CNTs can lead to synergetic enhancement of the mechanical performance of their polymer composites [11]. Furthermore, the unique electrical conductive network formed by the hybridization of graphene and CNTs lead to the increasing electrical conductivity [12].

Therefore, spinning polymer composites fibers by incorporation of CNTs-GO hybrids can be an excellent route to utilize their mechanical and electrical performance of their polymer composites.

Here, we report the preparation of PVA composite fibers with high content of MWCNTs-GO hybrids by gel spinning. For comparison, PVA composite fibers with high content of MWCNTs were also spun. The composites fibers were characterized by TGA, DSC, SEM, XL-2 yarn strength tester and electrical conductivity measurement.

2. Experimental

2.1. Materials

The pristine MWCNTs (Nanocyl™ NC7000, average diameters 9.5 nm, lengths 1.5 μm, carbon purity 90% by TGA) were purchased from Sambreville Nanocyl S.A. (Belgium) and used as received. Graphite was purchased from Shanghai Yifan Graphite Co. Ltd. PVA (Mw≈114400, 99% hydrolyzed) was received from Anhui Wanwei Updated High-tech Material Industry Co., Ltd. All the other reagents were supplied from Sinopharm Chemical Reagent Co., Ltd. and used as received. All reagents are analytical grade and used as received.

2.2. Fiber spinning

MWCNTs were oxidized by the mixture of H_2SO_4 and HNO_3 (volume ratio: 3:1). GO was prepared using the modified Hummer's method [13].

2.2.1. PVA/MWCNTs spinning dope preparation. 800 mg PVA was dissolved into 10 mL DMSO at around 100℃. Meanwhile, 200 mg oxidized MWCNTs was dispersed in 40 mL water and sonicated in a bath sonicator (SK2200H, 53 kHz, 100 W) for 2 hours. Then MWCNTs dispersion was gradually added into the PVA/DMSO solution. This mixture was subjected to further stirring for 12 h at around 100℃. The desired PVA/MWCNTs spinning dope was obtained by removing excess water using vacuum distillation.

2.2.2. *PVA/MWCNTs/GO spinning dope preparation.* 750 mg PVA was dissolved into 10 mL DMSO at around 100℃. Meanwhile, 200 mg oxidized MWCNTs was dispersed in 40 mL water and 50 mg GO which dispersed in 10 mL water was added into MWCNTs dispersion. Then the mixture was sonicated in bath sonicator for 2 hours. At last, the MWCNTs-GO hybrids aqueous dispersion was gradually added into the PVA/DMSO solution at around 100℃. This mixture was subjected to further stirring for 12 h at around 100℃. The desired PVA/MWCNTs/GO spinning dope was obtained by removing excess water using vacuum distillation.

Fiber spinning was done using syringe pump and syringe needle. The spinning dope was injected at 20 mL/h through a 0.5 mm diameter needle into methanol bath at room temperature. The as spun fibers were drawn at room temperature using a drawing machine.

For comparison, pure PVA fibers, PVA composite fibers with 20 wt% oxidized MWCNTs (PVA/MWCNTs 20 wt%), PVA composite fibers with 20 wt% oxidized MWCNTs and 5 wt% GO (PVA/MWCNTs 20 wt%/GO 5 wt%) were spun.

2.3. Characterization

SEM images were acquired on a JEOL JSM-5600 LV instrument (Japan) operating at 5 KV. Diamond DSC is at a heating rate of 10℃/min under nitrogen flow. TGA data were obtained on a TG 209 F1 Iris thermo-gravimetric analyzer (Germany), from room temperature to 800℃, at a heating rate of 20℃/min, under continuous nitrogen flow. Tensile strength was measured on XL-2 yarn strength tester. The electrical conductivity of composite fibers was measured by a two- probe method with Keithley 4200 SCS.

3. Results and discussion

3.1. TGA analysis

Based on TGA analysis (figure 1), the content of nano-fillers in composite fibers were estimated from the residual weight at 800℃, which are about 19.2%, 26.3% for PVA/MWCNTs 20wt% composite fibers and PVA/MWCNTs 20wt%/GO 5wt% composite fibers. The TGA curves of oxidized MWCNTs and GO shows multi-stage weight loss from room temperature to 800℃. The weight loss before 200℃ is due to the loss of bound water which absorbed by functional groups [14]. The weight loss at the temperature between 200℃ and 500℃ can be attributed to the decomposition of the functional organic moieties attached to the defect sites of MWCNTs and GO during functionalization process [14, 15]. At the temperature higher than 500℃, the slight weight loss is often ascribed to the decomposition of MWCNTs [14, 16]. For PVA/MWCNTs 20 wt% composite fibers and PVA/MWCNTs 20 wt%/GO 5 wt% composite fibers, there is an obvious weight loss at the temperature between 250℃ and 500℃ which is due to the decomposition of functional groups and PVA molecules. The residual weight remains stable at 800℃, and the content of nano-fillers can be confirmed.

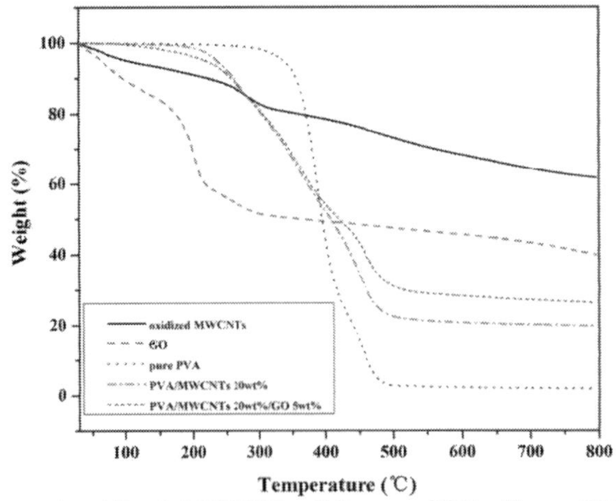

Figure 1. TGA curves of oxidized MWCNTs, GO, pure PVA fibers, PVA/MWCNTs 20 wt% composite fibers and PVA/MWCNTs 20wt%/GO 5wt% composite fibers.

3.2. DSC analysis

In order to understand the thermal properties of the composite fibers, DSC measurements were carried out on all fibers. As can be seen from figures 2(A) and (B), pure PVA fibers show obvious melting and crystallization peaks. However, for PVA/MWCNTs 20 wt%, PVA/MWCNTs 20 wt%/ GO 5 wt%, the melting and crystallization peaks decreased. This may be due to the reason that PVA chains are constrained in a tiny and fixed space by strong hydrogen bonding interaction between PVA chains and nano-fillers [17]. Compared to PVA/MWCNTs 20 wt%, the melting and crystallization peaks of PVA/MWCNTs 20 wt%/ GO 5 wt% is smaller. The major factor is that the introduction of GO enhanced the hydrogen bonding interaction between PVA chains and nano-fillers.

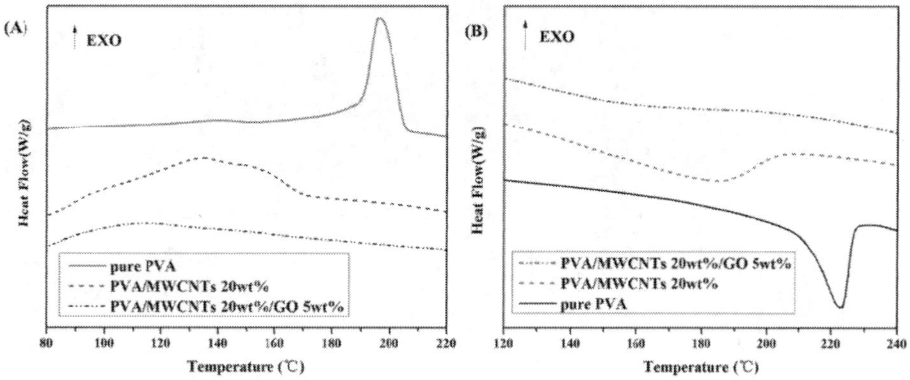

Figure 2. DSC curves of PVA composite fibers. (A) cooling curves and (B) melting curves.

3.3. Morphology characterizations

To investigate dispersion of MWCNTs and GO in PVA composite fibers, SEM was used to observe the cross-section morphology of as-spun fibers. As can be seen from figure 3(B), there are no obvious aggregates in PVA/MWCNTs 20 wt% composite fibers. This can be attributed to strong hydrogen bonding interaction between oxidized MWCNTs and PVA chains which damages the strong van der Walls interaction among MWCNTs. Compared to PVA/MWCNTs 20 wt% composite fibers, the dispersibility of MWCNTs in PVA/MWCNTs 20 wt%/ GO 5 wt% composite fibers improved slightly. The main reason is that GO serves as a superior dispersant to disperse MWCNTs in PVA composite fibers through supramolecular interactions [10]. The cross-section shape of the PVA/MWCNTs 20 wt% composite fibers is circle. However, that of PVA/MWCNTs 20 wt%/GO 5 wt% composite fibers turned into irregular circle after the introduction of GO. This may be due to the untight stack of GO and oxidized MWCNTs in composite fibers.

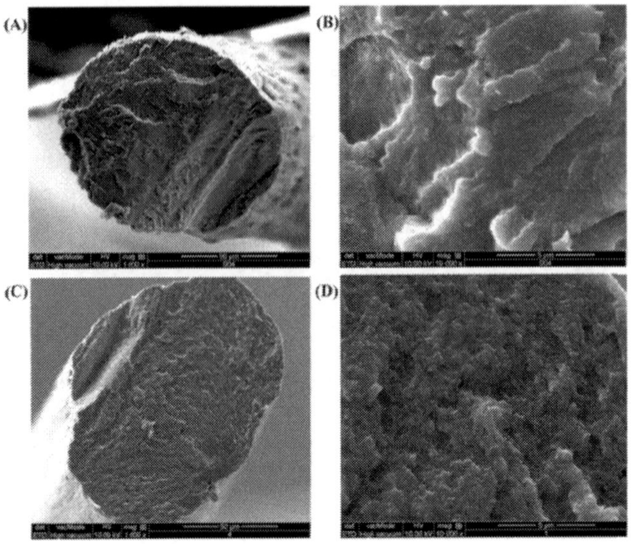

Figure 3. SEM images of the cross-section of (A), (B) PVA/MWCNTs 20wt% composite fibers, (C) and (D) PVA/MWCNTs 20wt%/GO 5wt% composite fibers.

3.4. Tensile strength

The mechanical properties of PVA composite fibers were shown in table 1. The tensile strength and Young's modulus are 0.21 GPa and 10.98 GPa, respectively, for PVA/MWCNTs 20 wt% composite fibers. The value for PVA/MWCNTs 20 wt%/ GO 5 wt% composite fibers spun in similar conditions increased by 38% and 67%, respectively, after the addition of GO. They are 0.29 GPa, 18.31 GPa, respectively. And the elongation for PVA/MWCNTs 20 wt% composite fibers is 20.52%, while that of PVA/MWCNTs 20 wt%/GO 5 wt% composite fibers is 2.74%. The major factor responsible for these differences is the interfacial interactions between the nano-fillers and PVA molecules and synergistic effect through hybridization of MWCNTs and GO [18]. There are a lot of carboxyl and hydroxyl groups on GO sheets, which formed hydrogen bonds with PVA molecules. The hydrogen bonding greatly enhances interfacial interactions and hinders the slippage of nano-fillers and PVA molecules, which leads to high tensile strength and low strain to break.

Table 1. Mechanical properties of gel-spun fibers.

Sample	Tensile strength (GPa)	Young's modulus (GPa)	Elogation (%)
PVA/MWCNTs 20wt%	0.21	10.98	20.52
PVA/MWCNTs 20wt%/GO 5wt%	0.29	18.31	2.74

3.5. Electrical property of the composite fibers

PVA/MWCNTs 20 wt% composite fibers were found to be poorly conductive despite the high content of MWCNTs they contains. Oxidation treatment which can bring in large damages to the MWCNTs must be the main reason. Moreover, the agglomerates also can reduce the electrical conductivity [18]. The electrical conductivity of PVA/MWCNTs 20 wt% composite fibers at room temperature is about 7.3×10^{-5} S/m. That of PVA/MWCNTs 20 wt%/ GO 5 wt% composite fibers increased to 5.3×10^{-2} S/m. The obvious improvement, by three orders of magnitude, is ascribed to the better dispersion of MWCNTs in PVA/MWCNTs 20 wt%/ GO 5 wt% composite fibers and synergistic effect through hybridization of MWCNTs and GO. GO can serve as a superior dispersant to disperse CNTs in polymer matrix through supramolecular interactions. And the introduction of GO allows better intertube contacts. These all contribute to the increasing electrical conductivity.

4. Conclusion

In summary, PVA composite fibers with high content of MWCNTs-GO hybrids were prepared by gel spinning. The total content of MWCNTs-GO hybrids in the PVA composite fibers is up to 25 wt%. The DSC measurement shows that the melting and crystallization peaks decreased due to strong hydrogen bonding interaction between PVA and nano-fillers which constrain the motion of PVA chains. After the introduction of GO, the dispersibility of MWCNTs in composite fibers improved slightly. The tensile strength and Young's modulus increase by 38% and 67%, respectively, due to the increased hydrogen bonding interaction and synergistic effect through hybridization of MWCNTs and GO. More significantly, the electrical conductivity of PVA/MWCNTs/GO composite fibers enhanced by three orders of magnitude with the addition of GO.

Acknowledgments

The authors would like to acknowledge the research center for analysis & measurement of Donghua University. This work is supported by Innovation Program of Shanghai Municipal Education Commission (13ZZ052) and Fundamental Research Program of Science and Technology Commission of Shanghai Municipality (13NM1401502).

References

[1] Iijima S 1991 Helical microtubules of graphitic carbon *Nature* **354** 56-8
[2] Cheng Q, Bao J, Park J, Liang Z, Zhang C and Wang B 2009 High mechanical performance composite conductor: Multi-walled carbon nanotube sheet/bismaleimide nanocomposites *Adv Funct Mater* **19** 3219-25
[3] Lee D H, Lee J A, Lee W J and Kim S O 2011 Flexible field emission of nitrogen-doped carbon nanotubes/reduced graphene hybrid films *Small* 795-100
[4] Chen Y, D. Shaw T, Bai X D, Wang E G, Lund C, Lu W M and Chung D D L 2001 Hydrogen storage in aligned carbon nanotubes *Appl Phys Lett* **78** 2128-30
[5] Castro M, Lu J, Bruzaud S, Kumar B and Feller J F 2009 Carbon nanotubes/poly(ε-caprolactone) composite vapour sensors *Carbon* **47** 1930-42
[6] Shin M K, Lee B, Kim S H, LeeJ A, Spinks G M, Gambhir S, Wallace G G, Kozlov M E, Baughman R H and Kim S J 2012 Synergistic toughening of composite fibres by self-alignment of reduced graphene oxide and carbon nanotubes *Nat Commun* **3** 1-8
[7] Mercader C, Denis-Lutard V, Jestin S, Maugey M, Derre A, Zakri C and Poulin P 2012 Scalable process for the spinning of PVA-carbon nanotube composite fibers *J Appl Polym Sci* **125** E191-6
[8] Vigolo B, Penicaud A, Coulon C, Sauder C, Pailler R, Journet C, Bernier P and Poulin P 2000 Macroscopic fibers and ribbons of oriented carbon nanotubes *Science* **290** 1331-4
[9] Xue P, Park K H, Tao X M, Chen W and Cheng X Y2007Electrically conductive yarns based on PVA/carbon nanotubes*Compos Struct* **78** 271-7
[10] Zhang C, Ren L, Wang X and Liu T 2010 Graphene oxide-assisted dispersion of pristine multiwalled carbon nanotubes in aqueous media *J Phys Chem C* **114** 11435-40
[11] Qiu L, Yang X, Gou X, Yang W, Ma Z-F, Wallace G G and Li D 2010 Dispersing carbon nanotubes with graphene oxide in water and synergistic effects between graphene derivatives *Chem Eur J* **16** 10653-8
[12] Chen M T, Tao T, Zhang L, Gao W and Li C Z 2013 Highly conductive and stretchable polymer composites based on graphene/MWCNT network *Chem Commun* **49**1612-4
[13] James D K and Tour 2012 Graphene: Powder, flakes, ribbons, and sheets *Acc Chem Res* **46** 2307-18

[14] Datsyuk V, Kalyva M, Papagelis K, Parthenios J, Tasis D, Siokou A, KallitsisI and Galiotis C 2008 Chemical oxidation of multiwalled carbon nanotubes *Carbon* **46** 833-40
[15] Mallakpour S and Zadehnazari A 2013 One-pot synthesis of glucose functionalized multi-walled carbon nanotubes: Dispersion in hydroxylated poly(amide-imide) composites and their thermo-mechanical properties *Polymer* **54** 6329-38
[16] Zhang L, Hashimoto Y, Taishi T and Ni Q Q 2011 Mild hydrothermal treatment to prepare highly dispersed multi-walled carbon nanotubes *Appl Surf Sci* **257** 1845-9
[17] Kou L and Gao C 2013 Bioinspired design and macroscopic assembly of poly(vinyl alcohol)-coated graphene into kilometers-long fibers *Nanoscale* **5** 4370-8
[18] Wang R, Sun J, Gao L, Xu C and Zhang J 2011 Fibrous nanocomposites of carbon nanotubes and graphene-oxide with synergetic mechanical and actuative performance *Chem Commun* **47** 8650-2

The study of UHMWPEF surface modification with plasma-induced polymerization

Yu-Fang Zhang[1], Qing-Xiu Jia, Xin Wang and Pei-Ran Zhang
Beijing Key Laboratory of Clothing Materials R&D and Assessment School of Material Science and Engineering Beijing Institute of Fashion Technology Beijing 100029, China

E-mail: zhang-y-f@163.com

Abstract. In order to improve the surface activity levels of the ultrahigh molecular weight polyethylene fiber (UHMWPEF), as well as enhancing the interface strength of the UHMWPEF based composite materials, the method of plasma-induced polymerization was applied to modify the UHMWPEF surface. In this study, the plasma's power, time, pressure and the grafting monomer concentration were introduced. Also, through a well-conducted comparison and analysis of the grafting rate, fabric surface functional groups and the microcosmic morphology, the most suitable plasma modification process was discovered and determined. The mechanics performance of hybrid composites with the modified UHMWPEF and unidirectional carbon fiber cloth (CF) was tested to reveal that, compared with the unmodified composites, the tensile strength and the laminar shear strength could be improved.

1. Introduction

The wettability performance of UHMWPEF surface is generally poor because of its chemical inert behavior, non-polarity, highly orientation and crystallinity of the surface. Thus, when producing the UHMWPEF based composite materials, the bond strength between fiber and matrix resin is small, which would affect the mechanical properties of the UHMWPEF composites, especially the laminar shear strength, transverse tensile strength and fracture toughness etc. This disadvantage of the UHMWPEF hindered the progress in application of structural materials [1]. In order to increase the interface adhesion of UHMWPEF, as well as taking full advantages of its mechanical properties, many modification methods have been researched in terms of improving the wettability performance. For instance the chemical etching method, the radiation induced grafting method and plasma-induced polymerization, etc. In this paper, low temperature plasma treatment method was applied to modify the properties of UHMWPEF, the variation of the percentage of the grafting polymer of UHMWPEF surface with plasma treatment parameters have been explored and the surfaces' functional groups and the microcosmic morphology is also investigated. In the experiment, the hand-made UHMWPEF/CF composite board was prepared for testing the tensile and shear properties of composite materials, and

[1] Address for correspondence: Yu-Fang Zhang, Beijing Key Laboratory of Clothing Materials R&D and Assessment School of Material Science and Engineering Beijing Institute of Fashion Technology Beijing 100029, China. E-mail: zhang-y-f@163.com.

the relationship between the mechanical properties and the grafting monomer concentration is discussed in detail.

2. Experimental section

2.1. Experimental materials and equipment
In this experiment, the fabric of UHMWPE we used was provided by the Beijing Tongyizhong Specialty Fiber Company. The fabric is of 300 g/m^2, and the density of warp and weft is 20 beam/in. The unidirectional carbon fiber sheet and ANKO-300s two-component epoxy adhesive was sourced from Mitsubishi Co, the carbon fiber is also of 300 g/m^2. Other materials are the analytical reagents (AR).

HD-2 the plasma surface-treatment apparatus was manufactured by Changzhoushitai plasma Co. The Soxhlet Extractor was self-produced.

2.2. Specimen preparation

2.2.1. Fiber cloth pretreatment. The UHMWPE fiber cloth was cut into 50mm×50mm samples, firstly they were dipped into the n-heptane for the period of 24 hours, then soaked in a Soxhlet Extractor with acetone extraction for 4 hours, to remove the impurities from the surface of the fiber. After extraction, the fabrics were dried up at 60°C in vacuo, until the weight remained constant.

2.2.2. Plasma treatment. The pre-treated UHMWPEF samples were immersed in acrylamide monomer solvent which was dissolved in N, N - dimethyl formamide (DMF). After soaking entirely, the samples were taken out to dry up. Then the samples were placed into the low temperature plasma processing chamber, where the carrier gas was oxygen. Thorough selection of plasma processing power, time, pressure and the grafting monomer concentration was made, then, the plasma equipment was shut down after achieving processing conditions, and samples were taken out. The homopolymer and impurities on the fabric surface were removed, as described in section 2.2.2, to dehydrate the samples until constant weight was reached at 60 ℃ in the vacuum oven.

2.2.3. UHMWPEF/CF composite material preparation. According to $(CF)_0(UHMWPE)(CF)_0$ layering method, the Plasma-induced polymerization treated UHMWPE fabrics and carbon fiber sheets were coated with ANKO-300s two-component epoxy adhesive by roller coating. Then the resin-impregnated composite product solidified at room temperature. The dosage of ANKO-300s adhesive is 500 g/m^2.

2.3. The performance tests and method

2.3.1. The calculation of grafting rate. The grafting rate of monomer on UHMWPEF surface is defined as:

$$\text{Grafting rate} = (W_1 - W_0)/W_0 \times 100\%$$

Where W_0 is the weight of UHMWPEF before processing, and W_1 is the weight of UHMWPEF after plasma treatment.

2.3.2. Tensile properties tests. The tensile property of UHMWPEF/CF composite was measured by the directional fiber reinforced plastic tensile performance test method (GB/T1447-2005) [2]. The width of the test specimen was 12.5 mm and the length was 230 mm. The sample length direction is the same as carbon cloth direction of composite surface.

2.3.3. Shear properties tests. The interlaminar shear property was measured by the fiber reinforced plastic interlayer shear strength test method (GB/T1450.1-2005) [3]. The length of the test specimen was 25 mm and the width was 6.5 mm.

2.3.4. Surface analysis. The functional groups spectra of the plasma-treated and untreated UHMPEF surface were analysed by Nicolet 670 type FT-IR. The microstructure of the plasma-treated and untreated UHMPEF surface was investigated by JEOL JMS – 6360 LV scanning electron microscopy (SEM).

3. Results and discussion

3.1. The influence factors of plasma modified UHMWPEF fabric

3.1.1. The effects of grafting monomer concentration. In the experiment, the plasma processing power was 140 W; the pressure was 40 Pa; the processing time was 3min, and the concentration of grafting acrylamide monomer was 0-2 mol/l. Figure 1 presents the variation of the grafting rates of UHMPEF surface with monomer concentration, which shows that the grafting rates increase gradually with further increase in monomer concentration. This is because in the plasma processing chamber, monomers vapour by glow discharging process which generated free radicals mixture of small molecules. These active free radicals reached the surface of the fabric, and reacted fast with monomer on the fabric surface to the generated active free radical macromolecules [4, 5]. As the monomer concentration increases, the more the chances of active free radicals were generated by monomer vapour, and the higher the probability that monomer polymerization was induced. Therefore, the grafting rate increases accordingly. In addition, at the interface of gaseous phase and liquid monomer of the plasma, high-energy electrons were mixed with organic solvent DMF, which produced ionic active free radicals, which would induce monomer polymerization. When this kind of ionic active free radical diffused and contacted with the liquid monomer, it would cause further polymerization of acrylamide monomer. That is the reason of the existence of solvent effect in the plasma induced polymerization [6, 7].

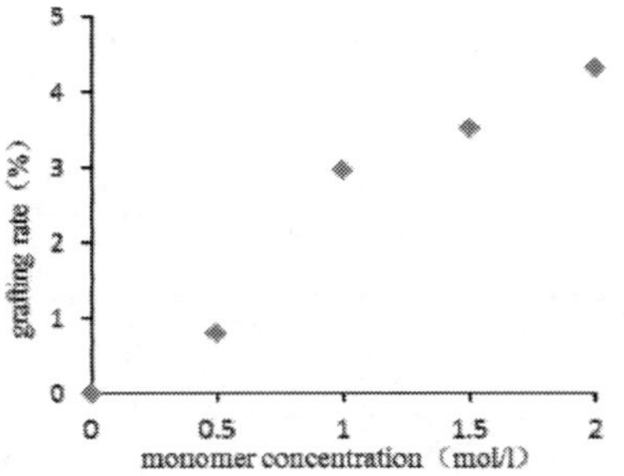

Figure 1. The variation of grafting rate with monomer concentration.

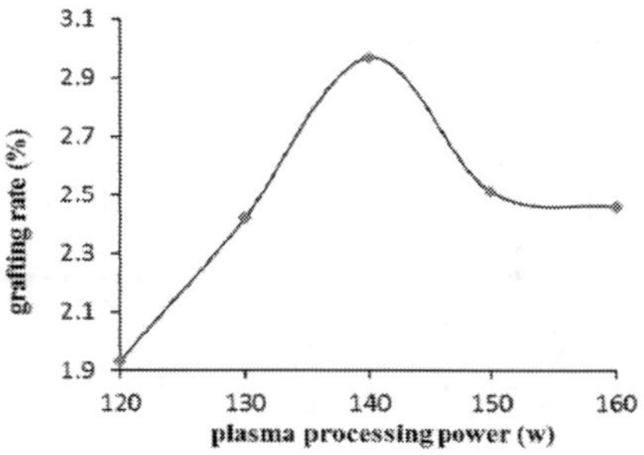

Figure 2. The variation of grafting rate with plasma processing power.

3.1.2. The influence of plasma processing power. In the use of plasma to the modified materials, the degree of the carrier gas ionization is different as the plasma output power changes, thus the plasma concentration inside the chamber is different. In the experiment, the concentration of grafting acrylamide monomer was 1.0 mol/l; the pressure was 40 Pa; the processing time was 3 min, and the plasma processing power was 120-160 W. The variation of the grafting rates of UHMPEF surface with processing power was displayed in figure 2. It is evident that the change of the power largely influenced the result of fabric modification. On one hand, if the processing power was too large, the high energy particles in plasma were increased in number, which would cause strikes to the material surface, neutralising some of the active groups on the material surface, and blocking the insertion of polar groups, thus not leading to graft polymerization [8]. On the other hand, when the processing power was insufficient, the carrier gas was not stimulated to produce plasma, so it was impossible to form active free radicals and to trigger reaction.

3.1.3. The influence of the plasma processing time. In the plasma modification, the processing time is a very important parameter. In this study, monomer concentration was 1.0 mol/l; processing power was 140 W; the pressure was 40 Pa. The variation of the grafting rate of UHMPEF surface together with processing time was displayed in figure 3. The UHMWPEF grafting rates show a maximum value at a plasma treatment time of 3 min and then gradually decrease with further increase in plasma-treatment time. This is mainly because at the beginning of the treatment, the active free radicals concentration increases sharply with the increase in processing time. After a certain period of time, the amount of monomer vapour decreases, making it difficult to form active free radicals [9]. Thus, the grafting rate of the fabric surface initially increases and as the processing time continues, active free radicals cause monomers to polymerize. Then, polymerization results in a decrease of free radical concentration, so the grafting rate declines. Therefore, when conducting the plasma optimization process, it is necessary to determine the appropriate processing time.

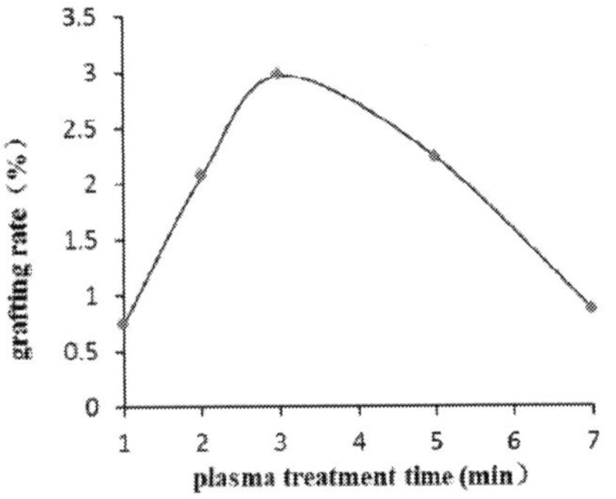

Figure 3. The variation of grafting rate with plasma treatment time.

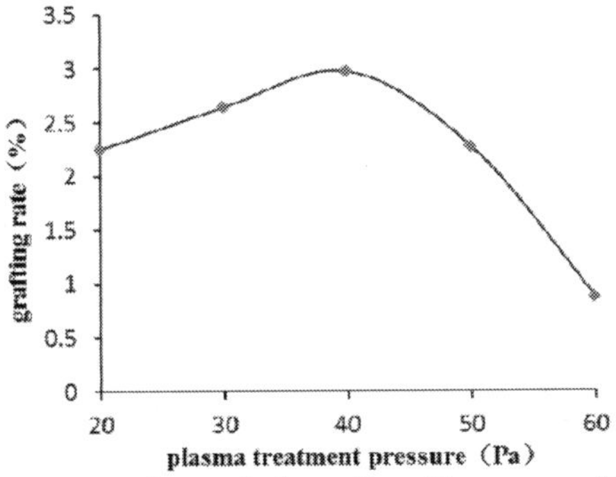

Figure 4. The variation of grafting rate with plasma treatment pressure.

3.1.4. The influence of plasma processing pressure. The conditions were changed to: monomer concentration was 1.0mol/l; the processing power was 140 W; the processing time for 3 min and pressure was 20-60 Pa. The variation of the grafting rate of UHMPEF surface with processing pressure was displayed in figure 4. The grafting rates show a maximum value at a plasma treatment pressure of 40Pa and then gradually decrease with further increase in pressure. This is mainly because the pressure directly affects the gas density in the reaction chamber. As the pressure increases, the plasma density in the chamber increases. The number of active free radicals increases too, and so does the grafting rate. When the pressure is too large, part of the particles carrying the energy of the plasma will collide with the other particles, resulting in energy transfer and lower plasma flight speed, and thus the plasma impact on the fabric surface will be slower. In addition, in the high-frequency AC field, heat quantity by gas molecules vibration poses a great influence on the UHMWPE fiber. Because of the UHMWPE fiber with low glass transition temperature and melting point, in plasma treatment high pressure would likely cause fiber fracture or contraction. Therefore the optimal pressure range is 20-40 Pa.

To sum up, using low temperature oxygen plasma treatment UHMWPE fabric, the most suitable process conditions are: the plasma power is 140 W, the pressure is 40 Pa together with the processing time set to 3 min.

3.2. Surface analysis

3.2.1. Infrared spectroscopy test results and analysis. Figure 5 presents FT-IR spectra of the UHMPE fabric surface. Figure 5(a) is the spectrum from an untreated fiber surface and figure 5(b) is for a surface plasma-treated where the monomer concentration is1.0 mol/l. There was a difference in these two spectra. It is known that amide groups are formed on the UHMPE fiber surface by oxygen-plasma induced polymerization treatment. Although it is difficult to analyze each chemical group quantitatively from these spectra, it is possible to understand the surface grafting polymerization state from the ratio of carbonyl peak and NH_2- peak area against the internal standard peak area. As showed in figure 5, the peak at 2916 cm^{-1}, 2849 cm^{-1}, 1471 cm^{-1} ,717 cm^{-1} is thought to be characteristics of the UHMWPE fiber -CH_2- peak, and the peak at 3359 cm^{-1} is assigned to the -NH_2 of acrylamide, the peak at 1658 cm^{-1} is assigned to the C=O vibration peak. It is shown that acrylamide has been grafted to the UHMWPEF surface.

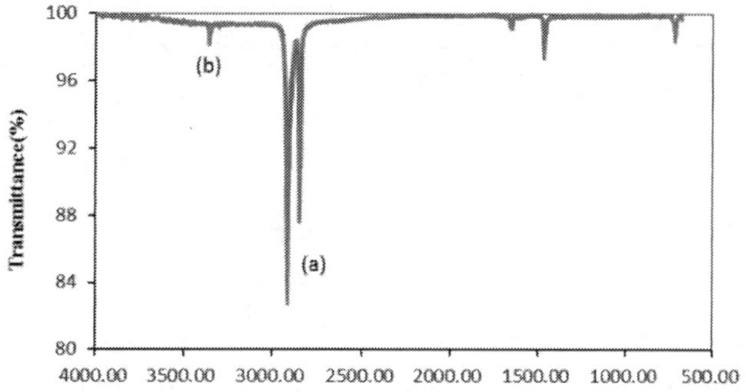

Figure 5. The FT-IR spectra of UHMWPEF surface: (a) in red, the spectrum of an untreated UHMPEF surface and (b) in blue, the spectrum of plasma treated UHMPEF surface (plasma output power: 140 W, pressure: 40 Pa, treatment time: 3min, monomer concentration: 1.0 mol/l).

3.2.2. The SEM test results and analysis. Figure 6 shows the scanning electron micrographs of fiber surfaces. The uniform bamboo-like cracks on the untreated fiber surface are shown in figure 6(a). After plasma processing, the UHMWPE fiber surface became coarse, and the uniform bamboo-like cracks appear covered by an additional coating layer (see figure 6(b)). It is shown that monomer was grafted onto the fiber surface.

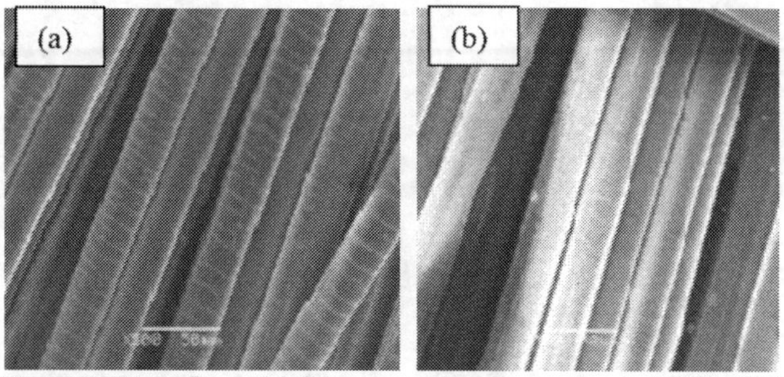

Figure 6. SEM of UHMWPEF surface: (a) the spectrum of an untreated UHMPEF surface and (b) the spectrum of plasma treated UHMPEF surface (plasma output power: 140 W, pressure: 40 Pa, treatment time: 3min, monomer concentration: 1.0 mol/l).

3.3. The mechanical performances of UHMWPEF/CF composites

3.3.1. Tensile property. Figure 7 shows the variation of the tensile strength of modified UHMWPEF/CF composites with grafting monomer concentration. It is clear that with the increase of grafting monomer concentration, the tensile strengths of the composites on $(CF)_0(UHMWPE)(CF)_0$ layers increased. It is because the plasma induced chemical modification can obviously improve the wettability of UHMWPEF surface, increasing the adhesion between the fiber and resin matrix.

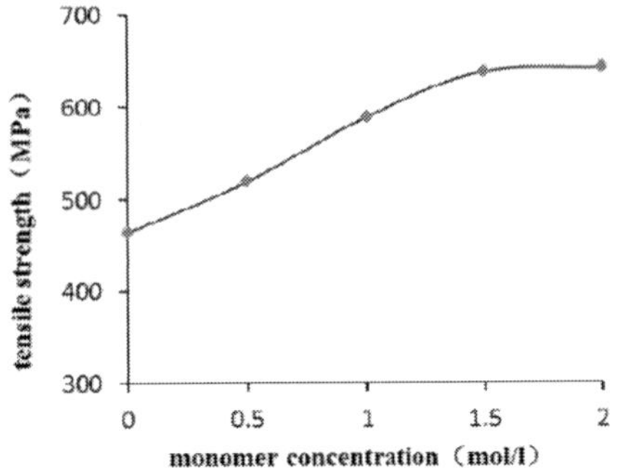

Figure 7. The variation of the tensile strengths with grafting monomer concentration.

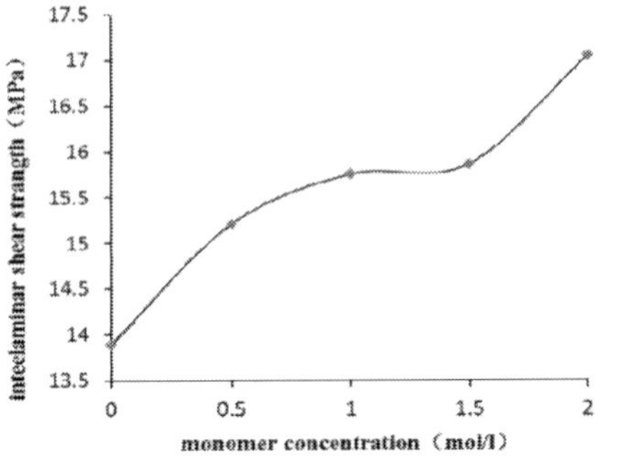

Figure 8. The variation of the Interlayer shear strengths with grafting monomer concentration.

3.3.2. Interlayer shear strengths. Figure 8 shows the variation of the interlayer shear strengths of modifying-UHMWPEF/CF composites with grafting monomer concentration. Same as the grafting rate of UHMWPEF surface, the interlayer shear strengths increase with the increase in grafting monomer concentration. As we know that the breakdown of the interlayer shear of the composite material mainly depends on the adhesion between the fiber and resin matrix, fiber and fiber, and different fiber layers. The plasma-induced grafting polymerization introduces active polymer mixture on the fiber surface by reaction, and this active mixture improves the adhesion between the UHMPE fiber and the epoxy adhesive by chemical bonding. This chemical bonding plays an important role in improving the interfacial adhesion between the UHMWPEF and the epoxy adhesive. On the other side, the oxygen plasma also brings micro-pitting on the fiber surface by etching, and this micro-pitting by mechanical interlocking improves the interfacial adhesion. Therefore the interlayer shear strengths of the composites would increase accordingly.

4. Conclusion

(1) Using low-temperature oxygen plasma induced acrylamide monomer polymerization on the UHMWPE fabric surface, the grafting rates of fabric surface with different processing parameters state that, the most apt condition for plasma treatment is the power 140 W, the time 3 min, and the pressure 40 Pa. As the grafting monomer concentrations increase, the grafting rates of fabric surface increase accordingly.
(2) Using SEM and FT-IR analysis the UHMWPEF surface, it is shown that the low temperature plasma treatment can induce acrylamide-DMF solvent polymerization on the UHMWPEF surface.
(3) The wettability of UHMWPEF surface enhanced largely by low temperature plasma treating. When it recombines with the unidirectional carbon fiber cloth on $(CF)_0(UHMWPE)(CF)_0$ layering, where the grafting monomer concentration is 2.0 mol/l, which compared with the untreated UHMWPEF, the tensile strength of the composite could be increased by 38.2%, and the interlayer shear strength can be increased by 22.7%.

Acknowledgments
In this paper, by the Beijing Natural Science Fundation Project, Beijing municipal education commission of science and technology plan to support key projects, serial number: KZ201510012013.

References
[1] Haixia M 2013 Structure and properties of special ultra-high relative molecular mass polyethylene for fiber *J. Synthetic Resin and Plastics* **30** 60
[2] The national bureau of standards of the People's Republic of China 2005 The tensile performance test methods of fiber-reinforced plastics composites GB/T1447-2005
[3] The national bureau of standards of the People's Republic of China 2005 The interlaminar shear strength test methods of fibre-reinforced plastic composites GB/T1450.1-2005
[4] Simionescu B C, Loanca M, Ioan S, et al. 1981 Plasma-induced Polymerization *J. Polymer Bulletin* **4** 415-9
[5] Simionescu C I and Simionescu B C 1992 The mechanism of plasma-induced polymerization C *Makromolekulare Chemie: Macromolecular Symposia* **54** 595-8
[6] Osada Y, Takase M and Iriyama Y 1983 Effects and role of the solvents on the plasma-initiated solution polymerization of vinyl monomers *J. Polymer Journal* **15** 81-6
[7] Kuzuya M, Kawaguchi T, Yanagihara Y, et al. 1986 Mechanism of plasma-initiated polymerization: Concerning the nature of solvent effects on the lifelike polymerization of water-soluble vinyl monomers *J. Polymer Science: Polymer Chemistry* **24** 707-13
[8] Helang Huang, Lidan Xie and Xiaoning Lu 2006 Effects of low temperature plasma treating on bonding strength of bamboo strip surface *J. Journal of Nanjing Forestry University (Natural science Edition)* **30** 23-6
[9] Yue Wu, Hongsheng Zhang and Zhiyun Xue 2001 Plasma grafting treatment of UHMWPE Fibers *J. Journal of Functional Polymers* **14** 76-80

Effect of coupling agent on nano-ZnO modification and antibacterial activity of ZnO/HDPE nanocomposite films

Y N Li[1], W M Xu and G Q Zhang

School of Mechanical Engineering, Wuhan Polytechnic University, Wuhan, 430023, P.R. China

E-mail: myllyn@126.com

Abstract. Commercial zinc oxide nanoparticles were modified by silane coupling agent of KH550 and KH560, respectively. The obtained nanocomposite was characterized by scanning electron microscopy (SEM), Thermogravimetric analyses (TGA) and apparent sedimentation stability (Sapp). Then, nano-ZnO/high-density polyethylene (HDPE) composite films were prepared via melt blending and a hot compression-molding process. The antibacterial testing of the films against E. coli and S. aureus was carried out via plate counting. SEM, TGA and Sapp suggest the modifier of silane changed the surface hydrophilicity and effectively broke the agglomerations of nanoparticles. Antibacterial testing results indicated the antibacterial rate of the films increased with increasing nano-ZnO content. At a low doped content of 0.2 wt %, the types of coupling agent influenced the antibacterial property with that of KH560-modified nano-ZnO/HDPE KH550-modified nano-ZnO/HDPE unmodified nano-ZnO/HDPE films. When the dosage nano-ZnO was over 0.2 wt %, the differences of antibacterial activity resulted by modification was concealed by the strong antibacterial activity with near 100% of the nano-ZnO/HDPE composite films.

1. Introduction

When Sawai and his colleagues found that ZnO powders had antibacterial activities against some bacteria strains in 1995 [1, 2], more and more researchers have embarked on the studies on ZnO nanoparticles as an antibacterial agent. In recent years, polymer/nano-ZnO composite [3-5] has attracted great attention because of its remarkable desirable properties, including mechanical strength, photostabilization, antibacterial activity, etc.

At the same time, it is well known that, to enhance the dispersion of nanoparticles in polymer matrix, the surface modification for ZnO nanoparticles is a necessity [6]. Silane coupling agents are often used to treat the zinc oxide filler due to their unique bifunctional structure with one end capable of reacting with the silanol groups on ZnO surface and the other end compatible with the polymer [7]. And researches showed that the types and amounts of silane coupling agents influence the properties of nanocomposites, such as morphology, processability, mechanical and barrier properties [8-10].

However, there are few detailed articles so far on the coupling agent on the antibacterial activity of the ZnO reinforced polymer to our knowledge. In this work, ZnO treated with aminoethyl aminopropyl-trimethoxy silane (KH550) and γ -aminopropyl-triethoxy silane (KH560) respectively

[1] Address for correspondence: Y N Li, School of Mechanical Engineering, Wuhan Polytechnic University, Wuhan, 430023, P.R. China. E-mail: myllyn@126.com.

filled high density polyethylene (HDPE) nanocomposite films were prepared via melt blending and hot compression molding process. The surface treatment of the ZnO nanoparticles was characterized by SEM, TGA and sedimentation experiment. Moreover, the effect of coupling agent as well as the ZnO content on the antibacterial properties of HDPE films was investigated.

2. Experimental

2.1. Materials
The HDPE resin (5021D) was from China National Offshore Oil Corporation in Beijing. The ZnO nanoparticles (around 50 nm average diameter, from Nanjing Haitai nano materials Co., China) were commercially available. Surface modifier, silane coupling agent of 3- aminopropyl-trimethoxysilane (KH550) andγ-glycidoxypropyl-trimethoxysilane (KH560) was bought from Shuguang Chemicals Company of Nanjing, China. The other reagents were analytical grade and were bought from chemical store.

2.2. Modification of nano-ZnO
Nano-ZnO particles (10g) and ethanol (100 g) were mixed with ultrasonator (KQ-300DV, China) for 30min. The modifier (0.5 g), pre-hydrolysed for 1h in ethanol/water (1/10, wt) solution, was slowly dripped in the system. After 1h reaction at a given temperature, the mixture was filtrated to remove the solvent and then get dry solid particles and grind them into powders.

2.3. Preparation of nano-ZnO/HDPE composite films
The master batch ((ZnO/HDPE) wt.=5:100) was fabricated firstly through HARRK internal mixer (PolylabRC.300P, Germany). The nano-ZnO/HDPE composite particles were prepared by mixing master batch and HDPE with a single-screw extruder (Brabender PLE330, Germany) and a granulator. Then the composite particles were made to a sheet via a two-roll mill (SK-160B, China). Test specimens (about 120 um) were compression-molded with the sheet by a press machine (SL-45, China).

2.4. Particle characterization
The observation of the nano-ZnO morphology was carried out using a scanning electron microscope (SEM, JSM-6380, JEOL Company, Japan). The sample was prepared by dropping the solution of nano-ZnO in ethanol on the electric conductive adhesive and making it evaporated at room temperature.

Thermogravimetric analyses (TGA) were performed with a TGA instrument (TGA Q500, TA Company, American) to study surface modification of nano-ZnO. The nano-ZnO powder, which had been dried on 120℃ for 6 h, was heated under nitrogen from room temperature to 800℃ at a ramping rate of 10℃/min. The weight loss of the nano-ZnO powder as a function of temperature was recorded.

The apparent sedimentation stability (S_{app}) was assessed by measuring the sedimentation speed of ZnO dispersion in water mixture with the aging time [11]. 10 wt.% ZnO nanoparticles-containing water dispersions were prepared and located in a glass tube of 0.5 cm inner diameter and 20 cm length with a scale and a cover. Then, the sedimentation stability of ZnO particles in the mixture was monitored at a given aging time, as equation (1):

$$S_{app}(\%) = \frac{l_{(t)}}{l_0} \times 100 \quad (1)$$

where, l_0 is the length of initial opaque dispersion and $l_{(t)}$ is the length of sediment part at a given time t in the tube, respectively.

2.5. Antibacterial tests

The antibacterial test of the composite films against *Escherichia coli* (ATCC10536, *E. coli*) and *Staphylococcus aureus* (ATCC6538, *S. aureus*) was according to the reference [12]. The antibacterial ratio was calculated using equation (2):

$$R = \frac{A-B}{A} \times 100\% \tag{2}$$

where R, antibacterial ratio (%); A, the CFU of the blank sample; B, the CFU of prepared samples.

3. Results and discussion

3.1. *Characterization of ZnO nanoparticles*

3.1.1. SEM. Scanning electron micrographs of original ZnO and modified ZnO nanoparticles are showed in figure 1. Figure 1(a) reveals that the original commercial ZnO nanoparticles are aggregated severely in ethanol solution. This should be attributed to the high surface energy of nanoparticles. From figures 1(b) and 1(c), it can be found that the dispersion of nanoparticles was greatly improved. That suggests that the modifier of silane effectively broke the agglomerations. However, the SEM pictures show no difference of the effect of the type of silane on the dispersion of nano-ZnO.

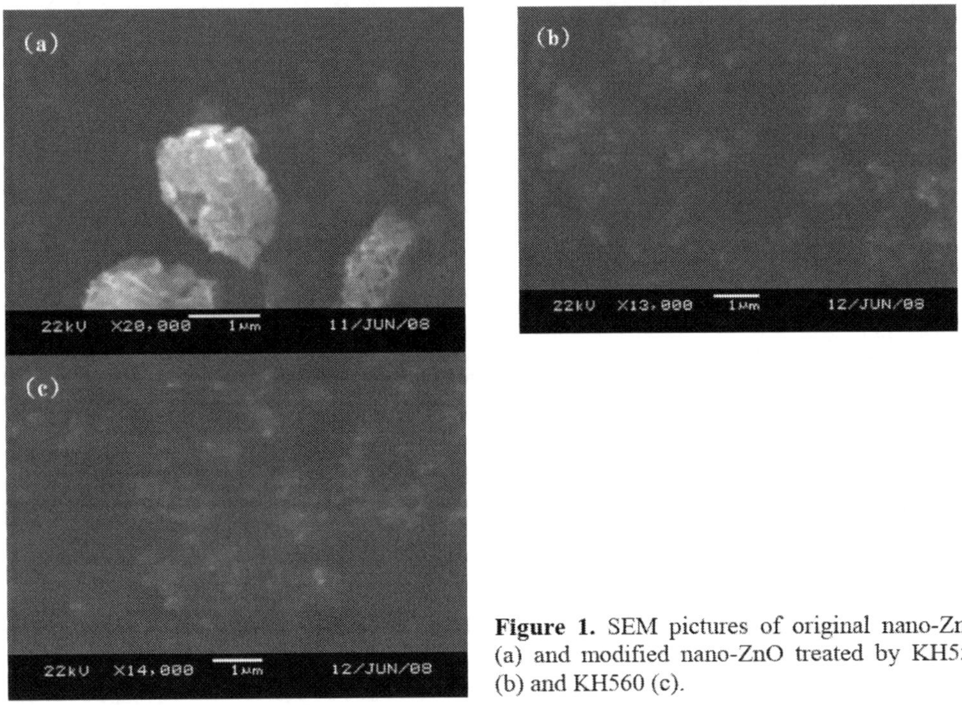

Figure 1. SEM pictures of original nano-ZnO (a) and modified nano-ZnO treated by KH550 (b) and KH560 (c).

3.1.2. Thermogravimetric analyses of nano-ZnO. The surface modification on the nano-ZnO was also characterized by TGA. It can be seen from figure 2 that the unmodified nano-ZnO has a weight loss (about 5wt%) around 100℃, which relates to the elimination of physically absorbed water on the surface [13]. The nano-size fillers possess a large surface area which makes it easy to pick up the moisture in a typical laboratory environment even after they are dried at 120℃ for 6 h. The weight loss above 150℃ for original nano-ZnO may be due to the chemically bonded water and residual organics from the process of the nano-ZnO synthesis [7]. On the other hand, it is found that the nano-

ZnO picks up less water after surface modification because of the changing of the surface hydrophilicity. The modified nano-ZnO begins to lose weight sharply after 200 ℃, which is contributed to the debonding and degradation of grafted silane functional group on the surface [14].

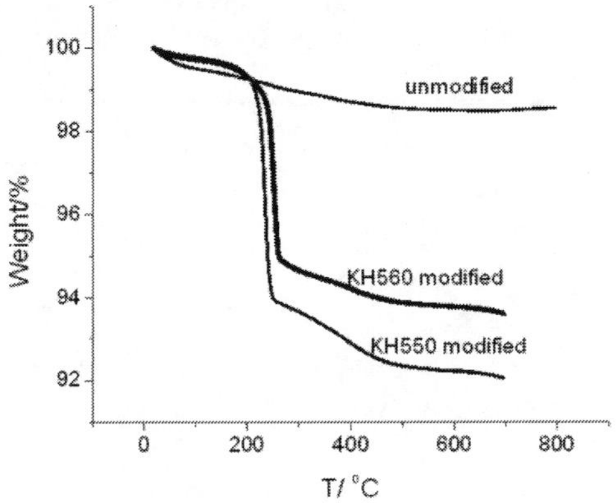

Figure 2. TGA graphs of nano-ZnO.

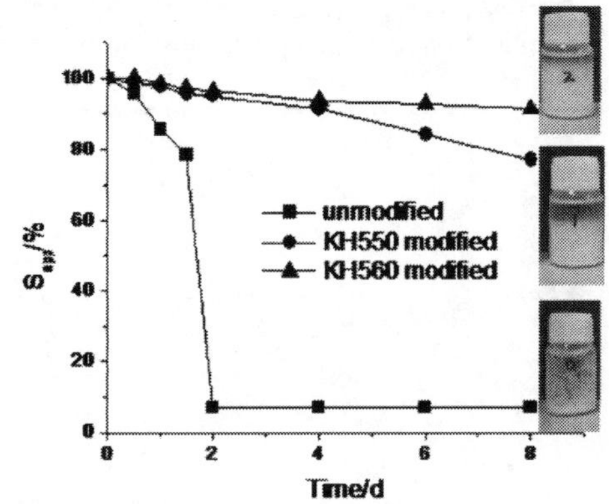

Figure 3. Sedimentation stability curves of nano-ZnO dispersed in water.

3.1.3. Sedimentation stability of nano-suspensions. To check the effect of surface modification, the stability of modified zinc oxide nanoparticle dispersion in water was compared with that of original zinc oxide nanoparticle (figure 3). The results show a remarkable difference. As shown in figure 3, original ZnO nanoparticle completely precipitated after 2 days. On the contrary, modified zinc oxide keeps a relatively stable colloidal dispersion in water for more than a week.

To check the effect of surface modification, the stability of modified zinc oxide nanoparticle dispersion in water was compared with that of original zinc oxide nanoparticle (figure 3). The results show a remarkable difference. As shown in figure 3, original ZnO nanoparticle completely precipitated after 2 days. On the contrary, modified zinc oxide keeps a relatively stable colloidal dispersion in water for more than a week.

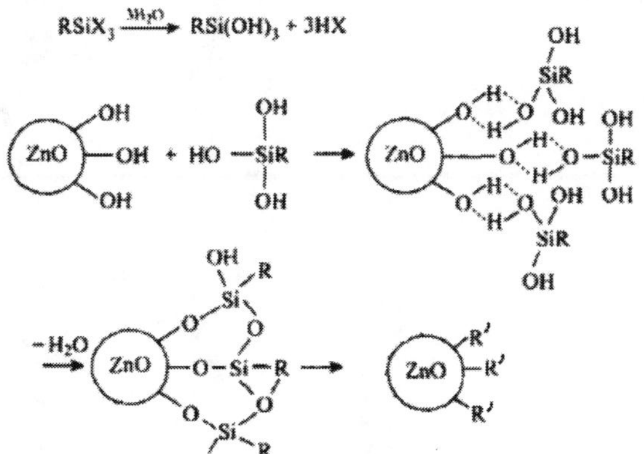

Figure 4. Schematic process for modification of silane coupling agents on ZnO nanoparticles.

The modification process is described as a hydrolysis and condensation reaction between the ZnO nanoparticle surface and silane coupling agents in a polar medium as shown in figure 4. Conventional inorganic nanoparticles own some hydroxyl groups (–OH) on the surface due to the impact of water or moisture. The hydroxyl groups of nano-ZnO particle surface interact with silanol group (Si-OH) of coupling agent and form poly (zinc silane) complex on the surface of ZnO nanoparticles. Consequently, the oligomers are initially grafted or anchored on the surface of the nanoparticle through one or several spots. The other terminal of silane is organic chain, which fulfill steric hindrance between inorganic nanoparticles [15]. This indicates that the grafting silane chain reduces the aggregation of ZnO nanoparticles (shown in figure 1) and increases the compatibility with organic solvent.

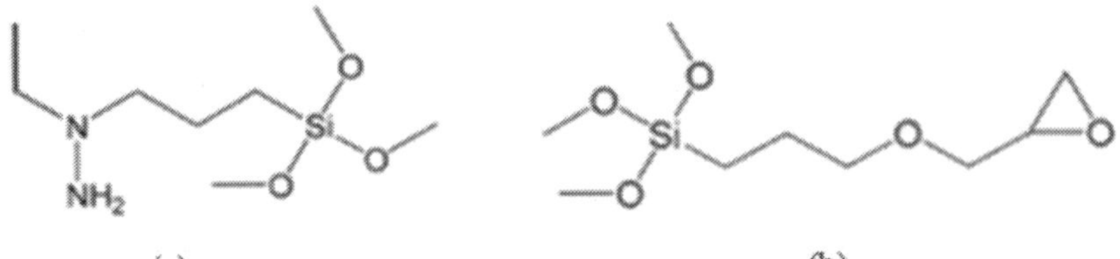

(a) (b)

Figure 5. Chemical structure of KH550 (a) and KH560 (b).

In addition, the chemical structures of silane coupling agents are shown in figure 5. Compared with epoxy group on KH560, the amino group on KH550 is still hydrophilic [7]. On the other hand, the amino group in the silane is easily to form hydrogen bonding with hydroxyl groups on the ZnO surface [16], which also influences the proceeding of reaction between ZnO and hydrolyzed silane. Those may lead to the lower S_{app} after 4d and visually significantly more precipitation at 8 d of the KH550-modified nano-ZnO due to the slightly worse effect of surface modification of KH550 in contrast with KH560.

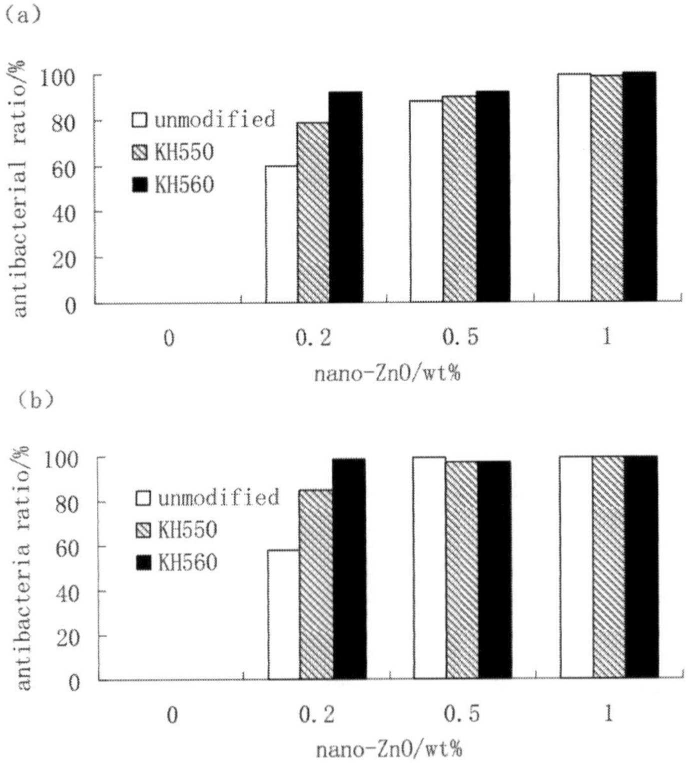

Figure 6. Antibacterial activity of nano-ZnO/HDPE composite films, (a) *E. coli*; (b) *S. aureus*.

3.2. Antibacterial property

The antibacterial activity of nano-ZnO/HDPE composite films against *E. coli* and *S. aureus* is shown in figure 6. It can be seen that the antibacterial ratio increased with the increase of nano-ZnO content, while the HDPE films had no any bacterial effect in this research with the antibacterial rate of 0%. When the concentration of nano-ZnO was 0.5wt %, the antibacterial ratio exceeded 90% against both *E. coli* and *S. aureus*. The films showed a stable strong antibacterial rate of near 100% when the content of nano-ZnO was over 0.5 wt %. That indicates that the composite films of nano-ZnO/HDPE possessed of excellent antibacterial property and the bacterial growth was effectively inhibited when the additon of nano-ZnO was only 0.5 wt %.

It was also found that for the low doped nano-ZnO with the content of 0.2wt %, the modification of nano-ZnO dramatically improved the antibacterial activity of the composite films against both *E. coli* and *S. aureus* compared with the unmodified nano-ZnO-doped HDPE films. In which, the antibacterial activity of KH560-modified nano-ZnO/HDPE films was better than that of KH550-modified nano-ZnO/HDPE films. The films showed a high antibacterial rate of near 100% against both *E. coli* and *S. aureus* when the concentration of ZnO nanoparticles was above 0.5 wt % with no influence by the surface modification of nano-ZnO and the type of coupling agents.

It is well known that nonpolar and hydrophobic HDPE resin lacks a strong interaction with polar and hydrophilic ZnO nanoparticles. Silane treatment improves the dispersability of nanoparticles in polymer matrix resulted from the reduction of the nanoparticle aggregation and enhancement of the interaction between HDPE and nano-ZnO [12, 17]. Therefore, at a relatively low doped content of modified nano-ZnO, such as 0.2 wt %, the aggregation of nano-ZnO in the HDPE matrix decreased after it was modified by the silane coupling agent. Moreover, from the result of sedimentation stability of nano-suspensions, the nanoparticles treated by KH560 would show better dispersability in the HDPE matrix than treated by KH550. Hence, the antibacterial activity of KH560-modified nano-ZnO/HDPE films was the best at a nano-ZnO content of 0.2 wt %.When the content of nano-ZnO was over 0.2 wt %, the differences of antibacterial activity resulted by modification was concealed by the strong antibacterial activity of the nano-ZnO/HDPE composite films.

4. Conclusions

- The modifier of silane changed the surface hydrophilicity and effectively broke the agglomerations of nanoparticles from SEM images and TGA curve. The effect of surface modification of KH560 was better than KH550 from the result of sedimentation experiment.
- The antibacterial rate of nano-ZnO/HDPE increased with increasing content of nano-ZnO. The composite films of nano-ZnO/HDPE possessed excellent antibacterial properties and that the bacterial growth was effectively inhibited when the dosage of nano-ZnO was only 0.5 wt %.
- The effect of surface modification influenced the antibacterial property due to dispersability of nanoparticles in HDPE matrix. The antibacterial activity of KH560-modified nano-ZnO/HDPE films was the best compared with KH550-modified nano-ZnO/HDPE and unmodified nano-ZnO/HDPE films at a low nano-ZnO content of 0.2 wt %. When the dosage nano-ZnO was over 0.2 wt %, the differences of antibacterial activity resulted by modification disappear due to the strong antibacterial activity with near 100% of the nano-ZnO/HDPE composite films.

Acknowledgment

The authors thank Wuhan Polytechnic University Scientific Research Funding (2013d13) and Research Funding of Hubei Provincial Education Department (D20151703) for the financial support.

References

[1] Sawai J, Igarashi H, Hashimoto A, Kokugan T, Shimizu M and Chem J 1995 Evaluation of growth inhibitory effect of ceramics powder slurry on bacteria by conductance method *Eng. JPN.* **28** 288-93

[2] Sawai J, Saito I, Kanou F, Igarashi H, Hashimoto A, Kokugan T, Shimizu M and Chem J 1995 Mutagenicity test of ceramic powder which have grown inhibitory effect on bacteria *Eng. JPN.* **28** 352-4

[3] Sato M, Kawata A, Morito S, Sato Y and Yamaguchi I 2008 Preparation and properties of polymer/zinc oxide nanocomposites using functionalized zinc oxide quantum dots *Eur. Polym. J.* **44** 3430-8

[4] Li Shanghua, Toprak Muhammet S, Jo Yun Suk, Jon Dobson, Do Kyung Kim and Mamoun Muhammed 2007 Bulk synthesis of transparent and homogeneous polymeric hybrid materials with ZnO quantum dots and PMMA *Adv. Mater.* **19** 4347-52

[5] Tang E, Liu H, Sun L, Zheng E and Cheng G 2007 Fabrication of zinc oxide/poly (styrene) grafted nanocomposite latex and its dispersion *Eur. Polym. J.* **43** 4210-8

[6] Hong R Y, Qian J Z and Cao J X 2006 Synthesis and characterization of PMMA grafted ZnO nanoparticles *Powder Technol.* **163** 160-8

[7] Sun Y Y, Zhang Z Q and Wong C P 2005 Study on mono-dispersed nano-size silica by surface modification for underfill applications *J. Colloid Interf. Sci.* **292** 436-44

[8] Sun S S, Li C Z, Zhang L, Du H L and Burnell-Gray J S 2006 Effects of surface modification of fumed silica on interfacial structures and mechanical properties of poly (vinyl chloride) composites *Eur. Polym. J.* **42** 1643-52

[9] Tian M, Liang W, Rao G, Zhang L and Guo C 2005 Surface modification of fibrillar silicate and its reinforcing mechanism on FS/rubber composites *Compos. Sci. Technol.* **65** 1129-38

[10] Douce J, Boilot J, Biteau J, Scodellaro L and Jimenez A 2004 Effect of filler size and surface condition of nano-sized silica particles in polysiloxane coatings *Thin Solid Films* **466** 114-22

[11] Shim J W, Kim J W, Han S H, Chang I S, Kim H K, Kang H H, Lee O S and Suh K D 2002 Zinc oxide/ polymethylmethacrylate composite microspheres by in situ suspension polymerization and their morphological study *Colloid Surf A.* **207** 105-11

[12] Li S and Li Y 2010 Mechanical and antibacterial properties of modified nano-ZnO/HDPE composite films in low doped content of nano-ZnO *J. Appl. Polym. Sci.* **116** 2965-9

[13] Dugas V and Chevalier Y 2003 Surface hydroxylation and silane grafting on fumed and thermal silica *J. Colloid Interface Sci.* **264** 354-61

[14] Li X, Wang G and Li X 2005 Surface modification of nano-SiO2 particles using polyaniline *Surf. Coat. Tech.* **197** 56-60

[15] Tang E, Cheng G, Ma X, Pang X and Zhao Q 2006 Surface modification of zinc oxide nanoparticle by PMAA and its dispersion in aqueous system *Appl. Surf. Sci.* **252** 5227-32

[16] Piers A S and Rochester C H. Infrared study of the adsorption of 1- aminopropyltrialkoxysilanes on silica at the solid/liquid interface *J. Colloid Interface Sci.* **174** 97-103

[17] Rong M Z, Zhang M Q, Pan S L and Friedrich K J 2004 Interfacial effects in polypropylene-silica nanocomposites *J Appl Polym Sci.* **92** 1771-81

Repair of filament wound composite pipes

Ramin Amali[1] and Heather Arnall

University of the West of England, Coldharbour Lane, Bristol, BS16 1QY, UK

Email: Ramin2.Amali@uwe.ac.uk

Abstract. Filament wound pipes are used in a wide variety of industries, due to the advantages composites have over metal pipes, such as a high strength to weight ratio, and resistance against frost, corrosion and heat. Composite pipes require minimal maintenance to ensure they are safe. Any damage occurring in composite pipes could lead to failure; therefore all damage should be assessed through NDT. If it is decided that the damage makes the pipe unsafe then a decision needs to be made whether to repair or replace the pipe. Repairing a composite pipe can be quicker, easier and cheaper than replacing it and can restore the strength of the pipe effectively. This investigation looks at the repair process and the parameters involved in determining the strength of the pipe following repair through the use of over 150 models in FEA software, Abaqus. Parameters considered include the pipe diameter and thickness, damage removal size and wrap width and thickness. It was found that if the pipe is thin-walled then it can be assumed that the pipe's thickness has no effect on the FOS following repair. Formulas were created to predict the FOS following repair for varying pipe diameters, damage sizes and wrap thickness. Formulas were also created to determine the wrap width required for varying wrap thicknesses and damage sizes.

1. Introduction

Filament wound pipes are readily used in a wide variety of industries, one example being the oil and gas industry. Composite pipes hold advantages over steel pipes due to being lighter, stronger and stiffer meaning less supports are required. They are also able to take greater loads and are resistant to frost, corrosion and heat, with minimal maintenance required. However if a filament wound pipe becomes damaged, there is the option to replace or repair. Damage mostly occurs at joints, fittings, bends, tees and reducers, with 49% being due to defective installation, 49% due to overloading of material due to shortcomings in design and 1% due to material defects [1]. Composites' strength also degrades with temperature elevations and moisture with the decrease in strength can be as high as 50-90% [2]. Damaged area should be removed with circular or oval shapes to avoid stress concentrations, but if this is not possible square or rectangular cut out can be used with rounded corners. A composite wrap can be used to repair pipes with diameters between 0.1 m and 1.42 m. It is 0.292 m long and needs to overlap the defect by 51mm on each side, if there is a large defect then multiple wraps can be used overlapped [3]. It is difficult to know the repaired pipe's new strength due to the number of parameter involved (pipe diameter and thickness, damage size, and the length and thickness of the wrap). Previous studies have considered the effect of the variation of individual parameters, but none have considered the effect of the parameters combined. This investigation considers these parameters

[1] Address for correspondence: Ramin Amali, University of the West of England, Coldharbour Lane, Bristol, BS16 1QY, UK. Email: Ramin2.Amali@uwe.ac.uk

and the effect they have on the factor of safety (FOS) of a damaged and repaired pipe through the use of FEA software. A total of over 150 models are analysed and the stresses and FOS are used to create formulas to predict the FOS following repair.

2. Literature review

Rousseau et al researched the influence of the winding pattern on the damage behaviour of filament wound glass/epoxy pipes under pressure with a [±55°] layup. The winding pattern was varied through different degrees of interweaving to find that interweaving should be minimised [4]. Meanwhile, surface cracks were researched by Tarakçioglu et al. by considering cracks of varying notch aspect ratios, notch to thickness ratios and winding angles under internal pressure. For the ±55° pipes they found that damage occurred at the notch tips of the pipe and then delamination occurred. The surface crack has the biggest effect on the burst strength for the ±75° winding angle [5]. Gemi et al. considered the progressive fatigue failure behaviour of glass/epoxy with a winding angle of $[\pm 75]_2$ under internal pressure. They found damage can be divided into three main steps, which starts with whitening and matrix cracking. The matrix cracking causes a leakage path to start forming from the inner side to the outer surface. Once the crack has reached the outside of the pipe failure occurs with catastrophic failure. They also concluded that if the applied load is high then the leakage and final failure coincides [6]. More recently Rafiee's work predicts the strength of GRP pipes with sand as a core material. The failure is modelled layer by layer, and once damaged the layer the mechanical properties are replaced by degraded properties [7]. Arikan considered inclined surface cracks, with varying crack angles, under static internal pressure in $[\pm 55°]_3$ filament wound pipes. It was concluded that as the crack angle increased the burst pressure increases and the effect of the crack angle decreases [8]. Günaydın et al. considered the effects of composite repair patches and number of layers on the fatigue behaviour of surface-notched composite pipes. They found that out of the amount of layers tested the maximum was the most effective at preventing fatigue failure [9]. Repairing impact damaged composite pipes was investigated by Kara et al, who investigated the amount of layers required to effectively repair the pipe, finding that 6 layers are required [10].

3. FEA modelling of composite pipe

Abaqus FEA software was used to model the pipe when undamaged, with the damage removed and following repair. In order for longitudinal stress to be considered one end of the pipe is closed. The damage and wrap is modelled through the use of partitions with air, pipe and wrap properties being applied to the relevant regions. These properties are given in Table 1.

Table 1. Material properties of glass fibre and air.

	Glass fibre	Air	Units
Young's Modulus	-	0.01	MPa
Poisson's Ratio	0.3	0.03	-
long. modulus E1	120	-	GPa
trans. modulus E2	8	-	GPa
shear modulus G12	6	-	GPa
long. tension Xt	1800	-	MPa
long. comp. Xc	-1200	-	MPa
trans.tension Yt	80	-	MPa
trans. comp. Yc	-200	-	MPa
Shear, S	150	-	MPa

Figure 1 shows the design of the pipe modelled in Abaqus, with length of 1 m and pressure of 6 MPa applied. The pipe has a fibre orientation of $[54.75\ -54.75]_4$, whilst the wrap has a fibre orientation of [0 90 0 90] s. A mesh study was carried out to consider the amount of elements required to correctly analyse the nominal and maximum stresses. Through considering the nominal stress, it was found that 6,600 elements were required to correctly analyse the nominal stress. This is equivalent to a node every 7 mm, but more elements were required for the maximum stress. Therefore the mesh was refined around the hole. Through the refinement mesh study it was found that 15 elements per line were required for refinement. This brings the total amount of elements to 8, 860 elements, which is shown in Figure 2. This mesh is structured with the elements being quad-dominated.

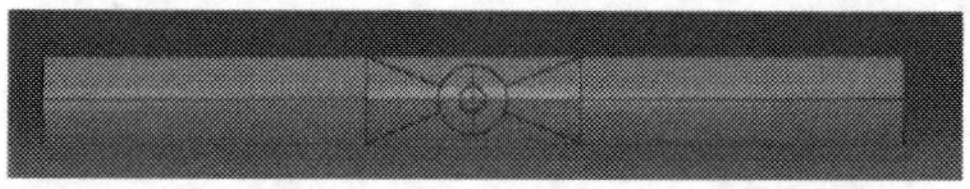

Figure 1. Design of pipe model.

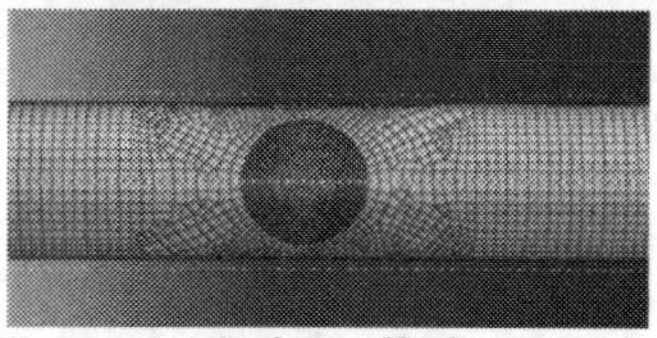

Figure 2. Mesh applied to models.

4. Effect of pipe diameter on FOS

Pipe diameter varies dramatically with applications therefore it is important to understand how pipe diameter affects effective pipe repair. In this investigation the FOS of the pipes prior and following damage and following repair are considered to obtain an understanding of the effect of removing damage and repairing a pipe for different size pipes: 0.1 m, 0.3 m and 0.5 m. This was done through the use of 9 initial FEA models. First to be considered was the effect of pipe diameter on the damage removal FOS, with the results shown in Figure 3, where FOS reduction factor is as defined in equation (1). These results were then used to create equation (2).

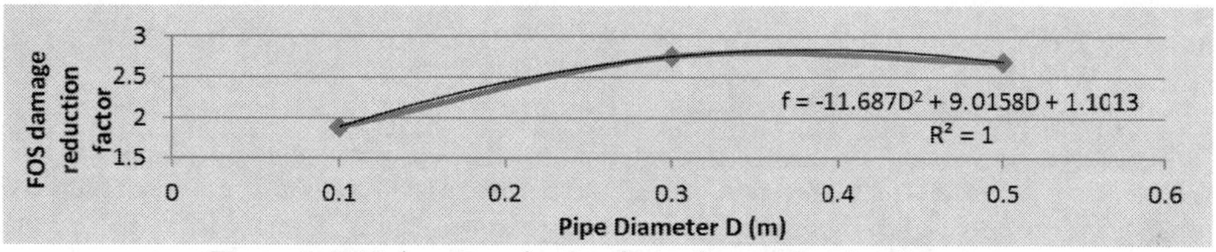

Figure 3. FOS damage reduction factor for varying pipe diameters.

$$FOS\ reduction\ (or\ gain) factor\ (f) = \frac{FOS_{prior}}{FOS_{after}} \tag{1}$$

$$FOS_{after\ damage\ removed} = \frac{FOS_{prior\ damage}}{-11.687D^2+9.0158D+1.1013} \tag{2}$$

Next the repair of damage was considered, by considering the relationship between FOS before repair and following. From the results equation (3) was produced to predict the FOS following repair if the FOS after damage removal is known. Often the middle stage of repair is not of interest; therefore results already obtained were used to consider the relationship before damage and after repair for varying pipe diameters to create equation (4).

$$FOS_{after\ repair} = \frac{FOS_{after\ damage\ removed}}{3.1873D^2-2.1499D+0.708} \tag{3}$$

$$FOS_{after\ repair} = \frac{FOS_{prior\ damage}}{2.7236D^2-1.2208D+1.0848} \tag{4}$$

These three equations are accurate for pipes with a diameter of up to 0.5 m. In order to determine the accuracy of these formulas an additional two pipe diameters (0.05 m and 0.2 m) were analyzed. It was found that the maximum percentage difference between FEA results and that predicted by the formulas was less than 5%, giving the formulas an accuracy of 95%.

5. Effect of pipe thickness on FOS

In this investigation thin-walled pipes are considered, this means that the thickness should be less than 10% of the radius of the pipe. As with the previous investigation the three stages (undamaged, damaged and repaired) are considered for the three pipes, giving a total of 9 models to be analysed. Six of the models were analysed to obtain an understanding of the effect of the thickness of the pipe on the FOS when removing damage. These results were used to create equation (5) to predict the FOS following damage removal for varying pipe thicknesses. To consider the repair of damage, the FOS following damage removal results were used and an additional three model (of varying thicknesses with wraps) should be analysed. These results can then be used to create equation (6) to predict the FOS following the wrap being applied.

$$FOS_{after\ damage\ removed} = \frac{FOS_{Prior\ damage}}{0.0174T^2 - 0.178T + 2.1234} \quad (5)$$

$$FOS_{after\ repair} = \frac{FOS_{after\ damage\ removed}}{-0.0066T^2 + 0.0669T + 0.449} \quad (6)$$

When considering the effect of the pipe diameter it was mentioned that often the FOS after damage but before repair is often not of interest. Therefore Figure 4 shows the overall reduction factor from, comparing before damage and after repair and this was used to create equation (7).

$$FOS_{after\ repair} = \frac{FOS_{prior\ damage}}{-0.0012T^2 + 0.01T + 1.01T + 1.0109} \quad (7)$$

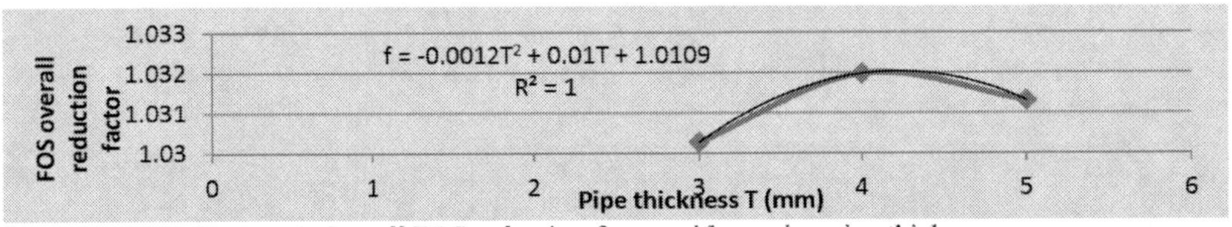

Figure 4. Overall FOS reduction factor with varying pipe thicknesses.

From Figure 4 it can be seen that the FOS overall reduction was constant despite the thickness of the pipe at approximately 1.03. Therefore it should be considered whether the FOS overall reduction factor is constant if the pipe is thin walled. This was done through additional pipe thicknesses of 3 mm and 15 mm to model six additional FEA models. From the results it was found that the overall FOS reduction factor was 1.03 and 0.9963 for the 3 mm and 15 mm pipes respectively. This is a difference of 0.04, for a thickness difference of 12 mm. Therefore it can be stated that the thickness of the pipes has no impact on the final FOS for thin walled pipes.

To gain an idea of how accurately the equations created can predict FOS, two additional thicknesses were considered (2 mm and 3.5 mm). These were modelled in Abaqus and the FEA results were compared with results obtained from the equations. The repaired FOS can be accurately obtained from the formulas created. In fact it was found that this formula had an accuracy of 99%. However the formulas did not predict the damaged FOS very well, with only an accuracy of 66%. These formulas are valid for pipes which are thin walled.

6. Effect of damage removal size on FOS

Damage comes in all different sizes and as little as suitable area surrounding the damage should be removed. Therefore it is required to investigate into the effect of different damage sizes. In this study, three initial studies of varying damage sizes are considered: 30 mm, 50 mm and 70 mm, through 7 models analysed in Abaqus. To consider the removal of damage, four of the models were analysed and the minimum FOS was taken from each model. These models were of the pipe prior to damage and after damage removal. These FOS were then used to obtain equation (8). The effect of repairing damage was considered to obtain equation (9).

$$FOS_{after\ damage\ removed} = \frac{FOS_{prior\ damage}}{-222.32d^2 + 27.228d + 0.839} \quad (8)$$

$$FOS_{after\ repair} = \frac{FOS_{after\ damage\ removed}}{117.05d^2 - 14.232d + 1.1695} \tag{9}$$

From all the results collected the overall gain or loss of strength can be considered, by looking at before damage and after repair. From these results were used to obtain Figure 5 and equation (10). It was noticed that the overall FOS gain factor varies very little for different damage diameters.

$$FOS_{after\ repair} = \frac{FOS_{prior\ damage}}{2.6216d^2 - 0.2097d + 1.2382} \tag{10}$$

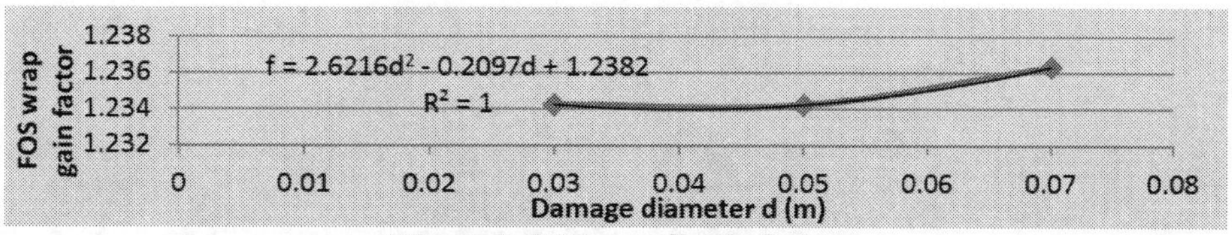

Figure 5. FOS overall gain factor.

To determine the accuracy of the formulas created in this section, an additional scenario was considered with damage of 40mm under the same conditions as previously. Two additional models are produced: before damage, following damage and following repair. The results from FEA are then compared with the FOS predicted from the formulas. It was found that there is a good comparison between FEA and the formulas created. The formulas can accurately predict the FOS once repaired, with a reliability of 99.98%. However there was a percentage difference of 38% between formulas and FEA for predicting before repair. These formulas are valid for damage sizes between 0.03m and 0.07 m.

7. Effect of damage removal shape on FOS

It is not always possible to remove damage with a circular hole, due to the geometry and the size of the damage and component. In these cases, a square cut out with rounded corners should be used instead. From a previous study the worst case scenario of a square cut out was considered and it was found that the best angle to remove damage is at 45° [11]. Therefore, an investigation was carried out to determine the difference in using a square cut out at 45° with a circular cut out. To make the shapes comparable the area between the cut out shapes were kept constant, as shown in Table 2, and the FOS following damage removal. Figure 6 and Figure 7 shows the results obtained. As expected as the damaged area increases, the FOS decreases. By using a square cut out the FOS is dramatically less than using a circular cut out. For each scenario it was found that the FOS following damage removal is 60% less if a square cut out rather than a circular cut out. It can be seen that the FOS following repair is approximately the same whether a circle or square is used to remove damage, with the square cut out FOS being only 0.5% less than the circular cut out. Therefore it can be stated that as long as a suitable wrap is used then it does not matter what shape is used to remove the damage.

8. Effect of wrap thickness

An investigation was carried out into the effect of varying the wrap thickness between 0.3 mm and 3.5 mm. A total of 9 models were considered and analysed using FEA software and from the results the minimum FOS was noted for each wrap thickness and used to create

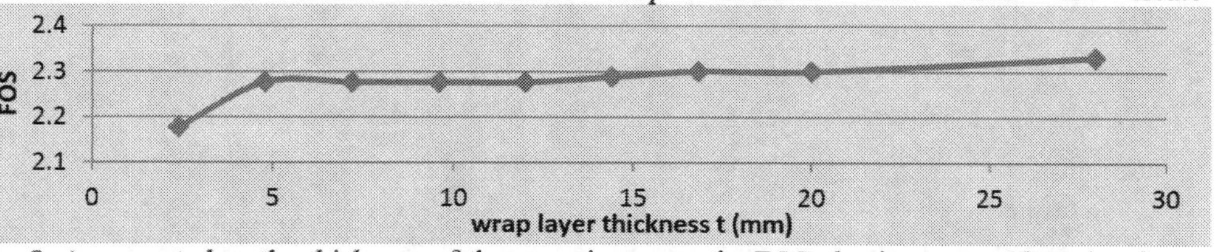

Figure 8. As expected as the thickness of the wrap increases the FOS also increases. There is a large

jump in the FOS between 2.4 and 4.8 mm, therefore a minimum thickness of 4.8 mm is suggested in this scenario.

Table 2. Damage Removal Shape Investigation Scenarios.

Scenario	Area (m^2)	Circular cut out diameter (m)	Square cut out side length (m)
1	7.069E-4	0.03	0.0266
2	1.9635E-3	0.05	0.0433
3	3.8485E-3	0.07	0.062

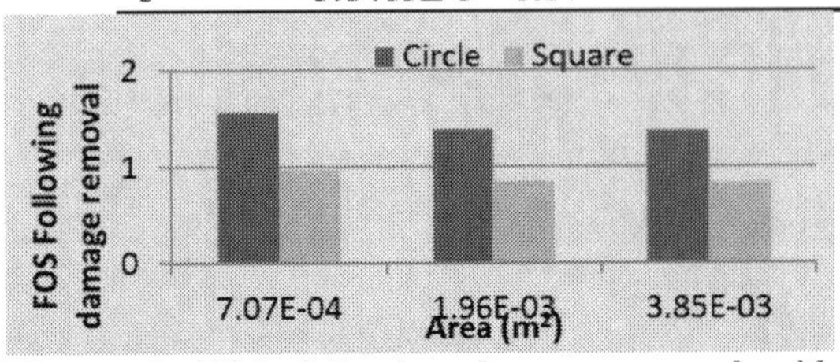

Figure 6. FOS following damage removal with varying cut out shapes.

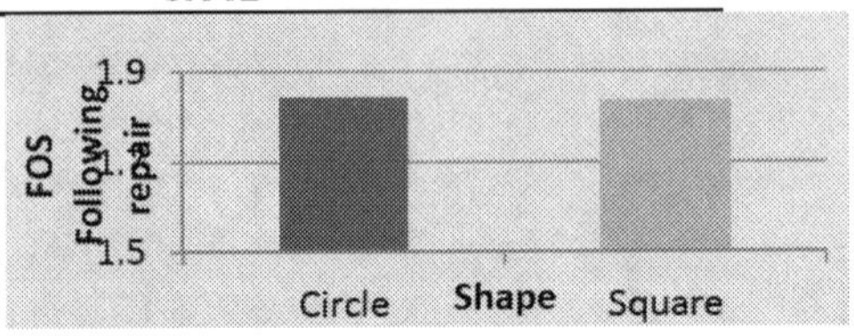

Figure 7. FOS following repair for varying cut out shapes.

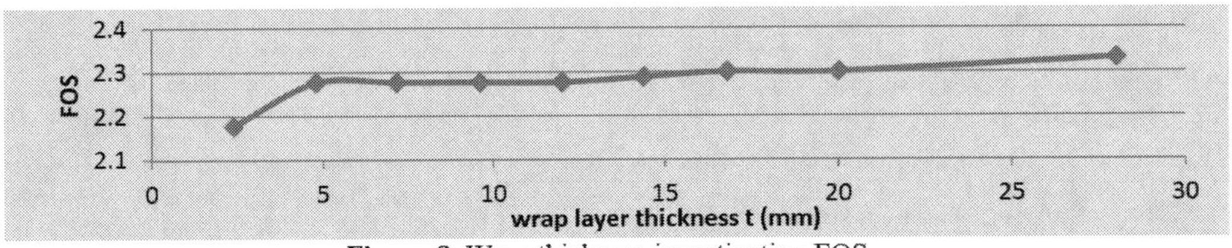

Figure 8. Wrap thickness investigation FOS.

9. Investigation into wrap width

Wrap manufacturers recommend at least 51mm of wrap either side of the damage, their standard wrap width is 292 mm. This section looks at whether this much wrap is actually needed either side of the damage. If the wrap is too short then it's possible that the wrap will not be able to deal with the stress concentrations caused by the hole and risk failure. But a wrap width of too much will waste material, money and time. Therefore three wrap thicknesses (13 mm, 8 mm, and 4 mm) and three damage sizes (30 mm, 50 mm and 70 mm) were considered in an extensive trial and error method of 70 models to determine the optimum wrap width for each scenario. The results can be seen in Figure 9, which is valid for wrap thicknesses between 4 mm and 13 mm and damage sizes between 0.03 m and 0.07 m.

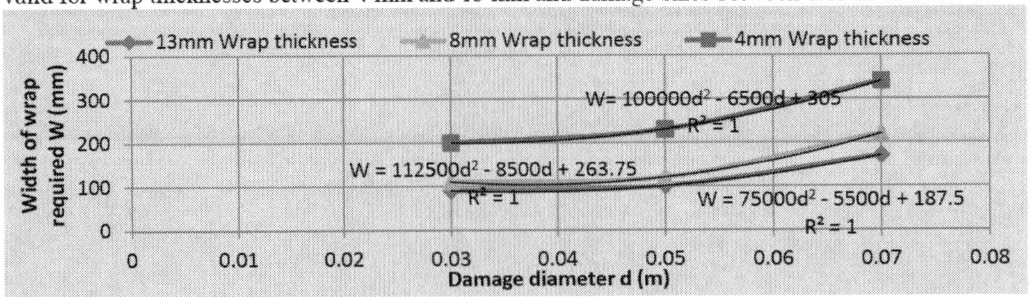

Figure 9. Wrap width investigation- Wrap width against damage size for varying wrap thicknesses.

10. Conclusion

In this investigation the parameters involved in repairing filament wound pipes are studied. The first parameter to be considered was the pipe diameter to create formulas to predict FOS following damage and FOS after repair, with a reliability of 95%. It was found that if the pipe is thin-walled then it can be assumed the pipe thickness has no impact on the FOS following repair.

The damage size was considered and formulas were created to estimate the FOS of the pipe when damaged and repaired. When checked with an additional scenario it was found that the FOS following repair could be predicted with an accuracy of 99.98% with the formulas created. However it was also found that the FOS after damage removal and before repair could only be predicted with an accuracy of 72%. However this FOS is not commonly of interest as the pipe will be repaired straight after damage removal.

The wrap was then investigated by looking at thickness and width. As expected it was found that as the wrap thickness increased the FOS following repair increased. An extensive trial and error method was used to investigate the wrap width for nine scenarios with varying damage sizes and wrap thicknesses. From this equations were created to predict the FOS before and after repair with an accuracy of 98%. It was also found that if T/D is constant then the wrap width required is the same length.

Through more than 150 models this investigation has provided useful equations to predict the FOS following damage and repair, which will save time and correct repair methods in industry. However experimental testing should be carried out to confirm findings.

References

[1] Dynaflow Research Group 2008 *FRP Pipe Failures & Lessons to be Learned* (Paulin Webinar)
[2] Li M 2000 *Temperature and Moisture Effects on Composite Materials for Wind Turbine Blades* (Montana: Montana State University-Bozeman)
[3] Alexander C 2007 *Guidelines for Repairing Damaged Pipelines using Composite Materials* (Texas: Stress Engineering Services Inc.)
[4] Rousseau J, Perreux D and Verdier N 1999 The influence of winding patterns on the damage behaviour of filament-wound pipes *Composites Science and Technology* **59** 1439-49
[5] Tarakcioglu N, Akdemir A and Avci A 2001 Strength of filament wound GRP pipes with surface crack *Composites Part B: Engineering* **32** 131-8
[6] Gemi L, Tarakcioglu N, Akdemir A and Sahin O 2009 Progressive fatigue failure behavior of glass/epoxy $(\pm 75)_2$ filament-wound pipes under pure internal pressure *Materials & Design* **30** 4293-8
[7] Rafiee R 2013 Experimental and theoretical investigations on the failure of filament wound GRP pipes *Composites Part B: Engineering* **45** 257-67
[8] Arikan H 2010 Failure analysis of $(\pm 55°)_3$ filament wound composite pipes with an inclined surface crack under static internal pressure *Composite Structures* **92** 182-7
[9] Gunaydin B, Daghan B and Avci A 2013 Fatigue behavior of surface-notched composite pipes repaired by composite patches *International Journal of Damage Mechanics* **22** 490-8
[10] Kara M, Uyaner M and Avci A 2015 Repairing impact damaged fiber reinforced composite pipes by external wrapping with composite patches *Composite Structures* **123** 1-8
[11] Amali R, Arnall H 2014 *Stress Concentration Due to the Removal of Damage in Composite Bridges* (England) (FRP Bridge Conference 2014)

Research on the defect types judgment in wind turbine blades using ultrasonic NDT

Suwei Li[1], Kezhong Shi, Kun Yang and Jianzhong Xu

Institute of Engineering Thermophysics, Chinese Academy of Sciences, Beijing, China

Email: lisuwei@iet.cn

Abstract. Beacuse of the complex internal structure of wind turbine blades, there are many types of defects. There will be important significance to determine the type of defects to assess the blade residual life and feedback on the design and production process. At present the type of defects found by ultrasonic nondestructive testing (NDT) can't be determined. Physical models of several typical defects are constructed. The sound intensity reflection coefficients of different physical models are calculated seperately, and the calculation method of the sound intensity reflection coefficient of actual defect is presented. Combined the calculated coefficients with the defect information tested, a method of recognize the type of defects is presented. Actual specimen testing is used to verify the effectiveness of the method.

1. Introduction

Wind energy is an important renewable energy. It is still a period of rapid development of wind power for China. Life and safety of wind power generator affects the pace of the utilization and development of wind power, and is the important factor for rapid and healthy development. Wind turbine blades as one of the wind turbine's foundation and key components, its service life and safety directly affects the life of the wind turbine and its security situation. In the process of the blades' production and operation, there will be structural and composite materials damages because of the production process and operation conditions. The damages will expansion under the effect of complex alternating load during operation, and then weak parts will appear. Eventually the life and operation safety of the blade will be reduced [1].

In order to guarantee the life and the safety of wind turbine blades, it is necessary to carry out the defect diagnosis. These defects in the blades can not be found out using conventional quality inspection. Nondestructive testing (NDT) is a kind of testing method that can find internal defects without destroying the structure of objects. Ultrasonic NDT technology has the advantage of larger penetration, greater sensitivity for plane type defects, more accurate judgments for depth and size, more safety operation, and so on. We can try to use the NDT method to search for defects in wind turbine blades.

There are many types of defects in wind turbine blades. Firstly, the defect types judgment in finished wind turbine blades can feedback on the design and production process to improve the quality of the product. Secondly, the defect type judgment can help to assess the blade residual life, because

[1] Address for correspondence: Suwei Li, Institute of Engineering Thermophysics, Chinese Academy of Sciences, Beijing, China. Email: lisuwei@iet.cn.

different defect types will have different effects on the blade in the process of operation. So the defect type judgment in wind turbine blades is very important. The judgment can also be carried out with ultrasonic NDT method. According to the literature study, the present applications of ultrasonic NDT in wind turbine blades or glass fiber reinforced plastics are if the the defects can be found [2], or how to find defects [3]. No research on the defect type judgment is published yet. Different types of defects may have similar ultrasonic signals in the process of NDT on wind turbine blades. How to use the echo signal of defect to determine the defect type is the article research content. Physical models of several typical defects are constructed. The sound intensity reflection coefficients of different physical models are calculated separately, and the calculation method of the sound intensity reflection coefficient of actual defect is presented. Combined the calculated coefficients with the defect information tested, a method of recognize the type of defects is presented. Actual specimen testing is used to verify the effectiveness of the method.

2. Analysis of several typical defects in blade

Defects are often generated due to production process in wind turbine blade manufacturing process. Inclusions, lack of structural adhesive glue and other defects are generated due to the randomness of human action. Delamination, gas holes and other defects are generated in the infusion process. With the comprehensive effect of load, vibration and environmental factors like sheat, acid and alkali, the initial defects will be extended, and new defects or damages like delamination and fracture will be generated.

Delamination defects are generated between the fiber laminates. It's generated in the interlayer by the low energy impact in the production process and operation conditions. When composite material containing delamination defect is under external load, local buckling will occur, causing delamination growth, resulting early damage of laminates, and the bearing capacity is greatly reduced [4]. There is gap between laminates containing delamination defect [2].

Gas holes are generated by the bubbles in resin in the production process, causing local fiber not saturated. Gas holes include the pores in glass fiber cloth layer bundle, the pores between bundles, and the pores between the glass fiber clothes. The presence of gas hole will reduce the fatigue performance of wind turbine blades [5].

Inclusion defects are generated because of the foreign body in layer or perfusion resin. The presence of inclusion will reduce the mechanical properties of wind turbine blades [5].

Structural adhesive lacked defects are generated when the structural adhesive does not fill between the girder and shell or between the two shells of the trailing/front edge. The scope of contact between two structures can not be completely filled with structual adhesive. The defects directly affect the internal force transfer between the bearing force structures in the blades, reducing the load bearing ability of blades [6].

Through analysis of the typical defects above, there are two kinds of interface with different properties in the damage surface.

3. Judgment Method

3.1. Analysis of the principle of ultrasonic nondestructive testing

The frequency range of ultrasonic used in NDT is generally between 0.25 MHz~15 MHz. The ultrasonic wave can spread in a certain speed and direction in the material, when it meet heterointerface with different acoustic impedance, such as defect or the bottom surface of the measured object, a reflection will be produced. The common method for detection is pulse echo testing at present. Pulse oscillator sends out voltage to the probe, while ultrasonic pulse generated by the probe propagates through the coupling agent into the material. When the ultrasonic wave meets the defects, partial reflection wave transfers along the original way and returns to the probe. The probe transforms the wave into electric pulse, and the pulse is amplified and displayed on the screen by the instrument. The principle of ultrasonic NDT is shown in figure 1. According to the location and amplitude of the reflected wave of defect showed on the screen, the position and the approximate size of defects can be determined [7].

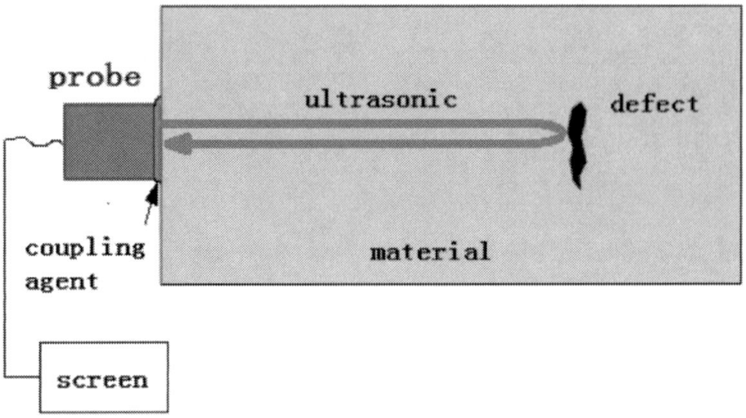

Figure 1. Principle of ultrasonic NDT.

Through analysis of typical defects of blades, in the delamination, gas hole, inclusion and Structural adhesive lacked defects, there are different acoustic impedance interfaces, such as interfaces between the composite material and air or composite materials and impurities. This interface can reflect ultrasonic wave.

Combined with the principle of ultrasonic and the generalization of blade defects, the blade defects above can be simplified to two physical models, glass fiber reinforced plastics-air (GFRP-Air) and glass fiber reinforced plastic-impurity (GFRP-Impurity), shown in figures 2 and 3, respectively. The delamination, gas hole, Structural adhesive lacked defects are simplified to GFRP-Air physical model, and the inclusion defect is simplified to GFRP-impurity physical model.

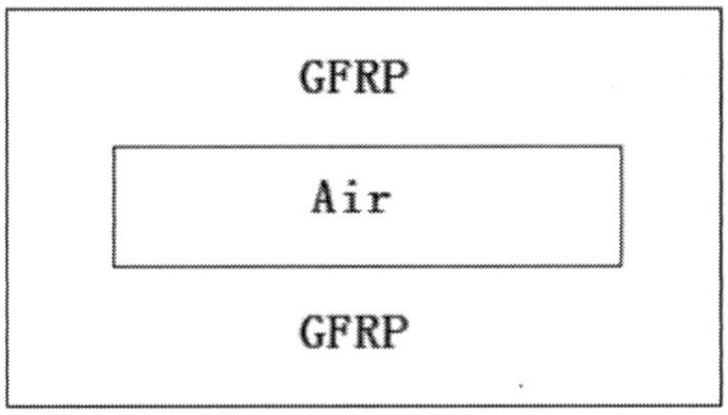

Figure 2. GFRP-Air physical model.

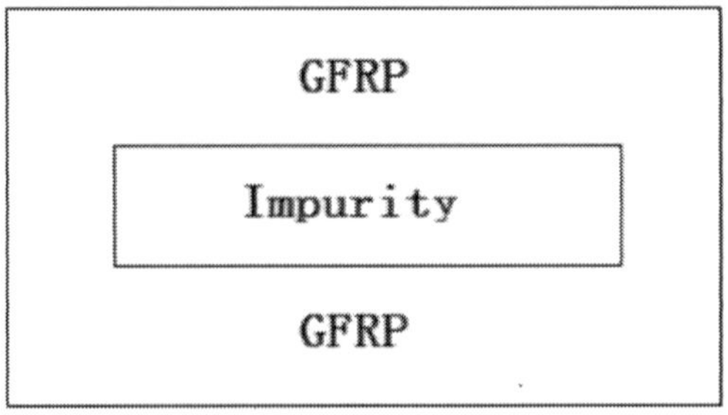

Figure 3. GFRP-Impurity physical model.

Ultrasonic NDT relies on the identification of reflection wave from the defect interface. The intensity of the sound wave reflection and transmission related to material on either side of the interface. The intensity can be reflected by the sound intensity reflection coefficient and transmission coefficient. Sound intensity reflection coefficient R and the transmission coefficient T can be expressed by equations (1) and (2).

$$R = \frac{I_r}{I_o} = \left(\frac{P_r}{P_o}\right)^2 = \left(\frac{Z_2 - Z_1}{Z_2 + Z_1}\right)^2 \tag{1}$$

$$T = \frac{I_t}{I_o} = \left(\frac{P_t^2}{2Z_2}\right) \Big/ \left(\frac{P_o^2}{2Z_1}\right) = \frac{4Z_1 Z_2}{(Z_1 + Z_2)^2} \tag{2}$$

I_o refers to the sound intensity of incident wave, I_r refers to the sound intensity of reflection wave, I_t refers to the sound intensity of transmission wave, Z_1 refers to the acoustic impedance of acoustic outgoing materials, Z_2 refers to the acoustic impedance of acoustic incoming materials, P_o refers to the sound pressure of incident wave, P_r refers to the sound pressure of reflection wave, P_t refers to the sound pressure of transmission wave. Acoustic impedance Z can be expressed by equation (3).

$$Z = \rho V_L \tag{3}$$

ρ refers to the density of the material, V_L refers to the longitudinal wave velocity.

From the equations (1) and (3) it can be seen that there will be different material interface in different physical models, and different material interface will produce different sound intensity reflection coefficient. When a material is fixed as the premise of GFRP, the sound intensity reflection coefficient of each defect interface is unique. So we can determine the corresponding physical model of defect according to the sound intensity reflection coefficient.

In order to determine the physical model corresponding to the actual defects, the two sound intensity reflection coefficients of physical model interface and actual defect interface, and compare the two coefficients with each other.

3.2. *The sound intensity reflection coefficient of physical model interface*

The structure of GFRP with actual defects can be shown in figure 4. The material around the defect is made by glass cloth layer and resin layer alternately. The GFRP in physical models in figures 2 and 3 will be simplified to homogeneous material.

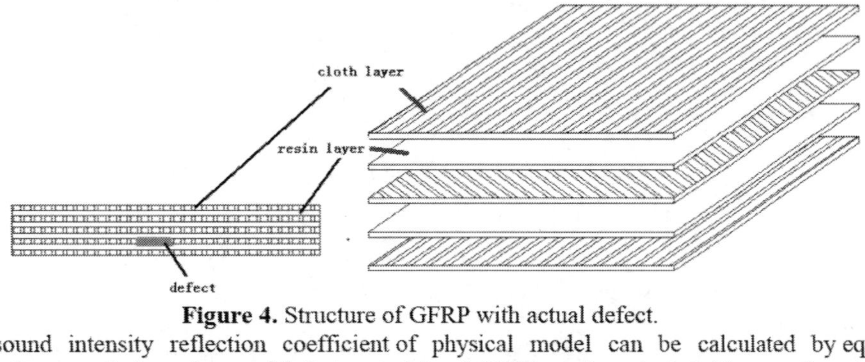

Figure 4. Structure of GFRP with actual defect.

The sound intensity reflection coefficient of physical model can be calculated by equation (1), while the acoustic impedance of defect material and GFRP can be calculated by equations (4) and (5) separately.

$$Z_1 = \rho_1 V_{L1} \tag{4}$$

$$Z_2 = \overline{\rho_2}\,\overline{V_{L2}} \tag{5}$$

ρ_1 refers to the density of defect material, V_{L1} refers to the longitudinal wave velocity of defect material, $\overline{\rho_2}$ refers to the average density of GFRP, $\overline{V_{L2}}$ refers to the average longitudinal wave velocity.

3.3. The sound intensity reflection coefficient of actual defect interface

Similar to the simplification of the physical model of GFRP, in the actual defects detection process, each layer echoes before the defect echo will be seen as attenuation of acoustic waves as shown in figure 5.

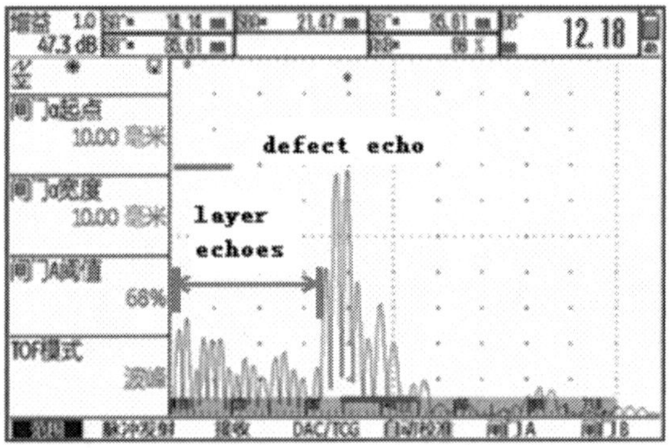

Figure 5. Actual defect detection echo sketch.

It's still need to rely on equation (1) to calculate the sound intensity reflection coefficient of actual defect interface. When using ultrasonic testing, sound pressure is directly proportional to the screen wave height.

$$P_1 / P_2 = H_1 / H_2 \tag{6}$$

P_1 refers to the sound pressure of position 1, P_2 refers to the sound pressure of poisition 2, H_1 refers to the screen wave height of position 1, H_2 refers to the screen wave height of position 2.

By the relationship between sound pressure and the screen wave height from equation (6), the sound intensity reflection coefficient of actual defect interface can be expressed through the screen wave height.

$$R_1 = \frac{I_{r-1}}{I_{o-1}} = \left(\frac{P_{r-1}}{P_{o-1}}\right)^2 = \left(\frac{H_{r-1}}{H_{o-1}}\right)^2 \tag{7}$$

I_{o-1} refers to the defect sound intensity of incident wave, I_{r-1} refers to the defect sound intensity of reflection wave, P_{o-1} refers to the defect sound pressure of incident wave, P_{r-1} refers to the defect sound pressure of reflection wave, H_{o-1} refers to the defect screen wave height of incident wave, H_{r-1} refers to the defect screen wave height of reflection wave.

Reflecting type ultrasonic NDT carries out detection by seeing the echo of target on the screen. If it meets the defect, there will be the reflection wave echo in the defect position on the screen. The defect screen wave height of reflection wave H_{r-1} can be read directly from the screen. But the defect screen wave height of incident wave H_{o-1} is virtual, which is unable to read from the screen, needing conversion with other parameters.

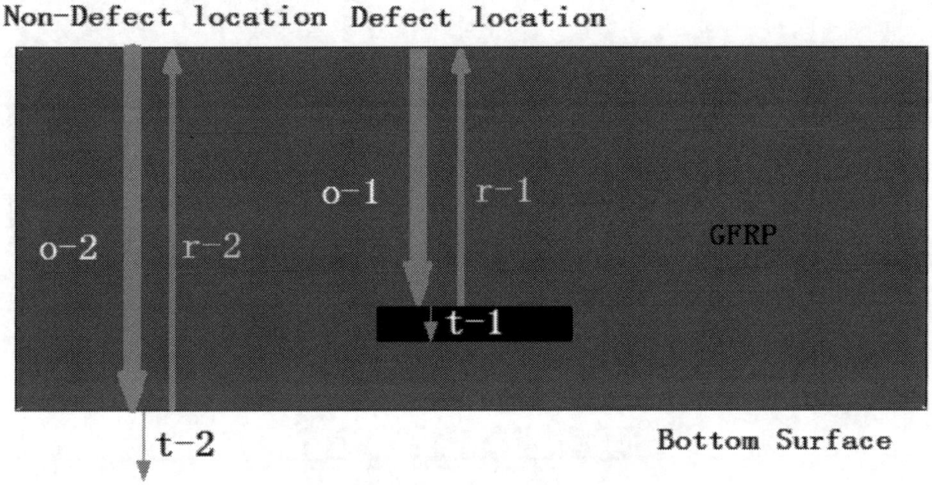

Figure 6. Schematic diagram of defect testing.

The schematic diagram of defect testing is shown in figure 6. In the location of defect testing, o-1 is the incident wave, in the GFRP - defect interface, part of the incident wave reflects back as reflection wave r-1, and the other part transmit into the defect as transmission wave t-1. In the location of non-defect testing, o-2 is the incident wave, in the GFRP-air interface, part of the incident wave reflects back reflection wave r-2, and the other part transmits into air transmission wave t-2. The incident wave height H_{o-1} in defect position needs conversion with the incident wave height H_{o-2} in non-defect position by considering as material attenuation from defect to bottom surface. And then H_{o-2} can be got by the reflection wave height H_{r-2} in non-defect position with the help of sound intensity reflection coefficient of GFRP-air interface R_2. And the reflection wave height H_{r-2} in non-defect position can be got from the ultrasonic instrument's screen when scanning non-defect position.

Basing on the derivation above, incident wave height H_{o-2} can be got by equation (7), showing as equation (8).

$$H_{o-2} = \frac{H_{r-2}}{\sqrt{R_2}} \tag{8}$$

The relationship among sound pressure, wave height and decibel is known as follows.

$$decibel = 20\lg\frac{P_1}{P_2} = 20\lg\frac{H_1}{H_2} \tag{9}$$

The incident wave height of defect interface can be got by equation (9), shown as equation (10):

$$H_{o-1} = H_{o-2} \times 10^{\frac{decibel}{20}} \tag{10}$$

The sound intensity reflection coefficient of actual defect interface can be got combined with equations (7), (8) and (10).

$$R_1 = \left(\frac{H_{r-1}}{H_{o-1}}\right)^2 = \left(\frac{H_{r-1} \times \sqrt{R_2}}{H_{r-2} \times 10^{\frac{decibel}{20}}}\right)^2 \tag{11}$$

3.4. *Judgment of the defect types*

By comparing the the sound reflection coefficient of actual defect interface and physical model interface, the type of physical model of the defect can be determined. If it's corresponding to the GFRP-Impurity model, the defect type is inclusion. If it's corresponding to the GFRP-air model, the type of defect will be determined according to the position and size. If defect appears in the GFRP layer, with continuous and large size, it's delamination defect. If defect appears in the GFRP layer, with discontinuous and small size, it's bubble defect. If defect appears out of the GFRP layer, it is judged as structural adhesive lacked defect.

4. Experiment and analysis of ultrasonic NDT

In order to verify this method, a specimen with pre-setting defect is prepared to carry out actual test. Thickness of the GFRP specimen as shown in figure 7 is 20 mm.

Figure 7. GFRP specimen being tested.

Firstly the sound intensity reflection coefficients of both GFRP-Air and GFRP-Impurity physical model interface will be calculated.

In the GFRP-Air physical model, ultrasonic transmits into air from GFRP. The average density of GFRP is 1.8×10^3 Kg/m^3, the average longitudinal velocity of GFRP is 3000 m/s, the density of the air is 1.29 Kg/m^3, and the longitudinal velocity of air is 340 m/s. Using equations (1) and (3), the sound intensity reflection coefficient from GFRP to air is 0.998.

Teflon is used as the inclusion in GFRP-Inclusion physical model in this test, because in the process of wind turbine,teflon is most likely as the mixed material. The density of teflon is 2.2×10^3 Kg/m^3, longitudinal velocity of teflon is 1350 m/s. By calculating, we can get the sound intensity reflection coefficient from GFRP to teflon is 0.084.

Next step is to calculate the interface sound interface reflection coefficient of actual defect. Ultrasonic NDT on the specimen is needed.

In the defect location, echo image is shown in figure 8, showing there is a defect in 15 mm in depth. The defect reflection wave height H_{r-15} is 86% of the screen height in 51.5 dB gain conditions.

In the non-defect location, echo image is shown in figure 9, showing the bottom surface is in 15 mm in depth. The bottom surface reflection wave height H_{r-20} is 58% of the screen height in 51.5 dB gain conditions.

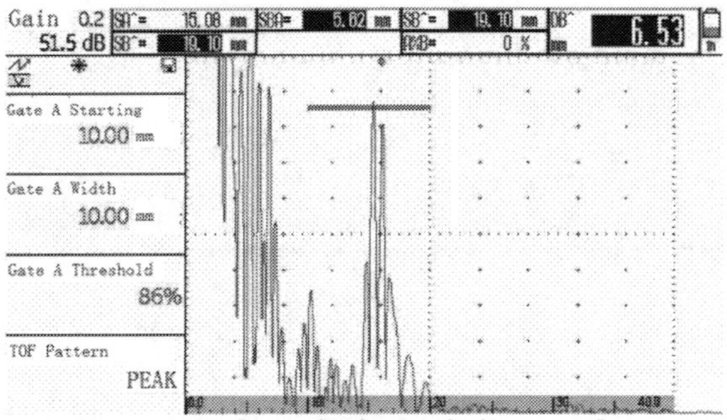

Figure 8. Defect echo image being tested.

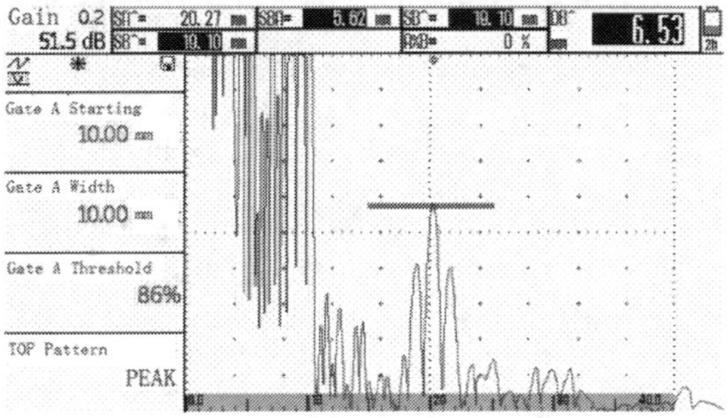

Figure 9. Bottom surface echo image being tested.

The incident wave in 20 mm depth is attenuated from the incident wave in 15 mm depth. In order to determine the attenuation of GFRP materials, another GFRP specimen is prepared to measure ultrasonic decibel values in different depth. And the curve shown in figure 10 is fitted by the decibel values.

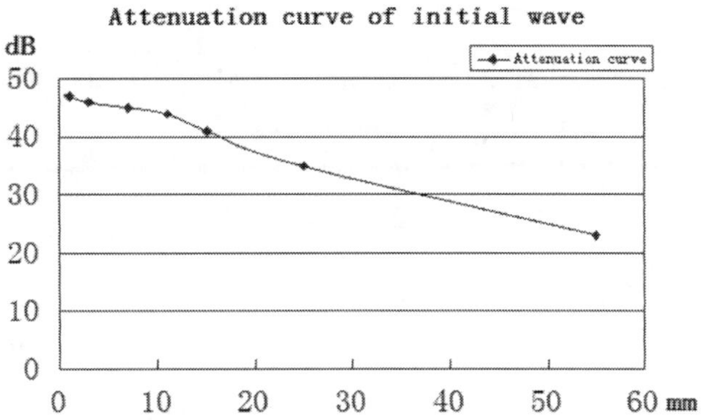

Figure 10. Attenuation curve of initial wave

Though the image in figure 10, we can get the decibel in 15 mm depth is 41 dB, the decibel in 20 mm depth is 37.5 dB, and the attention decibel is 3.5 dB.

Using the bottom surface reflection wave height H_{r-20}, the defect reflection wave height H_{r-15}, the attention decibel, and the equation (11), we get the sound interface reflection coefficient of defect in 15 mm is 0.982. By comparison with the sound interface reflection coefficient of

IOP Conf. Series: Materials Science and Engineering **87** (2015) 012056 doi:10.1088/1757-899X/87/1/012056

physical models, considering the influence of error, physical model of the actual defect can be determined as the GFRP-Air model.

Though testing range of the defect is about 30 mm*30 mm, and 15 mm in depth. The defect appears in the GFRP layer, with continuous and large size, so it's delamination defect belonging to the GFRP-Air model.

The selected specimen has a flat bottom hold in 15mm depth, used to simulate the delamination defect. Through the analysis above and comparison with practical specimen, this method is proved to be effective.

Using this method, through actual specimens of delamination, gas hole, inclusion and structural adhesive lacked defects can also be determined the types of defects.

5. Conclusion

- Due to the complexity of the design, production and operation of wind turbine blades, there may lead to many defects. It has good application prospect to use ultrasonic NDT on the flaw detection of blades.
- By establishing and analyzing physical models of typical defects, comparing with actual test of defect specimen, it is proved judgment of type of the typical defects can be realized, using the analysis of sound intensity reflection coefficient in ultrasonic flaw detection of wind turbine blades.
- In actual detection, we can calculate the actual sound intensity reflection coefficient, compare it with physical models of typical defects, confirm the type of models and determine the type of defect combining the information such as location and size.
- Through the calculation of sound intensity reflection coefficient, we can also guide the manufacture of defect simulating specimen, getting specimens as close as possible to the physical model and actual defects.

References

[1] Sun J L 2014 J. *Fiber Reinforced Plastics/Composites* **4** 81-4
[2] Amenabar A, Mendikute A and Lopez-A 2011 *J. Composites: Part B Engineering* **45** 1298-305
[3] Raišutis R, Jasiūnienė E and Žukauskas E 2008 *J. Ultragarsas (Ultrasound)* **63** 7-11
[4] Fang Y F, Liu W B, Zhang L 2010 *J. Experimental Technology and Management* **9** 49-51
[5] He J, Du L and Wang D S 2013 *J. Fiber Reinforced Plastics* **1** 1-7
[6] Yue D H, Li X L and Zhang H J 2011 *J. Infrared Technology* **33** 614-7
[7] The Chinese Society of NDT 2011 *M. Ultrasonic testing g* **2** (Beijing) 45-46

Preparation and multiple exciton generation of hydrogenated nanocrystalline silicon film prepared at different powers

Donghua Fan[1, 3] and Wenjin Long[2]

[1]Department of Applied Physics and Materials, Wuyi University, Jiangmen 529020, China
[2]Department of Applied Physics and Materials, Wuyi University, Jiangmen 529020, China

E-mail: donghua_fan@126.com

Abstract. The nc-Si:H films prepared by PECVD under different powers was characterized by Raman spectra, XRD pattern, and transmission measurements. The XRD pattern displays that the nc-Si:H nanocrystallites films is successfully prepared on quartz glass substrates. The average grain sizes of the prepared samples at different powers were estimated from full width at half maximum of (111) peak. The optical band gap of nc-Si:H films indicate that the optical band gap can be modulated by power. In the end, a simple statistical model was used to explain the effect of multiple exciton generation in the prepared nc-Si:H semiconductor material. The results indicate that multiple exciton generation efficiency in nc-Si:H film is related with both the size of quantum dots and incident photon energy.

1. Introduction

Currently, researches on the thin-film solar cells with low cost and high efficiency have aroused the increasing interest. Multiple exciton generation (MEG), as an important effect in nano semiconductor, can be used to enhance the efficiency of solar cell devices because absorbing a high energy photon can create the multiple electron-hole pairs [1-3]. So far, many experimental researches have been shown the existence of highly efficient MEG in semiconductor nanostructures or quantum dots, including, PbSe [4, 5], PbS [6, 7], Si [8] and so on. In addition, MEG in semiconductor nanocrystals has been explained by some theoretics, such as virtual exciton generation approach, coherent multiexciton model, atomistic model, and so on [9-11]. Su et al. had well explained the MEG effects in PbSe quantum dots by a simple statistical model based on the framework of Fermi statistical theory and impact ionization [12].

It is well konwn that hydrogenated nanocrystalline silicon (nc-Si:H) is a two-phase mixed materials, i.e. nanocrystals embedded in the amorphous silicon matix. This structure means that it has several inherent advantages, such as the strong optical absorption, high carrier mobility, tunable bandgap and good stability. Therefore, nc-Si:H film has received great attention for its potential application in photovoltaic cells. In addition, the existing Si quantum dots in the amorphous silicon matix also could yield the MEG effects, which could prompt the formation of electron-hole pairs to improve the

[3] Address for correspondence: Donghua Fan, Department of Applied Physics and Materials, Wuyi University, Jiangmen 529020, China. E-mail: donghua_fan@126.com.

efficiency of photovoltaic cells. However, there is a limited research on MEG in the nc-Si:H films. In this manuscript, the nc-Si:H films have been prepared by hydrogen diluted silane PECVD methods at different powers. We have investigated in details on the influence of the powers on the film properties and MEG.

2. Experimental set-up

The nc-Si:H thin films were prepared on the quartz glass substrates by PEVCD system with 13.56 MHz radio- frequency. During deposition, the pressure and substrate temperature were maintained at 100 Pa and 200°C, respectively. Silane (SiH_4) and hydrogen (H_2) was used as the source gases, where the hydrogen dilution rate ($SiH_4/(SiH_4+H_2)$) was kept constant at 1%. Keeping all these parameters constant, we grew the samples on the glass substrates with the various powers of 60, 70, 80, and 90 W, respectively. The structural properties of the synthesized samples were determined using the X-ray diffraction (XRD, X'Pert Pro with Cu Ka of 1.5406 Å). Raman spectra were recorded by a Jobin Yvon LabRAM HR UV-NIR micro-Raman system under 488 nm laser excitation. Transmission spectra were measured by using spectrophotometer (UV-2550).

3. Results and discussion

To investigate the structural properties of the nc-Si:H thin films prepared at different powers, we carried out XRD measurement. Figure 1(a) shows XRD pattern of the nc-Si:H thin films fabricated at 80 W. It displays two diffraction peaks at 2θ = 28.2 o and 47.2o, which is corresponding to the (111) and (220) crystallographic orientations of c-Si, respectively [13]. The broadening of diffraction peak related with c-Si phase indicates that a number of silicon nanocrystalline embed the prepared thin film. The strongest XRD peak intensity for the (111) plane shows that the nanocrystallites have preferentially grown along the (111) direction. The average grain sizes of the prepared samples at different powers were estimated from full width at half maximum of (111) peak using well-known Debye-Scherrer equation. Figure 1(b) shows the calculated average crystallite sizes between 6.2 and 8.5 nm with the increase of powers.

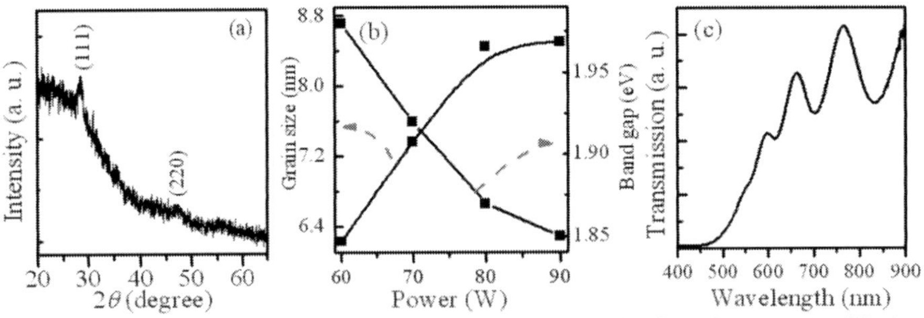

Figure 1. (a) and (c) XRD pattern of the nc-Si:H thin films fabricated at 80 W. (b) the power-dependent grain size and band gap.

Figure 1(c) shows the measured transmission spectra for the samples prepared at 80 W. If we igore the reflected light, the absorption coefficient can be calculated by the equation , where α, T and d is absorption coefficient, transmision and the film's thickness, respectively. The optical band gap Eg of the prepared nc-Si:H films can be estimated by Tayc's methods [14]. Figure 1(b) also shows the optical band gap E_g of the samples prepared at different powers. With the increase of power from 60 to 90 W, The optical gap of nanocrystalline silicon rise in gradual. The optical properties of nc-Si:H films can be greatly affected by many factors, including the ratio of volume fraction (c-Si, a-Si, and voids), surface effect, hydrogen content, quantum size effect, and so on [13]. It is still lacking a complete microscopic understanding of band gap depended on the microstructure in the films. However, the present observations clearly indicate that the optical band gap can be modulated by power.

According to Yukawa's meson theory, the collision between two high-energy nuclei will produce many elementary particles. In order to study this process of these elementary particles' generation, Fermi built a statistical model [15]. And MEG process is similar to it in some aspect because the multiple electron–hole pairs in the nanometer semiconductor are also generated by impact ionization after collisions. Therefore we can refer to Fermi statistical model to establish a new statistical model of MEG in the nano semiconductor materials.

The absorption of a single photon of high-energy hv > Eg (h is the Planck constant, v photon frequency and QDs band gap Eg) will generate an electron–hole pair (or an exciton) in semiconductor. And when hv > 2Eg, the nascent photogenerated charge carriers will generate the second exciton by collision with other bound electrons, which is so-called impact ionization [16]. If the new generated charge carriers have enough kinetic energy. It will generate more excitons. In the process, the regional volume of incident photon to release its energy is incident photon energy relaxation volume Ω. According to the effective mass approximation theory, the electron-hole pair kinetic energy can be shown as a function of the velocity v. The relationship between the band gap Eg of Si and should be content to the function. The semiconductor band gap and quantum size must have close relationship. According to the Brus equation [17, 18], we can establish an empirical equation by fitting the experimental observation, as shown in figure 2(a). It can be expressed as,

$$E_g = 1.69 + \frac{3.98}{d^{1.36}} - \frac{0.19}{d} \tag{1}$$

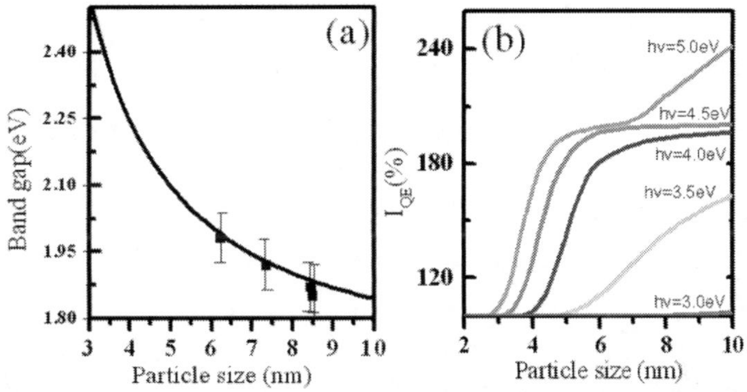

Figure 2. (a) Thedenpendence of band gab E_g on the particle size. (b) the I_{QE} at different particle size and photon energy.

Where the fitted value 1.69 is the band gap of amorphous silicon, and d is the diameter of Si quamtum dots. The fitting parameter -1.36 is in agreement with results of -1.37 [19]. With the increasing size of Si quantum dots, the band gap of nc-Si:H films will decrease.

The electron and hole release its energy in the crystal lattice whose mean free path is l_p ($l_p = 0.466nm$ in Si crystal).Then in t_c, the maximum energy relaxation length $l_{MAX} = \sqrt{l_p v t_c / 2}$. Therefore, the maximum photon relaxation volume relaxation volume V_{pr} can be expressed as $V_{pr} = \pi l_{MAX}{}^3 / 6 = \pi (l_p v t_c / 2)^{3/2} / 6$. The electron and hole can't move out nano semiconductor, so Ω should be $\Omega = \min(V_{pr}, V)$. V is the volume of nano semiconductor.In the nano semiconductor materials of volume V, when it absorbs a high-energy photon of energy hv and then produces n independent quasi-free particles [20, 21], the momentum of the n particles are $\bar{P}_1 \cdot \bar{P}_2 \cdot \bar{P}_3 \cdots \bar{P}_n$. Then the probability p(n) and the energy state density S(n) of the end equilibrium state can be described as [12],

$$p(n) = (\frac{\Omega}{V})^n ; \qquad S(n) = 2\times(\frac{V}{8\pi^3\hbar^3})^n \frac{dQ(E_{ks})}{dE_{ks}} \tag{2}$$

In equation (2), the reduced Planck constant $\hbar$ can be expressed as, $\hbar = h/2\pi$ =1.055 (unit: $10^{-34} J\cdot s$). $Q(E_{ks})$ is 3n- dimensional momentum space volume of the system total kinetic energy of the . According to the relationship between the total kinetic energy and particles mass $m_1, m_2, m_3 ... m_n$, the $S(n)$ in equation (2) can be expressed as [12],

$$S(n) = 2\times\frac{(m_1 m_2 m_3 \cdots m_n)^{3/2} V^n}{2^{3n/2}\pi^{3n/2}\hbar^{3n}} \frac{E_{ks}^{3n/2-1}}{(3n/2-1)!} \tag{3}$$

When nano semiconductor whose band gapE_g is absorbs photon energy of $h\nu$ and produces n particles, and the total kinetic energy of the particles can be expressed as:

$$E_{ks} = h\nu - \frac{nE_g}{2} \tag{4}$$

Therefore, the statistical weight can be written as:

$$\omega(n) = p(n)\times S(n) = 2\times\frac{(m_e m_h)^{3n/4}\Omega^n}{2^{3n/2}\pi^{3n/2}\hbar^{3n}} \frac{(h\nu - nE_g/2)^{3n/2-1}}{(3n/2-1)!} \tag{5}$$

When the semiconductor nanoparticle whose band gap $E_g$absorbs a single photon whose energy is $h\nu$, it can produce electron-hole pairs whose maximum number is N , $N = [h\nu/E_g]$. So nano semiconductor materials absorbs a high-energy photon and produces particles whose relative probability$W(n)$ can be calculated by the equation as follows [12]:

$$W(n) = \omega(n) / \sum_{n=2,even}^{2N} \omega(n), \qquad (n = 2, \cdots 2N) \tag{6}$$

According to equation (6) and statistical theory, nano semiconductor absorbs a high-energy photon and produces electron-hole pairs whose statistical average< N_{sa} > can be expressed as [12]:

$$\langle N_{sa}\rangle = \frac{1}{2}\sum_{n=2,even}^{2N} nW(n) \tag{7}$$

Nano semiconductor absorbs a singer photon [18] and produces the number of electron-hole pairs in its inner. This is the definition of MEG efficiency I_{QE} effect in the semiconductor materials. And it can be expressed as:

$$I_{QE} = \langle N_{sa}\rangle \times 100\% \tag{8}$$

According to the above equations, we can draw the following curve when it absorbs photons with different energy, i.e. 3 eV, 3.5 eV, 4 eV, 4.5 eV, and 6 eV. Figure 2(b) showed the relationship of MEG efficiency IQE and particle size. For a given incident light, IQE firstly is increasing with the rise of Si quantum dot diameter. In addition, it can be observed that for a given quantum dots, the larger the photo energy, the higher the MEG efficiency. The present results indicate that IQE in nc-Si:H film is related with both the size of quantum dots and incident photon energy.

4. Summary

In summary, we have investigated the influence of the power on microstructure and optical properties of the nc-Si:H films prepared by PECVD though Raman spectra, XRD pattern, and transmission measurements in detailes. The strongest XRD peak intensity for the (111) plane shows that the nanocrystallites have preferentially grown along the (111) direction. The average grain sizes of the prepared samples at different powers were estimated from full width at half maximum of (111) peak using well-known Debye-Scherrer equation. The optical band gap of nc-Si:H films indicate that the optical band gap can be modulated by power. In the end, a simple statistical model was used to explain MEG in the prepared nc-Si:H semiconductor material. The results indicate that IQE in nc-Si:H film is related with both the size of quantum dots and incident photon energy.

Acknowledgment
This work is partly supported by the Natural Science Foundation of Guangdong Province, China (Grant No. S2012010010017), the (key) Project of DEGP (No. 2013KJCX0181) and the Innovation Project of Department of Education of Guangdong Province (No. 2014KTSCX131).

References

[1] Beard M C and Ellingson R J 2008 *Laser Photonics Rev.* **2** 377-99
[2] Conibeer G 2007 *Mater. Today* **10** 42-50
[3] Nozik A J 2002 *Physica E* **14** 115-20
[4] Schaller R D and Klimov V I 2004 *Phys. Rev. Lett.* **92** 186601-4
[5] Ji M B, Park S, Connor S T, Mokari T, Cui Y and Gaffney K J, 2009 *Nano Lett.* **9** 1217-22
[6] Ellingson R J, Beard M C, Johnson J C, Yu P R, Micic O I, Nozik A J, Shabaev A and Efros A L 2005 *Nano Lett.* **5** 865-71
[7] Schaller R D, Sykora M, Pietryga J M and Klimov V I 2006 *Nano Lett.* **6** 424-9
[8] Beard M C, Knutsen K P, Yu P R, Luther J M, Song Q, Metzger W K, Ellingson R J and Nozik A J 2007 *Nano Lett.* **7** 2506-12
[9] Schaller R D, AgranovichV M and Klimov V I 2005 *Nat. Phys.* **1** 189-94
[10] Shabaev A, Efros A L and Nozik A J 2006 *Nano Lett.* **6** 2856-63
[11] Franceschetti A, An J M and Zunger A 2006 *Nano Lett.* **6** 2191-5
[12] Su W A and Shen W Z, 2012 *Solid State Commun.* **152** 798-801
[13] Parashar A, Kumar S, Gope J, Rauthan C M S, Dixit P N and Hashmi S A, 2010 *Sol. Energy Mater. Sol. Cells* **94** 892-9
[14] Tauc J, Grigorovici R and Vancu A 1966 *Phys. Stat. Solid. (b)* **15** 627-37
[15] Fermi E 1950 *Prog. Theor. Phys.* **5** 570-83
[16] Tauc J 1959 *J. Phys. Chem. Solids* **8** 219-23
[17] Brus L 1986 *IEEE J. Quan. Elec.* **22** 1909-14
[18] Brus L 1986 *J. Phys. Chem.* **90** 2555-60
[19] Wang L W and Zunger A, 1994 *J. Phys. Chem.* **98** 2158-65
[20] Hu Y Z, Lindberg M and Koch S W 1990 *Phys. Rev. B* **42** 1713-23
[21] Yoffe A D 1993 *Adv. Phys.* **42** 173-266

IOP Conf. Series: Materials Science and Engineering **87** (2015) 012058 doi:10.1088/1757-899X/87/1/012058

Synthesis and characterization of well-defined PVBCz-*b*-PDMAEMA multifunctional block copolymer prepared via ATRP

Tengfei Mao, Yanzi Gou[1] and Jun Wang

Science and Technology on Advanced Ceramic Fibers and Composites Laboratory, National University of Defence Technology, Changsha 410073, China

E-mail: y.gou2012@hotmail.com

Abstract. A functional polymer containing carbazole unit, PVBCz, was successfully prepared via atom transfer radical polymerization (ATRP) of 9-(4-vinylbenzyl)-9H-carbazole (VBCz). The controlled features of ATRP were confirmed by the linear increment of the molecular weight with the monomer conversion, while the molecular weight distribution was relatively narrow ($M_w/M_n \leq 1.33$). The block copolymer PVBCz-*b*-PDMAEMA was synthesized via ATRP, using PVBCz-Br as a macro-initiator and DMAEMA as the second monomer. ^{1}H-NMR, GPC and FT-IR characterization confirmed the successful synthesis of PVBCz-*b*-PDMAEMA. The resulting copolymer exhibited both fluorescence from PVBCz segments and thermo-responsiveness from PDMAEMA segments. Moreover, the fluorescent intensity of PVBCz-*b*-PDMAEMA in aqueous solutions increased as the temperature rose, which could be attributed to the coupling of two different functions.

1. Introduction

Atom transfer radical polymerization (ATRP) [1-3], which uses simple organic halides as initiators and transition metal complexes as catalysts, combines the superiority of conventional free-radical polymerization and living polymerization. This method has allowed the preparation of a variety of unique macromolecular architectures, such as block copolymers, graft copolymers, star polymers and hyperbranched polymers with predetermined molecular weights, narrow molecular weight distributions and various functional groups. These unique advantages make the ATRP method a powerful tool to synthesize functional polymers [4].

Carbazole compounds are an important series of nitrogen-containing heterocyclic compounds with a special rigid fused ring structure. The structure has a large conjugated system and a strong intra-molecular electron transfer, and it's susceptible to modification to introduce a variety of functional groups. Polymers containing a carbazole ring in the main or side chains are of considerable scientific and industrial interest due to their unique properties, such as high charge-carrier abilities, hole-transporting and fluorescent and photoconductive features [5-10]. Therefore, carbazole-containing polymers are considered promising organic photoelectric functional materials [11].

[1] Address for correspondence: Yanzi Gou, Science and Technology on Advanced Ceramic Fibers and Composites Laboratory, National University of Defence Technology, Changsha 410073, China. E-mail: y.gou2012@hotmail.com.

Poly [(dimethylamino) ethyl methacrylate] (PDMAEMA) is a typical so-called ''stimuli-responsive'' or ''smart'' polymer, due to its remarkable responsive behavior to environmental stimuli such as temperature, pH and ionic strength etc. [12]. The temperature-sensitive solubility of PDMAEMA in aqueous media also renders it potentially applicable in various fields such as drug delivery systems and sensors [13].

The multifunctional block copolymer can be obtained by copolymerization of carbazole-containing monomer and DMAEMA via ATRP. The carbazole segments exhibit the photoelectric function, and PDMAEMA segments exhibit the stimuli-responsive function. In addition, it's expected to achieve functional coupling, which can render the copolymer potentially applicable in sensors, smart optoelectronic functional materials. However, to the best of our knowledge, scientific and applied research in this area has rarely been publicly reported before.

In this paper, we report the synthesis of the well-defined homopolymer [(poly (9-(4-vinylbenzyl)-9H-carbazole) (PVBCz)] and the block copolymer PVBCz-*b*-PDMAEMA via ATRP. Meanwhile, the optical properties of polymers and their aqueous solutions were also investigated.

2. Experiment

2.1. Materials

1-(chloromethyl)-4-ethenyl-benzen-was purchased from J&K and passed through a short neutral alumina column to remove inhibitors prior to use, followed by final purification via distillation in vacuum. Copper (I) bromide (CuBr) (J & K, 99%) was purified by stirring overnight in acetic acid. After filtration, it was washed with ethanol, ether and then dried in vacuum. (Dimethylamino) ethyl methacrylate (DMAEMA) (J & K, 98%) was purified through a neutral alumina column to remove the inhibitor. Cyclohexanone (AR, Sinopharm) was dried with calcium hydride overnight and distilled in vacuum. Ethyl 2-bromoisobutyrate (EBiB), *N, N, N', N', N''*-pentamethyldiethylenetriamine (PMDETA) (98%), carbazole (99%) were purchased from J&K and used as received. Tetrahydrofuran (THF) (AR grade) was purchased from Sinopharm. Anhydrous THF was obtained by distillation under argon from sodium benzophenoneketyl. Deuterated chloroform ($CDCl_3$) with 1% v/v tetramethylsilane (TMS) (D, 99.8%) was purchased from Cambridge Isotope Laboratories. All other reagents were supplied as analytic pure and used without further purification.

2.2. Instrument

^{1}H-NMR spectra were recorded on an Agilent 400/54 NMR instrument. Monomer conversion was determined by integration of relative peaks in the ^{1}H-NMR spectra. Element analyses were obtained with a PerkinElmer PE-2400-II analyzer. The FT-IR spectra were recorded on a Perkin-Elmer 2000 FT-IR spectrometer. Molecular weight and polydispersity (PDi) were measured using Wyatt gel permeation chromatography (GPC) equipped with a Dawn Heleos-II detector and calibrated using PS standard samples. THF was used as the eluent at a flow rate of 1mL/min, operated at 25°C. Fluorescence (FL) spectra were recorded with a Horiba JY FL-3 spectrometer.

2.3. Synthetic procedures

2.3.1. Synthesis of 9-(4-vinylbenzyl)-9H-carbazole (VBCz). 9-(4-Vinylbenzyl)-9H-carbazole (VBCz) was synthesized according to the reference [14]. A mixture of carbazole (5.00 g, 30.0 mmol), sodium hydroxide (1.20 g, 37.5 mmol) and tetra-n-butylammonium bromide (TBAB) (200 mg, 0.62 mmol) in DMF 100 mL was stirred for 2 h. Then 1-(chloromethyl)-4-ethenyl-benzen (5.50 g, 36.0 mmol) was added dropwise, and the reaction mixture was continually stirred for 20 h at ambient temperature (scheme 1). The resulting mixture was poured into a large excess of deionized water. The resultant white precipitate was filtered and dried in vacuum at 70°C. The crude product was purified by recrystallization from acetone to give 5.40 g (63.5%) of white crystals. ^{1}H-NMR ($CDCl_3$, 400 MHz): δ (ppm) 5.17-5.22 (d, 1H, $C\boldsymbol{H_2}$=C), 5.51 (s, 2H, N-$C\boldsymbol{H_2}$), 5.65 (d, 1H, $C\boldsymbol{H_2}$=C), 6.65 (m, 1H, CH_2=$C\boldsymbol{H}$),

7.05-7.13 (d, 2H, Ar-***H***), 7.20-7.45 (m, 8H, Ar-***H***), 8.10-8.15 (d, 2H, Ar-***H***). Anal. Calcd for $C_{21}H_{17}N$: C, 89.01; H, 6.05; % N, 4.94. Found: C, 88.85; H, 5.96; N, 4.92.

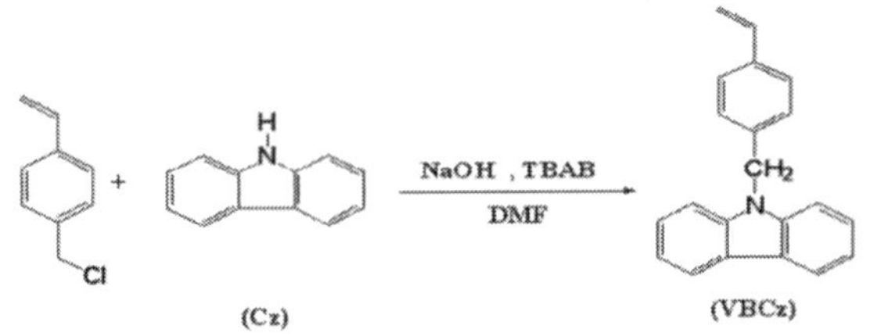

Scheme 1.

2.3.2. Synthesis of PVBCz-Br Macro-initiator (Homogeneous ATRP). A typical ATRP [15] was carried out as follows. VBCz (2.27 g, 8.0 mmol), PMEDTA (16.7 μL, 0.08 mmol) and EBiB (5.9 μL, 0.04 mmol) and anisole (internal NMR standard, 0.5 mL) were charged into a dry Schlenk flask with cyclohexanone (10.0 mL) as solvent. The flask subjected to three freeze-pump-thaw cycles to remove the air inside completely. The solution was then transferred via cannula under nitrogen into another Schlenk flask, previously evacuated and filled with nitrogen, containing Cu(I) Br (0.006 g, 0.04 mmol) and a stir bar. The reaction temperature was adjusted at to 90°C under constant stirring (t = 0). Samples were extracted periodically using a degassed syringe for molecular weight (GPC) and conversion (^{1}H-NMR) analysis. At the end of polymerization, the reaction mixture was diluted with 10 mL of THF, and air was bubbled through for 30 min. The mixture was passed through a short neutral alumina column and subsequently washed with THF to remove the copper complex catalyst. The resulting polymer solution was concentrated by rotary evaporation (ca. 10 mL), followed by being precipitated into methanol (ca. 200 mL). The white precipitate was filtered and dried under vacuum overnight at 50°C. The crude polymers were purified by a Soxhlet extractor with ethanol to further remove the remaining VBCz (scheme 2a).

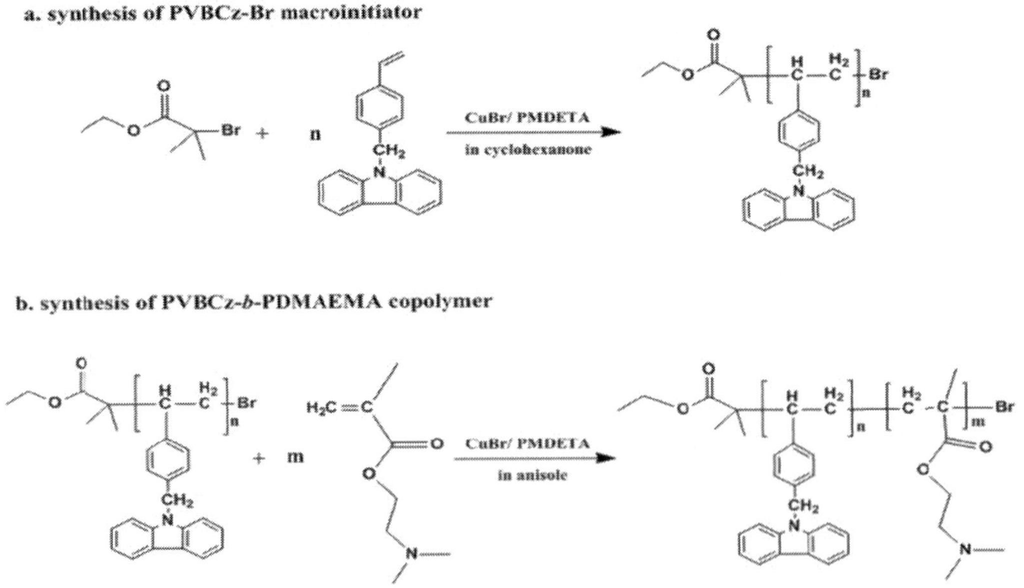

Scheme 2.

2.3.3. Synthesis of PVBCz-b-PDMAEMA copolymer. A PVBCz-*b*-PDMAEMA copolymer was synthesized using the PVBCz-Br macro-initiator via ATRP. The procedure was the same as discussed in the previous section, except that a macro-initiator (0.89 g, M_n=2.573×10^4, 0.05 mmol) and solvent (anisole, 10.0 mL) were added instead to the flask initially. Once the macro-initiator was dissolved, the second monomer (DMAEMA, 3.57 g, 23.0 mmol) and PMDETA (21.0 μL, 0.10 mmol) were added. The flask was subjected to three freeze-pump-thaw cycles. The solution was then transferred via cannula under nitrogen into another Schlenk flask, previously evacuated and filled with nitrogen, containing Cu (I) Br (0.007 g, 0.05 mmol) and a stir bar. Polymerization was performed at 60°C for 12 h. At the end of polymerization, the reaction mixture was diluted with 10 mL of THF, and air was bubbled through for 30 min. The mixture was passed through a short neutral alumina column and subsequently washed with THF to remove the copper complex. The copolymers were obtained by subsequent concentration, precipitation into *n*-hexane, filtration and finally drying under vacuum overnight at 40°C (scheme 2b).

3. Results and discussion

3.1. Kinetic studies of homogeneous ATRP

Figure 1 shows the kinetic plots of the ATRP of VBCz. The linearity of the semi-logarithmic plot of ln ($[M]_0/[M]_t$) vs time indicates that theATRP was first-order with respect to the monomer and that the concentration of the radicals remains constant during ATRP. Figure 2 shows the number average molecular weights ($M_{n,GPC}$) and the polydispersity indexes (PDi, M_w/M_n) as a function of monomer conversion. $M_{n,th}$ was calculated by the conversion and monomer/initiator ratio.

The reaction time, conversion, number average molecular weights (M_n), weight average molecular weights (M_w) and polydispersity indexes (PDi, M_w/M_n) are listed in table 1. The polydispersity indexes (PDi) of PVBCz are relatively narrow (1.12-1.33). All of these results suggest that the ATRP of VBCz using EBiB as an initiator and CuBr/PMDETA as a catalyst is a "living" polymerization process.

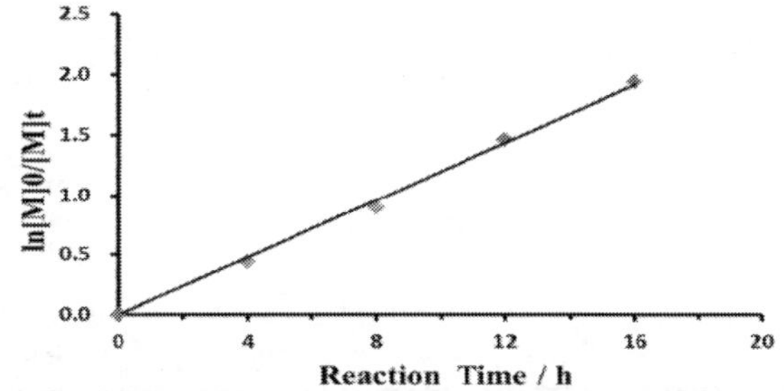

Figure 1. Semi-logarithmic plots of ln ($[M]_0/[M]_t$) *vs.* reaction time for the ATRP of VBCz.

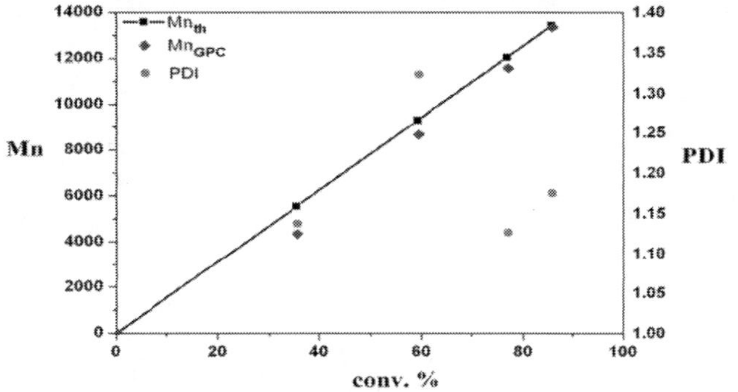

Figure 2. Plots of $M_{n,GPC}$, PDi and $M_{n,th}$ *vs*. monomer conversion for the ATRP of VBCz.

Table 1. Reaction time, M_n, M_w, PDi and and monomer conversion of PVBCz made by ATRP.

Reaction time (h)	Conversion (%)	M_n (g/ mol)	M_w(g/ mol)	PDi
4	35.6	4300	4900	1.14
8	59.4	8700	11500	1.32
12	76.9	11600	13000	1.13
16	85.7	13100	15300	1.18

3.2. Structural characterization of PVBCz and PVBCz-b-PDMAEMA

3.2.1. ^{1}H-NMR Characterization. Figure 3 gives the ^{1}H-NMR spectra of VBCz and PVBCz. The peaks at δ = 5.18 and 5.65 ppm belonging to the unsaturated C=H in the VBCz monomer disappear. However, the appearance of new peaks at δ = 1-2 ppm is obvious, which is attributed to saturated C-H in PVBCz.

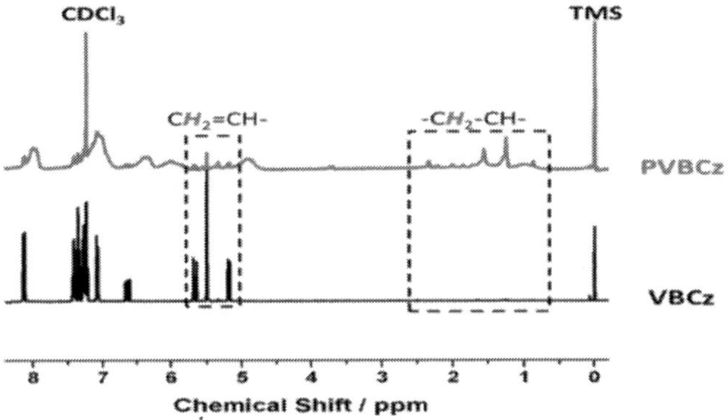

Figure 3. ^{1}H-NMR spectra of VBCz and PVBCz.

Figure 4 gives the ^{1}H-NMR spectra of PVBCz, PDMAEMA and PVBCz-*b*-PDMAEMA. In the ^{1}H-NMR spectrum of PVBCz-*b*-PDMAEMA, proton signals at δ = 2.27, 2.58 and 4.05ppm coming from N-methyl and methylene groups in PDMAEMA are observed [16]. Meanwhile, aromatic signals of PVBCz at 7.0–8.2 ppm can also be observed. These results confirmed the ATRP copolymerization of DMAEMA monomers initiated by the PVBCz-Br macro-initiator.

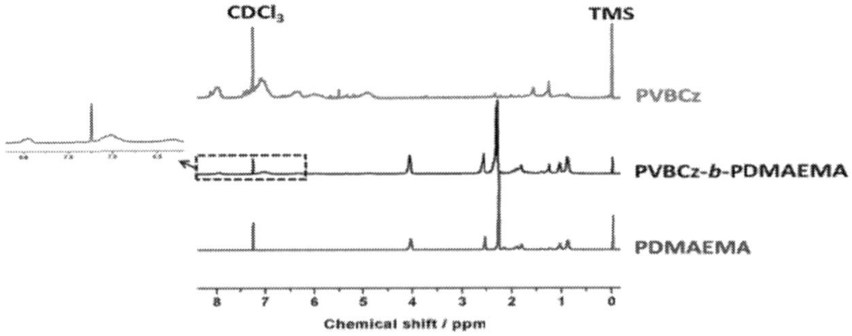

Figure 4. ^{1}H-NMR spectra of PVBCz, PDMAEMA and PVBCz-*b*-PDMAEMA.

3.2.2. GPC Characterization. The successful chain extension from a preformed polymer chain that acts as the macro-initiator is considered as an important criterion of the controlled/living nature ofATRP [17]. The PVBCz-Br macro-initiator (M_n = 25700 g/mol, PDi = 1.33) prepared via the ATRP of VBCz was employed for the chain extension experiment, using DMAEMA as the second monomer. The results are shown in figure 5 and table 2. M_n increased from 25700 to 68900 g/mol, while the molecular weight distributions still kept relatively narrow (PDi=1.11). This further confirmed the successful synthesis of the block copolymer PVBCz-*b*-PDMAEMA., which also indicated the living properties of ATRP.

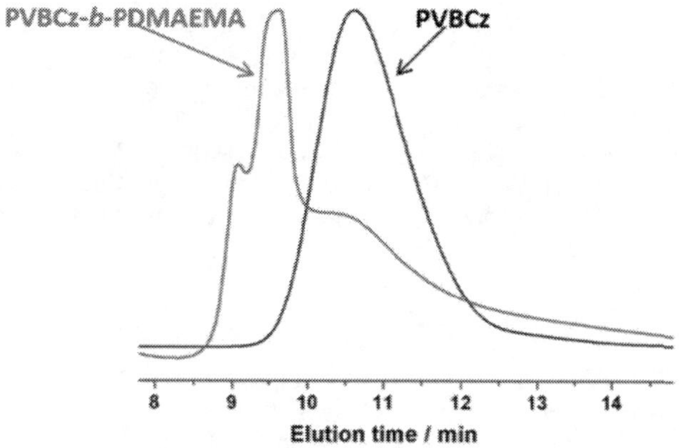

Figure 5. GPC traces of PVBCz and PVBCz-*b*-PDMAEMA.

Table 2. Molecular weight parameters of PVBCz and PVBCz-*b*-PDMAEMA.

Sample	$\overline{DP}$	$M_n(\times10^4)$	$M_w(\times10^4)$	PDi
PVBCz	91	2.573	3.427	1.33
PVBCz-*b*-PDMAEMA	(91+274)	6.890	7.670	1.11

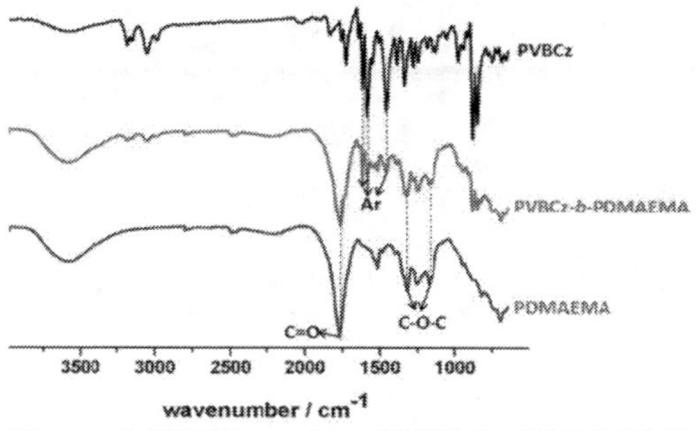

Figure 6. FT-IR spectra of PVBCz, PDMAEMA and PVBCz-*b*-PDMAEMA.

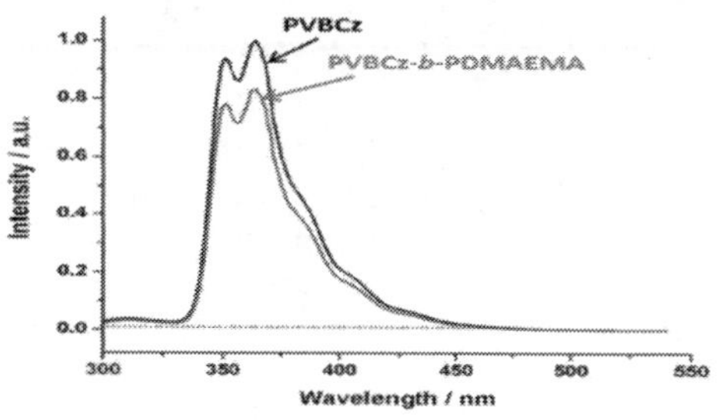

Figure 7. The fluorescence spectra of PVBCz and PVBCz-*b*-PDMAEMA in THF solution at ambient temperature. The concentration of the sample was 1 mg/mL. The excitation wavelength was 280 nm.

3.2.3. FT-IR Characterization. The FT-IR spectra of PVBCz, PDMAEMA and PVBCz-*b*-PDMAEMA are given in figure 6. As shown in the FT-IR spectra, the characteristic absorption bands at 1600, 1580 and 1450 cm^{-1} for the aromatic groups in the chain of PVBCz can be observed in both PVBCz and PVBCz-*b*-PDMAEMA [18]. In addition, after copolymerization with DMAEMA using PVBCz-Br as the macro-initiator, the characteristic absorption peaks at 1750 cm^{-1} (C=O stretching vibration) and 1300, 1100 cm^{-1} (C-O-C stretching vibration) [18] in figure 6 indicate the presence of the PDMAEMA block for PVBCz-*b*-PDMAEMA.

3.3. Optical properties of PVBCz and PVBCz-b-PDMAEMA

The fluorescence (FL) spectra of PVBCz and PVBCz-*b*-PDMAEMA in the THF solution at ambient temperature are shown in figure 7. The FL intensity of PVBCz is stronger than the intensity of PVBCz-*b*-PDMAEMA under the same mass concentration due to the existence of more carbazole moieties in PVBCz than in PVBCz-*b*-PDMAEMA.

Figure 8 shows the fluorescence changes of PVBCz-*b*-PDMAEMA in solvent mixtures (THF/water, the water volume fraction changed), and figure 9 exhibits the fluorescence change following temperature of PVBCz-*b*-PDMAEMA in aqueous solution. Both PVBCz and PDMAEMA are soluble in THF. PVBCz is hydrophobic. PDMAEMA is a classical thermo-responsive polymer, which is hydrophilic below the lower critical solution temperature (LCST). As long as the temperature exceeds the LCST, the polymer chains begin to collapse from coil to globule in aqueous solutions [19, 20]. In other words, it becomes hydrophobic above LCST. Due to the longer hydrophilic PDMAEMA segments below LCST, PVBCz-*b*-PDMAEMA could be dispersed in the aqueous medium.

As shown in figure 8, with the solution's water fraction increasing, the FL intensity decreased. The FL intensity of the THF solution is almost 10 times higher than that of the aqueous solution. As shown in figure 9, the FL intensity increased as the temperature of the solution rose. When the temperature was below 35°C, the FL intensity changed mildly. The FL intensity presented dramatical change as the temperature reached 40°C.

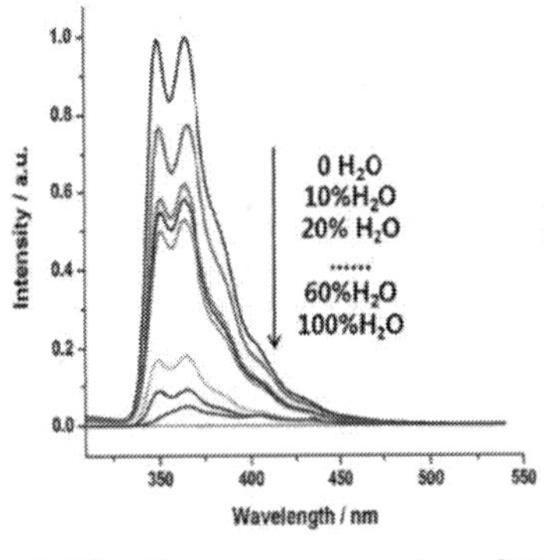

Figure 8. The fluorescence spectra of PVBCz-*b*-PDMAEMA in solvent mixtures (THF/water) at ambient temperature. The concentration of the sample was 1 mg/mL. The excitation wavelength was 280 nm.

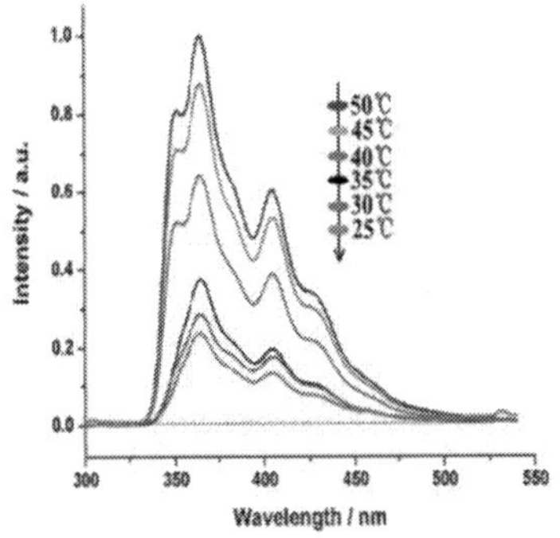

Figure 9. The fluorescence spectra of PVBCz-*b*-PDMAEMA in aqueous solution with variable temperature. The concentration of the sample was 1 mg/mL. The excitation wavelength was 280 nm.

3.4. Structure–property relationship

The mechanism of the unique optical properties of PVBCz-*b*-PDMAEMA is given in scheme 3. For linear conjugated polymers, the FL emission can be quenched by the π–π stacking of the conjugated segments during self-assembly easily (so-called "aggregation-caused quenching" (ACQ)) [21, 22]. In the THF solution, both polymer chains kept coiled. As the water fraction increased, the PVBCz chains aggregated to form solid-like cores while the PDMAEMA segments formed the shell. The FL intensity decreased due to the aggregation of PVBCz. As the temperature rose, the PDMAEMA chains also collapsed from coil to globule. The PVBCz cores were well wrapped in the interior of the micelles by PDMAEMA. Therefore, the aggregation of PVBCz cores was prevented effectually by the PDMAEMA shell, which is the reason why the FL intensity increased when the temperature was above the LCST.

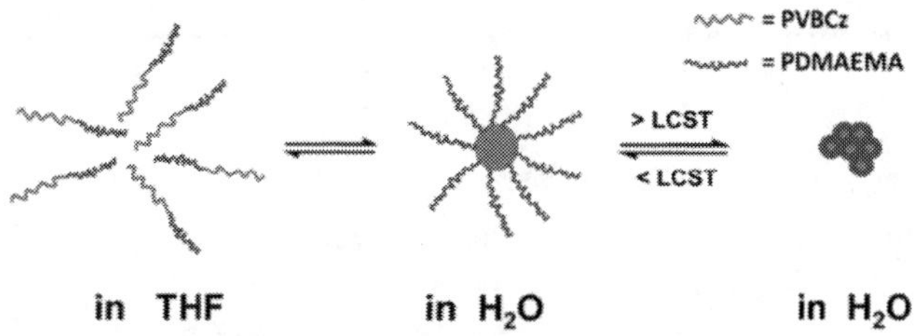

Scheme 3.

4. Conclusions

In this work, the functional polymers PVBCz and PVBCz-*b*-PDMAEMA were prepared via homopolymerization of 9-(4-vinylbenzyl)-9H-carbazole (VBCz) and copolymerization with DMAEMA initiated by macro-initiator PVBCz-Br, respectively. The controlled features of ATRP homopolymerization were confirmed by the kinetic study. The resulting PVBCz and PVBCz-*b*-PDMAEMA in the THF solution both exhibited fluorescence emission mainly at 350-400nm (λex=280 nm). The fluorescent intensity of PVBCz-*b*-PDMAEMA in the THF/water mixture solution decreased as the water fraction increased. The fluorescent intensity of PVBCz-*b*-PDMAEMA in the aqueous solution increased as the temperature rose to the LCST of the polymer, which could be attributed to the coupling of two different functional segments. Future work will focus on the self-assembly of the copolymers. Morphologic study is ongoing.

References

[1] Wang J and Matyjaszewski K 1995 *J. Am. Chem. Soc.* **117** 5614-5
[2] Percec V and Barboiu B 1995 *Macromolecules* 28 7970-2
[3] Kato M, Kamigaito M and Sawamoto M 1996 *Macromolecules* 28 1721-3
[4] Coessens V, Pintauer T and Matyjaszewski K 2001 *Prog. Polym. Sci.* **26** 337-77
[5] Grazulevicius J V, Strohriegl P and Pielichowski K. 2003 *Prog. Polym. Sci.* **28** 1297-353
[6] Zhang Y, Wang L and Sasabe H 1996 *Appl. Phys. Lett.* **69** 728-30
[7] Krotkus S, Kazlauskas K and Miasojedovas 2012 A *Phys. Chem. C* **116** 7561-72
[8] Moerner W E and Silence S M 1994 *Chem. Rev.* **94** 127-55
[9] Loiseau F, Campagna S and Hamaudaine A 2005 *J. Am. Chem. Soc.* **127** 11352-63
[10] Li W, Zhang X and Meetsma A 2008 *Organometallics* **27** 2052-7
[11] Romero D, Schaer M and M. Leclerc 1996 *Synth Met.* **80** 271-7
[12] Chen S J, Du F and Wu Z Q 1997 *Polym. Prepr.* **38** 534-7
[13] Cho S, John M and Lee H 1997 *J. Polym. Sci. Part B: Polym. Phys.* **35** 595-600
[14] Liu J, Cheng Z and Zhu X 2011 *Langmuir* **27** 12684-92
[15] Ladmiral V, Mantovani G and Haddleton D M 2006 *J. Am. Chem. Soc.* **128** 4823-30
[16] Zhang X, Xia J and Matyjaszewski K 1998 *Macromolecules* **31** 5167-9
[17] Zhang W, Yan Y and Zhu X 2008 *J. Euro. Polym.* **44** 3300-5
[18] Cheng Z, Zhu X and Neoh K G 2005 *Macromolecules* **38** 7187-92
[19] Fournier D, Hoogenboom R and Schubert U S 2007 *Macromolecules* **40** 915-20
[20] Plamper F A, Schmalz M and Ballauff A 2007 *J. Am. Chem. Soc.* **129** 14538-9
[21] Lin S T, Tung Y C and Chen W C 2008 *J. Mater. Chem*, **18** 3985-92
[22] Lin S T, Fuchise K and Sakai R 2009 *Soft Matter* **5** 3761-70

IOP Conf. Series: Materials Science and Engineering **87** (2015) 012059 doi:10.1088/1757-899X/87/1/012059

Finite element analysis when orthogonal cutting of hybrid composite CFRP/Ti

Jinyang Xu[1] and Mohamed El Mansori

MSMP Laboratoire, Arts et Métiers ParisTech, Rue Saint Dominique BP508, Châlons-en-Champagne 51006, France

E-mail: jinyang.xu@hotmail.com; jinyang.xu@ensam.eu

Abstract. Hybrid composite, especially CFRP/Ti stack, is usually considered as an innovative structural configuration for manufacturing the key load-bearing components in modern aerospace industry. This paper originally proposed an FE model to simulate the total chip formation process dominated the hybrid cutting operation. The hybrid composite model was established based on three physical constituents, i.e., Ti constituent, interface and CFRP constituent. Different constitutive models and damage criteria were introduced to replicate the interrelated cutting behaviour of the stack material. The CFRP/Ti interface was modelled as a third phase through the concept of cohesive zone (CZ). Particular attention was made on the comparative studies of the influence of different cutting-sequence strategies on the machining responses induced in hybrid stack cutting. The numerical results emphasized the pivotal role of cutting-sequence strategy on the various machining induced responses including cutting-force generation, machined surface quality and induced interface damage.

1. Introduction

Hybrid CFRP/Ti stacks are identified as an innovative structural configuration in modern aerospace industry, which has received a wide range of applications due to their enhanced characteristics and improved structural functions favouring energy saving. The composite-to-metal alliance commonly provides superior mechanical/physical properties for the aerospace components subjected to high thermo-mechanical stresses in their service. The CFRP/Ti coupling is considered a promising candidate to override conventional single-metal or single-composite applications in the aerospace field. A typical application is the usage of wing-fuselage connections in the Boeing 787 Dreamliner.

Prior to their post applications, the hybrid CFRP/Ti stacks are usually required to be manufactured in near-to-net shape in order to achieve dimensional tolerance and assembly performance. However, machining of such hybrid composite stack is characterized as the most challenging and difficult task in modern manufacturing sectors due to the extremely poor machinability of each constituent and their disparate natures. For instance, the CFRP exhibits anisotropic and abrasive nature, which always causes serious induced damage, and excessive tool wear [1-5]. In contrast, Ti alloy has low thermal conductivity and shows strong affinity to most used tool materials, which commonly results in severe adhesion wear, high concentrated cutting temperature, etc. [6-9]. To improve the machinability of CFRP/Ti stacks, various experimental studies have been conducted in the past few decades and have led to a better comprehension of the machining phenomena induced in cutting [10-12]. However, conventional experimental methodology used in CFRP/Ti stack cutting usually shows highly cost and

[1] Address for correspondence: Jinyang Xu, Arts et Métiers ParisTech, Rue Saint Dominique BP508, Châlons-en-Champagne 51006, France. Tel.: (+33) 0326699167; Fax: (+33) 0326699197 ; E-mail: jinyang.xu@hotmail.com.

time consuming. The number of parameters to control, the exhaustive material characterisation and the time-consuming procedure to determine various machining responses restricts its studies. In contrast, the numerical approach should be a promising tool that capable of offering feasible investigations of the machining results. Compared to tremendous scientific works dealing with single-metal and single-composite cutting modelling, nearly rare literatures exist on numerical simulation of hybrid CFRP/Ti cutting. The reason may be caused due to the complexity of various numerical parameters to control and the difficulty encountered in the interface behaviour modelling.

The interest of improving the knowledge concerning hybrid composite cutting has motivated the current work to develop an original FE model. In order to achieve the objective, a macro-mechanical model was established by using Abaqus/Explicit code allowing the simulation of the complete chip formation during the CFRP/Ti cutting process. The workpiece model consists of three constituents, i.e., the Ti layer, interface layer and CFRP layer, and each layer was developed individually and then assembled to conduct the FE computation. Several aspects of the cutting phenomena activated in different cutting-sequence strategies when cutting CFRP/Ti stacks were studied. The paper highlighted the significant role of cutting-sequence strategy in affecting the final machining responses.

2. Numerical model for CFRP/Ti cutting

2.1. *Description of OC model*

For present simulations, an original FE model of hybrid CFRP/Ti stack cutting was developed in Abaqus/Explicit code. To model 2D orthogonal cutting simulation, the quadrilateral elements with reduced integration were considered. The proposed OC model comprises four fundamental parts, i.e., the tool part, CFRP part, interface part and Ti part as shown graphically in figure 1.

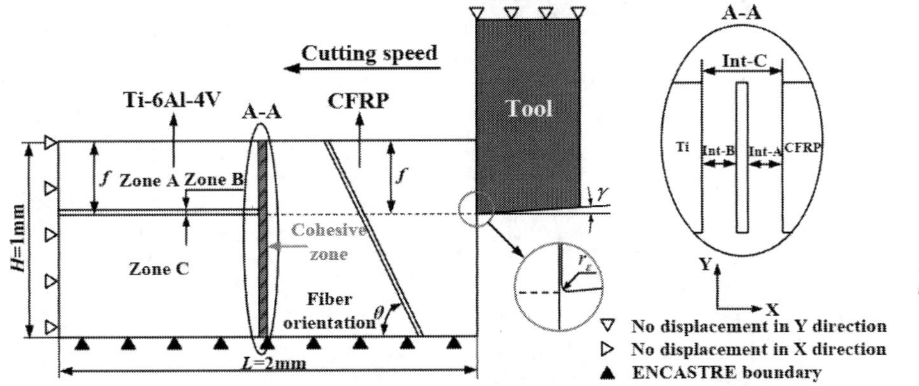

Figure 1. Scheme of OC model for hybrid CFRP/Ti stack cutting.

The total dimensions of the OC model were set as 2 mm × 1 mm ($L\times H$) and sufficient cutting length (around 1 mm) was made for both the Ti phase and CFRP phase in order to reach steady cutting conditions. The cutting tool used in the FE computation was modelled as a rigid body and imposed by a velocity to fulfil the cutting process. The tool geometries were defined by rake and clearance angles (α = 0° and γ = 7°) and nose radius (r_ε = 2 μm) as depicted in figure 1. The Ti part was assumed as fully isotropic and homogeneous material. A four-node plane-stress thermally coupled quadrilateral element type CPS4RT which has better convergence properties was implemented for the whole set of Ti elements. The interface part connecting Ti phase and CFRP phase was modelled by using the cohesive element functionally available in Abaqus/Explicit code. A triangular traction-separation cohesive formulation with linear softening is used to represent the mechanical response of the interface layer. The CFRP part was modelled as equivalent homogeneous material (EHM) by using 4-node plane-stress linearly interpolated elements with reduced integration and automatic hourglass control.

The kinematic contact algorithm was assigned to the contact pairs in order to avoid element penetration. Friction properties ruled in tool-work contact pairs were described by using Coulomb criterion with a constant friction coefficient equal to 0.3. Fixed displacements were applied on both the bottom and left edges of the stacked model. The bottom edge of the OC model was restrained in all directions (ENCASTRE) while the left side was constrained to move in the X direction, as shown in figure 1.

2.2. *Failure damage criteria*

In FE modelling, failure damage criteria play a pivotal role in influencing the rupture phenomena of the simulated work material and determining the finally separated chip morphology during the cutting process. The disparate nurtures of each stacked constituent make it more complex/difficult to simulate the chip removal process by using single failure damage criterion. For instance, the Ti part commonly exhibits isotropic and elastic-plastic behaviour, while the CFRP part shows anisotropic performance and heterogeneity strongly varied with fibre orientation (θ). The Ti phase and CFRP phase are required to implement different damage criteria to fulfil the chip separation process controlling the hybrid stack cutting. For Ti layer, the studied Ti alloy was Ti6Al4V and its properties were summarized as follows: density (4430 kg/m3), Young's modulus (113 GPa), Poisson's ratio (0.342), thermal expansion coefficient ($9.1\times10^{-6o}C^{-1}$), melting temperature (T_m = 1680°C), room temperature (T_r = 25°C), thermal conductivity (7.0 W/(m•°C)) and specific heat (546 J/(kg•°C)). The metal is assumed to suffer high plastic deformation, plane stress, strain rate and pronounced temperature effects during the cutting operation. In such case, the widely-used Johnson-Cook (JC) constitutive model and JC damage law [13, 14] were introduced to replicate the machining behaviour and cutting performance for Ti part. The expression of the JC constitutive equation is shown as follows:

$$\bar{\sigma}=\left(A+B\bar{\varepsilon}^n\right)\left(1+C\ln\frac{\dot{\bar{\varepsilon}}}{\dot{\bar{\varepsilon}}_0}\right)\left(\left[1-\left(\frac{T-T_r}{T_m-T_r}\right)^m\right]\right) \tag{1}$$

Where $\bar{\sigma}$ is the equivalent flow stress, $\bar{\varepsilon}$ is the equivalent plastic strain, $\dot{\bar{\varepsilon}}$ is the equivalent plastic strain rate, $\dot{\bar{\varepsilon}}_0$ is the reference equivalent plastic strain rate, T is the workpiece temperature, T_m is the material melting temperature, and T_r is the room temperature. A, B, C, m and n are material constants. The Johnson-Cook damage law was then presented as follows.

$$\bar{\varepsilon}_i=\left[D_1+D_2\exp\left(D_3\frac{P}{\bar{\sigma}}\right)\right]\left(1+D_4\ln\frac{\dot{\bar{\varepsilon}}}{\dot{\bar{\varepsilon}}_0}\right)\left[1+D_5\left(\frac{T-T_r}{T_m-T_r}\right)\right] \tag{2}$$

$$\omega=\sum\frac{\Delta\bar{\varepsilon}}{\bar{\varepsilon}_i} \tag{3}$$

In which, $\bar{\varepsilon}_i$ is plastic strain at damage initiation, P is hydrostatic pressure, $P/\bar{\sigma}$ is stress triaxiality, $\dot{\bar{\varepsilon}}$ is plastic strain rate, $\dot{\bar{\varepsilon}}_0$ is reference strain rate, T is the workpiece temperature, T_m is the material melting temperature, and T_r is the room temperature, D_1-D_5 are damage parameters, ω is the damage parameter and $\Delta\bar{\varepsilon}$ is the plastic strain increment.

Note that the damage is assumed to happen when damage parameter ω =1. Afterwards, the damage evolution was controlled by using effective plastic displacement in the FE calculation. The implementation of damage initiation and evolution ensured the complete chip separation. The input parameters for both JC constitutive model and JC damage criteria are summarized as follows: A = 1098 MPa, B = 1092 MPa, C = 0.014, m = 1.1, n = 0.93, $\dot{\bar{\varepsilon}}_0$=1 s^{-1}; D_1 = -0.09, D_2 = 0.25, D_3 = -0.5, D_4 = 0.014, D_5 = 3.87, respectively [15, 16].

For CFRP layer, it was modelled as isotropic but locally homogeneous. The constitutive behaviour and damage criteria of the composite phase were defined by using Hashin damage criteria [17] available in Abaqus/Explicit code. The studied CFRP material was carbon/epoxy T300/914 laminate and its material properties were summarized as follows: longitudinal modulus (E_1 = 136.6 GPa), transverse modulus (E_2= 9.6 GPa), in-plane shear modulus (G_{12}= 5.2 GPa), major Poisson's ratio (v_{12}

IOP Conf. Series: Materials Science and Engineering **87** (2015) 012059 doi:10.1088/1757-899X/87/1/012059

= 0.29), longitudinal tensile strength (X_T = 1500 MPa), longitudinal compressive strength (Y_T = 27 MPa), transverse compressive strength (Y_C = 200 MPa), in-plane shear strength (S = 80 GPa) [18].

The interface part was modelled through the concept of cohesive zone. Traction-separation law with linear softening was utilized to simulate the mechanical behaviour of the cohesive interaction. Damage initiation and evolution were controlled by using BK damage law available in Abaqus/Explicit. The relevant input parameters for interface zone modelling are summarized as follows: K_{nn}=2.0GPa, $K_{ss}=K_{tt}$ =1.5GPa, t_n^f=60MPa, $t_s^f=t_t^f$=80MPa, G_n^C =0.78N/mm, and $G_s^C=G_t^C$ =1.36N/mm.

2.3. *Validation of the stack model*

For validation purpose, each stacked constituent (i.e., Ti model and CFRP model) was verified carefully through the comparisons with experimental results from open literatures due to the significant lacking of experimental studies concerning orthogonal cutting of hybrid CFRP/Ti stacks. Both Ti-layer model and CFRP-layer model were validated in terms of cutting-force comparison by refer to experimental results gained by Umbrello [19] and Iliescu et al. [20], respectively. During the validation work, the cutting conditions were selected the same used in literatures. Figure 2 shows the comparison of simulated and experimental results for both Ti-layer validation and CFRP-layer validation. It can be seen that the established models produced the consistent results with the experimental observations from open literatures [19, 20] which indicated the accurate development of the proposed numerical models. When each model was refined and improved correctly, they were assembled to perform the FE simulation of hybrid CFRP/Ti cutting.

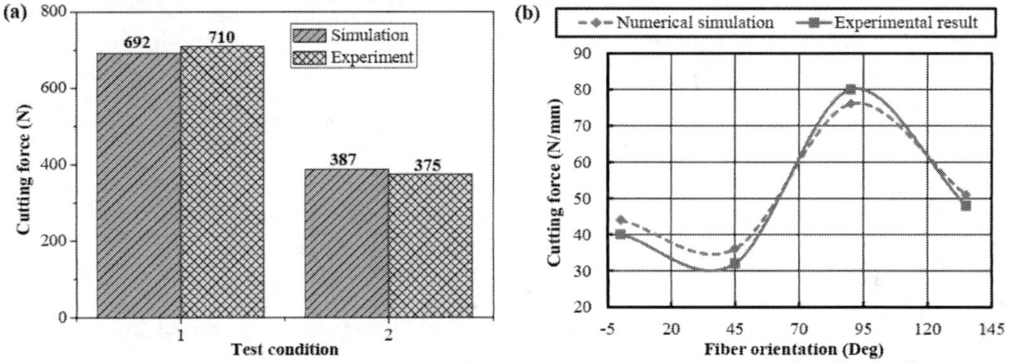

Figure 2. (a) Validation of Ti-layer model with open literature [19] and (b) validation of CFRP-layer model with open literature [20].

3. Results and discussion

In hybrid CFRP/Ti machining, typically two different cutting-sequence strategies (Ti→CFRP and CFRP→Ti) exist from the aspect of tool-entry and tool-exit throughout the chip removal process. In this section, a comparative studied focused on the different cutting-sequence strategies were performed in terms of force generation, machined surface quality and induced damage formation.

3.1. *Cutting-force investigation*

The cutting force generation (F_c) was studied and compared *vs.* cutting speed (v_c) and feed rate (f), and the obtained numerical results are shown in figure 3. In each cutting-sequence simulation, the F_c magnitudes were predicted based on individual layer cutting (i.e., Ti-phase cutting and CFRP-phase cutting) and repeated three times in order to minimize the measurement error.

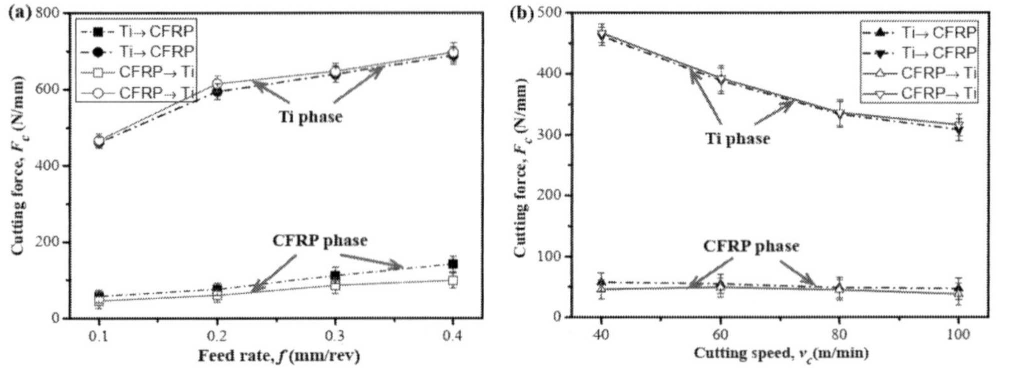

Figure 3. Comparison of simulated F_c under different cutting-sequence strategies: (a) F_c *vs.* f ($v_c = 40$ m/min and $\theta = 0°$); (b) F_c *vs.* v_c ($f = 0.1$ mm/rev and $\theta = 0°$).

As illustrated in the above figure 3, it was noticeable that the cutting-sequence strategy exhibited crucial role in affecting the cutting-force generation, irrespective of the used cutting parameters. In Ti→CFRP cutting sequence, the predicted F_c in CFRP-phase cutting yielded higher values than its counterpart produced in CFRP→Ti cutting sequence. Similarly, when operated in CFRP-Ti cutting sequence, the generated cutting force of Ti-phase cutting also achieved higher magnitudes than that obtained in Ti-CFRP cutting sequence. Such physical phenomena indicated that the identical-phase cutting which operated last in one specified cutting sequence typically promoted higher cutting-force generations than that operated initially in another cutting sequence, i.e., higher Ti-force generation in CFRP→Ti cutting sequence and higher CFRP-force generation in Ti→CFRP cutting sequence as depicted in figure 3. The key mechanisms controlling the physical phenomena can be attributed to favourable effects of chip adhesions on tool rake face that contributing to the increase of cutting resistance and hence the higher force generation.

In addition, feed rate (f) was observed to have significant influences on the cutting-force generations in such manner that a small increase of f typically resulted in great elevation of F_c in both Ti-phase cutting and CFRP-phase cutting. In contrast, the cutting speed (v_c) was found to have negative effects on the force generation, i.e., increased v_c favoured the reduction of force generation in hybrid stack cutting. The reason could be explained by the fact that increased v_c typically promoted high cutting-heat generation, softened the machined workpiece and hence resulted in less cutting resistance for further chip separation removal.

3.2. Machined surface morphology and induced damage

To clarify the practical influence of cutting-sequence strategy on hybrid cutting process, an inspection was made through the observations of the machined surface morphology *vs.* different cutting-sequence strategies under the fixed cutting conditions ($v_c = 40$ m/min, $f = 0.2$ mm/rev and $\theta = 0°$). Figures 4 and 5 then show the comparative numerical observations.

It can be seen that in hybrid CFRP/Ti cutting, disparate chip morphologies were generated during the entire chip separation process. Such phenomena were caused due to the dissimilar natures of the composite and metallic phases. In CFRP-phase cutting, the resected chip appeared "dust" like shape due to the brittle-fracture predominant chip-separation mode while in Ti-phase cutting, the chip shape was generated principally in the form of "continuous" appearance due to the dominated plastic-deformation mode. Moreover, the chip adhesion on tool rake face was also observed in the two cutting-sequence machining. The evidence strongly interpreted the phenomena of different force generations of identical-phase cutting under different cutting-sequence strategies (as illustrated in Section 3.1).

In addition, it was noticeable that the machined surface finish produced in CFRP→Ti cutting sequence seemed to be much smoother than that operated in Ti→CFRP cutting sequence. Moreover, only minor crack damage was detected beneath the machined CFRP surface in CFRP→Ti cutting sequence as shown in figure 4. In contrast, the Ti→CFRP cutting sequence commonly promoted serious crack imperfects concerning the CFRP-phase machined surface. Furthermore, the CFRP surface generated in Ti→CFRP cutting sequence appeared extremely uneven and rough as compared to that obtained in CFRP→Ti cutting sequence. Such phenomena could be attributed to the serious Ti-chip adhesion on tool rake face that generating severe scratching effects on the machined composite surface.

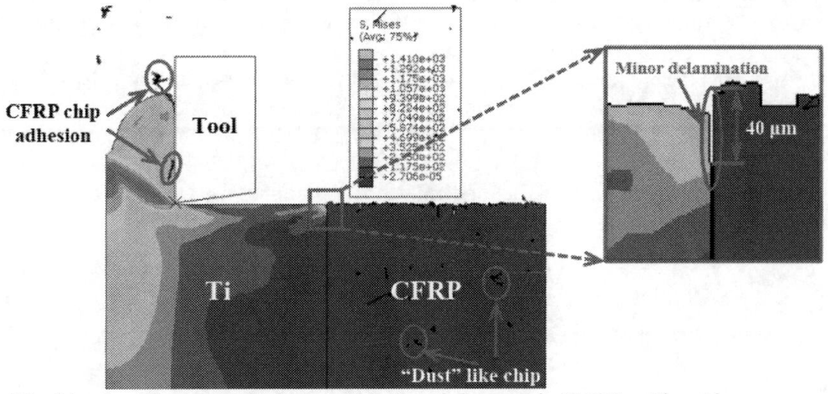

Figure 4. Machined surface morphology and induced damage in CFRP→Ti cutting sequence (v_c = 40 m/min, f = 0.2 mm/rev and θ = 0°).

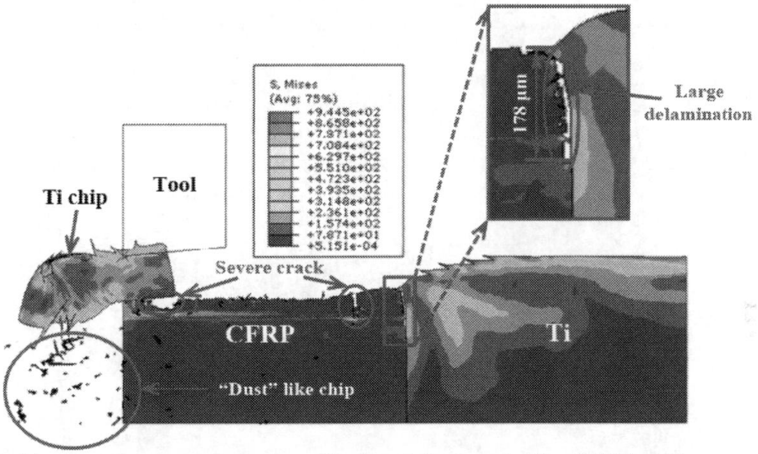

Figure 5. Machined surface morphology and induced damage in Ti→CFRP cutting sequence (v_c = 40 m/min, f = 0.2 mm/rev and θ = 0°).

With regard to the interface cutting, the damage mode occurred at the bi-material interface commonly exists in the form of delamination imperfection. In Ti→CFRP cutting sequence, typically a large length of delamination damage was produced focused on the bi-material interface. The predicted delamination length (d_m = 178 μm) was approximately 4.5 times longer than that (d_m = 40 μm) generated in CFRP→Ti cutting sequence. The reason can be explained by the fact that in CFRP→Ti cutting sequence the Ti alloy can act as the role of supporting plate in preventing CFRP laminate inflection and ensuring the stability of the tool-work interaction during the cutting operation. As a result, low extent of interface delamination was promoted. In contrast, when operated in Ti→CFRP cutting sequence, the CFRP phase could not act the role of supporting plate when tool firstly cut into the Ti phase and subsequently the interface region due to its brittle nature of the fibre/matrix system. Consequently, high extent of delamination damage was produced focused on the bi-material interface.

4. Conclusions

This paper developed an original FE model to simulate the chip formation process when cutting hybrid CFRP/Ti stacks. Different damage criteria were introduced to model the cutting behaviour of the composite phase and metallic phase. Special FE analyses were made focused on the comparative studies of different cutting-sequence strategies and also their influences on hybrid cutting operation. Based on the acquired FE results, some key conclusions can be drawn as follows:

- The use of the cohesive zone successfully replicated the delamination phenomenon activated among the CFRP/Ti interface. The numerical model allowed the reliable prediction of the key phenomena generated in hybrid CFRP/Ti cutting.
- Feed rate had significant positive impact on the cutting-force generation when cutting hybrid composite (CFRP/Ti) stacks. However, the impact arising from cutting speed was negative due to the softening effects of cutting temperature on the work material when cutting speed increased.
- The numerical results highlighted the favourable effects of CFRP→Ti cutting sequence on reducing the force generation, minimizing the interface delamination, and consequently promoting excellent machined surface morphology when orthogonal cutting of hybrid CFRP/Ti stacks.

Acknowledgment

The authors would like to cordially thank the China Scholarship Council (CSC) for its financial support (Contract No. 201306230091).

References

[1] Xu J, An Q and Chen M 2014 *Compos. Struct.* **117** 71-82
[2] Xu J, An Q, Cai X and Chen M 2013 *Int. J. Precis. Eng. Manuf.* **14** 1687-1696
[3] An Q, Cai X, Xu J and Chen M 2014 *Int. J. Abr. Technol.* **6** 183-196
[4] An Q, Xu J, Cai X and Chen M 2013 *Adv. Mater. Res.* **797**: 155-160
[5] Wei Y, Xu J, Cai X, An Q and Chen M **2013** *Key. Eng. Mater.* **589-590**: 173-178
[6] Xu J, Zheng X, An Q and Chen M 2012 *J. Shanghai Jiao Tong Uni.* **46** 1037-1042
[7] Liu Z, An Q, Xu J, Chen M and Han S 2013 *Wear* **305** 249-259
[8] Xu J, Liu Z, An Q and Chen M 2012 *Adv. Mater. Res.* **497** 30-34
[9] Xu J, Liu Z, Guo G and Chen M 2013 *Int. J. Adv. Manuf. Technol.* **67** 517-533
[10] Isbilir O and Ghassemieh E 2013 *Mach. Sci. Technol.* **17** 380-409
[11] Kim D and Ramulu M 2004 *Compos. Struct.* **63** 101-114
[12] Park KH, Beal A, Kim D, Kwon P and Lantrip J 2011 *Wear* **271** 2826-2835
[13] Johnson G R and Cook W H 1983 *Proc. of the 7th Int. Symposium on Ballistics* pp 541-547
[14] Johnson G R and Cook W H 1985 *Eng. Fract. Mech.* **21** 31-48
[15] Xi Y, Bermingham M, Wang G and Dargusch M 2013 *J. Manuf. Sci. Eng. – Trans. ASME* **135** 061014
[16] Lesuer D 1999 *Livermore: University of California, Lawrence Livermore National Laboratory* pp 1-36
[17] Hashin Z and Rotem A 1973 *J. Compos. Mater.* **7** 448-464
[18] Santiuste C, Miguélez H and Soldani X 2011 *Compos. Struct.* **93** 2706-2713
[19] Umbrello D 2008 *Mater. Process.Technol.* **196** 79-87
[20] Iliescu D, Gehin D, Iordanoff I, Girot F and Gutiérrez M E 2010 *Compos. Sci. Technol.* **70** 73-80

IOP Conf. Series: Materials Science and Engineering **87** (2015) 012060 doi:10.1088/1757-899X/87/1/012060

Light-curing polymers for laser plasma generation

E Y Loktionov[1], Y S Protasov, Y Y Protasov and V D Telekh

State Lab for Photon Energetics, Bauman Moscow State Technical University, 5-1 2nd Baumanskaya Street, Moscow 105005, Russia

E-mail: stcpe@bmstu.ru

Abstract. Solid rather than liquid media are used in pulsed laser plasma generators despite sophisticated transportation and dosing system need for a long-term operation. Liquid media could be more preferable due to transfer and dosing (down to 10^{-14} L) being well developed, but plasma generation of those results in intense droplet formation and kinetic energy losses. Combination of liquids transportation advantages and solids plasma generation efficiency might resolve this trade-off. Liquid-to-solid transition can be induced by cooling down to sublimation temperature, thermo-, photo- or electron induced polymerization (curing). Light cured polymers seem to be very useful as active media for plasma generators, since they can be solidified very fast (ca. 30 ms) just before impact. We considered experimentally several UV-curing polymer and mixtures ablation regimes and supply schemes for laser plasma generation. The best results were obtained for liquid polymer at high-power pulsed irradiation matching curing optimum wavelength.

1. Introduction

Application of light-curing polymers (LCP) [1] in laser ablation has a number of advantages for practical applications, which are related to high-intensity laser-matter interaction. Usually laser irradiation of solid surfaces is used to generate plasma flows because of high volumetric energy density in the material, while losses, in most cases, are related only heat dissipation. A specific feature of liquids laser irradiation is related to their linear absorption coefficient being high only within certain spectral intervals. Intense evaporation of liquid causes formation of droplets, as well as part of ablation plume and shock waves kinetic energy dissipation in liquid volume [2]. Although, using liquids is advantageous for working medium supply to the active zone in laser thrusters and technological setups both for arranging transport and precise dosing and for injection of different additives [3]. Liquids solidification just before laser irradiation makes real a combination of the advantages typical of these two aggregation states at different stages of operation. Using LCP for laser plasma generators was first suggested in [4].

It is known that the efficiency of laser ablation can be characterized by mass consumption $\Delta m/E$, momentum coupling coefficient $C_m \approx \Delta m \langle v \rangle / E$, and other parameters [5]. To eliminate drawbacks of using liquids, which are mainly related to losses of mechanical energy as a result of splashing [6], a number of authors proposed forming thin liquid films [7] and isolated droplets [8], filling porous target matrices with a liquid [9], and increasing viscosity [10]. However, the results of these studies show that the spatial confinement of the evaporation zone and increased liquid viscosity lead to an increase

[1] State Lab for Photon Energetics, Bauman Moscow State Technical University, 5-1 2nd Baumanskaya Street, Moscow 105005, Russia. E-mail: stcpe@bmstu.ru.

in the laser ablation efficiency by only 1-1.5 orders of magnitude. So, the absolute values of the discussed parameters remain small compared to those in the case of solid targets [11]. For instance, upon the film thickness decrease (or increase in the liquid viscosity), the ablation plume mass averaged velocity increases but remains within the interval $<v>\sim 30\text{-}80$ m/s. Although the momentum coupling coefficient is relatively large in the case of liquids, $C_m \sim (0.2\text{-}1)\ 10^{-2}$ Ns/J, laser energy conversion efficiency defined as $\eta \approx C_m <v>/2$ is small, on the order of $(1\text{-}20)\ 10^{-4}$.

Studies of laser-matter interaction efficiency for both solid and liquid states of the same material have been conducted only for water and ice [12]. It was demonstrated that the momentum coupling coefficient was 6-7 times higher in the case of water. However, due to the opposite ratio of mass averaged velocities of the plume, the energy efficiency was about the same in the both cases (4-6%).

Obviously, cooling a liquid down to the phase transition temperature can be used for its solidification. The latter can also take place as a result of polymerization, wherein photopolymerization occurs faster and is more controllable compared to thermal polymerization. Using special photoinitiators that bleach the reaction products for the incident laser radiation allows achieving UV-IR photopolymerization of layers with a thickness of ~1 mm [13]. Processes similar to photopolymerization, which usually takes place in the near-surface layer with a thickness of $\sim 10^{-6}$-10^{-4} m, can be initiated at a depth of up to 20 mm by exposing the corresponding substance to an electron beam [14].

Currently, LCPs are used as special optical and dental glues, inks for UV printing, and in stereolithography. The rate of photopolymerization can be regulated by changing the concentration and chemical composition of photoinitiators, or the wavelength and intensity of the radiation initiating the reaction. Solidification time of ca. 30 ms was achieved in UV curing ink jet printing technology at 365 nm irradiation ~ 10-30 mW/cm^2, which is much lower than the laser ablation threshold.

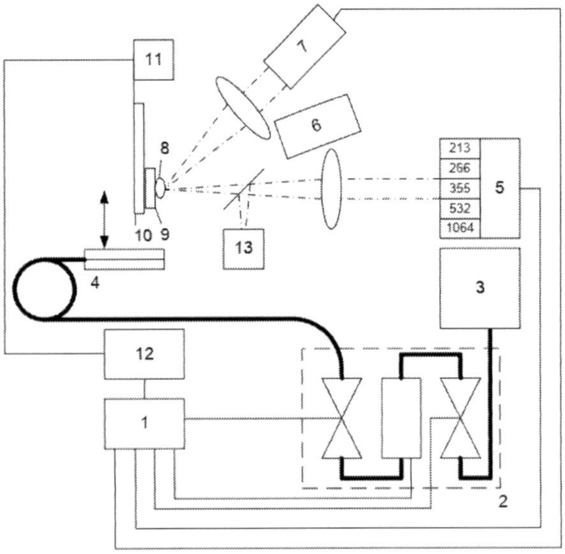

Figure 1. Experimental setup (1 – pulse generator, 2 – piezopump, 3 – reservoir, 4 – ceramic ferrule, 5 – nanosecond laser, 6 – microscope, 7 – diode laser, 8 – LCP droplet, 9 – glass slip, 10 – force sensor, 11 – signal amplifier, 12 – oscilloscope, 13 - calorimeter).

2. Experimental setup

We report the results of studying the efficiency of the laser action on acrylic UV-ink films (SPC-0659Y, Mimaki) in the liquid and cured states. The LCP was deposited on a coverslip substrate by means of a dosing dropper (Finnpipette Digital). The droplets had a nominal volume of ~0.5 µL. The glass substrate was weighed on analytical scales (CAS CAUW-120D) with an accuracy of 10 µg at all stages of the investigation. The targets were placed on a PVDF-film calibrated force sensor (Images SI PZ-02) connected to a digital oscilloscope (Tektronix 2024B). The target was cured by exposing it for ca. 2 s to a 405 nm diode laser (Lasever LSR405NL) 0.8 W/cm^2 radiation, an optimal wavelength for the studied LCP curing is 365±15 nm. For the laser ablation we used five harmonics (1064 nm, 532 nm, 355 nm, 266 nm, and 213 nm) of nanosecond (18 and 12 ns) Nd:YAG lasers (Lotis TII LS-2147 and Solar LQ929 – figure 1). The focal spot size (~0.02 mm^2) was measured by means of an optical microscope. Optical characteristics of the LCP were studied with an SF-2000 spectrophotometer in the interval from 190 to 1100 nm. The Raman spectra were measured using a Raman Systems R-3000 complex (785 nm, 200–2700 cm^{-1}).

A quasi-continuous LCP supply system sample has been designed using piezo micropump (Burkert 7616), driven by pulse generator (Stanford Research Systems DG645). A ZrO_2 ceramic ferrule for optical fibers (230 μm capillary – Thorlabs CF230) was used to produce LCP droplets at its tip.

3. Results

Solidification dynamics of LCPs was investigated in a number of applied studies [15]. The methods of diagnostics used in those studies were based either on light scattering by polymer chains or on the measurement of IR absorption. Our analysis of Raman spectra revealed some differences for the liquid and cured states: the intensity of the Rayleigh scattering was, as expected, higher in cured LCP, and the spectrum contained characteristic peaks at 610, 828, and 913 cm^{-1} (similar lines are known for polymethylmethacrylate (PMMA)). The luminescence quantum yields at ~240 nm were also different for the liquid and solidified states.

Characteristic time dependences of the recoil force are presented at figure 2. In contrast to the case of water and ice [12], momentum coupling coefficient for liquid LCP is high not due to an amplitude, but due to a longer duration. Laser irradiation of a liquid produced a large number of droplets with a characteristic size of ~0.2 mm, this substantially increased the mass flow rate and reduced the mass averaged velocity.

Specific characteristics of the laser ablation efficiency averaged over 10–30 pulses are shown in figure 3. Almost all the presented dependences reveal the existence of optimal values. Data for 1064 nm are not shown because they are not representative due to a low value of the spectral absorption coefficient. In this case, either the glass substrate was damaged, or the underlying layer of the force sensor coating was ablated. It should be taken into consideration that the laser ablation thresholds for the liquid and solidified phases at different wavelengths are substantially different; i.e., the relative regimes of laser action W/W_a [16] (W_a is the laser ablation threshold) are also different. Since laser irradiation was carried out under the condition $\ln(W/W_a)>>2$, we assume that the difference of the regimes did not influence the results significantly.

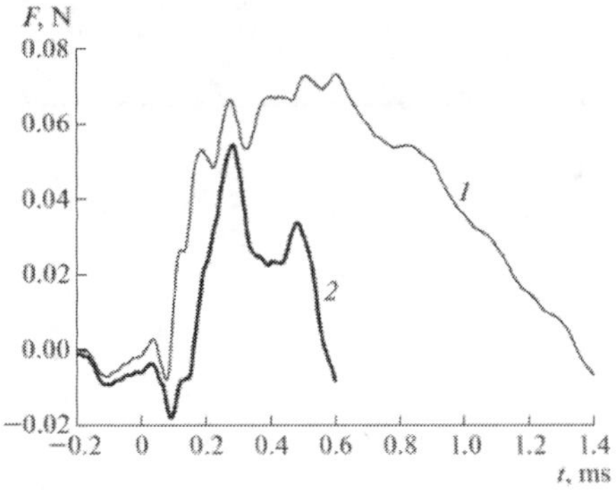

Figure 2. Recoil force at liquid (1) and cured (2) LCP.

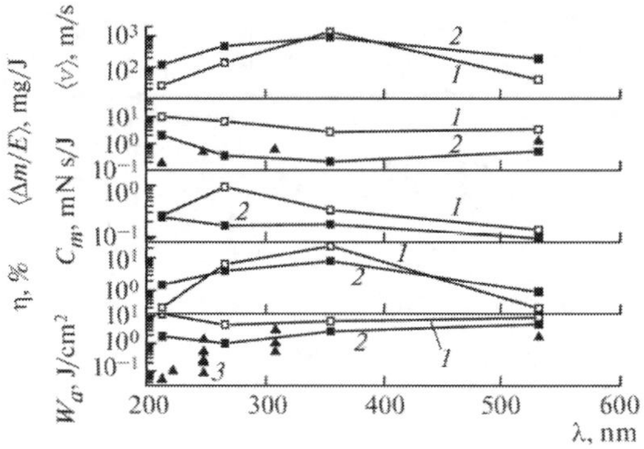

Figure 3. Laser ablation efficiency of a LCP ((1) liquid state, (2) cured state, and (3) reference data for nanosecond laser ablation of PMMA under ambient conditions.

Minimum ablation threshold W_a was observed at 266 nm. In most cases, the ablation thresholds of polymers decrease with increasing photon energy (it can be seen from figure 3 that our data on the laser action on solidified LCP at 266 nm are in good agreement with the data available in the literature for the laser action on PMMA). An increase in the ablation threshold at 213 nm in our experiments could be related to intense luminescence observed in this spectral interval, which leads to large energy losses. For this wavelength, the ablation threshold of liquid LCP reached a maximal value and the difference between the thresholds for the liquid and cured states was the largest.

A minimal specific mass flow was observed under target irradiation at 355 nm - close to the LCP's polymerization optimum wavelength. When a liquid LCP was exposed to such radiation (above a certain threshold, see below), it was cured during the nanosecond pulse (polymerization rate is proportional to light intensity square root), which led to a substantial increase in the mass averaged velocity of ejected plume and laser energy conversion efficiency. The data on momentum coupling coefficient in the case of the laser action on PMMA in the UV-NIR spectral range are not available in the literature (the value of ~0.04 mN s/J was obtained for 10.6 μm in [17]; an estimate of <v> from this work yields ~320 m/s and η~0.64%). The general tendency for polymers is that specific recoil momentum increases inversely proportional to the wavelength, which could be drastic due to the exothermal photochemical reactions initiated when quanta energy exceeds the energies of inter- and intramolecular bonds (usually <345 nm) [18]. With the exception of the laser action at 355 nm, energy efficiency of laser ablation of solidified LCP was higher than that of liquid LCP, which also agrees with the known data.

We investigated the dependence of the discussed characteristics of the laser ablation efficiency on the energy density of the radiation at 355 nm. Irradiation at this wavelength is of particular interest, because it initiates photochemical processes. It can be seen from figure 4 that there are two regimes of the laser action. One regime leads to macroscopic polymerization of the LCP by pulsed laser radiation, while the other does not. In the region of low energies of the radiation (below ~40 J/cm^2), the discussed characteristics of the laser ablation efficiency of liquid LCP substantially exceed those for cured ones, which is not typical for irradiation at other wavelengths and known tendencies mentioned above. In the region of high energies, this difference becomes much smaller. However, the initially liquid LCP still has an advantage. The dependence of the momentum coupling coefficient on laser fluence reveals two inflection points at which the minimum corresponds to the transition from high-energy regime to a low-energy one [19], while the maximum corresponds to a optimal value for the high-energy regime [20]. Other characteristics of laser ablation efficiency in high-energy region remain approximately equal for material in the initially liquid and cured states.

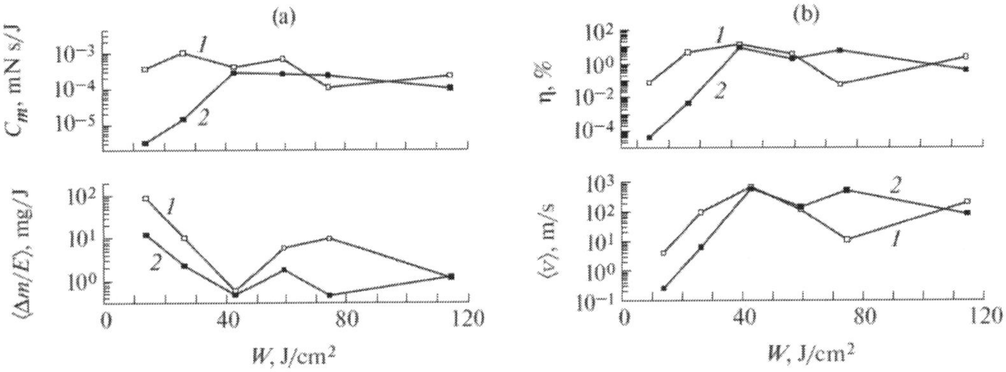

Figure 4. Dependence of LCP ablation efficiency on laser fluence at 355 nm (legend as in figure 3).

At LCP supply system tests irradiation of a ferrule tip and droplets at an intermediate surface were considered. Unlike plastic and metal analogues, been molten or cracked, ceramic ferrules were resistant to thermal and shock loads at laser irradiation. The main issue at tip irradiation is a balance of added and removed material, if the latter is smaller, supply capillary is blocked by cured LCP. There was no blockage observed at 355 nm (of a liquid) and 1064 nm. We also tested LCP solutions (down to 10% vol.) in ethanol, isopropanol, nitromethane, and those kept curing. Although, polymerization rate decreased proportionally to LCP fraction, this reduces capillary blockage risk.

4. Conclusions

Laser ablation efficiency of the same material was studied in liquid and solidified states. The proposed approach allows combining the advantages of transporting and precisely dosing (down to ~10^{-12} L) liquid working media with a high efficiency of laser action on solid substances. Using liquids allows, if necessary, easily changing chemical composition of the working medium by introducing absorbing, catalytic [21], energy, and ionizing [22] additives, creating heterogeneous [23], multilayer, and spatially structured targets, which may find a broad range of applications in laser and plasma chemistry. Pulsed irradiation with sufficient intensity and at the wavelength corresponding to optimal photopolymerization leads to fast solidification of LCP and increases laser ablation efficiency. LCP solutions could be used as buffers to fix metal nanoparticles solutions been produced by laser ablation at non-curing wavelengths.

Acknowledgments

This work was performed at the facility "Beam-M" (RFMEFI59014X0001), supported by the State Task of the Russian Ministry of Education and Science, grant MK-203.2014.8 and the Russian Foundation for Basic Research (grants 13-08-01391 and 14-08-01087).

References

[1] Ravve A 2006 *Light-Associated Reactions of Synthetic Polymers* (New York: Springer Science) p369
[2] Sinko J E and Pakhomov A V 2008 *Proc. AIP* **997** 195-208
[3] Phipps C, Birkan M, Bohn W, Eckel H-A, Horisawa H, Lippert T, Michaelis M M, Rezunkov Y, Sasoh A, Schall W, Scharring S and Sinko J 2010 *Journal of Propulsion and Power* **26** 609-37
[4] Loktionov E Y 2014 *Journal of Applied Spectroscopy* **81** 305-8
[5] Loktionov E Y, Ovchinnikov A V, Protasov Y S, Protasov Y Y and Sitnikov D S 2012 *Journal of Applied Spectroscopy* **79** 104-12
[6] Fardel R, Urech L, Lippert T, Phipps C, Fitz-Gerald J and Wokaun A 2009 *Applied Physics A* **94** 657-65
[7] Wang B, Li L, Tang Z-p and Cai J 2010 *AIP Conference Proceedings* **1230** 243-53
[8] Li X-q, Hong Y-j and He G-q 2010 *Journal of Propulsion Technology* **31** 105-10
[9] S. Uchida, Efficient Laser Ablation (Inst. Las. Technol., Osaka, 2006), Fin. Rep. No. AOARD 044033
[10] Choi S, Han T-h, Gojani A and Yoh J 2010 *Applied Physics A* **98** 147-51
[11] Loktionov E Y, Protasov Y S and Protasov Y Y 2013 *Optics and Spectroscopy* **115** 758-63
[12] Sinko J, Mukundarajan V, Porter S, Kodgis L, Kemp C, Lassiter J, Lin J and Pakhomov A V 2006 *Proc. SPIE* **6261** 626131-12
[13] Decker C and Bianchi C 2005 *Journal of Materials Science* **40** 5491-7
[14] Clark E 2009 An Innovative Way to Print: Ultra Violet and Electron Beam Curing (Rochester: Rochester Institute of Technology) p14
[15] Fouassier J-P 1995 *Photoinitiation, Photopolymerization, and Photocuring: Fundamentals and Applications* (Munich: Hanser-Gardner Publications) p387
[16] Loktionov E Y, Ovchinnikov A V, Protasov Y Y and Sitnikov D S 2010 *Technical Physics Letters* **36** 588-91
[17] Mori K and Sasoh A 2011 *International Journal of Aerospace Innovations* **3** 51-57
[18] Urech L, Lippert T, Phipps C R and Wokaun A 2007 *Applied Surface Science* **253** 7646-50

[19] Loktionov E Y, Ovchinnikov A V, Protasov Y S, Protasov Y Y and Sitnikov D S 2013 *High Temperature* **51** 867-9
[20] Loktionov E Y, Ovchinnikov A V, Protasov Y Y and Sitnikov D S 2010 *Doklady Physics* **55** 446-9
[21] Loktionov E Y and Protasov Y Y 2010 *Doklady Physics* **55** 313-6
[22] Loktionov E Y and Protasov Y Y 2011 *High Temperature* **49** 36-44
[23] Loktionov E Y and Protasov Y Y 2010 *Instruments and Experimental Techniques* **53** 601-6

IOP Conf. Series: Materials Science and Engineering **87** (2015) 012061 doi:10.1088/1757-899X/87/1/012061

Effect of EMA and antioxidants on properties of thermoplastic starch blown films

P Threepopnatkul[1,3], C Kulsetthanchalee[2], A Sittattrakul[1] and E Kaewjinda[1]

[1]Department of Materials Science and Engineering, Faculty of Engineering and Industrial Technology, Silpakorn University, Nakhon Pathom 73000, Thailand
[2]Event and Exhibition Design, Suan Dusit Rajabhat University, Bangkok 10300, Thailand

E-mail: poonsubt@yahoo.com

Abstract. The objectives of this study were to investigate the effect of poly(ethylene-*co*-methyl acrylate) (EMA) at 10, 30 and 50 wt% on the morphological properties, moisture sorption, water vapor permeability and biodegradability of thermoplastic starch (TPS). Urea and formamide were used as a mixed plasticizer. In addition, the effect of antioxidants namely, 3,5-di-*tert*-butyl-4-hydroxyhydrocinnamate (DTBH), butylated hydroxytoluene (BHT) and bis(octadecyl)hydroxylamine (BOH) at 1 wt% on the properties of TPS/EMA film was investigated. TPS/EMA films were produced by a blown film molding machine and characterized by scanning electron microscropy, moisture sorption, water vapor permeability and biodegradability measurement. Results found that the increment of EMA content in the TPS matrix could improve the water sorption, water vapor permeability and biodegradability properties of TPS/EMA films. For biodegradation, the weight loss of the blended films was directly proportional to TPS content. Regarding the antioxidants effect, the water vapor permeability of TPS/EMA films containing DTBH was higher than the one with BOH and BHT. However, the antioxidants contributed little to the biodegradability of TPS/EMA films and had no effect on the moisture sorption of TPS/EMA films.

1. Introduction

Thermoplastic starch (TPS) is a biodegradable polymer which is a promising material due to its low cost and native biodegradability [1]. Kiatkamjornwong S. et al. is focused on modifying the structure of starch [2] and Euaphasate N, et al. improved its ability to blend with other petroleum polymers to reduce its hydrophilicity and improve its mechanical properties [3]. The disadvantage of this approach is the poor compatibility between hydrophilic starch and hydrophobic polymers [4].

One way to enhance compatibility and to incorporate starch in petroleum polymer is to use a compatibilizer containing functional groups such as poly(ethylene-*co*-methyl acrylate), or EMA that can form hydrogen bonds with starch. Even though numerous works have reported on TPS blended with polymers such as low density polyethylene (LDPE) [1, 3-6], poly(lactic) acid (PLA) [7] and ethylene-vinyl acetate copolymer (EVA) [4], no one has studied EMA on the properties of blown TPS films. This research focuses on the effect of EMA on the properties of TPS film. EMA was used as a

[3] Address for correspondence: P Threepopnatkul, Department of Materials Science and Engineering, Faculty of Engineering and Industrial Technology, Silpakorn University, Nakhon Pathom 73000, Thailand. E-mail: poonsubt@yahoo.com.

IOP Conf. Series: Materials Science and Engineering **87** (2015) 012061 doi:10.1088/1757-899X/87/1/012061

compatibilizer in order to improve processability, moisture sorption, water vapor permeation and biodegradability of blown TPS films. An extensive study of these blends with the addition of various types of antioxidants was also performed on the properties such as moisture sorption, water vapor permeation and biodegradability of TPS/EMA blown films.

2. Experimental

2.1. Materials

Native cassava starch was purchased from a local market. Poly (ethylene-*co*-methyl acrylate) (EMA) copolymer with 20% methyl acrylate (MA) contents was purchased from Dupont. Plasticizers (Urea and formamide) were obtained from Ajax Finechem Pty Ltd. The ratio of plasticizers (urea/formamide) was 2:1 wt/wt. Three antioxidants, namely 3, 5-di-*tert*-butyl-4-hydroxyl hydrocinnamate (DTBH), butylated hydroxytoluene (BHT) and bis(octadecyl)hydroxylamine (BOHA) were obtained from Behnmeyer which chemical structures were listed in table 1.

Table 1. The chemical structures of EMA and antioxidants.

Materials	Chemical structure
EMA	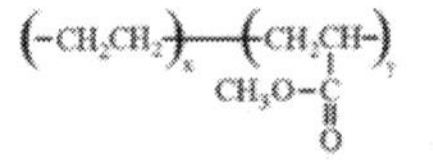
DTBH	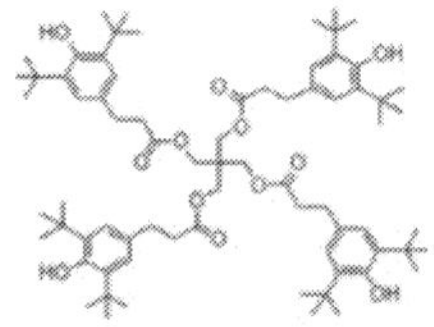
BHT	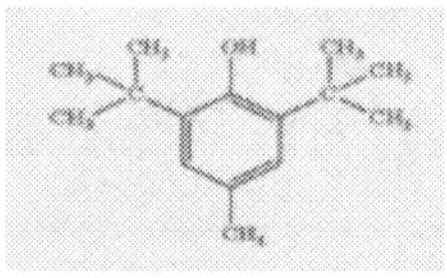
BOHA	

2.2. Compounding and preparation of films

Before blending the starch and plasticizers, urea and formamide were mixed in a high-speed mixer (100 rpm, 1 min). The ratio of plasticizers and starch was 30/70 wt/wt. EMA was blended into the plasticized starch in 10, 30 and 50 wt%. In this study, 1 wt% of antioxidants was added to the mixtures containing 10 wt% of EMA. The mixtures were mixed in a high-speed mixer (1000 rpm, 10 min). The mixtures were blended in twin screw extruder. The temperature profile along the extruder barrel was 50 -125 °C. Extruded strands were cut into pellets for further processing. The plasticized starch and EMA samples were referred to by number after TPS/EMA. The sample compositions of TPS/EMA films are listed in table 2.

Films were molded by blown film extruder with a screw diameter of 25 mm and L/D ratio of 30. The temperature profile of the extruder barrel was 70-115°C to obtain film thickness of approximately 0.2 mm. The films were conditioned for 3 days in closed chambers with saturated NaCl salt solutions (75% relative humidity) prior to any measurement.

Table 2. The sample codes and compositions of blended films.

Sample	Dry starch (wt%)	Urea/Fomamide (2:1) (wt%)	EMA (wt%)	Antioxidants
TPS	70	30	-	-
TPS/EMA1	63	27	10	-
TPS/EMA2	49	21	30	-
TPS/EMA3	35	15	50	-
TPS/EMA11	63	27	10	1% DTBH
TPS/EMA12	63	27	10	1% BHT
TPS/EMA13	63	27	10	1% BOHA

2.3. *Characterization and testing*

Scanning electron microscope (SEM) - The fracture surface of TPS/EMA film was carried out with SEM to analyze the morphological properties.

Differential scanning calorimeter (DSC) - Thermal analysis was conducted using Mettler Toledo DSC. Nitrogen gas was purged at a rate of 20 ml/min during experiments. Scanning was carried out in a heated-cooled-heated pattern at temperature ranging from -60 to 120°C at 10°C/min.

Moisture sorption (MS) - Moisture sorption measurements were performed under 75% relative humidity (saturated NaCl) at 25°C. Films were prepared in dimensions of 30×50 mm. Specimens were dried and weighted the original dry value (W_o). The percentage moisture uptake (M_t) as the following equation (1):

$$M_t = \frac{W_h - W_0}{W_0} \times 100 \quad (1)$$

where W_h is the weight of humid specimens. Moisture sorption data were fitted according to a Peleg model as the following equation (2).

$$M_t = M_0 + \frac{t}{k_1 + k_2 t} \quad (2)$$

where M_t denote the moisture content at any time (%), M_0 denote the initial moisture content (%), k_1 and k_2 are the Peleg rate and capacity constant, respectively, and t is time (hr).

Water vapor permeability (WVP) - Water vapor permeability tests were performed according to ASTM method E96 (1996) [8].

Biodegradability - The biodegradation of the blends was measured by soil burial. The specimens were placed in a box of 60×45 cm wide and 10 cm deep. Films were covered with soil. Water was initially added to the site and then again after for a week. The films remained at the site through the test period time with a sample removed each time in order to track their biodegradation. Measurements were performed to track the biodegradation of the samples weight loss.

3. Results and discussion

3.1. *Blends morphology*

The structures of the TPS and TPS/EMA films are illustrated in figure 1. The fracture surface of the TPS film showed a smoother surface which could be related to the complete destruction [9] as shown in figure 1a. The result showed the aggregated globules of EMA in the blended films. The higher the amount of EMA, the larger the aggregation of EMA was highly dispersed in the TPS matrix as shown in figures 1b-sd. The distinct interfaces and apparent cracks between starch and EMA were sign of the lack of adhesion and the immiscibility at the interface between the two polymers [10]. When 1 wt% of different types of antioxidants were added, no visible difference could be detected on the fracture surface of the TPS/EMA films, as shown in figures 1e-1g.

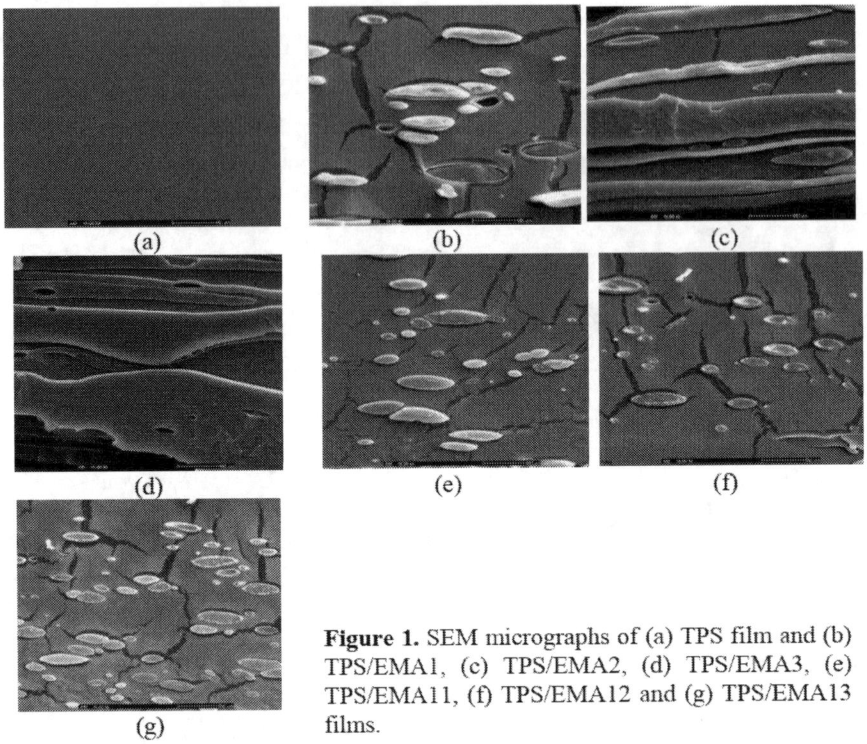

Figure 1. SEM micrographs of (a) TPS film and (b) TPS/EMA1, (c) TPS/EMA2, (d) TPS/EMA3, (e) TPS/EMA11, (f) TPS/EMA12 and (g) TPS/EMA13 films.

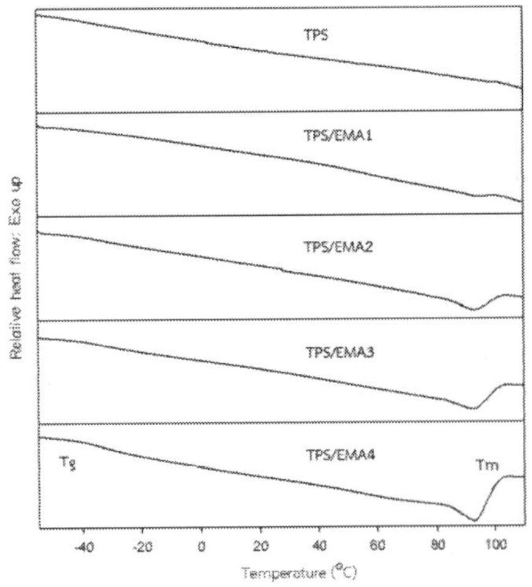

Figure 2. DSC thermograms of TPS and TPS/EMA films at various EMA content.

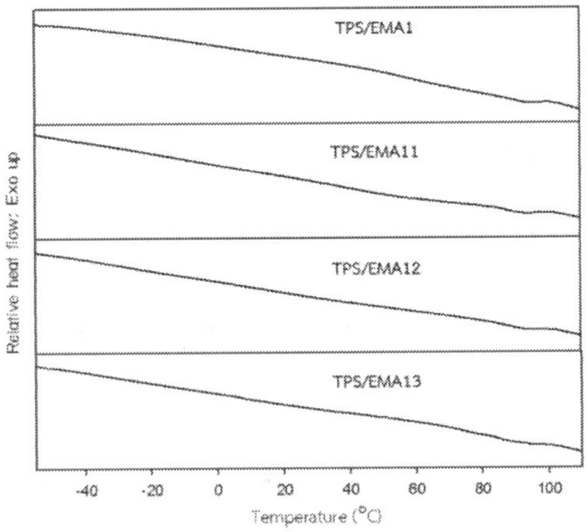

Figure 3. DSC thermograms of TPS/EMA films with various antioxidants.

3.2. Differential scanning calorimeter (DSC)

The differential scanning calorimetry thermograms of the TPS/EMA blended films are shown in figure 2. Results indicated that the glass transition temperature (T_g) of TPS could not be detected by the DSC technique, because starch granules were destroyed, which changed their structure from crystalline to amorphous when plasticizers were added. The melting temperature (T_m) of TPS was clearly unidentified. In the TPS/EMA films, T_g and T_m (-35.28°C and 92.53°C) which stand out were possibly related to the addition of EMA. The melting points and glass transition temperature of EMA had no significant differences when various EMA content were in the TPS/EMA films. Only a slight change in the blended films (1°C) compared with the one of neat EMA was observed. It could be concluded the EMA itself is least miscible with starch even with the methyl acrylate segment in the structure as its compatibilizer [6]. Figure 3 showed DSCs of TPS/EMA films containing antioxidant. The blended films had no significant difference compared with TPS/EMA1.

3.3. *Moisture sorption*

As TPS and TPS/EMA films with the influence of antioxidants are hydrophilic substance, the investigation of their moisture sorption is required. Figure 4 illustrates the effect of starch amount on moisture sorption of TPS/EMA films. EMA film did not demonstrate the moisture sorption. However, the moisture sorption rate and capacity of the TPS/EMA films were proportional to the starch amount, due to high water sorption capacity of the starch. As a function of time, the moisture sorption was relatively higher at the initial stages, and lesser amounts of moisture were absorbed in the later stages. Then, the moisture content of the films reached a plateau indicating that they became equilibrated with the surrounding relative humidity [3].

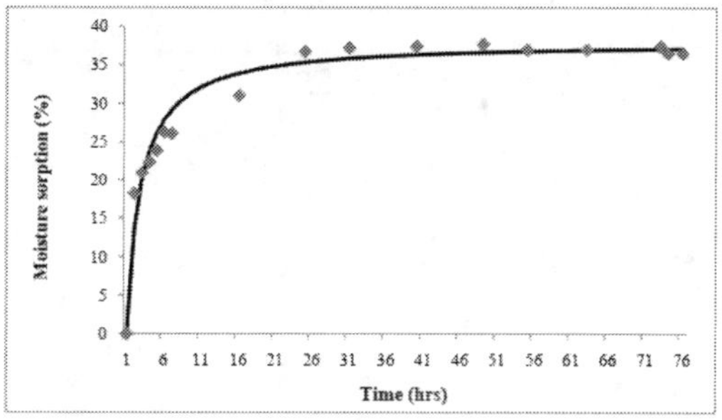

(a)

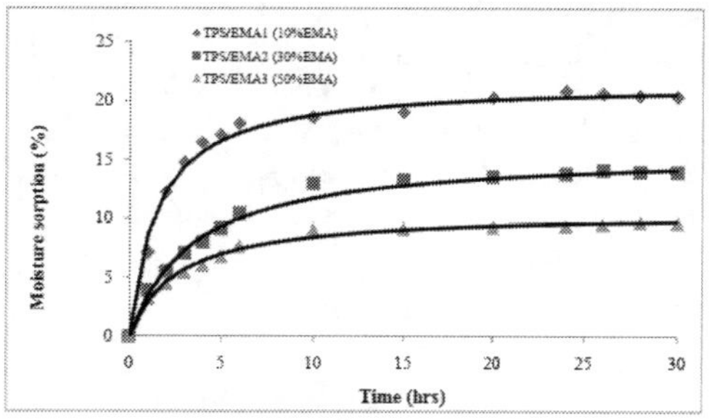

(b)

Figure 4. Moisture sorption of a) TPS and b) TPS/EMA films.

The moisture content data were fitted using Peleg's model to investigate moisture sorption behavior of the films. k_1 is related to the initial moisture absorption rate; k_2 is related to maximum moisture absorption capacity. k_1 and k_2 value are inversely proportional to the initial moisture absorption rate and the maximum moisture absorption capacity, respectively [11]. As seen in table 3, both k_1 and k_2 of TPS films were the lowest values. Therefore, TPS film possessed the highest initial moisture absorption rate and the maximum moisture absorption capacity. In TPS/EMA film, incremental increase of EMA contents tended to lower the absorption rate and capacity. When various types of antioxidants were added, the result of initial absorption rate and maximum moisture absorption capacities was not significantly different, as shown in table 3.

Table 3. Peleg constant from curve fitting using Peleg Model of films.

Blended films	**75% RH, 25°C**		
	k_1	k_2	r^2
TPS	0.0479	0.0263	0.9996
TPS/EMA1	0.0697	0.0463	0.9989
TPS/EMA2	0.2209	0.0633	0.9991
TPS/EMA3	0.2448	0.0946	0.9987
TPS/EMA11	0.0700	0.0483	0.9985
TPS/EMA12	0.0723	0.0445	0.9965
TPS/EMA13	0.0717	0.0464	0.9990

3.4. Water vapor permeability (WVP)

Table 4 illustrates water vapor permeability of TPS and TPS/EMA films with different types of antioxidants were investigated at 25°C. The results showed that both the EMA content and antioxidants had significant effects on the WVP of the films. The water vapor permeability decreased when the EMA content increased. These results suggested that the blended films show aggregated globules of EMA. Increasing the amount of EMA leads to larger aggregate parts of immiscible blend in the starch matrix, as shown in figures 1b-1d. These layers had acted as barrier from water penetration.

The incorporation of antioxidants into TPS/EMA film caused the films to possess difference water vapor permeability. Due to the abundance of hydroxyl groups in TPS matrix, there could be interaction, i.e. hydrogen bonding among those of starch and phenolic group from BOHA or the hydroxyl amine group from BOHA and BHT. The difference in WVP of the TPS/EMA films with different antioxidants established intermolecular interactions between the TPS network and those functional groups in antioxidants. Therefore, the availability of hydroxyl group to form hydrophilic bonding with water would be limited and subsequently lead to different degree of affinity of TPS/EMA film towards water. The hydroxyl group of BHT could likely act as hydrogen donor and thereby hydrogen bonds could be formed between antioxidants and starch molecules. This coincided with the study of Jongjareonraka, who worked on the fish skin gelatin film incorporated with BHT [12]. A large antioxidant molecule, such as DTBH, could experience the steric hindrance effect and thus make bigger holes between starch molecules. As a result, the TPS/EMA film with DTBH possessed relatively higher WVP than other antioxidants. The linear molecule with hydroxyl amine group, such as BOHA, could strongly established the interaction with the hydroxyl group of starch molecules and therefore cause relatively low WVP.

Table 4. WVP of TPS and TPS/EMA films at various EMA content and antioxidants.

Film samples	WVPx10^{-10} (g/(m.s.Pa))
TPS	8.65 ± 0.10
TPS/EMA1	7.25 ± 0.04
TPS/EMA2	1.17 ± 0.04
TPS/EMA3	0.69 ± 0.02
TPS/EMA11	6.01 ± 0.04
TPS/EMA12	5.69 ± 0.06
TPS/EMA13	4.98 ± 0.16

3.5. Biodegradability

Numerous works on biodegradable polymers have indicated that microorganism did consume starch and creating pores, increasing surface area of polymers/starch blends. This provides possibility for degradation. Starch consumption by microorganisms results in blend weight loss. Figure 5 illustrates difference of TPS/EMA3 film before and after soil burial. As seen in figure 6a, in the blended films with 10 wt% of EMA, the weight loss is substantial. The biodegradation rate seemed to diminish for the blended films with 30 and 50 wt% EMA.

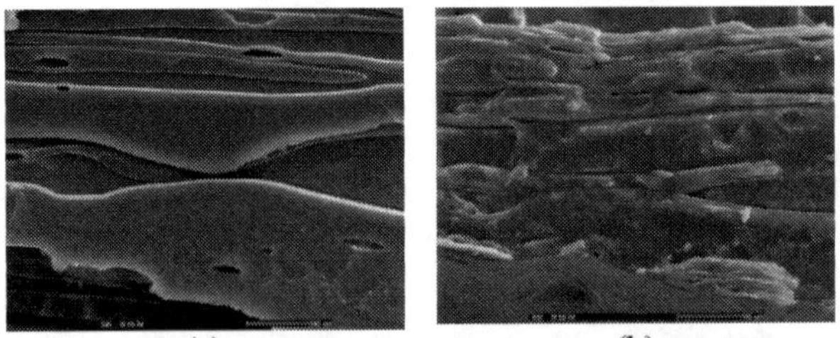

(a) (b)

Figure 5. SEM micrographs of TPS/EMA3 film (a) before and (b) after soil burial.

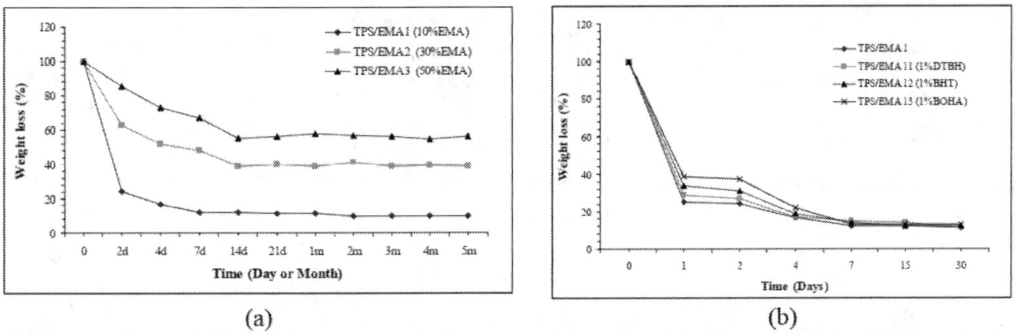

Figure 6. Effect of (a) EMA content and (b) antioxidants on weight loss of TPS/EMA films during soil burial period.

As shown in figure 6b, for the TPS/EMA films containing antioxidants, the antioxidants had little effect on weight loss. For the first seven days, the weight loss of TPS/EMA blend film with antioxidant took place at a lesser amount than TPS/EMA films. This could be explained by the polarity effect of TPS. As mentioned above, the hydroxyl groups or hydroxyl amine group of antioxidants could interact with starch; thus, TPS/EMA films with antioxidants could provide more hydrophobic properties, which would delay the degradation of TPS/EMA films.

4. Conclusions

EMA was to improve processability, although the EMA and TPS blended films were immiscible blends as confirmed by SEM and DSC. Moisture sorption and water vapor permeability in blended films were improved with the incremental increase of EMA content. EMA could act as a barrier against water permeation. In regard to biodegradability of TPS/EMA films, the weight loss of blended films depended on TPS content, and we assumed that starch consumption caused the weight loss. Regarding the antioxidant effect, the TPS/EMA blended films with added antioxidants, for example, DTBH, BHT or BOHA, interacted well with the starch matrix. A large antioxidant molecule, as in DTBH, could experience the steric hindrance effect, thus making larger holes between starch molecules. The TPS/EMA film with DTBH possessed relatively higher water vapor permeability than BHT and BOHA. Adding antioxidants would have little effect to weight loss of blended films.

Acknowledgment

The authors gratefully acknowledge the financial support and materials of the following: SURDI (Silpakorn University Research and Development Institute, Thailand) and Pacific Color Co., Ltd.

References

[1] Rodriguez-Gonzalez F J, Ramsay B A and Favis B D 2003 *Polymer* **44** 1517-26
[2] Kiatkamjornwong S, Thakeow P and Sonsuk M 2001 *Polym. Degrad. Stabil.* **73** 363-75
[3] Euaphasate N, Prachayawasin P, Uasopon S and Methacanon P 2008 *J. Met. Mat. Miner* **18** 103
[4] Da Roz A L, Ferreira A M, Yamaji F M and Carvalho A J F 2012 *Carbohyd. Polym* **90** 34-40
[5] Bikiaris D, Prinos J, Koutsopoulos K, Vouroutzis N, Pavlidou E, Frangi N and Panayiotou C 1998 *Polym. Degrad. Stabil* **59** 287-91
[6] Prinos J, Bikiaris D, Theologidis S and Panayiotou C 1998 *Polym. Eng. Sci.* **38** 594
[7] Shirai M A, Grossmann M V E, Mali S, Yamashita F, Garcia P S and Muller C M O 2013 *Carbohyd. Polym.* **92** 19-22
[8] Ning W, Xingxiang Z, Na H and Shihe B 2009 *Carbohyd. Polym.* **76** 68-73
[9] Zullo R and Iannace S 2009 *Carbohyd. Polym.* **77** 376-83
[10] Pushpadass H A, Bhandari P and Hanna M A 2010 *Carbohyd. Polym* **82** 1082-9

[11] Turhan M, Sayar S and Gunasekaran S 2002 *J. Food Eng.* **53** 153-9
[12] Jongjareonraka A, Benjakula S, Visessanguanb W and Tanaka M 2008 *Food Hydrocolloids* **22** 449-58

A facile method of fabricating mechanical durable anti-icing coatings based on CeO_2 microparticles

Pengren Wang[1], Chaoyi Peng[1, 3], Binrui Wu[1], Zhiqing Yuan[2], Fubiao Yang[1] and Jingcheng Zeng[1]

[1]College of Aerospace Science and Engineering, National University of Defense Technology, Changsha, Hunan, 410073, People's Republic of China
[2]School of Packaging & Materials Engineering, Hunan University of Technology, Zhuzhou, Hunan, 412008, People's Republic of China

E-mail: 952639193@qq.com

Abstract. Compromising between hydrophobicity and mechanical durability may be a feasible approach to fabricating usable anti-icing coatings. This work improves the contact angle of current commercial anti-icing coatings applied to wind turbine blades dramatically and keeps relatively high mechanical durability. CeO_2 microparticles and diluent were mixed with fluorocarbon resin to fabricate high hydrophobic coatings on the glass fiber reinforced epoxy composite substrates. The proportion of CeO_2 microparticles and diluent influences the contact angles significantly. The optimum mass ratio of fluorocarbon resin to CeO_2 microparticles to diluent is 1:1.5:1, which leads to the highest contact angle close to 140°. The microscopy analysis shows that the CeO_2 microparticles form nano/microscale hierarchical structure on the surface of the coatings.

1. Introduction

Superhydrophobic surfaces with the water contact angle higher than 150° and the sliding angle less than 10° have attracted much attention in the past ten years [1-3]. There are many superhydrophobic surfaces in nature, and the best well-known example is the lotus leaf with the water contact angle of up to 160° [4, 5]. The SEM pictures of the lotus surface showed that the combination of micrometer/ nanometer-scale roughness and epicuticular wax led to the superhydrophobic properties [6].

Superhydrophobic surfaces can be used in a variety of fields such as anti-icing, waterproof, self-cleaning, etc. Fabricating superhydrophobic surfaces includes two strategies: the one is constructing nano/microscale hierarchical structure on the intrinsic hydrophobic substrate, the other is modifying rough surface microstructures using low surface energy materials [7-9]. The common methods include template method [10], etching method [11], brush-coating [12], electrospinning technique [13], sol-gel method [14], layer-layer assembly [15-17], electrochemical deposition [2, 18, 19] etc.

However, most methods either have complicated processes or need expensive materials and experimental apparatus [20]. Thus, this is not conducive to large-scale production. Besides, for hydrophobic coatings, their mechanical durability is usually very poor and easily abraded by external forces [21]. Recent studies show that rare-earth metal oxide has not only good hydrophobicity but also

[3] Address for correspondence: Chaoyi Peng, College of Aerospace Science and Engineering, National University of Defense Technology, Changsha, Hunan, 410073, People's Republic of China, Email: 952639193@qq.com

durability properties [22]. CeO_2 is one kind of rare-earth metal oxide and has never been utilized to fabricate anti-icing coatings. Within, we report a simple method to fabricate a mechanical durable anti-icing coating by mixing CeO_2 microparticles and diluent with fluorocarbon resin. This method is easily controlled, and the raw materials are cheap [12, 20]. Furthermore, the coatings have good mechanical durability and can be cured at room temperature.

2. Experimental

2.1. Materials and method

Anti-icing coatings were made from the mixture of fluorocarbon resin (Obtained from Zhuzhou Times New Material Technology CO., LTD, China) with CeO_2 microparticles (obtained from Ganzhou Goring High-tech material Corporation Limited, China) and diluent. Diluent is the mixture of dimethylbenzene and butyl acetate (the mass ratio of dimethylbenzene to butyl acetate is 6:4). The curing agent is isocyanate (the ratio of resin to curing agent is 5:1). The diameter of CeO_2 microparticles is 10 μm.

Took 10 g fluorocarbon resin in the beaker, then diluent, CeO_2 microparticles and isocyanate were added to it in order. Finally, the mixture was homogeneously mixed by an electromagnetic stirrer. Took a certain amount of the coatings to brush on the composite material substrates and cured at room temperature.

We mainly studied the effects of the amount of CeO_2 microparticles and diluent on the hydrophobic properties of the coatings. Firstly, we studied the effects of diluent. The experimental formulations are shown in table 1.

Table 1. The experimental formulations of different amount of the diluent.

No.	Fluorocarbon resin (g)	Isocyanate (g)	Diluent (g)	CeO_2 microparticles (g)
A1	10	2	2.5	5
A2	10	2	5	5
A3	10	2	10	5
A4	10	2	20	5
A5	10	2	30	5
A6	10	2	40	5
A7	10	2	50	5

After finding the optimal amount of diluent based on the influence rules of which on the hydrophobic properties of the coatings, we changed the amount of CeO_2 microparticles to study the effects on the hydrophobicity of the coatings. The experimental formulations are shown in table 2.

Table 2. The experimental formulations of different amount of CeO_2microparticles.

No.	Fluorocarbon resin (g)	Isocyanate (g)	Diluent (g)	CeO_2 microparticles (g)
B1	10	2	10	1
B2	10	2	10	3
B3	10	2	10	5
B4	10	2	10	8
B5	10	2	10	10
B6	10	2	10	15
B7	10	2	10	20

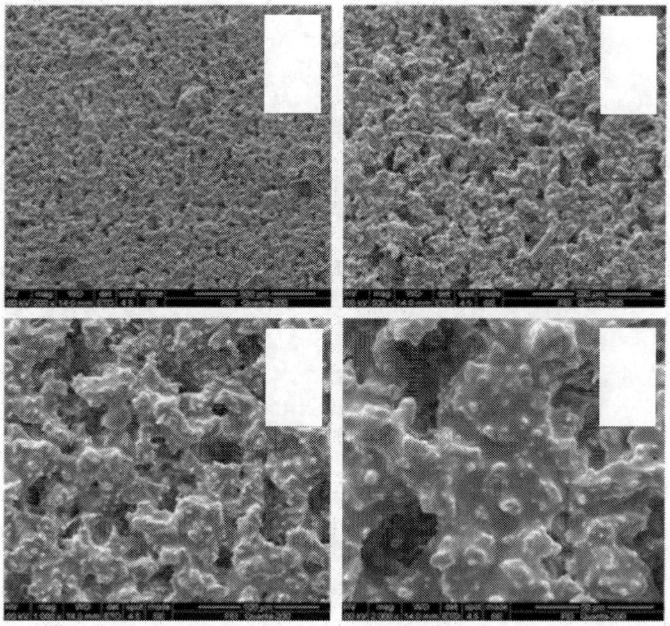

Figure 1. SEM pictures of anti-icing coatings whose mass ratio of fluorocarbon resin to CeO_2 microparticles to diluent was 1:0.5:0.5. The scales of SEM pictures were 500 μm, 200 μm, 100 μm and 50 μm for a, b, c and d, respectively.

Figure 2. SEM pictures of anti-icing coatings whose mass ratio of fluorocarbon resin to CeO_2 microparticles to diluent was 1:0.5:2. The scales of SEM pictures were 500 μm, 200 μm, 100 μm and 50 μm for a, b, c and d, respectively.

2.2. Characterization
The contact angle was measured by Standard Contact Angle Meter SL-200B (SHANGHAI SOLON INFORMATION TECHNOLOGY CO., LTD, China) in this work. A pipette was used to inject a water-drop on the surfaces of the coatings and the droplet formed a contact angle with the solid surface. The instrument was focalized to make the image of droplet showing on the screen clearly. After measuring the contact angle, the results were corrected by Elliptic Curve Method. Five different points of every sample were measured, and their averages were taken as the final results.

A scanning electron microscopy (SEM) of Quanta-200 (FEI Company, USA) was utilized to verify the nano/microscale hierarchical structures of the coatings. The samples were gold-sputtered before scanning. And then they were fixed on the sample plate to take photos under the vacuum condition.

3. Results and discussion

3.1. Surface morphology of coatings
Figures 1 and 2 display the results of SEM pictures of the obtained anti-icing coatings (the scales of SEM pictures are 500 μm, 200 μm, 100 μm and 50 μm, respectively). The mass ratio of fluorocarbon resin to CeO_2 microparticles to diluents was 1:0.5:0.5. Figure 1 displays that the surface presented to be porous and the diameter of the micropore was about 10μm. The main substance of the surface was the cured rock-shaped resin and some bulges comprised of CeO_2 microparticles were distributed on it. But the whole regularity was not good. The possible reason was the amount of resin was still large, and the CeO_2 microparticles were resin-clad.

In figure 2, the amount of diluent increased. The mass ratio of fluorocarbon resin to CeO_2 microparticles to diluent was 1:0.5:2. From the pictures, the surface still presented to be porous. An amount of chrysanthemum-shaped bulges were distributed on it. Nanoscale lamellar structures appeared on the tops of the chrysanthemum-shaped bulges. The combination of them constructed nano/microscale hierarchical structure.

From figures 1 and 2, the porous structure of the coatings surface and the bulges comprised of CeO_2 microparticles became clearer and clearer with the increase of the amount of diluent. When diluent increased, the resin in per unit volume of the coatings became less, the CeO_2 microparticles would not be resin-clad and could well subside on the surface to comprise nano/microscale hierarchical structure.

From figures 1 and 2, we also found that the CeO_2 microparticles had different effects on the hydrophobicity of the coatings under the condition of two kinds of amounts of diluent. So we fixed the mass ratio of fluorocarbon resin to diluent to be 1:2, then changed the amount of CeO_2 microparticles and observed the coatings microstructure. Under this condition, figure 3 shows the SEM pictures of anti-icing coatings whose mass ratio of fluorocarbon resin to CeO_2 microparticles was 1:0.3 and 1:1, respectively.

Figure 3 shows that the microstructures of these two samples were very similar. They both presented to be rough and had flower-shaped bulges comprised of CeO_2 microparticles. The results demonstrate that the change of the amount of CeO_2 microparticles has little effects on the coatings microstructure under this condition.

3.2. Hydrophobic properties of coatings
The contact angles (CA) of the anti-icing coatings are shown in tables 3 and 4.The results in table 3 indicate the influence rules of diluent on the hydrophobicity of the coatings. It shows that when the mass ratio of fluorocarbon resin to diluent was bigger than 1:1, the CA was less than 70°. While the mass ratio was 1:1, the CA significantly increased to 137.88° which indicated the coatings already had good hydrophobic properties. It is because when the ratio was bigger than 1:1, the surface of the substrate was mainly covered by resin, so the CeO_2 microparticles could not construct rough surface. But when the mass ratio was less than 1:1, the amount of resin decreased. So the sedimentary CeO_2 microparticles which have hydrophobic properties formed the bulges and constructed the porous

chrysanthemum-shaped microstructure (shown in figure 2). As a result, the CA increased significantly. From table 3 we also found that when the mass ratio of fluorocarbon resin to diluent was less than 1:1, keeping adding diluent had little effects on the CA.

Figure 3. SEM pictures of anti-icing coatings. For a and b, the mass ratio of fluorocarbon resin to CeO_2 microparticles to diluent was 1:0.3:2. For c and d, the mass ratio was 1:1:2. The scales of SEM pictures were 50 μm, 20 μm, 50 μm and 20 μm for a, b, c and d, respectively.

Figure 4. Water drop deposited on the anti-icing coatings surface: CA is 139.93.

Table 4 displays the influence rules of CeO_2 microparticles on hydrophobic properties of the coatings under the condition of high diluents concentration. The highest CA in table 4 is close to 140° of which the mass ratio of fluorocarbon resin to CeO_2 microparticles to diluent is 1:1.5:1(showed in figure 4). The influence of CeO_2 microparticles on the hydrophobic properties is not significant. In fact, when the amount of diluent is large, the CeO_2 microparticles can be easily homogenized in the coatings to construct nano/microscale hierarchical structure, so changing the amount of CeO_2 microparticles has little effects. Also, the CA became less with the decrease of CeO_2 microparticles. This is because when the particles decrease, there are fewer bulges on the coatings surface to construct nano/microscale hierarchical structure.

Table 3. Contact angles of anti-icing coatings (the mass ratio of fluorocarbon resin to CeO_2 microparticles is 1:0.5).

No.	Mass ratio of resin to diluent	CA (°)
A1	1 : 0.25	69.03
A2	1 : 0.5	65.40
A3	1 : 1	137.88
A4	1 : 2	136.45
A5	1 : 3	134.19
A6	1 : 4	138.62
A7	1 : 5	133.06

Table 4. Contact angles of anti-icing coatings (the mass ratio of fluorocarbon resin to diluent is 1:1).

No.	Mass ratio of resin to CeO_2 microparticles	CA (°)
B1	1 : 0.1	130.71
B2	1 : 0.3	131.93
B3	1 : 0.5	134.52
B4	1 : 0.7	128.11
B5	1 : 1	131.91
B6	1 : 1.5	139.93
B7	1 : 2	138.82

Recently, contact angles of most commercial anti-icing coatings for wind turbine blades are about 110° and these coatings are usually very weak to resist mechanical contact. In this work, we obtained coatings of which the highest water contact angle was close to 140°. Besides, these coatings also had good mechanical durability in theory.

4. Conclusions

This work presented a facile method of fabricating mechanical durable anti-icing coatings based on CeO_2 microparticles that could be cured at room temperature. The SEM pictures indicate that the CeO_2 microparticles construct nano/microscale hierarchical structure on the surface of the coatings. The optimum mass ratio of fluorocarbon resin to CeO_2 microparticles to diluent is 1:1.5:1, which leads to the highest contact angle close to 140°, about 30° higher than commercial anti-icing coatings. Furthermore, these coatings can well resist mechanical contact theoretically. This method is simple and conducive to large-scale production. It will promote the hydrophobic surfaces to practical application in the future

References

[1] Wang S and Jiang L 2007 Definition of superhydrophobic states *Advanced Materials* **19** 3423-4

[2] Darmanin T, Givenchy E T D, Amigoni S, et al. 2013 Superhydrophobic surfaces by electrochemical processes *Advanced Materials* **25**1378-94
[3] Erica Ueda and Pavel A Levkin 2013 Emerging applications of superhydrophilic-superhydrophobic Micropatterns *Advanced Materials* **25** 1234-47
[4] Deng X, Mannen L, Butt H J, et al. 2012 Candle soot as a template for a transparent robust superamphiphobic coating *Science* **335** 67-70
[5] Yao X, Song Y and Jiang L 2011 Applications of bio-inspired special wettable surfaces *Adv Mater* **23** 719-34
[6] Barthlott W, Neinhuis C, Jetter R, et al. 1996 Waterlily, poppy, or sycamore: on the systematic position of Nelumbo *FLORA* **191** 169-74
[7] Ma M and Hill R M 2006 Superhydrophobic surfaces *Current Opinion in Colloid & Interface Science* **11** 193-202
[8] Feng L, Zhang Z, Mai Z, et al. 2004 A super−hydrophobic and super−oleophilic coating mesh film for the separation of oil and water *Angewandte Chemie (International ed. in English)* **43** 2012-4
[9] Wu W, Wang X, Liu X, et al. 2009 Spray-coated fluorine-free superhydrophobic coatings with easy repairability and applicability *ACS Appl Mater Interfaces* **1** 1656-61
[10] Feng L, Zhang Y, Xi J, et al. 2008 Petal effect: A superhydrophobic state with high adhesive force *Langmuir* **24** 4114-9
[11] Öner D and McCarthy T J 2000 Ultrahydrophobic surfaces, effects of topography length scales on wettability *Langmuir* **16** 7777-82
[12] Seyedmehdi S A, Zhang H and Zhu J 2013 Fabrication of superhydrophobic coatings based on nanoparticles and fluoropolyurethane *Journal of Applied Polymer Science* **128** 4136–40
[13] Pisuchpen T, Chaim-ngoen N, Intasanta N, et al. 2011 Tuning hydrophobicity and water adhesion by electrospinning and silanization *Langmuir: the ACS Journal of Surfaces and Colloids* **27** 3654-61
[14] Berendjchi A, Khajavi R, et al. 2011 Fabrication of superhudrophobic and antibacterial surface on cotton fabric by doped silica-based sols with nanoparticles of copper *Nanoscale Res Lett* **6** 594
[15] Zhang M, Wang S, Wang C, et al. 2012 A facile method to fabricate superhydrophobic cotton fabrics *Applied Surface Science* **261** 561-6
[16] Yang H, Dou X, Feng Y, et al. 2013 Self-assembled biomimetic superhydrophobic hierarchical arrays *J Colloid Interface Sci* **405** 51-57
[17] Zhao Y, Xu Z, Wang X, et al. 2013 Superhydrophobic and UV-blocking cotton fabrics prepared by layer-by-layer assembly of organic UV absorber intercalated layered double hydroxides *Applied Surface Science* **286** 364-70
[18] Peng S, Tian D, Miao X, et al. 2013 Designing robust alumina nanowires-on-nanopores structures: superhydrophobic surfaces with slippery or sticky water adhesion *Journal of Colloid and Interface Science* **409** 18-24
[19] Valipour Motlagh N, Birjandi F C, Sargolzaei J, et al. 2013 Durable, superhydrophobic, superoleophobic and corrosion resistant coating on the stainless steel surface using a scalable method *Applied Surface Science* **283** 636-47
[20] Tang Y, Yang J, Yin L, et al. 2014 Fabrication of superhydrophobic polyurethane/MoS_2 nanocomposite coatings with wear-resistance *Colloids and Surfaces A: Physicochemical and Engineering Aspects* **459** 261-6
[21] Deng X, Mammen L, Zhao Y, et al. 2011 Transparent, Thermally Stable and Mechanically Robust Superhydrophobic Surfaces Made from Porous Silica Capsules *Advanced Materials* **23** 2962–5
[22] Azimi G, Dhiman R, Kwon H M, et al. 2013 Hydrophobicity of rare-earth oxide ceramics *Nature Materials* **12** 315-20

IOP Conf. Series: Materials Science and Engineering **87** (2015) 012063 doi:10.1088/1757-899X/87/1/012063

Numerical investigation of the geometrical effects on UHMWPE flow characteristics in small aperture spinneret orifice

Shengxue Qin[1], Zhanhuan Zhang[1], Xingming Xu[2], Jie Liu[1,3] and Hong Liu[1]

[1]College of Mechanical and Electrical Engineering Shandong University of Science and Technology Qingdao, China

[2]College of Mechanical and Electrical Engineering Tai-an Campus of Shandong University of Science and Technology Tai-an, China

E-mail: 15275177928@139.com

Abstract. Due to high viscosity of the ultra-high molecular weight polyethylene (UHMWPE), it is difficult for the melt UHMWPE to flow through the small aperture spinneret orifice in the melt spinning forming process. The geometrical parameters of the spinneret orifice become critical to the melt spinning process. Based on the theory of polymer rheology, the finite element model of UHMWPE melt spinning had been developed by using POLYFLOW, and the length-to-diameter ratio and taper angle of the spinneret orifice effects on the UHMWPE melt flow characters were discussed. The results show that suitable length-to-diameter ratio and taper angle are helpful for the compactness and flow stability of the melt.

1. Introduction

Ultra-high molecular weight polyethylene (UHMWPE), with weight average molecular weights in excess of 2×106 kg kmol-1 [1], is a thermoplastic material with an excellent set of properties, such as good abrasion resistance, the highest impact toughness of any plastic at cryogenic temperature, good corrosion resistance and environmental stress-crack resistance, a low coefficient of surface friction, and noise- and shock-abatement properties. Unfortunately, such high-molecular weight polyethylenes are either difficult or impossible to process in the melt due to their high melt viscosities, and much effort has been directed towards finding suitable processing methods for polyethylenes and, in particular, UHMWPE [2].

At present, the popular preparation method of UHMWPE fiber is gel spinning which use naphthalene, paraffin oil and other hydrocarbons as solvents. In gel spinning process, UHMWPE was prepared into half dilute solution [3], which was extruded through spinneret orifice and drawn to form gel fibers. Then gel fibers were ultra drawn to more than 30 times with a certain temperature after extracting and drying [4]. This gel spinning method is not very strict to the size of molecular weight, and it is an important way to obtain UHMWPE fiber with high strength and high modulus. Smith and Lemstra [5-7] developed the first commercial gel-spinning process. But this process involves significant quantities of solvents requiring removal and recycling, so the environment pollution

[3] Address for correspondence: Jie Liu, College of Mechanical and Electrical Engineering Shandong University of Science and Technology Qingdao, China. E-mail: 15275177928@139.com.

IOP Conf. Series: Materials Science and Engineering **87** (2015) 012063 doi:10.1088/1757-899X/87/1/012063

became a big problem for UHMWPE production when using gel spinning.

Melt spinning, as a new forming method to product UHMWPE fiber was getting more and more attention in the world. The polymer melt flows through spinneret orifice of the screw extruder quantitatively with metering pump, and shot into the air in the form of trickles, and then cooled into silk in the spinning channel [8]. No solvents and precipitation agent are needed during the forming process. There are many advantages in this forming method such as low cost, energy conservation, environmental protection, high production efficiency etc.

However, the unreasonable design of spinneret hole in the spinning process will lead to the failure of the molding because of large viscosity of UHMWPE. Therefore, reasonable design of spinneret hole is one of the key technologies of UHMWPE melt spinning forming based on the flow field analysis in the spinneret hole. The finite element model of UHMWPE melt spinning was established according to the polymer rheology theory and melt spinning process theory using POLYFLOW software. The spinning forming process of the internal melt was simulated with different structural parameters of spinneret hole, and the distribution of flow field in the hole was studied, which provided the theory basis for reasonable design of spinneret hole.

2. Numerical modelling and conditions

2.1. Geometry and FEM model

Spinneret orifice usually consists of guiding region, tapered region and small aperture orifice region. There D and L represent the diameter and length of the spinneret orifice respectively. In this paper, spinneret orifices with and W/O guiding hole were used as shown in figure 1, and α represents the taper angle. As the geometrical model of the spinneret orifice is axial symmetric, then half of axial cross section was selected for calculating. The ANSYS Polyflow software was used for numerical simulation, which provides advanced fluid dynamics technology for solving tasks in the polymer processing. As the models are axial symmetric, half of axial cross section was selected as the research object, and the finite element models meshed with quad elements are shown in figure 2.

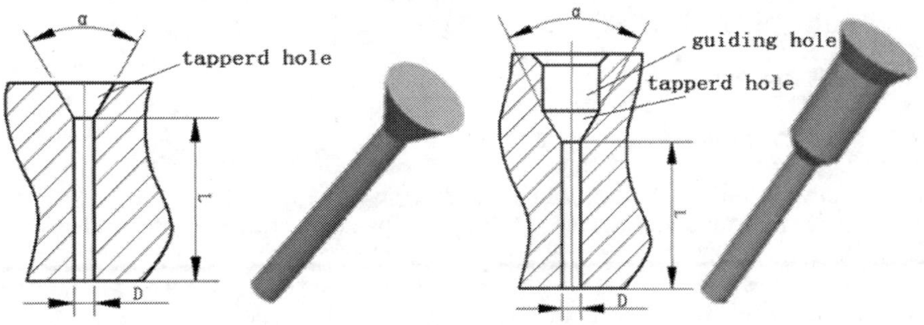

(a) (b)

Figure 1. Geometry model of the spinneret orifices. (a) W/O guiding hole; (b) With guiding hole.

Figure 2. Finite element models.

2.2. *Polymer melt rheology*

In extrusion molding process of polymer, the melt flow follows the law of mass conservation, the law of momentum conservation and the law of energy conservation, which are described respectively by continuity equation, motion equation and energy equation [9, 10].Melt flow behavior is described by the constitutive equation which reflects the relationship between flow stress and the motion parameters in the process of polymer melt flow. The common constitutive equations of the generalized Newtonian fluid include Bird-Carreau law, Power law, Cross law and so on. Power law [11] fluid is a common material model, the equation is expressed as:

$$\eta = \eta_0(\lambda \dot{\gamma})^{(n-1)} \tag{1}$$

where η_0 is zero shear viscosity, λ is relaxation time, $\dot{\gamma}$ is the shear rate, n is the non-Newtonian index.

Correlate the continuity equation, energy equation, motion equation and constitutive equation, and impose specific boundary conditions and initial conditions, the flow process of polymer melt can be solved. However, it is difficult to obtain analytic solutions of the above equations as a result of multivariate of high-order nonlinear partial differential equations, and numerical method is adopted usually. The finite element is widely adopted as a numerical method to solve the above equations and the distribution of velocity and pressure can be obtained, which provides the necessary theoretical basis for the polymer molding.

The UHMWPE used in this study was modified by adding nanometer montmorillonoid, and the modified UHMWPE parameters are shown in table 1 [12].

Table 1. Material parameters.

Density	$0.97g/cm^3$
Zero shear viscosity	13000Pa·s
Coefficient of heat conduction	$5\times10^5 W/(m^2\cdot K)$
Model index	0.275
Relaxation time	0.1s

3. Results and discussion

3.1. *Length-to-diameter ratio effect on flow characters*

For investigating the length-to-diameter ratio effect on the UHMWPE flow characters, two different models of spinneret orifices with and without guiding holes, as shown in figure 1, were used. The length-to-diameter ratio changes from 5 to 10. The spinneret orifice geometrical and processes parameters are shown in table 2.

Figure 3 shows flow velocity contours of the two cases as length-to-diameter ratio equals to 5. It can be seen from the figure that the velocity changes obviously in the taper region, especially at the conjunction region of the tapered hole and orifice as the UHMWPE melt is pushed from big space in a small aperture orifice. While the unstable flow phenomenon did not occur in the taper region as the UHMWPE melt has high melt viscosities. Then the flow velocity shows a stable state in the length direction after the melt flows into the orifice. In radius direction, the velocity gradient is obvious at the region closed to the inner wall of the orifice due to the friction between the melt with the orifice, and the flow velocity attached to the inner wall of the orifice is reduced to zero m/s.

Table 2. geometrical and process parameters.

Diameter of orifice	2mm
Taper angle	60°
Length-to-diameter ratio	5,6,7,8,9,10
Inlet melt temperature	473K
Volumetric flow	$3.5325\times10^{-8}m^3/s$.

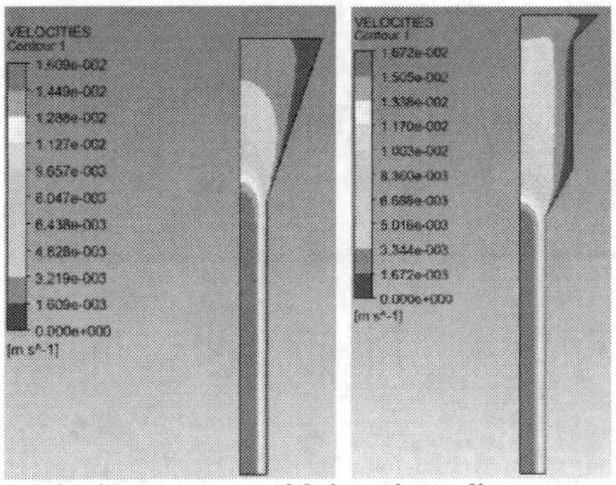

Figure 3. Flow velocity contours with length-to-diameter ratio equals to 5.

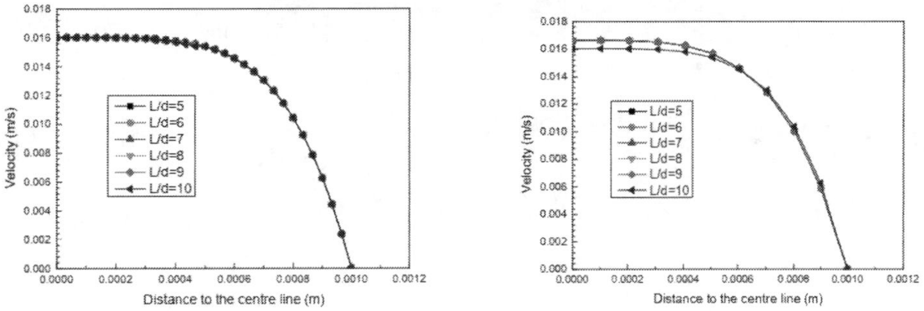

(a) (b)

Figure 4. Flow velocity at the exit with different length-diameter ratios. (a) W/o guiding hole; (b) whit guiding hole.

Figure 4 shows flow velocity plot at the mold exit in radius direction with different length-to-diameter ratios. It can be seen that the velocity gradient of the both cases of W/O and with guiding hole become steep from the point about 0.6 mm to the centre line due to the internal friction between the flow layers and the friction between the melt with the orifice. Figure 4(b) shows the flow velocity distribution in the spinneret orifice with guiding-hole have steeper gradient than the w/o guiding-hole cases of length-to-diameter ratios equals 5 to 9. And the velocity gradients are not affected by the length-to-diameter ratio obviously just as the same with the cases without guiding holes. While the velocity gradient could be decreased with larger length-to-diameter ratio as shown in figure 4(b). It may be that enough length of the orifice could make the UHMWPE melt has enough time to fully relax, which guaranteed the stability of the velocity distribution at the exit and orientation of material.

Figure 5 shows the pressure contours with length-diameter ratio equals to 5. It can be seen that the pressure decrease gradually from inlet to exit and the maximum pressure occurs in the region closed to the orifice's wall at inlet. Figure 6 shows the pressure distribution in length direction with different length-to-diameter ratios of the spinneret orifices with and without guiding holes. It can be seen from figure 6 that the curves have same gradient, and higher pressure according to larger length-to-diameter ratio value. Under the condition of constant origin flow rate, higher pressure was generated duo to the higher resistance induced by the high melt viscosities flow in the orifice. The pressure fluctuate more obviously in the guiding hole region as the

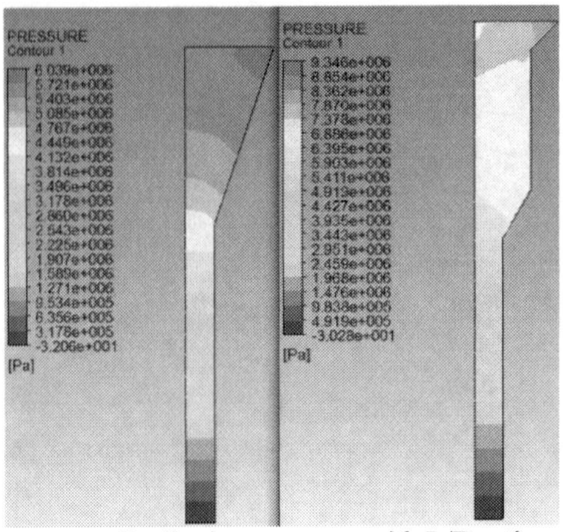

Figure 5. Pressure distribution contours with L/D ratio equals to 5.

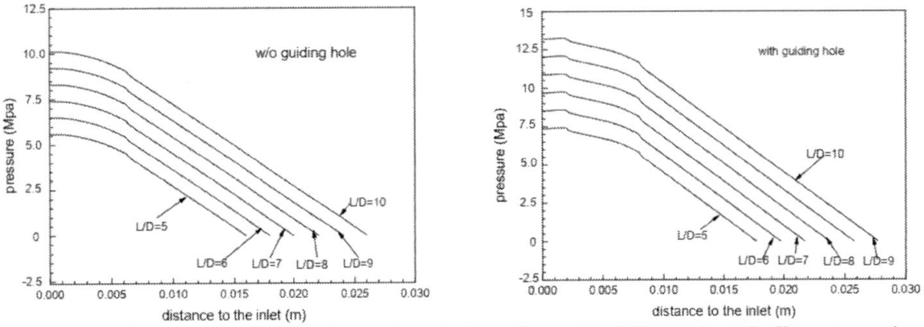

Figure 6. The pressure distribution in length direction with different length-diameter ratios.

Although higher flow pressure in orifice is helpful for improving compactness of the melt and decreasing the end breakage rate of spinning, the bigger length-to-diameter ratio means more difficulty to extrude the melt.

3.2. Taper angles influence on UHMWPE flow characters

From the flow results of the orifices with and w/o guiding hole, it can be seen that the orifice w/o guiding hole has advantages in producing and keeping more stable flow velocity at exit. So the spinneret orifice without guiding hole was used for investigating the taper angle's effect on the flow character. The process parameters are the same with different length-to-diameter ratios cases and the geometrical parameters are shown in table 3.

Figure 7 shows flow velocity distribution at orifice's exit in the radius direction with different taper angles. It can be seen that the velocity decreased gradually from the center to the wall, and the velocity gradient in the wall was larger. Taper angle of guide-hole increased from 60° to 120°, the convergence behavior of the melt flow increased significantly, the tensile deformation and shear deformation increased, which will leaded to instable flow. While in this study, unstable flow phenomenon was not appear in the transitional area of entrance due to the high viscosity and low flow rate of the UHMWPE melt.

Table 3. Geometrical parameters.

orifice diameter (mm)	L/D	Taper angle			
2	10	60°	75°	90°	120°

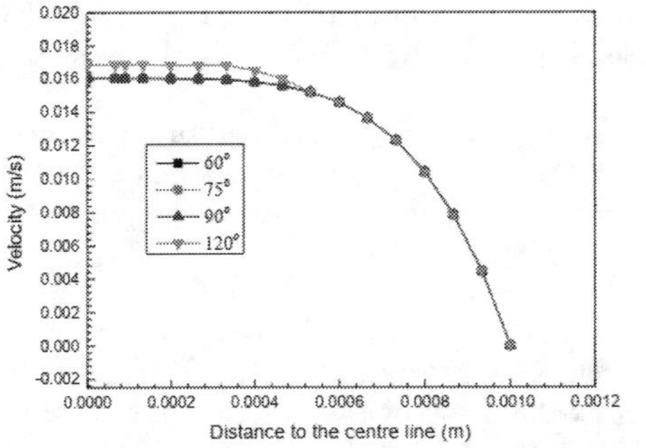

Figure 7. Flow velocity at the exit with different taper angles.

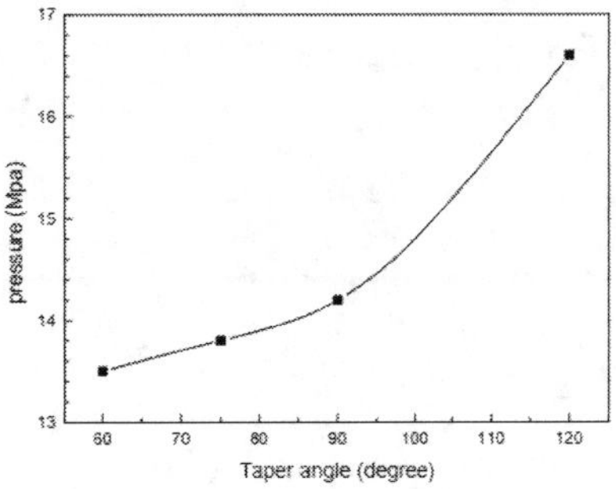

Figure 8. The maximum pressure changes with different taper angles.

Figure 8 shows the maximum pressure with different taper angles. The maximum pressure increases with the taper angle increase due to the convergence behavior induced by the "inlet effect". The pressure increases drastically as the taper angle between 90° and 120° as shown in figure 8. Although higher pressure is helpful to obtain more compactness fiber, too high pressure could induce more unstable flow velocity. Smaller taper angle is helpful to get more stable flow but could make the mold machining harder. So the taper angle should be chosen from 60° to 90°.

4. Conclusions

- The simulation results showed that velocity gradient existed in the radial direction when UHMWPE melt flowed in the spinneret hole, and the velocity decreased gradually from the center to the wall due to the inlet infect. The melt velocity changed obviously in the conjunction region of tapered hole and small aperture orifice. For the orifices with certain diameter, especially in the cases of mold with guiding hole, the melt in longer orifice could get more stable flow velocity in radius direction for the melt has longer relaxation time.
- The maximum pressure and pressure in the orifice were both increased with length-to-diameter ratio and taper angle increase. But too big taper angle may induce the flow velocity unstable, and too big length-to-diameter ratio will make it difficult to extrude the melt. The proper geometrical parameters suggested based the simulation results are that the length-to-diameter ratio should be chosen from 8 to 10, and the taper angle should be chosen from 60° to 90°.

Acknowledgment
This work is supported by the Qingdao Science and Technology program of basic projects (12-1-4-6-(6)-jch) and Research Foundation for new introduced doctor of Shandong University of Science and Technology.

References

[1] Termonia Y, Meakin P and Smith P 1985 Theoretical study of the influence of molecular weight on the maximum tensile strength of polymer fibers *Macro-molecules* **18** 2246-52
[2] Whitehouse C, Liu M L and Gao P 1999 Cold extrusion and in situ formation of self-blends of UHMWPE Part 1: Processability and thermal characterization *Polymer* **40** 1421-31
[3] EI-Domiaty A, EI-Fadaly M and Es. Nassef A 2002 Wear characteristics of ultrahigh molecular weight polyethylene (UHMWPE) *Journal of Materials Engineering and Performance* **11** 577-83
[4] Chen Chengsi, Hu Kaibo, Chen Jianfeng, et al. 2007 Industrialization of high strength polyethylene fibers and applications of their composites *Plastics* **1** 86-8
[5] Paul Smith and Piet J. Lemstra 1979 Ultrahigh-strength polyethylene filaments by solution

spinning/drawing, 2a) influence of solvent on the drawability *Makromol. Chem.* **180** 2983-6
[6] Smith P, Lemstra P J, Pijpers J P L and Kiel M 1981 A Ultra-drawing of high molecular weight polyethylene cast from solution, III. Morphology and structure *Colloid & Polymer Sci.* **259** 1070-80
[7] Smith Paul, Piet J. Lemstra and Jacques P. L. 1982 Pijpers tensile strength of highly oriented polyethylene, II. effect of molecular weight distribution *Journal of Polymer Science: Polymer Physics Edition* **20** 2229-41
[8] Guo Hongzhi, Lin Peijie, Wang Yanping, et al. 2011 Molding method of PTFE fibers *Synthetic Technology & Application* **26** 28-32
[9] Gan Xuehui, Ma Xiaojian, Liu Qiang, et al. 2011 Characteristics of the melt flow in a complex spinning micropore *Mechanical Science and Technology for Aerospace Engineering* **30** 1300-3
[10] Liu Hesheng, Tu Zhigang, Xiong Honghuai, et al. 2002 Viscoelastic constitutive equations of polymer melts *Polymer Materials Science and Engineering* **18** 22-5
[11] Yang Xiaodong, Liu Baochen, Liu Chuntai, et al. 2010 Modified cross viscosity model for polymer melt *Polymer Materials Science and Engineering* **26** 172-4
[12] Zhen Wanqing 2011 The preparation of UHMWPE fiber by melt spinning process (China: Shandong University of Science and Technology)

Effect of peppermint and citronella essential oils on properties of fish skin gelatin edible films

S Yanwong and P Threepopnatkul[1]

Department of Materials Science and Engineering, Faculty of Engineering and Industrial Technology, Silpakorn University, Nakhon Pathom 73000, Thailand

E-mail: poonsubt@yahoo.com

Abstract. Fish skin gelatin films incorporated with peppermint and citronella essential oils at difference concentrations (10, 20 and 30% w/w) were prepared by solution casting. Addition of peppermint oil contributed to a significant decrease of tensile strength and Young's modulus, while the percent elongation at break showed an obvious increase except at 30% w/w. On the other hand, addition of citronella oils promoted a great increase of tensile strength and young's modulus, but an intense decrease of the percent elongation at break. At the predetermined content, the film incorporated with citronella oils outperformed the one with peppermint oils in term of water vapor transmission and solubility in water. Thermal properties of gelatin films with citronella oils exhibited an enhancement in heat stability, while the one with peppermint oils showed slight decrease in heat stability. The additions with both of essential oils exhibited excellent antibacterial properties against both *Staphylococcus aureus* and *Escherichia coli*.

1. Introduction

Gelatin is one kind of protein that is hydrolyzed from collagen. The gelatin is derived from bones or skin such as beef, chicken, pork, and fish. The edible gelatin films were prepared from bovine and porcine skin. They possess high strength, low deformation and high water vapor permeability compared with the one from fish skin as reported by Sobral [1].

Gelatin can be applied widely in many fields, such as food, pharmaceuticals, cosmetic manufacturing and packaging. Packaging made of gelatin can be edible. Typically most gelatins have hygroscopic properties and poor antibacterial properties [2]. Therefore our work is determined to improve gelatin by incorporation with essential oils for packaging application. This helps prolong the food freshness and reduce the perishable rate.

Aguirre A [3] studied the triticale protein films containing the oregano essential oil and found that the properties of the films were improved against the Gram positive bacteria *Staphylococcus aureus* and Gram negative bacteria (*Escherichia coli* and *Pseudomonas aeruginosa*). Gholamreza K [4] reported the Bovine gelatin films incorporated with Zataria multiflora essential oil having excellent antibacterial properties against both Gram-positive and Gram-negative bacteria. Mehraj A [5] reported the gelatin film from the skin of unicorn leatherjacket containing the bergamot and lemongrass oil at different concentrations, concluding that the addition of lemongrass oil in films contributed to lower

[1] Address for correspondence: P Threepopnatkul, Department of Materials Science and Engineering, Faculty of Engineering and Industrial Technology, Silpakorn University, Nakhon Pathom 73000, Thailand. E-mail: poonsubt@yahoo.com.

water vapor permeability, and exhibited antibacterial properties against Gram-positive bacteria (*S. aureus and L. monocytogenes*) and Gram-negative bacteria (*E. coli and S. typhimurium*). Djamel D [6] studied the components of peppermint oil and its antibacterial properties against *E. coli* O157:H7 and *S. aureus* CECT 4459 in minced meat. They found that a significant drop in both bacterial growths in minced meat. Pires C. and Bárbara T [7, 8] showed the major components of citronella essential oil were Δ^3-Carene (22.5%), β-Citronellal (11.9%) and β-Citronellol (9.00%). They reported that the citronella essential oil could be the substance against *E. coli* O157:H7 bacteria.

Even though a lot of work has been reported on film blended with various essential oils [3-8], no one has studied the effect of peppermint and citronella oil on the properties of the fish skin gelatin films. Therefore, this research is to investigate the incorporation of two types essential oils i.e., peppermint and citronella oils in fish skin gelatin film, at different amounts to improve the properties of water vapor resistance and antibacterial activity. In addition, the effect of essential oils on the mechanical, thermal properties and morphology of fish skin gelatin films was also studied.

2. Experimental

2.1. *Materials*

Fish skin gelatin powder (~ 250 blooms), glycerol was used as a plasticizer, tween 80 was used as a surfactant between gelatin and essential oils, citronella essential oil and peppermint essential oil were purchased from Chemipan Co., Ltd., Thailand.

2.2. *Preparation of gelatin film solution casting*

The gelatin powder was dissolved in distilled water to gain the protein concentration of 5% (w/v) and then heated at 60°C for 30 min under continuous stirring. After that, the glycerol was added at concentration 30% (w/w) of gelatin content. Then, each essential oil was mixed with tween 80 at 25% (w/w) of essential oil content. After that, the mixture was added into the gelatin solution at concentrations of 10%, 20%, and 30% (w/w) of gelatin content, respectively. The gelatin-essential oil solution was homogenized at 5,000 rpm for 3 minutes by using a homogenizer. For preparing the film, the gelatin-essential oil solution was cast onto a glass tray and conditioned at 40°C for 24 h to obtain gelatin-essential oil films.

2.3. *Film characterization*

2.3.1. Mechanical properties. The mechanical properties of the film samples were measured by using a Universal Testing Machine. Films were cut into strips with a dimension of 150×25.4 mm, then placed at 25°C for 48 h and 50 ± 5% relative humidity (RH) before testing. The test was carried out by the ASTM D 882 standard method, with the initial grip separation of 100 mm, a crosshead speed of 10 mm/min, and load cell of 5 kN. The final value represented the average of six samples.

2.3.2. Film thickness. Film thickness was measured using a micrometer (SM-112, Teclock Corporation, Japan). The final value represented the average of five random measurements taken at different locations of the film.

2.3.3. Water vapor transmission rates. Water vapor transmission rate (WVTR) was measured using a modified ASTM E96 method. The glass bottles containing dried silica gel (0% RH) were closed at the mouth of the bottle by the film samples, placed at 25°C and 75% RH in desiccator. The weight of the glass bottles was measured at 1 h intervals during 24 h. The result demonstrated the relationship between the weight change (g) and time (h). WVTR of the film was calculated by equation (1):

$$\text{WVTR (g/m}^2\text{.h)} = G/(tA) = (G/t)/A \quad (1)$$

Where G is weight change (from the straight line), g; t is time, h; G/t is calculated from slope of the straight line, g/h; and A is test area (bottle mouth area), m^2.

2.3.4. Solubility in water. Film solubility in water was measured according to the method proposed by Gontard and Wang et al. [9, 10], and the samples were cut into small pieces and the films were dried in oven at 100°C for 24 h to a constant weight. Each dried film sample was immersed in 50 ml of distilled water for 24 h. After 24 h, all of the samples were removed from the solutions and the films were dried at 100°C for 24 h. The final weight was recorded and the solubility was calculated by equation (2):

$$\text{Solubility} = (\text{Initial Weight} - \text{Final Weight})/\text{Initial Weight}) \times 100\% \quad (2)$$

2.3.5. Thermo gravimetric analysis (TGA). Samples of approximately 5 mg were scanned using a thermo-gravimetric analyzer (TGA: TGA/DSC 1, Mettler Toledo, Switzerland) from 50 to 600°C at a rate of 10°C/min under a nitrogen flow of 30 mL/min. Weight losses of samples were measured as a function of temperature.

2.3.6. Antibacterial activity test using colony counting. Bacteria reduction method was used to investigate antibacterial efficacy of films. All of the film samples were cut into circular shape (with a diameter of 1.2 cm). In the first step, the plastic film was placed on the bottom of each bottle. Each prepared sample was placed on the plastic film. The active bacterial solution, McFarland Standard No. as 0.5, was dropped on the sample and covered them by plastic film again. After 24 h in an incubator, phosphate buffered saline (PBS) was added to each bottle. And 0.1 ml of each solution was sampled and diluted by PBS for 10 times. In each dilution, the solution was dropped into the prepared agar plate and kept in an incubator at 37°C for 18 h. Finally, the colony of bacteria in each plate was calculated.

2.3.7. Statistical analysis. Analysis of variance (ANOVA) was used to compare the mean differences of the samples and the differences between the means were determined by the least significant difference test at $p < 0.05$. The statistical software package (SPSS ver. 17.0, SPSS Inc. Chicago, IL, USA) was used to analyze the experimental data.

3. Results and discussion

3.1. Mechanical properties and thickness

Table 1 demonstrated the properties of tensile strength (TS), percent elongation at break (%EB), Young's modulus (YM) and the thickness of fish skin gelatin films with peppermint and citronella oils at various contents (10, 20 and 30% w/w). From the measurements, the thickness of gelatin films was found to have no differences when either peppermint or citronella essential oils were presented at the contents of 10% to 30% w/w compared with the pure gelatin film. However, the fish skin gelatin incorporated with 50% w/w root essential oils indicated the increase in film thickness because the interaction of essential oils between peptide chains could reduce the ordered alignment of gelatin chains and the compact network could not be developed [11].

The addition of peppermint oil into the gelatin films at the contents of 10% to 30% w/w demonstrated a significant decrease ($p < 0.05$) in TS compared with the pure gelatin film due to the effect of plasticized behavior of peppermint oil. This could reduce or even hinder the interaction between gelatin chains, thus diminishing the TS of the gelatin films [12].

On the other hand, the EB of gelatin films demonstrated an intense increase ($p < 0.05$) from 10% to 20% w/w of peppermint oils due to plasticizing effect. This led to the chains slippage during film stretching. However, the addition of peppermint oils at 30% w/w caused the EB reduction to 8.26 % due to the fact that an excessive amount of peppermint oils led to more interaction reduction between gelatin chains and even easy rupture [5]. Moreover, the presence of excess essential oils in gelatin matrix could interfere in the interactions between polymer chains and reduce the flexibility of the gelatin matrix [13].

In term of Young's modulus, YM, the addition of peppermint oils in gelatin films with concentrations of 10% to 20% w/w contributed to a significant decrease *($p < 0.05$)* in YM from 1777 to 1621 MPa respectively due to plasticizing effect. Polymer chains could be able to move, making the film flexible according to Zinoviadou et al. [14]. Their work was conducted on the addition of oregano oil and found that the YM of whey protein isolate films was decreased.

The addition of the citronella oils into the gelatin films at 10% promoted a reduction in TS and YM but an increase in EB. However, TS of fish gelatin film showed a significant increase from 27.78 MPa at 10%w/w concentration to 31.15 MPa at 30% w/w, while a significant decrease in EB from 15.33% at 10% w/w to 4.68 % at 30%w/w citronella oils was observed. These could be explained by the presence of β-citronellal in citronella oils. Such component could interact with proteins which promoted their cross-linking in gelatin film, resulting in the rigidity behavior in gelatin film [15]. This led to an increase in both TS and YM of fish skin gelatin films.

Table 1. Mechanical properties and thickness of fish skin gelatin films containing peppermint and citronella essential oils at 10, 20 and 30% w/w.

Essential oils	Content (%)	TS (MPa)	EB (%)	YM (MPa)	Thickness (mm)
Pure	-	33.46 ± 0.40^a	3.44 ± 0.40^a	1896.78 ± 30.14^a	0.1228 ± 0.0025^a
Peppermint	10	30.38 ± 0.72^b	7.38 ± 0.97^b	1777.58 ± 44.21^b	0.1233 ± 0.0021^a
	20	28.38 ± 0.69^c	12.61 ± 0.92^c	1621.05 ± 34.16^c	0.1241 ± 0.0025^a
	30	25.59 ± 0.74^d	8.26 ± 0.59^d	1696.76 ± 42.53^d	0.1249 ± 0.0017^a
Citronella	10	27.78 ± 0.84^e	15.33 ± 0.54^e	1522.41 ± 33.87^e	0.1235 ± 0.0018^a
	20	29.60 ± 0.78^f	13.10 ± 0.66^f	1647.33 ± 25.36^f	0.1245 ± 0.0027^a
	30	31.15 ± 0.46^g	4.68 ± 0.77^g	1791.56 ± 37.82^g	0.1254 ± 0.0021^a

[a]The values are presented as means ± standard deviation.
[b]Different letters in the same column indicate significant differences (*$p < 0.05$*), and the same letters indicate no significant differences.

Table 2. Water vapor transmission rates (WVTR) and solubility in water of fish skin gelatin films incorporated with peppermint and citronella essential oils at 10, 20 and 30% w/w.

Essential oils	Content (%)	WVTR (g/m^2.day)	Solubility in water (%)
Pure	-	65.13 ± 1.22^a	61.72 ± 1.12^a
Peppermint	10	59.82 ± 1.17^b	66.39 ± 0.88^b
	20	56.51 ± 1.75^c	70.13 ± 1.01^c
	30	53.86 ± 0.57^d	78.55 ± 1.28^d
Citronella	10	56.68 ± 0.99^e	67.78 ± 0.92^e
	20	53.03 ± 1.25^f	64.49 ± 0.69^f
	30	43.92 ± 1.30^g	62.28 ± 0.77^g

[a]The values are presented as means ± standard deviation.
[b]Different letters in the same column indicate significant differences (*$p < 0.05$*), and the same letters indicate no significant differences.

3.2. Water vapor transmission rates (WVTR)

WVTR of fish skin gelatin films containing peppermint and citronella essential oils at various contents is shown in table 2. Incorporations of peppermint oils and citronella oils at 10% ~ 30% w/w in gelatin films caused a significant decrease (*$p < 0.05$*) of WVTR from 59.82 to 53.86 g/m^2h and from 56.68 to 43.92 g/m^2h, respectively. These could be due to the different hydrophobicity of its components in peppermint [7] and citronella [8] essential oils, for instance, monoterpenes, a highly hydrophobic component [16]. Different amounts of monoterpenes could influence the hydrophobicity. In other words, monoterpenes is the governing factor in determining the hydrophobicity of gelatin film. As a result, the water vapor transmission through the gelatin films would be different between the films with peppermint and citronella oils. With the same concentration of essential oils, the fish skin gelatin film incorporated with citronella oils possessed lower WVTR than its counterpart with peppermint oils due to hydrophobicity of its component.

3.3. Solubility in water

The water solubility of fish skin gelatin films incorporated with peppermint and citronella oils at 10, 20 and 30% w/w is shown in table 2. The increase of peppermint oil from 10% to 30% w/w caused a significant increase ($p < 0.05$) of films solubility in water. The presence of peppermint oil in gelatin matrix may reduce the interaction between gelatin chains, which led to the weakening of the interactions between gelatin films and allowed to an increase of its leaching out [15].

The addition of citronella oil into gelatin films at concentration of 10% led to an increase of water solubility of films in comparison with pure gelatin film. Water solubility of fish skin gelatin films was decreased with the increasing amount of citronella oils from 10% to 30% w/w. These results were supported by the fact that the cross-linking between protein chains by β-citronella component caused a significant decrease of solubility in water of films.

3.4. Thermo gravimetric analysis (TGA)

The results of degradation temperatures (T_d), weight loss (Δw) and residue of fish skin gelatin films incorporated with peppermint and citronella oils at different contents (10, 20 and 30% w/w) are shown in table 3. All film samples showed three stages of degradation temperatures (T_d) and weight loss (Δw). The first stage at the range of up to 100 °C, T_{d1} and the Δw_1 of all films may be due to the loss of free water or other volatile compounds from the gelatin film. Gelatin films containing peppermint and citronella oils at any content (10, 20 and 30%) outperformed pure gelatin film on the weight loss due to the low ability in water absorption by the hydrophobic components in both peppermint and citronella essential oils [17].

Table 3. Thermal degradation temperatures (T_d, °C) and weight loss (Δw, %) of fish skin gelatin films incorporated with peppermint and citronella essential oils at 10, 20 and 30% w/w.

Essential oils types	Content (%)	Δ_1		Δ_2		Δ_3		Residue (%)
		$T_{d1,onset}$	Δw_1	$T_{d2,onset}$	Δw_2	$T_{d3,onset}$	Δw_3	
Pure	-	77.12	8.05	220.08	25.24	295.42	48.28	20.73
Peppermint	10	80.43	7.98	222.89	25.90	294.80	48.51	17.23
	20	79.32	6.47	228.18	25.71	292.52	51.29	15.97
	30	74.25	2.72	226.12	25.44	291.85	56.83	14.07
Citronella	10	69.27	6.43	227.75	25.66	294.63	46.19	21.54
	20	74.35	6.36	228.90	24.13	302.12	44.79	22.92
	30	71.20	2.53	233.59	24.01	302.82	42.35	24.45

The second stage, T_{d2} and the Δw_2, of all films indicated the loss of glycerol component and small protein molecule in gelatin films. The work was reported by Hoque et al. [18] that the degradation temperature in the range of 196.30–216.71°C appeared in cuttlefish skin gelatin film.

The third stage, T_{d3} and the Δw_3, of all film samples referred to the degradation of the larger molecules of protein in gelatin films [17]. The addition of peppermint oil at 10% to 30% w/w into gelatin films demonstrated caused a slight decrease of T_{d3} from 294.80 to 291.85°C and an increase of Δw_3 from 48.51 to 56.83%, respectively. The addition of peppermint oil at all contents led to the results of T_{d3} to have the lower temperature compared with pure gelatin film. In contrast, Δw_3 of fish skin gelatin film at all contents of peppermint oil was higher than that of the pure gelatin film. These could be explained by the plasticizing effect of peppermint oil. These may interfere in the interaction between protein chains in the gelatin films. On the other hands, the addition of citronella oil from 10% to30% w/w into gelatin films contributed to an increase of T_{d3} from 294.63 to 302.82 °C but a decrease of Δw_3 from 46.19 to 42.35%, respectively. This could be explained by the presence of β-citronellal, a component in citronella oil, which might cause the cross-linking between protein chains in gelatin films. Therefore, the gelatin film incorporated with citronella oil exhibited an improvement in the thermal stability.

3.5. Antibacterial activity test using colony counting

Antibacterial activities of fish skin gelatin films incorporated with peppermint and citronella oils at 10, 20 and 30% w/w are shown in table 4. As per our studies, pure gelatin film or the fish skin gelatin film without essential oil showed no activity against both *E. coli* and *S. aureus*. For both essential oils, the gelatin films incorporated with either peppermint or citronella oils at 30% showed a significant increase of antibacterial activity against both *E. coli* and *S. aureus* compared with its counterpart incorporated with either essential oil at 10%.

According to the report of Zivanovic et al. [19], the phospholipids of cellular membrane were attacked by the phenolic components of essential oils. Consequently, an increase in permeability and loss of cytoplasm took place. Antibacterial activity occurred when the phenolic components found in the essential oil interacted with such enzymes localized in the cell wall, which caused catastrophic incidents to the membranes. Most of the essential oils comprised phenolic components. Therefore, the gelatin films incorporated with peppermint and citronella essential oils exhibited activities against both *E. coli* and *S. aureus*. However, gelatin films incorporated with peppermint essential oils or citronella essential oils are more effective to *S. aureus* (Gram-positive) than to *E. coli* (Gram-negative) due to the presence of lipopolysaccharides, since the outer cell wall of Gram-negative bacteria was composed of lipopolysaccharides and could act as the protective membranes against the phenolic components from essential oils [20].

As for the films with the same amount of essential oil, the gelatin film with peppermint oil exhibited better antibacterial activity than its counterpart with citronella oil for both types of microorganisms since a higher amount of phenolic components existed in the peppermint oil [7] than in the citronella oil [8].

Table 4. Antibacterial reduction of fish skin gelatin films incorporated with peppermint and citronella essential oils at different concentrations.

Essential oils types	Content (%)	% Antibacterial Reduction	
		E. coli	*S. aureus*
		Gram negative	Gram positive
Pure	-	0.00	0.00
Peppermint	10	81.17	84.13
	20	82.63	85.07
	30	84.02	85.52
Citronella	10	79.97	83.92
	20	81.91	84.47
	30	83.59	85.27

4. Conclusion

The peppermint essential oil can act as plasticizer in gelatin films. When increasing the amount of peppermint oil, the mechanical and thermal properties of fish gelatin films are likely to decrease but solubility in water is increased. β-Citronellal component in citronella essential oils can act as the cross-linker between gelatin chains and will increase the mechanical and thermal properties of fish gelatin films, but the solubility in water will decrease when the amount of citronella oils increases. The outcome of antibacterial properties and water vapor permeability of gelatin films depends on the amount of phenolic and hydrophobic component in each essential oil. The fish skin gelatin film with citronella oil shows to be more effective in term of WVP than the film with peppermint oil because of the crosslinking of gelatin chains by β-Citronellal component in citronella oil. Incorporations of either peppermint or citronella oil into the fish skin gelatin films could improve the antibacterial activity against both *E. coli* (Gram-negative) and *S. aureus* (Gram-positive). But, the gelatin film with peppermint oil outperforms its counterpart with citronella oil for antibacterial activity due to its phenolic component.

References

[1] Sobral P J A, Menegalli F C, Hubinger M D and Roques M A 2001 Mechanical, water vapor barrier and thermal properties of gelatin based edible films *Food Hydrocolloids* **15** 423-32

[2] Gomez G and Perez M 2009 Fish gelatin: A renewable material for developing active biodegradable films *Trends in Food Science and Technology* **20** 3-16

[3] Aguirre A, Borneo R and Leon A E 2013 Antimicrobial, mechanical and barrier properties of triticale protein films incorporated with oregano essential oil *Food Bioscience* **1** 2-9

[4] Gholamreza K, Aytak R, Seyed M and Amin P 2014 Effects of essential oil on the water binding capacity, physicomechanical properties, antioxidant and antibacterial activity of gelatin films *Food Science and Technology* **57** 556-61

[5] Mehraj A, Soottawat B, Thummanoon P and Tri W A 2012 Physico-mechanical and antimicrobial properties of gelatin film from the skin of unicorn leatherjacket incorporated with essential oils *Food Hydrocolloids* **28** 189-99

[6] Djamel D, Mohammed A, Javier Y, Lamia I, Diego G and Pedro R 2012 Antioxidant and antibacterial effects of lavandula and mentha essential oils in minced beef inoculated with E. coli O157:H7 and S. aureus during storage at abuse refrigeration temperature *Meat Science* **92** 667-74

[7] Pires C, Ramos C, Teixeira B, Batista I, Nunes M L and Marques A 2013 Hake proteins edible films incorporated with essential oils: Physical, mechanical, antioxidant and antibacterial properties *Food Hydrocolloids* **30** 224-31

[8] Bárbara T, António M, Cristina R, Nuno R N, José M F, Jorge A S and Maria L N 2013 Chemical composition and antibacterial and antioxidant properties of commercial essential oils *Industrial Crops and Products* **43** 587-95

[9] Gontard N and Guilbert S 1994 Bio-packaging: Technology and properties of edible and/or biodegradable material of agricultural origin (Food Packaging and Preservation) Ed M Mthlouthi pp. 159-81

[10] Wang L, Liu L, Holmes J, Kerry J F and Kerry J P 2007 Assessment of film forming potential and properties of protein and polysaccharide-based biopolymer films *International Journal of Food Science and Technology* **42** 1128-38

[11] Phakawat T, Soottawat B and Thummanoon P 2013 Physico-chemical properties, morphology and antioxidant activity of film from fish skin gelatin incorporated with root essential oils *Journal of Food Engineering* **117** 350-60

[12] Avena-Bustillos R J, Chiou B and Olsen C W 2011 Gelation, oxygen permeability and mechanical properties of mammalian and fish gelatin films *Journal of Food Science* **76** 519-24

[13] Seyed F H, Masoud R, Mojgan Z and Farhid F 2015 Bio-based composite edible films containing origanum vulgare L. essential oil *Industrial Crops and Products* **67** 403-13

[14] Zinoviadou K, Koutsoumanis K P and Biliaderis C 2009 Physico chemical properties of whey protein isolate films containing oregano oil and their antimicrobial action against spoilage flora of fresh beef *Meat Science* **82** 338-45

[15] Hernández-Muñoz P, Villalobos R and Chiralt A 2004 Effect of cross-linking using aldehydes on properties of glutenin-rich films *Food Hydrocolloids* **18** 403-11

[16] Ikeda R M and Stanley W L 1962 The monoterpene hydrocarbon composition of some essential oils *Journal of Food Science* **27** 455-8

[17] Phakawat T, Soottawat B and Thummanoon P 2012 Properties and antioxidant activity of fish skin gelatin film incorporated with citrus essential oils *Food Chemistry* **134** 1571-9

[18] Hoque M S, Benjakul S and Prodpran T 2011 Effects of partial hydrolysis and plasticizer content on the properties of film from cuttlefish (Sepia pharaonis) skin gelatin *Food Hydrocolloids* **25** 82-90
[19] Zivanovic S, Chi S and Draughon A F 2005 Antimicrobial activity of chitosan films enriched with essentialoils *Journal of Food Science* **70** 45-51
[20] Burt S 2004 Essential oils: Their antibacterial properties and potential applications in foods, a review *International Journal of Food Microbiology* **94** 223-53

Terpene sensor array with bridge-type resistors by CMOS technology

Sung Pil Lee[1]

Department of Electronic Engineering, Kyungnam University, Kyungnam, Changwon 631-701, Korea

E-mail: sensors@kyungnam.ac.kr

Abstract. The interaction of terpene gas with the sensing element in the sensor array can cause changes in electrical properties because of a charge transfer and the polymer chain structure. Resistive type interdigited electrode sensor arrays covered with a mixture of molecularly imprinted polymer (MIP)/ conductive polymer (CP) were designed and fabricated to detect terpene gases. MIP coated on CP (MOC) type showed markedly higher sensitivity compared to mixture of MIP and CP (MMC) type. The gas detection patterns by PCA were used to get higher selectivity of multicomponent chemical media.

1. Introduction

Terpene is an oxygenated hydrocarbon, and is beneficial to humans in diverse ways including stabilization of people and ozone removal. A few types of terpene with such benefits are being utilized as commercial air fresheners. While terpene possesses unique scent based on its types, research is being delayed because there are many terpene types with similar characteristics. The major terpene detection methods that have been mainly researched involve high performance liquid chromatography (HPLC) [1], gas chromatography mass spectrometry (GC/MS) [2], and quartz crystal microbalance (QCM) [3, 4]. In addition, studies that use molecular imprinted polymers (MIP) to fabricate QCM or resistant terpene sensors are being actively performed. With high selectivity and superior molecular imprinting capacity, MIP [5, 6] is being widely applied in various fields. MIP is formed of cross-linker, functional monomers, and template, and functional monomers around the template are arranged at the binding site in advance for binding. To maintain the arrangement of the template bound function monomers, the cross-linker and polymerization initiator are added for polymerization. By removing the template from the polymers in the next step, 3-dimensional MIP with special pores in accordance to the template is synthesized. Template removed-MIP is utilized as a detection substance of sensors using the characteristic of simple rebinding between templates.

A conductive polymer (CP), which has π-conjugated electrons spread along the polymer backbone and has a delocalized electron structure after doping, is the solution to reducing the electrical resistance in resistive-type sensors [7]. The conductivity increases with increasing polyaniline content in the range 20 – 60 %. Polyaniline is classified based on its oxidation state, leucoemeraldine, emeraldine, or pernigraniline [8]. $y=1$ is called leucoemeraldine base, $y=0.5$ is emeraldine base, and

[1] Address for correspondence: Sung Pil Lee, Department of Electronic Engineering, Kyungnam University, Kyungnam, Changwon 631-701, Korea. Tel.: 82-55-2492642; Email: sensors@kyungnam.ac.kr.

y=0 is pernigraniline, respectively. Unlike most other polyaromatics, the polyaniline in the fully oxidized state is not conductive. Polyaniline becomes conducting when the moderately oxidized states, in particular emeraldine base, are protonated and charge carriers are generated. It is this process, generally called protonic acid doping, which makes polyaniline unique. The conduction mechanism is believed to involve polarons. In the case of protonated emeraldine, a delocalized poly (semiquinone radical cation) is the polaronic carrier. For intellectualization of the gas sensors or ion sensors, multi-functionality and miniaturization is essential. The purpose of fabricating sensor arrays in gas sensors or ion sensor system is for simultaneous detection of target substances of various types. In this study, the sensitivity of sensors was compared according to the fabrication method of resistant terpene sensors using conductive polymers and MIP.

2. Experimental

For the sensor array, a terpene sensor, temperature sensor, and analytical elements were designed within the same package [9]. Figure 1 shows the layout and sensing areas of the sensor array. A conductive polymer was coated on top of three resistance areas to create a bridge-type, and the substance that detects terpene was coated only on the top of an element used as a sensor. Four bridge resistance sensors were arranged to allow respond to each terpene. A general p-n junction was employed for the diode temperature sensor. The CMOS 0.5 μm rule was applied using the Cadence and Autocad with 30 bonding pads. To create a diode temperature sensor, a P-N junction was formed via ion implant and photolithography processes. After formation of the electrode pattern with poly silicon, a hole was connected with gold electrodes and bonded to the pad with aluminum.

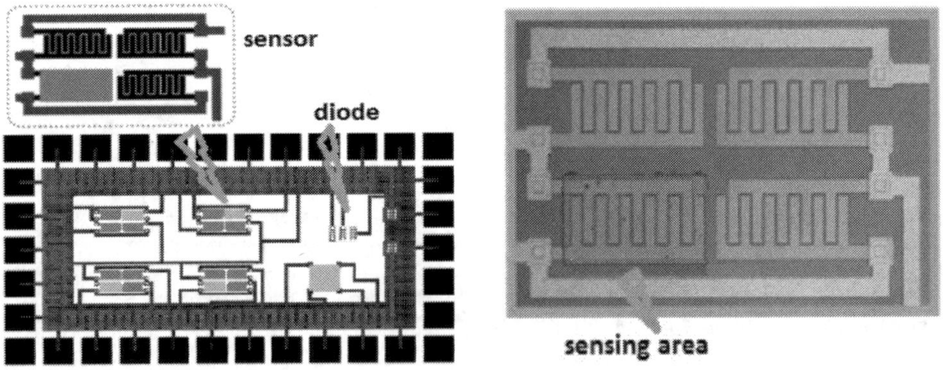

(a) (b)

Figure 1. Terpene sensor module with resistive bridge; (a) layout of sensor array and (b) photograph of sensing area.

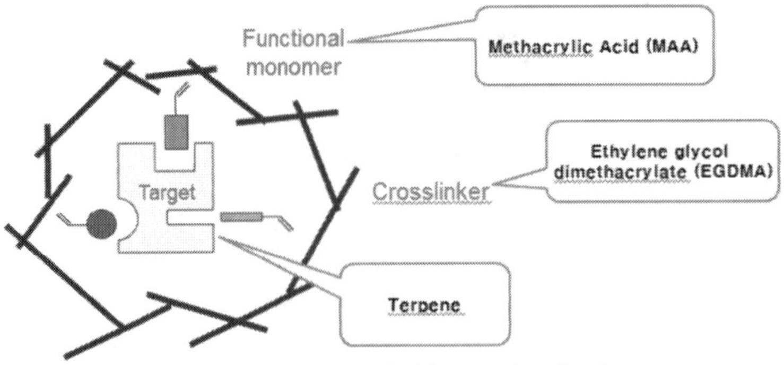

Figure 2. Formation of molecular imprinted polymers.

The methacrylic acid (MAA) as the functional monomer was positioned spatially around the template and their positions were fixed by copolymerization with cross-linking (ethylene glycol dimethacrylate, EGDMA). 2, 2'-azobisisobutyronitrile (AIBN) was selected as the polymerization initiator as shown in figure 2 [4]. The solution was purged with nitrogen gas for 10 minutes to remove any oxygen content as previously reported [10].

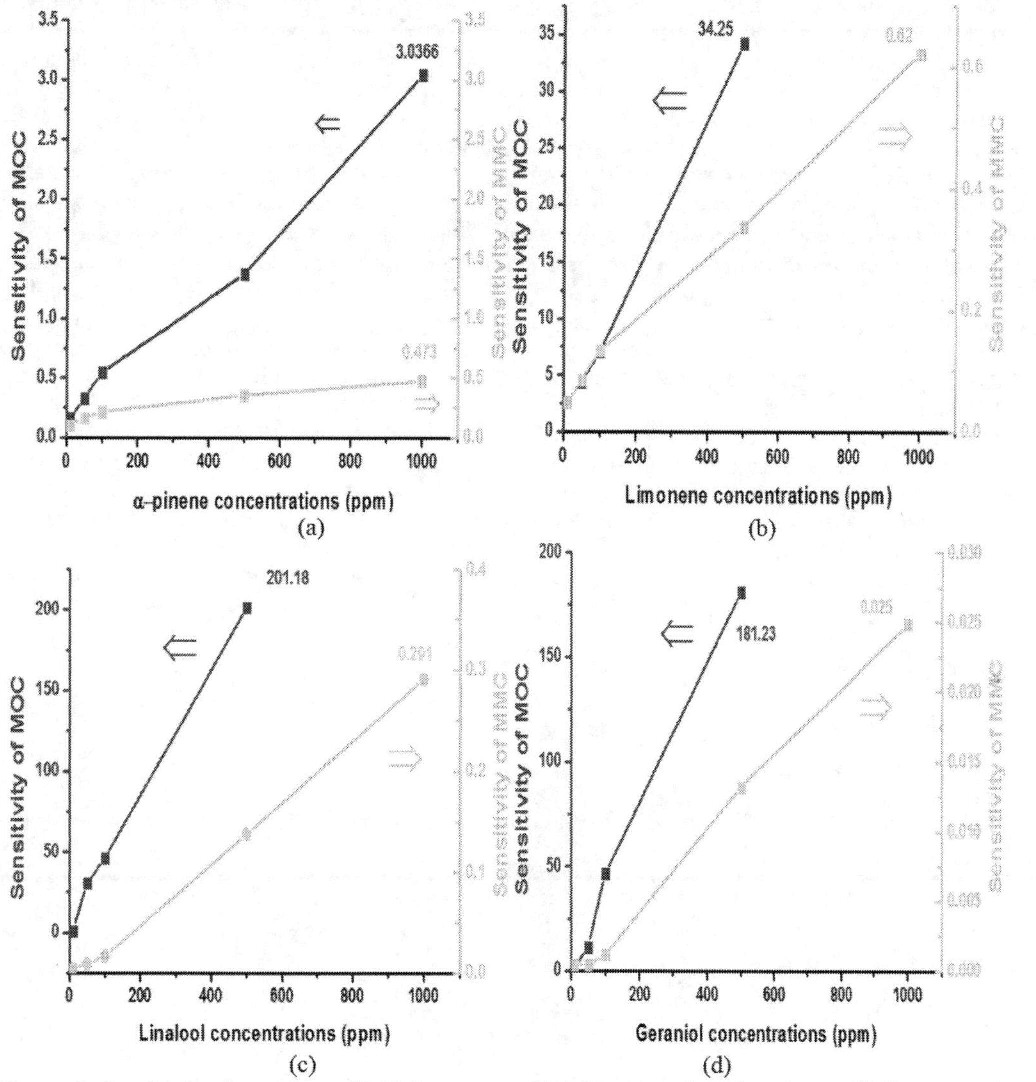

Figure 3. Sensitivity comparison of MOC sensor and MMC sensor for (a) α-pinene, (b) limonene, (c) linalool and (d) geraniol gas.

3. Results and discussion

Figure 3 shows a comparison of the characteristics of sensitivity to α-pinene and limonene in sensor arrays formed by MMC and MOC. The MMC sensor exhibited 0.473 sensitivity at the α-pinene concentration of 1000 ppm, whereas the MOC sensor showed much higher sensitivity (3.04). Regarding the response to limonene as presented, the MMC sensor exhibited sensitivity of 0.62 at

1000 ppm, while MOC sensor showed a much higher sensitivity of 34.25, even at 500 ppm. The sensor sensitivity is defined as the difference between electrical resistance before terpene gas injection and sensor resistance after gas adsorption. As for the response to linalool demonstrated in figure 3(c), the MMC sensor showed sensitivity of 0.29 at 1000 ppm, whereas the MOC sensor revealed a much higher sensitivity of 201.18, even at 500 ppm. Similarly, in response to geraniol (figure 3(d)), the MMC sensor showed the lowest sensitivity of 0.025 at 1000 ppm, but the MOC sensor exhibited a relatively high sensitivity of 181.23 at 500 ppm.

There are two possible conduction mechanisms of terpene sensors using MIP and conductive polymers. In the first, a charge transfer is induced between the cavity and the template through terpene gas adsorption, and in the second, the electrical characteristics are altered through the change of the polymer chain structure. Ryu et al. [8] reported that charge carrier mobility, parallel to the polyaniline chain, is a major causative factor of increases in electrical conductivity. In other words, the parallel conductivity ($\sigma_{\parallel}$) dramatically changes with the increase of the draw ratio, while the vertical structural conductivity ($\sigma_{\perp}$) hardly changes. Here, the parallel direction is the direction in which the film stretches, and the vertical direction is the direction perpendicular to the direction the film stretches. The MIP is synthesized via copolymerization of functional monomers in the presence of template molecules. After polymerization, monomers are separated from the template molecules. Template molecules are divided after monomer polymerization. Template molecule extraction from the completely formed polymer-nets creates a cavity that forms an arrangement in its functional group size, shape, and spatial arrangement.

Figure 4 demonstrates the selectivity for each terpene exhibited by the MOC-type sensor arrays. Sensor array was individually exposed to different terpene gases at the same concentration, and were normalized for inter-comparison according to the sensitivity of each sensor. Terpene is a substance that is biosynthesized from isoprene (molecular structure C_5H_8), and its basic structure retains $(C_5H_8)n$, a multi-structure of (C_5H_8). Limonene is monoterpene like α-pinene, but generally has a higher sensitivity for the same concentration. This is because limonene is monocyclic while α-pinene is bicyclic. With similar structures, the $C_{10}H_{16}$ group of limonene and α-pinene share similar structures and relatively low selectivity.

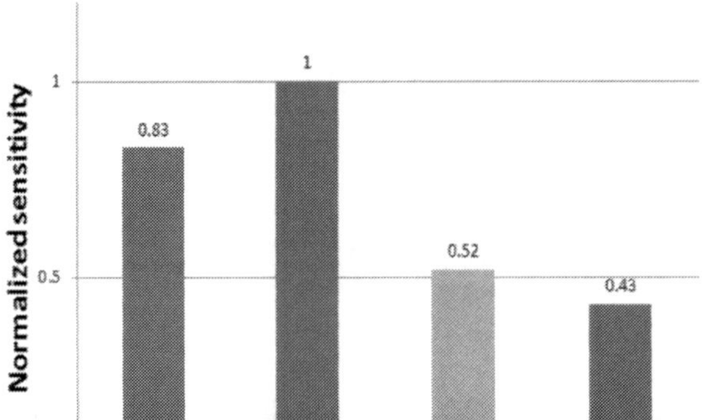

Figure 4. Selectivity of MOC type sensor array for limonene gas.

4. Conclusions

Using MIP molecular imprinting characteristics and conductive polymer, MIP mixed- or coated-resistant terpene sensor arrays were fabricated. The MOC-type sensor showed markedly higher sensitivity compared to the MMC-type sensor. At the terpene concentration of 500 ppm, the MOC sensor exhibited about 6.5 and 100-times higher sensitivity for α-pinene and limonene, respectively, compared to the MMC sensor. In the investigation of selectivity for each terpene using the sensor array, limonene and α-pinene possess identical chemical formulas with similar structures. We need to investigate to get the sensitivity and selectivity for more terpene gases.

Acknowledgment
The author wishes to acknowledge the financial support of Kyungnam University made in the program year of 2013.

References

[1] Onyeike E, Ikewuchi J, Ikewuchi C and Uwakwe A 2010 *The Pacific Journal of Science and Technology* **11** 480-2
[2] Yermakov A, Khlaifat A, Qutob H, Abramovich R and Khomyakov Y 2010 *Chemical Sciences Journal* **7** 1-10
[3] Tsuru N, Kikuchi M, Kawaguchi H and Shiratori S 2006 *Thin Solid Films* **499** 380-5
[4] Kikuchi M, Tsuru N and Shiratori S 2006 *Science and Technology of Advanced Materials* **7** 156-61
[5] Vasapollo G, Sole R, Mergola L, Lazzoi M, Scardino A, Scorrano S and Mele G 2011 *Int. J. Mol. Sci.* **12** 5908-45
[6] Yan H and Row K 2006 *Int. J. Mol. Sci.* **7** 155-78
[7] Skotheim T 1986 *Handbook of Conducting Polymers* (Dekker New York)
[8] Ryu K, Chang S, Kang S, Oh E and Yo C 1999 *Bull. Korean Chem. Soc.* **20** 333-6
[9] Lee S 2010 *Electronic Materials Letter* **6** 1-5
[10] Jung J and Lee S 2012 *Journal of Sensor Science and Technology* **21** 345-51

Grafting of poly (lactic acid) with maleic anhydride using supercritical carbon dioxide

R Khankrua[1, 4], S Pivsa-Art[2], H Hiroyuki[3] and S Suttiruengwong[1, 5]

[1]Department of Materials Science and Engineering, Faculty of Engineering and Industrial Technology, Silpakorn University, Sanamchandra Palace Campus, Nakhon Pathom 73000, Thailand
[2]Department of Material and Metallurgical Engineering, Faculty of Engineering, Rajamangala University of Technology Thanyaburi, Klong 6, Thanyaburi, Pathumthani 12110, Thailand
[3]Kyoto Institute of Technology, Hashigami-cho, Matsugasaki, Kyoto City 606-8585, Japan

E-mail: k.rattikarn_su@hotmail.com (R Khankrua); suttiruengwong_s@su.ac.th (S Suttiruengwong)

Abstract. The aim of this work was to modify poly lactic acid (PLA) via free radical grafting with maleic anhydride (MA) by using supercritical carbon dioxide ($SCCO_2$). Benzoyl peroxide (BPO) was used as an initiator. The solubility of MA in $SCCO_2$ was first determined to estimate the suitable grafting conditions and equilibrium. From the solubility study of MA in $SCCO_2$, it was found that the solubility of MA in $SCCO_2$ increased with the increasing pressure and dissolution time. PLA films were first prepared by compression molding. The ratio of MA to BPO was 2:1. The reaction temperature and pressure were 70°C and 100 bar respectively. The grafting reaction and the degree of grafting were characterized by nuclear magnetic resonance (NMR) spectroscopy and titration, respectively. Scanning electron microscope (SEM) technique and contact angle were used to confirm the changes in physical properties of PLA film grafted MA. NMR spectrum indicated that the grafting of MA onto PLA was successively achieved. Degree of grafting by using $SCCO_2$ was as high as 0.98%. This provided rather high grafting degree compared with other processes. SEM pictures showed the rough surface structure on modified PLA film. In addition, contact angle results showed an improvement of the hydrophilicity by maleic anhydride grafting onto polymers.

1. Introduction

Poly (lactic acid) (PLA) is one of the most well-known biodegradable polymers with high mechanical strength and modulus. It is now available largely in a commercial scale and a wide range of products. To expand the application window of PLA, modification on PLA properties has been carried out

[4] Address for correspondence: R Khankrua, Department of Materials Science and Engineering, Faculty of Engineering and Industrial Technology, Silpakorn University, Sanamchandra Palace Campus, Nakhon Pathom 73000, Thailand. E-mail: k.rattikarn_su@hotmail.com.
[5] Address for correspondence: S Suttiruengwong, Department of Materials Science and Engineering, Faculty of Engineering and Industrial Technology, Silpakorn University, Sanamchandra Palace Campus, Nakhon Pathom 73000, Thailand. E-mail: suttiruengwong_s@su.ac.th.

extensively for various applications and purposes. These include the blending of PLA with other polymers, composites and copolymers. The chemical modification of PLA can be performed by grafting desired molecules onto PLA chains in order to improve surface energy of polymer and its hydrophilicity to obtain polymer with suitable functional groups for the benefit of printing and coating applications, adhesion with polar polymers, metal and fibers. Some conventional grafting methods such as radiation [1], melt process [2, 3] and solution method [4] have been reported. However, all these methods have drawbacks, for example, in the melt process, most biodegradable polymers are polyesters and vulnerable to severe thermal degradation during processing [5], thus scarifying some properties like the mechanical and thermal properties. In the case of the solution process, the removal of organic solvent from the final product is required.

Until recently, the chemical modification of polymer using supercritical fluid, namely supercritical carbon dioxide ($SCCO_2$), has been explored for many polyolefins due to the ease of using organic solvents and mild processing conditions. High density like liquid and good diffusivity like gas are combined as a powerful solvent. Faster dissolution rate and higher solubility of solutes in supercritical fluid can be achieved. The supercritical fluid most commonly used is carbon dioxide due to its nontoxicity, low-cost, environmental benign, moderate critical conditions (T_c = 31.1 °C, P_c = 7.386 MPa) [6], but low polarity. The polarity of $SCCO_2$ can be improved using an entrainer, and in many cases ethanol is chosen.

Supercritical carbon dioxide is used in several applications such as separation processing, drug delivery [7, 8], chromatography, synthesis, purification and extraction, providing efficient processes with increased yields and selectivity, as comparable with conventional techniques [9, 10]. Moreover, it can be used for polymerization, polymer processing, plasticizer in polymer system and impregnating agent to make grafted polymer [11]. Grafting chemicals onto polyolefin using $SCCO_2$ has been carried out by many groups. The major advantage of this method is easy application for polymers in the solid state such as powder, sheet and film. J. H. Hayes and T. J. McCarthy [12] studied free radical grafting of maleic anhydride onto poly(4-methyl-1- pentene) (PMP) film using $SCCO_2$. It was found that the optimum temperature for the reactions was 125°C, which allowed for rapid reaction without significant sample deformation. In addition, the optimal molar ration of MAto DCP was found to be 3/2 for optimizing the degree of maleation. T. Liu and his coworkers [13] grafted maleic anhydride onto polypropylene in the solid state by using $SCCO_2$. The results showed the advantage of using $SCCO_2$ as an environmentally benign solvent and swelling agent for carrying out the grafting process of malic anhydride onto PP solid state. Moreover, this process produced much higher grafting contents than the conventional melt process did. A Galia and his colleagues [14] modified polypropylene by grafting of MA onto iPP sheet using $SCCO_2$ method. The result was successfully achieved by controlling the concentration of inorganic peroxide, MA and the density of supercritical fluid. For the modification of PLA, a few investigations have been performed to graft chemicals onto PLA. T. M. Quynh and his coworkers [15] prepared the crosslinked stereo blend of PLLA/PDLA by pretreatment with $SCCO_2$ with triallyl isocyanurate (TAIC) being used as a crosslinker. The result showed that the brittleness of blend was very much reduced by using $SCCO_2$. However, after radiation treatments, toughness and tensile strength of blend films can be significantly improved by simultaneous combination of $SCCO_2$ and radiation treatments.

Therefore, the objective of this work was to modify PLA film by radical grafting with maleic anhydride in $SCCO_2$. The solubility of MA in $SCCO_2$ was studied to estimate the suitable grafting condition. Benzoyl peroxide was used as initiator in this work.

2. Experimental

2.1. Materials

Poly (lactic acid) (PLA) 2003D was purchased from NatureWork ® LLC, USA. Maleic anhydride (MA) and benzoyl peroxide (BPO) initiator were purchased from Sigma Aldrich.

2.2. The solubility of MA in $SCCO_2$

To ensure the reaction in supercritical phase, the measurements of solubility of MA at different pressures and dissolution time were required. The solubility of MA in $SCCO_2$ was measured using the set-up of supercritical fluid process as shown in figure 1.

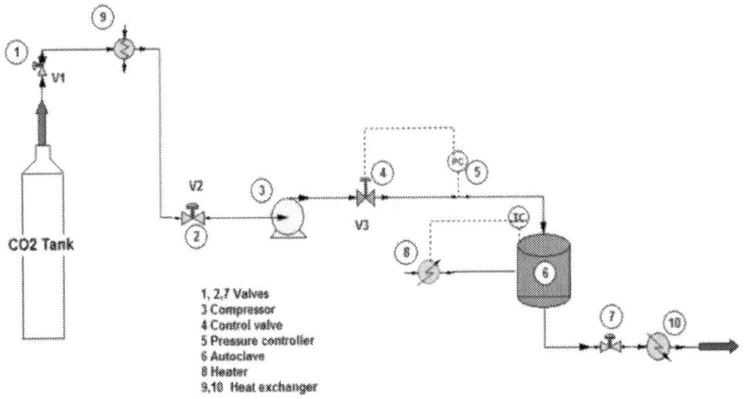

Figure 1. $SCCO_2$ process.

First, the MA (1 g) was placed in the reactor (Re248-D150L230, CHAREON TUT Co., Ltd, Thailand, No. 6 in figure 1). The temperature of reactor was set at 40±1oC. Then the reactor was closed and gradually carbon dioxide was introduced. Carbon dioxide was pumped (No. 3, SITEC, SITEC-Sieber /engineering AG, Switzerland) into the reactor until the desired pressure was reached. The studied pressures were 80, 90 and 100 bars and the dissolution time was varied for 6, 12 and 24 hours. After that, the system fluid was purged slowly. The purged MA in high pressure carbon dioxide was carefully collected with methanol. Then the concentration of MA in methanol was measured by using UV-Vis spectrometer (T80+ UV/Vis spectrometer, PG Instruments Ltd., Bangkok High LAB Co., Ltd) to confirm the results with the gravimetric method. The rest of MA in the reactor was removed from the reactor and then weighted. The solubility of MA in SCCO2 can be calculated from the decrease in MA weight according to equation (1). The amount of carbon dioxide in the reactor was calculated from the known volume and carbon dioxide density. The values of carbon dioxide density were taken from The National Institute of Standards and Technology (NIST) standard reference database [16].

$$\%\text{Solublity} = \frac{\text{Mass of soluble MA}}{\left(\text{Mass of soluble MA} + \text{Mass of CO}_2\right)} \times 100 \qquad (1)$$

2.3. Grafting reaction in $SCCO_2$

After the measurement of the MA solubility in $SCCO_2$, the suitable condition for grafting reaction was chosen. The film sample was prepared by compression molding (Labtech, LP-S-50, Labtech Engineering Co., Ltd.) at 170°C. The film thickness was around 0.2-0.3 mm. The film was cut into the sample of 50x30 mm. Then film sample, with the ratio of MA to BPO being 2/1, was placed in the reactor. The reactor was sealed tightly. The temperature of the reactor was set at 40°C, and then carbon dioxide was filled into the reactor. After that, the temperature was increased to 70°C to allow thermal decomposition of peroxide. Finally, the pressure was increased to the critical point and the reaction was kept for 3 hours. After the reaction, carbon dioxide was vented out. Let the system cool down and the sample was removed from the reactor for further characterizations.

2.4. Characterization and testing

IOP Conf. Series: Materials Science and Engineering **87** (2015) 012066 doi:10.1088/1757-899X/87/1/012066

For NMR analysis, PLA grafted MA sample was first dissolved in $CDCl_3$. The ^1H NMR spectrum was obtained using Bruker Avance 300 NMR Spectrometer.

The graft degree of maleated PLA can be determined by back-titration using phenolphthalein as indictor. First, PLA grafted MA sample was dissolved in chloroform, and then precipitated in methanol to remove the unreacted MA. And 1 g of sample was completely dissolved in 100 ml of chloroform. Then an excess of 0.02 N KOH in ethanol solution was added which made MA thoroughly react after that, 0.02 N HCl in ethanol solution was titrated against KOH. Finally, KOH ethanol solution was added again to the titration end point [17]. Degree of grafting (G%) was calculated from the consumed quantity of KOH, by the following equation:

$$G\% = \frac{\left(V_{KOH}N_{KOH} - V_{HCl}N_{HCl} - M_0\right)}{2W} \times 98.06 \times 100\% \tag{2}$$

where V is the volume (L), N is the normality (mol/L), M_0 is the blank value, and W is the specimen weight (g). The sample was determined for three times and the average degree of grafting was reported.

The morphology of cross section and surface of film samples was investigated by SEM technique. The samples were cooled down in the liquid nitrogen and then fractured. Then, the sample films were rinsed with water to remove the unreacted MA in the sample. They were gold-coated and studied under a working voltage of 15 kV with CamScan, MX 2000 SEM.

As for determining the contact angle determination, the sheet samples were prepared by compression molding (Labtech, LP-S-50, Labtech Engineering Co., Ltd.). The specimens were held on the slide glass and were conditioned at room temperature. Then, water droplets were dripped by a micropipette on the sample surface. The contact angle was measured with Goniometer (Model 200, Rame-Hart Instrument).

3. Results and discussion

3.1. Solubility of maleic anhydride in $SCCO_2$

The solubility of MA in $SCCO_2$ at pressures between 80-100 bar was shown in figure 2. To ensure the equilibrium for any constant pressure, the solubility of MA was measured at different dissolution time. It was found that the solubility of MA in $SCCO_2$ reached the maximum at 24 h for the pressure of 80 and 90 bar whereas at the pressure of 100 bar, the solubility of MA in $SCCO_2$ did not depend on the dissolution time. The solubility of MA in this pressure range lied between 0.35-0.63%. The pressure of 100 bar was chosen for further experiments due to its highest solvent power and independence of dissolution time.

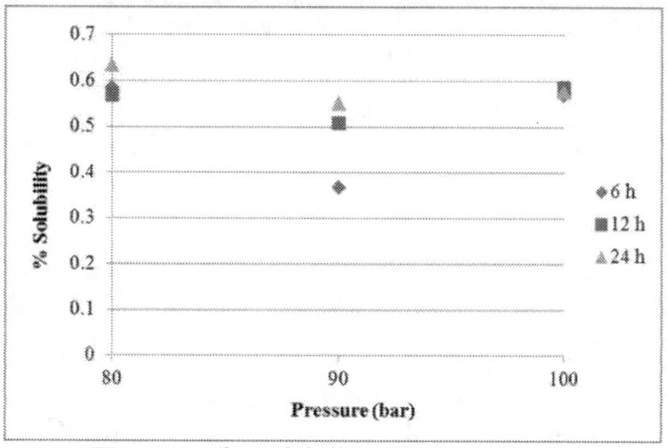

Figure 2. The solubility of MA in $SCCO_2$.

When considering the system consisting of peroxide, the decomposition temperature should be taken into account. The decomposition temperature of BPO is at least 70°C with a given half life time of 92 minutes [18]. Therefore, the reaction temperature should be increased to 70°C, and the dissolution time was then sufficiently reduced to 3 h.

3.2. Characterization of grafted PLA

After reaction, the sample was removed from the reactor and kept in the desiccator at room temperature for 24 h to remove CO_2 residue. The sample was washed with methanol. Then, all samples were characterized by ^{1}H NMR spectroscopy.

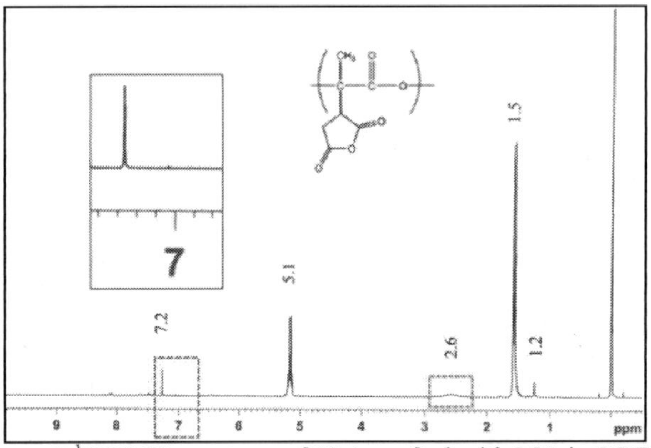

Figure 3 ^{1}H NMR spectrum of PLA grafted with MA in $SCCO_2$.

From figure 3, a signal at 7.2 ppm was the characteristic of $CDCl_3$. The signals appeared at 1.2-1.5 and 5.1 ppm corresponded to $-CH_3$ and –H of PLA, respectively. The succinyl functional groups were generally found in the region of 2-3 ppm [19]. A signal at around 2.6 ppm was assigned to the succinyl group of MA on PLA. The results indicated that graft of MA onto PLA was successfully achieved. However, an weak signal was also observed at around 7.0 ppm which was the characteristic signal of MA [20]. This indicated that there was the residual unreacted MA in sample. However, this problem can be solved by washing film with methanol. The signal at 7.0 ppm disappeared after washing sample film (NMR spectrum was not shown). It indicated that the unreacted MA was removed.

The MA grafting yield was determined by the titration method. The sample was first precipitated to ensure that the unreacted MA was removed. The average degree of grafting of maleated PLA was 0.976 %. It could be seen that grafting by using $SCCO_2$ provided higher grafting content than that by the solution method [19, 21] and melt method [22]. This may be due to the fact that when the temperature of reaction increased to 70 °C, higher than glass transition temperature of PLA (55-60°C), $SCCO_2$ can swell polymer [14] as well as have powerful solvent for the reaction. Therefore polymer chain had higher mobility and higher free volume as well as became softer. Consequently, this provides high diffusion of MA and BPO into polymer, resulting in the high degree of grafting.

3.3. Morphology of samples

The morphology of the cross section and surface of film samples was characterized by SEM technique as shown in figure 4. Compared with PLA grafted with MA, it was clearly observed that pure PLA film presented a smoother morphology (see figure 4(b)). PLA grafted with MA presented a rougher morphology and some solid crystal particles were impregnated in polymer matrix (see figure 4(d)). These particles sizes were around 10-20 μm which was the same size as MA (the SEM picture of MA crystals was not shown). Thus, this particle might be the MA crystals, which penetrated into the polymer film during the process and then the recrystallization of the finely dispersed MA within the polymer occurred upon reducing the pressure, thus allowing the encapsulation of MA crystals. This result agreed with NMR result as discussed previously. However, the sample film after rinsing with water was also examined and the result showed that a little bit of MA crystal still remained in the film (see figure 4(g)). This could be from the recrystallized MA tightly impregnated in PLA matrix. When considering the surface of the film, it could be seen that the surface of PLA grafted MA sample was rough and wavy (see figure 4(c)) and it became more pronounced in figures 4(e) and 4(h). This result was similar to the work reported by Liu Z. and co-workers [23]. This phenomenon was explained for grafted polypropylene after treating with $SCCO_2$ by the different initiation rates of amorphous and crystalline sites, partly arising from different swelling rates in amorphous and crystalline regions.

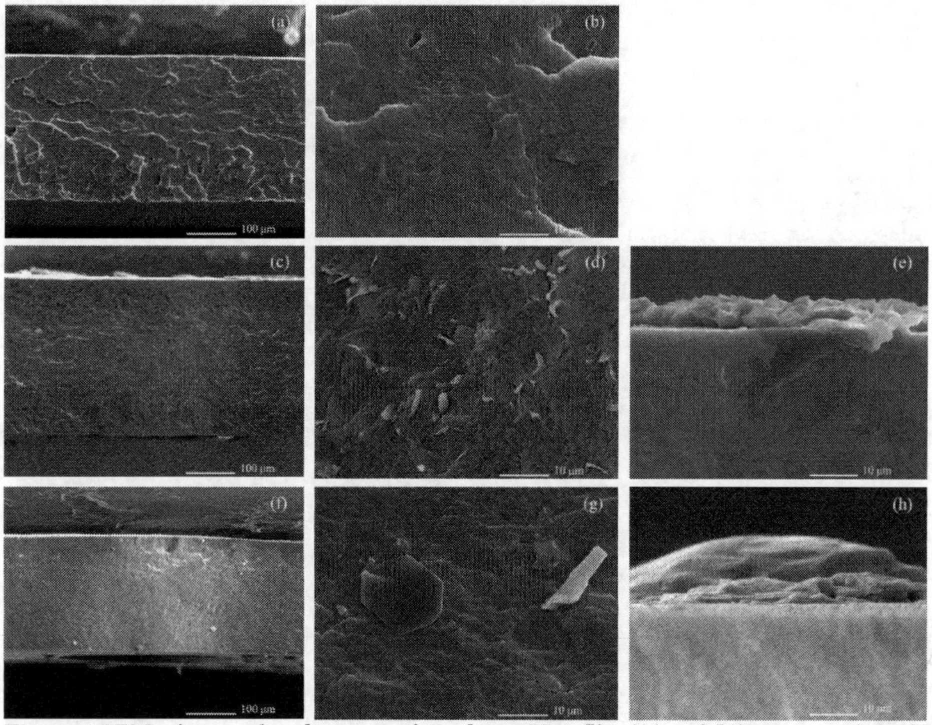

Figure 4 SEM micrographs of cross section of pure PLA film ((a) and (b)), cross section of PLA grafted with MA ((c) and (d)), surface of PLA grafted with MA ((e)), cross section of PLA grafted MA film after rinse ((f) and (g)) and surface of PLA grafted with MA after rinse ((h)).

3.4. *Water contact angle*

The average values of contact angle for pure PLA film and PLA grafted with MA were shown in table 1.

Table 1. The contact angle of samples.

Sample	Contact angle (°)
Pure PLA film	100.1
PLA grafted with MA	86.1

It was observed that the contact angle of polymers grafted with MA was lower compared with pure polymers. The decrease in contact angle was mainly ascribed to the modified surface of polymer films as shown in SEM results and it corresponded to the improvement of the hydrophilicity by maleic anhydride grafted onto polymers. This result was in good agreement with the work of D. Muenprasat and her coworkers that modified PLA with maleic anhydride through free radical solution reaction can improve the hydrophilic property of PLA [24].

4. Conclusion

The modification of PLA film by free radical grafting with MA is successfully prepared in $SCCO_2$. Pressure of CO_2 and the dissolution time affect the solubility of MA in $SCCO_2$. Thus, the reaction is carried out at 100 bar with 3 h reaction time for effective grafting reaction. The average degree of grafting is 0.976 %, which is relatively high compared with both solution and melt processes. The high grafting degree is required for tailoring most of the hydrophobic biodegradable polymers. In addition, PLA film grafted with MA shows better hydrophilicity as confirmed by contact angle. In terms of industrial production, the easy-to-setup supercritical fluid technology and high effectiveness are very attractive and can provide alternatives and versatile products. For the further work, the modification of as-received PLA resin will be conducted together with the elucidation of the relationship between molecular weight and the loading concentration of peroxides.

Acknowledgments

This work was supported by a grant from Department of Materials Science and Engineering, Faculty of Engineering and Industrial Technology, Silpakorn University. The authors would like to gratefully thank National Research Council Thailand (NRCT) for its financial support. The authors are also greatly thankful to Miss Faungchat Thammarakcharoen for conducting the contact angle testing at National Metal and Materials Technology Center (MTEC).

References

[1] Leonard D J, Pick L T, Farrar D F, Dickson G R, Orr J F and Buchanan F J 2009 *J. Biomed. Mater. Res A* **89A** 567-74
[2] Zhang J-F and Sun X 2004 *Biomacromolecules* **5** 1446-51
[3] Detyothin S, Selke S E M, Narayan R, Rubino M, and Auras R 2013 *Polym. Degrad. Stabil* **98** 2697-708.
[4] Plackett D 2004 *J. Polym. Environ.* **12** 131-8
[5] Khankrua R, Pivsa-Art S, Hiroyuki H and Suttiruengwong S 2014 *Polym. Degrad. Stabil* **108** 232-40
[6] Cooper A I 2000 *J. Mater. Chem.* **10** 207-34
[7] Subramaniam B, Rajewski R A and Snavely K 1997 *J. Pharm. Sci.* **86** 885-90
[8] Duarte A R C, Coimbra P, De Sousa H C and Duarte C M M 2004 *J. Chem. Eng. Data* **49** 449-52
[9] Lucien F P and Foster N R 2000 *J. Supercrit. Fluid* **17** 111-34
[10] Foster N, Mammucari R, Dehghani F, Barrett A, Bezanehtak K, Coen E, Combes G, Meure L, Ng A, Regtop H L and Tandya A 2003 *Ind. Eng. Chem. Res.* **42** 6476-93
[11] Kunita M H, Guilherme M R, Filho L C, Muniz E C, Franceschi E, Dariva C and Rubira A F 2011 *J. Colloid. Interf. Sci.* **361** 331-7
[12] Hayes H J and McCarthy T J 1998 *Macromolecules* **31** 4813-9
[13] Liu T, Hu G-H, Tong G-S, Zhao L, Cao G-P and Yuan W-K 2005 *Ind. Eng. Chem. Res.* **44** 4292-9
[14] Galia A, De Gregorio R, Spadaro G, Scialdone O and Filardo G 2004 *Macromolecules* **37** 4580-9
[15] Quynh T M, Mitomo H, Zhao L and Asai S 2008 *Carbohyd Polym* **72** 673-81
[16] Available at: http://webbook.nist.gov/chemistry/, March 25, 2015
[17] Chen C, Peng S, Fei B, Zhuang Y, Dong L, Feng Z, Chen S and Xia H 2003 *J. Appl. Polym. Sci.* **88** 659-68
[18] Available at: http://www.dowcorning.com, March 25, 2015

[19] Hwang S W, Lee S B, Lee C K, Lee J Y, Shim J K, Selke S E M, Soto-Valdez H, Matuana L, Rubino M and Auras R 2012 *Polym. Test.* **31** 333-44
[20] Available at: http://www.chemicalbook.com/, March 25, 2015
[21] Mani R, Bhattacharya M and Tang J 1999 *J. Polym. Sci. A1*. **37** 1693-1702
[22] Ma P, Jiang L, Ye T, Dong W and Chen M 2014 *Polymers*. **6** 1528-43
[23] Liu Z, Song L, Dai X, Yang G, Han B and Xu J 2002 *Polymer*. **43** 1183-8
[24] Muenprasat D, Suttiruengwong S and Tongpin C 2010 *Journal of Metals Materials and Minerals* **30** 189-92

Enhanced crystallization of poly (lactic acid) through reactive aliphatic bisamide

P Nanthananon[1], M Seadan[2], S Pivsa-Art[3] and S Suttiruengwong[1,4]

[1]Department of Materials Science and Engineering, Faculty of Engineering and Industrial Technology, Silpakorn University, Nakhon Pathom 73000, Thailand
[2]Department of Physics, Faculty of Science, Silpakorn University, Nakhon Pathom 73000, Thailand
[3]Department of Materials and Metallurgical Engineering, Faculty of Engineering, Rajamangala University of Technology Thanyaburi, Klong 6, Thanyaburi, Pathumthani 12110, Thailand

E-mail: suttiruengwong_s@su.ac.th (S Suttiruengwong)

Abstract. The poor crystallization rate of poly (lactic acid) (PLA) is a major drawback in terms of controlling the properties of final products. To overcome this, a nucleating agent is normally applied. In this work, the aliphatic bisamide, N, N'-(1,3-propylene) bis (10-undecenamide) (PBU), having reactive functional groups is used as a crystallization promoter for PLA by adding PBU in various concentration (0.1-0.7 wt%) into PLA together with peroxide via reactive melt blending. The conventional ethylene bis-stearamide (EBS) is used for a comparison. The extruded samples are characterized for gel content and FT-IR spectroscopy. The crystallization behaviour and rate, and spherulites morphology are investigated by differential scanning calorimetry (DSC) and polarized optical microscopy (POM), respectively. It is found that the addition of PBU into PLA results in the dramatic increase in crystallinity and crystallization rate of PLA compared with neat PLA and PLA added EBS. The crystallinity increases to 24.9-28.3% higher than neat PLA under even cooling rate of 7°C/min. The addition of 0.7 wt% PBU shows the fastest crystallization rate with $t_{1/2}$ value isothermally crystallized at 130°C of only 6 min. POM images indicate the increase in the nucleation density and very fine spherulites of PLA added PBU, promoting the fast crystallization.

1. Introduction

Many solutions have been established to reduce non-biodegradable plastic waste disposal and the consumption of non-renewable petroleum raw materials in order to cope with sustainability issues for balancing the use of resources and the growth of population. In terms of plastics, using renewable bio-derived raw materials and/or biodegradable plastics can best address the problem. Poly (lactic acid) (PLA) is one of the most fashionable biodegradable polymers which exhibits excellent performance in renewability, biocompatibility, transparency, processability, good mechanical, and barrier properties [1]. Unfortunately, semi-crystalline PLA exhibits very slow crystallization rate from melt, leaving its

[4] Address for correspondence: S Suttiruengwong, Department of Materials Science and Engineering, Faculty of Engineering and Industrial Technology, Silpakorn University, Nakhon Pathom 73000, Thailand. E-mail: suttiruengwong_s@su.ac.th (S Suttiruengwong).

products (especially processed from injection molding) amorphous and hence poor barrier property and thermal resistance, which allow to be used for only some certain applications. To cope with this drawback, there are many methods to increase the crytallinity of PLA such as post-annealing [2], in-mold annealing, imposing shear flow [3], low crosslinking structure [4], or adding nucleation agents (NA). Many kinds of NA, both inorganic and organic nucleating agents, have been reported in the literatures and even in some marketed products. These include talc [5], nano-$CaCO_3$ [6], modified carbon black [7, 8], carbon nanotubes (CNTs) [9], organically modified layered silicate (OMLS) [10], multi-wall carbon nanotube (MWCNT) [9], zinc citrate complex (ZnCC) [1], bisurea compound [11], cellulose nanofibers [12], hydrazide compounds [13], N,N'-ethylene bis-stearamide (EBS) [14], and N,N'-ethylenebis(-12hydroxystearamide) (EBH) [14],. Among organic NA, aliphatic amides such as EBS and EBH have been extensively used as nucleating agents for PLA to form crystallisable products such as thermo-forming process [14]. Forming the crystalline with one-step processing is easily achievable through melt blending. The addition of bisamide with reactive moiety would in some extent help promoting the crystallization together with possibility of *in situ* grafting or crosslink to form nuclei in the crystallization process. The influence of reactive aliphatic amides, which has double bonds at chain ends, on the crystallization of the PLA has not yet been reported. Therefore, in this study, the effect of reactive aliphatic amide on crystallization behaviour and rate of PLA and also its spherulites morphology are investigated. The small amount of reactive PBU (0.1, 0.3, 0.5 and 0.7 wt%) is introduced into PLA with peroxide via reactive melt blending, and EBS is selected for a comparison.

2. Experimental

2.1. Materials
PLA resin (3052D grade) was purchased from NatureWork®, Thailand. The N, N'- (l, 3-propylene) bis (10-undecenamide) (PBU) (T_m = 118°C, T_c = 111°C) was synthesized from 10-undecenoyl chloride and 1, 3-diaminopropane based on the procedure described our previous work. N, N'-ethylene bis-stearamide (EBS) (T_m = 144°C, T_c = 139°C) was purchased from Sigma-Aldrich Co,.Ltd, Thailand. Peroxide (Perkadox® 14-40B-pd) was purchased from AkzoNobel, Thailand.

2.2. Sample preparations
The different weight ratios of PLA blended with reactive PBU or EBS were prepared by using a twin-screw extruder (SHJ-25, China) with the screw speed of 80 rpm and the temperature profile of 100/110/130/160/180/185/190/200°C. Before mixing, PLA resins were dried at 65°C for 10 hr under vacuum. The blend samples were labelled as PLA/rPBUx, where r referred to reactive blends with peroxide 0.5 phr and x represented the weight ratio of PBU (0, 0.3, 0.5, 0.7 wt%). For a comparison, pure PLA and also PLA blended with EBS (1.0 wt%) samples were prepared by using the identical procedure, then labelled as neat PLA and PLA/EBS1.0, respectively.

2.3. Measurements
Differential scanning calorimetry (DSC) was carried out on a METTLER TOLEDO, DSC1 STAR instrument under a nitrogen atmosphere. To study the non-isothermal crystallization, the samples were first heated from 30 to 200°C at a rate of 7°C/min and held on for 3 min. the samples were then cooled down to 30°C and reheated to 200°C at rate of 7°C/min to study the melting behaviour of the samples. For the investigation of isothermal crystallization, the samples were initially heated up to 200°C at 10°C/min and kept for 3 min, then rapidly cooled (20°C /min) to 130°C and kept for a predetermined time. Spherulite morphologies of the samples were observed by OLYMPUS U-CMAD3 polarized optical microscopy (POM) equipped with digital camera and METTLER TOLEDO FP82HT hot stage. The samples were placed between two cover glasses and melted at 200°C for 3 min. Then they were quickly cooled to 130°C for isothermal crystallization. After the completion of crystallization, the spherulite morphologies were recorded. Gel content was measured by the weight remaining after dissolving the fine samples in chloroform using the following equation (1):

$$\text{Gel content (\%)} = (\frac{W_g}{W_0}) \times 100 \quad (1)$$

where W_0 is the original weight (dry) of the PLA/rPBUx samples, W_g is the weight remaining (dry gel component) of the PLA/rPBUx samples after being dissolved in chloroform at 40°C for 16 hr. Fourier transform infrared (FT-IR) spectroscopy was performed at ambient temperature on a Vertex 70, German spectrometer. Before testing, pure PLA, PBU and dry gel component of PLA/rPBU0.7 samples were mixed with potassium bromide (KBr) into flakes for FTIR measurements. To investigate the chemical reaction of PLA and reactive PBU, the FT-IR spectra of the samples were recorded in the wavenumber range of 400 - 4000 cm^{-1}.

Table 1. Thermal transition data in non-isothermal crystallization at 7°C/min for several samples.

Samples	T_{cc} (°C)	ΔH_{cc} (J/g)	T_m (°C)	ΔH_m (J/g)	T_c (°C)	ΔH_c (J/g)	X_c^a (%)
neat PLA	109.6	27.09	147.8, 154.1	27.19	-	-	0.11
PLA/rPBU0	108.7	28.80	147.9, 149.3	29.00	-	-	0.21
PLA/rPBU0.1	109.4	25.83	148.3, 154.4	26.11	-	-	0.30
PLA/rPBU0.3	98.2	3.25	148.2, 152.8	29.63	105.0	26.06	28.3
PLA/rPBU0.5	98.8	5.18	148.2, 152.7	28.39	103.0	24.28	24.9
PLA/rPBU0.7	101	4.48	148.9, 152.9	29.43	104.3	24.28	26.7
PLA/EBS1.0	115.9	26.27	149.7, 153.8	31.24	127.9	1.08	5.34

[a] %Crystallinity $= X_c = ((\Delta H_m - \Delta H_{cc}) / \Delta H_m^{\infty}) \times 100$; $\Delta H_m^{\infty} = 93.6$J/g [14].

3. Results and discussion

3.1. Non-isothermal melt crystallization behaviors

DSC analysis was used to study the effect of the amount of reactive PBU on the crystallization behavior of PLA compared with EBS (1.0 wt%). The results are shown in figure 1 and the interesting parameters such as cold crystallization temperature (T_{cc}), melting temperature (T_m), cold crystallization enthalpy (ΔH_{cc}), melting enthalpy (ΔH_m), crystallization temperature (T_c), crystallization enthalpy (ΔH_c), and %crystallinity (X_c) are shown in table 1. As seen in figure 1(a), in the cooling scan, no crystallization peak can be observed for neat PLA, PLA blended with only peroxide and even PLA added 0.1 wt% PBU. The cold crystallization peak appears at about 109°C for each sample in following heating scan (figure 1(b)). It indicates that they have low crystallization ability corresponding to their poor crystallinity, where the values are less than 1% as shown in table 1. With the addition of high amount of EBS (1.0 wt%), a small crystallization peak appears at 127.9°C in the cooling curve (figure 1(a)) and the X_c value increase to 5.34%. EBS can promote the crystallization of PLA due to the fact that the EBS crystallizes well at 118°C in the cooling process as known from its T_c. For the addition of higher amount of reactive PBU (0.3, 0.5 and 0.7 wt%) into PLA with peroxide, a large crystallization peak appears remarkably at approximately at 104°C and their crystallization enthalpy significantly increases, resulting in the higher X_c values of 24.9-28.3%. In addition, as seen from figure 1(b), the cold crystallization of the PLA/rPBU0.3-0.7 samples almost disappears in the heating scan; it implies that the crystallization of the samples is completed during cooling step, which is very desirable for practical use. From figure 1(b), two melting peaks are observed in the DSC heating scan for all samples. However, with the increase of reactive PBU contents, it shows the broad double peaks. Similar results are observed when adding EBS. This finding has also been reported by other groups [1, 14, 15]. Tang et al. [14] demonstrated that the double peaks of PLA samples corresponded to the mechanism based on melting of the two crystals formed in the cold crystallization stage and recrystallization stage, which is consistent with the mechanism of melt-recrystallization.

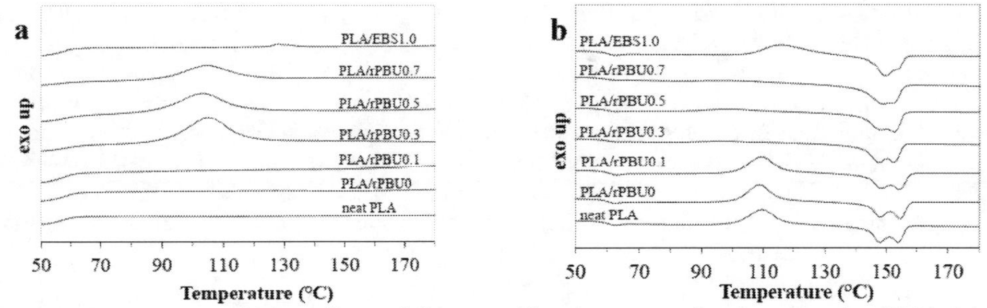

Figure 1. DSC curves of (a) cooling and (b) second heating scans of various samples (7°C/min).

3.2. Isothermal melt crystallization behaviors

The effect of reactive PBU compared with EBS on isothermal melts crystallization behaviours of PLA at 130°C are shown in figure 2(a). It is clearly seen that the crystallization rate of neat PLA is extremely slow, also by incorporating only peroxide without reactive PBU (PLA/rPBU0), which cannot crystallize within even 100 min. When 0.1 wt% reactive PBU is added, it shows a broad exothermal peak and PLA can crystallize within 60 min as well as the addition of 1.0 wt% EBS. With an increase of reactive PBU contents from 0.3 to 0.7 wt%, the sharper exothermal peak appears and PLA crystallizes much faster than the addition of 0.1 wt% PBU by 2-3 folds. For the initial crystallization of PLA/rPBU0.7 sample, its crystallization occurs almost immediately at the beginning of the isothermal crystallization at 130°C.

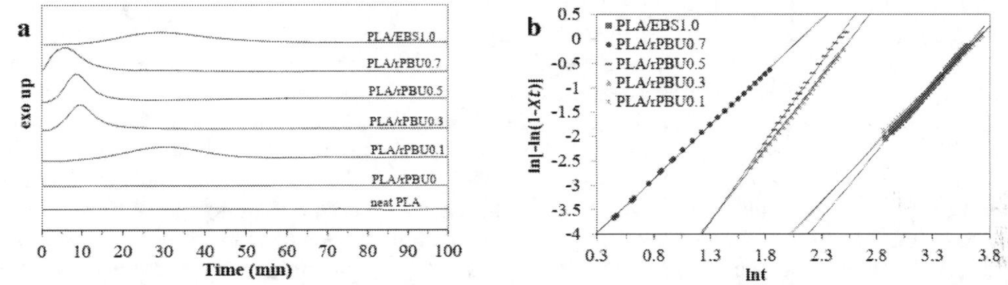

Figure 2. (a) DSC exotherms and (b) Avrami plots of samples isothermally crystallized at 130°C.

The Avrami equation is the most common method to describe the isothermal crystallization process of polymers, and its logarithmic form can be expressed as equation (2).

$$\ln[-\ln(1-X_t)] = \ln k + n \ln t \tag{2}$$

where X_t is the relative degree of crystallinity as a function of time, k is the crystallization rate constant, and n is the Avrami exponent, which is constant depending on the nature of nucleation and the dimension of crystal growth [1]. When ln[-ln(1-X_t)] is plotted against ln$t$ as shown in figure 2(b), the parameters n and k can be obtained from the slopes and intercepts of fitted lines, respectively. The value of crystallization half-time ($t_{1/2}$) is defined as the time required reaching 50% of the final crystallinity of the sample that can be determined from equation (3). n, k and $t_{1/2}$ values are listed in table 2.

$$t_{1/2} = \left(\frac{\ln 2}{k}\right)^{1/n} \tag{3}$$

The n values ranging between 2 and 3 are found. This suggests that two growth mechanisms are present: two-dimensional at the beginning of the crystallization process, with the formation of two-dimensional lamellar shape-like crystals and three-dimensional during further growth with the formation of spherulites [16]. As shown from figure 2(a), it implies that the crystallization time of neat PLA and the addition of peroxide samples are taken more than 100 min while the addition of 0.1 wt%

PBU in reactive blending has crystallization half-time ($t_{1/2}$) 34.53 min. On the basis of the value of k and $t_{1/2}$, the higher k and lower $t_{1/2}$ values is assumed as the higher crystallization rate. It is evident that the crystallization ability of PLA increases following the increase of reactive PBU content 0.3 to 0.7 wt%, results in much lower $t_{1/2}$ values from 10.53 to 6.05 min. As the $t_{1/2}$ value of 6.05 for PLA/rPBU0.7 is about one-fifth of $t_{1/2}$ for PLA/EBS1.0, the reactive PBU is supposed to accelerate the crystallization rate of PLA/EBS blended by five times. It can be concluded that PLA/rPBU0.7 sample is the most effective accelerating the crystallization rate of polymers. Due to the possible presence of fixed-structures in PLA melt, they act as the heterogeneous nuclei so that heterogeneous nucleation and crystal growth is induced and starts more rapidly under supercooled state.

Table 2. Kinetic parameters of several samples isothermally crystallized at 130°C.

Samples	n	k(min^{-n})	$t_{1/2}$(min)
PLA/rPBU0.1	2.4	1.7×10^{-4}	34.53
PLA/rPBU0.3	3.0	5.9×10^{-4}	10.53
PLA/rPBU0.5	3.3	3.9×10^{-4}	9.67
PLA/rPBU0.7	2.3	1.1×10^{-2}	6.05
PLA/EBS1.0	2.7	5.9×10^{-5}	32.23

3.3. Spherulite morphology

The spherulite images of neat PLA and its blends isothermally grown at 130°C are observed under a polarized optical microscope as depicted in figure 3. Normally, the rate of growth in the radial direction is constant until the spherulites meet [17]. However, each spherulite growth of neat PLA cannot meet to form the grain boundary in 10 hr due to the number of the nuclei of neat PLA is quite few and far away from each other, consequently obtaining large spherulites. The different size of the spherulite of neat PLA as shown in figure 3(a) indicates that the nucleation process is not formed at the same time following by an asynchronous crystallization. For the incorporation only peroxide into PLA (PLA/rPBU0), the spherulites are smaller and denser than the neat PLA as showed in figure 3(b), indicating that chains shortened by peroxide would increase its crystallization ability to some extent [4]. For adding small amount of PBU (0.1 wt%), it shows slight increase nuclei more than neat PLA; the spherulites can meet each other easier and then displays smaller crystals than neat PLA. With the further incorporation of PBU (0.3 to 0.7 wt%) as demonstrated in figure 3 (d-f), the spherulites appeared to be very fine throughout the recorded film area. The addition of the higher PBU could lead to higher degree of reactions, forming local crosslinks, which possibly act as nucleation sites to promote the crystallization process. Having a large number of nuclei, the crystallization will proceed very quickly to the limit of the space with small spherulites, thereby enhancing the overall rate of crystallization.

Figure 3. POM images of (a) neat PLA, (b) PLA/rPBU0, (c) PLA/rPBU0.1, (d) PLA/rPBU0.3, (e) PLA/rPBU0.5, (f) PLA/rPBU0.7 and (g) PLA/EBS1.0 isothermally crystallized at 130°C

This result is consistent with the DSC results in figure 2(a) and supported by the smaller $t_{1/2}$ values of their blends. Upon addition 1.0%wt of EBS, the spherulite size is also reduced compared with neat PLA but still bigger than the addition of reactive PBU (0.3 to 0.7 wt%). This indicated that PLA/EBS1.0 sample has less nucleated density than the reactive PBU blending samples even though EBS is added higher amounts.

3.4. Gel content and Infrared analysis

To elucidate the mechanism of reactive PBU, the gel contents and FT-IR of residual gels are measured. The gel content of the PLA/rPBU samples containing 0.5 phr of peroxide as a function of PBU content are shown in figure 4. It shows that the low gel content of the reactive blended samples starts at a very low concentration of PBU (0.3 wt%). The gel content increases from 0.5 to 16% (possible low cross-linking structure) with the increase of PBU contents from 0.3 to 0.7 wt%. After that the gel of PLA blended with 0.7 wt% PBU in the presence of peroxide is investigated using FT-IR spectroscopy technique, compared with PLA and PBU samples as shown in figure 5. Considering the gel spectrum, after the chemical reaction with PBU in the presence of peroxide, the peaks at 1757 cm^{-1} and 1184 cm^{-1} which belong to the C=O stretching and C–O–C stretching of PLA, respectively [4], and two new peaks can be observed at 3317 (N–H stretching of PBU) and 1637 cm^{-1} (C=O of amide, PBU). However, the peaks at 3093 (=C–H stretching of PBU) and 917 cm^{-1} (C=CH_2 bending of PBU) are disappear, which clearly confirms the chemical reaction between PLA and PBU.

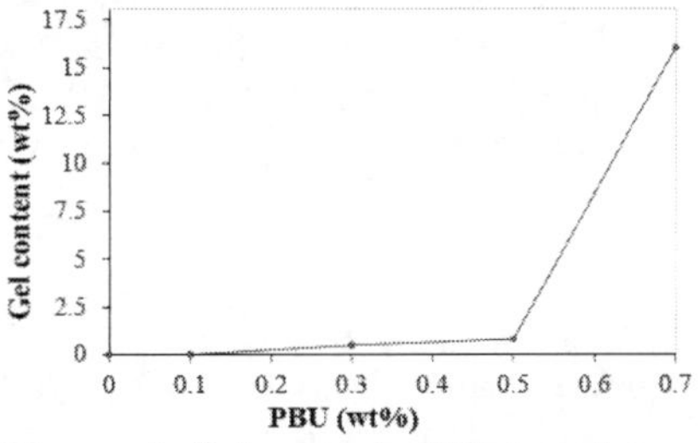

Figure 4. Gel content of PLA and peroxide blends as a function of PBU content.

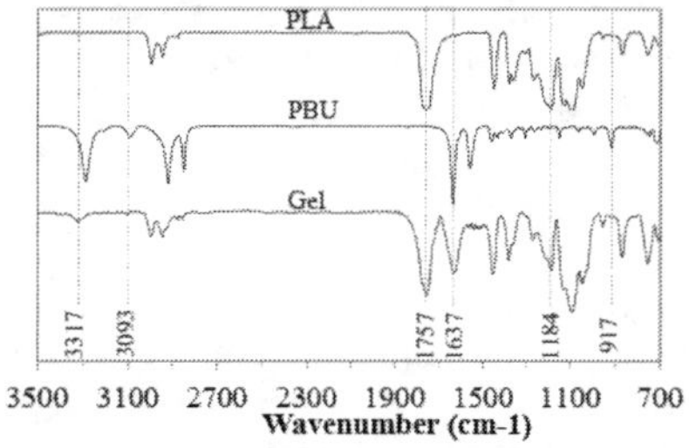

Figure 5. FT-IR spectra of PLA, PBU and Gel (PLA blended with 0.7 wt% PBU in the presence of peroxide).

Once the peroxide is added to the PLA/PBU melt blends, peroxide decomposed into the R_1O• and •OR_1O• radicals [4]. As the concentration of PLA is higher than PBU, then, the R_1O• or •OR_1O• radicals abstract hydrogen from PLA, which create the radicals at chiral centres in PLA chains [4]. After that, the double bonds in PBU are easily broken to produce radicals and then combine with the PLA radicals, and the chemical fixed-structure between PLA chains could take place as shown in scheme 1.

3.5. Proposed mechanism of reactive-induced crystallization

The reason for the effectiveness of reactive PBU over the EBS as induced crystallization for PLA remains unclear. With the incorporation of PBU, We propose that the nucleation of PLA is initiated by the fixed-structures between PLA chains so that the fixed-structures could have acted as nucleating centres, thereby inducing crystal growth. In addition, this reactive-induced crystallization process can aid the increased number of nuclei, thus leading to faster crystallization. These results were similarly reported by Y. Zhang et al. [4], With the addition of TAIC, the low crosslinking bundles are formed (not complete crosslinking network) and can act as precursor to promote the crystallization. During crystallization, the chains that surround the crosslinking bundles can be absorbed to stack in lamellas to forms spherulite [4]. It could imply that the fixed-structures can play the role as well as the low crosslinking bundles to induce the crystallization of PLA.

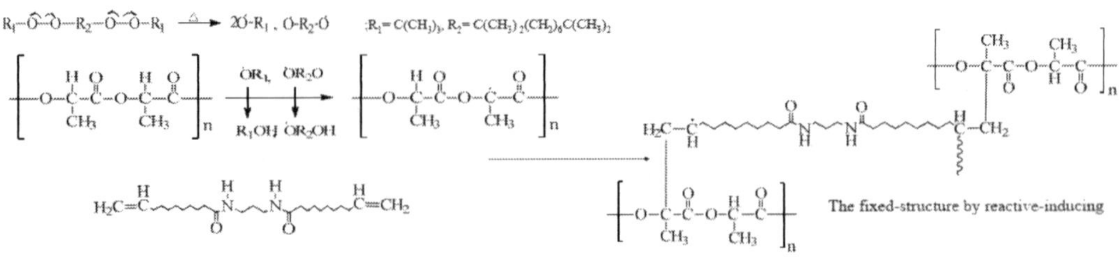

Scheme 1. A possible reaction of free radical reaction between PBU and PLA chains.

4. Conclusion

The local crosslinked structures could be achieved via the reactive melt blending of PLA with reactive bisamide (PBU) (for the small quantity of 0.3 to 0.7 wt%) incorporated with peroxide as an initiator. DSC results suggest that the fixed-structures created by partial local crosslinked structures can well promote and accelerate the crystallization of PLA while using peroxide alone does not affect the crystallization of PLA. With the incorporation of 0.3 to 0.7 wt% of reactive PBU, the crystallinity increases to 24.9-28.3%, which is higher than neat PLA and PLA added EBS almost by 260 and 6 times, respectively. The addition of 0.7 wt% PBU shows the fastest crystallization rate where the $t_{1/2}$ value isothermally crystallized at 130°C is only 6 min while the overall crystallization of neat PLA takes more than 100 min. The POM analysis confirms that a large number of nuclei are effectively involved on inducing the crystallization of PLA added PBU to be faster than PLA added EBS. Furthermore, the spherulites of the PLA/rPBU blends are smaller and denser compared to those of neat PLA and EBS. Thus, it can be summarized that the fixed-structures induced by reactive PBU can effectively induce crystallization of PLA as well.

Acknowledgments

The authors would like to gratefully thank National Research Council Thailand (NRCT) and Department of Materials Science and Engineering, Faculty of Engineering and Industrial Technology, and Silpakorn University, Thailand for financial supports and also Mr Chao Sripethdee for conducting the polarized optical microscopy (POM) testing.

References

[1] Song P, Chen G, Wei Z, Chang Y, Zhang W and Liang J 2012 *Polymer* **53** 4300-9
[2] Wang Y, Li M and Shen C 2011 *Mater. Lett.* **65** 3525-8
[3] Tang H, Chen J-B, Wang Y, Xu J-Z, Hsiao BS, Zhong G-J and Li Z-M 2012 *Biomacromolecules* **13** 3858-67
[4] Zhang Y, Wang C, Du H, Li X, Mi D, Zhang X and Wang T 2014 *Mater. Lett.* **117** 171-4
[5] Battegazzore D, Bocchini S and Frache A 2011 *Polymer Lett.* **5** 849-58
[6] Liang J-Z, Zhou L, Tang C-Y and Tsui C-P 2013 *Compos. B Eng.* **45** 1646-50
[7] Su Z, Li Q, Liu Y, Hu G-h and Wu C 2009 *J. Polymer Sci. B Polymer Phys.* **47** 1971-80
[8] Su Z, Li Q, Liu Y, Guo W and Wu C 2010 *Polymer Eng. Sci.* 1658-66
[9] Quan H, Zhang S-J, Qiao J-L and Zhang L-Y 2012 *Polymer* **53** 4547-52
[10] Ando Y, Sato H, Shinzawa H, Okamoto M, Noda I and Ozaki Y 2012 *Vib. Spectros.* **60** 158-62
[11] Ke T and Sun X 2003 *J. Appl. Polymer Sci.* **89** 1203-10
[12] Pei A, Zhou Q and Berglund LA 2010 *Compos. Sci. Tech.* **70** 815-21
[13] Cai Y, Yan S, Yin J, Fan Y and Chen X 2010 *J. Appl. Polymer Sci.* **121** 1408-16
[14] Tang Z, Zhang C, Liu X and Zhu J 2012 *J. Appl. Polymer Sci.* **125** 1108-15
[15] Xing Q, Zhang X, Dong X, Liu G and Wang D 2012 *Polymer* **53** 2306-14
[16] Salamone J C 1996 *Polymeric Materials Encyclopedia* vol 9 (Taylor & Francis) p 6825
[17] Sperling L H 1992 *Introduction to physical polymer science* 2nd (New York: John Wiley & Sons Inc.) p 236

Micromechanical analysis on anisotropy of structured magneto-rheological elastomer

R Li[1], Z Zhang[1], S W Chen[2] and X J Wang[1, 3]

[1]Chongqing univ of Posts & Telecommunications, Chongqing 400065, China
[2]Chongqing univ of Sci & Technol, Academy Math & Phys, Chongqing 400030, China

E-mail: xjwang@iamt.ac.cn

Abstract. This paper investigates the equivalent elastic modulus of structured magneto-rheological elastomer (MRE) in the absence of magnetic field. We assume that both matrix and ferromagnetic particles are linear elastic materials, and ferromagnetic particles are embedded in matrix with layer-like structure. The structured composite could be divided into matrix layer and reinforced layer, in which the reinforced layer is composed of matrix and the homogenously distributed ferromagnetic particles in matrix. The equivalent elastic modulus of reinforced layer is analysed by the Mori-Tanaka method. Finite Element Method (FEM) is also carried out to illustrate the relationship between the elastic modulus and the volume fraction of ferromagnetic particles. The results show that the anisotropy of elastic modulus becomes noticeable, as the volume fraction of particles increases.

1. Introduction

Magneto-rheological elastomer is comprised of polymer matrix and the micron-scale ferromagnetic particles inside it. The ferromagnetic particles are magnetized in the presence of magnetic field, and the MRE presents macroscopic behavior that the elastic modulus alters with the magnetic field. Compared with the magnetorheological fluids, the merits of MREs are their controllable stiffness, chemical and physical stability, and no sealing problems that make them attractive in semi-active vibration control [1, 2].

The elastic modulus of MRE can contribute to the magneto-induced modulus and zero-field modulus in the absence of magnetic field. The relations among magneto-induced modulus, ferromagnetic particle fraction, and magnetic field strength, have been investigated extensively [3]. In comparison with MRE whose particles are uniformly distributed, the MRE with aligned particles has a higher magneto-induced modulus [4]. Hence, it is very common to prepare the MREs where the magnetic particles are aligned orderly using magnetic field during curing process, resulting in anisotropic MREs.

However the research on the zero-field modulus is still inadequate, since only the empirical equation is utilized to evaluate the zero-field modulus of MRE in previous work [5, 6], and the particles distribution is merely taken in to consideration, which leads to an untrustworthy result. Many models are proposed in these years to help us understand the equivalent material properties, such as

[3] Address for correspondence: X J Wang, Chongqing univ of Posts & Telecommunications, Chongqing 400065, China. E-mail: xjwang@iamt.ac.cn.

method of upper and lower boundary from Voigt and Reuss [7, 8], Eshelby model [9] and Mori-Tanaka method [10]. Although those methods are helpful to evaluate the properties of composite like MRE, some errors worsen the theoretical result compared with empirical one. For example, the interaction between reinforced phases of composite is uncertain. Therefore, representative volume unit (RVU) [6] is set up for evaluation with the help of FEM software, and the results from FEM are proved to agree well with the experimental results, which can predict the properties of composites.

Considering the anisotropic property of the structured MREs, this paper assumes that magnetic particles are organized in layer in the presence of magnetic field, and MREs are a kind of anisotropic materials that stack the reinforced layer and matrix layer together. To investigate the zero-field modulus of the anisotropic MRE with different volume fraction of ferromagnetic particles, Mori-Tanaka method is carried out to build the theoretical model and RVU is founded and meshed to analyze the zero-field modulus of MRE by finite element theory.

2. Theoretical model

As shown in figures 1(a) and 1(b), the magnetic particles in MRE are orderly distributed in the presence of magnetic field, which make it become a kind of anisotropic material like fiber-reinforced composites.

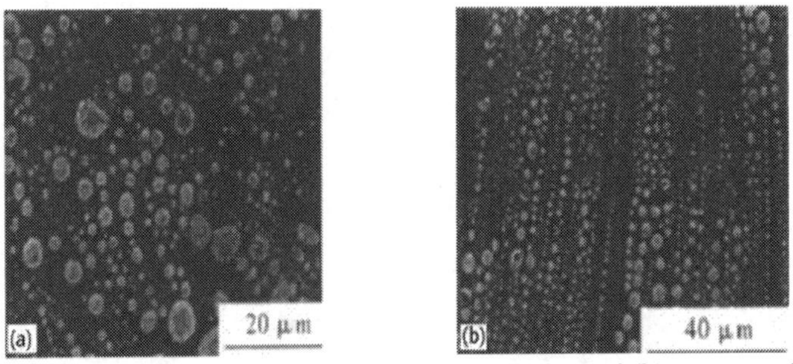

Figure 1. Microscopic structure of MRE. (a) Uniformed (b) Structured [1].

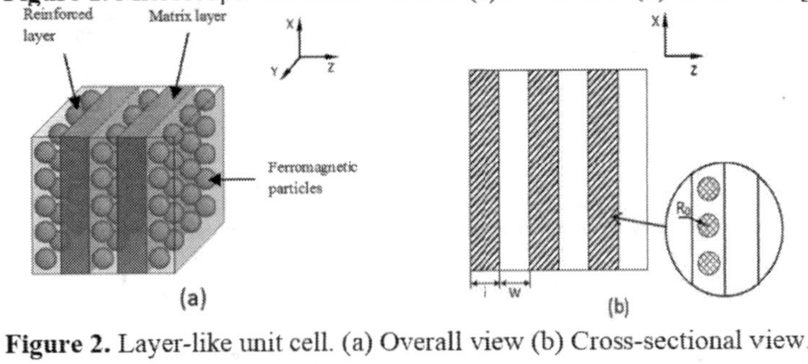

Figure 2. Layer-like unit cell. (a) Overall view (b) Cross-sectional view.

It is assumed that both matrix and ferromagnetic particles are isotropic linear elastic materials, and ferromagnetic particles are embedded in matrix with layer-like structure. Figure 2 shows that the composite is divided into two parts, i.e. reinforced layer and matrix layer. The reinforced layer is composed of matrix and the homogenously distributed particles inside it.

In figure 2, l and w represent the thickness of matrix layer and reinforced layer respectively. R_0 stands for the radius of ferromagnetic particles. In the following sections, we firstly analyze the equivalent elastic modulus of the reinforced layer. Then we obtain the equivalent elastic modulus of the MREs changed with particle volume fractions, at last we discuss the anisotropic property of the structured MREs in different deformation directions.

2.1. Equivalent modulus of reinforced layer

Given that particles in reinforced layer are seamlessly embedded in the matrix, and they are uniformly dispersed, the reinforced layer can be regarded as two phase composites. According to the relationship between stress and strain, the arithmetic mean symbol is defined as $\langle\cdot\rangle = \frac{1}{V}\oint \cdot\, dV$. Inside the reinforced layer, there are two phases: matrix phase and inclusion phase, where the average stress tensor $\langle\sigma_{ij}\rangle$ and the average strain tensor $\langle\varepsilon_{ij}\rangle$ of each unit cell satisfy the following equations:

$$\langle\sigma_{ij}\rangle = \overline{C}_{ijkl}\langle\varepsilon_{ij}\rangle, \tag{1}$$

$$\langle\sigma_{ij}\rangle_r = C^r_{ijkl}\langle\varepsilon_{ij}\rangle_r. \tag{2}$$

Where $\overline{C}_{ijkl}$ is the equivalent elastic tensor of a unit cell, r stands for the r phase of material. According to the definition of arithmetic mean symbol and the structure of reinforced layer, the average stress tensor of a unit cell can be also described as equation (3), which is the sum of average stress of each phase.

$$\langle\sigma_{ij}\rangle = \frac{1}{V}\oint \sigma_{ij} dV = \sum_{r=0}^{N-1}\frac{V_r}{V}\frac{1}{V_r}\int_{V_r}\sigma_{ij} dV = \sum_{r=0}^{N-1} c_r\langle\sigma_{ij}\rangle_r. \tag{3}$$

Where V is the total volume, V_r is the relative volume of phase r, $c_r = V_r/V$ is the volume fraction that phase r accounts for. The local relation between each phase is:

$$\langle\varepsilon_{ij}\rangle = B^r_{ijkl}\overline{\varepsilon}_{kl}. \tag{4}$$

where $\overline{\varepsilon}_{kl}$ is the uniform strain on the unit cell boundary, $\boldsymbol{B}$ is the local tensor. The equations (1)- (4) reduce to the equivalent elastic modulus :

$$\overline{C}_{ijkl} = C^0_{ijkl} + c_r(C^r_{ijkl} - CS^0_{ijkl})B^1_{klmn}. \tag{5}$$

According to Mori-Tanaka method, given that the strain around the certain inclusion is the average strain of matrix, and the average strain of ferromagnetic particles is expressed as:

$$\boldsymbol{B} = (\boldsymbol{I} + \boldsymbol{P}\delta\boldsymbol{C})^{-1}[c_1\boldsymbol{I} + c_1(\boldsymbol{I} + \boldsymbol{P}\delta\boldsymbol{C})^{-1}] \tag{6}$$

Where $\boldsymbol{I}$ is the fourth-rank tensor, $\delta\boldsymbol{C} = \boldsymbol{C_1} - \boldsymbol{C_0}$ is the difference between elastic modulus of inclusions $\boldsymbol{C_1}$ and elastic modulus of matrix $\boldsymbol{C_0}$. Tensor $\boldsymbol{P}$ is related with the inclusion shape. When the inclusions are chosen as spherical shape, equation (7) is derived as:

$$P_{ijkl} = \frac{1}{3G_0(4G_0+3K_0)}\delta_{ij}\delta_{kl} + \frac{3(2G_0+K_0)}{10G_0(4G_0+3K_0)}(\delta_{ij}\delta_{kl} + \delta_{ij}\delta_{kl} - \frac{2}{3}\delta_{ij}\delta_{kl}) \tag{7}$$

Where G_0, K_0 are shear modulus and volume modulus of matrix phase, respectively.δ_{ij} is Kronecker tensor. By considering equations (6) and (7), the equivalent modulus of reinforced layer satisfies the following equation:

$$\overline{\boldsymbol{C}} = \boldsymbol{C_0} + c_1[(\boldsymbol{C_1} - \boldsymbol{C_0})^1 + c_0\boldsymbol{P_1}]^{-1} \tag{8}$$

Where $\boldsymbol{P}_r$ is the r phase's $\boldsymbol{P}$ tensor, by substituting equation (7) into equation (8), it can be concluded that reinforced layer is isotropic, while the equivalent volume modulus and equivalent shear modulus are expressed as:

$$\overline{G} = G_0 + \frac{10G_0(4G_0+3K_0)(G_1-G_0)}{10G_0(4G_0+3K_0)+12c_0(3G_0+K_0)(G_1-G_0)}c_1, \tag{9}$$

$$\overline{K} = K_0 + \frac{3G_0(4G_0+3K_0)(K_1-K_0)}{3G_0(4G_0+3K_0)+9c_0(K_1-K_0)}c_1. \tag{10}$$

2.2. Equivalent modulus of MRE

The reinforced layer and matrix layer are made up of isotropic materials. Assuming that MRE is transverse isotropic, the five independent transverse elastic constants of elastic modulus are described as equations (12)-(16). Along the *x* or *z* axis stretching the MRE uniaxially, the displacement is Δl. Therefore, the resultant force generated by two layers is as:

$$P = A\overline{\sigma}_{11} = A_p\sigma_p + A_m\sigma_m, \tag{11}$$

Where A_p is the cross-sectional area of reinforced layer, A_m is the cross-sectional area of matrix. σ_p is the normal stress of reinforced layer, σ_m is the normal stress of matrix layer. Considering that the two layers have the same normal strain $\varepsilon_{11} = \Delta l/l$, according to the equation (11), the vertical Young's modulus of MRE $\overline{E}_1, \overline{E}_2$ are expressed as

$$\overline{E}_1 = \overline{E}_2 = \overline{E}_p c_p + E_m(1 - c_p) = E_m(1 - c_p) + \frac{9\overline{G}c_p\overline{K}}{3\overline{K}+\overline{G}}, \tag{12}$$

Where $c_p = A_p/(A_{pm} + A_p)$ represents the volume fraction of reinforced layer. Similarly, along the *y* axis stretches the MRE, assuming that both layers share the same normal stress, so the horizontal Young's modulus $\overline{E}_3$ can be obtained:

$$\overline{E}_3 = \frac{1}{(1-c_p)+c_p(E_m/\overline{E}_p)}. \tag{13}$$

Poisson coefficient ν affects the axial deformation when the material suffers from a uniaxial stretch. Assuming that the material is stretched along the direction which is perpendicular to the *y* axis, the total deformation of the elementary cell comes from the sum of vertical deformation of two elementary layers. Hence, the Poisson's ratio is derived as:

$$\overline{\nu}_{21} = \frac{\overline{E}_2}{\overline{E}_1}\overline{\nu}_{21} = \frac{\overline{E}_3}{\overline{E}_1}\overline{\nu}_{31} = (1 - c_p)\nu_m + c_p\nu_p, \tag{14}$$

When determining the shear modulus, given that a shear is along the *x-y*, and the shear stress of reinforced layer and matrix layer are the same, the *x-y* shear modulus is derived as:

$$\overline{G}_{12} = \frac{G_m\overline{G}_p}{(1-c_p)\overline{G}_p+G_m c_p}. \tag{15}$$

When a shear is along the opposite side, the reinforced layer and matrix layer have the same shear strain. So, the *x-z* shear modulus is derived as:

$$\overline{G}_{13} = \overline{G}_{23} = G_m(1 - c_p) + \overline{G}_p c_p. \tag{16}$$

3. FEM simulation

To further evaluate the theoretical model, the FEM is employed to obtain the elastic modulus of structured MRE. The model of a unit cell is shown as figure 3. When the composite suffers from the stretch along the *i* direction, a fixed constraint is applied on S_{efgh}, S_{dfgc} and S_{adef}, and the boundary condition on S_{abcd} is a prescribed displacement in the *i* direction, then we have:

$$\overline{E}_i = \frac{\overline{\sigma}_i}{\overline{\varepsilon}_i} = \frac{\int \sigma_i dV}{\int \varepsilon_i dV}, \tag{17}$$

$$\overline{\nu}_{ij} = \frac{\overline{\varepsilon}_{ii}}{\overline{\varepsilon}_{jj}} = \frac{\int \varepsilon_{ii} dV}{\int \varepsilon_{jj} dV}, \tag{18}$$

$$\overline{G}_{ij} = \frac{\overline{\sigma}_{ij}}{2\overline{\varepsilon}_{ij}} = \frac{\int \sigma_{ij} dV}{2\int \varepsilon_{ij} dV}, \tag{19}$$

When the composite suffers a shear force on the *i-j* plane, the equivalent shear modulus of MRE satisfies equation (19). In summary, by varying the volume fraction and layer distribution in the finite element model of composite element, the stress and strain of composite element model in different working conditions are acquired. Hence the equivalent elastic modulus is derived by introducing equations (17)-(19).

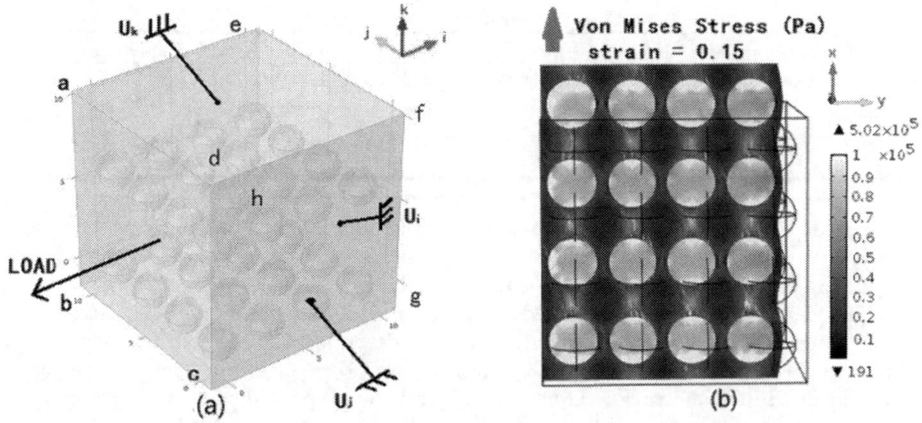

Figure 3. Finite element model of composite cell ($w/l = 1$),where the particles are aligned in layer-like form and cubic model represents a micro part of the material, and thus derives the equivalent elastic modulus of the MRE (a)Model (b)Von mises stress of model's cross section.

4. Results

The relation between particles proportion and anisotropy of MRE in layer-like structure is studied, with the results from FEM and theoretical calculation. Table 1 reveals the simulation parameters while figure 4 demonstrates the results of FEM simulation and model calculations.

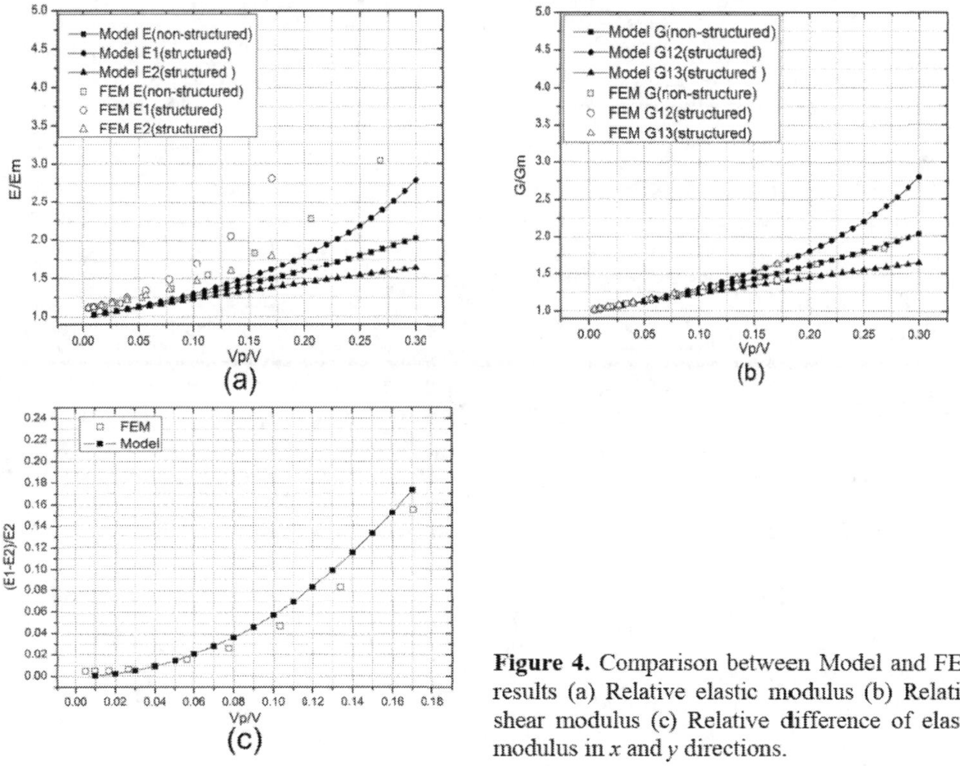

Figure 4. Comparison between Model and FEM results (a) Relative elastic modulus (b) Relative shear modulus (c) Relative difference of elastic modulus in x and y directions.

Table 1. Simulation parameter.

Parameter	Value	Parameter	Value (μm)
Particle Young's modulus E_p	200 (Gpa)	Reinforced layer width l	1.5
Particle Poisson's ratio v_p	0.29	Matrix layer width w	[0-1.5]
Matrix Young's modulus E_m	1 (Mpa)	Spheres radius R_0	[0.3-1.3]
Matrix Poisson's ratio v_m	0.49	Particle spacing l_0	1.5

Figure 4(a) shows that E/E_m varies with the particles volume fraction. Both of E_1, E_2 increase with the particles volume fraction. Due to the present theoretical model does not consider the interaction among neighboring particles, the theoretical solution for equivalent Young's modulus is lower than the FEM simulation results. However, both methods show the equivalent Young's modulus changing with particle volume fraction in the same trend. Figure 4(b) describes the relationship between shear modulus and particles volume fraction. The shear modulus of G_{12}, G_{13} has the same increasing trend with elastic modulus. In the shearing condition, the space between particles are fixed, which reduces the interaction among particles. Hence, the theoretical solutions for shear modulus well agree with the FEM solution.

Figure 4(c) illustrates that the difference between the Young's modulus E_1 in the x direction and the Young's modulus E_2 in the y direction. The result indicates that the elastic modulus along the reinforced direction increases significantly, as the particles proportion increases.

5. Conclusions

- In the same particle volume fraction, structured MRE has an enhanced Young's modulus along the direction of axis x, and the Young's modulus decreases along the y axis.
- As the particle percentage increases, the anisotropy of MRE becomes more noticeable.
- Solutions of equivalent shear modulus match well with the finite element solutions. Both solutions of equivalent Young's modulus have the same trend, but the errors still exist and the theoretical solutions are slightly lower.

Acknowledgments

This work was supported by the National Nature Science Foundation of People's Republic of China (Project NO. 11372366). Project is also supported by Chongqing Science Foundation for Distinguished Young Scholars (cstc2014jcyjjq40004), Chongqing Excellent Talents Plan for Universities and Nature Science Fund of Chongqing (cstc2012jjA40035).

References

[1] Fuchs A, Zhang Q, Elkins J, et al. 2007 J. *Appl. Polym. Sci.* **105** 2497-508a
[2] Li Y C, Li J C, Li W H and Samali B 2013 *Smart. Mater. Struct.* **22** 035005(12pp)
[3] Coquelle E, Bossis G, Szabo D, Giulier F 2006 *J. Mater. Sci.* **41** 5941-53
[4] Chen L, Gong X L and Li W H 2007 *Smart. Mater. Struct.* **16** 2645-50
[5] Ivaneyko D, Toshechevikov V P, Saphiannikova, et al. 2011 *Macromolecular Theory and Simulations* **20** 411-24
[6] Ivaneyk D,Toshchevikov V, Saphiannikova M and Heinrich G 2012 *Condensed Matter Physics* **15** 1-12
[7] Voigt W 1889 *Wied Ann* **38** 573-87
[8] Reuss A 1929 *ZAMM-Journal of Applied Mathematics and Mechanics* **9** 49-58
[9] Eshelby J D 1957 *Proceedings of the Royal Society of London. Series A. Mathematical and Physical Sciences* **241** 376-96
[10] Mori T and Tanaka K 1973 *Acta metall.* **21** 571-4

Impact property enhancement of poly (lactic acid) with different flexible copolymers

N Likittanaprasong[1], M Seadan[2] and S Suttiruengwong[1, 3]

[1]Department of Materials Science and Engineering, Faculty of Engineering and Industrial Technology, Silpakorn University, Sanamchandra Palace Campus, Nakhon Pathom 73000, Thailand
[2]Department of Physics, Faculty of Science, Silpakorn University, Sanamchandra Palace Campus, Nakhon Pathom 73000, Thailand

E-mail: suttiruengwong_s@su.ac.th

Abstract. The objective of this work was to improve the impact property of Poly (lactic acid) (PLA) by blending with different copolymers. Six flexible copolymers, namely, acrylonitrile butadiene styrene (ABS) powder, Biomax, polybutyrate adipate co-terephthalate (PBAT), polyether block amide (PEBAX), ethylene-vinyl acetate (EVA) and ethylene acrylic elastomer (EAE), with loading less than 20wt% were used and compared. The rheological, mechanical and morphological properties of samples were investigated by melt flow index, tensile testing, impact testing and scanning electron microscope (SEM), respectively. It was found that PLA added 20wt% EAE showed the highest impact strength (59.5 kJ/m^2), which was 22 times higher than neat PLA. The elongation at break was also increased by 12 folds compared to neat PLA. The SEM images showed good interface and distribution for PLA containing 20wt% EAE, 15 phr Biomax and 20 wt% PEBAX.

1. Introduction

Poly (lactic acid) (PLA) has emerged as a promising choice for replacing the existing commodity and industrial products. In terms of downstream manufacture of PLA products, the cost of PLA has dropped against the increasing demands. Therefore, many PLA-based products such as film wrap, glass, plate, spoon and fork have readily been marketed. The important features of PLA include high mechanical properties, compostability, transparency, while some weaknesses also exist such as low heat resistance, brittleness due to slow crystallization [1, 2]. To date, many types of polymers regardless of being biodegradable or not, have been reported for increasing impact properties of PLA [3]. The impact improvement was possible through first; the addition of copolymers such as OnCap™ BIO [4], random aliphatic copolyesters [4], ethylene/n-butyl acrylate/glycidyl methacrylate (EBA-GMA) terpolymer with zinc ion-containing ionomer [5], poly (butylene succinate) (PBS) [6], poly(-R-3-hydroxybutyrate) [7], polyurethane elastomer (PU) [8], poly (ethylene glycol) (PEG) [9]), and second; using *in situ* co-polymerization (e.g., c-caprolactone and L-lactide (c-CL/L-LA) [10], poly (β-

[3] Address for correspondence: S Suttiruengwong, Department of Materials Science and Engineering, Faculty of Engineering and Industrial Technology, Silpakorn University, Sanamchandra Palace Campus, Nakhon Pathom 73000, Thailand. E-mail: suttiruengwong_s@su.ac.th.

hydroxybutyrate-co-β-hydroxyvalerate) [11], ethylene-co-vinyl acetate [12], P[CL-co-LA] random copolyesters [13].

Sin L T et al. [14], used Poly(lactic acid) with Blendex 338 or Ultra High Rubber ABS as an impact modifier. They reported that PLA blended with Blendex 338 (at the amount of 20 wt%) showed the notched impact strength increased from 26.7 J/m to 518 J/m and the elongation at break increased from 10 % to 218 %. Krishnaswamy R et al. [15] demonstrated that blending Biomax strong 120 (other types of ethylene copolymer) at the amount of 10 wt% into PLA could improve the impact strength in a similar fashion. The impact strength increased from 0.2 t.lb/inch to 5.0 t.lb/inch. Afrifah K A et al. [16] used ethylene/acrylate copolymer as an impact modifier for blending with PLA and they found that the blends showed a decrease in tensile strength and modulus with the increasing impact modifier contents. In contrast, the ductility, elongation at break, and energy to break increased significantly. Although a lot of literatures have been reported the use of the above mentioned polymers for impact modifiers, there has not yet been attempt to systematically compare their performance in terms of mechanical properties. Therefore, in this research, six flexible copolymers; namely ABS powder (core shell (rubber core/ABS shell)), ethylene copolymer (Biomax), polybutyrate adipate co-terephthalate (PBAT), Polyether block amide (PEBAX), ethylene-vinyl acetate (EVA) and ethylene acrylic elastomer (EAE) were selected to melt-blending with PLA in order to study the effect of their types and compositions on impact properties together with tensile properties and physical appearance. The performance of toughened PLA was discussed and compared.

2. Experimental

2.1. Materials

PLA 2003D was purchased from NatureWorks LLC (USA). ABS Powder (rubber core/ABS shell, AP6200) was kindly supplied by IRPC, (Thailand). Biomax (Ethylene copolymer) and Ethylene Acrylic Elastomer (EAE, Vamac®) were kindly provided by Chemical Innovation Company Limited (Thailand). Ecoflex® (PBAT) was purchased from BASF (Germany). Polyamide elastomer; PEBAX was purchased from Arkema (France). Ethylene vinyl acetate (Escorene™ Ultra EVA copolymer, EVA) was purchased from Exxonmobil Chemical (USA).

2.2. Sample preparation

PLA resin was first dried at 60°C for 12 hr. Copolymers were then weighted according to table 1 and dried mixed before transferring to an internal mixer, Model MX 105-D40L50, Chareon tut CO. LTD., Thailand. Melt blending was performed at the temperature of 190°C and the rotor speed of 60 rpm.

Table 1. PLA formulations and blend compositions.

Sample formulation	PLA (g)	ABS powder (g)	Biomax (g)	PBAT (g)	PEBAX (g)	EVA (g)	EAE (g)
neat PLA	100						
PLA/ABS powder 10 phr	100	10					
PLA/ABS powder 15 phr	100	15					
PLA/Biomax 10 phr	100		10				
PLA/Biomax 15 phr	100		15				
PLA90/PBAT10	90			10			
PLA80/PBAT20	80			20			
PLA90/PEBAX10	90				10		
PLA80/PEBAX20	80				20		
PLA90/EVA10	90					10	
PLA80/EVA20	80					20	
PLA90/EAE10	90						10
PLA80/EAE20	80						20

The mixing time was 10 min. The test specimens were prepared using a compression molding machine (Chareon tut CO. LTD., Thailand). Compression molding was carried out at 190°C for 6 min to obtain micro tensile and impact specimen.

2.3. Mechanical properties testing
Tensile and impact tests were carried out according to ASTM D 638 and ASTM D256, respectively. Ten specimens were used for each sample. Tensile testing was measured using a Universal Testing Machine (Instron, Model 5969, Instron Engineering Corporation, USA). Testing speed was 10 mm/min. Impact testing was measured using a Zwick model Pendulum impact tester B5102.202.

2.4. Morphology
The morphology of the PLA/copolymers specimens was observed using SEM (CamScan, MX 2000, UK). The fracture surfaces of impact testing were coated with gold before SEM examinations.

3. Results and discussion

3.1. Torque and melt flow index (MFI)
From table 2, it was found that with the addition of all copolymer showed a decrease in the final torque values compared with neat PLA. PLA containing 20 wt% EVA showed the lowest final torque value at 7.34 Nm, which half of the value of neat PLA. These results were confirmed by the MFI results, and the highest MFI value was observed for the addition of 20 wt% EVA in PLA.

Table 2. Rheological properties of PLA and its blends containing different copolymers.

Sample	Final torque (Nm)	MFI 190°C (g/10min)
neat PLA	14.77	4.7±0.5
PLA/ABS powder 10 phr	13.37	4.8±0.5
PLA/ABS powder 15 phr	13.35	4.8±0.5
PLA/Biomax 10 phr	11.29	5.5±0.4
PLA/Biomax 15 phr	11.43	5.6±0.4
PLA90/PBAT10	11.41	6.4±0.7
PLA80/PBAT20	15.03	6.9±0.5
PLA90/PEBAX10	9.51	7.0±0.9
PLA80/PEBAX20	8.40	9.1±0.5
PLA90/EVA10	11.27	5.5±0.1
PLA80/EVA20	7.34	10.9±0.1
PLA90/EAE10	11.35	8.2±0.1
PLA80/EAE20	13.07	5.8±0.1

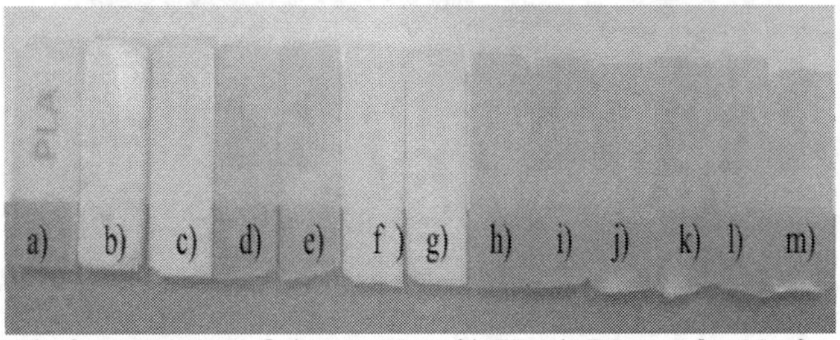

Figure 1. Physical appearance of a) neat PLA, b) PLA/ABS powder 10 phr, c) PLA/ABS powder 15 phr, d) PLA/Biomax 10 phr, e) PLA/Biomax 15 phr, f) PLA90/PBAT10, g) PLA80/PBAT20, h)PLA90/PEBAX10, i) PLA80/PEBAX20, j) PLA90/EVA10, k) PLA80/EVA20, l) PLA90/EAE10, m) PLA80/EAE20.

3.2. Physical appearance of blends

Figure 1 showed the physical appearance of polymer blended with copolymers. All added copolymer samples lost their optical property. When ABS powder and PBAT were added into PLA, the specimens became less clear. Samples b) and c) (PLA/ABS powder 10 and 15 phr) were whitened and opaque as well as sample f) and g) (PLA90/PBAT10 and PLA80/PBAT20). In case of samples d) and e) (PLA/Biomax 10 and 15 phr), they were clearer than sample h) and i) (PLA90/PEBAX10 and PLA80/PEBAX20, samples j) and k) (PLA90/EVA10 and PLA80/EVA20), and sample l) and m) (PLA90/EAE10 and PLA80/EAE20).

Table 3. Impact properties of PLA and the blends containing different copolymers.

Sample	Izod impact strength (IS) (kJ/m^2)
neat PLA	2.7±0.3
PLA/ABS powder 10 phr	8.6±0.7
PLA/ABS powder 15 phr	8.5±1.2
PLA/Biomax 10 phr	26.2±3.7
PLA/Biomax 15 phr	38.6±2.3
PLA90/PBAT10	5.6±0.8
PLA80/PBAT20	9.4±0.6
PLA90/PEBAX10	9.0±0.8
PLA80/PEBAX20	35.1±7.5
PLA90/EVA10	14.7 2.8
PLA80/EVA20	11.6±2.5
PLA90/EAE10	13.0±1.8
PLA80/EAE20	59.5±2.1

3.3. Impact properties and SEM of specimens

The izod impact strengths of neat PLA and its blends with copolymer were shown in table 3. With the addition of ABS powder (10 and 15 phr), it showed a slight increase in the impact strengths about 8.6 kJ/m^2 (3 times) compared with neat PLA. When adding Biomax 10 and 15 phr, the impact strengths of 26.2 and 38.6 kJ/m^2, respectively, which were higher than those of neat PLA by 9.7 and 14.3 times, respectively. With the incorporation of 10 phr of PBAT or PEBEX, the impact strengths were also 2 and 3 times higher than those of neat PLA, respectively. When increasing the amount of PBAT or PEBAX to 20 phr, the impact strengths were 4 and 13 times higher than neat PLA, respectively. In the case of the addition of EVA (10 to 20 wt%), in the blend, a decrease in impact strengths value was found, which is only 5 and 4 times compared with that of neat PLA, respectively. Surprisingly, when 20 wt% EAE was added, it showed the highest impact strength (59.5 kJ/m^2), which was higher than

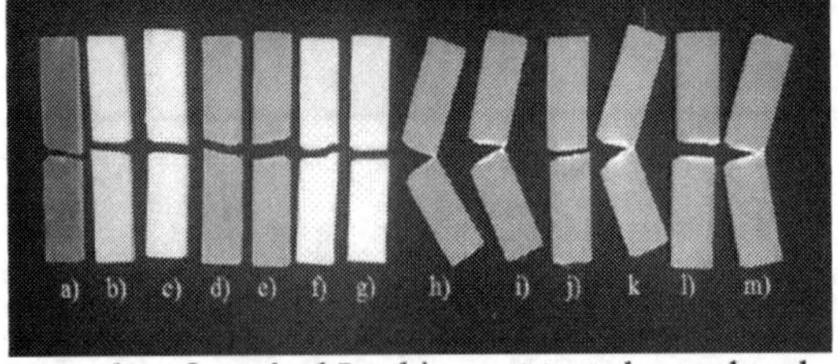

Figure 2. Photographs of notched Izod impact tested samples showing fracture modes. a) neat PLA, b) PLA/ABS powder 10 phr, c) PLA/ABS powder 15 phr, d) PLA/Biomax 10 phr, e) PLA/Biomax 15 phr, f) PLA90/PBAT10, g) PLA80/PBAT20, h)PLA90/PEBAX10, i) PLA80/PEBAX20, j) PLA90/EVA10, k) PLA80/EVA20, l) PLA90/EAE10, m) PLA80/EAE20.

that of neat PLA by 20 folds. The addition of elastomer EAE showed the dramatic improvement of impact toughness to PLA matrix, hence allowing energy dissipation mechanisms into PLA and retarding crack initiation and propagation.

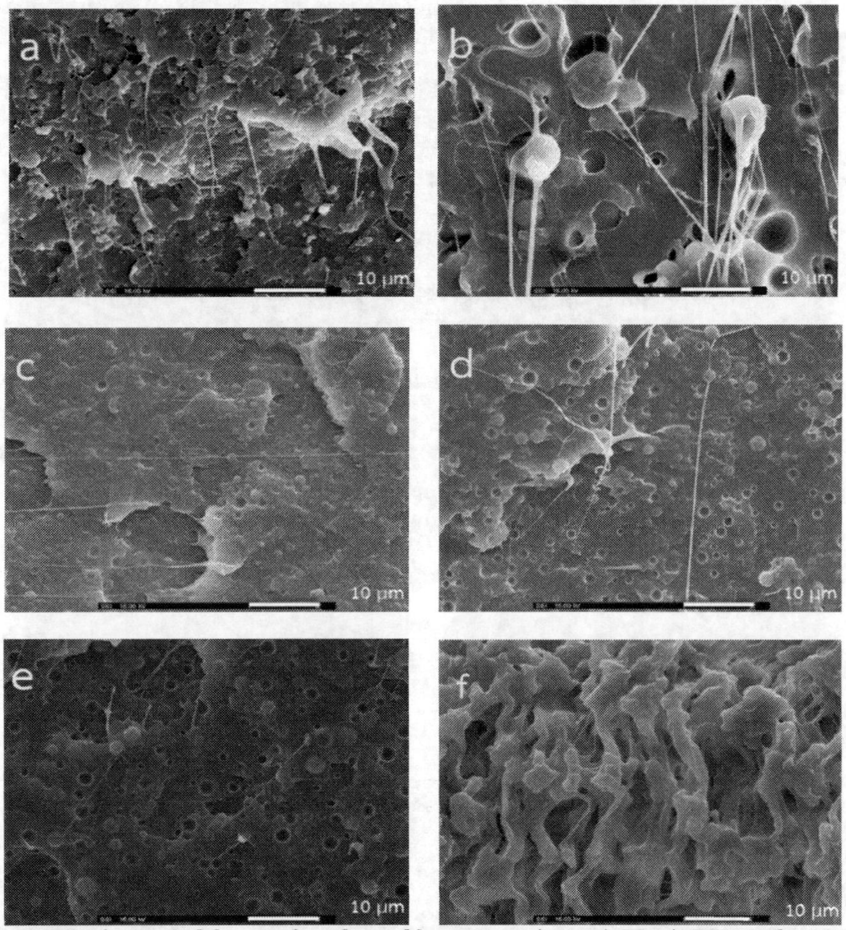

Figure 3. SEM images of fractured surface of impact specimen a) PLA/ABS powder 10 phr, b) PLA/Biomax 10 phr, c) PLA90/PBAT10, d) PLA90/PEBAX10, e) PLA90/EVA10, f) PLA90/EAE10.

From figure 2, it was demonstrated that complete breakage could be obtained with samples a), b), c), d), e), f), g), j) and l), whereas the partial breakage of specimens was found in samples h), i), k) and m). Stress whitening near the notched tip was observed due to crazing or micro cracks, and it showed that the matrix could bear most of the stress [17]. It is common that the continuous transfer of energy is an important factor for impact resistance. In adding dispersed phases into the polymer matrix, the good distribution of the dispersed phase, smaller particle sizes and good interfacial adhesion are key factors to determining the optimum performance of materials. Fractured surface of impact specimens corresponding to the same concentration of copolymers are shown in figures 3a)-3f). At the same amounts of copolymer (10 wt%), the SEM image of PLA added 10 phr ABS powder samples showed good dispersion of ABS powder throughout PLA matrix. In addition, separated particles of ABS powder phase could still be observed. The dispersion phases did not retain the spherical shapes, but partially adhered to PLA matrix. This might provide the positive impact property as seen in figure 3(a).

In figure 3(b), the particle phases in the matrix could retain the spherical shapes, copolymers particle elongation and had good interphase between modifier and the matrix. In figure 3(c), impact modifier particle sizes were regularly dispersed in PLA matrix. The particle sizes of dispersed phase were 3 μm. In this case, it showed the partial interfacial adhesion between PLA and PBAT matrix. In the case of PLA added PEBAX (in figure 3(d)), it was found that the adhesion between PEBAX and PLA was prominent and the distribution of tiny particle sizes in the matrix could be seen as the same as that in the case of Biomax. In figure 3(e), the addition of EVA demonstrated the superior distribution of particles in the matrix, but the particles of EVA were not-adhered to PLA matrix. In addition, dispersed phase particles were not spherical and do not change shape when applied force. In figure 3(f), broken fibrils from polymer blend between PLA with 10 wt% of EAE could still be observed. Particles of EAE were well dispersed in PLA matrix. There were many fibrils on fractured surface of the specimen. This indicated the toughening mechanism of polymer matrix. When force was applied to the samples by impact testing, EAE would adsorb force and the extension of particle took place. These results were attributed to the higher elongation at break which will be discussed later.

3.4. Tensile properties

The modulus, tensile strength and elongation at break were displayed in table 4. For all added copolymers, the increase in their contents led to the decrease in the modulus and tensile strengths, because copolymer was flexible than PLA. The decreased tensile strength probably resulted from the presence copolymers inserted in PLA matrix [18, 19]. The increase in elongation at break probably resulted from the samples were more ductile. For the addition of 20 wt% of PBAT, PEBAX and EVA, the elongation at break increased to over 35% (from 8.7% of neat PLA), whereas the tensile strength decreased by 30% when compared to neat PLA. PLA added 20 wt% PBAT showed the highest elongation at break value. It also appeared necking phenomenon which demonstrated the ductile behaviour of the polymer blends.

Table 4. Tensile properties of PLA and the blend.

Sample	Modulus [MPa]	Tensile Strength [MPa]	Elongation at break [%]
neat PLA	1619.2±115.1	91.6±8.3	8.7±1.4
PLA/ABS powder 10 phr	1517.8±59.2	76.0±2.8	39.4±10.5
PLA/ABS powder 15 phr	1231.8±11.2	57.0±4.3	41.1±7.6
PLA/Biomax 10 phr	1247.8±67.6	63.0±3.2	19.6±3.4
PLA/Biomax 15 phr	1179.6±48.3	57.5±3.2	44.1±23.6
PLA90/PBAT10	1287.1±70.5	85.4±2.2	96.7±40.6
PLA80/PBAT20	1117.3±96.5	65.2±4.9	185.9±34.0
PLA90/PEBAX10	1294.4±46.9	83.1±6.3	52.1±22.0
PLA80/PEBAX20	1042.8±49.5	66.0±2.4	110.1±10.1
PLA90/EVA10	1082.1 ± 51.2	52.6 ± 3.7	33.5 ± 7.1
PLA80/EVA20	934.5 ± 53.3	39.7 ± 1.7	96.5 ±32.4
PLA90/EAE10	1293.3±134.1	74.8±5.6	32.4±3.5
PLA80/EAE20	1030.8±105.6	57.8±2.9	104.1±53.2

4. Conclusion

This work aimed to study and compare the effect of six flexible copolymers on the impact property. The impact results show that the impact strength is significantly improved in all PLA added copolymer samples when compared to neat PLA. Especially, PLA with the addition of 20 wt% EAE, shows superior performance on impact strength improvement which its impact strength was 22 times higher than neat PLA and this is confirmed by SEM images of fractured surfaces from impact testing, which show good interface and distribution between PLA matrix and copolymer for PLA containing 20 wt% EAE, 15 phr Biomax and 20 wt% PEBAX. In addition, the elongation at break of all samples was greatly increased compared to neat PLA. Therefore, in summary, it indicates that the addition of 20 wt% EAE is the most effective thermoplastic elastomer for improving impact resistance of PLA.

Acknowledgments

The authors would like to gratefully thank National Research Council Thailand (NRCT) and this work was supported by a grant from Department of Materials Science and Engineering, Faculty of Engineering and Industrial Technology, and Silpakorn University, Thailand for financial supports, IRPC public company limited for impact modifier (ABS powder), Chemical Innovation Company Limited for impact modifier (Biomax and EAE).

References

[1] Rasal R M, Janorkar A V and Hirt D E 2010 *Progr Polymer Sci* **35** 338-56
[2] Matuana L M, Faruk O and Diaz C A 2009 *Biosci. Rep.* **100** 5947-54
[3] Jamshidian M, Tehrany E A, Imran M, Jacquot M and Desobry S 2010 *Compr. Rev. Food. Sci. F.* **9** 552-571
[4] Odent J, et al. 2012 *Eur Polymer J* **48** 331-40
[5] Song W, et al. 2012 *Polymer* **53** 2476-84
[6] Jompang L, et al. 2013 *Energy Procedia* **34** 493-9
[7] Yoon J-S, et al. 2000 *Eur Polymer J* **36** 435-42
[8] Imre B, et al. 2013 *Eur Polymer J* **49** 3104-13
[9] Choi K-M, et al. 2013 *Eur Polymer J* **49** 2356-64
[10] Hiljanen-Vainio M, Karjalainen T and Seppälä J 1996 *J Appl Polymer Sci* **59** 1281-8
[11] Ma P, et al. 2013 *Eur Polymer J* **49** 1523-31
[12] Ma P, et al. 2012 *Eur Polymer J* **48** 146-54
[13] Odent J, et al. 2013 *Eur Polymer J* **49** 914-22
[14] Lee Tin Sin A R R and Rahman W A W A 2012 *Handbook on the Google Book* p 210
[15] Available at: http://www.metabolix.com/sites/default/files/PHA_PLA-Slideshow.pdf
[16] Afrifah K A and Matuana L M 2010 *Macromol Mater Eng* **295** 802-11
[17] Liang J Z and Li R K Y 2000 *J Appl Polymer Sci* **77** 409-17
[18] Mengeloglu F, Matuana L M and King J A 2000 *J Vinyl Addit Techn* **6** 153-7
[19] Park D, et al. 2011 *Macromol Res* **19** 105-12

A study on reactive blending of (poly lactic acid) and poly (butylene succinate co adipate)

C Bureepukdee[1], S Suttireungwong[1, 3] and M Seadan[2]

[1]Department of Materials Science and Engineering, Faculty of Engineering and Industrials Technology, Silpakorn University, Sanamchandra Palace Campus, Nakhon Pathom 73000, Thailand
[2]Department of Physics, Faculty of Science, Silpakorn University, Sanamchandra Palace Campus, Nakhon Pathom 73000, Thailand

E-mail: suttireungwong_s@su.ac.th

Abstract. This research aims to study the blending of Polylactic acid (PLA) and Polybutylene succinate co adipate (PBSA) in order to understand the role of peroxide in free radical reaction on the compatibilization between these two biodegradable polyesters. Various ratios of PLA/PBSA blends with and without reactive agents were prepared in the twin screw extruder. Two types of peroxides, Di (tert-butylperoxyisopropyl) benzene (DTBP) and 2, 5-Dimethyl-2, 5-(t-butylperoxy) hexane (DTBH), were used with various concentrations to compare. From the torques measurement, DTBP was more reactive with PLA and PBSA than DTBH. PLA and PBSA 80:20, 60:40, 50:50, 40:60, and 20:80% by weight were melt-blended in a twin screw extruder. The reactive polymer blends were also prepared for the same ratios of the blends with addition of 0.08 and 0.1 phr of DTBP. The mechanical, thermal, rheological, and morphological properties were investigated. The impact strengths of the non-reactive blend increased with the increasing in PBSA content. The optimal impact strength was obtained at 40%wt of PBSA with 0.1 phr of DTBP. Adding 0.08 and 0.1 phr of DTBP led to the co-continuous phase morphology of PLA/PBSA blends. The per cent crystallinity of PLA increased when blended with PBSA. PBSA might induce the crystallization of PLA.

1. Introduction

In the last two decades, biodegradable plastics based on renewable and non-renewable resources have been recognized as a solution to address the environmental pollution caused by the increase of non-degradable plastic wastes and the threat of oil depletion. Poly (lactic acid) (PLA) is one of the most popular biodegradable plastics being utilized in various applications due to its high tensile strength, good heat stability during processing, an excellent gloss and clarity. However, there are some drawbacks, including its brittleness and poor heat resistance when applied to the temperature above its glass transition temperature.

The simple approach to overcome the brittleness of PLA is the blending with others ductile biodegradable polymers, such as Poly (butylene succinate) (PBS), Poly (butylene succinate co adipate) (PBSA), Poly (caprolactone) (PCL), Poly (butylene adipate co butylene terephthalate) (PBAT). These choices of polymers have been readily studied. An important factor in polymer blends is the compatibility. Unfortunately, the above mentioned polymers and PLA are incompatible. PBSA is a promising biodegradable polymer because of its lower crystallinity and more polymer chain flexible so it is appropriate for various applications [1]. PBSA is commercially available aliphatic polyester

[3] Address for correspondence: S Suttireungwong, Department of Materials Science and Engineering, Faculty of Engineering and Industrials Technology, Silpakorn University, Sanamchandra Palace Campus, Nakhon Pathom 73000, Thailand. E-mail: suttireungwong_s@su.ac.th.

with high flexibility, excellent impact strength, melt processibility, thermal, chemical resistance. Besides, PBSA has a low melting point of 90°C, which is more readily bio-degradable than others [2]. However, blending PLA with PBSA is not compatible so it is necessary to improve the compatibility between two phases. Sangmook Lee and Jae Wook Lee [2] investigated thermal properties and mechanical properties of binary blend of PLA and PBSA. They found that polymer blends had lower thermal stability than pure polymers. The modulus and tensile strength of polymer blends decreased when the content of PBSA increased. On the other hand, the impact strength of polymer blends was seriously increased higher than the rule of mixing.

Compatibilization through the in stitu formation of compatibilizer in polymer blends has become increasingly important alternative to replace the method of adding block or graph copolymers [3]. Vincent Ojijo et al. [4] blended PLA/PBSA in a presence of triphenyl phosphide (TPP); they found that the torque values were increased after adding the TPP. When increased the TPP content, the dispersed phase size was reduced and PLA and PBSA were linked so they were improved the compatibility. Deyun Ji et al. [5] prepared the polymer blends of PLA/PBS in a reactive blending by using dicumyl peroxide (DCP) as a reactive agent. Some cross linked/blanched structures were formed according to the rheological measurement and gel fraction results. PLA-PBS copolymers of the reactive acted as a compatibilizer for PLA and PBS phase so the compatibility between PLA and PBS phase were improved [5].

The blended between PLA and PBSA is incompatible so the properties of PLA/PBSA blends could be deteriorated [4]. From the previous research, adding additives (e.g. triphenyl phosphide, dicumyl peroxide) could improve the compatibility and dispersity of polymer blends [4, 5]. Moreover there is no research that used the peroxides as the reactive agent for PLA/PBSA blends and it was found that PBS which had the same functional groups as PBSA could react with peroxide so it might indicate that PBSA can also reacts with peroxides. Thus the properties of PLA/PBSA might improve with a presence of Peroxides. The effect of Di (tert-butylperoxyisopropyl) benzene (DTBP) to PLA/PBSA blends has not been reported. Therefore, in this study, it aimed to study the properties of polymer blends (PLA/PBSA) in non-reactive and reactive blends. Peroxide (DTBP) was used as a reactive agent for PLA/PBSA blends. The mechanical, rheological, thermal, and morphological of non-reactive and reactive polymer blends were investigated.

2. Experimental

2.1. Materials

Poly (lactic acid) (PLA) was purchased from Nature Works, LLC, USA and Poly (butylene succinate co adipate) (PBSA grade; Bionolle® 3001D) was purchased from Showa High Polymer Co., Ltd., Japan. Both of polymers were in the form of pellet dried at 60°C for 5 hours before processing. Di (tert-butylperoxyisopropyl) benzene (DTBP) (Perkadox14s) having 40% of peroxide and 2, 5-Dimethyl-2, 5-(t-butylperoxy) hexane (DTBH) (Luperox 101 XL45) having 50% of peroxide were purchased from Akzo Nobel and Arkema Inc., respectively.

2.2. Melt-blend of PLA or PBSA with reactive agents

To investigate the reactivity between polymers and two types of peroxides (DTBP and DTBH), PLA and PBSA were melt-blended with each peroxide at 190° in an internal mixer with screw speed 60 rpm. The torque values were detected in 10 minutes. The concentrations of peroxide varied from 0.1 to 1 phr.

2.3. Preparation of polymer blends.

The ratios of PLA: PBSA 80:20, 60:40, 50:50, 40:60, and 20:80 were weighted and melt-blended in a twin screw extruder. The temperature profile from feeding zone to die were 130/140/150/160/170/180/190°C and the screw speed was set at 80 rpm. The reactive blends were prepared by blending both of polymers with peroxide selected from 2.2, in this case0.08 and 0.1 phr of DTBP were used in the twin screw extruder with the same temperature profile as non-reactive PLA/PBSA blends. The symbol PBSA20 referred to 20%wt PBSA in the blend. The concentrations of Di (tert-butylperoxyisopropyl) benzene (DTBP) and 2, 5-Dimethyl-2, 5-(t-butylperoxy) hexane (DTBH) were written as Pxx and Lxx respectively (where xx is the concentration).

2.4. Melt flow index
Melt Flow Index (MFI) of polymer blends and neat polymers were investigated according to ASTM D1238. PLA/PBSA blends were characterized at 190°C, using 2.8 kN of load.

2.5. Fourier Transform Infrared Spectroscopy(FT-IR)
Fourier transformed infrared spectra of samples were detected over the range 400-4000 cm^{-1}, using Vertex70, Bruker, Germany. The samples were mixed with KBr and laminated.

2.6. Morphology
The morphology of non-reactive and reactive polymer blends was studied by Scanning Electron Microscope (SEM, made by Hitachi). The specimens were prepared from fractured surface of the impact testing specimens.

2.7. Mechanical properties
The modulus, elongation at break, and tensile strength of polymer blends were determined using Instron Crop, model 5969 with a load cell of 5 kN, according to ASTM D638-typeIV with rate at 10 mm/min at room temperature. For Impact testing, the specimens were tested according to ASTM D256 and using RESIL IMPACTOR, PIANEZZA-TORINO, Italy with a load 2.75 Jules.

2.8. Thermal Properties
Differential Scanning Calorimeter (DSC) was used to investigate the glass transition temperature and melting temperature of polymer blends. The samples were tested in the heat-cool-heat mode under the nitrogen atmosphere, with the heating and cooling rate at 5°C/min. The temperature was scanned from -60° to 200°C. The crystallinity of polymer blends were calculated from followed equation [6].

$$\%X_c = \frac{(\Delta H_m - \Delta H_{cc})}{\Delta H_f{}^\circ} x \frac{100}{fraction} \quad (1)$$

Where ΔH_m is the enthalpy of fusion and ΔH_{cc} is the enthalpy of crystallization which obtained from the second heating of DSC. $\Delta H_f{}^\circ$ is the enthalpy of completed crystallization and the fraction means PLA or PBSA weight ratios of the polymer blend.

3. Results and Discussions

3.1. Effect of peroxide types on torque values of neat PLA and PBSA
PLA and PBSA were dried-mixed with DTBP (P) and DTBH (L) and transferred to an internal mixer at the same mixing temperature and rotor speed to investigate the reactivity of each polymer and reactive agent. Figure 1 showed the torque values of PLA after adding DTBP and DTBH, The torque values of PLA increased when the concentration of DTBP and DTBH reached 1 phr. For the same amount of peroxides, DTBP was more reactive with PLA than DTBH in agreement with final torques. Torques of PBSA mixed with peroxides was shown in figure 2, the torque values of PBSA increased with increased amount of peroxide. At 0.1 phr of peroxides, the torque value was lowest and it was highest when added 1 phr of peroxides for both DTBP and DTBH. The increase in torque values indicated an increase in melt viscosity. It could be speculated that there were possible crosslinking reactions during the melt-blending [5]. When compared DTBP and DTBH, DTBP was more reactive with PBSA than DTBH due to the higher final torque value as seen in figure 2. Thus, 0.1 phr of DTBP was selected as a reactive agent of PLA/PBSA blends.

3.2. PLA/PBSA blends

3.2.1. Melt Flow Index (MFI). Table 1 showed Melt Flow Index (MFI) of polymer blends in non-reactive and reactive blends. The MFI of the blends lied between neat PLA and PBSA as shown in table 1. After adding DTBP at 0.08 and 0.1 phr, MFI of reactive polymer blends decreased, implying the increase in viscosity of polymer blends. Peroxides could act as the crosslink promoter as observed in the increase in torque values.

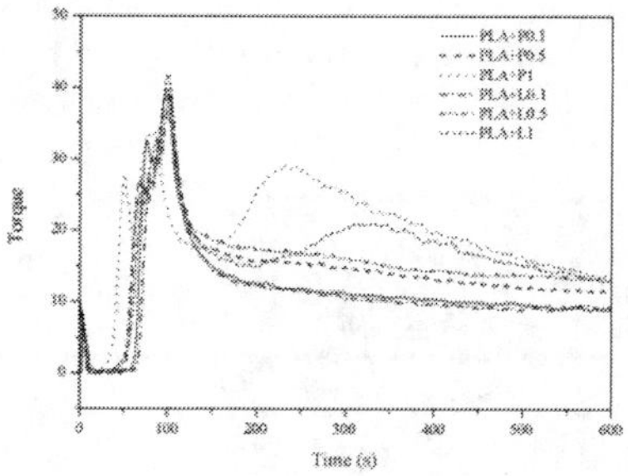

Figure 1. Torque-time of PLA mixed with peroxides (DTBP (P), DTBH (L)) with various contents.

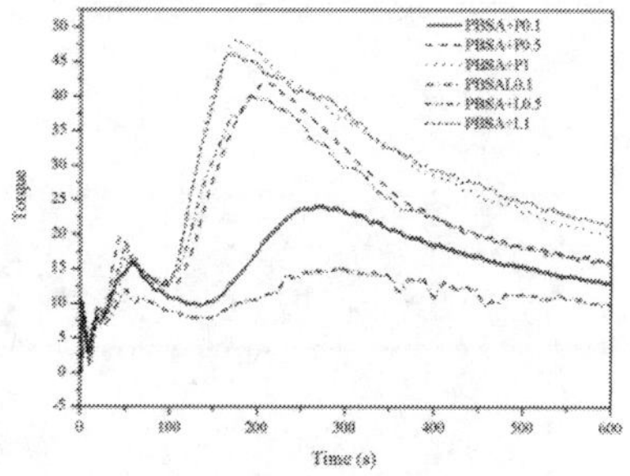

Figure 2. Torque –time of PBSA mixed with peroxides (DTBP (P), DTBH (L)) various contents.

Table 1. Melt flow index of polymer blends.

Formulation	MFI(g/10min)		
	non-reactive	DTBP 0.08phr	DTBP 0.1phr
Neat PLA	4.13±0.33	-	-
PBSA20	3.92±0.47	3.24±0.29	2.58±0.37
PBSA40	4.12±0.59	1.88±0.23	1.50±0.33
PBSA50	3.89±0.48	1.55±0.14	0.96±0.31
PBSA60	3.46±0.45	1.09±0.17	0.65±0.19
PBSA80	3.92±0.39	0.46±0.20	0.25±0.09
Neat PBSA	2.55±0.18	-	-

3.2.2. Fourier Transformation Infrared Spectroscopy (FT-IR). As seen in the figure 3, FT-IR spectra of PLA showed peaks at 2997.29 and 2945.91 cm^{-1} which was corresponded to $–CH_2$ stretching vibration and peaks at 1752.92 cm^{-1} and 1183.08 cm^{-1} were assigned to –C=O and C-O stretching vibration respectively. FTIR spectra of PBSA located at 1744.42 and 1181.73 cm^{-1} were assigned to C=O and C-O respectively. For the blends, the characteristic peaks of PLA and PBSA were appeared about 2900, 2800, and 1700 cm^{-1} respectively. Besides, it was found that the peaks of $–CH_2$, C=O, and C-O of blends in a presence of DTBP shifted towards lower wavenumber. It might suggest that the interaction between PLA and PBSA existed upon blending. However, very low degree of possible crosslinking reaction was difficult to be detected by FTIR due to peaks overlapping.

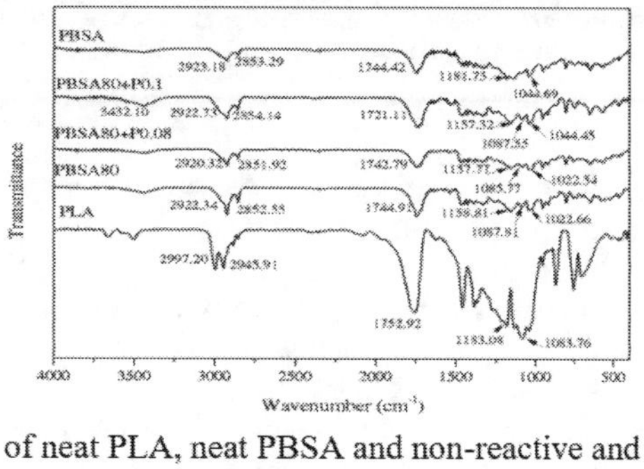

Figure 3. FT-IR spectra of neat PLA, neat PBSA and non-reactive and reactive PLA/PBSA blends.

3.2.3. Morphology. Figure 4 showed SEM micrographs of non-reactive and reactive polymer blends from the fractured surface of impact specimens. Without adding DTBP, as shown in figures 4(a-e), the dispersed phase particles were larger in sizes (the particle size was 30 microns) and the particle sizes of dispersed phase increased with PBSA contents. It indicated that PLA and PBSA in non-reactive blends were not compatible. However, at 80%wt of PBSA, the separation between two phases was not clearly observed. When adding 0.08 and 0.1 phr of DTBP, the morphology of the blends changed as seen in figures 4 (a'-e') and figures 4 (a''-e''). It was found that co-continuous phase occurred upon adding the DTBP in 40%wt of PBSA. In this case, PBSA became the matrix because the addition of peroxides resulted in higher viscosity of PBSA as compared to PLA so PLA was sheared easily and became the small dispersed phase. The increase of peroxides caused the better compatibility between PLA and PBSA.

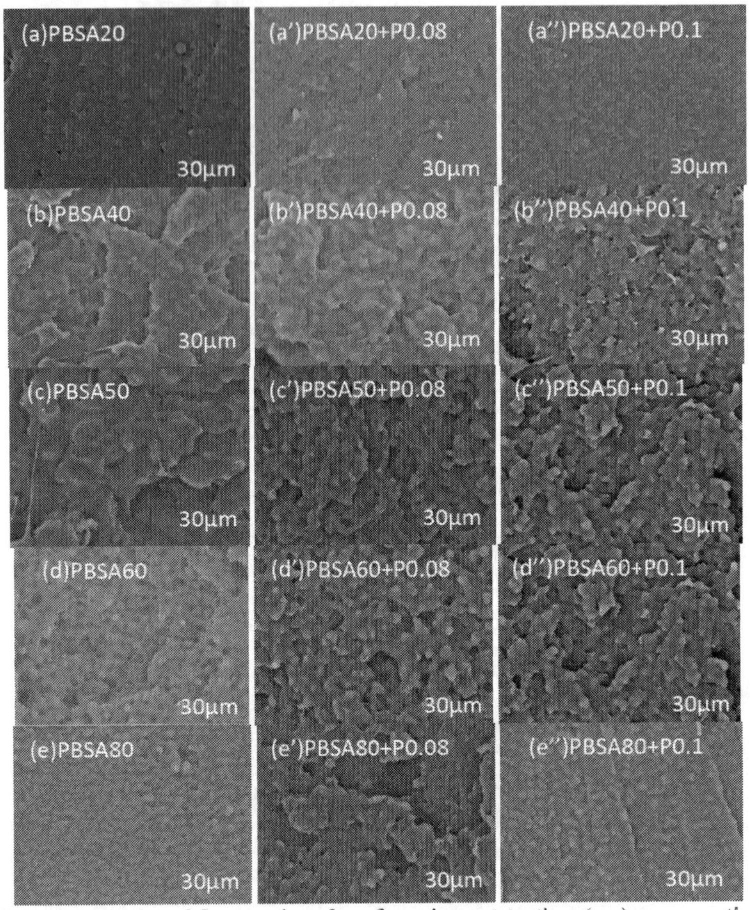

Figure 4. SEM micrographs of fractured surface from impact testing (a-e) non-reactive PLA/PBSA blends, (a'-e') adding 0.08 phr of DTBP, and (a''-e'') P LA/PBSA with adding 0.1 phr 0f DTBP. (Under 2000x magnification).

3.2.4. Mechanical properties. As seen in figure 5, the impact strengths of non-reactive polymer blends remained almost unchanged. When PBSA content was 80%wt, the impact strength increased dramatically and was even higher than the line of mixing rule. It indicated that there were better compatibility and dispersity in PLA/PBSA blends as seen in SEM micrographs (discussed in 3.2.3). The improvement of impact strength of polymer blends could be more pronounced when adding 0.08 and 0.1 phr of DTBP. Especially, when adding 0.1 phr of DTBP for 40%wt. of PBSA, it was found that the impact strength reached 35 kJ/m^2 (7 times higher than that of non-reactive blends at the same ratio). DTBP acted as an initiator for the possible partial crosslink reaction between PLA and PBSA, hence improving interfacial adhesion between the polymers. The impact strength increased with PBSA contents. At 40%wt. of PBSA with 0.1 phr of DTBP, the inversion phase might occur as the PBSA had higher viscosity than PLA and could become the matrix. Thus, PBSA40 with a presence of 0.1 phr of DTBP had the similar impact strength as neat PBSA and its impact strength was higher than line of mixing rule.

The modulus of polymer blends decreased when PBSA contents increased as seen in figure 6 and the tensile strength showed the similar trends (figure 7) because PBSA is tougher than PLA. In addition, it was found that adding 0.08 and 0.1 phr of DTBP caused the modulus and tensile strength of polymer blends to drop. The elongation at break of polymer blends increased when PBSA contents increased as shown in figure 8. For both non-reactive and reactive blends, the elongations at break were higher than the line of mixing rule. This implied that PBSA could improve the brittleness of PLA because of the elastic characteristic in PBSA. However, it was found that the elongation at break of polymer blends slightly decreased when adding 0.1 phr of DTBP.

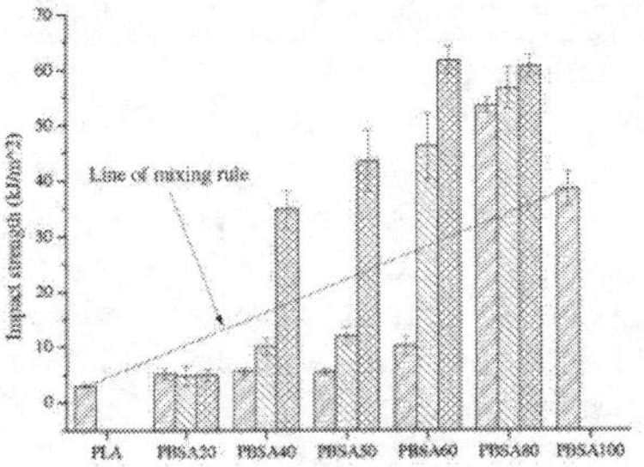

Figure 5. Impact strength of polymer blends.

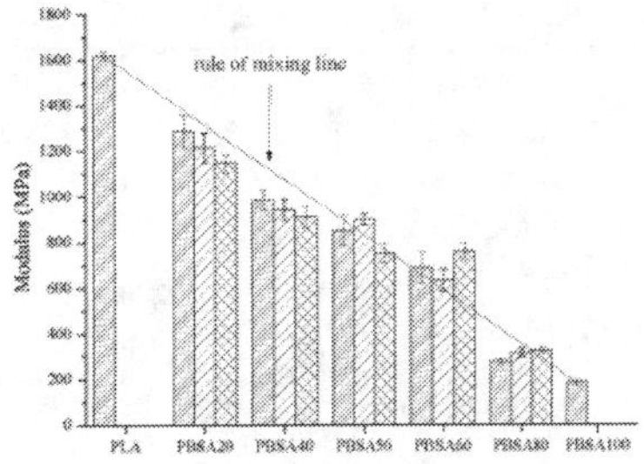

Figure 6. Modulus of polymer blends.

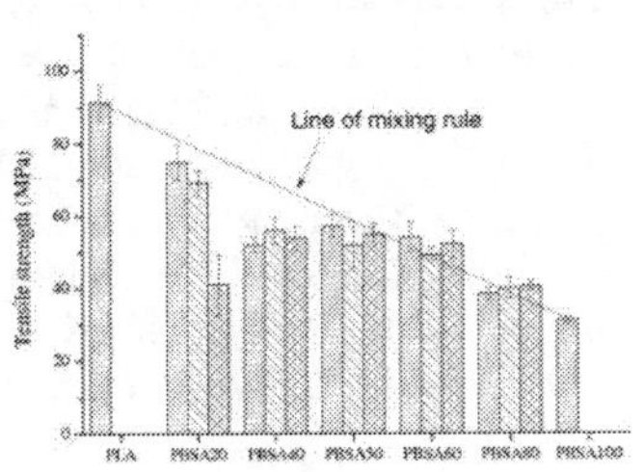

Figure 7. Tensile strength of polymer blends.

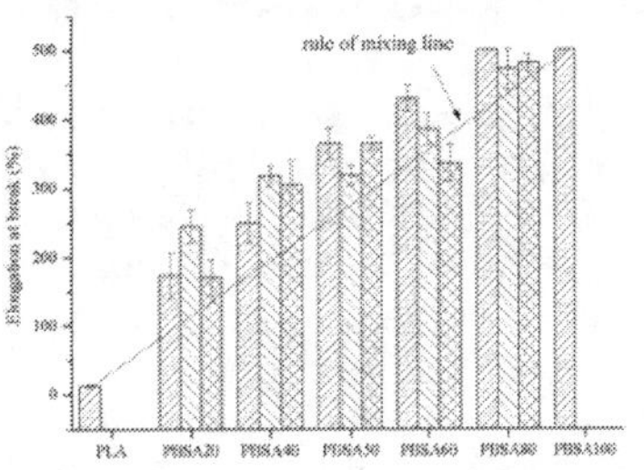

Figure 8. Elongation at break of polymer blends.

Non-reactive, 0.08 phr, and 0.1phr of DTBP

Table 2. Thermal properties of polymer blends obtained from 2nd heating of DSC thermograms.

Formulation	T_g (°C)		T_c (°C)		T_m (°C)				H_m(J/g)		%X_c	
	PBSA	PLA	PBSA	PLA	PBSA		PLA		PBSA	PLA	PBSA	PLA
					T_{m1}	T_{m2}	T_{m1}	T_{m2}				
neat PLA	-	61.91	-	-	-	-	147.4	153.1	-	28.26	-	4.108
PBSA20	-50.69	60.6	-	-	-	91.57	146.6	152.6	11.09	27.48	39.05	19.44
PBSA40	-44.85	62.03	59.45	-	82.33	92.28	146.9	152.5	16.39	17.6	28.86	12.22
PBSA50	-44.3	60.37	59.09	-	82.06	92.52	147.3	152.7	17.96	12.18	25.3	5.763
PBSA60	-42.77	59.76	59.26	-	81.98	92.5	147.5	-	23.09	10.98	27.1	7.5
PBSA80	-43.16	56.89	59.42	-	81.84	92.27	146.4	151.6	26.45	5.71	23.28	6.129
PBSA20+P0.08	-	60.45	65.49	-	-	91.06	145.2	152.9	10.29	26.53	36.23	18.8
PBSA40+P0.08	-45.63	60.24	64.73	-	85.05	91.62	146.1	152.9	14.47	17.94	25.48	11.52
PBSA50+P0.08	-43.26	60.44	65.73	-	86.18	92.09	146.4	153.3	17	11.93	23.94	7.806
PBSA60+P0.08	-44.21	59.67	65.67	-	85.21	92.11	146.1	153.1	20.86	14.02	24.48	11.18
PBSA80+P0.08	-43.42	-	68.43	-	-	88.11	144.7	152.3	46.47	4.66	40.91	8.118
PBSA20+P0.1	-	60.42	65.79	-	-	91.04	145.7	153.1	6.63	21.84	23.35	11.68
PBSA40+P0.1	-45.33	60.33	65.38	-	85.23	91.48	146	152.9	15.64	15.63	27.54	7.258
PBSA50+P0.1	-43.86	60.01	67.04	-	87.18	-	145.9	153	17.78	9.95	25.04	6.473
PBSA60+P0.1	-43.68	60.16	67.25	-	87.95	-	145.7	152.9	19.83	7.59	23.27	8.161
PBSA80+P0.1	-43.16	-	68.57	-	88.95	-	144.8	152.3	25.69	4.04	22.61	9.14
neat PBSA	-43.16	-	60.04	-	82.98	92.58	-	-	51.8	-	36.48	-

*Heat of fusion of PLA and PBSA are 93 J/g [6] and 142 J/g [7], respectively.

3.2.5. Thermal properties. Table 2 showed the results of DSC from second heating scans. It was found that the glass transition temperature of PBSA and PLA were about -43°C and 61°C respectively and there were no differences of the results after blending them together. It suggested that PLA/PBSA blends were immiscible. When considering the cool crystallization temperatures (T_c), T_c of all blends with 0.08 and 0.1 phr increased, especially when PBSA contents increased. However, PLA did not show the cool crystallization temperature because of its slow crystallization rate. The melting temperature(T_m) of PBSA was 92°C. After PBSA was blended with PLA, there were no differences of T_m between pure polymer and polymer blends. Moreover, PBSA showed two peaks of melting temperature corresponding to re-melting of newly formed crystalline [2]. PLA showed T_m which remained constant for both non-reactive and reactive blends. The addition of DTBP did not affect the melting temperature of polymer blends. The percent crystallinity (%X_c) of PLA in non-reactive blends was found to be increased. It might be due to the fact that PBSA was able to crystallize easily so it might induce the crystallization of PLA. Adding 0.08 and 0.1 phr of DTBP caused the decrease of the percent crystallinity of PLA in PLA/PBSA blends because possible crosslink might hinder the crystallization of PLA. The percent crystallinity of PBSA could not be accurately measured because the heat of fusion of PBSA overlapped with the heat of cold crystallization of PLA.

4. Conclusion

In this research, it aimed to study the properties of polymer blends in non-reactive and reactive blending. The effect of two types of peroxides and amount of peroxides on torque values was analyzed and it was found that DTBP was more reactive with PLA and PBSA than DTBH under specified blending condition as DTBP revealed higher torque value. For the impact testing, adding 0.1 phr of DTBP into PBSA40 could increase the impact strength dramatically. The interfacial adhesion and viscosity of PLA/PBSA blends could be improved. SEM micrographs showed the morphology close to co-continuous phase with a presence of DTBP. Furthermore, the elongation at break of polymer blends was improved when PBSA contents increased for both non-reactive and reactive blends. For thermal properties, blending the PLA with PBSA could increase the percent crystallinity of PLA. The percent crystallinity of PLA in reactive blends decreased when compared with non-reactive blends for all ratios.

Acknowledgments

Authors would like to thanks the Department of Materials Science and Engineering, Faculty of engineering and industrial, Silpakorn University for financial support and Nation Metal and Materials Technology Center (MTEC), Thailand for impact test measurement.

References

[1] Eslami H and Kamal M 2011 *App. Polym. Sci.* **127** 2290-306
[2] Lee S and Lee W 2005 *Korea-Aust. Rheol. J.* **17** 71-7
[3] Mani R, Bhattchara M and Tang J 1998 *J Polymer Sci Polymer Chem* **37** 1693-720
[4] Ojijo V, Ray S and Sadiku R 2013 *App. Mat. Interf.* **5** 11-7
[5] Ji D 2013 *App. Polym. Sci* **10** 1-8
[6] Rungsima H and Nattakarn H 2013 *Energy Procedia* **34** 871-9
[7] Available from: www.showadenko.us/file/bionolle/Bionelle-Flyer.pdf

IOP Conf. Series: Materials Science and Engineering **87** (2015) 012071 doi:10.1088/1757-899X/87/1/012071

A phenomenological intra-laminar plasticity model for FRP composite materials

Yinhua Zhou[1], Chi Hou, Wenzhi Wang, Meiying Zhao and Xiaopeng Wan

School of Aeronautics, Northwestern Polytechnical University, Xi'an, China

E-mail: zhouyh_123@163.com

Abstract. The nonlinearity of fibre-reinforced polymer (FRP) composites have significant effects on the analysis of composite structures. This article proposes a phenomenological intra-laminar plasticity model to represent the nonlinearity of FRP composite materials. Based on the model presented by Ladeveze et al., the plastic potential and hardening functions are improved to give a more rational description of phenomenological nonlinearity behavior. A four-parameter hardening model is built to capture important features of the hardening curve and consequently gives the good matching of the experiments. Within the frame of plasticity theory, the detailed constitutive model, the numerical algorithm and the derivation of the tangent stiffness matrix are presented in this study to improve model robustness. This phenomenological model achieved excellent agreement between the experimental and simulation results in element scale respectively for glass fibre-reinforced polymer (GFRP) and carbon fibre-reinforced polymer (CFRP). Moreover, the model is capable of simulating the nonlinear phenomenon of laminates, and good agreement is achieved in nearly all cases.

1. Introduction

FRP composite materials are widely used in aircrafts and many other industries, due to their high strength, stiffness and good fatigue resistance. Generally, composite materials exhibit noticeable nonlinear behavior before failure. Experimental investigations reported in public literature shows that many FRP laminates with multiple-angle layups exhibit obvious nonlinear or irreversible deformations before destructive failure. Another well-known feature is that most FRP composite materials show significant nonlinearity in shear. O'Higgins [1] studied the damage initiation and growth in CFRP HTA/6376 and GFRP S2FM94 composite materials. In his research, nonlinear performances were observed, relating to not only the stress-strain response in shear, but also to the tensile responses of laminates with either $[\pm22.5^\circ]_{2s}$, $[\pm35^\circ]_{2s}$, $[\pm55^\circ]_{2s}$ angle-ply layups or $[45^\circ]_8$ off-ply layup. H. Koerber [2] found similar phenomena in the off-ply laminates of carbon-epoxy GFRP IM7-8552. T. J. Vogler [3] also observed nonlinear behavior in the shear and transverse compression of unidirectional AS4/PEEK and investigated their interaction both experimentally and numerically.

Plasticity is not the only effect concerning nonlinearity. Additionally, material continuum degradation due to matrix micro-cracks or fibre/matrix debonding is another important feature inducing a soft decline; it is phenomenologically the same as nonlinearity. Consequently, many inelastic models have been proposed. Ladeveze [4] presented a widely used two-dimensional (2D)

[1] Address for correspondence: Yinhua Zhou, School of Aeronautics, Northwestern Polytechnical University, Xi'an, China. E-mail: zhouyh_123@163.com.

inelastic model including the combination of plasticity and the continuum damage mechanics (CDM) model. This model performed very well in simulations of composite structures, such as jointed fibre metal laminate structures [5] and composite bolted joints. More complexly, Comanho et al. [6] proposed a new three-dimensional (3D) transversely isotropic elastic-plastic constitutive model for unidirectional FRPs under the framework of the plasticity theory.

Unlike previous inelastic models, an attempt is made here to simplify the model describing the nonlinearity of composites. Although FRP composite materials generally have obvious non-linear behavior due to many physical reasons, such as damage accumulation resulting from fibre or matrix micro-cracks, fibre/matrix interface debonding and the nonlinearity of each individual constituents [7], disregarding the physical origins of the nonlinearity, a plasticity model is proposed to describe the nonlinear phenomenon. In addition, the hardening function is improved to make the plasticity model more rational.

2. Intra-laminar plastic model

2.1. The constitutive model

Nonlinearity is plastic deformation that occurs under mechanical loading. The model proposed here contains plasticity, so it predicts stress-strain behavior using plastic strain. The constitutive relations are given by:

$$\boldsymbol{\sigma} = \mathbf{C}^e : \boldsymbol{\varepsilon}^e \tag{1}$$

where $\mathbf{C}^e$ is the fourth order tensors of unidirectional composite laminates, and superscript **e** represents elasticity. $\boldsymbol{\varepsilon}^e$ is the elastic strain tensor. $\boldsymbol{\sigma}$ is the stress tensor, and the symbol (:) denotes a double dot product of two tensors.

For the elastic-plastic constitutive model presented in this article, the total strain, $\boldsymbol{\varepsilon}$, is a sum of the elastic strain $\boldsymbol{\varepsilon}^e$ and plastic strain $\boldsymbol{\varepsilon}^p$:

$$\boldsymbol{\varepsilon} = \boldsymbol{\varepsilon}^e + \boldsymbol{\varepsilon}^p \tag{2}$$

The yield domain, F, is then defined as a function of the effective stress:

$$F = F^p(\boldsymbol{\sigma}) - R_0 - R(\tilde{\varepsilon}_p) \tag{3}$$

where $F^p(\boldsymbol{\sigma}) = \sqrt{\sigma_{12}^2 + a^2\sigma_{22}^2} + bsign(-\sigma_{22})$, and is called the plastic potential. R_0 is the initial threshold, defined as $R(\tilde{\varepsilon}^p = 0)$, and it is evaluated experimentally. $R(\tilde{\varepsilon}^p)$ is the hardening function, and a^2 is the shear-transverse plasticity coupling property. Empirically, the compensate parameter for transverse compressive stresses, b is given as:

$$b = \frac{\sigma_{2t,p}^0}{\left(\sigma_{2t,p}^0 + \sigma_{2c,p}^0\right)} \tag{4}$$

where $\sigma_{2t,p}^0$ and $\sigma_{2t,p}^0$ are the initial yielding stress for transverse tension and transverse compression, respectively.

Hardening is assumed to be isotropic, so $R(\tilde{\varepsilon}^p)$ is obtained using the power-law relation as:

$$R(\tilde{\varepsilon}^p) = R_\infty(1 - \eta e^{\beta\tilde{\varepsilon}^p})(1 - e^{\mu\tilde{\varepsilon}^p}) \tag{5}$$

where β , η and μ are determined to fit the experimental hardening curve. Note that $\tilde{\varepsilon}^p$ is the equivalent effective plastic strain, while $\bar{\boldsymbol{\varepsilon}}^p$ is the effective plastic strain tensor. The effective plastic strain rate $d\bar{\boldsymbol{\varepsilon}}^p$ is given as:

$$d\boldsymbol{\varepsilon}^p = d\lambda \frac{\partial F}{\partial \boldsymbol{\sigma}} \tag{6}$$

Where λ is the plastic multiplier. The principle of plastic work equivalence is defined as [8]:

$$F^p(\boldsymbol{\sigma})d\tilde{\varepsilon}^p = \boldsymbol{\sigma} : d\boldsymbol{\varepsilon}^p \tag{7}$$

Substituting equations (3) and (6) into equation (7), the following relation is derived:

$$d\tilde{\varepsilon}^p = d\lambda \tag{8}$$

2.2. Numerical integration algorithm for plasticity

For a given load step, an elastic predictor stress at the beginning of an increment is used to evaluate whether the current load step is in the elastic or plastic domain. The yield function, F, will be less or equal to zero if no plasticity is occurring during the current increment. If F is greater than zero, the correct plastic strains have to be determined to update stress. The back Euler implicit integration algorithm is employed to compute the plastic strains [7]. The integration scheme is implicit in terms of plasticity parameter $\Delta\lambda_{n+1}$.

$$\boldsymbol{\varepsilon}_{n+1} = \boldsymbol{\varepsilon}_n + \Delta\boldsymbol{\varepsilon}_{n+1} \tag{9-a}$$

$$\boldsymbol{\varepsilon}^p_{n+1} = \boldsymbol{\varepsilon}^p_n + \Delta\boldsymbol{\varepsilon}^p_{n+1} = \boldsymbol{\varepsilon}^p_n + \Delta\lambda_{n+1}\frac{\partial \mathbf{F}_{n+1}}{\partial \bar{\boldsymbol{\sigma}}_{n+1}} \tag{9-b}$$

$$\tilde{\varepsilon}^p_{n+1} = \tilde{\varepsilon}^p_n + \Delta\lambda_{n+1} \tag{9-c}$$

$$\bar{\boldsymbol{\sigma}}_{n+1} = \boldsymbol{\sigma}^{trial} - \Delta\boldsymbol{\sigma} = \mathbf{C}^e : (\boldsymbol{\varepsilon}_{n+1} - \boldsymbol{\varepsilon}^p_{n+1}) \tag{9-d}$$

$$F_{n+1} = F(\bar{\boldsymbol{\sigma}}_{n+1}, \tilde{\varepsilon}^p_{n+1}) \tag{9-e}$$

The trial stress and plastic corrector are both in terms of the undamaged material stiffness tensor. The equation system equation (9) is solved iteratively using the Newton-Raphson algorithm, equations (10) and (11). For the numerical implementation in UMAT, it is assumed that the total strain, $\boldsymbol{\varepsilon}_{n+1}$, remains constant during the iteration, so that the initial conditions include $\Delta\boldsymbol{\varepsilon}_{n+1}$, $\boldsymbol{\varepsilon}_{n+1}$, $\tilde{\varepsilon}^p_n$, $\boldsymbol{\varepsilon}^p_n, \Delta\lambda_0 = 0$. Then, the aim of the Newton-Raphson iteration is to achieve proper $\Delta\lambda_{n+1}$; subsequently, $\tilde{\varepsilon}^p_{n+1}$ and $\boldsymbol{\varepsilon}^p_{n+1}$ make the yield domain, $F(\bar{\boldsymbol{\sigma}}_{n+1}, \tilde{\varepsilon}^p_{n+1}) \le TOL$, where $TOL \le 1\times10^{-5}$ is the error tolerance.

$$F^{(n)} + \left(\frac{dF}{d\Delta\lambda}\right)^{(n)} \delta\lambda^{(n)} = 0 \tag{10}$$

Where

$$\delta\lambda^{(n)} = \Delta\lambda_{n+1} - \Delta\lambda_n \tag{11}$$

2.3. The tangent stiffness matrix

The consistency condition for the yield domain is written as:

$$\frac{\partial F}{\partial \boldsymbol{\sigma}} : d\boldsymbol{\sigma} + \frac{\partial F}{\partial \tilde{\varepsilon}_p} d\tilde{\varepsilon}_p = 0 \tag{12}$$

The elastic strain increment is given by:

$$d\boldsymbol{\varepsilon}^e = d\boldsymbol{\varepsilon} - d\tilde{\boldsymbol{\varepsilon}}_p \tag{13}$$

Subsequently, the stress increment is:

$$d\boldsymbol{\sigma} = \mathbf{D} : d\boldsymbol{\varepsilon}^e \tag{14}$$

Substituting equations (6)-(8) into equation (12), the increment of the plastic multiplier is given as:

$$d\lambda = \frac{\frac{\partial F}{\partial \boldsymbol{\sigma}} : \mathbf{D} : d\boldsymbol{\varepsilon}}{\frac{\partial F}{\partial \boldsymbol{\sigma}} : \mathbf{D} : \frac{\partial F}{\partial \boldsymbol{\sigma}} - \frac{\partial F}{\partial \varepsilon_p}} \tag{15}$$

The tangent stiffness matrix D^{eq} is given by dividing the stress increment and the total strain increment:

$$\mathbf{D}^{eq} = \frac{d\boldsymbol{\sigma}}{d\boldsymbol{\varepsilon}} \tag{16}$$

After substituting equations (7), (8), (13), (14) and (15) into equation (16), the tangent stiffness matrix is:

$$\mathbf{D}^{eq} = \mathbf{D} - \mathbf{D} : \frac{\frac{\partial F}{\partial \boldsymbol{\sigma}} \otimes \frac{\partial F}{\partial \boldsymbol{\sigma}} : \mathbf{D}}{\frac{\partial F}{\partial \boldsymbol{\sigma}} : \mathbf{D} : \frac{\partial F}{\partial \boldsymbol{\sigma}} - \frac{\partial F}{\partial \varepsilon_p}} \tag{17}$$

Where $\otimes$ is the dyadic product.

3. Model verification and simulation results

3.1. Hardening curve

The parameter comparisons of the four-parameter plasticity model proposed in this paper are shown in figure 1. $R_\infty = 73.38$, $\eta = 0.6347$, $\beta = -61.51$ and $\mu = -871.2$ were obtained from data fitting, which matches the HTA/6376 hardening curve very well. For comparison, two additional alternative values for each parameter were given. The hardening stress here corresponds to $R(\tilde{\varepsilon}_p)$ in equation (3), not including the initial hardening stress (R_0). As we can see, each parameter plays its own irreplaceable role in the shape of the hardening curve. Physically, R_∞ is the ultimate hardening stress, which mainly controls the amplitude of the hardening curve. η is more related to the slope of the initial stage, playing the stretching-like role, while β influences the quantity of the hardening curve after the initial stage. Additionally, μ slightly adjusts the incline extent of the first half of the hardening curve.

It can be seen that the four-parameter model is capable of giving a more rational description of the hardening stress curve. Each parameter has its own role, which increases not only the understanding of the hardening curve, but also of the data fitting.

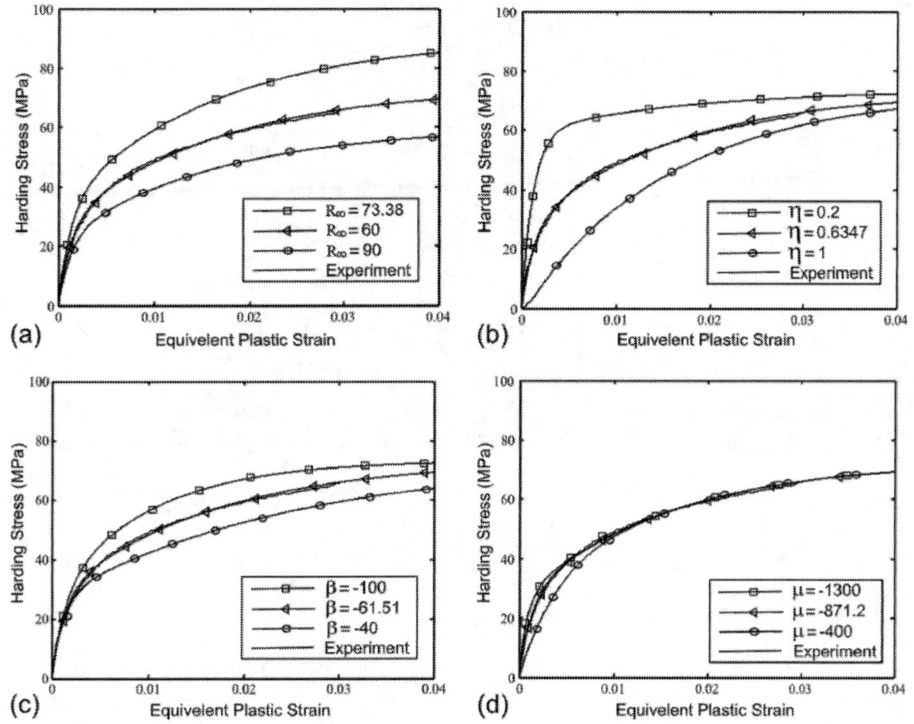

Figure 1. Parameter comparison for hardening model: (a) R_∞ (b) η (c) β (d) μ.

3.2. Stress-strain response

To validate the proposed plasticity model, comparisons were carried out at the element scale. Stress-strain curves from two advanced FRP composite material systems for glass, GFRP E-glass/MY750/HY917/DY063 [9] and CFRP HTA/6376 [1], were compared with the experiment's method.

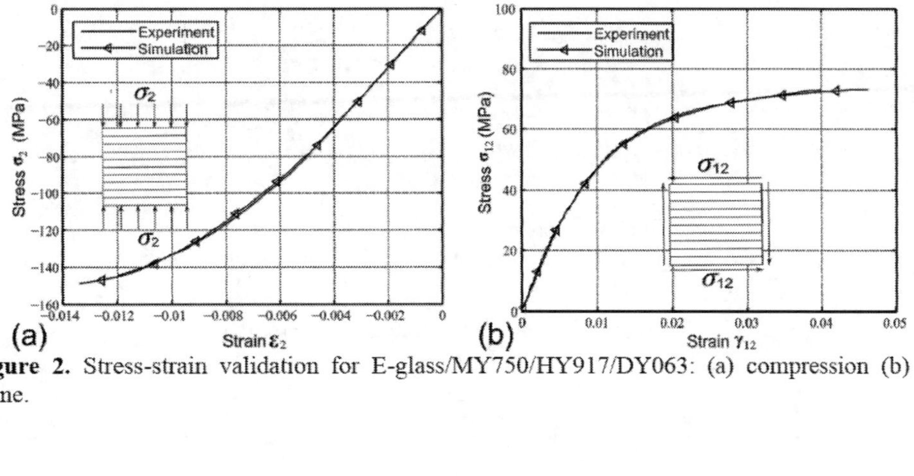

Figure 2. Stress-strain validation for E-glass/MY750/HY917/DY063: (a) compression (b) in-plane.

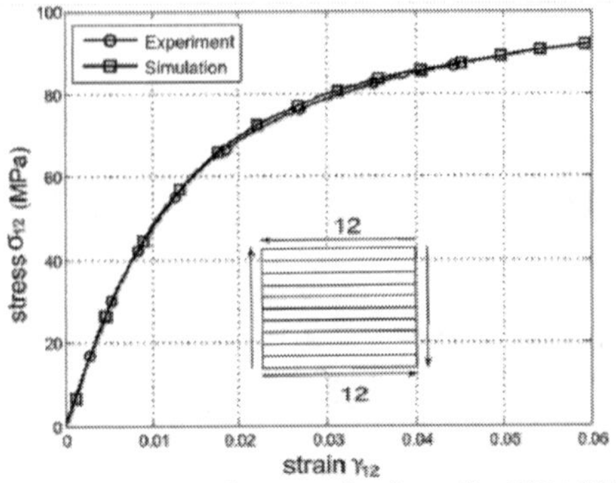

Figure 3. Stress-strain curve in shear for HTA/6376.

The numerical and experimental results of E-glass/MY750/HY917/DY063 (shown in figure 2) exhibit significantly nonlinear behavior under the transverse compression and in-plane shear. Figure 3 shows the plasticity hardening process in the shears of HTA/6376. The above simulation results in elementary scale match the experiment remarkably. This indicates that the plasticity model, including the four-parameter hardening model and the improved plastic potential, has the capability to simulate nonlinearity correctly.

3.3. Intra-laminar nonlinear response of laminates

3.3.1. Model details. Four types of selected laminate lay-ups were analyzed under uniaxial tensile load, which were respectively $[\pm 35^{\circ}]_{2s}$, $[45^{\circ}]_{8}$, $[\pm 55^{\circ}]_{2s}$ and $[\pm 67.5^{\circ}]_{2s}$ with a nominal ply thickness of 0.13 mm. Specially, $[45^{\circ}]_{8}$ is an off-axis layup. The typical stress-strain comparisons were made between experiments and simulations for these four laminates.

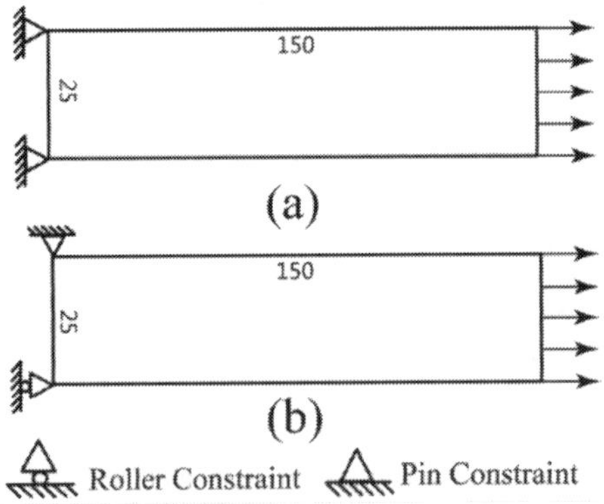

Figure 4. Schematic of specimen geometry and model constraints: (a) angle-ply (b) off-ply.

The axial stress in each of laminate specimen was calculated according to:

$$\sigma_x = \frac{P}{wt} \tag{18}$$

where P is the applied load, w is the measured specimen width, and t is the measured specimen thickness.

The axial strain in each of the laminates was calculated in the experiment as:

$$\varepsilon_x = \frac{\delta}{l_g - l_{off}} \tag{19}$$

where δ is the recorded extensometer displacement in the experiments, l_g is the extensometer gauge length, and l_{off} is the gauge length offset value recorded when the applied load was zero. In simulations, the axial strain is given as:

$$\varepsilon_x = \frac{u_x}{l} \tag{20}$$

where u_x is the applied displacement, and l is the geometric length of the finite element model in the loading direction.

As shown in figure 4, because of the asymmetry of off-ply laminates, the constraints for the finite element model of off-ply laminates is different from those of the angle-ply laminates. The constraint end of the angle-ply laminates is held stationary, while for off-ply laminates, the constraint in the vertical direction of the loading is freed at the holding end.

3.3.2. Simulation results. All simulations were implemented using implicit finite element model formulation. Abaqus (TM) solid elements C3D8R were chosen for the discretization. The minimum size of the mesh is 0.5 mm×0.5 mm×0.13 mm, a relatively fine mesh that is able to minimize mesh dependence. Since this study focuses on the plasticity of composites, laminate failure is not considered. However, the failure point is marked, as shown in figure 5.

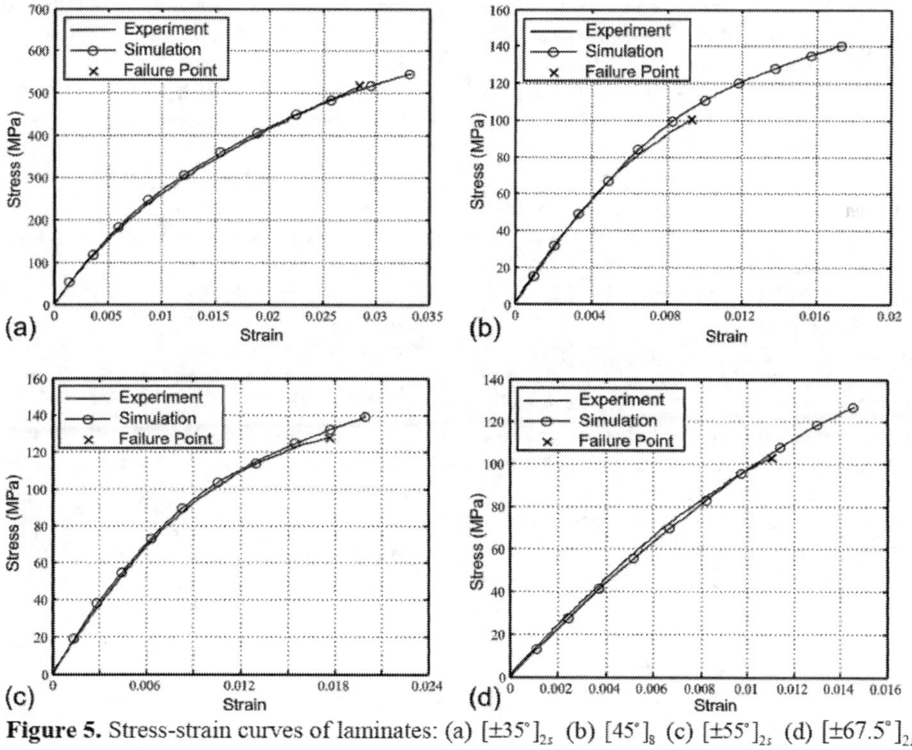

Figure 5. Stress-strain curves of laminates: (a) $[\pm35^\circ]_{2s}$ (b) $[45^\circ]_8$ (c) $[\pm55^\circ]_{2s}$ (d) $[\pm67.5^\circ]_{2s}$.

All four laminates exhibit obvious nonlinearity, as seen in figure 5. The $[\pm67.5^\circ]_{2s}$ laminate does not show as much plasticity as the other three laminates, due to the relatively higher shear stresses of the other three laminates. This indicates that shear nonlinearity plays a more important role in the CFRP HTA/6376 material system. There was very good agreement between experimental data and simulations in nearly all cases (figure 5). In only the 45° off axis tensile test, hardening behavior was slightly underestimated.

4. Conclusion
The proposed plasticity model is capable of simulating the nonlinearity of FRP composite materials under the frame of plasticity theory. Hardening behavior was successfully represented by the four-parameter hardening function, of which each parameter makes significant contributions. With the robustness added by the numerical algorithm and the detailed derivation of the tangent stiffness, the plasticity model in simulations achieved very good agreement with experiments.

Acknowledgment
The authors wish to acknowledge the financial support from the China Scholarship Council (CSC); the provision of research facilities from the Materials and Surface Science Institute (MSSI) at the University of Limerick are also gratefully appreciated.

References
[1] O'Higgins R M 2007 *An Experimental and Numerical Study of Damage Initiation and Growth in High Strength Glass and Carbon Fibre-Reinforced Composite Materials* (University of Limerick: College of Engineering)
[2] Koerber H, Xavier J and Camanho P P 2010 High strain rate characterisation of unidirectional carbon-epoxy IM7-8552 in transverse compression and in-plane shear using digital image correlation *Mechanics of Materials* **42** 1004-19
[3] Vogler T J and Kyriakides S 1999 Inelastic behavior of an AS4/PEEK composite under combined transverse compression and shear. Part I: experiments *International Journal of Plasticity* **15** 783-806
[4] Ladeveze P and E Ledantec 1992 Damage modeling of the elementary ply for laminated composites *Composites Science and Technology* **43** 257-67
[5] Frizzell R M 2010 *An Experimental and Numerical Investigation of Damage in Jointed Fibre Metal Laminate Structures* (University of Limerick)
[6] Vogler M, Rolfes R and Camanho P P 2013 Modeling the inelastic deformation and fracture of polymer composites - Part I: Plasticity model *Mechanics of Materials* **59** 50-64
[7] Chen J F, Morozov E V and Shankar K 2012 A combined elastoplastic damage model for progressive failure analysis of composite materials and structures *Composite Structures* **94** 3478-89
[8] Safaei M, et al. 2013 Evaluation of anisotropic constitutive models: Mixed anisotropic hardening and non-associated flow rule approach *International Journal of Mechanical Sciences* **73** 53-68
[9] Soden P D, Hinton M J and Kaddour A S 1998 Lamina properties, lay-up configurations and loading conditions for a range of fibre-reinforced composite laminates *Composites Science and Technology* **58** 1011-22

IOP Conf. Series: Materials Science and Engineering **87** (2015) 012072 doi:10.1088/1757-899X/87/1/012072

Synthesis of electroactive tetraaniline grafted polyethylenimine for tissue engineering

Shilei Dong[1], Lu Han[1,3], Muhang Cai[1], Luhai Li[1] and YanWei[1,2]

[1]BeijingEngineering Research Center of Printed Electronics, Beijing Institute of Graphic communication, Beijing 102600, China
[2]Department of Chemistry and Key Laboratory of Bioorganic Phosphorus Chemistry and Chemical Biology (Ministry of Education), Tsinghua University, Beijing 100084, China

E-mail: hanlu@bigc.edu.cn

Abstract. Tetraaniline grafted polyethylenimine (AT–PEI) was successfully synthesized in this study. Proton Nuclear Magnetic Resonance (1H NMR) Spectroscopy was used to determine the structure of carboxyl-capped aniline tetramer (AT-COOH) and AT–PEI. UV-Vis spectroscopy and Fourier transform infrared (FT-IR) spectroscopy were employed to characterize the absorption spectrum of the obtained AT–PEI samples. The morphology of AT–PEI copolymers in aqueous solution was determined by Scanning electron microscope (SEM). Moreover, AT–PEI copolymers demonstrated excellent solubility in aqueous solution and possessed electroactivity by cyclic voltammogram (CV) curves, which showed its potential application in the field of tissue engineering.

1. Introduction

Polyaniline is widely used in many fields such as metal anti-corrosion [1], gas sensor [2], and artificial muscle [3], owing to its good conductivity and reversible reoxidation. However, due to its rigid structure, polyaniline is generally insoluble in aqueous solution, which limits its practical application [4, 5]. As compared to polyaniline, aniline tetramer has well-defined structure, homogeneous molecular weight and various end groups for easy functionalization [6-8]. What is more important, it reserves almost the same electrical conductivity as polyaniline [9]. Recently, lots of literatures reported that the solubility, biocompatibility and biodegradability of aniline tetraaniline were dramatically improved by grafting with biocompatible molecule [10]. Wei and his colleagues [11] prepared electroactive and biocompatible triblock copolymer, i.e. (poly (L-glutamicacid)-graft-tetreaniline/poly (L-lysine)-graft-tetreaniline)$_n$, (PGA-g-TA/PLL-g-TA)$_n$, which could be used for bone tissue scaffold. Besides, they synthesized electroactive copolymer, i.e. PEG-aniline tetramer, which self-assembled into micelles in aqueous phase [12]. In addition, Ma [13] et al. fabricated nanofibrous electroactive scaffolds by grafting aniline tetramer to chitosan and using electricity spinning.

Polyethylenimine (PEI) is a cationic polymer containing amine groups, which can deliver genes into cells via endocytosis [14, 15]. Herein, aniline tetramer (AT) was conjugated with PEI (M_W=600),

[3] Address for correspondence: Lu Han, BeijingEngineering Research Center of Printed Electronics, Beijing Institute of Graphic communication, Beijing 102600, China. E-mail: hanlu@bigc.edu.cn.

which not only improve the solubility of AT, but also endow the synthesized copolymer, AT-PEI, with electroconductibility and gene transfection ability. Thus AT-PEI could be blended with hydrogel to form 3D scaffolds for facilitating gene delivery efficiency inside stem cells. In addition, Wei et al. [16] founded that, the poly (esteramide) s-graft-tetraaniline copolymers (PEA-g-TA) could promote adhesion and proliferation of mouse preosteoblastic MC3T3-E1 cells exposed to electrical stimulation. Therefore, AT-PEI may have the great potential application in stem cell gene therapy for bone and cartilage tissue defect.

2. Experimental section

2.1. Materials
N-phenyl-1, 4-phenylenediamine (aniline dimmer, AD, GR), succinic anhydride (GR) were Sigma-Aldrich (Steinheim, Germany). PEI (Mw=600, AR), N-(3-Dimethylaminopropyl)-N-ethylcarbodiimide hydrochloride (EDC·HCl, AR), N-hydroxysuccinimide (NHS, AR) were obtained from Aladdin Reagent Company (China). Dimethylformamide (DMF, AR), Dimethyl Sulphoxide (DMSO, AR) and Ammonium persulfate (AR) were received from Shanghai Reagent Corporation (China).

2.2. Synthesis of AT-COOH and AT-PEI copolymers
AT-COOH was prepared according to the reported method [17] with minor modification. Briefly, AD (1.84 g, 0.01 mol) was allowed to react with succinic anhydride (1.05 g, 10.5 mM) in DCM to obtain AD-COOH (figure 1). 0.01 mol AD and 0.01mol AD-COOH were dissolved in 130 ml of mixture solution containing 100 ml DMF, 15 ml HCl and 15ml deionized water. Ammonium persulfate (4.56 g, 0.2 mol) was dissolved in 50 ml of 1MHCl and added dropwise into the above AD suspension with stirring. After 4 h reaction, the obtained suspension was filtered under vacuum. The fully reduced leucoemeraldine (LM) of AT was prepared by reduction with $N_2H_4 \cdot H_2O$ and then doping with $NH_3 \cdot H_2O$ for overnight. The LMAT product was filtered under vacuum and washed thoroughly with deionized water and finally dried in vacuum at 45°C for 48 h.

The 0.928 g of AT (2 mM), 1.151 g of NHS (10 mM) and 1.917 g of EDC·HCl (10 mM) were dissolved in 20 ml DMF, which was distilled over CaH_2 before use. The mixture was stirred at room temperature for 24 h. PEI 600 (2.56 g, (C_2H_5N, 60 mM) was dissolved in 20 ml of DMSO: DMF (1:1). Then PEI solution was added dropwise into the above mixture with stirring for 24 h at RT. After that, the final product was washed using deionized water and subsequently anhydrous ether by centrifugation. The collected product was dried in vacuum at 45℃.

2.3. Characterization
^1H NMR spectra of AT and AT- PEI were obtained on a Bruker Avance 300 MHz NMR instrument using DMSO-d_6 as solvent. FT-IR spectra of AT and AT-PEI were recorded on a Nicolet 6700 FT-IR spectrometer (Thermo Scientific Instrument). The absorption spectra of AT, AT-COOH and AT-PEI were signified by UV-Vis spectrophotometer (UV-2501PC, Shimadzu). The morphology of AT-PEI copolymer was investigated via Scanning Electron Microscopy (SEM, Quanta 400, FEI) and the sample was prepared by depositing a drop of solution on a silicon slice. The particle size distributions were obtained by Dynamic light scattering (DLS, Malvern Zetasizer Nano ZS). Cyclic voltammetry (CV) of AT and AT-PEI were conducted on an Electrochemical Workstation (PGZ 402, VoltaLab, France) employing a three electrode system (DropSens, Spain) with a carbon disk as working electrode, and carbon and silver electrode as the auxiliary and reference electrodes, respectively.

3. Result and discussion

3.1. Characterization of AT-COOH and AT-PEI

In this work, AT-PEI was synthesized by amidization reaction between amine groups on PEI and carboxyl group of AT–COOH. The preparation procedure of AT-PEI was depicted in figure 1. In brief, succinic anhydride was reacted with N-phenyl-1, 4-phenylenediamine to obtain carboxyl-capped aniline dimmer (AD-COOH), which was then conjugated with AD to obtain AT-COOH by oxidative coupling. The AT-COOH was fully reduced by doping with $NH_3{\cdot}H_2O$. Subsequently, the carboxyl groups of AT-COOH were activated by NHS before grafting with PEI by amidization reaction.

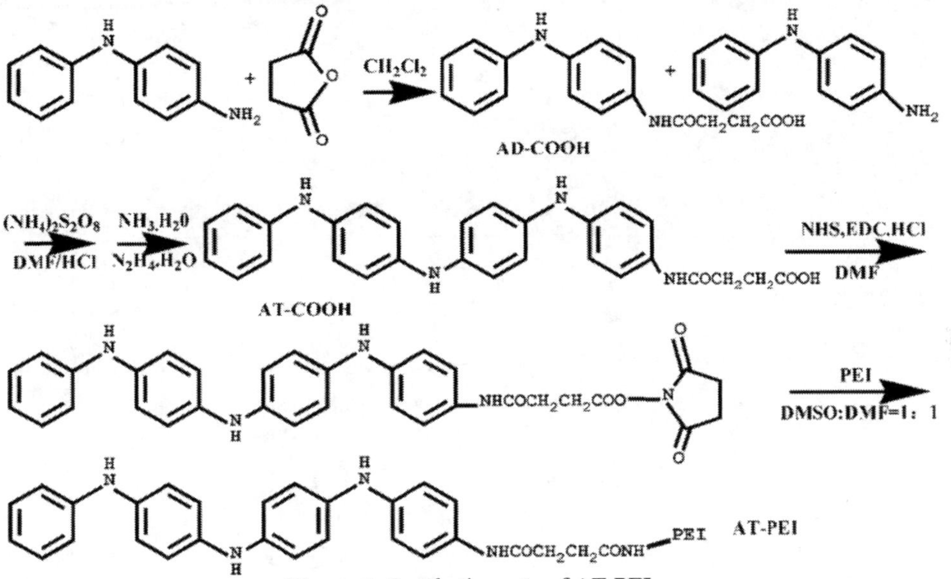

Figure 1. Synthetic route of AT-PEI.

^{1}H NMR and FT-IR spectra were used to characterize AT-PEI copolymer, the results of which was plotted in figures 2 and 3, respectively. As shown in figure 2, the proton signals between 7.94 and 6.63 ppm (multiplet) were ascribed to hydrogen of benzene ring. Proton signals at 3.9, 2.6 and 1.6 ppm were assigned to amino and alkyl groups of PEI, illustrating that AT was successfully grafted onto PEI molecule. Furthermore, the signals of –NH– groups of AT segment between 7.77 and 7.64 ppm are still present, indicating that the synthesized copolymer remained structural integrality of AT. As displayed in figure 3, the peaks at 1572 cm^{-1} and 1501 cm^{-1} were ascribed to the vibrations of quinoid and benzene rings, respectively. The stretching vibration of amide group (-CO-NH-) appeared at 1649 cm^{-1}. We also see that, the peak at 1720cm^{-1} corresponding to –COOH group was absent in the FT-IR curve of AT, demonstrated that AT was chemically grafted with PEI main chains by amidization reaction between carboxyl groups of AT and amine groups of PEI.

3.2. Self-assembling behavior of PEI-AT in water

According to self-assembling theory of the polyelectrolyte complexes [18], a hydrophilic cationic polymer, i.e. PEI in our study, fully stretched in aqueous solution due to the strong electrostatic repulsion. After conjugation with hydrophobic AT, the AT-PEI copolymer self-assembled into micelles with AT segments as nuclear and PEI molecular bone as a shell in aqueous phase. As shown in figure 4(A), nanoscale particles with the mean size of around 200 nm were observed by SEM, suggesting polyelectrolyte PEI-graft-AT formed nanoparticles owing to hydrophobic/hydrophilic interaction. However, the inhomogeneous size of particles may attribute to the broad range of molecular weight of PEI. The method of dynamic light scattering was further used to measure the hydrous size of assembled particles in water, as shown in figure 4(B). The mean diameter of 483 nm was detected with PDI of 0.463, demonstrating excellent dispersity. The discrepancy between the DLS

IOP Conf. Series: Materials Science and Engineering **87** (2015) 012072 doi:10.1088/1757-899X/87/1/012072

and SEM results was ascribed to that, high molecular weight, PEI expanded in the aqueous phase and shrunk after being dried.

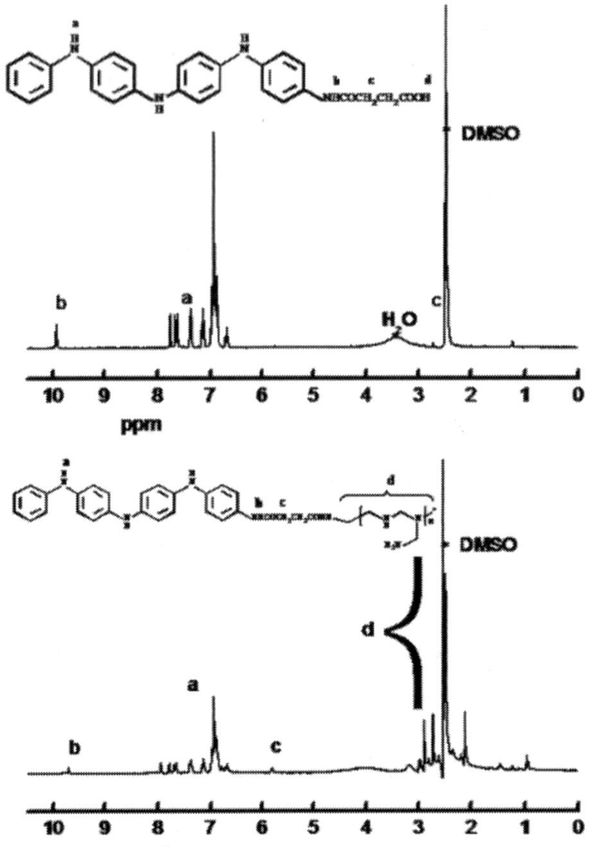

Figure 2. ^{1}H NMR spectrum of AT-COOH and AT-PEI.

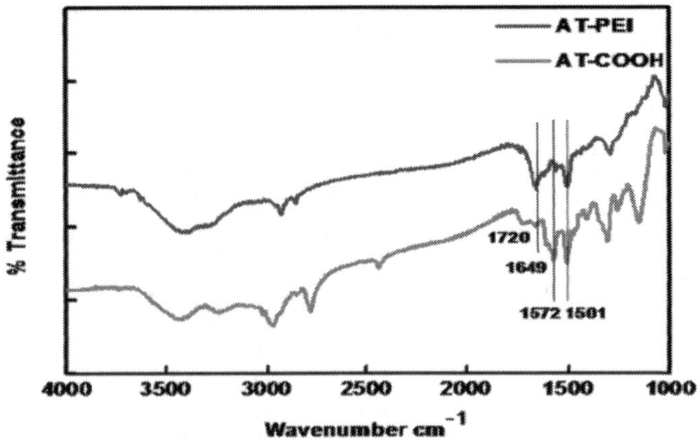

Figure 3. FTIR spectra of AT-COOH and AT-PEI.

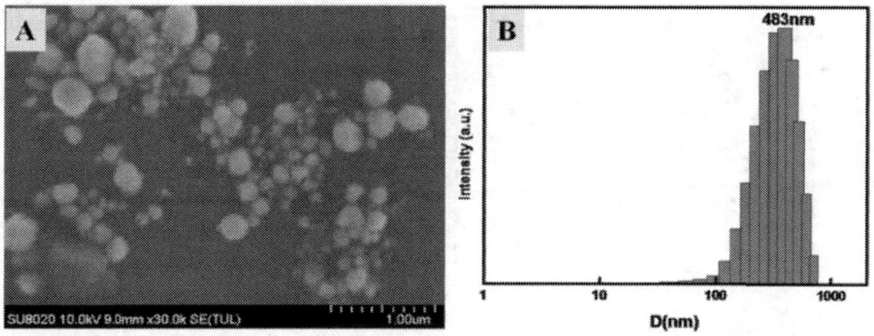

Figure 4. SEM image of spherical aggregates of AT-PEI (A) and Size distribution histograms of AT-PEI aggregate in water (B).

3.3. Electrochemical properties of AT-PEI

The electrochemical properties of synthesized products were characterized by UV-Vis spectrophotometer and electrochemical workstation. The UV-Vis spectra of AD-COOH, AT-COOH, and AT-PEI in DMF are shown in figure 5(A). The spectrum of AD-COOH exhibits an absorption peak at 298 nm, which was assigned to the absorption of benzene ring. The reduced state of AT-COOH and AT-PEI was observed only at 327 nm, which was ascribed to the π-π*transition of aromatic ring [19]. Figure 5(B) displayed the UV-Vis spectra of AT-PEI during in-situ oxidation by $(NH4)_2S_2O_8$. We can see that two absorption bands of AT-PEI were observed. Band was attributed to the absorption of conjugated phenyl ring. Band b was assigned to the benzenoid-to-quinoid excitonic transition with no blue-shift. Moreover, AT-PEI can only be tuned between fully reduced leucoemeraldine (LM) and emeraldine (EM) state under common oxidation conditions [20], as shown in figure 6. In figure 7, we observed only one oxidation peak at 0.559 V, which could be attributed to the transition of AT-PEI from LM to EM state of tetraaniline (AT) segment. This result demonstrated that, AT-PEI synthesized in our study was endowed with electrical conductivity form AT and excellent solubility because of PEI conjugation.

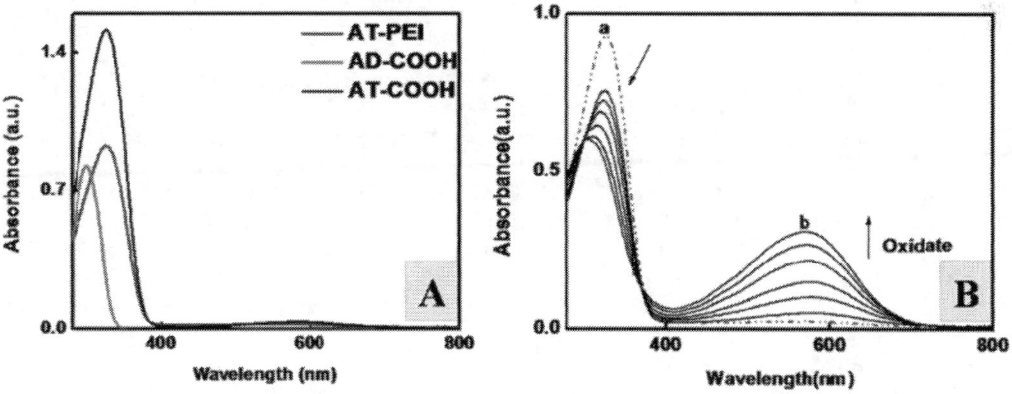

Figure 5. UV-Vis spectra of AD-COOH, AT-COOH and PEI-AT (A); AT-PEI(B) oxidized with ammonium persulfate dissolve in DMF.

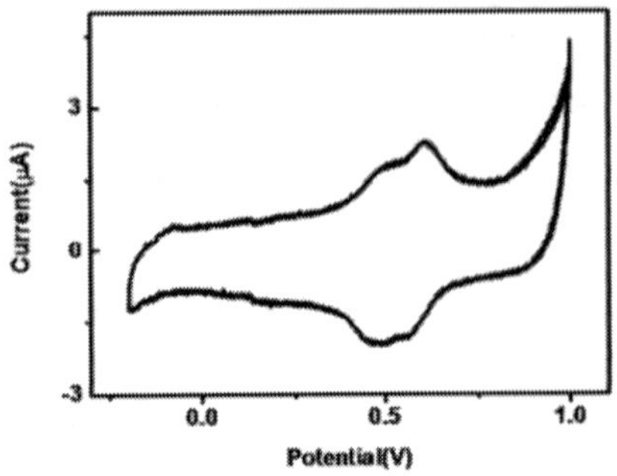

Figure 6. Cyclic voltammograms of AT-COOH and PEI-AT.

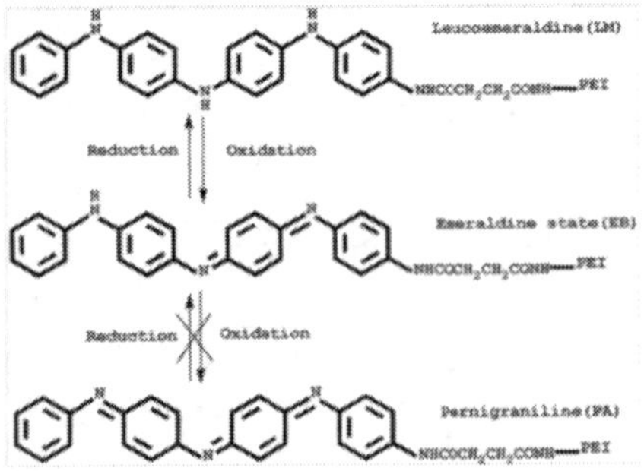

Figure 7. Proposed oxidation/reduction transitions of PEI-AT.

4. Conclusion

A novel electroactive AT-PEI copolymer was successfully synthesized by the amidization reaction between the carboxyl group of AT-COOH and amine group of PEI. The structure of AT-PEI was confirmed by ^{1}H NMR and FI-IR.UV-Vis and cyclic voltammetry was also employed to verify its electrochemical properties. The morphology of AT-PEI copolymer by SEM clearly revealed that, the copolymer could self-assemble into nanoparticles. AT-PEI showed one oxidation peak in 1.0 M sulfuric acid solution, suggesting AT-PEI still remained electroactivity in aqueous solution. All these results indicated that the electroactive copolymer, AT-PEI, might have potential applications in stem cell gene therapy for bone and cartilage tissue defect.

Acknowledgments

This work was financially supported in part by grants from the Natural Science Foundation of China (No. 31300820), Beijing Municipal Commission of Education Project (No. KM20140015001), a project founded by Beijing Institute of Graphic Communication (No. 27170115004/025), a grant from Beijing Municipal Commission of Education (No. 08150115/153) and Beijing Institute of Graphic Communication (No. 25000115006/009).

References

[1] Lu W, Elsenbaumer R L and Wessling B 1995 Corrosion protection of mild steel by coatings containing polyaniline *Synthetic Metals* **71** 2163-6

[2] Virji Huang, Kaner J, et al. 2004 Polyaniline nanofiber gas sensors: Examination of response mechanisms *Nano Letters* **4** 491-6

[3] Kaneto K, Kaneko M, Min Y, et al. 1995 Artificial muscle: Electromechanical actuators using polyaniline films *Synthetic Metals* **71** 2211-2

[4] Li S, Dong H and Cao Y 1989 Synthesis and characterization of soluble polyaniline *Synthetic Metals* **29** 329-36

[5] Huang J, Virji S, Weiller B H, et al. 2003 Polyaniline nanofibers: Facile synthesis and chemical sensors *Journal of the American Chemical Society* **125** 314-5

[6] Wei Z and Faul C F J 2008 Aniline oligomers–architecture, function and new opportunities for nanostructured materials *Macromolecular Rapid Communications* **29** 280-92

[7] Stejskal J and Trchová M 2012 Aniline oligomers versus polyaniline *Polymer International* **61** 240-51

[8] Mullen K and Wegner G 1998 *Electronic Materials: the Oligomer Approach* (New Jersey: Wiley)

[9] Lu F, Wudl F, Nowak M, et al. 1986 Phenyl-capped octaaniline (COA): An excellent model for polyaniline *J. Am. Chem. Soc.* **108** 8311-3

[10] Guo B, Lidija G and Ann-Christine A 2013 Biodegradable and electrically conducting polymers for biomedical applications *Progress in Polymer Science* **38** 1263-86
[11] Cui H, Liu Y, Cheng Y, et al. 2014 In vitro study of electroactive tetraaniline-containing thermosensitive hydrogels for cardiac tissue engineering *Biomacromolecules* **15** 1115-23
[12] Yang Z, Wang X, Yang Y, et al. 2010 Synthesis of electroactive tetraaniline-PEO-tetraaniline triblock copolymer and its self-assembled vesicle with acidity response *Langmuir* **26** 9386-92
[13] Ma X, Ge J, Li Y, et al. 2014 Nanofibrous electroactive scaffolds from a chitosan-grafted-aniline tetramer by electrospinning for tissue engineering *Rsc Advances* **26** 13652-61
[14] Pollard H, Remy J S, Loussouarn G, et al. 1998 Polyethylenimine but not cationic lipids promotes transgene delivery to the nucleus in mammalian cells *Journal of Biological Chemistry* **273** 7507-11
[15] Boussif O, Lezoualc'h F, Zanta M A, et al. 1995 A versatile vector for gene and oligonucleotide transfer into cells in culture and in vivo: polyethylenimine *Proceedings of the National Academy of Sciences* **92** 7297-301
[16] Cui H, Liu Y, Deng M, et al. 2012 Synthesis of biodegradable and electroactive tetraaniline grafted poly (ester amide) copolymers for bone tissue engineering *Biomacromolecules* **13** 2881-9
[17] Chen L, Yu Y, Mao H, et al. 2005 Synthesis of parent aniline tetramer and pentamer and redox properties *Materials Letters* **59** 2446-50
[18] Jiang M, Eisenberg A, Liu G, et al. 2006 *Macromolecular Self-Assembly* (Beijing: Science Press)
[19] Chen L. 2005 *Synthesis of Monodisperse Oligoanilines and Their Properties* (Changchun: Jilin University)
[20] Wang H, Guo P, Han Y 2006 Synthesis and surface morphology of tetraaniline–block–poly (L–lactate) diblock oligomers *Macromolecular Rapid Communications* **27** 63-68

Effect of peroxide and chain extender on mechanical properties and morphology of poly (butylene succinate)/poly (lactic acid) blends

W Cherykhunthod[1], M Seadan[2] and S Suttiruengwong[1,3]

[1]Department of Materials Science and Engineering, Faculty of Engineering and Industrial Technology, Silpakorn University, NakhonPathom, 73000, Thailand
[2]Department of Physics, Faculty of Science, Silpakorn University, Nakhon Pathom,73000, Thailand

E-mail: suttiruengwong_s@su.ac.th

Abstract. Poly (butylene succinate) (PBS) and poly (lactic acid) (PLA) are biodegradable polymers with high potential to replace commodity fossil-based polymers in a wide range of applications. However, these two polymers are immiscible in most ratios, but partially miscible when one of the two is a major phase. In this study, a one-step process in a twin-screw extruder was used to prepare the blends between poly (butylene succinate) (PBS) as a matrix and poly (lactic acid) (PLA) as a dispersed phase. To improve mechanical properties and morphology of blends, two reactive agents, peroxide (Perkadox) and multifunctional epoxide chain extender (Joncryl) were selected and compared. All samples were characterized for melt flow index (MFI), morphology, tensile, and impact properties. The results showed that the mechanical properties and morphology of PBS/PLA blends were improved when using both reactive agents. It was demonstrated that the increased mechanical properties resulted from good interfacial adhesion between PBS and finely dispersed PLA particles. The addition of 0.075 phr Perkadox to PBS/PLA (75:25 and 80:20) blends increased elongation at break by 7.2% and 38.4%, respectively compared with the blends without reactive agents. The results from gel content also revealed the graft copolymer existed at the interface when reactive agents were added. In the case of using multifunctional epoxide chain extender, the impact strength of the blends increased.

1. Introduction

In recent years, environmental pollution and the great depletion of non-renewable fossil resource have been increasingly concerned. The accumulation of non-biodegradable plastic wastes disposed daily leads to land and air pollution. One possible solution to this problem is to replace the commodity non-biodegradable fossil-based polymers with the renewable polymers or biodegradable polymers, which are readily susceptible to microbial action. Such biodegradable polymers are produced and processed to a wide spectrum of products, offering an alternative to fossil-based plastics [1-3]. Among biodegradable polymers, poly (butylene succinate) (PBS) is a popular choice due to its good

[3] Address for correspondence: S Suttiruengwong, Department of Materials Science and Engineering, Faculty of Engineering and Industrial Technology, Silpakorn University, NakhonPathom, 73000, Thailand. E-mail: suttiruengwong_s@su.ac.th.

mechanical strength and wide availability in an industrial scale. It is a renewable and fully-degradable plastic. It has high flexibility, impact strength, and thermal and chemical resistance but has low melt strength and is still expensive. Blending PBS with other biodegradable polyester can allow the fine-tuning of process ability of raw materials and properties of products. For most of the commercially biodegradable polymers, PLA is an interesting blend pair for PBS. PLA possesses high mechanical properties and is inexpensive. Moreover, PLA has recently been applied in various end-use applications [4-7].

A blend between PBS and PLA is partially miscible with the blend ratio of PBS/PLA up to 80/20 wt% [1]. PLA can offer strength and stability of PBS. However, the elongation at break of the blends decreases with the increasing PLA content [1, 6]. Mechanical properties of PBS/PLA blends strongly rely on the interfacial adhesion. Therefore, effective compatibilizers must be located at the interface for giving a low interfacial tension and a fine morphology of dispersed phase in the blend.

In order to achieve good compatibility of aliphatic polyesters, many methods such as non-reactive and reactive blending, copolymer synthesis and the addition of additives [8] were proposed. The superior compatibility of biodegradable polymer could be produced by adding reactive agents or grafting chemicals onto polymer backbone [9]. Many authors have reported the use of reactive agents in PBS/PLA blends such as dicumyl peroxide(DCP)[2],2,5-dimethyl-2,5-din (tertbutylperoxy) hexane [10], isocyanate [11], $Ti(OiPr)_4$ [12], $Ti(OBu)_4$ [12], and multifunctional epoxide [13]. The main advantages of reactive blending process include one-step process, cost-effectiveness and easy-to-handle features. The reactive can be selected relevant to polymers and processing parameters. Therefore, the aim of this work was to study the effect of two reactive agents on the mechanical properties and morphology of PBS/PLA (75:25 and 80:20). Two types of reactive agents, namely, peroxide (Perkadox 14-40B-PD) and multifunctional epoxide chain extender (Joncryl™ ADR-4368) with varying contents, were selected for comparison. Mechanical properties and morphology of PBS/PLA (75:25 and 80:20) were investigated and discussed. Gel content was determined to confirm the possible reactions.

2. Experimental

2.1. Materials

PBS (GS Pla FZ91PD) with MFI of 11 g/10 min (190°C, 2.16 kg) was purchased from the Mitsubishi Chemical Corporation (Japan). PLA (4043D) with a MFI of 2 g/10 min (190°C, 2.16 kg) was purchased from Nature Works LLC (USA). Di (tert-butylperoxyisopropyl) benzene (Perkadox 14-40B-PD), was purchased from Akzo Nobel (Netherlands).The peroxide initiator was absorbed on silica and calcium carbonate powder having reactive content of about 40%. The chain extender (Joncryl™ ADR-4368) was purchased from BASF The Chemical Co. (Thailand).The chain extender in solid flake had high epoxy functionality.

2.2. Preparation of PBS/PLA blends

PBS and PLA pellets were first dried at 60°C for 5 hours before further use. The ratios of PBS and PLA were 75:25 and 80:20 by weight. PBS, PLA and reactive agents were weighted and premixed prior to transferring to a twin-screw extruder (SHJ-25, China). The extrusion temperature was controlled with temperature profile being 100/130/140/150/160/170/180/195°C from feed zone to die exit. The screw speed was set at 100 rpm. The extruder was cooled in the water bath, then granulated and dried at 60°C for 12 hours. Two reactive agents systems, free radical reaction (Perkadox) and condensation reaction (Joncryl), were used for comparison. Reactive agents at various loading per 100 g of PBS/PLA blends were 0.0025, 0.0050, and 0.0075 phr for Perkadox system and 0.5, 1.0, and 1.5 phr for Joncryl system, respectively. PBS/PLA blend with Perkadox was referred to as Px and Joncryl as Jx, where x was the amount of reactive agents per 100g of blends. For instance, P0.0025 refers to a blend with 0.0025 phr of perkadox. A non-reactive PBS/PLA blend was prepared by using the identical procedure for comparison and denoted as PBS/PLA.

2.3. Characterization

2.2.1. Melt flow index. Melt flow index (MFI) of all samples was performed on a melt flow indexer (Intro D7053, KAYENESS) operating at 190°C with 2.16 kg load according to ASTM D1238.

2.2.2. Morphology. Scanning electron microscope (SEM) (Camscan MX-2000) was used to characterize the phase morphology and compatibility of PBS/PLA blends. The fracture surface was coated with spluttered gold and an accelerating voltage of 15 kV was used to produce the SEM photographs.

2.2.3. Mechanical properties. Notched Izod impact strength was tested according to ASTM D256 (ZwickB5102.202, Izod type). Tensile properties were measured using Universal Testing Machine (Instron 5969, Instron) according to ASTM D638, specimen type I. The specimens were tested at a cross-head speed of 50 mm/min.

2.2.4. Gel content analysis. To investigate the change in physical properties due to possible reactions, gel content test was measured using 1 g of PBS/PLA blends (with various reactive agents), which was thoroughly shaken with 20 ml dichloromethane and then left to rest at room temperature for 18 hours. PBS and PLA were removed from extraction residues. The residues were filtered through a 120-mesh sieve. Finally, the obtained residues were dried in an oven before calculating gel contents in each sample and performing FTIR analysis. The gel content (%) was calculated via equation (1).

$$\%\text{gel content}=\frac{W_1-W_2}{W_3}\times 100 \qquad (1)$$

Where W_1 denoted the weight of the sample and the cage after extraction and drying, W_2 was the weight of cage, and W_3: referred to weight of original sample.

3. Results and discussion

3.1. Melt flow index

From figure 1, MFI value of PBS/PLA (80:20) blend was similar to that of PBS/PLA (75:25) blend. MFI value of PBS/PLA (75:25) blends was 6.92 g/10min whereas PBS/PLA blends with reactive agents (P0.0025, P0.0050, P0.0075, J0.5, J1.0, and J1.5) expressed lower MFI values. Interestingly, MFI value was significantly decreased in the case of the blend with Perkadox (P) 0.0075 phr consisting of higher PLA content (grey bar, PBS/PLA(75:25)) (from 6.92 to 1.42 g/10 min), while the addition of Joncryl (J) 1.5 phr in the same composition of PBS/PLA(75:25) decreased the MFI value to 1.65 g/10 min. This indicated that some reactions in the blend with Perkadox and Joncryl were likely to occur.

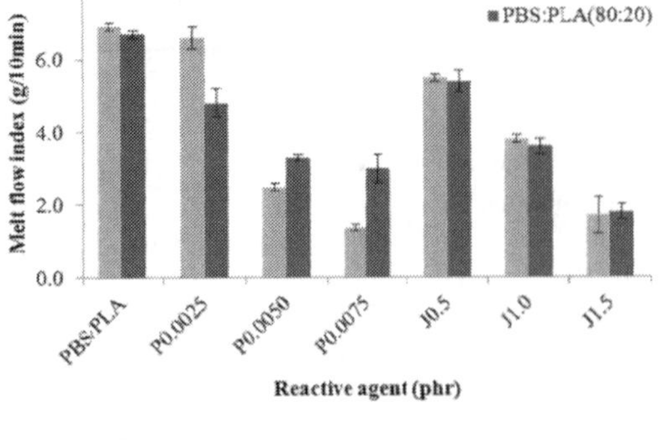

Figure 1. MFI of PBS/PLA blends at various contents of reactive agents.

3.2. Morphology

SEM micrographs of PBS/PLA (75:25) blend were shown in figure 2. The dispersion of PLA particles within the PBS matrix was less than 1 μm. The voids between PBS matrix and dispersed PLA particles could be observed in figure 2(a), indicating the poor interface between two polymers. In the case of PBS/PLA (75:25) blend at low contents of Perkadox between 0.0025-0.0075 phr (figures 2(b), 2(c) and 2(d), respectively), PLA particles were finely and uniformly dispersed in PBS matrix and showed a better interfacial adhesion than the blend without Perkadox. Considering the addition of Joncryl between 0.5-1.5 phr (figures 2(e)-2(g) for PBS/PLA (75:25) blend, PLA particles were dispersed in PBS matrix with fine particle sizes. Compared with the systems consisting of Perkadox with Joncryl, Perkadox system demonstrated a better interfacial adhesion, whereas Joncryl system only provided a fine dispersion phase. Similar results were also observed for SEM micrographs of PBS/PLA (80:20) blend at various contents of reactive agents in figure 3. Therefore, the addition of two reactive agents, namely, Perkadox and Joncryl, to PBS/PLA blends is likely to improve the morphology compared with non-reactive PBS/PLA blend.

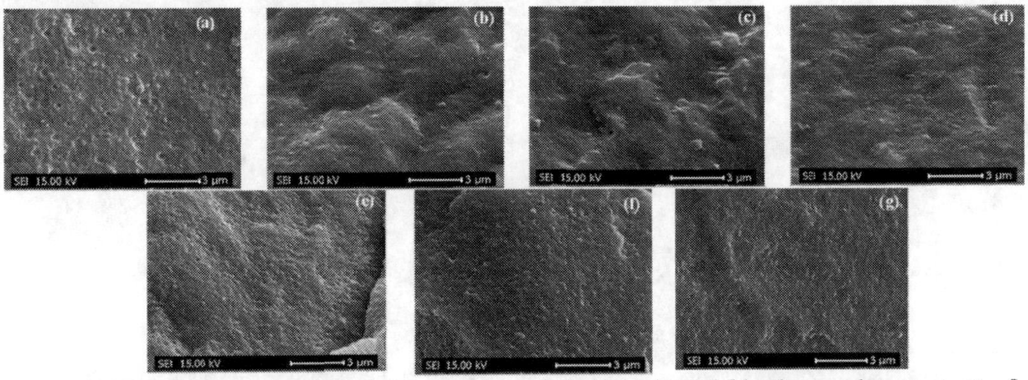

Figure 2. SEM micrographs of the fracture surface of PBS/PLA(75:25) blends at various contents of reactive agents(a) non-reactive agent, (b) Perkadox 0.0025 phr, (c) Perkadox 0.0075 phr, (d) Perkadox 0.0100 phr, (e) Joncryl 0.5 phr, (f) Joncryl 1.0 phr, and (g) Joncryl 1.5 phr.

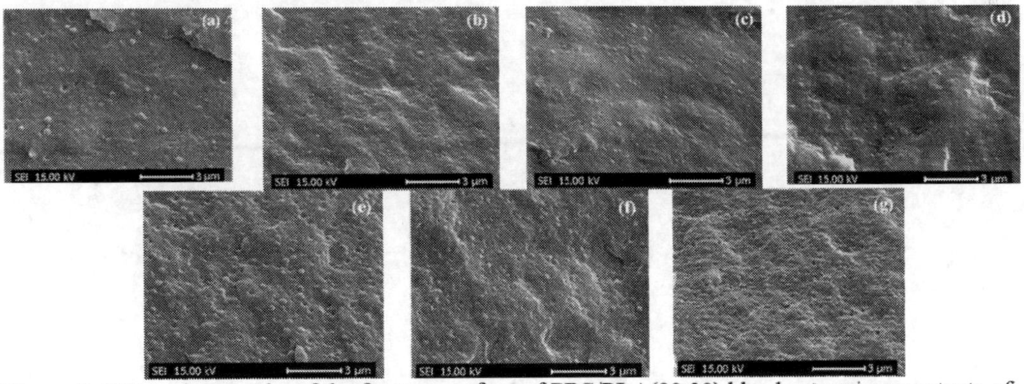

Figure 3. SEM micrographs of the fracture surface of PBS/PLA(80:20) blends at various contents of reactive agents(a) non-reactive agent , (b) Perkadox 0.0025 phr, (c) Perkadox 0.0075 phr, (d) Perkadox 0.0100 phr, (e) Joncryl 0.5 phr, (f) Joncryl 1.0 phr, and (g) Joncryl 1.5 phr.

3.3. Mechanical properties

According to two selected reactive systems, the effect of reactive agents on the mechanical properties of PBS/PLA blends was investigated. The impact strength was shown in figure 4. From figure 4, the impact strengths of PBS/PLA 75:25 and 80:20 blends were 5.98 and 6.97 kJ/m^2 respectively. The impact strength increased with the increasing Perkadox and Joncryl contents. The reactive agents systems, having Perkadox content between 0.0025 and 0.0075 phr, and having Joncryl content between 0.5 and 1.5 phr, clearly improved the impact strengths of PBS/PLA blends. This could be due to the free radical reaction of peroxide (Perkadox) with PBS/PLA to form the long branched structures or lightly crosslinked structures, whereas multifunctional epoxides of Joncryl could react with PBS/PLA providing the long linear of reconnected PBS/PLA chain. Therefore, they exhibited the difference in the improvement of impact strength compared with non-reactive agents.

The modulus, tensile strength, and elongation at break were characterized as shown in figures 5-7. Modulus, tensile strength, and elongation at break increased with the increasing Perkadox contents. From figure 5, modulus of PBS/PLA 75:25 and 80:20 blends were 653 and 636 MPa, respectively. A Perkadox system improved the modulus, tensile strength, and elongation at break of PBS/PLA blends, whereas for the Joncryl system, modulus remained almost unchanged with the increasing Joncryl contents while tensile strength and elongation at break decreased. In the case of the addition of Joncryl, the modulus, tensile strength, and elongation at break of PBS/PLA blends were unaffected.

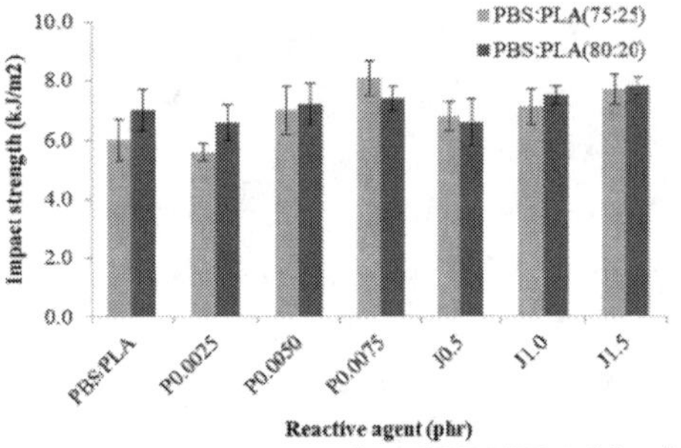

Figure 4. Impact strengths of PBS/PLA blends at various contents of reactive agents.

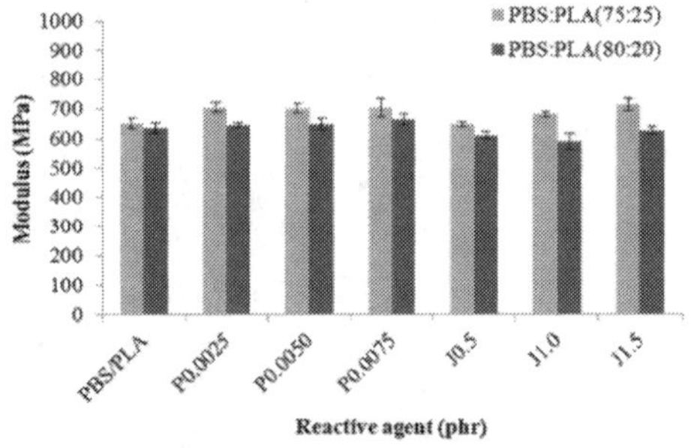

Figure 5. Modulus of PBS/PLA blends at various contents of reactive agents.

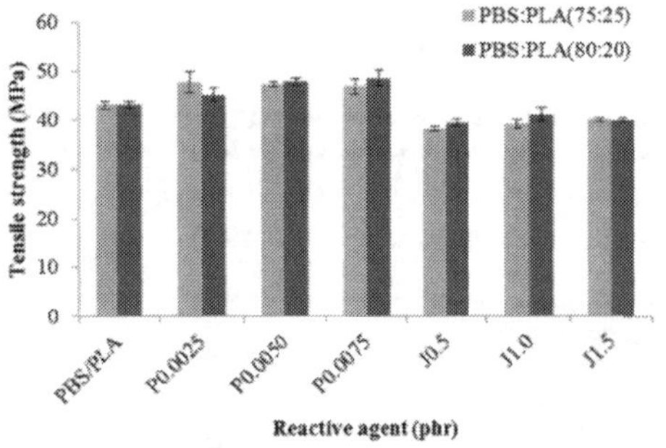

Figure 6. Tensile strength of PBS/PLA blends at various contents of reactive agents.

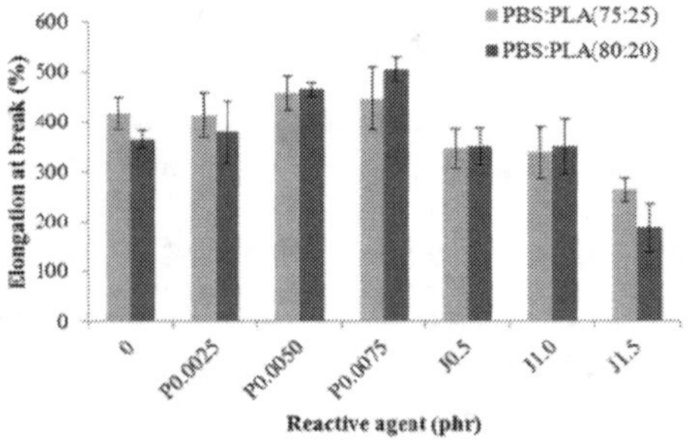

Figure 7. Elongation at break of PBS/PLA blends at various contents of reactive agents.

3.4. Gel content tests

It was evident that with the addition of Perkadox or Joncryl, the melt viscosity changed significantly. This implied possible reactions such as crosslink or branching. Gel content analysis was then performed to prove the physical changes of PBS/PLA blends with reactive agents. Any changes in chemical structures can result in the poor dissolution in the good solvent. Figure 8 represented the gel contents of PBS/PLA blends at various contents of reactive agents. Crosslinked polymers would give more gel contents in polymer blends, which were obtained after removing PBS and PLA by dichloromethane. The gel was not dissolved in dichloromethane. All of the blends showed low gel content (less than 10%), but it increased with the added amount of reactive agents. A small gel fraction could arise from possible crosslinked phase, branching structures and some graft copolymers in the blends.

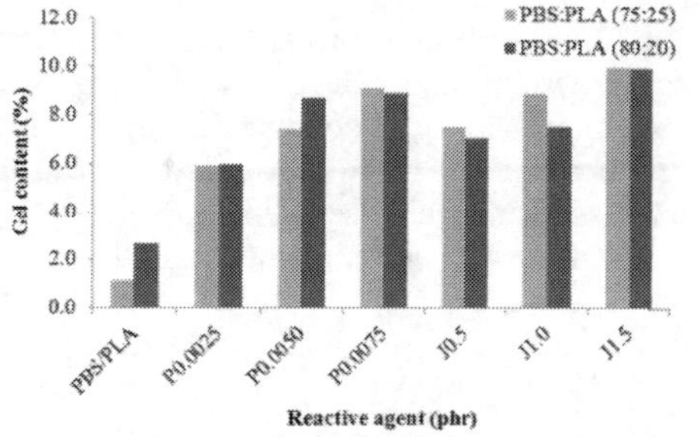

Figure 8. Gel contents of PBS/PLA blends at various contents of reactive agents.

The gel fractions of PBS/PLA blend from the extraction method were characterized by Fourier Transform Infrared Spectroscopy (FTIR), and compared with the PBS and PLA as shown in figures 9 and 10. The FTIR spectrum of PBS showed the characteristic at peak 1739cm^{-1} and 1170 cm^{-1} corresponding to carbonyl (C=O) groups and C-O stretching, respectively. The FTIR characteristic peaks of PLA were about 3157 cm^{-1} and 1406 cm^{-1} assigning to –OH stretch, and C-O, respectively. The FTIR spectra of the gel fractions from the reactive blends showed both characteristics of PBS and PLA. It indicated that the gel fraction contained a small portion of possible PBS-g-PLA (graft copolymer). This graft copolymer might be located at the interface of PBS and PLA, which led to the better interfacial adhesion between two polymers when reactive agents were added. Consequently, mechanical properties were improved.

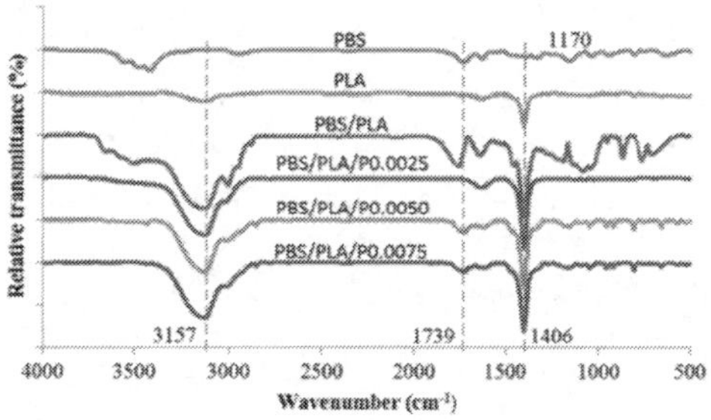

(a)

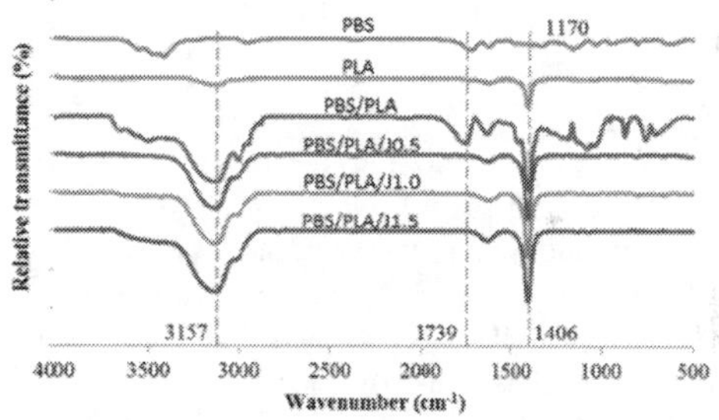

(b)

Figure 9. FTIR spectra of PBS/PLA (75/25) blend at various contents of (a) Perkadox and (b) Joncryl.

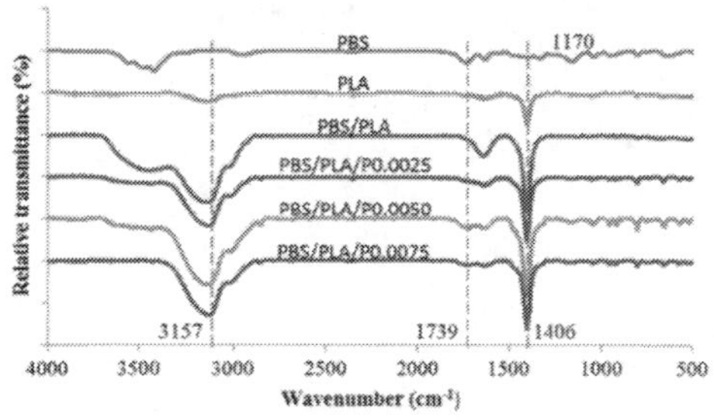

(a)

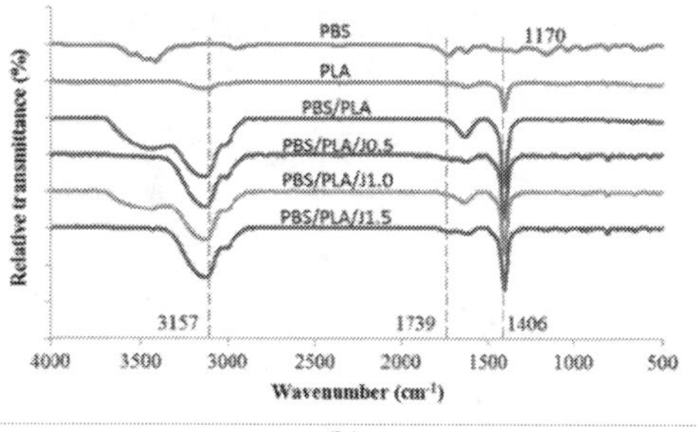

(b)

Figure 10. FTIR spectra of PBS/PLA (80/20) blend at various contents of (a) Perkadox and (b) Joncryl.

4. Conclusion

The addition of two reactive agents, Perkadox and Joncryl, into PBS/PLA blends improves the mechanical properties and morphology. The peroxide is likely to react with PBS/PLA via free radical reaction to form the long branched structures or lightly crosslinked structures whereas multifunctional epoxides (Joncryl) is prone to react with PBS/PLA to form the long linear of reconnected PBS/PLA chain. Therefore, they exhibit the differences in the improvement of impact strength and morphology compared with non-reactive agents. The results from gel content test and FTIR spectra reveal the formation of graft-copolymer at the interface of PBS and PLA when reactive agents are added. The slightly crosslinked polymers could be formed as suggested by low gel content. In addition, the addition of both reactive agents in PBS/PLA blends also improves the impact strength and morphology. Therefore, the reactive agents have the great potential to be used in PBS/PLA blends.

Acknowledgments
The authors would like to gratefully thank National Research Council Thailand (NRCT) and Department of Materials Science and Engineering, Faculty of Engineering and Industrial Technology, and Silpakorn University, Thailand for financial supports.

References

[1] Bhatia A, Gupta R K, Bhattacharya S N and Choi H J 2007 *Korea-Aust. Rheol. J* **19** 125-31
[2] Wang R, Wang S, Zhang Y, Wan C and Ma P 2009 *Polym. Eng. Sci.* 26-33
[3] Sotho K 2013 *Properties of Poly (Butylene Succinate) Base Blends via Reactive Extrusion* (Silpakorn University: Faculty of Engineering and Industrial Technology)
[4] Corporation M C GS Pla properties sheet. Available at: https://www.pttpm-techinfo.com/DownloadProduct.aspx, 2013
[5] LLC N 2010 Material Safety Data Sheet: 4043D. Available at: www.natureworksllc.com/media/Technical_Resources/Technical_Data_Sheets/TechnicalDataSheet_4032D_films_pdf, 2015
[6] Yokohara T and Yamaguchi M 2008 *Eur. Polym. J.* **44** 677-85
[7] Wilcox-Barsalou M 2010 Two concepts of Bioplastics–Degradability and Renewability. Available at: plasticsandtheplanet.com/archives/343, October 18, 2010
[8] Bicerano J 2010 A Practical Guide to Polymeric Compatibilizers for Polymer blends. Available at: www.plas2006.com/uploadfile/20063112235119.doc, 2010
[9] Ma P, Hristova-Bogaerds G D, Lemstra P J, Zhang Y and Wang S 2012 *Macromol. Mater. Eng.* **297** 402-10
[10] Coltelli M-B, Bronco S and Chinea C 2010 *Polym. Degrad. Stabil.* **95** 332-4
[11] Yang S-L, Wu Z-H, Yang W and Yang M-B 2008 *Polym. Test.* **27** 957-63
[12] Moura I, Nogueira R, Bounor-Legare V and Machado A V 2012 *Mater. Chem. Phys.* **134** 103-10
[13] Al-Itry R, Lamnawar K and Maazouz A 2002 *Polym. Degrad. Stabil.* **97** 1898-914

Chitosan-g-lactide copolymers for fabrication of 3D scaffolds for tissue engineering

T S Demina[1, 5], D S Zaytseva-Zotova[2], P S Timashev[3], V N Bagratashvili[3], K N Bardakova[3], Ch Sevrin[4], E A Svidchenko[1], N M Surin[1], E A Markvicheva[2], Ch Grandfils[4] and T A Akopova[1]

[1]Enikolopov Institute of Synthetic Polymer Materials, Russian Academy of Sciences, 70, Profsoyuznaya Str., 117393, Moscow, Russia
[2]Shemyakin & Ovchinnikov Institute of Bioorganic Chemistry, Russian Academy of Sciences, Miklukho-Maklaya Str., 16/10, Moscow, 117997, Russia
[3]Institute on Laser and Information Technologies, Russian Academy of Sciences, 2, Pionerskaya Str., Moscow (Troitsk), 142190, Russia
[4]Interfacultary Research Centre on Biomaterials (CEIB), University of Liège, Chemistry Institute, B6C, Allée du 6 août 11, B-4000 Liege (Sart-Tilman), Belgium

E-mail: detans@gmail.com

Abstract. Chitosan-g-oligo (L, D-lactide) copolymers were synthesized and assessed to fabricate a number of 3D scaffolds using a variety of technologies such as oil/water emulsion evaporation technique, freeze-drying and two-photon photopolymerization. Solid-state copolymerization method allowed us to graft up to 160 wt-% of oligolactide onto chitosan backbone via chitosan amino group acetylation with substitution degree reaching up to 0.41. Grafting of hydrophobic oligolactide side chains with polymerization degree up to 10 results in chitosan amphiphilic properties. The synthesized chitosan-g-lactide copolymers were used to design 3D scaffolds for tissue engineering such as spherical microparticles and macroporous hydrogels.

1. Introduction

Fabrication of 3D polymeric scaffolds that could serve as temporary supporting structures for growing cells and tissues plays an important role in developing of tissue engineering. These scaffolds should meet a number of design criteria related to material properties and processing conditions. The material should be biocompatible and bioresorbable with the same rate as new tissue form. Scaffolds should possess an optimum structure, e.g. surface topography, porosity and pore interconnectivity to provide efficient transfer of cells and nutrients. At the same time, the used polymer should be processable into scaffolds with the structure and properties tailored for the intended application. To reach a compromise between polymer properties and fabrication technologies as well as to enrich the material with special functions, a modification of polymer chemical structure is widely proposed.

[5] Address for correspondence: T S Demina, Enikolopov Institute of Synthetic Polymer Materials, Russian Academy of Sciences, 70, Profsoyuznaya str., 117393, Moscow, Russia. E-mail: detans@gmail.com.

Among a number of polymers used for biomedical applications chitosan, a linear polysaccharide derived from naturally occurring chitin, is widely investigated for scaffold fabrication, wound healing, drug delivery systems. Due to its biocompatibility, degradability by certain enzymes and positive charge of the macromolecule, chitosan is considered as a promising material for various biomedical applications. However, its unsuitability for processing in the molten state and solubility in limited range of solvents narrows a number of technologies to fabricate based on scaffolds and to modify its chemical structure.

This work aims to synthesize chitosan-g-oligo (L, D-lactide) in environmentally friendly solid-state mode and to prove the copolymer's processability with a range of 3D scaffold fabrication methods, such as oil/water emulsion evaporation technique, freeze-drying and laser microstereolithography based on two-photon polymerization. Chitosan-g-oligo (L, D-lactide) copolymers were synthesized through the solid-state reactive blending of chitosan and L, D-lactide mixtures by co-extrusion at low temperature to promote molecular level of mixing of these two components. Chemical reactions under shear deformation of the solid material induced by applying external mechanical energy in solvent-free mode provide high process effectiveness and ecological benefits. Grafting of hydrophobic oligolactide side chains on hydrophilic chitosan backbone results in copolymer's amphiphilic nature that allows us to extend a range of applicable processing technologies and to control other polymer properties, such as biocompatibility, degradation rate, mechanical characteristic, etc.

2. Materials and methods

2.1. Materials

Chitosan (Mw 60 kDa; acetylation degree 0.1) was prepared by mechanochemical solid-state synthesis as previously reported [1]. L, D-lactide was purchased from PuracBiochem (The Netherlands). The chitosan/L, D-lactide blends (marked as CL samples) were processed at lactide molar ratio of 0.5, 1, or 3 per chitosan unit in a semi-industrial co-rotating twin-screw extruder (ZE 40A×40D UTS, Berstorff, Germany) specially designed for powerful dispersion of solids. The process was run on 300-g batches at temperature range from 90 to 120°C [2]. The synthesis conditions and main macromolecular characteristics of synthesized chitosan-g-(L, D-lactide) copolymers are listed in table 1. Poly (D, L-lactide) (PDLLA) (Mw 76 kDa) used for microsphere preparation as a matrix polymer in oil phase was synthesized in CEIB. All other chemicals were of analytical grade and used without further purification.

Table 1. List of chitosan-g-poly (D, L-lactide) batches: conditions of treatment and the copolymer's characteristics.

Sample	Conditions of synthesis		Macromolecular features of the copolymer batches			Degree of chitosan amino group substitution	Average polymerization degree of the side lactide chains
	Lactide/ chitosan molar and weight ratio, mol (g/g)	Temperature (°C)	Relative amount of the reacted lactide (%)	Grafting degree (%)	Molar content of lactic acid units		
CL-0.5	0.5/(31:69)	120	57.1	26	0.58	0.19	3.0
CL-1	1/(47:53)	90	85.3	76	1.69	0.41	4.1
CL-3	3/(73:27)	120	59.3	160	3.58	0.37	9.7

2.2. 3D scaffolds fabrication and characterization

2.2.1. *Spherical microparticles via oil/water emulsion evaporation technology.* Microspheres were prepared by single oil/water emulsion solvent evaporation technique. Two different approaches to chitosan-g-oligo (L, D-lactide) copolymer application have been realized. As a first one, an oil phase obtained by dissolving of PDLLA in the solvent mixture (CH_2Cl_2: acetone = 9:1 v/v) at concentration of 8 wt-% was rapidly added to the aqueous phase contained 2 wt-% solution of the CL sample in mQ. The oil/water dispersion was mechanically stirred by four-blade propeller at 500 rpm for 1.5 hrs. The system contained as a water phase the 2 wt-% solution of non-modified chitosan in 2% acetic acid was prepared for comparative purpose. The second approach to CL application was based on their additional to an oil phase. Thus, the oil phase contained 30% of CL sample and 70% of PDLLA have been stirred within water phase consisted of mQ without any emulsifier. Temperature mode during the process was keep as 15°C within first 15 min of agitation then increased up to 30°C in order to promote the evaporation of organic solvents from the oil phase and to transform liquid-in-liquid dispersion into solid-in-liquid one. The microspheres were washed several times with mQ, settled by standard sieves with apertures of 500, 250, 125 μm and freeze-dried. Morphology of the prepared microspheres was studied by scanning electron microscopy (SEM) using JEOL-840M (Japan).

2.2.2. Macroporous hydrogels via freeze-drying technology. Water solutions of the chitosan-g-oligo(L,D-lactide) copolymers (1.5 wt-%) were poured into plastic tubes (15×25 mm) and frozen at -15°C, then freeze-dried. To cross-link the polymer and to convert the obtained sponges into insoluble form, the samples were heated in oven at 150°C for 3 hours. The control sample made of non-modified chitosan was prepared from 1.5% chitosan solution in 0.1 nHCl.

Morphology of the hydrogels in dried and swollen states were evaluated by SEM using Jeol JSM-5300 LV (Japan) and confocal laser scanning microscopy (CLSM) using Nikon TE-2000 inverted microscope equipped with an EZ-C1 confocal laser (Japan), respectively. Zeta-potential measurements were performed using PALS Zeta Potential Analyzer V. 3.54 (Brookhaven Instruments Corp., USA). Adsorption of bovine serum albumin (BSA), a major component of fetal bovine serum (FBS) used for cell cultivation, was measured by Bradford protein assay. Degradation of the samples into sterile solution of lysozyme (2 mg/mL) in PBS (pH 7.4) at 37°C was measured during 1 month by monitoring of weight loss. An ability of the hydrogels to support adhesion and growth of mouse fibroblasts L929 cell line was assessed using optical microscopy (Reichert Microstar 1820E, Germany) and MTT colorimetric assay.

2.2.3. Microstructured 3D scaffolds via two photon-induced polymerization (2PP). To fabricate microstructured 3D scaffolds 20 wt% CL solutions were prepared in mQ, then 100 mg of the prepared solution was mixed with 200 μL of 1 wt % Irgacure 2959 (BASF Kaisten AG) and 20 mg of poly (ethylene glycol) diacrylate (PEG-DA, Sigma-Aldrich) for 1.5 hours at RT. 2PP fabrication of scaffolds was performed using: ytterbium solid-state femtosecond laser TeMa 1053/100 system (Avesta Project, Russia), which delivers 80 fs pulses at an 69.7 MHz repetition rate. The experimental setup is similar to one that was previous described [3]. Structures were composed of two layers of hollow cylinder arrays. There was a 50 nm offset of cylinder layers, that's why the pores of the scaffold are hierarchically reduced from the top and bottom layer to the scaffold interior. The crosslinking process was followed by structures washing with mQ, 2 wt % acetic acid and aqueous ammonia to remove the uncross-linked material. The electronic absorption spectra of 1% solutions of non-modified chitosan and chitosan-g-oligo (L, D-lactide) copolymer were recorded on a Shimadzu UV 2501 PC spectrophotometer. The measurements were performed in quartz cuvettes with an optical path length of 10 mm.

3. Results and discussion

3.1. Microspheres

Polymeric microspheres are attractive microcarriers for tissue engineering and drug delivery systems. However, synthetic polymers traditionally used for their fabrication, such as polylactide, poly (lactide-co-glycolide), etc., suffer from hydrophobicity and a lack of functionalities needed to promote cell adhesion and proliferation, that limits its usage to design scaffolds for tissue regeneration. Surface modification of the preformed polylactide microparticles through physical sorption or chemical grafting of bioadhesive polymer has been proposed earlier [4, 5]. This last methodology is particularly tedious, because involving several steps after microparticle preparation to generate active functional groups to their surface before grafting the biomacromolecules. In order to simplify the above procedure but also to avoid non-degradable amphiphilic copolymer typically used to stabilize microparticles during their fabrication, we proposed to use chitosan-g-oligo (L, D-lactide) copolymers for preparation of polylactide-based microspheres.

Grafting of hydrophobic oligo (L, D-lactide) branch chains onto hydrophilic chitosan backbone allowed to obtain copolymers with amphiphilic properties that was clearly demonstrated by samples compatibility with organic solvents. Whilst non-modified chitosan were not dissolved and did not swell in chloroform at all, grafting of lactide chains allowed CL copolymers to swell in chlorinated solvents. At the same time the copolymers remained chitosan's ability to dissolve in diluted acids. Due to copolymer amphiphilic properties they have been applied for fabrication of spherical microparticles by an oil/water solvent evaporation technique using two different strategies. The first one allowed to replace classically used synthetic emulsifier PVA and to enrich the microparticle surface by chitosan moieties, which could positively affect cell attachment and growth on microparticle surface. The second approach relied on CL swelling in organic solvents which allowed to use it for oil phase modification. Therefore, the microparticles contained 30% of the CL sample and 70% of non-modified PDLLA have been prepared. Application of the CL samples in oil phase could provide chitosan moieties distribution over microsphere core and to avoid use of any emulsifier in aqueous phase.

In the frame of first option, water solubility of CL samples allowed us to directly dissolve copolymers in aqueous phase of the oil/water dispersion and use them as emulsifiers for microsphere preparation. As seen in figure 1, use of CL copolymers as the emulsifier in water phase allowed us to prepare spherical microparticles with smooth surface. The type of chitosan-based emulsifier drastically affected a total yield and size distribution of the prepared microparticles. The total microparticle yield was increased when degree of lactide grafting onto chitosan in the CL samples was increased. While application of non-modified chitosan acetic solution led to formation of particles with total yield of 26%, the use of CL-3 sample allowed us to enhance it up to 75% of PDLLA dissolved in oil phase. Better ability of the CL samples to stabilize microparticles during its preparation also appeared as a decrease of the particle size. Thus, the use of CL-0.5 and CL-3 allowed us to obtain the microparticles of 250 μm and 125 μm, respectively. Another approach dealt with 30% replacement of PDLLA in oil phase by the CL sample. It was proposed to fabricate microparticles without any emulsifier in water phase. Analysis of total yield and size distribution of the formed microparticles showed a tendency similar to water phase modification approach. Increase of lactic acid units per chitosan one among the CL-0.5, CL-1 and CL-3 samples led to increase of total microparticle yield: 21, 49 and 80 wt-%, respectively. The use of CL in oil phase gave a rise to smooth microspheres as well. SEM observation

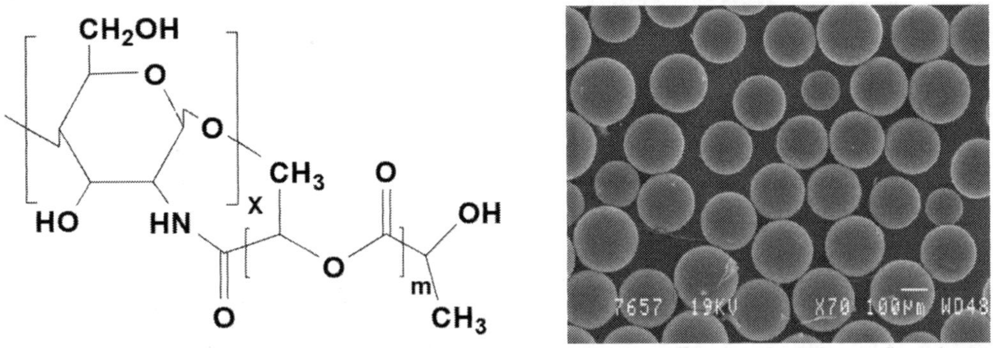

Figure 1. Chemical structure of chitosan-g-oligo (L, D-lactide) copolymer unit (left panel) and SEM image of microspheres made of CL-1 copolymer via oil/water emulsion evaporation technology (right panel).

of the cross-sectioned samples showed that the application of copolymers in oil phase led to formation of highly porous microparticles, an observation which can be explained by the amphiphilic nature of these copolymers.

We suppose that smooth surface could be promising to promote initial cell adhesion, whereas the macroporous internal structure of the microspheres provided the benefits during the microcarrier degradation, i.e. larger surface area for cell anchorage and facilitated diffusion of nutrients and penetration of cells and new tissue. Thus, chitosan-g-oligo (L, D-lactide) copolymers were successfully used for fabrication of microspheres possessed desired size and morphology to serve as drug/cell carriers for biomedical applications.

3.2. Macroporous hydrogels

Freeze-drying technology is a widely used approach for fabrication of macroporous 3D scaffolds for tissue engineering that could serve not only as templates to guide cell growth, but provides void volume for vasculatization, unhindered cell penetration and transfer of nutrients, oxygen and waste products. Freeze-drying allows to prepare macroporous structures as a result of formation of "walls" enclosing empty areas where polycrystals of frozen solvent (e.g. ice in the case of aqueous solutions) originally resided. The application of chitosan-based hydrogels for pharmaceutical applications has been already described in several reviews [6, 7]. Development of a new class of biocompatible and biodegradable hydrogels made of modified chitosan allows to control scaffold properties.

SEM observations of vertically and horizontally cross-sectioned slices of freeze-dried CL samples showed interconnected porous network with a pore mean size in a range of 70-140 μm. Control sample made of non-modified chitosan swell only under forced submergence, while the CL samples swelled up to the limit of their capacity within 1-2 min reaching equilibrium swelling degree up to 43 wt-%. The study of 3D architecture of the samples in swollen state using CLSM showed that their topography was similar to one in dried state, but the pore size had an increase in ~20 % (figure 2). An applicability of the macroporous chitosan-g-oligo (L, D-lactide) hydrogels as scaffolds for tissue engineering was evaluated *in vitro*. Protein adsorption, enzyme degradation rate and mouse fibroblast L929 cell line growth in the non-modified chitosan and CL-based hydrogels have been tested. Before cell/hydrogel interaction several phenomena usually take place, namely surface hydration and protein adsorption. In this research, we evaluated zeta-potential and adsorption of BSA on the control and CL hydrogels. We found that all hydrogels were positively charged, probably, due to the chitosan's cationic nature. The substitution of chitosan amino-groups led to decrease of zeta-potential. Since BSA was negatively charged, its adsorption should be higher on positively charged hydrogel samples. BSA adsorption profiles for various chitosan-based hydrogels were in a good agreement with zeta-potential measurements.

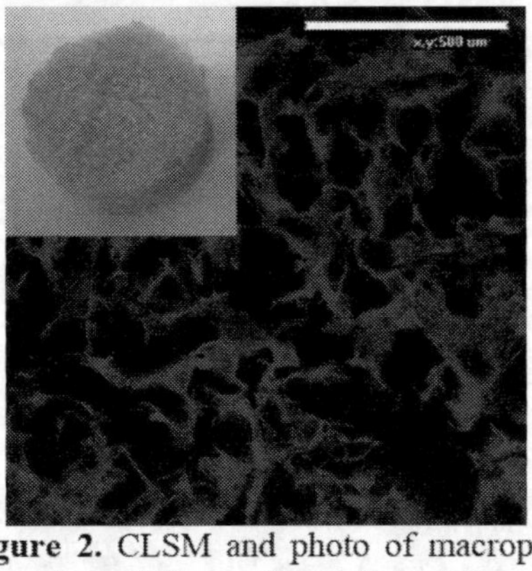

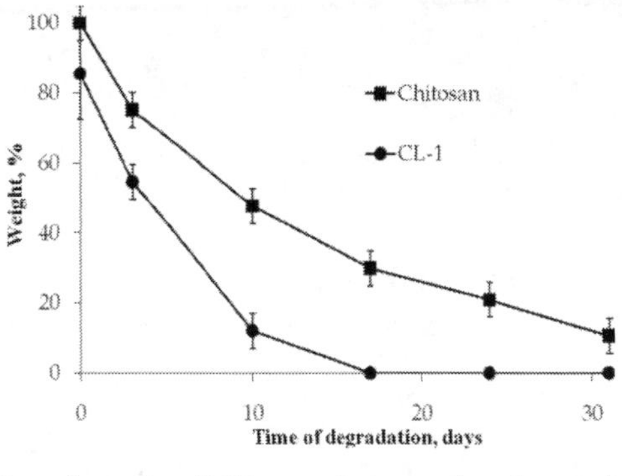

Figure 2. CLSM and photo of macroporous hydrogel made of CL copolymer using freeze-drying technology (left panel) and degradation of CL-1 hydrogels in PBS containing 2 mg/ml of lysozyme at 37°C (right panel).

An important scaffolds characteristic that could affect performance of cell growth, tissue regeneration and host response is a degradation rate. Ideally, scaffolds should degrade in a controllable manner with the same rate as new tissue form, thus, gradually transfer the structural and functional roles to the newly formed tissues. For the degradation study we selected lysozyme that is an enzyme present in the human body and is also produced by macrophages during wound healing. Monitoring of hydrogel's weight loss after incubation in lysozyme solution for various time intervals showed that CL samples degraded two times faster when hydrogels made of non-modified chitosan. It should be also stressed that during the degradation process hydrogels were not fallen into pieces but were found to be porous, very thin and fragile after degradation study. The study of ability the prepared hydrogels to support cell adhesion and growth by MTT-assay showed that a relative viability of L929 in the CL-based hydrogels was up to 80% higher than that in non-modified chitosan hydrogels. Optical observations showed that hydrogel's pore size was large enough for the cells to penetrate. The cells were seen to be mainly attached to the wall borders when to the wall surface. Thus, chitosan-g-oligo (L, D-lactide) copolymers were successfully used for fabrication of macroporous 3D hydrogels possessed enhanced biodegradation rate and the ability to support growth of animal cells.

3.3. 3D microstructures with well-defined architecture

Fabrication of 3D scaffolds through laser microstereolithography technique based on two-photon absorption polymerization (2PP) allows to achieve well-defined architecture with high resolution and reproducibility. The manufacturing of 3D objects by 2PP is based on a spatially controlled cross-linking process in a small volume at a computer-controlled beam focus. In contrast with single-photon polymerisation 2PP is excited by the nearly simultaneous absorption of two photons with low intensities that together provide enough energy to initiate cross-linking [8, 9]. This leads to high resolution (up to 200-400 nm) and possibility to carry out the scaffold formation without local overheating. Polymeric scaffolds of well-defined 3D architecture obtained by 2PP are considered as promising biomimetic constructions for tissue engineering [10].

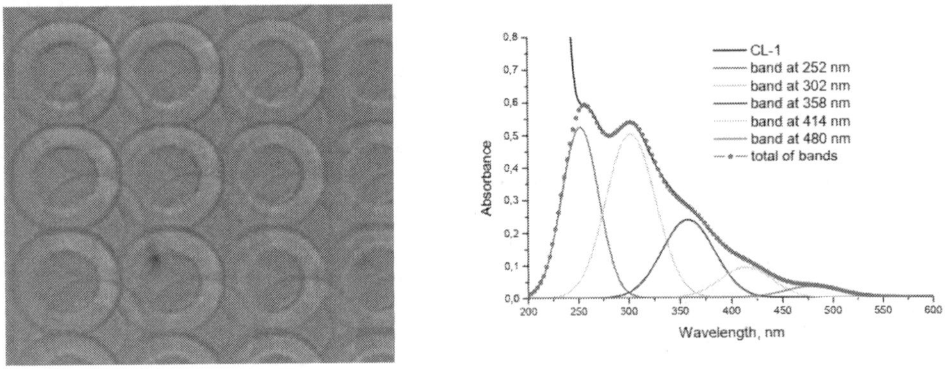

Figure 3. Optical micrographs of 3D microstructures fabricated by two-photon induced polymerization (left panel) and electron absorption spectra of CL-1 sample (right panel).

Here, we investigated the possibility to fabricate 3D scaffolds by 2PP from CL samples. It was found that photosensitive compositions based on CL samples could be successfully subjected to structuring under femtosecond laser beam. To reveal a difference in electron absorption spectra of non-modified chitosan and CL-1 solutions UV-Vis spectroscopy was carried out. Spectra of 1 wt-% non-modified chitosan solution in 0.1 n HCl was in a range of 200-420 nm and was consisted of three intensive bands having a maximum at 252, 299 and 346 nm and the bands intensity ratio as 1/1/0.4, while the chitosan-g-oligo (L, D-lactide) had two additional long wavelength bands with low intensities at 414 and 480 nm (figure 3). The CL sample showed a short wavelength band (< 200 nm) related to lactic acid. No significant differences were revealed in absorption spectra of both non-modified chitosan and chitosan-g-oligo (L, D-lactide) copolymer diluted solutions at concentration range from 0.1 to 1 wt-% that indicated the absence of an aggregation effect. However, analysis of absorption spectra showed that CL samples have particles with larger size than non-modified chitosan.

The microscopic observations (figure 3) of structures obtained by 2PP from CL samples demonstrate regular shape and definition, which shows applicability of this method for fabrication of CL-based microstructured objects. Although the simplicity of the presented structure, the objects of any shape could by produced with this technique.

The work was supported by Russian Foundation for Basic Research (project No. 15-02-06233 and No. 14-04-01861). The authors thank University of Liege for financial support of T S Demina.

4. Conclusions

Several types of 3D scaffolds, such as spherical microparticles and macroporous hydrogels, based on synthesized chitosan-g-oligo (L, D-lactide) have been successfully fabricated using a variety of technologies. Grafting of oligo (L, D-lactide) branch chains onto chitosan backbone allowed to broaden a number of technologies for fabrication of chitosan-based scaffolds and to improve materials properties.

References

[1] Akopova et al. 2011 *Focus on Chitosan Research* eds A N Ferguson and A G O'Neill (N.Y.: Nova Science Publishers) pp 223-54
[2] Demina et al. 2013 *Chemistry for Sustainable Development* **21** 631-7 (only in Russian)
[3] Koroleva et al. 2015 PLoS One **10** e0118164
[4] Messai et al. 2005 *Macromolecular Chemistry and Physics* **206** 1665
[5] Lao et al. 2008 *Colloids and Surfaces B: Biointerfaces* **66** 218-25
[6] Berger et al. 2004 *Eur. J. Pharm. Biopharm.* **57** 19-34
[7] Peppas et al. 2004 *Adv. Drug Delivery Rev.* **56** 1529-31
[8] Lee et al. 2008 *Prog. Polym. Sci.* **33** 631-81
[9] Melchels et al. 2010 *Biomaterials* **31** 6121-30
[10] Raimondi et al. 2012 J Appl Biomater Biomech **10** 55-65

Research of morphology structure and properties of bamboo charcoal acrylic fiber

Yongjiu Zhang[1] and Aifen Feng

Hebei University of Science and Technology, Shijiazhuang, Hebei

E-mail: zhangyongjiu1958@sohu.com

Abstract. In order to understand the properties of bamboo charcoal acrylic fiber, the tensile properties, friction properties and hygroscopicity of it, the bamboo charcoal acrylic fiber and the ordinary acrylic fiber were tested, compared and analyzed. The burning behaviors of the two kinds of fibers were observed by burning test, and their cross-sectional and longitudinal morphology was observed with scanning electron microscope (SEM). The SEM pictures showed that there are the uneven sizes of microspores on the surface of bamboo charcoal acrylic fiber and in it. It was found that the friction coefficients of the bamboo charcoal acrylic fiber are smaller and its tensile and moisture absorption are better than those of the ordinary acrylic fiber. However, there are no obvious differences of the burning behaviors between the two fibers.

1. Introduction

Acrylic fiber has obtained the considerable development and has become an indispensable part of synthetic fibers since its industrialization production in the 1950s [1]. In China's acrylic fiber market, the conventional acrylic fiber is saturated, but large quantities of differentiated and functional acrylic fibers, such as composite, superfine, noncircular cross-section, antibacterial fibers and more, are needed to import [2]. Therefore, it is necessary to develop functional acrylic fibers in order to adapt to the market demand.

The advent of bamboo charcoal fiber meets the people's increasing demand for functional textiles, and it has been paid attention to in the textile industry widely. It is a new kind of fiber following original bamboo fiber and bamboo pulp fiber made from bamboo and enjoys "Black Diamond" reputation [3]. The fabrics manufactured from bamboo charcoal fiber has good adsorption, deodorization, moisture absorption, heat storage, warm, antibacterial property, mould proof and anti-ultraviolet. Also they can emit far infrared rays and release negative ions. Therefore they can be used to made high-grade underwear, sportswear, coats, jackets and others [4].

Bamboo charcoal acrylic fiber is a new type of functional fiber and developed by Jilin Chemical Fiber Group CO., LTD. It has good heat storage and heat insulation functions. The fabrics made from bamboo charcoal acrylic fiber and general acrylic fiber were tested by Swiss Company SGS. The results show that the temperature of the fabric made from the bamboo charcoal acrylic fiber was 18.5℃ higher than that of the fabric made from the general acrylic fiber when the fabrics were irradiated by the certain heat source. This shows that the fabric made from the bamboo charcoal

[1] Address for correspondence: Yongjiu Zhang, Hebei University of Science and Technology, Shijiazhuang, Hebei. E-mail: zhangyongjiu1958@sohu.com.

acrylic fiber can reflect human body's heat radiation, store heat energy and prevent heat loss. Therefore, the fabric can meet the requirements of winter clothing with beauty and warm. The fabrics made from the blended yarns of the bamboo charcoal acrylic fiber with viscose, bamboo fiber, Tencel, cotton fiber have hygroscopic heat efficiency with the average temperature up to 3.7℃ and the highest up to 6.7℃. This efficiency makes people feel warmer and more comfortable. The fiber can also radiate far infrared rays which produce resonance with the human body cell molecular and go into the skin and subcutaneous tissue to improve the body's microcirculation, promote its metabolism and delay its fatigue. The normal emission rate of the fiber's far infrared rays is up to 0.87 and is durable. The fiber has good warm retention property, air permeability and comfort, and is suitable for underwear, sweaters, wadding, socks, gloves, scarves, blankets, bedding, and so on [5].

Because the bamboo charcoal acrylic fiber is a new type of fiber, the research on its properties is now in the initial stage. As is known to all, fiber properties have important influences on the textile processing, dyeing and finishing. Therefore it is necessary to research its characteristics.

2. Experimental details

2.1. Materials

1.67dtex bamboo charcoal acrylic fiber and normal acrylic fiber from Jilin Chemical Fiber Group CO., LTD.

2.2. Methods

2.2.1. Burning performance. The test is according to the textile industry standard FZ/T 01057.2-2007 "Test method for identification of textile fibers—part2:_Burning behavior". Use tweezers to clip a small bunch (about 50 ~ 100mg) fiber, and carefully observe the fiber in the proximity of alcohol lamp flame and in flames, the burning state of the fiber being removed from the flame, the generated order during its burning and its burning residue characteristic.

2.2.2. Microstructures. According to the textile industry standard FZ/T 01057.2-2007 "Test method for identification of textile fibers—part 3: Microscopy", the cross-sectional and longitudinal morphology of the fibers was observed by Hitachi S450 scanning electron microscope made in Japan. The monofilament and sampled thin cross-section of fibers are respectively stuck on the object loading platform by the conductive glue, and the platform is put into a vacuum tank and is evacuated. Under the conditions of the scanning electron microscope with test voltage of 15 kV, the magnification of 1500 times and the resolution of 1.0nm, the morphology of the fibers are detected to observe by adjusting brightness and contrast being the best, and the most suitable pictures are selected.

2.2.3. Mechanical properties. Fibers can deform and even be destroyed under a variety of external forces such as tensile, bending, torsion, compression, friction and so on in textile processing. The characteristics of fibers bearing all kinds of external forces are called mechanical properties [6, 7]. The mechanical properties tested include the breaking tenacity and friction performance of fibers. The samples had been conditioned at the temperature of 20℃ and the relative humidity of 41% for 24h, and then were tested.

Breaking Tenacity: The instrument of YG004N electronic single fiber strength tester is used. The breaking tenacity of the fibers was tested at the drawing speed of 100mm/min and the length of the specimen of 15 mm according to the national standard "GB/T14337-2008 Testing Method for Tensile Properties of Man-Made Staple Fibers". Each fiber was tested 20 times and the results were averaged

Friction Performance: The test instrument of type Y151 friction coefficient tester is used to test the friction coefficient of the fibers. The winch is used to be measured [8]. The friction coefficients of the two kinds of acrylic fibers respectively with metal roller, fiber roller and rubber roller were tested. The test is under the conditions of which the speed of friction roller is 30 rpm, the line speed of 75 cm/min, and pre-tension is set to 100 mg. Each hanging single fiber was demanded to repeat the measurement operation for 2~3 times. Each roller was tested for 6 of each fiber and 5 rollers of the same types are used to obtain 30 values which were recorded. Based on the values, the friction coefficients of each fiber are calculated according to the formula of the friction coefficient.

2.2.4. Moisture absorption performance. The test is according to the national standard "GB6503-2008Testing method for moisture regain of man-made fibres". The moisture regains of two fibers were tested by using typeY802A eight-basket constant temperature oven at the working voltage of 220 V and the heating power of 2.75 KW. After the fibers had been conditioned at the temperature of 20℃ and the relative humidity of 41% for 24 h, 10 g fiber for each was weighed and then put into the 110℃ oven to expose for 2 h.

3. Results and discussion

3.1. Burning performance analysis

The burning characteristics of two kinds of acrylic fibers are as shown in table 1.

Table 1. The burning behaviors of the fibers.

Fiber	Close to the flame	In the flame	After leaving the fire	Smell during burning	Combustion residue
Bamboo charcoal acrylic	Contracting	Burning with yellow flame and a lot of black smoke	Continue to burn rapidly	Slight smell of burning paper	Crisp, black and fragile lumps
Normal acrylic	Contracting and slight melting	Melting and burning with white and very bright, and a little of black smoke	Continue to burn slowly	Pungent smell	Crisp, black and fragile lumps

The burning behaviors of the bamboo charcoal acrylic fiber are similar to those of the normal acrylic fiber without a significant difference as shown in table 1. The bamboo charcoal acrylic fiber contracted when it was close to the flame; it was continuing to burn after it had left the flame with a burning speed more rapidly than the normal acrylic fiber; and there was a lot of black smoke to produce and a slight smell of burning paper during the burning process. A lot of black smoke by the bamboo charcoal acrylic fiber maybe because there is some bamboo charcoal powder to add into the fiber. The burning residues of two fibers are crisp, black and fragile.

3.2. Morphological analysis

Figure 1 shows the scanning electron microscopy images of the cross-section and longitudinal surface morphology of the bamboo charcoal acrylic fiber and the normal acrylic fiber.

There is much honeycomb microspore in the cross-section of the bamboo charcoal acrylic fiber with irregular section being similar to the round or oval as shown in figure 1. The longitudinal surface of the fiber has many grooves and is slightly rough. Therefore, in the process of spinning, the cohesion between the fibers will be increased, the fibers are not easily slip, and the yarn strength will increase. Because of the porous structure of the surface and internal of the fiber, the yarns and fabrics made from it not only have good moisture absorption and moisture absorption properties, thermal properties and good air permeability, but also have a certain absorption performance. The light undergarments, such as T-shirts, underwear, sportswear and casual wear, made from the fiber can quickly absorb skin moisture and sweat from the body's skin and spread into the surrounding air quickly because of its porous structure. Therefore to wear the light undergarments made from the fiber can keep the skin dry, and allow the wearer to keep dry and comfortable. The weight products, such as quilts, warm clothing, coats, socks and gloves for winter, can prevent the invasion of cold air and are warm and light because there is a lot of static air in the fiber's microspore.

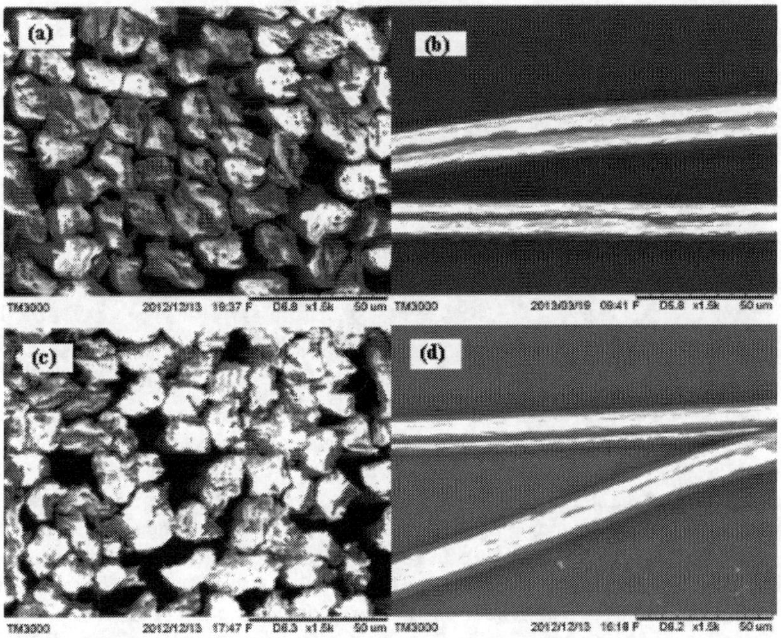

Figure 1. Scanning electron micrographs of fibers: (a) Cross-section morphology of bamboo charcoal acrylic fiber, (b) Longitudinal surface morphology of bamboo charcoal acrylic fiber, (c) Cross-section morphology of normal acrylic fiber, (d) Longitudinal surface morphology of normal acrylic fiber.

3.3. Mechanical properties' analysis

Table 2 shows the results of the breaking strength of the fibers.

Table 2. Tensile properties of fibers.

fiber	fineness/dtex	Breaking strength/CN	Breaking Elongation/mm	Breaking Tenacity/cN • dtex^{-1}	Rate of Breaking Elongation/%
Bamboo charcoal acrylic	1.67	5.72	4.05	3.4	27.02
Normal acrylic	1.67	4.64	4.25	2.7	25.07

The breaking strength and elongation of the bamboo charcoal acrylic fiber are larger than that of the normal acrylic fiber as shown in table 2. This may be because the nanometer bamboo charcoal powder to add into the fiber causes the internal microstructure of the fiber changed, which influences the tensile properties of the fiber.

Fiber friction performance affects not only spinning and weaving processing, but also the handle and style of finished products. What is more, friction will lead to fiber wear and deformation, and produce mass transfer, heat and electrostatic phenomena [9, 10].

Table 3 shows the friction coefficients of two fibers. The friction coefficients of the bamboo charcoal acrylic fiber are a little smaller than those of the normal acrylic fiber. The differences between the static friction coefficient and the dynamic friction coefficient of each fiber are very small. Because the cross section of the normal acrylic fiber is round, and its surface is smooth, the contact area is smaller when relatively sliding, which causes its friction coefficients slightly higher. The surface of the bamboo charcoal acrylic fiber with bamboo charcoal powder is rougher, so the contact area is larger when relatively sliding, which causes the friction coefficients can reduce.

Table 3. Friction coefficients of two fibers.

Fiber	Static friction coefficient μ_s	Dynamic friction coefficient μ_d	μ_s-μ_d
Bamboo charcoal acrylic to metal bar	0.3165	0.2762	0.0403
Bamboo charcoal acrylic to its bar	0.3342	0.2838	0.0504
Bamboo charcoal acrylic to rubber bar	0.5124	0.4288	0.0836
Normal acrylic to metal bar	0.3252	0.2838	0.0414
Normal acrylic to its bar	0.3434	0.2917	0.0517
Normal acrylic to rubber bar	0.5459	0.4413	0.1046

The static friction coefficients of two fibers are respectively higher than their dynamic frictional coefficients. Fiber's dynamic and static friction coefficients and the differences between them affect its handle. The smaller they are, the softer they are, and vice versa. Therefore, the bamboo charcoal acrylic fiber compared to ordinary acrylic fiber is softer according to the results.

Fiber's friction coefficients not only influence its handle, but also affect its spinning process and its yarn quality. The larger the friction coefficient among fibers is, especially the static one, the larger the cohesion among fibers. Larger cohesion contributes to the fiber aggregation in the process, prevents fiber diffusion, and improves the yarn's quality and strength [11]. The static friction coefficients of the bamboo charcoal acrylic fiber are larger than its corresponding dynamic friction coefficients, which is beneficial to the fiber winding, pressurization, anti-adhesion etc. The friction coefficients of the fiber to its bar are larger than those of it to the metal bar but smaller than those of it to the rubber bar. Large friction coefficient between the fiber and the machine parts is good at holding the strip grip by the jaw in the drawing process. However, in the opening and carding process, the fiber transfer will be difficult and form the strikes repeatedly, which results in fiber damage and causes the wear of yarns and machine parts. Therefore, in order to ensure the spinning smoothly, proper measures should be taken to reduce the friction coefficient between the fiber and the parts.

3.4. Moisture absorption analysis

Usually, the ability for fiber to absorb water from the atmosphere or Release water to the atmosphere is called hygroscopicity [6]. The level of the moisture absorption of textile materials usually is represented by the moisture regain. The larger the moisture regain of a textile material is, the better its moisture absorption is.

Based on the results, the moisture regain of the bamboo charcoal acrylic fiber is higher than the normal acrylic fiber's, respectively 1.71% and 1.52%. There is a lot of microspore on the surface of bamboo charcoal acrylic fiber and in it which makes it have good moisture absorption and liberation, warmth retention property and air permeability. In addition, the bamboo charcoal particles in the fiber have the adsorption function which makes the fiber has better moisture absorption and increase its moisture regain. The increased moisture regain of the fiber not only helps its textile processing and reduces static electricity, but also improves the wearing comfort being suitable for work clothing, sportswear, underwear and other clothing.

4. Conclusions

There is a lot of microspore with many sizes distributed unevenly on the surface of the bamboo charcoal acrylic fiber and in it. This structure can increase the cohesion among the fibers and improve the quality of yarn, fabric moisture permeability, air permeability, puff performance, coverage, and warmth. Therefore the bamboo charcoal acrylic fiber can be used to produce sportswear, work clothes and other products whose comfort will be improved. The breaking strength and breaking elongation of bamboo charcoal acrylic fiber are higher than the normal acrylic fiber's, so its durability is better too. The friction coefficients of the bamboo charcoal acrylic fiber are a little smaller than the normal acrylic fiber's, so its handle feels softer. The moisture regain of the bamboo charcoal acrylic fiber is larger than the normal bright acrylic fiber's, so it is easy to process in spinning and weaving, and its products are comfortable to wear.

References

[1] Lu Dong, Li Qingshan and Yang Xiuzhen 2006 Prensent situation of China acrylic fiber industry *Synthetic Fiber* **6** 28-29
[2] Jin Lichen 2007 Development of China acrylic fiber industry *China Synthetic Fiber Industry* **30** 56-59
[3] Wang Xianfeng, Pan Fukui, Luo Jiali, etc. 2006 Performance and application of bamboo charcoal fiber *Shandong Textile Science & Technology* **6** 54-56
[4] Li Wang, Li Xiaorong and Cao Qiuling 2006 Bamboo charcoal and its development and application in textiles *China Textile Leader* **2** 36-7
[5] Jilin chemical fiber: Vigorously promote lai ma and heat storage fibers http://www.tteb.com:2012-8-24
[6] Jiang Yaoxing and Qin Feng 2005 *An Introduction to Textile* (Beijing: China Textile & Apparel Press)
[7] Xufen Yu 2004 *Experimental Technology of Textile Materials* (Beijing: China Textile & Apparel Press)
[8] Zhao Shujing 2007 *Textile Material Experiment Tutorial* (Beijing: China Textile & Apparel Press)
[9] Bowden F P and Tabor D 1964 *The Friction and Lubrication of Solids, Part II* (London: Oxford University Press)
[10] Rabinowicz E *Friction and Wear of Materials* 1965 (New York: Wiley)
[11] Li Ruizhou, Zheng Yuansheng and Ao Limin 2010 Research of silver plated fiber friction property and spinning oil *Cotton Textile Technology* **38** 432-3

Study of silicon carbide/graphite double coating polyester woven fabric EMW absorbing property

Yuanjun Liu and Xiaoming Zhao[1]

School of Textiles, Tianjin Polytechnic University, Tianjin 300387, China

E-mail: texzhao@163.com

Abstract. In this paper, polyester woven fabric was selected as the base fabric. Silicon carbide and graphite were the underlying and surface layer absorbing agents, respectively. The influence of coating thickness of the silicon carbide and graphite absorbents on the dielectric constant was discussed.

1. Introduction

Microwave absorptive fabrics are widely studied not only in the military field for stealth technology, but also in civilian sphere, with regards to pollution by EM radiation deriving from various electronic apparatuses [1-3]. In past years, the properties of intrinsic conductive fabrics or fabrics coated with absorbers, such as carbon fiber fabric, nonwoven polyacrylonitrile and poly (ethylene terephthalate) impregnated with polyaniline, glass fibers coated with multi-walled carbon nanotubes, microwave absorptive fabrics, epoxycotton fabric coated with barium ferrite, doped polyaniline, etc. have been investigated [4-10].

In this paper, silicon carbide and graphite with high absorption and low-cost production were the underlying and surface layer absorbing agents, respectively. The influence of coating thickness on the dielectric constant was discussed. Silicon carbide/graphite carbide double coating polyester woven fabric absorbing materials with the best wave absorption performance were prepared.

2. Experimental procedure

2.1. Materials and instruments

Polyester woven cotton fabrics (weave fabrics) used for this work were provided by the YOUNGOR Co., Ltd. (Zhejiang, China). All other reagents were purchased from Tianjin Chemical Reagent Co., Ltd. (Tianjin, China).

2.2. Fabric coating preparation process

- Add a weighed amount of ethanol, the diluent, to the beaker with the epoxy resin, and then stir continuously with the electric mixer for 30 min until the mixture is uniformly mixed;
- Add the absorbing agent powder and curing agent to the mixture; stirring for 15 min to ensure the mixture is uniformly mixed.

[1] Address for correspondence: Xiaoming Zhao, 399 Binshui West Road, Xiqing District, Tianjin 300387, P. R. China, Tel.: 86-22-83955239; Fax: 86-22-83955287; E-mail: texzhao@163.com.

- Fix the plain weave fabric woven onto the coating machine, then pour the prepared coating liquid on the fabric, making the first layer (bottom layer).
- After completion of the underlying coating, dry for 0.5 hours at 60℃ until the coating is slightly dry, and then repeat (4) for the second layer (surface layer) coating; place the coated fabric in the oven to dry at 60℃ for 3 hours to obtain double layer absorbing coating fabric. The absorbing coating fabric structure model is shown in figure 1.

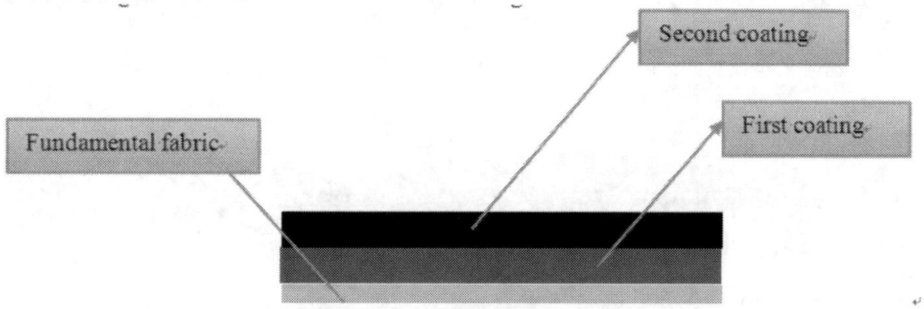

Figure 1. Absorbing coating fabric structure model.

2.3. Measurement of the permittivity

The composites were die-pressed to form cylindrical toroidal specimens with a 7.0 mm outer diameter and a 3.0 mm inner diameter. The measurements of EMW loss property for the specimens were carried out using a PNA 3629D vector network analyzer in the 30–6000 MHz range. The ε' and ε'' of complex permittivity correlate polarization and loss.

3. Results and discussion

3.1. The influence of silicon carbide coating thickness on the dielectric constant

In order to explore the influence of silicon carbide coating thickness on the dielectric constant, the woven fabrics had been selected as fundamental fabric, a series of different coating thicknesses of silicon carbide of the silicon carbide /graphite layer coating woven fabric were made. Sample specifications are shown in table 1.

Table 1. Sample specification of different silicon carbide coating thickness.

Number	Under layer thickness (mm)	Surface layer thickness (mm)
1	0.25	0.50
2	0.50	0.50
3	0.75	0.50
4	1.00	0.50
5	1.25	0.50

Figures 2-4 show the influence of different silicon carbide coating thicknesses on the dielectric constant with frequency change. When the content of the absorbent is constant, the coating thickness determines the amount of absorbent contained in the coating, thereby affecting the extent of the absorption of the electromagnetic wave into the coated fabric. As seen from the figure, the real part of the dielectric constant decreases with increasing frequency from the figure 2. At $f > 10^4$ Hz, when the silicon carbide coating thickness increases, the real part of the dielectric constant of the coated fabric increases, at 10 Hz $< f < 10^4$ Hz; when the silicon carbide coating thickness is 1.00mm, the dielectric constant has the largest real part, so polarization is strongest. Over the entire frequency range, when the silicon carbide coating thickness is 1.00 mm, the loss tangent and the imaginary part of the dielectric constant is maximum, so the loss is strongest. The loss tangent and permittivity imaginary part curves of other thickness nearly coincide, at $f > 10^3$ Hz. Thus, when silicon carbide coating thickness is 1.00mm, the coated fabrics have the best absorbing properties.

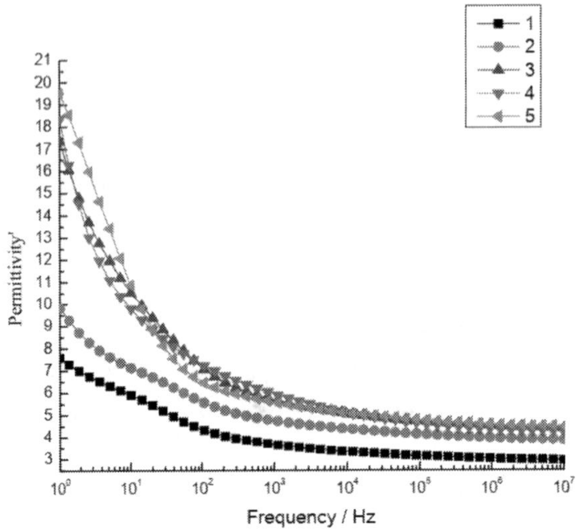

Figure 2. The influence of silicon carbide coating thickness on the real part of the dielectric constant.

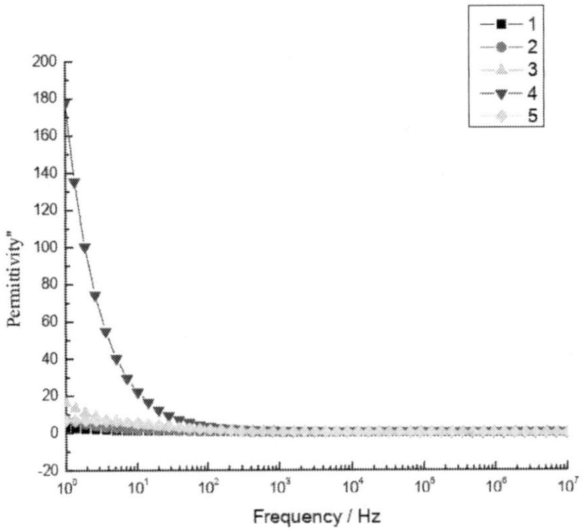

Figure 3. The influence of silicon carbide coating thickness on the imaginary part of permittivity.

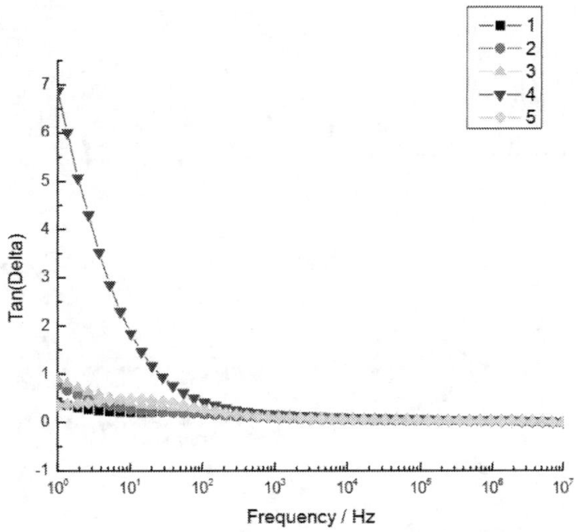

Figure 4. The influence of silicon carbide coating thickness on the loss tangent.

3.2. The influence on the graphite coating thickness of the dielectric constant

In order to explore the influence of graphite coating thickness on the dielectric constant, the woven fabrics had been selected as fundamental fabric, a range of different graphite coating thickness weave fabrics of silicon carbide/graphite were prepared. Sample specifications are shown in table 2.

Table 2. Sample specification of different graphite coating thickness.

Number	Under layer thickness (mm)	Surface layer thickness (mm)
1	1.00	0.25
2	1.00	0.50
3	1.00	0.75
4	1.00	1.00
5	1.00	1.25

Figures 5-7 reflect the dielectric constant of different graphite coating thicknesses with frequency change. When the content of the absorbent is constant, the coating thickness determines the amount of absorbent contained in the coating, thereby affecting the extent of the absorption of the electromagnetic wave into the coated fabric. As can be seen from the figure, when frequency increases, the real part of the dielectric constant decreases. At $f > 10$ Hz, as graphite coating thickness increases, the real part of the dielectric constant of the coated fabric showed an increased fluctuation trend. When the graphite coating thickness is 0.25 mm, the coated fabrics have the minimum real part of the dielectric constant. When the graphite coating thickness of 1.25 mm, the coated fabrics have the largest real part of the dielectric constant. When the graphite coating thickness is 1.00 mm, the loss tangent and the imaginary part of the dielectric constant are maximum, so the loss is strongest. The curves of other graphite coating thicknesses coincide approximately at $f > 10^3$ Hz. The curve of several different thicknesses of the permittivity imaginary part and the loss tangent nearly coincide and change slowly; the value trend is a constant. Thus, a graphite coating thickness of 1.00 mm has the best absorbing property.

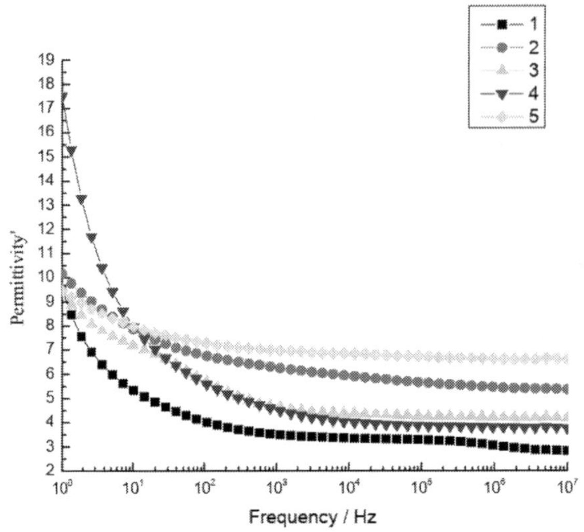

Figure 5. The influence of graphite coating thickness on the real part of the dielectric constant.

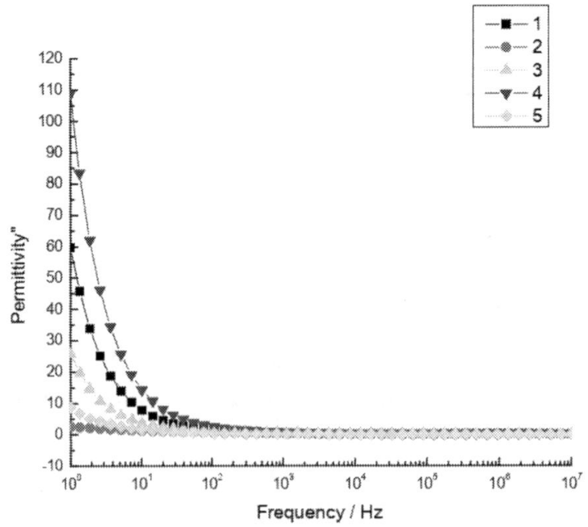

Figure 6. The influence of graphite coating thickness on the imaginary part of permittivity.

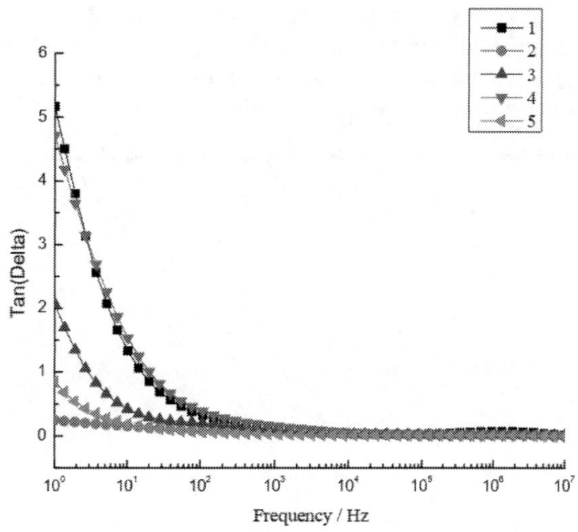

Figure 7. The influence of graphite coating thickness on the loss tangent.

4. Conclusions

At lower frequency bands (f <10^3 Hz), coating thickness has a great influence on the dielectric constant. At higher frequencies (10^3 Hz <f <10^6 Hz), the curves of the loss tangent and the imaginary part of the dielectric constant show a gentle trend, almost a constant. When the silicon carbide coating layer (under layer) has a thickness of 1.00 mm and the graphite coating (surface layer) has a thickness of 1.00 mm, the coated fabrics have the best absorbing performance.

Acknowledgment

This work was supported by the National Natural Science Fundation of China under Grant 51206122.

References

[1] Kim S S, Han D H and Choowave S B 1994 Absorbing properties of sintered Ni–Zn ferrite *IEEE Transactions on Magnetics* **30** 4554-6
[2] Yang Y L, Gupta M C, Dudley K L and Lawrence R W 2005 Novel carbon nanotube–polystyrene foam composites for electromagnetic interference shielding *Nano Letters* **5** 2131-4
[3] Zhang X Z and Sun W EMW 2010 Absorbing properties of double-layer cementitious composites containing Mn–Zn ferrite *Cement and Concrete Composites* **32** 726-30
[4] Folgueras L D C, Nohara E L, Faez R and Rezende M C 2007 Dielectric EMW absorbing material processed by impregnation of carbon fiber fabric with polyaniline *Materials Research* **10** 95-9
[5] Kim S H, Oh K W and Bahk J H 2004 Electrochemically synthesized polypyrrole and Cu-plated nylon/spandex for electrotherapeutic pad electrode *J. Appl. Polym. Sci* **91** 4064-71
[6] Hashisho Z, Rood M J, Barot S and Bernhard J 2009 Role of functional groups on the EMW attenuation and electric resistivity of activated carbon fiber cloth *Carbon* **47** 1814-23
[7] Lee S E, Park K Y, Oh K S and Kim C G 2009 The use of carbon/ dielectric fiber woven fabrics as filters for electromagnetic radiation *Carbon* **47** 1896-904
[8] Folgueras L D C and Rezende M C 2007 Hybrid multilayer structures for use as EMW absorbing material *Proceedings of the 2007 SBMO/IEEE MTT-S International EMW & Optoelectronics Conference* **1-2** 483-7

[9] Park K Y, Lee S E, Kim C G and Han J H 2007 Application of MWNT-added glass fabric/epoxy composites to electromagnetic wave shielding enclosures *Composite Structures* **81** 401-6
[10] Aksit A C, Onar N, Ebeoglugil M F, Birlik I, Celik E and Ozdemir I 2009 Electromagnetic and electrical properties of coated cotton fabric with barium ferrite doped polyaniline film *Journal of Applied Polymer Science* **113** 358-66

IOP Conf. Series: Materials Science and Engineering **87** (2015) 012077 doi:10.1088/1757-899X/87/1/012077

Improvement of Gd123 superconductor bulks with the additions of $BaFe_{12}O_{19}$

Yufeng Zhang[1, 3], Liqi Peng[1], Wenli Zhou[1], Xiaojuan Zhou[1], Lingling Jia[1] and Mitsuru Izumi[2]

[1]Department of Mathematics and Physics, Shanghai University of Electric Power, 28 Xuehai Road, New Pudong District, Shanghai 201300, P.R. China
[2]Lab of Applied Physics, Tokyo University of Marine Science and Technology, Tokyo 135-8533, Japan

Email: 2009000018@shiep.edu.cn

Abstract. The flux pinning performance of the superconductors is important for the application of the Gd123 bulk superconductors. The study shows that to introduce the secondary phases into the Gd123 bulk matrix can enhance the flux pinning performance. In this article, by using top-seeding melt texture growth process method, single domain $GdBa_2Cu_3O_{7-\delta}$ superconductor bulks doping with the different amounts of $BaFe_{12}O_{19}$ (0.0mol% to 0.8mol%) were successfully achieved. The property and micro-structure have also been investigated. The result shows that there is an obvious improvement on J_C with 0.2mol% $BaFe_{12}O_{19}$ addition. The fine distribution and smaller size of Gd211 particles appear in the micro-structure which may result in the enhancement of J_C. At the same time, $BaFe_{12}O_{19}$ may also form an effective pinning center to increase J_C.

1. Introduction

$REBa_2Cu_3O_{7-\delta}$ (RE denotes a rare earth element) bulk superconductors can be applied in the fields of superconducting motor and superconducting magnetic energy storage [1, 2]. Improving the flux pinning is the key to achieve higher critical currents density (J_C), in order to improve the flux pinning so many experiments have been done, and the experiments reveals that various nano-sized particles which meet the coherence lengths of RE123 (2-4 nm) can form the pinning centers to prevent thermally activated giant flux motion [3-5], thereby improving the flux pinning performance. According to the previous studies, RE_2BaCuO_5 (RE211) and RE_2Ba_4CuMOy (M=Nb, Mo, Zr and Hf) can act as strong pinning centers, improving the critical currents density (J_C) [6, 7]. Recently, Beizhan Li et al. reported that $BaHfO_3$ and $BaTiO_3$ can form the pinning centers in the Gd123 bulk superconducting matrix and enhance the properties of the Gd123 bulk superconductors [8, 9].

In this article, we get Gd123 bulk superconductors by using a top-seeding melt textured growth process, $BaFe_{12}O_{19}$ addition will be added into the bulk as second phase. The superconductivity and surface morphology of the Gd123 bulk superconductors have also been investigated.

2. Experimental

[3] Address for correspondence: Yufeng Zhang, Department of Mathematics and Physics, Shanghai University of Electric Power, 28 Xuehai Road, New Pudong District, Shanghai 201300, Email:2009000018@shiep.edu.cn

Commercial powders of Gd123 (99.9%), Gd211 (99.9%), Ag_2O (99.9%), Pt (99.9%) were using as starting material for the top-seeding melt texture growth. The Gd123 and Gd211 powders were made as a molar ratio of Gd123:Gd211=1:0.4. To improve the mechanical properties and reduce the coarsening of Gd211 particles, 10 wt% Ag_2O and 0.5 wt% Pt were also added. $BaFe_{12}O_{19}$ (99.9%) was introduced into Gd123 as the second phase. The final composition of the bulk were Gd123+0.4Gd211+ x$BaFe_{12}O_{19}$ (x=0.0 mol%, 0.2 mol%, 0.4 mol%, 0.8 mol%)+10 wt%Ag_2O+0.5 wt%Pt. The mixed powders were pressed into pellets, each pellet with the size of 20mm in diameter and 10mm in thickness. MgO crystal covered by Nd123 thin film was put on the top of pellet as seed to nucleate the growth of the Gd123 bulk. Then the pellets were put placed into the conventional box. We set the temperature to about 83℃ above the peritectic temperature (Tp) of Gd123, and stay at that temperature about 1h, then we cool the temperature to 2℃ above the Tp and slowly decreased by 25℃ with 0.25℃/h. During the growth, Gd123 can decompose to Gd211 phase and Liquid phase. It is reported that the lower the number of the Gd211 phase, the higher the nucleation sites. More Gd211 particles will also suppress the growth of the single domain Gd123 bulk. The ratio of Gd123:Gd211=1:0.4 is suitable [10, 11].

To investigate the influence of the $BaFe_{12}O_{19}$ addition on the Gd123 bulk superconductors, magnetization measurements should have been done. Generally, several rectangular tablets of 2 mm×2 mm×1 mm were cut from each bulk, we usually take these tablets from under the seed position and near the boundary of the bulk. Superconducting properties were measured by a SOUID magnetometer, dc magnetization was investigated with a magnetic field parallel to the c axis. And J_C was deduced by the extended bean's critical state model [12].

3. Results and discussion

Figure 1 shows the photograph getting from the top surface of Gd123 bulk with 0.4 mol% $BaFe_{12}O_{19}$.The final Gd123 bulk superconductors can be the size with the diameter of 17 mm and thickness of 8 mm.We can see the Nd123 seed remain stable and no obvious melting phenomenon can be found. The single-domain Gd123 superconductor bulks have been successfully achieved.

Figure 1. Surface appearance of the Gd123 bulk doping with 0.4 mol% $BaFe_{12}O_{19}$.

Figure 2 shows the critical currents density J_C position under the seed site of Gd123 bulk. The enhancement of J_C can be obvious found in the Gd123 bulk superconductor with the addition of 0.2mol% $BaFe_{12}O_{19}$ during the whole field. Another peak value appears in the bulks with 0.2 mol% $BaFe_{12}O_{19}$. Due to the similar size of Gd^{3+} and Ba^{2+}, $Gd_{1+y}Ba_{2-y}Cu_3O_7$ appears in the bulks with 0.2 mol% $BaFe_{12}O_{19}$ which may enhance the peak effect and result in the improvement on critical currents density J_C under the middle field. The superconducting transition temperature, T_C, may be suppressed as shown in figure 3. Owing to the push effect of Gd211, the numbers of Gd211 particles will increase from the seed site to the edge of the bulk. So that, the J_C under the self-field exhibits an increase along the *a* growth direction from the seed to the edge [11, 13]. The result is shown in figure 4. With the addition of 0.4mol% $BaFe_{12}O_{19}$, J_C decreases.

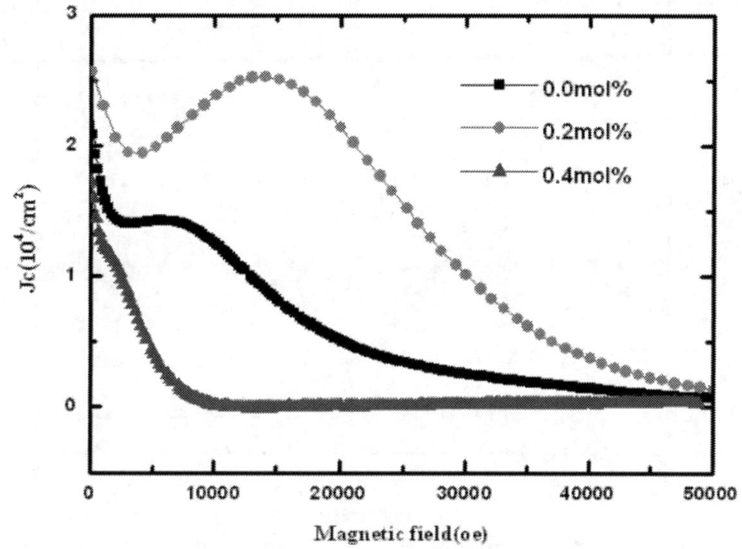

Figure 2. Applied magnetic field dependence of J_C positioned under the seed site for bulks with the addition of 0.0, 0.2, 0.4mol%$BaFe_{12}O_{19}$.

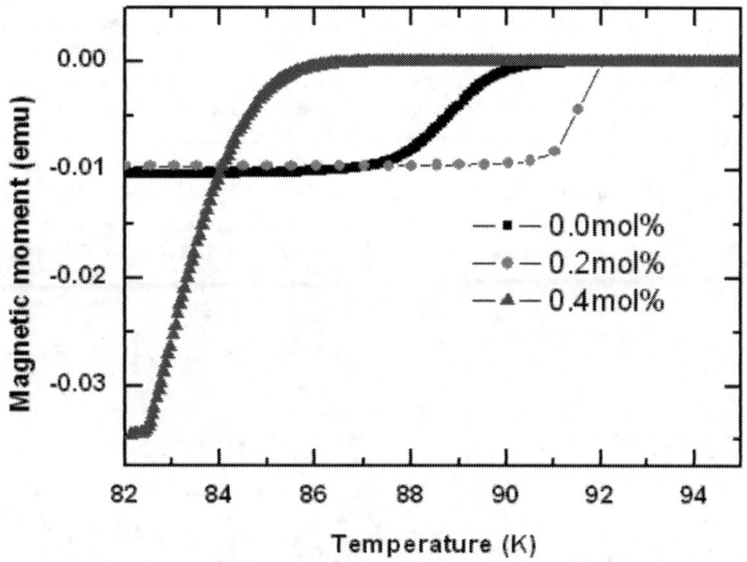

Figure 3. The superconducting transition curve, T_C with the additions of 0.0mol%, 0.2mol%, 0.4mol% $BaFe_{12}O_{19}$.

Figure 3 shows the superconducting transition temperature with 0.0 mol%, 0.2 mol%, 0.4 mol% $BaFe_{12}O_{19}$ addition. The T_C stays around 91K when Gd123 bulk superconductor with no $BaFe_{12}O_{19}$ addition. There is a slight increase on *Tc* when Gd123 bulk superconductor with the addition of

IOP Conf. Series: Materials Science and Engineering **87** (2015) 012077 doi:10.1088/1757-899X/87/1/012077

0.2mol% $BaFe_{12}O_{19}$, *Tc* increases to around 92K. For the addition of 0.4mol% $BaFe_{12}O_{19}$, T_C decreases to 86K, which may decrease J_C of the bulks as shown in figure 3. The result suggested that Gd123 bulk doping with 0.2mol% $BaFe_{12}O_{19}$ can improve the J_C without destroying the superconductivity.

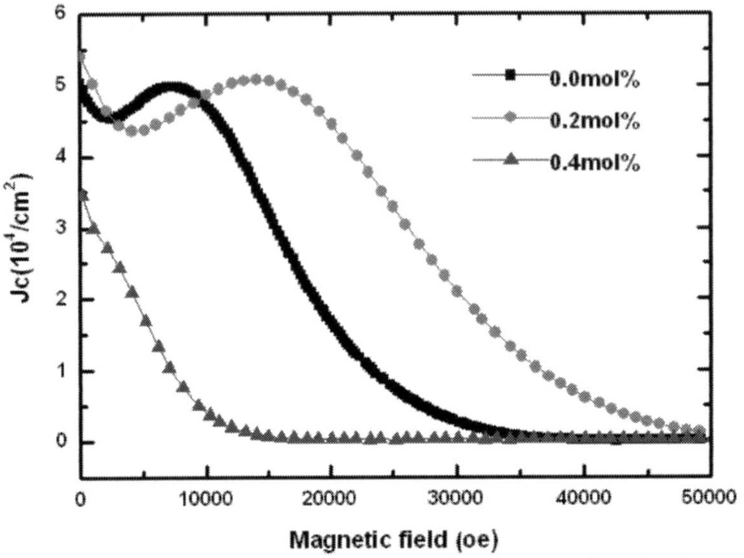

Figure 4. Applied magnetic field dependence of J_C near the edge of bulks with the additions of 0.0, 0.2, 0.4mol% $BaFe_{12}O_{19}$.

Figure 4 shows the critical currents density J_C near the edge of the Gd123 bulks. For 0.2 mol% $BaFe_{12}O_{19}$ addition, there is an obvious enhancement than that in the bulk with 0.0mol% $BaFe_{12}O_{19}$, the highest J_C can reach 55000 A/cm^2 at zero-field, and we can see another peak value in J_C at 1.5T, the value can reach 51000 A/cm^2. It indicates that $BaFe_{12}O_{19}$ additions can almostly improve the critical currents J_C in the whole field. However, with the addition of 0.4 mol% $BaFe_{12}O_{19}$, the J_C decreases comparing with that without doping $BaFe_{12}O_{19}$. It illuminated the excessive addition may affect the superconductivity properties to decrease J_C. And the J_C positioned near the edge is higher than that positioned under the seed site, the inhomogeneous Gd211 and $BaFe_{12}O_{19}$ distribution from the seed to the edge may cause the difference in the J_C.

Figure 5. Scanning electron micr-graphs of the Gd123 bulks superconductors with 0.0mol%, 0.2mol% addition.

Figure 5 shows the scanning electron micrographs of Gd123 bulks with 0.0 mol%, 0.2 mol% addition. For the addition of 0.0 mol% $BaFe_{12}O_{19}$, the large mean particle size of Gd211 appeared in Gd123 bulks superconductors, so that the critical currents density J_C is a lower value. With the increase of the doping amount x=0.2 mol%, the mean size of Gd211 particles were smaller than that with 0.0mol% addition. So that the JC increases with these smaller size of Gd211 particles, which is consist with the previous reports [14, 15]. And we found tiny Fe around Gd211 particles with the doping amount x=0.2mol%, we suggested that $BaFe_{12}O_{19}$ may form effective magnetic flux centers, make an effect on the distribution and the size of Gd211 particles at the same time, so as to enhance the performance of J_C.

4. Conclusion

Gd123 bulk superconductors doping with different $BaFe_{12}O_{19}$ were successfully obtained by using a top-seeding melt texture growth method. We have also investigated the effect of $BaFe_{12}O_{19}$ on the Gd123 bulk superconductor. For the additions of 0.2 mol% $BaFe_{12}O_{19}$, the critical currents density J_C (positioned near the edge of bulk) is higher than that with 0.0 mol% $BaFe_{12}O_{19}$ under the middle and high field. The J_C can reach 55000 A/cm^2 under the self-field. The superconducting transition temperature T_C can stay 92K. The smaller size of Gd211 particles was observed by the microstructure scanning, which enhance J_C. The result indicates that doping with the proper $BaFe_{12}O_{19}$ can be an effective second phase for the enhancement of the Gd123 superconductor bulks

Acknowledgments

Financially supported by the National Natural Science Foundation of China (NSFC, Grant No.11004129), Scientific Research Foundation for the Returned Overseas Chinese Scholars, State Education Ministry (SRF for ROCS, SEM), Innovation Program of Shanghai Municipal Education Commission (Grant No. 11YZ197, 12ZZ174), Shanghai University Scientific and Cultivation for Outstanding Young Teachers in Special Fund (Grant No. sdl10005), "Chen Guang" project of Shanghai Educational Development Foundation (No.12CG63).

References

[1] Tomita M and Murakami M 2003 *Nature* **421** 517-20
[2] Murakami M 2007 *Int. J. Appl. Ceram. Technol* **4** 225-41
[3] Larbalestier D, Gurevich A, Feldmann D M and Polyanskii A 2001 *Nature* **414** 368-77
[4] Haugan T, Barnes P N, Wheeler R, Meisenkothen F and Sumption M 2004 *Nature* **430** 867-70
[5] Muralidhar M and Murakami M 1998 *Physica C* **309** 39-42
[6] Cardwell D A, Babu N H, Shi Y H and Iida K 2006 *Supercond. Sci. Technol* **19** S510
[7] Babu N H, Reddy E S and Cardwell D A 2003 *Supercond. Sci. Technol* **16** L44
[8] Li Beizhan, Xu Kun, Hara Shogo, Zhou Difan, Zheng Yufeng and Izumi Mistsuru 2012 *Physica C* **475** 51-6
[9] Xu Kun, Zhou Difan, Li Beizhan, Hara Shogo, Deng Zigang and Izumi Mitsuru 2015 *Physica C: Superconductivity and Its Applications* **510** 54-6
[10] Zhou Difan, Xu Kun, Hara Shogo, Li Beizhan, Deng Zigang, Tsuzuki Keita and Izumi Mitsuru 2012 *Supercond. Sci. Technol.* **25** 025022
[11] Desgardin G, Monot I and Raveau B 1999 *Supercond. Sci. Technol.* **12** R115
[12] Chen D X and Goldfarb R B1989 *J. Appl. Phys.* **66** 2489
[13] Cardwell D A, Shi Y-H, Hari Babu N, Pathak S K, Dennis A R and Iida K 2010 *Supercond. Sci. Technol.* **23** 034008
[14] Zhang Y F, Izumi M, Xu Y and Kimura Y 2010 *J. Phys*: Conference Series **234** 012052
[15] Zhang Y F, Izumi M, Kimura Y and Xu Y, 2009 *Physica C: Superconductivity and Its Applications* **469** 1169-71

Effect of natural fiber types and sodium silicate coated on natural fiber mat/PLA composites: Tensile properties and rate of fire propagation

C Thongpin[1], J Srimuk, N hipkam and P Wachirapong

Department of Materials Science and Engineering, Faculty of Engineering and Industrial Technology, Silpakorn University, Sanamchandra Palace Campus, Nakornpathom, 73000, Thailand

E-mail: CHANCHAI_C@su.ac.th; cmaterials@hotmail.com

Abstract. In this study, 3 types of natural fibres, i.e. jute, sisal and abaca, were plain weaved to fibre mat. Before weaving, the fibres were treated with 5% NaOH to remove hemi cellulose and lignin. The weaving was performed by hand using square wooden block fit with nails for weaving using one and two types of natural fibres as weft and warp fibre to produce natural fibre mat. The fibre mat was also impregnated in sodium silicate solution extracted from rich husk ash. The pH of the solution was adjusted to pH 7 using H_2SO_4 before impregnation. After predetermined time, sodium silicate was gelled and deposited on the mat. The fabric mat and sodium silicate coated mat were then impregnated with PLA solution to produce prepreg. Dried pepreg was laminated with PLA sheet using compressing moulding machine to obtain natural fibre mat/PLA composite. The composite containing abaca aligned in longitudinal direction with respect to tension force enhanced Young's modulus more than 300 %. Fibre mat composites with abaca aligned in longitudinal direction also showed tensile strength enhancement nearly 400 % higher than neat PLA. After coating with sodium silicate, the tensile modulus of the composites was found slightly increased. The silicate coating was disadvantage on tensile strength of the composite due to the effect of sodium hydroxide solution that was used as solvent for silicate extraction from rice husk ash. However, sodium silicate could retard rate of fire propagation about 50 % compare to neat PLA and about 10% reduction compared to fibre mat composites without sodium silicate coated fibre mat.

1. Introduction

Natural fiber reinforced composites have been paid high attention to extend their applications all over the world. Composites base on natural fiber and biopolymer [1-4] have received much attention as they are environmental friendly. The applications of natural fiber reinforced biopolymer mostly are on non structural part such as some part in the building or automotive parts [5]. Generally short natural fibers are added to bioplastic, as matrix, due to its processing ability using general thermoplastic processing techniques such as twin screw extruder and injection molding machine. The fiber length is one of the factor controlling mechanical properties of composites [6-9]. Fiber added for compounding

[1] Address for correspondence: C Thongpin, Department of Materials Science and Engineering, Faculty of Engineering and Industrial Technology, Silpakorn University, Sanamchandra Palace Campus, Nakornpathom, 73000, Thailand. E-mail: CHANCHAI_C@su.ac.th; cmaterials@hotmail.com.

the composites was normally in the range of 0.1-13 mm. The length which is longer than 20 mm may cause agglomeration during processing. The content of fiber incorporated into matrix was also limited to the highest content of 30-40% by weight. The fiber could be degraded physically, length shortening, during processing [10]. Misra and coworker [11] studied laminated biocomposites which prepared from stacking 4 layers of poly (lactic acid) sheet and 3 layers of kenaf fibers alternatively. They were laminated by using compression molding machine. The fiber content used was highest at 40 % content. They found that the flexural modulus was found about 4 times higher than neat PLA. Khonder et al. [12] prepared unidirectional fiber composite using microbraid yarns wrapped around the metal frame and followed with compression mold. The microbraid yarns prepared from jute fibers being braided around with polymer matrix yarn. Memon et al. [13] had also adapted the same technique to fabricate unidirectional composite of jute yarn hybrid with PLA yarn. The composites were prepared by using compression molding machine. They prepared composites with fiber volume fraction of 37-53%. The impregnation of PLA into the fibers was dependent upon temperature used for compression. Skrifvars and coworker [14] studied thermoplastic composites manufactured using PLA/hemp co-wrapped bybrid yarn prepreg. The manufacturing was carried out by wrapping PLA filaments around a core that prepared from a 400 twists/m of 25 tex hemp yarn and 18 tex PLA yarn. They vary hemp yarn in the range of 10-45% by weight. The prepreg was prepared by compression molding. They found that Young's modulus and tensile strength were increased about 2 times of neat PLA. The fiber mat made from woven and knitting has also been paid attention [15-25]. They are advantage in term of mechanical performance. Among them, Rajini and coworker [25] studied unidirectional natural fiber unsaturated polyeter composites having banana and kenaf fibers and hybrid fibers, i.e. banana fiber hybrid with kenaf fiber, mat. The mat was prepared by weaving fiber and hybrid fiber in different pattern. The composites were prepared by hand lay-up technique. They report that the mat prepared by plain weaving hybrid composites can improve mechanical strength better than other pattern. Maranon and coworker [17] prepared laminate composite from bamboo fabric reinforcing in PLA. They found that composite with fiber in weft direction is greater than warp direction. Khan and coworkers [21] prepared woven jute fabric reinforced PLA composite using also unbalanced plain weave. In this research they found composites in warp direction are higher than those in weft direction. They also reported that tensile strength and tensile modulus were improved about 103% and 211%, respectively.

Prepreg is a class of premix composites or compound of composite that consists of fiber impregnated with polymer resin. Brouwer explained the possibility of preparation natural fiber prepreg for automotive parts [22]. Thermosets, thermoplastic and biopolymer matrix have potential for prepreg preparation. Flammability and thermal properties of natural fiber composites have been also very important for automotive and other applications [23, 24].

It was widely known that natural fibres are various in their properties [25, 26]. Their properties were also dependent upon their size and morphology [26-31]. In this study, we selected three types of natural fibers which are abaca jute and sisal fibers. Jute fiber is from as bast of Corchorus plant whereas abaca and sisal are from leaf of banana and Agave sisalana leaves respectively. These three types of fibers are commonly available in Thailand. This preliminary work was intended to prepare natural fiber mat by plain weave using the same size of fiber in warp and weft direction. Type of fibers used in weaving could be the same type and hybrid of fiber types. The prepreg was prepared impregnation of fiber mat in PLA solution. This study was also aimed to improve flame resistance of natural fiber mat by coating silica on fiber via so-gel process where the silicate solution was extracted from rice husk ash.

2. Materials and method

2.1. Materials

Abaca, sisal and jute fibers were supplied by OTOP centre from Hup Kraphung Market, Prachuap Khiri Khan Province, south of Thailand. PLA 4043D was purchased from NatureWork LLC. Rice Hush ash was kindly supplied from local electric power plant.

IOP Conf. Series: Materials Science and Engineering **87** (2015) 012078 doi:10.1088/1757-899X/87/1/012078

2.2. Method

2.2.1. Preparation of fiber mat. Before plain weaving, all fibers used in this study were treated with 5% NaOH at 80°C for 4 hours. The fibers were cooled to room temperature before neutralization with 5% HCl to pH 7. The fiber was left overnight for thoroughly neutralization. Water was squeezed out, before being washed several times with distilled water. Treated fibers were dried in a hot air oven at 60 °C until totally dried. The dried fiber was then aligned and plain weaved in the square wooded block with the dimension of 30x30 cm in figure 1. The balance fiber mat, of 56x56 (56 yarns in warp and 56 yarns in weft directions per 150 mm) and average weight of 230 g/m^2, was obtained. The fiber mat and hybrid fibers mat was designed and designated as stated in table 1. In order to study the effect of silicate coating, fiber mat was impregnated in sodium silicate solution in which its pH was adjusted in order to gel sodium silicate within predetermined time. The sodium silicate solution was obtained from extraction of rice husk ash, 10 grams with 300 ml 1 *M* NaOH at 100-110°C for 16 hours. After predetermined time sodium silicate gelled and was deposited on the fiber mat surface with average weight gained about 20% by weight of dried fiber mat.

Table 1. Code represent fiber types used for weaving to natural fiber mat. A=Abaca, J=Jute and S=Sisal.

Warp / Weft	Sisal Fibre	Abaca Fibre	Jute Fibre
Sisal Fibre	S + S	S + A	S + J
Abaca Fibre	A + S	A + A	A + J
Jute Fibre	J + S	J + A	J + J

Figure 1. Wooden block used for plain weave to produce natural fiber mat.

2.2.2. Molding of natural fiber mat/PLA composites. The preparation of fiber mat composites was conducted in 2 steps. Firstly, the fiber mat obtained from section 2.2.1 was impregnated in PLA solution prepared from dissolution 20 g of PLA in 100 ml chloroform. Fiber mat was then left to dry in air. 20 grams of dried PLA was compressed in a compression molding machine at 170°C with the hydraulic pressure of 500 psi giving rise to PLA sheet with the thickness of 0.8-1 mm. The PLA sheet was cut into the same size as prepreg and laminated with PLA sheet using compression mold at 170°C with the pressure of 500 psi. The composite sheet was obtained with the thickness of 1-2 mm. Weight fraction of fabrics in composites was about 25 % in all fabricated composite.

2.2.3. Mechanical measurements. Tensile test was performed in warp direction (as longitudinal direction) according to ASTM D 638 using Universal Testing Machine (Instron 3300). The specimens for tensile were cut into dumbbell shape with 50 mm length and 35 mm gauge length. The test was performed at the cross head speed of 5 mm/min. At least 15 specimens were used for investigation.

2.2.4. Fire resistance. Flame rating was performed according to UL-94 flammability standard. The specimen was cut using laser cutter (Laser Engraving Cutting Machine model 6040 N) to a rectangular shape with the length, width and thickness of 125 mm, 13 mm and 0.8 mm respectively. The specimen was clamped vertically after conditioned in 50% RH at 25°C for 48 hours. Blue flame of Bunsen burner was applied to the edge of the specimen for 10 minutes. The flaming time was recorded with respect to 10 mm length on the specimen. Plastic drip was allowed to fall on surgical cotton that placed at 30 cm. below the burning specimen. Burning rate was calculated from burning length divided by flaming time.

3. Result and discussion

3.1. Stress-strain behavior

Plain weave prepared in this study was balanced plain weave where fibers in warp and weft directions have the same size. Warp direction was tensioned as longitudinal direction (parallel) with respect to tensile force. Weft fibre was in transverse direction (perpendicular) with respect to tensile force. The first letter is short abbreviation of fibre such as A which stood for abaca fibre, J stood for jute fibre and S stood for sisal fibre. For labelling the composite system, the first alphabet was designated for fibres in transverse direction and the second one was designated for fibres in longitudinal direction. The fiber mat composites were prepared by compression prepreg with PLA sheet. The content of fabric was kept at the content of 20% by weight.

Stress-strain curves of PLA reinforced with natural fiber mat and hybrid natural fiber mat composites is shown in figure 2. It was shown that natural fiber mat/PLA composites are brittle than neat PLA which has strain extension at 4.13%.

Table 2. Percentage weight gained after sol-gel process.

Fiber Fabric system	A+A	J+A	S+A	A+J	J+J	S+J	A+S	J+S	S+S
% weight gained	28.76	10.46	9.18	10.29	18.17	19.68	9.42	8.64	9.60

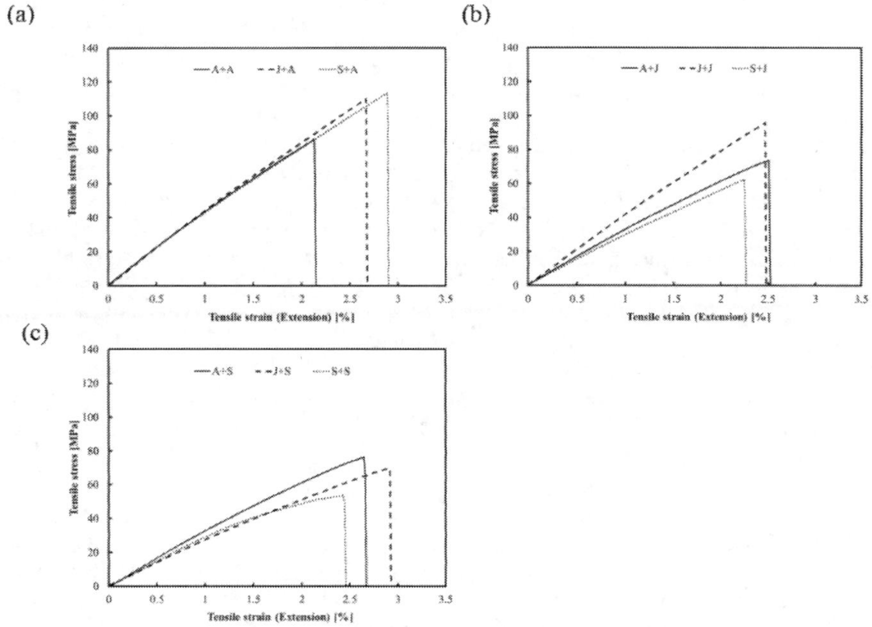

Figure 2. Tensile stress-strain of natural fiber mat/PLA and hybrid natural fiber mat/PLA with (a) abaca fiber (second alphabet) A, (b) jute fiber, J, and (c) sisal, S in longitudinal direction.

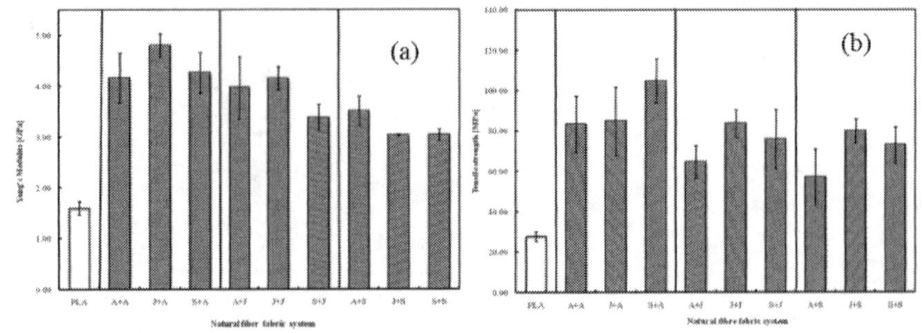

Figure 3. Young's modulus (a) and tensile strength (b) of natural fiber mat/PLA and hybrid natural fiber mat/PLA composites.

Figures 3(a)-3(c) show Young's modulus and tensile strength of neat PLA, natural fibre mat/PLA compared with hybrid natural fibre mat/PLA composites. It was shown that Young's modulus and tensile strength of natural fibre mat /PLA and hybrid natural fibre mat /PLA were improved about 200-400% dependent on fiber system used for reinforcement.

Young's Modulus of the composites having abaca fiber aligned in longitudinal direction was found to be achieved about 258-299% compared to neat PLA, figure 3(a). It could also be seen that if the composite have abaca fiber aligned in longitudinal direction and jute fiber aligned in transverse, the composite would be enhanced about 299%. This result showed the synergism of two fiber types where network structure of abaca fiber [27] induced good penetration of PLA, seen by fracture surface examined by SEM, in figure 4(a), led to good adhesion at the interface. Jute fiber also has similar structure as abaca which jute is produced from stem. The small filament of jute, as shown in figure 5, could be responsible for its higher modulus [32]. The finding in this present study was found to have modulus achievement very much higher than those reported by Memon [13] where they found the achievement in modulus about 95% at optimum moulding temperature. Although the content of fabric in this study was lower, 20% by weight. This could be by the impregnation of PLA into fabric prepared in this study and the network structure of abaca, as mentioned previously.

The natural fibre mat having jute fibre aligned in longitudinal direction was found to achieve slightly lower modulus than abaca. SEM micrograph of fractured surface, shown in figure 4(b) confirms the poorer adhesion between jute fiber bundle and PLA and this resulted in lower in modulus compared to composite reinforced with fiber mat having abaca fiber aligned in longitudinal direction. Not only the poor adhesion but jute fiber also has low cellulose content than abaca fiber [33]. Sisal fiber also shows poor adhesion at the interface, seen figure 4(c). It was explained by Smole et al. [33] that sisal fiber is fairly coarse and inflexible. Although sisal contains percentage of cellulose similar to jute fiber, its tensile properties are not uniform along its length. Sisal fibre also has lower Young's modulus than jute fibre [27]. These could be the reasons that affect natural fiber mat composite having sisal fiber in longitudinal direction to have inferior modulus than abaca and jute fibers.

Tensile strength of natural fibre mat composites are shown in figure 3(b). It was generally seen that natural fibre mat/PLA composites having abaca fibre in longitudinal direction shows tensile strength achievement about 306–385%. The network structure and good adhesion at the interface was responsible for high tensile strength. In fibre mat composite system having jute fibre aligned in longitudinal direction, their tensile strength achievement was found in the range of 237-307% compared to neat PLA, which were slightly lower than that of abaca fibre system. As explained earlier, jute showed adhesion between fibre and matrix inferior to abaca fibre. This was indicated by SEM micrograph in figure 4(b). Moreover, Jute has cellulose content (61-71%) lower than that of abaca fiber (76.6%) [33]. It could be another reason for their tensile strength. Again natural fiber mat/PLA composites having sisal fiber aligned in longitudinal direction generally showed the lowest tensile strength, 210-293% achievement which resulted from non uniform tensile properties along fiber length and also from the cellulose content in the fiber (70%) [33].

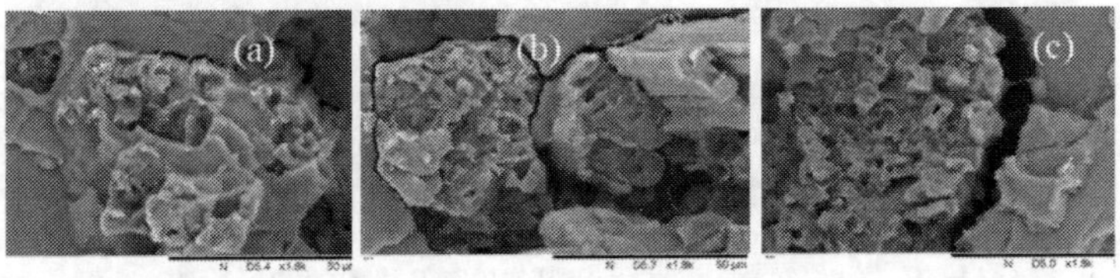

Figure 4. Fractured surfaces of natural fiber/PLA mat composites focusing on longitudinal fiber fractured showing (a) abaca fiber, (b) jute fiber and (c) sisal fiber.

3.2. *Effect of Sol-gel Sodium silicate on fabric surface*

The fiber mat that was coated with sol-gel of sodium silicate solution shows weight gained after impregnated in pH adjusted sodium silicate solution for predetermined time. The coated fabric was then dried at 80°C in a hot air oven until dry. The weight gained was tabulated in table 2. Percentage weight gained was varied depending on fiber and fabric structure. As abaca fiber has network structure, it could absorb sodium silicate solution and hence sol-gel occurred much more than other fiber fabric systems.

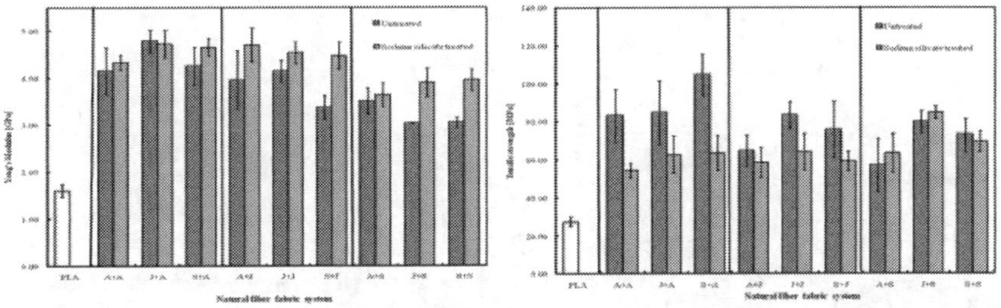

Figure 5. Natural fiber mat/PLA composites with various fiber mat system (a) Young's Modulus and (b) tensile strength compared between sodium silicate untreated and treated fiber mat.

Tensile properties, i.e. Young's modulus, tensile strength of fiber mat/PLA composite compared with sodium silicate treated fiber mat/PLA composites are shown in figures 5(a) and 5(b). The silicate coated on the fiber mat caused Young's modulus of the composites to be increased due to silicate stiffness. However, natural fiber fabric having abaca aligned in longitudinal direction showed small increment in Young's modulus. Due to the network structure of abaca caused absorption of sodium silicate solution more than jute and sisal fibers. Zhang [34] found that under long time NaOH treatment, natural fiber could be swollen and fragmented. It can affect the mechanical behavior of the fiber. For sisal fiber, the fiber morphology is quite different from abaca and jute. It could be seen from figure 6. The diameter of sisal fiber seems to be larger than that of abaca and jute fibers. Silicate could gel only on the outer surface of fiber fabric. The content of silicate coated on the fabric shown in table 2 also showed the lowest weight gained hence less fiber swollen and fragmented. Its mechanical properties could still be sustained.

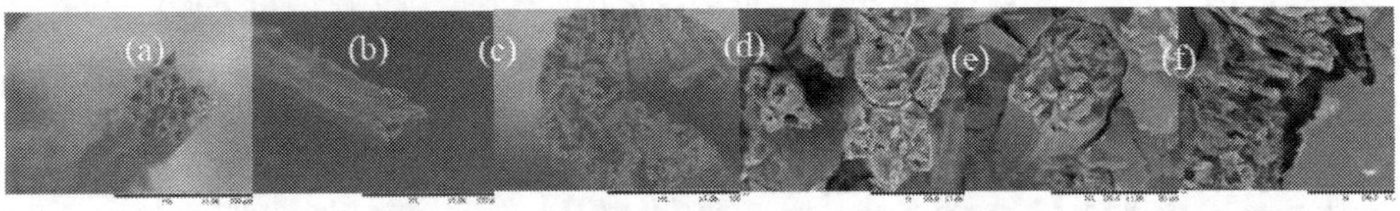

Figure 6. Virgin (a) abaca fiber, (b) jute fiber and (c) sisal fiber; fractured surface of silicate treated natural fiber mat (d) abaca, (e) jute and (f) sisal aligned in longitudinal direction.

3.3. Fire resistance

Figure 7 shows average rate of flame propagation obtained from 10 specimens of each composite compared to neat PLA, using UL-94. PLA was burning very fast with dripping and the dripping carry on flaming. Whereas natural fiber fabric/PLA composites decreased rate of fire propagation evaluated in term of fire propagate distance per second about 40%. The decreased in fire propagation rate could be caused by lower in amount of polymer and also by the presence of natural fibers. Dripping was fond decreased due to the fabric protecting molten plastic to be dripped. Type of fibers did not show significant difference of the fire propagation. After treatment, sodium silicate was deposited on the fabric surface. The improvement was about 17% compared to untreated fabric. This was due to silicate which is widely known for their heat resistance [35-36].

4. Conclusions

Plain weave of natural fibre fabric was prepared using both one type of fibre and hybrid natural fibres. The fabric prepreg was prepared by impregnation the fabric in PLA solution. Prepreg was laminated with PLA sheet using compression moulding machine. The content of the fibre fabric was fixed at 20% by weight. It was found from this work that fabric having abaca aligned in longitudinal direction

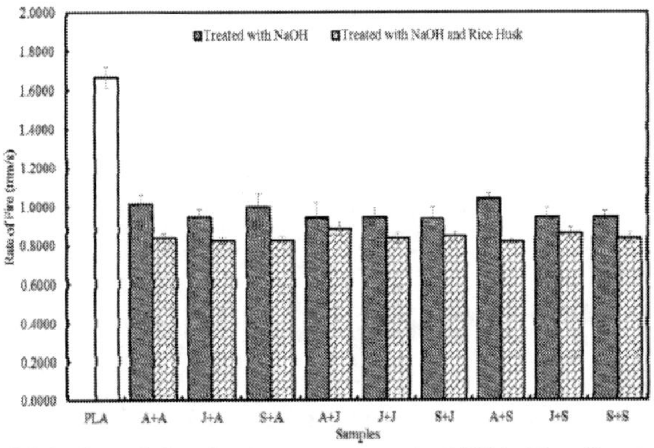

Figure 7. Rate of fire propagating of natural fiber mat/PLA, hybrid natural fiber mat /PLA and sodium silicate treated fabrics composites specimen average from 10 specimens.

exhibited modulus improvement up to 399%. Tensile strength was found 385% improvement. The achievement was dependent upon type of fibre aligned in transverse direction. It was found that fabric having abaca aligned in longitudinal and jute aligned in transverse direction exhibited highest Young's modulus. Tensile strength was also found showed the same trend. The morphology of fibre (network structure and size of fibre bundle are the factors affecting mechanical properties of natural fibre fabric/PLA composite as it control matrix and fabric adhesion at the interface. This work tried to apply sodium silicate extracted from rice husk silica as thermal insulator and hence flame retarder. It was found that after impregnation fabric in sodium silicate solution and allowed to gel for predetermined time sodium silicate was deposited on the fabric. The weight of silicate gained was dependent upon morphology of fibre. Fabric having abaca aligned in longitudinal direction was found to gain the highest silicate deposition. But the effect of alkaline on fibre resulted in slightly increased of modulus. Although less sodium silicate was deposited on the surface of sisal but it was not as swollen as abaca and jute fibre. The mechanical properties of fabric containing sisal aligned in longitudinal direction then clearly exhibited significantly increment in both Young's modulus and tensile strength. Flammability of composites is crucial for their applications. The fire propagation rate of natural fibre fabric composite was found decrease about 40%. Whereas fabric treated with sodium silicate was found to improve flammability about 50% compared to neat PLA. This finding could give an idea of applying natural resources for composite materials and gave the some basic information for future researches.

References

[1] Weyenberg I, Ivens J, Coster A, Kino B, Baetens E and Verpoest I 2003 Influence of processing and chemical treatment of flax fibres on their composites *Composites Science and Technology* **63** 1241-6

[2] Ying-Chen Z, Hong-Yan W and Yi-Ping Q 2010 Morphology and properties of hybrid composites based on polypropylene/polylactic acid blend and bamboo fiber *Bioresourse Technology* **101** 7944-50

[3] Nassiopoulos E and Njuguna J 2015 Thermo-mechanical performance of poly(lactic acid)/flax fibre-reinforced biocomposites *Materials & Design* **66** 473-85

[4] Oksman K, Skrifvars M and Selin J F 2003 Natural fibres as reinforcement in polylactic acid (PLA) composites *Composites Science and Technology* **63** 1317-24

[5] Cristaldi G, Latteri A, Recca G and Cicala G 2010 *Composites based on natural firbres fabrics in Woven fabric engineering* ed Polona Dobnik Dubrovski (online: In. Tech. Europe) chapter 17 pp 317-342.

[6] Thumm A and Dickson A R 2013 The influence of fibre length and damage on the mechanical performance of polypropylene/wood pulp composites *Composites: Part A* **26** 14-52

[7] Santos P A, Spinace M, Fermoselli K and De Paoli M A 2007 Polyamide-6/vegetal fibre composite prepared by extrusion and injection molding *Composite: Part A* **38** 2404-11

[8] Migneaut S, Koubaa A, Erchiqui F, Chaala A, Enhlund K and Wolcot M P 2009 Effects of processing method and fiber size on the structure and properties of wood-plastic composites *Composites: Part A* **40** 80-5

[9] Phelps J H, El Rahman A, Kunc V and Tucker III C L 2013 A model for fiber length attrition in injection-molded long-fiber composites *Composites: Part A* **51** 11-21

[10] Jayaraman K 2003 Manufacturing sisal-polypropylene composites with minimum fiber degradation *Composite Science and Technology* **63** 367-74

[11] Huda M S, Drzal L T, Mohanty A K and Misra M 2008 Effect of fiber surface-treatments on the properties of laminated biocomposites from poly (lactic acid) (PLA) and kenaf fibers *Composites Science and Technology* **68** 424-32

[12] Khondker O A, Ishiaku U S, Nakai A and Hamada H 2006 A novel processing technique for thermoplastic manufacturing of unidirectional composites reinforced with jute yarn *Composites: Part A* **37** 2274-84

[13] Memon A and Nakai A 2013 Fabrication and mechanical properties of jute spun yarn/PLA unidirection composite by compression molding *Energy Procedia* **34** 830-8

[14] Baghaei B, Skrifvars M and Berglin L 2013 Manufacture and characterisation of thermoplastic composites made from PLA/hemp co-wrapped hybrid yarn prepregs *Composites: Part A* **50** 93-101

[15] Graupner N, Hermann A S and Mussug J 2009 Natural and man-made cellulose fibre-reinforced poly (lactic acid) (PLA) composites: An overview about mechanical characteristics and application areas *Composites: Part A* **40** 810-21

[16] Kannan T G, Wu C M and Cheng K B 2012 Effect ofdifferent knitted structure on the mechanical properties and damage behavior of Flax/PLA (poly lactic acid) double covered uncoomingled yarn composites *Composites Part B* **43** 2836-42

[17] Porras A and Maranon A 2012 Development and characterization of a laminate composite material from polylactic acid (PLA) and woven bamboo fabric *Composites: Part B* **27** 82-8

[18] Misnon M I, Islam M M, Epaarachchi J A and Lau K 2014 Potentiality of utilising natural textile materials for engineering composites applications *Materials and Design* **59** 359-68

[19] Song Y S, Lee J T, Ji D S, Kim M W and Lee S H 2012 Viscoelastic and thermal behavior of woven fiber reinforced poly (lactic acid) composites *Composites: Part B* **43** 856-60

[20] Gu Y, Tan X, Yang Z, Li M and Zhang Z 2014 Hot compaction and mechanical properties of ramie fabric/epoxy composite fabricated using vaccuum assisted resin infusion molding *Materials and Design* **56** 852-61

[21] Khan G M, Terano M, Gafur M A and Alam M S 2013 Studies on the mechnical properties of woven jute fabric reinforced poly (L-lactic acid) composites *Journal of King Saud University-Engineering Sciences* (in press)

[22] Jayaramadu J, Siva Mohan Reddy G, Varaprasad K and Sadiku E R 2013 Structure and properties of poly (lactic acid)/ *Sterculia urens* uniaxial fabric biocomposites *Carbohydrate Polymers* **94** 822-8

[23] Yan L, Chouw N and Jayaraman K 2015 Effect of UV and water spraying on the mechanical properties of flax fabric reinforced polymer composites used for civil engineering *Materials and Design* **71** 17-25

[24] Tausif M, Ahmad F, Hussain U, Basit A and Hussain T 2015 A Comparative study of mechanical and comfort properties of bamboo viscose as an eco-friendly alternative to conventional cotton fibre in polyester blended knitted fabric *Journal of Cleaner Production* **80** 110-5

[25] Alavudeen A, Rajini N, Kartiheyarn S, Thiruchitrambalan M and Venkateshwaren N 2015 Machanical properties of banana/kenaf fibre-reinforced hybrid polyester composites: Effect of woven fabric abd random orientation *Materiala and Design* **66** 246-57

[26] Faruk O, Bledzki A, Fink H and Sain M 2012 Biocomposites reinforced with natural fibers: 2000-2010 *Progress in Polymer Science* **37** 1552-96

[27] Abiola O, Kupolati W, Sadiku E and Ndambuki J M 2014 Utilisation of natural fibre as modifier in bituminous mixes: A review *Construction and Building Materials* **54** 305-12

[28] Ernestin M, Fidelis A, Vitorino T, Pereiraa C, Gomes O, Silva F and Filho R 2013 The effect of fiber morphology on the tensile strength of natural fibers *Journal of Materials Research Technology* **2** 149-57

[29] Munawar S S, Umemura K and Kawai S 2007 Characterization of the morpjology, physical and mechanical properties of seven nonwood plant fiber bundles *Journal of Wood Science* **53** 108-13

[30] Ausias G, Bourmaud A, Corooler G and Baley C 2013 Study of the fibre morphology stability in polypropylene-flax composites *Polymer Degradation and Stability* **98** 1216-24

[31] Peltola H, Paakkonen E, Jetsu P and Heinemann S 2014 Wood based PLA and PP composites: Effect of fibre type and matrix polymer on fibre morphology, dispersion and composite properties *Composites Part A: Applied Science and Manufacturing* **61** 13-22.

[32] Bledzki A K, Mamun A A and Faruk O 2007 Abaca fiber reinforced PP composites and comparison with jute and flax fiber PP composites express *Polymer Letters* **1** 755-62

[33] Smole M S, Hribernik S K S and Kleinshek Kreze T 2013 *Plant Fibers for Textile and Technical Applications in Advanced in Agrophysical Research* ed Stanislaw Grundas and Andrzej Stepniewski (online: In. Tech. Europe) chapter 15

[34] Zhang S, Wang W C, Li F X and Yu J Y 2013 Swelling and dissolution of cellulose in NaOH aqueous solvent systems *Cellulose Chem. Technol.* **47** 671-9

[35] Thongpin C, Juntum J, Sa-ngun-moo R and Sombatsompop N 2010 Thermal stability of PVC with gamma-APS-g-MMT and zeolite stabilizers by TGA technique *Journal of Thermoplastic Composite Materials* **23** 435-45

[36] Gao D, Li R, Lv B, Ma J, Tian F and Zhang J 2015 Flammability, thermal and physical-mechanical properties of cationic polymer/montmorillonite composite on cotton fabric *Composite Part B* **77** 329-37

IOP Conf. Series: Materials Science and Engineering **87** (2015) 012079 doi:10.1088/1757-899X/87/1/012079

Solid state synthesis of chitosan and its unsaturated derivatives for laser microfabrication of 3D scaffolds

T A Akopova[1, 3], T S Demina[1], V N Bagratashvili[2], K N Bardakova[2], M M Novikov[2], I I Selezneva[2], A V Istomin[1], E A Svidchenko[1], G V Cherkaev[1], N M Surin[1] and P S Timashev[2]

[1]Enikolopov Institute of Synthetic Polymer Materials, Russian Academy of Sciences, 70 Profsoyuznaya Str., Moscow 117393, Russia
[2]Institute on Laser and Information Technologies, Russian Academy of Sciences, 2 Pionerskaya Str., Moscow 142190, Troitsk, Russia

E-mail: akopova@ispm.ru

Abstract. Chitosans with various degrees of deacetylation and molecular weights and their allyl substituted derivatives were obtained through a solvent-free reaction under shear deformation in an extruder. Structure and physical-chemical analysis of the samples were carried out using nuclear magnetic resonance (NMR), ultraviolet (UV) and infrared radiation (IR) spectroscopy. Photosensitive materials based on the synthesized polymers were successfully used for microfabrication of 3D well-defined architectonic structures by laser stereolithography. Study on the metabolic activity of NCTC L929 cultured in the presence of the cured chitosan extracts indicates that the engineered biomaterials could support adhesion, spreading and growth of adherent-dependent cells, and thus could be considered as biocompatible scaffolds.

1. Introduction

3D models of various cell cultures that are performed in a volume of polymer matrices of well-defined architectonics have been considered as promising biomimetic constructions for tissue engineering in recent years [1, 2]. Natural polysaccharides, including chitosan and its derivatives, are attractive polymers to create scaffolds of that kind due to high cell affinity alongside with biodegradation capability [3]. These scaffolds should possess an optimum microarchitecture to provide efficient transfer of nutrients and cell growth and proliferation. Laser microstereolithography technique based on two-photon absorption polymerization allows to induce a cross-linking spatial process in a small volume at a beam focus. Thus formed scaffolds possess well-managed reproducibility of the 3D structure with high resolution up to 200 nm [4, 5]. Hydrogels or scaffolds for tissue engineering have been developed through two-photon-induced cross-linking of the photosensitive polymer compositions based on gelatin methacrylate [6], polylactic [7] and hialuronic [8] acids. Chitosan – natural cationic polyelectrolyte which has a free primary amino group almost in each of polymeric units, has been applied early via "guest-host" scheme to design 3D triacrylate-based structures through two-photon polymerization [9].

[3] Address for correspondence: T A Akopova, Enikolopov Institute of Synthetic Polymer Materials, Russian Academy of Sciences, 70 Profsoyuznaya Str., Moscow 117393, Russia. E-mail: akopova@ispm.ru.

To increase the reactivity of the functional groups of the polymers, which are subjected to cross-linking under laser radiation, it is promising to introduce unsaturated groups into their structure. This study reports the solid-state synthesis of chitosan and its unsaturated ether – allyl chitosan – for the purpose of 3D structures fabrication by laser stereolithography.

2. Materials and methods

2.1. Solid state synthesis of chitosan and its allyl substituted derivatives

The reactive blends of allyl bromide and chitosan characterized by the predetermined ratios of the components (0.5 mol of alkylation agent per 1 unit of chitosan) were processed in a semi-industrial co-rotating twin-screw extruder (Berstorff ZE40, Germany) in the absence of any liquid dispersion media and catalyst at -5°C. The extruder is equipped with processing elements which performed compression and shear deformation of the substances in a thin layer. Plastic flow realized at such conditions leads to intense chemical interaction of the blend components up to quantitative yield of the products [10]. Chitosan was synthesized through solid-state alkaline deacetylation of crab chitin (Vostok-Bor, Russia) in accordance with a published procedure [11]. Two samples having different content of residual N-acetylated units, 10 and 30 mol%, and comparatively low molecular weight of approximately 50-90 kDa, were obtained. Allyl bromide (99%) was purchased from Sigma-Aldrich. The products were purified with isopropyl alcohol to remove unreacted monomer before characterization.

2.2. IR spectroscopy

Infrared spectra were recorded on a Bruker Vertex 70 spectrometer and treated using the program Bruker Opus (version 7.0). All spectra were normalized with respect to the composite stretching band at 1060 cm^{-1} of the C–O pyranose cycle used as an internal standard. The following major bands in the IR-spectrum of allyl-modified chitosan were observed: c.a. 3400 (sh) cm^{-1} – O-H stretching; 3365, 3357 cm^{-1} – N-H asymmetrical and symmetrical stretching, correspondingly; 1651 cm^{-1} – Amide-I band; 1590 cm^{-1} – NH_2-groups – deformational mode; 1500-1200 cm^{-1} region – multiple peaks typical for C-C and C-N mixed vibrations; c.a. 1150 cm^{-1} – C-O-C stretching in glycoside bridge; c.a. 1070 (sh), 1060 and 1055 (sh) – several overlapped bands of C-O-C and C-O(H) stretching of pyranose ring. Only small changes in relative intensity and the shape of complex bands in the spectrum of modified product could be observed as compared with the spectrum of initial chitosan.

2.3. UV-Vis spectroscopy

Electronic absorption spectra of 1% polymer solution in 0.1 M HCl were recorded in the 200-420 nm region on a Shimadzu UV 2501 PC spectrophotometer. The measurements were performed in quartz cuvettes with an optical path length of 10 mm.

2.4. NMR analysis

^{1}H (300 MHz), ^{13}C (75.5 MHz) and ^{13}C-{^{1}H} APT (attached proton test) NMR spectra were recorded on a Bruker Avance II 300 spectrometer at 25 and 70°C in D_2O solution doped with HCl. Signal ascriptions to give a ratio of initial and allyl-substituted pyranose cycles as well as a content of residual N-acetylated units are shown.

2.5. Scaffold fabrication

For scaffold fabrication, 20 wt% highly viscous polymer solutions were prepared in 4 wt% acetic acid. In order to prepare photosensitive material, 100 mg of polymer solution was mixed with 200 mkL 1 wt% Irgacure 2959 (BASF Kaisten AG) as well as 20 mg poly(ethylene glycol) diacrylate (PEG-DA, Sigma-Aldrich) for 90 min at ambient temperature.

3. Results

3.1. Polymers characterization

^1H (300 MHz) spectrum of allyl chitosan sample is shown in figure 1. And 8-10 mol% of allyl-modified units and similar content of residual N-acetylated units were found in accordance with the data. Chemical shift of carbon in allyl methylene group (53 ppm) in ^{13}C-{^1H} APT NMR spectra indicates its bonding with nitrogen atom, therefore N-alcylation mainly takes place at the given conditions of the solid state synthesis. Since relatively low degree of amino-groups substitution (about 20 mol% in total) was found, chitosan modified with hydrophobic allyl moieties retains solubility in acidic aqueous media which is characteristic of initial polymer.

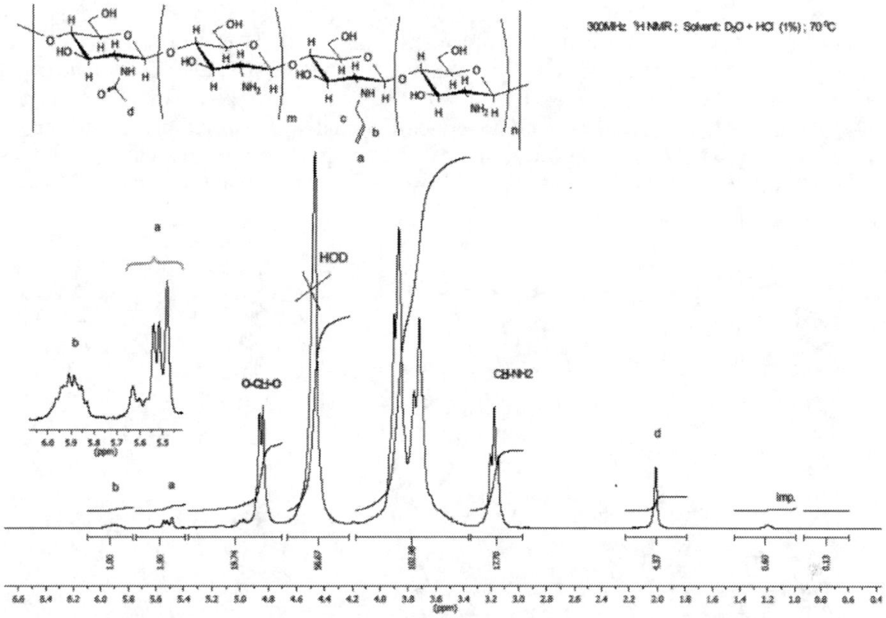

Figure 1. ^1H NMR (300 MHz) spectrum of allyl chitosan in D_2O solution doped with 1% HCl.

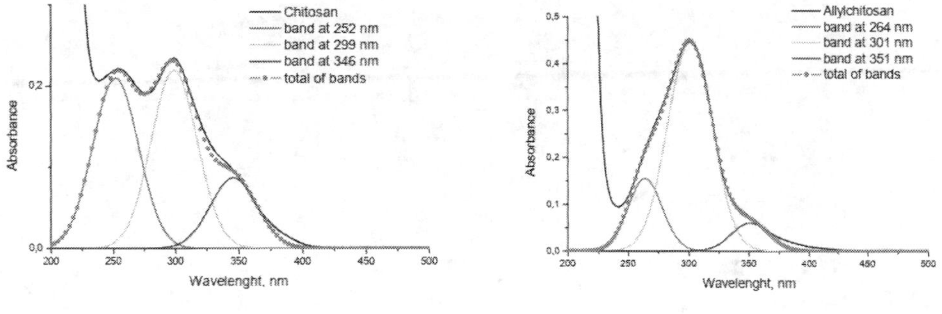

Figure 2. Electronic absorption spectra of 1% polymer solution in 0.1 M HCl in a range of 200-420 nm: initial chitosan (left) and allyl-modified chitosan (right) samples.

Electronic absorption spectra of the initial and modified chitosan samples are shown in figure 2. The spectrum of allyl chitosan sample exhibits one strong band at 301 nm and two less intense bands at 264 and 351 nm (with intensity ratio of 1/0.3/0.1, correspondingly). The intensity of the main absorption band (at 301 nm) of allyl chitosan is increased two fold in comparison with initial chitosan sample, while the intensity of other two bands is significantly decreased. No significant difference is revealed in the absorption spectra of both chitosan and allyl chitosan diluted solutions at the concentration range from 0.1 to 1 wt%, which indicates the absence of an aggregation effect.

3.2. 3D microfabrication
Concentrates of the samples were subjected to structuring by a microstereolithography setting developed in ILIT RAS and equipped with a laser TEMA-1053/100 (Avesta-Project, Russia) and an Epiplan 20× microscope objective (Zeiss, Oberkochen, Germany). Second harmonic of a femtosecond laser pulses (80 fs, 69.7 MHz, 1050 nm) was used for polymer structuring via two-photon absorption (2 PP) method in accordance with a published procedure [12]. Single laser absorption (1 PP) scaffolds were fabricated with LS 120 stereolithography setting (ILIT RAS, Russia), equipped with He-Cd laser with a wavelength of 325 nm and intensity 10 mW. The scanning speed was 1.5 m/sec, a beam diameter - 0.10 mm. After fabrication, unpolymerized material was washed away with mQ, 2 wt% acetic acid and aqueous ammonia. In figure 3, microscopic images demonstrate that the matrices are of a regular shape structures, and specified partial resolution is transmitted without distortion that indicates the uniformity of cross-linking.

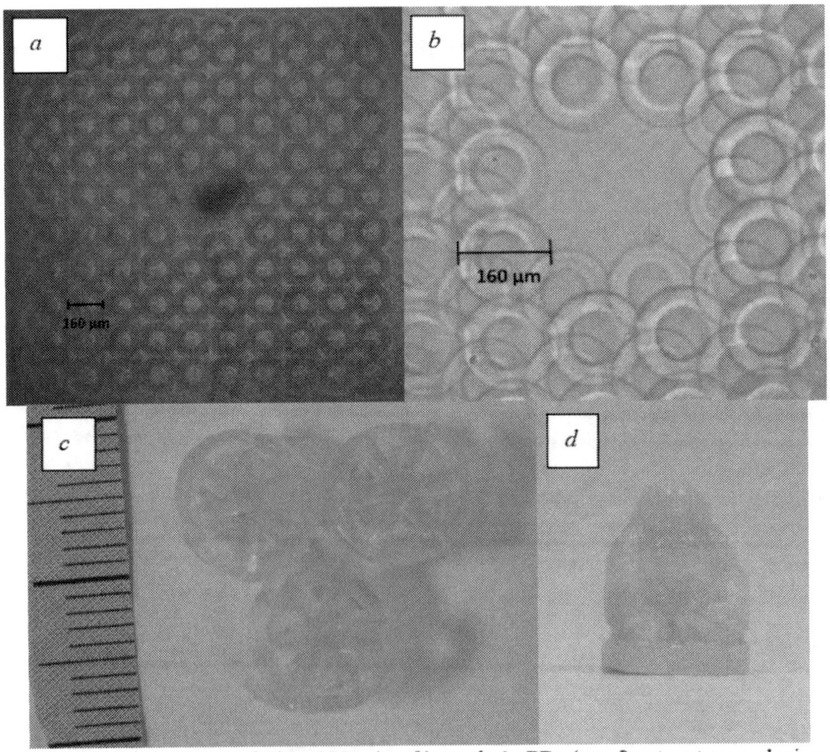

Figure 3. Microscopic images of the 2PP (*a*, *b*) and 1 PP (*c*, *d*) structures designed on a microstereolithography setting using the photosensitive chitosan-based compositions.

3.3. Study of biocompatibility
Study on the metabolic activity of NCTC L929 cultured in the presence of the engineered chitosan extracts indicates that the engineered biomaterials could support adhesion, spreading and growth of adherent-dependent cells, and thus could be considered as biocompatible scaffolds (see figure 4).

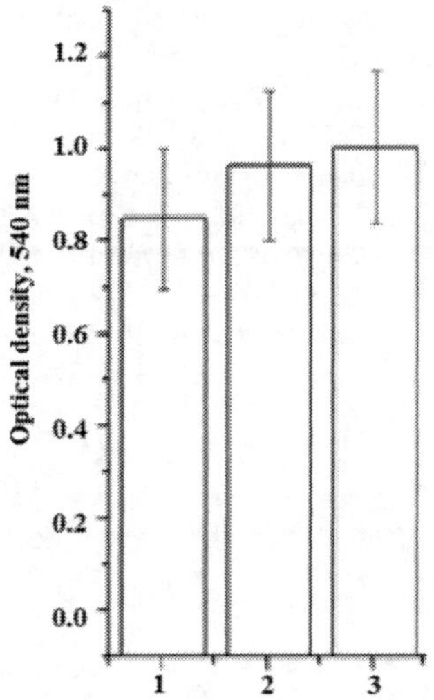

Figure 4. Metabolic activity of NCTC L929 cultured for 48 hours in 3-day extracts of: 1– chitosan + Irgacure 2959 (0.92%), 2 – chitosan + Irgacure 2959 + PEG-DA, 3 – control. Results of MTT-test are expressed as mean ± standard deviation for three replicates.

4. Conclusion

The laser microstereolithography techniques based on non-linear absorption of femtosecond laser pulses allows to fabricate structures of photosensitive materials with a resolution from tens of micrometers down to hundreds of nanometers. These 3D engineering approaches offer numerous opportunities for tissue engineering biomaterial scaffolds that mimic the complex architecture of native tissues and provide mechanical support for the cells. In this study, both 2PP and 1PP techniques are applied to develop precisely defined biocompatible 3D chitosan scaffolds. Chitosan with various degrees of deacetylation and molecular weights and their allyl substituted derivatives are obtained through a solvent-free reaction under shear deformation in the semi-industrial extruder. Photosensitive materials based on the chitosan and its unsaturated derivatives are successfully used for microfabrication of 3D structures by laser stereolithography methods. The study on the cytotoxicity of the fabricated chitosan-based structure reveals no significant effects on animal cell viability.

Acknowledgments

This work was supported by the Russian Foundation for Basic Research (Grant No. 14-29-07234-ofi_m) and Russian Scientific Foundation (Grant No. 14-25-01422).

References

[1] Aubin H, Nichol J W, Hutson C B, et al. 2010 Directed 3D cell alignment and elongation in microengineered hydrogels *Biomaterials* **31** 6941-51

[2] Chen A A, Tsang V L, Albrecht D R and Bhatia S N 2007 *BioMEMS and Biomedical Nanotechnology* (Therapeutic Micro/Nanotechnology vol 3) eds M Ferrari, T Desai and S N Bhatia (Berlin: Springer) Chapter 2 pp 23-38

[3] Prabaharan M and Tiwari A 2010 *Polysaccharides: Development, Properties and Applications* ed A Tiwari (New York: Nova Science Publishers Inc) pp 151-76

[4] Ovsianikov A, Mironov V, Stampfl J and Liska R 2012 Engineering 3D cell-culturematrices: Multiphoton processing technologies for biological and tissue engineering applications *Expert Rev. Med. Devices* **9** 613-33

[5] Correa D S, Cardoso M R, Tribuzi V, Misoguti L and Mendonca C R 2012 Femtosecond laser

in polymeric materials: Microfabrication of doped structures and micromachining *IEEE J Sel Topics Quantum Electron* **18** 176-86
[6] Ovsianikov A, Deiwick A, Van Vlierberghe S, et al. 2011 *Materials* **4** 288-99
[7] Gill A and Claeyssens F 2011 *3D Cell Cult* **695** 309-21
[8] Burdick J, Chung C, Jia X, Randolph M A and Langer R 2005 *Biomacromolecules* **6** 386-91
[9] Correa D S, Tayalia P, Cosendey G, et al. 2009 *J Nanosci Nanotech* **9** 5845-9
[10] Zharov A A 1994 *High Pressure Chemistry and Physics of Polymers* ed A L Kovarskii (London, Tokyo: CRC Press) Chapter 7 pp 268-301
[11] Rogovina S Z, Akopova T A and Vikhoreva G A 1998 *J Appl Polym Sci* **70** 927-33
[12] Ovsianikov A, Gaidukeviciute A, Chichkov B N, et al. 2008 *Laser Chem* Article **2008** 493059

Investigation and modeling of moisture sorption behaviour of rice starch/carboxymethyl chitosan blend films

R Suriyatem[1], C Rachtanapun[2], P Raviyan[1], P Intipunya[1] and P Rachtanapun[3, 4]

[1]Division of Food Science and Technology, Chiang Mai University, Chiang Mai 50100, Thailand
[2]Department of Food Science and Technology, Faculty of Agro-Industry, Kasetsart University, Bangkok 10900, Thailand
[3]Division of Packaging Technology, Chiang Mai University, Chiang Mai 50100, Thailand

E-mail: pornchai.r@cmu.ac.th

Abstract. The biopolymer films from rice starch (RS) and carboxymethyl chitosan (CMCh) were developed by solution casting. The effect of the ratios of rice starch to CMCh (100:0, 88:12, 67:33, 50:50, 33:67, 12:88 and 0:100) on water barrier properties and moisture sorption isotherm of blend films was studied. Water vapor permeability of rice starch film and CMCh film were 4.8 and 9.1 g.mm/m^2.mHg.day, respectively, while those of the RS/CMCh blend films ranged between 5.0 and 9.1 g.mm/m^2.mHg.day. The sorption isotherm of RS/CMCh blend films was determined at 25°C. The sorption behaviour of RS/CMCh blend films could be categorized as type II and type III isotherms. The highest equilibrium moisture content (63.5 g water/100 g dry solid) was obtained in the CMCh film at a_w of 0.87. For further application of sorption isotherm data, the moisture sorption characteristic of the films can be predicted using empirical models. Lewicki, Peleg, Guggenheim-Anderson-deBoer (GAB), Brunauer–Emmett–Teller (BET), and Oswin models were tested to fit the experimental data. The Peleg equation showed the best fit to the experimental data of the RS/CMCh blend films with $r^2 \approx 0.997$ and the lowest %RMS of 6.2-44.3.

1. Introduction

Natural biopolymers are the main materials used to prepare edible films. These edible films can improve handling properties, extend shelf life of food products, and help reduce the environmental pollution caused by synthetic polymer packaging waste [1]. Edible films can be prepared from polysaccharides, proteins, and lipids. Among these biopolymers, starch is one of the most attractive materials because of its good film forming capability, biocompatibility, relative inexpensiveness, renewability and abundance. Rice starch is made from a rice by-product after rice processing. The major components of rice starch—amylose and amylopectin—are attractive raw materials to be utilized as barriers in packaging materials [2]. Although films of polysaccharides have overall suitable mechanical and optical properties [3], their application is limited by their efficient barrier against low polarity [2]. Therefore, there is a current trend to combine different biomaterials for food packaging

[4] Address for correspondence: P Rachtanapun, Division of Packaging Technology, Chiang Mai University, Chiang Mai 50100, Thailand. E-mail: pornchai.r@cmu.ac.th.

and coatings in order to overcome the shortcomings [3]. Starch film is often blended with another biopolymer to create a new biocomposite with the more overall benefits than when used individually [2, 3].

Chitosan is a linear polysaccharide containing β-(1-4)-linked N-acetyl-D-glucosamine and D-glucosamine units [4]. It is nontoxic and biodegradable. Chitosan is extracted from chitin, which is abundantly available in nature. Chitosan exhibits excellent biocompatibility, biodegradability, non-toxicity, adsorption properties, etc. [4]. Thus, it has a wide range of applications, including utilization in food processing, agriculture, biomaterials, biotechnology, biomedicine, and pharmaceuticals [4]. However, its limitation of only dissolving in acidic water affects the finished product by causing a bad odor [5]. Carboxymethyl chitosan (CMCh) is an etherified chitosan, which can be used in various industries including food and non-food. It is an interesting choice for use in polymer blending because of its dissolvability in cold water. There are a few studies that have been carried out to investigate the preparation and properties of blends between starch and CMCh [6].

To date, no research has presented the moisture sorption characteristics of rice starch/ carboxymethyl chitosan (RS/CMCh) blend film. The objectives of this study were (1) to determine the moisture sorption isotherm including their relation to water vapour permeability and (2) to investigate the sorption model and determine the best fitted model.

2. Methodology

2.1. Materials

Rice starch was purchased from Thai Flour Industry Company Ltd. (Bangkok, Thailand). Shrimp chitosan flakes (molecular weight ranging between 900,000-1,300,000 Dalton, degree of deacetylation=98%) were purchased from Taming Enterprises (Thailand). Monochloroacetic acid was purchased from Sigma-Aldrich (Steinheim, Germany). Sodium hydroxide, glacial acetic acid, isopropanol, ethanol and methanol were purchased from Lab-scan Co., Ltd. (Bangkok, Thailand). Glycerol was purchased from Union Science Co., Ltd. (Chiang Mai, Thailand). All chemicals used in the experiment were AR grade or the equivalent.

2.2. Synthesis of carboxymethyl chitosan (CMCh)

Chitosan was synthesized into CMCh according to the method propounded by Chen and Park [7] and Rachtanapun and Suriyatem [8] with slight modifications. Chitosan flakes were grounded into powder using a hammer mill to obtain particle size smaller than 60 mesh. Chitosan powder (25 g) was suspended in solution of NaOH:isopropanol:distilled-water (50 g:400 mL:100 mL) with continuous stirring at 50°C for 1 h using a hotplate stirrer (IKA C-MAG HS 7, USA). The solution of monochloroacetic acid (etherifying agent) and isopropanol (50 g: 50 mL) was added into the mixture and continuously stirred at 50°C for 4 h. Next, the mixture was separated into two phases—liquid phase and solid phase—and the liquid phase was removed. The solid phase was suspended in methanol, neutralized with glacial acetic acid, and filtered. The residue was washed with 500 mL of 70% (v/v) ethanol five times; followed by washing with 500 mL of 95% (v/v) ethanol to remove undesirable products. The last product, CMCh, was dried in an oven at 55°C for 18 h and stored in a sealed container.

2.3. Film preparation

Film forming solutions (5% w/v) with different rice starch (RS) to CMCh ratios were prepared to yield different RS/CMCh blended films (table 1). RS was dispersed in distilled water and constantly stirred at 90°C for 5 min. CMCh was dissolved in distilled water at 80°C for 10 min. The RS and CMCh solution were mixed and glycerol (25% wt of solid content) was added as a plasticizer. The mixture was degasified, cooled to 25°C, casted on a silicone plate, and then dried at 25°C for 36 h. The films were peeled and kept in an aluminium foil bag until use.

2.4. Water vapor permeability (WVP)

WVP was investigated using the ASTM E96-93. The cups containing dried silica gel were covered with the circular specimens and sealed with paraffin wax. Sealed cups were weighed and kept in a desiccator with saturated solution of $NaNO_2$ to provide 65% RH at 25°C. The cups were re-weighed daily for seven days. WVP was calculated by equation (1) [9]:

$$WVP = \frac{WVTR}{\Delta P} xL \quad (1)$$

where L is the thickness of the film specimen and ΔP is the partial pressure difference of water vapor across the film. The WVTR (water vapor transmission rate) was acquired from the calculation of the slope of the linear relationship between time (independent variable) and weight gain (dependent variable) divided by the area of the mouth of the test cup. All measurements were performed in triplicate.

2.5. Moisture sorption isotherm

The film specimen, approximately 1.0 g in weight and 20 mm x 20 mm in size, was dried in an oven at 55°C for 24 h and preconditioned in a desiccator containing dried silica gel for seven days. The film specimen was placed in the desiccators containing various saturated salt solutions. The specimen was weighed every 24 h until equilibrium was achieved; this occured when the change in weight did not exceed 0.1% for three consecutive weight measurements. Equilibrium moisture content was calculated by equation (2) [10]:

$$EMC = \{\frac{We}{Wi}(Mi+1)-1\}x100 \quad (2)$$

where EMC is the equilibrium moisture content (g water/ 100 g dry solid); We is the equilibrium weight of the film specimen (g); Wi is the initial weight of the film specimen (g); and Mi is the initial moisture content of the film specimen (g/g).

2.6. Moisture sorption isotherm model

Five isotherm models from the literature were selected to fit the experimental data of adsorption isotherms of RS/CMCh blend films. The models are characterized as given in table 1 [11-16].

Table 1. List of nomenclatures and models describing the moisture sorption isotherms.

Nomenclature		Model	Mathematical expression
a_w	water activity	Lewicki	$M = [F/(1-a_w)^G] - [F/(1+a_w^H)]$
RH	relative humidity in desiccator	Peleg	$M = A.a_w^B + C.a_w^D$
A, B, C, D	constants of the Peleg model	GAB (Guggenheim–Anderson–deBoer)	$M = (M_0Cka_w)/[(1-a_w)(1+(C-1)ka_w)]$
C, k	constants of the GAB model	BET (Brunauer–Emmett–Teller)	$M = [(M_0C_1a_w]/[(1-a_w)(1+(C_1-1)a_w)]$
C_1	constant of the BET model	Oswin	$M = C_1[a_w/(1-a_w)]^{C2}$
F, G, H	constants of the Lewicki model	M, Equilibrium moisture content; M_0, monolayer moisture content; A, B, C, C_1, C_2, D, F, G, H, and k constants specific to individual mathematical expression.	

2.7. *Statistical analysis*
All data was analyzed by an analysis of variance (ANOVA) and Duncan's multiple range test (DMRT; $P \leq 0.05$) using the SPSS version 11.

3. Results and discussion

3.1. *Water vapor permeability*
The various RS/CMCh film-forming solutions were prepared and casted into seven different blend films using the ratios between the two main solids—rice starch and carboxymethyl chitosan—as shown in table 2. The WVPs of the blend films were measured at a vapor pressure difference of 0/65% across the film and are listed in table 2. The results presented that CMCh film (RS/CMCh-07) and the film containing 88% CMCh (RS/CMCh-06) provided around two times higher WVP than rice starch film (RS/CMCh-01). This result could be explained by the higher hydrophilicity of CMCh material (NH_3+ and COO- groups) than rice starch (OH- groups) [2]. The other RS/CMCh blend films, which composed 12–67% CMCh, showed no significantly different WVPs, but all were lower than that of CMCh film. These results are similar to previous researches by Wu et al. [17] for pullulan-CMCh blend film and Bourtoom and Chinnan [2] for rice starch-chitosan blend film. They presented that WVPs of the films increased with an increase of chitosan [2] or CMCh content [17]. This tendency is due to the higher hydrophillicity of NH_3+ groups by chitosan in the films [2] or the bulkier side groups of chitosan and CMCh that caused the increased free-volume of the blend matrix [17]. Likewise, Zhong et al. [1] studied the physicochemical properties of chitosan and kudzu starch composite film, and they found that the WVP declined when starch content increased.

Table 2. Composition of the film-forming solutions and WVP of RS/CMCh blend films.

Film code	Composition (g/100g solid)		glycerol	WVP
	RS	CMCh	(%w/w solid)	($g.mm/m^2.mHg.\ day$)
RS/CMCh-01	100	0	25	4.800 ± 0.342 a
RS/CMCh-02	88	12	25	5.459 ± 0.492 a
RS/CMCh-03	67	33	25	5.059 ± 0.426 a
RS/CMCh-04	50	50	25	5.368 ± 0.628 a
RS/CMCh-05	33	67	25	5.985 ± 0.767 a
RS/CMCh-06	12	88	25	9.078 ± 1.909 b
RS/CMCh-07	0	100	25	9.068 ± 0.669 b

3.2. *Sorption Isotherm*
Moisture sorption isotherm curve (a_w *versus* EMC) of RS/CMCh blend films are presented in figure 1a. The isotherm curve of rice starch film is classified as a sigmoid-shaped curve type II (figure 1a), which is a behaviour of a glassy polymer-organic on a water vapour system [18]. This curve type is typically obtained for the soluble material and displays an asymptotic trend as a_w tends toward 1 [16, 19]. The isotherm is more rapid in the initial stages (a_w 0.00–0.20), and a lesser amount of moisture is adsorbed while a_w increased toward 0.56. The EMC of film was dramatically raised above a_w 0.85. However, it can be verified that the isotherm curve of CMCh film shown as type III displays the crystalline components (figure 1a) [10]. The more RH surrounding, the more pronounced the effect. The EMC of the film soared after a_w 0.77. By analysing the curve of other blend films, it (except the RS/CMCh-02) can be classified as the type III isotherm (figure 1a), which is similar to that of CMCh film. This is possibly due to the effect of increasing the proportion of CMCh in the blend films. However, the isotherm curve of RS/CMCh-02 can be categorized as type II (figure 1a) because it contains a small amount of CMCh. The results indicate that the sorption isotherm characteristic is dependent on the composition of the main material. In this study, CMCh has more influence on the moisture sorption characteristic of the blend than rice starch. This is possibly because of its bulkiness and polar side chain. Similar results were found in a previous study for the moisture isotherm of rice starch/carboxymethyl cellulose from durian rind blend films [16]. Furthermore, these observations are

in agreement with those of carboxymethyl cellulose from papaya peel/corn flour blend films [10], cassava flour film [19] and cassava starch/carboxymethyl cellulose blend films [9].

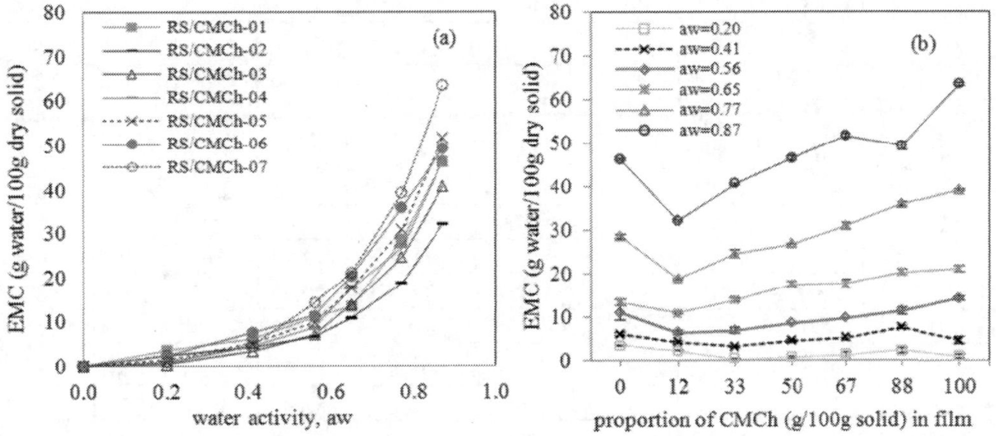

Figure 1. (a) Moisture sorption isotherm curves of RS/CMCh blend films and (b) equilibrium moisture content *versus* proportion of CMCh in RS/CMCh blend film at various a_w. (mean values ± standard error; in triplicate).

Figure 1b exhibits the moisture sorption isotherm curve of RS/CMCh blend films in terms of proportion of CMCh *versus* EMC. The difference between the sorption capacities of each film can be more obviously seen at higher a_w. The CMCh film provided a significant higher EMC than the rice starch. This result is in agreement with the previous works for the blend films between starch and derivatives from carboxymethylation [9, 16]. In this case, the EMC of the blend film decreased with a decrease of the CMCh composition (12–100 g/ 100 g solid). Interestingly, when small amounts of CMCh were employed in the films (12–33 g/ 100 g solid), the EMC of the blend film dropped and it was lower than that of the films of both main materials.

3.3. Fitting of sorption isotherm models to experimental data

Experimental sorption isotherm data of the RS/CMCh blend films were fitted to Lewicki, Peleg, GAB, BET and Oswin's equations. Table 3 lists sorption isotherm model constants including coefficient of determination (r^2) and percentage root mean square error (%RMS), which are the criteria used to estimate the fit of each model. All models showed a good fit to the experimental data ($r^2 > 0.99$).

The Lewicki model was followed to predict the moisture sorption data at a high range of a_w and it fitted well with the data at high humidity [11]. This model predicted that the amount of water tented toward infinity as a_w reached 1.0 [11]. Therefore, the Lewicki model is considered to be a suitable choice to predict the moisture adsorption of materials at different environmental humidities. The %RMS value was in the range between 12.5 and 184.7%.

The Peleg model was applied to predict both sigmoid and non-sigmoid isotherms, and it can fit comparably to or better than the GAB model [12]. However, its constants have no physical meaning. In this case, the range of %RMS was the lowest (6.2–44.3%) among all test models.

The GAB and BET models are the most accepted model for foods and edible materials [15]. The GAB equation gave the best fit for many kinds of the fruits, meats and vegetables examined [20]. The BET model is the most significant isotherm model for the explanation of multilayer sorption isotherms, especially for sigmoidal type II isotherm characteristics [21]. Both models allow for the evaluation of monolayer moisture content (M_0) values. The M_0 is utilized to measure the number of sorption sites, and indicates the maximum moisture content which can be adsorbed in a single layer of 1 g of dry film [9]. The M_0 calculated from the GAB and BET equations for the RS/CMCh blend films ranged between 8.6–220.6 g/ 100 g (dry basis) and 4.6–9.4 g/ 100 g (dry basis), respectively. Results

Table 3. Sorption isotherm model constant, coefficient of determination (r^2) and percentage of root mean square error (%RMS) for RS/CMCh blend films.

Sample ∨	Lewicki					BET			
Parameter >	F	G	H	r^2	RMS(%)	m0	C1	r^2	RMS(%)
RS/CMCh-01	5.091	1.197	0.620	0.996	13.66	8.19	1.01	0.996	19.705
RS/CMCh-02	8.660	0.729	6.006	0.999	14.45	4.58	1.99	0.997	10.320
RS/CMCh-03	8.403	0.840	4.560	0.995	184.75	6.12	1.22	0.992	191.315
RS/CMCh-04	10.424	0.801	4.221	0.997	79.56	6.82	1.50	0.994	95.344
RS/CMCh-05	11.628	0.804	4.697	0.998	35.87	7.64	1.44	0.995	43.471
RS/CMCh-06	92.331	0.089	4.642	0.998	12.54	7.17	3.61	0.981	37.995
RS/CMCh-07	15.249	0.776	4.427	0.996	95.78	9.42	1.48	0.992	108.248
Sample ∨	GAB					Oswin			
Parameter >	m0	C	k	r^2	RMS(%)	C1	C2	r^2	RMS(%)
RS/CMCh-01	8.575	0.924	0.995	0.996	20.188	8.198	1.000	0.996	19.742
RS/CMCh-02	9.521	0.547	0.917	0.999	18.708	6.719	0.840	0.998	14.621
RS/CMCh-03	129.939	0.045	0.805	0.998	131.039	7.591	0.900	0.993	242.370
RS/CMCh-04	96.668	0.074	0.802	0.998	53.004	9.322	0.862	0.996	125.362
RS/CMCh-05	120.540	0.064	0.804	0.999	16.770	10.169	0.874	0.996	61.527
RS/CMCh-06	220.594	0.061	0.706	0.995	11.457	15.188	0.662	0.990	76.266
RS/CMCh-07	163.894	0.061	0.795	0.998	63.869	13.014	0.855	0.994	146.791
Sample ∨	Peleg								
Parameter >	A	B	C	D	r^2	RMS(%)			
RS/CMCh-01	12.280	0.794	106.864	6.781	0.998	10.550			
RS/CMCh-02	26.325	3.703	26.325	3.703	0.992	44.286			
RS/CMCh-03	9.286	1.855	63.859	4.661	1.000	24.910			
RS/CMCh-04	38.196	3.716	38.196	3.716	0.997	33.440			
RS/CMCh-05	14.477	1.453	82.824	5.244	1.000	6.150			
RS/CMCh-06	76.496	3.397	2.458	0.061	0.997	11.534			
RS/CMCh-07	57.644	2.738	62.053	6.766	0.999	11.420			

represented that the GAB model provided higher M_0 than that of the BET for each RS/CMCh blend film. These observations are consistent with the previous results for carboxymethyl cellulose from papaya peel/ corn flour blend films [10] and rice starch/ carboxymethyl cellulose from durian rind blend films [16]. According to the BET model, the M_0 of the blend films showed the relation to the equilibrium moisture content described previously (figure 1). The CMCh content decreased (100–12 g/ 100 g solid), the EMC and the M_0 tended to decrease. The decrease in M_0 may reveal a reduction in the number of active sites, carboxymethyl groups. For the GAB equation, the C parameter is correlated to the difference of the magnitude in the upper layers and in the monolayer [9]. The C values obtained from RS/CMCh blend films were in the range of 0.04–0.92. The %RMS of the GAB and the BET model ranged between 11.5–131.0% and 10.3–191.3%, respectively.

The Oswin model allows good descriptions of the moisture isotherms all over the entire range of water activity [14]. This model is used for predicting sigmoid shape curve and is able to provide a better fit for protein and starchy foods [14]. Unfortunately, in this case, the maximum %RMS (14.6-242.4%) was found for the Oswin model.

Therefore, the Peleg model is the best estimator for predicting the EMC of RS/CMCh blend films, followed by the GAB, Lewicki, BET and Oswin models, respectively. This is in agreement with the previous study for rice starch/carboxymethyl cellulose from durian rind [16]. However, Suppakul et al. [19] found that the GAB model was the most optimal (among the Peleg, Lewiki and Oswin models) to fit the sorption isotherm for cassava flour film. Furthermore, other researches revealed that the GAB model was found to be the best fit for cassava starch based films blended with gelatin and carboxymethyl cellulose [9] and carboxymethyl rice starch films [22].

Figure 2 reveals the experimental *versus* predicted moisture sorption contents by the Lewicki, Peleg, GAB, BET and Oswin models for the RS/CMCh blend films. The achieved points are placed on the diagonal for low and intermediate a_w levels (0.20–0.77). This observation indicated low interaction occurring between the film compositions, according to their separation in the independent phases as appeared during film drying [19]. The point increased rapidly on the diagonal, as an effect of the interaction between the water molecules and the polar groups of the film, at a high level of a_w (0.85) [19]. These results pointed out that these four types of predicting models can be used to predict the moisture content of the RS/CMCh blend films at a_w 0.20–0.87.

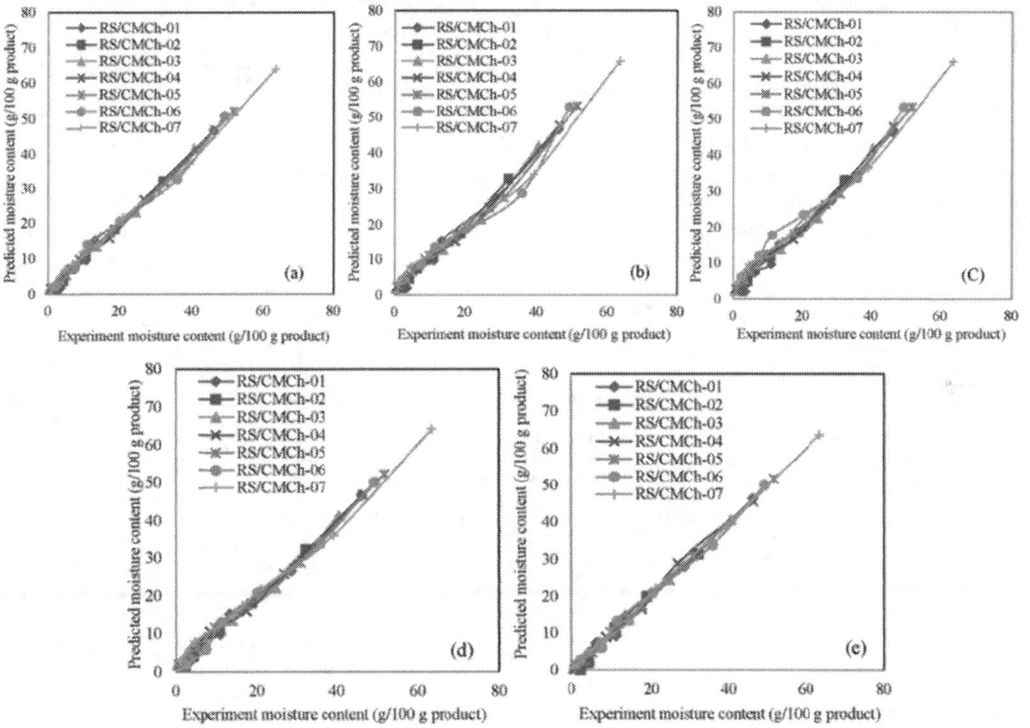

Figure 2. Comparison between experimental and predicted data of moisture content by (a) model, (b) BET model, (c) Peleg model for RS/CMCh blend fils.

4. Conclusion

Hydrophilicity of carboxymethyl chitosan and its proportion in the blend films were the interrelated factors in an investigation of water resistance and moisture sorption characteristics for the blend film composed of rice starch and carboxymethyl chitosan. The Peleg, GAB, Lewicki, BET and Oswin models were beneficial to fit moisture sorption isotherm data. The Peleg model was found to give the best prediction of sorption isotherm of RS/CMCh blend films at 25°C. These blend films possess a potential use as an edible coating and food packaging using moisture-resistant and/or antioxidative modification.

Acknowledgments
The authors gratefully acknowledge the Thailand Research Fund through the Royal Golden Jubilee Ph.D. Program as well as Chiang Mai University for the financial support.

References

[1] Zhong Y, Li Y and Zhao Y 2012 *J. Food Sci.* **77** E280-6
[2] Bourtoom T and Chinnan M S 2008 *LWT-Food Sci. Technol.* **41** 1633-41
[3] Al-Hassan A A and Norziah M H 2012 *Food Hydrocolloid.* **26** 108-17
[4] Ravi Kumar M N V 2000 *React. Funct. Polym.* **46** 1-27
[5] Ferreira C O, Nunes C A, Delgadillo I and Lopes-da-Silva J A 2009 *Food Res. Int.* **42** 807-13
[6] Duan B, Sun P, Wang X and Yang C 2011 *Starch - Stärke* **63** 528-35
[7] Chen X-G and Park H-J 2003 *Carbohyd. Polym.* **53** 355-9
[8] Rachtanapun P and Suriyatem R 2012 *J. Agri. Sci. Tech. A* **2** 50-57
[9] Tongdeesoontorn W, Mauer L J, Wongruong S and Rachtanapun P 2009 *As. J. Food Ag-Ind.* **2** 501-14
[10] Rachtanapun P and Thondeesuoontorn W 2009 *As. J. Food Ag-Ind.* **2** 791-801
[11] Lewicki P P 1998 *J. Food. Process. Eng.* **21** 127-44
[12] Peleg M 1993 *J. Food. Process. Eng.* **16** 21-37
[13] Brunauer S, Emmett P H and Teller E 1938 *J. Am. Chem. Soc.* **60** 309-19
[14] Oswin C R 1946 *J. Soc. Chem. Ind-L.* **65** 419-21
[15] van den Berg C and Bruin S 1981 *Water activity and its estimation in food systems: theoretical aspects* (Water Activity: Influences on Food Quality) ed L B R Stewart and F George (New York: Academic Press) p 1
[16] Suriyatem R and Rachtanapun P 2013 *Appl. Mech. Mater.* **431** 32-36
[17] Wu J, Zhong F, Li Y, Shoemaker C F and Xia W 2013 *Food Hydrocolloid.* **30** 82-91
[18] Rosa G S, Moraes M A and Pinto L A A 2010 *LWT-Food Sci. Technol.* **43** 415-20
[19] Suppakul P, Chalernsook B, Ratisuthawat B, Prapasitthi S and Munchukangwan N 2013 *LWT-Food Sci. Technol.* **50** 290-7
[20] Lomauro C J, Bakshi A S and Labuza T P 1985 *LWT-Food Sci. Technol.* **18** 111-7
[21] Siripatrawan U and Jantawat P 2006 *Food Sci. Technol. Int.* **12** 459-65
[22] Rachtanapun P and Tongdeesoontorn W 2011 *Chiang Mai J. Sci.* **38** 380-8

Value added of mulberry paper waste by carboxymethylation for preparation a packaging film

P Rachtanapun[1,2,3], S Kumthai[1], N Mulkarat[1], N Pintajam[1] and R Suriyatem[3]

[1]Division of Packaging Technology, Chiang Mai University, Chiang Mai 50100, Thailand
[2]Materials Science Research Center, Chiang Mai University 50200, Thailand
[3]Division of Food Science and Technology, Chiang Mai University, Chiang Mai 50100, Thailand

E-mail: pornchai.r@cmu.ac.th

Abstract. Cellulose from mulberry paper waste was converted to carboxymethyl cellulose (CMCm) by etherification using chloroacetic acid and various sodium hydroxide (NaOH) concentrations (30-60 g/100 mL). The degree of substitution of the various CMCm materials produced ranged between 0.33-0.45. The chemical structure of cellulose and the synthesized CMCm materials was characterized using Fourier transform infrared spectroscopy. CMCm films were cast and tested. The tensile strength (TS) and percent elongation at break (EB) of the films were investigated. The highest TS (32.58 MPa) and EB (2.39%) were found using the 50g /100 mL NaOH-synthesized CMCm film. The effect of glycerol as a plasticizer on TS and EB of this CMCm film was also evaluated. Increasing the glycerol content decreased TS and increased EB of the CMCm films.

1. Introduction

Mulberry paper, or *saa* paper as it is known in Thailand, is widely used to make lanterns, book covers, decorative products, greeting cards, office equipment, packaging and a variety of other items [1]. Mulberry paper is made from the bark of the mulberry tree, a plentiful fibrous resource [2]. Mulberry trees are extensively grown in many countries of Asia, including Thailand, China, India, Laos and Myanmar. In northern Thailand, many industries produce mulberry paper and mulberry paper products for both domestic consumption and export. The waste from the paper production is discarded. This waste is both natural and a superlative cellulose source [3].

Cellulose normally appears in the cell walls of plants. It consists of a linear chain of glucose units joined together by β-glycosidic bonds between the C-4 of one sugar unit and the anomeric C-1 of the second sugar. Cellulose can be utilized in many forms, including as its original unmodified fibers or with derivatives as carboxymethyl cellulose (CMC) [4, 5]. CMC is a copolymer of two units: ß-D-glucose and ß-D-glucopyranose 2-O-(carboxymethyl)-monosodium salt that are connected *via* β-1, 4-glycosidic bonds. Due to its simple and inexpensive production, CMC is used extensively in food and non-food industrial applications [6]. CMC is obtained *via* carboxymethylation between cellulose and

[3] Address for correspondence: P Rachtanapun, Division of Packaging Technology, Chiang Mai University, Chiang Mai 50100, Thailand. E-mail: pornchai.r@cmu.ac.th.

chloroacetic acid (an etherifying agent) in an alkaline condition. The first step is an alkalization procedure as in equation (1), where NaOH reacts with the hydroxyl groups of the cellulose.

$$\text{Cellulose-OH} + \text{NaOH} \rightarrow \text{Cellulose-ONa} + H_2O \quad (1)$$

In the second step, etherification occurs as in equation (2) to obtain CMC and a side reaction as in equation (3), which results in sodium glycolate [4].

$$\text{Cellulose-ONa} + \text{Cl-CH}_2\text{-CO-ONa} \rightarrow \text{Cellulose-O-CH}_2\text{-COONa} \rightarrow \text{NaCl} \quad (2)$$
$$\text{NaOH} + \text{Cl-CH}_2\text{-CO-ONa} \rightarrow \text{HO-CH}_2\text{-CO-ONa} + \text{NaCl} \quad (3)$$

According to published reports, CMC has been produced from a variety of agricultural waste cellulose sources, including cotton linters [7], sugar beet pulp [8], cashew tree gum [9], cavendish banana pseudo stem [10], sago waste [11], papaya peel [12], Mimisa pigra peel [6], durian rind [4] and cotton gin waste [13]. However, the production and characterization of CMC film from mulberry paper waste (CMCm) has not previously been reported.

This study evaluated the influence of varying NaOH concentrations for CMC synthesis from mulberry paper waste on the degree of substitution and chemical structure of CMCm. The study also investigated the effects of NaOH concentrations on the mechanical properties, including tensile strength (TS) and percent elongation at break (EB), of the synthesized CMCm films. In addition, the effect of the amount of plasticizer on the CMCm films, including on TS and EB, was determined.

2. Methodology

2.1. Materials
Mulberry paper waste was purchased from Bankradassa (Chiang Mai, Thailand). All chemicals used in the preparation and analysis of carboxymethyl cellulose (CMC) were AR grade or the equivalent. Isopropanol, methanol, ethanol, glycerol, distilled water and commercial grade carboxymethyl cellulose (CMCc) were purchased from Northern Chemical Co., Ltd. (Chiang Mai, Thailand). Monochloroacetic acid was purchased from Sigma-Aldrich (Steinheim, Germany). Sodium hydroxide and glacial acetic acid were purchased from RCI Lab-scan Co., Ltd. (Bangkok, Thailand).

2.2. Synthesis of carboxymethyl cellulose and charaterization
Carboxymethyl cellulose (CMCm) from mulberry paper waste was synthesized according to the procedure described by Rachtanapun and Rattanapanone [6], with slight modification. Cellulose from mulberry paper waste (15 g), 750 mL of isopropanol and 150 mL of various concentrations (30, 40, 50 and 60 g/100 mL) of NaOH were mixed in a beaker for 90 min. Monochloroacetic acid (MCA, 22.5 g) was introduced and the mixture was continuously stirred for 90 min. The beaker was covered with aluminum foil and placed in a hot air oven at 55°C for 3.5 h, where the mixture separated into a liquid and solid phase. The liquid phase was removed by filtration. Then, the solid phase was isolated by precipitation into 100 mL of methanol (70% v/v) and neutralized with glacial acetic acid. The neutralized product was filtered and washed 5 times by soaking in 300 mL of ethanol (70% v/v) for 10 min and filtering. The product (CMCm) was then washed again with 300 mL of absolute methanol, filtered and dried at 55°C for 12 h. The CMCm was stored in a PE bag in a dry place.

The DS was determined using the USP XXXII method described for Crosscarmellose sodium in two steps – titration and collection of residue on ignition – as previously described by Rachtanapun and Rattanapanone [6] and Rachtanapun et al. [4]

2.3. Infrared spectroscopy (IR)
The functional groups of the cellulose and CMCm samples were determined using an infrared spectrophotometer (Bruker, Tensor 27, Germany). Pellets were made from cellulose and the CMCm samples (~2 mg) and KBr. Transmission levels measured for wavelengths of 4000-400 cm^{-1}.

2.4. Film preparation
CMC (3 g) was dissolved in 100 mL of distilled water and constantly stirred at 80°C for 10 min. The resulting solution was cooled to around 20-25°C and cast on Mylar® film using a tape casting machine)DP 150, Japan (at 20 rpm set to produce a film thickness of 30 μm. The casted solution was left to dry at room temperature for 20 h, producing the CMC film. The film was peeled and kept in a dry place.

2.5. Mechanical properties
The TS and EB of the film were determined using an Instron Universal Testing Machine Model 1000 (H1K-S, England) with the procedure according to ASTM D882-80a. The film specimen was cut into 15 x 140 mm rectangles. The initial grip separation and crosshead speed was set at 100 mm and 20 mm/min, respectively. Before testing, each film was pre-conditioned at 65±2% relative humidity for 24 h at 27±2°C. Ten specimens were measured.

2.6. Film preparation with glycerol as plasticizer
From the mechanical property tests, we selected the most promising film to further test with plasticizer. Various amounts (0.1, 0.2, 0.3 and 0.4 mL) of glycerol were introduced in the film solution to overcome film brittleness. The method of film preparation presented above and subjected to the mechanical property tests as described earlier.

3. Results and discussion

3.1. Characterization of CMCm materials
CMCm materials were synthesized *via* carboxymethylation between alkalized cellulose from mulberry paper waste and an etherifying agent. Various alkali concentrations (30, 40, 50 and 60 g/100 mL) during synthesis yielded four different CMCm as CMCm30, CMCm40, CMCm50 and CMCm60, respectively.

DS measurements indicate the water solubility of CMC, with increasing DS values indicating increasing hydrophillicity. CMC is completely soluble at DS values above ~0.4. Below ~0.4, CMC is swellable, but insoluble [8]. The DS values of the CMCm produced here ranged from 0.33–0.45 (table 1). The DS value increased when the NaOH concentration increased from 30 to 40 g/100 mL, reaching its maximum DS of 0.45 from CMCm40. Beyond 40 g/100 mL NaOH, the DS decreased. This result can be explained by the reactions of CMC synthesis, as described in Rachtanapun and Rattanapanone [6] and Rachtanapun et al. [4] The process of carboxymetylation in the experiment yielded two main products. The first product is a desired CMC obtained from the reaction of cellulose hydroxyl and NaMCA. The secound product is a sodium glycolate as a by-product formed from the conversion of NaMCA and NaOH [14]. The results of this study imply that if the alkaline level is higher than 40 g/100 mL, the latter reaction is favored, therefore reducing the DS value of the obtained CMC. Previous studies of CMC from sugar beet pulp [8], sago waste [11], banana pseudo stem [10], Mimosa pigra peel [6] and durian rind [4] reported similar DS results.

Table 1. Degree of substitution (DS) of CMCm.

Sample	DS
CMCm30	0.4473
CMCm40	0.4540
CMCm50	0.3521
CMCm60	0.3320

The IR spectrum of CMC materials normally show the typical absorptions of the cellulose backbone, including the presence of the carboxymethyl ether group at 1605 cm^{-1} [11]. The frequency of the absorption bands of the representative spectrum of cellulose from mulberry paper (figure 1a) and CMCm synthesized with various NaOH concentrations (figures 1b-1e) were similar. This result indicated that cellulose and CMCm contained similar functional groups. A broad band at 3100-3700 cm^{-1} is due to O–H stretching. The band at 2900 and 1600 cm^{-1} is due to a C–H stretching and C–O stretching vibration, respectively. The bands around 1450 and 1000-1200 cm^{-1} are due to $-CH_2$ scissoring and –O– stretching, respectively. The band at 1060 cm^{-1} is due to $-CH-O-CH_2$ stretching [15]. The absorbent bands in each CMCm spectrum were similar, but differed from those of cellulose. The differences were noticed at 1600, 1420 and 1100 cm^{-1}. This phenomenon confirmed that carboxymethylation took place on the cellulose molecules. The carboxymethyl substituent for the absorption bands were COO–, $-CH_2$ and –O–, respectively [4, 6, 10]. Similar results have been reported previously [4, 6, 11].

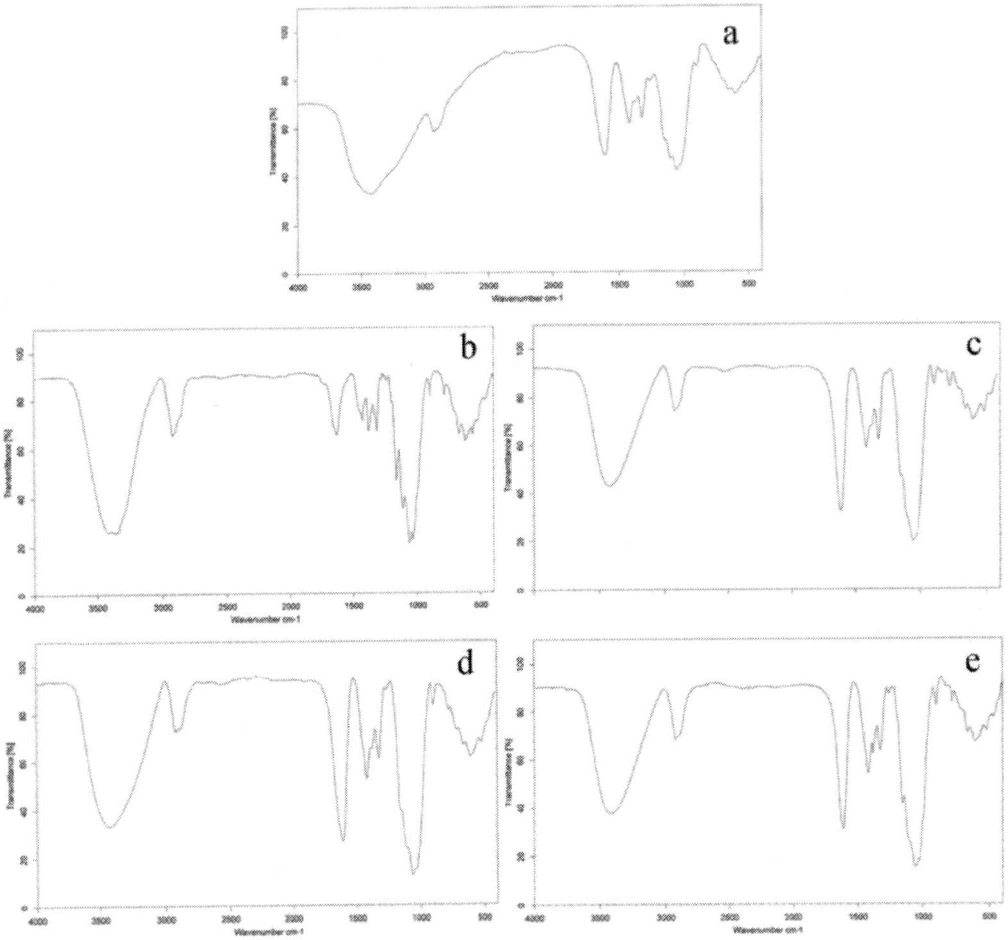

Figure 1. IR spectra of cellulose from mulberry paper waste (a) and CMCm synthesized from various NaOH concentrations: 30 (b), 40 (c), 50 (d) and 60 (e) g/100 mL NaOH.

3.2. CMCm films: effect of alkaline level of CMC synthesis on mechanical properties

The effect of the various NaOH concentrations in synthesizing CMCm on the mechanical properties – strength and flexibility – of the resultant films are reported below.

Different CMCm films were prepared. They appeared white and translucent, because of residual insoluble cellulose (figures 2a-2d). A commercial grade CMC (CMCc) film was also prepared, but was transparent (figure 2e). The CMCm60 wrinkled after drying, and as a result could not be formed into a film (figure 2d). The CMCm60 film forming solution flowed faster than the others when casted. This occurred because the hydrolysis reaction in the cellulose chain took place when a high level of alkaline was utilized, and the intermolecular hydrogen bonds decreased with the increasing basic concentration [16]. As a result, CMCm60 film could not be tested for mechanical properties.

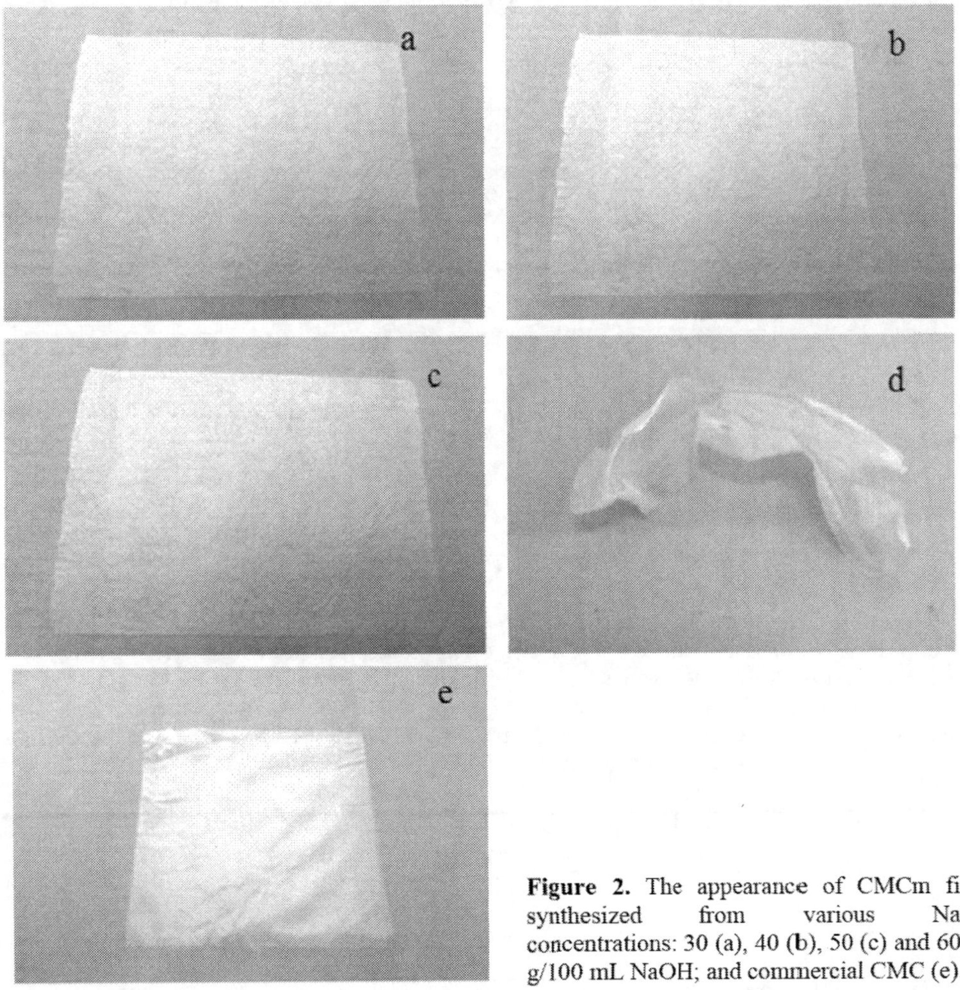

Figure 2. The appearance of CMCm films synthesized from various NaOH concentrations: 30 (a), 40 (b), 50 (c) and 60 (d) g/100 mL NaOH; and commercial CMC (e).

Figure 3a shows the TS of the CMCm films. The TS of the films increased with increasing NaOH concentrations (30-50 g/100 mL) used in CMC synthesis. The CMCm50 showed the highest TS value of the CMCm films, and compared favorably with the commercial grade film. The increasing TS of the CMCm films may be a result of more combination between the swellable part (fibers) and soluble component (desired CMC) of the CMCm materials. The addition of fibers reinforced the tensile properties of the films, increasing their TS [17]. However, our result is contrary with previous reports for CMC films from durian rind [4] and Mimosa pigra peel [6], where the TS correlated directly with the DS value of the CMC materials. An increasing DS due to the substitution of carboxymethyl groups increased the ionic character and intermolecular force between the polymer chains [6, 10]. Despite this, the DS of CMC materials and the TS of CMC films decreased when too high of a NaOH concentration was employed, because of sodium glycolate formation as a by-product and polymer degradation. This leads to low CMC content, which lowers the intermolecular forces [18].

The EB of the polymer is a measure of its flexibility. In this study, the EB of the CMCm films increased with increasing NaOH concentrations (figure 3b), similar to the TS results. This result agrees with the previous report of CMC films from Mimosa pigra peel [6]. In contrast, the EB of CMC films from durian rind did not change significantly with different levels of NaOH in CMC synthesis.[4] The CMCm50 showed the highest EB value, higher than that of the CMCc film (figure 3b).

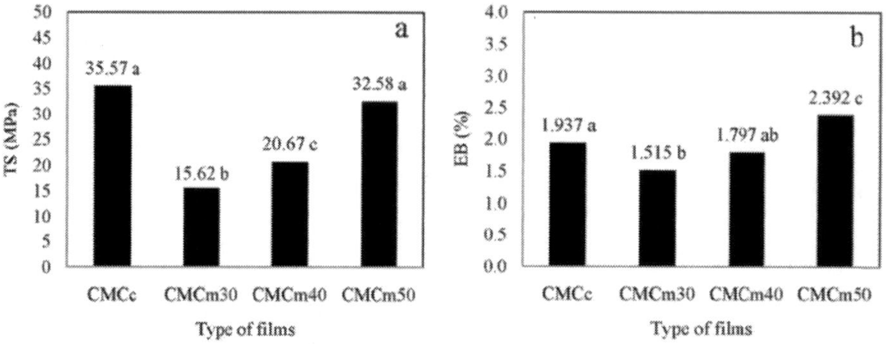

Figure 3. Mechanical properties, (a) TS and (b) EB, of the CMCm films synthesized from various NaOH concentrations compared with commercial CMC.

From our experiment, the CMCm50 film provided the best mechanical properties. However, as prepared, it was too brittle to use for packaging. Given its underlying mechanical advantages, we introduced glycerol as a plasticizer in an effort to overcome its brittleness, as reported in the next section.

3.3. CMCm50 films: effect of a plasticizer on mechanical properties

Various glycerol concentrations as a plasticizer were introduced into the film forming solutions of CMCm50 in an effort to improve its mechanical properties before casting. Figures 4a and 4b present the TS and EB of the CMCm50-plastisized films, respectively. Increasing glycerol content in the CMCm-plasticized films decreased the TS and increased the EB. The glycerol increased the flexibility of the films due to its ability to reduce internal hydrogen bonding between polymer chains with increasing of a space between the polymers [6, 19]. The increasing flexibility of hydrophilic films with increasing plasticizer content has been reported previously [6, 12, 17, 20].

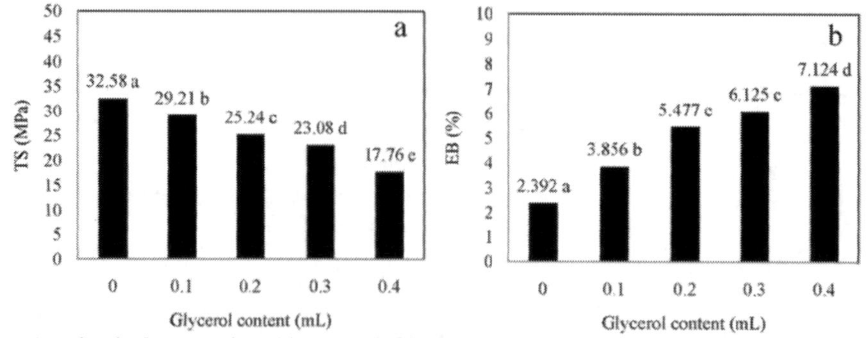

Figure 4. Mechanical properties, (a) TS and (b) EB, of the CMCm50 films plasticized with several amounts of plasticizer.

IOP Conf. Series: Materials Science and Engineering **87** (2015) 012081 doi:10.1088/1757-899X/87/1/012081

4. Conclusions

Mulberry paper waste was successfully employed as a raw material to produce carboxymethyl cellulose using MCA as an etherifying agent and isopropanol as a medium under various alkaline conditions. The DS value of the CMCm materials decreased with increasing NaOH concentrations. The CMCm synthesized with 40 g/100 mL NaOH produced the highest DS value of 0.45. As shown in this study, the NaOH concentration used in CMC synthesis directly correlated with the mechanical properties of the CMCm. The CMCm film derived from 50 g/100 mL NaOH had the best mechanical properties. This film was chosen to improve its flexibility by adding glycerol as a plasticizer. The amount of plasticizer, in turn, strongly affected the mechanical properties of the film. The CMCm materials produced here are new biopolymers with potentially broad industrial applications, including as film forming agents, binding agents, gelling agents and/or emulsifying agents.

Acknowledgments

The authors gratefully acknowledge the New Research Project of Chiang Mai University, the Thai Research Fund in Funding Research Project for Undergraduate Students (IRPUS) and the National Research University Project under Thailand's Office of the Higher Education Commission for their financial support. We also thank Mr. Prem Thongchai, the Department of Packaging Technology, Faculty of Agro-industry, Chiang Mai University.

References

[1] Ithisoponakul S, Memon A, Pramoonmak S, Montip L and Nateechai P 2011 *Proc. Int. Conf. on 9th Eco-Energy and Materials Science and Engineering Symposium*, (Chiang Rai: Thailand).
[2] Li R, Fei J, Cai Y, Li Y, Feng J and Yao J 2009 *Carbohyd. Polym.* **76** 94-99
[3] Punsuvon V and Wansanug K 1997 *Proc. Int. Conf. on 36th Kasetsart University Annual Conference*, (Bangkok: Thailand).
[4] Rachtanapun P, Luangkamin S, Tanprasert K and Suriyatem R 2012 *LWT-Food Sci. Technol.* **48** 52-58
[5] Majewicz T G, Erazo-Majewicz P E and Podlas T J 2002 *Cellulose Ethers* (Encyclopedia of polymer science and technology) ed (New York: John Wiley & Sons, Inc.)
[6] Rachtanapun P and Rattanapanone N 2011 *J. Appl. Polym. Sci.* **122** 3218-3226
[7] Xiquan L, Tingzhu Q and Shaoqui Q 1990 *Acta Polym.* **41** 220
[8] Togrul H and Arslan N 2003 *Carbohyd. Polym.* **54** 73-82
[9] Silva D A, Paula R C M, Feitosa J P A, Brito A C F, Miciel J S and Paula H C B 2004 *Carbohyd. Polym.* **58** 136-171
[10] Adinugraha M P, Marseno D W and Hayadi 2005 *Carbohyd. Polym.* **62** 164-169
[11] Pushpamalar V, Langford S J, Ahmad M and Lim Y Y 2006 *Carbohyd. Polym.* **64** 312-318
[12] Rachtanapun P, Kumthai S, Yagi N and Uthaiyod R 2007 *Proc. Int. Conf. on 45th Kasetsart University Annual Conference*, (Bangkok, Thailand) pp 790-799.
[13] Haleem N, Arshad M, Shahid M and Tahir M A 2014 *Carbohyd. Polym.* **113** 249-255
[14] Kirk R E and Othmer D F 1967 *Cellulose* (Encyclopedia of chemical technology) ed (New York: John Wiley & Sons, Inc.) p 593
[15] Kondo T 1997 *Cellulose* **4** 281-292
[16] Toth T, Borsa J and Takacs E 2003 *Radiat. Phys. Chem.* **67** 513-515
[17] Dias A B, Müller C M O, Larotonda F D S and Laurindo J B 2011 *LWT-Food Sci. Technol.* **44** 535-542
[18] Barai B K, Singhal R S and Kulkarni P R 1997 *Carbohyd. Polym.* **32** 229-231
[19] Lieberman E R and Gilbert S G 1973 *J. Polym. Sci. Pol. Sym.* **41** 33-43
[20] Shaw N B, Monahan F J, O'Riordan E D and O'Sullivan M 2002 *J. Food Eng.* **51** 299-304

Synthesis and characterization of carbon fibers functionalized with poly (glycidyl methacrylate) via atom transfer radical polymerization

Yongwei Wu, Lei Xiong[1], Xiaokang Qin, Zhengyue Wang, Bei Ding, Huan Ren and Xiaolong Pi

School of Material Science and Engineering, Nanchang Hangkong University, Nanchang 330063, China

E-mail: x_lei81@163.com

Abstract. In this work, polyacrylonitrile (PAN)-based carbon fibers (CF) were chemically modified with poly (glycidyl methacrylate) (PGMA) via atom transfer radical polymerization (ATRP) to improve the interaction between the CF and polymer matrix. The FT-IR, TGA, and XPS were used to determine the chemical structure of the resulting products and the quantities of PGMA chains grafted from the CF surface. The experimental results confirm that the CF surface was functionalized and glycidyl methacrylate was graft-polymerized onto the CF, and the grafting content of polymer could reach 10.2%.

1. Introduction

In the past decade, the carbon fibers (CF) have attracted considerable attention because of their high mechanical properties, excellent high thermal stability and light weight, which have been used in a variety of fields [1-4]. Even though the carbon fibers possess these excellent properties, they are poor wetted by resins because of their smooth surface and inertness characteristics, resulting in a low interlaminar shear strength of composites [5-7].

Therefore, various methods have been developed for surface modification of CF with functional groups or polymers mainly to overcome this problem and improve the compatibility between CF and polymer matrix [8-10]. These methods include plasma treatment, electrochemical method, wet chemical, surface grating, thermal treatment and so on [11-15].

Among these methods, the functionalization of the CF with polymer is an effective method. It can be accomplished by many different approaches, including conventional free radical polymerization, cationic, anionic, esterification reaction and controlled radical polymerization [16, 17]. In those methods, the ATRP has become an important method to modify the CF surface with polymer chains because of its good controllability over the molecular weight and molecular weight distribution of the grafting polymers. Furthermore, the experimental condition of ATRP is not stringent, and a large numbers of monomers can be used in this method [18-20]. Herein we present a method of grafting

[1] Address for correspondence: Lei Xiong, School of Material Science and Engineering, Nanchang Hangkong University, Nanchang 330063, China. Tel.: 86-791-83863025; Fax: 86-791-86453203; E-mail: x_lei81@163.com.

poly (glycidyl methacrylate) onto the CF surface via ATRP. The introduction of polymers onto the CF surface can improve the compatibility between CF and resins, which results in high interfacial properties of the CF reinforced composites.

2. Experimental

2.1. Materials

The PAN-based carbon fibers were purchased from Hexcel Corporation, and the diameter of fiber is between 5 and 10 μm. The carbon fibers were cleaned by ethanol and acetone in turn, and then were dried for overnight at 60°C. Glycidyl methacrylate (GMA) was purchased from Shanghai Reagents Company and was purified under reduced pressure to remove the polymerization inhibitor. Thionyl chloride, 2-bromoisobutyryl bromide (BIBB), N, N, N′, N″, N″-pentamethyldiethylenetriamine (PMDETA), N, N-dimethylformamide (DMF), and diethylenetriamine (DETA) were purchased from Shanghai Chemical Company. All other reagents used in this study were obtained from commercial sources and used as received.

2.2. Immobilizing the ATRP initiators onto the carbon fibers

The carbon fibers were first treated with a 70% nitric acid that was stirred for 1.5h at 115°C. Then the resulted CF was washed with diluted water and dried for 24 h at 70°C. In the next step, the acid-treated carbon fibers were reacted with excess thionyl chloride at 70°C for 36 h. The resulted acyl-derivated carbon fibers (CF-COCl) were washed with tetrahydrofuran (THF) and dried for overnight at 40°C. The CF-COCl was then put in an excess amount of DETA at 80°C for 36h, and washed with ethanol and dried at 90°C for 24 h. The resulted products were amine treated carbon fibers (CF-NH_2). Finally, the CF-NH_2 was reacted with BIBB at room temperature. After 48 h, the resulted products were washed with a solution of acetone/distilled water (1:1 v/v). The black final product (termed CF-Br) was dried at 60°C for overnight.

2.3. Synthesis of carbon fibers grafted with PGMA via ATRP

The carbon fibers grafted with PGMA were prepared by using ATRP method. The CF-Br, GMA, CuBr, PMDETA and DMF were added into a dried flask. The flask was degassed and sealed under vacuum. The sealed flask was then immersed in an oil bath at 80°C for 6 h. The reaction mixture was then intensively washed with ethanol and acetone (sequentially) and dried in an oven at 25°C for overnight. The final products were the carbon fibers grafted with PGMA (CF-PGMA). Figure 1 shows the procedure of grafting PGMA chains onto the CF via ATRP.

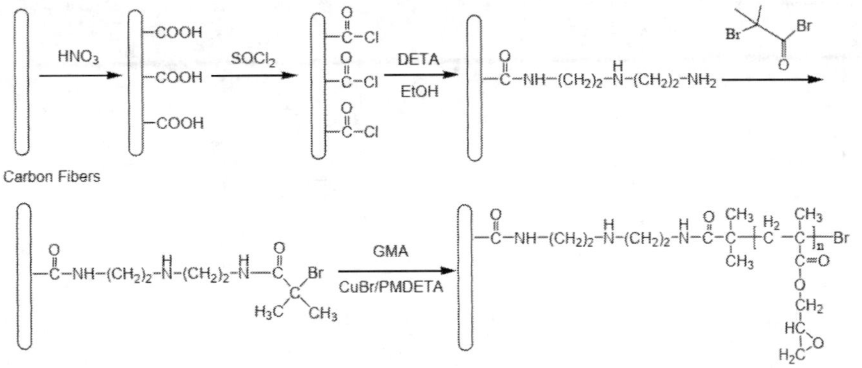

Figure 1. Synthesis of CF-PGMA via ATRP.

2.4. Measurements

IOP Conf. Series: Materials Science and Engineering **87** (2015) 012082 doi:10.1088/1757-899X/87/1/012082

Fourier transform infrared (FT-IR) spectra of the pure and functionalized CF in KBr pellets were obtained with Perkin-Elmer spectrometer at room temperature. The X-ray photoelectron spectroscopy (XPS) (monochromatized Mg (Kα) radiation) was used to characterize surface chemical compositions of different CF. Thermogravimetric analyses (TGA) were carried out using a Shimadzu TG-50 system under N_2 flow. The samples were heated from 25°C to 800°C at a rate of 20°C·min^{-1}.

3. Results and discussion

3.1. Characterization of carbon fibers fixed with ATRP initiators

The carboxyl groups of the carbon fibers were treated with thionyl chloride, DETA, and 2-bromoisobutyryl bromide, then the ATRP initiators were introduced on the CF surface. The formation of CF-Br is detected by FT-IR and TGA. Figure 2 shows the FT-IR spectra of pure CF and CF-Br. Compared with the FT-IR spectrum of the pure CF, the spectrum of CF-Br clearly shows a new peak at 1720 cm^{-1} which is ascribed C=O stretching band. The occurrence of C=O peak indicates that the acylamide groups were formed through the reaction between CF-NH_2 and BIBB. The absorption at 2855 cm^{-1} and 2920 cm^{-1} ascribed to C-H of grafting organic compounds. The FT-IR results confirm that the synthesis of CF-Br is successful.

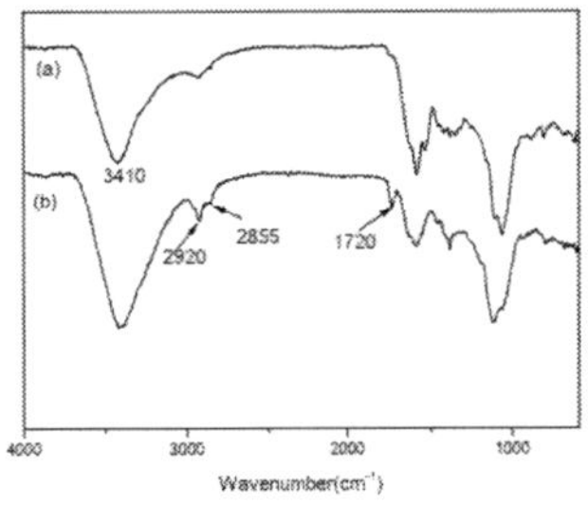

Figure 2. The FT-IR spectra of (a) pure CF and (b) CF-Br.

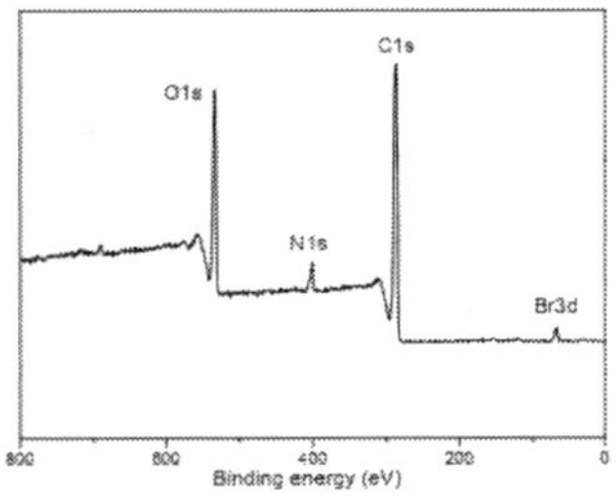

Figure 3. XPS spectrum of CF-Br.

The FT-IR results can also be confirmed by the results of XPS. Figure 3 shows the XPS spectrum of CF-Br. In this spectrum, the peaks at 530.9, 399.2, 285.3 and 69.6 eV are assigned to O1s, N1s, C1s, and Br3d, respectively. The appearance of Br3d peak in the spectrum shows that the ATRP initiators were grafted onto the CF by the reaction between CF-NH_2 and BIBB.

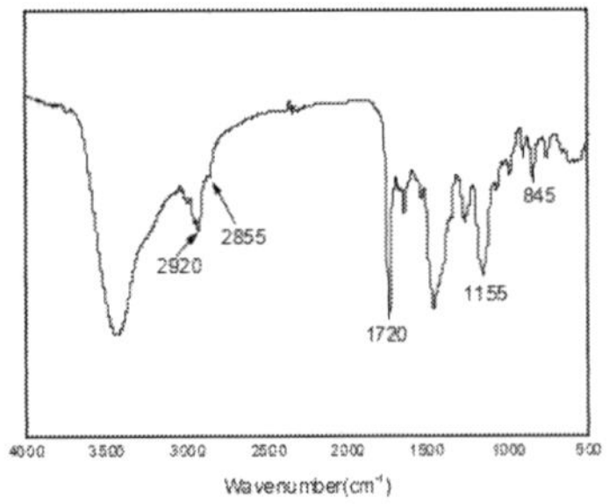

Figure 4. The FT-IR spectrum of CF-PGMA.

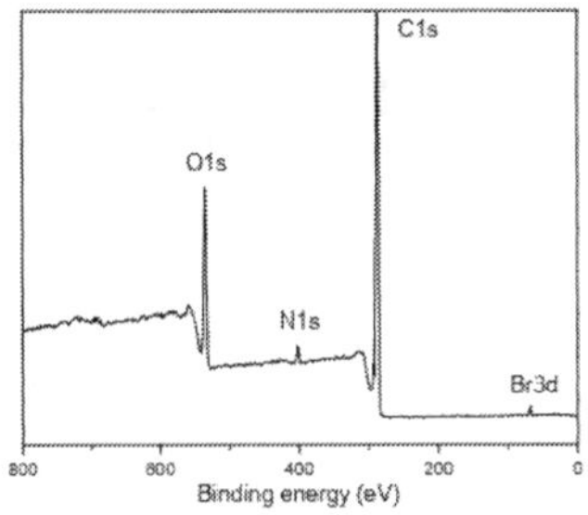

Figure 5. XPS spectrum of CF-PGMA.

3.2. Characterization of carbon fibers grafted with PGMA

After the immobilization of ATRP initiators, the monomer of GMA can be graft-polymerized onto the CF surface via ATRP. The CF functionalized with PGMA was characterized by means of FT-IR, XPS, and TGA. Figure 4 presents the FT-IR results of CF-PGMA. In this spectrum, it can be found that there is a strong absorption band at 1720 cm^{-1} ascribed to the ester carbonyl (C=O), which is associated with the grafting PGMA. The bands at 2920 cm^{-1} and 1155 cm^{-1} are assigned to the C-H and C-O-C, respectively. The spectrum of CF-PGMA also shows a new band at 845 cm^{-1}, which is ascribed to the epoxy groups. The FT-IR results demonstrate that the synthesis of the CF grafted with PGMA chains was successful.

XPS was used to determine the chemical compositions of the functionalized CF surface. Figure 5 shows the XPS spectrum of CF-PGMA. As compared with the spectrum of CF-Br, the spectrum of CF-PGMA shows that the peaks corresponding to O1s, Br3d and N1s become weak. On the other hand, the peak for C1s grows strong. The XPS results further suggest that the synthesis of CF-PGMA was successful.

Thermal degradation of pure CF, CF-Br and CF-PGMA were investigated by heating up from 25°C to 800°C under N_2 flow with a heating rate of 20°C·min^{-1}. The TGA results of different CF are shown in figure 6. Pure CF (figure 6a) did not present significant weight loss between room temperature and 800°C. In contrast, the CF-Br (figure 6b) shows a gradual decomposition as the temperature rises. From the TG curves it can be estimated that the weight loss in the range 30-800°C is about 5.6%, which may be ascribed to the decomposition of the ATRP initiators immobilized onto the carbon fiber surface. The TGA curve of CF-PGMA (figure 6c) is much different to the pure CF and CF-Br. It can be found that this curve shows a major decomposition near 300°C because of the thermal decomposition of PGMA, which was grafted from the CF-Br surface. According to the TGA data, the weight fraction of PGMA on the CF surface is approximately 10.2%.

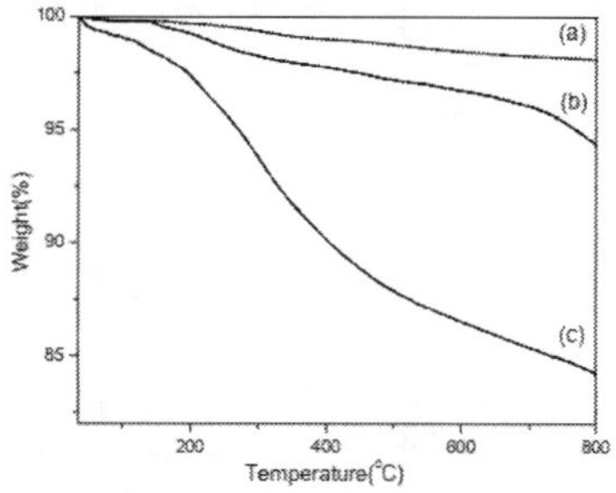

Figure 6. TGA curves of (a) pure CF, (b) CF-Br and (c) CF-PGMA.

4. Conclusion

To improve the interfacial properties between the CF and polymer resins, the monomer of glycidyl methacrylate (GMA) was graft-polymerized on the CF surface via ATRP method in this paper. The functionalized CF was characterized in terms of the XPS, FT-IR, and TGA. The experimental results confirmed that the occurrence of PGMA functionalization and the content of grafting polymer on the CF surface is about 10.2%. Therefore, this is an effective method to the preparation of the CF functionalized with polymer. It provides a potential approach for high performance composites developing.

Acknowledgment

The research work was supported by the the National Natural Science Foundation of China (No. 51463017), Jiang'xi Educational Committee (GJJ12423) and Science Foundation of Aeronautics of China (2014ZF56018).

References

[1] Fitzer E 1989 *Carbon* **27** 621–45
[2] Hung K B, Li J, Fan Q and Chen Z H 2008 *Compos. Part A* **39** 1133-40
[3] Yue Z R, Jiang W, Wang L, Gardner S D and Pittman Jr C U 1999 *Carbon* **37** 1785-96

[4] Pamula E and Rouxhet P G 2003 *Carbon* **41** 1905-15
[5] Edie D D 1998 *Carbon* **36** 345-62
[6] Dvir H, Jopp J and Gottlieb M 2006 *J. Colliod. Interface. Sci.* **304** 58-66
[7] Park S J and Kim B J 2005 *Mater. Sci. Eng. A* **408** 269-73
[8] Severini F, Formaro L, Pegoraro M and Posca L 2002 *Carbon* **40** 735-41
[9] Qian H, Bismarck A, Greenhalgh E S and Shaffer M 2010 *Compos. A* **41** 1107-14
[10] Xu B, Wan X S and Lu Y 2006 *Appl. Surf. Sci.* **253** 2695-9
[11] Montes-Moran M A, Martinez-Alonso A and Tascon J M 2001 *Compos. A* **32** 361-71
[12] Severini F, Formaro L, Pegoraro M and Posca L 2002 *Carbon* **40** 735-41
[13] Fukunaga A and Ueda S 2000 *Compos. Sci. Tech.* **60** 249-54
[14] Cao H, Huang Y, Zhang Z and Sun J 2005 *Compos. Sci. Technol.* **65** 1655-62
[15] Lee W H, Lee J G and Reucroft P J 2001 *Appl. Surf. Sci.* **171** 136-142
[16] Edmondson S, Osborne V L, Huck W T S 2004 *Chem. Soc. Rev.* **33** 14-22
[17] Ohno K, Morinaga T, Koh K, Tsujii Y and Fukuda T 2005 *Macromolecules* **38** 2137-42
[18] Matyjaszewski K and Xia J 2001 *Chem. Rev.* **101** 2921-90
[19] Braunecker W A and Matyjasezewski K 2007 *Prog. Polym. Sci.* **32** 93-146
[20] Zhou L L, Yuan W Z, Yuan J Y and Hong X Y 2008 *Mater. Lett.* **62** 1372 -5

IOP Conf. Series: Materials Science and Engineering **87** (2015) 012083 doi:10.1088/1757-899X/87/1/012083

Electrocaloric effect of lead-free inorganic (1-x) [$Bi_{0.5}$ $(Na_{0.84}K_{0.16})_{0.5}$]$TiO_3$+x$BaTiO_3$+0.3wt%$Nb_2O_5$ ceramics

Jong-Dae Han[1], Ju-Hyun Yoo[1,4], Lark-Hoon Hwang[1], Yeong-Ho Jeong[2] and Sung-Ill Lee[3]

[1]Department of Electrical Engineering, Semyung University, Jecheon 390-711, Republic of Korea
[2]Department of Electrical Engineering, Korea National University of Transportation, Chungju 380-702, Republic of Korea
[3]Department of Safety Engineering, Korea National University of Transportation, Chungju 380-702, Republic of Korea

E-mail: juhyun57@semyung.ac.kr

Abstract. In this work, in order to develop the ceramics with an excellent electrocaloric effect, (1-x) [$Bi_{0.5}(Na_{0.84}K_{0.16})_{0.5}$]$TiO_3$+x$BaTiO_3$+0.3wt%$Nb_2O_5$ [(1-x)BNKT-xBT] ceramics (x= 0, 0.04) were fabricated by conventional solid state reaction method. To investigate the electrocaloric effect of the ceramics, P-E hysteresis loops were measured in various temperature ranges from room temperature to 200°C. The adiabatic temperature change ΔT of these ceramics was measured using indirect method, which was calculated using the Maxwell's relations. The optimal value of adiabatic temperature change ΔT was obtained as 0.31°C at 165°C under applied electric fields 45 kV/cm in x =0 ceramics.

1. Introduction

Over the past several years, the demand for various kinds of cooling devices has been rapidly increased according to industrial development [1-3]. These cooling devices can cause the various environmental problems because of the usages of Freon gas. To solve the environmental problems, a large number of efforts have been made in environment friendly solid state cooling technologies. One of the technologies is the electrocaloric effect (abbreviated as ECE) [4]. The ECE is a change in temperature and/or entropy of a ferroelectric material upon the application or withdrawal of an electric field under adiabatic conditions. The first research based on ECE materials was carried out in 1930s. However, the ECE was difficult in using as cooling device because the adiabatic temperature change is very small. Generally, it was well known that the material with ferroelectric characteristic have a phase transition temperature (Curie Temperature: T_c) changing from ferroelectric phase to paraelectric phase, where the entropy change is very large. The ferroelectric materials based on $Pb(Zr,Ti)O_3$ showed the outstanding adiabatic temperature change [5, 6]. For example, Mischenko et al. reported that the adiabatic temperature change of over 12°C was obtained at PZT thin film. However, the toxicity of PbO in lead based ceramics has caused serious environmental pollution due to the volatilization of PbO during sintering process. Therefore, it is necessary to develop environmentally friendly lead-free

[4] Address for correspondence: Ju-Hyun Yoo, Department of Electrical Engineering, Semyung University, Jecheon 390-711, Republic of Korea. E-mail: juhyun57@semyung.ac.kr.

materials for the replacement of lead based ceramics. It was well known that the lead-free ceramics based on Ba(Na,K)TiO_3 have excellent ferroelectric properties and low phase transition temperature [7, 8].

Accordingly, in this study, in order to develop the ceramics with an excellent electrocaloric effect, (1-x)[$Bi_{0.5}(Na_{0.84}K_{0.16})_{0.5}$]$TiO_3$+x$BaTiO_3$+0.3wt%$Nb_2O_5$ ceramics (x=0, 0.04) were fabricated by conventional solid state reaction method. To investigate ECE, P-E hysteresis loop was measured in various temperature ranges. The adiabatic temperature change ΔT was calculated using Maxwell's relations.

2. Experimental

The specimens were manufactured by using a conventional mixed oxide process. The compositions used in this study were as follows;

$$(1\text{-}x)[Bi_{0.5}(Na_{0.84}K_{0.16})_{0.5}]TiO_3 + xBaTiO_3 + 0.3wt\%Nb_2O_5 \ (x = 0, 0.04) \quad (1)$$

The raw materials, Bi_2O_3 (99%), Na_2CO_3 (99.5%), K_2CO_3 (99.5%), TiO_2 (99%), $BaCO_3$ (99%), Nb_2O_5 (99.9%) were weighed by mol ratio to synthesize the given composition and the powder ball-milled for 24 h using a zirconia ball in acetone. After drying, they were calcined at 800°C for 1 h. Then, the calcined powder remixed with CuO and ZnO for reducing the sintering temperature of the ceramics and dried again. A polyvinyl alcohol (PVA: 5wt% aqueous solution) was subsequently added to the dried powder, the well mixed powder were pressed into a disk (Φ17 mm x 1 mm) at under pressure of 15 MPa. After burned out at 600°C for 3 h, specimens were sintered at 1050°C for 5 h in air. The surfaces of specimen were electroded by silver electrode and fired at 600°C for 10min. The crystal structure and microstructure of sample was observed by X-ray diffraction (XRD) and scanning electron microscope (SEM), respectively. Bulk density was calculated by Archimedes method in distilled water. P-E hysteresis loops were measured by using Sawyer-Tower circuit for range of temperature from near room temperature to 200°C and temperature change (ΔT) were calculated by indirect measurement method using Maxwell's relation.

3. Results and discussion

Figure 1 shows the temperature dependences of dielectric constants as a function of x content. As can be seen figure 1, depolarization temperature of all ceramics was observed in the vicinity of 150°C. It was conformed that all ceramics showed phase transition from ferroelectric to anti-ferroelectric at vicinity of depolarization temperature. With increasing x content, the depolarization temperature was slightly shifted to right side.

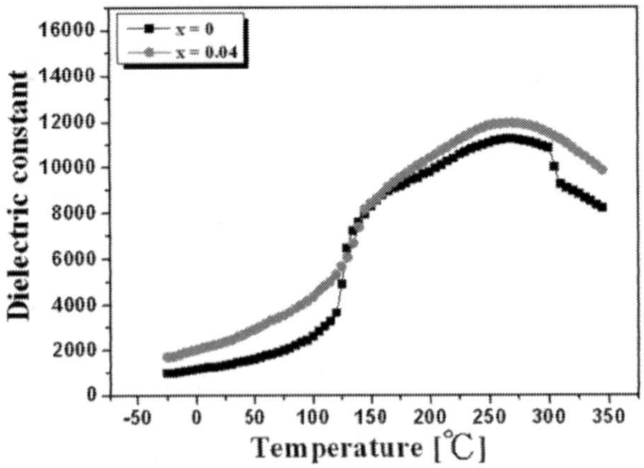

Figure 1. Temperature dependences of dielectric constants as a function of x content for (1-x) BNKT-xBT ceramics.

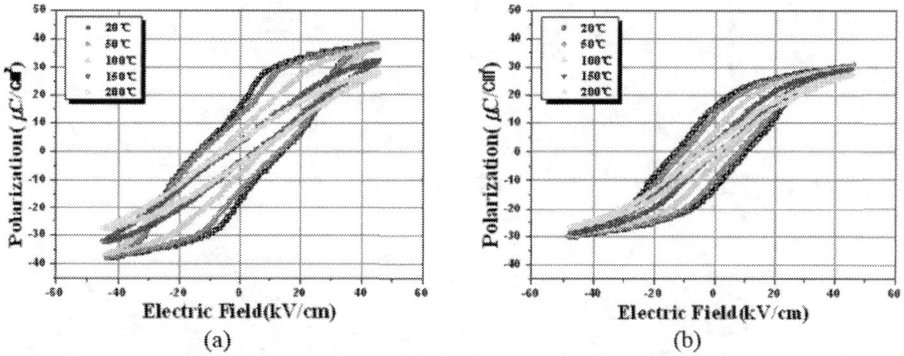

Figure 2. The P-E hysteresis loops of (1-x) BNKT-xBT ceramics as a function of variation temperature. (a) x=0, (b) x=0.04.

In order to calculate the adiabatic temperature change ΔT, the P-E hysteresis loops of all ceramics were measured at temperature range from room temperature to 200°C. The P-E hysteresis loops of all ceramics exhibited tendency to become slim with further increasing temperature. These results can be attributed to the depolarization temperature as shown in figure 1. All ceramics became anti-ferroelectric phase at more then depolarization temperature.

The Polarization versus Temperature curves (abbreviated as P-T curves) were obtained to calculate the adiabatic temperature change ΔT from P-E hysteresis loops in figure 2. The P-T curves were rapidly decreased in the vicinity of 150°C. It was corresponded to the depolarization temperature in figure 1.

In this experiment, ECE of (1-x)BNKT-xBT ceramics were measured using indirect method, and the adiabatic temperature changes ΔT were calculated by Maxwell's relation as

$$\Delta T = -T\int_{E_1}^{E_2} \frac{1}{C_E(T,E)}\left(\frac{\partial P}{\partial T}\right)_E dE \tag{2}$$

where $C_E(T,E)$ is the heat capacity [$J/(m^3K)$] at constant electric field, and E_1 and E_2 are the initial and final applied electric fields, respectively. $(\partial P/\partial T)_E$ is the pyroelectric coefficient of these ceramics. The value of $(\partial P/\partial T)_E$ was calculated from P-T curves fourth-order polynomial fitting in figure 3. As can be seen figure 4, the optimal values of ΔT was obtained about 0.32°C in x=0 ceramics under applied electric fields 45 kV/cm. These results can be attributed to the increase of entropy at depolarization temperature changing from the ferroelectric phase to anti-ferroelectric phase.

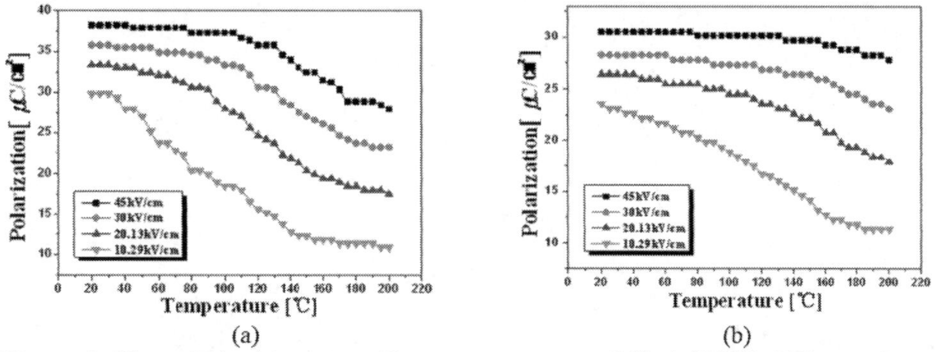

Figure 3. The polarizations versus Temperature curves of (1-x) BNKT-xBT ceramics as a function of applied electric fields. (a) x=0, (b) x=0.04

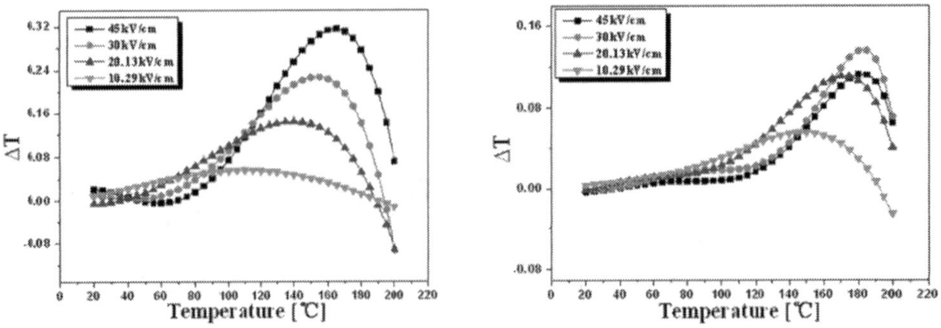

(a) (b)

Figure 4. The adiabatic temperature changes ΔT of (1-x) BNKT-xBT ceramics as a function of applied electric fields. (a) x=0, (b) x=0.04.

4. Conclusions

In this study, the electrocaloric effect in lead-free (1-x) $[Bi_{0.5}(Na_{0.84}K_{0.16})_{0.5}]TiO_3$+$xBaTiO_3$+0.3wt%$Nb_2O_5$ ceramics was investigated using indirect method. The results obtained from the experiment are as follows;

- The depolarization temperature T_d was observed to near 160°C in x=0, 0.04 ceramics. With increasing x content, the T_d was slightly shifted to right side.
- The optimal ΔT of these ceramics was obtained as 0.31°C at 165°C under applied electric fields 45 kV/cm in x =0 ceramics.

References

[1] Kim Y S, Yoo J H, Jeong Y H and Lee J Y 2013 *J. KIEEE* **26** 801-5
[2] X. Jiang, L. Luo, B. Wang, W. Li and H. Chen 2014 *Ceram. Int* **40** 2627-34
[3] Cao W P, Li W L, Xu D, Hou Y F, Wang W and Fei W D 2014 *Ceram. Int.* **40** 9273-8
[4] Wang J, Yang T, Chen S, Li G, Zhang Q and Yao X 2013 *J. Alloys Compd* **550** 561-3
[5] Qiu J H and Jiang Q 2008 *Phys. Lett. A* **372** 7191-5
[6] Bai T, Zheng G P and Shi S Q 2011 *Mater. Res. Bull* **46** 1866-9
[7] Kaddoussi H, Gagou Y, Lahmar A, Belhadi J, Allouche B, Dellis J L, Courty M, Khemakhem H, El Marssi M 2015 *Solid State Commun* **201** 64-7
[8] Chen H, Ren T L, Wu X M, Yang Y and Liu L T 2009 *Appl. Phys. Lett.* **94** 182902-4

Electrocaloric effect of inorganic low temperature sintering $(Pb_{0.88}La_{0.08})(Zr_{0.65}Ti_{0.35})O_3$ ceramics

Cheol-Min Ra[1], Ju-Hyun Yoo[1, 3], Seung-Hun Chol[1] and Jie-Young Lee[2]

[1]Semyung University, Jecheon 390-711, Republic of Korea
[2]Department of Computer Science, Semyung University, Jecheon 390-711, Republic of Korea

E-mail: juhyun57@semyung.ac.kr

Abstract. In this study, in order to develop the composition ceramics with the excellent electrocaloric properties, $(Pb_{0.88}La_{0.08})(Zr_{0.65}Ti_{0.35})O_3$ ceramics were fabricated by the conventional solid-state method. Electrocaloric effects of inorganic $(Pb_{0.88}La_{0.08})(Zr_{0.65}Ti_{0.35})O_3$ ferroelectric ceramics were investigated and discussed using the characteristics of P-E hysteresis loops at wide temperature range from room temperature to 220℃. The temperature change ΔT due to the electrocaloric effect was calculated by Maxwell's relations. The temperature change ΔT of ceramics sintered at 950℃ reached the maximum of ~0.19 at 190℃ under applied electric field of 30 kV/cm.

1. Introduction

In recent years, temperature of the earth has been gradually increased because of its global warming. As a result, the use of refrigerators and air conditioners has been increasing. Therefore, the usages of Freon gas have been also increasing. It is well known that Freon gas is the major cause of the destruction of ozone layer due to Greenhouse effect. Accordingly, in order to protect the global environment, the substitute material for Freon gas is urgent [1]. However, perfect substitute materials for Freon gas have not yet been found. Electrocaloric (abbreviated as EC) effect as a new refrigeration technologies has been received the attention of many researchers. Unlike other solid-state refrigeration technologies, ferroelectric cooler based on the principle of EC effect could be small size and environmentally friendly [2-4]. The EC effect is a change of temperature in a dielectric material by the change in entropy on application of electric field [5, 6]. It is well known that change of entropy in ferroelectric material is maximum near the phase transition temperature because a high change in polarization in vicinity of the transition temperature [7, 8]. In order to utilize the high EC effect at room temperature, research on ferroelectric with low curie temperature has been carried out. PZT based ferroelectric shows outstanding EC effect. For example, Mischenko et al. observed refrigeration temperature change of ~12k in PZT thin films. Lu et al. also observed that thin films of PLZT and polymer exhibit a refrigeration temperature change of 20K and 40K under an electric field of 120 kV/mm, 160 kV/mm, respectively [9, 10].

[3] Address for correspondence: Ju-Hyun Yoo, Semyung University, Jecheon 390-711, Republic of Korea. E-mail: juhyun57@semyung.ac.kr.

IOP Conf. Series: Materials Science and Engineering **87** (2015) 012084 doi:10.1088/1757-899X/87/1/012084

In this work, in order to develop ferroelectric ceramics with outstanding EC effect, $(Pb_{0.88}La_{0.08})(Zr_{0.65}Ti_{0.35})O_3$ ceramics were prepared by a conventional solid-state reaction method and then ferroelectric and ECE properties were investigated by using characteristics of P-E hysteresis loops to measure EC effect.

2. Experimental procedure

The compositions used in this study were as follow;

$$(Pb_{0.88}La_{0.08})(Ti_{0.65}Zr_{0.35})O_3 + \text{sintering aid } (Bi_2O_3 + CuO + Li_2O_3) \qquad (1)$$

The $(Pb_{0.88}La_{0.08})(Ti_{0.65}Zr_{0.35})O_3$ ceramics was fabricated by using a conventional mixed oxide process. The raw materials, PbO (99.5%), La_2O_3 (99.99%), TiO_2 (99%), ZrO_2 (99%) compositions were weighed by mol ratio synthesize the given composition and the milled for 24h using a zirconia ball in acetone. After drying, they were calcined at 850℃ for 2 h and then, the calcined powder remixed with sintering aid and dried again. A polyvinyl alcohol (PVA: 5 wt% aqueous solution) was subsequently added to the dried powder, the well mixed powder were into a disk (Φ17 mm) at under pressure of 15 Mpa. After burned out at 600℃ for 3 h, specimens were sintered at 950℃ and 970℃ for 4 h in air. The surfaces of specimen were electrode by silver electrode and fired at 600℃ for 10min. P-E hysteresis loops were measured by using Sawyer-Tower circuit for range of temperature from near room temperature to 220℃ and temperature change (ΔT) were calculated by indirect measurement method using Maxwell's relation.

3. Results and discussion

Figure 1 shows the temperature dependence of dielectric constant of specimens sintered at 950℃ and 970℃, respectively. Curie temperature (T_c) of all the specimens was observed in the vicinity of about 190℃. It was conformed that Tc with sintering temperature significantly unchanged.

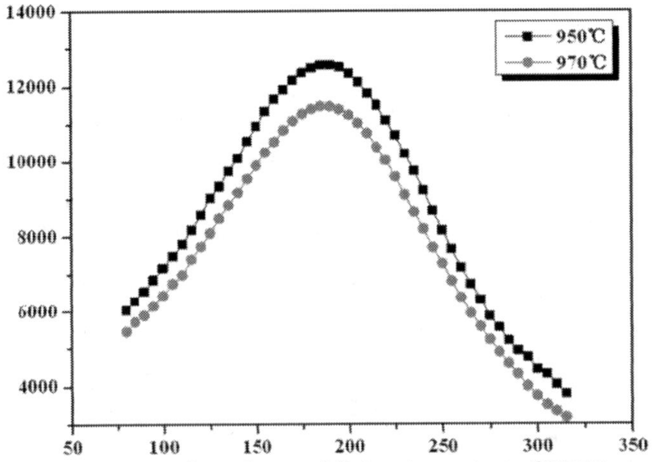

Figure 1. Temperature dependence of dielectric constant of PLZT ceramics.

Figure 2 shows P-E hysteresis loops of PLZT ceramics with electric field at a range of temperature from 20℃ to 220℃. The specimen sintered at 950℃ showed the maximum values of E_c=9.135 $_p$C/N, P_r=10.98 μC/cm^2. Coercive electric field (E_c) and Remnant polarization (P_r) were decreased as a function of increasing temperature. In particular, as shown in figure 2, it can be seen that P-E hysteresis loops in the vicinity of 180℃ rapidly were slimed due to the phase transition from ferroelectric phase to paraelectric phase was occurred at T_c.

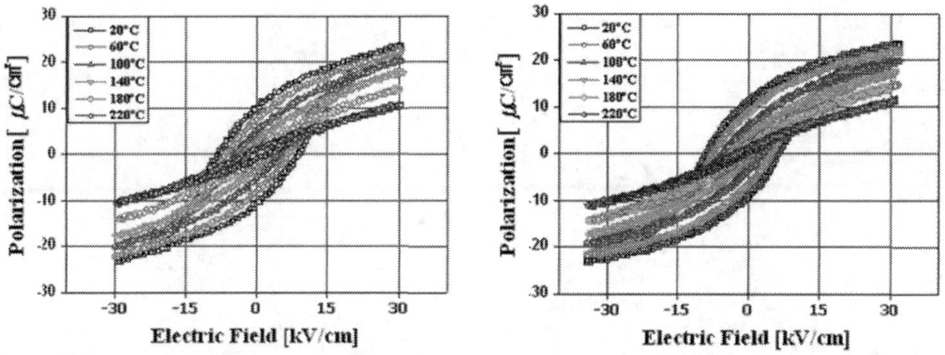

(a) (b)

Figure 2. P-E hysteresis loops of PLZT ceramics as a function of electric fields at different temperature (a=950℃, b=970℃).

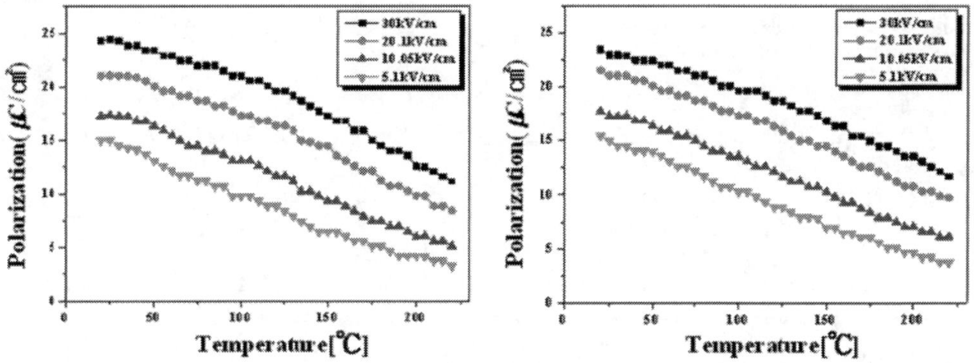

(a) (b)

Figure 3. The polarization versus Temperature of PLZT ceramics as a function of temperature change under applied electric fields (a=950℃, b=970℃).

Figure 3 presents the changes in polarization with temperature under applied electric fields. As can be seen figure 3, the remnant polarization decreased with increasing temperature. In particular, decrease at range of temperature from 100℃ to 220℃ is larger than that at range of different temperature. These results show that the phase transition was occurred from ferroelectric phase to paraelectric phase at Tc.

Figure 4 shows temperature variation (ΔT) with temperature under applied electric fields. In this work, the EC effect of PLZT ceramics was measured by indirect method, and ΔT calculated by Maxwell's relation.

$$\Delta T = -T\int_{E_1}^{E_2} \frac{1}{C_E(T,E)}\left(\frac{\partial P}{\partial T}\right)_E dE \quad (2)$$

Heat capacity of specimens is approximately constant for ferroelectric ceramics at 2.74 x 106 (J/m3K).

As shown in figure 4(a), ΔT of ceramic sintered at 950℃ showed the maximum value of ~0.19℃ under application of 30kV/cm at 190℃, while ceramics sintered at 970℃ exhibited the temperature variation of ~0.15℃. As a result, the maximum value of ΔT was attributed to the increase of entropy because the phase transition from ferroelectric phase to paraelectric phase was occurred in the vicinity of T_c.

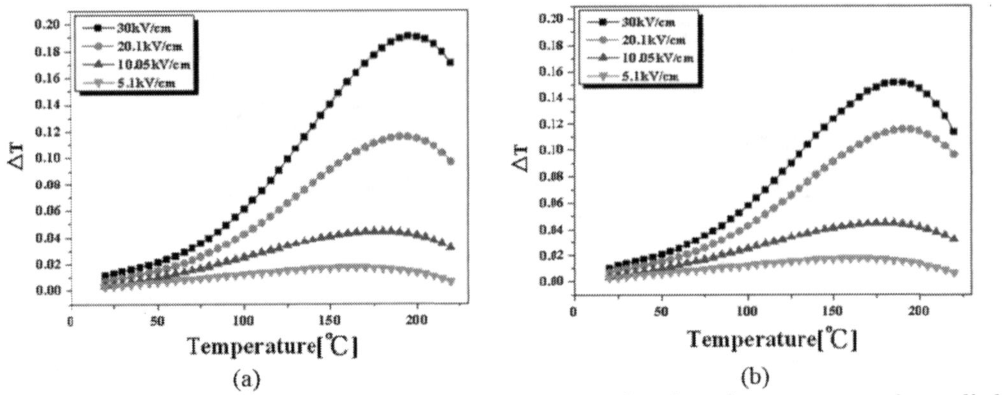

Figure 4. Temperature variation (ΔT) of PLZT ceramics as a function of temperature under applied electric fields (a=950℃, b=970℃).

4. Conclusions

In this study, the electrocaloric effects of $(Pb_{0.88}La_{0.08})(Zr_{0.65}Ti_{0.35})O_3$ ferroelectric ceramics were investigated and discussed.

- Curie temperature (T_c) of PLZT ceramics was observed in the vicinity of 190℃.
- Coercive electric field (E_c) and Remnant polarization (Pr) were rapidly decreased at T_c. Due to the phase transition from ferroelectric phase to paraelectric phase.
- Temperature variation (ΔT) was calculated by Maxwell's relations. ΔT of ceramic sintered at 950℃ showed the maximum value of ~0.19℃ with application of 30 kV/cm at 190℃.

References

[1] Xiao D Q, Wang Y C, Zhang R L, Peng S Q, Zhu J G and Yang B 1998 *Mater. Chem. Phys.* **57** 182-5
[2] Bai Y, Zheng G P, Ding K, Qiao L, Shi S Q and Guo D 2011 *J. Appl. Phys.* **110** 094103
[3] Rožič B, Kosec M, Ursic H, Holc J, Malič B, Zhang Q M, Blinc R, Pirc R and Kutnjak Z 2011 *J. Appl. Phys.* **110** 064118
[4] Zheng X C, Zheng G P, Lin Z and Jang Z Y 2012 *J. Electroceram.* **28** 20-6
[5] Bai Y, Zheng G P and Shi S Q 2011 *Mater. Res. Bull.* **46** 1866-9
[6] Rožič B, Zhang Q M, Kutnjak Z, Li X, Furman E, Gorny L J, Lin M, Malič B, Kosec M, Blinc R, Pirc R 2010 *Appl. Phys. Lett.* **97** 162904
[7] Maiwa H and Kim S H 2013 *Ceram. Int.* **39** S497-S500
[8] Wang J, Yang T, Chen S, Li G, Zhang Q and Yao X 2013 *J. Alloys Compd.* **550** 561-3
[9] Singh G, Tiwari V S 2012 *J. Alloys. Compd.* **523** 30-5
[10] Saranya D, Chaudhuri A R, Parui J and Krupanidhi S B 2009 *Bull. Mater. Sci.* **32** 259-62

Effect of compatibilizing agents on the interface and mechanical behaviour of polypropylene/hemp bast fiber biocomposites

M Boruvka[1], P Lenfeld, P Brdlik and L Behalek

Technical University of Liberec, Department of Engineering Technology, Studentska 2, 46117, Liberec, Czech Republic

E-mail: martin.boruvka@tul.cz

Abstract. During the last years automotive industry has given a lot of attention to the biobased polymers that are sustainable and eco-friendly. Nevertheless fully green composites are currently too expensive for most applications. A viable solution and logical starting point at this material revolution lies in reinforced synthetic thermoplastics based on plant derived biodegradable fibers. Plant fibers (PF´s) have potential to reduce weight of composite vehicle parts up to 40% compared with the main automotive composites filler, glass fibers (GF´s). Production of GF´s composites is much more energy intensive and polluting compared with growing, harvesting and preparing of PF´s. The main disadvantage of PF´s lies in combination of non-polar hydrophobic polymer matrix and polar hydrophilic fibers. This combination creates poor interface with low adhesion of both components. That implies poor wettability of fibres by polymer matrix and low mechanical properties of biocomposites. Therefore specific compatibilizing agents (Struktol SA1012, Fusabond P353, Smart + Luperox) were used in order to enhance compatibility between reinforcement and matrix. In this paper sets of biocomposite compounds were prepared by twin screw extrusion considering different type and weight percentage (wt. %) of compatibilizing agents, hemp bast fibres (HBF´s) within ratio 20 (wt. %) and polypropylene (PP) THERMOFIL PP E020M matrix. Resulting compounds were than injection molded and tested samples were characterized by means of scanning electron microscopy (SEM) and mechanical testing.

1. Introduction

Producing the conventional composites is energy intensive and polluting, while their durability is often seen as an advantage, on the other hand it is also their biggest disadvantage. Glass, carbon, and aramid fibre reinforced polymers do not degrade naturally, it is hard to dispose them of and difficult to recycle them [1]. Glass fibres compared to aramid and carbon fibres are the most widely used reinforcement due to their low cost and relatively good mechanical properties. However, these fibres have serious drawbacks as indicate table 1 [2]. But to meet European Union directive about end-of-life of vehicles (ELV), which requires that 95% of all new vehicles should be recyclable by 2015, there is strong need to find new and more suitable solutions. Thermoplastics offer some solution, as the resin could be

[1] Address for correspondence: M Boruvka, Technical University of Liberec, Department of Engineering Technology, Studentska 2, 46117, Liberec, Czech Republic. E-mail: martin.boruvka@tul.cz.

thermally recycled [1]. Variety of automotive components based on environmentally friendly reinforced fibre polymers are now manufactured instead of glass [2]. There are already companies (Lear Corp., Johnson Controls Inc., Ford, etc.) using hemp fibers as reinforcement in number of automotive components like door trim panels and trunk trims [3].

Table 1. Comparison between natural and glass fibres [2].

	Natural fibres	**Glass fibres**
Density	Low	Twice that of natural fibres
Cost	Low	Low, but higher than NF
Renewability	Yes	No
Recyclability	Yes	No
Energy consumption	Low	High
Distribution	Wide	Wide
CO_2 neutral	Yes	No
Abrasion to machines	No	Yes
Health risk when inhaled	No	Yes
Disposal	Biodegradable	Not biodegradable

The modern fibre corps production oriented on high yield requires huge amounts of water, pesticides, fungicides, and herbicides. Those factors, with the other agricultural techniques, like monorops growth without crop rotation and tillage are contributing to soil degradation. In consequence this could lead to disruption of ecological balance. Nevertheless, fibre crops as a renewable resource of PF´s could be important for world environment, economy and sustainable future [4].

Compared to the other fibre crops, production of Hemp (Cannabis sativa L.) is easy to achieve organically. Therefore many of the ecological problems in chemical farming could be avoided [5]. Hemp, considered as one of the oldest crops known to man and has been cultivated at least 12,000 years [4]. During 19th century heavy taxes was imposed on industrial hemp that essentially legislated the industry out of existence [6]. Recently interest has been renewed, possibility of agro-ecological cultivation and hemp products sustainability are the main drivers for a future expansion of this multifunctional crop. The economical value of hemp growing can be related to the possibility of very large number of end-use applications (figure 1). Traditionally, hemp had a very important position in crop rotations because it restores nutrients to the soil, thus provide beneficial effect on following crops [5]. Gorchs et al. [7] reported increased yield of wheat grown after hemp rotation. Moreover, very high yield compared with many other crops, up to 25 tons (t) above ground dry matter per hectare (20 $t{\cdot}ha^{-1}$ stem dry matter), depending on environmental conditions and agronomy [8]. Other benefit of hemp is that it can be grown on polluted areas and can contribute to phytodepuration (reduction of contaminants in waste water by means of biological and physicochemical processes driven by aquatic photosynthetic organism) [9]. Despite of many positive features of hemp, there still barriers which are needed to be overcome: improvement of mechanical and biological processing for fiber separation; innovation of harvesting and agro-production systems; legislation related to progressive lowering of the legal delta-9-tetrahydrocannabinol (THC) limit content in plants, seeds availability limitation and so on [5].

The main disadvantage of those biocomposites is combination of non-polar hydrophobic polymer matrix and polar hydrophilic fibers. This combination creates poor interface with low adhesion of both components. That implies poor wettability of fibres by polymer matrix and low mechanical properties of biocomposite. Due to this it is necessary to modify HBF´s surface properties prior to using them as reinforcement to ensure good dispersion and mechanical properties. In this paper, chemical modification of the fiber-matrix interphase with different compatibilizing agents for comparison was used. Generally, coupling agents has two functions to ensure proper stress transfer at the interface between fiber and matrix. The first function is to reaction with surface OH groups of cellulose and the second is to react with functional groups of the matrix [1].

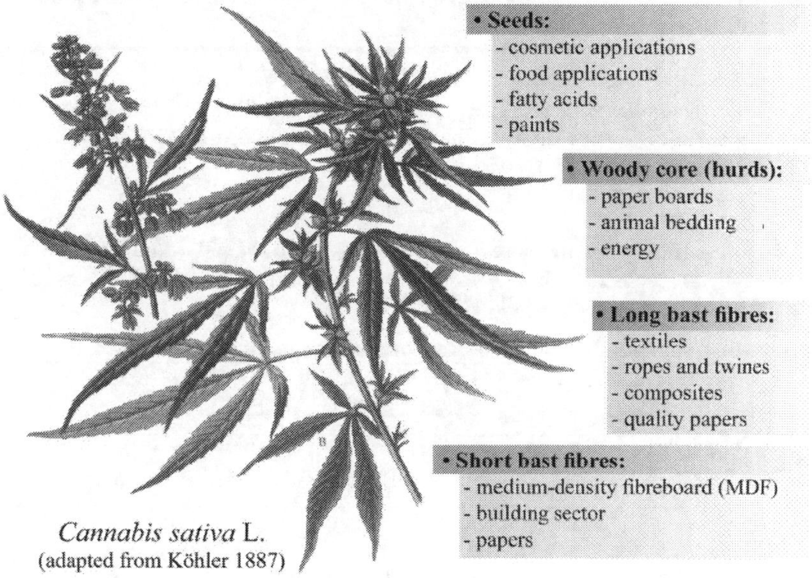

Figure 1. Main uses of the hemp products.

2. Experimental

2.1. Materials

Hemp bast fibres were purchased from Atelier Johanna (CZ). Coupling agents were purchased as fallow: Fusabond P353 - DuPont (USA), Struktol SA1012 - Struktol Company of America (USA), Smart + Luperox - DoW Corning Corporation (USA). Matrix polypropylene THERMOFIL PP E020M was purchased from Sumika Polymer Compounds (EU).

2.2. Preparation of hemp bast fibers (HBF)

For the composite material preparation, fibres were milled at the shear mill (RETSCH SM 300) (n = 3000 min^{-1}) with trapezoidal holes of 0.75 mm to modify their length. Subsequently were fibres washed with deionized water to remove organic impurities and greases. Milled fibres were than dried. An optical microscope (Leica DM 2005P) was used to analyze the length of fibres (at least 30 individual fibres were observed). Range of fibres lengths were within 0.5 mm up to 2 mm.

2.3. Compounding procedure

For fabrication composite pellets was used twin screw extruder (ZAMAK EHP-2x130di) flowed by water bath and pelletizer. Temperature profile of extrusion line was se to 150°C up to 180°C. Coupling agents and matrix were dosed directly to hopper. HBF was dosed directly into the melting chamber of extruder in the recommended front position by external device, working on the gravimetric principle. The reason for dosing fibres in the front parts of the extruder (near granulation head) is to prevent excessive shear stress of fibres during compounding melt composite and thus their damage or thermal degradation.

IOP Conf. Series: Materials Science and Engineering **87** (2015) 012085 doi:10.1088/1757-899X/87/1/012085

2.4. Drying

Due to the fact that in the previous operation extruded string passed through a water bath, it was necessary to dry granulated composite compound. For drying (80°C, 6 hours) was used stationary laboratory oven (Venticell). Moisture would adversely affect workability and properties (dimensional changes) of the final molded part, for this reason it was necessary to reduce the moisture content to the lowest possible level.

2.5. Injection molding

Further than were injection molded testing specimens according to ISO 527. Based on the experimental tests on injection molding machine (ARBURG 270S 400-100) was chosen increasing temperature profile (160°C up to 180°C) of the melting chamber.

2.6. Mechanical testing

The mechanical properties were measured according to international standards (table 2). Testing machines for measuring mechanical properties were used as fallow: Tensile - TIRAtest 2300, Impact - CEAST Resil 5.5 and Flexural - Hounsfield H10KT.

Table 2. Summarization of mechanical properties.

	Coupling agents:			1,5% SMART + 0,2% LUPEROX		4% FUSABOND P353		3% STRUKTOL SA 1012	
	MATRIX / FIBRE		PP	PP	PP + 20% HBF	PP	PP + 20% HBF	PP	PP + 20% HBF
Tensile Modulus of Elasticity (1 mm/min)	ISO 527/ 1B/1	MPa	1596	1419	2143	1395	1917	1337	1992
Yield Strength (Ultimate Strength) (50 mm/min)	ISO 527/ 1B/50	MPa	26,7	29,7	37,3	27,7	34,9	27,6	33,7
Nominal Strain at Fracture (50 mm/min)	ISO 527/ 1B/50	%	39,4	38,0	8,9	30,5	7,1	20,3	5,8
Charpy Impact Energy (+23°C)	ISO 179-1/1eA	kJ/m^2	8,8	10,1	5,6	7,7	4,7	8,3	4,5
Charpy Impact Energy (-35°C)	ISO 179-1/1eA	kJ/m^2	3,8	3,4	3,6	3,3	3,0	3,1	2,9
Flexural Modulus of Elasticity (2 mm/min)	ISO 178	MPa	1287	1427	2064	1305	1929	1193	1859
Flexural Strength (2 mm/min)	ISO 178	MPa	34,8	39,2	49,5	35,9	47,3	34,7	44,3

2.7. Scanning electron microscope (SEM) examination

Micrographs of PP/HBF composite fractures were taken on scanning electron microscope Tescan Vega XMU. To investigate adhesion of fibres and interface improvement according to used coupling agents were taken fractures from Charpy impact energy measurements under low temperatures (-35°C).

3. Results and discussion

The effect on interface and mechanical properties of resulted PP/HBF biocomposites were investigated due to use of compatibilizing agents based on different bonding principles. The concentrations of compatibilizing agents were used due to producer's recommendations and experimental tests.

Fusabond P353 (F) is one of the most commonly used compounding agents. This additive is on the basis of maleic anhydride (MA) which is by the producer grafted (g) with PP chains. Grafting is usually carried out by radical mechanism, as an initiator there is peroxide group which by the reaction with the PP chain creates free radical which subsequently reacts with the PP molecule at creation compound MA-g-PP. PP connected on the MA is during compoundation physically implemented into PP matrix by which there is complete interlacing of the PP matrix with natural fibres.

Struktol SA 1012 (S) is compatibilizer and processing additive designed to deliver better physical properties and dispersion. Struktol has a similar function as additive already mentioned above. There is difference between free radical creations, which enable connection of the polar and non-polar composite components. Covalent bond is created by the ion mechanism initiated by the additive hydrolysis.

Smart + Luperox (SL) is two component liquid additive based on patent invention No. WO 2011/083045 A1. The above mentioned coupling technologies functionalize polypropylene using grafting that is accompanied by degradation of the polymer by β-chain scission. Such degradation results in a decrease of the viscosity of the material to be processed. Furthermore, this degradation results in a polymer having inferior performance compared to the starting material. Compared to this, PP, the unsaturated silane and the compound capable of generating free radical sites (peroxide) in the PP are mixed together before twin screw extrusion process at ambient temperature. The hydrolysable groups, for example silyl-alkoxy groups, present in the silane moieties grafted to the PP react in the presence of ambient moisture with hydroxyl groups on HBF surface.

3.1. Mechanical properties

The mechanical properties are generally affected by the volume fraction of the added reinforcement (HBF), the dispersion of the HBF in the matrix and the improvement of interaction between the HBF and the matrix by compatibilizing agents. Summarization of mechanical properties is in table 2.

A maximum tensile modulus of 2143 MPa (34% increase compared to pure PP control sample) was achieved for PP/HBF treated with SL compatibilizing agent. The results for tensile strength showed with same compatibilizing agent (SL) best improvement of 37.3 MPa (40% increase compared to control sample). The strain at break was reduced significantly for all PP/HBF reinforced composites. The largest decrease of 85% was achieved for S compatibilizing agent and lowest decrease of 77% for SL compatibilizing agent. The reason is that stiff HBF caused substantial local stress concentrations and failure at strain. The impact strength of PP/HBF reinforced composites also significantly decreased for booth temperatures and all compatibilizing agents. The highest impact strength at 23°C of 5.6 kJ/m^2 (36% decreasing) was achieved again for SL compatibilizing agent and lowest 4.5 kJ/m^2 (49% decrease) for S compatibilizing agent. The only little energy is absorbed to break samples at low temperatures (-35°C) for all PP/HBF composites. This decrease for booth temperatures may be attributed to the fact that the notch serves as the stress concentration zone and under sudden impact, fracture is very easily propagated. The booth flexural modulus and flexural strength increased for all PP/HBF treated with compatibilizing agents. The highest values attained for flexural modulus (2064 MPa, 60% increase) and strength (49.5 MPa, 42% increase) showed SL compatibilizing agent. The lowest values attained for flexural modulus (1859 MPa, 44% increase) and strength (44.3 MPa, 27% increase) showed S compatibilizing agent.

3.2. Interfacial properties

The SEM micrographs of Charpy impact fractures of PP/HBF composites treated with compatibilizing agents under low temperatures (-35°C) are shown at figure 2. To improvement of mechanical properties with addition of compatibilizing agents were confirmed above. As can be seen all three micrographs showed embedded HBF in PP matrix with very tight interface. That implies significant improve of HBF wettability and interface adhesion of both components.

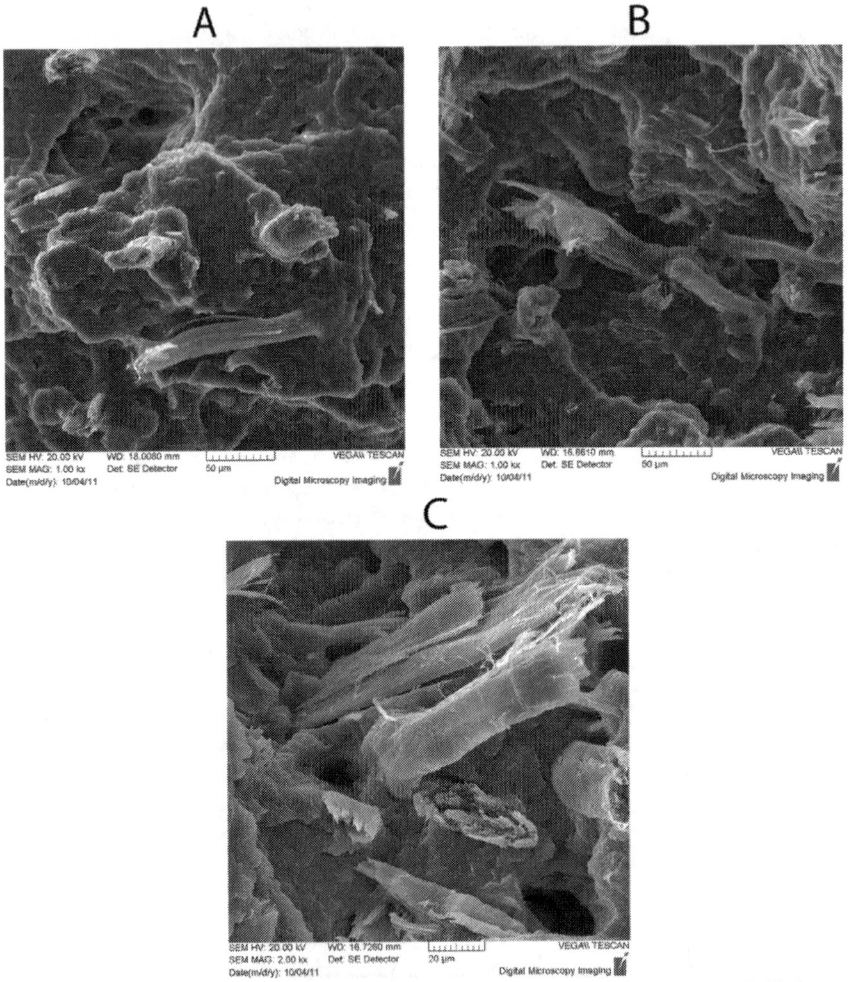

Figure 2. SEM micrographs of PP/HBF composite fractures treated with compatibilizing agents: A) Smart + Luperox, B) Fusabond P353, C) Struktol SA1012.

4. Conclusion

In this paper, PP/HBF composites modified with compatibilizing agents based on different bonding principles have been investigated due to processing and industrial applications such as injection molding. Hemp bast fibres have properties that make them a suitable eco-friendly replacement compared to glass fibres. The economical value of hemp growing can be related to the possibility of very large number of end-use applications and possibility of organic growing can be related to environmental and sustainable production. These composites have shown promising mechanical properties. However, impact strength represents the lowest mechanical property affected by compatibilizing agents. It was found that Smart + Luperox compatibilizing agent based on silane has shown best improvement in mechanical properties. The lowest mechanical properties were found for Smart SA 1012 compatibilizing agent. In conclusion it can be postulated that, all used compatibilizing agents can serve as a value added and cost effective improvement in biocomposite material processing based on HBF.

Acknowledgments
This paper was written with research project SGS 21005 support.

References

[1] March G 2003 Next step for automotive materials *Materials Today* **6** 36-43
[2] Wambua P, Ivens P and Verpoest I 2003 Natural fibres: Can they replace glass in fibre reinforced plastics? *Composites Science and Technology* **63** 1259-64
[3] Beldzki A K, Sperber V E and Faruk O 2002 Natural and wood fibre reinforcement in polymers *Rapra Review Reports* (Shawbury, U.K.: Rapra Technology Ltd.) **13** 144
[4] Mohanty A K, Misra M and Drzal L T, 2005 *Natural Fibers, Biopolymers, and Biocomposites* (Boca Raton: Taylor & Francis) p 896
[5] Müssig J 2010 *Industrial Application of Natural Fibres: Structure, Properties, and Technical Applications* (Chichester, West Sussex, U.K: Wiley) p 560
[6] Roulac J 1997 *Hemp horizons: The Comeback of the World's Most Promising Plant* (White River Junction: Chelsea Green Publishing) p 211
[7] Gorchs G, Llovers J and Comas J 2000 Effect of hemp (Cannabis sativaL.) in a crop rotation hemp–wheat in the humid cool areas of north-eastern Spain *Proceedings of the Conference on Crop Development for the Cool and Wet Regions of Europe* (Italy: Pordenone) pp 581-9
[8] Struik P C, Amaducci S, Bullard M J, Stutterheim N C, Venturi G and Cromack H T H 2000 Agronomy of fibre hemp (Cannabis sativa L.) in Europe *Industrial Crops and Products* **11** 107-18
[9] Kozlowski J, Baraniecki P, Grabowska L and Mankowski J 1995 Recultivation of degraded areas through cultivation of hemp *Proceedings of the First International Symposium 'Biorohstoff Hanf'* (Frankfurt am Main, nova-Institut Köln/Hürth) pp 259-67

IOP Conf. Series: Materials Science and Engineering **87** (2015) 012086 doi:10.1088/1757-899X/87/1/012086

Research on the property improvement of PVC using red mud in industrial waste residue

Xiaopeng Nie[1], Xingang Li and Songxian Shuai

Guizhou Construction Science Research & Design Institute Limited Company of CSCEC, 4 Ganping Road, Guiyang, 550006, China

E-mail: niexiaopeng2009@163.com

Abstract. Red mud is a red solid power waste that is discharged in the aluminium refinery industry during production. It is a strong alkali and can be categorized as polluting industrial residue. How to make comprehensive use of red mud has become a worldwide issue. In this paper, we put red mud into PVC (polyvinyl chloride polymer), taking advantage of the complicated chemical properties of red mud derived from the Bayer process. The results are compared with silica fume, coal ash and calcium carbonate under the same experimental conditions, which shows that improvement of PVC plastication can be achieved by adding red mud.

1. Introduction

Red mud is waste when refined alumina from bauxite. According to aluminum smelting methods which were divided into Bayer process red mud, Sintering red mud and joint method red mud. With the developing of the aluminum industry, the world produces almost 50 million tons of red mud annually. However, the main method for dealing with red mud is pilling up in the world, so how to comprehensively utilize red mud is a world problem [1-4]. Because the red mud dealing with alkaline solution, the alkali content is very high, and the pH content of red mud is 15%~30%, from 10.0 to 12.5. Because we don't make full use of the amount of red mud, plenty of land it occupies will lead to alkaline soil, polluted groundwater and harm to health. In order to resolve the problem of environment pollution, scientific scholars research the physical and chemical properties of red mud in the every country [5-7].

We found that there are two ways to deal with red mud: one is to recycle valued metals from the red mud [8-10]; the other is as a common mineral resources [11-14].

Earlier research about recycling iron from red mud is from the 1950s, and mang refining technology has been patented. Laszlo et al.[15] tested the red mud content many metallic oxide such as Fe_2O_3 ,Al_2O_3 and TiO_2 which respectively is 66%(mass fraction, the same below), 13.3% and 6.1%. Ercagt et al. [16] retrieved Fe,TiO_2 and Al_2O_3 from smelting red mud in the electric-arc furnace. There has been not only research on retrieving metal, but also on recycling rare earth metals from red mud. Orhsenkuhnu [17] studied rare earth metals recycling under different acid soaking and leaching methods.

Study on red mud as a mineral resource, and put it in the cement. At present, research mainly

[1] Address for correspondence: Xiaopeng Nie, Guizhou Construction Science Research & Design Institute Limited Company of CSCEC, 4 Ganping Road, Guiyang, 550006, China. E-mail: niexiaopeng2009@163.com.

produces cement high in iron oxide, special cement and ordinary portland cement. Tsakiridis et al. have researched producing portland cement technology by adding 3.5% red mud in raw material. With this method, the cement possesses higher early strength than ordinary portland cement in mechanics performance. Except for red mud as raw material produce cement, Sglavo et al. [18] studied red mud mixes with clay calcining ceramics. Additionally, red mud can remove impurities; Gupta V K et al. [19] used H_2O_2 to treat Bayer process red mud to remove organic matter on the surface. Altundogan S et al. [20] studied red mud adsorption of As^{5+} and As^{3+}.

This paper mainly researches the effect of Bayer process red mud on PVC organic material.

2. Experiments

2.1. Equipment

The heat & cool mixer model is JZRL10/25. It is made of fuxin haoteer plastic machinery CO., LTD.

A ZJL-200A torque rheometer controlled by computer is made of Chuang chun Intelligent Apparatus CO., LTD.

2.2. Material

This test used many organic materials, such as PVC, stabilizer, Yellow foaming agent etc. Those materials types are listed in table 1.

Table 1. The type of raw material.

name	type	Place of production	remark
PVC	Tianyuan SG-8	Yibin Sichuan	
stabilizer	Shengtailong GB-01	Shenyang Liaoning	
Yellow foaming agent	Joysun NH600A	Shanghai	
ACR	Ruifeng LP-901	Zibo Shandong	
Processing agent KW58	Joysun KW58	Shanghai	

In order to find the results, sub-nanometer fly ash, fly ash, calcium carbonate and micro-spheres were used as fillers [21].

There are many fillers, and they will lead to different test results, so it is necessary to analyze their material microcosmic structure and chemistry aspects. The physical properties of red mud are listed in table 2.

Table 2. Physical properties of red mud.

Producing method	Density kg/cm^3	Content (%)		Melting temperature (℃)	Plastic index	Dry induction coefficient (khoo)
		≥0.2 mm	≤0.008 mm			
Bayer process red mud	3.3~3.4	1.8~2.6	4.3~5.4	1350~1370	3.5~6.0	<1.2

After analysis and testing by the Institute of Geochemistry Chinese Academy of Science, the main chemistry components that the aluminum industry discharged bayer process red mud in Guizhou are shown in table 3.

Table 3. The main chemistry component of Bayer process red mud (wt %).

SiO_2	CaO	Al_2O_3	Fe_2O_3	MgO	Na_2O	TiO_2	Ignition loss
13.35	20.11	33.34	4.16	0.36	5.00	5.10	15.79

Fly-ash: most pellets are globular (over 98%), withelectron microscope found Nanoparticles adhere to parts of the ball granule, developing framboidal texture, as shown in figure 1.

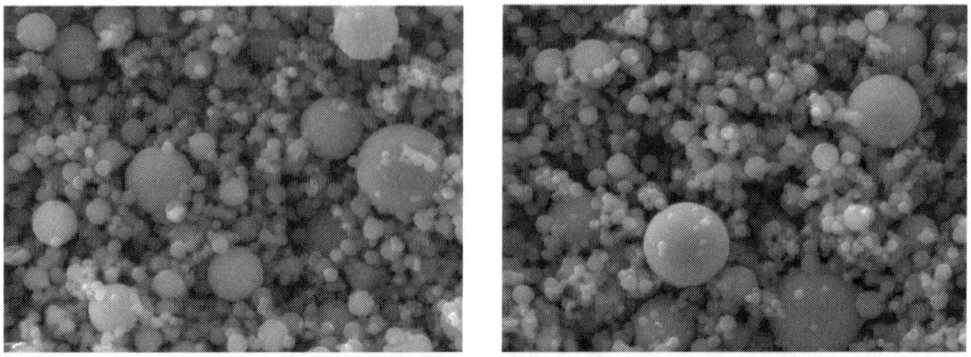

Figure 1. Fly-ash in the SEM photograph.

Figure 2. Sub-nanometer silicon fume in the SEM photograph.

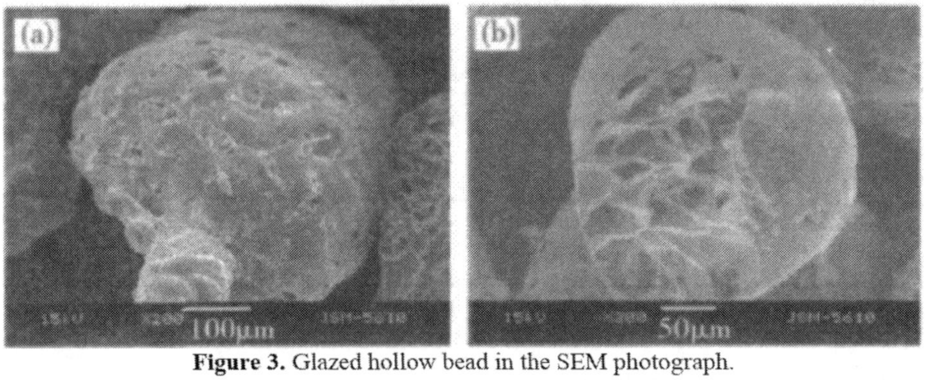

Figure 3. Glazed hollow bead in the SEM photograph.

Its chemistry component includes CaO_2 SiO_2 MgO Fe_2O_3 KCl Ti, etc.

Sub-nanometer silicon fume: most of the particles are angular, under the SEM discovered: Nanoparticles adhere to parts of the ball granule, as shown in figure 2.

Light calcium carbonate: it is an important inorganic material that light calcium carbonate was made by chemistry method. It is also called precipitated calcium carbonate, and it has fine grain size, high whiteness, good crystal form, lower prices, etc. [22].

Glazed hollow bead: it is an acidic glassy lava mineral with its internal porous surface closed. It is made of SiO_2 Al_2O_3 CaO. The SEM picture is shown in figure 3.

2.3. Testing method

The materials compounding as table 4.

Table 4. The basic testing formula.

Raw material	Copies(Volume ratio)	Raw material	Copies(Volume ratio)
PVC	75	ACR	3
filling	68	stabilizer	3
Yellow foaming agent	1		

The temperature was set at 180℃, the affection of the rotor speed is 30 r/min, and the quantity of total material is 65 g.

3. Results and discussion

These figures show that under the same experiment, in figure 4 the curve of torque is a line parallel X-axis in the process of test, it proved adding this material PVC did not plastify. But the curve of torque in figure 5 shown PVC mixed with red mud would plastify, so the red mud as adding possess character that promotes PVC plasticized in this test condition.

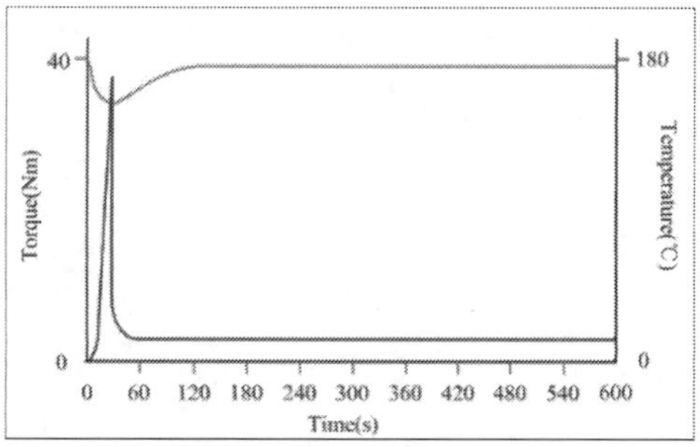

Figure 4. Testing curve.

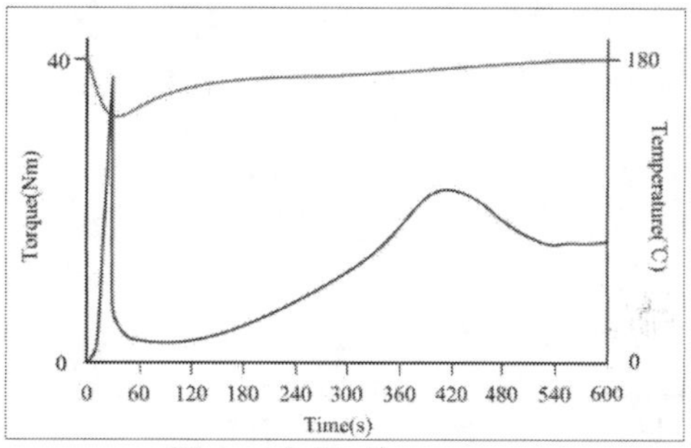

Figure 5. Testing curve.

4. Conclusions

We made the follow conclusions:

- As red mud contains rich chemical elements, it will help plasticized polyvinyl chloride plastified during maxing.
- Mixed with appropriately mount of red mud will instead of parties processing adding in PVC.

Acknowledgments

This paper was financially supported by the research project: research on optimize and integrated application of plastic form board and support systems (CSCEC-2014-Z-12)

References

[1] Zhu J, Lan J K 2008 *J Conservation and Utilization of Mineral Resource* **2** 52-4
[2] He B Q, Zhou G H and Xue Y L 2001 *J. Light Metals* **2** 24-6
[3] Zhou W X, Xie Y J and Liu B J 2002 *J. Concert.* **1** 37-40
[4] Dong F Z, Liu X Z and Yao D 2004 *J. Utilization of Mineral Resource* **6** 37-9
[5] Brunori C, Cremisini C, Massanisso P, et al. 2005 *J. J. Hazard. Master* **117** 55-63
[6] Paramguru R K, Rath P C and Misra V N 2005 *J. Miner. Process .Extr. Metall. Rev.* **26** 1-29

[7] Gu H N and Wang N, et al. *J Acta Mieralogica Sinca* **29** 397
[8] Smirnov D I and Molchanova T V 1997 *J. Hydrometallurgy* **3** 249-59
[9] Ochsenkuhn P, Maria T and Hatzilyberis K S J 2002 *J Industrial and Engineering Chemistry Research* **23** 5794-801
[10] Dobos S G, Horvath G and Felfoldi Z 1974 *J. Desalination* **12** 151-9
[11] Vincenzo M S and Renzo C Bauxite 2000 *Journal of the European Ceramic Society* **1** 235-44
[12] Vincenzo M S and Renzo C Bauxite 2000 *Journal of the European Ceramic Society* **6** 245-52
[13] Maneesh S and Upadhayay S N 1997 *J. Cement and Concrete Research* **7** 1037-46
[14] Laszlo T and Jozsef K 1957 *Journal of Hazardous Materials* **16** 460-5
[15] Ercagt E and Apak R Furnace 1997 *Journal of Chemical Technology and Biotechnology* **8** 103-8
[16] Orhsenkuhnu P Bauxaline 2002 *J. Light Metals* **1** 125-31
[17] Vangelatosa I, Angelopoulos G N and Boufounos D 2009 *Journal of Hazardous Materials* **1** 473-8
[18] Tsakiridis P E and Agarzini Leonardou S 2004 *Journal of Hazardous Materials* **8** 103-10
[19] Gupta V K, Gupte M and Sharma S 2001 *J. Water* **5** 1125-34
[20] Altundogan H S, Altundogan S, Tmen F, et al. 2002 *J. Waste Management* **22** 357-63
[21] Piga F P and Stoppa L 1993 *Journal of Management* **11** 5-9
[22] Zhang D, Yang H T and Shen S Y 2001 *J. Non-Metallic Mines* **4** 27-8

IOP Conf. Series: Materials Science and Engineering **87** (2015) 012087 doi:10.1088/1757-899X/87/1/012087

Synthesis and characterization of a sphere-like modified chitosan and acrylate resin composite for organics absorbency

S S Xin, Y H Wang, Q R Li[1], Q Zhang and X P Wang

Key Laboratory of Applied Chemistry, Yanshan University, Qinhuangdao 066004, Hebei Province, China

E-mail: liqiurong63@aliyun.cn

Abstract. In this study, the chitosan (deacetylation degree > 95%) was modified with vinyltriethoxysilane (A151) and became hydrophobic. The modified chitosan and acrylate resin composite can be synthesized by butyl methacrylate (BMA) , butyl acrylate (BA), poly vinyl alcoho (PVA), N,N'-methylene bisacrylamide (MBA), benzoyl peroxide (BPO), and ethyl acetate under microwave irradiation. The optimal synthetic condition was as follows: the molar ratio of BA and BMA was 1.5:1, the dosage of ethyl acetate, PVA, MBA, BPO and modified chitosan were 50 wt.%, 10 wt.%, 1.5 wt.%, 2.0 wt.% and 1.0 wt.% of monomers, respectively. The adsorption capacity of the composite for $CHCl_3$ and CCl_4 were approximate to 53 g/g and 44 g/g, respectively. The organics absorbency and regeneration of the samples were also tested, and the samples were characterized by analysis of the scanning electron microscope and simultaneous thermo gravimetric/differential thermal.

1. Introduction

The exhausted oils from industrial activities or operational maintenance of machines are usually treated by specialized firms while the disposal of waste containing oils at low concentration occurs in conventional sewage treatment systems and causes environmental damages. In order to reduce the polluting power of oily wastes, they might be treated separately before discharge [1]. In recent years, many research methods are chosen methyl acrylates as raw material; however, the adsorption capacities of acrylate resins are very low. To solve this problem, some natural materials have attracted the attention of many researchers. Asha Srinivasan researched fungus mucor roux rich with chitosan for removal oil from water [2]. H.H. Sokker, etc. reported an adsorbent, which was chitosan gelatin based on polyacrylamide compound, and showed a high adsorption capacity of crude oil from aqueous solution [3]. S.M. Sidik, etc. investigated oil palm leaves modified by lauric acid to adsorb crude oil and the maximum adsorption capacity was 1176±12.92 mg/g at 303K [4]. Chitosan is a partially acetylated glucosamine biopolymer with many useful features such as hydrophilicity, biocompatibility and biodegradability. Recently, chitosan has been also used as natural templates for the fabrication of functional inorganic oxides [5, 6]. In addition, the amino groups on the chitosan also can be used for further functionalization, such as various drugs, specific binding sites, or other functional groups [7, 8]. The aim of this study was to synthesize modified chitosan and acrylate resin composite and research

[1] Address for correspondence: Q R Li, Key Laboratory of Applied Chemistry, Yanshan University, Qinhuangdao 066004, Hebei Province, China. Tel.: +86 335 8387744; Fax: +86 335 8061569; E-mail address: liqiurong63@aliyun.com.

organics adsorption performance of composite resin, because chitosan combined with methyl acrylate and butyl acrylate resins by hydrogen bond, which improved the performance of the compound material.

2. Experimental

2.1. Materials

All regents were all of analytical grade. The chitosan (deacetylation degree > 95%) was obtained from Jinan Haidebei Marine Biological Engineering Co., LTD. Benzoperoxide (BPO) is recrystallized from ethanol, dried at 60℃, and stored in a desiccator. Deionized water was used in all the experiments. Two gram of chitosan, a certain amount of A151 and 50 mL deionized water were added into a round bottom flask. The mixed solution was stirred at a certain temperature by microwave radiation. After the reaction, the sample was washed several times with deionized water and dried in the oven at 80℃, which is modified chitosan for next experiment.

2.2. Synthesis of composite

PVA was dissolved in two-neck flask under stirring at 90℃. Then the mixture included the monomers (BMA and BA), pore-forming agent (ethyl acetate), initiator (BPO), cross-linker (MBA) and hydrophobic chitosan, were injected into the PVA aqueous solution under stirring in the presence of nitrogen. The reaction mixture was kept at 80℃for 80 min by microwave irradiation (300W). The sample was washed with ethanol and deionized water, respectively. Finally, the sample was dried in vacuum oven at 60℃for 12 h.

2.3. Characterization

The IR spectra were collected using a Fourier transform infrared spectroscopy (FTIR) spectrophotometer (Nicolet iS10). The surface morphology of the samples characterized by scanning electron microscopy (SEM) was performed by S-4800 Field Emission SEM (Hitachi, Japan). Thermo gravimetric (TG) and differential thermal analyses (DTA) were carried out simultaneously using a DTG-60A instrument (Shimadzu Corporation, Japan).

2.4. Oil adsorption experiments

All tests were performed at 25℃. The absorbent was immersed in organics solution for 12 h. The sorption capacity of the absorbent was expressed in grams of the pollutant (oil or organics) absorbed by per gram of the absorbent (g/g):

$$q_{eq} = \frac{w_e - (w_o + w_w)}{w_o} \tag{1}$$

Where q_{eq} (g/g) is the sorption capacity of the absorbent, w_e (g) is the weight of the swollen state of the sample at equilibrium, w_o(g) is the weight of the dry state of the sample (initial weight), and w_w is the weight of adsorbed water (g). In the pure oil or organics medium without any water, w_w is equal to zero.

2.5. Hydrophobic test

Equation (2) is Washburn's equation that presents the relation between a squared height of the penetrating liquid and the penetrating time [9]).

$$h^2=r\gamma t(\cos\theta)/2\eta \tag{2}$$

where h (m) is the water level in tube, r (m) the average geometric radius of filling powders' gap, γ (N/m) is the surface tension of liquid, t (s) is the penetrative time, θ is the contact angles of powders, η ($p_a \cdot s$) is the viscosity of solvent.

3. Results and discussion

3.1. Hydrophobic test of modified chitosan

The chitosan powder was modified by different amounts of A151 (4, 10, 12 and 20% of the chitosan powder weight). Figure 1A showed that the original chitosan powder was hydrophilicity. After modified by a small amount of A151, the hydrophobic capability of chitosan powder was improved significantly. When A151 was 20 wt. % of the chitosan powder, the slope of curve was approximate to horizontal, it illustrated that the modified chitosan was completely hydrophobic.

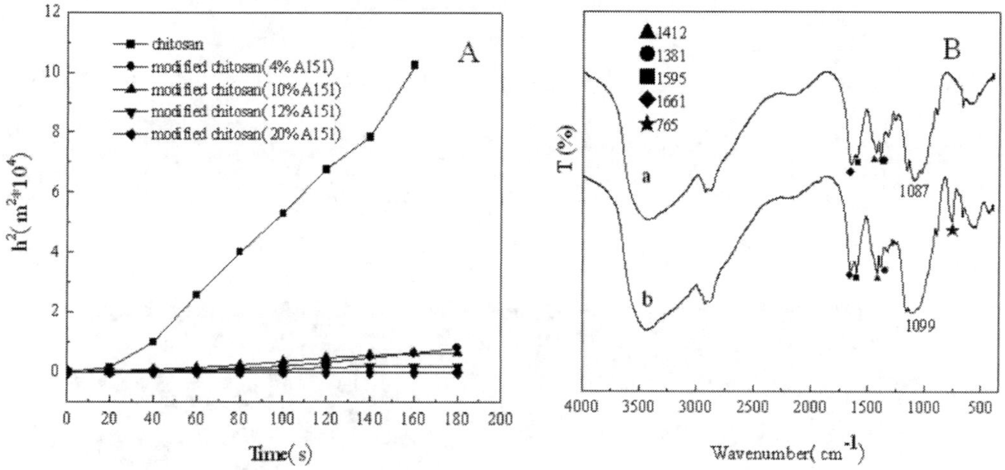

Figure 1. A: Dynamic hydrophobic test of modified chitosan; B: The FTIR spectra of chitosan (a) and modified chitosan (b).

3.2. Characterizations

Figure 1B showed the FTIR spectra of chitosan and modified chitosan. Figures 1A and 1B showed the characteristic peak at 1661 cm^{-1} corresponds to C=O stretching vibration of secondary amide, the peaks at 1381 cm^{-1} and 1412 cm^{-1} are assigned to plane rocking and twisting of $-CH_3$. In figure 1b, the peak at 765 cm^{-1} are due to C-H surface rocking vibration of $-(CH_2)_n-$ group($n<4$), the strong and broad peak at 1099 cm^{-1} nearby can be assigned to C-O stretching vibration of alkyl ether. The sharp peak at 1087 cm^{-1} is assigned to C-O stretching vibration of chitosan (figure 1A), the peak at 1595 cm^{-1} is belong to C=C stretching vibration of vinyltriethoxysilane (figure 1B).

Figure 2A shows a schematic illustration of the binding mechanisms of composite resin. The surface morphologies of acrylate resin and composite resin were observed by FESEM in figure 2B and 2C, and figure 2B showed that the surface of acrylate resin was very smooth, unlike acrylate resin; composite resin (figure 2C) exhibited a rough surface, which indicated that modified chitosan had penetrated into the inner macropore of resin by combining with acrylate resin by hydrogen bond. The real photos of composite before and after adsorption organics are presented sphere-like structure in figures 2D and 2E, and there are many holes on the surface of the composite. The characterization of composite should make oil absorbency increased.

The weight loss of samples due to thermal degradation was obtained from the thermo gravimetric analysis. As shown in figure 3, it represented the TG–DTA curves of acrylate resin and composite. The resin decomposed in two consecutive steps. The first mass loss occurred in the temperature range of 67–290℃ with the loss of about 20% based on the weight of resin, the second mass loss took place in the temperature range of 290–394℃ with 80% loss (figure 3A). However, the first mass loss of the

IOP Conf. Series: Materials Science and Engineering **87** (2015) 012087 doi:10.1088/1757-899X/87/1/012087

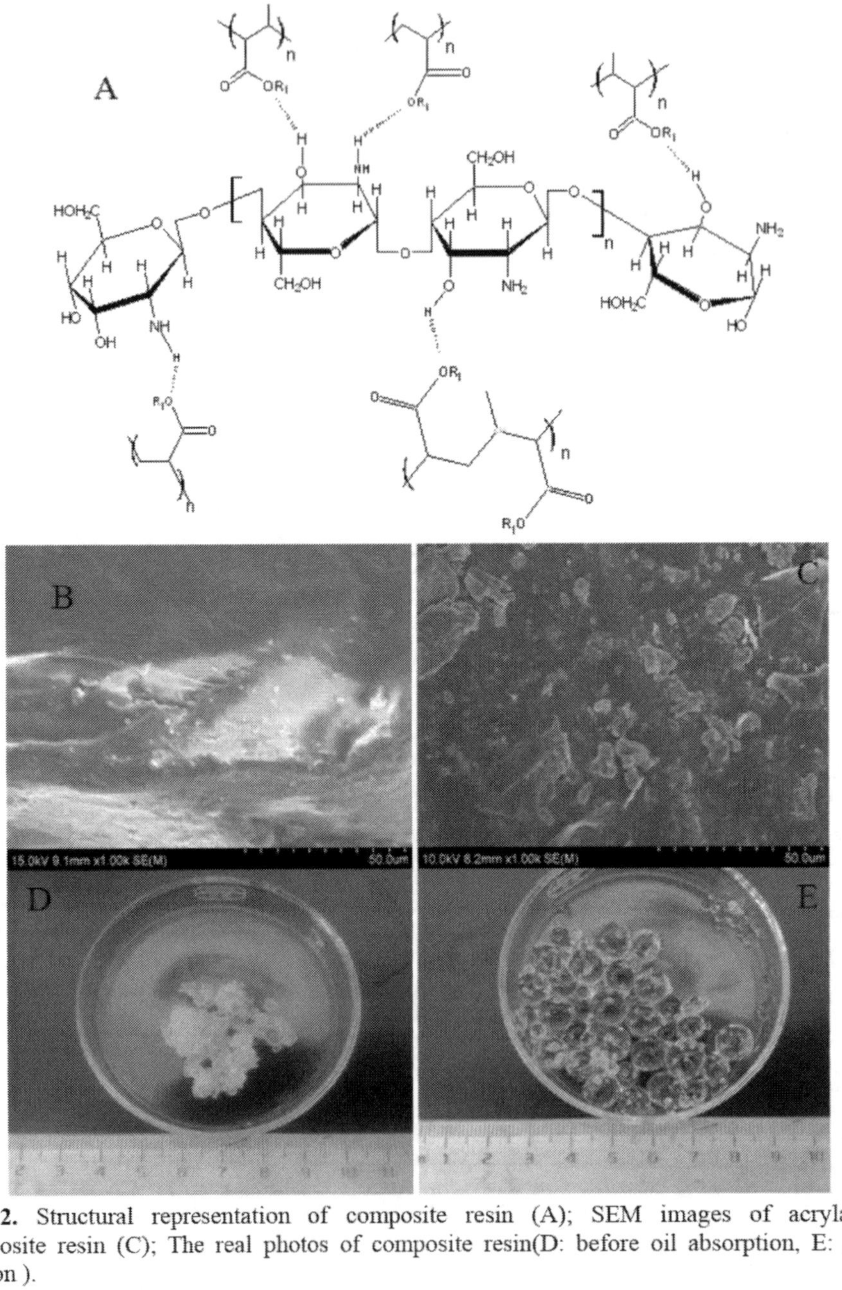

Figure 2. Structural representation of composite resin (A); SEM images of acrylate resin (B)composite resin (C); The real photos of composite resin(D: before oil absorption, E: After oil absorption).

composite resin occurred in the temperature range of 148–242℃ with the loss of 5% based on the weight of resin, the second mass loss took place in the temperature range of 242–409℃ with 90% loss (figure 3B). The thermostability of the composite was more than that of acrylate resin. The DTA plot showed one distinct endothermic peaks, which at 357℃ corresponded to vaporization of acrylate resin, however the DTA plot of the composite showed one distinct exothermic peak (figure 3B), because of the crosslinking of modified chitosan and acrylate resin, the forces and structures of compound resin were changed.

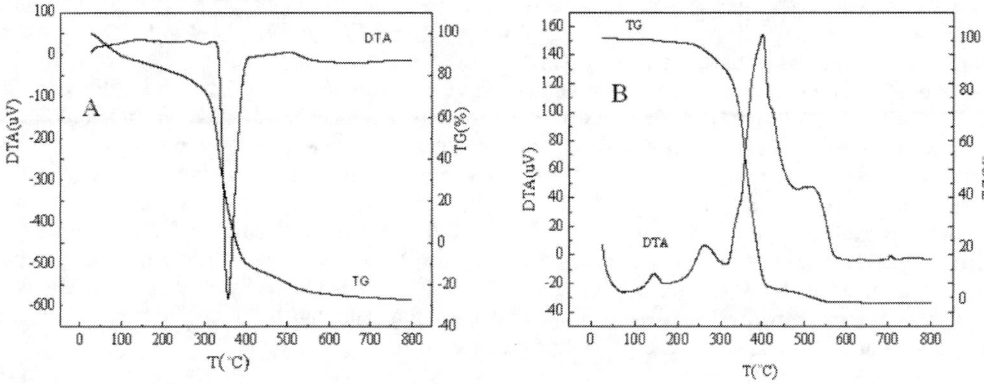

Figure 3. TG–DTA curves of acrylate resin and the composite resin (A: acrylate resin; B: composite resin).

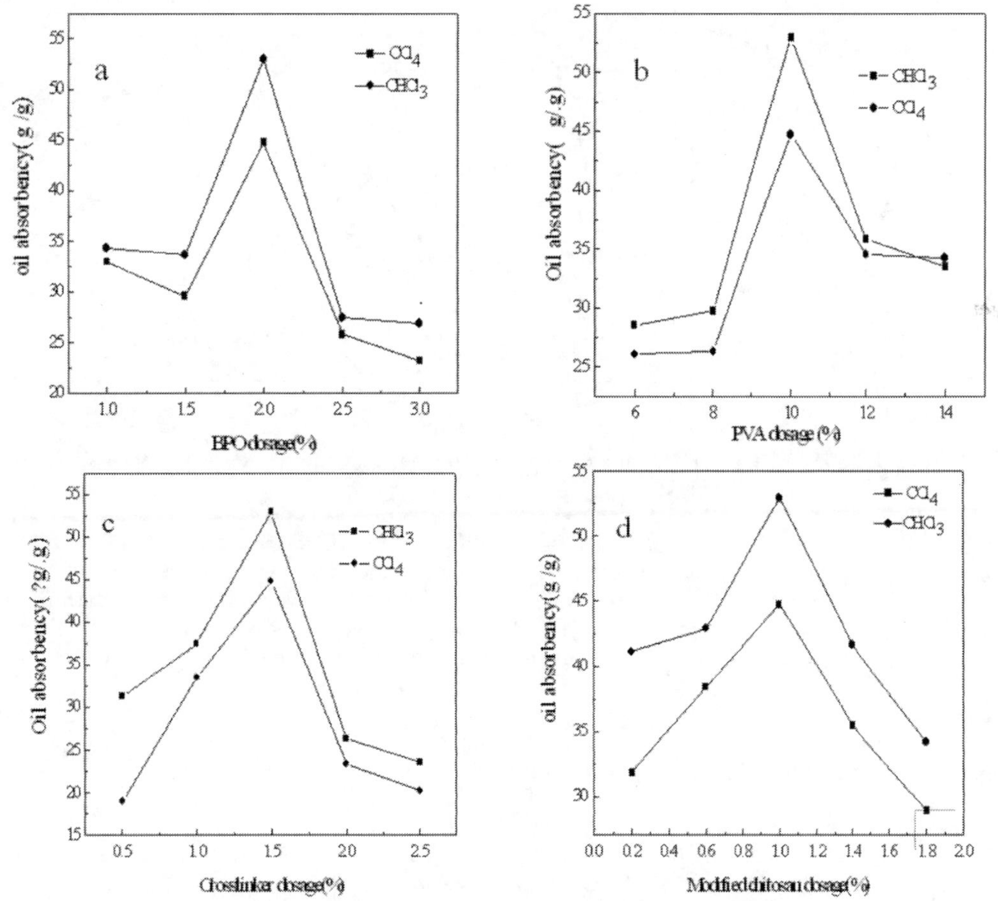

Figure 4. Effect of some factors on oil absorbency.

3.3. Oil absorptivity and renewability of composite resin

Although oil absorption mechanism of oil-absorptive resin and mechanism of superabsorbent are essentially similar, but superabsorbent resins absorb water by using hydrogen bonds, and oil-absorptive resin can only absorb oil or organics by Van der Waals force between the molecules. This difference explained the reason that the oil-absorptive resins have lower absorptive ability. Hydrophobic chitosan and acrylate resin composite will change the molecular structure of the original resin network, which can improve the oil absorptive capacity and mechanical strength. In addition, oil absorption resin would achieve oil-absorption and oil-retention properties by its molecular network structures. The effects of initiator, dispersant, cross-linker and modified chitosan were shown in figure 4, results showed that the optimum synthesis condition is as follows: the dosages of cross-linker, initiator, dispersant and modified chitosan are 1.5 wt.%, 2.0 wt.%, 10 wt.% and 1.0 wt.% of monomers, respectively. Desorption of the composite was done by distillation method, then the sample was put into organics again, and then the absorption capacity was calculated. The reusability of composite was shown in figure 5. From figure 5, we can see that the composite can be repeatedly used, and the regenerative composite has good adsorption performance.

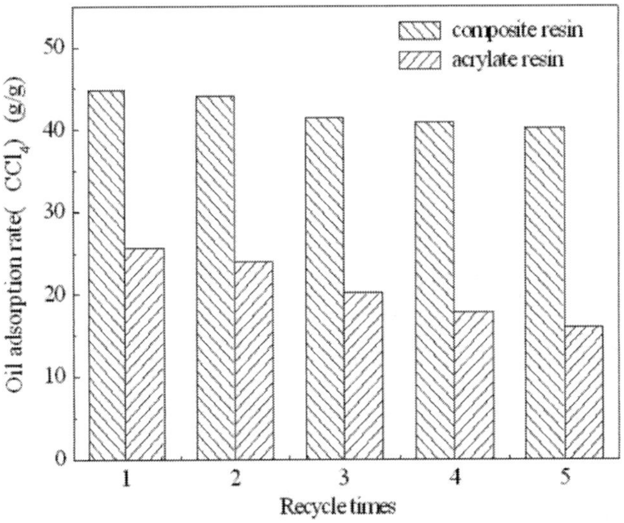

Figure 5. Oil absorbency of acrylate resin and compound resin with the reused times at 298K.

4. Conclusions

A sphere-like modified chitosan and acrylate resin composite as efficient sorbent material for oil and organics was synthesized by suspension polymerization under microwave irradiation. From the above-mentioned results, it can be concluded that the organics adsorption performance of the composite is better than that of acrylate resin. Through the thermo gravimetric analysis, thermal decomposition and heat stability of the as-prepared sample were improved distinctly.

Acknowledgments

The authors are grateful to the Natural Science Foundation of Hebei Province (B2014203143) and the key laboratory of applied chemistry in Hebei province for support during this study.

References

[1] Lim T-T and Huang X 2006 *J J Hazard Mater.* **137** 820
[2] Srinivasan A andViraraghavan T 2010 *J J Hazard Mater.* **175** 695
[3] Sokker H H, El-Sawy N M, Hassan M A and El-Anadouli B E 2011 *J J Hazard Mater.* **190** 359

[4] Sidik S M, Jalil A A, Triwahyono S, Adam S H, Satar M A H and Hameed B H 2012 *J Chem.Eng.J.* **203** 9
[5] Kadib A E, Molvinger K, Cacciaguerra T, Bousmina M and Brunel D 2011 *J Micropor Mesopor Mat.* **142** 301
[6] Thakre D, Jagtap S, Sakhare N, Labhsetwar N, Meshram S and Rayalu S 2010 *J Chem.Eng.J.* **158** *315*
[7] Qu J, Liu G, Wang Y and Hong R 2010 *J Adv Powder Tech.* **21** 461
[8] Huacai G, Wan P and Dengke L 2006 *J Carbohyd Polym.* **66** 372
[9] Kirdponpattara S, Phisalaphong M and Newby B-m Z 2013 *J J Colloid Interf Sci.* **397** 169

Reinforcement of epoxy-based composites by magnetically aligned multi walled carbon nanotube

Xianjuan Xu[1], Song Lin[2], Moyu Li[3], Wusheng Li[1], Xiaolong Jia[1, 4, 5], Qing Cai[1] and Xiaoping Yang[1, 4]

[1]State Key Laboratory of Organic-Inorganic Composites, College of Materials Science and Engineering, Beijing University of Chemical Technology, Beijing 100029, P. R. China
[2]Aerospace Research Institute of Materials and Processing Technology, Beijing 100076, China
[3]China Petroleum Engineering Co., Ltd., Beijing Company, Beijing 100085, China
[4]Changzhou Institute of Advanced Materials, Beijing University of Chemical Technology, Jiangsu 213164, P. R. China

E-mail: jiaxl@mail.buct.edu.cn

Abstract. Multi walled carbon nanotubes decorated with ferriferrous oxide nanoparticle (MWCNTs-Fe_3O_4) complex was used as an effective reinforcement in the polymer composites. The MWCNTs-Fe_3O_4 with various grafting contents of Fe_3O_4 nanoparticles were successfully prepared by combining in situ atom transfer radical polymerization (ATRP) and coprecipitation process, which was characterized with Fourier transform infrared spectroscopy (FT-IR) and transmission electron microscope (TEM). The MWCNTs-Fe_3O_4 complex showed the strong magnetic response behavior, which could be easily aligned in an external magnetic field. The alignment state of MWCNTs-Fe_3O_4 complex could be modulated by adjusting the intensity of external magnetic field, grafting content of Fe_3O_4 nanoparticles and viscosity of the solvent. Moreover, with the addition of MWCNTs-Fe_3O_4, tensile strength and modulus of epoxy composites were enhanced by 12.3 and 10.9 %, respectively, which was due to the reinforcing effect of the aligned MWCNTs-Fe_3O_4 within magnetic field.

1. Introduction

MWCNTs have been the focus of intense research and widely applied in many different fields such as composites and biotechnology due to their unique structure-dependent electrical, mechanical and optical properties. Especially in some applications where high-performance lightweight structural materials are required, and they can be developed by adding aligned MWCNTs to fiber reinforced composites. With aligned MWCNTs, the lateral mechanical performance of fiber reinforced composites can be substantially increased [1]. However, MWCNTs are easy to cluster, bundle together and entangle, causing many defects in the fiber reinforced composites, and limiting the performance of MWCNTs on polymer matrices. Besides, poor interface property resulting from weak interactions with

[5] Address for correspondence: Xiaolong Jia, State Key Laboratory of Organic-Inorganic Composites, College of Materials Science and Engineering, Beijing University of Chemical Technology, Beijing 100029, P. R. China. E-mail: jiaxl@mail.buct.edu.cn.

the matrix also limit the efficiency of MWCNTs in fiber reinforced composites [2]. Furthermore, the alignment of MWCNTs in the fiber reinforced composites is difficult to achieve. Thus, the main challenges are to improve the dispersion, interface property and alignment of MWCNTs in the polymer matrix when preparing these nanocomposites.

There are several methods to improve the dispersion and interface property of MWCNTs in polymer matrices, such as employing optimum physical blending, in situ polymerization and chemical functionalization. Among these methods, in situ polymerization such as ATRP has been proved an efficient way to improve the dispersion and interface property of MWCNTs [3]. And the obtained MWCNTs-polymer composites exhibit excellent performance due to the MWCNTs' unique mechanical, surface and multi-functional properties, and strong interactions with the matrix. What's more, there are several techniques to achieve and characterize the alignment of MWCNTs in polymer matrices, such as employing flow field, magnetic field, electrical spinning field and orientation tensor description. Among these techniques, magnetic field is an important way because the alignment of MWCNTs can be adjusted and it can be easily applied in fiber reinforced composites by using a external magnetic field [4, 5]. Besides, the orientation tensor description provides an effective way to embody all MWCNTs-Fe_3O_4 directions in the selected region, which has been successfully used to describe the orientation states of short fibers. The second moments of the probability distribution function would be adopted to numerically characterize the alignment of MWCNTs, which are called orientation tensors. Via the attachment of magnetic nanoparticles onto MWCNTs, the alignment of MWCNTs can be easily achieved in weak magnetic field [6]. Recently, several methods have been reported for the attachment of magnetic nanoparticles onto MWCNTs. Miguel A. Correa-Duarte et al. obtained the coating of MWCNTs with magnetic nanoparticles by combining the polymer wrapping and layer-by-layer (LbL) assembly techniques [7]. Via the employment of polymer onto the surface of MWCNTs, the attachment between the magnetic nanoparticles and MWCNTs is stronger, and the grafting content of the magnetic nanoparticles can be much higher, meaning MWCNTs can be aligned in weak magnetic field which is more practical in industry [8]. Nevertheless, compared to ATRP, the interaction between polymer and MWCNTs by polymer wrapping process is much weaker, meaning the polymer and magnetic nanoparticles can be detached. Besides, the grafting content of the magnetic nanoparticles cannot be easily controlled.

In this contribution, we supposed to prepare MWCNTs- Fe_3O_4 complex and control the alignment of MWCNTs under weak magnetic fields, based on ATRP of polymers and coprecipitation process of magnetic nanoparticles onto the surface of the MWCNTs, which were characterized by FT-IR, SEM and TEM. The method combines the ATRP and coprecipitation process techniques allowing the noncovalent attachment of nanoparticles onto MWCNTs. Via this method, the interface property, dispersion and orientation of obtained MWCNTs- Fe_3O_4 complex could be largely promoted, and the Fe_3O_4 nanoparticles could be efficiently grafted onto the surface of MWCNTs [9]. Furthermore, the content of Fe_3O_4 nanoparticles was high and can be numerically controlled. The alignment of MWCNTs was achieved in the presence of an external magnetic field, upon grafting of magnetic nanoparticles onto the MWCNTs surface, and the alignment was numerically characterized via orientation tensor description. Based on the results, the mechanism of the alignment formation was proposed, and the effect of the intensity of magnetic field and the grafting content of magnetic particles was discussed.

2. Experimental section

2.1. Materials

MWCNTs-COOH (purity: ≥95%, diameter: 50-80 nm, length: 10-20 μm) were purchased from Chengdu Organic Chemicals Co. Ltd. (China). Glycidylmethacrylate monomer (GMA) (purity: ≥97%) was obtained from SIGMA-ALDRICH Company. Silane coupling agent, KH550 (purity: ≥98%), is obtained from Alfa-Aesar Company. CuBr is obtained from Alfa Aesar Company. N, N, N', N'', N''-pentmethyldiethylenetriamine (PMDETA) is obtained from TCI Shanghai Co. LTD.

2-bromo-2-methylpropionyl bromide (Bibb) is obtained from SIGMA-ALDRICH Company. Triethylamine is obtained from Beijing Tong Guang Fine Chemical Company. $FeCl_3 \cdot 6H_2O$ and Fe $Cl_2 \cdot 4H_2O$ was obtained from Tianjin Fuchen Chemical Reagents Factory. Tetramethylammonium hydroxide (TMAOH) (purity: ~25 %) was obtained from Tianjin Guangfu Fine Chemical Research Institute. All other chemicals and reagents were purchased from Beijing Chemical Works (China) and used directly.

2.2. Synthesis of MWCNTs-Fe_3O_4 complex

MWCNTs-NH_2 was prepared by mixing KH550 with MWCNT-COOH suspension under bath sonication foras following our previous works. The dried MWCNTs-NH_2 was dispersed in tetrahydrofuran (THF) and triethylamine by bath sonication for 0.5 h. Subsequently, Bibb mixed with THF was added into the dispersion. Then, the mixture was slowly stirred and allowed to react for 24 h at 30°C. The Bibb grafted MWCNTs (MWCNTs-Br) were obtained by centrifugation, washing with absolute THF several times and drying in vacuum. Then, MWCNTs-Br was dispersed in N, N-Dimethylformamide (DMF) by bath sonication for 0.5 h under nitrogen and subsequently mixed with GMA, PMDETA and CuBr. The resulting solution was further stirred for 24 h at 30°C. MWCNTs-PGMA was obtained by centrifugation, washing with absolute ethyl alcohol several times and drying in vacuum. The dried MWCNTs-PGMA was dispersed with de-ionized water and TMAOH by bath sonication for 1 h under nitrogen. Subsequently, Fe^{2+}/Fe^{3+} aqueous solution was added dropwise till the pH of the final mixture is in the range of 11–13. Finally, the resultant product, MWCNTs- Fe_3O_4 complex, was obtained by centrifugation, washing with de-ionized water several times and drying in vacuum. The final product prepared using MWCNTs-PGMA, TMAOH, $FeCl_3 \cdot 6H_2O$ and $FeCl_2 \cdot 4H_2O$ was named as MWCNT-Fe_3O_4.

2.3. Characterizations

Changes of the functional groups on the surface of MWCNTs were detected with a FT-IR (Nicolet 670). Morphologies of various MWCNTs were observed by TEM (JEM100CX), respectively. The alignment of the sample was conducted by a magnet at 0.1 T. The morphology and aligned feature of as-prepared samples were also characterized using SEM (S4700) and optical microscope (IX51-21PH).

3. Results and discussion

3.1. Characterization of MWCNTs-Fe_3O_4 complex

Figure 1 shows the FT-IR spectra of MWCNTs-COOH, MWCNTs-NH_2, MWCNTs-Br, MWCNTs-PGMA and various MWCNTs-Fe_3O_4. Compared to MWCNTs-COOH, an additional peak at 1119 cm^{-1} in the spectrum of MWCNTs-NH_2 was corresponded to Si-O-C_2H_5 bending vibration, which meant that the coupling agent KH550 had been grafted onto MWCNT. This grafting of KH550 onto MWCNTs was further confirmed by characteristic peak of Si-C group at 830 cm^{-1}. In the spectrum of MWCNTs-Br, the absorption peak at 666 cm^{-1} was attributed to characteristic peak of -Br group, indicating the changes of functional groups on the surface of MWCNTs due to the bromization. The strong peaks at 1635 cm^{-1} and 1726 cm^{-1} appeared in the spectra of MWCNTs-PGMA, which were originated from the characteristic C=O stretching vibration of carboxylic groups and ester groups corresponding to the carbonyl groups of PGMA polymer. Noticeably, two obvious adsorption peaks at 1149 cm^{-1} and 1261 cm^{-1} were all originated from asymmetrical and symmetrical stretching peak of -CH_3 group and the absorption peak at 902 cm^{-1} was ascribed to the epoxide groups from PGMA polymer. The peaks at 580 and was resulted from the stretching vibration due to the interactions of Fe-O-Fe in Fe_3O_4. These characteristic peaks in the FT-IR spectra manifested that Fe_3O_4 particles were truly grafted onto the surface of MWCNTs.

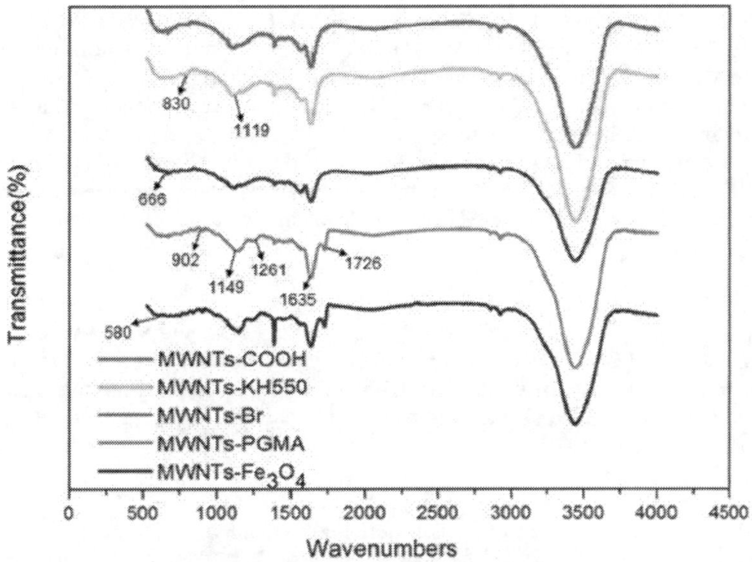

Figure 1. FT-IR spectra of MWCNTs-COOH, MWCNTs-NH_2, MWCNTs-Br, MWCNTs-PGMA, MWCNTs-Fe_3O_4.

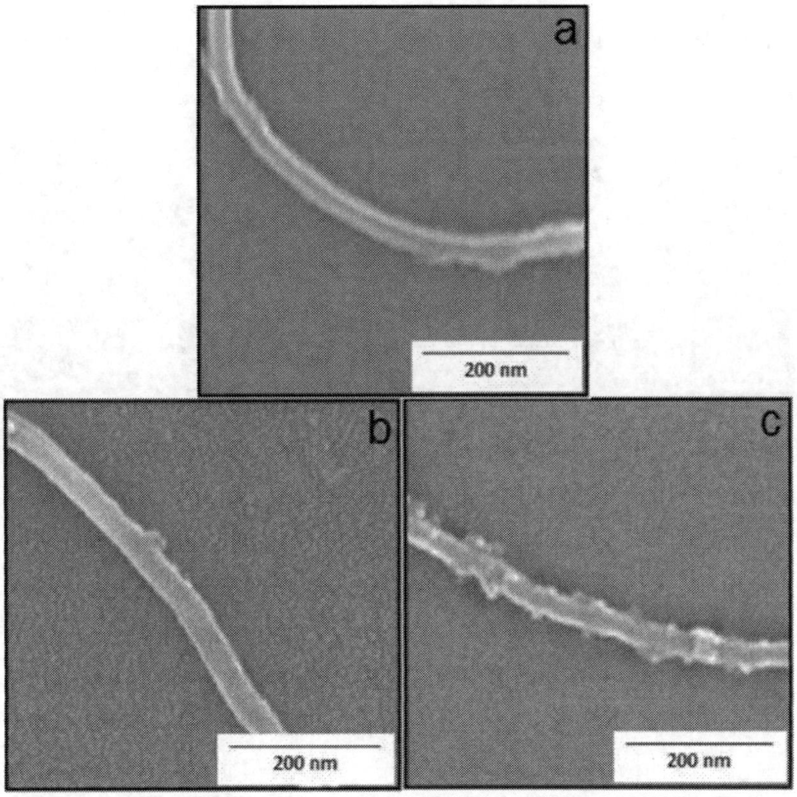

Figure 2. SEM of MWCNTs-COOH, MWCNTs-PGMA, MWCNTs-Fe_3O_4.

IOP Conf. Series: Materials Science and Engineering **87** (2015) 012088 doi:10.1088/1757-899X/87/1/012088

Figure 2 shows SEM images of MWCNTs-COOH, MWCNTs-PGMA and MWCNTs-Fe_3O_4. As shown in figure 2(a), the surface of the MWCNTs-COOH was very smooth, while the surface of MWCNTs-PGMA looked much rougher with a layer of the coating, confirming that PGMA was successfully grafted onto MWCNTs. For MWCNTs- Fe_3O_4, a large amount of nanoparticles were found to be uniformly distributed on the surface of MWCNTs as showed in figure 2(c), which indicated that the magnetic Fe_3O_4 nanoparticles were deposited onto MWCNTs. The uniform distribution of Fe_3O_4 nanoparticles on the surface of MWCNTs was attributed to the electrostatic repulsion between the positively-charged nanoparticles as well as the impeding effect on the aggregation of nanoparticles by the introduction of TMAOH.

3.2. Observation of MWCNTs-Fe_3O_4 alignment

Figure 3 shows SEM images of MWCNTs-Fe_3O_4 in the magnetic field with various intensities. As showed in figure 3(a), MWCNTs-Fe3O4 was partially aligned at 0.08 T along the axis of the magnetic field instead of being randomly distributed. MWCNTs- Fe_3O_4 was partially stretched by the magnetic force between MWCNTs-Fe_3O_4 complexes and magnets, but the magnetic force at the intensity of 0.08 T was not strong enough to make all the MWCNTs-Fe_3O_4 well-aligned along the axis direction of magnetic field. When the intensity of magnetic field increased to 0.1 T, most of MWCNTs- Fe_3O_4 was obviously aligned along the axis direction of magnetic field as showed in figure 3(b), which indicated that intensity increasing of magnetic field was beneficial to enhancing alignment degree of MWCNTs due to the increasing of the magnetic force applied on MWCNTs. However, it was unexpected that MWCNTs- Fe_3O_4 was found to be much less aligned at 0.12 T than at 0.1 T along the axis direction of magnetic field and started to form a number of clusters. This was because the magnetic force applied on MWCNTs was so strong that the whole MWCNTs-Fe_3O_4 could move and pack together. Therefore, these observations revealed that the alignment state of MWCNTs-Fe_3O_4 was sensitive to the intensity of magnetic field. Only if the intensity value of magnetic field was in a proper range to balance the moving state of MWCNTs, the optimum alignment of MWCNTs-Fe_3O_4 could be achieved.

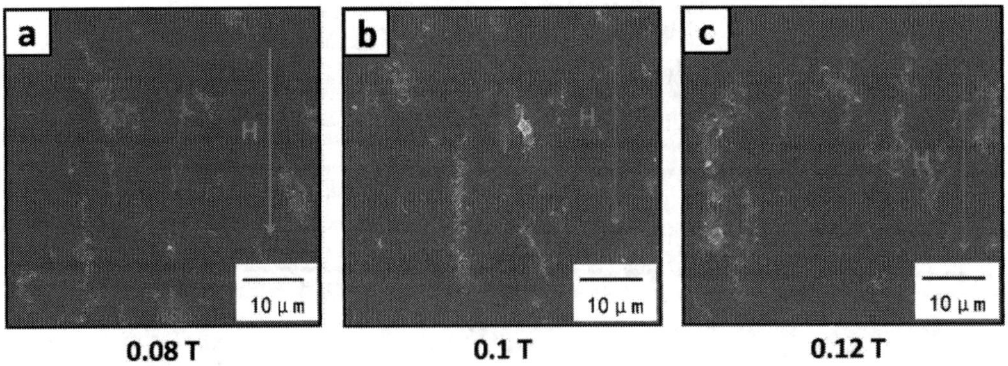

Figure 3. SEM images of MWCNTs-Fe_3O_4 in the magnetic field with various intensities.

3.3. Reinforcing effect of aligned MWCNTs-Fe_3O_4 on epoxy composites

With the aim of evaluating the advantage of aligned MWCNT-Fe_3O_4 in the application of reinforcement, MWCNT-Fe_3O_4 was selected to prepare epoxy-based composites and their mechanical properties were tested. Figure 4 shows tensile properties of pristine epoxy and epoxy composites with 0.5 wt.% of MWCNT-Fe_3O_4 without and with the application of external magnetic field. In comparison with the data of pristine epoxy, epoxy composites with MWCNT-Fe_3O_4, prepared with or without external magnetic field, apparently demonstrated higher tensile and flexural properties, showing the reinforcing effect of MWCNT. With the addition of 0.5 wt.% of MWCNT-Fe_3O_4 into epoxy in the absence of magnetic field, the tensile strength and modulus of resulting composite reached 75.1 ± 5.2 MPa and 2.1 ± 0.1 GPa, respectively. The application of magnetic field could further enhance the values by 12.3 and 10.9%, respectively. These enhancements were attributed to the directional alignment of MWCNT-Fe3O4 in the epoxy matrix along the length direction of specimens since the magnetic field was set parallel to the length direction. The aligned MWCNTs were suggested able to bear more loads and subsequently enhance effective stress transfer in the composites by taking full advantage of the superior mechanical strength and modulus of CNTs themselves. This finding provided a practical possibility of applying magnetic CNTs as reinforcement for resins, in which, the CNTs could be effectively aligned by applying the proper magnetic field to achieve high performance composites.

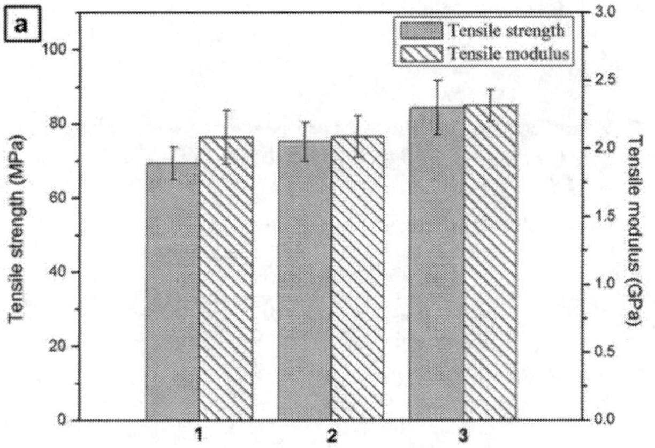

Figure 4. Tensile properties of (1#) pristine epoxy and epoxy composites with 0.5 wt.% of MWCNT-Fe_3O_4 prepared (2#) without and (3#) with the application of external magnetic field.

4. Conclusions

The nanoscale complex composed of one-dimensional multiwalled carbon nanotubes and zero-dimensional ferriferrous oxide nanoparticles (MWCNTs-Fe_3O_4) were successfully prepared by combining atom transfer radical polymerization (ATRP) and coprecipitation process. The alignment state of MWCNTs-Fe_3O_4 complex could be modulated by adjusting the intensity of the external magnetic field and the grafting content of Fe_3O_4 nanoparticles on MWCNTs. The corresponding preferential alignment degree of MWCNTs-Fe_3O_4 was effectively and quantitatively characterized by employing an orientation tensor description in a selected region, which provided a feasible way to optimize the distribution state and exploit the whole performance of MWCNTs in various applications. Furthermore, the detailed studies to elucidate the effect of aligned MWCNTs-Fe_3O_4 complex on properties of fiber reinforced composites were underway.

Acknowledgment

The authors are very pleased to acknowledge financial support from the National Natural Science Foundation of China (Grant No. U1362205, 5157031049), Natural Science Foundation of Jiangsu Province (Grant No. BK20140271), National High Technology Research and Development Program of China (Grant No. 2012AA03A203), Beijing Youth Talent Plan (Grant No. YETP0492) and 2014 Open Project of State Key Laboratory of Organic-Inorganic Composites.

References

[1] Chen Z Q, Dai X J, Magniez K, Lamb P R, Leal DRDC, Fox B L, et al. 2013 Improving the mechanical properties of epoxy using multiwalled carbon nanotubes functionalized by a novel plasma treatment *Compos Part A* **45** 145-52

[2] Lubineau G and Rahaman A 2012 A review of strategies for improving the degradation properties of laminated continuous-fiber/epoxy composites with carbon based nanoreinforcements *Carbon* **50** 2377-95

[3] Homenick C M, Lawson G and Adronov A 2007 Polymer grafting of carbon nanotubes using living free–radical polymerization *Polymer Reviews* **47** 265-90

[4] Gao C, Li W, Morimoto H, et al. 2006 Magnetic carbon nanotubes: Synthesis by electrostatic self-assembly approach and application in biomanipulations *The Journal of Physical Chemistry B* **110** 7213-20

[5] Steinert B W and Dean D R 2009 Magnetic field alignment and electrical properties of solution cast PET–carbon nanotube composite films *Polymer* **50** 898-904

[6] Sen R, Zhao B, Perea D, et al. 2004 Preparation of single-walled carbon nanotube reinforced polystyrene and polyurethane nanofibers and membranes by electrospinning *Nano Letters* **4** 459-64

[7] Correa-Duarte M A, Grzelczak M, Salgueiriño-Maceira V, et al. 2005 Alignment of carbon nanotubes under low magnetic fields through attachment of magnetic nanoparticles *The Journal of Physical Chemistry B* **109** 19060-3

[8] Ma P C, Siddiqui N A, Marom G, et al. 2010 Dispersion and functionalization of carbon nanotubes for polymer-based nanocomposites: A review *Composites Part A: Applied Science and Manufacturing* **41** 1345-67

[9] Jia X, Li G, Liu B, et al. 2013 Multiscale reinforcement and interfacial strengthening on epoxy-based composites by silica nanoparticle-multiwalled carbon nanotube complex *Composites Part A: Applied Science and Manufacturing* **48** 101-9

Novel giant magnetostrictive material current sensor

Yan Shen[1], Zhao Liu, Qiuyan Lin, Jinming Ge, Guoqing Zhang and Wenbin Yu

Department of Electrical Engineering, Harbin Institute of Technology, Harbin 150080, China

E-mail: yanshen@hit.edu.cn

Abstract. Because of the shortcomings of the traditional inductive current sensor and optical current sensor, this paper proposes a new type of current sensor that uses a giant magnetostrictive material (GMM) to monitor and control power lines. This paper introduces operating principle and structure design of a GMM current sensor. To eliminate the frequency-doubled effect and to obtain good linearity, we set the bias magnetic field to 11.53 kA/m and the prestress force to 6 MPa. The strains of the 100- and 200-mm GMM sticks under the same magnetic field were compared; the results showed that the 100-mm stick had a larger strain. The magnetic field interference during a single-phase measurement of power lines was also studied. Finally, we analyzed the device sensitivity and discussed its influencing factors. The sensitivity reached 4×10^{-9} m^2/A.

1. Introduction

Using traditional current sensors to obtain an aperiodic current is difficult; the use of these traditional sensors can also be accompanied by the serious phenomenon of magnetic saturation based on electromagnetic induction p [1]. Optical current sensors have therefore become the latest trend in current research [2, 3]. Optical current sensors, which compared with the traditional current sensors, enjoy the advantages of wide dynamic measurement range, no transient magnetic saturation, and strong ability to avoid electromagnetic interference [2]. Currently, however, the a magneto-optic glass is primarily used in sensing head of an optical current sensor, which is subject to the Faraday effects, leading to issues such as birefringence and temperature drift [3].

Two methods are thus clearly necessary so that these sensors can be improved: one method involves the complete replacement of the optical sensor technology, and the other method involves the application of a new functional material, e.g., a giant magnetostrictive material (GMM), which has been developed for years since 1970. In a GMM, the material size changes with the magnetic field intensity [4]. Compared with a normal magnetostrictive material, the GMM has the advantages of large magnetostriction coefficient, high-temperature adaptation, wide linearity range, smart response, and robust stability [5].

Owing to the many advantages of GMM, the application of this material involves a significant breakthrough for magnetism—e.g., sound-transform products [6]. Many of the early GMM applications involved long-range transmission sonar [7]. Currently, the applications for GMM include high-precision-speed micro-displacement actuators and large power low-frequency sonar systems,

[1] Address for correspondence: Yan Shen, Department of Electrical Engineering, Harbin Institute of Technology, Harbin 150080, China. E-mail: yanshen@hit.edu.cn.

among others [8]. In this paper, we report a type of current sensor which designed using a GMM and its application to monitoring of power lines. This sensor can convert the change in the magnetic field near a power line to a displacement of an output rod. When a fault occurs in the power line, such as a short circuit, which causes a dramatic change in the magnetic field, the output rod will trigger the circuit breaker to protect the line and adjacent equipment and prevent a large-scale blackout. The advantages of this device include the following: no outside energy source, no noise source, high speed, high precision, simple structure, and low cost.

2. Structure and operating principle

2.1. Structure of the GMM current sensor

The structure of a GMM applicated current sensor is shown in figure 1. The current sensor consists of an output rod, a permanent magnet, a prestress mechanism, and a GMM stick. These devices are fixed in place by a shell, a rubber ring, a permanent magnet frame, and a pedestal.

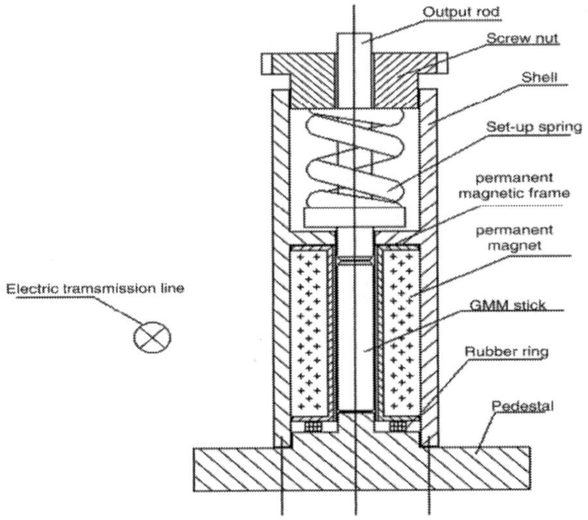

Figure 1. Basic structure of the GMM current sensor.

2.2. Operating principle of the GMM sensor

Figure 1 shows that the GMM stick, shell, output rod, and pedestal form a closed magnetic path. A magnetic field alternately superimposes a biased magnetic field on the magnet. We place the device next to power lines, which causes the GMM to expand. The size of the GMM stick thus vibrates with a set response frequency because of the superimposed field, and the displacement resulting from a strain is transferred to others which called output rod. The displacement can acquiring or monitoring data, and it also can be directly applied as a protector. A set-up spring and a screw nut are included in the prestress mechanism comprises. This mechanism reduces the gap between the GMM stick and the output rod and amplifies the strain of the GMM stick that in order to achieve a satisfactory output. In addition, it makes the energy conversion of the device more efficiency.

3. Bias magnetic field and prestress

3.1. Bias magnetic field analysis

One property of the magnetostrictive effect is that its relative strain depends on the value of the magnetic field and not depends on its direction; thus, this strain can be expressed as $\Delta l / l = \psi(B^2)$, where l is the strain and B is the magnetic field strength. Therefore this function can be considered as a quadratic function $\Delta l / l \infty B^2$. According to the Hooke's Law, Tm is determined to be proportional to the strain; thus, $Tm = \gamma B^2$, where γ is the proportionality factor and Tm is magnetostrictive force. Once an alternating current with an angular frequency flows by a power line, the size of the ferromagnet would be also changes under the influence of the alternating magnetic induction intensity, i.e., $B = B_m \cdot \cos \omega t$, where B_m is the peak value of B and ω is the angular frequency. Thus, we obtain

$$Tm = \gamma B^2 = \gamma B_m^{\ 2} \cdot \cos^2 \omega t = (\gamma B_m^{\ 2} / 2) + (\gamma B_m^{\ 2} / 2) \cos 2\omega t \qquad (1)$$

The frequency of the Tm is two times that of the input signal, a phenomenon that is called the frequency-doubled effect [3]. We ofen use a permanent magnet to induced a biased magnetic field B_p to eliminate this effect,; thus

$$Tm = \gamma(B_p^{\ 2} + B_m^{\ 2} / 2) + 2\gamma B_p B_m \cdot \cos \omega t + \gamma(B_m^{\ 2} / 2) \cos 2\omega t \qquad (2)$$

In equation (2), $\gamma(B_m^{\ 2} / 2) \cos 2\omega t$ is the distortion part. This distortion part can be neglected when we set $2B_p \gg B_m / 2$;Namely, that effect can be eliminated in the case of that we set a more biased magnetic field than the magnetic field of the power line.

Furthermore, although the magnetostrictive effect of the GMM stick is non-linear, a large section in the characteristic curve exists that can be treated as non-linear because of the shape of the curve. Therefore, a biased magnetic field is setted can also make the operating point turn to the linear part.

3.2. *Prestress analysis and biased magnetic field setting*

The GMM characteristic curves would be various under different prestress settings. Once under axial prestress, the GMM can have higher energy transfer efficiency and a larger displacement output; thus, we can control the relationship curves between the magnetostriction coefficient ε and magnetic field H by choosing the prestress value. The experimental magnetostrictive characteristic curves obtained using a GMM stick (model Terfenol-D with 10 mm diameter) are shown in figure 2 (P1 indicates that the prestress is 1 MPa) [9].

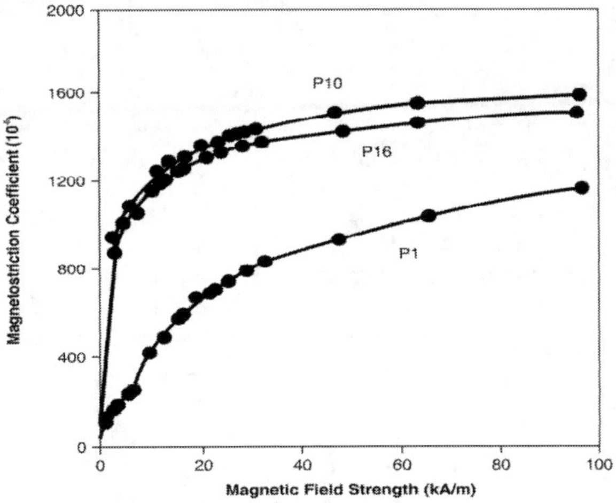

Figure 2. Experimental magnetostrictive characteristic curve.

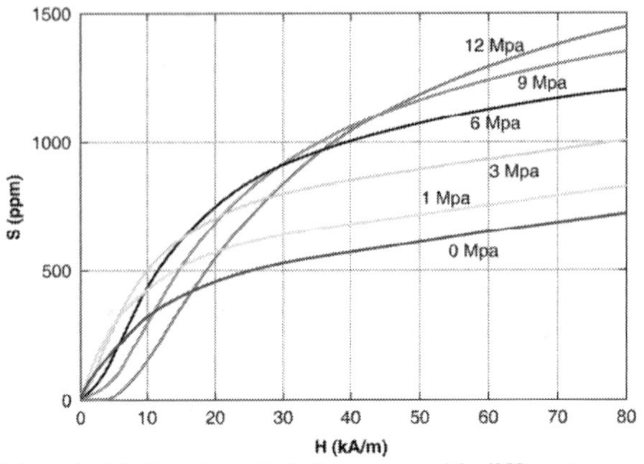

Figure 3. Magnetostrictive characteristic curves with different prestress values.

With respect to the experimental curves, a numerical fitting method using a magnetostrictive model of an exponential function plus a polynomial is used to obtain the following equation:

$$\varepsilon = 165.776Pe^{-2.1612P/H} + (12 - P)(0.5928 + 3.655H - 0.1342H^2 + 2.8667\times10^{-3}H^3 \\ -3.532\times10^{-5}H^4 + 2.57\times10^{-7}H^5 - 1.0899\times10^{-9}H^6 + 2.4719\times10^{-12}H^7 - 2.3259\times10^{-15}H^8) \quad (3)$$

Here, ε is the magnetostriction coefficient, H is the magnetic field strength, and P is the prestress value. We can obtain the characteristic curves under different prestress values by using this equation, as shown in figure 3.

Figure 3 shows that once the prestress reaches 6 MPa, the curve could exhibit a good linearity and a considerable magnetostriction coefficient when the magnetic field is in the range of 10–20 kA/m. Therefore, we choose 6 MPa as the prestress value. If we want the prestress force coould be applicable to all working conditions, we should choose the screw nut and set-up spring as the prestress unit to clamp down the material to create the stress.

Using the characteristic curves, we set the biased magnetic field to be 10 kA/m. Then, according to equation (3), with $B_p = 10$ and $B_m = 1.53$, we decided that $2B_p \gg B_m / 2$. Additionally, if we set the biased magnetic field to 10 kA/m, the operating point would in the middle of the linear part of the curve.

4. Material standard selection

A Terfenol-D ($T_{b0.3}D_{y0.7}Fe_{1.95}$) GMM is selected in this study. We chose 10 mm as the diameter of the GMM stick and compared two sticks with lengths of 100 and 200 mm to determine which had a larger strain.

Because it is not parallel between the power line and the GMM stick, the magnetic field strength is not the same throughout the stick. Thus, the strain which applied on the stick is also not an average value. Therefore, calculating the average magnetic field strength on the GMM stick is essential to determine the strain on the entire stick. By assuming that the gap between the power line and the GMM stick is 0.1 m, the distance between the center of the stick and some points is x, and the length of the stick is y, the average distance between the power line and every point on the stick is

$$r_{av} = \left(\int_0^{0.5y} \sqrt{0.1^2 + x^2 dx}\right) / 0.5y \quad (4)$$

When the length of the GMM stick is 100 mm, the average distance between the power line and the points on the stick is $r_{av(100mm)}=(\int_0^{0.05}\sqrt{0.1^2+x^2}dx)/0.05=0.104(\mathrm{m})$ according to equation (4). When the length is 200 mm, the average distance is $r_{av(200mm)}=(\int_0^{0.1}\sqrt{0.1^2+x^2}dx)/0.1=0.1148(\mathrm{m})$.

We calculated the strain of the 100- and 200-mm standard GMM sticks when the current that flows through the power line is 1000 A, 3000 A, and 40 kA; the results are listed in table 1.

From the results, it is a larger magnetostrictive coefficient in the 100-mm standard Terfenol-D stick.

Table 1. Comprehensive results.

Length	Item	1000 A		3000 A		40 kA	
		Largest strain	Minimum strain	Largest strain	Minimum strain	Largest strain	Minimum strain
100 mm	Field strength (kA/m)	11.53		14.59		71.2	
	Coefficient (ppm)	540	208	619	353	1381	688.4
	Strain (mm)	0.054	0.028	0.062	0.035	0.138	0.069
200 mm	Field strength (kA/m)	11.386		14.158		65.45	
	Coefficient (ppm)	527	188	602	312	1335	655
	Strain (mm)	0.105	0.038	0.120	0.062	0.267	0.133

5. Magnetic field interference analysis

The primary factor that influences the precision is the magnetic field interference in practical application. The other two phase currents will interfere with the measured phase; thus, the interference must be analyzed when the device is moved to different positions.

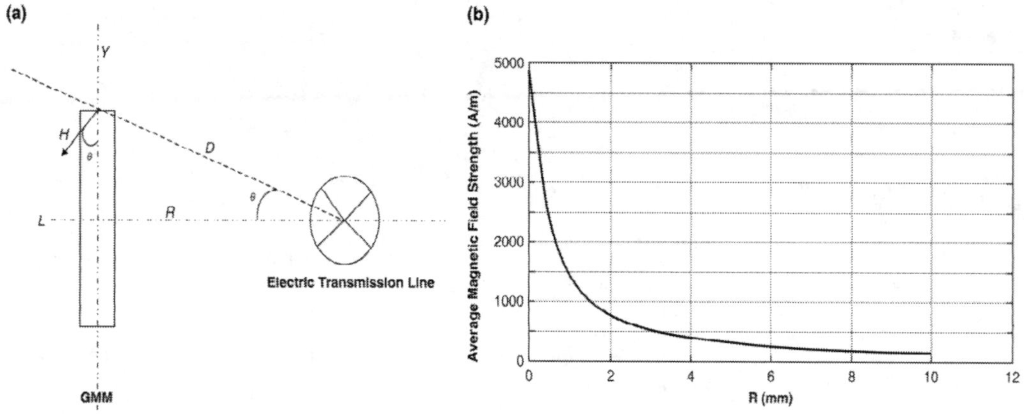

Figure 4. Average magnetic field strength analysis without interference. (a) Relative position of the GMM stick and the power line. (b) Average magnetic field strength curve.

5.1. No magnetic field interference

The relative positions of the power line and the GMM stick are shown in figure 4(a), and the average magnetic field strength curve that with no interference is shown in figure 4(b). The average magnetic field strength equation is

$$\overline{H_i} = I \arctan(0.5L / R) / \pi L \tag{5}$$

In this equation, L is the length of the GMM, I is the current on the power line, and R is the distance between the power line and the. GMM

Figure 4(b) shows that the average magnetic field strength exponentially decreases as R increases.

5.2. Interference from phase A

The relative positions of the power line and the GMM stick are shown in figure 5(a), and we can see that phases A, B, and C are aligned in a row with the same distance. The average magnetic field strength curves when phase A is measured are shown in figure 5(b). Distance R_0 is 1 m, and R_x and R_y denote the horizontal and vertical distances, respectively, between the phase A and GMM stick. We can conclude that when the GMM stick at a same height as the power line ($R_y = 0$), the average magnetic field strength curve approximates the condition without interference. However, when $R_y \neq 0$, the distortion of the curve is obvious, particularly in positions closer to the phase A line. The error in the magnetic field strength when phase A is measured originates from the other two phases when $R_y = 0$, as shown in figure 5(c); the relative error varies from 0.05 to 0.54.

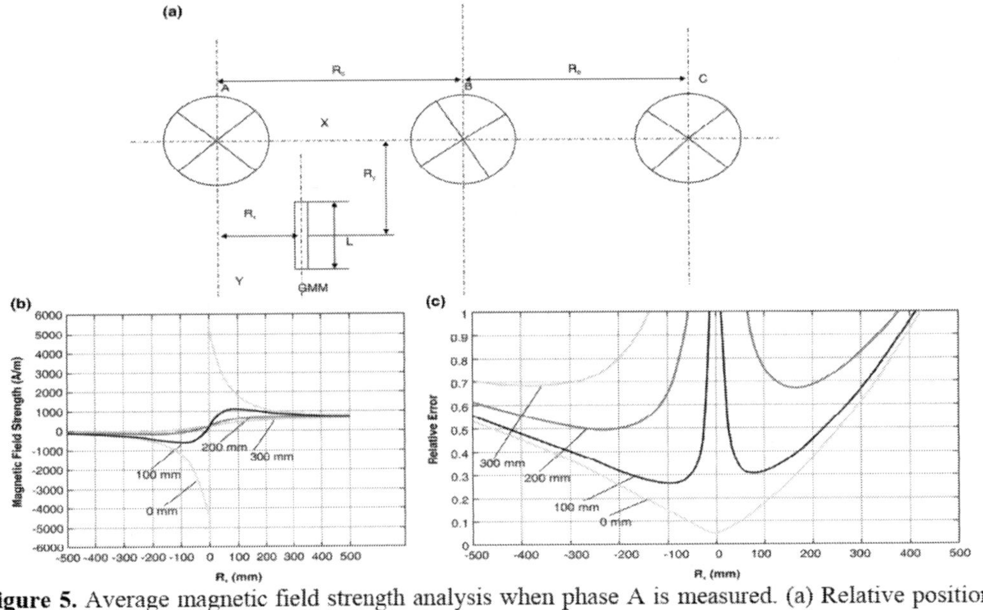

Figure 5. Average magnetic field strength analysis when phase A is measured. (a) Relative positions of the GMM stick and power line. (b) Average magnetic field strength curves. (c) Relative error curves.

Because phases A and C are symmetrical, the interference when phase C is measured is the same as that for phase A; therefore, additional analysis of phase C is not required.

5.3. Interference from phase B

The relative positions of the GMM stick and the power line are shown in figure 6(a), and figure 6(b) shows the average magnetic field strength curves of phase B. We can draw the same conclusion as that

IOP Conf. Series: Materials Science and Engineering **87** (2015) 012089 doi:10.1088/1757-899X/87/1/012089

drawn for phase A. The relative error curves are shown in figure 6(c), which clearly shows that the relative error when phase B is measured is much smaller because phase B is in the middle of the three phases, and the interference from phases A and C neutralizes each other. When R_y = 0, the relative error varies from 0 to 0.65, and when R_y is fixed, with the distance $|R_x|$ increases, the error also increases.

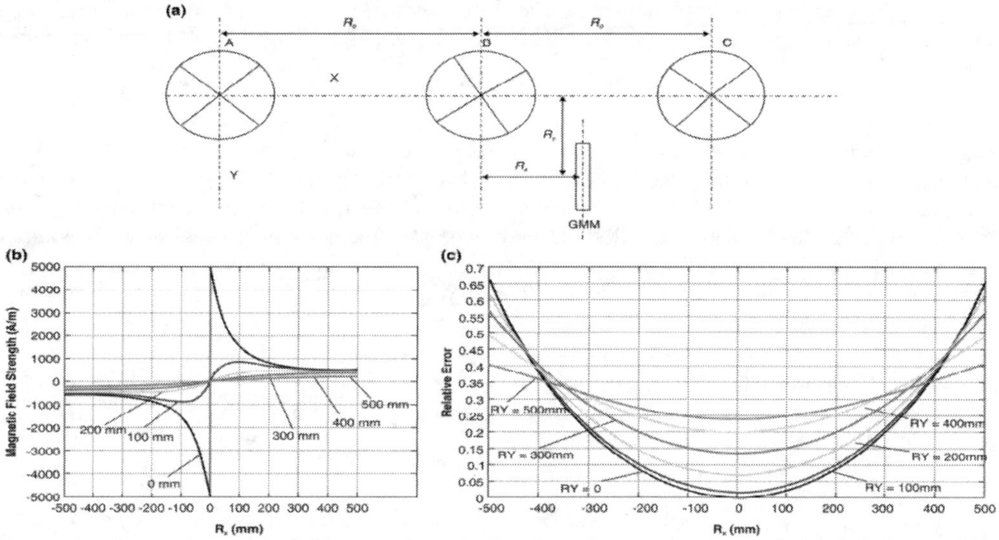

Figure 6. Average magnetic field strength analysis when phase B is measured. (a) Relative positions of the GMM stick and power line. (b) Average magnetic field strength curves. (c) Relative error curves.

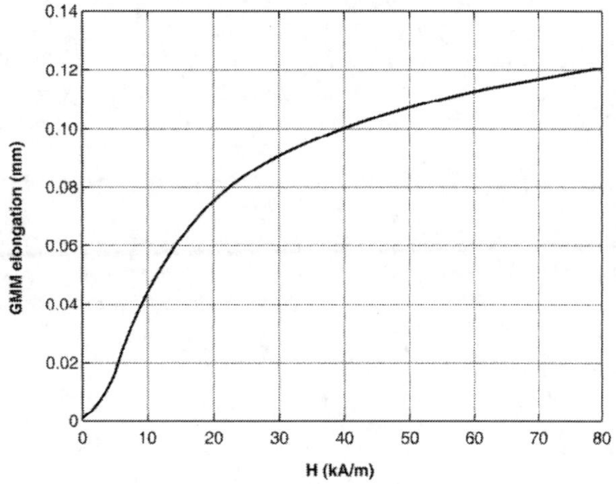

Figure 7. GMM characteristic curve.

6. Device sensitivity analysis

Sensitivity is indicated by the differential relationship of the variables; the derivative reflects the influence of the changes in the design variables on the target function [10]. We analyzed the sensitivity of the device to confirm the influence of the magnetic field on the output and set the strength of the coupling using follow-up subsystems to complete the design. To calculate the sensitivity and to investigate the factors that affect it, we plotted the GMM characteristic curve, as shown in figure 7. By calculating the GMM output sensitivity, we can obtain the derivative of equation (3) and employ it at the operating point. The slope is 4×10^{-9} m^2/A. This high sensitivity can be applied to the monitoring of power lines. Because the biased magnetic field and the prestress force are already set, the biggest influencing factor is the temperature. GMM is a ferromagnetic material; thus, magnetic hysteresis loss and eddy-current loss are inevitable. The heat caused by the losses will lead to a temperature rise, and the GMM will incur additional elongation, which lowers the device precision.

7. Conclusions

Because traditional inductive and optical current sensors have some shortcomings, this paper proposes a new type of current sensor that uses a giant magnetostrictive material (GMM) to monitor and control power lines. We also analyzed the biased magnetic field and set it to 10 kA/m, which eliminated the frequency-doubled effect and also moved the operating point to the linear part of the curve. In addition, the prestress force was 6 MPa, which led to good linearity and a high magnetostriction coefficient. Furthermore, we studied the magnetic field interference when a single-phase power line was measured, and the results showed that when the GMM stick was at the same height as the power lines, the lowest relative error was achieved. In addition, the relative error when phases A and C were measured varied from 0.05 to 0.54 compared with that of phase B, which varied from 0 to 0.65. The error increased as the distance increased. Finally, the sensitivity of the device was determined to be 4×10^{-9} m^2/A; the factors that influenced this value have been discussed.

Acknowledgments

This work was supported in part by the National Natural Science Foundation of China (No. 11004040) and in part by the Natural Science Foundation of Heilongjiang Province (E201237).

References

[1] Huang Y 2010 *Coastal Enterprises and Science & Technology* **4** 123-5
[2] Wang J, Guo Z, Zhang G, Shen Y and Yu W 2012 *Power System Technology* **36** 37-41
[3] Wang X 2008 *Electric Power Science and Engineering* **24** 34-6
[4] Li K, Xu J, Yang H, Yuan Y, Xu J, Yu D and Zhang S 2004 *China Rare Earth* **25** 51-6
[5] Shen Y, Rong H, Zhang G, Yu W, Guo Z and Lu Y 2013 *Applied Mechanics and Materials* **437** 710-5
[6] Wang B and Yan R 2004 Application and devices *Journal of Hebei University of Technology* **33**, 16-22
[7] Wang J, Cai G, Gao L, Xu Y and Zhao Y 2007 *High Voltage Engineering* **33** 67-71
[8] Li M 2005 *Development & Innovation of Machinery & Electrical Products* **18** 35-7
[9] Guo X, Gao Y and Wang J 2006 *Science & Technology of Baotou Steel (Group) Corporation* **32** 27-41
[10] Yan L, Chen X and Wang Z 2005 *Aeronautical Computer Technique* **35** 1-6

Effect of reactive agent and transesterification catalyst on properties of PLA/PBAT blends

S Pitivut[1], S Suttiruengwong[1,3] and M Seadan[2]

[1]Department of Materials Science and Engineering, Faculty of Engineering and Industrial Technology, Silpakorn University, Sanamchandra Palace Campus, Nakhon Pathom 73000, Thailand
[2]Department of Physics, Faculty of Science, Silpakorn University, Sanamchandra Palace Campus, Nakhon Pathom 73000, Thailand

E-mail: suttiruengwong_s@su.ac.th

Abstract. This research aimed to study the properties of poly (lactic acid) (PLA)/poly (butylene adipate-co-terephthalate) (PBAT) blends with two different reactive systems: free radical reaction through peroxide (Perkadox) and transesterification catalyst (tetrabutyl titanate; TBT). Two blends composed of PLA as a matrix phase with the composition of 80 and 70 percent by weight. PLA/PBAT blends with Perkadox were prepared in twin screw extruder, whereas PLA/PBAT blends with TBT were prepared in an internal mixer. The morphology of the blends was investigated by scanning electron microscope (SEM). Tensile and impact testings of the blends were reported. In case of the blends with Perkadox, SEM micrographs revealed that the size of particles was substantially reduced when adding more Perkadox. Young's modulus and the tensile strength of all blend ratios were insignificantly changed, whereas the elongation at break was decreased when compared to non-reactive blends due to the possible crosslinking reaction as observed from melt flow index (MFI) values. When adding Perkadox, the impact strength of PLA/PBAT (80/20) remained almost unchanged. However, the impact strength of PLA/PBAT (70/30) was enhanced, increasing to 110% for 0.05 phr Perkadox. In case of the blends with TBT, SEM micrographs showed the decrease in the particle size of PBAT phase when adding TBT. Young's modulus and the tensile strength of all blend ratios were not different, but the elongation at break was improved when adding TBT owing to the transesterification reaction. For PLA/PBAT (80/20), the elongation at break was increased by 39%, whereas the elongation at break was increased by 15% for PLA/PLA (70/30). The impact strength of all blend ratios unaffected.

1. Introduction

Since the discovery of bioplastics, the growth and availability of some bio-based and/or biodegradable plastics manufacturers have become key factors to compete with existing non-biodegradable fossil-based plastics. These marketed bioplastics include poly (lactic acid) (PLA) under trade name, Ingeo^{TM} by NatureWorks, Sulzer Polylactide by Sulzer Chemtech, ECOPLAN by SK Chemicals, poly(butylene

[3] Address for correspondence: S Suttiruengwong, Department of Materials Science and Engineering, Faculty of Engineering and Industrial Technology, Silpakorn University, Sanamchandra Palace Campus, Nakhon Pathom 73000, Thailand. E-mail: suttiruengwong_s@su.ac.th.

adipate-co-terephthalate) (PBAT) provided by BASF (Ecoflex®), Biocosafe 2003 by Xinfu Pharmaceutical Co. Ltd. and poly(butylene succinate) under trade name Bionolle (Showa Denko) or GsPLA (Mitsubishi Chemical). The combinations of these biodegradable polymers are also readily available as compound resins in a wide range of applications such as Ecovio® by BASF, Bioflex® (FKur). Among the bioplastics mentioned, the blends of PLA and PBAT have been extensively explored due to their complete biodegradable feature and mechanical balance.

PLA possesses high modulus and tensile strength, but it is brittle, resulting in limited applications [1, 2]. To improve its brittleness, many methods such as copolymerization, plasticization, blending with flexible biodegradable polymers (e.g. poly(butylene adipate-co-terephthalate)) [1, 2] have been explored. PBAT has been extensively investigated as an appropriate blend pair for enhancing brittleness of PLA as PBAT has high toughness and is fully biodegradable [1, 2]. However, PLA shows complete phase separation with PBAT, resulting in limited improvement in the toughness by physical blending [3, 4]. For such a system, depending on the required properties of the blends and products, whether PLA or PBAT becomes a major phase, the blend is less problematic.

The addition of reactive agents can improve compatibility between PLA and PBAT in blends; for instance, dicumyl peroxide (DCP) [5, 6], tetrabutyl titanate (TBT, $Ti(BuO_4)$) [4, 7, 8], bioxazoline [9], phthalic anhydride [9], acetyl tributyl citrate [10, 11], and glycidyl methacrylate(GMA) [12, 13].

Reactive agents can promote in-situ copolymerization through the free radical reaction using peroxide initiator. Coltelli M-B, et al. [5] studied reactive compatibilization between PLA and PBAT with various 2, 5-Dimethyl-2, 5-di (tert-butylperoxy) hexane contents. The results showed that mechanical properties (modulus and tensile strength) of the blends were improved, but the elongation at break was decreased due to the crosslinking reaction. The occurrence of crosslink was also confirmed by torque, melt volume-flow rate (MVR) and size-exclusion chromatography (SEC). Furthermore, the particle size of dispersed phase in reactive blends was lower than non-reactive blend due to improving interfacial adhesion between PLA and PBAT through free radical reaction. Nishida M, et al. [6] studied the effect of dialkyl peroxide on PLA/PBAT blends. The results showed that the elongation at break and the impact strength were improved when adding dialkyl peroxide at a ratio of PLA to PBAT to dialkyl peroxide being 60:40:1. It indicated that dialkyl peroxide resulted in improving ductility of the blends due to the crosslinking reaction. From these findings, peroxide could be used to modify the morphology and mechanical properties of PLA/PBAT blend.

Alternative to the free radical reaction is the transesterification reaction. Both reactions are reactive blending, so they are easy to process and low cost. Coltelli M-B, et al. studied the reaction time of transesterification for blending PLA, PBAT and TBT. The results showed that the particle size of dispersed phase in the blends with TBT was smaller than that of the blend without TBT and the elongation at break was improved due to the transesterification reaction. Moreover, the reaction time affected various properties of the blends. Lin S, et al. [4] studied PLA/PBAT with various TBT contents. The results showed that the tensile strength, the elongation at break and the impact strength were improved when adding TBT to the blends, particularly, PLA/PBAT blend with incorporation of TBT 0.5 wt.%. The compatibility between PLA and PBAT confirmed by SEM was also enhanced due to the transesterification reaction. Moreover, for dynamic mechanical analysis (DMA) results, storage modulus and glass transition temperature were improved. Girdthep S, et al. [8] studied PLA/PBAT (70/30) blends with TBT and kaolinite. TBT was added as compatibilizer to improve interfacial affinity between PLA and PBAT via transesterification. Kaolinite was added to increase the interlamellar space of the silicate layers. It was found that mechanical properties of PLA/PBAT blends were enhanced. Tensile strength was dropped and elongation at break and modulus were increased at TBT 0.5 phr, an optimal TBT concentration. SEM results showed that adding TBT could improve dispersion of PBAT in PLA phase. However, the reaction time of transesterification reaction is usually longer than that of the free radical reaction.

Therefore, this work aimed to study the two reactive systems involving free radical reaction using peroxide and transesterification reaction using TBT as a reactive agent for PLA/PBAT blends, where

PLA was used as a major phase with the composition of 80 and 70 percent by weight. The influences of both systems on the morphological and mechanical properties were separately discussed.

2. Experimental

2.1. Materials
PLA (4043D grade) was purchased from NatureWorks LLC, USA. PBAT (Ecoflex®) was purchased from BASF Chemical Company, Germany. Di (tert-butylperoxyisopropyl) benzene (Perkadox14s), having 40% of peroxide, was supplied by Akzo Nobel, Netherlands. Tetrabutyl titanate (TBT) was purchased from Sigma-Aldrich, USA.

2.2. Preparation of PLA/PBAT blends
PLA and PBAT were dried at 60°C for 5 hours prior to use. The weight ratios of PLA/PBAT were 80/20 and 70/30. The polymers and reactive agent (Perkadox) were weighted, premixed, and then fed into twin screw extruder using temperature 120/130/150/160/170/180/190/190°C and screw speed of 80 rpm. The amount of Perkadox (0.01, 0.03, 0.05 phr) was varied. Due to a slow reaction time, the experiments for transesterification catalyst were performed using an internal mixer to follow the reaction as a function of torques. The polymers and transesterification catalyst (TBT) were weighted, premixed, and added to internal mixer at 185°C and rotor speed of 60 rpm for 15 minutes. The amount of TBT (0.3, 0.5, 0.1 phr) was varied. The formulations consisting of Perkadox and TBT were referred as pX and TX, respectively and X denoted the quantity of reactive agents in phr.

2.3. Testing
Melt flow index (MFI) of all samples was tested at 190°C/2.16 kg, according to ASTM D1238. The fractured surface morphology of PLA/PBAT blends was examined by SEM (CamScan, MX-2000, UK). Mechanical properties were evaluated by tensile and impact testing. The tensile and impact specimens were prepared by compression moulding at 190°C. Tensile testing was carried out by universal testing machine (Instron, Model 5969, USA) equipped with a 5 kN load cell at a crosshead speed of 10 mm/min, according to ASTM D638. The specimen was type V. Notched Izod impact testing was tested according ASTM D256 (Zwick model pendulum impact tester B5102.202)

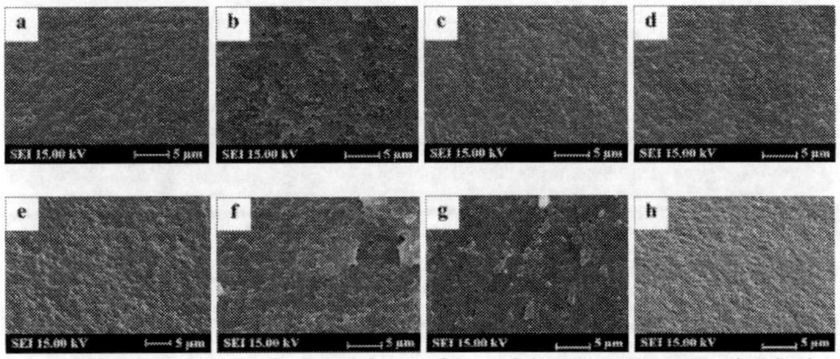

Figure 1. SEM micrographs of fractured surface (a) PLA/PBAT (80/20) (b) PLA/PBAT (80/20)+p0.01 (c) PLA/PBAT (80/20)+p0.03 (d) PLA/PBAT (80/20) + p0.05 (e) PLA/PBAT (70/30) (f) PLA/PBAT (70/30)+p0.01 (g) PLA/PBAT (70/30)+p0.03 (h) PLA/PBAT (70/30)+p0.05.

3. Results and discussion

3.1. PLA/PBAT blends with perkadox

3.1.1. Morphology. In case of PLA/PBAT (70/30) (figure 1(e)), PBAT particle sizes were larger than those of PLA/PBAT (80/20) (figure 1(a)). Upon increasing Perkadox contents, PBAT particles were still finely dispersed in PLA matrix and the size of PBAT particles of both blend ratios became smaller. In case of PLA/PBAT (80/20), the size of PBAT particles reduced from about 0.6 to 0.2-0.4 μm. In case of PLA/PBAT (70/30), the size of PBAT particles also reduced from about 0.8 to 0.2-0.4 μm.

3.1.2. Melt flow index. MFI value of PLA/PBAT lied between that of their neat ones. For PLA/PBAT with Perkadox, the MFI values tended to decrease with the increase of Perkadox contents. This result could be observed for both blend ratios. The addition of Perkadox to the blends could cause lower MFI values since crosslinks could evolve between two polymers by Perkadox. However, with the same amount of Perkadox, PBAT contents increased in blends, so PLA/PBAT (70/30) led to higher MFI values.

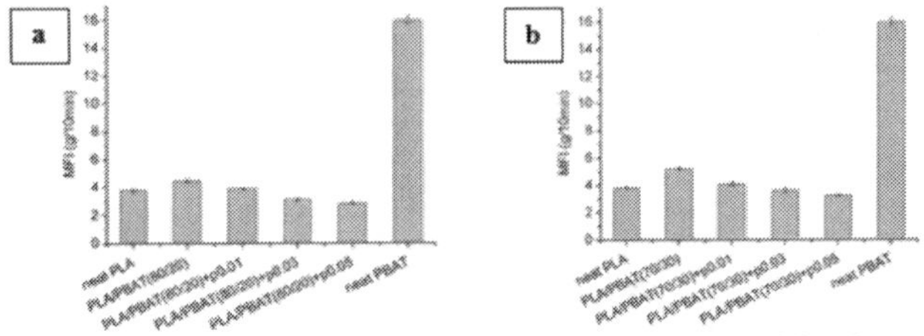

Figure 2. Melt flow index of neat PLA, neat PBAT and PLA/PBAT blends at various amounts of Perkadox (a) PLA/PBAT (80/20) blends and (b) PLA/PBAT (70/30) blends.

3.1.3. Tensile testing. Upon increasing PBAT contents in blends, Young's modulus and tensile strength decreased, but the elongation at break of the blends increased as demonstrated in figures 3(a) and 3(b), respectively. When the content of Perkadox in PLA/PBAT (70/30) increased, Young's modulus and tensile strength of the blends remained unchanged, whereas the elongation at break was slightly dropped when compared to PLA/PBAT without Perkadox probably due to the crosslinking reaction evolved by free radical reaction as indicated by the decrease in MFI values. As a result, blends with Perkadox were stiffer and more brittle than those without Perkadox. This finding was in good agreement with the elongation at break of PLA/PBAT (75/25) added 2, 5-Dimethyl-2, 5-di (tert-butylperoxy) hexane reported by Coltelli M-B, et al. [5]. For PLA/PBAT (80/20), when adding Perkadox, Young's modulus and tensile strength of blends were almost unchanged, but the elongation at break was slightly increased when compared to PLA/PBAT without Perkadox. The results could suggest that the interfacial adhesion between PLA and PBAT phases was improved.

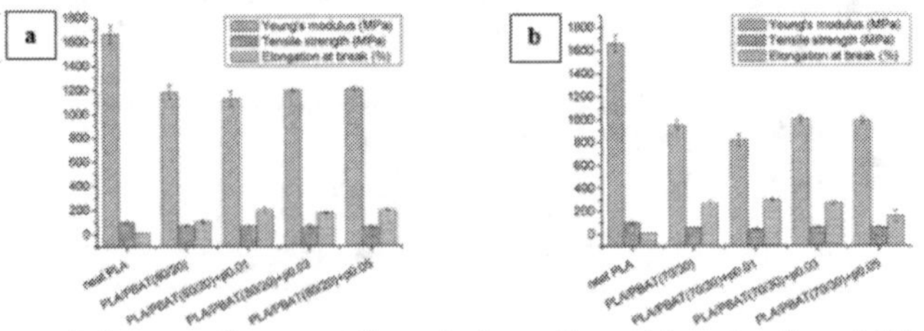

Figure 3. Young's modulus, tensile strength and elongation at break of neat PLA and PLA/PBAT blends (a) PLA/PBAT (80/20) blends and (b) PLA/PBAT (70/30) blends.

3.1.4. Impact testing. As expected, the impact strength of PLA/PBAT blend was higher than that of neat PLA due to the flexibility of chemical structure of PBAT. The addition of PBAT resulted in energy dissipation from PLA phase to PBAT phase because of the occurrence of very little adhesion between two phases [14]. The impact strength of PLA/PBAT (70/30) was almost two times higher than that of PLA/PBAT (80/20). The impact strength of PLA/PBAT (80/20) blends was insignificantly changed when adding greater amount of Perkadox (figure 4(a)). This result was similar to that reported by Nishida M, et al. [6]. They found that PLA/PBAT (80/20) with dialkyl peroxide 1 wt.% had the similar impact strength to that of PLA/PBAT(80/20). However, the impact strength of PLA/PBAT (70/30) blends increased with the increasing amount of Perkadox. This may be due to the improvement of interfacial adhesion between PLA and PBAT phases (figure 4(b)). PLA/PBAT (70/30) + p0.05 had the highest impact strength of all samples. Its impact strength value was increased by about 110% (from 12.5 to 26.2 kJ/m^2) when compared to non-reactive blend. The improvement of the impact strength by adding as high as 1 wt.% diakyl peroxide of 133% (from 10.0 to 23.3 kJ/m^2) has been reported by other group [6]. Based on this finding, Perkadox was effective even at the very small amount.

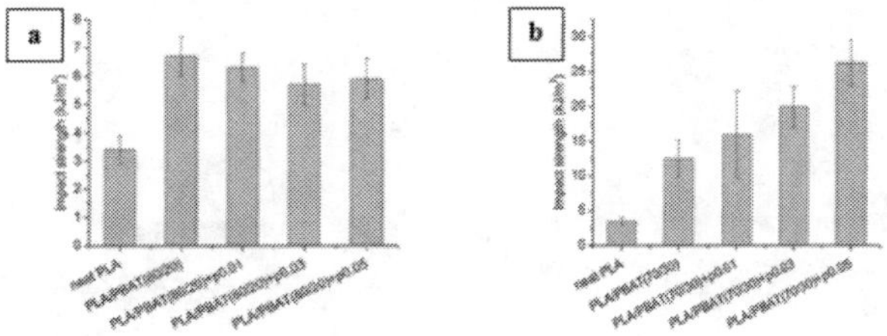

Figure 4. Impact strength of neat PLA and PLA/PBAT blends (a) PLA/PBAT (80/20) blends and (b) PLA/PBAT (70/30) blends.

3.2. PLA/PBAT blends with TBT

3.2.1. Morphology. SEM images of PLA/PBAT (80/20) blends (figures 5(a) and 5(b)) demonstrated more uniform PBAT phase than PLA/PBAT (70/30) blends (figures 5(c) and 5(d)). Adding TBT to both blend ratios, PBAT particles were smaller than the blends without TBT, which was in good agreement with Coltelli M-B, et al. [7]. They found that when adding TBT, the size of PBAT particles reduced from 0.95 to 0.75 µm (for PLA/PBAT (75/25) at 200°C and the mixing time 10 minutes in an internal mixer).

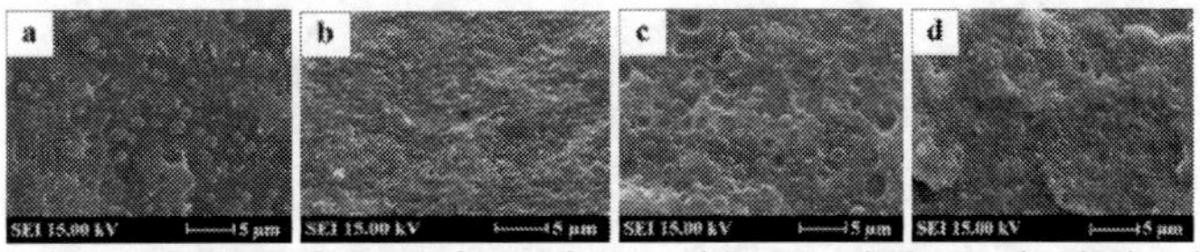

Figure 5. SEM micrographs of fractured surfaces. (a) PLA/PBAT (80/20), (b) PLA/PBAT (80/20)+T0.3, (c) PLA/PBAT(70/30) and (d) PLA/PBAT (70/30)+T0.3.

3.2.2. Melt flow index. From figure 6(a), the MFI values of PLA/PBAT (80/20) blends increased when increasing an amount of TBT, while those of PLA/PBAT (70/30) blends, however, remained almost constant (figure 6(b)). This suggested that the transesterification reaction did not change the viscosity of the blend, implying that the effect of TBT on the molecular weight was very small.

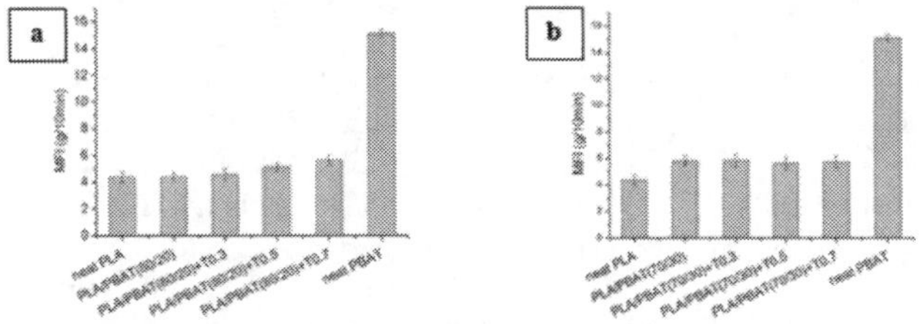

Figure 6. Melt flow index of PLA/PBAT blends with and without TBT. (a) PLA/PBAT (80/20) blends and (b) PLA/PBAT (70/30) blends.

IOP Conf. Series: Materials Science and Engineering **87** (2015) 012090 doi:10.1088/1757-899X/87/1/012090

3.2.3. Final torques. To ensure the complete reaction, torques of molten polymer blends were recorded in 15 minutes. The final torques were then taken to analyze the effect of TBT on the viscosity of blends. The final torques of PLA/PBAT blends with TBT were higher than that of PLA/PBAT blends without TBT. The results showed that the transesterification reaction could take place after 15 minutes (data not shown). Coltelli M-B [7] demonstrated that the reaction time of TBT in molten PLA/PBAT blend was around 20 minutes for the temperature of 200°C. For PLA/PBAT (80/20) blends, PLA/PBAT (80/20) + T0.3 had the highest final torque. For PLA/PBAT (70/30) blends, PLA/PBAT (70/30) + T0.5 had the highest final torque.

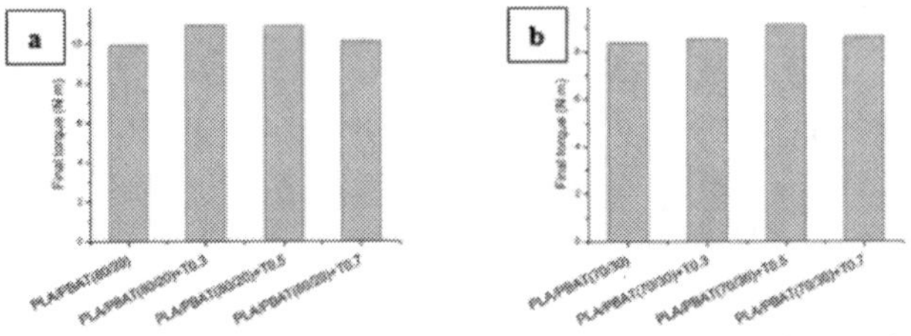

Figure 7. Final torques of PLA/PBAT blends with and without TBT (a) PLA/PBAT (80/20) blends and (b) PLA/PBAT (70/30) blends.

3.2.4. Tensile testing. Yong's modulus and tensile strength of PLA/PBAT blends with TBT were insignificantly different when compared to PLA/PBAT blends without TBT for both blend ratios. However, the elongation at break increased, particularly, PLA/PBAT (80/20) + T0.3 (from 188 to 262%) and PLA/PBAT (70/30) +T0.5 (from 191 to 221%) (figure 8(a)-(b)). These results indicated that the transesterification between PLA and PBAT reaction could occur [7] as also suggested in the final torque (figure 7). Girdthep S, et al. [11] found that the tensile strength of PLA/PBAT with and without TBT was not different, Young's modulus and the tensile strength were improved after the addition of TBT, particularly TBT 0.5 phr. In case of PLA/PBAT (80/20) with TBT less than 0.5 phr could improve the elongation at break of the blend, and the elongation at break of this blend was similar to that of PLA/PBAT (70/30) +T0.5.

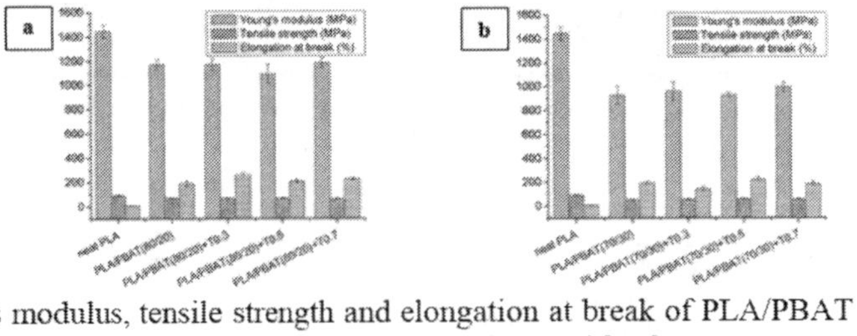

Figure 8. Young's modulus, tensile strength and elongation at break of PLA/PBAT with and without TBT (a) PLA/PBAT (80/20) blends and (b) PLA/PBAT (70/30) blends.

3.2.5. Impact testing. The impact strength of PLA/PBAT(80/20) blends dropped when increasing TBT content (figure 9(a)), which was in agreement with SEM images. In case of PLA/PBAT(70/30), the impact strength was not improved (figure 9(b)), which was in agreement with SEM images. The particle size of PBAT of PLA/PBAT(80/20)+T0.3 was larger than that of PLA/PBAT(80/20) without TBT (figure 10(a)-(b)). For PLA/PBAT(70/30), the particle sizes of PBAT of PLA/PBAT with and without TBT were not different. There were the cavities that were occurred by disengaging of dispersed phase from matrix (figure 10(c)-(d)). Furthermore, the interfacial adhesion between PLA and PBAT phases was poor despite adding TBT.

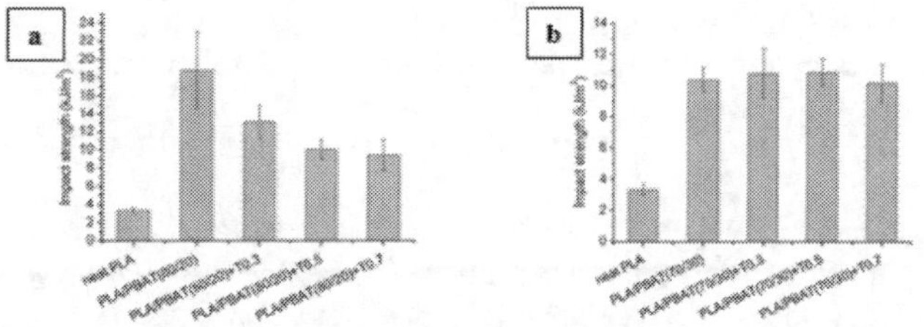

Figure 9. Impact strength of PLA/PBAT blends without and with TBT concentration (a) PLA/PBAT (80/20) blends and (b) PLA/PBAT (70/30) blends.

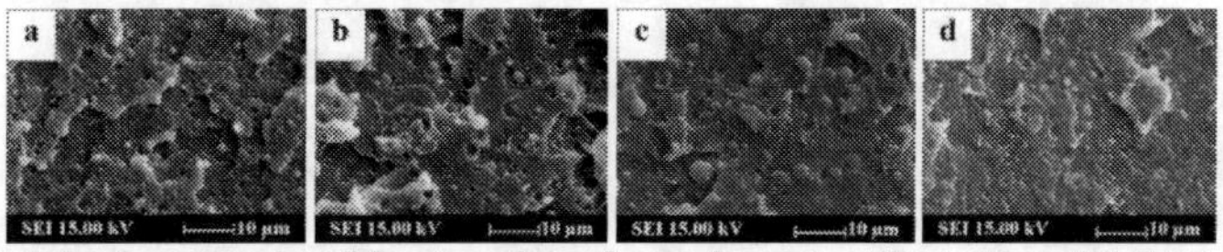

Figure 10. SEM micrograph of fractured surfaces from tensile testing (a) PLA/PBAT(80/20) (b) PLA/PBAT(80/20)+T0.3 (c) PLA/PBAT(70/30) (d) PLA/PBAT(70/30)+T0.3.

4. Conclusion

In case of PLA/PBAT blends with and without Perkadox, PLA, PBAT and Perkadox were blended in twin screw extruder. From SEM results, the particle sizes of PBAT phases of the blends with Perkadox were reduced from those of the blends without Perkadox for all blend ratios due to the improvement of interfacial adhesion between PLA and PBAT phases. The MFI values of all blend ratios were decreased when adding more Perkadox thanks to the crosslinking reaction. Although Young's modulus and the tensile strength of all blend ratios remained unchanged, the elongation at break tended to decrease when adding more Perkadox. The impact strength of PLA/PBAT (80/20) was unchanged even though the impact strength PLA/PBAT (70/30) was improved, increasing by 110% when adding Perkadox 0.5 phr. In case of PLA/PBAT blends with and without TBT, SEM results revealed that the particle sizes of PBAT phases of the blends with TBT were reduced when compared to those of the blends without TBT for all blend ratios due to the improvement of interfacial adhesion between PLA and PBAT phases through the transesterification reaction. Although Young's modulus and the tensile strength of all blend ratios were unchanged, the elongation at break tended to increase when adding TBT. The elongation at break of PLA/PBAT (80/20) was increased by 39% when adding TBT 0.3 phr while the elongation at break of PLA/PBAT (70/30) was increased by 15% when adding TBT 0.5 phr. The impact strength of all blend ratios was not improved.

Acknowledgments

The authors would like to gratefully thank Department of Materials Science and Engineering, Faculty of Engineering and Industrial Technology, Silpakorn University, Thailand for financial supports.

References

[1] Jiang L, Wolcott M P and Zhang J 2006 *Biomacromolecules* **7** 199-207
[2] Yeh J T, Tsou C H, Huang C Y, Chen K N, Wu C S and Chai W L 2010 *J. Appl. Polym. Sci.* **116** 680-7
[3] Dong W, Zou B, Yan Y, Ma P and Chen M 2013 *Int. J. Mol. Sci* **14** 20189-203..
[4] Lin S, Guo W, Chunyin C, Jianli M and Biaobing W 2012 *Mater. Design.* **36** 604-8
[5] Coltelli M B, Bronco S and Carlos Chinea C 2010 *Polym. Degrad. Stabil.* **95** 332-41
[6] Masahiro N, Hiroki I, Hiroki W, Norio F and Hiroaki I 2012 *Engng. Trans.* **60** 171-84
[7] Coltelli M B, Toncelli C, Ciardelli F and Bronco S 2011 *Polym. Degrad. Stabil.* **96** 982-90
[8] Girdthep S, Worajittiphon P, Molly R, Leejarkpai T and Punyodom W 2015 *Polym. Int.* **64** 203-11

[9] Dong W, Zhou B, Ma P, Liu W, Zhou X, Shi D, Ni Z and Chen M 2013 *Polym. Int.* **62** 1783-90
[10] Coltelli M B, Maggiore I D, Bertoldo M, Signori F, Bronco S and Ciardelli F 2008 *J. App. Polym. Sci.* **110** 1250-62
[11] Quero E, Muller A J, Signori F, Coltelli M B and Bronco S 2012 *Macromol. Chem. Physic.* **213** 36-48
[12] Kumar M, Mohanty S, Nayak S K and Parvaiz M R 2010 *Bioresource. Technol.* **101** 8406-15
[13] Al-Itry R, Lamnawar K and Maazouz A 2012 *Polym. Degrad. Stabil.* **97** 1898-914
[14] Farsetti S, Cioni B and Lazzeri A 2011 *Macromol. Symp.* **301** 82-9

IOP Conf. Series: Materials Science and Engineering **87** (2015) 012091 doi:10.1088/1757-899X/87/1/012091

Preparation of hyperbranched polymer via single electron transfer living radical polymerization

X Zhang, X H Chen, J Li, X X Cao and C J Cheng[1]

School of Chemistry and Chemical Engineering, Jiangxi Science & Technology Normal University, Nanchang, Jiangxi, 330013, P.R. China

E-mail: chengcj530@163.com

Abstract. α-Trichloromethyl benzyl alcohol is converted to α-trichloromethyl benzyl methacrylate in 62% yield under esterification conditions. The ester proves to be a good inimer to prepare hyperbranched polymer by utilizing SET-LRP approach. The monomer conversion is about 72% at 60°C for 6 h.

1. Introduction

Hyperbranched polymers have attracted wide attention in recent years due to their unique architectures with dense end groups on their periphery. Compared with their linear counterparts, the outstanding structures of these kinds of polymers result in outstanding properties: soluble in solvents though they are highly branched; low viscosity in solution due to their compact shape [1-3]. Therefore, the polymers have found a variety of practical or potential applications in many fields such as coatings [4], drug delivery [5, 6], optical and electrical polymers [1], etc. Synthesis of hyperbranched polymers usually involves three methods: one is copolymerization of commercially available mono-vinyl monomers with small amount of bi- or multi-vinyl monomers; the second is polymerization of bi- or multi-vinyl monomers in the presence of excess initiator, which can guarantee low molecular weight of the polymers and avoid crosslinking by controlling feedstocks; the third is self-condensing vinyl polymerization via use of a species called inimer [7].

Single electron transfer living radical polymerization (SET-LRP) is one of living radical polymerization approaches developed by Percec and coworkers [8-10]. In comparison with other methods, SET-LRP shows some unique advantages such as mild reaction conditions, rapid polymerization rate, low catalyst loading, and high molecular mass of the polymer products [11].

α-Trichloromethyl benzyl alcohol is a functional small molecule, which can be easily synthesized by dichloro-carbene method. This molecule has been successfully used as an initiator of SET-LRP reactions in our previous work [12]. Herein, as a continuation of our previous work, an inimer is synthesized from α-trichloromethyl benzyl alcohol and further utilized to prepare hyperbranched polymer.

2. Experimental

[1] Address for correspondence: C J Cheng, School of Chemistry and Chemical Engineering, Jiangxi Science & Technology Normal University, Nanchang, Jiangxi, 330013, P.R. China. E-mail: chengcj530@163.com.

2.1. Reagents and instruments
NMR spectra were measured by BrukerAV 400MHz nuclear magnetic resonance meter.
Benzenaldehyde, chloroform, tetrabutylammonium bromide (TBAB), methacryloyl chloride, triethyl amine (TEA), dichloromethane, petroleum ether (PE), ethyl acetate (EA), dimethyl sulfoxide (DMSO), and 2, 2'-bipyridyl (BPY) were all AR grade purity and used as received. Before use, aqueous sodium hydroxide was utilized to treat methyl methacrylate (MMA) to remove possible inhibitor.

2.2. Synthesis of methacrylate ester (2) of α-trichloromethyl benzyl alcohol
Compound 1 was synthesized in moderate yield from benzaldehyde and chloroform according to literature [12]. Compound 1 (6.77 g, 30 mmol) and triethyl amine (6.07 g, 60 mmol) were dissolved in dichloromethane (30 mL), and the resulting solution was cooled to -5°C. Then, methacryloyl chloride (2.09 g, 20 mmol) in dichloromethane (10 mL) was added dropwise. After the addition was completed, the mixture was stirred at -5°C for 12 h. The solid was removed by filtration, and the filtrate was washed with water, saturated Na_2CO_3, and brine, dried with anhydrous sodium sulfate. After concentrated in vacuo, the crude product was purified on silica gel chromatography using PE/EA (10: 1, v : v) as the eluant to give an oily product 2 (3.64 g, 62%). ^{1}H NMR (400 MHz, $CDCl_3$): δ 7.68-7.65 (m, 2H), 7.45-7.38 (m, 3H), 6.44 (s, 1H), 6.38 (s, 1H), 5.75-5,74 (m, 1H), 2.04 (s, 3H); ^{13}C NMR (100 MHz, $CDCl_3$): δ 164.9, 135.3, 133.2, 129.8, 129.6, 128.0, 99.5, 83.5, 83.2, 82.7, 82.3, 18.2.

2.3. Preparation of hyperbranched polymer by using MMA and the inimer (2)
To a flask were added copper power (0.05 g, 0.2 mmol), MMA (5.0 g, 50 mmol), 2, 2'-bipyridyl (0.156 g, 1 mmol), α-trichloromethyl benzyl methacrylate (**2**) (1.48 g, 5 mmol), and DMSO (10 mL). The mixture was stirred at 60°C for 6 h under N_2 atmosphere. Crude hyperbranched polymer was given by precipitation and subsequent filtration. Further purification on neutral Al_2O_3 column chromatography gives a white solid. $M_{n,GPC}$ = 90 kDa, PDI = 1.25.

3. Results and discussion

3.1. Synthesis and characterization of the inimer α-trichloromethyl benzyl methacrylate (2)
It is known that multi-halo-compounds such as $CHCl_3$, CCl_4, $CHBr_3$ can be used as an atom transfer radical polymerization (ATRP) or SET-LRP initiator [13]. Our previous work proved that α-trichloromethyl benzyl alcohol (1) can also be a SET-LRP initiator as its structure was similar to chloroform. Thus, compound 1 can further react with methacryloyl chloride to form α-trichloromethyl benzyl methacrylate (2) (scheme 1). As the reactant methacryloyl chloride and the product 2 are all very reactive, the reaction must be conducted at low temperature to avoid some side reactions such addition or polymerization.

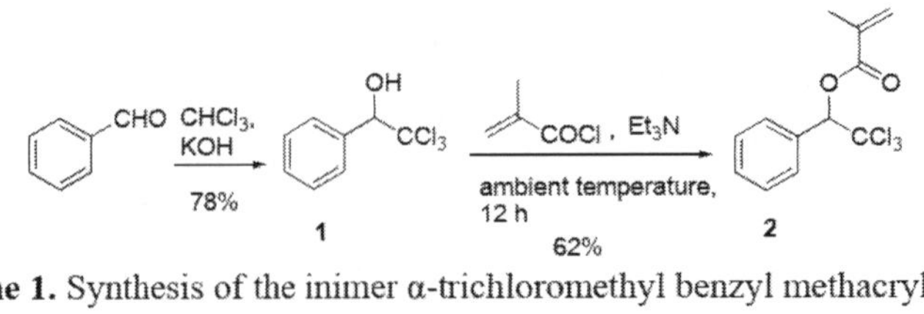

Scheme 1. Synthesis of the inimer α-trichloromethyl benzyl methacrylate (2).

^{1}H NMR and ^{13}C NMR spectra of the inimer α-trichloromethyl benzyl methacrylate (2) were shown in figure 1, together with those of α-trichloromethyl benzyl alcohol (1) for comparison.

3.2. Preparation of hyperbranched polymer from the inimer α-trichloromethyl benzyl methacrylate (2)
In the compound α-trichloromethyl benzyl methacrylate (2), the trichloromethyl group can be SET-LRP initiator, and methacryloyl group be polymeric monomer. Therefore, compound 2 is able to be an inimer. Thus, with Cu(0) as the catalyst, BPY as the ligand, DMSO as the solvent, the hyperbranched

IOP Conf. Series: Materials Science and Engineering **87** (2015) 012091 doi:10.1088/1757-899X/87/1/012091

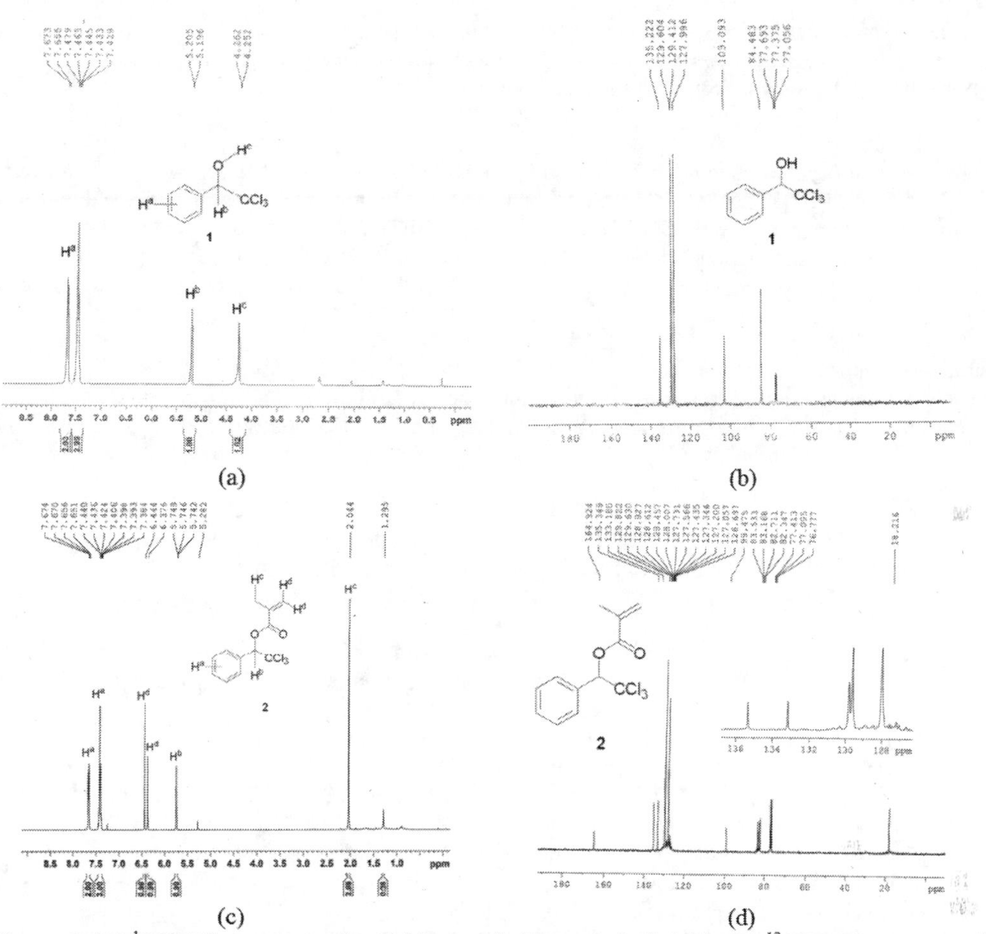

Figure 1. (A) ^{1}H NMR spectrum of α-trichloromethyl benzyl alcohol (1); (B) ^{13}C NMR spectrum of α-trichloromethyl benzyl alcohol (1); (C) ^{1}H NMR spectrum of the inimer α-trichloromethyl benzyl methacrylate (2); (D) ^{13}C NMR spectrum of the inimer α-trichloromethyl benzyl methacrylate (2).

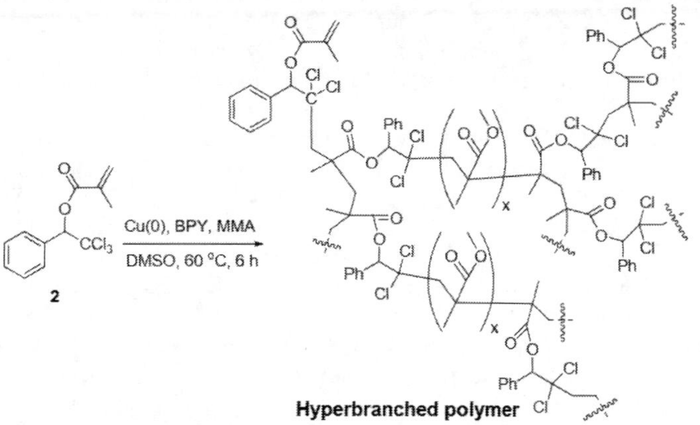

Scheme 2. Preparation of hyperbranched polymer by SET-LRP.

polymer is prepared under SET-LRP conditions (scheme 2). Under the conditions, i.e., at 60°C for 6 h, the hyperbranched polymer is easily obtained, with monomers conversion of about 72%. Further optimization of the reaction conditions is still underway.

4. Conclusions

A new inimer α-trichloromethyl benzyl methacrylate is designed and synthesized in 62% yield from α-trichloromethyl benzyl alcohol. The intermediate and the inimer are well characterized by NMR spectra. The inimer is then self-polymerized under SET-LRP conditions to form the desired hyperbranched polymer. The polymer is characterized by GPC to obtain M_n of about 90 kDa, PDI = 1.25. Further studies on the characterization of the polymer by NMR, DSC, viscosity is under further investigation.

Acknowledgment

The authors acknowledge financial support from the Natural Science Foundation of China (No. 21264008) and the 8th Foundation of Creativity and Study for College Students in Jiangxi Science Technology Normal University.

References

[1] Wu W B, Tang R L, Li Q Q and Li Z 2015 Functional hyperbranched polymers with advanced optical, electrical and magnetic properties *Chem. Soc. Rev.* (in press)

[2] Hult A, Johansson M and Malmström E 1999 Hyperbranched polymers *Adv. Polym. Sci.* **143** 1-34

[3] Rikkou-Kalourkoti M, Matyjaszewski K and Patrickios C S 2012 Synthesis, characterization and thermolysis of hyperbranched homo- and amphiphilic co-polymers prepared using an inimer bearing a thermolyzable acylal group *Macromolecules* **45** 1313-20

[4] Mehta P N 2006 Hyperbranched polymers: Unique design tool for coatings *Surface Coatings International Part B: Coatings Transactions* **89** 333-42

[5] Paleos C M, Tsiourvas D, Sideratou Z and Tziveleka L A 2010 Drug delivery using multifunctional dendrimers and hyperbranched polymers *Expert Opin Drug Deliv.* **7** 1387-98

[6] Zill A, Rutz A L, Kohman R E, Alkilany A M, Murphy C J, Kong H and Zimmerman S C 2011 Clickable polyglycerol hyperbranched polymers and their application to gold nanoparticles and acid-labile nanocarriers *Chem. Commun.* **47** 1279-81

[7] Tsarevsky N V, Huang J Y and Matyjaszewski K 2009 Synthesis of hyperbranched degradable polymers by atom transfer radical (co) polymerization of inimers with ester or disulfide groups *J. Polym. Sci. A: Polym. Chem.* **47** 6839-51

[8] Percec V, Guliashvili T, Popov A V and Ramirez-Castillo E 2005 Catalytic effect of dimethyl sulfoxide in the Cu(0)/ tris (2-dimethylaminoethyl) amine-catalyzed living radical polymerization of methyl methacrylate at 0–90°C initiated with CH_3CHClI as a model compound for α, ω-di (iodo) poly (vinyl chloride) chain ends *J. Polym. Sci. A: Polym. Chem.* **43** 1935-47

[9] Levere M E, Nguyen N H, Leng X F and Percec V 2013 Visualization of the crucial step in SET-LRP *Polym. Chem.* **4** 1635-47

[10] Zhang Q, Zhang Z B, Wang W X, Cheng Z P, Zhu J, Zhou N C, Zhang W, Wu Z Q and Zhu X L 2011 In Situ Cu(0) Catalyzed SET-LRP: The first attempt *J. Polym. Sci. A: Polym. Chem.* **49** 4694–700

[11] Lu G L, Li Y J, Guo H, Du W Y and Huang X Y 2013 SET-LRP synthesis of novel polyallene-based well-defined amphiphilic graft copolymers in acetone *Polym. Chem.* **4** 3132-9

[12] Bai X X, Hu Y, Zhang X, Ai L L and Cheng C J 2014 Single electron transfer living radical polymerization via a new initiator *IOP Conf. Series: Materials Science and Engineering* **62** 012008

[13] Matyjaszewski K and Xia J H 2001 Atom transfer radical polymerization *Chem. Rev.* **101** 2921-90

A NH_4Ac-assisted solvothermal fabrication and properties of size-controlled cobalt ferrite nanospheres

Aiguo Yan[1,3] **and Haomei Liu**[2]

[1]School of Materials Science and Engineering, Xiamen University of Technology, Xiamen, Fujian 361024, P.R. China
[2]Personnel Department, Xiamen University of Technology, Xiamen, Fujian 361024, P.R. China

E-mail: yanaiguod4@163.com

Abstract. Monodisperse cobalt ferrite nanospheres with the mean size 100 nm have been synthesized via a simple solvothermal method by using NH_4Ac as protective agent. The sizes of the particles can be controlled from 20 to 200 nm by changing reaction time or protective agents. Magnetic and microwave absorption properties of Co ferrite nanospheres with the different sizes were investigated in details. Magnetic studies revealed that the saturation magnetization (Ms) and the coercivity (Hc) increased with the increase in particle size, that is, the Ms and Hc are 53.5 $A{\cdot}m^2/kg$ and 14.6×10^3 $A{\cdot}m^{-1}$ (100 nm), 68.9 $A{\cdot}m^2/kg$ and 49.9×10^3 $A{\cdot}m^{-1}$ (125 nm) and 72.5 $A{\cdot}m^2/kg$ and 54.5×10^3 $A{\cdot}m^{-1}$ (200 nm), respectively. Electromagnetic and resulting microwave adsorption properties showed that the maximum magnitudes of the dips are −11.96 dB (at 9.00 GHz for 100 nm), −23.02 dB (at 5.55 GHz for 125nm), and −32.79 dB (at 10.47 GHz for 200 nm), respectively.

1. Introduction

As an important functional material, cobalt ferrite exhibits excellent properties such as high coercivity, moderate magnetization, high Curie temperature, large magneto-crystalline anisotropy, excellent chemical stability, and mechanical hardness [1-3]. Cobalt ferrite thus is realized as a promising candidate material that used for high density information storage device [4], ferrofluids technology [5], biomedical drug delivery [6], and magneto-optical devices [7], etc. On the other hand, as the physical and chemical properties of nanomaterials are strongly depended on their morphology and size, the relationship between size and property over cobalt ferrite materials are widely investigated. For example, when the size of cobalt ferrite nanoparticle is decreased to 5 nm, cobalt ferrite becomes superparamagnetic at room temperature, which makes it highly useful in cell-separation and drug delivery in biomedicine field [8]. Therefore, synthesis of monodisperse cobalt ferrite nanoparticles with adjustable sizes is great significance.

Up to now, considerable efforts have been done on the shape-controlled synthesis of cobalt ferrite crystals. Among them, one-dimensioned cobalt ferrites nanoparticles such as nanowires [9], nanorods

[3] Address for correspondence: Aiguo Yan, School of Materials Science and Engineering, Xiamen University of Technology, Xiamen, Fujian 361024, P.R. China. E-mail: yanaiguod4@163.com.

IOP Conf. Series: Materials Science and Engineering **87** (2015) 012092 doi:10.1088/1757-899X/87/1/012092

[10], and three-dimensioned cobalt ferrites [11], such as nanospheres and nanocubes have been successfully synthesized. Recently, cobalt ferrite nanoparticles with the diameter of particles from 2.7 nm to 17 nm have been synthesized through the "Self-propagating High temperature Synthesis technique (SHS)" [12]. However, the controlled synthesis of cobalt ferrite nanoparticles with monodisperse particle size is still a highly sophisticated challenge. Herein, we demonstrated a one-step, economical and efficient solvothermal method for the large-scale fabrication of monodisperse cobalt ferrite solid nanospheres. In this route, NH_4Ac was used as protective agent, and the size and inner-structure of these particles could be readily controlled by simply adjusting reactive time or the nature of the protective agent. The size-dependent properties of magnetic, electromagnetic and microwave adsorption properties were investigated.

2. Experimental section

$FeCl_3 \cdot 6H_2O$ (2.0 mmol), $CoCl_2$ (1.0 mmol), and $NH_4 \cdot Ac$ (15.0 mmol) was dissolved in 30.0 ml ethylene glycol. The mixture was magnetically stirred to form clear solution at room temperature. Then, the mixture was sealed in a teflon-lined stainless-steel autoclave (50 ml capacity) and heated at 180℃ for 24 hours (h). After reaction, the autoclave was cooled down to room temperature naturally. At last, the product was centrifuged at 2000 rpm for 10 min, washed (with deionized water and 99.5% ethanol), and dried (at 60°C for 6 h) to obtain cobalt ferrite powder.

The obtained samples were characterized on a Rigaku Dmax-2000 X-ray powder diffractometer (XRD) with Cu Kα radiation (λ=1.5418 Å). The operation voltage and current were kept at 40 kV and 40 mA, respectively. The size and morphology of the as-synthesized products were determined at 20 kV by a XL30 S-FEG scanning electron microscope (SEM) equipped with an Energy-dispersive X-ray spectroscopy (EDS). Magnetic properties of the samples were carried out by using a LDJ-9600 vibrating sample magnetometer (VSM). The complex permittivity and permeability and resulting microwave adsorption property have been studied by HP-8720ET vector network analyzer.

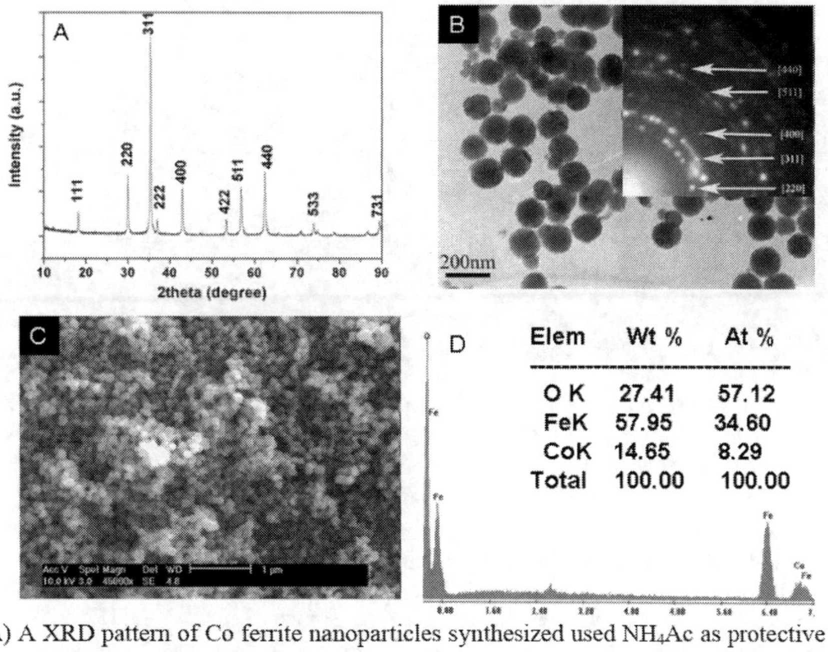

Figure 1. (A) A XRD pattern of Co ferrite nanoparticles synthesized used NH_4Ac as protective agents at 180℃ for 24 h. (B) A TEM image of the same batch of sample. The inset shows the SAED pattern taken from an individual nanosphere. (C) A SEM images of the same batch of sample. (D) An EDS pattern of the same batch of sample.

IOP Conf. Series: Materials Science and Engineering **87** (2015) 012092 doi:10.1088/1757-899X/87/1/012092

3. Result and discussion

Figure 1A shows the XRD pattern of the product prepared at 180℃ for 24 h. All the peaks could be readily identified as the pure cubic phase [space group: Fd3m (227)] with cell constants a = 8.394 °A (JCPDS 22-1086), indicating that the as-prepared product shows high purity. So, well crystallized and pure phase Cobalt ferrite crystalline was successfully synthesized. Then, the morphology and sizes of prepared Cobalt ferrites are characterized by transmission electron microscope analyses. Figure 1B shows a typical TEM image of the Co ferrite nanospheres, which illustrates that the cobalt ferrite nanoparticles prepared by the solvothermal method are monodisperse with monodisperse nanoparticles about 100 nm. The inset is a corresponding selected area electron diffraction pattern, which further confirmed that the nanopartilce has well crystallinity. The labeled diffraction rings could be indexed to the spinel structure of cobalt ferrites, which are consistent with XRD result. Figure 1C shows the SEM image of the product observed in large area. It is shown all the particles are sphere like with monodisperse particle size about 100 nm. Then, the chemical compositions of sample have also been investigated by EDS, which is shown in figure 1D. The as-prepared particles contain Fe, Co and O and there is no other contamination element detected. So, high purity Cobalt ferrets is successfully obtained.

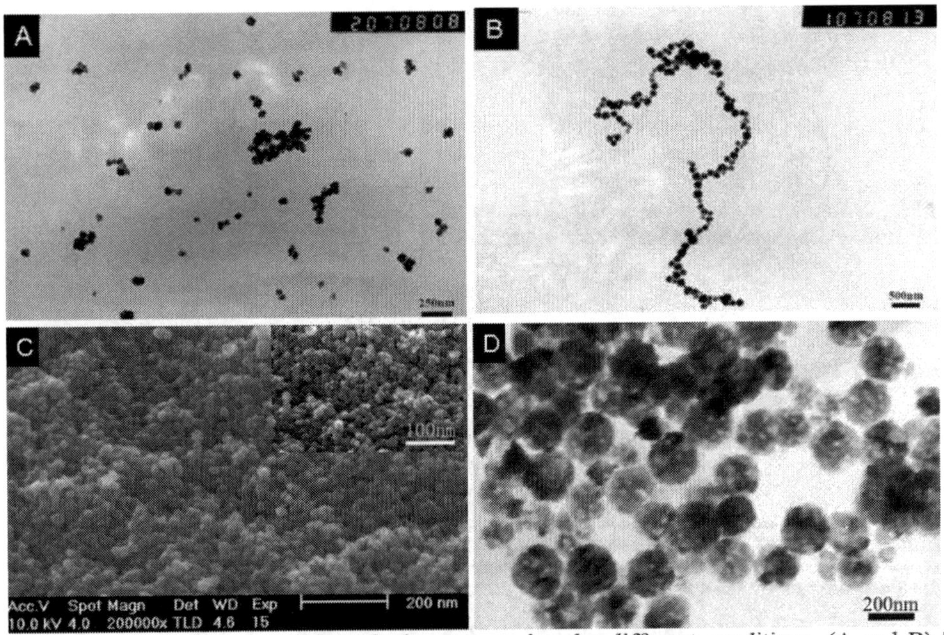

Figure 2. TEM or SEM images of Co ferrites prepared under different conditions. (A and B) The same as in figure 1, except that growth time was changed from 24 h to 12 and 48 h, respectively. (C and D) The same as in figure 1, except that the protective agent of NH_4Ac was substituted with NaAc and the mixed protective agents of 15.0 mmol NaAc and of 3.0 mmol SDS, respectively. The inset in (C) shows the higher magnification SEM image obtained from a selected area of the figure 2(C).

The morphology and sizes of the final product were found to strongly depend on reaction conditions and the nature of the protective agent. For example, when the reaction time was reduced from 24 to 12 h, the average sizes of cobalt ferrite decreased from 100 nm to 75 nm and the shape of the as-prepared was not monodisperse nanospheres (as shown in figure 2A). However, if the reaction time was prolonged from 24 to 48 h, the sizes of monodisperse Co ferrite spheres were increased to about 125 nm and the nanospheres self-assembled (figure 2B). Besides that, the sizes of nanoparticles prepared in this condition are also not as monodisperse as the product prepared at reaction time of 24 h.

So, the 24 h is an optimized reaction time for preparing monodisperse Co ferrite nanospheres.

We have also tried to use the different protective reagent to prepare Co ferrite nanoparticles under similar reaction conditions. As shown in figure 2C, when 15.0 mmol of the NH_4Ac was substituted with equal NaAc, Co ferrite nanoparticles with average size about 20 nm is obtained with reaction time of 24 h. The inset in figure 2(C) is a higher magnification SEM image obtained from a selected area of the figure 2(C) and reveals that these nanospheres were monodisperse. Furthermore, when some surface agent is added, such as sodium dodecyl sulfates (SDS), only nanoparticles with sizes about 200 nm could be obtained (mixture of 15.0 mmol NaAc and of 3.0 mmol SDS, reaction time 24 h).

In the following study, the relationship between sizes and magnetic properties was studied by using a vibrating sample magnetometer (VSM). Figure 3 shows the typical hysteresis loops of the different sizes Co ferrites nanocrymaterials at 300 K. It can be seen from figure 3 that both of the saturation magnetization (Ms) and the coercivity (Hc) gradually increases with as the particle size increases. The values of Ms and Hc of 100 nm as-synthesized sample are respectively 53.5 $A\cdot m^2/kg$ and 14.6×10^3 $A\cdot m^{-1}$, while the two corresponding values are 68.9 $A\cdot m^2/kg$ and 49.9×10^3 $A\cdot m^{-1}$ (125 nm) and 72.5 $A\cdot m^2/kg$ and 54.5×10^3 $A\cdot m^{-1}$ (200 nm).

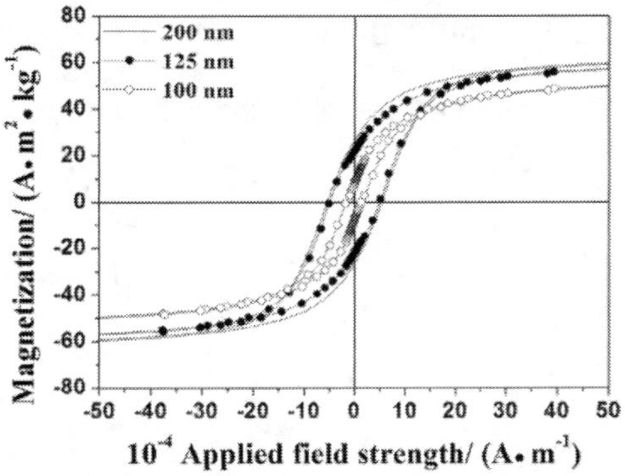

Figure 3. Magnetic hysteresis loops of Co ferrite nanospheres with the different sizes at 300 K.

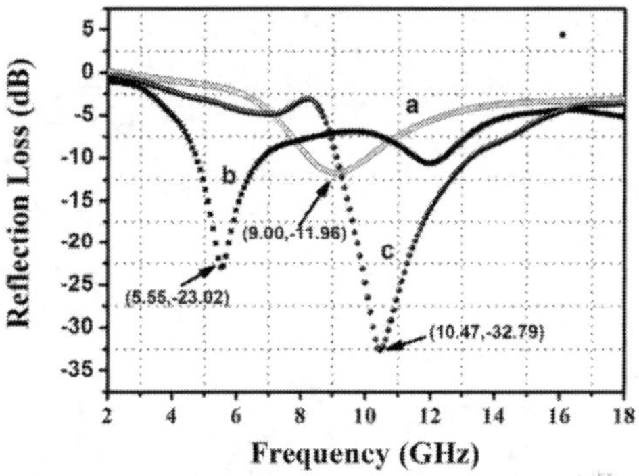

Figure 4. Absorption characteristics of different average sizes of Co ferrite nanospheres at radar band from 2 to 18 GHz: (a) 100 nm, (b) 125 nm, and (c) 200 nm.

The relationship between sizes and microwave absorption properties calculated from a computer simulation using the values of electromagnetic parameters (ε', ε'', μ' and μ'') and other parameters, such as the speed of light (c) and the thickness of the samples (t) was also investigated. The real and imaginary parts of complex permittivity (ε', ε'') and permeability (μ', μ'') of Co ferrite particles were measured using an 8720ET vector network analyzer working at the 2–18 GHz band (the test samples were firstly mixed together as-synthesized cobalt ferrites and paraffine with the mass ratio of 3:1 and then molded into the 3.1 mm-thickness hollow pipe of rectangular cavities). The results are shown in figure S (Supporting Information), in which the letters "a", "b" and "c" correspond to particle size of about 100, 125 and 200 nm, respectively. The calculation process referred to the work of Yusoff [13]. Figure 4 shows that an enhanced microwave absorption properties as the particle size of Co ferrites increases. Figures 4(a), 4(b) and 4(c) also correspond to the particle sizes of Co ferrites with 100, 125 and 200 nm, respectively. Figure 4(a) shows that there is only one wide and shallow dip around 9.00 GHz and the reflection loss minimum or the dip in plot which is equivalent to the occurrence of minimal reflection of the microwave absorber is −11.96 dB at 9.00 GHz. The number, magnitudes and situations of the dips vary as the sizes of the as-prepared Co ferrites increases. Two dips can be observed in the different frequency ranges for 125 samples (figure 4(b)) and 200 nm one (figure 4(c)). The maximum magnitudes of the dips are −23.02 dB at 5.55 GHz for 125nm one, and −32.79 dB at

10.47 GHz for 200 nm one, respectively. Further study is needed to fully probe the mechanism leading to the difference of microwave absorption properties of Co ferrites nanoparticles with the different sizes.

4. Conclusion

In conclusion, monodisperse cobalt ferrite nanospheres have been successfully synthesized via a simple NH_4Ac-assisted solvothermal method. The sizes of these particles can be controlled by changing some experimental parameters, such as reaction time and the nature of protective reagents. Magnetic studies revealed that the saturation magnetization and the coercivity increased with the increase in particle size (the values of Ms and Hc are 53.5 $A{\cdot}m^2/kg$ and 14.6×10^3 $A{\cdot}m^{-1}$ for 100 nm cobalt ferrite nanospheres, 68.9 $A{\cdot}m^2/kg$ and 49.9×10^3 $A{\cdot}m^{-1}$ for 125 nm one and 72.5 $A{\cdot}m^2/kg$ and 54.5×10^3 $A{\cdot}m^{-1}$ for 200 nm one). Microwave absorption properties studies revealed that the maximum magnitudes of the dips are −11.96 dB at 9.00 GHz for 100 nm cobalt ferrite spheres, −23.02 dB at 5.55 GHz for 125nm one and −32.79 dB at 10.47 GHz for 200 nm one, respectively.

Acknowledgments

The support for this project was given by Hunan Province Natural Science Foundation (No. 12JJ6049), Science and technology project of Hunan Province (No. 2011RS4056), Hunan Province Department of education platform project (No. 11K022), Natural Science Project of Xiamen University of Technology (No. YKJ13033R), and National Fund pre-research Project of Xiamen University of Technology（No. XYK201411）is gratefully acknowledged.

Appendix

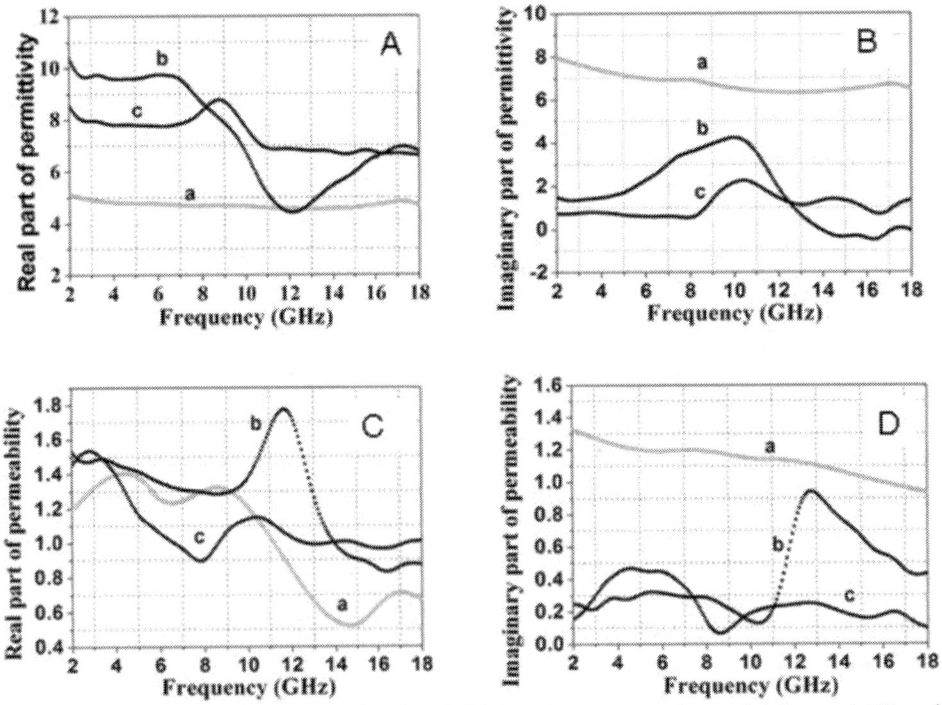

Figure S. Frequency dependence on (A) real and (B) imaginary parts of permittivity, and (C) real and (D) imaginary parts of permeability for different average sizes of Co ferrite nanospheres. (a) 100 nm, (b) 125 nm, and (c) 200 nm.

References

[1] Peddis D, Cannas C, Piccaluga G, Agostinelli E and Fiorani D 2010 *Nanotechnology* **21** 125705
[2] Comes R, Liu H, Khokhlov M, Kasica R, Lu J and Wo S A 2012 *NanoLett.* **12** 2367-73
[3] Peddis D, Yaacoub N, Ferretti M, Martinelli A, Piccaluga G, Musinu A, Cannas C, Navarra G, Greneche J M and Fiorani D 2011 *J. Phys. Condens. Matter* **23** 426004
[4] Sugimoto M 1999 *J. Am. Ceram. Soc.* **2** 269-80
[5] Pileni M P 2001 *Adv. Funct. Mater.* **5** 323-36
[6] Liu C, Zou B, Rondinone A J and Zhang Z J 2000 *J. Phys. Chem. B* **6** 1141-5
[7] Jamon D, Donatini F, Siblini A, Royer F, Perzynski R, Cabuil V and Neveu S 2009 *J. Magn. Magn. Mater.* **321** 1148-54
[8] Loo L, Guenther R H, Lommel S A and Franzen S 2007 *J. Am. Chem. Soc.* **129** 11111-7
[9] Ji G B, Tang S L, Xu B L, Gu B X and Du Y W 2003 *Chem. Phys. Lett.* **379** 484-9
[10] Ji G B, Tang S L, Ren S K, Zhang F M, Gu B X and Du Y W 2004 *J. Crys. Growth* **270** 156-61
[11] Song O and Zhang Z J 2004 *J. Am. Chem. Soc.* **126** 6164-8
[12] de Biasi R S, Figueiredo A B S, Fernandes A A R and Larica C 2007 *Solid State Communs.* **144** 15-17
[13] Yusoff A N, Abdullah M H, Ahmad S H, Jusoh S F, Mansor A A and Hamid S A A 2002 *J. Appl. Phys.* **92** 876-82

IOP Conf. Series: Materials Science and Engineering **87** (2015) 012093 doi:10.1088/1757-899X/87/1/012093

Preparation and application of poly 3, 4-ethylenedioxythiophene (PEDOT) nanofibers in the pre-treatment of samples before the determination of elements in children fingernails

J L Qiu[1], X J Kang[1, 3], L Ma[1], W Y Huang[2] and Q Y Ge[1, 4]
[1]Key Laboratory of Child Development and Learning Science, Ministry of Education, Research Centre for Learning Science, University of Southeast, Nanjing, 210096, China
[2]School of Sport, Normal University of Jiangxi, Nanchang, 331000, China

E-mail: xjkang64@163.com; qinyuge@seu.edu.cn

Abstract. Solid phase extraction (SPE) has been used widely for sample preparation in the analytical process. Many efforts have focused on developing novel adsorbents to enrich and purify the analytes effectively. In this study, poly-3, 4-ethylenedioxythiophene (PEDOT) nanofiber was prepared and used as the SPE adsorbent. The fiber performed good in extraction of metal ions, Cd, Sn, Hg, Pb, Al, and As, with the extraction recoveries ranged from 53.9% to 99.6% in the wet digested samples of fingernails. A PEDOT namofibers SPE column coupled with ICP-MS was established for assay of elements in fingernails. The levels of elements (Cd, Sn, Hg, Pb, Al, and As) in the fingernails of 77 healthy Chinese children (6-7 and 10-11 years) were determined. Independent t test shows that no significance has been found between boys and girls. On the contrary, there was obvious difference on the levels of most elements between the two grade groups.

1. Introduction

Biological information from the subject can be obtained from fingernails. Children fingernails have been recognized as biochemical indicators, which have prominent advantages for monitoring children physical health. Compared to blood and urine, fingernails grow at a rate of approximately 1-1.4 mm per week, thus, the analysis and detection of fingernails has certain advantages, including easier sampling, transfer, and storage [1].

Elements in fingernails indicate exposure to hazardous material during the previous 2 to 18 months. Fingernails analysis is useful for detecting the levels of elements in the body, which keep information on the nutrition, habitat, and other environmental conditions.

Before determination, fingernails must be wet digested. However, direct determination of metal ions at trace levels in digested samples may be limited due to their low concentrations and matrix

[3] Address for correspondence: X J Kang, Key Laboratory of Child Development and Learning Science, Ministry of Education, Research Centre for Learning Science, University of Southeast, Nanjing, 210096, China. E-mail: xjkang64@163.com.
[4] Q Y Ge, Key Laboratory of Child Development and Learning Science, Ministry of Education, Research Centre for Learning Science, University of Southeast, Nanjing, 210096, China. qinyuge@seu.edu.cn.

IOP Conf. Series: Materials Science and Engineering **87** (2015) 012093 doi:10.1088/1757-899X/87/1/012093

interferences. Pre-concentration and separation methods have been routinely used to eliminate matrix effects and cope with low metal levels. Solid-phase extraction (SPE) is a cost-effective technique which is used not only to extract traces of targets from samples, but also to remove the interfering components of the complex matrices. In recent years, conductive polymers have attracted great interest for the extraction of target compounds due to a variety of advantages. The main advantage of using conducting polymers in SPE is that the charge of the sorbent can readily be controlled by oxidation and reduction of the polymers [2].

In this paper, poly 3, 4-ethylenedioxythiophene (PEDOT) nanofibers were prepared and used as the SPE absorbents. The fibers were prepared by electrostatic spinning of polystyrene as a template and in situ polymerization. Owing to their large surface area to volume ratio, electrospun nanofibers facilitate the miniaturization of SPE when they are packed into a mini column as the sorbent beds. Packed-fiber solid-phase extraction (PFSPE) has been successfully applied in various sample matrices. However, most of the published work has concentrated on the extraction of organic compounds [3]. Its application for the extraction of metal ions is still limited. In this work, coupled with this PFSPE column, the levels of elements in the fingernails of 77 healthy Chinese school children (6-7 and 10-11 years) were determined by Inductively Coupled Plasma Mass Spectrometer. The data is analyzed to find out whether sex and grade (age) are the factors influencing the content of elements in children' bodies.

2. Materials and methods

2.1. Sample collection

77 healthy Chinese school children (6-7 and 10-11 years) were recruited from pupils living in Nanchang city, Jiangxi province. Fingernail samples were collected using stainless steel nail clippers and sealed in labeled polyethylene zip lock bags. Informed consent was obtained from each surveyed participant following guidelines approved by the school and parents. The age range spanned 6-11years.

2.2. Equipment and reagents

Inductively Coupled Plasma Mass Spectrometer (ICP-MS), model agilent 7700, was used for analytical determinations. Double distilled water and guarantee reagents, HNO_3 and H_2O_2, were used to prepare the solutions. Methanol (chromatographic purity) was from Chengdu Xindu area of the industrial development zone (Chengdu, China). Dimethylformamide (DMF), Ferric chloride, Ethanol and Tetrahydrofuran (THF) were from Nanjing Chemical Reagent Company (Jiangsu Province, China). Buffer solution (Borax/boric acid, pH 7.5-8.0) were prepared using water as the solvent. Polystyrene (PS, Mw=185,000) was from Shanghai Institute of Chemical Reagent (Shanghai, China). 3, 4 – ethylene dioxythiophene monomers was from Suzhou Yake Reagent Company (Jiangsu Province, China).

2.3. Preparation of solid-phase extraction column packed with PEDOT nanofibers

The polystyrene (PS) was added into a mixed solvent composed of dimethylformamide and tetrahydrofuran (4: 6, v/v) and then stirred slowly at room temperature for 10 h. Therefore, a 10% (w/v) solution of polystyrene was prepared. After dissolving totally, as can be seen in figure1 the solution was removed quickly and put in the syringe with the stainless steel needle connected with the positive pole of 15 KV high voltage power. The collection screen which collected to the negative power was placed at a distance of about 20 cm from the stainless steel needle. Until now, a fall mesh was shown in the collection screen.

After PS nanofibers were gained, it would be firstly cleaned with ethanol and water. 3, 4 - ethylene dioxythiophene monomer (EDOT) was dissolved with a small amount of ethanol and then added into 50 ml distilled water. The PS nanofibers were put into the solution mentioned above and mixed for 15 minutes. 50 ml of 0.25 g/ml ferric chloride solution was added into the solution mentioned above and then placed in 60°C water bath for 24 hours [4]. Lastly, the PEDOT nanofibers should be dried after being washed with ethanol and distilled water. The solid phase extraction column used in this study was prepared by filling 5 mg PEDOT nanofibers in the thin bottom of the column.

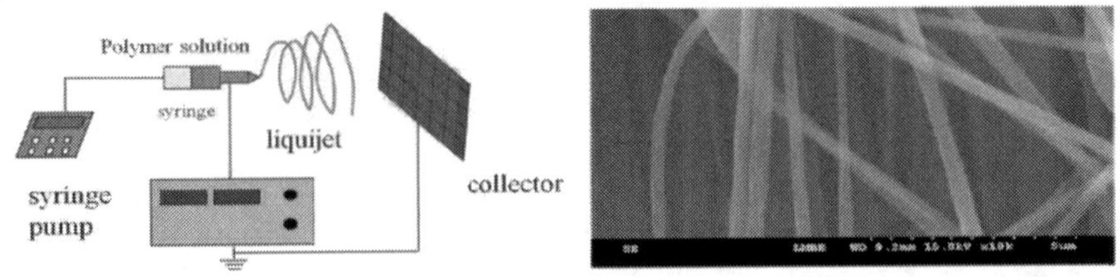

(a) (b)

Figure 1.The schematic view for electrostatic spinning apparatus (a) and SEM graph of polystyrene nanofibers (b).

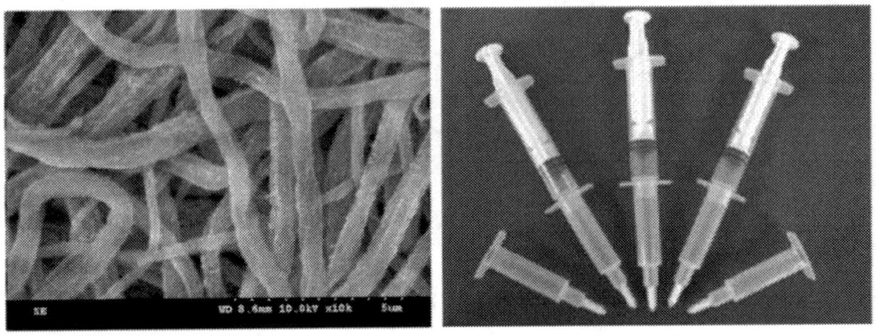

(a) (b)

Figure 2. The graph of PEDOT nanofibers (a) and the solid phase extraction column (b).

2.4. Sample pre-treatment and analysis

The fingernail samples were immersed in the mixture of cleanser and deionised water and then placed in an ultrasonic bath for 10 min. After sonication, the solution was discarded and the nails were rinsed thoroughly with deionised water (18 MΩ), then dried at 60°C for 2 h. Following cleaning, 10mg nail sample was accurately weighed into Teflon beaker, 5 ml of 65% HNO_3 and 2 ml H_2O_2 were added into each beaker, the beakers were capped and put on the far infrared electric heating plate. After heating at 100°C for 0.5 h, open the lips and continue heating at 150°C until the nitric acid is steamed off completely [5].

The columns in figure 2 and the liquid storage cartridge were washed with 0.1 mol/L nitric acid and deionized water for cleaning up the interfering ions. 2 mol/L sodium hydroxide aqueous solution and deionized water were used sequentially to activate the nanofibers. Then, the remaining digested liquid for each fingernail sample along with the water for washing the beaker was transferred into the column, the pH value of which was adjusted to 7.5-8.0 by the buffer solution. After loading and elution, the column was desorbed with 0.1 mol/ L nitric acid solution in a small volume (1 ml).

The concentrations of elements were assayed in triplicates by use of ICP-MS with a flow injection system. A series of standards were prepared for instrumental calibration by diluting in commercial standards containing 1000 μg/L of the metals. Standard and blank samples were determined in the same manner as the samples.

3. Results and discussion

The extraction recovery for the target objects is an important criteria for a material used as adsorbent. In this study, the extraction recoveries of metal ions by PEDOT nanofibers solid phase column were presented in table 1. As can be seen, most of the metal ions have high extraction recoveries.

Table 1. The extraction recoveries for the metal ions.

Metals	Cd	Sn	Hg	Pb	Al	As
Extraction recovery (%)	93.8	85.3	88.7	99.6	53.9	65.3

Table 2 presents the results obtained from the monitoring of elements in the fingernails of boys and girls living in Nanchang city. Independent samples t test shows that no obvious effect has been found on the levels of elements except Cd. Also, boys spend more time playing outside than girls, so will be more exposed to pollutants. This may result in the high levels of Cd in the boy group. However, the mean levels of Sn, Al and Pb were lower in boy group. Girls like eating unhealthy snack food containing these metals may be one of the reasons for this phenomenon.

Table 2. Descriptive statistics of metal levels (μg/g) in fingernail samples of different sex pupil groups along with independent t-test (P = 0.05).

Metal	Boys (n=35)		Girls (n=42)		t-value at P=0.05	p-value
	Mean	Range	Mean	Range		
Cd	0.16±0.05	0.0-1.4	0.04±0.01	0.0-0.2	0.01	0.01*
Sn	0.58±0.13	0.0-4.3	1.0±0.6	1.0±0.6	0.50	N.S
Hg	0.09±0.009	0.0-0.2	0.09±0.01	0.09±0.01	0.93	N.S
Pb	4.79±0.81	0.0-18.8	5.2±1.2	5.2±1.2	0.79	N.S
Al	48.2±9.0	0.0-239.6	52.2±12.0	0.0-483.7	0.79	N.S
As	0.29±0.05	0.0-0.1	0.28±0.05	0.0-1.4	0.89	N.S

[a]Indicate significant difference level at P=0.05.

Table 3 compares element levels of fingernail samples (μg ± SE) in two grade (age) groups. It can be seen from P values, grade (age) has significant correlation with the levels of most elements. However, the trend for the levels of elements between the two groups is not the same because the factors affect the contents of elements are varied [6]. As can be seen, the levels of Hg and Cd were no obvious change. However, the differences for the levels of remain elements were remarkable. Currently, there is no hypothesis can explain these significant differences. The factors influenced the contents of elements are various and complex. It may be resulted from varied diet and living habits.

Table 3. Descriptive statistics of metal levels (μg/g) in fingernail samples of different age groups along with independent t-test (P = 0.05).

Metal	Junior Group (n=39)		Senior Group (n=38)		t-value at P=0.05	p-value
	Mean	Range	Mean	Range		
Cd	0.1±0.04	0.0-1.4	0.1±0.02	0.0-0.8	0.02	N.S
Sn	1.2±0.6	0.0-23.6	0.4±0.1	0.0-0.3	1.25	0.02*
Hg	0.1±0.01	0.01-0.23	0.08±0.01	0.0-0.2	1.61	N.S
Pb	3.5±0.7	0.0-24.1	6.6±1.3	0.0-44.7	-2.15	0.04*
Al	73.4±14.0	0.0-483.7	26.7±3.4	0.0-86.1	3.21	0.002*
As	0.37±0.05	0.0-1.4	0.2±0.04	0.0-1.0	2.38	0.02*

[a]Indicate significant difference level at P = 0.05.

4. Conclusion

In summary, a convenient method was developed to prepare the PEDOT nanofibers, and it was successfully used as absorbents of SPE for the elements enrichment and purity. The levels of metals in fingernail samples between boys and girls were found to be no significant difference. Whereas, age has significantly influences on the levels of metals between two grade groups.

Acknowledgments
This work was supported by the National Basic Research Program (2012CB933302), the National Instrumental Research Program (2014YQ06077303) and Suzhou Science and Technology Department Foundation (Grant No. ZXG201441).The authors also acknowledge the help of Nanjing Institute of Product Quality Inspection, special thanks to the school authorities in Nanchang, teachers, children and parents for their voluntary participation.

References
[1] Parsons P 2007 Blood lead levels in children aged 5 –9 years living in Cartagena, Colombia *J. Total Environ.* **372** 707-16
[2] Chen L Q, Wang Y, Qu J S, Deng J J and Kang X J 2015 Nanofiber packed solid-phase extraction for the analysis of plasma, urine, and saliva with high-performance liquid chromatography *J. Instrumentation Science and Technology.* **43** 170-80
[3] Zhang Y Y, Kang X J, Chen L Q and Pan C 2008 *J. Anal. Bioanal. Che.* **391** 2189-97
[4] Tian T, Zhen S L, Ye B F, Qu B, Zhao Y J, Kang X J and Gu Z Z 2013 Poly-3,4-ethylenedioxythiophene nanoclusters for high effective solid phase extraction *J. Chromatography A.* **275** 17-22
[5] Molina V, Casana M and Odriguez B M 2015 Biomonitoring of arsenic, cadmium, lead, manganese and mercury in urine and hair of childrenliving near mining and industrial areas *J. Chemosphere* **124** 83-91
[6] Ondimu K 2000 Severe anaemia during pregnancy in Kisumu District of Kenya, prevalence and risk factors *J. Health Care Quality Qssurance.* **12** 230-3

Microstructure-property relationships in alumina trihydrate filled poly (methyl methacrylate) composite materials

Ruoyu Zhang[1]

Mechanical Engineering Department, Imperial College London, London, United Kingdom

E-mail: r.zhang13@imperial.ac.uk

Abstract. The mechanical properties (Young's modulus and fracture toughness) of composite made from a poly (methyl methacrylate) (PMMA) matrix filled with alumina trihydrate (ATH) are reported. The experiments were performed using flexural tests and single edge notched bend (SENB) tests. The composites samples were tested at a range of filler volume fractions (34.7%, 39.4% and 44.4%) and mean filler diameters (8 μm, 15 μm and 25 μm). The data of Young's modulus agreed well with the results of Lielens model and finite element analysis (FEA) model.

1. Introduction

Particle-filled polymer composites are important among engineering materials due to their ability to provide both economic and technical advantages. The work here is concerned with ATH filled PMMA composites at high filler volume fractions. These composites have been widely studied [1-3]. One of the major concerns is the relationship between the microstructure and the mechanical properties of these composites. The elastic behavior of particle-filled polymers is well understood and several models have been developed [4, 5] to describe the relationship between Young's modulus and the addition of filler. However, the effects of filler on the fracture toughness are not well understood. Generally, as the modulus increases, the fracture toughness decreases. In this work, fracture toughness tests were performed on composites with different filler volume fractions (34.7%, 39.4% and 44.4%) and mean particle diameters (8 μm, 15 μm and 25 μm). In addition, by comparing with the experimental results, the reliability of FEA modelling results is verified.

2. Experiments and methods

2.1. Materials

ATH/PMMA composites used in this work were provided by E.I. DuPont Nemours & Co. (Inc.) in the form of sheets (250 mm x250 mm x12 mm). The particles were mixed with the matrix using a four-blade marine-type propeller and the sheets were manufactured using casting process. These samples were all prepared using a proprietary adhesion promoting agent as part of the matrix composition. Each composite is formulated using the same type and amount of adhesion promoter and crosslinker. Table 1 shows the manufacturer's (DuPont) specifications. Particles with a 15 μm mean diameter were

[1] Address for correspondence: Ruoyu Zhang, Mechanical Engineering Department, Imperial College London, London, United Kingdom. E-mail: r.zhang13@imperial.ac.uk.

IOP Conf. Series: Materials Science and Engineering **87** (2015) 012094 doi:10.1088/1757-899X/87/1/012094

used for the manufacturing of composites with codes 1, 3 and 5. For these composites, the filler volume fractions were 34.7%, 39.4% and 44.4%, respectively. Keeping the same filler volume fraction as 3 (i.e. 39.4%), 8 μm particles were used for manufacturing composite 2, while 25 μm particles were used in composite 4.

Table 1. Particle size and volume fraction of the ATH/PMMA composites.

Composite Code	Volume Fraction ATH Filler (%)	Mean Particle Size of ATH Filler
1	34.7	15
2	39.4	8
3	39.4	15
4	39.4	25
5	44.4	15

2.2. *Experimental*

The flexural tests were performed according to ASTM D790M-10 [6]. The specimens used were simple rectangular beam specimens, with dimensions of 12x12x200 (mm^3) in accordance with the standards. The loading span L was 160 mm. The crosshead rate was 4.5 mm/min.

The SENB tests were performed according to ISO 13586 [7]. The specimens were rectangular beams, of dimensions 10x20x100 (mm^3). A 10mm notch was machined in each specimen and then sharpened using a razor blade. The tests were performed at a cross-head motion rate of 10mm/min, while the compliance correction test was performed at 1mm/min.

2.3. *Analytical model*

The results of Lielens model were used to verify the experimental results. Like most analytical models, Lielens model is based on the following equation [4]:

$$\boldsymbol{C} = \boldsymbol{C_m} + \emptyset_f(\boldsymbol{C_f} - \boldsymbol{C_m})\boldsymbol{A} \tag{1}$$

where $\boldsymbol{C}$ are the elastic stiffness tensors, while the subscripts m and f represent the matrix and fillers, respectively. $\boldsymbol{A}$ is the strain concentration factor [4] and $\emptyset_f$ is the particle volume fraction. $\boldsymbol{C}^{-1}$, or compliance tensor, $\boldsymbol{D}$, can be expressed as:

$$\boldsymbol{D} = \begin{bmatrix} \frac{1}{E_{11}} & -\frac{v_{12}}{E_{11}} & -\frac{v_{12}}{E_{11}} & 0 & 0 & 0 \\ -\frac{v_{12}}{E_{11}} & \frac{1}{E_{11}} & -\frac{v_{23}}{E_{22}} & 0 & 0 & 0 \\ -\frac{v_{12}}{E_{11}} & -\frac{v_{23}}{E_{22}} & \frac{1}{E_{22}} & 0 & 0 & 0 \\ 0 & 0 & 0 & \frac{11v_{23}}{E_{22}} & 0 & 0 \\ 0 & 0 & 0 & 0 & \frac{1}{2\mu_{12}} & 0 \\ 0 & 0 & 0 & 0 & 0 & \frac{1}{2\mu_{12}} \end{bmatrix}$$

where E_{11} is longitudinal Young's modulus, E_{22} is transverse Young's modulus, μ_{12} longitudinal shear modulus, v_{12} is longitudinal Poisson's ratio and v_{23} is transverse Poisson's ratio. Hence, if $\boldsymbol{A}$ is obtained, the mechanical properties of the composite can be calculated accordingly.

Lielens model is a bounding method, in which the reference materials for upper and lower bound are filler and matrix, respectively. Using the rule of mixtures between the upper and lower bound, Lielens et al. proposed a 'convenient' strain-concentration factor $\boldsymbol{M_L}$, which can be calculated according to:

$$M_L = \left[(1 - F_L)(M_{lower})^{-1} + F_L\left(M_{upper}\right)^{-1}\right]^{-1} \quad (2)$$

where $M_{lower} = \left[I + S_m C_m^{-1}(C_m - C_f)\right]^{-1}$, $M_{upper} = \left[I + S_f C_f^{-1}(C_m - C_f)\right]^{-1}$ and $F_L = \emptyset_f + \emptyset_f^2/2$. S_m and S_f are the Eshelby's tensors of the matrix and the particle, respectively.
The strain concentration factor of Lielens' model can then be calculated according to:

$$A_L = M_L\left[(1 - \emptyset_f)I + \emptyset_f M_L\right] \quad (3)$$

2.4. Finite element analysis

The FEA modelling was performed using Abaqus [8], and the microstructure geometries of the composites were converted from SEM images using a Matlab python proposed by Tarleton et al. [9]. In a binary image, the position of a pixel can be defined by its coordinate (i, j). By defining a black pixel (matrix) as f_{ij}=0 and a white one (filler) as f_{ij}=1, the position of a white pixel can be identified according to its four neighbours. By calculating a value, $k_{ij}=f_{i+1j}+f_{i-1j}+f_{ij+1}+f_{ij-1}$ (see figure 1), if k_{ij}=4, the pixel lies in the bulk of a particle. If k_{ij}=3, the particle lies on a vertical or horizontal boundary of a particle. If k_{ij}=2, the pixel lies on a diagonal boundary or a corner. If $k_{ij}=1$ or $k_{ij}=0$, the pixel will be removed from the microstructure to simplify the microstructure. The position of each pixel is then input into Abaqus for the creation of the 2-D FEA microstructure.

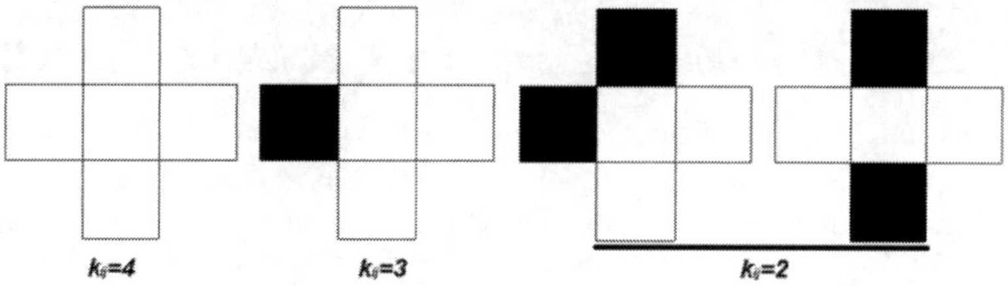

Figure 1. The identification of the position of a white pixel.

3. Results and discussion

Figure 2 shows the G_{IC} values as a function of filler volume fraction (A) and particle diameter (B). It can be seen that higher filler content lead to lower fracture .This can be explained by that as more particles were added, the deformation of the matrix around the crack tip was more limited and hence, less energy can be absorbed during the crack growth. In addition, due to the high filler volume fractions of the composites, further increase in the filler content can lead to the agglomeration of the particles. The agglomerated particles induce stress concentrations, which can initiate cracks and make

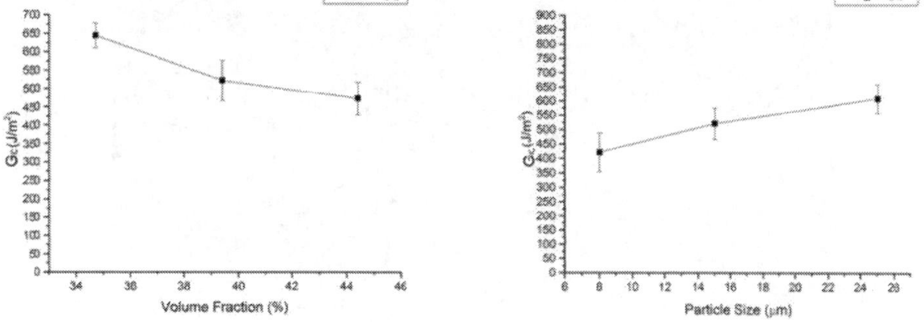

(A) (B)

Figure 2. GIC values as function of (A) filler volume fraction and (B) particle diameter.

the crack extend until the cracks are larger than the critical size that cause failure, and hence, reduce the fracture toughness of the composites.

On the other hand, larger particles lead to higher G_{IC} values. This is because that in the composites filled with smaller particles, the crack propagated in the matrix was deflected by the particles slightly, while in the composites filled with larger particles, the crack propagated through both the matrix and the particles. This conclusion is supported by the images shown in figure 3, in which a smoother fracture surface can be observed in the composite filled with 8μm particles than in the one filled with 25 μm particles.

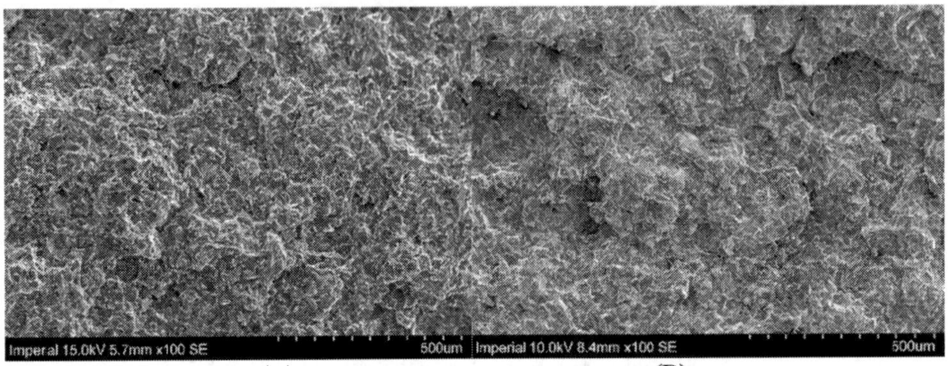

(A) (B)

Figure 3. Fracture surface of composite filled with (A) 8 μm particles and (B) 25 μm particles.

An image-converted FEA microstructure is shown in figure 4. A tensile simulation was performed on the microstructure of each composite to get the Young's modulus. The results of FEA modelling agreed well with the results of Lielens model (figure 5), which means the image based FEA geometry can represent the microstructure of the composites and hence, could be useful in the future work of fracture toughness modelling.

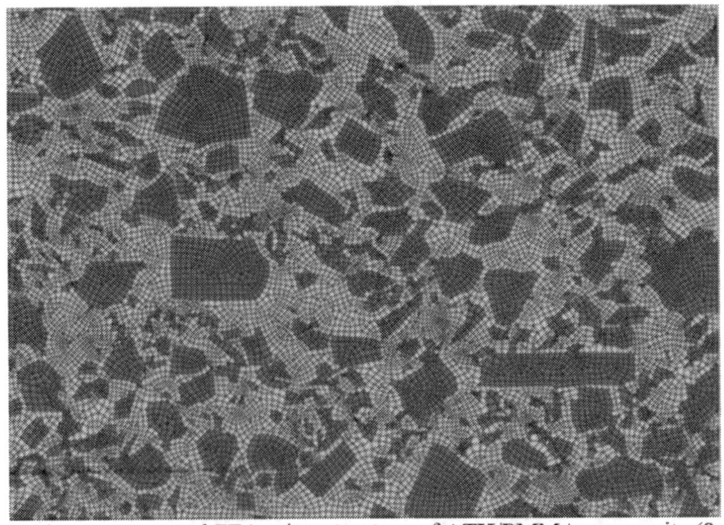

Figure 4. SEM image converted FEA microstructure of ATH/PMMA composite (8 μm, 39.4%).

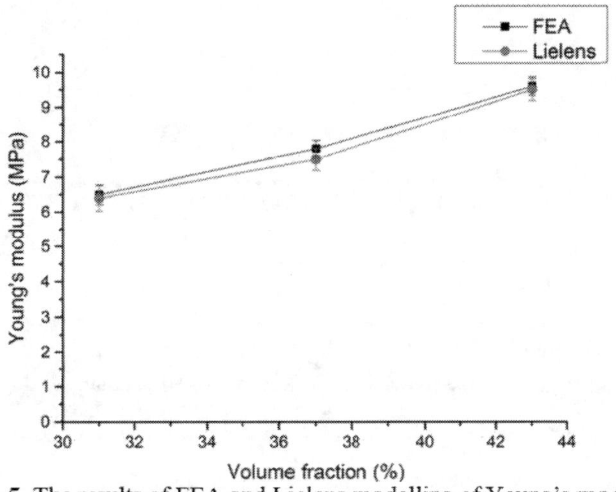

Figure 5. The results of FEA and Lielens modelling of Young's modulus.

4. Conclusion and future work

G_{IC} values of composites increase with the increasing particle size, while higher filler content means lower fracture toughness. The results of FEA modeling and Lielens modeling agreed well with each other, which means that the image converted FEA geometries can represent the microstructure of the composites. Therefore, in the future, this method will be used in the modelling of fracture toughness.

References

[1] Stapountzi O A 2008 *Stiffness and Fracture Properties of Alumina Trihydrate Ffilled Poly (Methyl Methacrylate) Composites* (Imperial College London: PhD thesis)

[2] Nie S 2005 *A Mechanical Study of the Damage Mechanicas of Acrylate Particulate Composites under Thermomechanical Loading* (State University of New York at Buffalo: PhD dissertwation)

[3] Tarleton E, Charalambides M N and Leppard C 2012 Image-based modelling of binary composites *Computational Materials Science* **64** 183-186.

[4] Lielens G, Pirotte P, Couniot A, Dupret F and Keunings R 1998 Prediction of thermo-mechanical properties for compression moulded composites *Composites Part A: Applied Science and Manufacturing* **29** 63-70

[5] Mori T and Tanaka K 1973 Average stress in matrix and average elastic energy of materials with misfitting inclusions *Acta Metallurgica* **21** 571-4

[6] American Society for Testing and Materials D790M-10n 2010 *Standard Test Methods for Flexural Properties of Unreinforced and Reinforced Plastics and Electrical Insulating Materials*

[7] ISO 13586-2000 2000 *Plastics-determination of fracture toughness (GIC and KIC)-Linear elastic fracture mechanics (LEFM) approach* (Gevena: International Organization of Standardization)

[8] ABAQUS 2012 *Version 6.12 Documentation* (USA: Dassault Systemes Simulia Corp. Providence, RI)

[9] Tarleton E, Charalambides M N and Leppard C 2012 Image-based modelling of binary composites *Computational Materials Science* **64** 183-6

Study on moisture absorption and sweat discharge of honeycomb polyester fiber

Aifen Feng[1] **and Yongjiu Zhang**

Hebei University of Science and Technology, Shijiazhuang, Hebei, China

E-mail: fengaf774@sohu.com

Abstract. The moisture absorption and liberation properties of honeycomb polyester fiber were studied in order to understand its moisture absorption and sweat discharge. Through testing moisture absorption and liberation regains of honeycomb polyester fiber and normal polyester fiber in standard atmospheric conditions, their moisture absorption and liberation curves were depicted, and the regression equations of moisture regains to time during their reaching the balance of moisture absorption and moisture liberation were obtained according to the curves. Their moisture absorption and liberation rate curves were analyzed and the regression equations of the rates to time were obtained. The results shows that the moisture regain of honeycomb polyester fiber is much bigger than the normal polyester fiber's, and the initial moisture absorption and moisture liberation rates of the former are much higher than the latter's, so that the moisture absorbance and sweat discharge of honeycomb polyester fiber are excellent.

1. Introduction

With the continuous improvement of living standards, the functions and comfort of clothing have been paid more and more attention. The fabrics made from polyester fiber have good elasticity, wrinkle resistance, shape retention, excellent wash-and-wear performance and durability, and so on, so that it is widely used in all kinds of apparel fabrics. However, because polyester fiber is poor in moisture absorption, its clothing makes wearer to feel hot and sticky, produces static electricity easily which results in clothing absorbing dust and clinging to the body, and has poor comfort. Therefore, it is one of the research hotspots in man-made fiber and textile industry in recent years that hygroscopic polyester fiber is given the certain hydrophobic property. The polyester fibers with moisture absorption performance emerged as the times require under this kind of background [1].

In the early1980's, Teijin Company began to research and develop moisture absorption and perspiration polyester and produced the microporous hollow fiber. Coolmax fibers are functional made by DuPont Co R & D. They are not round, but are slightly oblong in cross-section with grooves running lengthwise along the threads. They are manufactured in either a tetrachannel or hexachannel style. The series of closely spaced channels creates capillary action that wicks moisture through the core and out to a wider area on the surface of the fabric which increases evaporation [2, 3]. Later, a lot of fiber producers in Taiwan have also developed a variety of moisture absorption and perspiration fibers, such as Technofine, Topcool and Coolplus [4]. Internationally, many fiber production

[1] Address for correspondence: Aifen Feng, Hebei University of Science and Technology, Shijiazhuang, Hebei, China. E-mail: fengaf774@sohu.com.

enterprises and companies have developed a variety of moisture absorption and perspiration fibers, such as "Ekslive" (Toyobo Company), "Aerocool" (South Korean Hyosung Corp.), and "Wellkey-MA" (Teijin Company). A lot of moisture absorption and perspiration fibers with having good properties have been developed in China, such as Coolbst, Cooldry and Coolnice.

Honeycomb polyester fiber with microporous structures is developed through the methods of the chemical and physical modification by Shangyu Hongqiang color polyester Co. Ltd. in order to improve the handle, moisture permeability and dyeing property of polyester fiber. It is also called wicking fiber or breathing fiber. There are honeycomb microporous structures on its surface and in the interior (figure 1), it is quick drying and fast in heat dissipation. Not only can it be dyed by cationic dyes under the conditions of normal temperature and pressure, but also its dyeing time is short and its dye-uptake is high, which can save energy and decrease the cost. The fiber is soft and fluffy like cotton and has good drape and air permeability is good, which improves the comfort of ordinary polyester fibers. It can be widely used for underwear, sportswear, casual wear, athletic wear, T-Shirts, shirts, pants, sports wear, suit, dress, sock lining etc. [5], and will be used in other fields, such as home furnishing, agriculture, health care and other areas, especially in military uniform. The natural conditions of military training, exercises and combat are hard, especially in hot areas or summer, and sweat and heat will not only affect one's health, but also may affect one's mood, therefore if they wear the clothing of moisture absorption and sweat discharge, the problems would be solved or improved [5-7].

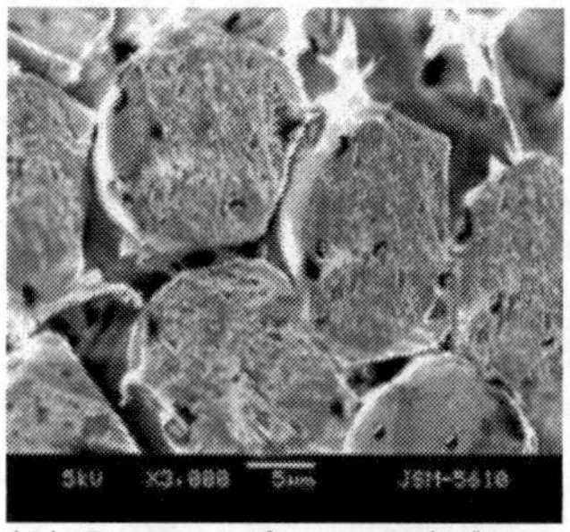

(A) Cross-section morphology

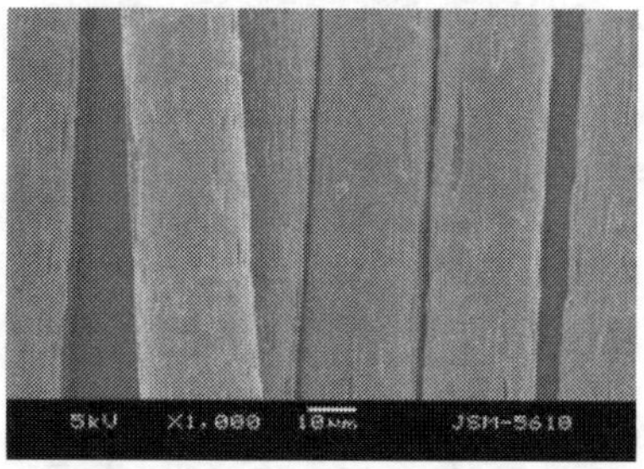

(B) Longitudinal surface morphology

Figure 1. SEM pictures of honeycomb polyester fiber.

2. Experimental materials and methods

2.1. Materials

1.33dtex honeycomb polyester fiber and 1.33dtex normal polyester fiber from Shangyu Hongqiang Color Polyester Co. Ltd.

2.2. Principles and methods

2.2.1. Principles. The speed of fiber absorbing moisture or releasing moisture is indicated by the variation of the moisture regain of unit mass fiber under the standard atmospheric conditions as shown in (1) [8, 9]:

$$v = \left| \frac{\mathrm{d}w}{\mathrm{d}t} \right| \tag{1}$$

Where:
w=the moisture regain of fiber absorbing moisture or releasing moisture at any time,
t =the time for fiber absorbing moisture or releasing moisture.
The principle of the moisture absorption rate is in the same as the moisture desorption's, but they are

opposite. Through the equation (1) for the integral operation is integral, the regression equation of moisture regain to time is:

$$w = a + be^{-ct} \quad (2)$$

Where: *a, b, c*= constant.

That is the curve of the moisture regain to time is an exponential curve during the process of the balance of fiber's moisture absorption and desorption.

By the equation (1) is known, the regression equation of fiber moisture absorption rate and desorption rate is:

$$v = \left| -bce^{-ct} \right| \quad (3)$$

2.2.2. Apparatus. MP-200A electronic balance with the accuracy of 0.001 g from Yuhua Instrument Co. ltd. , Y802A eight basket oven (Changzhou Textile Instruments Factory), glass dishes, YG715B constant temperature and humidity box (Ningbo Textile Instruments Factory).

2.2.3. Conditioning and Methods. Tests are according to the national standard "GB6503-2008: Testing method for moisture regain of man-made fibers".

Moisture absorption: the 1 g fiber is weighed after conditioning it for more than 24 h under standard atmosphere, 20±2℃ and 65±5% relative humidity, and then put in the oven to dry at the temperature of 105℃ for 1h and then weighed in the oven. Afterwards, the fiber is weighed once every 10 min. until it loses no more then 0.05% of its mass at 10 min. intervals to stop drying and the latter mass is as the fiber's dry weight. The fiber (the sample of constant mass) is carried out of the oven and weighed quickly in the conditioning room. The mass is the initial one (G_0). The fiber is put in the glass dish, and it is weighed once every 5min and the mass (G_t) is recorded until it increases less than 0.05% of its mass at 5 min intervals. The moisture regain at every weighted time for the fiber is calculated for drawing the moisture regain curve. The calculation equation of the fiber's moisture regain (Wt) is as follows:

$$W = (W_t - W_0) / W_0 \times 100\% \quad (4)$$

Where: G_0= dry mass, G_t=fiber mass at the time t.

Moisture liberation: The 1 g fiber is weighed at the standard atmosphere and put in the constant temperature and humidity box of 20℃ and 100% relative humidity for 96 h until the fiber researches moisture saturation. Then the fiber is removed out of the box and put quickly in the standard atmosphere. The measurement method is in the same as the moisture absorption's. Finally, the fiber which has balanced is dried and weighed. The moisture regains at the time t is calculated and the moisture liberation curve is drawn.

3. Results and discussion

3.1. Moister regains and liberation curves and discussion

3.1.1. Moister regains and liberation curves. The curves drawn according to the results of the moisture regain of the honeycomb polyester and the normal polyester fiber are as shown in figures 2 and 3.

It is known from figure 2 that the moisture regain of the honeycomb polyester fiber at the moment of the moisture absorption balance is much larger than the normal polyester fiber's, respectively 1.19% and 0.21%, the former is 5.67 times of the latter. However, the moisture absorption rates of both fibers are similar in the trends. The fibers' moisture absorption rates are relatively quick in the initial stage of their moisture absorption, but they slow down obviously with the increase of the moisture absorption time. The speed of the moisture absorption of the honeycomb polyester fiber has slowed at the time 25 min. to 40 min., but the normal polyester fiber's has slowed at the time 10 min. to 25 min., afterwards their moisture absorption were gradually tends to balance.

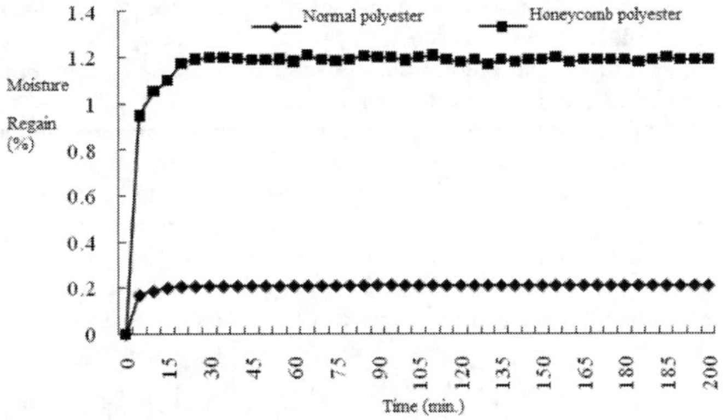

Figure 2. Moisture absorption curves.

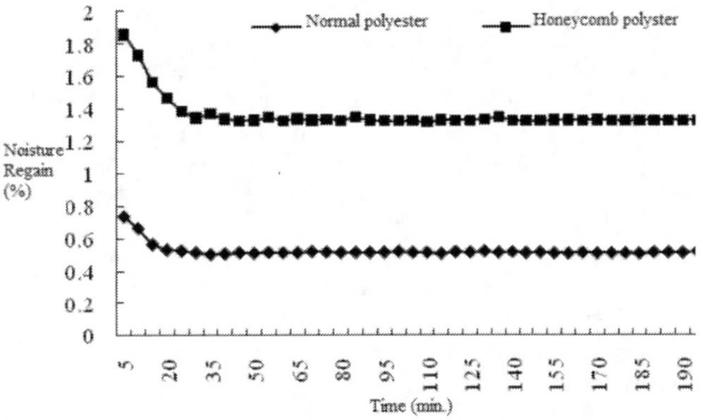

Figure 3. Moisture liberation curves.

The moisture regain of the honeycomb polyester fiber is much larger than the normal polyester fiber's after they have been processed in the constant temperature and humidity box as shown in figure 3. The fibers released moisture quickly in the initial stage, but they do slowly with the extension of time. The moisture regains of two fibers are respectively 1.33% and 0.51% after they have been at the moisture liberation balance. The moisture regains of the fibers at the moisture absorption balance are higher than the corresponding ones at the moisture liberation balance, respectively 11.7% and 142.88%, which are caused by the hygroscopic hysteresis of the fibers [10]. However, the moisture regains of honeycomb polyester fiber at the moisture absorption balance and the moisture liberation balance are respectively higher of 466.67 and 160.78% than those of the normal polyester fiber.

3.1.2. Analysis of moister absorption and liberation. (1) Moister absorption and liberation: The capacity of the moisture absorption and desorption of fibers depends on their chemical and physical structures. The gaseous water from the skin surface evaporating is first absorbed by fibers and then released in the air through the fibers' surfaces. The capillary effect caused by the microporous on the surfaces of honeycomb fibers and in them and the interspaces among them gap makes the liquid water from the skins be adsorbed by the fibers, diffuse and evaporate. These two kinds of interaction make the moisture and water migration. These properties are not only relative to the fiber surface charges and electrical property [11], but also relative to fiber's specific surface area and internal voids. The larger the surface area of a fiber is, the more the number of molecules on its surface is, so its surface energy is larger and its surface adsorption capability is stronger. Therefore the ability that the fiber surface adsorbs water molecules is stronger, and its moisture absorption is better [10].

The specific surface area of honeycomb polyester fiber with honeycomb microporous structure is 170 times that of cotton fiber, and its microporous are connective each other and with the atmosphere. So that its moisture absorption and desorption are better than theses of normal polyester fiber and it has capillary wicking effect and higher moisture regain.

(2) Moisture regains at the balances of moisture absorption and desorption: Because the fiber molecules and water molecules attract each other and bind when the fiber absorbs water, the kinetic energy of water molecules will reduce and convert into heat energy to be released so then the heat will be generated [10]. In the initial stage of moisture absorption, the steam pressure is high in the atmosphere, and water molecules enter into the fiber, which leads to the fiber moisture regain larger , at the same time, the temperature of the fiber increases because of the effect of moisture absorption and heat release. More molecules and higher temperature make the steam pressure of the fiber become larger gradually. When the steam pressure of the fiber is close to that of the air, the fiber is in the moment equilibrium of moisture absorption and stops absorbing moisture. However, the fiber will release moisture when the steam pressure of the fiber is larger than that of the air. When the heat generated by moisture-absorption quickly transfers to the air, the fiber will begin to absorb moisture until the steam pressure and temperature of the fiber are in the same as those in the atmospheric conditions when the fiber moisture absorption is in the state of equilibrium. Because the moisture and temperature in the fiber change its steam pressure, its moisture regain can fluctuate. When the fiber is dry, its steam pressure is low because there is little moisture in it, so that its moisture regain does not fluctuate in the initial stage of moisture absorption. This is because though initially a little moisture enters into the fiber to cause its temperature higher, but it is not enough to make the steam pressure of the fiber in the same as that in the air [12]. The principles of the fiber's moisture absorption and desorption are same, but their processes are opposite.

The balanced moisture regains of the moisture absorption and desorption of honeycomb polyester fiber fluctuate very slightly, but the normal polyester fiber's do not basically as shown in figures 2 and 3. Because the moisture regain of the former fiber is larger than that of the latter fiber, the steam pressure of the former fiber would have some change, but the steam pressure of the latter fiber would not have any change basically.

3.2. Establishment of regression equations of moisture regains and rates of moisture absorption and liberation of two fibers

The rates of the moisture absorption and liberation of the fiber during the processes are always in change. When the results were done curve fitting, the regression equations of the moisture regains of two fibers's moisture absorption and liberation to time have been obtained as shown in table 1. Also, the curves of their regression equations have been obtained as shown in figures 4 and 5.

The rate equations and curves of two fibers' moisture absorption and liberation have been obtained by using the origin75 software to analyze the data, such as shown in table 2 and figure 6. The rates of two fiber's moisture absorption and liberation with time decreased exponentially.

Table 1. Regression equations of fibers' Moisture absorption and liberation to time.

Sample	Moisture absorption	Moisture liberation
Honeycomb polyester fiber	w=0.01189-0.01183e$^{-0.28623t}$	w=0.01328+0.00882e$^{-0.0913t}$
Normal polyester fiber	w=0.0021-0.00209e$^{-0.30258t}$	w=0.00512+0.00454e$^{-0.13285t}$

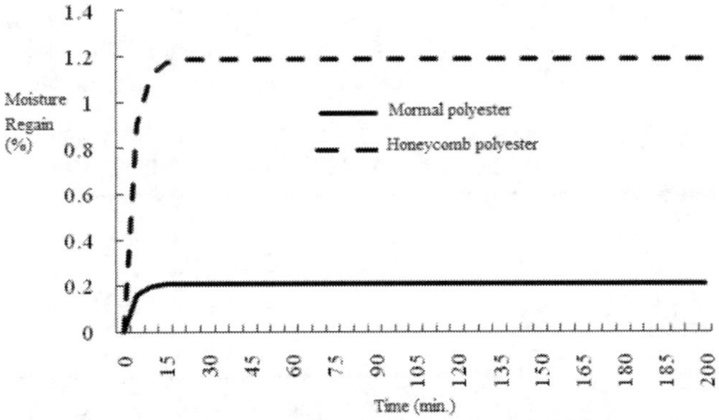

Figure 4. Moisture absorption regression curves of two fibers.

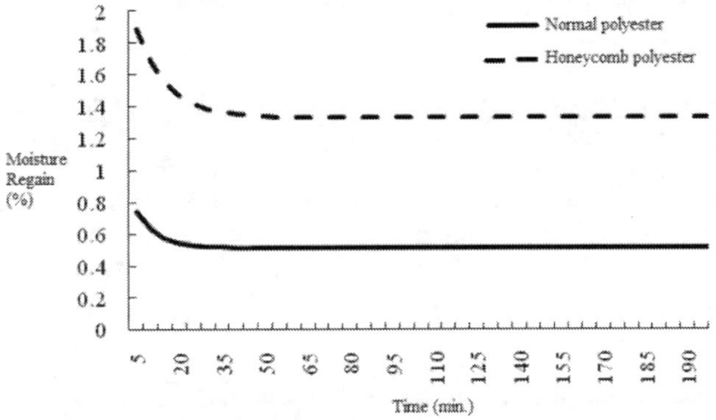

Figure 5. Moisture liberation regression curves of two fibers.

In the initial stage of moisture absorption, the moisture absorption rate of the honeycomb polyester fiber is faster, but the rate rapidly decreases with the time. The absorption rate from 10 min. is obviously slow and is gradually approaching the moisture absorption rate of the normal polyester fiber. The curves of the moisture absorption rates of two fibers after 20 min. begin to coincide and gradually trend to 0, which the moisture absorption of two fibers is in the state of balance. In the stage of moisture liberation, the moisture liberation rate of the honeycomb polyester fiber is faster than that of normal polyester fiber, but they have a similar trend. Two fibers' rates from 10min. are obviously slow, but the rate of the former is obviously faster than that of the letter. The curves of the moisture liberation rates of two fibers from 30 min. begin to coincide and gradually trend to 0, which the moisture liberation of two fibers is in the state of balance. Therefore, the honeycomb polyester fiber has excellent moisture absorption and desorption performance.

Table 2. Regression equations of rates of fibers' moisture absorption and liberation to time.

Sample	Moisture absorption	Moisture liberation
Honeycomb polyester fiber	w=0.00339e$^{-0.28623t}$	w=0.00081e$^{-0.0913t}$
Normal polyester fiber	w=0.00063e$^{-0.30258t}$	w=0.0006e$^{-0.13285t}$

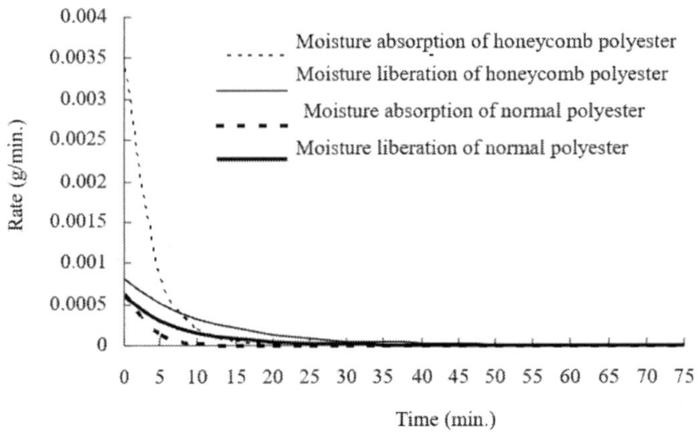

Figure 6. Rate curves of moisture absorption and liberation of two fibers.

4. Conclusions

The honeycomb polyester fiber has excellent moisture absorption and liberation. Its initial moisture absorption and liberation rates are much faster than those of the normal polyester fiber. The garments made of the honeycomb polyester fiber can absorb the sweat from human bodies and has good moisture wicking properties. There are a lot of microspore connected each other on the surface of honeycomb polyester fiber and in it, which increases its specific surface area. The moisture regain of the fiber is much higher than that of the normal polyester fiber, which is in favor of textile processing and increases the comfort of garments.

References

[1] Gu Chaoying 2005 Development and application of moisture absorption and liberation polyester fibers *China Textile Leader* **7** 57-62

[2] Tang Xin and Lin Jianhua 2007 Research on properties of moisture absorption & perspiration elimination fiber and fabric and some problems in dyeing & finishing *Melliand China* **1** 77-9

[3] Zhang Jin 2007 Moisture absorption & perspiration elimination fibers-Coolmax and Coolplus *Silk Textile Technology Overseas* **4** 38-40

[4] Hua Xinghong, Hou Cuiling and Wang Huaping, et al. 2005 The development of moisture absorption and sweat releasing synthetic fiber *Synthetic Fiber in China* **1** 44-6

[5] Qi Sumei, Xu Yinglian and Ye Qilin 2011 The effect of microstructure of honeycomb cationic modification polyester on the properties *Journal of Zhejiang Sci-Tech University* **28** 44-9

[6] Huang Yanli 2011 The performances of a new type of honeycomb micropore modified polyester fabric *Shanghai Textile Science & Techonology* **39** 52-3

[7] Liu Yulei and Meng Jiaguang 2009 Classification and application of moisture absorption and perspiration fabric *Progress in Textile Science & Technology* **5** 27-30

[8] Zhu Jinzhong, Su Yuheng, Mao Huixian, et al. 2010 Study on moisture absorption of some cellulose fibers *Journal of Tianjin Polytechnic University* **29** 29-32

[9] Li Ruqin and Wang Yu 1991 Measuremnet of moisture absorption rate of textile materials Journal of *Textile Research* **12** 35-7

[10] Yu Weidong and Chu Caiyuan 2009 *Textile Physcis* (Shanghai: Donghua University Press) 2nd Edi

[11] Ji Peizhen 1994 Moisture absorption and desorption of polyester and polyamide fibre modifed by sericin *Journal of Zhenjiang Institute of Silk Textiles* **11** 23-6

[12] Zhang Zhen, Zaho Hansheng and Chen Mingzhen 2010 Comparison of moisture absorption and liberation properties of bamboo fiber and regenerated bamboo fiber *Knitting Industries* **7** 15-7

Study on Pores and Crack Sensitivity of Ni- based Composite Coating by Laser Cladding

Tiangang Zhang[1, 2, 3] **and Ronglu Sun**[1, 2]

[1]School of Mechanical Engineering, Tianjin Polytechnic University, Tianjin 300387, China
[2]Tianjin Area Major Laboratory of Advanced Mechatronics Equipment Technology, Tianjin 300387, China

[3]E-mail: 113099506@qq.com

Abstract: This study aims to explore the types and mechanism of pores and crack on Ni-based composite coating of laser cladding. The results show that pores are generally found on the surface and bottom of cladding layer, and the gas comes from the cladding or the interaction between composites during the cladding process. There are two major categories of cracks, vertical crack and horizontal ones, which is mainly related to the low melting eutectic in the cladding area, and also relevant to coating composition, contents of inclusion and laser specific energy.

1. Introduction

Laser cladding technique is a new surface modification technology. Surface coating material and the substrate material melt quickly and solid to form metallurgical bond cladding layer, thereby improving the wear resistance, corrosion resistance and high temperature oxidation resistance, and has broad application prospects [1-4]. But presence of pores and cracks on the cladding layer affects the quality and performance of the layer. Therefore, further study of laser cladding should focus on the types and mechanism and influencing factors, so that the promotion of industrial technology has practical significance.

This paper analyzes the form and mechanism of pores and cracks on Ti alloy laser cladding Ni-based MoS2 of Ni package and investigates the effects of laser specific energy and the ratio of the coating material on the cladding layer crack sensitivity. We have determined the main factors affecting its formation, thus we can effectively control and eliminate the cladding defects, and ensure the quality of laser cladding process.

2. Materials and methods

TC4 alloy substrate materials was used in the study, and its chemical composition as shown in table 1, with the sample size of 60 mm x 30 mm x 10 mm. Samples were cut and grinded on the surface before cladding to each mayonnaise roughness of 0.2 um, its surface oxide layer was removed with metallurgical sandpaper, and then cleaned with acetone. Ni60 self-fluxing alloy powder and MoS2 of

[3] Address for correspondence: Tiangang Zhang, School of Mechanical Engineering, Tianjin Polytechnic University, Tianjin 300387, China. E-mail: 113099506@qq.com.

Ni package powder were used as cladding material, and their chemical composition and physical properties are shown in table 1. 0% to 80% of MoS2 of Ni package powder was added to Ni-based self-fluxing alloys, then fully grinded and mixed. Cladding material was pre-set on the surface of the substrate sample using bonding method, and the preset coating had a thickness of 1.0 mm.

YLS-4000 fiber laser was used to perform cladding, with laser beam spot diameter of 4 mm, the protective gas N_2 was 5 MPa. Laser power was 1.5 ~ 2.5 kW, and scanning speed was 6 ~ 12 mm/ s.

The samples were cut along the direction perpendicular to the scanning direction. The cross-section of cladding layer was made into a metallurgical samples, and then etched with a mixed solution of hydrofluoric acid, nitric acid and water with the composition ratio of 7 ml: 3 ml: 50 ml, etching time 10 ~ 30 s. OLYMPUSGX-51 optical microscope was used to observe pore morphology and distribution on cross-sectional area of the cladding layer, SZ660 stereo microscope was used to observe and measure the number of cracks, S-3400N scanning electron microscopy (SEM) and energy dispersive spectroscopy XFlash5030 (EDS) was used to analyze cladding cracks and tissues.

Table 1. Chemical composition and physical parameters of experi mental materials.

Experimental materials	Composition (quality fraction %)	Powder density /μm	Melting point/°C	Coefficient of thermal expansion/10^{-6}°C
TC4 alloy	6.01Al,3.84V,0.1C,0.3Fe, Allowance for Ti	-	1660	7.89
Ni60	16Cr,3.3B,4.5Si,0.9C,≤8.0 Fe,Allowance for Ni	140~250	1027	13.4
Ni-MoS2	14.6Mo,9.7S, Allowance for Ni	200	1185	10.7

3. Results and discussion

3.1. Pore types and mechanism

Pores are caused the gas that fails to escape during solidification and remains in the liquid metal [5]. The size, number and distribution of pores are important to characterize the quality of the macro-cladding layer. It not only affects the density of the cladding layer, but also affects the corrosion protection coating capabilities, in addition, there are parts of the cladding holes could easily lead to stress concentration, can easily lead to microcracks. Therefore, controlling the pore rate is one of the important measures to improve the density and mechanical properties of corrosion resistance.

Besides produced by laser cladding process, gas can also be produced by interaction of components. Crystal water contained in the cladding material itself to stomatal formation can not be ignored. Wherein the gas generated in the reaction bath, to go through bubble nucleation growth and escape three processes. In the molten pool, there is a bubble and a stable existence. After the gas around the yuan will spread to the bubble in the bubble to grow away when the bubble grows to a certain critical size in molten pool under the action of buoyancy. Gradually rise and escape away. But not all of the gas can escape completely, when the pool of solidification rate is greater than the escape velocity of the bubble, bubbles left in the formation of pores in the cladding layer.

Figure 1 is low times in different parts of the cladding layer morphology. As can be seen, the distribution of pores of different cladding, mostly located in the binding region. Surface porosity is close to the large pore surface of the cladding layer of a single distribution, which is due to the protection of N_2 gas flow rate is too large, severe turbulence is involved in the bath, the solubility of N_2 in the phase transition of a sudden drop in the surface layer remaining in the cladding layer. Such pore size is usually greater than 30 AIDS m, when the protective gas is involved in too much mayonnaise diameter of 1 to 5 m, and with micro-crack initiation. Its causes may be decomposed to form MoS_2 Ni package of free sulfur when more than 400℃, and oxygen combine to generate SO_2 tiny air mass, $2MoS_2 + 7O_2 \rightarrow 2MoO_3 + 4SO_2$, can not be discharged in the pool said the rapid solidification process may also be a cladding powder material moisture in the air, the powder in the water under the laser beam vaporization, residue in the cladding layer.

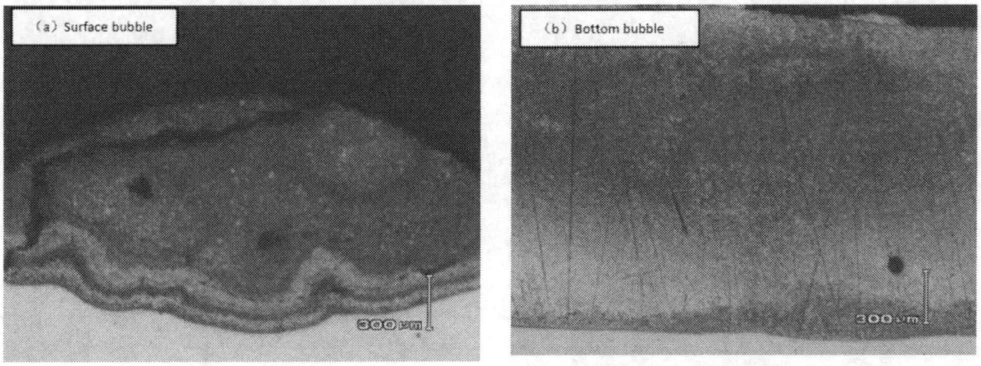

Figure 1. Morphology at different parts of the cladding layer.

3.2. Crack morphology and mechanism

Crack is a serious flaw in laser cladding process due to the combined effect of plastic reserve, stress, gas content and gas in the coating. It directly affects the quality of the cladding layer, shortenes product life. To improve the quality of the cladding layer, to reduce and eliminate the crack is the key to improve the laser cladding technology, so the cladding crack research has been important in laser cladding technique subject.

Depending on the shape, cracks can be divided into the inner longitudinal and transverse, shown in figure 2. Longitudinal cracks were seen on both sides of the cladding layer, starting from the top downwards, discontinuous distribution, and sometimes there is a small crack on both sides of microcracks cloud. Transverse cracks straight smooth, continuous distribution, mostly originated in the combined area of 10 to 60 AIDS bottom cladding layer and the substrate m away from the junction.

In order to understand the distribution of the elements in and around the cracks, energy spectrum analysis was performed and results are shown in figure 3. The data in table 2 show that Si, S C content at the crack were higher than normal value, indicating that segregation Si, S, C and other elements, resulting in increased cracking susceptibility. In the cladding layer crystallization process, the first solidified metal pure, secondary crystallization containing Si, C, S push slag and other low melting point to hard phase grain boundaries, constituting a discontinuous film, reduced hard bonding strength of the matrix phase. Then cracks are formed because the stresses generated in the cooling. The above analysis shows that cracks are closely relevant to internal cladding eutectic.

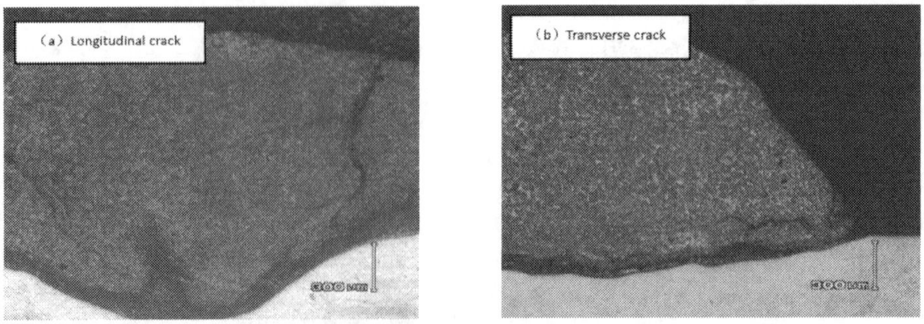

Figure 2. Crack shape of cladding layer.

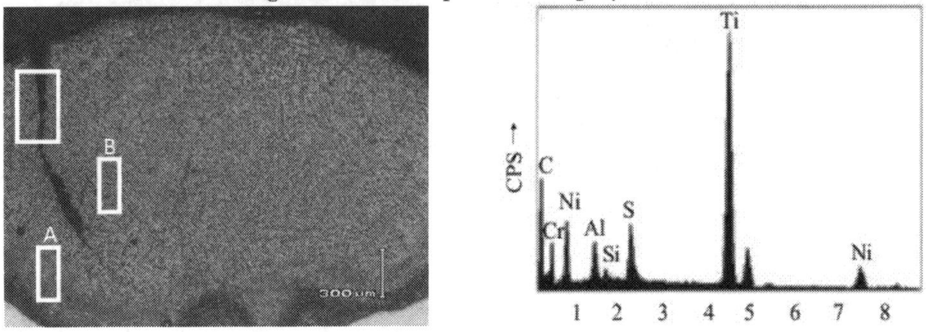

Figure 3. Crack morphology and its energy spectrum analysis.

Table 2. Comparison of element contents of crack elements and alloy (except oxygen and fluorine elements).

Element	Ti	Al	Ni	Cr	Si	C	S
Crack place	47.68	2.11	15.27	1.02	0.34	30.44	3.14
Alloy A	52.96	4.39	22.97	0.95	0.26	11.15	2.58
Alloy B	55.97	3.59	20.64	1.18	0.13	12.23	1.58

3.3. Cracking susceptibility factors

Cracks on the cross-section of the cladding layer were colored to examine the number of cracks in a certain length, so that cracking rate can be obtained. Factors affecting Cracking susceptibility was obtained through exploring the relationship between laser specific energy, coating composition and the crack rate, and results are shown in figure 4. The results show that with the increase of MoS2 of Ni package, crack rate showed an increasing trend after the initial fall, when MoS2 content increased to 20%, the rate tends to a minimum. With increase of the laser energy, which prolongs the existence bath time, the internal organization of the cladding layer has a relatively long time to nucleation and growth, forming a thick dendrites and granular tissue, increased sensitivity to crack, the crack rate increases

Analysis shows that factors contributing to reduced residual tensile stress will reduce cracking tendency, and factors contributing to increased r stress will raise cracking tendency. From table 1, the linear expansion coefficient of the cladding material is significantly higher than that of the linear expansion coefficient of the base material, while the temperature decrease is much larger than the base of the cladding layer, resulting a bigger contraction than the base of the cladding layer. When the

IOP Conf. Series: Materials Science and Engineering **87** (2015) 012096 doi:10.1088/1757-899X/87/1/012096

tensile stress is greater than the ultimate strength of the cladding material, the cracks will form in the cladding. Therefore, physical properties of the material itself cracking are fundamental to susceptibility of the cladding layer.

Factors contributing to reduce the inclusion within the cladding holes and other defects can suppress the crack formation and expansion. Results from previous studies [6] shows that the low melting point inclusions at grain boundaries coarse hard phase will form a liquid film, the liquid film to form pores in the shrinking process or rupture, becomes a source of cracks. Weld pool solidification process, because no time left in the bottom of the cladding layer inclusions floating particles and holes and other factors, led to the cladding surface crack initiation expansion. Therefore, the number of inclusions and porosity content directly affect the cladding cracking susceptibility. Those who help to increase the toughness of plastic cladding factors will reduce the tendency to crack. Kwok Wing-keung, etc. [7] studied the microstructure and properties of Ni-based composite coating by laser cladding, noting that strengthening mechanisms of laser cladding particulate reinforced metal matrix composite coating is mainly fine grain strengthening hard phase dispersion strengthening and discretion - Ni solid solution strengthening. When the hard phase when TiC precipitates in the petaloid large grain, greatly increases the brittleness of the material and with said diamond fine particulate dispersed in solid solution, not only significantly improves the strength of the alloy, and toughness of the alloy can be substantially maintained. Laser than can determine the status and distribution of the coating microstructure and, therefore, the laser process parameters have an impact on the cladding cracking susceptibility. From the above results and analysis, the formation of cracks in laser cladding layer is the result of many factors, the influence of laser cladding cracking susceptibility

The above results and analysis show that cracks on laser cladding layer is the interaction of many factors, which are coating composites, pores and laser specific energy. There are many ways to reduce crack sensitivity. One is to optimize powdered ingredients, control the number of reinforcing phase, and try to make them evenly distributed in the liquid and solid; the other is to lengthen cooling time to reduce stress and therefore reduce cracks through pre-heating and post-cladding processing [8] We can also use addictives with suitable additives linear expansion coefficient as the transition layer to reduce internal stress in the coating, so that cladding cracking susceptibility is minimized [9].

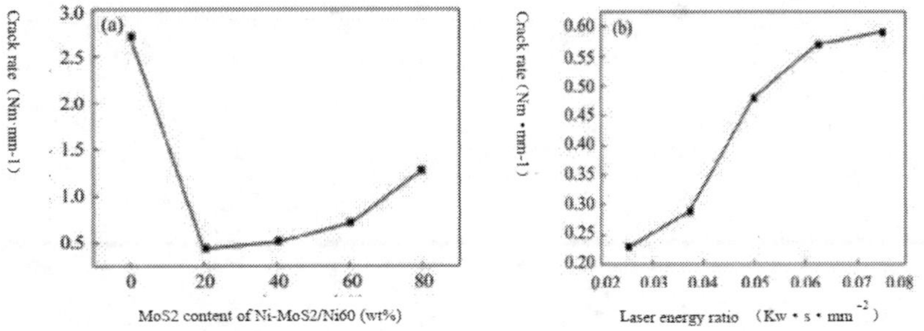

Figure 4. Relation between crack rate and composition of coating and the laser energy.

4. Conclusions

- Pores on laser cladding Ni-based composite coating are mainly on the surface and the bottom. Pores on the surface are produced by protective gas N_2 gas involved, but pores on the bottom are produced when cladding layer is oxidized o to SO_2 and vaporized crystal water fails to escape.
- Cracks can be divided into vertical and horizontal ones. Horizontal cracks are found on both sides, extending from the top to bottom, and the source is the conjunction of cladding layer and the substrate, extending parallel to the binding region. Crack is closely related to low melting eutectic cladding. Segregation of Si, S, C and other elements can reduce the combined strength of the coating, results in crack susceptibility, and produces cracks under cooling shrinkage stress.

- Coating composition, pores and laser specific energy can have a certain effect on the crack sensitivity ratio, wherein the coating composition has the greatest impact. There are several ways to reduce cladding crack susceptibility: one is to optimize allocation of powdered ingredients, and the other is to eliminate residual stress through pre-heating and post-cladding processing. In addition, gradient cladding can also be used.

References

[1] Paul C P, Alemohanmmad H, Toyserkani E, et al. 2007 Cladding ofWC-1 2 Co on low carbon steel using a pulsed Nd: YAG laser *Material Science and Engineering* **464** 170-6

[2] Lei Y W, Tang Y and Niu W 2010 A new theoretical model for highpower laser clad TiC/ NiCrBSiC composite coatings on Ti6Al4Valloys *Optics and Laser in Engineering* **48** 899-905

[3] Zhang Y K, Chen J F, Lei W N, et al. 2008 Effect of laser surface melting on friction and wear behavior of AM50 magnesiumalloy *Surface & Coatings Technology* **202** 3175-9

[4] Li Yongquan and Xiefa Qin 2013 Laser Cladding Cr/ Ni-base alloy composite coating and corrosion research *Thermal Processing* **42** 133-5

[5] Yu L, Nakata K and Liao J. 2009 Weld porosity in fibre laser weld of thixomolded heat resistant Mg alloys *Science and Technology* **14** 554-8

[6] Yu Ting, Deng Qilin, Zhang Wei, et al. 2012 NiCrBSi laser cladding alloy coating crack formation mechanism *Journal of Shanghai Jiaotong University* **7** 1043-8

[7] Kwok Wing-keung, Sunrong Lu, Wen Leiyi, et al. 2009 Laser cladding situ TiC particle reinforced Ni-based composite coating microstructure and properties *Materials and Heat Treatment* **3** 178-81

[8] Puja Kadolkar and Narendra B Dahotre 2002 Variation of structure with input energy during laser surface engineering of ceramic coatings on aluminum alloys *Applied Surface Science* **199** 222-33

[9] David Maohuai, Dong Yuan, Zhang Yuan and Liu Zefu 2008 Influence of laser cladding net additions for crack control *Journal of Tianjin University* **5** 553-7

IOP Conf. Series: Materials Science and Engineering **87** (2015) 012097 doi:10.1088/1757-899X/87/1/012097

Effect of substrate preheating temperature and coating thickness on residual stress in plasma sprayed hydroxyapatite coating

Dapei Tang[1]

School of Mechanics and Engineering, Southwest Jiaotong University, 610031, Chengdu, China

E-mail: dapeitang@163.com

Abstract. A thermal-mechanical coupling model was developed based on thermal-elastic-plastic theory according the special process of plasma spraying Hydroxyapatite (HA) coating upon Ti-6Al-4V substrate. On the one hand, the classical Fourier transient heat conduction equation was modified by introducing the effect item of deformation on temperature, on the other hand, the Johnson-Cook model, suitable for high temperature and high strain rate conditions, was used as constitutive equation after considering temperature softening effect, strain hardening effect and strain rate reinforcement effect. Based on the above coupling model, the residual stress field within the HA coating was simulated by using finite element method (FEM). Meanwhile, the substrate preheating temperature and coating thickness on the influence of residual stress components were calculated, respectively. The failure modes of coating were also preliminary analyzed. In addition, in order to verify the reliability of calculation, the material removal measurement technique was applied to determine the residual stress of HA coating near the interface. Some important conclusions are obtained.

1. Introduction

Hydroxyapatite (HA) is a biological active material with nearly the same chemical composition as natural bone, and can be used to guide bone regeneration [1]. One of the most important clinical applications of HA is as a coating on bone or tooth implants [2], especially plasma sprayed HA coating applied on Ti-6Al-4V substrate. These implants combine the mechanical advantages of substrate with the excellent biocompatibility and bioactivity of HA. However, It has been reported that failure often occurs at the HA/substrate interface during the surgery or after implantation [3]. One of the major causes of failure is the existence of high residual stresses within the implant. Therefore, the overall understanding of the residual stress is crucial in failure analysis of these composite materials. Moreover, modeling and simulation of the residual stress field will be also useful for guiding the design process of HA coating.

Several techniques are usually employed for the measurement of residual stress in HA coating. These include specimen curvature [4], Raman spectroscopy [5], X ray diffraction [6], nanoindentation [7], high-energy synchrotron [8], and material removal method [9]. Unfortunately, the residual stress

[1] Address for correspondence: Dapei Tang, School of Mechanics and Engineering, Southwest Jiaotong University, 610031, Chengdu, China. E-mail: dapeitang@163.com.

measured by these methods may vary over at least two orders of magnitude. Besides, some test results are tensile in nature, but others are compressive in nature. These contradictory results are in dispute. There are multiple reasons causing these test results inaccurate or conflicting. Firstly, the in-plane stress state or biaxial stress state is often assumed in these tests techniques [10], i.e. the stress component perpendicular to interface and the shear stress are ignored. The thickness of coating is thinner than that of substrate, but the thickness of the coating/substrate system is not thin. Therefore, the above simplified assumption seems to be inappropriate. Secondly, each of the aforesaid techniques has its own advantages and disadvantages. For example, The X-ray diffraction is limited by the shallow penetration of the X-ray beam, therefore only the top surface average residual stress can be obtained; the accuracy of Raman spectroscopy depends on the resolution of Raman frequency shift, since this shift is apparently small, uncertainty of residual stress result might be incurred. In fact, there is not any unique standard method for evaluation of residual stress, it is nearly impossible to compare one reported data with another.

In order to obtain the complete and accurate value and state of the residual stress, numerical simulation technology is worth to adopt. Up to now, there are few numerical researches undertaken on residual stress of HA coating [8, 11]. Previous investigations only considered the effect of temperature on deformation, not considered the influence of reverse, i.e. without studied the thermo-mechanical coupled effect. In addition, the influences of substrate preheating temperature and coating thickness on residual stress were also less simulated. This paper intends to continue these work.

2. Thermal-mechanical coupling model

It is well known that the classical Fourier heat conduction equation is derived under the assumption of "unit volume constant", which ignores the effect of strain field on the temperature field. However, in plasma spraying process, the change of temperature is sudden, and its change rate is great, volume change caused by the sudden deformation would affect the heat transfer. Therefore, Fourier heat conduction equation need to be modified in order to reflect the influence of deformation on temperature. The modified Fourier heat conduction equation according to the first law of thermodynamics is as follows:

$$\lambda_t T_{,ii} + Q = \rho c_v \dot{T} + (3\lambda + 2\mu)\alpha T_0 \dot{\varepsilon}_{ii} \tag{1}$$

Where T is temperature, ε is strain, Q is heat source density, λ_t is thermal conductivity coefficient, ρ is density, c_v is specific heat , α is thermal expansion coefficient, λ and μ are Lame constant. The last item of equation (1) shows the coupling between deformation and temperature.

Johnson-Cook model [12] is adopted to describe the material constitutive under the high temperature and high strain rate in the paper (seen in equation (2)), which considers temperature softening effect, strain hardening effect and strain rate reinforcement effect.

$$\sigma = \left[A + B\left(\varepsilon^{pl}\right)^n\right]\left[1 + C\ln\left(\frac{\dot{\varepsilon}^{pl}}{\dot{\varepsilon}_0}\right)\right]\left[1 - \left(\frac{T - T_r}{T_m - T_r}\right)^m\right] \tag{2}$$

Where σ is stress, ε^{pl} is plastic strain, T is temperature, T_m is melting point, T_r is room temperature. Five Johnson-Cook model constants by compression experiment are as follows: A=1000MPa, B=780 MPa, C=0.033, m=1.02 and n=0.47.

3. Finite element simulation

A simple cylindrical Ti-6Al-4V substrate with HA coating on top surface of the sample is selected as the calculating geometry model, shown in figure 1(a). The xoz plane is the interface between HA coating and Ti-6Al-4V substrate, the y axis is the axis of symmetry. Because the model is symmetrical, so only for half a symmetry plane is simulated by FEM, seen in figure 1(b), x and y directions are radial and axial direction, respectively. The radius of the sample is 5 mm; the substrate thickness t_s is 3 mm. The initial temperature of HA coating is considered to be its melting point temperature. The substrate material is preheated before HA coating sprayed upon it. Both the coating thickness t_c and the substrate preheating temperature T_s are chosen as variable parameters, in order to study their effects on residual stress.

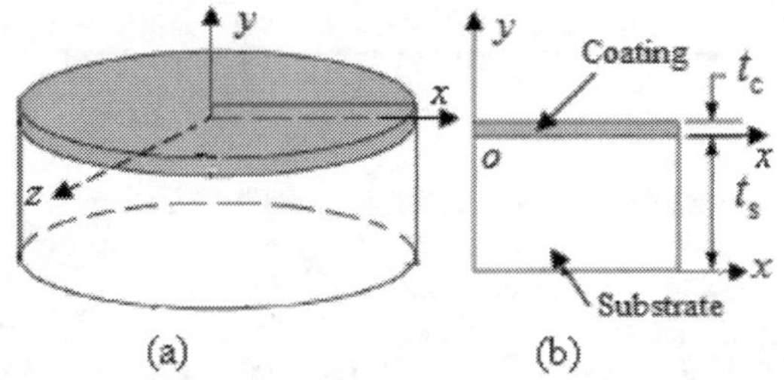

Figure 1. Geometry schematic used in FE simulation.

After the deposition process, heats are lost from every surface of the coating and substrate to the ambient air by convection. The heat transfer coefficient is 16 W/(m^2•K). Materials' properties used in simulation are summarized in table 1.

Table 1. Materials' properties.

Properties	HA	Ti-6Al-4V
Density(kg·m^{-3})	3100	4500
Thermal Conductivity(W·m^{-1}·K^{-1})	0.72	7.2
Specific Heat(J·(kg·K)$^{-1}$)	2500	560
Young's Modulus(GPa)	16	115
Poisson Ratio	0.23	0.32
CTE(×10^{-6}K^{-1})	11.5	8.9

4. Results and discussion

4.1. *Fundamental analysis of residual stress field*

In this section, the coating thickness t_c and initial substrate preheating temperature T_s are given the following parameter values respectively: t_c=100μm, T_s=600℃. Figures 2(a) and 2(b) show the typical contour of von Mises residual stress in the whole region of the sample and in the local region around the free edge of coating and substrate interface, respectively. For all specimens (including other thickness, other preheating temperature, etc), modeling result shows that, von Mises stress in the interface is larger than in other locations, and there is a significant stress concentration near the free

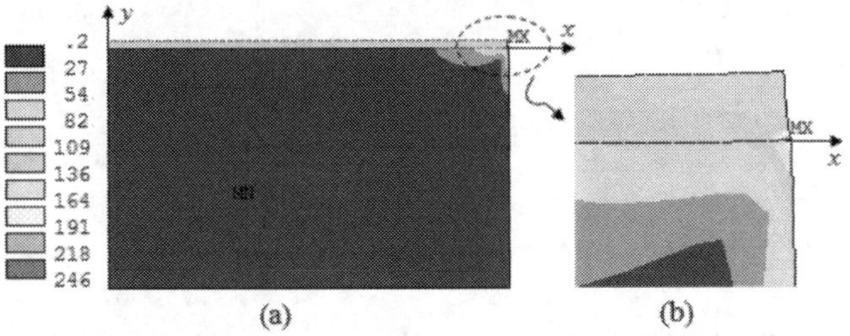

Figure 2. Typical contour of von Mises stress (MPa): (a) the whole region of the sample; (b) the local region near the edge of the coating and substrate interface.

edge of the interface. It has also been reported that failure often occurs at the HA/substrate interface rather than at HA/ bone interface from a direct shear/ bending loading [13]. Accordingly, it can well explain that why failure always occurs at the HA/substrate interface during the clinical use of implants [14].

Figures 3(a)–3(d) shows the distributions of radial residual stress, axial residual stress, shear residual stress, and von Mises residual stress of the coating and substrate at the interface respectively. From figure 3(a), it can be seen that, for the most part locations in far away from the interface edge, the radial residual stress of coating (denoted as σ_x(Coating)) is tensile stress, the radial residual stress of substrate (denoted as σ_x(Substrate)) is compressive stress, and they does not change significantly in these locations. However, in the local positions near the interface edge, there is an obvious stress concentration. Figure 3(b) shows that, for the most part locations in far away from the interface edge, the axial residual stress of coating is not large, and the axial residual stress of substrate is compressive stress, however, near the interface edge, the axial residual stresses of coating and substrate change from compressive stress to tensile stress, and there is a remarkable stress concentration. The tensile axial stress, acting in the direction normal to the interface, may likely tear the coating apart from the substrate. In places far from interface edge, shear residual stresses (seen in figure 3(c)) of coating and substrate are fairly small, but near the interface edge, both the absolute value of them increase abruptly, they are easy to cause coating delamination or dislocation. It is also shown from von Mises residual stresses (seen in figure 3(d)) that, a remarkable stress concentration exists at the edge of the interface, and it may cause the spallation or crack of the coating.

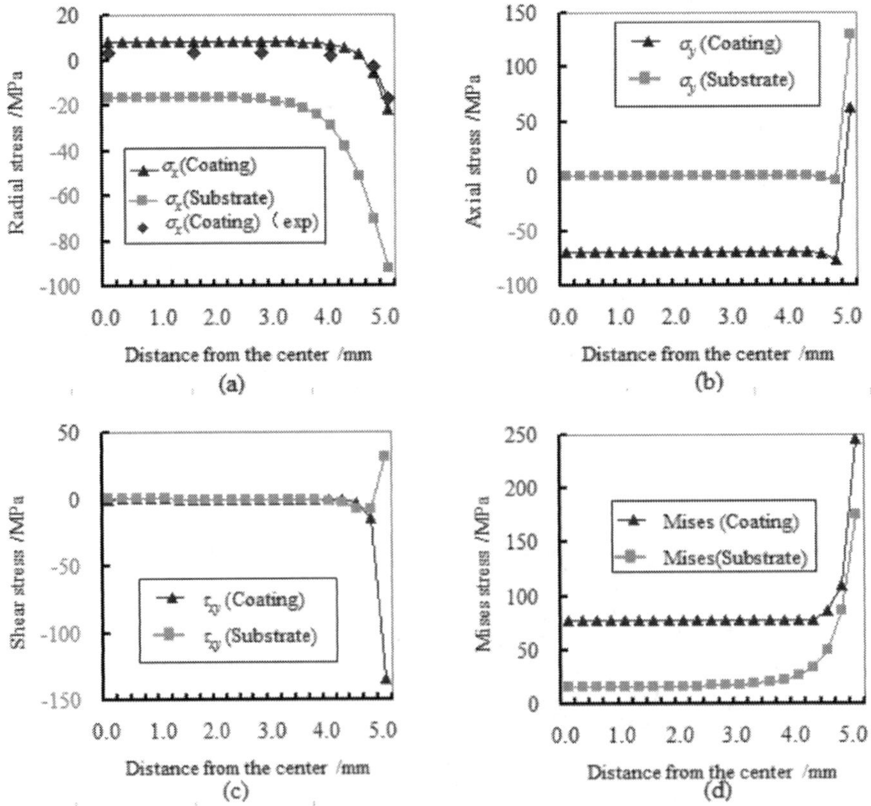

Figure 3. The distributions of residual stress components of coating and substrate at the interface respectively.

4.2. Effect of substrate preheating temperature

In this section, the coating thickness t_c is still 100μm, but the initial substrate preheating temperature T_s is used as a variable parameter between 25 to 900℃. The rest of parameters are the same as the section 3. Figure 4 shows the effect of substrate preheating temperature on the maximum of single residual stress component in stress concentration area of interface, i.e. the maximum of radial residual stress, the maximum of axial residual stress, the maximum of shear residual stress, and the maximum of Mises residual stress, denoted as σ_{x0}, σ_{y0}, τ_{xy0}, and $Mises_0$, respectively. It can be seen that, all absolute values of these stress components decrease with the rise of preheating temperature, especially the decreases of σ_{x0}, τ_{xy0}, and $Mises_0$ are most obvious. The reason is that, the temperature difference between coating and substrate is reduced if the substrate has been preheated to higher temperature before the HA coating sprayed upon it. Therefore, the substrate preheating has significant influence on residual stress. It should be noted that σ_{x0} would changes from compressive stress to tensile stress while preheating temperature over 700℃, which perhaps more likely to cause coating failure. Therefore, it is important to advice that the substrate is preheated to higher temperature, but not more than 700℃.

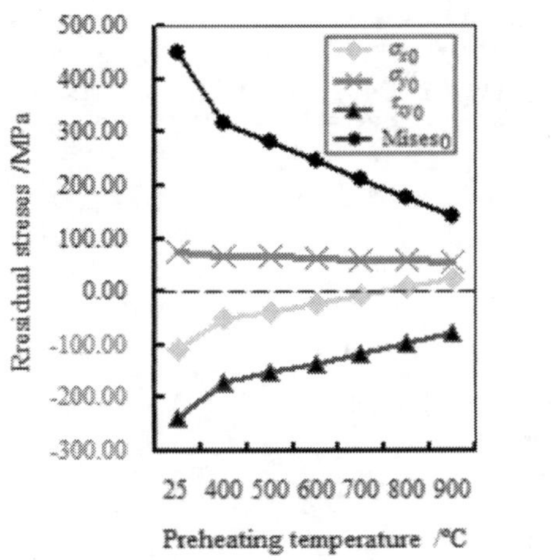

Figure 4. Effect of substrate preheating temperature on residual stresses of coating.

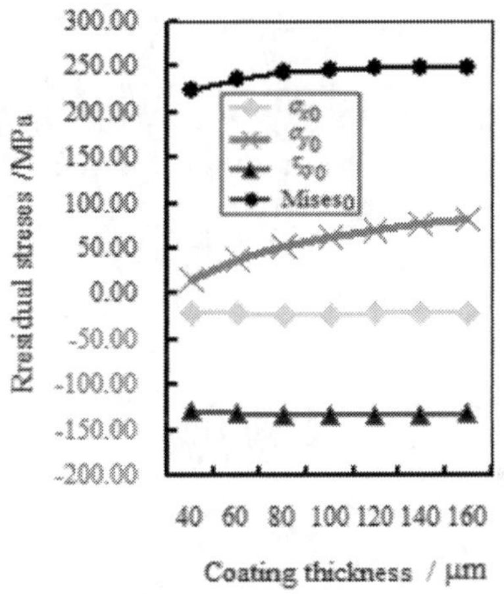

Figure 5. Effect of coating thickness on residual stresses of coating.

4.3. Effect of coating thickness

In this section, the initial substrate preheating temperature T_s=600℃, the coating thickness t_c is used as a variable parameter between 40 to 160μm. The rest of parameters are the same as the section 3. Figure 5 shows the effect of coating thickness on σ_{x0}, σ_{y0}, τ_{xy0}, and $Mises_0$. It can be seen that, in addition to the changes of σ_{x0} and τ_{xy0} are not obvious, both σ_{y0} and $Mises_0$ increase with the increasing of coating thickness. So, in order to decrease the level of residual stresses and control the stress concentration in critical regions, the coating should not be very thick.

4.4. Comparison of experimental and numerical results

Figure 3(a) shows the radial residual stresses of coating at the interface by calculation (denoted as σ_x (Coating)) and by experiment (denoted as σ_x (Coating)(exp)). The latter is measured by using the material removal method. From figure 3(a), it can be seen that, the absolute value of the data by experiment is smaller than by calculation, which may be due to a small part of the residual stress in coating has been released after removing the constraint of substrate. The distribution trend of them is the same, so the calculation result and experimental result is basically coincident.

5. Conclusion

There is a remarkable stress concentration near the interface edge. It is helpful to reduce the residual stress by increasing properly the substrate preheating temperature. The residual stress increases with the increasing of coating thickness. The calculated result by FEM is consistent with the experimental result.

References

[1] Kuo Y H, Sung C L and Chung S S 2013 *Surface and Coatings Technology* **231** 337-45
[2] Tanvir A H S and Jankowski A F 2011 *Thin Solid Films* **520** 1516-9
[3] Achariya R N, Yuichi O and Yoshiharu M 2014 *Journal of the Mechanical Behavior of Biomedical Materials* **36** 98-108
[4] Tsui Y C, Doyle C and Clyne T W 1998 *Biomaterials* **19** 2015-29
[5] Sergo V , Sbaizero O and Clarke D R 1997 *Biomaterials* **18** 477-82
[6] Yang Y C and Chang E 2001 *Biomaterials* **22** 1827-36
[7] Arjun D and Anoop K M 2014 *Ceramics International* **40** 1263-9
[8] Paul F, Bruno C and Pierre M 2005 *IEEE Transactions on Biomedical Engineering* **52** 1161-6
[9] Yang Y C 2011 *Surface & Coatings Technology* **205** 5165-71
[10] Yang Y C, Chang E, Hwang B H and Lee S Y 2000 *Biomaterials* **21** 1327-37
[11] Luigi B, Ilaria C and Franco M 2008 *Simulation Modelling Practice and Theory* **16** 971-87
[12] Johnson G R and Cook W H 1983 *Proceedings of 7th International Symposium on Ballistics* (Netherlands) pp 541-47
[13] Li H, Khor K A and Cheang P 2007 *Engineering Fracture Mechanic* **74** 1894-903
[14] Yang Y C and Yang C Y 2013 *Ceramics International* **39** 6509-16

IOP Conf. Series: Materials Science and Engineering **87** (2015) 012098 doi:10.1088/1757-899X/87/1/012098

Electrospinning synthesis of spinel $Li_4Ti_5O_{12}$ and its characterization

Xiao Li[1], Kai Yang[2, 3], Fei Gao[2], Chen Hu[2], Hao Liu[2], Dahe Li[2] and LinaWang[2]

[1]University of Science and Technology Beijing, 30 Xueyuan Road, Haidian District, Beijing, 100083, PR China

[2]China Electric Power Research Institute, Haidian District, Beijing, 100192, PR China

E-mail:yangkai@epri.sgcc.com.cn

Abstract. The electrode material $Li_4Ti_5O_{12}$ (LTO) with excellent electrochemical properties was synthesized by an electrospinning method using $Ti(OCH(CH_3)_2)_4$ and CH_3COOLi as raw materials. The effects of the solution viscosity and calcination conditions on the structure and electrochemical performance of LTO were investigated by scanning electron microscopy (SEM), X-ray diffractometry (XRD) and charge-discharge test. The optimal synthesis condition is 11 cP for the viscosity of the precursor solution, 800℃ for 12h with the heating rate of 1℃·min^{-1} for the calcination of precursor filaments. The product has a narrow diameter distribution and small amounts of impurities. The discharge capacity of LTO sample is 199.51 mAh·g^{-1}.

1. Introduction

Lithium-ion batteries (LIBs) are widely used in small mobile electronic devices for its high energy density, high power density and other characteristics. Currently, people focus on the superior LIBs which are suitable for electric vehicles and smart grid [1]. As a new anode material, spinel $Li_4Ti_5O_{12}$ (LTO) has a very flat charge-discharge plateau at ~1.55V (vs Li^+/Li) [2], which prevents the growth of lithium dendrites and thereby improves the safety of LIBs. Moreover, it has the "zero-strain" characteristic during Li^+ insertion/extraction process, thus makes it has a good stability in the long-term cyclic usage. Therefore, it becomes one of the most promising anode materials [3]. Different methods are used for synthesizing LTO, such as solid-state method, sol-gel method and hydrothermal method, etc. The hard agglomeration is a problem of easy occurring in LTO powders, which leads to the increase of preparation cost. As a new method for preparing submicro- or nano- materials, electrospinning is simple and efficient. Materials prepared by electrospinning don not easily form hard agglomerates due to electrostatic repulsion effect. In this paper, LTO submicron fibers were synthesized by an electrospinning method, and the effect of synthesis conditions on the structure of LTO and its electrochemical performance were investigated.

2. Experimental details

2.1. Synthesis of the spine LTO

[3] Address for correspondence: Kai Yang, China Electric Power Research Institute, Haidian District, Beijing, 100192, PR China. E-mail:yangkai@epri.sgcc.com.cn.

The precursor solution for electrospinning was prepared from lithium acetate, tetraisopropyl titanate, acetic acid and different doses of polyvinyl pyrrolidone (PVP) (1 g, 2 g, 3 g and 4 g). The viscosities of the resulting solution were 11cP, 38cP, 74cP and 117cP with different doses of PVP (Brookfield viscometer at a temperature of 21℃ using spindle No. 3 at 100 rpm). A certain amount of the solution was loaded into a plastic syringe and subjected to electrospinning. The flow rate was 1.5 ml•h-1 and the distance between the tip of the needle and collector was 10 cm, the collector drum speed was 800~1000 rpm, experiment voltages was 12 kV [4]. Then LTO samples were obtained under different calcination conditions. LTO samples with the PVP dosage of 1 g, 2 g, 3 g and 4 g were marked for LTO-11cP, LTO-38cP, LTO-74cP and LTO-117cP, respectively.

2.2. *Morphology and structure*

X-ray diffractometry (XRD) measurements were performed on a Rigaku UltimaIV instrument with Cu Ka radiation (10°<2θ< 90°). The surface morphologies of the fibers were studied by scanning electron microscopy (Hitachi X−650).

2.3. *Preparation of electrode films and the button cell assembly*

The electrodes were prepared by mixing 80 wt.% active materials with 10 wt.% carbon black (CB) and 10 wt.% polyvinylidene fluoride (PVDF) dissolved in N-methyl-2-pyrrolidone (NMP) onto a Cu foil. The slurry was coated on a copper current collector and dried for 12 h at 80℃ in order to remove the NMP solvent. The resulting electrode foil was rollpressed and punched to a circular disc. The electrode films were preserved in argon-filled dry box (Unilab, MB 200B) in which oxygen and moisture contents were maintained less than 1 ppm. The coin cells were fabricated with LTO, metallic lithium anode, 1 M $LiPF_6$ in 1:1 ethylene carbonate (EC) and diethylene carbonate (DEC) electrolyte, and Celgard polypropylene separator.

2.4. *Electrochemical measurements*

The cell was cycled between 1.0 and 2.5 V [5] versus Li/Li^+ electrode at 24℃. The charge–discharge test was performed by using a multichannel cell cycling system (Maccor S4000).

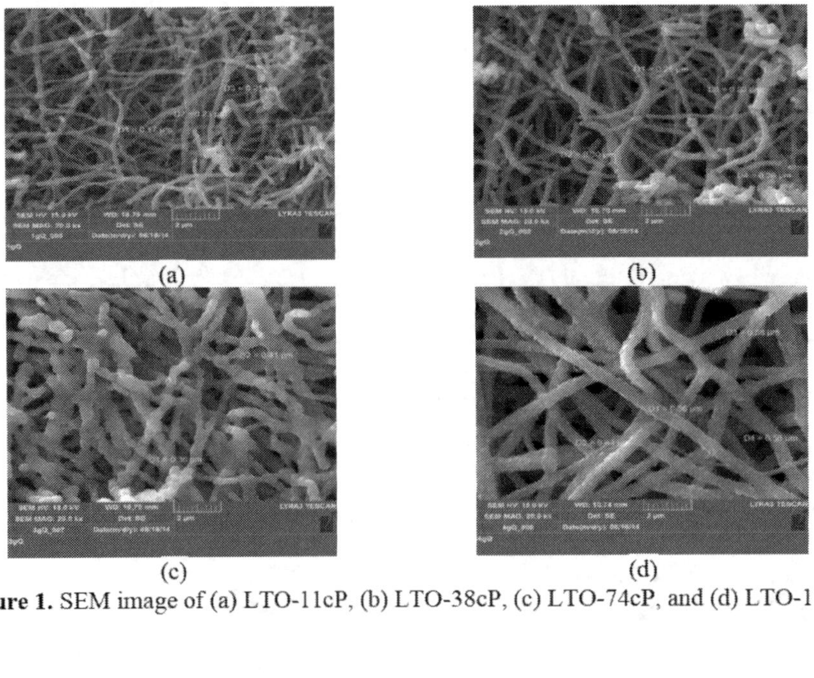

(a) (b) (c) (d)

Figure 1. SEM image of (a) LTO-11cP, (b) LTO-38cP, (c) LTO-74cP, and (d) LTO-117cP.

3. Results and discussion

3.1. Morphology and structure

Figure 1 shows the SEM images of LTO submicron fibers obtained after the calcination of electrospun precursor at 800℃ for 12 h in air. As shown in figure 1, the fiber diameters of LTO-11cP, 38cP, 74cP and 117cP were about 0.17~0.23 μm, 0.23~0.30 μm, 0.36~0.41 μm and 0.44~0.58 μm, which indicating that the viscosity of LTO precursor solution dramatically affects the morphology of LTO submicron fiber. When the viscosity of the solution increases, it tends to resist the bending-instability occurring in the solution injection process, thereby reducing the spinning paths. It means the tensile force of the solution decreases which causes the formation of thicker fibers.

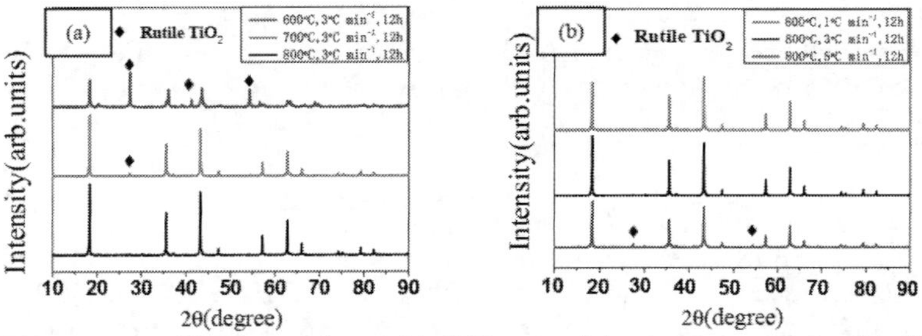

Figure 2. XRD patterns of LTO samples at (a) different calcination temperature, (b) different heating rates.

Figure 2 shows the XRD patterns of samples which were obtained under different calcination conditions. The sharp diffraction peaks of LTO samples are consistent with the spinel LTO (JCPDS PDF 49-0207), which means it is a single-phase cubic material with an Fd3m space group. From figure 2, it can be seen that there are some diffraction peaks that corresponds to the existence of rutile TiO_2 (JCPDS PDF 26-1198) for samples at the calcination temperature of 600, 700℃ (figure 2(a)) as well as the heating rates of 3, 5℃·min^{-1} at 800℃ (figure 2(b)). In addition, the peak intensities of LTO sample at 1℃·min^{-1} are obviously stronger than those of LTO samples at 3, 5℃·min^{-1} in the figure 2(b). In summary, the optimum calcination conditions are 1℃·min^{-1} heating rate, 800℃, holding time 12 h.

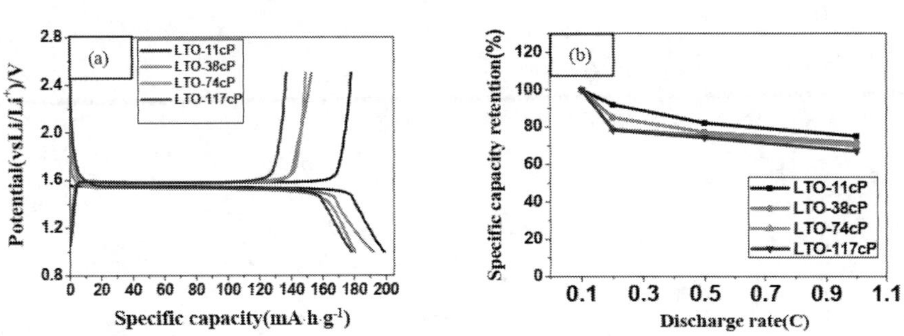

Figure 3. The charge-discharge curves of (a) LTO-11cP, (b) LTO-38cP, (c) LTO-74cP, and (d) LTO-117cP at 0.1, 0.2, 0.5, and 1.0 C rate.

3.2. Electrochemical test

Figure 3(a) shows the charge-discharge curves of LTO samples at different rates. The LTO samples have the c.a. 1.55V voltage platform and the discharge capacities of LTO-11cP, LTO-38cP and LTO-74cP samples at 0.1C rate are 199.51, 192.79, 182.51, and 178.51 mAh·g^{-1}, respectively. Figure 3(b) reveals the obvious influence of LTO fiber diameter on the rate capability of LTO samples. The rate capability of LTO-11cP (c.a. 0.17~0.23 μm) is comparatively better than other LTO samples which can be seen in figure 3 (b), its specific capacity retention reaches 75% at 1.0 C discharge rate.

4. Conclusion

In the paper, LTO precursor fiber was prepared by an electrospinning technology and then LTO submicron fiber was synthesized by solid state sintered method. The proper viscosity of LTO precursor solution was 11 cP and the optimum conditions are heating rate of 1℃·min^{-1}, calcination temperature of 800℃, calcination time of 12 h. The discharge capacity of LTO-11Cp is 199.51 mAh·g^{-1} at 0.1C rate and it possesses good rate capability.

Acknowledgment

The work was financially by the science and technology project of State Grid Corporation of China (DG71-13-037).

References

[1] Zhao L, Hu Y S, Li H, Wang Z X, Chen L Q 2011 Porous $Li_4Ti_5O_{12}$ coated with N-doped carbon from ionic liquids for li-ion batteries *Advanced Materials* **23** 1385-8.

[2] Li J, Jin Y L, Zhang X G, Yang H 2007 Microwave solid-state synthesis of spinel $Li_4Ti_5O_{12}$ nanocrystallites as anode material for lithium-ion batteries *Solid State Ionics* **178** 1590-4

[3] Sandhya C P, Bibin John and Gouri C 2013 Synthesis and electrochemical characterisation of electrospun lithium titanate ultrafine fibres *Journal of Materials Science* **48** 5827-32

[4] Hao Y J, Lai Q Y, Lu J Z, Wang H L, Chen Y D and Ji X Y 2006 Synthesis and characterization of spinel $Li_4Ti_5O_{12}$ anode material by oxalic acid-assisted sol-gel method *Journal of Power Sources* **158** 1358-64

[5] Xu R, Li J R, Tan A, Tang Z L and Zhang Z T 2011 Novel lithium titanate hydrate nanotubes with outstanding rate capabilities and long cycle life *Journal of Power Sources* **196** 2283-8

[6] Wang Y S, Wang X Y, An H F, Wang G B, Pei F, Zheng L P, Liu X M, Chen Q Q, Liu L and Jiang L L 2010 Preparation and electrochemical properties of spinel $Li_4Ti_5O_{12}$ by hydrothermal method *The Chinese Journal of Nonferrous Metals* **12** 2366-71

Physical properties of inorganic PMW-PNN-PZT ceramics

Sang-Hoon Sin[1], Ju-hyun Yoo[1, 3], Yong-Jin Kim[1], Sam-ki Baek[1], Jun-Soo Ha[1], Chung-Han No[1], Hyun-Seon Song[1] and Dong-Chan Shin[2]

[1]Semyung University, Sinwol-dong Jecheon Chungbuk 390-711, Republic of Korea
[2]Dong-IL Electronics Chungju 380-921, Republic of Korea

E-mail: Juhyun57@semyung.ac.kr

Abstract. In this work, inorganic $Pb(Mg_{1/2}W_{1/2})_{0.03}(Ni_{1/3}Nb_{2/3})_x(Zr_{0.5}Ti_{0.5})_{0.97-x}O_3$ (x= 0.02~ 0.12) composition ceramics were fabricated by the conventional solid state reaction method. And then their microstructure and ferroelectric properties were investigated according to the amount of PNN substitution. Small amounts of Li_2CO_3 and $CaCO_3$ were used in order to decrease the sintering temperature of the ceramics. The 0.10 mol PNN-substituted PMW-PNN-PZT ceramics sintered at 920℃ showed the excellent physical properties of piezoelectric constant (d_{33}), electromechanical coupling factor (k_p), mechanical quality coefficient (Q_m), and dielectric constant of 566 pC/N, 0.61, 73, and 2183, respectively.

1. Introduction

Recently, PZT-based ceramics have been widely used for applications such as transformer, and multilayer actuator [1-4]. Because PZT-based ceramics including $Pb(Ni_{1/3}Nb_{2/3})O_3$-$Pb(Zr,Ti)O_3$ and $Pb(Mg_{1/3}Nb_{2/3})O_3$-$Pb(Zr,Ti)O_3$ show an excellent piezoelectric properties at MPB(Morphotropic Phase Boundary), these three-component perovskite compounds have been actively studied [5]. However, a main problem encountered in manufacturing these ceramics is the volatilization of PbO during sintering process. High temperature sintering of PZT-based ceramics brings about serious problems such as environmental pollution and compositional fluctuations due to the volatilization of the lead oxide. Therefore, it is necessary to improve the physical properties of the ceramics using low temperature sintering. Many researchers have been tried to lower the sintering temperature of PZT-based ceramics without degrading the properties of the materials. In order to lower sintering temperature of the ceramics, sintering aids as additives are widely used [6-8]. Sintering aids used in this experiment were Li_2CO_3 (melting point=732℃) [9] and $CaCO_3$ (melting point=825℃) [10]. In this study, Pb $(Mg_{1/2}W_{1/2})_{0.03}(Ni_{1/3}Nb_{2/3})_x(Zr_{0.5}Ti_{0.5})_{0.97-x}O_3$ were prepared as a base composition . And also, in order to develop piezoelectric ceramics with excellent piezoelectric and dielectric properties, the amount of PNN substitution was changed.

[3] Address for correspondence: Ju-hyun Yoo, Semyung University, Sinwol-dong Jecheon Chungbuk 390-711, Republic of Korea. E-mail: Juhyun57@semyung.ac.kr.

2. Experimental procedure

The specimens were prepared by a conventional mixed oxide process. The compositions used in this study were as follows.

0.9925 $Pb(Mg_{1/2}W_{1/2})_{0.03}(Ni_{1/3}Nb_{2/3})_x(Zr_{0.5}Ti_{0.5})_{0.97-x}O_3$+0.0075$BiFeO_3$ (x= 0.02, 0.04, 0.06, 0.08, 0.10, 0.12)

The raw materials of PbO, MgO, WO_3, NiO, Nb_2O_5, ZrO_2 and TiO_2 were weighed by mole ratio and mixed together by ZrO_2 balls using the acetone as the medium. After drying, they were calcined at 850℃ for 2h. The calcined powder and $BiFeO_3$ was milled again with the additives, Li_2CO_3 (0.2 wt%) and $CaCO_3$ (0.25 wt%) for 24 h. The powders were pressed into disk of diameter of 14Φ and 2mm in thickness under pressure of 1,000 kg/cm^2. Burned out at 600℃ for 3h, and then sintered at 920℃ for 1 h 30 min. For measuring the piezoelectric properties, after sintering the piezoelectric ceramics were lapped into 1 mm thickness and then metallized with Ag paste at 600℃ for 15 min. Specimens were poled under a dc field of 3.5 kv/mm in silicon oil at 120℃ for 30 min. All electrical measurements were performed about 24h after poling. The crystal structure and microstructure of specimens were analyzed through scanning electron microscopy (SEM) and X-ray diffraction (XRD), respectively. For measuring the dielectric properties, capacitance was measured at 1 kHz using an LCR meter (ANDO AG 4304) and dielectric constant was calculated. For piezoelectric properties, the resonant and anti-resonant frequencies were measured by an Impedance analyzer (Agilent 4294) in conformity to IEEE standard and then the electromechanical coupling factor and mechanical quality factor were calculated. Piezoelectric constants were obtained using d_{33} meter (APC YE2730A).

3. Results and discussion

Figure 1 shows density of specimens as a function of PNN substitution. With increasing PNN substitution, the density was gradually increased. The maximum value of 7.85 g/cm^3 at x=0.12 was appeared. It is considered that the densification of density can be improved by sintering aids.

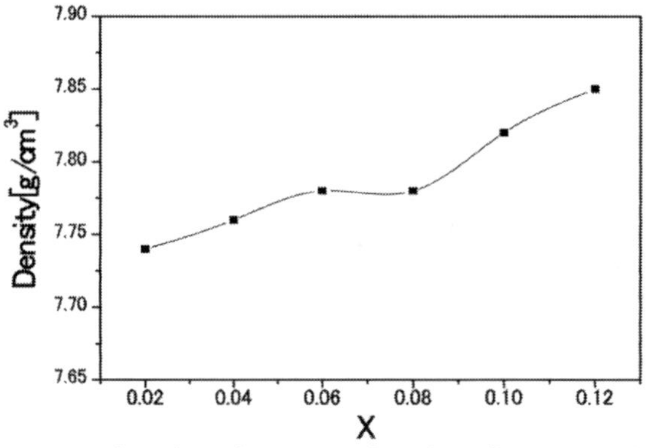

Figure 1. Density of specimens as a function of PNN substitution.

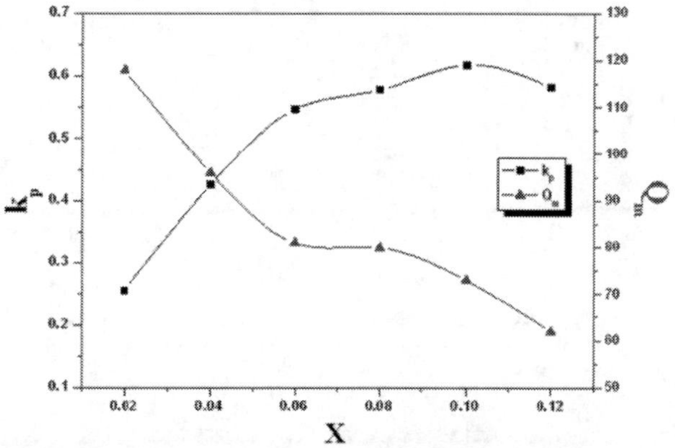

Figure 2. k_p and Q_m with the amount of PNN substitution.

Figure 2 shows electromechanical coupling factor (k_p) and mechanical quality factor (Q_m) with the amount of PNN substitution. k_p increases with the increase of PNN substitution, and reaches the maximum value of 0.618 at x=0.10. Subsequently, k_p was decreased at x=0.12. This phenomenon is due to the excess substitution of PNN. The variation of Q_m showed the opposite properties to piezoelectric coefficient (d_{33}), electromechanical coupling factor (k_p) and dielectric constant (ε_r). When the PNN substitution is x=0.10, the Q_m was reached 73 and then decreased with the increase of PNN substitution. Generally, PZT -based ceramics show the decreased Q_m value at MPB region.

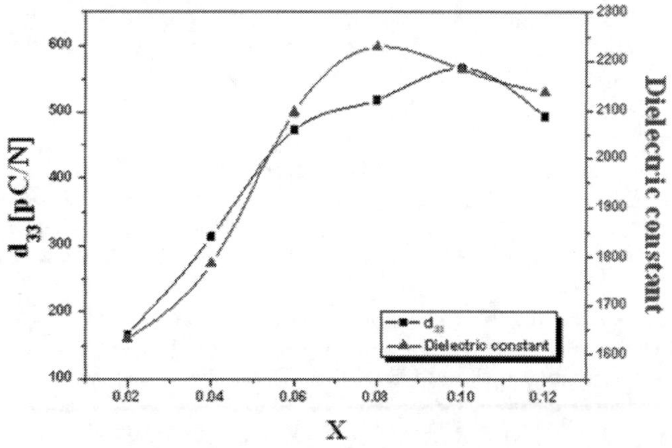

Figure 3. Piezoelectric coefficient (d_{33}) and dielectric constant (ε_r) of specimens as a function of PNN substitution.

Figure 3 shows piezoelectric coefficient (d_{33}) and dielectric constant (ε_r) of specimens as a function of PNN substitution, measured at room temperature. As results in figure 2, piezoelectric coefficient (d_{33}) increases with the increase of PNN substitution. The variation of d_{33} showed the maximum value of 566pC/N at x=0.10 PNN substitution. In addition, these results are because of appearance of MPB at x=0.10 and the feasibility of domain wall movement can be increased owing to increased polarization efficiency at MPB. Dielectric constant (ε_r) of specimens was increased with the increase of PNN substitution, reached the maximum value of 2,229 at x=0.08, and then decreased with the further increase of PNN substitution. This result can be explained from the improvement of sinterability and increase of density.

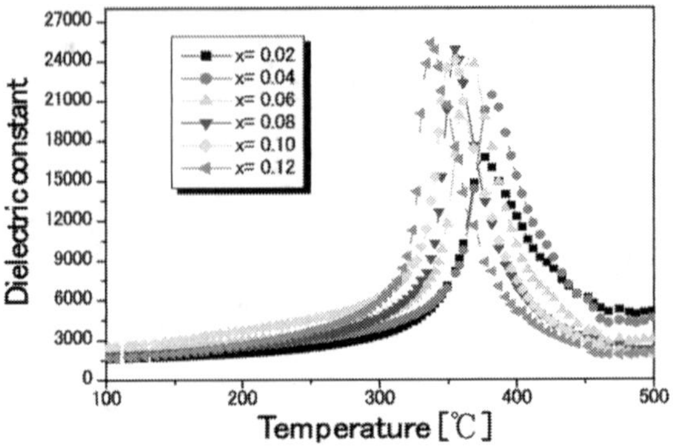

Figure 4. Temperature dependence of dielectric constant as a function of PNN substitution.

Figure 4 shows the temperature dependence of dielectric constant as a function of PNN substitution. Curie temperature (T_c) was decreased with the increase of PNN substitution. It shifted from 377℃ to 337℃. In addition, when the PNN substitution amount was x= 0.10, Curie temperature (T_c) was 355℃.

4. Conclusions

In this study, in order to develop the ceramics with excellent piezoelectric properties, $0.9925[Pb(Mg_{1/2}W_{1/2})+(Ni_{1/3}Nb_{2/3})_x(Zr_{0.5}Ti_{0.5})_{0.97-x}]O_3 + 0.0075\ BiFeO_3$ ceramics were investigated as a function of PNN substitution. The results obtained from the experiment are as follows:

- When x= 0.10, the maximum values of k_p= 0.61, ε_r= 2183, and d_{33}= 566 pC/N were shown.
- As amount of PNN substitution increased, Curie temperature (Tc) shifted to low temperature range. When the x = 0.10, Curie temperature(Tc) was 355℃.

References

[1] Jeong Y H, Yoo K G and Lee J H 2009 *J Electroceram* **23** 387-91
[2] Chao X, Yang Z, Dong M and Zhang Y 2011 *J Magn Magn Mater* **323** 2012-6
[3] Shin T H, Ha J Y, Song H C, Yoon S J, Park H H and Choi J W 2013 *Ceram Int* **39** 1327-33
[4] Ha J Y, Choi J W, Kang C Y, Choi D J, Kim H J and Yoon S J 2005 *Mater Chem Phys* **90** 396-400
[5] Yoon K H, Kim K S and Choi B S 2001 *Journal of the Korean Institute of Electrical and Electronic Material Engineers* **14** 20-5
[6] Yoo J H, Kim K J, Jeong Y H and Lee S H 2007 *Journal of the Korean Institute of Electrical and Electronic Material Engineers* **20** 766-70
[7] Lee M W, Kim S J, Yoon M S, Ryu S L and Kweon S Y 2008 *Journal of the Korean Institute of Electrical and Electronic Material Engineers* **21** 1077-82
[8] Lee K S, Lee L H, Yoo J H and Ryu S L 2007 *Journal of the Korean Institute of Electrical and Electronic Material Engineers* **20** 1034-8
[9] Hu T, Jantunen H, Deleniv A, Leppavuori S and Gevorgian S 2004 *J. Am. Ceram. Soc* **87** 578-83
[10] Licht S, Wu H J, Hettige C, Wang B, Asercion J and Stuart J 2012 *Electronic Supplementary Information (Chem. Comm.)* **1** 1-20

2015 Global Conference on Polymer and Composite Materials (PCM 2015) IOP Publishing
IOP Conf. Series: Materials Science and Engineering **87** (2015) 012100 doi:10.1088/1757-899X/87/1/012100

Impact comparative study of phone carcasses behavior by FEM

Cărăuşu Constantin[1], **Simona Plăvănescu and Nedelcu Dumitru**

Technical University "Gheorghe Asachi" of Iasi, Romania, Blvd. Mangeron, No. 59A, 700050 Iasi, Romania

E-mail: c_carausu@yahoo.com

Abstract. A constant concern of scientific research is based on plastics replace with biodegradable materials that reduce the adverse impact of waste on the environment. A biodegradable material that arouses interest lately is Arboform which is made of lignin, a component of wood and woody plants. Replacing plastic with Arboform in carrying components of products requires technical and economic studies on the implications of such replacement. Numerical simulation methods are a fast and economical way of analyzing the behavior of a product in various mechanical, thermal, electromagnetic and so on. The paper presents comparative results of numerical simulation using the software package SolidWorks impact behavior through the "Drop Test" of half shells made of High Density Polyethylene (HDPE) and of the Arboform LV3 Nature. Simulation watched the half-carcass behavior in three cases of accidental impact, "head", "corner" and the "back side". We analyzed the size and location of the maximum voltage and maximum deformation resulting from impact. Simulations have shown for all three cases a maximum voltage increase when using Arboform to use PEDH 93% for impact "forward" and "corner" and only 48.77% "back side" impact. If the maximum displacement, it increasing from carcasses of Arboform 4% for impact "head" and 6% for impact "corner", but fell by 2.7% for the "back side" impact. The significant increase of stress can be attributed to the higher density of Arboform to PEDH, which led to different weights of the two half-carcasses.

1. Introduction

The use of plastics is increasing and in a continuous diversification. If in 2007 the production of plastics was 260 million tons, in 2011 it had reached 280 million tons and was estimated an increase of 4% annually until 2016 [1]. New manufacturing technologies, such as 3D printing, have been implemented.

Many uses of plastics refer to the consumer goods industry characterized by a high volume production and reduced life cycle. Many of the plastics used are made from petroleum products. A direct result of this trait is the generation of large amounts of waste which represent a complex socio-economic and political problem with direct implications on the environment [2].

One solution to reduce this problem is the use of biodegradable materials [3]. It results from the use of biomass in the composition of some material.

[1] Address for correspondence: Cărăuşu Constantin, Technical University "Gheorghe Asachi" of Iasi, Romania, Blvd. Mangeron, No. 59A, 700050 Iasi, Romania. E-mail: c_carausu@yahoo.com.

IOP Conf. Series: Materials Science and Engineering **87** (2015) 012100 doi:10.1088/1757-899X/87/1/012100

"Liquid wood" made from lignin - a component of biomass represented by wood and other woody plants - is an alternative replacement of petroleum products manufactured from plastic. Arboform, trade name of the new material, presents a number of properties that makes it suitable to manufacture a wide range of products [4] such as covers for mobile phones.

To check the behavior in exploitation, mobile phone carcasses are subject to a number of mechanical stresses (accidentally) including the impact with a flat surface. To verify this test, carcass behavior is subject to the so-called "drop test" which, consists in observing the consequences to a certain speed impact between the carcass and a plan.

Drop test can be simulated using finite element software packages.

In the following is presented the simulation results using "Drop Test" from SolidWorks software [5], the test is done on a mobile phone half-carcass made of High Density Polyethylene (HDPE) and, respectively, Arboform LV3 Nature.

2. Methodology

The 3D model of half-carcass made in SolidWorks is presented in figure 1.

The properties of the two materials, PHDE and Arboform are shown in Table 1. Data for PHDE were taken from the database of the SolidWorks package, and those relating to Arboform are taken from [4, 5]. It is noted that the value of longitudinal and transverse elastic modulus of Arboform is double that of HDPE. Arboform's density is bigger with 29% than that of the HDPE.

Figure 2 presents the discretized model of the half-carcass. Was used a uniform mesh with tetrahedral elements, being involved 63.195 nodes and 73.195 elements. The goal of the simulations was to determine the stress and strain on the half-carcasses impacted with the plan in three situations: a. the impact of the half-carcasses "head"; b. the impact with a corner; c. at the back side. Figure 2 shows the meshed model of the half-carcass. The impact speed was set at 20 m/s. Weight of half-carcass made from HDPE was calculated at 6.9 g and for Arboform 9.8 g.

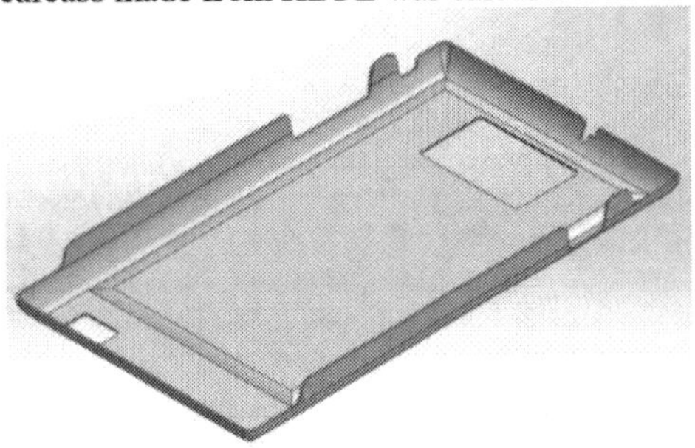

Figure 1. Solidworks half-carcass model.

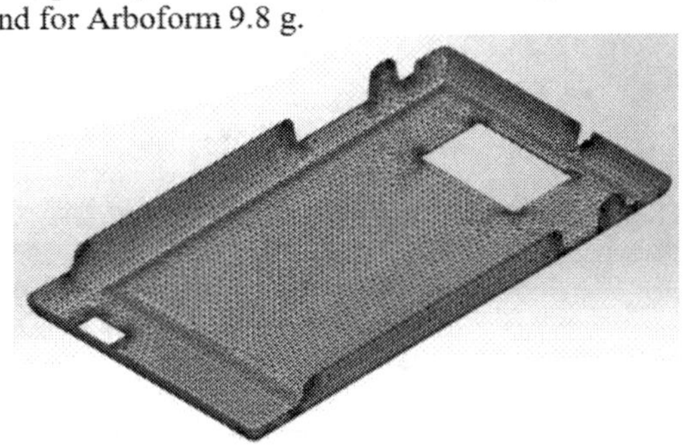

Figure 2. Discretized model of the half-carcass.

Table 1. Fizico-mechanical properties of Polyethylene High Density (PEHD) and Arboform.

Material	Elastic Modulus [N/m^2]	Poisson Ratio	Shear Modulus [N/m^2]	Tensile Strength [N/m^2]	Density [Kg/m^3]
PEHD	$1.07.10^9$	0.4101	$377.2.10^6$	$22.1.10^6$	952
Arboform	$2.22.10^9$	0.38	$804.3.10^6$	$19.2.10^6$	1350

3. Results and discussion

3.1. "Head" impact of the half-carcass

3.1.1. For HDPE

In figure 3 is presented the equivalent stress state over von Mises criterion for HDPE half-carcass to "Drop test".

The maximum stress of 119.5 MPa was located as shown in figure 3(b) on the half-carcass inside part.

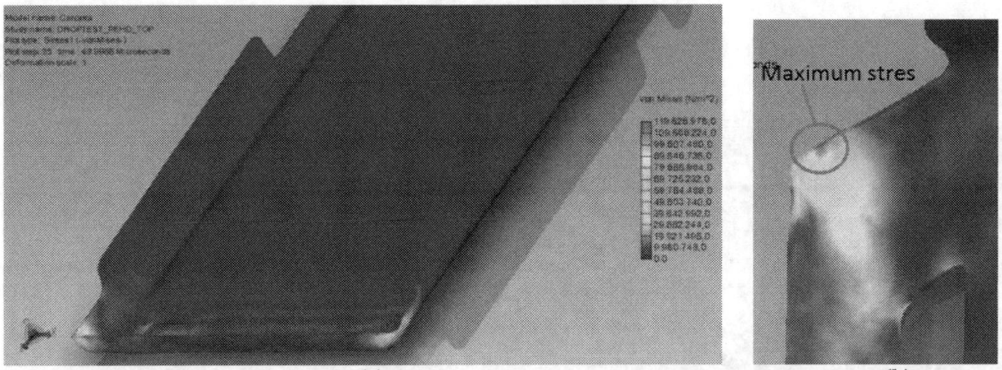

(a) (b)

Figure 3. (a) Stress state for "Drop test" "in head" of HDPE half-carcass. (b) maximum stress detail.

Figure 4. Deflection state for "Drop test" "in head" of HDPE half-carcass.

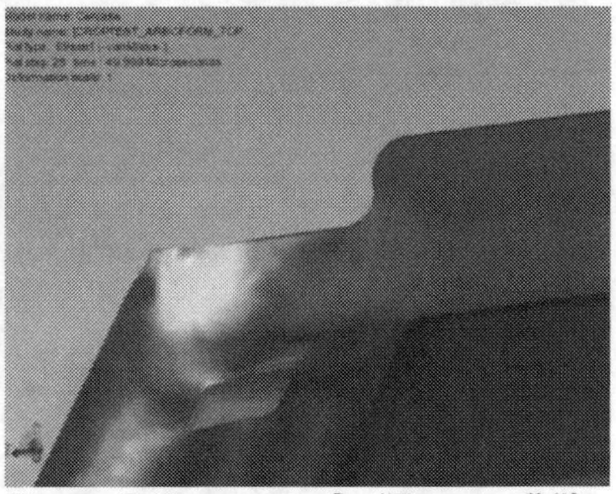

Figure 5. Stress state for "Drop test" "in head" of Arboform half-carcass.

The state of deformation is shown in figure 4. The maximum impact deflection at "head" for HDPE half-carcass is 1.23 mm.

3.1.2. For Arboform

For "head" impact of the Arboform half-carcass, locations of maximum stresses and maximum strains are similar with those of the HDPE half-carcass. The values of these parameters are: maximum stress 231 MPa and maximum deformation 1.28 mm. It is noted that the stress distribution in the impact area is different from that of HDPE half-carcass (figure 5) by the fact that the impact edge stresses are reduced.

In case of "head" impact test for the Arboform half-carcass, maximum stress increased by 93% from that of HDPE, while the maximum displacement increased only by 4%.

3.2. "Corner" impact of the half-carcass

3.2.1. For HDPE

In Figure 6 is represented the equivalent stress state over von Mises criterion for HDPE half-carcass to "Drop test".

The maximum stress of 148.3 Mpa was located as shown in figure 6(b) on the half-carcass inside part. Maximum stress is with 24.1% bigger than the "head" impact.

The state of deformation is shown in figure 7. The maximum impact deflection in "corner" for HDPE half-carcass is 1.42 mm. Maximum stress is with 1% bigger than the "head" impact.

3.2.2. For Arboform

As in the case of a, For "corner" impact of the Arboform half-carcass, locations of maximum stresses (figure 8) and maximum strains (figure 9) are similar with those of the HDPE half-carcass. The values of these parameters are: maximum stress 286 MPa and maximum deformation 1.5 mm.

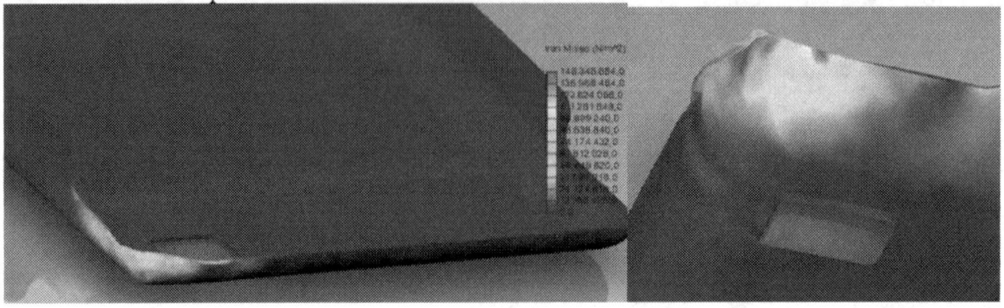

(a) (b)

Figure 6. (a). Stress state for "Drop test" "in corner" of HDPE half-carcass. (b). maximum stress detail.

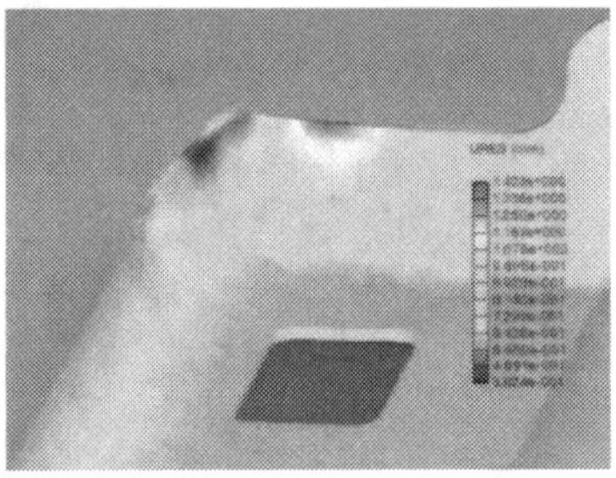

Figure 7. Deflection state for "Drop test" "in corner" of HDPE half-carcass.

Figure 8. Stress state for "Drop test" "in corner" of Arboform half-carcass.

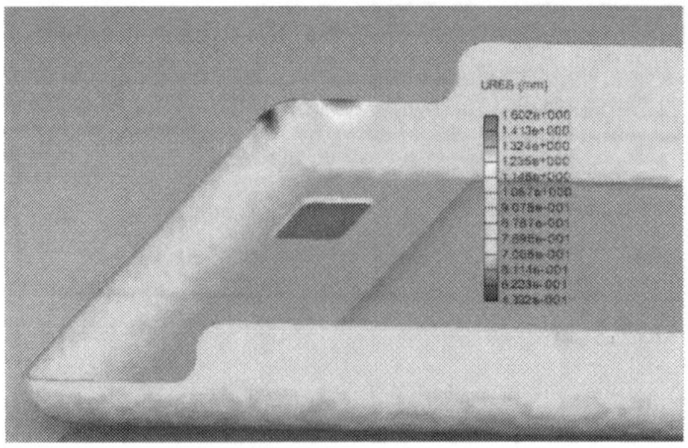

Figure 9. Deflection state for "Drop test" "in corner" of Arboform half-carcass.

Comparing the behavior of the two materials resouts that in case of "head" impact for Arboform half-carcase maximum stress increased by 93% from that of HDPE (similar increase in case a), while the maximum displacement increased by only 6% (higher than the growth of case a).

3.3. "Back side" impact of the half-carcass

3.3.1. For HDPE

In figure 10 is presented the equivalent stress state over von Mises criterion for HDPE half-carcass to "Drop test".

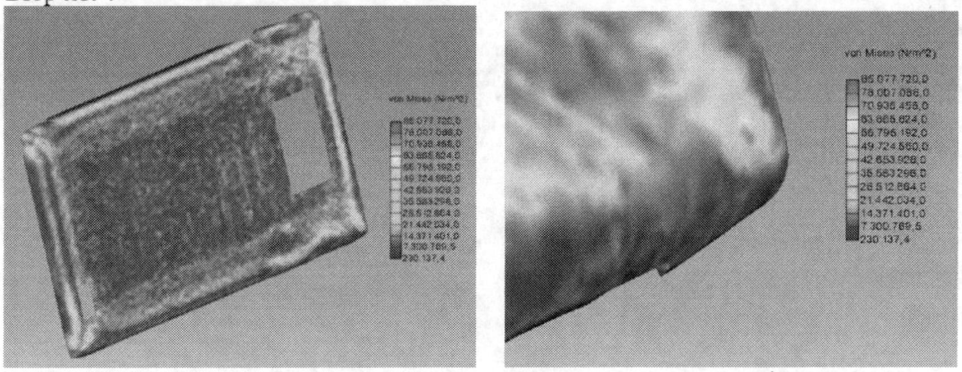

a. b.

Figure 10. (a) Stress state for "Drop test" on the " back side " of HDPE half-carcass. (b) maximum stress detail.

As indicated the figure 10(b), maximum stress of 85.1 MPa was located in the corners of the half-carcass with higher cropping. On the ribs from the back side of half-carcass high stresses are located. Maximum stress is with 28.8% smaller than the "head" impact.

The state of deformation is shown in figure 11. The maximum impact deflections are located in the edges of the half-carcass recesses. The biggest release presents the maximum deformation of 1.5 mm. Increasing the maximum deformation in relation to that of the impact of "head" is with 21.95%.

3.3.2. For Arboform

For the Arboform half-carcass case the state of deformation is shown in figure 12. The maximum stress (126.2 Mpa) is located in the half-carcass corner area (figure 12(b)). Compared to the HDPE case, maximum stress for Arboform is higher by 48.3%.

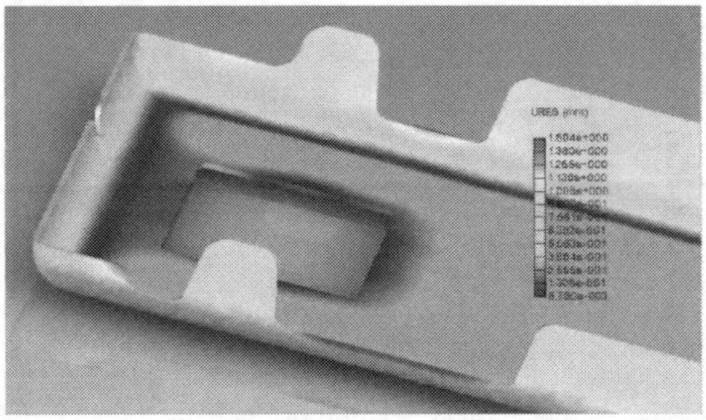

Figure 11. Deflection state for "Drop test" on the "back side" of HDPE half-carcass.

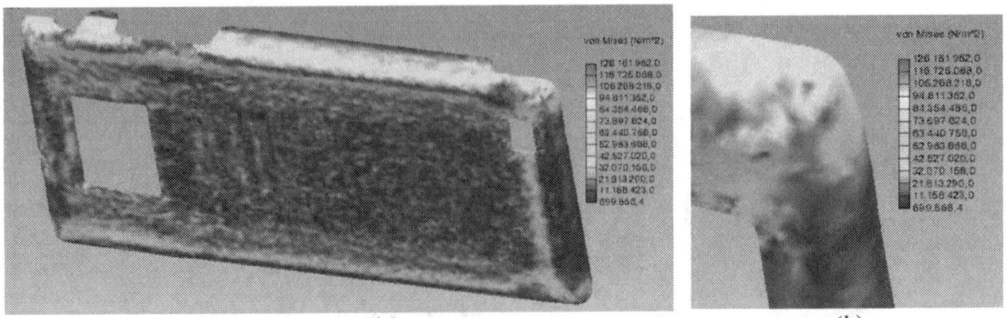

(a) (b)

Figure 12. (a) Stress state for "Drop test" on the "back side" of Arboform half-carcass. (b) maximum stress detail.

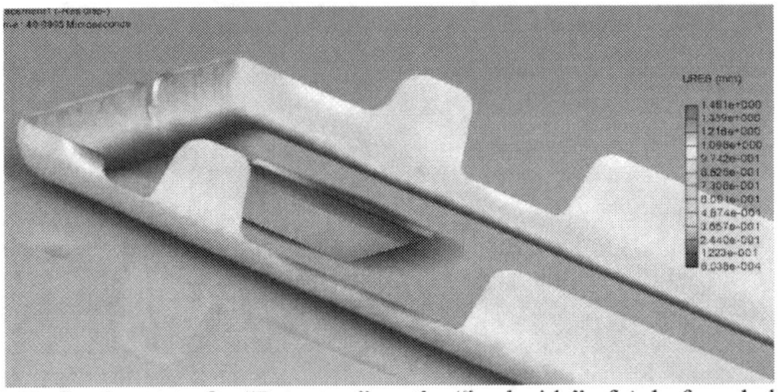

Figure 13. Deflection state for "Drop test" on the "back side" of Arboform half-carcass.

Deflection state for "back side" impact of Arboform half-carcass is gived in figure 13. The maximum displacement is 1.46 mm, which is lower by 2.6% compared to the HDPE case.

4. Conclusions

To reduce the impact of plastic waste (based on petroleum products) on the environment, there is a tendency to replace them with biodegradable materials. Concerns in this area are manifested mainly on consumer products with a relatively low life cycle because, they generate a huge amount of waste. Among these are mobile phones parts. Most of these phones have carcasses made of plastics.

Arboform is a biodegradable material made from lignin, a component of wood and woody plants. The properties posed by this material recommend it as a potential substitute for plastics in the mobile phones carcasses manufacture. For this purpose are imposed research regarding the behavior of various Arboform tests faced by mobile phones.

A accidental test at which can be subjected a mobile phone is the impact with a plan. To verify this, the so-called Drop test is made. It can be performed under real conditions or numerical simulation by using specialized finite element packages.

In the study was performed a numerical simulation of the half-carcass mobile phone impact with a plan using SolidWorks software package. It was considered a half-carcass made from HDPE and, respectively Arboform. The "drop test" was performed in three circumstances depending on the side that hits the plan: "head", "corner" and the "back side" of the half-carcass.

Compared for the two materials used were monitored/ followed maximum values of stress and deformations and also their localization on the half-carcass.

Simulations have shown for all three cases a maximum stress increase when Arboform was used, towards to use of HDPE when the "head" impact, respectively the "corner" impact is with 93% bigger and only 48.77% on the "back side" impact.

In case of maximum displacement, this one increase with 4% for Arboform half-carcass for "head" impact and 6% for "corner" impact, but decrease with 2.7% for the "back side" impact of the half-carcass.

The significant increase of stress can be attributed to the higher density of Arboform toward to HDPE, which led to different weights of the two half-carcasses. In case of assembled phones, the difference in weight between the two solutions, HDPE *vs* Arboform, is reduced and is expected an impact behavior better of the Arboform half-carcass.

Simulation models closer to the real part is a solution but that leads to important consumer of computer resources and time. Another solution that would save time and computer resources is the location on specific area on the body half-carcass of masses (weights) to replace the phone's electronic components/elements.

References

[1] OECD Environment Directorate *Material case study* http://www.oecd.org/environment/waste/46156 805.pdf
[2] Srikanth Pilla 2011 Engineering applications of bioplastics and biocomposites-An overview *Handbook of Bioplastics and Biocomposites Engineering Applications* (Willey)
[3] Ying Jian Chen 2014 Bioplastics and their role in achieving global sustainability *Journal of Chemical and Pharmaceutical Research* **6** 226-31
[4] Nedelcu D, Ciofu C, Lohan N M, 2013 Microindentation and differential scanning calorimetry of "liquid wood" *Composites Part B: Engineering* **55** 11-5
[5] http://www.tecnaro.de/english/daten.htm?section=arboform
[6] Drop test analysis, http://help.solidworks.com/2012/English/SolidWorks/cworks/Drop_Test_Studies.htm?id=d4e82ba0ff8e4881bcc4fa447e4eea15#Pg0

Adsorption of the water molecule on monolayer graphene surface has effect on its optical properties

Y F Peng[1, 2, 3], J Wang[1, 2, 3], Z S Lu[1, 2, 3] and X Y Han[1, 2, 3, 4]

[1]Henan Normal University College of Physics and Electronic Engineering, Xinxiang, Henan 453007, China
[2]Engineering Laboratory of Optoelectronic Technology and Advanced Manufacturing of Henan Province, Henan 453007, China
[3]Key Laboratory of Electromagnetic Wave Detection of Henan Province, Henan 453007, China

E-mail: yufengp@htu.cn

Abstract. The adsorption of water molecules on the surface of a monolayer graphene can be studied with the Materials Studio software and be applied density function theory from first principles. By studying the interaction of graphene with water molecule, it uses DFT (density function theory) with the PBE-GGA (the generalized gradient approximation of Perdew-Burke-Ernzerhof) and Periodic plane model, on the one hand working out the adsorption energy, and on the other hand getting related optical properties. It is shown that a single water molecule on graphene has very small adsorption energy, mainly owning to the van der Waals interactions. Graphene has high hydrophobic; adsorbed water molecule has little effect on the electronic structure of the graphene. The optical properties of the graphene have changed after the adsorption.

1. Introduction

Graphene is a single layer of graphite as the basic unit. Unique two-dimensional structure gives it attractive physical properties, such as excellent optical properties and special band structure [1]. In recent years, graphene becomes popular with the large number of researchers. For adsorption, graphene is an excellent material due to its several useful properties [2]. Because of its essentially two-dimensional structure, it almost completely exposed to the environment, leading to being significantly influenced by surrounding environment. There are numbers of literature in the study of alkali metal atoms [3], inert gas, and hydrocarbon [4], etc. adsorb on graphene. Whether in daily life or scientific researches, water plays a crucial role in the surface. The adsorption of water on the surfaces of metals has been studied in lubrication, corrosion, electrochemical and catalytic in the industrial production due to its important role. All kinds of researches are in progress and are going deep into smaller scale gradually; finally focus on the atomic scale [5].

In the experiment, a major problem exists in the calculation of a single water molecule adsorbs on the surface: Water molecules are so easy to form clusters that limit the study of adsorption.

[4] Address for correspondence: X Y Han, Henan Normal University College of Physics and Electronic Engineering, Xinxiang, Henan 453007, China. E-mail: yufengp@htu.cn.

IOP Conf. Series: Materials Science and Engineering **87** (2015) 012101 doi:10.1088/1757-899X/87/1/012101

Theoretically, distribution and hydrogen bonding of water molecules may cause difficulties in the calculation of water molecules interact with the material surface. To calculate the electronic structure and various properties, density functional theory is used in usual.

In recent years, the studies of water molecules adsorption on graphene surface are increasing, examples are as follows: Jie Ma studied the adsorption and diffusion of water on graphene from first principles [6], P. Cabrera San Felix, researches the structure of water on the surface of graphite [7], Gabor Nagy studied the water structure at the graphite surface by STM measurements [8] etc.. On the influence of the adsorption of water molecules on graphene, the article illustrates a hydrophobic related theory about graphene was written by Leenaerts O [9].

Based on the results of their study, Materials Studio software simulation calculation is applied in adsorption of water molecules on the monolayer graphene surface, using the first principles density functional theory with the generalized gradient approximation and periodic plane model. In this paper, the interaction between different water molecules and graphene is primarily a van der Waals interaction. The influence on monolayer graphene adsorbed by water molecules is explained and the corresponding optical properties are calculated.

2. Computational details

All computations have been done using the CASTEP (Cambridge Serial Total Package) programs in Materials Studio and using DFT (Density Functional Theory) to describe the exchange-correlation potential. Moreover, these use a plane-wave basis set with ultra soft pseudo potentials. PBE version of GGA (the General Gradient Approximation) was used rather than PW91[10], which mainly describing gas phase water adsorption on metal surfaces. Under the condition of the gas phase study, it is common to select the B3LYP version of GGA. Here, the paper doesn't need to accurately describe approximate degenerate of many clusters of water molecules of isomers. So researches don't need pay attention to the differences of these methods in detail, but focus on the selected part.

Monolayer Graphene crystals model is a two-dimensional hexagonal grid with repetition structure. Considering the boundary conditions and computing time, typically graphene is replaced by a hexagonal 4 x 4 x 1 supercell sample model in simulation calculation, and it have used a vacuum gap of 15 Å (see figure 1). Numerical test results show that the structure can not only satisfy the accuracy of all calculation, but also won't make a too large amount of calculation.

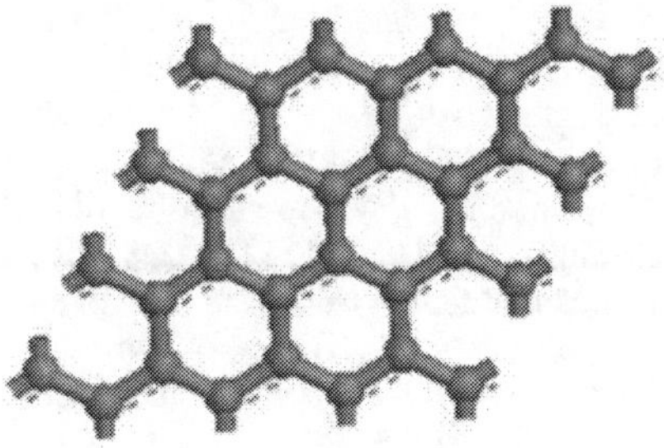

Figure 1. Monolayer graphene.

Because water molecules relatively far from the plane, it makes no difference whether use a metallic or insulator treatment. Self-consistent iteration method (SCF) was used in the energy calculation with 1000 iteration steps, and ultra soft pseudo potential in Reciprocal space used BFGS algorithm.

In geometry optimization, system total energy convergence value is 1×10^{-6} eV/atom, the force on the atom is 0.03 eV, maximum stress is 0.05 Gpa, the maximal displacement of 0.01 angstrom, and cut-off energy is 310.0 eV. Self-consistent calculation of irreducible Brillion k point $3\times3\times2$ is produced by the program automatically, with BFGS algorithm geometry optimization and surface relaxation.

Water molecules adsorbed on the surface at random. In order to simulate the adsorption of water molecules on the graphene in reality, combining with numerical simulation conditions, and respectively adding one water molecule, two water molecule and three water molecules in random location and random direction to realize geometry optimization. Following calculations have been done on the basis of optimized structure.

Several different optimized models' vertical view can be seen as follows (in figure 2).

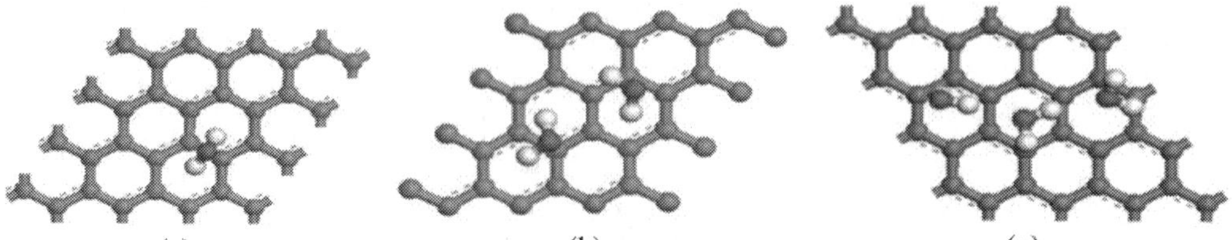

(a) (b) (c)

Figure 2. (a) 1 H_2O; (b) 2 H_2O; (c) 3 H_2O.

1. One water molecule model is shown in figure 2(a): Water molecule present bridge site; the oxygen atom is just above the middle of the carbon-carbon bond; two hydrogen molecules are symmetrically distribute on both sides of the carbon-carbon bond, and the height of the water molecules from graphene is 3.26358 Å.

2. Two water molecule model is shown in figure 2(b): Both of two water molecules present bridge site; two oxygen atoms are located in two carbon-carbon bonds in different carbon rings; length between water molecules is 3.430 Å; and the heights of the water molecules from graphene are 3.04383 Å and 2.94343 Å.

3. Three water molecule model is shown in figure 2(c): Three water molecules are located above three different carbon rings. One of a water molecule hydrogen atoms pointing down; lengths between water molecules are 2.961 Å and 2.955 Å. It became a small cluster of water molecule, conforming to liquid water, and the heights of the water molecules from graphene are 3.24965 Å, 3.42859 Å and 3.8967 Å.

3. Results and discussion

Optimization of the lattice constant of graphene is 2.4638 Å, being in good agreement with the theoretical value of 2.46 Å. The water molecule's O-H bond length in the first model, after optimized is about 0.9903 Å and O-H bond angle is 104.534°; The water molecules' O-H bond lengths in the second model after optimized are about 0.9700 and 0.96941 Å ,and O-H bonds angles are 104.067°, 104.745°; The water molecules' O-H bond lengths in the third model after optimized are about 0.9864, 0.9691, 0.9696 Å, and O-H bonds angles are 104.305°, 104.938°, 105.301°; conforming to the actual weather conditions of water molecule bond length of 0.97 and bond angle of 104.00°.

Thus, it can be seen that Supercell geometry optimization and water molecules geometry optimization's results are accord with the theory. The method to select the model is properly and the calculation method is feasible.

3.1. Adsorption energy

Adsorption energy can reflect the strength of the interaction between the graphene and the water molecules, defining as equation (1):

$$E_{ad}=E_{total}-(E_{H20}+E_{gra}) \qquad (1)$$

In the equation, E_{ad} is the adsorption energy; Etotal is system total energy of a single water molecule adsorbed on the monolayer graphene; E_{H20} is water molecule energy; E_{gra} is the monolayer graphene energy.

IOP Conf. Series: Materials Science and Engineering **87** (2015) 012101 doi:10.1088/1757-899X/87/1/012101

The energy of one water molecular adsorption on graphene Ead is -0.0213952 eV; the energy of adsorption of two water molecular on graphene Ead is -0.0203484 eV in average; the energy of adsorption of three water molecular on graphene Ead is -0.0244898 eV in average.

The results of optimized structure and adsorption energy can be analyzed as follows: In the first two models, water molecule only interacts with graphene surface and the interaction between water molecules is so small that can be ignored when calculating. In the third model, located on the edge of the periodic structure will result in the interaction between two water molecules, leading to water molecules form changed after the optimization. Water molecules form clusters and adsorption energy in equation (1) will be an error.

Calculation of adsorption energy use PBE in GGA includes van der Waals interactions. It can be seen that the interaction of water molecules and graphene is so small that Van der Waals interaction plays a main role. Because the C-C bond between carbon atoms is highly saturated in graphene, making the interaction of water molecules and graphene is very weak, that id the structure of graphene is high hydrophobicity structure.

Observing the data, the water molecules average adsorption energy of different models are very similar to each other, which showing that the number of adsorbed water molecules have no effect on the average adsorption energy. The conclusion is in accordance with and the water adsorption on the graphite surface [11].

3.2. Density of states

In the figure 3, monolayer has a summit value is 57.68912 electrons/eV in the point of -265.65 eV, and a fragment of summits exist between -19.4235 eV to 4.232958 eV. In this section, a summit value 34.22898 electrons/eV exists at the point of -0.04977 eV. For the model of adsorption, summits height and scope of the original image is few changes [12]. Compared with the monolayer graphene, a water molecule adsorbed on graphene cause a small height (3.932914 electrons/eV) adding at the point of 504.888 eV. Two or three water molecules adsorption on graphene follows the same rule.

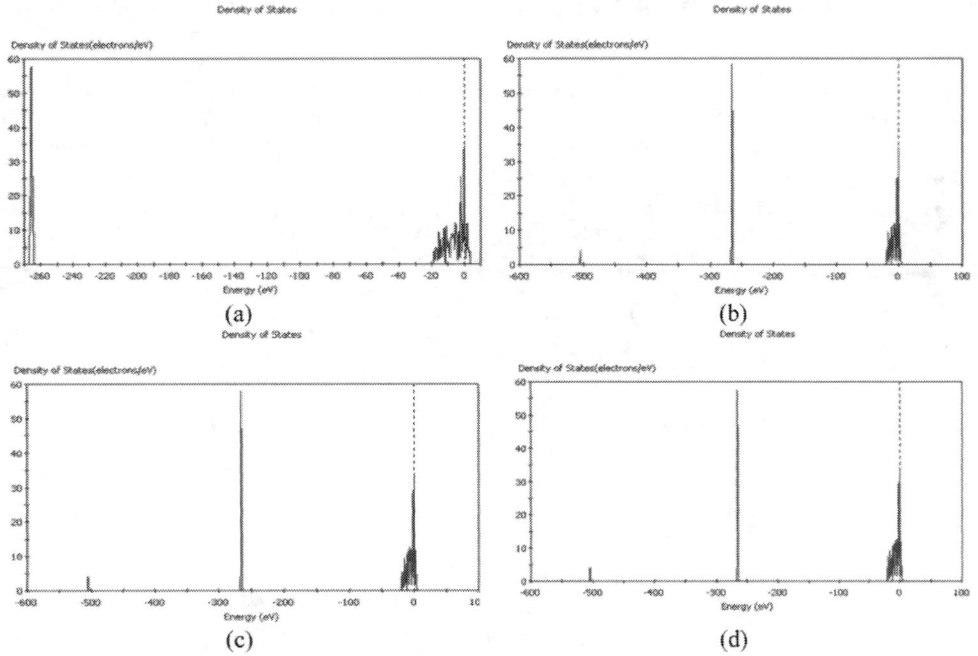

Figure 3. Density of state. (a) Graphene; (b) 1 H_2O; (c)2 H_2O; (d) 3 H_2O.

Figure 3(a) monolayer graphene; (b) A water molecule adsorption on graphene; (c) Two water molecule adsorption on graphene; (d) Three water molecules adsorption on graphene [13]. Only the valence electrons were considered without the inner electrons.

As shown in figure 4, water molecules state of density. It can be seen clearly that the extra wave was due to the superposition of a part of the water molecules density of state.

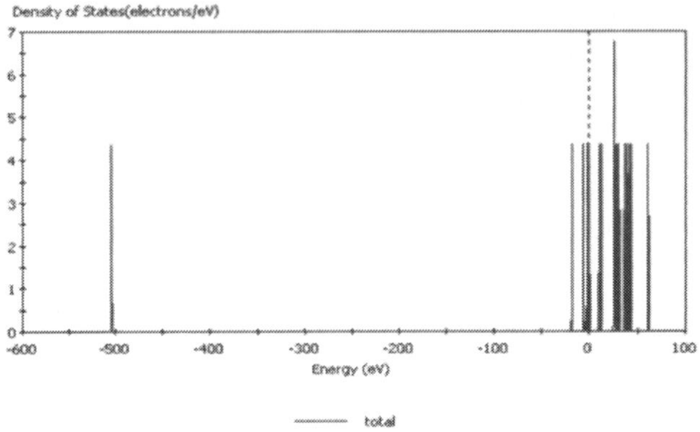

Figure 4. Density of state.

3.3. Band structure

Band structure of graphene is shown in figure 5. Because graphene is a two-dimension crystal with hexagonal symmetric structure consisting of a bipartite lattice of two triangular sub-lattices, four valence electrons are tied to the C nucleus tightly and each atom is tied to its three nearest numbers via σ bonds which are a result of sp2 hybridization of 2s, 2px and 2py orbitals for the three valence electrons. Near the Fermi surface level, 2pz orbitals make the main contribution, which is π bond.

Graphene energy gap is zero. The conduction band and valence band intersect at the Dirac point, there are two atoms per unit cell and hence there are two electrons per unit cell. When the lower π band is completely filled full, leaving the upper π^* band empty completely, Fermi surface is just located in the intersection point. Fermi surface E=0. Under the Fermi surface, electronic state is the corresponding with πorbitals bonding state; above the Fermi surface energy level, electronic state is the corresponding π^* orbitals anti-bonding.

After the adsorption of water molecules, the band structure of graphene does not change, that is water molecules has little influence on the band structure of graphene.

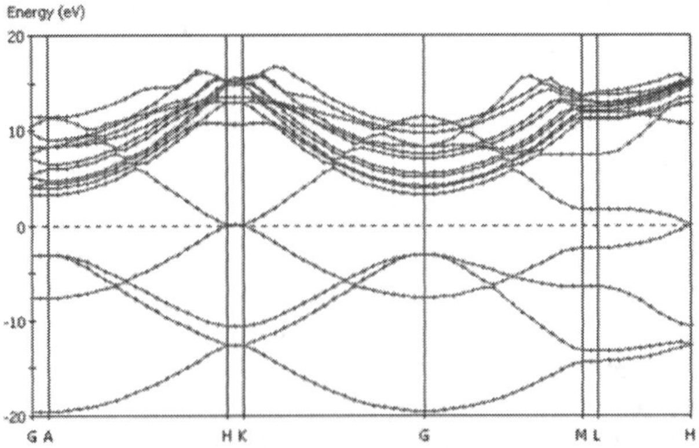

Figure 5. Band structure.

3.4. Optical properties

In this paper, the calculation of optical properties is under polarized light, and parameters are as follows: Graphene has zero band gap; the scissors operator is 0 eV; the polarization direction of the polarized light is (1 0 0).

Limitations: the calculation of optical properties in CASTEP based on the electronic structure, but it doesn't contain the excited states, so it has some influence on the accuracy of optical computing. For example, the transition of the electric field inside the material is considered no attenuation while the transition is calculating, but in fact not the case.

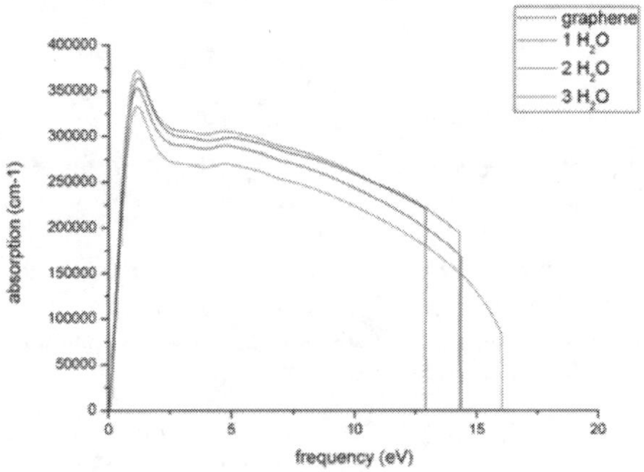

Figure 6. Absorption.

Figure 6 is the absorption rate. The absorption coefficient is usually referred to the material energy attenuation when the electromagnetic waves pass through the unit thickness of material [14].

Absorption spectrums are calculated by CASTEP is only considered about intrinsic absorption. Seeing the curves in the figure, on the same wave band, the absorption rate is arranged as follows: one water molecular adsorption on graphene > two water molecular adsorption on graphene > three water molecular adsorption on graphene.

Absorption coefficient equation is as follows:

$$\alpha(\omega) = \frac{\omega \varepsilon_2(\omega)}{nc} \tag{2}$$

According to the relationship of quantum mechanics energy equation:

$$E=hv=hck=pc,\ h=6.626\times10^{-34},\ 1\ eV=2.41796\times1101\ Hz$$

As is known to all, the wavelength of visible light is 390 nm to 780 nm; the frequency of the corresponding is 3.84×1014 to 7.70×1014 Hz, that is 1.6 eV to 3.2 eV. In the figure, light absorption in the range of 0 ~ 17 eV; a summit located in 2 eV and the absorption coefficient is 0 when the frequency is greater than 17 eV.

When light through the model, in visible band graphene shows favourable nonlinear optical absorption properties, the adsorption of one water molecular model is a little smaller than pure graphene. Owing to the zero gap of graphene and its wide wave band adsorption, graphene is easy to saturate to light. Therefore, graphene has low light saturation flux, so it has prospects for practical application such as manufacturing optical photonic crystals, fiber laser and laser switch in the area of the optical field.

Figure 7 is light conductivity, calculating by $\sigma=\sigma1+i\sigma2$, the phenomenon of the conductance change of the semiconductor material is caused by illumination change is called photoconductive effect.

This parameter is used to describe the optical properties of the metal, but the CASTEP expand the scope of it to insulators and semiconductors. As for the metal, the main difference in calculating process is the real component is closely related to an internal band when considering Conversion relationship in optical spectroscopy while the other does not consider these factors.

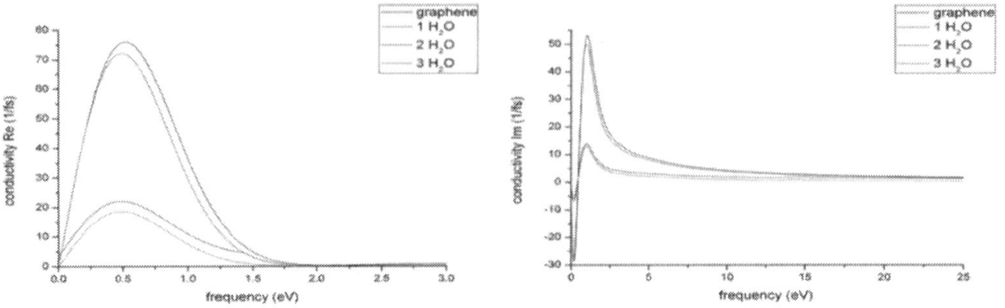

(a) (b)

Figure 7. Conductivity. (a) conductivity Re; (b) conductivity Im.

In figure 7(a), the light conductivity of the graphene in the real number part is very strong in the low-frequency part, but will be weak in the high-frequency part. The summit value 76 of graphene located in 1eV. When one, two and three water molecular adsorption on graphene, the summit value drops to 73, 21 and 19 respectively. In the low-frequency part of the imaginary, the light conductivity value of the graphene is bigger than others, then dropping to 0 in the high-frequency part. The summit value of pure graphene is 51 at 2 eV, and dropping to 50, 15, 15 when one, two and three water molecular adsorption on it, then they are all quickly dropping to near zero. It shows that the optical conductivity of graphene is very big in infrared wave band, but it decreased after adsorbing water molecular.

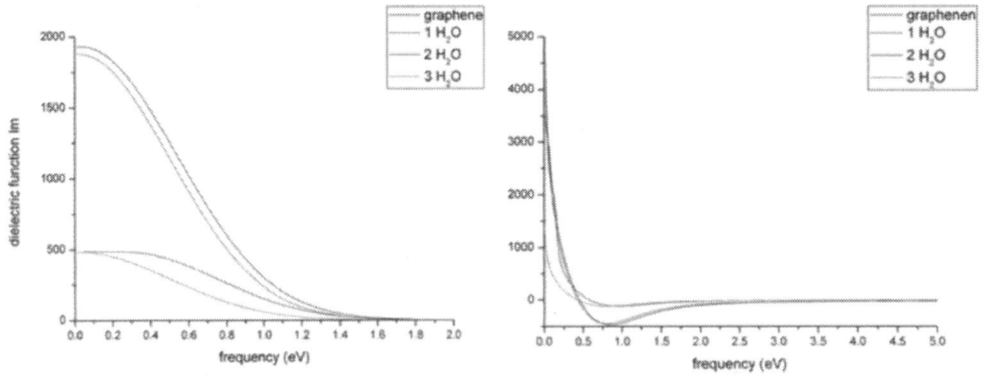

(a) (b)

Figure 8. (a) Dielectric function Re; (b) dielectric function Im.

Figure 8 is the dielectric function.

Imaginary part of a dielectric function expresses the transition information between band gaps, and reflects the solid band structure and various kinds of spectral information. It depends on the size of the electron transition energy of conduction band and valence band. In fact, it is the electronic absorption spectrum of the transition between the bands.

Dielectric function spectrum peaks are formed by electron transition from the valence band to the conduction band; transitions satisfy to selection rule which corresponding to the width value of band gap. The position of the absorption peak was not corresponding with band difference well, because of the relaxation effect when the transition energy absorption by electronic in the process of transition, leading to the deviation. In band structure, there are many levels have contributed to the same peak by directly or indirectly transition, so we can't simply use two energy levels difference to calculate, and can't simply think crystal absorption spectrum is caused by a single transition.

Imaginary part of dielectric function equation (3) as follows:

$$\varepsilon(\vec{q} \to O_u, h\omega) = \frac{2e^4\pi}{\vec{\Omega}\varepsilon_0} \sum_{k,v,c} |< \psi_k^c | u \cdot r | \psi_k^v >|^2 \delta(E_k^c - E_k^v - E) \tag{3}$$

In equation (3), q is a vector in the First Brillouin zone, and the physical meaning is the change of electronic before and after the transition. Because the photon momentum is far smaller than the electron momentum, the electron momentum remains the same approximately. So q tends to be 0, and Ou is a zero vector, and the Ω refers to the original cell volume.

Graphene has a zero band gap, and its spectrum produced by the electron transition between energy levels. In the real part, the curve of one water molecule adsorption on graphene is coincident with pure graphene, and going down with energy's growth in the low energy part. The minimum value is -500 at 0.8 eV, growing with energy, arriving at value 0 at 2.5 eV, saturated. Adsorption of two or three water molecules change smoothly. In the imaginary part, curve monotonous gently fall; Dielectric function of graphene and one water molecule adsorption model changes more than the adsorption of two or three water molecules. That is the more water molecular adsorption on graphene, the more energy is loss.

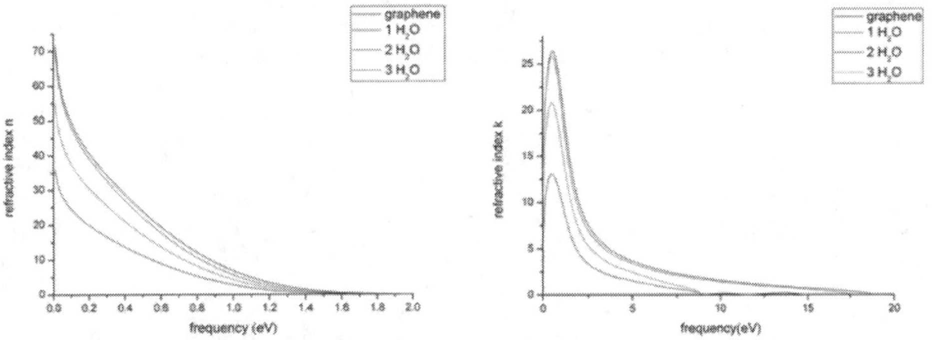

(a) (b)

Figure 9. (a) Reflective index n; (b) reflective index k.

Figure 9 is the refractive index n and k.

The equation of refractive index n and k is as follows:

$$\mathrm{n}(\omega) = \frac{1}{\sqrt{2}}\{(\varepsilon_1^2 + \varepsilon_2^2)^{1/2} + \varepsilon_1\}^{1/2}$$
$$\mathrm{k}(\omega) = \frac{1}{\sqrt{2}}\{(\varepsilon_1^2 + \varepsilon_2^2)^{1/2} - \varepsilon_1\}^{1/2} \tag{4}$$

The refractive index is the light attenuation in the crystal. In figure 9, all curves have no peak, and attenuation rate drop to zero in low energy part soon. At about 2eV, graphene refractive index k is 27, and the refractive index value is almost zero in the high energy region (more than 15 eV). It shows that graphene absorption is very weak for high-frequency electromagnetic wave. Analysis of the figure shows that only one water molecule adsorption on graphene has little effect on transparency.

Figure 10 is a loss function.

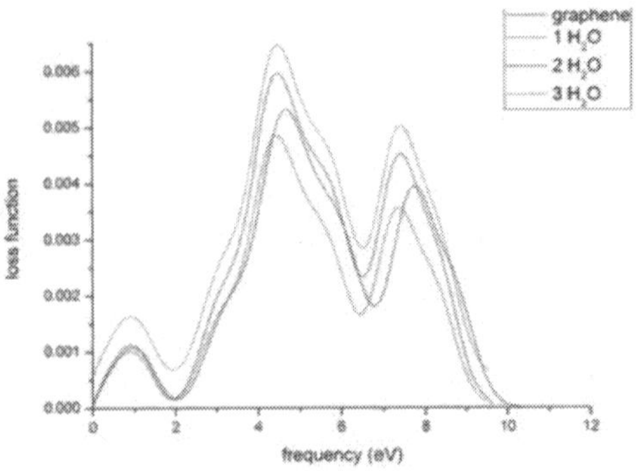

Figure 10. Loss function.

The energy loss function in theory is Im(- 1/ε), it is deduced by experiments Optical energy loss function Im[−1/ε(ω)]. If q ≠0, energy loss function is Im[−1/ε(q,ω)].

In figure 10, the energy loss maximum value is at about 5 eV; in addition, there are secondary energy loss values at about 1 eV and 8 eV. For graphene, every C atom is tied to each other via σ bonds that lie in the graphene plane with angles of 120°, 2s, $2p_x$ and $2p_y$ orbitals form to sp^2 hybridization. The electron is in the $2p_z$ orbital that is orthogonal to the monolayer graphene. Plasma excitation peaks at 5 eV and 8 eV are caused by π and σelectrons transit to excited states, but plasma excitation peak at 1 eV is caused by secondary plasma excitation of σelectrons.

The main peak of energy loss of water molecular adsorption on graphene is at 4.3 eV, and second peaks of energy loss are at 1 eV and 7.5 eV. In figure 10, the more adsorption of water molecules, the bigger the energy loss.

All above is optical properties simulation model of graphene and different water molecular adsorption on graphene calculated by CASTEP. CASTEP provides the calculation of polarization system, but only in the same spins conversion. But in reality, due to the complexity of the incident light, geometrical structure on the material surface is not idea; and because the polarization effect exist in the crystal structure, causing the dielectric coefficient may not isotropic. These factors will hinder us to predict the optical properties of the material, leading to a random error.

4. Conclusions

Based on the first principles calculation of plane wave ultra-soft pseudo-potential technology, it was performed by the Density Function Theory (DFT).The plane wave function was studied on different number of water molecules adsorption on the monolayer graphene, electronic structure total energy shows that the nature of the adsorption is physical adsorption, and the adsorption is very weak. The density of states are analyzed, and optical properties including absorption rate, dielectric function, refractive index n and k, loss function are calculated.

Acknowledgments

The authors would like to thank the reviewers for their useful comments. This study was supported by Engineering Laboratory of Optoelectronic Technology and Advanced Manufacturing of Henan Province.

References

[1] Zheng Z X, Du Y L, Feng Q L 2012 Facile method to prepare Pd/ graphene-polyaniline nanocomposite and used as new electrode material for electrochemical sensing *Journal of Molecular Catalysis A Chemical* **80-86** 353-4

[2] Salavagione Horacio J, Díez-Pascual Ana M and Lázaro Eduardo 2014 Chemical sensors based on polymer composites with carbon nanotubes and graphene: The role of the polymer *Journal of Materials Chemistry A 2* **35** 14289-328

[3] Giovannetti G, Khomyakov P A, Brocks G, et al. 2008 Doping graphene with metal contacts *Phys. Rev. Lett* **101** 026803-6

[4] He M C and Zhao J 2013 Methane adsorption on graphite (0001) films: A first-principles study *Chinese Physics B* **1** 452-6

[5] Falk K, Sedlmeier F, Joly L, Netz R R and Bocquet L 2010 *Nano Lett.* **10** 4067

[6] Jie Ma, Angelos Michaelides and Dario Alfe 2011 Adsorption and diffusion of water on grapheme from first principles *Phys. View B* **84** 033402

[7] Cabrera Sanfelix P, Holloway S and Kolasinski K W 2003 The structure of water on the (0001) surface of graphite *Elsevier Science B.V.* **532-535** 166-172

[8] Gabor Nagy 1995 Water structure at the graphite (0001) surface by STM measurements *Journal of Electroanalytical Chemistry* **409** 19-23

[9] Leenaerts O, Partoens B and Peeters F M 2009 Water on grapheme: Hydrophobic and dipole moment using density functional theory *Phys. Rev B* **79** 235440

[10] Shamir N, Tiferet E and Zalkind S 2006 *Surf. Sci.* **600** 657-64

[11] Birgenau R J, Taub H and Dutta P 1980 *Ordering in Two Dimensions* Edi S K Sinha (The Netherlands, Amsterdam: Elsevier)

[12] Hammer B, Hansen L B and Norkov J K 1999 Improved adsorption energetics with density-functional theory using revised perdew-burke-ernzerth of functionals *Phys Rev B* **59** 7413-21

[13] Tian X F, Wang H, Xiao H X, et al. 2014 Adsorption of water on UO2 (111) surface: Density functional theory calculations *Computational Materials Science* **91** 364-71

[14] Yuan F, Lu S X, Xu W G, et al. 2014 LSDA + U calculations of the electronic and optical properties of rutile TiO_2 (110) vs (011)-2×1 *Computational Materials Science* **90** 1-6

IOP Conf. Series: Materials Science and Engineering **87** (2015) 012102 doi:10.1088/1757-899X/87/1/012102

Fabrication and performances of Al/CuO nano composite films for ignition application

Yong Li[1], Yun Gao[2], Xin Jia[1], Bin Zhou[1, 3] and Rui-Qi Shen[1]

[1]Nanjing University of Science & Technology, Jiangsu, 210094, China
[2]Changzhou Sunan Blasting Demolition Co., Ltd., Jiangsu, 213016, China

E-mail: zhoubin8266@sina.com.cn

Abstract. In an effort to explore the application possibility of composite films in ignition field, Al/CuO was fabricated on semiconductor bridge (SCB) chip by ion beam sputtering technique. Surface morphology and elemental composition of the composite films were analysed by scanning electron microscope (SEM) and energy dispersive spectrometer (EDS). Spatial size and duration of the products was detected with the open-air combustion experiment. The results showed that the prepared composite films surface is smooth, flat, and uniform. Element weight ratio meets the design requirements. And the chemical reaction of the Al/CuO nCFs improves output performances of ignition chip.

1. Introduction

Semiconductor bridge (SCB) has been widely used in military and civilian field since invented by Sandia back in 1986 because of the advantages of rapidly function, low energy requirement, and good safety. When current flows through SCB, bridge material goes through heating, melting, vaporization and plasma generation to ignite explosives [1-3]. But, the output energy relies on the stimulation energy and may not be sufficient to provide reliable ignition when there is any gap between the ignition chip and explosives [4]. D A Benson [5] and B Martinez-Tovar [6] attempted to deposit a metal layer onto SCB to improve its output performance. However, without any further reaction, the metal layer cannot release enough energy.

The nano energetic multilayer films (nEMFs) systems can be initiated by self-propagating exothermic reaction to produce a large quantity of heat with a small thermal pulse or an electrical stimulus [7-11]. Among those nEMFs system, the Al/CuO nEMFs with the exothermic reaction have been widely studied in recent years [12]. K J Bolbaum [13, 14] and M Petrantoni [15, 16] have reported the preparation and reaction of Al/CuO nEMFs. N Amini-Manesh [17, 18] studied the self-propagating reaction speed of Al/CuO on a different substrate. Those work indicated a possible application of Al/CuO nEMFs in ignition field because of good reactivity and easy integration.

Theoretically, the igniter integrated SCB with Al/CuO composite films may inherit the advantages of SCB, and also has a good output performance. In this paper, Al/CuO composite films have been prepared by ion beam sputtering technique. The surface morphology, elemental composition, and open-air combustion performances were described.

[3] Address for correspondence: Bin Zhou, Nanjing University of Science & Technology, Jiangsu, 210094, China. E-mail: zhoubin8266@sina.com.cn.

2. Fabrication

The fabrication process of SCB chip bases on a standard complementary metal-oxide semiconductor (CMOS) procedure. A silica layer forms onto a silicon substrate by thermal oxidation. Then heavily doping to form a 2 μm thick polysilicon layer, where the doping element is phosphorus and doping concentration is about 10^{20} atom/cm^3. And last, metal layers are deposited to form contact pads. The shape and size of the bridge is decided by contact pads and polysilicon layer. The structure of SCB chip is shown in figure 1.

Figure 1. Schematic of SCB chip structure.

Based on the prepared SCB wafer, the Al/CuO-SCB chip fabrication process is a combination of lithography, development and ion beam sputtering techniques. During the process, two masks are demanded. One determines the size of the insulating layer, and the other determines the size of the Al/CuO. An insulating layer is deposited onto the bridge and contact pads to avoid any current flows through composite films. There is an adhesive layer between the insulating layer and copper oxide. The whole fabrication process is shown in figure 2, fabrication conditions listed in table 1, and structure of the chip is shown in figure 3.

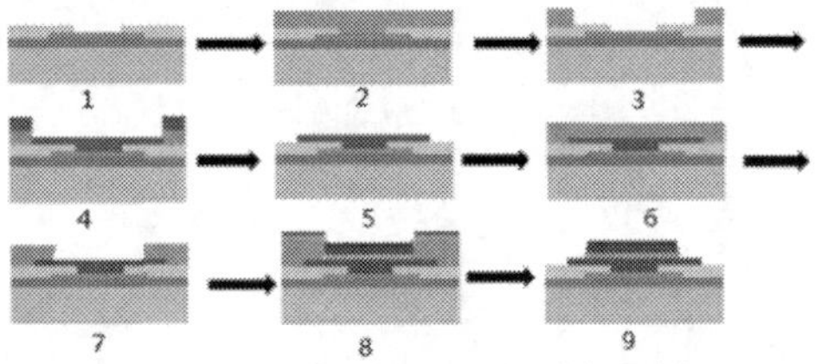

Figure 2. Fabrication process of the chip.

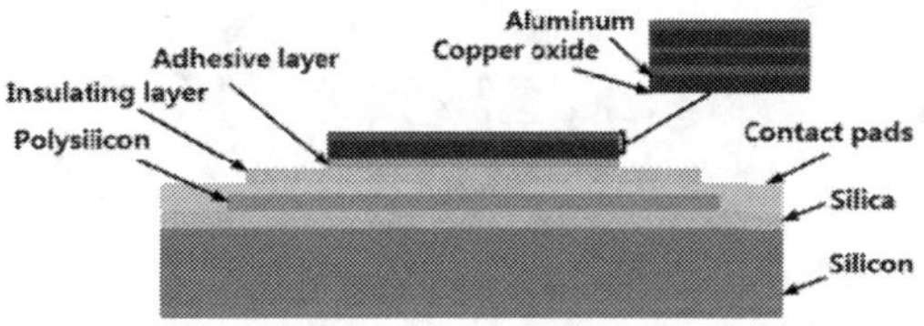

Figure 3. Al/CuO-SCB chip structure.

Table 1. Fabrication conditions.

Base pressure	Ion energy	Sputtering current	Sputtering speed		Monolayer thickness		Total thickness
			Al	CuO	Al	CuO	
<5×10-4 Pa	650 eV	70 mA	10 nm/min	7 nm/min	100 nm	200 nm	2.4 μm

3. Characterization

Surface morphology and elemental composition are characterized by scanning electron microscope (SEM) and energy dispersive spectrometer (EDS). As shown in figure 4, the film surface is smooth, flat, and void-free defect.

Figure 4. Surface morphology of Al/CuO composite films.

Elemental analysis results are shown in figure 5 and listed in table 2. The selected area contains Al/CuO composite films, insulating layer and contact pads. The results include plenty of aluminium, copper and oxygen, and also a small amount of silicon, titanium, and gold. The atomic percentage ratio of aluminium and copper oxide is 26.96:37.86. While the designed film thickness ratio is 1:2, leads to a mole ratio of 2:3. So the prepared film system meets the design requirement.

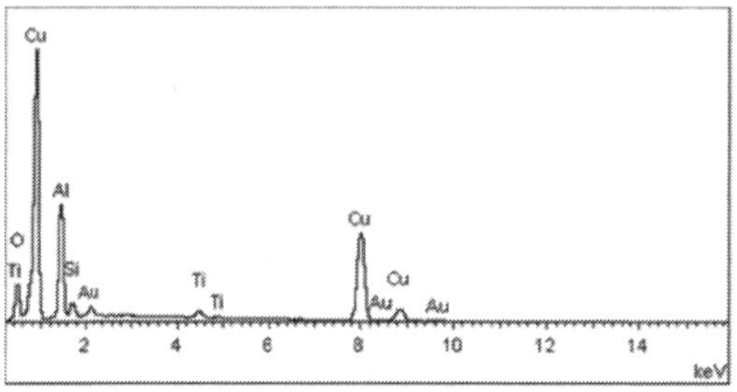

Figure 5. Element analysis results from SEM and EDS.

Table 2. Element analysis results from SEM and EDS.

Element	Weight percentage	Atomic percentage
O	12.33	30.14
Al	18.60	26.96
Si	2.27	3.16
Ti	1.35	1.10
Cu	61.50	37.86
Au	3.96	0.79
Total	100.00	100.01

4. Open-air combustion

In the experiment, Al/CuO-SCB chip is mounted into a ceramic plug with epoxy. The foot wires of ceramic plug and contact pads are connected with bonding wires by ultrasonic bonding technology. Then bonding wires are covered and protected by conductive adhesive. The packaged igniters are shown in figure 6.

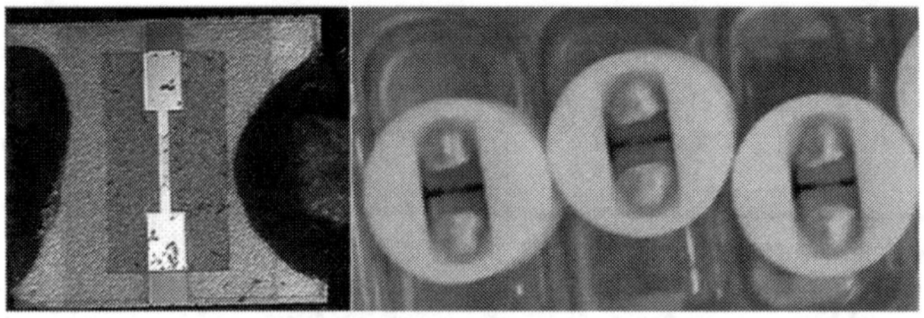

Figure 6. Photos of packaged igniters.

The stimulus resource is a capacitor discharge unit (CDU), voltage and current variations during the function process are recorded by an oscilloscope, and products conditions are recorded by a high-speed camera. The experiment circuit diagram is shown in figure 7.

Under the stimulus conditions of 47 μF 50 V, products status of SCB and Al/CuO-SCB are shown in figure 8. In the frames, the bright core is plasma generated by polysilicon. Plasma spatial size reaches maximum at 40 μs, and then begins to dissipate. Start from 80 μs, the products spatial size of

Al/CuO-SCB is larger than that of SCB. The products mainly are the combustion of the Al/CuO composite films.

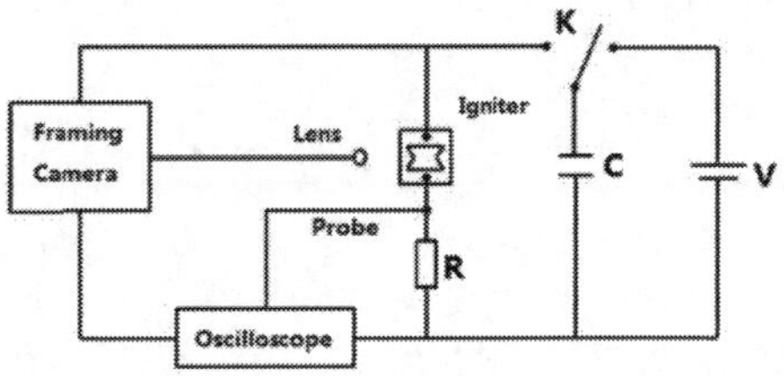

Figure 7. Experiment circuit diagram.

The voltage and current variations of those two samples are shown in figure 9. Voltage and current variations of Al/CuO-SCB are almost the same as that of the SCB, which indicates a good insulating effect of the insulating layer. Generally the time that current drops to zero is the function time. As shown in figure 9, the function time of SCB is 69.8 μs, and Al/CuO-SCB is 89.1 μs. Further demonstrate the products after 100μs in figure 8 is the combustion of Al/CuO composite films.

So, when current flows through Al/CuO-SCB chip, the polysilicon layer goes through heating, melt, vaporization and plasma generation. During the function process, polysilicon heat the Al/CuO composite films form below. Then the films are ejected by the shock wave generated with plasma growth, and combustion in the air to release large quantity of energy.

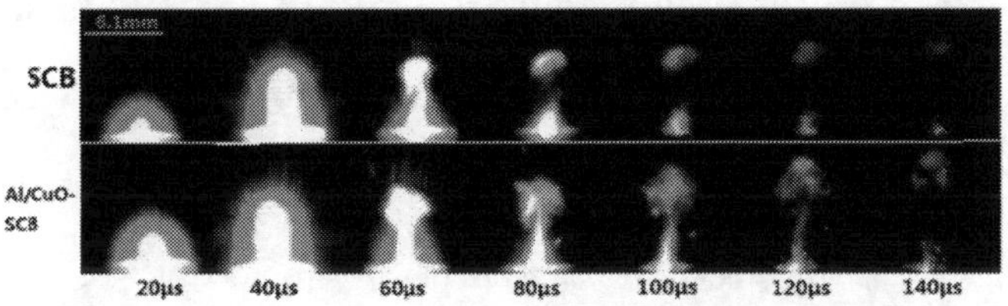

Figure 8. Products variations of SCB and Al/CuO-SCB.

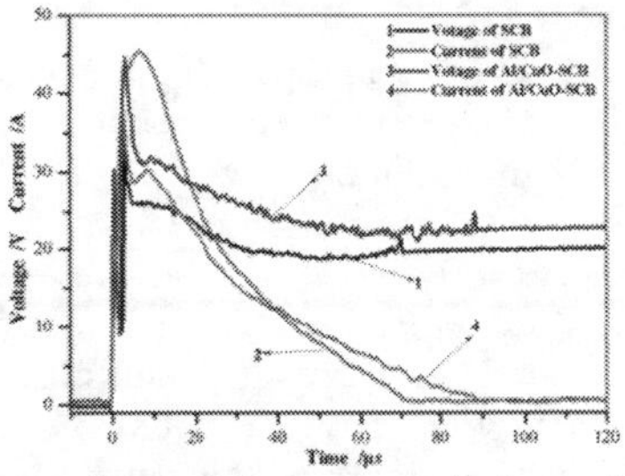

Figure 9. Voltage and current variations of samples.

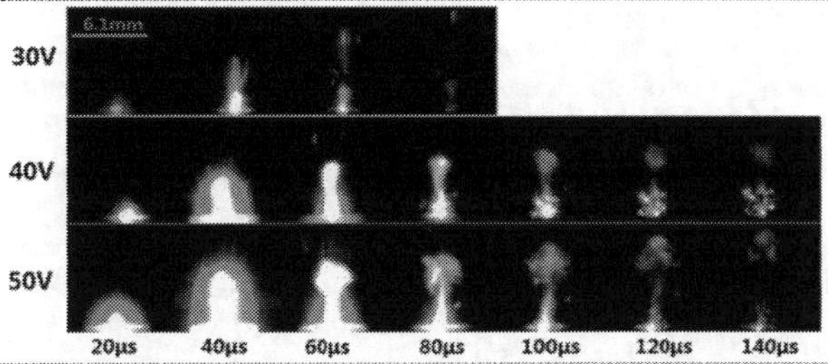

Figure 10. Products status variations with respect to time.

Figure 10 illustrates the products status versus time of Al/CuO-SCB with different stimulus energy. As it shows, spatial size and duration of products are all increase with voltage. At 50 V, the height of combustion flame is higher than 6 mm.

5. Conclusions

In this paper, Al/CuO multilayer composite films are deposited onto SCB chip to form a new kind of igniter by employing the ion-beam sputtering technique. The prepared films surface is flat, smooth and uniform. Element analysis indicates that the atomic weight ratio of aluminium and copper oxide meets the design requirement. In the open-air combustion experiment, characteristic parameters variations indicate that current only flows through polysilicon layer. Polysilicon goes through heat, melt, vaporization and plasma generation and heat the composite films from below. The composite films are ejected from ceramic plug by the shock wave generated by plasma growth, and combustion in the air to release a large quantity of energy. The products spatial size and durations of Al/CuO-SCB is much larger than that of the SCB, which demonstrates higher output energy of Al/CuO-SCB.

References

[1] Hollander Jr. 1968 Semiconductor explosive igniter *US: USP3366055*
[2] Willis K E and Whang D 1995 *Semiconductor Bridge Technologies* **95** 25-48
[3] Bickes R W and Schwarz A C 1987 Semiconductor bridge (SCB) igniter *US: USP4708060*
[4] Baginski T A and Parker T S 2005 Electro-explosive device with laminate bridge *US: USP 2005/0115435 A1*
[5] Benson D A, Bickes R W and Blewer R S 1990 Tungsten bridge for the low energy ignition of explosive and energetic material *US: USP4976200*
[6] Martinez-Tovar B and Foster M C 2008 Titanium semiconductor bridge igniter *US: USP2008/0017063A1*
[7] Rogachev A S and A S Mukasyan 2010 Combustion of heterogeneous nanostructural systems *Combustion Explosion and Shock Waves* **46** 243-66
[8] Shteinberg A 2011 Thermal analysis of high-temperature fast reactions in energetic materials *Journal of Thermal Analysis and Calorimetry* **106** 39-46
[9] Qiu X, Tang R, Liu R, et al. 2011 A micro initiator realized by reactive Ni/Al nanolaminates for MEMS applications *IEEE* **137** 1665-8
[10] Petrantoni M, Rossi C, Salvagnac L, et al. 2011 Multilayered Al/CuO thermite formation by reactive magnetron sputtering: Nano versus micro *Journal of Applied Physic***s** **108** 084323
[11] Zhou X, Shen R, Ye Y, et al. 2011 Influence of Al/CuO reactive multilayer films additives on exploding foil initiator *Journal of Applied Physic* **110** 094505
[12] Zhang K, Rossi C and Ardila Rodriguez G A 2007 Development of a nano-Al/CuO based energetic material on silicon substrate *Applied Physics Letters* **91** 113-7
[13] Bolbaum K J, M E Reiss, et al. 2003 Deposition and characterization of a self-propagating CuOx/Al thermite reaction in a multilayer foil geometry *Journal of Applied Physics* **94** 2915-22
[14] Bolbaum K J, Wagner A J, Plizko J M, et al. 2003 Investigating the reaction path and growth kinetics in CuOx/Al multilayer foils *Journal of Applied Physics* **94** 2923-9
[15] Petrantoni M, Rossi C, Conédéra V, et al. 2010 Synthesis process of nanowired Al/CuO thermite *Journal of Physics and Chemistry of Solids* **71** 80-3
[16] Petrantoni M, Rossi C, Salvagnac L, et al. 2010 Multilayered Al/CuO thermite formation by reactive magnetron sputtering: Nano versus micro *Journal of Applied Physic***s** **108** 084323
[17] Manesha N A, Basu S and Kumar R 2010 Experimental flame speed in multi-layered nano-energetic materials *Combustion and Flame* **157** 4476-80
[18] Manesha N A, Basu S and Kumar R 2011 Modeling of a reacting nanofilm on a composite *Energy* **36** 1688-97

IOP Conf. Series: Materials Science and Engineering **87** (2015) 012103 doi:10.1088/1757-899X/87/1/012103

Influence of deposited CNTs on the surface of carbon fiber by ultrasonically assisted electrophoretic deposition

J J Jiang[1], F Liu, C Deng, L C Fang and D J Li

Shaanxi Engineering Research Center for Digital Manufacturing Technology, Northwestern Polytechnical University, Xi'an 710072, P.R. China

E-mail: jianjun@nwpu.edu.cn

Abstract. The surface property of carbon fiber directly affects the interface performance between carbon fiber and matrix. To improve the surface property of carbon fiber, we proposed a simple method to prepare carbon nanotubes /carbon fiber hybrid fiber via ultrasonically assisted electrophoretic deposition (EPD). Surface morphologies and surface functional group of carbon fibers were examined by atomic force microscopy (AFM), scanning electron microscopy (SEM) and fourier transform infrared spectrometer (FTIR), respectively. The results show that the deposition of carbon nanotubes changed the surface morphologies of carbon fibers and introduced polar groups (C=O and C-O) to carbon fiber surface. Comparing the results with EPD-only, ultrasonically assisted EPD increased the uniformity of carbon nanotubes coatings whereas only sparse and not uniformly deposition formed without ultrasonic.

1. Introduction

Carbon fiber (CF) reinforce composites with their favorable mechanical properties has widely used in aviation industry. Mechanical properties of carbon fiber reinforce composite is closely related to its interfacial bond strength of carbon fiber and matrix. However, the interface performance related to the carbon fiber surface properties, the interface bonding of untreated carbon fiber that has large surface inertia and resin matrix is weak which affects the carbon fiber composite material's excellent performance into full play [1, 2]. As a result, a lot of studies have been devoted to the surface treatment of CF to improve surface properties of CF, such as electrochemical method [3], sizing process [4], plasma treatment [5], preparation of carbon nanotubes (CNTs)/CF hybrid fiber [6]. Among these methods, CNTs with their excellent properties have caused enough attention on hybrid fiber. Grafting CNTs onto fiber surface is a practical method to improve fiber surface area, forming mechanical interlocking on the interface between fiber and matrix, which may improve stress transfer and interfacial properties [7].

Recently, many grafting methods have been proposed such as chemical vapor deposition (CVD) [8] and electrophoretic deposition (EPD) [9]. Both methods reported to deposited CNTs on the fiber surface successfully. However, high temperatures and predeposited catalysts for CVD, seriously limits the practical application of this technique for the fabrication of CNTs/CF hybrid fiber. EPD is known to be one of the most promising manipulation techniques to produce large-scale reinforcement of

[1] Address for correspondence: J J Jiang, Shaanxi Engineering Research Center for Digital Manufacturing Technology, Northwestern Polytechnical University, Xi'an 710072, P.R. China. E-mail: jianjun@nwpu.edu.cn.

IOP Conf. Series: Materials Science and Engineering **87** (2015) 012103 doi:10.1088/1757-899X/87/1/012103

nano-particles in composite applications, which have several advantages over other surface coating process, such as simple process, homogeneity of the deposited films and easy control of the deposited thickness [10]. All these characteristics can be used for the preparation of CNTs/CF hybrid fiber to avoid these problems.

Most research reported improved properties of composites. However, limited attention has been paid to the influence of deposited CNTs on the surface of carbon fibers. In this study, a facile route is proposed to prepare carbon nanotubes/ carbon fiber hybrid fiber via electrophoretic deposition (EPD) without complex chemical reactions. Special emphasis has been put on the influence of CNTs coatings on carbon fiber by ultrasonically assisted electrophoretic deposition on its surface property. Surface roughness and surface morphologies of the fibers were examined using atomic force microscopy (AFM) and scanning electron microscopy (SEM), changes in surface chemistry were studied using fourier transform infrared spectrometer (FTIR).

2. Experimental

2.1. Materials and processing

PAN-based CF tow (T700, 12 K and diameter about 7 um) was purchased from Toray Industries, Japan. Carbon nanotubes (95% purity, diameter about 10–20 nm, length 10–30 um) were purchased from Nanjing Xianfeng. Port. Co., Ltd. The epoxy resin was bought from Yueyang Chemical Reagent Co., Ltd. of China. The curing agent was supplied by Shanghai jingchun Chemical Reagent Co., Ltd. of China.

The CF tow was refluxed in acetone at 80°C for 72 h, then washed with deionized water repeatedly and dried under vacuum at 80°C for 3 h to remove sizing agent (labeling as desized CF).

The carbon nanotubes was refluxed in a 3:1 (v: v) mixture of concentrated sulfuric acid and nitric at 100℃ for 4h to introduce carboxylic acid groups, then diluted with deionized water and suction filtering, deionized water washing to neutral.

2.2. Preparation of CNTs/CF hybrid fiber

A schematic of the EPD setup is shown in figure 1. For the EPD of CNTs onto the CF, the CF tow was used as the deposition electrode, and two graphite plates were used as the counter electrodes. In order to obtain 0.05 mg/mL CNTs dispersions, we disperse CNTs in deionized water by ultrasonic. The EPD process was carried out at constant voltages of 20 V for 20 min; the sonicator (60 W, 40 KHz) used to provide ultrasonic during ultrasonically assisted EPD. After EPD process, sonicator was switched off and the samples were dried in air at room temperature.

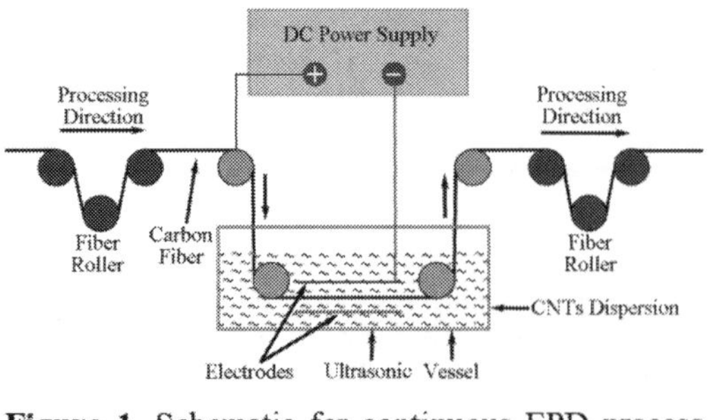

Figure 1. Schematic for continuous EPD process of CNTs onto carbon fiber.

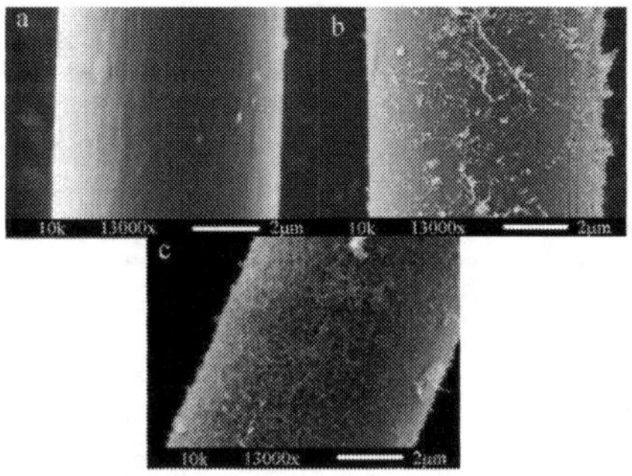

Figure 2. SEM images of the surface of carbon fiber; (a) desized; (b) CNTs deposited without ultrasonic; (c) CNTs deposited with ultrasonic.

IOP Conf. Series: Materials Science and Engineering **87** (2015) 012103 doi:10.1088/1757-899X/87/1/012103

2.3. *Characterizations*

2.3.1. Scanning electron microscopy. Surface morphologies of the CF was characterized using scanning electron microscope at 10 kV (JOEL, JCM-6000). Fiber samples were cut with scissors to observe a transverse cross-section for SEM observation

2.3.2. Atomic force microscopy. Surface roughness of CF was characterized by atomic force microscopy (Park, XE-100). All AFM images of CF used the tapping mode to obtain with the scan area of 4 mm×4 mm.

2.3.3. *Fourier transform infrared spectrometer.* Fourier transform infrared spectra (FTIR) were conducted on Nicolet 20DXB 60,000 spectrophotometer with the range of 450–4000 cm^{-1} to evaluate the chemical structures using a KBr pellet.

3. Results and discussion

3.1. *Surface roughness and surface morphologies of carbon fiber*

Figure 2 shows typical SEM micrographs of the surface of CF at different process stages. Initially, the surface of desized CF was rough and there are some grooves on the surface of CF (figure 2a). Figures 2(b) and 2(c) shows that the surface of CF has been deposited CNTs. It also shows that EPD process can prepare CNTs/CF hybrid fiber successfully. Meanwhile, homogeneous CNTs films due to ultrasonic can be seen from the SEM images of CNTs/CF hybrid fiber, the surface of CF with CNTs deposited showed some slight concave pits and convex hills without ultrasonic (figure 2b), whereas the surface of the CF showed much more convex hills with ultrasonic and homogeneously cover the whole surface (figure 2c). It appears that ultrasonic during the EPD process can increase the quality and quantity of depositing CNTs.

In order to have better understanding of the surfaces, AFM studies have been carried out. Figure 3 shows the increased roughness of CF surface with the CNTs deposition. It can be found that desized CF (figure 3a) shows a smooth surface with streaks, as shown in figures 3b and 3c), the introduction of CNTs can increase the surface roughness of CF obviously. In addition, the surface roughness increased when introduce ultrasonic to deposition of CNTs on the CF surface (figures 3b and 3c). The increase of surface roughness is beneficial to enhance mechanical interlocking between the fiber and matrix [9].

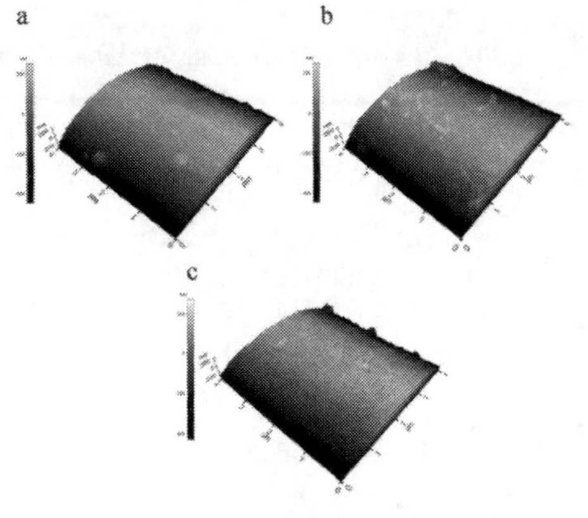

Figure 3. AFM images of CF (a) desized; (b) CNTs deposited without ultrasonic; (c) CNTs deposited with ultrasonic.

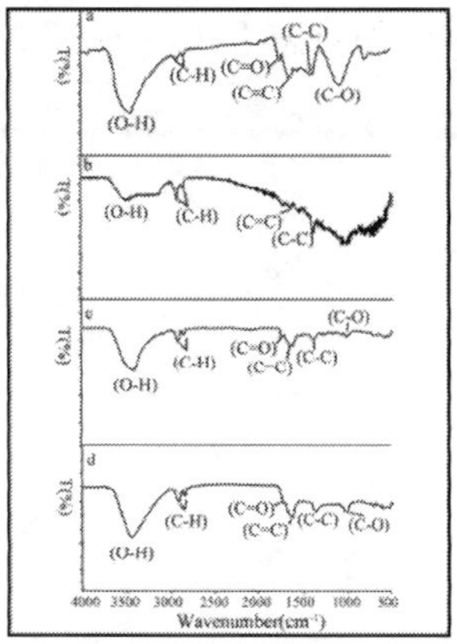

Figure 4. FTIR spectra of (a) CNTs; (b) desized CF; (c) CNTs deposited without ultrasonic; (d) CNTs deposited with ultrasonic.

3.2. Surface functional group of carbon fiber

The FTIR was used to determine the functionalization during EPD process, as shown in figure 4. The spectrum of CNTs-COOH showed peaks at 1731 cm^{-1}, 1628 cm^{-1}, and 1042 cm^{-1}, which are assigned to C=O stretching vibration, C=C deformation vibration, C-O stretching vibration (figure 4a) [11], respectively. For the desized CF, four peaks located at 3468 cm^{-1}, 2925 cm^{-1}, 1644 cm^{-1}, 1384 cm^{-1}in the FTIR spectrum are observed (figure 4b), which are allocated to ν(O-H), ν(C-H), ν(C=C), ν(C-C). After depositing CNTs on the desized CF without ultrasonic, ν(C=O) at 1731 cm^{-1} are visible (figure 4c), which means carboxyl are formed as a result of CNTs were successfully deposited on the surface of CF. For depositing CNTs on the desized CF with ultrasonic, the broader peaks of ν(C=O) at 1731 cm^{-1} (figure 4d) indicate that the amount of carbon nanotubes were increased by ultrasonically assisted EPD process.

4. Conclusions

In summary, carbon nanotubes was successfully coated onto carbon fiber surfaces via ultrasonically assisted electrophoretic deposition, which avoids complex chemical reactions, therefore provides great potential for industrial applications. SEM and AFM analysis showed that carbon nanotubes film with high roughness is homogeneously formed on carbon fibers. Comparing the results with EPD-only, EPD with ultrasonic increased the amount and homogeneity of CNTs coatings, this comes from ultrasonic improves the dispersion of CNTs and inhibits water electrolysis reaction.

References

[1] Chand S 2000 Review carbon fibers for composites *J. Mater. Sci.* **35** 1303-13

[2] Akbar D and Gungor U E 2014 Study of high radio frequency plasma discharge effects on carbon fiber using raman spectroscopy *Surf. Coat. Technol.* **240** 233-42

[3] Cao H L, Huang Y D, Zhang Z Q and Sun J T 2005 Uniform modification of carbon fibers surface in 3-D fabrics using intermittent electrochemical treatment *Compos. Sci. Technol.* 65 1655-62

[4] Liu J, Ge H, Chen J, Wang D and Liu H 2012 The preparation of emulsion type sizing agent for carbon fiber and the properties of carbon fiber/vinyl ester resin composites *J. Appl. Polym. Sci.* **124** 864-72

[5] Xie J F, Xin D W and Cao H Y 2011 Improving carbon fiber adhesion to polyimide with atmospheric pressure plasma treatment *Surf. Coat. Technol.* **206** 191-201

[6] Bekyarova E, Thostenson E T and Yu A 2007 Multiscale carbon nanotube-carbon fiber reinforcement for advanced epoxy composites *Langmuir.* **23** 3970-4

[7] Thostenson E, Li W, Wang D, Ren Z and Chou T 2002 Carbon nanotube/carbon fiber hybrid multiscale composites *J. Appl. Phys.* **91** 6034-7

[8] Greef N D, Zhang L and Magrez A 2015 Direct growth of carbon nanotubes on carbon fibers: Effect of the CVD parameters on the degradation of mechanical properties of carbon fibers *Diamond. Relat. Mater.* **51** 39-48

[9] Guo J H, Lu C X and An F 2012 Effect of electrophoretically deposited carbon nanotubes on the interface of carbon fiber reinforced epoxy composite *J. Mater. Sci.* **47** 2831-6

[10] Subramanian P, Jonsson J N, Lesniewski A, Wang Q and Li M S 2014 Preparation of reduced graphene oxide-$Ni(OH)_2$ composites by electrophoretic deposition: application for nonenzymatic glucose sensing *J. Mater. Chem.* **2** 5525-33

[11] Byrappa K, Dayananda A S, Sajan C P and Basavalingu B 2008 Hydrothermal preparation of ZnO: CNT and TiO_2: CNT composites and their photocatalytic applications *J. Mater. Sci.* **43** 2348-55

IOP Conf. Series: Materials Science and Engineering **87** (2015) 012104 doi:10.1088/1757-899X/87/1/012104

Density function theory study on structures and electronic properties of $2Meq_2AlOPh$ and its derivatives

Yanju Ji[1], Xiaocun Liu, Gang Fu, Ying Chen, Fengxiang Wang and Junqing Zhao

School of Science, Shandong Jianzhu University, Jinan 250101, China

E-mail: jiyanju@sdjzu.edu.cn

Abstract. Structures and electronic properties of $2Meq_2AlOC_6H_4R$ (R=CN, Cl, H, OMe, NMe_2) have been studied in the scheme of density function theory with the B3LYP and the 6-31G (d) basis set. The substitutions of the electron-withdrawing groups (CN, Cl) increase the band gap while the substitutions of the electron-donating groups (OMe, NMe_2) decrease the gap. The substitutions of electron-withdrawing groups enhance the electron injection ability. In addition, the vertical ionization energies (IE_V), the vertical electronic affinities (EA_V) and reorganization energies of $2Meq_2AlOPh$ and its derivatives are calculated in the work. All complexes are better electron transport materials.

1. Introduction

Since the discovery of Organic light-emitting device (OLEDs) by Tang C W [1] and Burroughes J H [2], OLEDs receive much attention due to its unique properties [3-8]. Tris (8-hydroxyquinoline) aluminum (Alq_3) has been a popular compound because of its green emission, good electron mobility, and thermal stability up to 350°C [9]. Numerous studies, both theoretical and experimental, have been carried out on the derivatives of Alq_3 [10-15]. As an efficient green emitter with emission maximum in the range of 526 nm, Alq_3 could meet the requirement of green OLEDs. Until now, however, blue electroluminescent materials derived from organometallic complexes are rare. Leung et al. [16] synthesized a high-efficiency blue dye bis (2-methyl-8-quinolinolato) aluminum (III) hydroxide complex ($2MAlq_2OH$). Another complex, bis (2-methyl-8-quinolinolato) (phenolate) aluminum (III) ($2Meq_2AlOPh$), showed a higher chemical stability [17, 18].

In general, an OLED consists of a series of sequentially-deposited organic thin films sandwiched between two electrodes. Electrons which are injected from cathode should fill in the lowest unoccupied molecular orbital (LUMO) while holes injected from the anode should concentrate on the highest occupied molecular orbital (HOMO). The electrons can recombine the holes to form electron-hole pairs to emit light. The information about the ease of the electron injection can be given by discussing the injection energy from cathode to the complex.

In the present study, we report the theoretical study on the geometry structures and electronic structures of $2Meq_2AlOC_6H_4R$ (R=CN, Cl, H, OMe, NMe_2). Comparison of charge transfer properties of the five complexes was carried out.

[1] Address for correspondence: Yanju Ji, School of Science, Shandong Jianzhu University, Jinan 250101, China. E-mail: jiyanju@sdjzu.edu.cn.

2. Calculation method and models

The molecular models used in our calculations are given in figure 1, where the labels A-C represent the three different ligands. The geometries of $2Meq_2AlOC_6H_4R$ (R=CN, Cl, H, OMe, NMe_2) in their neutral, cationic, and anionic states were fully optimized using Becke's three-parameter hybrid functional [19] combined with the Lee, Yang and Parr (LYP) correlation functional [20], denoted B3LYP [21] with the 6-31G(d) [22-25] basis, which have given reliable results for organic materials [26-28]. The frontier orbital energies and molecular orbital maps of HOMO and LUMO for complexes were also given using B3LYP with 6-31G (d). All the calculations results have been performed using Gaussian 03 package [29].

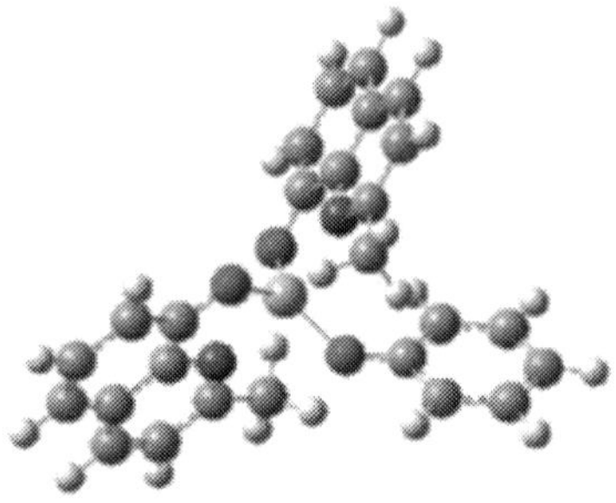

a

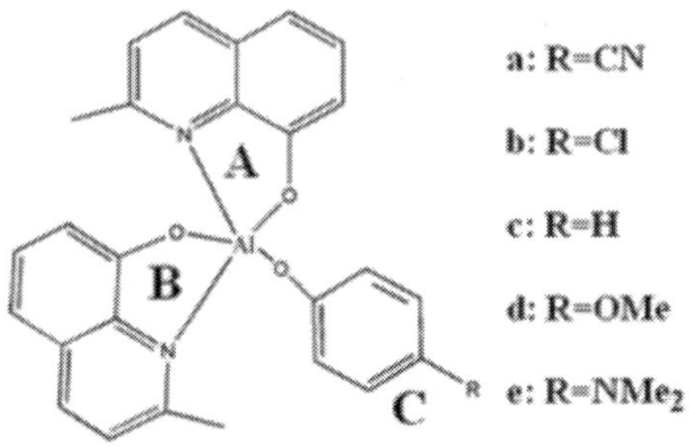

b

Figure 1. (a) The geometry of $2Meq_2AlOPh$. (b) Calculation models of $2Meq_2AlOPh$ and its derivatives with labels A-C for three ligands.

3. Results and discussion

3.1. Molecular geometries

Some optimized bond lengths for $2Meq_2AlOPh$ and its derivatives are presented in table 1. In comparison with complex c ($2Meq_2AlOPh$), there are slightly geometrical changes in the four derivatives. The Al-N_A bond is predicted to be negligible shortened in complex a while lengthened in complexes b, d, and e. The Al-N_B, Al-O_A, Al-O_B bonds are predicted to be shorted while the Al-O_C bonds are lengthened in complexes a and b. The bond lengths of Al-N and Al-O of complex d are almost similar as in e.

Table 1. Selected optimized bond lengths in angstrom (Å) for $2Meq_2AlOPh$ and its derivatives at the B3LYP/6-31G* level

Parameters	a	b	c	d	e
Al-N_A	2.086	2.089	2.087	2.090	2.091
Al-N_B	2.073	2.075	2.076	2.078	2.078
Al-O_A	1.815	1.817	1.820	1.820	1.820
Al-O_B	1.815	1.818	1.820	1.821	1.822
Al-O_C	1.776	1.768	1.764	1.761	1.761

3.2. Frontier molecular orbitals

The energies for the frontier molecular orbital energies and energy gaps of complexes a-e are shown in figure 2 and the molecular orbital maps of HOMO and LUMO for complexes b-d are shown in figure 3. The LUMO levels in complexes a-e are not significantly changed while the HOMOs are changed significantly. We reason that due to the molecular orbital maps of HOMO and LUMO for the complexes. The LUMOs are mostly localized on A and B but rarely on C while the HOMOs are localized mostly on C. The metal Al only plays a role in supporting and restricting ligand. It doesn't contribute to HOMO and LUMO.

The substitutions of electron-withdrawing groups (CN, Cl) increase the gap, while the substitutions of an electron-donating group (OMe, NMe_2) decrease the band gap. Complex a shows the highest HOMO-LUMO gap of 3.74eV and complex e shows the lowest HOMO-LUMO gap of 2.87 eV. A gradual decrease in the HOMO-LUMO gap comes from electron-withdrawing groups to electron-donating groups, a>b>c>d>e. As already known, the magnitudes of the HOMO-LUMO gap correlate with the emission properties. The smaller the gap is, the longer the emission wavelength is. Thus, the substitutions of electron-withdrawing groups (CN, Cl) show blue shifts, while the substitutions of electron-donating group (OMe, NMe_2) show red shifts.

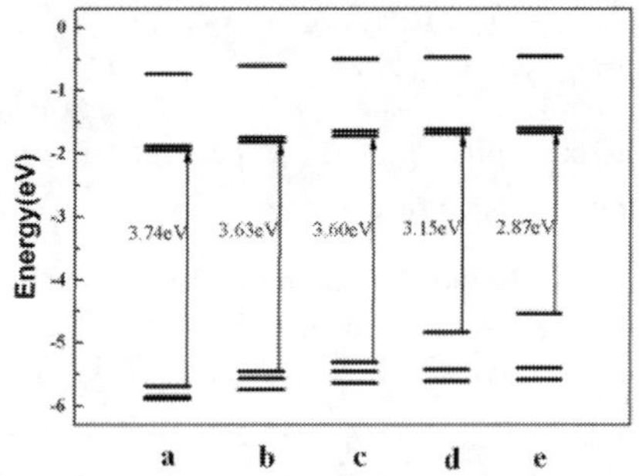

Figure 2. HOMO-LUMO gap for complexes a-e.

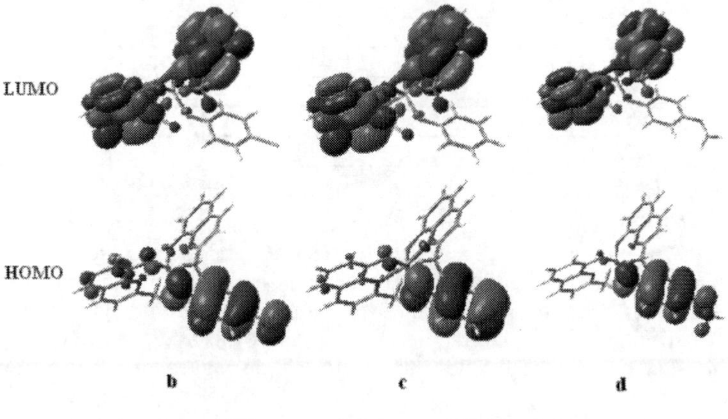

Figure 3. Molecular orbital maps of HOMO and LUMO for complexes b-d.

3.3. *Electron injection*

The information about the ease of the electron injection can be given by discussing the injection energy (E_{inj}) from a cathode (as in Al) to the complex. The E_{inj} is calculated as follow:

$$E_{inj}=E_{LUMO}-E_{Al} \quad (1)$$

where E_{LUMO} are the LUMO energies, E_{Al} is the work function of Al.

Obviously, a low LUMO energy is favorable for enhancing the electron injection ability. Since the LUMO energies (eV) of complexes a-e are -1.95, -1.82, -1.71, -1.68, -1.66 eV and the work function of Al is -4.3 eV [30], the injection energies are 2.35, 2.48, 2.59, 2.62 and 2.64 eV. For the five complexes, the order of the calculated E_{injs} is a<b<c<d<e. The substitutions of electron-withdrawing groups (CN, Cl) enhance the electron injection ability.

3.4. Reorganization energies

The charge transfer rate in the organic solid can be accounted for by the Marcus theory [31] via the following equation:

$$K_{h/e} = \frac{V_{h/e}}{\hbar} \times \left(\frac{\pi}{\lambda_{h/e} kT} \right)^{\frac{1}{2}} \times \exp\left(-\frac{\lambda_{h/e}}{4kT} \right) \tag{2}$$

where k and ħ, respectively, refer to the Boltzmann and Plank constant, T the temperature, $V_{h/e}$ the electronic couplings between adjacent molecules and $\lambda_{h/e}$ the reorganization energies for hole or electron carriers.

Obviously, the electron transfer rate is determined by two major parameters: $V_{h/e}$ and $\lambda_{h/e}$. It has reported that $V_{h/e}$ shows a narrow range of values [32, 33], therefore, the charge transfer rate is dominated by the reorganization energies $\lambda_{h/e}$. The smaller the reorganization energies, the better the carrier transfer properties. The hole/electron reorganization energies are as

$$\lambda_h = IE_V\text{-}HEP;\ \lambda e = EEP\text{-}\ EA_V \tag{3}$$

where HEP is the hole extraction potential and EEP is the electron extraction potential.

The vertical ionization energies (IE_V), the vertical electronic affinities (EA_V) and reorganization energies of $2Meq_2AlOPh$ and its derivatives are shown in table 2.

Table 2. Vertical ionization energies, vertical electronic affinities, hole/electron extraction potentials (HEP/EEP) and hole/electron reorganization energies (λ_h/λ_e) (in eV) for complexes a-e at the DFT/B3LYP/6-31G* level.

	a	b	c	d	e
IE_V	6.917	6.703	6.617	6.344	6.086
EA_V	0.693	0.553	0.448	0.418	0.407
HEP	6.797	6.530	6.461	5.667	5.128
EEP	0.937	0.792	0.667	0.685	0.649
λ_h	0.120	0.173	0.156	0.677	0.956
λ_e	0.244	0.239	0.219	0.267	0.242

The calculated vertical ionization energy and vertical electronic affinity for $2Meq_2AlOPh$ are 6.617 eV and 0.448 eV. In line with the change in their HOMO and LUMO, substitutions of electron-withdrawing groups increase the vertical ionization energy and vertical electronic affinity, while the substitutions of electron-donating groups decrease them. Both the vertical ionization energy and vertical electronic affinity for all of the complexes are in following order: a>b>c>d>e.

The reorganization energies of the derivatives are a little higher than that of $2Meq_2AlOPh$ for electron while the hole reorganization energies among all complexes are quite different. The hole reorganization energy of complex a is smaller than that of complex c. But the hole reorganization energies of complexes d and e are quite large. Based on equation (2), the smaller the reorganization energies, the better the carrier transfer properties. The calculated electron reorganization energies of derivatives are higher than that of $2Meq_2AlOPh$, but the effect is not significant. As a result, $2Meq_2AlOPh$ and its derivatives are better electron transport materials.

4. Conclusions

The geometries and electronic structures of $2Meq_2AlOC_6H_4R$ (R=CN, Cl, H, OMe, NMe_2) were fully optimized using B3LYP with 6-31G (d) set. The structures of derivatives change slightly. The LUMOs are mostly localized on A and B but rarely on C, while the HOMOs are localized mostly on C and the metal Al doesn't contribute to HOMO and LUMO. The substitutions of electron-withdrawing groups increase the gap and enhance the electron injection ability. The vertical ionization energies, the vertical electronic affinities and reorganization energies of $2Meq_2AlOPh$ and its derivatives are given. All complexes are better charge transport materials.

Acknowledgment
This work was supported by the National Natural Science Foundation of China under Grant Nos 11005070, 51042002 and 11075097, and the Natural Science Foundation of Shandong Province, China (ZR2009FM031).

References

[1] Tang C W and VanSlyke S A 1987 *Appl. Phys. Lett.* **51** 913
[2] Burroughes J H, Jones C A and Friend R H 1988 *Nature* **335** 137
[3] Burroughes J H, Bradley D D C, Brown A R, Marks R N, Mackay K, Friend R H, Burns P L and Homes A B 1990 *Nature* **347** 539
[4] Forrest S R 1989 *IEEE Circuits Device Mag.* **5** 33
[5] Adachi C, Tokito S, Tsutsui T and Saito S 1988 *Jpn. J. Appl. Phys.* **27** L269/L713
[6] Goetzberger A, Luther J and Willeke G 2002 *Solar Energy Mater. Solar Cells* **74** 1
[7] Choi J H, Kim K H, Choi S J and Lee H H 2006 *Nanotechnology* **17** 2246
[8] Sun Y, Giebink N C, Kanno H, Ma B, Thompson M E and Forrest S R 2006 *Nature* **440** 908
[9] Chen C H and Shi J 1998 *Coord. Chem. Rev.* **171** 161
[10] Zhang J and Frenking G 2004 *Chem. Phys. Lett.* **394** 120
[11] Gahungu G and Zhang J 2005 *J. Phys. Chem.* B **109** 17762
[12] Wang W, Dong S, Yin S, Yang J and Lu J 2008 *J. Mol. Struct.* (*THEOCHEM*) **867** 116
[13] Wu C C, Lin Y T, Wong K T, Chen R T and Chen Y Y 2004 *Adv. Mater.* **16** 61
[14] Chan L H, Lee R H, Hsieh C F, Yeh H C and Chen C T 2002 *J. Am. Chem. Soc.* **124** 6469
[15] Irfan A, Cui R H and Zhang J P 2009 *Theor. Chem. Acc.* **122** 275
[16] Leung L M, Lo W Y, So S K, Lee K M and Choi W K 2000 *J. Am. Chem. Soc.* **122** 5640
[17] Qiu Y, Shao Y, Zhang D and Hong X 2000 *Jpn. J. Appl. Phys.* **39** 1151
[18] Shen Z, Burrows P E, Bulovic V, Forrest S R and Thompson M E 1997 *Science* **276** 2009
[19] Becke A D 1993 *J. Chem. Phys.* **98** 5648
[20] Lee C, Yang W and Parr R G 1988 *Phys. Rev.* B **37** 785
[21] Stephens P J, Devlin F J, Chabalowski C F and Frisch M J 1994 *J. Phys. Chem.* **98** 11623
[22] Hariharan P C and Pople J A 1974 *Mol. Phys.* **27** 209
[23] Gordon M S 1980 *Chem. Phys. Lett.* **76** 163
[24] Frisch M J, Pople J A and Binkley J S 1984 *J. Chem. Phys.* **80** 3265
[25] Foresman J B, Head-Gordon M, Pople J A and Frisch M J 1992 *J. Phys. Chem.* **96** 135
[26] Zhang R Q, Lu W C, Lee C S, Hung L S and Lee S T 2002 *J. Chem. Phys.* **116** 8827
[27] Irfan A, Cui R and Zhang J 2010 *J. Mol. Struct.* (*THEOCHEM*) **956** 61
[28] Gahungu G, Zhang J, Ntakarutimana V and Gahungu N 2010 *J. Phys. Chem.* A **114** 652
[29] Frisch M J et al. 2003 *GAUSSIAN 03 Revision B. 03* (Pittsburgh: Gaussian Inc.)
[30] Michaelon H B 1997 *J. Appl. Phys.* **48** 4729
[31] Marcus R A and Sutin N 1985 *Biochim Biophys Acta* **811** 265
[32] Nelsen S F, Trieber D A, Ismagilov R F and Teki Y 2001 *J. Am. Chem. Soc.* **123** 5684
[33] Nelsen S F and Blomgren F 2001 *J. Org. Chem.* **66** 6551

The energy saving research of the flat tire vulcanization process

X Tang[1, 2], J-Y He[1], J Y Zhang[1, 2], Y An[1], W-M Yang[1] and J Tan[1, 2, 3]

[1]Beijing University of Chemical Technology, Beijing 100029, China
[2]The National Engineering Laboratory of Tire Design and Manufacturing Process, Beijing 100029, China

E-mail: 18612948131@163.com

Abstract. Vulcanization was an important step in tire production process, directly determining the mechanical properties of the tire which was the most energy-intensive link. The traditional vulcanization process of a flat tire did not consider the post curing effect, which may result in a waste of energy. Early finite element simulation showed that the traditional process caused excessive vulcanization. In order to find ways to improve the curing process, in the paper effect of process parameters on the flat tire vulcanization had been studied. Curing parameters included heating temperature, curing time and preheating temperature. Corresponding to these three factors, the three-dimensional model was established by ABAQUS finite element software to carry out three groups of simulation. Result showed that heat source temperature on the maximum temperature was the largest, effect of preheating temperature was the lowest; heat source temperature on the curing degree was the largest, effect of the curing time was lowest. According to the data analysis the optimal process parameters had been raised.

1. Introduction

The developments of automobile industry put forward higher requirements for tire safety and comfort. The tire breakthrough was not only made in the structure design but also in tire curing process. The design of tire curing process could not stay in experience or half experience, but by simulating the actual vulcanization process and calculating vulcanization degree of rubber, the reasonable curing conditions would be determined to improve overall tire performance [1]. Tire had complex structure and components, with various parts of different thickness, in curing process some parts would over cure and some parts would under cure. So understanding tire curing process had the vital significance for curing process optimization. By improving the structure, changing material and curing process, the best mechanical properties of tire were obtained with the lowest heat, so as to achieve the purpose of energy saving and low carbon. In the factory tire curing temperature was generally measured and curing degree was calculated. Compared with computer simulation, actual measurement method was time-consuming, and did not consider the influence of curing degree after vulcanization, conservative process would result in a waste of energy. Vulcanization was the most energy-intensive link in tire

[3] Address for correspondence: J Tan, Beijing University of Chemical Technology, Beijing 100029, China, The National Engineering Laboratory of Tire Design and Manufacturing Process, Beijing 100029, China

IOP Conf. Series: Materials Science and Engineering **87** (2015) 012105 doi:10.1088/1757-899X/87/1/012105

manufacturing. Through the computer finite element simulation, curing temperature and curing time were adjusted according to the simulation results, ultimately achieving the purpose of energy saving.

In this paper, with 255/30R22 tire as example, the finite element model was established by using the software of ABAQUS, the three groups of simulation was respectively carried, changing the heat source temperature, curing time and preheating temperature. Reaction heat was considered in three sets of simulations. Subroutine HETVAL was loaded by curing reaction heat. The temperature fields at sampling points were obtained. The subroutine UVARM was compiled to get curing degree field [2]. The establishment of finite element model, the simulation of temperature field and the calculation of vulcanization degree referred to articles [3-5].

In order to study the effect of sulfuration parameters on rubber vulcanization, the three comparison simulation had been designed. In the first group the vulcanization time was changed, set as 780 s, 720 s and 660 s. In the second group steam temperature was changed, set as 180℃, 170℃ and 170℃. In the third group preheating temperature or initial setting temperature was set as 150℃, 140℃ and 130℃.

2. The finite element simulation

2.1. The finite element model

Throughout the development of numerical simulation of tire curing it could be found that the heat transfer properties of anisotropic composite materials were not considered. In general, the mechanical properties of composites were anisotropic, and its thermal properties were also anisotropic [6]. The two-dimensional axisymmetric model and the three-dimensional model were as figure 1. In this thesis, curing temperature field and cure degree field of 255/30R22 tire had been researched. Figure 2 was sampling point diagram.

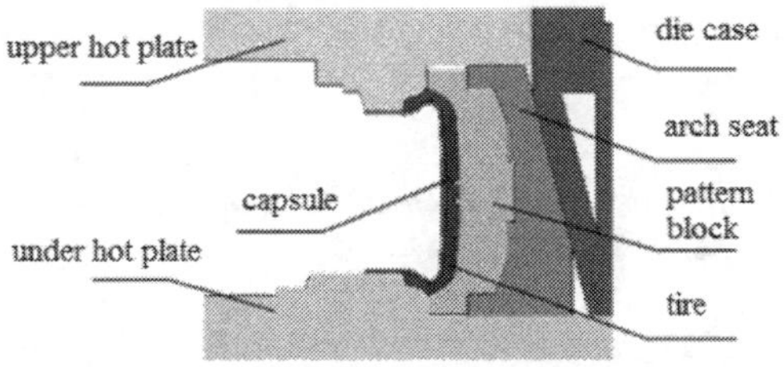

(a) The two-dimensional axisymmetric model

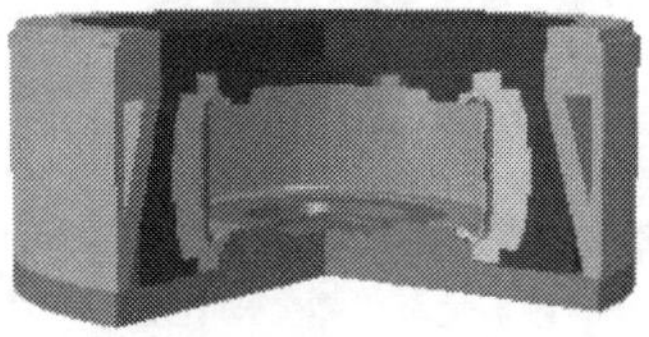

(b) 3 d model

Figure 1. Finite element model of tire vulcanization.

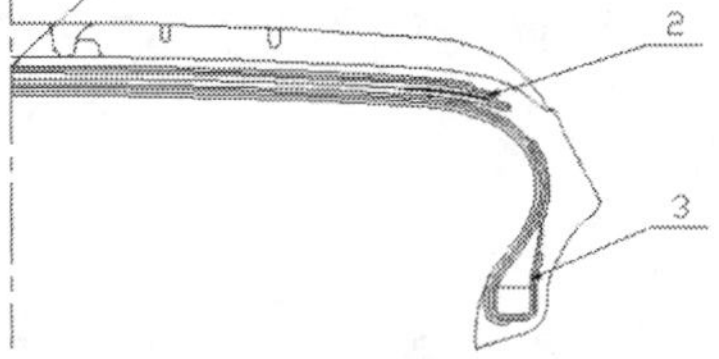

Figure 2. Sampling point diagram.
[a]1 point between tread and crown.
[b]2 point at tire shoulder.
[c]3 point at triangle rubber.

2.2. Vulcanization process

Although specific heat capacity and thermal conductivity of each rubber were similar, in order to ensure accuracy, change of thermal physical parameters with temperature was still considered [7]. According to the actual conditions of experiment, the initial temperature of the tire was set as 20℃, the initial temperatures of capsule and mold were 150℃. Vulcanization process was divided into five stages, continuing for 1540s. The boundary conditions of each phase were as follows:

The first phase was that the tire was in the mould. Upper hot plate and under hot plate were set as 180℃. This process lasted for 580s. The second stage was the tire was taken out from the mold. The entire tire was exposed to the air, the main way of heat dissipation was radiation and the radiation factor was 0.94.

3. Simulation results and analysis

3.1. The influence of curing time

Early by comparison of temperature simulation values and the measured temperature values at sampling points it showed that the simulation with reaction heat was accurate, simulation temperature curve and measured temperature curve were almost identical at each point. In this section the simulated values were no longer compared with the measured values. Figure 3 showed that temperature reduced by 5℃ as the vulcanization time reduced by 60s. In the previous stage of vulcanization and after cure stage temperature curves were the same at each point, just the highest temperature and its corresponding time were changed. Cure degree of the triangle rubber was the highest. As vulcanization time reduced by 60 s, cure degree at each point reduced by 0.3. In figure 4, G curve was the optimal cure degree. By comparison the vulcanization time could be optimized for 600 s.

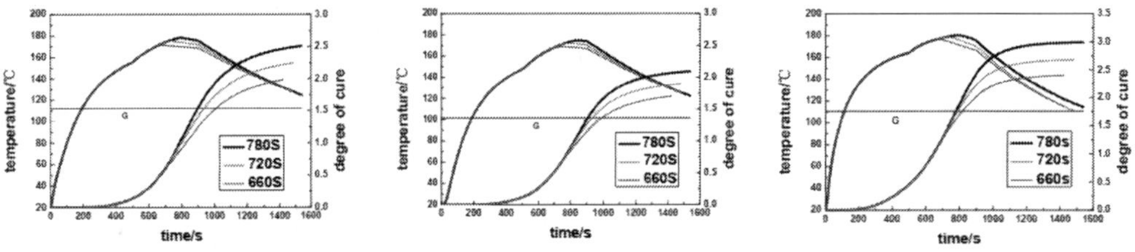

(a) 1 point (b) 2 point (c) 3 point

Figure 3. The temperature and vulcanization degree curves over time at the sampling points.

3.2. The influence of heat source temperature

Figure 4 showed that impact of the heat source change on initial curing was not obvious, it was because in the initial vulcanization stage the tire was heated by residual heat in the mould. The change of the heat source changed the highest temperature and leaded to curve translation. As heat source temperature reduced by10℃, the highest temperature at the sampling point reduced by 7℃. The effect of heat source on the cure degree was larger than the effect of curing time. As the heat source temperature reduced by10℃, vulcanization degree reduced by 1. Heat source temperature of the traditional process could be optimized for 170℃.

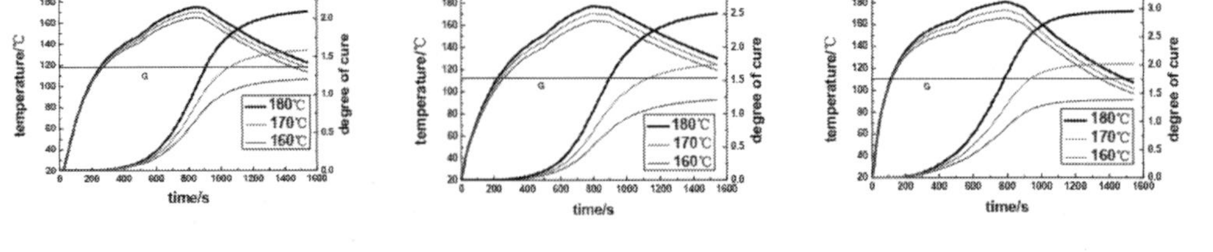

(a) 1 point (b) 2 point (c) 3 point

Figure 4. The temperature and vulcanization degree curves over time at the sampling points.

3.3. The influence of preheat temperature

Figure 5 showed that when preheat temperature reduced by 10℃, the temperature curve at each point only translated downward for a small distance. The impact of preheat temperature on the temperature curve was the weakest among three factors. But with time accumulated preheat temperature had large effect on cure degree, second only to the influence of the curing time. The optimal preheat temperature was 130℃.

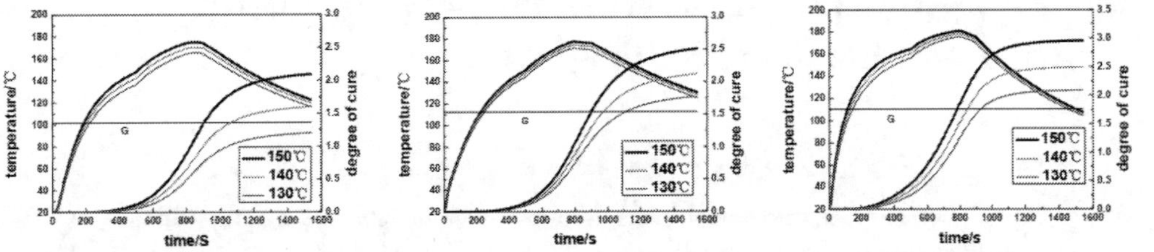

(a) 1point (b) 2point (c) 3point

Figure 5. The temperature and vulcanization degree curves over time at the sampling points.

4. Conclusions

Based on curing time, heat source temperature, preheating temperature, three groups of simulations had been carried out in the paper. The following conclusions could be drawn from temperature curves and cure degree curves:

- According to the influence on the highest temperature from high to low, factors were the heat source temperature, curing time, preheating temperature. According to the influence on vulcanization degree from high to low, factors were the heat source temperature, preheating temperature, curing time.
- The traditional crafts had optimizable space. By only adjusting the single factor, vulcanization time could be adjusted to the 600 s, heat source temperature could be adjusted to 170℃, preheating temperature could be adjusted to130℃.

References

[1] Wei Zhang, Tao Deng and Shugao Zhao 2006 The tire vulcanization process condition optimization Rubber Industry 26 225-7

[2] Eckert E R G and Drake R M 1972 *Analysis of Heat and Mass Transfer* (NewYork: McGraw-Hill) pp 1-49

[3] Shugao Zhao, Zhang Ping, Deng Tao, et al. 2001 The numerical simulation of tire vulcanization process Rubber Industry 21 617-22

[4] Xiuping Su 2000 *The Numerical Simulation of Tire Vulcanization Process and Cure Degree Analysis* (Qingdao: Qingdao Institute of Chemical Industry)

[5] Qingzhi Yang 1997 *Modern Rubber Technology* (Beijing: China Petrochemical Press)

[6] Benlong Li 2010 *Giant Radial Tire Vulcanizing Simulation Research* (China: Harbin Institute of Technology)

[7] Wang Yue 2009 *Tire Vulcanizing Temperature Field Finite Element Analysis* (Japan: Tohoku University)

The preparation of flame-retardant poly (ethylene terephthalate) fabric by ionic agent

F Xu[1,2], F X Zhang[3], Y S Zhang[1,2] and G X Zhang[1,2,4]

[1]College of Textile & Garments, Southwest University, Chongqing, China
[2]Chongqing Engineering Research Center of Biomaterial Fiber and Modern Textile, Chongqing, China
[3]School of Chemistry and Chemical Engineering, Southwest University, Chongqing, China

E-mail: zgx656472@sina.com

Abstract. In this paper, PET fabric was modified by the flame retardant HOP to improve flame retardance. The results showed that the flame retardance of PET fabric improved a lot. With the increase of the HOP concentration, the limiting oxygen index could increase to 34%, the horizontal burning length could decrease to 0 mm, the horizontal burning rate could decrease to 0 mm/s, and the vertical ignite time could increase to 5 s. The infrared spectroscopy showed the modification of HOP made PET fabric gain phosphate groups. Moreover, the breaking strength, stiffness and crease recovery angle were kept well.

1. Introduction

Poly (ethylene terephthalate) (PET) is wildly used in many fields, such as clothing, home furnishings, engineering domain and so on due to its many outstanding properties like good chemical stability and excellent mechanical properties [1]. However, PET fabric is combustible fabric and it may cause fire which will end up with bodily injury and financial loss [2]. Therefore, the flame retardant modification has great significance and there are many studies intending to impart PET fabric with flame retardance [3-5].

It is commonly to use intumescent flame retardant and phosphorus-based flame retardant to improve the flame retardance of PET fabric. The flame retardant modification by using intumescent flame retardant belongs to gaseous phase flame retardancy that has high flame retardant efficiency [6], while the flame retardant modification by using phosphorus-based flame retardant belongs to solid phase flame retardancy that reduces combustible gas [7]. Considering of the safety of some intumescent flame retardants, phosphorus-based flame retardants are mainly applied in fabric flame retardant.

The aim of this study was to impart high-efficiency flame retardance to PET fabrics using the flame retardant HOP. The flame retardant HOP can introduce phosphorus into PET macromolecular by chemical reaction in order to improve the flame retardance of PET fabric. This method for modifying PET fabric is not only rapid but also efficient.

[4] Address for correspondence: G X Zhang, College of textile & garments, Southwest University, Chongqing, China. E-mail: zgx656472@sina.com.

2. Experimental

2.1. Material
Polyester fabric (27.0tex×22.2tex/500×360) was bought from Chongqing market. HOP [self-synthesized, $(HO)_2OP\text{-}R\text{-}PO(OH)_2$, molecular weight is 392, which react with PET fabrics to generate -PO(OH)-PET] was a flame retardant agent.

2.2. Flame retarded modification of PET fabrics
Flame retardant agent HOP could be dissolved in water. HOP solutions with different concentrations ranging from 10 g/L to 30 g/L were made with distilled water.

The PET fabric (50 cm × 70 cm) was immersed into HOP solution for 5 minutes at room temperature and the bath ratio was 1:20. After PET fabric was soaked, the PET fabric was dried at 60°C for 30 min and baked in an oven at 125°C for 25 min. The baked PET fabric was then washed and dried at 60°C. This procedure was repeated thrice.

2.3. Test for flame retardant

2.3.1. Burning behavior. The limiting oxygen index (LOI) values of different treatments PET fabrics were measured according to an international standard ASTM D2863-2000 with M606B digital oxygen index apparatus from Qingdao Shanfang Instrument Co. Ltd., Shandong, China.

The horizontal burning behavior of PET fabrics was measured according to ASTM D 5132-2004, the standard test method for horizontal burning rate of polymeric materials used in occupant compartments of motor vehicles with the YG815D horizontal fabric FR tester from Nantong Sansi Electromechanical Science & Technology Co. Ltd., China.

The vertical ignite time of the PET fabrics treated by HOP were measured by the YG815B vertical fabric FR tester from Nantong Sansi Electromechanical Science & Technology Co. Ltd., China, according to ISO 6940:2004, Textile fabrics-Burning behavior-Determination of ease of ignition of vertically oriented specimens, MOD.

2.3.2. Characterization of the Modified PET Fabric Samples. Fourier transform infrared (FTIR) spectra of the fabric samples were acquired with a Spectrum GX spectrometer (PE Co., U.S.). The spectra were acquired over the range 2500-400 cm^{-1} using the reflection-absorption technique.

2.3.3. Mechanical properties. The tensile strength of fabric sample was measured according to ASTM 5035-2006 standard with an Electronic fabric tension tester, HD026N (Nantong Hongda Experiment Instruments Co. Ltd., China).

Based on ASTM D 1388-96 (2002), the standard test method for stiffness of fabrics, the stiffness of fabric was measured using a fabric stiffness tester (Changzhou, China).

Crease recovery angles of treated fabrics were measured using a wrinkle recovery tester (Wenzhou, China) according to AATCC Test Method 66-2008 for wrinkle recovery of woven fabrics: recovery angle.

Table 1. The limiting oxygen index, horizontal burning behavior and the time to ignite the fabric of the polyester treated by HOP.

Concentration of HOP (g/L)	0	10	20	30
The limiting oxygen index (%)	21	26	28	34
Hrizontal burning length (mm)	254	74.3	32.5	0
Hrizontal burning rate (mm/s)	5.5	4.5	3.7	0
The time to ignite the fabric (s)	1	1	2	5

3. Results and discussion

3.1. Flammability of PET fabric modified by HOP

Table 1 showed the limiting oxygen index, horizontal burning behavior and vertical ignite time of the original PET fabric and the modified PET fabrics.

From the table, it can be observed that the limiting oxygen index increased when the concentration of HOP increased. The limiting oxygen index of original PET fabric was 21%, and it could increase to 34% when the PET fabric was modified by 30 g/L HOP solution for three times. The horizontal burning rate decreased quickly. The horizontal burning rate of original PET fabric was 5.5 mm/s and it decreased to 0 mm/s when the PET fabric modified by HOP with 30 g/L concentration for three times. The horizontal burning lengths of original and modified PET fabrics decreased quickly with the increase of the HOP concentration. The horizontal burning length of original PET fabric was 254 mm/s, the fire spread fast and the sample were burned out. The horizontal burning length could decrease to 0 mm when the PET fabric was modified by 30 g/L HOP solution for three times. And the time for the modified PET fabric to be ignited by vertical method became longer. It cost 1 s to ignite the original PET fabric, and the ignite time increased to 5 s after PET fabric was modified by 30 g/L HOP for three times. After modified by flame retardant HOP, the PET fabric was difficult to be ignited. The flame retardance of PET fabric improved greatly.

3.2. Characterization of the modified PET fabric samples

Figure 1 was the infrared spectroscopy of PET fabric before and after modified by HOP. The concentration of HOP was 30 g/L and the process was repeated thrice. As shown in the figure, the modified PET fiber had stronger absorption peaks at 570 cm^{-1}, 600 cm^{-1}, 650 cm^{-1}, 1270 cm^{-1} and 1610 cm^{-1}. The moderately strong absorption peaks 570 cm^{-1}, 600 cm^{-1}, 650 cm^{-1} were assigned to the P-C single bond, which indicated that there was a dehydration reaction between HOP and PET macromolecular, and phosphorus was grafted onto PET macromolecular. At the absorption peak of 1270 cm^{-1}, the absorption of modified PET fiber had strengthened. It was assigned to the P=O exists.

At 1610 cm^{-1}, a characteristic absorption peak of carboxylate appeared. It showed that there were transesterification and hydrolysis reaction between HOP and PET macromolecular. From the modification process, although PET fabric was soaked in HOP for 25 min, the water on the surface of fabric would rapidly volatilize in 125°C. There were a few chances for PET fiber to be hydrolysis. Carboxyl was supposed to appear by transesterification.

HOP is a flame retardant agent containing phosphorus, the structure of HOP is $(HO)_2OP\text{-}R\text{-}PO(OH)_2$. HOP can react with PET fabrics to generate -PO(OH)-PET, as shown in scheme 1. Therefore, the modification of HOP made PET fabric gain phosphate groups, and the PET fabric could get flame-retardant phosphorus to improve the flame retardance.

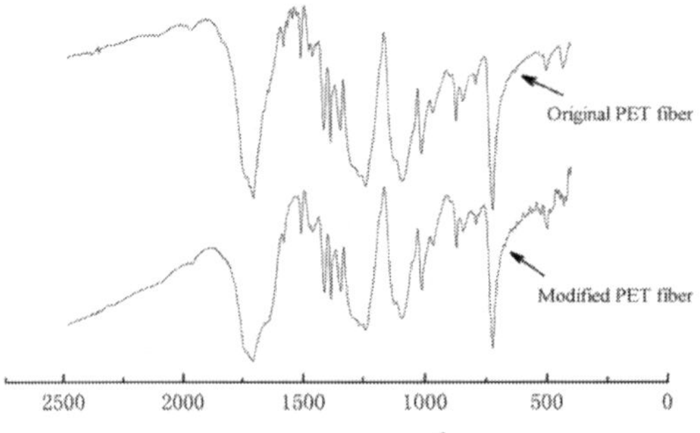

Figure 1. FTIR spectra of the original and modified PET fabrics.

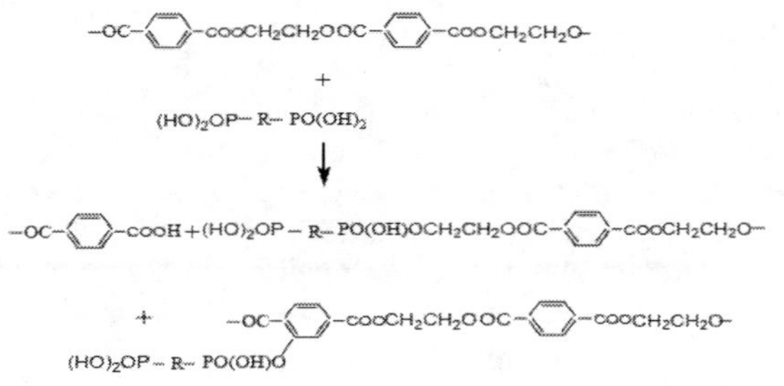

Scheme 1. The possible reaction equation of the modification.

3.3. Breaking strength of modified PET fabric

The relationship between the HOP concentrations and the breaking strengths of original and modified PET fabric was shown in figure 2. From figure 2, it can be seen that the breaking strengths decreased gradually as the HOP concentration increased. The breaking strength of the original PET fabric was 834.00 N, and the breaking strength decreased to 709.20 N when the PET fabric modified by 30 g/L HOP for three times. Overall, the breaking strength of the modified fabric was still high enough to be used in textile.

3.4. Stiffness of modified PET fabric

Figure 3 showed the relationship between the HOP concentrations and the bending lengths. According to figure 3, the bending length increased a little as the HOP concentration increased. The bending length of the original PET fabric was 24.02 mm, and the bending length increased to 26.28 mm when the PET fabric modified by 30 g/L HOP for three times. This indicated that the modification process had only a small effect on the stiffness of the PET fabric, and the limpness was kept well.

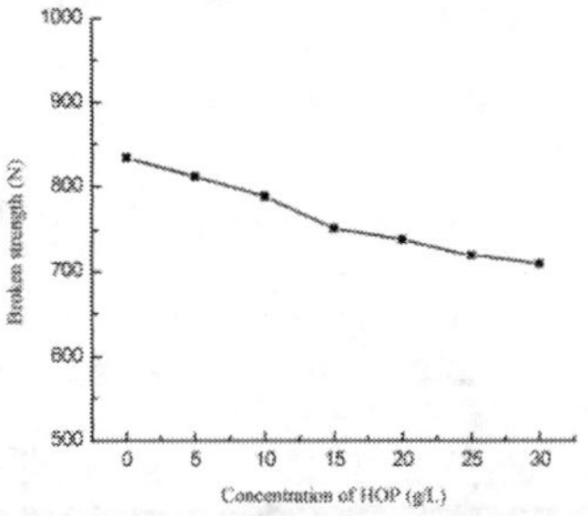

Figure 2. The relationship between the HOP concentrations and the breaking strengths.

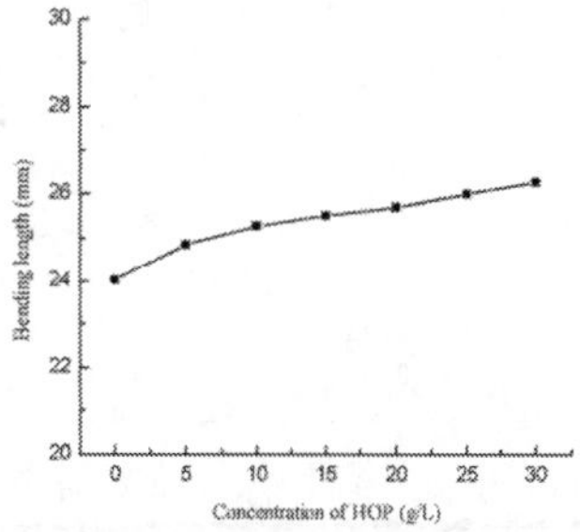

Figure 3. The relationship between the HOP concentrations and the bending lengths.

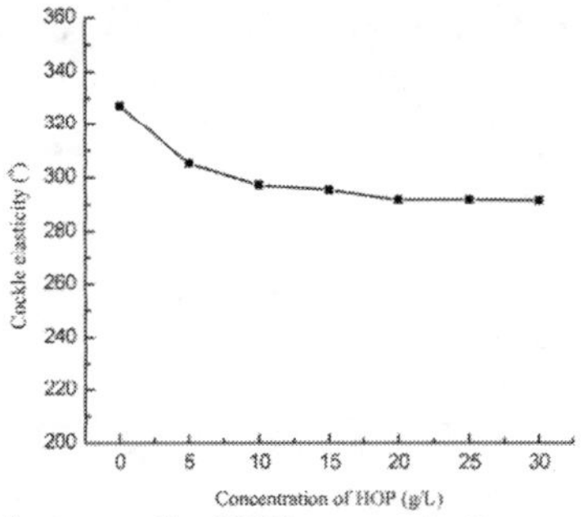

Figure 4. The relationship between the HOP concentrations and the crease recovery angles.

3.5. Crease recovery angle of modified PET fabric

Figure 4 showed the relationship between the HOP concentrations and the crease recovery angles. From figure 4, it can be seen that the crease recovery angle decreased a little as the HOP concentration increased. The crease recovery angle of the original PET fabric was 327.10°, and the crease recovery angle decreased to 291.50° when the PET fabric was modified by 30 g/L HOP for three times. The crease recovery angles decreased slightly, and they were still high enough actually.

4. Conclusions

The modification with flame retardant HOP could endow high flame retardance to PET fabric. The flame retardance of PET fabric improved a lot when the fabric was modified by 30 g/L HOP for three times. The modification of HOP made PET fabric gain phosphate groups to improve the flame retardance of PET fabric. When the concentration of HOP increased, the horizontal burning rate decreased rapidly, and the vertical ignite time increased quickly. When PET fabric was modified by 30 g/L HOP for three times, the horizontal burning rate decreased to 0 mm/s, and the vertical ignite time increased to 5 s. At the same time, the breaking strength, stiffness and crease recovery angle were kept well.

Acknowledgments

This work was supported by the Fundamental Research Funds for the Central Universities (Grant No. XDJK2013A021), the Chongqing City Graduate Research and Innovation Project (Grant No. CYS14059), and the Key Application and Development Project of the Chongqing Science & Technology Commission (Grant No. CSTC2014yykfB50002).

References

[1] Bessem K, et al. 2014 UV irradiation-assisted grafting of poly (ethylene terephthalate) fabrics *Colloid Surface A* **441** 606-13

[2] Chengshou Z, et al. 2008 A novel halogen-free flame retardant for glass-fiber-reinforced poly (ethylene terephthalate) *Polym Degrad Stabil* **93** 1188-93

[3] Horrocks A R 2011 Flame retardant challenges for textiles and fibres: New chemistry versus innovatory solutions *Polym Degrad Stabil* **96** 377-92

[4] Przemysław R and Grazyna J 2013 Influence synergetic effect of halloysite nanotubes and halogen-free flame-retardants on properties nitrile rubber composites *Thermochim. Acta.* **557** 24-30

[5] Lecoucet B, et al. 2013 A comprehensive study of the synergistic flame retardant mechanisms of halloysite in intumescent polypropylene *Polymer Degradation and Stability* **98** 2268-81

[6] Luying W, et al. 2011 Status and development of flame retardant polyester fabric *Synthetic Fiber in China* **40** 6-10

[7] Levchik V S and Well D 2005 Review flame redundancy of thermoplastic polyesters-a review of the recent literature *Polym Int* **54** 11-35

Study of a novel phosphorus-containing flame retardant for cotton fabric

W W Gao[1], Y H Lu[2, 3], F Xu[2, 3], G X Zhang[2, 3, 4] and F X Zhang[1, 5]

[1]Institute of Bioorganic and Medicinal Chemistry, College of Chemistry and Chemical Engineering, Southwest University, China
[2]College of Textile and Garment, Southwest University, China
[3]Chongqing Engineering Research Center of Biomaterial Fiber and Modern Textile, Chongqing, China

E-mail: zgx656472@sina.com.cn (G X Zhang); zhangfx656472@sina.com.cn (F X Zhang)

Abstract. In this paper, a high efficiency FR named HPA was applied to treat cotton fabric. The results of LOI values and vertical flammability test showed that HPA treated cotton fabric had the best flame retardancy (LOI value was 36.0%), when the FR concentration is 50 g/L, and cured at 180°C for 7 min. During the process of holding back the combustion, HPA behaves the excellent properties of FR for cotton fabric.

1. Introduction

Cotton fabric is one of the excellent natural materials and is one of the most widely used textile biopolymers in the world. However, cotton fabric is flammable easily [1]. Fires from cotton fibre textiles are causing many injuries, deaths and considerable financial losses [2]. So, flame retardant (FR) treatment of cotton fabric is extremely necessary in many applications for reducing their flammability [3]. FR exerts its flame retardancy by means of the heat absorbing, the covering effect, inhibition of chain reaction and gas dilution [4].

For cotton fabric, the most commonly used FRs are based on halogen and phosphorus compounds. The use of halogen FRs will generate poisonous substances such as dioxins when burnt at too low a temperature for normal commercial incinerators. Halogen FRs are bioaccumulation which is harmful to humans and the environment [5]. Therefore, the general trend is to substitute these FRs by non-halogenated alternatives. The mostly used phosphorus FRs are Proban and Pyrovatex CP [6]. They can react with the cotton fibre or form cross-linked structures on the fibre. These compounds can increase the formation of char, prevent the formation of levoglucosan and flammable volatiles, thus act as FRs for cellulose [7]. Proban and Pyrovatex CP, however, have the shortcoming of releasing formaldehyde which is known as a carcinogenic compound [8]. So, it is important to develop halogen-free and formaldehyde-free FRs for cotton fabrics.

[4] Address for correspondence: G X Zhang, College of Textile and Garment, Southwest University, China. E-mail: zgx656472@sina.com.cn.
[5] F X Zhang, Institute of Bioorganic and Medicinal Chemistry, College of Chemistry and Chemical Engineering, Southwest University, China. E-mail: zhangfx656472@sina.com.cn.

In this study, a phosphorus based FR named HPA was synthesized to treat cotton fabrics. It can behave excellent FR performance in the process of holding back the combustion.

2. Experimental

2.1. Materials
100% Cotton fabric, scoured and bleached, was obtained from the market (123.27 g/m^2) was a FR agent.

2.2. Methods

2.2.1. Preparation of HPA solution. HPA was dissolved in distilled water. The concentration of HPA was ranging from 1% to 6%.

2.2.2. The treatment of cotton fabric with HPA solution. The cotton fabric was impregnated in HPA solution with the bath ratio of 1:20 under 50°C thermostatic water bath for 20 min.

2.2.3. Curing the FR finished cotton fabric. The HPA finished cotton fabrics were padded through two dips and two nips to reach an average wet pickup of 120%. After the treatment, the fabric was cured in an automatic shaping and drying machine (HL-3842, Nantong Sansi Electromechanical Science & Technology Co. Ltd., China) with different temperatures from 120°C to 190°C. The cured time is ranging from 5 to 15 min. Then the treated sample was washed thoroughly.

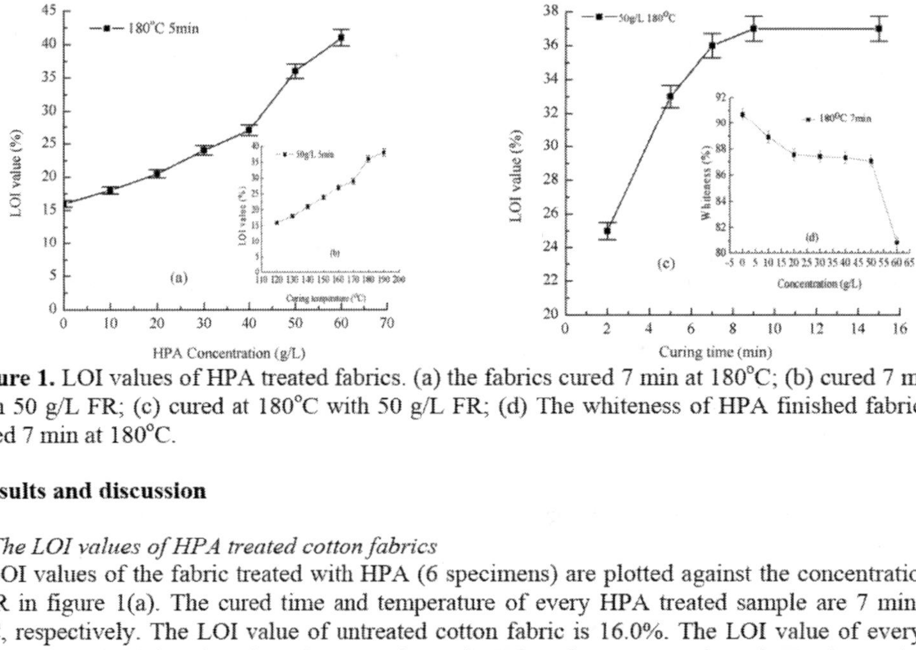

Figure 1. LOI values of HPA treated fabrics. (a) the fabrics cured 7 min at 180°C; (b) cured 7 min with 50 g/L FR; (c) cured at 180°C with 50 g/L FR; (d) The whiteness of HPA finished fabrics, cured 7 min at 180°C.

3. Results and discussion

3.1. The LOI values of HPA treated cotton fabrics
The LOI values of the fabric treated with HPA (6 specimens) are plotted against the concentration of the FR in figure 1(a). The cured time and temperature of every HPA treated sample are 7 min and 180°C, respectively. The LOI value of untreated cotton fabric is 16.0%. The LOI value of every FR treated samples is higher than that of untreated sample. When the concentration of HPA is 40 g/L, the LOI value is 27.0% which does not meet with the cotton fabric of FR standard LOI value of 28.0%. But the LOI values would reach 36.0% and 41.0%, respectively, when the concentration of HPA were 50 g/L and 60 g/L. Thus, the concentration of HPA should keep above 50 g/L. The curing temperature is also an important parameter to enhance the flame retardancy. From figure 1(b), the relationship between LOI value and curing temperature was exhibited. Besides, the cured time is 7 min, and 50 g/L

IOP Conf. Series: Materials Science and Engineering **87** (2015) 012107 doi:10.1088/1757-899X/87/1/012107

HPA is used for every specimen. When the curing temperature was 170°C, the LOI value was 29.0%. And LOI value would be up to 36.0%, if the curing temperature was 180°C. Therefore, the temperature must be at least 180°C.

The curing time is another important factor to increase the flame retardancy. The LOI values of HPA finished fabrics, cured at 180°C with 50 g/L FR were showed in figure 1(c). The LOI value increased a lot (from 25% to 37%) when the time expanded from 2 min to 9 min. If expanding the curing time continuously, the LOI value would not increase any more, kept 37%. So, the curing time of 7 min is the optimum. In addition, the whiteness of FR treated fabric cured 7 min at 180°C decreased with the increasing concentration of HPA in figure 1(d). The whiteness of control cotton fabric is 90.69%. When the concentration of HPA was 50 g/L, the whiteness of FR finished cotton fabric would decrease to 87.06%, this value is acceptable. The whiteness, however, would be 80.82%, if the concentration of FR was 60 g/L. It means that the concentration of 60 g/L is very high to damage the whiteness of the treated fabric.

From the results of LOI values and whiteness, there are three parameters, including FR concentration, curing temperature and time, affecting the flame retardancy of HPA treated cotton fabric. The optimum FR concentration is 50 g/L, the best curing temperature and time are 180°C and 7 min, respectively. HPA behaves excellent FR properties for cotton fabric. Because of the high temperature, the tensile strength, however, decreased to a certain degree (table 1).

Table 1. Tensile strength of HPA finished cotton fabric.

Samples	Tensile strength / (N)	
	Warp	Weft
Control cotton fabric	543.50	306.00
50g/L HPA treatment	460.00	263.50

3.2. The vertical flammability test of HPA treated cotton fabrics

The vertical flammability test test results were shown in table 2. The control fabric ignited and burned rapidly in the air, and the after flame time and afterglow time of control sample were 6 s and 11 s, respectively. After treated with 50 g/L HPA, there are lots of chars remained when removing the ignition source and the char length was 81 mm. And the treated sample has no after flame time and afterglow time, which means that after removing the ignition source, the flame extinguished immediately and no afterglow observed. Thus, the flame retardancy of cotton fabric treated with HPA improved greatly.

Table 2. The vertical flammability tests of control and HPA treated samples.

Samples	After flame time (s)	After glow time (s)	Char length (mm)
Control sample	6	11	0
50 g/L HPA treated sample	0	0	81

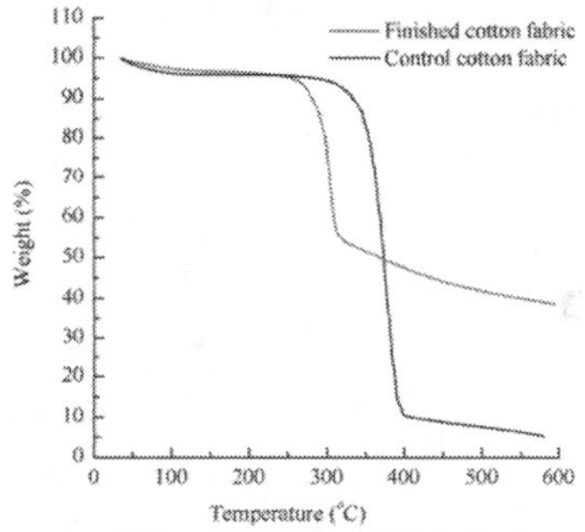

Figure 2. TG of the control and finished cotton fabrics.

3.3. The vertical flammability test of HPA treated cotton fabrics

Figure 2 showed the TG of the control and HPA finished samples. For the control sample, the weight decreased rapidly from 325 to 400°C with about 80% weight loss, and 5.3% ash was left until 580°C. The finished sample with FR, however, rapidly lost its weight ranged from 292.1 to 310.7°C with only about 46.73% weight loss and 38.64% residual was remained at 595°C. It was demonstrated that the decomposition temperature of the finished fabric was lower than that of the control fabric, and the amount of residue was much more than that of control sample at the end.

4. Results and discussion

During the process of holding back the combustion, HPA behaves the excellent properties of FR. From LOI values and vertical flammability tests, when the FR concentration is 50 g/L, and curied at 180°C for 7 min, the treated cotton fabric had the best flame retardancy and the LOI value was 36.0%. The treatment by HPA could protect cotton fabric from fire to a certain degree based on the TG results. HPA exerts its FR action on cotton fabric at a temperature well below that of the pyrolysis of these normal fabrics.

Acknowledgments

The authors would like to thank the Project No. XDJK2013A021, and this paper was supported by the Fundamental Research Funds for the Central Universities.

References

[1] Xie K L, Gao A Q and Zhang Y S 2013 Flame retardant finishing of cotton fabric based on synergistic compounds containing boron and nitrogen *Carbohyd. Polym.* **98** 706-10

[2] Lecouvet B, Sclavons M, Bailly C and Bourbigot S 2013 A comprehensive study of the synergistic flame retardant mechanisms of halloysite in intumescent polypropylene *Polym. Degrad. Stab.* **98** 2268-81

[3] Levchik S V and Weil E D 2004 Flame retardance of thermoplastic polyesters-a review of the recent literature *Polym. Int.* **54** 11-35

[4] Wang D Q, Tian L, Jiao L, Wang C X and Bi X Y 2011 The flame retardant finishing effect on physical properties of polyester fabric *Shandong Textile Science and Technology* **52** 4-6

[5] Smith R, Georlette P, Finberg I and Reznick G 1996 Development of environmentally friendly multifunctional flame retardants for commodity and engineering plastics *Polym. Degrad. Stab.* **54** 167-73

[6] Horrocks A R, Kandola B K, Davies P J, Zhang S and Padbury S A 2005 Developments in flame retardant textiles—a review *Polym. Degrad. Stab.* **88** 3-12

[7] Lessan F, Montazer M and Moghadam M B 2011 A novel durable flame-retardant cotton fabric using sodium hypophosphite, nano-TiO_2 and maleic acid *Thermochimica Acta.* **520** 48-54

[8] Yang H and Yang C Q 2007 Non-formaldehyde flame retardant finishing of the Nomex/Cotton blend fabric using a hydroxy-functional organophosphorus oligomer *J. Fir. Sci.* **25** 425-46

Evaluation of tensile strength of hybrid fiber (jute/gongura) reinforced hybrid polymer matrix composites

G Venkatachalam[1], A Gautham Shankar[2, 3], Vijay Kumar V[2], Chandan Byral R[2], Prabaharan G P.[2] and Dasarath Raghav[2]

[1]Associate Professor, School of Mechanical and Building Sciences, VIT University, Vellore, India
[2]Mechanical Engineering, School of Mechanical and Building Sciences, VIT University, Vellore, India

E-mail: g.venkatachalam@vit.ac.in

Abstract. The polymer matrix composites attract many industrial applications due to its light weight, less cost and easy for manufacturing. In this paper, an attempt is made to prepare and study of the tensile strength of hybrid (two natural) fibers reinforced hybrid (Natural + Synthetic) polymer matrix composites. The samples were prepared with hybrid reinforcement consists of two different fibers such as jute and Gongura and hybrid polymer consists of polyester and cashew nut shell resins. The hybrid composites tensile strength is evaluated to study the influence of various fiber parameters on mechanical strength. The parameters considered here are the duration of fiber treatment, the concentration of alkali in fiber treatment and nature of fiber content in the composites.

1. Introduction

Polymer Composites are materials having two or more chemically distinct constituents. The discontinuous phases are embedded in the continuous phase to form a material composite. The discontinuous phase is called as reinforcement, and the continuous phase are called as the matrix. If the matrix material is polymeric, then the composites are called as polymer composites (PMC) [1, 2]. The reinforcing phase can either be fibrous or non-fibrous, and if the fibers are derived from plants or natural resources, they are called natural-fibers. The fiber reinforced polymers (FRPs) consist of fibers of high strength and modulus embedded in or bonded to a matrix with a distinct interface between them [3]. In this form, both fibers and matrix retain their physical and chemical identities. In general, fibers are the principal load-carrying members, while the matrix keeps them at the desired location and orientation. The matrix also acts as a load transfer medium between the fibers, and protects them from environmental damage [4, 5].

Hybrid composites are materials that are made by combining two or more different types of fibers or different types of matrix. Viswanath et al. investigated the tensile behavior of coir fiber reinforced hybrid polymer matrix composites. The authors used CNSL and general purpose resin for getting hybrid polymer. Vimalanand et al. explored the tensile, and flexural strength of aloe vera fiber

[3] Address for correspondence: Mechanical Engineering, School of Mechanical and Building Sciences, VIT University, Vellore, India. E-mail: g.venkatachalam@vit.ac.in

reinforced hybrid polymer matrix composites. Venkatachalam et al. studied the flexural behavior of coir fiber reinforced hybrid polymer matrix composites. Authors [6-8] used CNSL and general purpose resin for getting hybrid polymers. An attempt is made in this work to have hybrid (two) fibers as reinforcement and hybrid polymer as the matrix. Taguchi's L9 orthogonal array was used for determining different combination of specimens. This material composition will attract engineering markets for the good environmental friendly feature and improved mechanical properties.

2. Experimental

Natural fibers are eco-friendly, available in abundant on earth and are cheaper. The jute fiber and the gongura fibers were chosen as reinforcement material. Jute is the cheapest and most abundantly available natural fibers. Gongura are cultivated for its fiber in south India. Polyester resin was used to bind fiber as a matrix in polymer composite. To improve the biodegradation of the polymer composite, about 25% of the natural resin extracted from cashew nut shell (cashew nut shell resin-CNSL) is mixed with polyester resin. Hand layup technique was followed for the preparation of composites. The fibers are knitted as shown in the figure 1. After the knitting of fiber in the mold, the resin mixture is poured into the mold cavity.

The molds were kept at room temperature for curing. Different composite specimens are prepared according to the composition specified in the table 1 and is presented in figure 2.

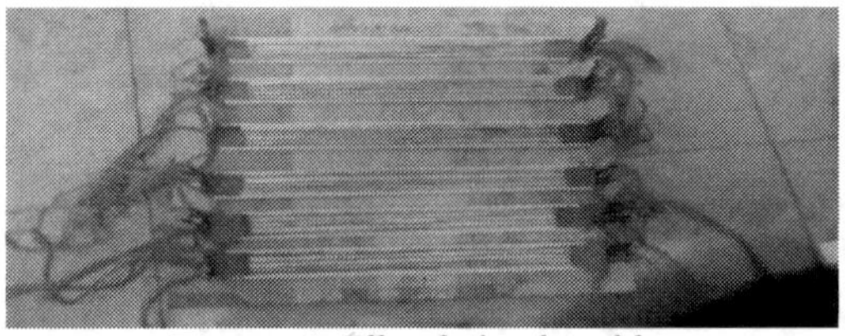

Figure 1. Fiber knitted mold.

Figure 2. Composite specimens.

Table 1. Specimen composition.

Sample No	Fiber Parameter		
	Duration of Fiber treatment (Hrs)	**Fiber treatment % of NaOH concentration**	**Fiber Hybridization-proportion of fibers**
1	6	5	3 Jute-1 Gongura
2	6	10	2 Jute-2 Gongura
3	6	15	1 Jute-3 Gongura
4	12	5	2 Jute-2 Gongura
5	12	10	1 Jute-3 Gongura
6	12	15	3 Jute-1 Gongura
7	24	5	1 Jute-3 Gongura
8	24	10	3 Jute-1 Gongura
9	24	15	2 Jute-2 Gongura

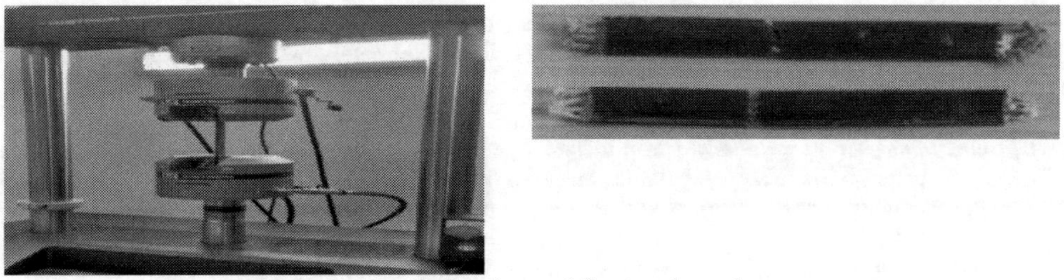

Figure 3. Tensile Testing and fractured specimens.

2.1. Tensile tests

After the curing process, the specimen is tested for its tensile strength that is shown in figure 3. Tensile testing is carried out using Instron machine according to ASTM D 638–76. The load vs displacement curves were obtained, and the modulus were calculated.

3. Results and discussions

3.1. Analysis of tensile strength

The results obtained from tensile testing of composites with hybrid fibers as reinforcement and hybrid polymers as matrix are given in table 2.

Table 2. Yield strength, ultimate tensile strength and young's modulus for the various compositions of specimens.

Sample No	Yield strength (MPa)	Ultimate Strength (MPa)	Young Modulus (MPa)
1	1.35	3.24	152.53
2	0.6	2.86	178.306
3	0.54	2.18	150.235
4	0.909	2.647	176.214
5	0.89	2.243	154.563
6	0.649	3.323	156.902
7	0.769	2.113	145.578
8	0.89	3.561	168.431
9	0.802	2.345	142.854

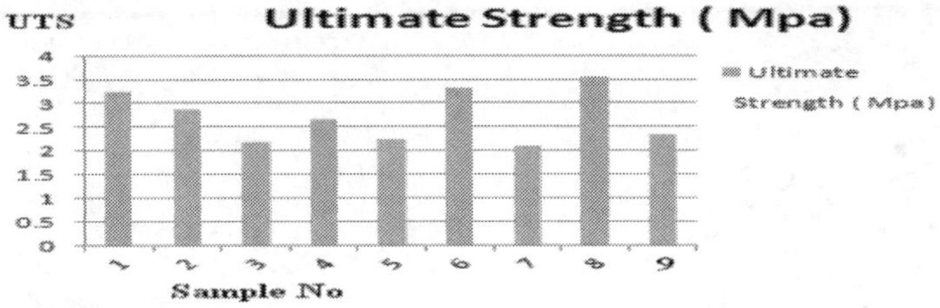

Figure 4. Comparative analysis of ultimate strength of composite specimens tested.

A comparison analysis of the ultimate strength of various specimens is plotted and shown in the figure 4. The study on ultimate strength of specimens showed that composites with the following composition exhibited good ultimate strength

- 3:1 proportion of Jute and Gongura fiber, treated for 6 Hours with 5% NaOH solution
- 3:1 proportion of Jute and Gongura fiber treated for 12 Hours with 15% NaOH solution

IOP Conf. Series: Materials Science and Engineering **87** (2015) 012108 doi:10.1088/1757-899X/87/1/012108

- 3:1 proportion of Jute and Gongura fiber treated for 24 Hours with 10% NaOH solution

Hence, it is decided to study the influence of each fiber parameters on tensile strength by ANOVA techniques to find the optimal material composition.

3.2. Optimization of Composite proportion using ANOVA Techniques

The ANOVA technique is used to evaluate the influence of fiber parameters on tensile strength. The Minitab software is used to carry out ANOVA.

3.2.1. Study on influence of various fiber parameters on ultimate strength, yield strength, and Young's modulus. The main effects plots, presented in figure 5, for ultimate strength shows that the duration of fiber treatment has little influence on ultimate strength and fiber treatment with 10% concentration exhibits higher ultimate strength. But the increase in fiber combination reduced the ultimate strength. The same is conferred in yield strength study also. In case of modulus of elasticity, the strength is good for the specimen with the fiber treatment for 12 hours, 10% concentration NaOH solution and medium fiber combinations.

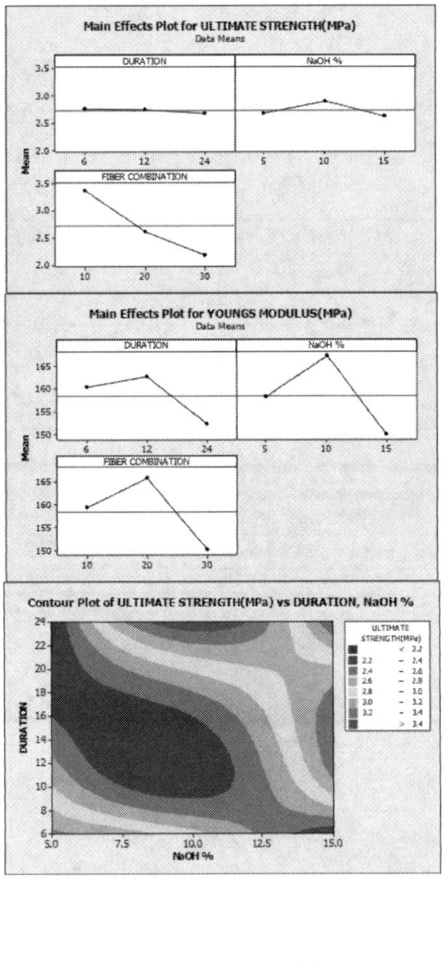

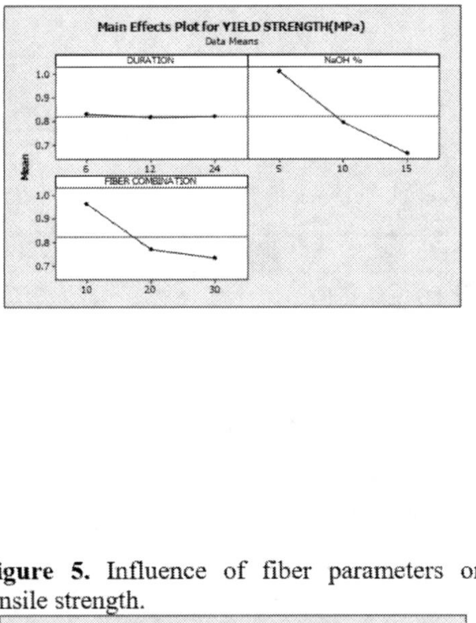

Figure 5. Influence of fiber parameters on tensile strength.

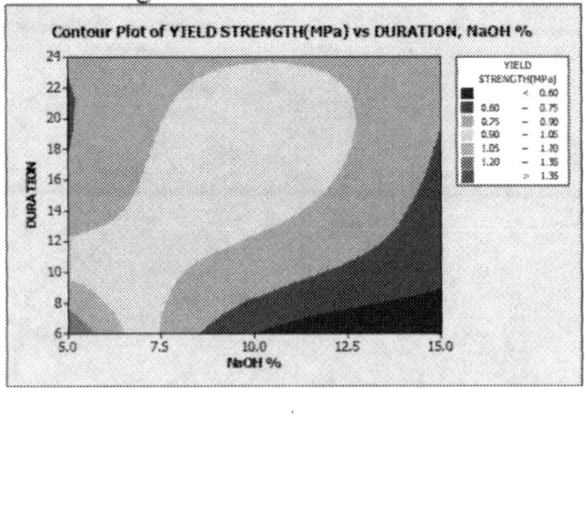

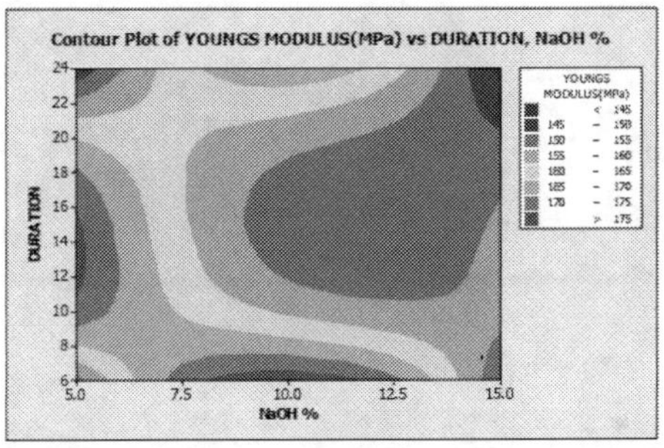

Figure 6. Contour plots of influences of and NaOH solution concentration on tensile strength.

3.2.2. Influence of treatment duration and NaOH solution concentration on tensile strength. The contour plots, shown in figure 6, are extracted to study the influences of various fiber treatment solution concentrations and duration of treatment on tensile strength. It was found from the plot that the tensile strength decreases with increase in treatment duration and alkali solution concentration.

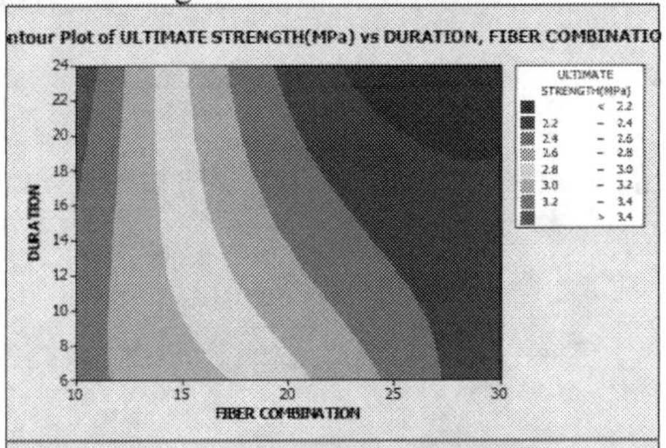

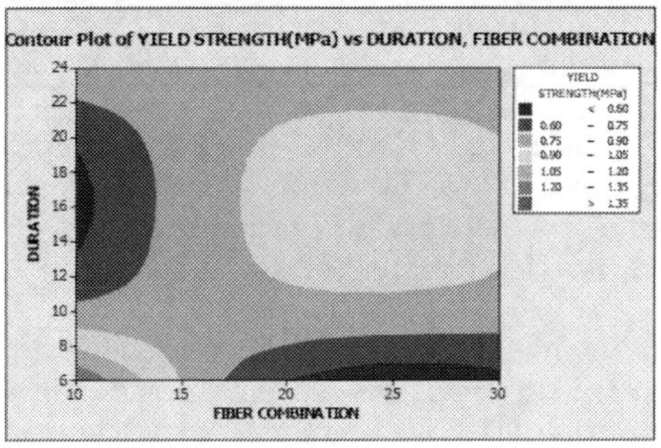

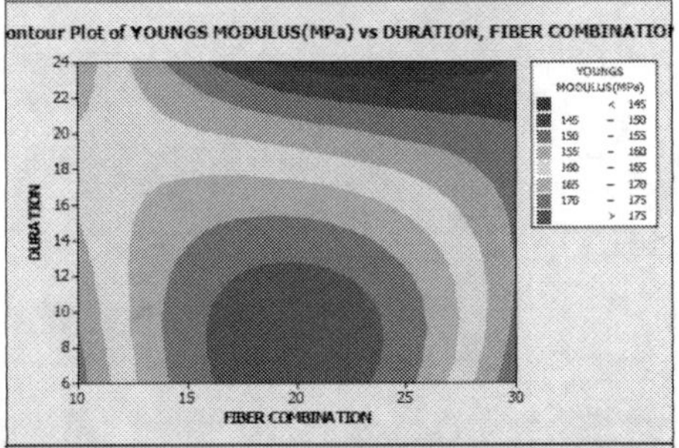

Figure 7. Influence of fiber treatment duration and fiber combination on tensile strength.

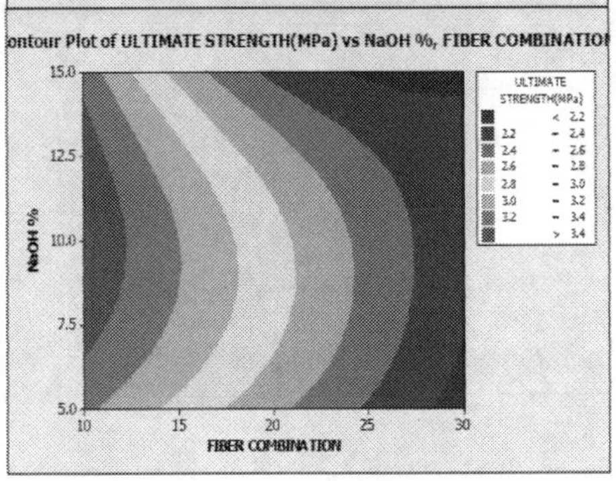

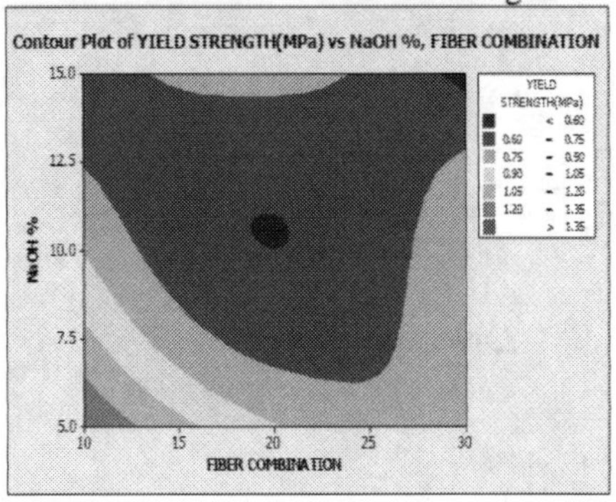

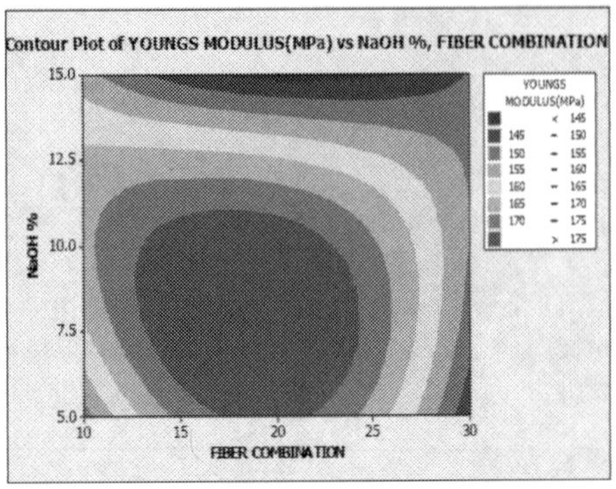

Figure 8. Influence of NaOH concentration and fiber combination on tensile strength.

3.2.3. Influence of fiber treatment duration and fiber combination on tensile strength. The contour plot, presented in figure 7, between fiber treatment duration and fiber combination, reveals that the increase in duration of treatment will increase the tensile strength whereas an increase in fiber combination proportions will decrease the tensile strength.

3.2.4. Influence of NaOH concentration and fiber combination on tensile strength. During the study on influence of NaOH concentration and fiber combination on tensile strength, it is found that increase in NaOH concentration will increase the tensile strength. An increase fiber concentration will reduce the tensile strength of the material that is shown in figure 8.

3.2.5. Regression equations. Regression equations were obtained via ANOVA analysis in MINITAB software with which we can find the approximate tensile strength of the composite with other fiber combination, alkali concentration and duration of treatment.

The regression equation is

- YIELD STRENGTH(mpa) = 1.40 - 0.00041 DURATION - 0.0346 naoh % - 0.0115 FIBER COMBINATION
- ULTIMATE STRENGTH(mpa) = 4.04 - 0.00491 DURATION - 0.0051 naoh % - 0.0598FIBER COMBINATION
- YOUNGS MODULUS(mpa) = 183 - 0.507 DURATION - 0.81 naoh %- 0.458 FIBER COMBINATION

4. Conclusion

As the present industry requirement is environment friendly materials, a new composite material with hybrid polymer and hybrid fiber is developed, and its properties are evaluated. It was concluded from the analysis that fiber treatment duration and concentration of alkali solution had to be chosen as 12 hours and 10% respectively, and hybridization of fiber should be low for the better tensile strength

References

[1] Laly A Pothan, Zachariah Oommen and Sabu Thomas 2003 Dynamic mechanical analysis of banana fiber reinforced polyester composites *Science and Technology* **63** 283-93

[2] Sapuan S M, Leenie A, Harimi M and Beng Y K 2006 Mechanical properties of woven banana fiber reinforced epoxy composites *Materials & Design* **27** 689-93

[3] Laly A Pothan and Sabu Thomas 2003 Polarity parameters and dynamic mechanical behavior of chemically modified banana fiber reinforced polyester composites *Science and Technology* **63** 1231-40

[4] Maries Idicula, Malhotra S K, Kuruvilla Joseph and Sabu Thomas 2005 Dynamic mechanical analysis of randomly oriented intimately mixed short banana/sisal hybrid fiber reinforced polyester composites *Science and Technology* **65** 1077-87

[5] Murali K, Mohan Rao and Mohana Rao K 2007 Extraction and tensile properties of natural fibers: Vakka, date and bamboo *Composite Structures* **77** 288-95

[6] Viswanath S, Venkatachalam G, Ragavender K and Ashka, 2014 Investigation on tensile behavior of coir fibre made hybrid polymer matrix composite *Porceedings of IMEC* (India: Tiruchirappalli) pp 311-4

[7] Vimalanand S, Venkatachalam G, Gautham A, Rohit A, Gupta G and Anantrao U S 2014, Investigations on tensile and flexural strengths aloevera fibre reinforced hybrib polymer matrix composite materials *Souvenir on INCCOM* (India: Thiruvananthapuram) p 5A6

[8] Venkatachalam G, Gupta A and Gautham Shankar A 2015 Flexural analysis of coir fibre reinforced hybrid polymer matrix composites *Composites: Mechanics, Computations, Applications* **6** 105-12

IOP Conf. Series: Materials Science and Engineering **87** (2015) 012109 doi:10.1088/1757-899X/87/1/012109

The effectiveness of combined gripping method in tensile testing of UHMWPE single yarn

H X Wang[1], P J Hazell, K Shankar, E V Morozov and J P Escobedo

School of Engineering and Information Technology, University of New South Wales Canberra, ACT 2600, Australia

E-mail: Hongxu.Wang@student.adfa.edu.au

Abstract. This paper presents the experimental study on the effectiveness of combined gripping method employed in the tensile testing of UHMWPE (Dyneema® SK75) single yarn. Seven different solutions including epoxy, acrylic, and ethyl cyanoacrylate adhesives were tested under quasi-static loadings in order to determine the most effective adhesive for bonding UHMWPE single yarn to aluminium sheets. The ethyl cyanoacrylate adhesive combined with polyolefin surface primer was found to be the best choice which could prevent yarn slippage and ensure the failure of yarn occurs in the gauge section. The single yarns were then tested at three strain rates of 3.3×10^{-5}, 3.3×10^{-3}, and 0.33 s^{-1}. The tensile strength, maximum strain, and Young's modulus were determined from the measured stress-strain curves and compared with the values from literature; the results showed these tensile properties of single yarn depend on strain rate over the range tested.

1. Introduction

Ultra high molecular weight polyethylene (UHMWPE) fibres are produced through gel-spinning and subsequent hot drawing. This process creates highly oriented molecular chains which exhibit outstanding specific tensile strength and stiffness [1]. In addition, their excellent chemical, fatigue, and abrasion resistance render them an attractive material for potential use in high-performance textiles and composites. Woven fabrics made of these fibres are increasingly used in impact-related applications, such as soft body armour and propulsion engine containment system, where large deformation and high energy absorption are required. These applications have stimulated a demand for numerical simulation of fabrics and in-depth study on the behaviour of yarns and their interaction, which requires experimentally verified material properties of single yarn, the basic load carrying component of a fabric. Although the quasi-static tensile properties for the single fibre form of the material are usually available from the manufacturers, these characteristics cannot be extrapolated and scaled up directly for a yarn consisting of many fibres. Previous study showed the tensile strength of a single Dyneema® SK76 fibre exceeded that of a yarn by about 20% [2]. Furthermore, the strain rates observed in impact applications are far beyond the order of magnitude of the strain rates in the quasi-static loadings [3]. Hence, the material properties of constitutive single yarn must be studied under experimental conditions similar to impact events.

[1] Address for correspondence: H X Wang, School of Engineering and Information Technology, University of New South Wales Canberra, ACT 2600, Australia. E-mail: Hongxu.Wang@student.adfa.edu.au.

It is generally accepted that a longitudinal strain wave, travelling at the speed of sound in the material, is generated in fabric yarns upon impact [4]. This wave stretches the yarns and causes the material primarily subjected to axial tension loading. Although there is an increasing demand for characterising the tensile properties of high performance fibres, how to grip the fibres tightly enough without introducing excessive stress concentration is a challenging task. The principal requirements of a suitable gripping method are preventing fibre slippage while maintaining the fibre tension required to bring them to failure, and ensuring the failure occurs in the gauge section such that the failure is not initiated by the grips [3]. Several different methods for gripping fibres during tensile test were explored. The direct clamping between flat jaws is the most common method of gripping specimens in material testing. This gripping mechanism is done by either an external or a self-clamping force that is almost uniform over the entire clamped length of the specimen, with friction acting on both sides of the specimen. Shim et al. [5] tested the dynamic tensile properties of Twaron® fibres using a specially designed direct clamping device. ASTM D3379 (Standard test method for tensile strength and Young's modulus for high-modulus single-filament materials, withdrawn in 1998 with no replacement) [6] introduced an adhesive gripping method which adheres the test fibres onto a mounting tab with a centre hole or a longitudinal slot of a fixed gauge length. The mounting tab is then gripped by the test machine, and the middle part of the tab is cut or burned away, leaving the fibre specimen free to be tested. This method is also specified in the ASTM C1557 (Standard test method for tensile strength and Young's modulus of fibers, current version 2014) [7]. This adhesive gripping method has been adopted in the quasi-static tensile testing of carbon fibres and glass fibres [8, 9], and the quasi-static tensile testing of poly(p-phenylene terephthalamide) (PPTA) fibres [10, 11]. There are some other methods using different gripping mechanisms were reported in the literature. For example, PET textile cable was secured by a progressive binding on the grips in dynamic tensile tests [12]; a capstan method which grips the yarn by winding it's ends around circular pins was used in quasi-static tensile tests of Twaron® single yarn [13].

The compression strength of UHMWPE fibres is only around 1% of the tensile strength along the fibre axis [1]. So when using a direct gripping method, insufficient pressure can lead to fibre slippage, while too much pressure would easily result in fibre crush by the rigid surfaces of the clamp. Moreover, UHMWPE fibres are difficult to grip using the standard specified adhesive method due to their poor adhesion characteristic which is attributed to the non-polar nature of the $(—CH_2—CH_2—)_n$ backbone and the low surface energy [14]. Therefore, it is even more difficult to grip UHMWPE fibres effectively in tensile testing. Several potential methods were experimented with. Tan et al. [13] improved the direct clamping device designed by Shim et al. [5]; the Twaron® fibres were sandwiched between two layers of pliable urethane foam in order to protect the fibres from large clamping forces and increase the contact area, and then clamped by a pair of flat grips. Kim et al. [10, 15] developed a method of direct gripping on single PPTA fibre utilising two poly methyl methacrylate (PMMA) blocks, and the clamping force of the blocks was controlled by a spring. This method was also employed for tensile testing of Dyneema® SK76 single fibre at different loading rates [16]. Zhu et al. [17, 18] designed a combined gripping method to reduce stress concentration and improve load transfer in grips. Both ends of a Kevlar® single yarn were sandwiched between two aluminium sheets and glued to them using epoxy adhesive. The aluminium face-sheets were then friction gripped between steel wedges that had serrated surfaces. Russell et al. [2] also used this combined gripping method for tensile testing of Dyneema® SK76 single fibre and single yarn. The fibre and yarn samples were sandwiched between two rubber sheets and bonded to them using cyanoacrylate adhesive. Besides, Russell et al. [2] wrapped the yarn around a semicircular anvil and adhered to it by cyanoacrylate adhesive in both quasi-static and dynamic tensile tests, however, the tensile strength determined from these two gripping methods showed obvious difference.

In this paper, the effectiveness of a combined gripping method similar to that proposed by Russell et al. [2] and Zhu et al. [17, 18] has been experimentally studied for clamping Dyneema® SK75 single yarn in the tensile testing. The suitable adhesive was selected from the trial test of seven different adhesives. After the adhesive was selected, the single yarn specimens were tested at three strain rates of 3.3×10^{-5}, 3.3×10^{-3}, and 0.33 s^{-1}. The tensile strength, maximum strain, and Young's modulus were calculated from the measured stress-strain curves and compared with the values from manufacturer and published elsewhere. Although the ultimate goal is the characterisation of yarn properties under high speed loadings mimicking the loading rates that are seen in an impact event, this initial study focuses on the assessment of the combined gripping method and the yarn properties taken under quasi-static tension loadings.

2. Experimental methodology

2.1. Specimen preparation

The as-received material was a roll of 4-harness satin weave fabric consisting of Dyneema® SK75 fibres (see figure 1). The warp yarns were randomly extracted from the fabric with care for tensile testing. The bulk density and linear density of the yarn are 0.97 g/cm^3 and 1500 dtex respectively. The cross-sectional area of the yarn was calculated as 0.155 mm^2 by dividing the linear density of the yarn by its bulk density. The yarn ends were sandwiched between cleaned aluminium plates and bonded to them using suitable adhesives, and a gauge length of 50 mm was left in the middle (see figure 2).

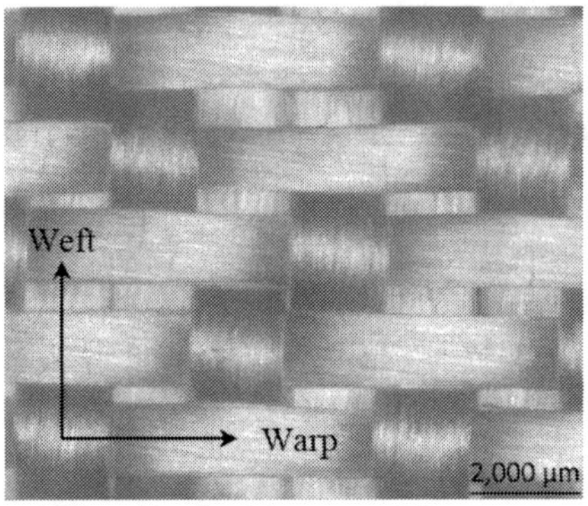

Figure 1. Optical micrograph showing the 4-harness satin weave pattern of Dyneema® fabric.

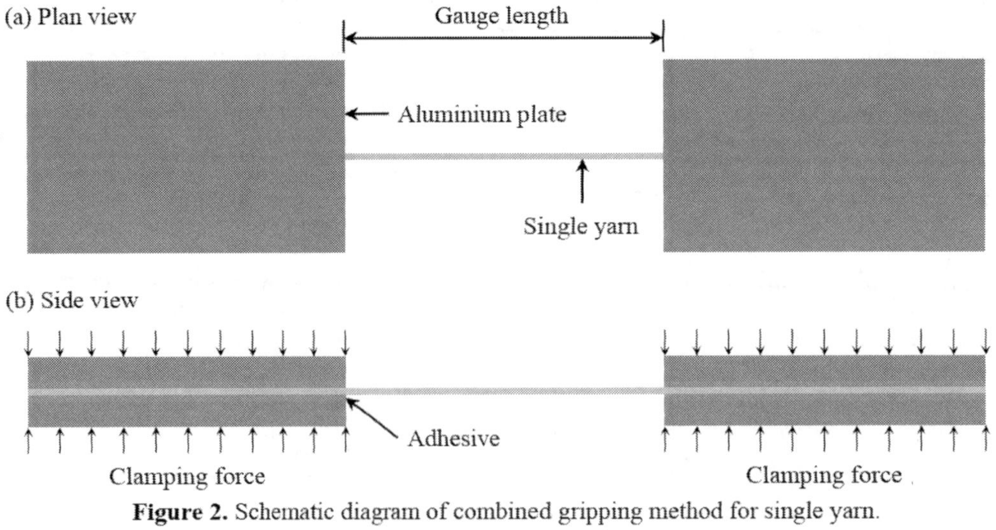

Figure 2. Schematic diagram of combined gripping method for single yarn.

Adhesive selection is crucial to producing a high strength bonding between the yarn specimen and aluminium sheet, because any yarn slippage will affect the force and displacement measurement.

There are many kinds of adhesives available in the market, and each one is better for use in different environments and with different material substrates. In order to find out the effective adhesive can be used in the tensile testing of UHMWPE single yarn, seven solutions were experimented with in this study. Table 1 lists the adhesive products and their chemical types which fall into three categories: epoxy, acrylic, and ethyl cyanoacrylate. Epoxy adhesive was successfully used for quasi-static and dynamic tensile testing of Kevlar® 49 single yarn [18], so two general-purpose high strength epoxy adhesives were tested in this study. Two acrylic-based adhesives, which are designed to bond low surface energy plastics including UHMWPE without special surface pre-treatment, were included in the test as well. Two ethyl cyanoacrylate adhesives were employed because this kind of adhesive was used for tensile tests of Dyneema® SK76 single fibre and yarn [2]. Except for using cyanoacrylate adhesives alone, an aliphatic amine primer was applied on the yarn ends to make the surface more suitable for bonding with corresponding cyanoacrylate adhesive. All the adhesives were used strictly according to the instruction of each product. More details of these adhesives can be found in the technical data sheet and material safety data sheet of the products.

Table 1. Adhesives tested for the combined gripping method.

Adhesive product	Chemical type
Epirez Episet Structural Adhesive 8242	Epoxy resin
Loctite 3801 Epoxy Adhesive	Epoxy resin
Loctite 3038 Structural Adhesive	Acrylic
3M Scotch-Weld Structural Plastic Adhesive DP-8005	Acrylic
Bostik Super Glue	Ethyl cyanoacrylate
Loctite 401 Instant Adhesive	Ethyl cyanoacrylate
Loctite 401 Instant Adhesive & Loctite 770 Polyolefin Primer	Ethyl cyanoacrylate

2.2. Experiment setup

The quasi-static tensile tests were conducted using a Shimadzu AG-X universal testing machine. The loading rate was controlled by the speed of the crosshead and the tensile force was measured by a load cell with a capacity of 100 kN. The loading speed was set at 0.1, 10, and 1000 mm/min for the specimen with a gauge length of 50 mm, corresponding to nominal strain rates of 3.3×10^{-5}, 3.3×10^{-3}, and 0.33 s^{-1}, respectively. The traditional strain-sensing devices, such as strain gauge and extensometer, may cause damage to the delicate fibres. Therefore, a non-contacting video extensometer, Shimadzu DVE-101, was employed to perform strain measurement by using a CCD camera to capture digital images of the test specimen. The image data was then processed to calculate the elongation of the gauge length. The aluminium face-sheets were gripped by steel wedges that had serrated surfaces. As the specimen was loaded, there was a uniform stretching of the yarn until failure.

3. Results and discussion

3.1. Effectiveness of adhesives

An effective adhesive for the combined gripping method is considered to be one which could prevent yarn slippage in the gripping region and insure the fibre failure occurs in the gauge section such that there is no fibre failure initiated by the grips. Of all the adhesives tested in the experiment, the ethyl cyanoacrylate-based adhesive combined with polyolefin surface primer was found to be the best solution that meets the abovementioned requirements. Figure 3 shows the tensile process of single yarn subjected to a loading speed of 10 mm/min using Loctite 401 Instant Adhesive and Loctite 770 Polyolefin Primer. The gauge length was divided into four equal portions using a marker pen before the test. It is obvious that the four portions keep fairly equal up to the moment just before failure occurs (time = 16s), which means the yarn was uniformly stretched in the entire gauge length and there was negligible yarn slippage in the gripping region.

Yarn pull-out from the gripping region was observed when the acrylic-based adhesives were employed, as illustrated in figure 4. The yarn was pulled out from the upper grip because the top portion of gauge length (L_1) was becoming much longer than the other portions as the test progressed. Besides, there was little fibre fracture occurred during the test. The cured acrylic adhesives were semi-rigid and cannot provide enough strength to hold the yarn. When the epoxy-based adhesives were used, there was no yarn pull-out observed in the gripping region up to failure, however, some fibres fractured in the gripping region as shown in figure 5. The reason for this phenomenon may be the viscosity of the used epoxy adhesives because it was not low enough to infiltrate all the fibres within the yarn.

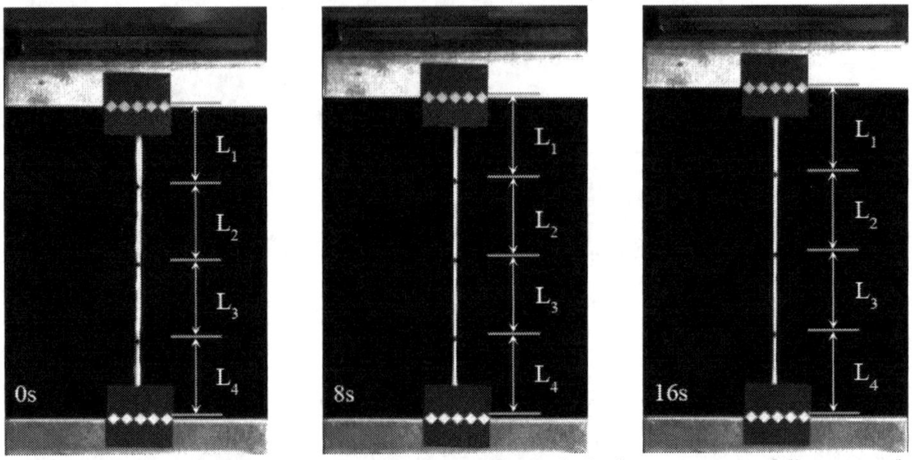

Figure 3. Video frames showing the uniform stretching of single yarn up to failure tested at the loading speed of 10 mm/min using Loctite 401 Instant Adhesive & Loctite 770 Polyolefin Primer.

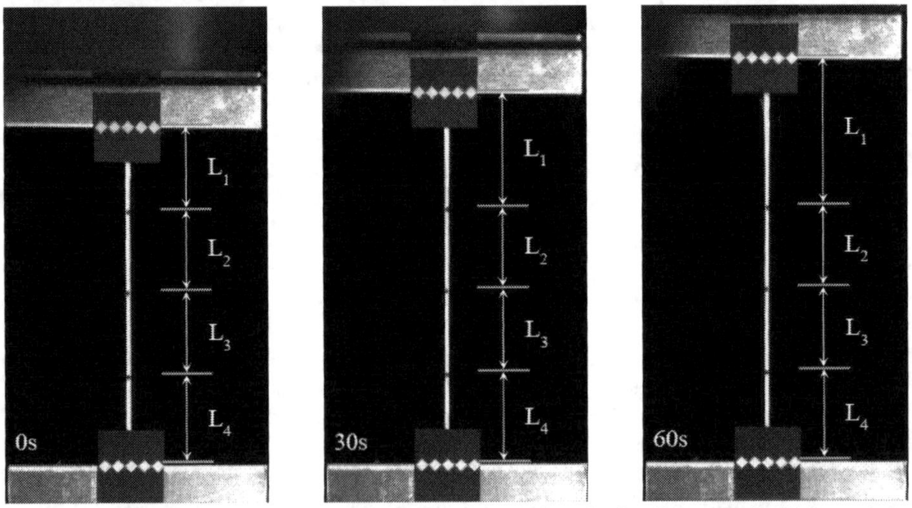

Figure 4. Video frames showing the pull-out of yarn tested at the loading speed of 10 mm/min using Loctite 3038 Structural Adhesive.

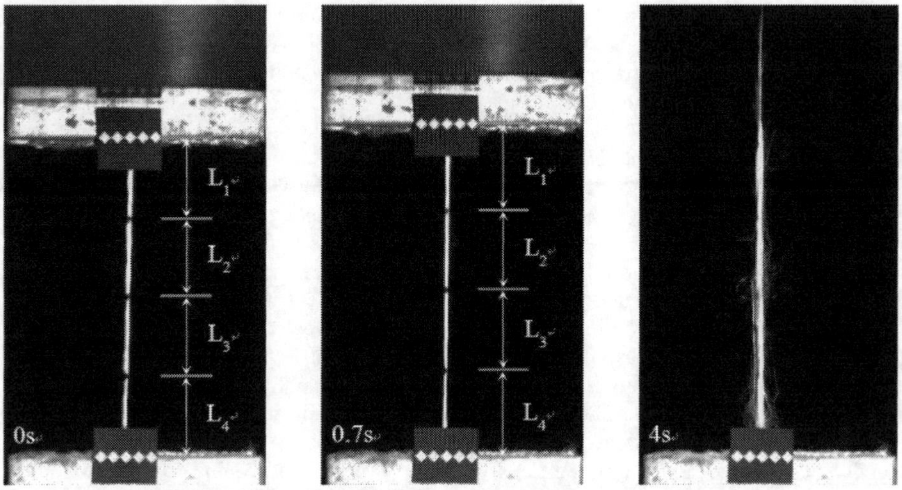

Figure 5. Video frames showing the fibre fracture in the gripping region tested at the loading speed of 1000 mm/min using Epirez Episet Structural Adhesive 8242.

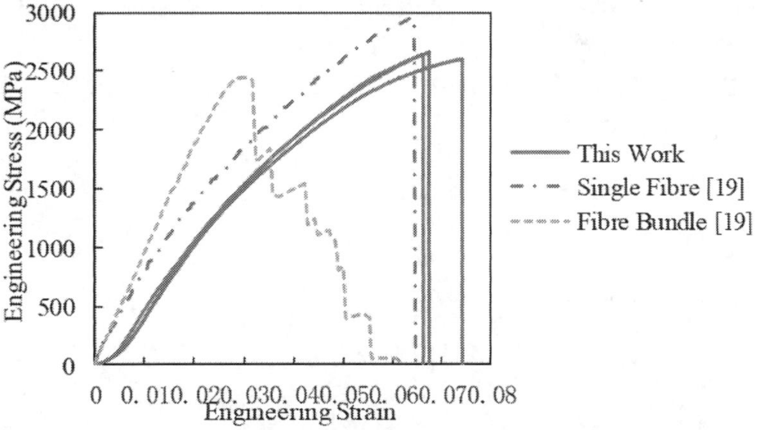

Figure 6. Stress-strain curves measured at strain rate of 3.3×10^{-5} s^{-1} and extracted from literature [19].

3.2. *Stress-strain curves*

After the suitable adhesive was selected, the single yarn samples were tested at strain rates of 3.3×10^{-5}, 3.3×10^{-3}, and 0.33 s^{-1}, respectively. For each strain rate, at least three repeated tests were carried out. The tensile behaviour at the strain rate of 3.3×10^{-5} s^{-1} in terms of the engineering stress and engineering strain is shown in figure 6. The curves show good repeatability which means the method and adhesive used in this test can produce stable and repeatable tensile behaviour of single yarn. The quasi-static stress-strain relationships of Dyneema® SK75 single fibre and fibre bundle tested by Kromm et al. [19] are also included in figure 6 for comparison, although the strain rates for the representative curves were not indicated explicitly in their paper. It is clear that the stress-strain curves of single yarn measured here are very similar to the curve of single fibre obtained by Kromm et al. [19]. The single yarn specimens tested in this work have an inherent residual crimp due to the weave structure of the fabric (see figure 1). In the initial region, the stress-strain curves show a relative large increase in strain for a very small increase in stress, and the tensile load essentially straightens the yarn by removing the crimp at the beginning. The curve of fibre bundle tested by Kromm et al. [19] shows a different behaviour, that is a stepped drop of stress after peak, which indicates the fibres in the bundle failed one by one. There can be a problem of not aligning all fibres in the bundle very well with the load applied in their work which leads to non-uniform stretching of the fibres. The method and adhesive used here could align the fibres in the yarn properly.

3.3. Tensile properties at different strain rates

In order to examine the strain rate effect on the tensile properties of single yarn, the tensile strength, maximum strain, and Young's modulus are calculated from the measured stress-strain relationships at different strain rates, as shown in table 2. The Young's modulus is determined from the linear slope after the initial decrimping region from the stress-strain curves. The results indicate the tensile properties of Dyneema® SK75 single yarn are strain rate dependent; the tensile strength and Young's modulus increase with strain rate while the maximum strain decreases.

It is instructive to compare these results with the values published in the literature [19] and given by the manufacturer [20] (see table 2). According to the results published by Kromm et al. [19], only the maximum strain of Dyneema® SK75 single fibre exhibits some sensitivity to strain rate while the tensile strength and Young's modulus show no dependency on strain rate in the investigated range of their study. When compared to the properties from the manufacturer, the measured tensile strength at low strain rate (3.3×10^{-5} s^{-1}) in this study is lower. The yarn bundling and fabric weaving process will cause damage to the fibres, which leads to the reduction in the tensile strength. It is common that the tensile strength of a single yarn is lower than that of a single fibre [2]. The maximum strains measured in this study are higher than the values from literature and manufacturer, and this is only because the presence of waviness within a yarn.

Table 2. Tensile properties of single yarn and single fibre at different strain rates.

Strain Rate (s^{-1})	Material Form	Tensile Strength (GPa)	Maximum Strain (%)	Young's Modulus (GPa)
3.3×10^{-5}	Yarn	2.63	6.94	65.69
3.3×10^{-3}	Yarn	3.04	5.19	78.37
0.33	Yarn	3.33	5.03	81.19
1.7×10^{-5} [19]	Fibre	2.4	5.88	56
1.7×10^{-4} [19]	Fibre	2.4	3.95	71
6.7×10^{-4} [19]	Fibre	2.4	4	71
1.7×10^{-2} [19]	Fibre	2.4	3.76	62
Manufacturer [20]	Fibre	3.3~3.9	3~4	109~132

4. Conclusion and future work

The combined gripping method was employed to perform tensile tests of single yarn under quasi-static loading conditions. The ethyl cyanoacrylate-based adhesive combined with polyolefin surface primer was determined to be the ideal choice for application to UHMWPE single yarn. Image analysis showed no yarn slippage from the aluminium sheets and the measured stress-strain curves showed good repeatability of the results. The tensile strength and Young's modulus of Dyneema® SK75 single yarn was found to increase with the strain rate while the maximum strain decrease. The present work is a step towards a complete experimental characterisation of Dyneema® SK75 single yarn, and dynamic tensile tests using this combined gripping method will be conducted in the near future.

Acknowledgments

The authors gratefully acknowledge Mr. D Sharp for his technical assistance during the experimental work reported in this paper. H Wang would like to thank UNSW Canberra and the China Scholarship Council for funding his PhD study.

References

[1] Marissen R 2011 Design with ultra strong polyethylene fibers *Materials Sciences and Applications* **2** 319-30
[2] Russell B P, Karthikeyan K, Deshpande V S and Fleck N A 2013 The high strain rate response of Ultra High Molecular-weight Polyethylene: From fibre to laminate *International Journal of Impact Engineering* **60** 1-9
[3] Farsi Dooraki B, Nemes J A and Bolduc M 2006 Study of parameters affecting the strength of yarns *J. Phys. IV France* **134** 1183-8
[4] Shahkarami A, Cepus E, Vaziri R and Poursartip A 2006 Material responses to ballistic impact, *Lightweight Ballistic Composites* ed A Bhatnagar (England: Woodhead Publishing) pp 72-100
[5] Shim V P W, Lim C T and Foo K J 2001 Dynamic mechanical properties of fabric armour, *International Journal of Impact Engineering* **25** 1-15
[6] ASTM International 1989 Standard test method for tensile strength and Young's Modulus for high-modulus single-filament materials (PA: West Conshohocken) ASTM D3379 (1989) e1
[7] ASTM International 2014 Standard test method for tensile strength and young's modulus of fibers (PA: West Conshohocken) ASTM C1557-14
[8] Pardini L C and Manhani L G B 2002 Influence of the testing gage length on the strength, young's modulus and Weibull modulus of carbon fibres and glass fibres *Materials research* **5** 411-20
[9] Ilankeeran P K, Mohite P M and Kamle S 2012 Axial tensile testing of single fibres *Modern Mechanical Engineering* **2** 151-6
[10] Kim J, Heckert N A, Leigh S, Kobayashi H, McDonough W, Rice K and Holmes G 2013 Effects of fiber gripping methods on the single fiber tensile test: I. Non-parametric statistical analysis *Journal of Materials Science* **48** 3623-37
[11] Lim J, Zheng J Q, Masters K and Chen W W 2011 Effects of gage length, loading rates, and damage on the strength of PPTA fibers *International Journal of Impact Engineering* **38** 219-27
[12] Guégan P, Othman R, Pasco F and Bruant R 2012 Dynamic tensile test of single PET textile cables *EPJ Web of Conferences* **26** 01003
[13] Tan V B C, Zeng X S and Shim V P W 2008 Characterization and constitutive modeling of aramid fibers at high strain rates *International Journal of Impact Engineering* **35** 1303-13
[14] Brennan A B, Arnold J J and Zamora M P 1995 Surface modification of ultra-high molecular weight polyethylene fibers by γ-radiation-induced grafting *Journal of Adhesion Science and Technology* **9** 1031-48
[15] Kim J H, Alan Heckert N, Leigh S D, Rhorer R L, Kobayashi H, McDonough W G, Rice K D and Holmes G A 2014 Statistical analysis of PPTA fiber strengths measured under high strain rate condition *Composites Science and Technology* **98** 93-9
[16] Sanborn B and Weerasooriya T 2015 Tensile properties of dyneema SK76 single fibers at multiple loading rates using a direct gripping method *Dynamic Behavior of Materials* vol 1 ed B Song, D Casem and J Kimberley (Berlin: Springer International Publishing) pp 1-4
[17] Zhu D, Mobasher B and Rajan S 2011 Dynamic tensile testing of Kevlar 49 Fabrics *Journal of Materials in Civil Engineering* **23** 230-9
[18] Zhu D, Mobasher B, Erni J, Bansal S and Rajan S D 2012 Strain rate and gage length effects on tensile behavior of Kevlar 49 single yarn *Composites Part A: Applied Science and Manufacturing* **43** 2021-9
[19] Kromm F X, Lorriot T, Coutand B, Harry R and Quenisset J M 2003 Tensile and creep properties of ultra high molecular weight PE fibres *Polymer Testing* **22** 463-70
[20] Dyneema D S M 2008 Dyneema® high-strength, high-modulus polyethylene fiber fact sheet *avaialble at : http://www.pelicanrope.com/pdfs2010/DYNEEMA_factsheet_UHMWPE.pdf*

IOP Conf. Series: Materials Science and Engineering **87** (2015) 012110 doi:10.1088/1757-899X/87/1/012110

Fabrication and properties of HDPE/CF/$CaCO_3$/PE-g-MAH quaternary composites

X-L Wang[1, 3], H Ming[2] and H Yin[1]

[1]College of Material Science and Engineering, Chongqing University of Technology, Chongqing 400054, China
[2]College of Mechanical and Electrical Engineering, Yangtze Normal University, Chongqing 408100, China

E-mail: wangxuanlun@cqut.edu.cn

Abstract. In this research, carbon fiber (CF) was taken as reinforcing filler, nano calcium carbonate ($CaCO_3$) was taken as toughener, maleic anhydride grafted polyethylene (PE-g-MAH) was taken as compatibilizer for high density polyethylene (HDPE) modification. Through orthogonal test, the influence of different amount of ingredient CF, $CaCO_3$ and PE-g-MAH on the mechanical properties of the HDPE composites was researched. The optimal composition of the quaternary composites with the good toughness and high strength was obtained.

1. Introduction

High density polyethylene (HDPE) is a kind of non-polar thermoplastic resin which has high crystalline degree. Comparing with engineering plastics product, the application of HDPE product is limited by its low strength, low hardness and low environmental stress crack resistance [1-3]. So that it cannot be widely used as structural material. CF is a kind of reinforcing material which has excellent properties [4, 5]. The adding of CF can obviously improve the tensile strength, bending strength and stiffness of HDPE, meanwhile the impact strength which displays the toughness of the material can have a sharp decrease. PE-g-MAH can improve the interfacial compatibility of CF and HDPE and thus improve the impact strength of the material. As an excellent toughener, nanometer scale inorganic rigid filler $CaCO_3$ can largely increase the toughness of material [6].

In this research, firstly, $CaCO_3$ was added into HDPE/CF blend system, the impact of $CaCO_3$ on mechanical properties of blend system was studied. Secondly, CF as reinforcing filler, $CaCO_3$ as toughener, and PE-g-MAH as compatibilizer were considered as three main factors which influence the mechanical properties of HDPE matrix material. In order to investigate the effect of the relative amount of the three components on the mechanical properties of blend system and thus get a HDPE composite with high strength and high toughness, an orthogonal test was designed.

2. Experimental

2.1. Materials

[3] Address for correspondence: Xuan-Lun Wang, College of Material Science and Engineering, Chongqing University of Technology, Chongqing 400054, China. E-mail: wangxuanlun@cqut.edu.cn.

High density polyethylene (60550AG-Ⅱ) supplied by PetroChina Lanzhou Petrochemical Company; Carbon Fiber (filament diameter 0.7 μm, length 5 mm) supplied by China Carbon Composite Materials Co., LTD; PE-g-MAH (2909, with a grafting ratio of 1.2%) supplied by Shenzhen HuaLin Chemical Co., LTD; Nanoscale $CaCO_3$ (average grain diameter 60-80 nm, activating rate larger than 90%) supplied by Shanghai Hongqing New Materials Co., LTD.

2.2. Experimental design

HDPE (100 phr/per hundred resin), CF (15 phr), $CaCO_3$ (varies from 0 to 15 phr, each 5 phr as a step) were mixed and four different blends were prepared, marked as $CaCO_3$0, $CaCO_3$5, $CaCO_3$10 and $CaCO_3$15, respectively.

Table 1. Factor levels designed in orthogonal experiment.

Level	Factor		
	A (CF/phr)	B ($CaCO_3$/phr)	C (PE-g-MAH/phr)
1	5	5	2
2	10	10	4
3	15	15	6

The orthogonal test was designed as follows: amount of CF (A), $CaCO_3$ (B) and PE-g-MAH (C) and they were taken as the factors which affected the mechanical properties of blend system. Each factor has three levels thus the orthogonal experiment L9 (3^4) was applied. The factors and level settings were shown in table 1. The scheme of orthogonal experimental design was shown in table 2. The optimum recipe can be found after nine groups of test.

Table 2. The orthogonal experiment design.

Test number	Factor			Sample number
	A/phr	B/phr	C/phr	
1	5	5	2	$A_1B_1C_1$
2	5	10	4	$A_1B_2C_2$
3	5	15	6	$A_1B_3C_3$
4	10	5	4	$A_2B_1C_2$
5	10	10	6	$A_2B_2C_3$
6	10	15	2	$A_2B_3C_1$
7	15	5	6	$A_3B_1C_3$
8	15	10	2	$A_3B_2C_1$
9	15	15	4	$A_3B_3C_2$

2.3. Composites fabrication procedure

Take ethyl alcohol as solvent to prepare 1 wt.% KH560 silane coupling agent solution. $CaCO_3$ was mixed with the solution in an environment away from light and soaked for an hour. Then the sample was filtered and washed by deionized water for three to four times, dried under 70°C and cooled for further use. According to the recipe in table 2, blend the dried materials on double-roller blending rolls. The order of adding materials should be: adding CF into blend system after milling HDPE on an open mill for 10mins. After the blend was well mixed, $CaCO_3$ was added into the system. Finally the compatilizer PE-g-MAH was added into the blend system.

2.4. Characterizations

An electronic universal testing machine was used for tensile strength test. The specimens of HDPE matrix composites for tensile test were prepared by injection molding. The method and standard of the test shall refer to China GB/T 1040.2-2006 Standard.

Izod impact tests were conducted by Izod impact test machine (XJU-5.5 type, Chengde Jinjian) under room temperature. The specimens of HDPE matrix composites for impact test were prepared by injection molding. Each value obtained represented the average of five samples.

An electronic universal testing machine was used for bending strength test. When running the test, all the stress was loaded at the same direction. The distance between the two fulcrums was 64 mm.

The liquid-nitrogen-frozen fractured HDPE matrix composites with different contents of CF, $CaCO_3$ and PE-g-MAH were studied using a JSM-6460LV SEM (JEOL). The SEM samples were etched by acetone for a given time and gold-sputtered prior to observation.

3. Results and discussion

3.1. The mechanical properties of HDPE/CF/$CaCO_3$ system

The effect of nano calcium carbonate loading on tensile strength and bending strength of CF (15 phr) strengthened HDPE was shown in figures 1 and 2. As shown, the adding of nano calcium carbonate lowered the stiffness of material. When the $CaCO_3$ was added to 15 phr, tensile strength and bending strength of material reached the minimum value at the same time (35.95 MPa and 40.11 MPa respectively). When compared with HDPE/CF binary system, the tensile strength and bending strength decreased to 82.64% and 91.45%, respectively. This suggests that when the amount of nano $CaCO_3$ is high, the combination between the different phases in the system is not so good which resulted in deterioration of materials' ability to resist deformation.

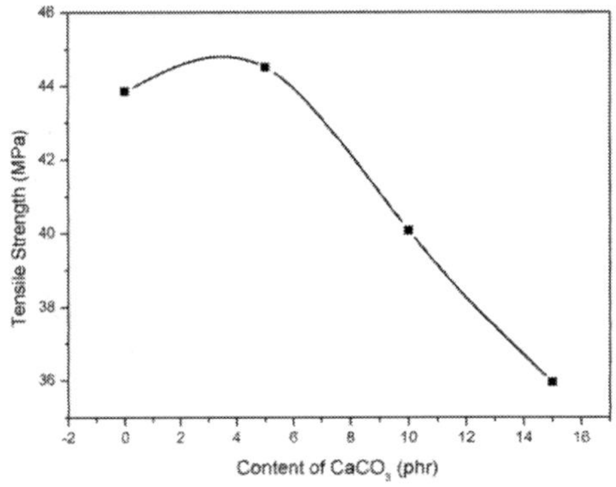

Figure 1. Effect of nano $CaCO_3$ on the tensile strength of the HDPE composites.

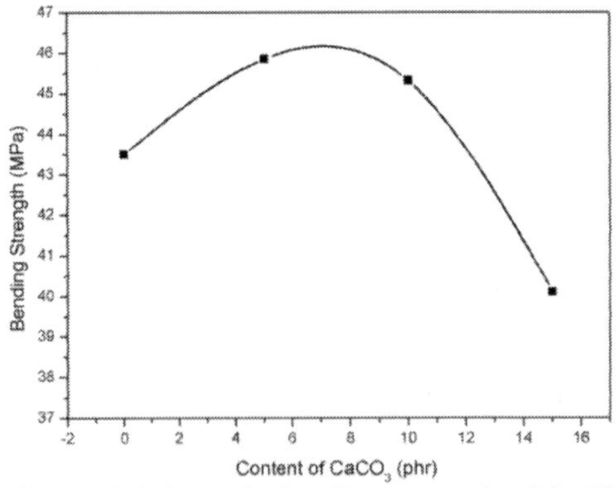

Figure 2. Effect of nano $CaCO_3$ on the bending strength of the HDPE composites.

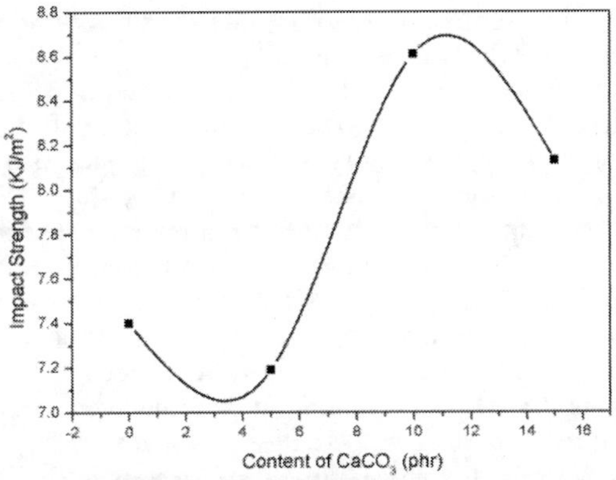

Figure 3. Effect of nano $CaCO_3$ on the impact strength of the HDPE composites.

When the amount of $CaCO_3$ is 10 phr, the tensile strength of HDPE/CF/$CaCO_3$ system (40.07 MPa) has decreased slightly comparing to HDPE/CF binary system (43.5 MPa), but it showed a 58.00% improvement when compared with HDPE (25.36 MPa); the bending strength reached 45.32 MPa, showed a slight increase comparing to HDPE/CF binary system.

The effect of nano $CaCO_3$ on the impact strength of the blend system was shown in figure 3. When the amount of $CaCO_3$ was 5 phr, the impact strength of system was 7.19 kJ/m^2, showing a slight decrease comparing with that of HDPE/CF binary system (7.4 kJ/m^2). With the increasing of $CaCO_3$ amount, the impact strength of system reached 8.61 kJ/m^2 when amount of $CaCO_3$ was 10 phr, showed a 16.35% increase comparing with that of HDPE/CF binary system. When the amount of $CaCO_3$ was 15 phr, the impact strength dropped to 8.13 kJ/m^2. As analyzed, when the amount of $CaCO_3$ was 10phr, polar CF and non-polar $CaCO_3$ could both be well contacted with matrix resin, the density of disperse increased. Among the system, $CaCO_3$ can be seen as a third phase which can transmit stress, CF played the role of reinforcing agent. In consequence, the impact strength of the composites was obviously improved. But the amount of $CaCO_3$ shall not be too much. Excessive $CaCO_3$ can make the interfacial properties between each component to be worse. $CaCO_3$, as a relatively plentiful phase, were likely to make direct contact with CF. As a result, stress cracking were more likely to occur when the material was under external impact.

3.2. Orthogonal test and range analysis

The orthogonal test was carried out according to table 2. The results are shown in table 3. The impact strength values in table 3 were range analyzed.

Table 3. Results of orthogonal test.

Test number	Impact strength kJ/m^2
$A_1B_1C_1$	6.806
$A_1B_2C_2$	7.698
$A_1B_3C_3$	6.612
$A_2B_1C_2$	4.266
$A_2B_2C_3$	6.422
$A_2B_3C_1$	6.314
$A_3B_1C_3$	8.851
$A_3B_2C_1$	6.106
$A_3B_3C_2$	5.872

The impact strength of quaternary blending system showed a decreasing trend when the adding amount of CF was low. The reason is speculated as follows: since the quantity of CF in each unit section was low, the average stress that CF in each unit section borne was high. Stress concentrated in the position of CF. As results, either CF was pulled out easily, or crack occurs in the interface between CF and matrix resin, both leaded to the sharp decrease of impact strength. When CF is further added to 15 phr, impact strength of quaternary blending system showed an increasing trend, the mechanism of such phenomenon is speculated as follows: since the quantity of CF in each unit section increased, along with the existence of compatilizer PE-g-MAH which improved the interfacial compatibility of CF with HDPE, CF could deliver more impact energy. When the material is impacted, the possibility of interfacial debonding dropped. Thus, the impact strength of material is improved.

With the increase of $CaCO_3$ amount, the impact strength of blending system rose at the beginning and then dropped down, similar to the trend in ternary system. The impact strength reached the peak when $CaCO_3$ amount is 10 phr. $CaCO_3$ can demonstrate the strengthen ability when its amount is low. However, $CaCO_3$, as polar filler, have low compatibility with non-polar matrix. Agglomeration occurs among $CaCO_3$ particles, the interfacial compatibility also became worse. These are the main reasons led to the sharp decrease of impact strength.

Table 4. Results of range analysis of impact strength in orthogonal test.

Mean impact strength of every level, kJ/m^2	Factors A/phr	B/phr	C/phr
k1	6.875	6.641	6.407
k2	5.667	6.742	5.945
k3	6.943	6.266	7.295
Range	1.086	0.476	1.350

When the amount of PE-g-MAH was 6 phr, the impact strength of quaternary system reached the maximum. When the amount of PE-g-MAH was low, it could not be well mixed with $CaCO_3$ and CF fillers, the interaction between four kinds of component was weak; the phase interfaces was obvious so the material is easy to break at the interfaces when impacted by external force. Thus the material showed low impact strength. When the addition of PE-g-MAH increased from 4 phr to 6 phr, the average impact strength of blend system climbed from 5.945 kJ/m^2 to 7.295 kJ/m^2. The increase rate reached 22.71%. The reason may be that the increase of compatilizer had strengthened the interaction between CF, $CaCO_3$ and matrix. That is, the ethylene chain segments grafted polyethylene affined well with the HDPE chain, at the same time, the grafted MAH group was strong chemistry bounded to the surface of CF and $CaCO_3$. PE-g-MAH showed the function of "bridge" among fillers and matrix and improved the surface condition among each component. So the impact strength of filled system had improved.

3.3. The determine of the optimum recipe

According to the results of range analysis in table 4, the maximum value of range of factor C (the amount of PE-g-MAH) is 1.350. The range reflects the degree of effect that the factor made to index. The results showed that among the quaternary system, the amount of PE-g-MAH has the greatest influence to the mechanical properties of blend system, i.e. the amount of PE-g-MAH is the main factor. From figure 3 we can conclude, the level of C (the amount of PE-g-MAH) should be C3.

Then, the range of factor A (the amount of CF) is 1.086. The results showed that the amount of PE-g-MAH has great influence to the mechanical properties of blend system. From table 4 we can conclude, the level of A (the amount of CF) should be A3. Then, the range of factor B (the amount of $CaCO_3$) is 0.476. The results showed that the amount of $CaCO_3$ has little influence to the mechanical properties of blend system. Thus the factor C is the secondary factor. Also we can conclude, the level of B (the amount of $CaCO_3$) should be B2.

From the above, the conclusion can be gotten the optimum recipe should be $A_3B_2C_3$. That is, when the amount of matrix resin HDPE was 100 phr, the amount of CF was 15 phr, the amount of $CaCO_3$

was 10 phr and the amount of PE-g-MAH was 6 phr, the impact strength of blend system may reach the maximum.

Recipe $A_3B_2C_3$ did not occur in the scheme of orthogonal test. So we did a further test with the blending system of the recipe $A_3B_2C_3$. The properties of recipe $A_3B_2C_3$ and pure HDPE were compared as table 5. The impact strength indicates the toughness of material had risen 18.74%, from 8.75 kJ/m^2 (pure HDPE) to 10.39 kJ/m^2 ($A_3B_2C_3$). This had experimentally verified the accuracy of influence each factor level made on the impact strength of material. Meanwhile, tensile strength and bending strength were also great. Tensile strength of recipe $A_3B_2C_3$ rose from 25.36 MPa (pure HDPE) to 42.17 MPa, which was a 66.29% increase. Bending strength of recipe $A_3B_2C_3$ rose from 21.48 MPa (pure HDPE) to 60.12 MPa, which was a 137.52% increase.

Table 5. Properties comparison of pure HDPE, HDPE/CF binary composites and $A_3B_2C_3$ quaternary composites.

	Pure HDPE	HDPE/CF (15 phr)	$A_3B_2C_3$
Impact strength (KJ/m^2)	8.75	7.41	10.39
Tensile strength (MPa)	25.36	43.51	42.17
Bending strength (MPa)	21.48	43.86	60.12
Vicat softening temperature (°C)	70.5	80.5	80.6
MFR (g/10 min, 2.16 Kg)	6.3	2.6	5.7

From the above results, we can conclude that recipe $A_3B_2C_3$ is the composite material which has obvious strengthening effect and best toughness obtained by the experiment. The reason maybe the compatibilizer PE-g-MAH improved the dispersion of nano $CaCO_3$ and CF. Furthermore, it enhanced the interface bonding between the fillers and HDPE matrix.

The heat resistance of $A_3B_2C_3$, HDPE and HDPE/CF were also compared in table 5. As shown, the vicat softening temperature of $A_3B_2C_3$ reached 80.6°C, showed 14.33% increase when compared with HDPE (70.5°C), showed a slightly increase when compared with HDPE/CF binary system. It was clear that $A_3B_2C_3$ has good heat resistance.

The melt flow rate (MFR) of $A_3B_2C_3$, HDPE and HDPE/CF were investigated. As shown, the MFR value of $A_3B_2C_3$ reached 5.7 g/10 min, showed slightly decrease when compared with HDPE (6.3 g/10 min). Meanwhile it showed a 142.69% increase when compared with HDPE/CF binary system (2.6 g/10 min) in which the amount of CF is 15 phr. It was clear that $A_3B_2C_3$ has better processing fluidity than the HDPE/CF binary system.

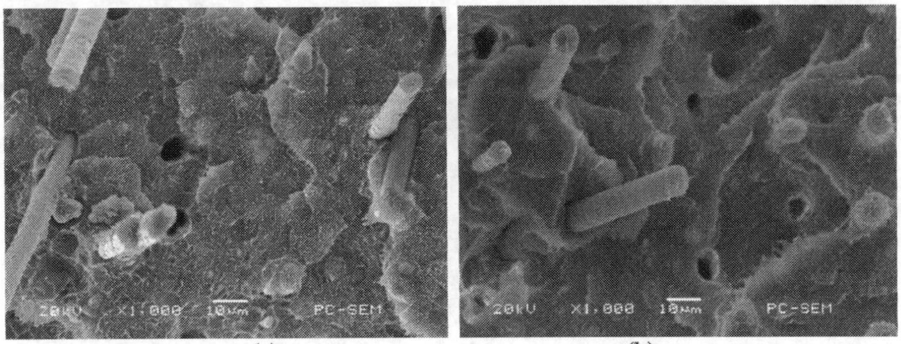

(a) (b)

Figure 4. SEM micrographs of HDPE/CF/$CaCO_3$ ternary composites (a); HDPE/CF/ $CaCO_3$/ PE-g-MAH quaternary composites (b).

From figures 4(a) and 4(b), we can see that CF and $CaCO_3$ have better dispersion in the HDPE/PE-g-MAH composite matrix than the pure HDPE matrix. Some hollow holes were discovered on the impact fractured surface. Materials broke at the same layer that meant there was no stress concentration inside material, at the break position it was uniformly forced and there were no large crack occurs at the break. As shown, CF and matrix resin were well wrapped, $CaCO_3$ were also well deposited in matrix resin. PE-g-MAH improved the interfacial adhesion between the fillers and the HDPE matrix. From above we can conclude that there were fine combinations among matrix resin and fillers.

4. Conclusions

In this study, the HDPE/CF/$CaCO_3$/PE-g-MAH quaternary composites were fabricated and their properties were characterized. CF as reinforcing filler, $CaCO_3$ as toughener, and PE-g-MAH as compatibilizer were considered as three main factors which influence the mechanical properties of HDPE matrix material. Orthogonal test and range analysis were designed and conducted. The optimum recipe obtained from the experiment was $A_3B_2C_3$, i.e. the amount of matrix resin HDPE was 100 phr, the amount of CF was 15 phr, the amount of $CaCO_3$ was 10 phr, and the amount of PE-g-MAH was 6 phr. Analyses were taken with the result of orthogonal test. The impact strength had risen 38.53%, from 7.5 kJ/m^2 (pure HDPE) to 10.39 kJ/m^2 ($A_3B_2C_3$). It had experimentally verified the accuracy of influence each factor level made on the impact strength of material. Meanwhile, tensile strength and bending strength were also great. The vicat softening temperature which indicates the HDPE/CF/$CaCO_3$/PE-g-MAH quaternary composites have good heat resistance and processing fluidity. From the micro morphology measurement conducted by SEM, it was concluded that all fillers in the composites were well dispersed and interface bonding was good.

Acknowledgments

The authors gratefully acknowledge the financial support of Chongqing Municipal Key Laboratory of Institutions of Higher Education for Mould Technology (Grant No. MT201507)

References

[1] Bartczak Z, Argon A S, Cohen R E and Weinberg M 1999 *Polymer* **40** 2347-65
[2] Lazzeri A, Zebarjad S M and Pracella M 2005 *Polymer* **46** 827-44
[3] Park S J and Kim B J 2005 *Mat Sci Eng A* **408** 269-73
[4] Severini F, Formaro L, Pegoraro M and Posca L 2002 *Carbon* **40** 735-41
[5] Li J, Huang Y, Xu Z and Wang Z 2005 *Mater Chem Phys* **94** 315-21
[6] Lin Y, Chen H, Chan C M and Wu J 2008 *Macromolecules* **41** 9204-13

About a permeability of graphene pores

Mikhail A Bubenchikov, Alexander I Potekaev, Alexey M Bubenchikov, Olesya V Usenko[1] and Anton V Ukolov

National Research Tomsk State University, 634050, 36 Lenin Avenue, Tomsk, Russia

E-mail: usenko.olesya@yandex.ru

Abstract. The present paper theoretically studies helium and helion passage through pores in a graphene sheet using classical molecular dynamics. It was found that, within the frame of the suggested model, for considerably large pores with the diameter of 0.75 nm, helium permeates much better than helion, which is quite an unexpected result in itself.

1. Introduction

In [1, 2] practical aspects of ^4He, ^3He and H_2 molecules' penetration through graphene pores are considered. In this paper, using numerical methods, we solve the problem of defining the permeability gap for CH_4, ^4He and ^3He within a graphene structure containing a hole in the crystal lattice. [3] reports and describes in detail the simulation model of energetically stable vacancies in graphene plates, as well as presents the results of modelling the process of separation for mixtures CO_2-CH_4, CO_2-N_2 with the help of these structures. Regarding the separation properties, a comparison with carbon nanotubes having similar diameters or the S/V ratio is made. The results obtained can be used in applications connected with adsorption of CO_2. [4] studies permeability of graphene sheets containing nanopores with the diameter of 0.32 nm - 0.64 nm. Components of a methane-butane mixture were separated. Unexpected results were obtained: in the case of large pores, butane penetrates much better than methane.

2. Molecular-kinetic description

In our theory, interaction between individual molecules is determined by the classical Lennard-Jones potential. Interaction with the structure, in turn, is described by the law of action independence as the sum of each structure molecule's impact onto a given test molecule. According to this law, if a material point is influenced by several forces, it acquires acceleration equal to the geometric sum of accelerations that it would acquire in case of those forces acting separately [5].

The total potential of interaction between the structure molecules and the test molecule can be written as the sum of individual potentials:

$$\Phi = 4\varepsilon \sum_{j=1}^{N_g} \left[\left(\frac{\sigma}{\rho_j} \right)^{12} - \left(\frac{\sigma}{\rho_j} \right)^{6} \right], \tag{1}$$

[1] Address for correspondence: Olesya V Usenko, National Research Tomsk State University, 634050, 36 Lenin Avenue, Tomsk, Russia. E-mail: usenko.olesya@yandex.ru.

where ρ_j is the distance between the considered molecules and the j-th molecule of the structure; σ and ε are the parameters of the interaction between the pairs of substances; σ is the repulsion radius for the molecular pairs; ε is the depth of the potential well for the corresponding pairs of identical molecules.

The values of the constants ε and σ, included in the Lennard-Jones potential, for homogeneous pairs of substances, are given in the table 1.

Table 1. The values of the constants ε and σ

Interacting molecules	Relative depth of the potential well (K)	Influence radius of the interaction potential (nm)
^3He — ^3He	$\varepsilon / k = 10{,}2$ [K]	$\sigma = 0{,}256$ [nm]
He — He	$\varepsilon / k = 10{,}2$ [K]	$\sigma = 0{,}228$ [nm]
CH_4 — CH_4	$\varepsilon / k = 148$ [K]	$\sigma = 0{,}38$ [nm]

If the system under study consists of heterogeneous molecules (atoms), for the parameters ε and σ the following averaging rules (the Lorentz-Berthelot mixing rules) are observed:

$$\sigma_{CH} = \frac{\sigma_{CC} + \sigma_{HH}}{2}, \quad \varepsilon_{CH} = (\varepsilon_{CC} \cdot \varepsilon_{HH})^{1/2} \tag{2}$$

It is known that the Lennard-Jones potential is attractive at long distances, has a minimum at the point $r = \sqrt[6]{2}\sigma$ and is strongly repulsive at short distances [6]. The view of the potential is shown in figure 1.

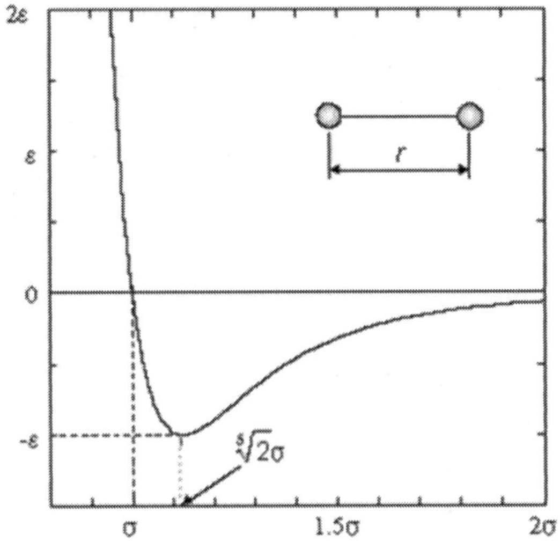

Figure 1. View of the Lennard-Jones potential.

In order to define the permeability gap for each of the gases, we take the average value of the parameter σ as the minimum.

Peculiarities of applying the method of classical molecular dynamics with nanofiltration objectives are introduced in [7-11].

Figures 2-4 show permeability gaps for methane, helium, and helion with 12 remote nodes.

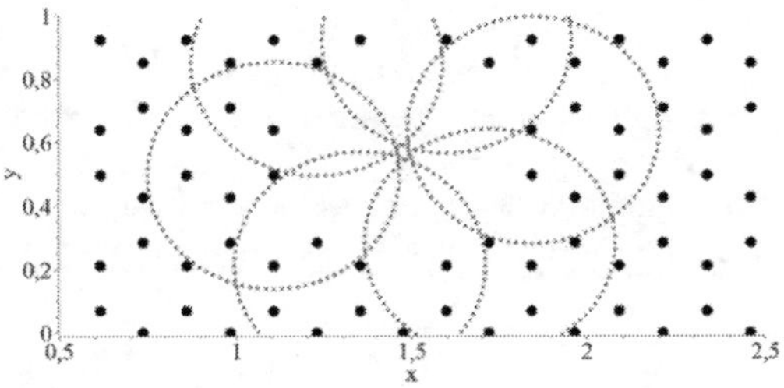

Figure 2. Permeability gap and fields of repulsive forces (hatched) within the area of 12 remote crystal nodes for methane molecules.

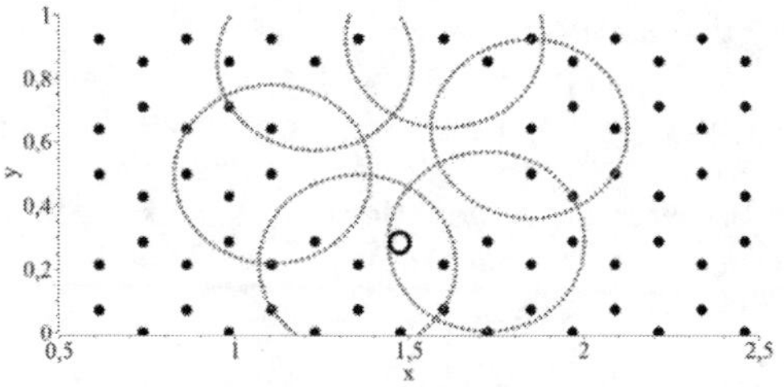

Figure 3. Permeability gap and fields of repulsive forces (hatched) within the area of 12 remote crystal nodes for helium molecules.

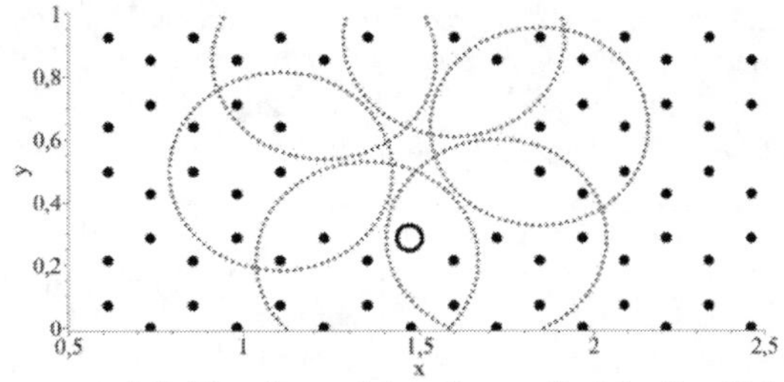

Figure 4. Permeability gap and fields of repulsive forces (hatched) within the area of 12 remote crystal nodes for helion molecules.

In order to verify these results a series of molecular-ballistic tests was conducted. At first, shots were made from the point of the remote interior node of the graphene plate, which is included neither in the permeability gap of helium, nor in that of helion: x_0 =1,5, y_0 =0,4. It was determined that molecules of both gases pass through the defect. Then shots were made from the position displaced from this node to the left by a quarter of the distance between carbon molecules in the graphene crystal structure (0.036 nm). For this position, helium test results appeared to be positive and those of helion - negative (see figures 5 and 6).

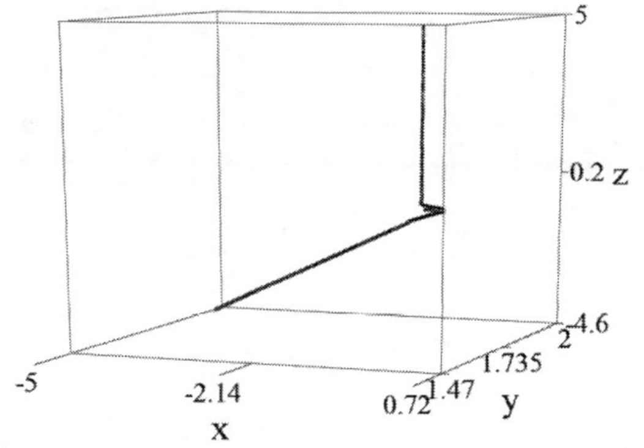

Figure 5. Passage of helium molecule through given pore in grapheme.

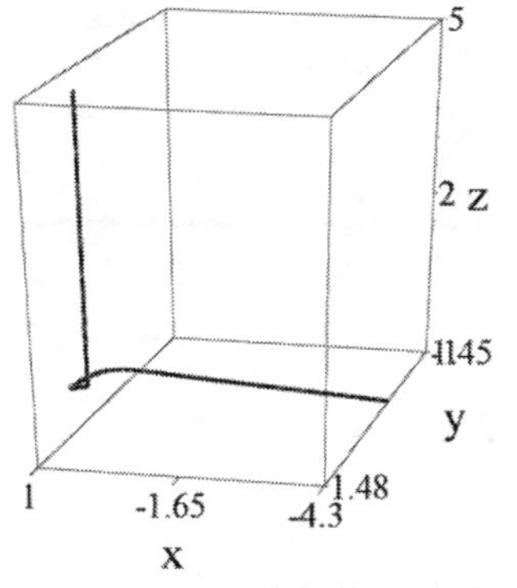

Figure 6. Passage of helion molecule through pore sized minus twelve carbon atoms.

As it can be seen from figure 6 the helion molecule, reaching the graphene plate, was repulsed and passed along.

Despite the fact that helion possesses a greater heat rate and a smaller mass than helium, a graphene plate does not show greater permeability for it.

IOP Conf. Series: Materials Science and Engineering **87** (2015) 012111 doi:10.1088/1757-899X/87/1/012111

3. Summary

We analyzed pores having zero areas of permeability gaps for all neutral molecules under study, including the helium isotope in question. Despite the simplicity of the mathematical model, which does not take account of carbon-carbon bonds within a graphene plate, as well as changing force fields on the edge of the pore, the calculations give correct relations between permeabilities of individual components. The calculations show that perforated graphene sheets can not be used for separation of a helion-helium mixture.

Acknowledgments

Funding was provided by Tomsk State University Competitiveness Improvement Program.

References

[1] Hauser A W, Schrier J and Schwerdtfeger P 2012 *Journal of Physical Chemistry* C **116** 10819-27
[2] Liu H, Dai S and Jiang D 2013 *Solid State Communications* **175** 101-5
[3] Furmaniak S, Terzyk A P, Gauden P A, Kowalczyk P and Harris P J 2014 *J. Phys Condens Matter* **26** 485006
[4] Nieszporek K and Drach M 2015 *Phys. Chem. Chem. Phys.* **17** 1018-24
[5] Tomilov E D 1966 *Theoretical Mechanics* (Tomsk: Tomsk State University)
[6] Fedorov A S, Sorokin P B, Avramov P and Ovchinnikov S G 2006 *Modelling of the properties of the electronic structure of a number of carbon and non-carbon nanoclusters and their interaction with light elements* (Novosibirsk: Publishing House of the Russian Academy of Sciences)
[7] Potekaev A I, Bubenchikov A M and Bubenchikov M A 2013 *Russian Physics Journal* **55** 1434-43
[8] Bubenchikov M A, Potekaev A I and Bubenchikov A M 2013 *Russian Physics Journal* **56** 341-8
[9] Bubenchikov M A, Potekaev A I and Bubenchikov A M 2013 *Russian Physics Journal* **56** 785-90
[10] Bubenchikov M A, Potekaev A I and Bubenchikov A M 2015 *Applied Mechanics and Materials* **698** 525-8
[11] Bubenchikov M A, Potekaev A I and Bubenchikov A M 2015 *Applied Mechanics and Materials* **698** 627-30

IOP Conf. Series: Materials Science and Engineering **87** (2015) 012112 doi:10.1088/1757-899X/87/1/012112

A study on properties of PLA/PBAT from blown film process

P Hongdilokkul[1], K Keeratipinit[1], S Chawthai[1], B Hararak[2], M Seadan[3] and S Suttiruengwong[1, 4]

[1]Department of Materials Science and Engineering, Faculty of Engineering and Industrial Technology, Silpakorn University, Sanamchandra Palace Campus, Nakhon Pathom 73000, Thailand
[2]National Metal and Material Technology Center (MTEC), National Science and Technology Development Agency, Pathum Thani, 12120, Thailand
[3]Department of Physics, Faculty of Science, Silpakorn University, Nakhon Pathom, 73000, Thailand

E-mail: suttiruengwong_s@su.ac.th

Abstract. The aim of this work was to study the properties of films based on PLA/PBAT blend prepared by the reactive compounding. PLA/PBAT blends were prepared at the weight ratio of 80:20 together with peroxide as a reactive agent in a twin screw extruder with temperature profile of 160/170/180/210/220/190/175/150°C from feed to die zone. All blended samples, neat PLA, and neat PBAT were characterized for morphology, mechanical and rheological properties. SEM micrographs showed finely dispersed phases of PBAT in PLA in all cases. The particle sizes of PBAT were around 1 μm. The results indicated that the drawability and toughness properties of PLA were greatly improved when blended with 20%wt PBAT. The interface adhesion, and mechanical properties of PLA/PBAT blends were also improved when adding a very small quantity of peroxide. PLA/PBAT blends were then used to produce films. The film characteristics and mechanical properties were examined. Tensile strength of films was significantly improved in the machine direction in PLA/PBAT/peroxide blends whereas the good optical transparent property were remained the same compared with neat PLA.

1. Introduction

A sustainable use of resources and the pollution prevention are important issues for generations to come. The needs of both renewable and non-renewable feedstocks management are vital. One of the most concerned feedstocks is a fossil, which is normally utilized for energy and as an intermediate for petrochemical industries including plastics. In terms of plastics, most of the conventional plastics are generally fossil-based and can be accumulated in the environment after usage as solid wastes. An alternative approach to replacing the non-biodegradable or/and fossil-based plastics is the use of biobased plastics (could be either biodegradable or not), which have been recognized for decades.

Among bio-based polyester [1, 2], polylactic acid (PLA) is very fashionable and one of the most explored bioplastics. Polylactic acid has high modulus and strength comparable to that of petroleum-

[4] Address for correspondence: S Suttiruengwong, Department of Materials Science and Engineering, Faculty of Engineering and Industrial Technology, Silpakorn University, Sanamchandra Palace Campus, Nakhon Pathom 73000, Thailand. E-mail: suttiruengwong_s@su.ac.th.

based plastic [3]. Its tensile strength and elastic modulus are comparable to poly (ethylene terephthalate) (PET) [4]. However, PLA is brittle, possesses low heat distortion temperature (HDT around 60°C), low water vapor, and gas barrier properties [5-7], which limited its applications.

Many methods have been investigated in the literature to improve the toughness of PLA such as plasticization [8], copolymerization [9], and blending with a variety of flexible polymers, elastomers or rubbers. Blending PLA with other biodegradable polymers presents a more practical and economic measure to obtain products with properties not currently attainable [10]. In this sense, improving the brittleness of PLA is considered to be economical as PLA is less expensive when compared to other biodegradable polyesters. PLA/Poly (butylene-adipate-co-terephthalate) (PBAT) blends have been extensively investigated by many researchers [11-17] due to the flexibility of PBAT. In terms of the high toughness PBAT, it is considered to be a good candidate for the improved toughness of PLA [18]. The elongation at break of PLA was increased from 4% to 200% by addition PBAT 20 wt % (Ecoflex®) [19]. However, PLA /PBAT blend are immiscible [19]. In situ compatibilization could be achieved in PLA/PBAT blends in a presence of glycidyl methacrylate (GMA) using dicumyl peroxide (DCP) as a free radical initiator [11]. It led to reducing PBAT domain size and enhanced in their interfacial adhesion [20]. To ensure the material performance and feasibility of production, most of the compounds should be tested for finished products such as films, injection-molded articles and thermoformed items. However, there is very few information on blown film processing of PLA compounded with PBAT. Therefore, in this study we reported the reactive blend between PLA and PBAT with fixed ratio of 80:20. Their rheological, mechanical and morphological properties on reactive compounded polymers were investigated whereas the mechanical and optical properties of the resulting films were fully examined.

2. Experimental

2.1. Materials

PLA (4043D grade) was purchased from Nature Works LLC, USA. PBAT (Ecoflex F blend C1200) was purchased from BASF Chemical Company, Germany. Di (tert-butylperoxyisopropyl) benzene (Paradox® 14-40B-pd) was commercially supplied by Akzo Nobel (Asia) Co., Ltd. (Shanghai, PR China).

Table 1. The composition of reactive blend and the non-reactive blend of PLA/PBAT blends.

Sample name	PLA	PBAT	Perkadox (phr)
Neat PLA	100	0	0
80/20	80	20	0
P0.002	80	20	0.002
P0.02	80	20	0.02
Neat PBAT	0	100	0

2.2. Blending and sample preparation

Before use, both PLA and PBAT resin were dried at 60°C for six h in an oven in order to remove all trace moisture and to prevent potential hydrolytic degradation during melt processing. PLA and PBAT were blended via a twin screw extruder (Labtech engineering, Thailand). The temperature profiles were set from feed zone to die at 160/170/180/210/220/190/175/150°C. The screw speed was maintained at 80 rpm. The weight ratios of PLA, PBAT and reactive agents in this study were shown in table 1. The extrudate was cooled in a water bath, pelletized and dried at 60°C for six h. The pellets were compression-molded into micro-tensile and impact specimens at 180°C for 4 min through a hot compression molding machine (Chareon tut CO., LTD). The films of blends were produced via the single screw extruder blown film (HAAKE); annular die with a 34 mm diameter. The temperatures were set at 160, 170 and 180°C and the temperature of die was set at 190°C. The blow-up ratio (BUR) of the bubble was 2.5:1. This setting produced a bubble with an average thickness of 0.03 mm.

2.3. Characterization and testing

2.3.1. Morphology. The micrographs of the blends were examined using a scanning electron microscope (SEM) operated at 15 KV. The specimens were fractured under cryogenic condition in liquid nitrogen and the fractured surfaces were sputtered with gold.

2.3.2. Rheological property. Shear Viscosity of sample melt were measured by using a capillary rheometer (ROSAND Rh 2200, Rosand (Bohlin Instrument)) at 190°C in a shear rate from 10-5000 s-1, according to ASTM D 3835.

2.3.3. Mechanical properties. Mechanical properties of the specimens were investigated by using tensile and impact testing. The tensile testing was carried out on an Instron machine (High Wycombe, UK) at 23°C, crosshead speed of 10 mm/min on at least five dumbbell-shaped specimens of each material typed V and gauge length of 17 mm, according to ASTM D 638. Tensile modulus, elongation at break, tensile strength and toughness were determined. The Izod impact testing was performed on 5-10 specimens of each material using the impact tester (GT-7016-A2, Gotech Testing Machines,Taiwan) with maximum hammer energy of 4 J. Shape of the specimens conformed to ASTM D256. The tensile testing of blown film sample was carried out according to ASTM D882 via a universal testing machine. The crosshead speed of the test was set at 10 mm/min. This test was recorded for both machine and transverse directions (MD and TD).

2.3.4. Optical properties. Optical properties of films were investigated by using Hazemeter (Haze-gard plus, BYK additives and instruments, USA), according to ASTM D1003.

3. Results and discussion

PLA/PBAT (80:20) with and without peroxide were first compounded in a twin screw extruder. Neat PLA and PBAT were also processed at the same conditions for comparison. The morphology, mechanical and rheological properties were then examined for further film processing.

3.1. Morphology of the blends

SEM micrographs of PLA and PLA/PBAT blends were shown in figure 1. The cryo-fractured surface of the neat PLA was fairly smooth (not shown here). In the case of PLA/PBAT 80/20 blend, PBAT phase was finely dispersed in PLA matrix. PBAT particles were less than 1 μm. They could be considered as a partially compatible blend at this ratio. For PLA/PBAT/Perkadox reactive blends with adding Perkadox as less as 0.002 phr, finer morphology of the blend was achieved (figures 1(b)-1(c)) and the particle size of dispersed phases was reduced. The in situ free radical reaction of peroxide improved the compatibility of PLA/PBAT blend, providing better interfacial adhesion between PLA and PBAT phases.

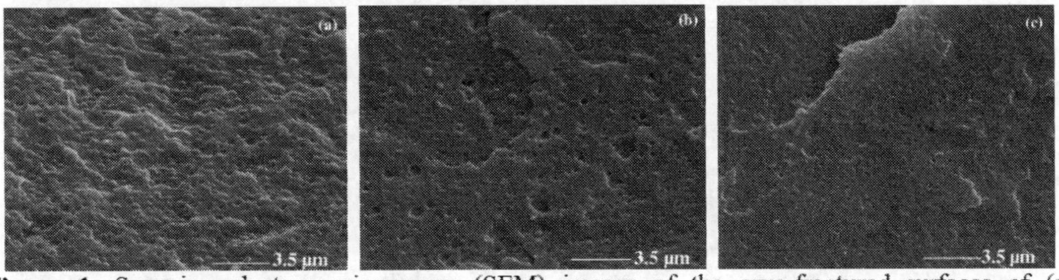

Figure 1. Scanning electron microscopy (SEM) images of the cryo-fractured surfaces of (a) PLA/PBAT 80/20, (b) P0.002 and (c) P0.02.

3.2. Rheological property of the blends

Rheological properties of neat PLA, PLA/PBAT blends, and neat PBAT were evaluated as shown in figure 2. All samples showed that shear viscosity were reduced with increasing shear rate, and this was generally corresponded to shear thinning effect of the polymer. For neat PLA, the effect of shear thinning was more pronounced where its shear viscosity was dropped more rapidly for the shear rate above 1000 s^{-1}. The shear viscosity at the shear rate between 100 and 5000 s^{-1} were presented in figure 3. At the shear rate of 100 s^{-1}, PLA showed the highest shear viscosity whereas at the shear rate of 5000 s^{-1} PLA exhibited the lowest shear viscosity. In the case of PLA/PBAT blend with and without peroxide, its viscosity was close to that of neat PLA for the shear rate of 100 s^{-1} while its viscosity was close to that of PBAT for shear rate of 5000 s^{-1}. The addition of PBAT 20 wt. % could change the rheology properties of PLA. In the case of the reactive blends of PLA/PBAT/Perkadox, upon increasing the peroxide contents, the shear viscosity of the blends were increased and higher peroxide content (P0.002) exhibited the highest shear viscosity. These results confirmed that there was the possible interaction at the interface between PLA and PBAT. However, when adding Perkadox 0.021 phr, the shear viscosity was decreased when compared to that of PLA/PBAT blend without peroxide probably due to chain scission from free radicals. At high shear rate region (5000 s^{-1}), the viscosity of all samples except neat PLA was very close. Thus, it could be concluded that PLA/PBAT blends remained processable when compared with neat PLA even at the high shear processing.

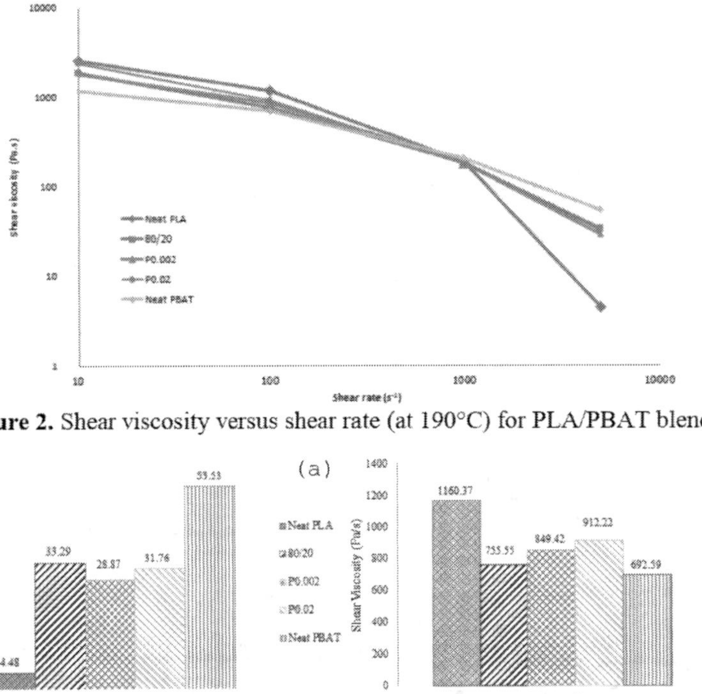

Figure 2. Shear viscosity versus shear rate (at 190°C) for PLA/PBAT blends.

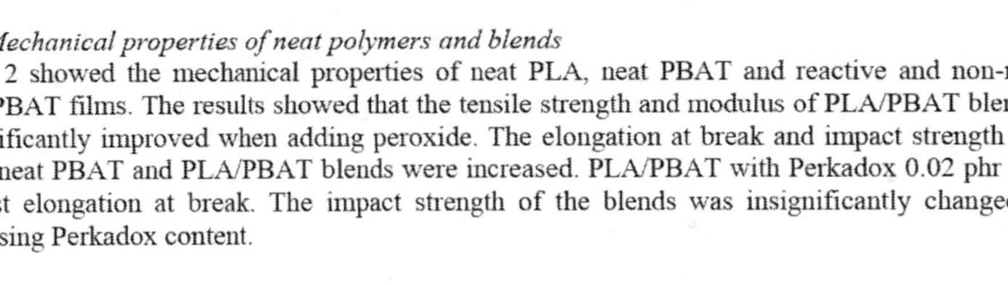

Figure 3. Shear viscosity of the blends at shear rate (a) 5000 s^{-1} and (b) 100 s^{-1}.

3.3. Mechanical properties of neat polymers and blends

Table 2 showed the mechanical properties of neat PLA, neat PBAT and reactive and non-reactive PLA/PBAT films. The results showed that the tensile strength and modulus of PLA/PBAT blend were insignificantly improved when adding peroxide. The elongation at break and impact strength of neat PLA, neat PBAT and PLA/PBAT blends were increased. PLA/PBAT with Perkadox 0.02 phr had the highest elongation at break. The impact strength of the blends was insignificantly changed when increasing Perkadox content.

Table 2. Mechanical properties of the blends.

Sample name	Tensile strength (MPa)	Modulus (MPa)	Elongation at break (%)	Impact strength (KJ/m^2)
Neat PLA	90.6±6.3	1470±101	10.6±2.1	3.3±0.4
80/20	66.1±4.6	1087±75	157.7±11.0	6.4±0.9
P0.002	67.2±5.3	1095±70	180.4±35.9	6.0±1.2
P0.02	63.6±2.5	1126±59	288.9±23.1	6.1±0.9
Neat PBAT	14.1±1.0	44±2	>300	Not break

Figure 4 demonstrated the stress-strain relationship of neat PLA and PLA/PBAT blends. The toughness of the samples could be determined by the area under the stress-strain curve. All blends had a characteristic of drawability curve. Neat PLA yielded the tensile strength of about 90 MPa and elongation at break of about 10 %. It showed very weak drawability and poor toughness property. The tensile strength and modulus of all blends decreased to about 60 MPa, but the elongation at break was increased to about 290%. However, PLA/PBAT with peroxide (P0.002 and P0.02) exhibited better strength tensile strength and modulus. The elongation at break was improved when adding peroxide to the blends. A lot of research of immiscible PLA blends with other polymers occurring the dispersed phase indicated that having defect lead to poor mechanical properties [21]. The improved drawability of reactive PLA/PBAT blends could correspond to the better interfacial adhesion between the two polymers as represented by fine dispersion morphology discussed earlier in SEM.

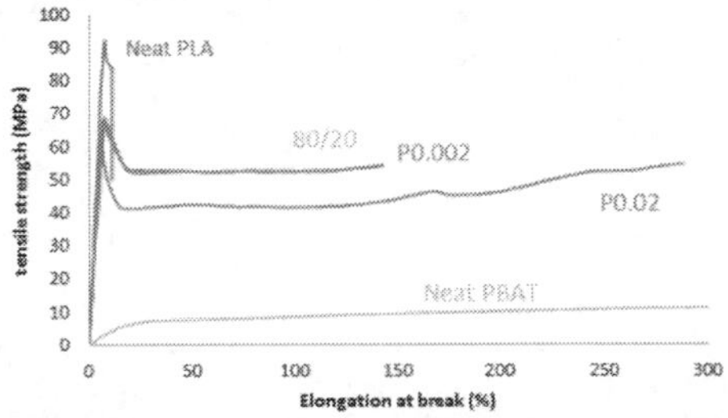

Figure 4. Stress-strain curves of the blends with peroxide contents.

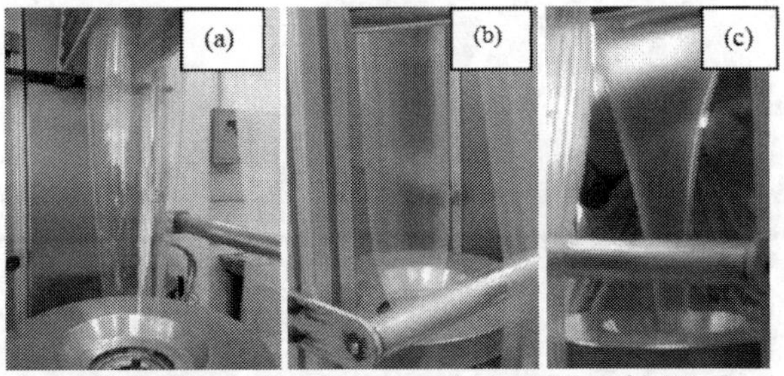

Figure 5. Comparison of the illustration between a) Neat PLA, b) P0.02 and c) Neat PBAT.

3.4. Blown film processing

The blown film machine was then used to convert the compounds (neat polymers and blends) into films. The physical appearance and film properties were evaluated. Figures 5(a)-(c) showed blown film processing of neat PLA, PBAT and the reactive blend of PLA/PBAT (P0.02). Blowing neat PLA was difficult, and the bubble shape was unstable due to low melt strength of PLA. The resulting film had a large variation of dimension and thickness. The film was transparent as PLA was amorphous and had low crystallization rate. In the case of neat PBAT film processing, it was easier to blow. The shape of PBAT bubble was dominated by a long neck champagne similar to that of HDPE bubble. However, the PBAT film was stick together after conveying into the nip roll. In the case of

PLA/PBAT blends with and without peroxide, the films could be processed more easily when compared to neat PLA. The bubble shape was tubular similar to that of LDPE blown film.

3.5. Mechanical properties of films

Table 3 showed mechanical properties of PLA/PBAT (80/20) and P0.02 films in both machine and transverse directions. In the machine direction, in the case of P0.02, the tensile strength, elongation at break and secant modulus were all improved and more than those of non-reactive PLA/PBAT (80/20). On the other hand, in the transverse direction, the tensile strength, elongations at break and secant modulus of P0.02 were decreased when compared with PLA/PBAT (80/20). Figure 6 showed the stress-strain curve of machine direction film samples.

Table 3. Mechanical properties of the films.

Sample name	Direction	Tensile strength (MPa)	Modulus (MPa)	Elongation at break (%)
Neat PLA	MD	51.8±3.7	4068±525	4.9±0.9
	TD	37.9±6.3	3813±367	5.1±1.2
80/20	MD	36.4±4.0	3274±130	87.7±29.9
	TD	40.5±6.9	2949±587	209.6±30.3
P0.002	MD	24.1±5.0	2497±391	123.1±27.2
	TD	42.3±5.9	3514±212	219.9±15.9
P0.02	MD	41.9±4.2	3515±90	199.1±11.2
	TD	26.1±2.6	3099±240	9.3±3.8
Neat PBAT	MD	16.38±1.1	287±12	>300
	TD	13.2±1.1	290±12	>300

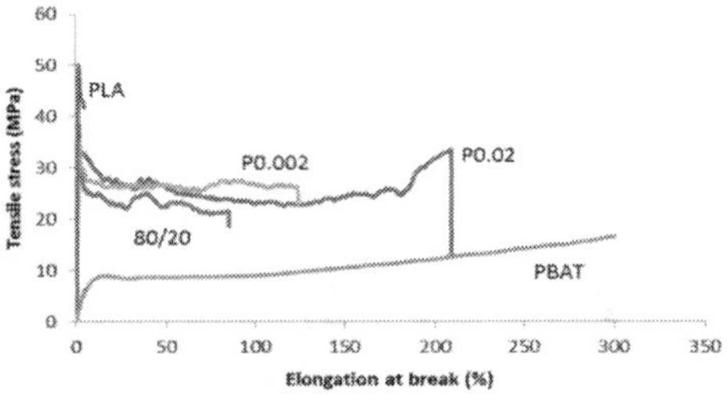

Figure 6. Stress-strain curve of the blends with film samples in machine direction.

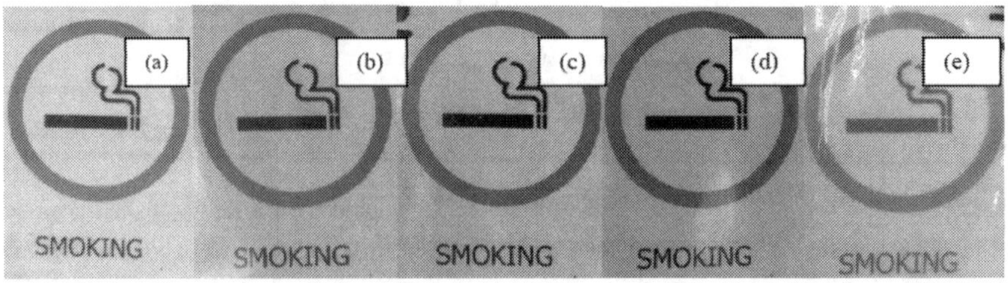

Figure 7. Clarity Comparison of a) neat PLA, b) 80/20, c) p0.002, d) p0.02, e) neat PBAT films.

3.6. Optical properties

Table 4 showed results from haze testing of neat polymers and the blends. Neat PLA had a low haze whereas PBAT had a high haze values that contrary to PBAT, PLA is transparent. All blends despite the different phase morphology (SEM), also showed low haze values indicated that upon adding

PBAT, the optical clarity (figure 7) remained unaffected due to finely dispersed phases of PBAT. The addition of peroxide did not influence the haze values.

Table 4. Haze values of films.

Sample name	Haze (%)
Neat PLA	0.46±0.16
80/20	6.27±0.72
P0.002	5.95±0.82
P0.02	5.49±0.77
Neat PBAT	61.2±1.31

4. Conclusions

PLA/PBAT blends were prepared at the weight ratio of 80:20 with two concentrations of peroxide (0.02 and 0.0002 phr) as a reactive agent. The improvement of the drawability obtained with 0.02 phr of peroxide was due to the improved interfacial adhesion between PLA and PBAT phases. Neat PLA, neat PBAT, and PLA/PBAT blends showed similar shear thinning effect of the polymer. The shear viscosity of neat PLA was dropped more rapidly at the shear rate above 1000 s^{-1}. Adding PBAT 20 wt. %, PBAT could change the rheological properties of PLA at high shear rate (above 1000 s^{-1}. All of the blends non-reactive and reactive PLA/PBAT blends had the same rheology properties. The tensile strength and modulus of PLA/PBAT blend were insignificantly changed when adding Perkadox. PLA/PBAT with 0.02 phr peroxide had the highest elongation at break (288.9%). The impact strength of the blends was insignificantly changed when increasing peroxide contents. PLA/PBAT with peroxide could improve drawability.

For blown film processing, neat PLA was difficult to keep the tubular profile shape due to low melt strength. The PLA/PBAT film with or without peroxide could be better processed by blown film. For mechanical properties of the films in the machine direction, in the case of P0.02, the tensile strength, elongation at break and secant modulus were improved more than those of PLA/PBAT (80/20). The optical properties of PLA had a low haze, but PBAT had a high haze.

Acknowledgments

This work is financially supported by the Department of Materials Science and Engineering, Faculty of Engineering and Industrial Technology, Silpakorn University. The National Metal and Materials Technology Centre (MTEC) was gratefully acknowledged for experimental works.

References

[1] Sodergard A and Stolt M 2002 Properties of lactic acid based polymers and their correlation with composition *Progress in Polymer Science* **27** 1123-63

[2] Iannace S, et al. 2001 Influence of crystal and amorphous phase morphology on hydrolytic degradation of PLLA subjected to different processing conditions *Polymer* **42** 3799-807

[3] Martin O and Averous L 2001 Poly (lactic acid): Plasticization and properties of biodegradable multiphase systems *Polymer* **42** 6209-19

[4] Auras R, Harte B and Selke S 2004 An overview of polylactides as packaging materials *Macromolecular Bioscience* **4** 835-64

[5] Petersen K, Nielsen P V and Olsen M B 2001 Physical and mechanical properties of biobased materials - Starch polylactate and polyhydroxybutyrate *Starch-Starke* **53** 356-61

[6] Bastioli C 2001 *Global status of the production of biobased packaging materials Starch-Starke* **53** 351-5

[7] Fortunati E, et al. 2010 Development and thermal behaviour of ternary PLA matrix composites *Polymer Degradation and Stability* **95** 2200-6

[8] Pillin I, Montrelay N and Grohens Y 2006 Thermo-mechanical characterization of plasticized PLA: Is the miscibility the only significant factor? *Polymer* **47** 4676-82

[9] Harada M, et al. 2007 Increased impact strength of biodegradable poly(lactic acid)/poly(butylene succinate) blend composites by using isocyanate as a reactive processing agent *Journal of Applied Polymer Science* **106** 1813-20

[10] Armentano I, et al. 2010 Biodegradable polymer matrix nanocomposites for tissue engineering: A review *Polymer Degradation and Stability* **95** 2126-46

[11] Kumar M, et al. 2010 Effect of glycidyl methacrylate (GMA) on the thermal, mechanical and morphological property of biodegradable PLA/PBAT blend and its nanocomposites *Bioresource Technology* **101** 8406-15

[12] Al-Itry R, Lamnawar K and Maazouz A 2012 Improvement of thermal stability, rheological and mechanical properties of PLA, PBAT and their blends by reactive extrusion with functionalized epoxy *Polymer Degradation and Stability* **97** 1898-914

[13] Nofar M, et al. 2015 Interfacial and rheological properties of PLA/PBAT and PLA/PBSA blends and their morphological stability under shear flow *Journal of Rheology* **59** 317-33

[14] Quero E, et al. 2012 Isothermal cold-crystallization of PLA/PBAT blends with and without the addition of acetyl tributyl citrate *Macromolecular Chemistry and Physics* **213** 36-48

[15] Al-Itry R, Lamnawar K and Maazouz A 2014 Reactive extrusion of PLA, PBAT with a multi-functional epoxide: Physico-chemical and rheological properties *European Polymer Journal* **58** 90-102

[16] Al-Itry R, Lamnawar K and Maazouz A 2014 Rheological, morphological, and interfacial properties of compatibilized PLA/PBAT blends *Rheologica Acta* **53** 501-17

[17] Signori F, Coltelli M B and Bronco S 2009 Thermal degradation of poly(lactic acid) (PLA) and poly(butylene adipate-co-terephthalate) (PBAT) and their blends upon melt processing *Polymer Degradation and Stability* **94** 74-82

[18] Li D D, Shentu B X and Weng Z X 2011 Morphology, rheology, and mechanical properties of polylactide/poly(ethylene-co-octene) blends *Journal of Macromolecular Science Part B-Physics* **50** 2050-9

[19] Jiang L, Wolcott M P and Zhang J W 2006 Study of biodegradable polyactide/poly(butylene adipate-co-terephthalate) blends *Biomacromolecules* **7** 199-207

[20] Ma P, et al. 2014 In-situ compatibilization of poly (lactic acid) and poly (butylene adipate-co-terephthalate) blends by using dicumyl peroxide as a free-radical initiator *Polymer Degradation and Stability* **102** 145-51

[21] Kowalczyk M, et al. 2014 Toughening of polylactide by blending with a novel random aliphatic-aromatic copolyester *European Polymer Journal* **59** 59-68

Mechanics performance test and feasibility analysis to replace the rigid sucker rod for 6K T300

Changhong Tong[1]

Institute of mechanical detection and fault diagnosis, Lanzhou City University, Lanzhou, 730070, China

E-mail: tch643@163.com

Abstract. A experiment plan was designed according to the working conditions of sucker rod and the requirements for pump depth in 3000 m in the oil field, the tensile strength for 6K T300 under a normal temperature and high temperature was measured by using universal testing machine, and then, the resistance to corrosion for a crude oil was verified by measuring the tensile strength for 6K T300 after crude oil immersion at a certain time, and the conclusions are that the material is sensitive relatively to corrosion of crude oil and that the tensile strength of the 6K T300 compared with similar products is lower, a proposal to the GH company that to meet the need of oil field production instead of the rigid rod the tensile strength and corrosion resistant for a crude of the T300 6 k materials have to do further efforts was pointed out.

1. Introduction

More than 90% of oil pumping wells are mechanical oil pumping in the world petroleum industry today, and over 80% of mechanical oil pumping wells are sucker-rod pumping wells [1]. Sucker rod can be divided into three grades which are C, D and K according to the API [2]. Grade C sucker rods is mainly applied to light load, middle load and weak corrosion oil wells, which is mainly carbon steel or chrome alloy steel, the tension strength is 794~965 MPa. Grade D sucker rods is mainly applied to middle load, heavy load and sour oil wells, which is mainly carbon steel or manganese steel, such as series of Gr-Mo, the tension strength is 620~794 MPa. Meanwhile, Grade K sucker rods is mainly applied to light load, middle load, low sulfur and medium corrosion oil wells at normal temperature, which is mainly nickel-molybdenum alloy steel, such as series of Ni-C-Mo or Ni-Cr, the tension strength is 588~794 MPa.

Performance of traditional rigid rod has many deficiencies. For example: high cost, heavy, easy corroding, pumping with higher energy consumption, difficult to install and maintain, etc. However, the carbon fibers have the advantage of high strength, lightweight, fatigue tolerance, abrasion resisting, and corrosion resistance and so on. The carbon fibers steady of rigid rod have been studied by the western literature in the age of twenty century ninetieth. The result is that carbon fiber composite is a new material which can solve the problem of oil exploitation of high water cut oil wells, deep well, ultra-deep well and corrosive wells [3-7]. Our domestic scholars made the same conclusion quite independently.

literature [8] points out that the coiled flexible rod of carbon fibers has higher specific strength and

[1] Address for correspondence: Changhong Tong, Institute of mechanical detection and fault diagnosis, Lanzhou City University, Lanzhou, 730070, China. E-mail: tch643@163.com.

specific modulus, higher fatigue strength, better wear resistance and higher corrosion resistance than common or fiber reinforced plastics rod, while it has good heat resistance, and it is more convenient to oil exploitation, the daily fluid production rate raised by over 30%, electricity consumption were visibly reduced and its working life were considerably extended. The domestic research on flexible continuous sucker rod of carbon fibers and the application of oil field in North China, Qinghai, Central Plains and Sheng Li have been described in the literature [9], it is pointed out that the flexible continuous sucker rod improved the pump efficiency greatly, saved lots of energy and reduces consumption. It is also concluded that using flexible continuous sucker rod is the development trends in future. Zhang [10] and co-workers studied phenolic resin-based carbon fiber as materials of flexible continuous sucker rod, indicating that it could replace the traditional rigid rod. These studies suggest that it is possible to replace rigid rod by using carbon fibers.

However, most of the above investigations were based on the imported carbon fibers; no study was carried out on domestic ones, this really renders the development and application of domestic carbon fibers. Herein, mechanical properties of 6K T300 (produced by GH Company) are tested; it is focused on the study of using carbon fibers based flexible continuous sucker rod to replace rigid rod. According to the measurements of tensile strength as well as oil corrosion resistance under low and high temperature, the carbon fibers based flexible continuous sucker rod (produced by GH Company) can replace traditional rigid rods.

2. The experiment plan design

The experiment aims are to verify the feasibility of using 6K T300 materials in the underground oil field pumping environment, and thus verify the mechanical properties which related to the underground pipeline pumping oil according to the environmental conditions.

2.1. *Working condition of pumping-oil*

- When the oil pumping rod of beam pumping unit works, if the beam swings upward, it will be driven the plunger rod of piston pump and oil to complete the pumping action – this is upstroke; if the beam swings down, the plunger rod of piston pump will move down according to its own weight – this is down stroke (empty return trip). Thus, the oil pumping rod must have sufficient tensile strength to withstand the load suspension point when pumping.
- The temperature will increase 3℃ when the oil well depth increase 100 m [1], if the pump setting depth at 3000 m, the temperature of oil and water mixture will be close to 100℃, therefore, the high temperature performance of 6 K T300 must be verified.
- The composition of the crude oil is complex, including the various mineralization, paraffin, acid, alkali, water and other ingredients, and even organic ingredients. The working conditions are very poor, it is needed to verify the corrosion resistance of carbon fiber in crude medium, and also tensile strength in corrosive conditions.

2.2. *Experiment scheme design*

According to the conditions (1), the tensile strength of 6 K T300 will be tested, then compared to that of traditional oil pumping rod to verify its feasibility.

According to the conditions (2),the temperature is 100℃ when the pump setting depth at 3000 m, the tensile strength of 6K T300 at different conditions (from room temperature to 100℃) will be measured to verify its high temperature properties.

According to the conditions (3), the 6K T300 materials will be soaked in crude oil for different time (from 3 to 360 days), then tensile strength of this materials will be tested to verify its corrosion resistant for the crude oil condition.

2.3. *Laboratory equipment*

Laboratory equipment: AG-IS-10 kN universal testing machine.

Force measurement precision: ±0.5%.

Sampling rate: 1.25 m/sec.
Test speed range: 0.005~1000 mm/min.

3. The experimental results

3.1. Test the tensile strength for 6K T300 *at the room temperature*

Production material of 6K T300 by GH company,mono filament d = 7 μm, a bunch of tensile diameter (6 k): D= $\sqrt{6000\times10d^2}=542$ μm, that is, D=0.542 mm. Use 1 kN universal testing machine to measure room temperature tensile strength for the 6K T300, the data can be seen in table 1.

Table 1. At 20℃ material tensile strength.

No.	Diameter of sample /mm	Original length /mm	Tensile strength /MPa	Lower yield Strength /MPa	Break elongation /%
1	0.542	100	567	383	17%
2		110	548	381	17%
3		57	537	372	16%
4		90	583	408	18%
5		71	544	383	21%
6		55	523	358	20%
7		77	541	365	15%
8		78	530	361	25%
9		82	562	388	20%
10		71	521	348	17%
Average	0.542	79	525	345	21%

According to the standard, 6 K tensile strength index should be 3.5 GPa forT300 material under the room temperature, apparently the material strength is low.

3.2. For 6K T300 tensile strength test at different temperature

For less than 100℃, use the heat water, heat preservation material in the container. By controlling the water temperature set temperature the material respectively measure tensile strength at 40℃, 60℃, 80℃, 98℃ temperature, the results are shown in table 2-5.

Table 2. At 40℃ material tensile strength.

No.	Diameter of sample /mm	Original length /mm	Tensile strength /MPa	Lower yield strength/MPa	Break elon-Gation /%
1	0.542	135	427	205	0.14
2		155	694	227	0.083
3		145	339	211	0.2
4		145	1057	245	0.08
5		140	920	230	0.114
Average	0.542	144	687	224	0.12

Table 3. At 60℃ material tensile strength.

No.	Diameter of sample/mm	Original length /mm	Tensile strength /MPa	Lower yield strength MPa	Break elongation/%
1	0.542	190	886	282	0.05
2		140	723	247	0.12
3		135	510	240	0.13
4		140	1437	294	0.05
5		140	431	216	0.07
Average	0.542	149	781	256	0.08

Table 4. At 80℃ material tensile strength.

No.	Diameter of sample /mm	Original length /mm	Tensile strength /MPa	Lower yield strength /MPa	Break elongation/%
1	0.542	180	418	231	0.08
2		160	445	227	0.1
3		145	626	363	0.09
4		150	439	259	0.08
5		155	565	313	0.08
Average	0.542	158	499	279	0.08

Table 5. At 98℃ material tensile strength.

No.	Diameter of sample /mm	Original length /mm	Tensile strength /MPa	Lower yield strength /MPa	Break elongation /%
1	0.542	155	255	209	0.16
2		160	310	227	0.15
3		155	310	221	0.18
4		160	339	220	0.18
5		155	439	210	0.17
Average	0.542	157	331	217	0.17

Results analysis:tensile strength of 6 k T300 around 60℃ is the highest, the tensile strength declined slightly at 60℃~98℃,but the reduction is not big, the average tensile strength of still keep 331 MPa until to 98℃, explain the T300 6 k material of high temperature stability is good.

3.3. Oil corrosion resistance experiment

To soak T300 6 k materials in crude oil medium for 3 days, 5 days, 10 days, 30 days, 60 days and 180 days, each time period sampling 5 groups, tensile test, respectively. The results are shown in table 6-11.

Table 6. The tensile strength on corrosion for 3 days.

No.	Diameter of sample /mm	Original length /mm	Tensile strength /MPa	Lower yield strength /MPa
1	0.542	155	189	165
2		80	157	129
3		100	200	173
4		100	201	174
5		80	191	168
Average	0.542	103	187	162

Table 7. The tensile strength on corrosion for 5 days.

No.	Diameter of sample /mm	Original length /mm	Tensile stre-Ngth /MPa	Lower yield strength /MPa
1	0.542	50	158	46
2		70	139	124
3		100	96	41
4		50	91	53
5		80	119	70
Average	0.542	80	121	67

Table 8. The tensile strength on corrosion for 10 days.

No.	Diameter of sample /mm	Original length /mm	Tensile stre-Ngth /MPa	Lower yield strength /MPa
1	0.542	100	86	57
2		70	97	88
3		70	78	52
4		80	82	68
5		60	96	75
Average	1.715	76	88	68

Table 9. The tensile strength on corrosion for 30 days.

No.	Diameter of sample mm	Original length /mm	Tensile stre-Ngth /MPa	Lower yield strength /MPa
1	0.542	90	81	53
2		70	87	79
3		70	71	53
4		80	78	62
5		60	86	75
Average	0.542	74	80.6	64.4

Table 10. The tensile strength on corrosion for 60 days.

No.	Diameter of sample mm	Original length /mm	Tensile stre-Ngth /MPa	Lower yield strength /MPa
1	0.542	80	76	52
2		70	83	75
3		70	78	55
4		80	79	61
5		60	81	72
Average	0.542	72	79.6	63

Table 11. The tensile strength on corrosion for 180 days.

No.	Diameter of sample mm	Original length /mm	Tensile stre-Ngth /MPa	Lower yield strength /MPa
1	0.542	70	78	53
2		70	76	65
3		70	79	62
4		80	69	61
5		60	81	70
Average	0.542	72	76.6	62.2

From the experiment, 6 K T300 severe corrosion in the oil is very severity, have a great influence on the material tensile strength, especially in the previous month, the tensile strength is down in a straight line, decline down is slow to a month later. Explain T300 6 k material's resistance to corrosion of crude oil is bad, need to improve.

4. Conclusion

- From the point of mechanical properties, tensile strength of T300 6 k room temperature tensile strength is an average of 525 MPa, less than 1/6 of the material standards (3.5GPa), to replace D grade Cr-Mo steel rigid rod, the material strength needs to be improved.
- 6 k T300 tensile strength in 100℃ or less is relatively stable, proving that the material high temperature resistant performance is good.
- 6 k T300 is highly affected by the crude oil corrosion on the tensile strength, to adapt to the underground oil production operation environment, must be to improve the resistance to corrosion of crude oil property.

In short, from the point of mechanical properties and corrosion resistance, GH Company'S 6 K T300 can't now replace the rigid rod.

References

[1] Xue C J 2003 Preparation and properties of heat resistance carbon fiber pultruded composites for continuous sucker rods *Journal of Beijing University of Petrochemical* **30** 55-8
[2] API 11B Standard for Sucker Rod *available at: http://www.docin.com/p-429208236.html*
[3] Tanner C J , Bender R E, Simson A K, et al. 1983 Oil well setup and pumping apparatus *USP,4416329*
[4] Hensley H N and Tanner C J 1984 Graphite composite tape in beam-pumped oil well *SPE13200* (Houston Texas: Society of Petroleum Engineers)
[5] Tanner C J, Bender R E, Simson A K, et al. 1986 Ribbon rod for use in oil well apparatus *USP,4563391*

[6] Lee J F and Hensley H N 1994 Space age material technology expends beam life capacity *SPE28523* (Houston Texas: Society of Petroleum Engineers)

[7] Foley W L and Hensley H N 1996 Ribbon rod-improvement sucker rod technology shows need to re-evaluate current artificial lift installations *SPE35708* (Houston Texas: Society of Petroleum Engineers)

[8] Zhu Bo, Cai Huasu and Sun Naiwu 2003 Development and application of carbon fiber reinforced plastic continuous sucker rod *Oil Machinery* **31** 29-31

[9] Mo Yimin and Pan Yunping 2003 Research and application of assemble process plan and assemble simulation system *Journal of Wuhan University of Science and Technology* **25** 41-3

[10] Zhang Qiang, Zhu Bo and Cai Huasu 2004 Investigation in the characteristics of carbon fiber continuous sucker rod with phenoli cresina smatrix *Oil Machinery* **32** 22-4

Preparation of $AgInS_2$ nanocrystals and their application as sensitizers for TiO_2 nanorod array photoelectrodes

Jianbo Yin[1,3] and Junhong Jia[2]

[1]State Key Laboratory of Advanced Processing and Recycling of Non–ferrous Metals, Lanzhou University of Technology, Lanzhou 730050, P. R. China
[2]State Key Laboratory of Solid Lubrication, Lanzhou Institute of Chemical Physics, Chinese Academy of Sciences, Lanzhou 730000, P. R. China

E-mail: jianbery@163.com

Abstract. This paper reports the controlled synthesis of $AgInS_2$ nanocrystals and their application as sensitizers for TiO_2 nanorod array photoelectrode. The analytical results of high resolution transmission electron microscope (TEM) and laser particle analyzer show that the as–prepared $AgInS_2$ nanoparticles are highly crystallized with an average diameter of 16.67 nm. The results of ultraviolet (UV)–visible absorption spectra indicate that the as–prepared $AgInS_2$ nanoparticles have strong absorption in the spectral range of 300 nm–700 nm. The photovoltaic detectable results of TiO_2 nanorod array photoelectrode sensitized by $AgInS_2$ nanocrystals reveal that the composite films red shift to visible light range with high photovoltaic conversion efficiency of 2.076%, which is 3 times higher than that of bare TiO_2 photoelectrodes.

1. Introduction

In recent years, the controllable synthesis of nanostructural materials remains a research topic of great interest due to their unique chemical and physical properties and potential applications in various areas such as energy storage, luminescence and photoelectricity [1-3]. Metal sulfide nanostructural materials have attracted more and more researchers' attention due to their excellent luminescence and photoelectricity properties [4], for example, the excellent photoluminescence property of ZnS [5], the potential application of Cu_2S and CdS in photovoltaic fields [6, 7], and the appropriate band-gap of In_2S_3 and Bi_2S_3 [8-10]. Ternary metal sulfide semiconductors, alternatives for appropriate band–gap energies, have been the focus of much recent research because of their unique optical properties and electrical properties as well as their potential applications in linear and nonlinear optical devices and solar cells [11-14]. $CuInS_2$ with band gap energy of 1.50 eV and absorption coefficient $1.0\times10^5\ cm^{-1}$, is considered to be ideal in solar cells [15-20]. However, few studies have been conducted on the photovoltaic characteristics of $AgInS_2$ so far, though its band gap energy is only 1.8 eV [21] and suitable for photovoltaic application. Yet, most researchers keep their eyes on the photocatalytic and luminescent properties. Tsuji and his coworkers reported photocatalytic characteristics of $AgInS_2$ and ZnS composite materials [22]; Torimoto and his partners studied their color–adjustable luminophore

[3] Address for correspondence: Jianbo Yin, to be contacted at State Key Laboratory of Recycling of Non–ferrous Metals, Lanzhou University of Technology, Lanzhou 730050, P. R. China. Tel.: +86 931 2973563; Fax: +86 931 2973563; E-mail: jianbery@163.com.

[23]; Hamanaka and his coworkers only studied the photoluminescence properties of $AgInS_2$ quantum dots [24], while ignored their photovoltaic characteristic; Cheng and his partners used $AgInS_2$ crystals as sensitizers on TiO_2 thin film solar cell, but only 0.05% low energy conversion efficiency was achieved [25].

In this paper, a high energy conversion efficiency of $AgInS_2$ nanocrystals as sensitizers is achieved and a systematic study of their photovoltaic application on TiO_2 nanorod arrays is discussed. The study results show that $AgInS_2$ nanocrystals are appropriate sensitizers.

2. Experimental

2.1. Chemicals

Indium (III) chloride tetrahydrate ($InCl_3{\cdot}4H_2O$), titanium tetrachloride ($TiCl_4$) and thiourea were purchased from Shanghai Chemical Reagent Corporation; thioglycolic acid, silver nitrate ($AgNO_3$), iodine, potassium iodide, hydrochloric acid, methanol anhydrous, ethanol absolute, hexane and acetonitrile were all obtained from Tianjin Chemical Reagent Corporation; Cetylamine (Chengdu) was purchased and used as–received.

2.2. Synthesis of $AgInS_2$ nanocrystals

Typical synthesis of $AgInS_2$ nanocrystals was shown as follows: $AgNO_3$ (1.0 mmol), $InCl_3{\cdot}4H_2O$ (1.0 mmol) and thiourea (2.0 mmol) were added to 20 mL cetylamine in a 50 mL three–necked bottle at 120°C under the protection of N_2. After reaction for 5 minutes for nucleation, the system was subsequently heated to 180°C and the reaction continued for 60 minutes under N_2 flow. Cooling naturally, the nanocrystals were separated by adding 100 mL absolute ethanol. The nanocrystals were washed by excess absolute ethanol for two more times and dispersed in a methanol solution of thioglycollic acid (120 mM) and tetramethylammonium hydroxide (150 mM) to exchange cetylamine molecules which were adsorbent on the surface of nanocrystals for 60 minutes to activate the optical activity. After being precipitated with an ethyl acetate–hexane (1/4, v/v) solution, the as–prepared samples were redispersed in methanol.

2.3. Preparation of $AgInS_2$ nanocrystal sensitive solar cell

TiO_2 nanorod arrays were prepared on fluorine doped tin dioxide (FTO) substrates (20×20 mm) by hydrothermal route and the process was similar to the literature [26]. In order to attach the linker molecules to the TiO_2 surface, TiO_2 nanorod array film was immersed in acetonitrile solution of thioglycolic acid (1.2 M) and sulfuric acid (0.15 M) for 12 h. After that, spin–coating method was used for sensitized TiO_2 nanorod arrays by $AgInS_2$ nanocrystals for 2 times, 5 times and 10 times (1000 R/min), respectively. Then, I_2/KI solution with a definite concentration was used as the electrolyte, and transparent platinum glass electrode as the cathode to prepare $AgInS_2$ nanocrystal sensitive solar cell.

2.4. Characterization

High resolution transmission electron microscope (HRTEM, Hitach H–7500, Japan) was used to study the microstructure of the $AgInS_2$ nanocrystals, and the average diameter of the $AgInS_2$ nanocrystals was studied by laser particle size analyzer (Zetasizer Nano 3600). Scanning electron microscope (FESEM, JSM–6701F, Japan) was utilized to study the microstructure of the thin films. The crystal structure measurements of the samples were performed by power X–ray diffraction (XRD, Rigaku D/MAX–2500, Japan) using Cu Kα radiation (λ=0.15418 nm) at the scanning speed of 1.2°/min. UV–vis absorption spectra were recorded using a LabTech (China) spectrophotometer within the wavelength range of 300–1000 nm. The photovoltaic conversion efficiency of the sensitized TiO_2 arrays was studied by solar simulator of CELL–S500 (China).

3. Results and discussions

Figure 1 shows the morphology and structure characterization of TiO_2 nanorod array film. Figure 1a shows the general SEM image of TiO_2 nanorod arrays, and it can be seen that they are uniform on FTO from the inset of which the thickness is observed to be 1.0 μm. Seen from the TEM image (figure 1b), the average diameter of the nanorods is about 120 nm. Figure 1c reveals the clear lattice fringes of TiO_2 with an interplanar distance of about 0.323 nm, which is a typical characteristic fringe spacing of the (110) plane of rutile, showing that the TiO_2 nanorods grow along the (001) plane.

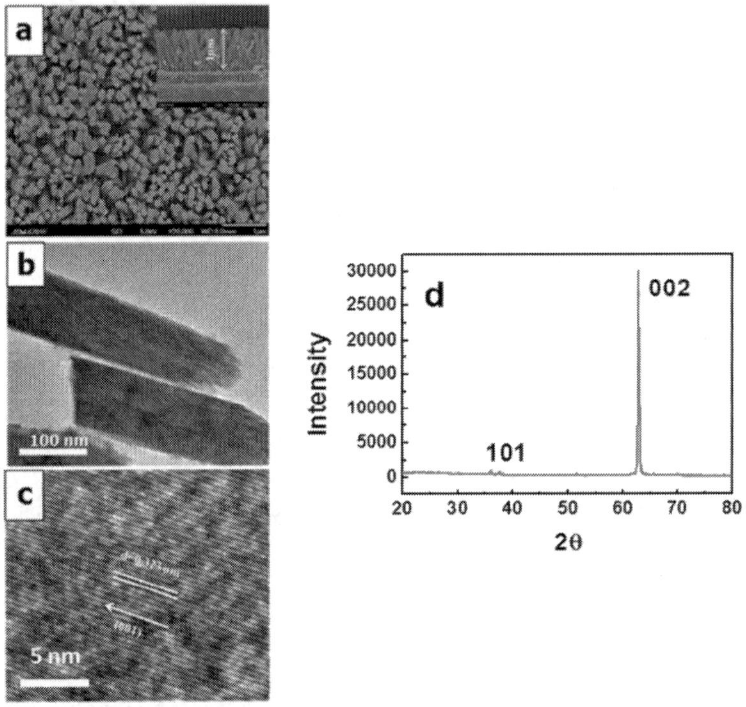

Figure 1. Characterization of TiO_2 arrays. (a) SEM image; (b) TEM image; (c) HRTEM; (d) XRD pattern.

Figure 1d displays the XRD patterns of TiO_2 nanorod arrays. All the diffraction peaks agree well with the tetragonal rutile phase (SG, P42/mmm; JCPDS No. 88–1175, a = b = 0.4517 nm and c = 0.2940 nm). Compared with the powder diffraction pattern, the (002) diffraction peak is significantly enhanced, and some diffraction peaks including (110), (111), and (211) are absent, which indicates that the nanorod arrays are highly oriented with respect to the substrate surface and the TiO_2 nanorods grow in the (001) direction with the growth axis parallel to the substrate surface. Moreover, these conclusions are also confirmed by HRTEM measurement.

Figure 2 shows characterization of $AgInS_2$ nanocrystals. Figure 2a displays the TEM image of monodispersed $AgInS_2$ nanocrystals with the average size range of 7–20 nm, which coincides with the characterization result of the laser particle size analyzer as shown in figure 2d where the average diameter is observed to be 16.67 nm. From the HRTEM image (figure 2b), it can be seen that these nanoparticles are well crystallized, and the inset of figure 3b reveals clear lattice fringes with an interplanar distance of about 0.356 nm which is a characteristic fringe spacing of the orthorhombic $AgInS_2$ crystal phase in the (120) crystal plane. The average particle size of the $AgInS_2$ nanocrystals is calculated to be about 16.5 nm according to Scherrer Equation which also matches the results of the

IOP Conf. Series: Materials Science and Engineering **87** (2015) 012114 doi:10.1088/1757-899X/87/1/012114

TEM and the laser particle size analyzer very well.

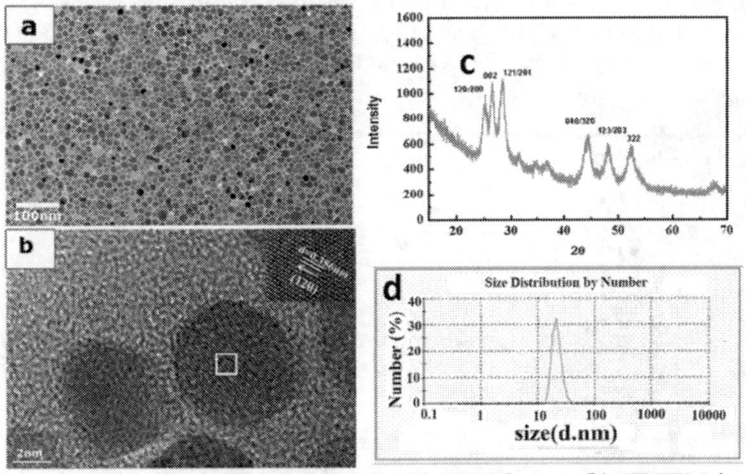

Figure 2. Characterization of $AgInS_2$ nanocrystals: (a) TEM image; (b) HRTEM image; (c) XRD pattern; (d) size distribution chart by number.

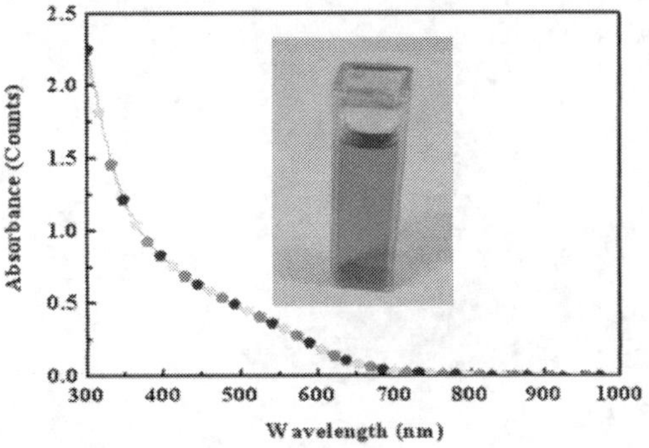

Figure 3. UV–vis absorption spectra of $AgInS_2$ nanocrystals.

The crystal structure of the as–synthesized products is examined by XRD. Figure 2c shows the XRD patterns of $AgInS_2$ nanocrystals, from which it can be seen that all the diffraction peaks can be assigned to the orthorhombic–phase $AgInS_2$, and this result has been also proven by the HRTEM (figure 2b).

The UV–vis absorption spectrum of $AgInS_2$ nanocrystals is shown in figure 3. It shows a wide absorption in the spectral range of 300 nm to 700 nm, and the wavelength of the absorption edge can be clearly identified at around 700 nm (corresponding to E_g of 1.8 eV). Since the E_g of $AgInS_2$ is reported to be 1.8 eV within the band energy range of 0.9–1.9 eV in which most of the solar energy can be absorbed, it can be one of the best solar cell sensitizers.

Figure 4 displays the SEM image of TiO_2 nanorod arrays sensitized by $AgInS_2$ nanocrystals for different spin–coating cycle times: 2, 5, and 10 times, respectively. The gap in the nanorod arrays is infilled slowly by $AgInS_2$ nanocrystals with the increase of spin–coating cycle times. After spin–coating for 2 times, few $AgInS_2$ nanocrystals can be observed under SEM (figure 4a), but after spin–coating for 10 times, very compact nanocomposite film is obtained, and the nanorod arrays can't be observed (figure 4c). The optimal spin-coating cycle is 5 times demonstrated by figures 4b and 5a and b, which indicates that they absorb a sufficient number of $AgInS_2$ nanocrystals with TiO_2 nanorod arrays being exposed on the surface. Figure 5c shows XRD patterns of the TiO_2 nanorod arrays sensitized by $AgInS_2$ nanocrystals. All the diffraction peaks can be attributed to the tetragonal rutile phase TiO_2 and orthorhombic $AgInS_2$ crystals. The peak in (002) plane is weakening compared with figure 1b because it is covered by $AgInS_2$ crystals. The peaks in (121), (040) and (323) crystal planes show that the $AgInS_2$ nanocrystals have no remarkable change.

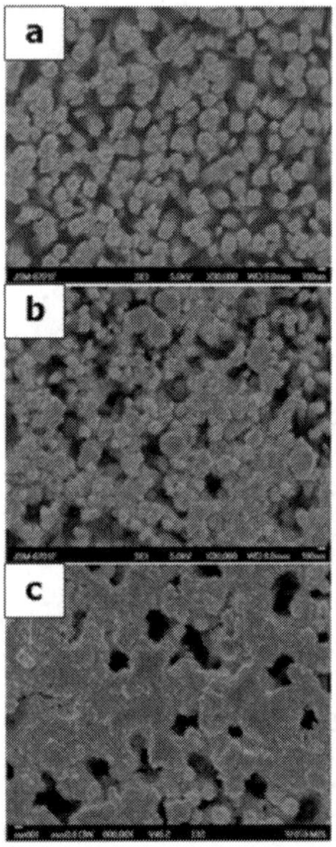

Figure 4. FESEM images of TiO_2 nanorod arrays for different spin–coating times. (a) for 2 times; (b) for for 5 times. (c) for 10 times.

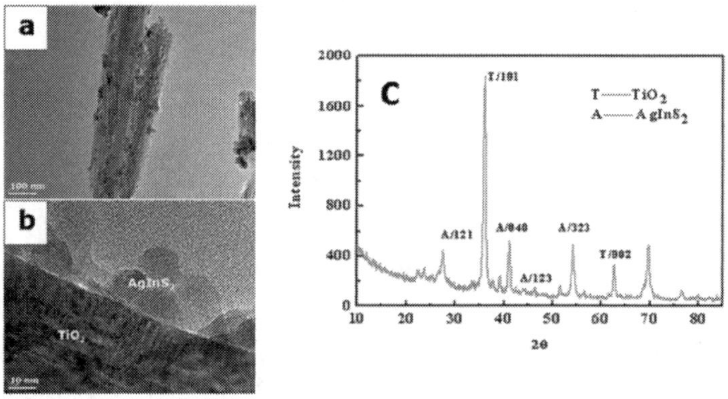

Figure 5. TEM images of TiO_2 nanorod arrays sensitized by $AgInS_2$ nanocrystals spin–coating for 5 times. (a) general image; (b) detailed image; (c) XRD pattern.

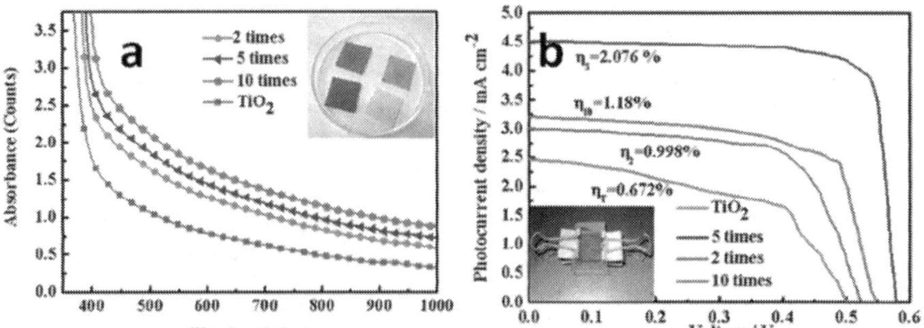

Figure 6. (a) UV–vis absorption spectra of TiO_2 nanorod arrays sensitized by $AgInS_2$ crystals; (b) the photocurrent density–voltage (J–V) curves of the TiO_2 solar cells sensitized by $AgInS_2$ nanocrystals for different spin–coating times under AM 1.5 illumination of 100 mW cm^{-2}.

The UV–vis absorption spectra of TiO_2 nanorod arrays sensitized by $AgInS_2$ nanocrystals are shown in figure 6a. It can be seen that the optical absorption of the TiO_2 nanorod arrays is in the ultraviolet region. However, after being sensitized by $AgInS_2$ nanocrystals, the absorption edges of the thin films can red shift to the visible region, and the red shift length increases with the sensitized times, indicating that $AgInS_2$ nanocrystals as the sensitizers for TiO_2 nanorod arrays expand the photoabsorption range.

Figure 6b shows the photocurrent density–voltage (J–V) curves of the solar cells based on the TiO_2 photoelectrodes sensitized by $AgInS_2$ nanocrystals with an active area of about 0.78 cm^2 under AM 1.5 illumination of 100 mW cm^{-2} for 20 times cycles. The photovoltaic parameters of the solar cells are summarized in table 1. It shows that the solar cell based on bare TiO_2 indicates a short–circuit photocurrent density (Jsc) of 2.474 mA cm^{-2}, an open–circuit voltage of 501 mV, a fill factor (FF) of 59.5%, and a conversion efficiency of 0.672%. The relatively low efficiency is ascribed to the low density uniform TiO_2 nanorod arrays thin film, which decreases the surface area and the effective quantity of the photo–generated electrons. The relatively high efficiency of thin film can be attributed to the sensitizers $AgInS_2$ nanocrystals on the surface of TiO_2 nanorods that absorb more solar energy and generate more photo–generated electrons than bare TiO_2 as shown in table 1 (Cell 2 and Cell 3). After being sensitized for 2 times, the energy conversion efficiency is 1.18%, and it almost improves twice up to 2.076% after being sensitized for 5 times and the FF is 70.8% which is the max one of all the values shown in table 1. The decrease of conversion efficiency of Cell 4 is due to a large amount of $AgInS_2$ nanocrystals being deposited on the TiO_2 film, which agglomerate into chunks and lead to a decrease of quantum efficiency.

Table 1. Photovoltaic parameters of the TiO_2 solar cells sensitized by $AgInS_2$ nanocrystals for different spin–coating times: 2 times (TiO_2–2), 5 times (TiO_2–5) and 10 times (TiO_2–10) under AM 1.5 illumination of 100 mW cm^{-2}.

Sample	Sensitized times	V_{oc} (mV)	J_{sc} (mA cm^{-2})	FF (%)	η (%)
Cell 1	0	501	2.474	59.5	0.672
Cell 2	2	546	3.218	67.1	0.998
Cell 3	5	576	4.513	70.8	2.076
Cell 4	10	525	3.013	65.4	1.180

4. Conclusion

Orthorhombic $AgInS_2$ nanocrystals are synthesized by a simple wet chemical route. The results of TEM and laser particle size analyzer show that the diameter of the $AgInS_2$ nanocrystals is 16.67 nm; and they have strong optical absorption in the spectral range of 300 nm–700 nm. The study results show that $AgInS_2$ nanocrystals are suitable for use as sensitizers for solar cells. The sensitized results show that the max energy conversion efficiency is 2.076 % after spin–coating for 5 times, and after that the energy conversion efficiency decreases with the increasing the spin–coating times.

References

[1] Bondarev V A 1997 Variational method for solving non-linear problems of unsteady-state heat conduction *Int. J. Heat Mass Transf.* **40** 3487
[2] Zhu J, Li Q, Bi W, et al. 2013 Ultra–rapid microwave–assisted synthesis of layered bultrath in birnessite $K_{0.17}MnO_2$ nanosheets for efficient energy storage *J. Mater. Chem. A.* **1** 8154-9
[3] Mishra S, Rajeswari R, Vijayan N, et al. 2013 Probing the structure, morphology and multifold blue absorption of a new red–emitting nanophosphor for LEDs *J. Mater. Chem. C.* **1** 5849-55
[4] Zhu Y P, Ren T Z and Yuan Z Y 2014 Mesoporous non–siliceous inorganic–organic hybrids: A promising platform for designing multifunctional materials *New J. Chem.* **38** 1905-22
[5] Mao B, Chuang C H, McCleese C, et al. 2014 Near–infrared emitting $AgInS_2$/ZnS nanocrystals *J.*

Phys. Chem. C. **118** 13883-9

[6] Zhu Y P, Li J, Ma T Y, et al. 2014 Sonochemistry–assisted synthesis and optical properties of mesoporous ZnS nanomaterials *J. Mater. Chem. A.* **2** 1093-101

[7] Zhu G, Pan L, T. Xu, et al. 2011 One–step synthesis of CdS sensitized TiO_2 photoanodes for quantum dot–sensitized solar cells by microwave assisted chemical bath deposition method *ACS Appl. Mater. Interf.* **3** 1472-8

[8] Palacios P, Aguilera P, Wahnon I, et al. 2009 Simple and generalized synthesis of semiconducting metal sulfide nanocrystals *Adv. Function. Mater.* **19** 1645-9

[9] Ye F, Wang C, Du G, et al. 2011 Large–scale synthesis of In_2S_3 nanosheets and their rechargeable lithium–ion battery *J. Mater. Chem.* **21** 17063-5

[10] Heeá Kim Y, Haká Lee J, Mină Park S, et al. 2010 Synthesis of shape–controlled β–In_2S_3 nanotubes through oriented attachment of nanoparticles *Chem. Commun.* **46** 2292-4

[11] Wu T, Zhou X, Zhang H, et al. 2010 Bi_2S_3 nanostructures: a new photocatalyst *Nano Research* **3** 379-86

[12] Chang S H, Chiu B C, Gao T L, et al. 2014 Selective synthesis of copper gallium sulfide ($CuGaS_2$) nanostructures of different sizes, crystal phases, and morphologies *Cryst. Eng. Comm.* 16 3323-30

[13] Yin J and Jia J 2014 Synthesis of Cu_3BiS_3 nanosheet films on TiO_2 nanorod arrays by a solvothermal route and their photoelectrochemical characteristics *Cryst. Eng. Comm.* **16** 2795-801

[14] Lin Z, Fei X and Ma Q 2014 $CuInS_2$ quantum dots@silica near–infrared fluorescent nanoprobe for cell imaging *New J. Chem.* **38** 90-6

[15] Yang F, Kuznietsov V and Lublow M 2013 Solution synthesis of high–quality $CuInS_2$ quantum dots as sensitizers for TiO_2 photoelectrodes *J. Mater. Chem. A.* **1** 6407-15

[16] Tang Y, Ng Y H, Yun J H, et al. 2014 Fabrication of a $CuInS_2$ photoelectrode using a single–step electrodeposition with controlled calcination atmosphere *RSC Adv.* **4** 3278-83

[17] Li Q, Zhai L, Zou C, et al. 2013 Wurtzite $CuInS_2$ and $CuIn_xGa_{1-x}S_2$ nanoribbons: Synthesis, optical and photoelectrical properties *Nanoscale* **5** 1638-48

[18] Guo K, Liu Z, Han J, et al. 2014 Hierarchical TiO_2–$CuInS_2$ core–shell nanoarrays for photoelectrochemical water splitting *Phys. Chem. Chem. Phys.* **16** 16204-13

[19] Yang J, Bao C, Zhang J, et al. 2013 In situ grown vertically oriented $CuInS_2$ nanosheets and their high catalytic activity as counter electrodes in dye–sensitized solar cells *Chem. Commun.* **49** 2028-30

[20] Gunawan, Septina W and Ikedal S, et al. 2014 Platinum and indium sulfide–modified $CuInS_2$ as efficient photocathodes for photoelectrochemical water splitting *Chem. Commun.* **50** 8941-3

[21] Cheng K W and Wang S C 2009 Effects of complex agents on the physical properties of Ag–In–S ternary semiconductor films using chemical bath deposition *Mater. Chem. Phys.* **115** 14-20

[22] Tsuji I, Kato H, Kobayashi H, et al. 2004 Photocatalytic H_2 evolution reaction from aqueous solutions over band structure–controlled (AgIn)xZn_2(1–x)S_2 solid solution photocatalysts with visible–light response and their surface nanostructures *J. Am. Chem. Soc.* **126** 13406-13

[23] Torimoto T, Adachi T, Okazaki K, et al. 2007 Facile synthesis of ZnS–$AgInS_2$ solid solution nanoparticles for a color–adjustable luminophore *J. Am. Chem. Soc.* **129** 12388-9

[24] Hamanaka Y, Ogawa T and Tsuzuki M 2011 Photoluminescence properties and its origin of $AgInS_2$ quantum dots with chalcopyrite structure *J. Phys. Chem. C* **115** 1786-92

[25] Cheng K, Law W C, Yong K T, et al. 2011 Synthesis of near–infrared silver–indium–sulfide ($AgInS_2$) quantum dots as heavy–metal free photosensitizer for solar cell applications *Chem. Phys. Lett.* **515** 254-7

[26] Liu B and Aydil E S 2009 Growth of oriented single–crystalline rutile TiO_2 nanorods on transparent conducting substrates for dye–sensitized solar cells *J. Am. Chem. Soc.* **131** 3985-90

IOP Conf. Series: Materials Science and Engineering **87** (2015) 012115 doi:10.1088/1757-899X/87/1/012115

Application of water borne paint on the tractor chassis

H C Kang[1,3], A K Cai[1], R Bi[2]; B Wang[2], C Q Ma[1], L F Yan[1], X Liu[1], B B Zhang[1] and X Y Shi[1]

[1] Technology and Materials Research Institute, YTO Group Corporation, Luoyang 471003, China
[2] Harbin Institute of Technology Shenzhen Graduate School, Shenzhen 518055, China

E-mail: hchk878@163.com

Abstract. This paper introduces the test and application of cleaner water borne paint coating mode on great wheel tractor chassis.

1. Introduction

At present, accounting for the world's most polluted cities PM2.5 more than half in China, the ecological environment is grim, haze and other natural disasters frequently occur, the number of cancer patients is dramatically increasing. It is the high time for China to develop a low carbon economy to implement the cleaner production. The HJ/T293-2006 Cleaner production standard–Automobile manufacturing (Painting) was released by China in 2006. Recently, water borne paint and short process have been used in almost of Chinese new automobile coating lines, controlling the generation of pollutants from the source of production [1]. As the international agricultural enterprises and Chinese automobiles, engineering machinery industry have entered, the competition is fierce for Chinese tractor industry. It is an inevitable trend that the tractor coating technology tends to economic goals [2].

As safety accidents caused by corrosion occurred continuously, so the great tractor chassis has a direct relationship with safety of the driving and drivers. The shape structure of tractor chassis is complicated and hulking. which includes four major series of more than 1000 varieties, and its maximum is 4500 mm×2600 mm×1600 mm, and maximum weight is 3800 kg. Compared with the other parts of a car, the diversity and complexity increase significantly in many aspects including the shape structure of the chassis. As a result, copying the existing water borne paint spraying technology and robotics technology doesn't work. Water Borne Paint has been investigated in this paper to apply

[3] Address for correspondence: Huichun Kang (1963—_): female, master of engineering, senior engineer, is skilled in coating process and equipment design and project management. E-mail: hchk878@163.com; Tel.: 0379-64969914.

on the tractor chassis".

2. The major contents of research

2.1. The research of main technical indicators of chassis painting with water borne paint

According to the requirements of technology advances in tractor manufacturing and the standard of advanced foreign counterparts in several aspects including environmental influence, salt spray resistance and adhesion, etc [3], it can be benchmarked to advanced counterparts. The main indications such as salt spray resistance was increased from 96 h to 240 h, paint film thickness was elevated by 15~25 um to 30~50 um.

2.2. Feasibility research of the chassis painting with water borne paint

Achieving clean paint, adopting high efficient robots, low consumption of airmix spraying and other new technology is the best choice, which whether or not be applied to the great wheel tractor chassis, it need to complete the following research:

2.2.1. Research of new airmix Spraying Application on Chassis. It is a compared experiment between the new airmix spraying and conventional air spraying mode. The laboratory test shows that utilization rate of airmix spraying way more than higher 40% than air spraying way; the axle test shows that the amount of paint of air spraying is the airmix spraying four times for unit film thickness. As shown in the table 1.

Table 1. Experimental data of different spraying ways of axle.

Serial number	Spraying way	Axle		
		The average thickness of paint film (μm)	Paint consumption (kg)	Paint consumption of unite of film thickness (kg/μm)
1	Airmix spraying	44.4	0.04	0.0025
2	Air spraying	57.8	0.61	0.0105

2.2.2. Research of water borne paint products meeting the chassis. For four sample water borne paints, their performances were test in laboratory. To make spraying test in production site, choosing B and C two sample from test data, as shown in the tables 2 and 3. Through the experiments of single, low-volume and batch spraying, it shows that the film properties and appearance of product B meet the requirements. The phenomenon that shrinks of paint happened in the testing ground. Combining with the airmix spraying mode, the prescriptive of water borne paint about thixtotropic was adjusted through commutating with B product technical staffs. It met the requirements of batch spray in the production site.

Table 2. Screening experiment.

...	Design indicators	A	B	C	D
Solids (%)	≥48	49.9	48.7	48.5	45.5
Viscosity (s)	≥60	154	106	62	134
Fineness	≤30 μm	30	28	26	35
Hardness	>0.4	0.61	0.5	0.69	0.53
Gloss (%)	≥80	80	80	96.2	67.2
Film color and appearance	Film formation uniform, pinhole-free, particle and Air bubbles defects	The shrink of the Paint film on surface	Standard template	Standard template	A little of pinhole on the surface of the paint film
Salt spray resistance (h)	240	144 Corrosion of paintfilm:1.2 mm, Large area of bubble and wrinkle	240	240	240 Mild foaming

Research Conclusions:

Applications of epoxy acrylic water-based paint on the chassis, the appearance of the paint film in Chassis was smooth, and the environment is improved obviously. So it can replace the oil paint for mass production. The application of water borne paint in the chassis also can reduce the emissions of atmospheric pollutants VOC from the source(from 133 g/table dropped to 10 g/table) to meet clean production standards. Research and development of epoxy acrylic water borne paint which is suitable for a variety of chassis material (the casting, mechanical parts, plastic parts, rubber parts) and a variety of the coated surface (primer coat, top coat, smooth surface) bottom and low temperature (80℃), and high corrosion resistance (240 h), it improves the coating's corrosion resistance and weather resistance. Then the coated surface adhesion of form a complete set is ensured, the spraying operation environment is improved, and the fire danger is also reduced. A lot of technical problems are resolved at the same time, which are the low resistant to high temperature of rubber-plastic part and the energy dissipation and large heat capacity of casting-forging.

2.2.3. Research of robot automatic spray technology on the chassis. In order to meet the requirements of the chassis in high quality and fast step coating (2 min) for different coated surfaces, variable flow technology of automatic airmix spray was developed which improves the efficiency of the coating uniformity, spraying and painting rate (from 30%-40% to 45%-60%); the coating thickness can reach to 40 μm±5 μm which can replace traditional manual air spraying, reduce equipment investment and covers of area [5]. Optimized automatic spraying area can be up to 85%-95% which achieved automatic spraying in high level, then reduced the dosage of artificial.

Conclusion of Experiment:

- For the tractors' chassis parts with large and complex shape, preliminary program was completed

Table 3. Production test.

Serial number	Spray form	Models (The great wheel tractor chassis)	B (HD—E0704—1)		C	
	…	…	Spraying quantity	Spraying effect	Spraying quantity	Spraying effect
1	Single spraying	Above 140 horsepower	4 batches 6 chassis	surface of paint film is smooth, no ills; the spraying environment is good, no odor	2 chassis	Surface of Paint film is good. problem: The color is greyish after spraying, the viscosity of paint cannot meet the technical requirements, so the on-site spray next step was canceled
2		Below 140 horsepower	12 batches 20 chassis	surface of paint film is smooth, no ills	…	…
3	Small batch spraying	Above 140 horsepower	230 chassis	surface of paint film is smooth, no ills	…	…
4		Below 140 horsepower	3 batches 113 chassis	surface of paint film is smooth, no ills	…	…
5	batch spraying	Below 140 horsepower	300 chassis	surface of paint film is smooth, no ills	…	…

by adopting robot simulation with off-line and brief 3D drawings. Robot spraying program can be optimized by adopting reverse modulus and online teaching [5, 6]. As the impact of catenary hanging workpiece, a small area of about 5 to 10 percent can not be sprayed, it needs manually fill spray in simulation process.

- Four P250iA/15 robots, two sides and bottom of the workpiece were sprayed by two robots, the surface of the workpiece was sprayed by other robots. Each robot the spraying flow is in an average of 360 cc/m, and the flow of the spray gun must be 500 cc/m from the perspective in protection of spray gun (i.e., the maximum flow of the spray gun for using is 70%). Spraying of the woorkpieces was completed in 2.0 min.
- In the condition of any robot unattended, the most of spraying work can be completed by

extending the production rhythm appropriately and using a robot on each side.

- Automatic identification of different artifacts with RFID technology, selecting the spraying process can meet the needs of the many kinds of mixed production line.
- Spraying robot, conveyor system and centralized paint supply were controlled by mutual interlock, providing a reliable guarantee for the equipment at the scene of production safety.

3. Factors of chassis with water borne paint

3.1. The viscosity effect

Because the water paint is pseudoplastic non-Newtonian fluid, so its viscosity can be influenced by the shear stress and temperature. The coating viscosity curve of water paint was always plotted in the lab by coating suppliers, while the flow time was determined by -4#cup as the construction basis on site paint mix. If construction viscosity of the water paint needs to be controlled precisely, test instrument of rotational viscometer is recommended because it can determine the different viscosity numbers under different shear stress.

The viscosity with rotational viscometer was used in laboratory, and the water addition proportion should be controlled in 5%~10%. According to the original paint viscosityin construction, deionized water index was controlled to be PH=7, and conductivity to be 220 s/cm. they were mixed well by agitator, then test quickly the flow time up to 40～50 s by 4-cup in construction, resulting in better painting result.

3.2. The effect of cleanness

The requirement of cleanness level of parts' surface for the water borne paints is higher than solvent paint, because water has higher surface tension, and faults such as attachment difficult for paint film and shrinkage cavity would be shown in the case of there was a contaminant as oil, etc .

Therefore, in water-based paint coating, two processes both are added up for the degreasing and washing. Small flow preflush was added before the first hot wash and predegreasing, and fresh water flush was added at the outlet, what is more, the degreaser must also be compatible with the water paint.

3.3. The effect of environment

The evaporation speed of the diluent of water paint evaporate is slower than solvent of solvent paint. The water paint has more stringent requirements for range of work temperature and humidity. The optimum spraying environment is listed as follows: the temperature is (23±3)℃[7], the humidity is (65±5)%, spraying booth keep slight positive pressure, spraying area settlement wind speed is 0.2~0.5 m/s.

The paint coat defects would be shown as orange peel if the surface drying is too fast, when the temperature is too high and humidity too low; the paint film is easily sagging under the action of gravity if the temperature is too low or humidity is too high; or in the water in the paint coat did not volatilize in time, which lead to defect as pinhole and heat rash after drying.

If the settlement wind speed in the spraying area is too high, it will take too much coating material, and paint film is easily sagging; if the settlement wind speed is too low, paint spray will fly in all directions, the equipment will be contaminated.

Air distribution system of air-conditioning in the spraying booth need configure temperature control, humidity control, and filtration device. For YTO chassis coating Plant, wetting device is the only device in winter according to the location and climate.

3.4. The effect of drying

The diluent of water paint is water, which vaporization heat is higher and is difficult to control. Normally, the baked temperature of water borne paint is higher than 100℃, it cannot quick heat like the oil paint after spraying and flowing flat, it needs self-volatilizing, or 3~5min flash dry at 60~80℃ (Absolute humidity is controlled at 7~10 g/kg, the wind speed of vent is controlled at 6~10 m/s), keeps the solid content of paint film in 80%~90%, then go into process the dry oven, so flash dry device should be set up.

If the time for flash dry is too short, the water in the paint coating cannot volatilize in time, lead to defect as pinhole and heat rash when high-temperature baking.

If the time for flash dry is too long, the surface of paint film will be too dry, paint coat defects would be shown as orange peel. The greatest weight of chassis is 3.8 tons, choice the 80℃ fast dry water paint, make water natural evaporation by prolonging the flow flat time, at the same time, chassis is heating slowly. It needs 8~10 min for the surface temperature of parts increasing to 80℃, so adding flash dry device is not needed. Water paint dry curves are shown in the figure 1.

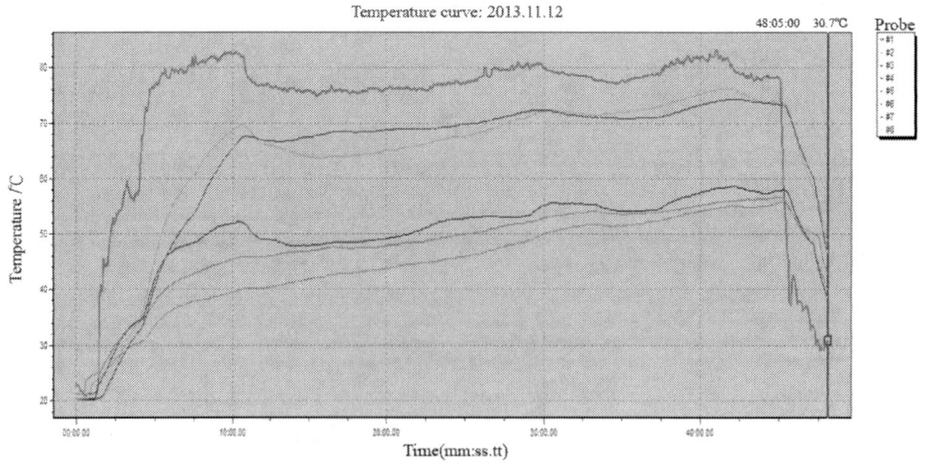

Figure 1. Water paint dry curves.

4. Equipment of chassis with water born paint and points of fire

4.1. Points of robot application

Robot airmix spraying with water paint, based on the off-line programming of the spraying track, according to different varieties and production tact, optimizing the track and the process parameters can be ensured quality and efficiency of spray. Through adjusting track and parameters, the got process parameter about 2min [8] of production tact as follows:

Spraying speed is 600~700 mm/s: After the parameters such as thickness of paint film, overlapping rate of spray span and times of spraying was determined, the productivity is only concerned to the spraying speed and the wide of spray span. Because each of the spray gun has an optimum running speed, in this case we used gas mixing spray gun, the proposed optimum spraying speed is 600～700 mm/s, then productivity is only concerned in spray span.

The paint quantity of spray gun is 400 cc/min [9] to get the optimum spray effect, the flow of airmix spraying should not exceed 500cc/min, in this case, spraying flow is 400 cc/min, the flow controlling pressure is 0.2 MPa, the atomization controlling pressure is 0.1 MPa, the thickness of paint film is 30μm.

The wide of spray span is 250~280 mm. As the spray span of the gun related the atomization, according to the text, the spray span of the airmix gun should be controlled in 250~280 mm, and the spray spancontrolling pressure is 0.1 MPa.

Overlapping rate is 50%: Considering the paint film character of chassis parts, selecting overlapping rate of spray spanat is 50%; for automobile panel, the overlapping rate of spray span is always 2/3 or 3/4.

Purge pressure is 0.6 MPa. Spraying robot is explosion-proof robot, before the start of each service; it needs purge to get rid of the flammable and explosive toxic gas in the robot's cavity. At this time it needs consume a large amount of compressed air 1250 L/min for each of machine: the pressure of compressed air should be keep more than 0.6 MPa, and compressed air need further treatment. The compressed air technical index as follows: the pressure is 0.6 MPa, the particle diameter: <5 um; the concentration: <5 mg/m3, the pressure dew point: -20℃; the oil content: <0.0005 mg/m3. If the above parameters cannot be kept, the work of the robot may be wrong.

4.2. Points of coatings supply system used

4.2.1. Control of temperature. Avoiding dry and crust of paint caused by water evaporation of paint bucket. Coatings supply system should have a temperature control system, to ensure the coating temperature stable at a certain range of 37±3℃ (related to the use of paint products). It can avoid great fluctuation and unreliability.

4.2.2. Low shear [10]. Water paint has poor dispersion stability against mechanical force. The solid dispersed particle in paint is compressed into solid particles when the flow rate changes dramatically in the pipeline, leading to paint film pitting. To prevent coating precipitation and leaving dead-angle, all the agitators and pumps should use low shear device, and pipelines and their joints should be smooth. The diameter of pipes should be larger than that of solvent paint to reduce the resistance. The main pipe: φ16 mm, the branch pipe: φ10 mm, the flow rate; ＞18 m/min.

4.2.3. Coating filtration [11]. 180~200 mesh filter bags were used in air spraying; 220 mesh filter bags were used in airmix spraying, pressure of Filtration diaphragm should be chosen 3:1.

4.2.4. Liquid level control. After water paint level drop, coating attached to the wall of the container leads to powder chipping easily because of water evaporation, and it can't be removed when adding new coating. The powder chipping put a bad affect upon the coating quality, so the requirement of liquid level control for water paint is higher than solvent paint. In paint and circulating buckets, the level of liquid level controller should be added.

4.2.5. Pipeline design and maintenance. It's better to adopt dual line circulation or three line circulation, the main pipes and branch pipes join into the recycle, so it isn't easily precipitate. The recycle of main pipe should be adopted, the branch pipe and spray gun every day should be to prevent clogging after spraying, main pipe should be cleaned thoroughly each six months.

4.3. Corrosion requirements of process equipment

As water paint has higher surface tension than oil paint and higher electrical conductivity, so it easily occurs electrolytic corrosion reaction. The phenomena will lead to oxidation and rusty of pipeline and coatings device, so the stainless steel or aluminizing plate are used in water borne paint equipment, such as coating supply system, spray room, flow flat room dry oven and coatings supply system.

4.4. Fire requirements

Though water paint is water-soluble, it has the same performance with solvent paint after paint film dried, so the contaminated device and grid should be clean in time to avoid the clog of ventilation system, which will influence the spray effect, or increase the difficulty of cleaning and maintenance and cause the fire risk. Recommended to setting fire extinguishing equipment in paint spraying room, leveling room and drying room. Paint mixing room and the combustible gas alarm device may not be provided.

5. Conclusions

Because the window of water paint construction is narrow, in circumstance of certain shear force, the viscosity with a greater influence on the environment temperature, it easily occurs problems in winter and summer. In winter and summer, so temperature and humidity should be controlled in mixing room, coatings supply system and painting room, otherwise it occurs the phenomenon of difficult transport which due to leveling and high viscosity.

The great tractor chassis adopts water born paint technology of robot airmix spraying, operating costs be equivalent to consumption of original paint, the film quality can meet or exceed the quality of the original paint, salt spray resistance also can be improved [12]. The investment on equipment increases 5%, so operation cost of equipment will be increased slightly. Application of water borne paint on the great tractor chassis is first line in Chinese agriculture industry [13], it embodies the social responsibility of YTO group that attaches great importance to environmental protection. In addition, it supplies a good demonstration effect to the domestic agricultural machinery industry for the application of water borne paint.

References

[1] Xing W P 2009 Application of water borne paint on automobile painting line *Shanghai Coatings* **8** 26-9
[2] Le H M 2010 Application omobile painting *Shanghai Coatings* **3** 40-3

[3] Wang X C, Li W G and Gong J B 2014 Enhance the responsibility of creating green painting workshop: to improve the design level of coating process *Modern Paint and Finishing China* **1**1-5

[4] Aznar A C, Pardini O R and Amalvy J I 2006 Glossy topcoat exterior paint formulations using water borne polyurethane/acrylic hybrid binders *Progress in Organic Coatings* **55** 43-9

[5] Chen H, Sheng W, Xi N, et al. 2002 Automated robot trajectory planning for spray painting of free-form surfaces in automotive manufacturing *Proc. ICRA'02. IEEE Int. Conf. on Robotics and Automation* vol 1 pp 450-5

[6] Vecellio A M 1985 Robot painting system for automobiles *US Patent* 4532148

[7] Liu S D, Ma C Q, Kang H C, et al. 2011 Analysis of the application of tractor chassis painting with water borne paint *Modern Paint and Finishing* **8** 50-2

[8] Arıkan S and Balkan T 2000 Process modeling, simulation, and paint thickness measurement for robotic spray painting *Journal of Robotic Systems* **17** 479-94

[9] Kang H C and Wang J J 2012 Large tractor chassis integrated system of automatic spray robot *Modern Painting* **7** 59-62

[10] Tian Y, Wang M and Zhang D M 2009 Construction of water borne paint coating line equipment requirements *Automotive Technology and Materials* **2** 27-9

[11] Yan J, Yang K F, Li G B and Pan Z L 2010 Application and development of environmental water brone paint *Painting Technology* **7** 113-5

[12] Kang H C, Xie Z. 2013 Clean production of coating process and equipment on large tractor chassis *Plating and Fishing* **1** 76-81

[13] Wang H, Lu H X and Xu C 2010 Application of water borne paint on automobile body painting *Painting Industry* **4** 22-5

A sensor device with specific recognition sites for formaldehyde based on molecular imprinting technique

Yumin Zhang[1, 2], Jin Zhang[1, 2], Zhongqi Zhu[1, 2] and Qingju Liu[2, 3]

[1] School of Physical Science and Technology, Yunnan University, Kunming 650091, China
[2] Yunnan Key Laboratory for Micro/Nano Materials & Technology, Yunnan University, Kunming 650091, China

E-mail: qjliu@ynu.edu.cn

Abstract. A novel gas sensor for the detection of formaldehyde was developed based on molecular imprinting technique (MIT) with formaldehyde as template. The structure of the molecular imprinting nanoparticles (MINs) is of orthogonal perovskite. The MINs with a good dispersion possess high surface-to-volume ratio. The particles are spherical and uniform, and then the MINs are printed onto an alumina tube to form a sensor device. At 92℃, the response to 1 ppm formaldehyde based on the sensor is 8, while the response to the other test gases is less than 3. The response time and recovery time are 40 s and 90 s, respectively.

1. Introduction

Formaldehyde is considered to be one of the most important industrial and commercial chemicals due to its chemical activity, high purity and relative cheapness [1-3]. It is well known that formaldehyde is classified as a mutagen and possible human carcinogen by both the US Environmental Protection Agency (EPA) and the World Health Organization (WHO) for its toxin, anaphylaxis and accumulation [4, 5]. Therefore, it is very necessary to monitor the concentration of formaldehyde in atmospheric environment quickly and accurately. Gas sensors based on semiconducting oxides are thought to be an effective means to monitor the gases because they are small, low cost and easy to use [6]. Peng et al. [7] developed a formaldehyde gas sensor comprising ZnO nanorods deposited on an ITO/glass substrate. The experimental results showed that the response of the nanorods to 110 ppm of formaldehyde was around 120 times higher with UV light irradiation than that without it. However, the minimal detection limit was just 1.8 ppm. Lee et al. [8] presented a gaseous formaldehyde sensor comprising a suspended silicon nitride microstructure with an integrated micro Pt heater, a thin-film NiO sensing layer, and Pt interdigitated electrodes. It was shown that the change in resistance varied linearly with the formaldehyde concentration in the range of 0–5 ppm. Lv et al. [9] developed a formaldehyde gas sensor incorporating a thin film of SnO_2-NiO nanometer polycrystalline composite deposited on a micro-hotplate (MHP). It was shown that the device was capable of detecting gaseous concentrations as low as 0.06 ppm given a MHP working temperature of 300℃. Moreover, the device

[3] Address for correspondence: Qingju Liu, Yunnan Key Laboratory for Micro/Nano Materials & Technology, Yunnan University, Kunming 650091, China. Tel.: +86 871 65035376; Fax: +86 871 65035376; E-mail: qjliu@ynu.edu.cn.

showed good selectivity in the presence of common interferents such as alcohol, toluene, α-pinene and acetone. Commonly used materials include ZnO [10-14], WO_3 [15], hybrid materials [16], NiO [17-20] and so on.

Molecular imprinting is a means to prepare specific binding sites for given molecules in appropriate solid matrices. In this approach, the shape and functionality of a template can be transcribed onto microporous materials. The configuration of the functional groups in the template may be memorized within the matrix. Today, this field has become dominated by the use of organic polymeric materials for chromatographic separation, enzyme-mimicking catalysts, chemical sensors, and biosensors [21-26]. In this paper, MIT were used to prepare specific binding sites for small organic molecules (such as formaldehyde). The response to 1 ppm formaldehyde based on the as-prepared sensor is 8 at 92℃, and the response to other test gases is less than 3 to the other test gases. The response time and recovery time are 40 s and 90 s, respectively.

2. Experimental

2.1. Preparation of Ag-$LaFeO_3$

Based on our previous experiments [27], the sample carbon nanotube (CNTs) modified Ag-$LaFeO_3$ with CNTs: Ag-$LaFeO_3$ = 0.75% weight ratio (called Sample A, namely, Sample A = 0.75%CNTs-Ag-$LaFeO_3$, the same below) was found to be the most promising for the sensing properties to formaldehyde gas. In this study, Sample A was further modified based on molecular imprinting technique. To prepare MINs, formaldehyde is used as template, acrylamide (AM) as functional monomer, azodiisobutyronitrile (AIBN) as initiator and Sample A as crosslinker. The molar ratio of template to functional monomer was defined as x = mol (formaldehyde): mol (AM) =1:1, 1:2, 1:3 and 1:4. And 1.0 mmol formaldehyde mixed with x mmol AM was treated by ultrasonic concussion for 30 min, and stood for 8 h as Solution A. Then, 1.0 mmol AIBN was dissolved in 20 mL ethanol and mixed with Solution A and Sample A. The final mixture was treated by ultrasonic concussion for 30 min, stirred at 50℃ for 12 h with the protection of nitrogen and circulating water and then dried. The MINs were finally prepared.

2.2. Fabrication of gas sensor

The prepared MINs were further mixed with distilled water and ground to form a paste, which was subsequently printed onto an alumina tube. There are two Au electrodes placed at the both end sides of the tube. The length of the alumina tube is 4 mm and the diameter is 1.2 mm. In order to improve their stability and repeatability, the gas sensors were aged at 150℃ for 170 h in air. The gas response β was defined as the ratio of the electrical resistance in gas (R_g) to that in air (R_a).

2.3. Characterization

The X-ray diffraction (XRD) patterns were obtained for the phase identification with a D/max23 diffractometer using Cu $K_{\alpha 1}$ radiation (λ=1.54056 Å), where the diffracted X-ray intensities were recorded as a function of 2θ. The accelerating voltage was 35 kV and the applied current was 25 mA, and the sample was scanned from 10° to 90° (2θ) in steps of 0.02°. The particle morphology of the sample was tested by transmission electron microscope (TEM, JEM-2100).

3. Results and discussion

Gas sensing properties of the MINs with x=1:2 is better than those with x=1:1, 1:3 and 1:4. This paper mainly discussed the MINs with x=1:2.

3.1. Structure and micro-morphology characterization

X-ray powder diffraction patterns of MINs are shown in figure 1. The patterns indicate that the structure of MINs is of orthogonal perovskite, which only includes one phase of $LaFeO_3$, because the amount of Ag and CNTs is too small to be detected and the template, functional monomer and initiator are removed, and only some functional groups on the crosslinker are left [28], which can't be detected either. The average crystallite size is estimated by means of Scherrer formula ($D=k\lambda/\beta Cos\theta$, $k=1$, where λ is the wavelength of X-ray, θ is the diffraction angle, and β is the true half-peak width.) through measuring the half-peak widths of the lines in the pattern. The average crystallite sizes for the MINs with x=1:1, 1:2, 1:3 and 1:4 are about 29 nm, 29 nm, 31 nm and 33 nm, respectively.

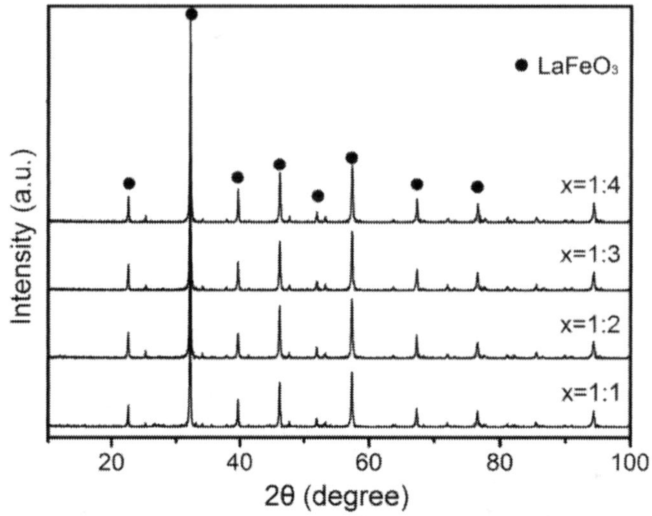

Figure 1. XRD patterns of MINs (1:4≤x≤1:1).

TEM images for the MINs with x=1:2 are shown in figure 2. For this sample, there are many spherically shaped particles which are uniform in size and well dispersed, and the particle size is in the range of 50 nm-90 nm. The specific surface area of the sample is big, which can adsorb formaldehyde more easily. Compared with the XRD results, TEM images reveal that there are various sizes of particles in the MINs. The big particles are composed of small crystallites.

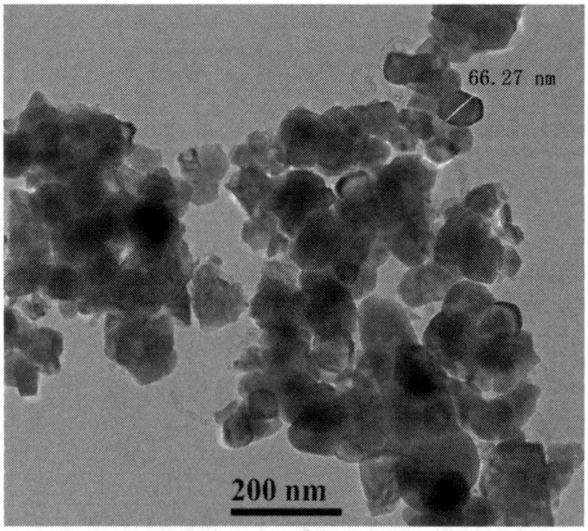

Figure 2. TEM images of the MINs with x=1:2.

3.2. Gas sensing properties

Figure 3 depicts the relationship between response and operating temperature of MINs (1:4≤x≤1:1) sensors to 1 ppm formaldehyde gas. As is shown, the sensor with x=1:2 exhibits higher response to formaldehyde gas at the operating temperature between 40℃ and 92℃, respectively. Taking the conception of sensing properties and cost conservation into consideration, the sensor with x=1:2 is chosen as an optimal element to do further study.

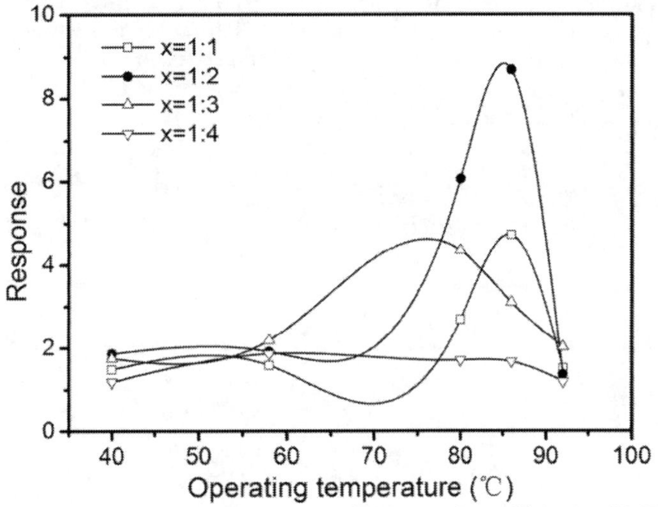

Figure 3. Response -operating temperature curves for 1 ppm formaldehyde of MINs (1:4≤x≤1:1).

To investigate the selectivity of MINs with x = 1:2, the responses of formaldehyde, methanol, toluene, acetone and gasoline were tested for 1 ppm at different operating temperature respectively. It can be found that the sensor behaves well in selectivity to formaldehyde in figure 4. The best response to formaldehyde is 8 at 92℃. To the other test gases, the highest response is lower than 3.

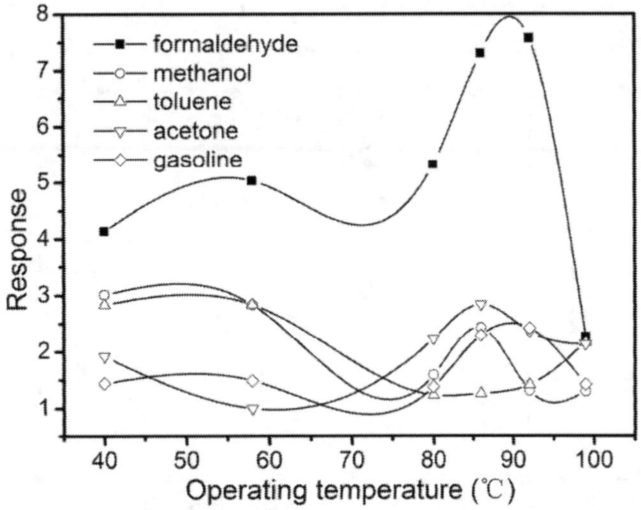

Figure 4. Response-operating temperature curves for 1 ppm different tested gases of the sensor based on the MINs with x=1:2.

Then, the relationship between response and formaldehyde concentration was investigated at 92℃ as illustrated in figure 5. It can be seen that the response increasing for MINs with x = 1:2 is almost linear with the concentration of the formaldehyde at the operating temperature of 92℃. This is because there are a lot of formaldehyde-adsorbing vacancies on the surface of the sensor. When the concentration of test formaldehyde increases, the quantity of adsorbed formaldehyde on the surface of the sensor increases unceasingly, the electronics that the sensor has got also increase gradually. Thus, for p-type semiconductors, the resistance value of the sensor also rises, eventually resulting in the increase of the response. The detailed sensing mechanism is discussed in Section 3.3. At the same time, gas concentration ranges from 0.5 ppm to 3 ppm, the element response with formaldehyde concentration in a linear increase also shows that the sensor can be used for a continuous real-time monitoring at a lower concentration of formaldehyde [29].

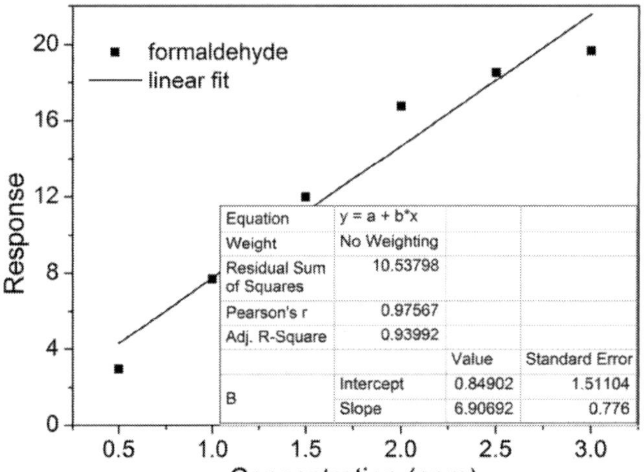

Figure 5. Variation in response to different concentration of formaldehyde gas at an operating temperature of 92℃ of the sensor based on the MINs with x=1:2.

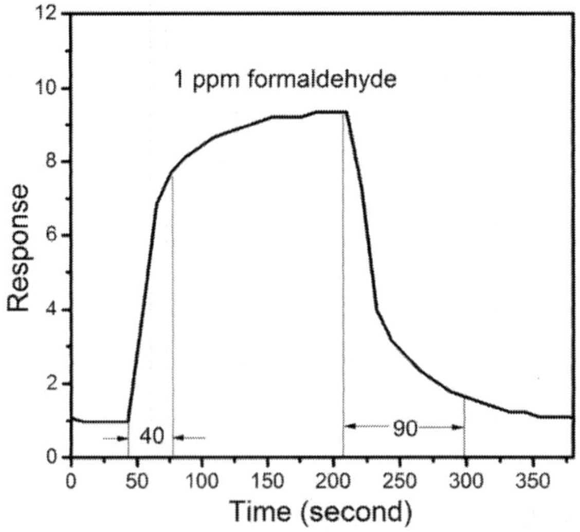

Figure 6. Response-recovery time curve of the sensor based on the MINs with x=1:2 at 92℃.

The response/recovery time are two key quantities for a sensor, which are respectively defined as the time required to reach 90% response/recovery/ when gas is in/out [30]. Figure 6 presents response-recovery characteristic for the sensor based on MINs with x=1:2 operated at 92℃. The resistance of the sensor increases sharply when formaldehyde flow is in but returns to its original state when the gas flow is out although with some delay. When formaldehyde is in, the sensor begins to response which takes 40 s, and when formaldehyde is out, the sensor begins to recover which takes 90 s. The response and recovery time are about 40 s and 90 s, respectively.

4. Conclusion

The selectivity of sensors can be improved by using molecular imprinting method theoretically. In this paper, the optimal amount of functional monomer-acrylamide is x=1:2. The structure of MINs is of orthogonal perovskite and the average grain size of the MINs with x=1:2 is about 29 nm. The MINs have good selectivity for low concentration formaldehyde. To 1 ppm formaldehyde, the sensitivity is 8 at the operating temperature 92℃, and the response time and recovery time are 40 s and 90 s, respectively. These finding may present a new feasible way for exploring $LaFeO_3$-based molecular imprinting sensing materials.

Acknowledgments

This work was supported by National Natural Science Foundation of China (No. 51262030), the Natural Science Foundation of Yunnan Province, China (NO. 2013FZ003).

References

[1] Li J, Dasgupta P K and Luke W 2005 *Anal. Chim. Acta.* **531** 51-68
[2] Wang Z, Bai Z, Yu H, Zhang J and Zhu T 2004 *Energ. Buildings* **36** 1299-308
[3] Yu C and Crump D 1998 *Build. Environ.* **33** 357-74
[4] Belpomme D, Irigaray P, Hardell L, Clapp R, Montagnier L, Epstein S. and Sasco A J 2007 *Environ. Res.* **105** 414-29
[5] Arts J H, Rennende M A and Heer C 2006 *Regul. Toxicol. Pharm.* **44** 144-60
[6] Barsan N, Koziej D and Weima U 2007 *Sensors and Actuators B* **121** 18-35
[7] Peng L, Zhao Q, Wang D, Zhai J, Wang P, Pang S and Xie T 2009 *Sens. Actuators B* **136** 80-5
[8] Lee C Y, Chiang C M, Wang Y H and Ma R H 2007 *Sens. Actuators B* **122** 503-10
[9] Lv P, Tang Z A, Yu J, Zhang F T, Wei G F, Huang Z X and Hu Y 2008 *Sens. Actuators B* **132** 74-80
[10] Bai Z, Xie C, Hu M and Zhang S 2008 *Physica E* **41** 235-9
[11] Chu X, Chen T, Zhang W, Zheng B and Shui H 2009 *Sens. Actuators B* **142** 49-54
[12] Han N, Wu X, Zhang D, Shen G, Liu H and Chen Y 2011 *Sens. Actuators B* **152** 324-9
[13] Zhang L, Zhao J, Lu H, Gong L, Li L, Zheng J, Li H and Zhu Z 2011 *Sens. Actuators B* **160** 364-70
[14] Zhang L, Zhao J, Zheng J, Li L and Zhu Z 2011 *Appl. Surf. Sci.* **258** 711-8
[15] Deng L, Ding X, Zeng D, Tian S, Li H and Xie C 2012 *Sens. Actuators B* **163** 260-6
[16] Xie C, Xiao L, Hu M, Bai Z, Xia X and Zeng D 2010 *Sens. Actuators B* **145** 457-63
[17] Lee C Y, Chiang C M, Wang Y H and Ma R H 2007 *Sens. Actuators B* **122** 503-10
[18] Dirksen J A, Duval K and Ring T A 2001 *Sens. Actuators B* **80** 106-15
[19] Hotovy I, Rehacek V, Siciliano P, Capone S and Spiess L 2002 *Thin Solid Film* **418** 9-15
[20] Wang Y H, Lee C Y, Lin C H and Fu L M 2008 *Microsyst. Technol.* **14** 995–1000
[21] Vlatakis G, Andersson L I, Muller R and Mosbach K 1993 *Nature* **361** 645-7
[22] Wulff G 1995 *Angew. Chem. Int. Edit.* **34** 1812-32
[23] Kriz D, Ramstrom O and Mosbach K 1997 *Anal. Chem.* **69** 345A-9A
[24] Ramstrom O and R J Ansell 1998 *Chirality* **10** 195-209
[25] Haupt K and Mosbach K 2000 *Chem. Rev.* **100** 2495-504
[26] Whitcombe M J and Vulfson E N 2001 *Adv. Mater.* **13** 467-78

[27] Zhang Y M, Zhang J, Chen J L, Zhu Z Q and Liu Q J 2014 *Sens. Actuators B* **195** 509-14
[28] Gao B, Lu J, Chen Z and Guo J 2009 *Polym* **50** 3275-84
[29] Zhang Y M, Lin Y T, Chen J L, Zhang J, Zhu Z Q and Liu Q J 2014 *Sens. Actuators B* **190** 171-6
[30] Zeng W, Liu T M and Liu D J 2010 *Physica B* **405** 4235-9

Size-dependent free vibrations of electrostatically pre-deformed functionally graded micro-cantilevers

Masoud Tahani[1, 3]**, Romesh C. Batra**[2] **and Amir R. Askari**[1]

[1]Department of Mechanical Engineering, Ferdowsi University of Mashhad, Mashhad, Iran

[2]Department of Engineering Science and Mechanics, Virginia Polytechnic Institute and State University, Blacksburg, VA 24061, USA

E-mail: mtahani@um.ac.ir

Abstract. Size-dependent free vibration analysis of cantilever-type resonant micro-sensors made of non-homogeneous functionally graded materials with two material phases is the objective of the present paper. For this aim, the modified couple stress theory together with the Euler-Bernoulli beam model is considered, and the size-dependent equation of motion which accounts for the non-linear and distributed electrostatic force is derived using the Hamilton's principle. The natural frequencies of the system are extracted using the Galerkin weighted residual method. It is found that the fundamental frequency of the system is decreased with an increase of applied voltage and becomes zero when the input voltage reaches the pull-in voltage of the system. The findings of the present paper are compared and validated by available results in the literature and excellent agreements between them are observed.

1. Introduction

Micron and sub-micron scale structures are frequently used in different applications such as micro-electro-mechanical systems (MEMS) nowadays. Because of their small size, low power consumption and the reliability of batch fabrications, there are lots of potential applications in engineering [1]. Resonant micro-sensors as one of the largest categories of these systems are extensively used in different applications such as signal filtering and chemical and mass sensing [2, 3]. The building blocks of these sensors, which employ the vibrational characteristics of the mechanical structure as the sensing method, are electrostatically actuated micro-cantilevers and doubly clamped micro-beams [4, 5]. In general, an electrostatically actuated micro-beam is an electrically conductive and elastic thin plate suspended over a stationary rigid electrode [6]. In resonant MEMS sensors, the movable electrode is deflected toward the fixed electrode and vibrates about its static deflection, where the direct current (DC) applied voltage is responsible for the static deflection and the oscillatory motion is occurred due to applying the alternating current (AC) voltage [7]. It is noteworthy that the resonance frequencies of micro-sensors in which the applied AC voltage is much smaller than the DC ones are controlled by the DC voltage [8]. Therefore, free vibration analysis of such structures is quiet useful and has application in determining design parameters of these devices.

[3] Address for correspondence: Masoud Tahani, Department of Mechanical Engineering, Ferdowsi University of Mashhad, Mashhad, Iran. E-mail: mtahani@um.ac.ir.

The applied DC voltage has an upper limit in which the non-linear electrostatic attraction overcomes the elastic restoring force of the beam. In this manner the movable part suddenly collapses toward the fixed substrate. This unstable behavior is called pull-in instability which is simultaneously observed by Nathanson et al. [9] and Taylor [10]. Also the upper limit of the DC voltage is called the pull-in voltage.

Recently, variety of experiments showed that the material mechanical behavior in small scales is size-dependent [11-14]. Size-dependent behavior is an intrinsic property of certain materials, which emerges when the characteristic size, e.g. the diameter or the thickness is comparable to the material length scale parameter. Material length scale parameter for a specific material can be determined using some typical experiments such as micro-torsion test [11], micro-bend test [12, 13] and micro/nano indentation test [14-16]. For example, the length scale parameter for single crystals of Al, Ag, Ni, polycrystalline Cu and ploy-synthetically twinned (PST) lamellar α_2-TiAl and γ-TiAl have been determined, respectively as: 2762 nm, 6233 nm, 4315 nm, 1120 nm, 74 nm and 49 nm [17].

The classical continuum mechanics cannot predict the size-dependent behavior of materials that occurs in micron and sub-micron scale structures. In 1960s Toupin [18], Koiter [19] and Mindlin [20, 21] proposed the classical couple stress elasticity theory based on the Cosserat continuum mechanics [22]. Based on this theory, besides the classical stress tensor, the couple stress components should be included to describe the manner of media. In comparison to the classical continuum mechanics, the couple stress theory has two additional parameters (high-order material length scales) other than two classical Lame's constants in constitutive equations for isotropic elastic materials. Recently, a modified version of this theory has been elaborated by Yang et al., in which constitutive equations involve only one additional internal material length scale parameter besides two classical material constants [23].

The modified couple stress theory (MCST) has been successfully utilized to predict the mechanical behavior of micro-structures in recent years. Here some of these works are reviewed. Park and Gao [24] showed that the bending rigidity predicted by the MCST is larger than that calculated by the classical theory (CT) and the difference between the deflections predicted by these two models is significant when the beam thickness is small. Kong et al. [25] investigated the size effect on natural frequencies of the Euler-Bernoulli micro-beam. Ke et al. [26] studied the thermal effect on the free vibration and buckling of micro-beams using the MCST and Timoshenko beam theory through the differential quadrature method (DQM). Rahaeifard et al. [27] investigated the size effect on the deflection and static pull-in voltage using the MCST. They could remove the gap between the experimental observations and the results of CT for the static pull-in voltage and calculated the silicon length scale parameter as $l = 0.592\ \mu\text{m}$. Kong [28] introduced an analytic approximate solution to static pull-in problem and calculated pull-in voltage and pull-in displacement based on the MCST using the Rayleigh–Ritz method. He found that pull-in voltage predicted by the MCST is 3.1 times greater than that predicted by the CT when the micro-beam thickness is equal to material length scale parameter. Furthermore, the normalized pull-in displacement is size-independent and equals to 0.448 and o.398 for cantilever and clamped-clamped micro-beams, respectively.

To achieve all material and economical requirements, functionally graded materials (FGMs) have been frequently employed and analyzed in the past years [29-41]. Recently these types of materials are extensively utilized in MEMS structures [42-45]. Therefore, many researchers are motivated to develop size-dependent mechanical models for functionally graded structures at micron and sub-micron scales using the MCST. Reddy [46] developed size-dependent non-linear Euler-Bernoulli and Timoshenko beam theories for micro-beams made of FGM with two material phases based on the MCST. He also presented some analytic solutions as well as finite element models for investigating free vibrations, bending, buckling and post-buckling responses in such systems. Ke and Wang [47] investigated dynamic buckling instability of functionally graded (FG) micro-beams utilizing the MCST and Timoshenko beam model. Ke et al. [48] also investigated non-linear free vibration of FG Timoshenko micro-beam using the MCST and solved the resulting equations through iterative DQM. Furthermore, Akgöz and Civalek [49] investigated free vibrations of tapered axially FG micro-beams based on the Euler-Bernoulli beam model and the MCST. They extracted the natural frequencies of vibrations using the Rayleigh–Ritz method.

Although many researchers have dealt with the mechanical behavior of micro-beams, the research effort devoted to free vibration analysis of electrostatically pre-deformed micro-beams are very limited. The objective of the present work is to establish a size-dependent electromechanical model for free vibration analysis of FG micro-cantilevers pre-formed by an electric field on the basis of MCST.

The rest of the paper is organized as follows. In sections two and three we present a brief review of the MCST and describe the present size-dependent electro-mechanical model for FG micro-beams, respectively. In section four we provide details of the solution procedure. In section five we compare and validate our findings with the available results in the literature. Also the rest of the section is devoted to a parametric study which indicates the significant effects of couple stress components on the natural frequencies of the system. The main concluding remarks of the present study are summarized in section six.

2. The modified couple stress theory

According to the MCST presented by Yang et al. [23] in 2002, both strain tensor (conjugated with the stress tensor) and curvature tensor (conjugated with the couple stress tensor) are included in the strain energy density. Based on this theory, the strain energy U in a deformed isotropic linear elastic material occupying region Π is given by

$$U = \frac{1}{2}\int_{\Pi} (\ddot{\boldsymbol{\sigma}} : \ddot{\boldsymbol{\varepsilon}} + \ddot{\mathbf{m}} : \ddot{\boldsymbol{\chi}})\, d\Pi \tag{1}$$

where $\ddot{\boldsymbol{\sigma}}$, $\ddot{\boldsymbol{\varepsilon}}$, $\ddot{\mathbf{m}}$ and $\ddot{\boldsymbol{\chi}}$ are the Cauchy stress, strain, deviatoric part of couple stress and symmetric curvature tensors, respectively. These tensors for cases with small slopes and deflections can be written as

$$\ddot{\boldsymbol{\sigma}} = \lambda \mathrm{tr}(\ddot{\boldsymbol{\varepsilon}})\ddot{\mathbf{I}} + 2\mu\ddot{\boldsymbol{\varepsilon}} \tag{2}$$

$$\ddot{\boldsymbol{\varepsilon}} = \frac{1}{2}\left[\nabla\mathbf{u} + (\nabla\mathbf{u})^{\mathrm{T}}\right] \tag{3}$$

$$\ddot{\mathbf{m}} = 2l^2\mu\,\ddot{\boldsymbol{\chi}} \tag{4}$$

$$\ddot{\boldsymbol{\chi}} = \frac{1}{2}\left[\nabla\boldsymbol{\theta} + (\nabla\boldsymbol{\theta})^{\mathrm{T}}\right] \tag{5}$$

where $\nabla = \mathbf{e}_x \frac{\partial}{\partial x} + \mathbf{e}_y \frac{\partial}{\partial y} + \mathbf{e}_z \frac{\partial}{\partial z}$, $\mathbf{u}$ is the displacement vector, λ and μ are Lame's constants (μ is also known as shear modulus), l is a material length scale parameter, $\ddot{\mathbf{I}}$ is the identity tensor and $\boldsymbol{\theta}$ is the rotation vector expressed as

$$\boldsymbol{\theta} = \frac{1}{2}\mathrm{curl}\,\mathbf{u} \tag{6}$$

Obviously, only one length scale parameter l involved in addition to two Lame's constants in the constitutive equation of the MCST. The material length scale parameter l is mathematically defined as the square of the ratio of the curvature modulus to the shear modulus. Also, it is physically treated as a material property characterizing the effect of couple stress components [50]. It is noted that both $\ddot{\boldsymbol{\sigma}}$ and $\ddot{\mathbf{m}}$, as respectively defined in equations (2) and (4) are symmetric due to the symmetry of $\ddot{\boldsymbol{\varepsilon}}$ and $\ddot{\boldsymbol{\chi}}$ given in equations (3) and (5), respectively.

3. The MCST formulation for FG Euler-Bernoulli micro-beam pre-deformed by an electric field

Figure 1 shows a schematic of clamped-free electrically actuated micro-beam under the action of electrostatic force. The thickness, length, width and density of micro-beam are h, L, b and ρ, respectively. The initial gap between the non-actuated beam and the stationary electrode is d. Also, x, y and z are the coordinates along the length, width and thickness, respectively, w is deflection of the beam and t is time.

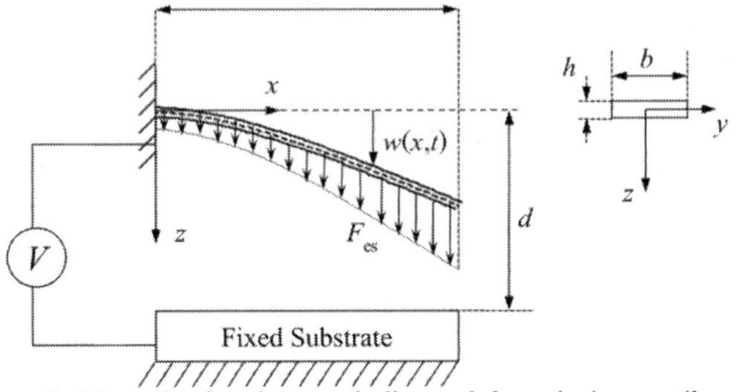

Figure 1. Schematic of an electrostatically pre-deformed micro-cantilever.

The electrostatic excitation by polarized DC voltage V without the effect of fringing field per unit length of the beam can be expressed as [4]:

$$F_{es} = \frac{\varepsilon b V^2}{2(d-w)^2} \tag{7}$$

where ε is the dielectric constant of the medium. It is noted that the fringing field does not have a sizable effect especially for the case of wide micro-beams [51].

According to the basic hypothesis of the Euler-Bernoulli beam model, the displacement field $(\tilde{u}, \tilde{w})$ of an arbitrary point on the micro-beam can be expressed as [52]:

$$\tilde{u} = -z\frac{\partial w}{\partial x}, \quad \tilde{w} = w(x,t) \tag{8}$$

where w is the transverse displacement of a point on the mid-plane of micro-beam (i.e. $z = 0$). For the displacement field introduced in equation (8), the strain components can be written as [52]:

$$\varepsilon_x = -z\frac{\partial^2 w}{\partial x^2}, \quad \varepsilon_y = \varepsilon_z = \varepsilon_{xy} = \varepsilon_{xz} = \varepsilon_{yz} = 0 \tag{9}$$

Substitution of equation (8) into equation (6) and the subsequent result into equation (5) yields

$$\chi_{xy} = -\frac{1}{2}\frac{\partial^2 w}{\partial x^2}, \quad \chi_x = \chi_y = \chi_z = \chi_{xz} = \chi_{yz} = 0 \tag{10}$$

For slender beam, the Poisson's effect is negligible [46], so by substituting equations (9) and (10) into equations (2) and (4), respectively, the non-zero components of the Cauchy stress and high-order couple stress can be determined as:

$$\sigma_x = -E(z)z\frac{\partial^2 w}{\partial x^2} \tag{11}$$

$$m_{xy} = -\mu(z) l^2(z) \frac{\partial^2 w}{\partial x^2} \tag{12}$$

where $E(z)$, $\mu(z)$ and $l(z)$ are the Young modulus, the shear modulus and the material length scale parameter of the micro-beam and vary continuously through the beam thickness according to the power-law [46, 53, 54]. We assume that the beam is made of a two-constituent FGM through the thickness. Hence, the Young modulus, the shear modulus, the material length scale parameter and the density of the beam take the form

$$E(z) = E_1 + \left(\frac{2z+h}{2h}\right)^n (E_2 - E_1) \tag{13a}$$

$$\mu(z) = \frac{E(z)}{2(1+\nu)} = \frac{1}{2(1+\nu)}\left[E_1 + \left(\frac{2z+h}{2h}\right)^n (E_2 - E_1)\right] \tag{13b}$$

$$l(z) = l_1 + \left(\frac{2z+h}{2h}\right)^n (l_2 - l_1) \tag{13c}$$

$$\rho(z) = \rho_1 + \left(\frac{2z+h}{2h}\right)^n (\rho_2 - \rho_1) \tag{13d}$$

where the parameters with 1 and 2 sub-scripts are referred to the properties of the material used in the bottom and top surfaces of the beam, respectively. Also n is the power-law index and ν is the Poisson's ratio of the beam. It is also assumed that the Poisson's ratio is a constant for the FGM [46]. It should be noted that by setting $n = 0$, the present formulation will be simplified to that of homogeneous micro-beams. Upon substitution of equations (9)-(12) into equation (1), the following result is obtained

$$U = \frac{1}{2}\left[(EI)_{\text{eq}} + (\mu A l^2)_{\text{eq}}\right] \int_0^L \left(\frac{\partial^2 w}{\partial x^2}\right)^2 dx \tag{14}$$

where $(EI)_{\text{eq}}$ and $(\mu A l^2)_{\text{eq}}$ are

$$(EI)_{\text{eq}} = \int_A z^2 E(z)\, dA \tag{15a}$$

$$(\mu A l^2)_{\text{eq}} = \frac{1}{2(1+\nu)} \int_A l^2(z) E(z)\, dA \tag{15b}$$

in which A is referred to the cross-sectional area of the micro-beam. The virtual work done by the electrostatic loading is

$$\delta W_{\text{es}} = \int_0^L F_{\text{es}}\, \delta w dx = \int_0^L \frac{\varepsilon b V^2}{2(d-w)^2} \delta w dx \tag{16}$$

The kinetic energy of the slender micro-beam with symmetric cross-section can be expressed as:

$$KE = \frac{1}{2}(\rho A)_{\text{eq}} \int_0^L \left(\frac{\partial w}{\partial t}\right)^2 dx \tag{17}$$

where $(\rho A)_{\text{eq}}$ is

$$(\rho A)_{\text{eq}} = \int_A \rho(z)\, dA \tag{18}$$

The Hamilton principle for an elastic body states [55]

$$\int_{t_i}^{t_f} (\delta KE - \delta U + \delta W_{\text{es}})\, dt = 0 \tag{19}$$

By substitution of equations (14), (16) and (17) into equation (19), the equation of motion is obtained as:

$$(\rho A)_{\text{eq}} \frac{\partial^2 w}{\partial t^2} + \left[(EI)_{\text{eq}} + \left(\mu A l^2\right)_{\text{eq}} \right] \frac{\partial^4 w}{\partial x^4} = \frac{\varepsilon b V^2}{2(d-w)^2} \tag{20}$$

The corresponding boundary conditions for clamped-free micro-beam are also determined as:

$$w(0,t) = 0\,, \quad \left.\frac{\partial w}{\partial x}\right|_{x=0,\,t} = 0 \tag{21a}$$

$$\left.\frac{\partial^2 w}{\partial x^2}\right|_{x=L,\,t} = 0\,, \quad \left.\frac{\partial^3 w}{\partial x^3}\right|_{x=L,\,t} = 0 \tag{21b}$$

It is to be noted that for free vibration analysis, no initial conditions are required. Also, for convenience, the following dimensionless variables are introduced

$$\hat{w} = \frac{w}{d}, \quad \hat{x} = \frac{x}{L}, \quad \hat{t} = \frac{t}{T} \tag{22}$$

where

$$T = \sqrt{\frac{(\rho A)_{\text{eq}} L^4}{(EI)_{\text{eq}}}} \tag{23}$$

Upon substitution of the dimensionless quantities given in equation (22) into equation (20) and dropping the hats, the following result will be obtained

$$\ddot{w} + (1+\alpha) w'''' = \frac{\beta}{(1-w)^2} \tag{24}$$

where dot and prime signs denote derivatives with respect to t and x, respectively. The dimensionless parameters of the system are also introduced as:

$$\alpha = \frac{\left(\mu A l^2\right)_{\text{eq}}}{(EI)_{\text{eq}}}, \quad \beta = \frac{\varepsilon b V^2 L^4}{2d^3 (EI)_{\text{eq}}} \tag{25}$$

The dimensionless boundary conditions also take the form

$$w(0,t) = w'(0,t) = w''(1,t) = w'''(1,t) = 0 \tag{26}$$

To obtain the free vibration equation of micro-beams pre-deformed by an electric field, the deflection is divided into two counterparts as:

$$w = w_s + \Delta w \tag{27}$$

where w_s denotes the static deflection of the micro-beam occurred due to the application of the electrical field and Δw describes the oscillating part of the deflection. Upon substitution of equation (27) into equation (24), utilizing the static equilibrium equation (i.e. equation (24) without the inertia terms) and retain linear terms in Δw, one can obtain

$$(1+\alpha)\frac{\partial^4 \Delta w}{\partial x^4} + \frac{\partial^2 \Delta w}{\partial t^2} - \frac{2\beta}{(1-w_s)^3}\Delta w = 0 \tag{28}$$

Equation (28) governs free vibration of the pre-deformed electrostatically actuated micro-beams. Assuming $\Delta w = \varphi_n(x)\exp(i\omega_n t)$, the following eigenvalue problem will be obtained

$$(1+\alpha)\frac{d^4 \varphi_n}{dx^4} - \left(\frac{2\beta}{(1-w_s)^3} + \omega_n^2\right)\varphi_n = 0 \tag{29}$$

where $i = \sqrt{-1}$. Also ω_n and φ_n denote the nth natural frequency and its associated mode-shape of the electrostatically pre-deformed micro-beam, respectively.

4. Solution procedure

4.1. Obtaining the static deflection

To find the natural frequencies of the present electrostatically pre-deformed micro-cantilever, the static deflection of the micro-beam under the applied electrostatic force will be required. The equation that governs the static equilibrium of the beam can be obtained by setting all time derivatives in equation (24) to zero. Doing so, one would get

$$(1+\alpha)w_s'''' = \frac{\beta}{(1-w_s)^2} \tag{30}$$

Due to the high non-linearity involved in equation (30), a closed-form solution for this equation cannot be found. Hence, an approximate solution will be developed here through the Galerkin weighted residual method. Based on this procedure, the nano/micro-beam deflection can be expressed as a linear combination of a complete set of linearly independent basis functions [55]. It is noted that these functions must satisfy all kinematic boundary conditions [55]. Therefore, linear and undamped mode shapes of the undeformed nano/micro-beam can be used as these basis functions. It is proved that using only the first mode for static, dynamic, vibrational and pull-in analysis of electrically actuated nano/micro-beams maybe very accurate [56-59]. Hence, the deflection of the beam based on one-mode solution can be expressed as:

$$w_s(x) = \bar{w}\varphi_1(x) \tag{31}$$

where $\bar{w}$ is an unknown parameter determined through the Galerkin procedure and $\varphi_1(x)$ is the first linear and undamped mode shape of the undeformed nano/micro-beam determined as [60]:

$$\varphi_1(x) = \cosh(\gamma_1 x) - \cos(\gamma_1 x) - \sigma_1\left[\sinh(\gamma_1 x) - \sin(\gamma_1 x)\right] \tag{32}$$

where the parameters γ_1 and σ_1 for clamped-free nano/micro-beam are given by [60]:

$$\gamma_1 = 1.8751, \quad \sigma_1 = 0.7341 \tag{33}$$

It should be noted that the mode-shapes of the present size-dependent FG micro-cantilever are as same as those of homogenous cantilevers introduced in equation (32) [61]. Because their governing boundary value differential equation of free vibrations, as well as their corresponding boundary conditions are the same. According to the Galerkin procedure, we multiply equation (30) by $\varphi_1(x)$, substitute equation (31) into the resulting equation, integrate the outcome from $x=0$ to 1 and obtain

$$\mathrm{K}\bar{w}-\beta\int_0^1\left\{\varphi_1(x)\left[1-\bar{w}\varphi_1(x)\right]^{-2}\right\}dx=0 \tag{34}$$

where

$$\mathrm{K}=(1+\alpha)\int_0^1\varphi_1\varphi_1''''dx \tag{35}$$

The deflection of the nano/micro-beam under the electrical load can be determined by solving the algebraic equation (34). In this paper, this equation is solved using the method presented by Tahani and Askari [58]. According to the iterative nature of this procedure [58], the non-linear equation (34) is rewritten in the following form

$$\mathrm{K}\bar{w}+\mathcal{N}(\bar{w})=0 \tag{36}$$

where

$$\mathcal{N}(\bar{w})=-\beta\int_0^1\left\{\varphi_1(x)\left[1-\bar{w}\varphi_1(x)\right]^{-2}\right\}dx \tag{37}$$

At the first step, the non-linear term in equation (34) (i.e., $\mathcal{N}(\bar{w})$) is calculated when $\bar{w}_0=0$ as:

$$\mathcal{N}_0=-\beta\int_0^1\left\{\varphi_1(x)\left[1-\bar{w}_0\varphi_1(x)\right]^{-2}\right\}dx=-\beta\int_0^1\varphi_1(x)\,dx \tag{38}$$

Substituting $\mathcal{N}_0$ from equation (38) into equation (36), yields the first estimation of the unknown parameter $\bar{w}$ (i.e. $\bar{w}_1$) as:

$$\bar{w}_1=\frac{-\mathcal{N}_0}{\mathrm{K}} \tag{39}$$

The next estimation of $\mathcal{N}(w)$ (i.e., $\mathcal{N}_1$) will be obtained by substituting $\bar{w}_1$ from equation (39) into equation (37) as:

$$\mathcal{N}_1=-\beta\int_0^1\left\{\varphi_1(x)\left[1-\bar{w}_1\varphi_1(x)\right]^{-2}\right\}dx \tag{40}$$

Substituting $\mathcal{N}_1$ from equation (40) into equation (36), leads to the second estimation of the unknown parameter $\bar{w}$ (i.e. $\bar{w}_2$). This iterative procedure is continued till the convergence achieves or pull-in happens. The convergence criterion is defined as:

$$\left|(\bar{w}_i-\bar{w}_{i-1})\bar{w}_i^{-1}\right|\le 10^{-6} \tag{41}$$

and the pull-in will happen if

$$\bar{w}_i\ge 1 \tag{42}$$

It is noted that the integral $\int_0^1 \left[\varphi_1 \left(1-\bar{w}_i \varphi_1\right)^{-2} \right] dx$ should be calculated numerically and repeated at the each step. After finding $\bar{w}$ with sufficient accuracy, the static deflection of the beam can be obtained through equation (31).

4.2. *Extracting the natural frequencies*

After finding the static deflection corresponding to the applied voltage V, using the Galerkin procedure, the associated ith dimensionless natural frequency of the pre-deformed micro-cantilever can be easily obtained from equation (29) as [62]:

$$\omega_i(V) = \left((1+\alpha)\int_0^1 \varphi_i \varphi_i'''' \, dx - 2\beta \int_0^1 \frac{\varphi_i^2}{(1-w_s)^3} dx \right)^{1/2} \tag{43}$$

The ith dimensional natural frequency of the system (Hz) can also be determined as:

$$f_i(V) = \frac{1}{2\pi T} \left((1+\alpha)\int_0^1 \varphi_i \varphi_i'''' \, dx - 2\beta \int_0^1 \frac{\varphi_i^2}{(1-w_s)^3} dx \right)^{1/2} \tag{44}$$

where the ith mode-shape of the micro-cantilever takes the form [60]

$$\varphi_i(x) = \cosh(\gamma_i x) - \cos(\gamma_i x) - \sigma_i \left[\sinh(\gamma_i x) - \sin(\gamma_i x) \right] \tag{45}$$

in which the values of γ_i and σ_i for the first four mode-shapes are given in table 1 [60]. It is to be noted that the present micro-beam's mode-shapes are normalized such that $\int_0^1 \varphi_i^2 \, dx = 1$. Also the ith dimensional natural frequency of the undeformed micro-cantilever (Hz) can also be extracted using equation (44) when $V = 0$ as:

$$f_i(0) = \frac{\gamma_i^2}{2\pi} \left\{ \left[(EI)_{eq} + (\mu A l^2)_{eq} \right] (\rho A)_{eq}^{-1} L^{-4} \right\}^{1/2} \tag{46}$$

which is a more comprehensive version of the exact expression derived by Asghari et al. [61] for the natural frequencies of undeformed micro-cantilevers. It is to be mentioned that, Asghari et al. [61] did not account for the variation of the length scale parameter of the FGM through its thickness. However, the present expression (i.e. equation (46)) accounts for the through thickness variation of the length scale parameter as one of the micro-beam properties. It is also noteworthy that this can be considered as verification for present analysis.

Table 1. The first four mode-shape parameters for clamped-free micro-beams [60].

Mode-shape parameters	$i=1$	$i=2$	$i=3$	$i=4$
γ_i	1.8751	4.6941	7.8548	10.9955
σ_i	0.7341	1.0185	0.9992	1.0000

5. Results and discussion

5.1. *Comparison and validation*

As discussed at the end of the previous section, the present solution procedure leads to the exact solution for size-dependent natural frequencies of undeformed FG micro-cantilevers. Since, according to the best of authors' knowledge, there are no published results in the open literature for size-dependent natural frequencies of FG electrostatically pre-deformed micro-beams, we neglect the effects of size and through thickness variation of the material properties (i.e. set $n = l_1 = l_2 = 0$) and compare our findings with those obtained by De and Aluru [63] in figure 2 for an 110-direction silicon micro-cantilever with properties listed in table 2. It is to be noted that the 110-direction silicon will be abbreviated to silicon hereinafter.

Table 2. Geometric and material properties of present silicon micro-cantilever.

L (μm)	b (μm)	h (μm)	E (GPa)	ν	ρ (kg/m^3)
80	10	0.5	169	0.33	2331

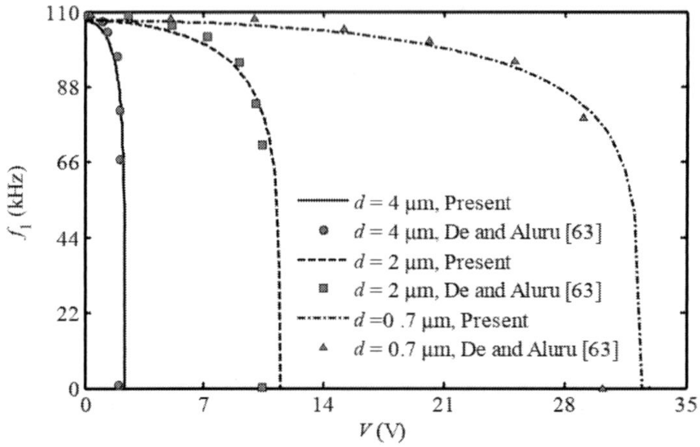

Figure 2. Variation of the fundamental natural frequency (kHz) versus the applied voltage (V) for micro-cantilevers with properties presented in table 2.

It is seen from figure 2 that the present findings agree well with those obtained by De and Aluru [63] except for near pull-in cases. It should be noted that De and Aluru [63] solved the coupled non-linear equations of motion developed based on the Lagrangian description of both mechanical and electrical domains using the relaxation and Newton schemes. However, the present paper employs the improved version of the simple and accurate electromechanical model which was introduced by Batra et al. [4] for homogenous materials. It should be mentioned here that the accuracy of this model in comparison to three-dimensional (3-D) finite element (FE) simulations was proven by Batra et al. [4]. According to the results of figure 2, by increasing the applied voltage, the fundamental natural frequency of the system decreases and suddenly drops to zero when the input voltage reaches the pull-in voltage of the system. To validate the accuracy of present results for near pull-in cases, the calculated pull-in voltages for silicon micro-cantilevers with material properties presented in table 2 and geometric properties listed in table 3 are compared with experimental observations reported by Osterberg [64] in figure 3. It is to be noted that the value of the length scale parameter is also set to $l = 0.592$ μm for preparing the results of figure 3 [27]. Based on the results of this figure and those of figure 2, one can conclude that the present procedure can provide reliable and high-accurate results for both far from and near pull-in cases. However, as it is seen from figure 2, the results reported by De and Aluru [63] are not accurate for cases under input voltage close to their pull-in voltage.

Table 3. Geometric properties of silicon micro-cantilevers tested by Osterberg [64].

The range for length (μm)	b (μm)	h (μm)	d (μm)
75-250	50	2.94	1.05

5.2. The effect of size on FG micro-cantilevers

This section investigates the effect of size on natural frequencies of micro-cantilevers made of FGM with two distinct material phases. It is assumed that the beam is made of FG poly-SiAg layer structure where silver (Ag) and silicon (Si) are used as material phase 1 and material phase 2, respectively. The material properties of both silicon and silver are listed in table 4. Also, it is assumed that geometric properties of the micro-cantilever are as same as those presented in table 2.

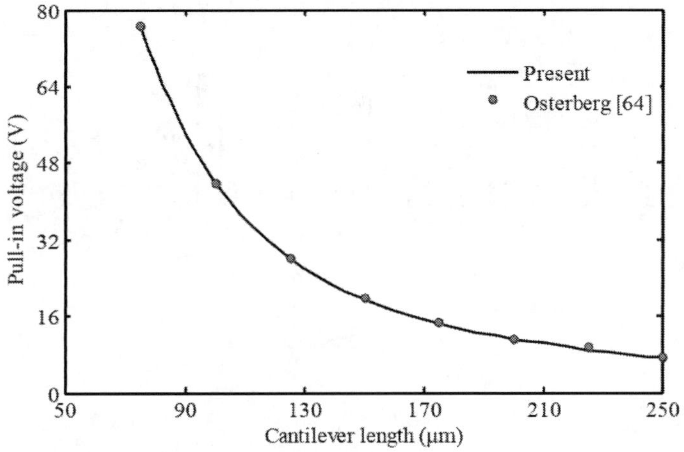

Figure 3. Comparison between present results and experimental observations reported by Osterberg [64] for pull-in voltages of silicon micro-cantilevers with geometric properties presented in table 3.

Table 4. Material properties of silver and silicon.

Material	E (GPa)	ν	ρ (kg/m^3)	l (μm)
1. Silver (Ag)	83	0.37	10490	6.233 [17]
2. Silicon (Si)	169	0.33	2331	0.592 [27]

Table 5 represents the first two natural frequencies of the undeformed micro-cantilevers made of both homogenous and non-homogenous FG materials obtained by both MCST and CT. According to the results of this table, by increasing the values of power-law index (i.e. n) the values of the natural frequencies lead to those of silver; because the volume fraction of silver increases with an increase of the power-law index value (i.e. n). It is noteworthy that the volume fraction of each material phase takes the form [46]

$$\mathrm{VF}_{\mathrm{Ag}}(z)=1-\left(\frac{2z+h}{2h}\right)^n,\quad \mathrm{VF}_{\mathrm{Si}}(z)=\left(\frac{2z+h}{2h}\right)^n \tag{47}$$

Based on equation (47), the properties of the FGM may lead to those of silicon and silver, when the power-law index leads to zero and infinity, respectively.

Table 5. The first two natural frequencies (kHz) for undeformed micro-cantilevers made of both homogenous and non-homogeneous FG materials.

Material	f_1 (MCST)	f_1 (CT)	f_2 (MCST)	f_2 (CT)
Silicon (Si)	288.8889	107.4354	1810.4502	673.2917
FGM, $n=0.1$	501.6917	89.5343	3144.0736	561.1063
FGM, $n=1$	822.7274	55.9488	5155.9866	350.6280
FGM, $n=10$	922.7817	40.3917	5783.0212	253.1326
Silver (Ag)	926.7893	35.4993	5808.1364	222.4721

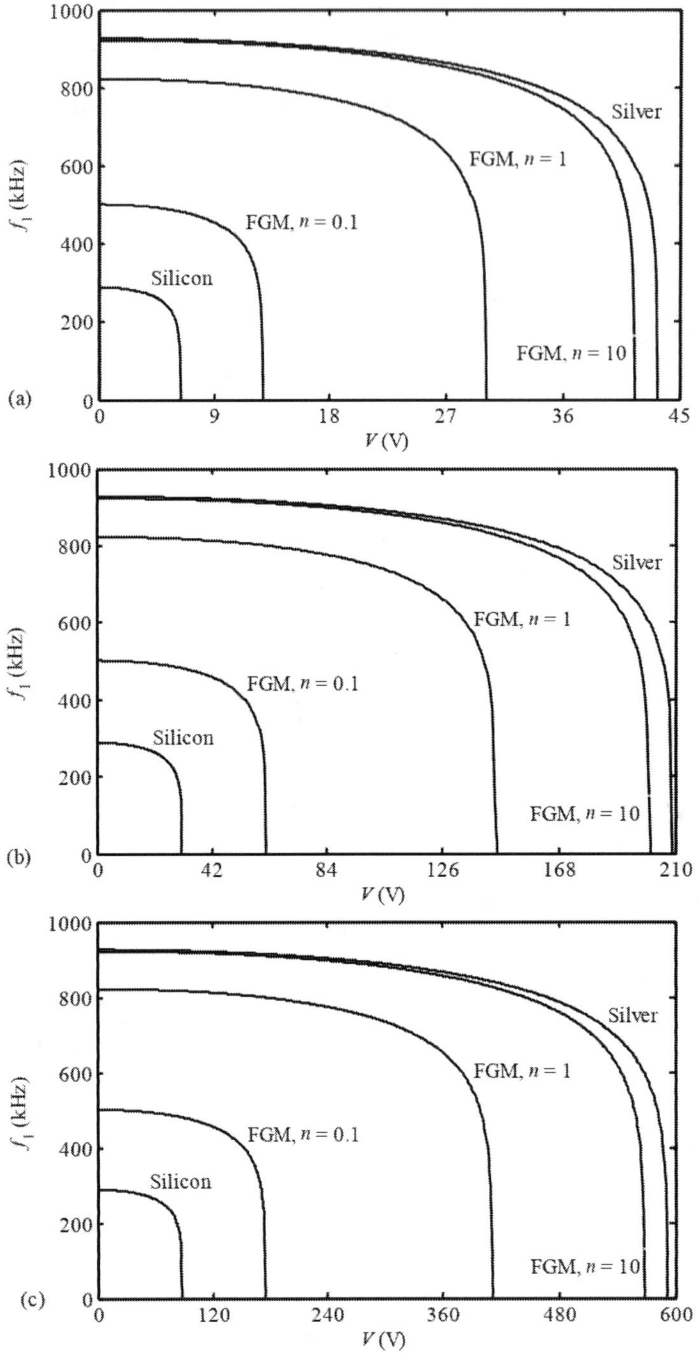

Figure 4. Variation of the fundamental natural frequency (kHz) versus the input voltage (V) for both FG and homogenous micro-cantilevers with geometric properties listed in Tabel2: (a) $d = 0.7\ \mu m$, (b) $d = 2\ \mu m$ and (c) $d = 4\ \mu m$.

As one can observe from table 5, neglecting the effect of size may lead to much more in-accurate results for silver than silicon, since the material length scale parameter of silver is much more than that of silicon. Because, based on equation (46), the effect of size may be more considerable for cases with

IOP Conf. Series: Materials Science and Engineering **87** (2015) 012117 doi:10.1088/1757-899X/87/1/012117

greater values of material length scale parameter. Furthermore, according to the results of table 5, accounting for the effect of couple stress components increases the bending rigidity of the micro-beam that results in increasing the natural frequencies of the system.

Variation of the fundamental natural frequency of present FG poly-SiAg, as well as silicon and silver micro-cantilevers versus the applied voltage is depicted in figure 4 for systems with different initial gaps. According to the results of this figure, increasing the initial gap of the system increases its pull-in voltage. Furthermore, increasing the power-law index increases the bending rigidity of the system which results in the increase of its pull-in voltage and natural frequencies. Because by increasing the value of power-law index, the properties of the FGM are more and more similar to those of silver that is stiffer than silicon in such dimensions. It should be noted that, although the Young modulus of silver is smaller than that of silicon, it is stiffer than silicon. Because its larger material length scale parameter plays a crucial role in the dimensions of present micro-cantilever. This is the concept of size-dependent behavior at micron and sub-micron scales which must be taken into account for designing MEMS.

6. Concluding remarks

Extracting the size-dependent natural frequencies of FG micro-cantilevers pre-deformed by an electric field was the objective of the present paper. To this end, a size-dependent Euler-Bernoulli beam model was employed and the equation of motion, which accounts for the through thickness variation of material properties and distributed non-linear electrostatic attraction, was derived using the Hamilton's principle. The eigenvalue equation which governs on the free vibrations of electrostatically pre-deformed micro-cantilevers was also obtained and solved using the Galerkin weighted residual method. The extracted natural frequencies, as well as the pull-in voltage of the system were compared and validated by available results in the literature and excellent agreements between them were observed. The results showed that the fundamental natural frequency of the system decreases with an increase of applied voltage and suddenly drops to zero when the applied voltage reaches the pull-in voltage of the system.

References

[1] Younis M I 2011 *MEMS Linear and Nonlinear Statics and Dynamics* (New York: Springer)
[2] Rebeiz G M 2003 *RF MEMS: Theory, Design, and Technology* (NewYork: John Wiley & Sons)
[3] Rhoads J F, Shaw S W and Turner K L 2006 *J. Micromech. Microeng.* **16** 890-9
[4] Batra R C, Porfiri M and Spinello D 2008 *J. Sound Vib.* **309** 600-12
[5] Younis M I, Abdel-Rahman E M and Nayfeh A H 2003 *J. Microelectromech. S.* **12** 672-80
[6] Pelesko J A 2002 *SIAM J. Appl. Math.* **62** 888-908
[7] Nayfeh A H and Younis M I 2005 *J. Micromech. Microeng.* **15** 1840-7
[8] Zook J D, Burns D W, Guckel H, Sniegowski J J, Engelstad R L and Feng Z 1992 *Sensor Actuat. A-Phys.* **35** 290-4
[9] Nathanson H C, Newell W E, Wickstrom R A and Davis J R 1967 *IEEE T. Electron. Dev.* **14** 117-33
[10] Taylor G I 1968 *Proc. of Roy. Soc. A.* **306** 423-34
[11] Fleck N A, Muller G M, Ashby M F and Hutchinson J W 1994 *Acta Metall. Mater.* **42** 475–87
[12] Stolken J S and Evans A G 1998 *Acta Mater.* **46** 5109-15
[13] Chong A C M and Lam D C C 1999 *J. Mater. Res.* **14** 4103-10
[14] McFarland A W and Colton J S 2005 *J. Micromech. Microeng.* **15** 1060-7
[15] Begley M R and Hutchinson J W 1998 *J. Mech. Phys. Solids* **46** 2049-68
[16] Nix W D and Gao H 1998 *J. Mech. Phys. Solids* **46** 411-25
[17] Zhao M, Slaughter W S, Li M and Mao S X 2003 *Acta Mater.* **51** 4461-9
[18] Toupin R A 1962 *Arch. Ration. Mech. Anal.* **11** 385-414
[19] Koiter W T 1964 *Proc, K. Ned. Akad. Wet. B.* **67** 17-44
[20] Mindlin R D 1964 *Arch. Ration. Mech. Anal.* **16** 51-78

[21] Mindlin R D 1965 *Int. J. Solids Struct.* **1** 417-38
[22] Cosserat E and Cosserat F 1909 *Theorie des Corps Deformables* (Paris: Hermann et Fils)
[23] Yang F, Chong A C M, Lam D C C and Tong P 2002 *Int. J. Solids Struct.* **39** 2731-43
[24] Park S K and Gao X L 2006 *J. Micromech. Microeng.* **16** 2355-9
[25] Kong S, Zhou S, Nie Z and Wang K 2008 *Int. J. Eng. Sci.* **46** 427-37
[26] Ke L L, Wang Y S and Wang Z D 2011 *Physica E* **43** 1387-93
[27] Rahaeifard M, Kahrobaiyan M H, Asghari M and Ahmadian M T 2011 *Sensor Actuat. A-Phys.* **171** 370-4
[28] Kong S 2013 *Appl. Math. Model.* **37 (12-13)** 7481-8
[29] Belabed Z, Houari M S A, Tounsi A, Mahmoud S R and Anwar Bég O 2014 *Compos. Part B-Eng.* **60** 274–83
[30] Benachour A, Tahar H D, Atmane H A, Tounsi A and Ahmed M S 2011 *Compos. Part B-Eng.* **42** 1386-94
[31] Ben-Oumrane S, Abedlouahed T, Ismail M, Mohamed B B, Mustapha M and El Abbas A B 2009 *Comp. Mater. Sci.* **44** 1344-50
[32] Bessaim A, Houari M S A, Tounsi A, Mahmoud S R and Adda Bedia E A 2013 *J. Sandw. Struct. Mater.* **15** 671-703
[33] Bouderba B, Houari M S A and Tounsi A 2013 *Steel Compos. Struct.* **14 (1)** 85-104
[34] Bouiadjra M B, Houari M S A and Tounsi A 2012 *J. Therm. Stresses* **35** 677-94
[35] Bourada M, Tounsi A, Houari M S A and Adda Bedia E A 2012 *J. Sandw. Struct. Mater.* **14** 5 - 33
[36] El Meiche N, Tounsi A, Ziane N, Mechab I and Adda.Bedia E A 2011 *Int. J. Mech. Sci.* **53** 237-47
[37] Fekrar A, Houari M S A, Tounsi A and Mahmoud S R 2014 *Meccanica* **49** 795-810
[38] Hebali H, Tounsi A, Houari M S A, Bessaim A and Adda Bedia E A 2014 *J. Eng. Mech. -ASCE* **140** 374-83
[39] Houari M S A, Tounsi A and Anwar Bég O 2013 *Int. J. Mech. Sci.* **76** 102-11
[40] Mahi A, Adda Bedia E A, Tounsi A and Mechab I 2010 *Compos. Struct.* **92** 1877-87
[41] Tounsi A, Houari M S A, Benyoucef S and Adda Bedia E A 2013 *Aerosp. Sci. Technol.* **24** 209-20
[42] Witvrouw A and Mehta A 2005 *Mater. Sci. Forum* **492-493** 255-60
[43] Gromova M, Mehta A, Baert K and Witvrouw A 2006 *Sensor Actuat. A-Phys.* **130–131** 403-10
[44] Hasanyan D J, Batra R C and Harutyunyan S 2008 *J. Therm. Stresses* **31** 1006–21
[45] Jia X L, Yang J, Kitipornchai S and Lim C W 2010 *Smart Mater. Struct.* **19** 115028 (115013pp)
[46] Reddy J N 2011 *J. Mech. Phys. Solids* **59** 2382-99
[47] Ke L L and Wang Y S 2011 *Compos. Struct.* **93** 343-50
[48] Ke L L, Wang Y S, Yang J and Kitipornchai S 2012 *Int. J. Eng. Sci.* **50** 256-67
[49] Akgöz B and Civalek Ö 2013 *Compos. Struct.* **98** 314-22
[50] Ma H M, Gao X L and Reddy J N 2008 *J. Mech. Phys. Solids* **56** 3379-91
[51] Chao P C P, Chiu C W and Liu T H 2008 *J. Micromech. Microeng.* **18** 115008
[52] Reddy J N 2007 Theory and analysis of elastic plates and shells (Philadelphia: Taylor & Francis)
[53] Sankar B V 2001 *Compos. Sci. Technol.* **61** 689-96
[54] Chakraborty A, Gopalakrishnan S and Reddy J N 2003 *Int. J. Mech. Sci.* **45** 519-39
[55] Reddy J N 2002 *Energy principles and variational methods in applied mechanics* (New York: John Wiley & Sons)
[56] Askari A R and Tahani M 2012 *Int. J. Aerosp. Lightweight Struct.* **2** 317-34
[57] Askari A R and Tahani M 2014 *Mech. Res. Commun.* **57** 34-39
[58] Tahani M and Askari A R 2014 *Physica E* **63** 151-9
[59] Askari A R and Tahani M 2015 *Appl. Math. Model.* **39** 934-46
[60] Balachandran B and Magrab E 2009 *Vibrations* (Toronto: Cengage Learning)

[61] Asghari M, Ahmadian M T, Kahrobaiyan M H and Rahaeifard M 2010 *Mater. Design* **31** 2324-9
[62] Rao S S 2007 *Vibration of Continuous Systems* (USA: John Wiley & Sons)
[63] De S K and Aluru N R 2004 *J. Microelectromech. S.* **13** 737 -58
[64] Osterberg P M 1995 *Electrostatically actuated microelectromechanical test structures for material property measurement* (Massachusetts Institute of Technology, Cambridge, Massachusetts, USA: PhD thesis)